MATHEMATICS OF QUANTIZATION
AND QUANTUM FIELDS

Unifying a range of topics that are currently scattered throughout the literature, this book offers a unique and definitive review of some of the basic mathematical aspects of quantization and quantum field theory. The authors present both elementary and more advanced subjects of quantum field theory in a mathematically consistent way, focusing on canonical commutation and anti-commutation relations. They begin with a discussion of the mathematical structures underlying free bosonic or fermionic fields, such as tensors, algebras, Fock spaces, and CCR and CAR representations (including their symplectic and orthogonal invariance). Applications of these topics to physical problems are discussed in later chapters. Although most of the book is devoted to free quantum fields, it also contains an exposition of two important aspects of interacting fields: the diagrammatic method and the Euclidean approach to constructive quantum field theory. With its in-depth coverage, this text is essential reading for graduate students and researchers in departments of mathematics and physics.

This title, first published in 2013, has been reissued as an Open Access publication on Cambridge Core.

JAN DEREZIŃSKI is a Professor in the Faculty of Physics at the University of Warsaw. His research interests cover various aspects of quantum physics and quantum field theory, especially from the rigourous point of view.

CHRISTIAN GÉRARD is a Professor at the Laboratoire de Mathématiques at Université Paris-Sud. He was previously Directeur de Recherches at CNRS. His research interests are the spectral and scattering theory in non-relativistic quantum mechanics and in quantum field theory.

CAMBRIDGE MONOGRAPHS ON MATHEMATICAL PHYSICS

General Editors: P. V. Landshoff, D. R. Nelson, S. Weinberg

[†] Issued as a paperback.

Mathematics of Quantization and Quantum Fields

JAN DEREZIŃSKI

University of Warsaw

CHRISTIAN GÉRARD

Université Paris-Sud

CAMBRIDGE
UNIVERSITY PRESS

CAMBRIDGE
UNIVERSITY PRESS

Shaftesbury Road, Cambridge CB2 8EA, United Kingdom

One Liberty Plaza, 20th Floor, New York, NY 10006, USA

477 Williamstown Road, Port Melbourne, VIC 3207, Australia

314–321, 3rd Floor, Plot 3, Splendor Forum, Jasola District Centre, New Delhi – 110025, India

103 Penang Road, #05–06/07, Visioncrest Commercial, Singapore 238467

Cambridge University Press is part of Cambridge University Press & Assessment,
a department of the University of Cambridge.

We share the University's mission to contribute to society through the pursuit of
education, learning and research at the highest international levels of excellence.

www.cambridge.org
Information on this title: www.cambridge.org/9781009290821

DOI: 10.1017/9781009290876

First published 2013
Reissued as OA 2022

A catalogue record for this publication is available from the British Library.

ISBN 978-1-009-29082-1 Hardback
ISBN 978-1-009-29083-8 Paperback

Since my high school years, I have kept in my memory the following verses:

Profesor Otto Gottlieb Schmock
Pracuje już dziesiąty rok
Nad dziełem co zadziwić ma świat:
Der Kaiser, Gott und Proletariat.

As I checked recently, it is a somewhat distorted fragment of a poem by Julian Tuwim from 1919. I think that it describes quite well the process of writing our book.

Jan Dereziński

Je dédie ce livre à mon pays.

Que diront tant de Ducs et tant d'hommes guerriers
Qui sont morts d'une plaie au combat les premiers,
Et pour la France ont souffert tant de labeurs extrêmes,
La voyant aujourd'hui détruire par soi-même?
Ils se repentiront d'avoir tant travaillé,
Assailli, défendu, guerroyé, bataillé,
Pour un peuple mutin divisé de courage
Qui perd en se jouant un si bel héritage.
        (Pierre de Ronsard, 1524–1585)

Christian Gérard

Contents

Introduction

Quantum fields and *quantization* are concepts that come from quantum physics, the most intriguing physical theory developed in the twentieth century. In our work we would like to describe in a coherent and comprehensive way basic aspects of their mathematical structure.

Most of our work is devoted to the simplest kinds of quantum fields and of quantization. We will mostly discuss mathematical aspects of *free* quantum fields. We will consider the quantization only on *linear* phase spaces. The reader will see that even within such a restricted scope the subject is rich, involves many concepts and has important applications, both to quantum theory and to pure mathematics.

A distinguished role in our work will be played by representations of the *canonical commutation* and *anti-commutation relations*. Let us briefly discuss the origin and the meaning of these concepts.

Let us start with *canonical commutation relations*, abbreviated commonly as the *CCR*. Since the early days of quantum mechanics it has been noted that the *position operator* x and the *momentum operator* $D = -\mathrm{i}\nabla$ satisfy the following commutation relation:

$$[x, D] = \mathrm{i}\mathbb{1}. \tag{1}$$

If we set $a^* = \frac{1}{\sqrt{2}}(x - \mathrm{i}D)$, $a = \frac{1}{\sqrt{2}}(x + \mathrm{i}D)$, called the *bosonic creation* and *annihilation operators*, we obtain

$$[a, a^*] = \mathbb{1}. \tag{2}$$

We easily see that (1) is equivalent to (2).

Strictly speaking, the identities (1) and (2) are ill defined because it is not clear how to interpret the commutator of unbounded operators. Weyl proposed replacing (1) by

$$\mathrm{e}^{\mathrm{i}\eta x}\mathrm{e}^{\mathrm{i}qD} = \mathrm{e}^{-\mathrm{i}q\eta}\mathrm{e}^{\mathrm{i}qD}\mathrm{e}^{\mathrm{i}\eta x}, \ \eta, q \in \mathbb{R}, \tag{3}$$

which has a clear mathematical meaning. (1) is often called the *CCR in the Heisenberg form* and (3) *in the Weyl form*.

It is natural to ask whether the commutation relations (1) determine the operators x and D uniquely up to unitary equivalence. If we assume that we are given two self-adjoint operators x and D acting irreducibly on a Hilbert space and satisfying (3), then the answer is positive, as proven by Stone and von Neumann.

Relations (1) and (2) involve a classical system with one degree of freedom. One can also generalize the CCR to systems with many degrees of freedom. Systems with a finite number of degrees of freedom appear e.g. in the quantum mechanical description of atoms or molecules, while systems with an infinite number of degrees of freedom are typical for quantum many-body physics and quantum field theory.

In the case of many degrees of freedom it is often useful to use a more abstract setting for the CCR. One can consider a family of self-adjoint operators $\phi_1, \phi_2, \ldots$ satisfying the relations

$$[\phi_j, \phi_k] = \mathrm{i}\omega_{jk}\mathbb{1}, \tag{4}$$

where ω_{jk} is an anti-symmetric matrix. Alternatively, one can consider the Weyl (exponentiated) form of (4) satisfied by the so-called *Weyl operators* $\exp\!\left(\mathrm{i}\sum_i y_i \phi_i\right)$, where y_i are real coefficients.

A typical example of CCR with many, possibly an infinite number of, degrees of freedom appears in the context of second quantization, where one introduces *bosonic creation* and *annihilation operators* a_i^*, a_j satisfying an extension of (2):

$$
\begin{aligned}
[a_i, a_j] &= [a_i^*, a_j^*] = 0, \\
[a_i, a_j^*] &= \delta_{ij}\mathbb{1}.
\end{aligned}
\tag{5}
$$

The Stone–von Neumann theorem can be extended to the case of regular CCR representations for a finite-dimensional symplectic matrix ω_{jk}. Note that in this case the relations (4) are invariant with respect to the symplectic group. This invariance is implemented by a projective unitary representation of the symplectic group. It can be expressed in terms of a representation of the two-fold covering of the symplectic group – the so-called *metaplectic representation*.

Symplectic invariance is also a characteristic feature of classical mechanics. In fact, one usually assumes that the phase space of a classical system is a symplectic manifold and its symmetries, including the time evolution, are described by symplectic transformations. One of the main aspects of the correspondence principle is the fact that the symplectic invariance plays an important role both in classical mechanics and in the context of canonical commutation relations.

The symplectic invariance of the CCR plays an important role in many problems of quantum theory and of partial differential equations. An interesting – and historically perhaps the first – non-trivial application of this invariance is due to Bogoliubov, who used it in the theory of superfluidity of the Bose gas; see Bogoliubov (1947b). Since then, applications of symplectic transformations to the study of bosonic systems often go in the physics literature under the name *Bogoliubov method.*

Let us now discuss the *canonical anti-commutation relations*, abbreviated commonly as the *CAR*. They are closely related to the so-called *Clifford relations*, which appeared in mathematics before quantum theory, in Clifford (1878). We say that operators $\phi_1, \ldots, \phi_n$ satisfy Clifford relations if

$$[\phi_i, \phi_j]_+ = 2g_{ij}\mathbb{1}, \tag{6}$$

where g_{ij} is a symmetric non-degenerate matrix and $[A, B]_+ := AB + BA$ denotes the anti-commutator of A and B. It is not difficult to show that if the representation (6) is irreducible, then it is unique up to a unitary equivalence for n even, and there are two inequivalent representations for n odd.

In quantum physics, CAR appeared in the description of fermions. If $a_1^*, \ldots, a_m^*$ are *fermionic creation* and $a_1, \ldots, a_m$ *fermionic annihilation operators*, then they satisfy

$$[a_i^*, a_j^*]_+ = 0, \quad [a_i, a_j]_+ = 0, \quad [a_i^*, a_j]_+ = \delta_{ij}\mathbb{1}.$$

If we set $\phi_{2j-1} := a_j^* + a_j$, $\phi_{2j} := \frac{1}{\mathrm{i}}(a_j^* - a_j)$, then they satisfy the relations (6) with $n = 2m$ and $g_{ij} = \delta_{ij}$. Besides, the operators ϕ_i are then self-adjoint.

Another family of operators satisfying the CAR in quantum physics are the *Pauli matrices* used in the description of spin $\frac{1}{2}$ particles. The *Dirac matrices* also satisfy Clifford relations, with g_{ij} equal to the Minkowski metric tensor.

Clearly, the relations (6) with $g_{ij} = \delta_{ij}$ are preserved by orthogonal transformations applied to $(\phi_1, \ldots, \phi_n)$. The orthogonal invariance of CAR is implemented by a projective unitary representation. It can be also expressed in terms of a representation of the double covering of the orthogonal group, called the *Pin group*.

The orthogonal invariance of CAR relations appears in many disguises in algebra, differential geometry and quantum physics. In quantum physics its applications are again often called the *Bogoliubov method*. A particularly interesting application of this method can be found in the theory of superconductivity and goes back to Bogoliubov (1958).

The notion of CCR and CAR representations is quite elementary in the case of a finite number of degrees of freedom. It becomes much deeper for an infinite number of degrees of freedom. In this case there exist many inequivalent CCR and CAR representations, a fact that was not recognized before the 1950s.

The most commonly used CCR and CAR representations are the so-called *Fock representations*, acting on *bosonic*, resp. *fermionic Fock spaces*. These spaces have a distinguished vector Ω called the *vacuum*, killed by annihilation operators and cyclic with respect to creation operators.

In the case of an infinite number of degrees of freedom, the symplectic or orthogonal invariance of representations of CCR, resp. CAR becomes much more subtle. In particular, not every symplectic, resp. orthogonal transformation is unitarily implementable on the Fock space. The *Shale*, resp. *Shale–Stinespring theorem* say that implementable symplectic, resp. orthogonal transformations belong to a relatively small group $Sp_j(\mathcal{Y})$, resp. $O_j(\mathcal{Y})$. Other interesting objects in the case of an infinite number of degrees of freedom are the analogs of the metaplectic and Pin representation.

CCR and CAR representations provide a convenient setting to describe various forms of *quantization*. By a quantization we usually mean a map that transforms a function on a classical phase space into an operator and has some good properties. Of course, this is not a precise definition – actually, there seems to be no generally accepted definition of the term "quantization". Clearly, some quantizations are better and more useful than others.

We describe a number of the most important and useful quantizations. In the case of CCR, they include the *Weyl, Wick, anti-Wick, x, D- and D, x-quantizations*. In the case of CAR, we discuss the *anti-symmetric, Wick and anti-Wick quantizations*. Among these quantizations, the Weyl, resp. the anti-symmetric quantization play a distinguished role, since they preserve the underlying symmetry of the CCR, resp. CAR – the symplectic, resp. orthogonal group. However, they are not very useful for an infinite number of degrees of freedom, in which case the Wick quantization is much better behaved. The x, D-quantization is a favorite tool in the *microlocal analysis* of partial differential equations.

The non-uniqueness of CCR or CAR representations for an infinite number of degrees of freedom is a motivation for adopting a purely algebraic point of view, without considering a particular representation. This leads to the use of operator algebras in the description of the CCR and CAR. This is easily done in the case of the CAR, where there exists an obvious candidate for the *CAR C^*-algebra* corresponding to a given Euclidean space. This algebra belongs to the well-known class of *uniformly hyper-finite algebras*, the so-called UHF(2^∞) algebra. We also have a natural *CAR W^*-algebra*. It has the structure of the well-known *injective type* II_1 *factor*.

In the case of the CCR, the choice of the corresponding C^*-algebra is less obvious. The most popular choice seems to be the C^*-algebra generated by the Weyl operators, called sometimes the *Weyl CCR algebra*. One can, however, argue that the Weyl CCR algebra is not very physical and that there are other more natural choices of the C^*-algebra of CCR.

Essentially all CCR and CAR representations used in practical computations belong to the so-called *quasi-free representations*. They appear naturally, e.g. in the description of thermal states of the Bose and Fermi gas. They have interesting mathematical properties from the point of view of operator algebras. In particular, they provide interesting and physically well motivated examples of *factors of type II and III*. They also give good illustrations for the *Tomita–Takesaki modular theory* and for the so-called *standard form of a W^*-algebra*.

The formalism of CCR and CAR representations gives a convenient language for many useful aspects of quantum field theory. This is especially true in the case of free quantum fields, where representations of the CCR and CAR constitute, in one form or another, a part of the standard language. More or less explicitly they are used in all textbooks on quantum field theory. Usually the authors first discuss quantum fields *classically*. In other words, they just describe algebraic relations satisfied by the fields without specifying their representation. In relativistic quantum field theory these relations are usually derived from some form of classical field equations, like the *Klein–Gordon equation* for bosonic fields and the *Dirac equation* for fermionic fields.

In the next step a representation of CCR or CAR relations on a Hilbert space is introduced. The choice of this representation usually depends on the dynamics and the temperature. At the *zero temperature*, it is usually the Fock representation determined by the requirement that the dynamics should be implemented by a self-adjoint, bounded from below *Hamiltonian*. At *positive temperatures* one usually chooses the *GNS representation* given by an appropriate *KMS state*.

Another related topic is the problem of the unitary implementability of various symmetries of a given theory, such as for example Lorentz transformations in relativistic models. If the generator of the dynamics depends on time, one can also ask whether there exists a time-dependent Hamiltonian that implements the dynamics.

Models of quantum field theory that appear in realistic applications are usually *interacting*, meaning that they cannot be derived from a linear transformation of the underlying phase space. Interacting models are usually described as formal perturbations of free ones. Various terms in perturbation expansions are graphically depicted with *diagrams*. The diagrammatic method is a standard tool for the perturbative computation of various physical quantities.

In the 1950s, mathematical physicists started to apply methods from spectral theory to construct rigorously interacting quantum field theory models. After a while, this subject became dominated by the so-called *Euclidean methods*. The main idea of these methods is to make the real time variable purely imaginary. The Euclidean point of view is nowadays often used as the basic one, at both zero and positive temperature.

Many concepts that we discuss in our work originated in quantum physics and have a strong physical motivation. We believe that our work (or at least some of its parts) can be useful in teaching some chapters of quantum physics. In fact, we believe that the mathematical style is often better suited to explaining some concepts of quantum theory than the style found in many physics textbooks.

Note, however, that the reader does not have to know physics at all in order to follow and, it is hoped, to appreciate our work. In our opinion, essentially all the concepts and results that we discuss are natural and appealing from the point of view of pure mathematics.

We expect that the reader is familiar and comfortable with a relatively broad spectrum of mathematics. We freely use various basic facts and concepts from linear algebra, real analysis, the theory of operators on Hilbert spaces, operator algebras and measure theory.

The theory of the CCR and CAR involves a large number of concepts coming from algebra, analysis and physics. Therefore, it is not surprising that the literature about this subject is very scattered, and uses various conventions, notations and terminology.

We have made an effort to introduce terminology and notation that is as consistent and transparent as possible. In particular, we tried to stress close analogies between the CCR and CAR. Therefore, we have tried to present both formalisms in a possibly parallel way. We make an effort to present many topics in their greatest mathematical generality. We believe that this way of presentation is efficient, especially for mathematically mature readers.

The literature devoted to topics contained in our book is quite large. Let us mention some of the monographs. The exposition of the C^*-algebraic approach to the CCR and CAR can be found in Bratteli–Robinson (1996). This monograph also provides extensive historical remarks. One could also consult an older monograph, Emch (1972). Modern exposition of the mathematical formalism of second quantization can be also found e.g. in Glimm–Jaffe (1987) and Baez–Segal–Zhou (1991). We would also like to mention the book by Neretin (1996), which describes infinite-dimensional metaplectic and Pin groups, and review articles by Varilly–Gracia-Bondia (1992, 1994). A very comprehensive article devoted to CAR C^*-algebras was written by Araki (1987). Introductions to Clifford algebras can be found in Lawson–Michelson (1989) and Trautman (2006).

The book can be naturally divided into four parts.

(1) Chapters 1, 2, 3, 4, 5 6 and 7 are mostly collections of basic mathematical facts and definitions, which we use in the remaining part of our work. Not all the mathematical formalism presented in these chapters is of equal importance for the main topic of work. Perhaps, most readers are advised to skip these chapters on the first reading, consulting them when needed.

(2) Chapters 8, 9, 10 and 11 are devoted to the canonical commutation relations. We discuss in particular various kinds of quantization of bosonic systems and the bosonic Fock representation. We describe the *metaplectic group* and its various infinite-dimensional generalizations.

(3) In Chaps. 12, 13, 14, 15 and 16 we develop the theory of canonical anti-commutation relations. It is to a large extent parallel to the previous chapters devoted to the CCR. We discuss, in particular, the fermionic Fock representation. As compared with the bosonic case, a bigger role is played by operator algebras. We give also a brief introduction to Clifford relations for an arbitrary signature. We discuss the *Pin* and *Spin groups* and their various infinite-dimensional generalizations.

(4) The common theme of the remaining part of the book, that is, Chaps. 17, 18, 19, 20, 21 and 22, is the concept of quantum dynamics – one-parameter unitary groups that describe the evolution of quantum systems. In all these chapters we treat the bosonic and fermionic cases in a parallel way, except for Chaps. 21 and 22, where we restrict ourselves to bosons.

In Chap. 17 we discuss *quasi-free states*. These usually arise as KMS states for a physical system equipped with a free dynamics. In Chaps. 18 and 19 we study quantization of free fields, first in the abstract context, then on a (possibly, curved) space-time. Chapters 20, 21 and 22 are devoted to interacting quantum field theory. In Chap. 20 we discuss in an abstract setting the method of *Feynman diagrams*. In Chap. 21 we describe the *Euclidean method*, used to construct interacting bosonic theories. In Chap. 22 we apply Euclidean methods to construct the so-called *space-cutoff* $P(\varphi)_2$ *model*.

Acknowledgement

The research of J. D. was supported in part by the National Science Center (NCN), grant No. 2011/01/B/ST1/04929.

1
Vector spaces

In this chapter we fix our terminology and notation, mostly related to (real and complex) linear algebra. We will consider only algebraic properties. Infinite-dimensional vector spaces will not be equipped with any topology.

Let us stress that using precise terminology and notation concerning linear algebra is very useful in describing various aspects of quantization and quantum fields. Even though the material of this chapter is elementary, the terminology and notation introduced in this chapter will play an important role throughout our work. In particular we should draw the reader's attention to the notion of the complex conjugate space (Subsect. 1.2.3), and of the holomorphic and anti-holomorphic subspaces (Subsect. 1.3.6).

Throughout the book $\mathbb{K}$ will denote either the field $\mathbb{R}$ or $\mathbb{C}$, all vector spaces being either real or complex, unless specified otherwise.

1.1 Elementary linear algebra

The material of this section is well known and elementary. Among other things, we discuss four basic kinds of structures, which will serve as the starting point for quantization:

(1) Symplectic spaces – classical phase spaces of neutral bosons,
(2) Euclidean spaces – classical phase spaces of neutral fermions,
(3) Charged symplectic spaces – classical phase spaces of charged bosons,
(4) Unitary spaces – classical phase spaces of charged fermions.

Throughout the section, $\mathcal{Y}, \mathcal{Y}_1, \mathcal{Y}_2, \mathcal{W}$ are vector spaces over $\mathbb{K}$.

1.1.1 Vector spaces and linear operators

Definition 1.1 *If $\mathcal{U} \subset \mathcal{Y}$, then* Span $\mathcal{U}$ *denotes the space of finite linear combinations of elements of $\mathcal{U}$.*

Definition 1.2 *$\mathcal{Y}_1 \oplus \mathcal{Y}_2$ denotes the* external direct sum *of $\mathcal{Y}_1$ and $\mathcal{Y}_2$, that is, the Cartesian product $\mathcal{Y}_1 \times \mathcal{Y}_2$ equipped with its vector space structure. If $\mathcal{Y}_1, \mathcal{Y}_2$ are subspaces of a vector space $\mathcal{Y}$ and $\mathcal{Y}_1 \cap \mathcal{Y}_2 = \{0\}$, then the same notation $\mathcal{Y}_1 \oplus \mathcal{Y}_2$ stands for the* internal direct sum *of $\mathcal{Y}_1$ and $\mathcal{Y}_2$, that is, $\mathcal{Y}_1 + \mathcal{Y}_2$ (which is a subspace of $\mathcal{Y}$).*

Definition 1.3 $L(\mathcal{Y}, \mathcal{W})$ *denotes the space of linear maps from* $\mathcal{Y}$ *to* $\mathcal{W}$*. We set* $L(\mathcal{Y}) := L(\mathcal{Y}, \mathcal{Y})$.

Definition 1.4 $L^{\mathrm{fd}}(\mathcal{Y}, \mathcal{W})$*, resp.* $L^{\mathrm{fd}}(\mathcal{Y})$ *denote the space of* finite-dimensional *(or* finite rank*) linear operators in* $L(\mathcal{Y}, \mathcal{W})$*, resp.* $L(\mathcal{Y})$.

Definition 1.5 *Let* $a_i \in L(\mathcal{Y}_i, \mathcal{W})$*,* $i = 1, 2$*. We say that* $a_1 \subset a_2$ *if* $\mathcal{Y}_1 \subset \mathcal{Y}_2$ *and* a_1 *is the restriction of* a_2 *to* $\mathcal{Y}_1$*, that is,* $a_2\big|_{\mathcal{Y}_1} = a_1$.

Definition 1.6 *If* $a \in L(\mathcal{Y}, \mathcal{W})$*, then* $\mathrm{Ker}\, a$ *denotes the* kernel *(or* null space*) of* a *and* $\mathrm{Ran}\, a$ *denotes its* range.

Definition 1.7 $\mathbb{1}_{\mathcal{Y}}$ *stands for the* identity *on* $\mathcal{Y}$.

1.1.2 2×2 *block matrices*

If $\mathcal{Y} = \mathcal{Y}_+ \oplus \mathcal{Y}_-$, every $r \in L(\mathcal{Y})$ can be written as a 2×2 block matrix. The following decomposition, possible if a is invertible, is often useful:

$$r = \begin{bmatrix} a & b \\ c & d \end{bmatrix} = \begin{bmatrix} \mathbb{1} & 0 \\ ca^{-1} & \mathbb{1} \end{bmatrix} \begin{bmatrix} a & 0 \\ 0 & d - ca^{-1}b \end{bmatrix} \begin{bmatrix} \mathbb{1} & a^{-1}b \\ 0 & \mathbb{1} \end{bmatrix}. \tag{1.1}$$

Here are some expressions for the inverse of r:

$$r^{-1} = \begin{bmatrix} \mathbb{1} & -a^{-1}b \\ 0 & \mathbb{1} \end{bmatrix} \begin{bmatrix} a^{-1} & 0 \\ 0 & (d - ca^{-1}b)^{-1} \end{bmatrix} \begin{bmatrix} \mathbb{1} & 0 \\ -ca^{-1} & \mathbb{1} \end{bmatrix} \tag{1.2}$$

$$= \begin{bmatrix} (a - bd^{-1}c)^{-1} & (c - db^{-1}a)^{-1} \\ (b - ac^{-1}d)^{-1} & (d - ca^{-1}b)^{-1} \end{bmatrix}. \tag{1.3}$$

If $\mathcal{Y}$ is finite-dimensional, then, using the decomposition (1.1), we obtain the following formulas for the determinant:

$$\begin{aligned} \det r &= \det a \det(d - ca^{-1}b) \\ &= \det c \det b \det(ac^{-1}db^{-1} - \mathbb{1}). \end{aligned} \tag{1.4}$$

1.1.3 *Duality*

Definition 1.8 *The* dual *of* $\mathcal{Y}$*, denoted by* $\mathcal{Y}^{\#}$*, is the space of* linear functionals *on* $\mathcal{Y}$*. Three kinds of notation for the action of* $v \in \mathcal{Y}^{\#}$ *on* $y \in \mathcal{Y}$ *will be used:*

(1) *the* bra–ket *notation* $\langle v|y \rangle = \langle y|v \rangle$,
(2) *the* simplified *notation* $v \cdot y = y \cdot v$,
(3) *the* functional *notation* $v(y)$.

There is a canonical injection $\mathcal{Y} \to \mathcal{Y}^{\#\#}$. We have $\mathcal{Y} = \mathcal{Y}^{\#\#}$ iff $\dim \mathcal{Y} < \infty$.

Definition 1.9 *If* $y \in \mathcal{Y}$*, we will sometimes write* $|y\rangle$ *for the operator*

$$\mathbb{K} \ni \lambda \mapsto |y\rangle \lambda := \lambda y \in \mathcal{Y}.$$

If $v \in \mathcal{Y}^{\#}$*, we will sometimes write* $\langle v|$ *instead of* v.

As an example of this notation, suppose that $y \in \mathcal{Y}$ and $v \in \mathcal{Y}^{\#}$ satisfy $\langle v|y \rangle = 1$. Then $|y\rangle\langle v|$ is the projection onto the space spanned by y along the kernel of v.

Definition 1.10 *Let* $(e_1, \ldots, e_n)$ *be a basis of a finite-dimensional space* $\mathcal{Y}$*. Then there exists a unique basis of* $\mathcal{Y}^{\#}$*,* $(e^1, \ldots, e^n)$*, called the* dual basis*, such that* $\langle e^i|e_j \rangle = \delta^i_j$.

1.1.4 Annihilator

Definition 1.11 *The* annihilator *of* $\mathcal{X} \subset \mathcal{Y}$ *is defined as*

$$\mathcal{X}^{\mathrm{an}} := \{ v \in \mathcal{Y}^{\#} \ : \ \langle v|y \rangle = 0, \ y \in \mathcal{X} \}.$$

The pre-annihilator *of* $\mathcal{V} \subset \mathcal{Y}^{\#}$ *is defined as*

$$\mathcal{V}_{\mathrm{an}} := \{ y \in \mathcal{Y} \ : \ \langle v|y \rangle = 0, \ v \in \mathcal{V} \}.$$

Note that

$$(\mathcal{X}^{\mathrm{an}})_{\mathrm{an}} = \mathrm{Span}\,\mathcal{X}, \quad (\mathcal{V}_{\mathrm{an}})^{\mathrm{an}} = \mathrm{Span}\,\mathcal{V}.$$

1.1.5 Transpose of an operator

Definition 1.12 *If* $a \in L(\mathcal{Y}_1, \mathcal{Y}_2)$*, then* $a^{\#}$ *will denote the* transpose *of* a*, that is, the operator in* $L(\mathcal{Y}_2^{\#}, \mathcal{Y}_1^{\#})$ *defined by*

$$\langle a^{\#} v|y \rangle := \langle v|ay \rangle, \quad v \in \mathcal{Y}_2^{\#}, \quad y \in \mathcal{Y}_1. \tag{1.5}$$

Note that a is bijective iff $a^{\#}$ is. We have $a^{\#\#} \in L(\mathcal{Y}_1^{\#\#}, \mathcal{Y}_2^{\#\#})$ and $a \subset a^{\#\#}$.

1.1.6 Dual pairs

Definition 1.13 *A* dual pair *is a pair* $(\mathcal{V}, \mathcal{Y})$ *of vector spaces equipped with a bilinear form*

$$(\mathcal{V}, \mathcal{Y}) \ni (v, y) \mapsto \langle v|y \rangle \in \mathbb{K}$$

such that

$$\langle v|y \rangle = 0, \ v \in \mathcal{V} \ \Rightarrow \ y = 0, \tag{1.6}$$
$$\langle v|y \rangle = 0, \ y \in \mathcal{Y} \ \Rightarrow \ v = 0. \tag{1.7}$$

Clearly, if $(\mathcal{V}, \mathcal{Y})$ is a dual pair, then so is $(\mathcal{Y}, \mathcal{V})$. If $\mathcal{Y}$ is finite-dimensional and $(\mathcal{V}, \mathcal{Y})$ is a dual pair, then $\mathcal{V}$ is naturally isomorphic to $\mathcal{Y}^{\#}$.

In general, $(\mathcal{V}, \mathcal{Y})$ is a dual pair iff $\mathcal{V}$ can be identified with a subspace of $\mathcal{Y}^{\#}$ (this automatically guarantees (1.7)) satisfying $\mathcal{V}_{\mathrm{an}} = \{0\}$ (this implies (1.6)).

1.1.7 Bilinear forms

Definition 1.14 *Elements of $L(\mathcal{Y}, \mathcal{Y}^{\#})$ will be called* bilinear forms.

Let $\nu \in L(\mathcal{Y}, \mathcal{Y}^{\#})$. Then ν determines a bilinear map on $\mathcal{Y}$:

$$\mathcal{Y} \times \mathcal{Y} \ni (y_1, y_2) \mapsto y_1 \cdot \nu y_2 = \langle y_1 | \nu y_2 \rangle \in \mathbb{K}. \tag{1.8}$$

Definition 1.15 *We say that ν is* non-degenerate *if* $\operatorname{Ker} \nu = 0$.

Definition 1.16 *We say that $r \in L(\mathcal{Y})$* preserves the form ν *if*

$$r^{\#} \nu r = \nu, \quad \text{i.e.} \quad (r y_1) \cdot \nu r y_2 = y_1 \cdot \nu y_2, \quad y_1, y_2 \in \mathcal{Y}.$$

We say that $a \in L(\mathcal{Y})$ infinitesimally preserves the form ν *if*

$$a^{\#} \nu + \nu a = 0, \quad \text{i.e.} \quad (a y_1) \cdot \nu y_2 = -y_1 \cdot \nu a y_2, \quad y_1, y_2 \in \mathcal{Y}.$$

Remark 1.17 *We will use three kinds of notation for bilinear forms:*

(1) *the* bra–ket notation $\langle y_1 | \nu y_2 \rangle$, *going back to Dirac,*
(2) *the* simplified notation $y_1 \cdot \nu y_2$,
(3) *the* functional notation $\nu(y_1, y_2)$.

Usually, we prefer the first two kinds of notation (both appear in (1.8)).

1.1.8 Symmetric forms

Definition 1.18 *We will say that $\nu \in L(\mathcal{Y}, \mathcal{Y}^{\#})$ is* symmetric *if*

$$\nu \subset \nu^{\#}, \quad \text{i.e.} \quad y_1 \cdot \nu y_2 = y_2 \cdot \nu y_1, \quad y_1, y_2 \in \mathcal{Y}.$$

The space of all symmetric elements of $L(\mathcal{Y}, \mathcal{Y}^{\#})$ will be denoted by $L_{\mathrm{s}}(\mathcal{Y}, \mathcal{Y}^{\#})$.

Let $\nu \in L_{\mathrm{s}}(\mathcal{Y}, \mathcal{Y}^{\#})$.

Definition 1.19 *A subspace $\mathcal{X} \subset \mathcal{Y}$ is called* isotropic *if*

$$y_1 \cdot \nu y_2 = 0, \quad y_1, y_2 \in \mathcal{X}.$$

Definition 1.20 *Let $\mathcal{Y}$ be a real vector space. ν is called* positive semi-definite *if $y \cdot \nu y \geq 0$ for $y \in \mathcal{Y}$. It is called* positive definite *if $y \cdot \nu y > 0$ for $y \neq 0$.*

A positive definite form is always non-degenerate.

Assume that ν is non-degenerate. Using that ν is symmetric and non-degenerate we see that $\langle v | y \rangle = 0$ for all $v \in \nu \mathcal{Y}$ implies $y = 0$. Thus $(\nu \mathcal{Y}, \mathcal{Y})$ is a dual pair and $\mathcal{Y}$ can be treated as a subspace of $(\nu \mathcal{Y})^{\#}$. Hence, ν^{-1}, a priori defined as a map from $\nu \mathcal{Y}$ to $\mathcal{Y}$, can be understood as a map from $\nu \mathcal{Y}$ to $(\nu \mathcal{Y})^{\#}$. We easily check that ν^{-1} is symmetric and non-degenerate. If ν is positive definite, then so is ν^{-1}.

Proposition 1.21 *Let $\mathcal{Y}$ be finite-dimensional. Then,*

(1) $\nu \in L_s(\mathcal{Y}, \mathcal{Y}^\#)$ *iff* $\nu^\# = \nu$.
(2) *If ν is non-degenerate, then $\nu\mathcal{Y} = \mathcal{Y}^\#$, so that $\nu^{-1} \in L_s(\mathcal{Y}^\#, \mathcal{Y})$ is a non-degenerate symmetric form.*

1.1.9 (Pseudo-)Euclidean spaces

Definition 1.22 *A couple $(\mathcal{Y}, \nu)$, where $\nu \in L_s(\mathcal{Y}, \mathcal{Y}^\#)$ is non-degenerate, is called a* pseudo-Euclidean space. *If $\mathcal{Y}$ is real and ν is positive definite, then $(\mathcal{Y}, \nu)$ is called a* Euclidean space. *In such a case we can define the norm of $y \in \mathcal{Y}$, denoted by $\|y\| := \sqrt{y \cdot \nu y}$. If $\mathcal{Y}$ is complete for this norm, it is called a* real Hilbert space.

Let $(\mathcal{Y}, \nu)$ be a pseudo-Euclidean space.

Definition 1.23 *If $\mathcal{X} \subset \mathcal{Y}$, then $\mathcal{X}^{\nu\perp}$ denotes the ν-orthogonal complement of $\mathcal{X}$:*

$$\mathcal{X}^{\nu\perp} := \{y \in \mathcal{Y} \ : \ y \cdot \nu x = 0, \ x \in \mathcal{X}\}.$$

Definition 1.24 *A symmetric form on a real space, especially if it is positive definite, is often called a* scalar product *and denoted $\langle y_1 | y_2 \rangle$ or $y_1 \cdot y_2$. In such a case, the orthogonal complement of $\mathcal{X}$ is denoted $\mathcal{X}^\perp$. For $x \in \mathcal{Y}$, $\langle x|$ will denote the following operator:*

$$\mathcal{Y} \ni y \mapsto \langle x|y := \langle x|y \rangle \in \mathbb{K}.$$

If $\langle x|x \rangle = 1$, then $|x\rangle\langle x|$ is the orthogonal projection onto x.

Most Euclidean spaces considered in our work will be real Hilbert spaces. Real Hilbert spaces will be further discussed in Subsect. 2.2.2.

1.1.10 Inertia of a symmetric form

Let $\mathcal{Y}$ be a finite-dimensional space equipped with a symmetric form ν. In the real case we can find a basis

$$(e_{1,+}, \ldots, e_{p,+}, e_{1,-}, \ldots, e_{q,-}, e_1, \ldots, e_r)$$

such that if

$$(e^{1,+}, \ldots, e^{p,+}, e^{1,-}, \ldots, e^{q,-}, e^1, \ldots, e^r)$$

is the dual basis in $\mathcal{Y}^\#$, then

$$\nu e_{j,+} = e^{j,+}, \qquad \nu e_{j,-} = -e^{j,-}, \qquad \nu e_j = 0.$$

The numbers (p, q) do not depend on the choice of the basis. ν is positive definite iff $q = r = 0$.

Definition 1.25 *We set* inert $\nu := p - q$.

In the complex case, we can find a basis

$$(e_{1,+}, \ldots, e_{p,+}, e_1, \ldots, e_r)$$

such that if

$$(e^{1,+}, \ldots, e^{p,+}, e^1, \ldots, e^r)$$

is the dual basis in $\mathcal{Y}^{\#}$, then

$$\nu e_{j,+} = e^{j,+}, \quad \nu e_j = 0.$$

The number p does not depend on the choice of the basis.

Definition 1.26 *We set* inert $\nu := p$.

1.1.11 Group $O(\mathcal{Y})$ and Lie algebra $o(\mathcal{Y})$

Let $(\mathcal{Y}, \nu)$ be a Euclidean space and $a \in L(\mathcal{Y})$.

Definition 1.27 *We say that*

$$
\begin{aligned}
a \text{ is isometric} \quad & \textit{if } a^{\#} \nu a = \nu, \\
a \text{ is orthogonal} \quad & \textit{if } a \textit{ is isometric and bijective}, \\
a \text{ is anti-self-adjoint} \quad & \textit{if } a^{\#} \nu = -\nu a, \\
a \text{ is self-adjoint} \quad & \textit{if } a^{\#} \nu = \nu a.
\end{aligned}
$$

The set of orthogonal elements in $L(\mathcal{Y})$ is a group for the operator composition, denoted by $O(\mathcal{Y})$. The set of anti-self-adjoint elements in $L(\mathcal{Y})$, denoted by $o(\mathcal{Y})$, is a Lie algebra, equipped with the commutator $[a, b]$.

Definition 1.28 *If $(\mathcal{Y}, \nu)$ is pseudo-Euclidean, we keep the same definitions, except we replace* isometric, orthogonal, anti-self-adjoint *and* self-adjoint *with* pseudo-isometric, pseudo-orthogonal, anti-pseudo-self-adjoint *and* pseudo-self-adjoint.

1.1.12 Anti-symmetric forms

Definition 1.29 *We will say that $\omega \in L(\mathcal{Y}, \mathcal{Y}^{\#})$ is* anti-symmetric *if*

$$-\omega \subset \omega^{\#}, \quad \textit{i.e.} \quad y_1 \cdot \omega y_2 = -y_2 \cdot \omega y_1, \quad y_1, y_2 \in \mathcal{Y}.$$

The space of all anti-symmetric elements of $L(\mathcal{Y}, \mathcal{Y}^{\#})$ will be denoted by $L_{\mathrm{a}}(\mathcal{Y}, \mathcal{Y}^{\#})$.

Let $\omega \in L_{\mathrm{a}}(\mathcal{Y}, \mathcal{Y}^{\#})$.

Definition 1.30 *A subspace $\mathcal{X} \subset \mathcal{Y}$ is called* isotropic *if*

$$y_1 \cdot \omega y_2 = 0, \quad y_1, y_2 \in \mathcal{X}.$$

A maximal isotropic subspace is called Lagrangian.

Definition 1.31 *A non-degenerate anti-symmetric bilinear form is called* symplectic.

If ω is symplectic, then $(\omega\mathcal{Y}, \mathcal{Y})$ is a dual pair and we can treat $\mathcal{Y}$ as a subspace of $(\omega\mathcal{Y})^{\#}$. We can also define a symplectic form $\omega^{-1} \in L_a(\omega\mathcal{Y}, \mathcal{Y}) \subset L_a(\omega\mathcal{Y}, (\omega\mathcal{Y})^{\#})$.

Proposition 1.32 *Let $\mathcal{Y}$ be finite-dimensional.*

(1) *ω is anti-symmetric iff $\omega^{\#} = -\omega$.*
(2) *An isotropic subspace $\mathcal{X}$ is Lagrangian iff $\dim \mathcal{X} = \frac{1}{2}\dim \mathcal{Y}$.*
(3) *If ω is symplectic, then $\omega\mathcal{Y} = \mathcal{Y}^{\#}$, so that $\omega^{-1} \in L_a(\mathcal{Y}^{\#}, \mathcal{Y})$ is a symplectic form.*

1.1.13 Symplectic spaces

Definition 1.33 *The pair $(\mathcal{Y}, \omega)$, where ω is a symplectic form on $\mathcal{Y}$, is called a symplectic space.*

Let $(\mathcal{Y}, \omega)$ be a symplectic space.

Definition 1.34 *The symplectic complement of $\mathcal{X} \subset \mathcal{Y}$ is defined as*

$$\mathcal{X}^{\omega\perp} := \{y \in \mathcal{Y} \ : \ y \cdot \omega x = 0, \ x \in \mathcal{X}\}.$$

Let $\mathcal{X}$ be a subspace of $\mathcal{Y}$. Note that $\mathcal{X}$ is isotropic iff $\mathcal{X}^{\omega\perp} \supset \mathcal{X}$ and it is Lagrangian iff $\mathcal{X}^{\omega\perp} = \mathcal{X}$.

Definition 1.35 *We say that $\mathcal{X}$ is co-isotropic if $\mathcal{X}^{\omega\perp} \subset \mathcal{X}$.*

If $\mathcal{X}$ is co-isotropic, then $\mathcal{X}/\mathcal{X}^{\omega\perp}$ is naturally a symplectic space.

Note that $\mathcal{X}$ is isotropic in $\mathcal{Y}$ iff $\mathcal{X}^{\mathrm{an}}$ is co-isotropic in $\mathcal{Y}^{\#}$.

1.1.14 Group $Sp(\mathcal{Y})$ and Lie algebra $sp(\mathcal{Y})$

Let $(\mathcal{Y}, \omega)$ be a symplectic space and $a \in L(\mathcal{Y})$.

Definition 1.36 *We say that*

$$a \text{ is symplectic} \quad if \quad a \text{ is bijective and } a^{\#}\omega a = \omega,$$
$$a \text{ is anti-symplectic} \quad if \quad a \text{ is bijective and } a^{\#}\omega a = -\omega,$$
$$a \text{ is infinitesimally symplectic} \quad if \quad a^{\#}\omega = -\omega a.$$

The set of symplectic elements in $L(\mathcal{Y})$ is a group for the operator composition denoted by $Sp(\mathcal{Y})$. The set of infinitesimally symplectic elements, denoted by $sp(\mathcal{Y})$, is a Lie algebra equipped with the commutator.

Proposition 1.37 *Assume that $\mathcal{Y}$ is finite-dimensional and $r \in L(\mathcal{Y})$. Then*

(1) *$r \in Sp(\mathcal{Y})$ iff $r^{\#}\omega r = \omega$.*
(2) *$r \in Sp(\mathcal{Y}, \omega)$ iff $r^{\#} \in Sp(\mathcal{Y}^{\#}, \omega^{-1})$.*
(3) *$r \in Sp(\mathcal{Y})$ implies $r^{-1} = \omega^{-1}r^{\#}\omega$.*

1.1.15 Involutions and super-spaces

Definition 1.38 $\epsilon \in L(\mathcal{Y})$ *is called an* involution *if* $\epsilon^2 = \mathbb{1}$.

Definition 1.39 *If* $\epsilon \in L(\mathcal{Y})$ *is an involution, we set* $\mathcal{Y}^{\pm\epsilon} := \mathrm{Ker}(\mathbb{1} \mp \epsilon)$.

Every involution determines a decomposition $\mathcal{Y} = \mathcal{Y}^\epsilon \oplus \mathcal{Y}^{-\epsilon}$, the operators $\frac{1}{2}(\mathbb{1} \pm \epsilon)$ being the projections onto $\mathcal{Y}^{\pm\epsilon}$ along $\mathcal{Y}^{\mp\epsilon}$.

Conversely, a decomposition $\mathcal{Y} = \mathcal{Y}_0 \oplus \mathcal{Y}_1$ determines an involution given by the matrix $\epsilon = \begin{bmatrix} \mathbb{1} & 0 \\ 0 & -\mathbb{1} \end{bmatrix}$.

Operators $a \in L(\mathcal{Y})$ commuting with ϵ are of the form $a = \begin{bmatrix} a_{00} & 0 \\ 0 & a_{11} \end{bmatrix}$.

Definition 1.40 *We say that* $(\mathcal{Y}, \epsilon)$ *is a* $\mathbb{Z}_2$-graded space *or a* super-space *if* ϵ *is an involution on* $\mathcal{Y}$. ϵ *is often called the* $\mathbb{Z}_2$-grading.

Definition 1.41 *In the context of super-spaces one often writes* $\mathcal{Y}_0$ *for* $\mathcal{Y}^\epsilon$ *and its elements are called* even. *One writes* $\mathcal{Y}_1$ *for* $\mathcal{Y}^{-\epsilon}$ *and its elements are called* odd. *Elements of* $\mathcal{Y}_0 \cup \mathcal{Y}_1$ *will be called* homogeneous *or* pure. *The operator* $p = 0 \oplus \mathbb{1}$ *is called the* parity, *so that* $\epsilon = (-\mathbb{1})^p$. *Sometimes, the parity of a homogeneous element* $y \in \mathcal{Y}$ *is denoted* $|y|$.

Remark 1.42 *The name "super-space" came into use under the influence of super-symmetric quantum field theory. The prefix "super" is often attached to mean "$\mathbb{Z}_2$-graded" in various contexts; see e.g. Subsects. 3.3.9 and 6.1.4.*

If $\mathcal{Y}$ has an additional structure, we will often assume that it is preserved by ϵ. For instance, we have the following terminology (see Subsect. 1.3.8):

Definition 1.43 $(\mathcal{Y}, \epsilon)$ *is a* super-Hilbert space *if* $\mathcal{Y}$ *is a Hilbert space and* ϵ *is a unitary involution; it is a* super-Kähler space *if* $\mathcal{Y}$ *is a Kähler space and* ϵ *is a symplectic and orthogonal (and hence complex linear) involution.*

Let $(\mathcal{Y}, \epsilon)$, $(\mathcal{W}, \varepsilon)$ be two super-spaces. The space of linear transformations from $\mathcal{Y}$ to $\mathcal{W}$, that is, $L(\mathcal{Y}, \mathcal{W})$, is itself naturally a super-space, with the grading given by

$$L(\mathcal{Y}, \mathcal{W}) \ni r \mapsto \varepsilon r \epsilon \in L(\mathcal{Y}, \mathcal{W}).$$

Written in the matrix notation, the decomposition of an element of $L(\mathcal{Y}, \mathcal{W})$ into its even and odd parts is

$$\begin{bmatrix} a_{00} & a_{01} \\ a_{10} & a_{11} \end{bmatrix} = \begin{bmatrix} a_{00} & 0 \\ 0 & a_{11} \end{bmatrix} + \begin{bmatrix} 0 & a_{01} \\ a_{10} & 0 \end{bmatrix}.$$

We can form other super-spaces in an obvious way, for example, $(\mathcal{Y} \oplus \mathcal{W}, \epsilon \oplus \varepsilon)$, $(\mathcal{Y} \otimes \mathcal{W}, \epsilon \otimes \varepsilon)$.

1.1.16 Conjugations on a symplectic space

Let $(\mathcal{Y}, \omega)$ be a symplectic space.

Definition 1.44 *A map $\tau \in L(\mathcal{Y})$ is called a* conjugation *if it is an anti-symplectic involution.*

Let $(\mathcal{V}, \mathcal{X})$ be a dual pair of vector spaces. Define $\omega \in L(\mathcal{V} \oplus \mathcal{X}, \mathcal{V}^{\#} \oplus \mathcal{X}^{\#})$ and $\tau \in L(\mathcal{V} \oplus \mathcal{X})$ by

$$\omega = \begin{bmatrix} 0 & \mathbb{1} \\ -\mathbb{1} & 0 \end{bmatrix}, \quad \tau = \begin{bmatrix} \mathbb{1} & 0 \\ 0 & -\mathbb{1} \end{bmatrix}. \tag{1.9}$$

In other words, for $(\eta_1, q_1), (\eta_2, q_2) \in \mathcal{V} \oplus \mathcal{X}$ we have

$$(\eta_1, q_1) \cdot \omega(\eta_2, q_2) = \eta_1 \cdot q_2 - \eta_2 \cdot q_1, \quad \tau(\eta_1, q_1) = (\eta_1, -q_1). \tag{1.10}$$

Then ω is a symplectic form on $\mathcal{V} \oplus \mathcal{X}$ and τ is a conjugation.

We can also define ω^{-1} and $\tau^{\#}$ on $\mathcal{V}^{\#} \oplus \mathcal{X}^{\#}$. We obtain a symplectic form and a conjugation:

$$\omega^{-1} = \begin{bmatrix} 0 & -\mathbb{1} \\ \mathbb{1} & 0 \end{bmatrix}, \quad \tau^{\#} = \begin{bmatrix} \mathbb{1} & 0 \\ 0 & -\mathbb{1} \end{bmatrix}, \tag{1.11}$$

or equivalently

$$(x_1, \xi_1) \cdot \omega^{-1}(x_2, \xi_2) = \xi_1 \cdot x_2 - \xi_2 \cdot x_1, \quad \tau^{\#}(x_1, \xi_1) = (x_1, -\xi_1). \tag{1.12}$$

We will see below that the above construction describes a general form of a symplectic space equipped with a conjugation.

Proposition 1.45 *Let τ be a conjugation on a symplectic space $\mathcal{Y}$. Then the spaces $\mathcal{Y}^{\pm\tau}$ are Lagrangian.*

Proof The spaces $\mathcal{Y}^{\pm\tau}$ are clearly isotropic. Since $\mathcal{Y} \simeq \mathcal{Y}^{\tau} \oplus \mathcal{Y}^{-\tau}$ we have $\mathcal{Y}^{\#} \simeq (\mathcal{Y}^{\tau})^{\#} \oplus (\mathcal{Y}^{-\tau})^{\#}$, and we can write ω as the matrix

$$\begin{bmatrix} 0 & a \\ -b & 0 \end{bmatrix},$$

where $a : \mathcal{Y}^{-\tau} \to (\mathcal{Y}^{\tau})^{\#}$ and $b : \mathcal{Y}^{\tau} \to (\mathcal{Y}^{-\tau})^{\#}$ are injective and

$$a^{\#}\big|_{\mathcal{Y}^{\tau}} = b, \quad b^{\#}\big|_{\mathcal{Y}^{-\tau}} = a.$$

If $\mathcal{Y}^{\tau} \subsetneq \mathcal{X}$, where $\mathcal{X}$ is isotropic, then there exists $e \notin \mathcal{Y}^{\tau}$ such that $y \cdot \omega e = 0$ for all $y \in \mathcal{Y}^{\tau}$. Then $(\mathbb{1} - \tau)e \neq 0$ and $y \cdot \omega(\mathbb{1} - \tau)e = y \cdot a(\mathbb{1} - \tau)e = 0$ for all $y \in \mathcal{Y}^{\tau}$, which contradicts the fact that a is injective. Hence $\mathcal{Y}^{\pm\tau}$ are Lagrangian. $\square$

Proposition 1.46 *Let $\mathcal{Y}$ be a symplectic space $\mathcal{Y}$ with a conjugation τ. We use the notation of the proof of Prop. 1.45. Set*

$$\mathcal{X} := \mathcal{Y}^{-\tau}, \quad \mathcal{V} := b\mathcal{Y}^{\tau}.$$

Then $(\mathcal{V}, \mathcal{X})$ *is a dual pair and* $b \oplus \mathbb{1}$ *sends bijectively* $\mathcal{Y} = \mathcal{Y}^\tau \oplus \mathcal{Y}^{-\tau}$ *onto* $\mathcal{V} \oplus \mathcal{X}$. *With this identification,* ω *and* τ *are given by (1.10).*

If in addition the dimension of $\mathcal{Y}$ *is finite, then* $\mathcal{V} = \mathcal{X}^\#$ *and we obtain a bijection of* $\mathcal{Y}$ *onto* $\mathcal{X}^\# \oplus \mathcal{X}$ *and of* $\mathcal{Y}^\#$ *onto* $\mathcal{X} \oplus \mathcal{X}^\#$.

Proof Clearly, $\mathcal{V} \subset \mathcal{X}^\#$. We need to show that $\mathcal{V}_{\mathrm{an}} = \{0\}$. Let $x \in \mathcal{V}_{\mathrm{an}}$. For any $y \in \mathcal{Y}^\tau$, we have

$$0 = \langle by|x \rangle = \langle y|b^\# x \rangle = \langle y|ax \rangle,$$

since $b^\#\big|_{\mathcal{Y}^{-\tau}} = a$. This implies that $ax = 0$, and hence $x = 0$, since a is injective. Therefore, $(\mathcal{V}, \mathcal{X})$ is a dual pair. $\square$

Theorem 1.47 *Let* $\mathcal{Y}$ *be a finite-dimensional symplectic space. There exists a conjugation in* $L(\mathcal{Y})$. *Consequently, there exists a vector space* $\mathcal{X}$ *such that* $\mathcal{Y}$ *is isomorphic to* $\mathcal{X}^\# \oplus \mathcal{X}$.

Proof Let f_1 be an arbitrary non-zero vector in $\mathcal{Y}$. Since ω is non-degenerate, we can find a vector e_1 such that $f_1 \cdot \omega e_1 = 1$. f_1 is not proportional to e_1, because ω is anti-symmetric. Let $\mathcal{Y}_1 = \{y \in \mathcal{Y} \ : \ y \cdot \omega f_1 = y \cdot \omega e_1 = 0\}$. Then $\dim \mathcal{Y}_1 = \dim \mathcal{Y} - 2$. We continue our construction in $\mathcal{Y}_1$, finding vectors f_2, e_2 etc.

In the end we set $\tau = \mathbb{1}$ on $\mathrm{Span}\{f_1, \dots, f_d\}$ and $\tau = -\mathbb{1}$ on $\mathrm{Span}\{e_1, \dots, e_d\}$. $\square$

1.2 Complex vector spaces

Throughout the section, $\mathcal{Z}, \mathcal{W}$ are complex vector spaces.

1.2.1 Anti-linear operators

Definition 1.48 *Let* a *be a map from* $\mathcal{Z}$ *to* $\mathcal{W}$. *We say that it is* anti-linear *if it is linear over* $\mathbb{R}$ *and* $ia = -ai$.

Definition 1.49 *Let* a *be anti-linear from* $\mathcal{Z}$ *to* $\mathcal{W}$. *The* transpose *of* a *is the operator in* $L(\mathcal{W}^\#, \mathcal{Z}^\#)$ *defined by*

$$\langle a^\# v|y \rangle := \overline{\langle v|ay \rangle}, \quad v \in \mathcal{Y}_2^\#, \quad y \in \mathcal{Y}_1. \tag{1.13}$$

Note that the transpose of an anti-linear operator is also anti-linear.

1.2.2 Internal conjugations

Definition 1.50 *An anti-linear map* χ *on* $\mathcal{Z}$ *such that* $\chi^2 = \mathbb{1}$ *is called an* (internal) conjugation. *The subspace* $\mathcal{Z}^\chi := \{z \in \mathcal{Z} \ : \ \chi z = z\}$ *is sometimes called a* real form *of* $\mathcal{Z}$. *According to an alternative terminology,* $\mathcal{Z}^\chi$ *is called the* real subspace *and* $\mathcal{Z}^{-\chi} := \{z \in \mathcal{Z} \ : \ \chi z = -z\}$ *the* imaginary subspace (for χ).

Definition 1.51 *Operators $a \in L(\mathcal{Z}, \mathcal{W})$ satisfying $a = \chi a \chi$ will be sometimes called* real *(for χ).*

Clearly, the space of real operators can be identified with $L(\mathcal{Z}^\chi, \mathcal{W}^\chi)$.

Sometimes, an internal conjugation will be denoted by $\overline{z}$ instead of χz. In such a case, if $a \in L(\mathcal{Z})$, we will write $\overline{a}$ for $\chi a \chi$.

1.2.3 Complex conjugate spaces

In this subsection we discuss the external approach to the complex conjugation. This is a very simple and elementary subject, which, however, can be a little confusing.

Definition 1.52 $\overline{\mathcal{Z}}$ *will denote a complex space equipped with an anti-linear isomorphism*

$$\mathcal{Z} \ni z \mapsto \overline{z} \in \overline{\mathcal{Z}}. \tag{1.14}$$

We will call $\overline{\mathcal{Z}}$ the space complex conjugate *to $\mathcal{Z}$. We will use the convention that the inverse of (1.14) is denoted by the same symbol, so that $\overline{\overline{z}} = z$, $z \in \mathcal{Z}$ and $\overline{\overline{\mathcal{Z}}} = \mathcal{Z}$.*

In practice, one often uses one of the following two concrete realizations of the complex conjugate space.

The first approach is the most canonical (it does not introduce additional structure). We set $\overline{\mathcal{Z}}$ to be equal to $\mathcal{Z}$ as a real vector space. The map $\mathcal{Z} \ni z \mapsto \overline{z} \in \overline{\mathcal{Z}}$ is just the identity. One defines the multiplication by $\lambda \in \mathbb{C}$ on $\overline{\mathcal{Z}}$ as

$$\overline{\lambda z} := \overline{\lambda} z, \quad z \in \mathcal{Z}, \, \lambda \in \mathbb{C}.$$

In the second approach, we choose $\overline{\mathcal{Z}} = \mathcal{Z}$ as complex vector spaces and we fix an internal conjugation χ. Then we set $\overline{z} := \chi z$. Thus we are back in the framework of Subsect. 1.2.2.

Definition 1.53 *If $a \in L(\mathcal{Z}, \mathcal{W})$, then one defines $\overline{a} \in L(\overline{\mathcal{Z}}, \overline{\mathcal{W}})$ by*

$$\overline{a}\,\overline{z} := \overline{az}. \tag{1.15}$$

The map $L(\mathcal{Z}, \mathcal{W}) \ni a \mapsto \overline{a} \in L(\overline{\mathcal{Z}}, \overline{\mathcal{W}})$ is an anti-linear isomorphism which allows us to identify $L(\overline{\mathcal{Z}}, \overline{\mathcal{W}})$ and $\overline{L(\mathcal{Z}, \mathcal{W})}$ as complex vector spaces.

Sometimes the notation $z \mapsto \overline{z}$ is inconvenient for typographical reasons, and we will denote the complex conjugation by a letter, e.g. χ. Thus $\chi : \mathcal{Z} \to \overline{\mathcal{Z}}$ is a fixed anti-linear map and we write χz for $\overline{z}$.

In particular, if $a \in L(\mathcal{Z}_1, \mathcal{Z}_2)$, and the conjugations $\mathcal{Z}_i \to \overline{\mathcal{Z}}_i$ are denoted by χ_i, then $\overline{a} = \chi_2 a \chi_1^{-1}$.

A typical situation when this alternative notation is more convenient is the following. Suppose that b is an anti-linear map from $\mathcal{Z}_1$ to $\mathcal{Z}_2$. Then, instead of b, it may be more convenient to use one of the following two *linear* maps:

$$b\chi_1^{-1} \in L(\overline{\mathcal{Z}}_1, \mathcal{Z}_2), \quad \text{or} \quad \chi_2 b \in L(\mathcal{Z}_1, \overline{\mathcal{Z}}_2). \tag{1.16}$$

Note that b is a conjugation on $\mathcal{Z}$ iff the linear map $a := b\chi^{-1} \in L(\overline{\mathcal{Z}}, \mathcal{Z})$ satisfies

$$\overline{a}a = \mathbb{1}.$$

1.2.4 Anti-linear functionals

If $\overline{w} \in \overline{\mathcal{Z}^\#}$, we let it act on $\overline{\mathcal{Z}}$ as

$$\langle \overline{w} | \overline{z} \rangle := \overline{\langle w | z \rangle}, \ z \in \mathcal{Z}.$$

This identifies $\overline{\mathcal{Z}^\#}$ with $\overline{\mathcal{Z}}^\#$. (This is a special case of (1.15) for $\mathcal{W} = \mathbb{C}$.)

Definition 1.54 *The* anti-dual of $\mathcal{Z}$ *is defined as*

$$\mathcal{Z}^* := \overline{\mathcal{Z}}^\#.$$

Thus $\mathcal{Z}^*$ is the space $\overline{L(\mathcal{Z}, \mathbb{C})}$ of anti-linear functionals on $\mathcal{Z}$. Several kinds of notation for the action of $w \in \mathcal{Z}^*$ on $z \in \mathcal{Z}$ will be used:

(1) *the* bra–ket notation $(z|w) = \langle \overline{z} | w \rangle = \langle w | \overline{z} \rangle$,
(2) *the* simplified notation $\overline{z} \cdot w = w \cdot \overline{z}$,
(3) *the* functional notation $w(z)$.

Since $\overline{\mathcal{Z}^\#} = \overline{\mathcal{Z}}^\#$, we see that $\mathcal{Z}^{**} = \mathcal{Z}^{\#\#}$, so that $\mathcal{Z} \subset \mathcal{Z}^{**}$ and in the finite-dimensional case $\mathcal{Z} = \mathcal{Z}^{**}$.

Remark 1.55 *We will consistently use the following convention. The round brackets in a pairing of two vectors will indicate that the expression depends anti-linearly on the first argument and linearly on the second argument. In the case of the angular brackets the dependence on both arguments will always be linear, in both the real and the complex case.*

1.2.5 Adjoint of an operator

Let $a \in L(\mathcal{Z}_1, \mathcal{Z}_2)$.

Definition 1.56 *We define the* adjoint of a, *denoted by $a^* \in L(\mathcal{Z}_2^*, \mathcal{Z}_1^*)$, by*

$$(a^* w_2 | z_1) := (w_2 | a z_1), \ \ w_2 \in \mathcal{Z}_2^*, \ \ z_1 \in \mathcal{Z}_1. \tag{1.17}$$

We see that

$$a^* = \overline{a}^\# = \overline{a^\#}, \ \ a \subset a^{**}. \tag{1.18}$$

Definition 1.57 *Let a be an anti-linear map from $\mathcal{Z}_1$ to $\mathcal{Z}_2$. The adjoint of a, instead of by (1.17), is defined by*

$$(z_1 | a^* w_2) = (w_2 | a z_1),$$

$$\text{or, equivalently,} \ \ \overline{(a^* w_2 | z_1)} = (w_2 | a z_1), \ \ w_2 \in \mathcal{Z}_2^*, \ \ z_1 \in \mathcal{Z}_1. \tag{1.19}$$

It is an anti-linear operator from $\mathcal{Z}_2^*$ to $\mathcal{Z}_1^*$ satisfying (1.18).

1.2.6 Anti-dual pairs

Definition 1.58 *An* anti-dual pair *is a pair* $(\mathcal{W}, \mathcal{Z})$ *of complex vector spaces equipped with a form*

$$(\mathcal{W}, \mathcal{Z}) \ni (w, z) \mapsto (w|z) \in \mathbb{C}$$

anti-linear in $\mathcal{W}$ *and linear in* $\mathcal{Z}$ *such that*

$$(w|z) = 0, \ w \in \mathcal{V} \ \Rightarrow \ z = 0,$$
$$(w|z) = 0, \ z \in \mathcal{Z} \ \Rightarrow \ w = 0.$$

Properties of anti-dual pairs are obvious analogs of the properties of dual pairs. For instance, if $\mathcal{Z}$ is finite-dimensional and $(\mathcal{W}, \mathcal{Z})$ is a dual pair, then $\mathcal{W}$ is naturally isomorphic to $\mathcal{Z}^*$.

1.2.7 Sesquilinear forms

Definition 1.59 *Elements of* $L(\mathcal{Z}, \mathcal{Z}^*)$ *will be called* sesquilinear forms.

Let $\beta \in L(\mathcal{Z}, \mathcal{Z}^*)$. β determines a map

$$\mathcal{Z} \times \mathcal{Z} \ni (z_1, z_2) \mapsto (z_1|\beta z_2) = \overline{z_1} \cdot \beta z_2 \in \mathbb{C} \tag{1.20}$$

anti-linear in the first argument and linear in the second argument.

Definition 1.60 *We say that* β *is* non-degenerate *if* $\mathrm{Ker}\,\beta = \{0\}$.

Definition 1.61 *An operator* $r \in L(\mathcal{Z})$ preserves β *if*

$$r^*\beta r = \beta, \text{ i.e. } \quad (rz_1|\beta r z_2) = (z_1|\beta z_2), \ z_1, z_2 \in \mathcal{Z}.$$

An operator $a \in L(\mathcal{Z})$ infinitesimally preserves β *if*

$$a^*\beta + \beta a = 0, \text{ i.e. } \quad (az_1|\beta z_2) = -(z_1|\beta a z_2), \ z_1, z_2 \in \mathcal{Z}.$$

Remark 1.62 *Note that we adopt the so-called* physicist's convention *for sesquilinear forms. A part of the mathematical community adopts the reverse convention: they assume sesquilinear forms to be linear in the first and anti-linear in the second argument.*

Remark 1.63 *We will use three kinds of notation for sesquilinear forms:*

(1) *the* bra–ket notation $(z_1|\beta z_2)$, *going back to Dirac,*
(2) *the* simplified notation $\overline{z}_1 \cdot \beta z_2$,
(3) *the* functional notation $\beta(\overline{z}_1, z_2)$.

Note that in all cases the notation indicates that the form is sesquilinear and not bilinear: by the use of round instead of angular brackets in the first case, and by the use of the bar in the remaining cases. Usually, we will prefer the first two notations, both given in (1.20).

1.2.8 Hermitian forms

Let $\beta \in L(\mathcal{Z}, \mathcal{Z}^*)$.

Definition 1.64 *We will say that*

$$\beta \text{ is Hermitian } \text{ if } \beta \subset \beta^*, \quad \text{i.e. } (z_2|\beta z_1) = \overline{(z_1|\beta z_2)}, \; z_1, z_2 \in \mathcal{Z},$$
$$\text{or equivalently } (z|\beta z) \in \mathbb{R}, \; z \in \mathcal{Z};$$

$$\beta \text{ is anti-Hermitian } \text{ if } \beta \subset -\beta^*, \quad \text{i.e. } (z_2|\beta z_1) = -\overline{(z_1|\beta z_2)}, \; z_1, z_2 \in \mathcal{Z},$$
$$\text{or equivalently } (z|\beta z) \in \mathrm{i}\mathbb{R}, \; z \in \mathcal{Z}.$$

Clearly, β is Hermitian iff $\mathrm{i}\beta$ is anti-Hermitian.

Definition 1.65 *The space of all Hermitian elements of $L(\mathcal{Z}, \mathcal{Z}^*)$ will be denoted $L_{\mathrm{h}}(\mathcal{Z}, \mathcal{Z}^*)$. Such operators are also called* Hermitian forms.

If $\mathcal{Z}$ is finite-dimensional then $\beta \in L_{\mathrm{h}}(\mathcal{Z}, \mathcal{Z}^*)$ iff $\beta^* = \beta$.

Definition 1.66 *A Hermitian form β is called* positive semi-definite *if $(z|\beta z) \geq 0$ for $z \in \mathcal{Z}$. It is called* positive definite *if $(z|\beta z) > 0$ for $z \neq 0$. A positive definite form is also often called a* scalar product.

Positive definite forms are always non-degenerate.

If $\beta \in L_{\mathrm{h}}(\mathcal{Z}, \mathcal{Z}^*)$ is non-degenerate, then $(\beta\mathcal{Z}, \mathcal{Z})$ is an anti-dual pair. Hence, we can define $\beta^{-1} \in L_{\mathrm{h}}(\beta\mathcal{Z}, \mathcal{Z}) \subset L_{\mathrm{h}}(\beta\mathcal{Z}, (\beta\mathcal{Z})^*)$. (Note that $\mathcal{Z} \subset (\beta\mathcal{Z})^*$.) The form β^{-1} is non-degenerate and is positive definite iff β is positive definite.

1.2.9 (Pseudo-)unitary spaces

Definition 1.67 *A couple $(\mathcal{Z}, \beta)$, where $\beta \in L_{\mathrm{h}}(\mathcal{Z}, \mathcal{Z}^*)$ is non-degenerate, is called a* pseudo-unitary space. *If β is positive definite, then $(\mathcal{Z}, \beta)$ is called a* unitary space. *In such a case we can define the norm of $z \in \mathcal{Z}$ denoted by $\|z\| := \sqrt{(y|\beta y)}$. If $\mathcal{Z}$ is complete for this norm, it is called a* Hilbert space.

Note that the notion of a pseudo-unitary space is closely related to that of a *charged symplectic space*, which is defined later, in Subsect. 1.2.11.

Let $(\mathcal{Z}, \beta)$ be a pseudo-unitary space.

Definition 1.68 *If $\mathcal{U} \subset \mathcal{Z}$, then $\mathcal{U}^{\beta\perp}$ denotes the β-orthogonal complement of $\mathcal{U}$:*

$$\mathcal{U}^{\beta\perp} := \{z \in \mathcal{Z} \; : \; (u|\beta z) = 0, \; u \in \mathcal{U}\}.$$

Definition 1.69 *Let $(\mathcal{Z}, \beta)$ be a unitary, pseudo-unitary, resp. charged symplectic space. Then $\overline{\mathcal{Z}}$ has a natural unitary, pseudo-unitary, resp. charged symplectic structure:*

$$(\overline{z}_1|\overline{\beta}\overline{z}_2) := \overline{(z_1|\beta z_2)}.$$

Definition 1.70 *A non-degenerate Hermitian form, especially if it is positive definite, is often called a* scalar product *and denoted* $(z_1|z_2)$ *or* $\overline{z}_1 \cdot z_2$*. In such a case, the orthogonal complement of* $\mathcal{U}$ *is denoted* $\mathcal{U}^\perp$*. For* $w \in \mathcal{Z}$*,* $(w|$ *will denote the following operator:*

$$\mathcal{Z} \ni z \mapsto (w|z := (w|z) \in \mathbb{C}.$$

For example, if $(w|w) = 1$, then $|w)(w|$ is the orthogonal projection onto w.

Most unitary spaces considered in our work will be (complex) Hilbert spaces. Hilbert spaces will be further discussed in Subsect. 2.2.2.

1.2.10 Group $U(\mathcal{Z})$ and Lie algebra $u(\mathcal{Z})$

Let $(\mathcal{Z}, \beta)$ be an unitary space and $a \in L(\mathcal{Z})$.

Definition 1.71 *We say that*

$$\begin{aligned}
&a \text{ is isometric} &&\text{if } a^*\beta a = \beta, \\
&a \text{ is unitary} &&\text{if } a \text{ is isometric and bijective,} \\
&a \text{ is self-adjoint} &&\text{if } a^*\beta = \beta a, \\
&a \text{ is anti-self-adjoint} &&\text{if } a^*\beta = -\beta a.
\end{aligned}$$

The set of unitary operators on $\mathcal{Z}$ *is a group for the operator composition denoted by* $U(\mathcal{Z})$*. The space of anti-self-adjoint operators on* $\mathcal{Z}$*, denoted by* $u(\mathcal{Z})$*, is a Lie algebra equipped with the usual commutator.*

Let b be an anti-linear operator on $\mathcal{Z}$.

Definition 1.72 *We say that*

$$\begin{aligned}
&b \text{ is anti-unitary} &&\text{if } b^*\beta b = \beta \text{ and } a \text{ is bijective,} \\
&b \text{ is a conjugation} &&\text{if it is an anti-unitary involution.}
\end{aligned}$$

Recall from Subsect. 1.2.3 that we sometimes use two alternative symbols for the complex conjugation: χ and the "bar".

Clearly, b is anti-unitary iff $\chi b : \mathcal{Z} \to \overline{\mathcal{Z}}$ is unitary.

If $\mathcal{Z}$ is a pseudo-unitary space, we can repeat Subsect. 1.2.10, replacing the terms isometric, unitary, anti-self-adjoint and self-adjoint with *pseudo-isometric, pseudo-unitary, anti-pseudo-self-adjoint* and *pseudo-self-adjoint*.

1.2.11 Charged symplectic spaces

Definition 1.73 *If* ω *is anti-Hermitian and non-degenerate, then* $(\mathcal{Z}, \omega)$ *is called a* charged symplectic space.

Note that the difference between a pseudo-unitary and charged symplectic space is minor (passing from β to $\omega = i\beta$ changes a pseudo-unitary space into a charged symplectic space). We will, however, more often use the framework of a charged symplectic space. The terminology in this case is somewhat different.

Let $(\mathcal{Z}, \omega)$ be a charged symplectic space and $a \in L(\mathcal{Z})$.

Definition 1.74 *We say that*

$$\begin{array}{ll} a \text{ preserves } \omega & \textit{if } a^*\omega a = \omega, \\ a \text{ anti-preserves } \omega & \textit{if } a^*\omega a = -\omega, \\ a \text{ is charged symplectic} & \textit{if } a \text{ preserves } \omega \text{ and is bijective}, \\ a \text{ is charged anti-symplectic} & \textit{if } a \text{ anti-preserves } \omega \text{ and is bijective}, \\ a \text{ is infinitesimally charged symplectic} & \textit{if } a^*\omega = -\omega a. \end{array}$$

The set of charged symplectic operators on $\mathcal{Z}$ is a group for the operator composition denoted by $ChSp(\mathcal{Z})$. The space of infinitesimally charged symplectic operators on $\mathcal{Z}$, denoted by $chsp(\mathcal{Z})$, is a Lie algebra equipped with the usual commutator.

Let a be an anti-linear operator on $\mathcal{Z}$.

Definition 1.75 *We say that*

$$\begin{array}{ll} a \text{ preserves } \omega & \textit{if } a^*\omega a = \omega, \text{ or } (z_1|\omega z_2) = \overline{(az_1|\omega az_2)}, \\ a \text{ anti-preserves } \omega & \textit{if } a^*\omega a = -\omega, \text{ or } (z_1|\omega z_2) = -\overline{(az_1|\omega az_2)}, \\ a \text{ is anti-charged symplectic} & \textit{if } a \text{ preserves } \omega \text{ and is bijective}, \\ a \text{ is anti-charged anti-symplectic} & \textit{if } a \text{ anti-preserves } \omega \text{ and is bijective}. \end{array}$$

Remark 1.76 *The terminology "charged symplectic space" is motivated by applications in quantum field theory: such spaces describe charged bosons.*

1.3 Complex structures

When we quantize a classical system, the phase space is often naturally equipped with more than one complex structure. Therefore, it is useful to develop this concept in more detail.

Besides complex structures, in this section we discuss the so-called (pseudo-) Kähler spaces, which can be described as (pseudo-)unitary spaces treated as real spaces.

1.3.1 Anti-involutions

Let $\mathcal{Y}$ be a vector space.

Definition 1.77 *We say that* $\mathrm{j} \in L(\mathcal{Y})$ *is an* anti-involution *if* $\mathrm{j}^2 = -\mathbb{1}$.

If $\mathcal{Y}$ is a real vector space with an anti-involution j, then $\mathcal{Y}$ can be naturally endowed with the structure of a complex space:

$$(\lambda + \mathrm{i}\mu)y := \lambda y + \mu \mathrm{j}y, \quad y \in \mathcal{Y}, \quad \lambda, \mu \in \mathbb{R}. \tag{1.21}$$

Therefore, anti-involutions on real spaces are often called *complex structures*.

Definition 1.78 $\mathcal{Y}$ *converted into a vector space over* $\mathbb{C}$ *with the multiplication* *(1.21) will be denoted* $\mathcal{Y}^{\mathbb{C}}$, *or by* $(\mathcal{Y}^{\mathbb{C}}, \mathrm{j})$ *if we need to specify the complex structure that we use. It will be called a* complex form *of* $\mathcal{Y}$.

Definition 1.79 *Conversely, any complex space* $\mathcal{W}$ *can be considered as a real vector space, called the* realification *of* $\mathcal{W}$ *and denoted* $\mathcal{W}_{\mathbb{R}}$. *It is equipped with an anti-involution* $\mathrm{j} \in L(\mathcal{W}_{\mathbb{R}})$ *(the multiplication by the complex number* i).

Let $\mathcal{Y}_1$, $\mathcal{Y}_2$ be real spaces with anti-involutions j_1, j_2. Then

$$L(\mathcal{Y}_1^{\mathbb{C}}, \mathcal{Y}_2^{\mathbb{C}}) = \{a \in L(\mathcal{Y}_1, \mathcal{Y}_2) \ : \ a\mathrm{j}_1 = \mathrm{j}_2 a\}.$$

1.3.2 Conjugations on a space with an anti-involution

Let $\mathcal{Y}$ be a vector space equipped with an anti-involution $\mathrm{j} \in L(\mathcal{Y})$.

Definition 1.80 *We say that* $\chi \in L(\mathcal{Y})$ *is a* conjugation *if it is an involution and* $\mathrm{j}\chi = -\chi\mathrm{j}$.

Recall that χ determines a decomposition $\mathcal{Y} = \mathcal{Y}^{\chi} \oplus \mathcal{Y}^{-\chi}$ (see Def. 1.39). Let us write $\mathcal{X} := \mathcal{Y}^{-\chi}$. Then $\mathrm{j}\mathcal{X} = \mathcal{Y}^{\chi}$. The map

$$\mathcal{Y} \ni y \mapsto \left(\mathrm{j}\frac{\mathbb{1} + \chi}{2}y, \frac{\mathbb{1} - \chi}{2}y\right) \in \mathcal{X} \oplus \mathcal{X} \tag{1.22}$$

is bijective. Thus $\mathcal{Y}$ can be identified with $\mathcal{X} \oplus \mathcal{X}$, so that

$$\mathrm{j} = \begin{bmatrix} 0 & -\mathbb{1} \\ \mathbb{1} & 0 \end{bmatrix}, \quad \chi = \begin{bmatrix} \mathbb{1} & 0 \\ 0 & -\mathbb{1} \end{bmatrix}.$$

$r \in L(\mathcal{X} \oplus \mathcal{X})$ commutes with j iff it is of the form

$$r = \begin{bmatrix} a & -b \\ b & a \end{bmatrix}, \tag{1.23}$$

for $a, b \in L(\mathcal{X})$.

r commutes with both j and χ iff

$$r = \begin{bmatrix} a & 0 \\ 0 & a \end{bmatrix}, \tag{1.24}$$

for $a \in L(\mathcal{X})$.

1.3.3 Complexification

Let $\mathcal{X}$ be a real vector space

Definition 1.81 *The* complexification *of* $\mathcal{X}$, *denoted by* $\mathbb{C}\mathcal{X}$, *is the complex vector space* $(\mathcal{X} \oplus \mathcal{X})^{\mathbb{C}}$, *equipped with the anti-involution given by* $\begin{bmatrix} 0 & -\mathbb{1} \\ \mathbb{1} & 0 \end{bmatrix}$, *which will be denoted simply by* i. $\mathbb{C}\mathcal{X}$ *is also equipped with the conjugation* χ *given by* $\begin{bmatrix} \mathbb{1} & 0 \\ 0 & -\mathbb{1} \end{bmatrix}$. *According to the convention in Subsect. 1.2.3, we will usually write* $\overline{z} := \chi z$, $z \in \mathbb{C}\mathcal{X}$.

Note that $L(\mathbb{C}\mathcal{X})$, in the representation $\mathcal{X} \oplus \mathcal{X}$, consists of matrices of the form (1.23).

Let $a \in L(\mathcal{X})$.

Definition 1.82 *We set*

$$a_{\mathbb{C}} := \begin{bmatrix} a & 0 \\ 0 & a \end{bmatrix}, \qquad a_{\overline{\mathbb{C}}} := \begin{bmatrix} a & 0 \\ 0 & -a \end{bmatrix}. \tag{1.25}$$

$a_{\mathbb{C}}$, resp. $a_{\overline{\mathbb{C}}}$, is the unique (complex) linear, resp. anti-linear extension of a to an operator on $\mathbb{C}\mathcal{X}$. Often, we simply write a instead of $a_{\mathbb{C}}$.

1.3.4 Complexification of a Euclidean space

Let $(\mathcal{X}, \nu)$ be a Euclidean space. Then the scalar product in $\mathcal{X}$ has two natural extensions to $\mathbb{C}\mathcal{X}$: if $w_i = (x_i + \mathrm{i}y_i) \in \mathbb{C}\mathcal{X}$, $i = 1, 2$, we can define the *bilinear form*

$$w_1 \cdot \nu_{\mathbb{C}} w_2 := x_1 \cdot \nu x_2 - y_1 \cdot \nu y_2 + \mathrm{i}x_1 \cdot \nu y_2 + \mathrm{i}y_1 \cdot \nu x_2$$

and the *sesquilinear form*

$$(w_1 | w_2) = \overline{w_1} \cdot \nu_{\mathbb{C}} w_2 := x_1 \cdot \nu x_2 + y_1 \cdot \nu y_2 + \mathrm{i}x_1 \cdot \nu y_2 - \mathrm{i}y_1 \cdot \nu x_2.$$

We will more often use the latter. It makes $\mathbb{C}\mathcal{X}$ into a unitary space. The canonical conjugation χ defined in Subsect. 1.3.3 is anti-unitary. We also see that if $r \in O(\mathcal{X})$, resp. $r \in o(\mathcal{X})$, then $r_{\mathbb{C}} \in U(\mathbb{C}\mathcal{X})$, resp. $r_{\mathbb{C}} \in u(\mathbb{C}\mathcal{X})$.

Assume now that $(\mathcal{W}, (\cdot|\cdot))$ is a unitary space and that χ is a conjugation on $\mathcal{X}$ in the sense of Subsect. 1.2.8. Let $\mathcal{X} := \mathcal{W}^{\chi}$ as in Subsect. 1.3.2. Then $\mathcal{X}$ equipped with $y_1 \cdot \nu y_2 := (y_1 | y_2)$ is a Euclidean space. The identification of $\mathcal{X} \oplus \mathcal{X} \simeq \mathbb{C}\mathcal{X}$ with $\mathcal{W}$ as complex spaces defined in Subsect. 1.3.2 is unitary from $(\mathbb{C}\mathcal{X}, (\cdot|\cdot))$ to $(\mathcal{W}, (\cdot|\cdot))$.

1.3.5 Complexification of a symplectic space

Let $(\mathcal{X}, \omega)$ be a symplectic space. Then $\mathbb{C}\mathcal{X}$ can be equipped with the non-degenerate anti-symmetric form ω defined for $w_i = (x_i + \mathrm{i}y_i) \in \mathbb{C}\mathcal{X}$, $i = 1, 2$, by

$$w_1 \cdot \omega_{\mathbb{C}} w_2 := x_1 \cdot \omega x_2 - y_1 \cdot \omega y_2 + \mathrm{i}x_1 \cdot \omega y_2 + \mathrm{i}y_1 \cdot \omega x_2,$$

as well as a charged symplectic form

$$\overline{w_1} \cdot \omega_{\mathbb{C}} w_2 := x_1 \cdot \omega x_2 + y_1 \cdot \omega y_2 + \mathrm{i}x_1 \cdot \omega y_2 - \mathrm{i}y_1 \cdot \omega x_2.$$

where $w_i = (x_i + \mathrm{i}y_i)$, $i = 1, 2$.

1.3.6 Holomorphic and anti-holomorphic subspaces

Assume that a real space $\mathcal{Y}$ is equipped with an anti-involution $j \in L(\mathcal{Y})$. Thus $(\mathbb{C}\mathcal{Y})_\mathbb{R}$ has two distinguished anti-involutions: the usual i, and also $j_\mathbb{C}$.

Definition 1.83 *Set*

$$\mathcal{Z} := \{y - ijy \; : \; y \in \mathcal{Y}\}.$$

$\mathcal{Z}$ *will be called the* holomorphic subspace of $\mathbb{C}\mathcal{Y}$.

$$\overline{\mathcal{Z}} := \{y + ijy \; : \; y \in \mathcal{Y}\}$$

will be called the anti-holomorphic subspace of $\mathbb{C}\mathcal{Y}$.

The corresponding projections are $\mathbb{1}_\mathcal{Z} := \frac{1}{2}(\mathbb{1} - ij_\mathbb{C})$ and $\mathbb{1}_{\overline{\mathcal{Z}}} := \frac{1}{2}(\mathbb{1} + ij_\mathbb{C})$. Clearly, $\mathbb{1} = \mathbb{1}_\mathcal{Z} + \mathbb{1}_{\overline{\mathcal{Z}}}$, and $\mathbb{C}\mathcal{Y} = \mathcal{Z} \oplus \overline{\mathcal{Z}}$. We have $\mathcal{Z} = \mathrm{Ker}(j_\mathbb{C} - i)$, $\overline{\mathcal{Z}} = \mathrm{Ker}(j_\mathbb{C} + i)$, thus on $\mathcal{Z}$ the complex structures i and $j_\mathbb{C}$ coincide, whereas on $\overline{\mathcal{Z}}$ they are opposite.

The canonical conjugation on $\mathbb{C}\mathcal{Y}$ is bijective from $\mathcal{Z}$ to $\overline{\mathcal{Z}}$, which shows that we can treat $(\overline{\mathcal{Z}}, i)$ as the conjugate vector space $\overline{(\mathcal{Z}, i)}$.

Using the decomposition

$$\mathbb{C}\mathcal{Y} = \mathcal{Z} \oplus \overline{\mathcal{Z}}, \tag{1.26}$$

we can write

$$i = \begin{bmatrix} i & 0 \\ 0 & i \end{bmatrix}, \quad j_\mathbb{C} = \begin{bmatrix} i & 0 \\ 0 & -i \end{bmatrix}.$$

The converse construction is as follows: Let $(\mathcal{Z}, i)$ be a complex vector space. Set

$$\mathrm{Re}(\mathcal{Z} \oplus \overline{\mathcal{Z}}) := \big\{(z, \overline{z}) \in \mathcal{Z} \oplus \overline{\mathcal{Z}} \; : \; z \in \mathcal{Z}\big\}.$$

Clearly, $\mathrm{Re}(\mathcal{Z} \oplus \overline{\mathcal{Z}})$ is a real vector space. It can be equipped with the anti-involution

$$j(z, \overline{z}) := (iz, \overline{iz}) = (iz, -i\overline{z}).$$

We identify $\mathbb{C}\mathrm{Re}(\mathcal{Z} \oplus \overline{\mathcal{Z}})$ with $\mathbb{C}\mathcal{Y} = \mathcal{Z} \oplus \overline{\mathcal{Z}}$ as follows: if $y_i = (z_i, \overline{z}_i) \in \mathcal{Y}$ for $i = 1, 2$, then

$$\mathbb{C}\mathcal{Y} \ni y_1 + iy_2 \mapsto (z_1 + iz_2, \overline{z}_1 + i\overline{z}_2) \in \mathcal{Z} \oplus \overline{\mathcal{Z}}. \tag{1.27}$$

With this identification we have

$$j_\mathbb{C} \simeq \begin{bmatrix} i & 0 \\ 0 & -i \end{bmatrix},$$

which shows that this is the converse construction.

$\mathcal{Z} \oplus \overline{\mathcal{Z}}$ is equipped with a conjugation

$$\epsilon(z_1, \overline{z}_2) := (\overline{z}_2, z_1).$$

Note that $\mathrm{Re}(\mathcal{Z} \oplus \overline{\mathcal{Z}})$ is the real subspace of $\mathcal{Z} \oplus \overline{\mathcal{Z}}$ for the conjugation ϵ. Clearly, under the identification (1.27), ϵ coincides with the usual complex conjugation on $\mathbb{C}\mathcal{Y}$.

Often it is convenient to identify the space $\mathcal{Z}$ with $\mathrm{Re}(\mathcal{Z} \oplus \overline{\mathcal{Z}}) = \mathcal{Y}$.

Definition 1.84 *For any $\lambda \neq 0$, we introduce an identification between a space with an anti-involution and the corresponding holomorphic space:*

$$\mathcal{Y} \ni y \mapsto T_\lambda y = \lambda \frac{\mathbb{1} - \mathrm{ij}}{2} y \in \mathcal{Z}. \tag{1.28}$$

The inverse map is

$$\mathcal{Z} \ni z \mapsto T_\lambda^{-1} z := \frac{1}{\lambda}(z + \overline{z}) \in \mathcal{Y}. \tag{1.29}$$

In the literature one can find at least two special cases of these identifications: for $\lambda = 1$ and for $\lambda = \sqrt{2}$. Each one has its own advantages. Note that in the bosonic case, we will typically use the identification $T_{\sqrt{2}}$, and in the fermionic case, the identification T_1. The arguments in favor of $T_{\sqrt{2}}$ will be given in Subsect. 1.3.9.

Let us discuss an argument in favor of T_1. Consider the natural projection from $\mathbb{C}\mathcal{Y}$ onto $\mathcal{Y}$:

$$\mathbb{C}\mathcal{Y} \ni w \mapsto \frac{w + \overline{w}}{2} + \mathrm{j}\frac{w - \overline{w}}{2\mathrm{i}} \in \mathcal{Y}. \tag{1.30}$$

Then

$$\mathcal{Z} \ni z \mapsto T_1^{-1} z = z + \overline{z} \in \mathcal{Y} \tag{1.31}$$

is the restriction of (1.30) to $\mathcal{Z}$.

T_1 appears naturally in the following context. Suppose that we have a function $\mathcal{Z} \ni z \mapsto F(z) \in \mathbb{C}$. One often prefers to move its domain onto $\mathcal{Y}$ by considering

$$\mathcal{Y} \ni (z, \overline{z}) \mapsto F(T_1(z, \overline{z})) = F(z). \tag{1.32}$$

Abusing notation, one can denote (1.32) by $F(z, \overline{z})$. This notation is especially common in the literature if F is not holomorphic.

Let us assume for a moment that $\mathcal{Y}$ is a complex space. We can realify $\mathcal{Y}$, and then complexify it, obtaining $\mathbb{C}\mathcal{Y}_\mathbb{R}$. Denote the original imaginary unit of $\mathcal{Y}$ by j. Introducing $\mathcal{Z}$ and identifying it with $\mathcal{Y}$ with help of T_1 we can write

$$\mathbb{C}\mathcal{Y}_\mathbb{R} \simeq \mathcal{Y} \oplus \overline{\mathcal{Y}}. \tag{1.33}$$

1.3.7 Operators on a space with an anti-involution

Let $\mathcal{Y}$ be a real space with an anti-involution j. Let $\mathcal{Z}, \overline{\mathcal{Z}}$ be the holomorphic and anti-holomorphic spaces defined in Subsect. 1.3.6. Let us collect the form of various operators on $\mathbb{C}\mathcal{Y}$ after the identification of $\mathbb{C}\mathcal{Y}$ with $\mathcal{Z} \oplus \overline{\mathcal{Z}}$.

We have

$$\epsilon = \begin{bmatrix} 0 & \chi \\ \overline{\chi} & 0 \end{bmatrix}, \quad j_{\mathbb{C}} = \begin{bmatrix} i & 0 \\ 0 & -i \end{bmatrix}, \quad i = \begin{bmatrix} i & 0 \\ 0 & i \end{bmatrix}.$$

where $\mathcal{Z} \ni z \mapsto \epsilon z := \overline{z} \in \overline{\mathcal{Z}}$.

An operator in $L(\mathbb{C}\mathcal{Y})$ is of the form

$$\begin{bmatrix} a & b \\ c & d \end{bmatrix},$$

where $a \in L(\mathcal{Z})$, $b \in L(\overline{\mathcal{Z}}, \mathcal{Z})$, $c \in L(\mathcal{Z}, \overline{\mathcal{Z}})$, $d \in L(\overline{\mathcal{Z}})$.

An operator in $L(\mathbb{C}\mathcal{Y})$ equal to $r_{\mathbb{C}}$ for some $r \in L(\mathcal{Y})$ is of the form

$$\begin{bmatrix} p & q \\ \overline{q} & \overline{p} \end{bmatrix},$$

where $p \in L(\mathcal{Z})$, $q \in L(\overline{\mathcal{Z}}, \mathcal{Z})$.

Finally an operator $L(\mathbb{C}\mathcal{Y})$ equal to $r_{\mathbb{C}}$ for $r \in L(\mathcal{Y}^{\mathbb{C}})$ (which means that $[r, j] = 0$) is of the form

$$\begin{bmatrix} p & 0 \\ 0 & \overline{p} \end{bmatrix},$$

for $p \in L(\mathcal{Z})$.

1.3.8 *(Pseudo-)Kähler spaces*

Let $(\mathcal{Y}, (\cdot|\cdot))$ be a (pseudo-)unitary space. Then $\mathcal{Y}_{\mathbb{R}}$ is a (pseudo-)Euclidean space for the scalar product

$$y_2 \cdot \nu y_1 := \mathrm{Re}(y_2|y_1), \tag{1.34}$$

a symplectic space for the symplectic form

$$y_2 \cdot \omega y_1 := \mathrm{Im}(y_2|y_1), \tag{1.35}$$

and has an anti-involution

$$jy := iy. \tag{1.36}$$

The name "*(pseudo-)Kähler space*" is used for a unitary space treated as a real space with the three structures (1.34), (1.35) and (1.36). Below we give a more precise definition:

Definition 1.85 *We say that a quadruple* $(\mathcal{Y}, \nu, \omega, j)$ *is a* pseudo-Kähler space *if*

(1) $\mathcal{Y}$ *is a real vector space,*
(2) ν *is a non-degenerate symmetric form,*
(3) ω *is a symplectic form,*
(4) j *is an anti-involution,*
(5) $\omega j = \nu$.

If in addition ν is positive definite, then we say that $(\mathcal{Y}, \nu, \omega, \mathrm{j})$ is a Kähler space.

Definition 1.86 *If $(\mathcal{Y}, \nu, \omega, \mathrm{j})$ is a (pseudo-)Kähler space, we set*

$$(y_1|y_2) := y_1 \cdot \nu y_2 + iy_1 \cdot \omega y_2. \qquad (1.37)$$

Then $(\mathcal{Y}^{\mathbb{C}}, (\cdot|\cdot))$ is a (pseudo-)unitary space.

Definition 1.87 *If a Kähler space $\mathcal{Y}$ is complete for the norm $(y \cdot \nu y)^{\frac{1}{2}}$, we say that $\mathcal{Y}$ is a* complete Kähler space. *In other words $\mathcal{Y}^{\mathbb{C}}$ equipped with $(\cdot|\cdot)$ is a Hilbert space.*

Two structures out of ν, ω, j determine the other. This is used in the following three definitions. In all of them $\mathcal{Y}$ is a real vector space, ω is a symplectic form and ν is a non-degenerate symmetric form.

Definition 1.88 (1) *We say that a pair (ω, j) is* pseudo-Kähler *if $\omega\mathrm{j}$ is symmetric. If in addition $\omega\mathrm{j}$ is positive definite, then we say that (ω, j) is* Kähler.

(2) *We say that a pair (ν, j) is* pseudo-Kähler *if $-\nu\mathrm{j}$ is a symplectic form. If in addition ν is positive definite, then we say that (ν, j) is* Kähler.

(3) *We say that a pair (ν, ω) is* pseudo-Kähler *if $\operatorname{Ran}\omega = \operatorname{Ran}\nu$ and $\omega^{-1}\nu$ is an anti-involution. If in addition ν is positive definite, we say that (ν, ω) is* Kähler.

The definitions (1) and (2) have other equivalent versions, as seen from the following theorem:

Theorem 1.89 (1) *Let $(\mathcal{Y}, \omega)$ be a symplectic space. Consider the following conditions:*

  (i) $\mathrm{j}^{\#}\omega\mathrm{j} = \omega$ *(j preserves ω),*
  (ii) $\mathrm{j}^{\#}\omega + \omega\mathrm{j} = 0$ *($\mathrm{j} \in sp(\mathcal{Y})$, or equivalently $\omega\mathrm{j}$ is symmetric),*
  (iii) $\mathrm{j}^2 = -\mathbb{1}$ *(j is an anti-involution).*

Then any pair of the conditions (i), (ii), (iii) implies the third condition and that the pair (ω, j) is pseudo-Kähler.

(2) *Let $(\mathcal{Y}, \nu)$ be a (pseudo-)Euclidean space. Consider the following conditions:*

  (i) $\mathrm{j}^{\#}\nu\mathrm{j} = \nu$ *(j is (pseudo-)isometric),*
  (ii) $\mathrm{j}^{\#}\nu + \nu\mathrm{j} = 0$ *($\mathrm{j} \in o(\mathcal{Y})$, or equivalently $\nu\mathrm{j}$ is anti-symmetric),*
  (iii) $\mathrm{j}^2 = -\mathbb{1}$ *(j is an anti-involution).*

Then any pair of the conditions (i), (ii), (iii) implies the third condition and that the pair (ν, j) is (pseudo-)Kähler.

1.3.9 Complexification of a (pseudo-)Kähler space

Let $(\mathcal{Y}, \nu, \omega, \mathrm{j})$ be a (pseudo-)Kähler space. We have seen that the space $\mathbb{C}\mathcal{Y}$ is equipped with

(1) the symmetric form $w_1 \cdot \nu_{\mathbb{C}} w_2$,
(2) the Hermitian form $(w_1|w_2) := \overline{w_1} \cdot \nu_{\mathbb{C}} w_2$,
(3) the symplectic form $w_1 \cdot \omega_{\mathbb{C}} w_2$, and
(4) the charged symplectic form $\overline{w_1} \cdot \omega_{\mathbb{C}} w_2$,

where $w_1, w_2 \in \mathbb{C}\mathcal{Y}$.

The spaces $\mathcal{Z}$ and $\overline{\mathcal{Z}}$ introduced in Subsect. 1.3.6 are isotropic for both bilinear forms $\nu_{\mathbb{C}}$ and $\omega_{\mathbb{C}}$ and are mutually orthogonal for both sesquilinear forms.

Let us concentrate on the (pseudo-)unitary structure on $\mathbb{C}\mathcal{Y}$ given by the form $(\cdot|\cdot)$. Using the fact that j is anti-self-adjoint for ν on $\mathcal{Y}$ we see that $j_{\mathbb{C}}$ is anti-self-adjoint for $(\cdot|\cdot)$ on $\mathbb{C}\mathcal{Y}$. Therefore, the projections $\mathbb{1}_{\mathcal{Z}}$ and $\mathbb{1}_{\overline{\mathcal{Z}}}$ are orthogonal projections and hence the spaces $\mathcal{Z}$ and $\overline{\mathcal{Z}}$ are orthogonal for $(\cdot|\cdot)$. The map $T_{\sqrt{2}}$, introduced in (1.29) is (pseudo-)unitary, if we interpret $\mathcal{Y}$ as a (pseudo-)unitary space $\mathcal{Y}^{\mathbb{C}}$ equipped with the scalar product (1.37). This is the main reason why the identification $T_{\sqrt{2}}$ is often used, at least for bosonic systems.

The converse construction is as follows. Let $\mathcal{Z}$ be a (pseudo-)unitary space. Set $\mathcal{Y} := \mathrm{Re}(\mathcal{Z} \oplus \overline{\mathcal{Z}})$. Recall from Subsect. 1.3.6 that $\mathcal{Z}$ is naturally isomorphic to the holomorphic space for $(\mathcal{Y}, j)$, where the anti-involution j is given by

$$j(z, \overline{z}) = (iz, \overline{iz}) = (iz, -i\overline{z}).$$

$\mathcal{Y}$ is equipped with the symmetric form

$$(z_1, \overline{z}_1) \cdot \nu(z_2, \overline{z}_2) := 2\mathrm{Re}(z_1|z_2),$$

and the symplectic form

$$(z_1, \overline{z}_1) \cdot \omega(z_2, \overline{z}_2) = 2\mathrm{Im}(z_1|z_2).$$

Then $(\mathcal{Y}, \nu, \omega, j)$ is a (pseudo-)Kähler space.

If we first take a (pseudo-)Kähler space $\mathcal{Y}$, take its holomorphic space $\mathcal{Z}$ equipped with its (pseudo-)unitary structure, and then go to the (pseudo-)Kähler space $\mathcal{Y} = \mathrm{Re}(\mathcal{Z} \oplus \overline{\mathcal{Z}})$ constructed as above, we return to the original structure.

If $\mathcal{Z}$ is complete, then the topological dual $\mathcal{Y}^{\#}$ can be identified with $\mathrm{Re}(\overline{\mathcal{Z}} \oplus \mathcal{Z})$ by setting

$$\langle (z, \overline{z})|(\overline{w}, w) \rangle := (z|w) + (\overline{z}|\overline{w}) = 2\mathrm{Re}(z|w).$$

With this identification we have

$$\omega(z, \overline{z}) = (-i\overline{z}, iz).$$

1.3.10 *Conjugations on a (pseudo-)Kähler space*

Proposition 1.90 *Let $(\mathcal{Y}, \nu, \omega, j)$ be a Kähler space. Let $\tau \in L(\mathcal{Y})$ be an involution. Then the following statements are equivalent:*

(1) τ *is anti-unitary on* $(\mathcal{Y}^{\mathbb{C}}, (\cdot|\cdot))$.
(2) $\tau \in O(\mathcal{Y}, \nu)$, $\tau j = -j\tau$.
(3) τ *is anti-symplectic,* $\tau j = -j\tau$.

Definition 1.91 *If the conditions of Prop. 1.90 are satisfied we say that τ is a conjugation of the Kähler space $\mathcal{Y}$.*

Def. 1.91 is consistent with the definitions of a conjugation on a complex space, a symplectic space and a (pseudo-)unitary space.

Assume that $\mathcal{Y}$ is a complete Kähler space with a conjugation τ. Let $\mathcal{X} := \mathcal{Y}^{-\tau}$, which is a real Hilbert space for ν. We can identify $\mathcal{Y}$ with $\mathcal{X} \oplus \mathcal{X}$ by (1.22), as in Subsect. 1.3.2. Having in mind applications to CCR representations (see Subsect. 8.2.7), we prefer, however, to describe a more general identification. We fix a bounded, positive and invertible operator c on $\mathcal{X}$. Then the map

$$\mathcal{Y} \ni y \mapsto \left((2c)^{-\frac{1}{2}} \mathrm{j} \frac{\mathbb{1}+\tau}{2} y, (2c)^{\frac{1}{2}} \frac{\mathbb{1}-\tau}{2} y \right) \in \mathcal{X} \oplus \mathcal{X} \qquad (1.38)$$

is bijective. With this identification we have

$$\tau = \begin{bmatrix} \mathbb{1} & 0 \\ 0 & -\mathbb{1} \end{bmatrix}, \quad \mathrm{j} = \begin{bmatrix} 0 & -(2c)^{-1} \\ 2c & 0 \end{bmatrix},$$

$$(x_1^+, x_1^-) \cdot \nu(x_2^+, x_2^-) = x_1^+ \cdot \nu 2c x_2^+ + x_1^- \cdot \nu(2c)^{-1} x_1^-,$$

$$(x_1^+, x_1^-) \cdot \omega(x_2^+, x_2^-) = x_1^+ \cdot \nu x_2^- - x_1^- \cdot \nu x_2^+, \quad (x_i^+, x_i^-) \in \mathcal{X} \oplus \mathcal{X}, \quad i = 1, 2.$$

1.3.11 Real representations of the group $U(1)$

Let $\mathcal{Y}$ be a real space. Consider the group $U(1) \simeq \mathbb{R}/2\pi\mathbb{Z}$ and its representation

$$U(1) \ni \theta \mapsto u_\theta \in L(\mathcal{Y}). \qquad (1.39)$$

Definition 1.92 *Let $n \in \{0, 1, \dots\}$. A representation (1.39) is called a charge n representation if there exists an anti-involution j_{ch} such that*

$$u_\theta = \cos(n\theta)\mathbb{1} + \sin(n\theta)\mathrm{j}_{\mathrm{ch}}, \quad \theta \in U(1). \qquad (1.40)$$

Proposition 1.93 *(1) If (1.39) is a charge 1 representation, then*

$$u_\theta y \neq y, \quad 0 \neq y \in \mathcal{Y}, \quad 0 \neq \theta \in U(1), \qquad (1.41)$$

and the operator j_{ch} in (1.40) coincides with $u_{\pi/2}$.
(2) If the representation (1.39) satisfies (1.41), then $u_{\pi/2}$ is an anti-involution.

Proof (2) Clearly, $u_\pi^2 = \mathbb{1}$. Therefore, u_π is diagonalizable and $\frac{1}{2}(\mathbb{1} \pm u_\pi)$ are the projections onto its eigenvalues ± 1. By (1.41), $\mathrm{Ker}(\mathbb{1} - u_\pi) = \{0\}$. Therefore, $u_\pi = -\mathbb{1}$. Now $u_{\pi/2}^2 = u_\pi = -\mathbb{1}$. $\qquad \square$

Proposition 1.94 *Assume that $\mathcal{Y}$ is either finite-dimensional or a real Hilbert space and the representation (1.39) is orthogonal. In both cases we suppose that the representation is strongly continuous. Then*

(1) $\mathcal{Y} = \bigoplus_{n=0}^{\infty} \mathcal{Y}_n$, *where $\mathcal{Y}_n$ are invariant and (1.39) restricted to $\mathcal{Y}_n$ is a charge n representation.*

(2) *The set of vectors $y \in \mathcal{Y}$ satisfying (1.41) equals $\mathcal{Y}_1$.*

Proof We can complexify $\mathcal{Y}$ and write that $u_{\theta,\mathbb{C}} = e^{\mathrm{i}\theta c}$ on $\mathbb{C}\mathcal{Y}$, for some operator c. Clearly, spec $c \subset \mathbb{Z}$. Then $\mathcal{Y}_n := \mathrm{Ran}\, 1\!\!1_{\{n,-n\}}(c) \cap \mathcal{Y}$. $\qquad\qquad\square$

Charge 1 representations are related to (pseudo-)Kähler structures.

Proposition 1.95 *Consider a charge 1 representation*

$$u_\theta = \cos(\theta)1\!\!1 + \sin(\theta)\mathrm{j_{ch}}, \quad \theta \in U(1). \tag{1.42}$$

(1) *If $\mathcal{Y}$ is a real Hilbert space and $u_\theta \in O(\mathcal{Y})$, $\theta \in U(1)$, then $\mathrm{j_{ch}}$ is a Kähler anti-involution.*

(2) *If $\mathcal{Y}$ is a symplectic space and $u_\theta \in Sp(\mathcal{Y})$, $\theta \in U(1)$, then $\mathrm{j_{ch}}$ is a pseudo-Kähler anti-involution.*

1.4 Groups and Lie algebras

In this section we fix terminology and notation concerning groups and Lie algebras, mostly consisting of linear or affine transformations.

Throughout the section, $\mathcal{Y}$ and $\mathcal{W}$ denote finite-dimensional spaces.

1.4.1 General linear group and Lie algebra

Definition 1.96 $GL(\mathcal{Y},\mathcal{W})$ *denotes the set of invertible elements in $L(\mathcal{Y},\mathcal{W})$. The* general linear group *of $\mathcal{Y}$ is defined as $GL(\mathcal{Y}) := GL(\mathcal{Y},\mathcal{Y})$.*

$$SL(\mathcal{Y}) := \big\{ r \in GL(\mathcal{Y}) \,:\, \det r = 1 \big\}$$

is its subgroup called the special linear group *of $\mathcal{Y}$.*

Definition 1.97 *The* general linear Lie algebra *of $\mathcal{Y}$ is denoted $gl(\mathcal{Y})$ and equals $L(\mathcal{Y})$ equipped with the bracket $[a,b] := ab - ba$.*

$$sl(\mathcal{Y}) := \big\{ a \in gl(\mathcal{Y}) \,:\, \mathrm{Tr}\, a = 0 \big\}$$

is its Lie sub-algebra called the special linear Lie algebra *of $\mathcal{Y}$.*

1.4.2 Homogeneous linear differential equations

Assume that $\mathbb{R} \ni t \mapsto a_t \in gl(\mathcal{Y})$ is continuous, and $t \geq s$.

Definition 1.98 *We define the* time-ordered exponential *by the following convergent series:*

$$\mathrm{Texp} \int_s^t a_u \,\mathrm{d}u := \sum_{n=0}^\infty \int \ldots \int_{t \geq u_n \geq \cdots \geq u_1 \geq s} a_{u_n} \cdots a_{u_1} \,\mathrm{d}u_n \ldots \mathrm{d}u_1.$$

For $y \in \mathcal{Y}$, $s \in \mathbb{R}$, there exists a unique solution of

$$\frac{\mathrm{d}}{\mathrm{d}t} y_t = a_t y_t, \quad y_s = y. \tag{1.43}$$

It can be expressed in terms of the time-ordered exponential as

$$y_t = \mathrm{Texp} \int_s^t a_u \, \mathrm{d}u \, y.$$

Clearly, if $a_t = a \in gl(\mathcal{Y})$ does not depend on t, we can use the usual exponential instead of the time-ordered exponential:

$$\mathrm{Texp} \int_s^t a \, \mathrm{d}u = \mathrm{e}^{(t-s)a}.$$

1.4.3 Affine transformations

Definition 1.99 $AL(\mathcal{Y}, \mathcal{W})$ *will denote* $\mathcal{W} \times L(\mathcal{Y}, \mathcal{W})$ *acting on* $\mathcal{Y}$ *as follows: if* $(w, a) \in AL(\mathcal{Y}, \mathcal{W})$ *and* $y \in \mathcal{Y}$, *then* $(w, a)y := w + ay$. *Elements of* $AL(\mathcal{Y}, \mathcal{W})$ *are called* affine *maps from* $\mathcal{Y}$ *to* $\mathcal{W}$. *We set* $AL(\mathcal{Y}) := AL(\mathcal{Y}, \mathcal{Y})$.

Definition 1.100 *If* $G \subset L(\mathcal{Y}, \mathcal{W})$, *we set* $AG := \mathcal{W} \times G$ *as a subset of* $AL(\mathcal{Y}, \mathcal{W})$.

In particular, if $G \subset L(\mathcal{Y})$ is a group, then so is AG. The multiplication in $AG(\mathcal{Y})$ is

$$(y_2, r_2)(y_1, r_1) = (y_2 + r_2 y_1, r_2 r_1).$$

Thus $AG(\mathcal{Y})$ is an example of a semi-direct product of $\mathcal{Y}$ and G, determined by the natural action of G on $\mathcal{Y}$, and is often denoted by $\mathcal{Y} \rtimes G$.

Definition 1.101 *The* general affine Lie algebra *of* $\mathcal{Y}$ *is* $agl(\mathcal{Y}) := \mathcal{Y} \times L(\mathcal{Y})$ *equipped with the bracket*

$$[(y_2, a_2), (y_1, a_1)] = (a_2 y_1 - a_1 y_2, a_2 a_1 - a_1 a_2).$$

Definition 1.102 *If* $g \subset gl(\mathcal{Y})$, *then we set* $ag := \mathcal{Y} \times g$ *as a subset of* $agl(\mathcal{Y})$.

Clearly, if g is a Lie algebra, then so is ag. It is an example of the semi-direct product of $\mathcal{Y}$ and g, determined by the natural action of g on $\mathcal{Y}$, and is often denoted by $\mathcal{Y} \rtimes g$.

1.4.4 Inhomogeneous linear differential equations

Consider a continuous function $\mathbb{R} \ni t \mapsto (w_t, a_t) \in agl(\mathcal{Y})$. Then, for $y \in \mathcal{Y}$, $s \in \mathbb{R}$, there exists a unique solution of

$$\frac{\mathrm{d}}{\mathrm{d}t} y_t = w_t + a_t y_t, \qquad y_s = y. \tag{1.44}$$

It can be written as

$$y_t = \int_s^t \left(\mathrm{Texp} \int_v^t a_u \, \mathrm{d}u \right) w_v \, \mathrm{d}v + \left(\mathrm{Texp} \int_s^t a_u \, \mathrm{d}u \right) y. \tag{1.45}$$

If $(w_t, a_t) = (a, w) \in agl(\mathcal{Y})$ does not depend on t, then (1.45) reduces to

$$y_t = a^{-1}(e^{(t-s)a} - \mathbb{1})w + e^{(t-s)a}y.$$

This motivates setting

$$e^{(w,a)} := (a^{-1}(e^a - \mathbb{1})w, e^a) \in AGL(\mathcal{Y}).$$

Note in particular that

$$e^{(0,a)} = (0, e^a), \quad e^{(w,0)} = (w, \mathbb{1}).$$

1.4.5 Exact sequences

Let $\pi : F \to G$, $\rho : G \to H$ be homomorphisms between groups.

Definition 1.103 *By saying that*

$$F \xrightarrow{\pi} G \xrightarrow{\rho} H \tag{1.46}$$

is an exact sequence, we mean that $\operatorname{Ran} \pi = \operatorname{Ker} \rho$.

Often, if they are obvious from the context, π, ρ are omitted from (1.46). The one-element group is often denoted by 1. Therefore,

$$1 \to F \to G \to H \to 1 \tag{1.47}$$

means that F is a normal subgroup of G and we have a natural isomorphism $H \simeq G/F$.

1.4.6 Cayley transform

Let $\mathcal{Y}$ be a vector space. Let $r \in L(\mathcal{Y})$ and $r + \mathbb{1}$ be invertible.

Definition 1.104 *We define the* Cayley transform *of* r *as*

$$\gamma := (\mathbb{1} - r)(\mathbb{1} + r)^{-1}.$$

Note that $\gamma + \mathbb{1}$ is again invertible and

$$r = (\mathbb{1} - \gamma)(\mathbb{1} + \gamma)^{-1}.$$

Hence the Cayley transform is an involution of

$$\{a \in L(\mathcal{Y}) \ : \ r + \mathbb{1} \text{ is invertible}\}. \tag{1.48}$$

Let r_1, r_2, r belong to (1.48) with $r = r_1 r_2$. Let $\gamma_1, \gamma_2, \gamma$ be their Cayley transforms. Then we have the identity

$$\mathbb{1} + \gamma = (\mathbb{1} + \gamma_2)(\mathbb{1} + \gamma_1\gamma_2)^{-1}(\mathbb{1} + \gamma_1). \tag{1.49}$$

Suppose that $\mathcal{Y}$ is a finite-dimensional symplectic space. Then the Cayley transform is a bijection of

$$\{r \in Sp(\mathcal{Y}) \ : \ r + \mathbb{1} \text{ is invertible}\}$$

onto

$$\{\gamma \in sp(\mathcal{Y}) \: : \: \gamma + \mathbb{1} \text{ is invertible}\}.$$

If $\mathcal{Y}$ is a Euclidean space, then the same is true with $Sp(\mathcal{Y})$, $sp(\mathcal{Y})$ replaced with $O(\mathcal{Y})$, $o(\mathcal{Y})$.

If $\mathcal{Y}$ is a unitary space, then the same is true with $Sp(\mathcal{Y})$, $sp(\mathcal{Y})$ replaced with $U(\mathcal{Y})$, $u(\mathcal{Y})$.

1.5 Notes

Most of the material in this section is a collection of concepts and facts from any basic linear algebra course, after a minor "cleaning up". The need for a particularly precise terminology in this area is especially important in differential geometry. Therefore, in the literature such concepts as Kähler, symplectic and complex structures typically appear in the context of differentiable manifolds; see e.g. Guillemin–Sternberg (1977). They are rarely considered in the (much simpler) context of linear algebra.

2

Operators in Hilbert spaces

In this chapter we recall basic properties of operators on topological vector spaces. We concentrate on Hilbert spaces, which play the central role in quantum physics.

2.1 Convergence and completeness

We start with a discussion of various topics related to convergence and completeness.

2.1.1 Nets

Nets are generalizations of sequences. In this subsection we briefly recall this useful concept.

Definition 2.1 *A* directed set *is a set I equipped with a partial order relation $\leq$ such that for any $i, j \in I$ there exists $k \in I$ such that $i \leq k$, $j \leq k$.*

We will often use the following directed set:

Definition 2.2 *Let I be a set. We denote by 2^I_{fin} the family of finite subsets of I. It becomes a directed set when we equip it with the inclusion.*

Definition 2.3 *Let $\mathcal{S}$ be a set. A* net *in $\mathcal{S}$ is a mapping from a directed set I to $\mathcal{S}$, denoted by $\{x_i\}_{i \in I}$.*

Definition 2.4 *A net $\{x_i\}_{i \in I}$ in a topological space $\mathcal{S}$* converges *to $x \in \mathcal{S}$ if for any neighborhood $\mathcal{N}$ of x there exists $i \in I$ such that if $i \leq j$ then $x_j \in \mathcal{N}$. We will write $x_i \to x$. If $\mathcal{S}$ is Hausdorff, then a net in $\mathcal{S}$ can have at most one limit and one can also write $\lim x_i = x$.*

Definition 2.5 *Let $\mathcal{X}$ be a topological space and $\mathcal{U} \subset \mathcal{X}$. Then $\mathcal{U}^{\text{cl}}$ will denote the* closure *of $\mathcal{U}$, which is equal to the set of limits of all convergent nets in $\mathcal{U}$.*

2.1.2 Functions

Definition 2.6 *Let $\mathcal{X}, \mathcal{Y}$ be sets. Then $c(\mathcal{X}, \mathcal{Y})$ is the set of all functions from $\mathcal{X}$ to $\mathcal{Y}$. Clearly, $c(\mathcal{X}, \mathbb{K})$, is a vector space over $\mathbb{K}$. We often write $c(\mathcal{X})$ for $c(\mathcal{X}, \mathbb{C})$. $f \in c(\mathcal{X}, \mathbb{K})$ is called* finitely supported *if $f^{-1}(\mathbb{K}\backslash\{0\})$ is finite. $c_{\text{c}}(\mathcal{X}, \mathbb{K})$*

denotes the space of finitely supported functions in $c(\mathcal{X}, \mathbb{K})$. If $x \in \mathcal{X}$, define $\delta_x \in c_c(\mathcal{X}, \mathbb{K})$ by $\delta_x(y) := \begin{cases} 1, & x = y, \\ 0, & x \neq y. \end{cases}$. Clearly, each element of $c_c(\mathcal{X}, \mathbb{K})$ can be written as a unique finite linear combination of $\{\delta_x \ : \ x \in \mathcal{X}\}$. Sometimes, it will be convenient to write x instead of δ_x.

Definition 2.7 *Let* $\mathcal{X}, \mathcal{Y}$ *be topological spaces. Then* $C(\mathcal{X}, \mathcal{Y})$ *is the set of all continuous functions from* $\mathcal{X}$ *to* $\mathcal{Y}$. *Clearly,* $C(\mathcal{X}, \mathbb{K})$ *is a vector space over* $\mathbb{K}$. *We often write* $C(\mathcal{X})$ *for* $C(\mathcal{X}, \mathbb{C})$. $C_c(\mathcal{X}, \mathbb{K})$ *denotes the set of* compactly supported *functions in* $C(\mathcal{X}, \mathbb{K})$.

We will use various styles of notation to introduce a function f with domain $\mathcal{X}$, such as $\mathcal{X} \ni x \mapsto f(x)$ or $\{f(x)\}_{x \in \mathcal{X}}$. Sometimes, we will simply write that *we are given a function* $f(x)$. This is possible, if we declared before that x is the *generic variable in* $\mathcal{X}$, or at least if it is clear from the context that x should be understood this way. Thus x is not a concrete element of $\mathcal{X}$, it is just a symbol for which we can substitute an arbitrary element of $\mathcal{X}$.

For example, the notation $[a_{ij}]$ is sometimes used for a matrix. Here, i is understood as the generic variable in $\{1, \ldots, n\}$ and j as the generic variable in $\{1, \ldots, m\}$, where n, m should be clear from the context. Thus $[a_{ij}]$ is an abbreviation for $\{1, \ldots, n\} \times \{1, \ldots, m\} \ni (i, j) \mapsto a_{i,j}$.

Generic variables are also used in some other situations, e.g. as a part of the notation for integration or differentiation.

2.1.3 Topological vector spaces

Let $\mathcal{E}$ be a topological vector space.

Definition 2.8 *If* $\mathcal{U} \subset \mathcal{E}$, *we will use the shorthand* $\mathrm{Span}^{\mathrm{cl}}(\mathcal{U})$ *for* $(\mathrm{Span}(\mathcal{U}))^{\mathrm{cl}}$.

Definition 2.9 *A net* $\{x_i\}_{i \in I}$ *in a topological vector space* $\mathcal{E}$ *is* Cauchy *if, for any neighborhood* $\mathcal{N}$ *of* 0, *there exists* $i \in I$ *such that if* $i \leq j, k$, *then* $x_j - x_k \in \mathcal{N}$. $\mathcal{E}$ *is* complete *if every Cauchy net is convergent.*

Proposition 2.10 *There exists a complete topological vector space containing* $\mathcal{E}$ *as a dense subspace. If* $\mathcal{E}_1$ *and* $\mathcal{E}_2$ *are two such complete spaces, then there exists a unique linear homeomorphism* $T : \mathcal{E}_1 \to \mathcal{E}_2$ *such that* $T\big|_{\mathcal{E}} = \mathbb{1}_{\mathcal{E}}$.

Definition 2.11 *The complete vector space, described in Prop. 2.10 uniquely up to isomorphism, is called the* completion *of* $\mathcal{E}$ *and denoted* $\mathcal{E}^{\mathrm{cpl}}$.

2.1.4 Infinite sums

Let $\mathcal{E}$ be a topological vector space and $\{x_i\}_{i \in I}$ a family of elements of $\mathcal{E}$.

Definition 2.12 *We say that the* series $\sum_{i \in I} x_i$ *is* convergent *if the net* $\left\{ \sum_{i \in J} x_i \right\}_{J \in 2^I_{\text{fin}}}$, *is convergent. The limit of the above net will be denoted by*

$$\sum_{i \in I} x_i.$$

Assume that $\mathcal{E}$ is a normed space.

Definition 2.13 *We say that the series* $\sum_{i \in I} x_i$ *is* absolutely convergent *if the numerical series* $\sum_{i \in I} \|x_i\|$ *is convergent.*

Proposition 2.14 (1) *For every absolutely convergent series, the set* $\{i : x_i \neq 0\}$ *is at most countable.*
(2) *Every absolutely convergent series in a Banach space is convergent.*
(3) *In a finite-dimensional space, a series is convergent iff it is absolutely convergent.*

2.1.5 Infinite products

Let $\{x_i\}_{i \in I}$ be a family in $\mathbb{C}$.

Definition 2.15 *First assume that $x_i \neq 0$ for all $i \in I$. In this case, the* infinite product $\prod_{i \in I} x_i$ *is called* convergent *if the net* $\left\{ \prod_{i \in J} x_i \right\}_{J \in 2^I_{\text{fin}}}$ *converges to a nonzero limit in $\mathbb{C}$. The limit will be denoted by*

$$\prod_{i \in I} x_i.$$

In the general case, one says that $\prod_{i \in I} x_i$ *is convergent if $I_0 = \{i \in I : x_i = 0\}$ is finite and the infinite product* $\prod_{i \in I \setminus I_0} x_i$ *is convergent in the above sense. If $I_0 \neq \emptyset$, one sets*

$$\prod_{i \in I} x_i := 0.$$

It is easy to see that the convergence of $\prod_{i \in I} x_i$ is equivalent to the convergence of $\sum_{i \in I} |x_i - 1|$. Therefore, if $\prod_{i \in I} x_i$ converges, then the set $\{i \in I : x_i \neq 1\}$ is at most countable and $x_i \to 1$.

2.2 Bounded and unbounded operators

2.2.1 Normed vector spaces

Let $\mathcal{H}, \mathcal{K}$ be normed spaces over $\mathbb{K} = \mathbb{R}$ or $\mathbb{C}$.

Definition 2.16 *We equip the complex conjugate space $\overline{\mathcal{H}}$ with the norm $\|\overline{\Phi}\| := \|\Phi\|$, $\Phi \in \mathcal{H}$.*

Definition 2.17 *$B(\mathcal{H}, \mathcal{K})$ denotes the space of bounded linear operators from $\mathcal{H}$ to $\mathcal{K}$. We set $B(\mathcal{H}) := B(\mathcal{H}, \mathcal{H})$. $\mathcal{H}^{\#} := B(\mathcal{H}, \mathbb{K})$ is the topological dual of $\mathcal{H}$ and $\mathcal{H}^{*} := B(\overline{\mathcal{H}}, \mathbb{K}) = \overline{\mathcal{H}^{\#}} = \overline{\mathcal{H}}^{\#}$ is the topological anti-dual of $\mathcal{H}$.*

Remark 2.18 *Note that the meaning of the symbol $\mathcal{H}^{\#}$, resp. $\mathcal{H}^{*}$ depends on the context: if we consider $\mathcal{H}$ as a vector space without a topology, it will denote the algebraic dual, resp. anti-dual. see Defs; 1.8, resp. 1.54. If $\mathcal{H}$ is considered together with its topology, it will denote the topological dual, resp. anti-dual.*

Definition 2.19 *By saying that A is a* linear operator from $\mathcal{H}$ to $\mathcal{K}$, *we will not necessarily mean that it is defined on the whole $\mathcal{H}$. We will just mean that there exists a subspace $\mathcal{D}$ of $\mathcal{H}$ such that $A \in L(\mathcal{D}, \mathcal{K})$. The space $\mathcal{D}$ will be called the* domain *of A and denoted $\mathrm{Dom}\,A$. The subspace $\mathrm{Gr}\,A := \{(\Phi, A\Phi) \ : \ \Phi \in \mathrm{Dom}\,A\} \subset \mathcal{H} \oplus \mathcal{K}$ is called the* graph *of A.*

Definition 2.20 *A linear operator A from $\mathcal{H}$ to $\mathcal{K}$ is* closed *if $\mathrm{Gr}\,A$ is closed in $\mathcal{H} \oplus \mathcal{K}$. It is called* closable *if it has a closed extension. Its minimal closed extension is called the* closure *of A and denoted by A^{cl}. $Cl(\mathcal{H}, \mathcal{K})$ will denote the set of closed, densely defined operators from $\mathcal{H}$ to $\mathcal{K}$.*

Proposition 2.21 *Let $A \in B(\mathcal{H}, \mathcal{K})$. Then A is closable as an operator from $\mathcal{H}^{\mathrm{cpl}}$ to $\mathcal{K}^{\mathrm{cpl}}$ and $A^{\mathrm{cl}} \in B(\mathcal{H}^{\mathrm{cpl}}, \mathcal{K}^{\mathrm{cpl}})$.*

Definition 2.22 *Let A be an operator on $\mathcal{H}$. We say that $z \in \mathbb{C}$ belongs to the* resolvent set *of A if $A - z\mathbb{1} : \mathrm{Dom}\,A \to \mathcal{H}$ is bijective and $(A - z\mathbb{1})^{-1} \in B(\mathcal{H})$. The resolvent set of A is denoted by $\mathrm{res}\,A$. The set $\mathrm{spec}\,A = \mathbb{C} \backslash \mathrm{res}\,A$ is called the* spectrum *of A.*

Definition 2.23 (1) *If A is an injective linear operator, then we set $\mathrm{Dom}\,A^{-1} := \mathrm{Ran}\,A$.*

(2) *If A, B are two linear operators, we set*

$$\mathrm{Dom}\,AB := \{\Phi \in \mathrm{Dom}\,B \ : \ B\Phi \in \mathrm{Dom}\,A\}.$$

(3) *If A, B are two linear operators on $\mathcal{H}$, their* commutator *and* anti-commutator *are the operators given by*

$$[A, B] := AB - BA, \ [A, B]_{+} := AB + BA, \ on \ \mathrm{Dom}\,AB \cap \mathrm{Dom}\,BA.$$

In the case that $\mathcal{H}$ is a Hilbert space, sometimes we will consider $[A, B]$, $[A, B]_{+}$ as quadratic forms on $\mathrm{Dom}\,A \cap \mathrm{Dom}\,A^{*} \cap \mathrm{Dom}\,B \cap \mathrm{Dom}\,B^{*}$. For example,

$$(\Phi|[A, B]\Psi) := (A^{*}\Phi|B\Psi) - (B^{*}\Phi|A\Psi).$$

2.2.2 Scalar product spaces

Let $\mathcal{H}$ be a unitary space (a complex space equipped with a scalar product). The scalar product of $\Phi, \Psi \in \mathcal{H}$ will be denoted by $(\Phi|\Psi)$ or $\overline{\Phi} \cdot \Psi$. Recall that a complete unitary space is called a complex Hilbert space, where one usually omits the word "complex". Note that if $\mathcal{H}$ is a Hilbert space, then $\overline{\mathcal{H}}$ equipped with the scalar product $(\overline{\Psi}|\overline{\Phi}) := \overline{(\Psi|\Phi)}$ is a Hilbert space as well, and the map $\mathcal{H} \ni \Phi \mapsto \overline{\Phi} \in \overline{\mathcal{H}}$ is anti-unitary (see Subsect. 1.2.10). The Riesz lemma says that $\mathcal{H}^* = \overline{\mathcal{H}}^{\#}$ is naturally isomorphic to $\mathcal{H}$. Sometimes, however, other identifications are convenient; see Subsect. 2.3.4.

In a Euclidean space (a real space equipped with a scalar product) we prefer to denote the scalar product by $\langle\Phi|\Psi\rangle$ or $\Phi \cdot \Psi$. Recall that a complete Euclidean space is called a real Hilbert space. If $\mathcal{H}$ is a real Hilbert space, the Riesz lemma says that $\mathcal{H}^{\#}$ is naturally isomorphic to $\mathcal{H}$.

Remark 2.24 *If we compare Def. 1.54 with this subsection, we see that $\overline{z} \cdot w$ or $(w|z)$ may stand for the pairing between vectors in two distinct spaces in an anti-dual pair, or for the scalar product of two vectors in the same Hilbert space.*

Analogously, if we compare Def. 1.8 with this subsection, we see that $v \cdot y$ or $\langle v|y \rangle$ may stand for the pairing within a dual pair, or for the scalar product in the same real Hilbert space.

There are more such ambiguous notations, whose exact meaning depends on the context; see e.g. Remark 2.18. These ambiguities should not cause any difficulties.

Remark 2.25 *As we see above, there are minor differences in the notation and terminology between real and complex Hilbert spaces. In what follows, we often discuss both cases at once. We then use the notation and terminology of complex Hilbert spaces, their modification to the real case being obvious.*

Definition 2.26 *Let $\mathcal{H}$ be a real or complex Hilbert space. A family of vectors $\{e_i\}_{i \in I}$ is called an* orthonormal system *if $(e_i|e_j) = \delta_{ij}$. If in addition $\mathrm{Span}^{\mathrm{cl}}\{e_i : i \in I\} = \mathcal{H}$, we say that it is an* orthonormal basis, *or an o.n. basis for brevity.*

Definition 2.27 *Let $\mathcal{H}$ be a topological vector space. We say that it is a* Hilbertizable space *if there exists a scalar product on $\mathcal{H}$ that generates its topology and $\mathcal{H}$ is complete in the corresponding norm.*

2.2.3 Operators on Hilbert spaces

In this subsection we discuss basic definitions concerning operators on complex and real Hilbert spaces. We try to be as close as possible to the usual terminology, fixing, however, some of its obvious flaws (see Remark 2.30).

We start with the complex case. Let $\mathcal{H}_1, \mathcal{H}_2, \mathcal{H}$ be complex Hilbert spaces. Let A be a densely defined operator from $\mathcal{H}_1$ to $\mathcal{H}_2$.

Definition 2.28 *The operator A^* from $\mathcal{H}_2$ to $\mathcal{H}_1$ defined by*

$$(\Phi_2, \Psi_1) \in \operatorname{Gr} A^* \Leftrightarrow (\Phi_2 | A\Phi_1) = (\Psi_1 | \Phi_1), \ \Phi_1 \in \operatorname{Dom} A,$$

is called the adjoint *of A. We set $A^\# := \overline{A}^* = \overline{A^*}$, which is an operator from $\overline{\mathcal{H}}_2$ to $\overline{\mathcal{H}}_1$.*

Note that A^* and $A^\#$ are automatically closed. Moreover, A is closable iff $\operatorname{Dom} A^*$, or $\operatorname{Dom} A^\#$ is dense. We then have $A^{**} = A^{\#\#} = A^{\mathrm{cl}}$.

If A is bounded, then so are A^* and $A^\#$. As an example of adjoints, consider $\Phi \in \mathcal{H}$ and let us note the identities $|\Phi)^* = (\Phi|$ (see Def. 1.70).

Definition 2.29 (1) *Densely defined operators on $\mathcal{H}$ satisfying $A \subset A^*$ are called* Hermitian.
(2) *Densely defined operators from $\overline{\mathcal{H}}$ to $\mathcal{H}$ satisfying $A \subset A^\#$ are called* symmetric.

Remark 2.30 *Note that, unfortunately, in a part of the literature the word "symmetric" is often used to denote Hermitian operators. This is an incorrect usage.*

Definition 2.31 (1) *Densely defined operators on $\mathcal{H}$ satisfying $A^* = A$ are called* self-adjoint *and those satisfying $A^* = -A$* anti-self-adjoint. *The set of bounded self-adjoint operators on $\mathcal{H}$ is denoted by $B_{\mathrm{h}}(\mathcal{H})$, and the set of all self-adjoint operators on $\mathcal{H}$ by $Cl_{\mathrm{h}}(\mathcal{H})$.*
(2) *The set of bounded symmetric, resp. anti-symmetric operators from $\overline{\mathcal{H}}$ to $\mathcal{H}$ is denoted $B_{\mathrm{s}}(\overline{\mathcal{H}}, \mathcal{H})$, resp. $B_{\mathrm{a}}(\overline{\mathcal{H}}, \mathcal{H})$. The set of all operators from $\overline{\mathcal{H}}$ to $\mathcal{H}$ satisfying $A = A^\#$, resp. $A = -A^\#$ is denoted $Cl_{\mathrm{s}}(\overline{\mathcal{H}}, \mathcal{H})$, resp. $Cl_{\mathrm{a}}(\overline{\mathcal{H}}, \mathcal{H})$.*

Self-adjoint and anti-self-adjoint operators are automatically closed. Likewise, operators in $Cl_{\mathrm{s}}(\overline{\mathcal{H}}, \mathcal{H})$ and $Cl_{\mathrm{a}}(\overline{\mathcal{H}}, \mathcal{H})$ are automatically closed.

A is anti-self-adjoint iff iA is self-adjoint.

Let us now consider the real case. Let $\mathcal{H}_1, \mathcal{H}_2, \mathcal{H}$ be real Hilbert spaces. Let A be a densely defined operator from $\mathcal{H}_1$ to $\mathcal{H}_2$.

Definition 2.32 *The operator $A^\#$ from $\mathcal{H}_2$ to $\mathcal{H}_1$ defined by*

$$(\Phi_2, \Psi_1) \in \operatorname{Gr} A^\# \Leftrightarrow \langle \Phi_2 | A\Phi_1 \rangle = \langle \Psi_1 | \Phi_1 \rangle, \ \Phi_1 \in \operatorname{Dom} A,$$

is called the adjoint *of A.*

Note that $A^\#$ is automatically closed. Moreover, A is closable iff $\operatorname{Dom} A^\#$ is dense and we then have $A^{\#\#} = A^{\mathrm{cl}}$.

If A is bounded, then so is $A^\#$. As an example of adjoints, consider $\Phi \in \mathcal{H}$ and let us note the identity $|\Phi)^\# = \langle \Phi|$ (see Def. 1.24).

Definition 2.33 *Densely defined operators on $\mathcal{H}$ satisfying $A \subset A^\#$ are called* symmetric.

Definition 2.34 *Densely defined operators on $\mathcal{H}$ satisfying $A^\# = A$, resp.
$A^\# = -A$ are called* self-adjoint, *resp.* anti-self-adjoint. *The set of bounded self-
adjoint, resp. anti-self-adjoint operators on $\mathcal{H}$ is denoted by $B_s(\mathcal{H})$, resp. $B_a(\mathcal{H})$.
The set of all self-adjoint, resp. anti-self-adjoint operators on $\mathcal{H}$ is denoted by
$Cl_s(\mathcal{H})$, resp. $Cl_a(\mathcal{H})$.*

Self-adjoint and anti-self-adjoint operators are automatically closed.

2.2.4 Product of a closed and a bounded operator

Proposition 2.35 *Let $G \in Cl(\mathcal{H}_1, \mathcal{H}_2)$, $H \in B(\mathcal{H}_2, \mathcal{H}_3)$. We define HG and
G^*H^* with their natural domains, as in Def. 2.23. Then HG is densely defined,
so that we can define its adjoint, and we have*

$$(HG)^* = G^*H^*. \tag{2.1}$$

Besides, G^*H^* is closed.

Proof By Def. 2.23,

$$\mathrm{Dom}\, HG = \mathrm{Dom}\, G, \tag{2.2}$$
$$\mathrm{Dom}\, G^*H^* = \{\Phi \in \mathcal{H}_3 : \ H^*\Phi \in \mathrm{Dom}\, G^*\}. \tag{2.3}$$

G is densely defined. By (2.2), so is HG. It immediately follows that

$$(HG)^* \supset G^*H^*.$$

Suppose that $\Psi \in \mathrm{Dom}(HG)^*$. This means that for some C

$$|(\Psi|HG\Phi)| \le C\|\Phi\|, \quad \Phi \in \mathrm{Dom}\, G.$$

Thus

$$|(H^*\Psi|G\Phi)| \le C\|\Phi\|, \quad \Phi \in \mathrm{Dom}\, G.$$

Hence, $H^*\Psi \in \mathrm{Dom}\, G^*$. Thus

$$(HG)^* \subset G^*H^*.$$

This ends the proof of (2.1). G^*H^* is closed as the adjoint of a densely defined
operator.   □

2.2.5 Compact operators

Let $\mathcal{H}_1, \mathcal{H}_2, \mathcal{H}$ be real or complex Hilbert spaces.

Definition 2.36 *We denote by $B_\infty(\mathcal{H}_1, \mathcal{H}_2)$ the space of compact operators from
$\mathcal{H}_1$ to $\mathcal{H}_2$ and set $B_\infty(\mathcal{H}) := B_\infty(\mathcal{H}, \mathcal{H})$.*

Proposition 2.37 *If $A \in B_\infty(\mathcal{H})$ is self-adjoint, then $\mathcal{H}$ has an o.n. basis $\{e_j\}_{j \in I}$ of eigenvectors of A for a family $\{\lambda_j\}_{j \in I}$ of real eigenvalues having 0 as its only possible accumulation point.*

2.2.6 Hilbert–Schmidt and trace-class operators

Let $\mathcal{H}_1, \mathcal{H}_2, \mathcal{H}$ be real or complex Hilbert spaces.

Definition 2.38 *$A \in B(\mathcal{H}_1, \mathcal{H}_2)$ is called* Hilbert–Schmidt *if $\mathrm{Tr}\, A^* A < \infty$. The space of Hilbert–Schmidt operators is denoted $B^2(\mathcal{H}_1, \mathcal{H}_2)$ and is a Hilbert space for the scalar product $\mathrm{Tr}\, B^* A$.*

Definition 2.39 *If $A \in B(\mathcal{H}_1, \mathcal{H}_2)$, then $|A| := \sqrt{A^* A}$ is called the* absolute value *of A. We say that A is* trace class *if $\mathrm{Tr}|A| < \infty$. The space of trace-class operators is denoted $B^1(\mathcal{H}_1, \mathcal{H}_2)$.*

Note the following proposition:

Proposition 2.40 *Let $A \in B^1(\mathcal{H})$ and $B_n \in B(\mathcal{H})$, with $B_n \to B$ weakly. Then $\mathrm{Tr}\, B_n A \to \mathrm{Tr}\, BA$.*

Definition 2.41 *Positive elements of $B^1(\mathcal{H})$ having trace 1 are called* density matrices.

Definition 2.42 *If $\beta > 0$ is a number, H a self-adjoint operator and $\mathrm{Tr}\, e^{-\beta H} < \infty$, then the density matrix*

$$e^{-\beta H} / \mathrm{Tr}\, e^{-\beta H}$$

is called the Gibbs density matrix *for the Hamiltonian H and inverse temperature β.*

Definition 2.43 *For $1 \leq p < \infty$, the* p-th Schatten ideal *is*

$$B^p(\mathcal{H}_1, \mathcal{H}_2) := \big\{ A \in B(\mathcal{H}_1, \mathcal{H}_2) \; : \; \mathrm{Tr}|A|^p < \infty \big\}.$$

2.2.7 Fredholm determinant

Let $\mathcal{H}$ be a real or complex Hilbert space.

Definition 2.44 *We denote by $\mathbb{1} + B^1(\mathcal{H})$ the set of operators of the form $\mathbb{1} + A$ with $A \in B^1(\mathcal{H})$. If $\mathcal{H}$ is a complex, resp. real Hilbert space, we set*

$$U_1(\mathcal{H}) := U(\mathcal{H}) \cap (\mathbb{1} + B^1(\mathcal{H})), \; \text{resp.} \; O_1(\mathcal{H}) := O(\mathcal{H}) \cap (\mathbb{1} + B^1(\mathcal{H})).$$

Theorem 2.45 *There exists a unique function $\mathbb{1} + B^1(\mathcal{H}) \ni R \mapsto \det R \in \mathbb{C}$ satisfying the following properties:*

(1) *If $\mathcal{H} = \mathcal{H}_1 \oplus \mathcal{H}_2$ with $\dim \mathcal{H}_1 < \infty$ and $R = R_1 \oplus \mathbb{1}$, then $\det R = \det R_1$, where $\det R_1$ is the usual determinant of the finite-dimensional operator R_1.*

(2) *$B^1(\mathcal{H}) \ni A \mapsto \det(\mathbb{1} + A)$ is continuous in the trace norm.*

Definition 2.46 *$\det R$ is called the* Fredholm determinant *of R.*

The following properties follow easily from Thm. 2.45:

Proposition 2.47 (1) *$\det R_1 R_2 = \det R_1 \det R_2$, $\det R^* = \overline{\det R}$.*
(2) *Let $A \in B^1(\mathcal{H})$. Then $\mathbb{1} + A$ is invertible iff $\det(\mathbb{1} + A) \neq 0$.*
(3) *If $\mathcal{H}$ is a complex, resp. real Hilbert space, then*

$$|\det R| = 1, \text{ for } R \in U_1(\mathcal{H}), \text{ resp. } \det R = \pm 1, \text{ for } R \in O_1(\mathcal{H}).$$

Definition 2.48 *Let $A \in B^2(\mathcal{H})$. The* regularized determinant *of $\mathbb{1} + A$ is*

$$\det_2(\mathbb{1} + A) := \det\big((\mathbb{1} + A)e^{-A}\big). \tag{2.4}$$

The regularized determinant can sometimes be used instead of the usual determinant.

Proposition 2.49 *Let $A \in B^2(\mathcal{H})$. Then $\mathbb{1} + A$ is invertible iff $\det_2(\mathbb{1} + A) \neq 0$.*

2.2.8 Derivatives

For functions on a vector space, one can distinguish several kinds of derivatives. In the following definition we recall the directional derivative, the Gâteaux derivative and the (most commonly used) Fréchet derivative.

Let $\mathcal{Y}$ be a real or complex vector space and G be a complex-valued function defined on a subset U of $\mathcal{Y}$. To define the directional derivative of G at a point $y_0 \in U$, U has to be *finitely open*, i.e. the intersection of U with any finite-dimensional subspace of $\mathcal{Y}$ should be open (for its canonical topology).

Definition 2.50 *Let $\mathcal{Y}$ be a real or complex normed space and G be a complex-valued function defined on a subset U of $\mathcal{Y}$.*

(1) *Assume that U is finitely open. We say that the* derivative of G in the direction of $y \in \mathcal{Y}$ at y_0 *exists if*

$$y \cdot \nabla G(y_0) := \frac{\mathrm{d}}{\mathrm{d}t} G(y_0 + ty)\big|_{t=0} \text{ exists.}$$

(Here t is a real parameter if $\mathcal{Y}$ is real, and complex if $\mathcal{Y}$ is complex.)
We say that G is Gâteaux differentiable *at y_0 if*

$$\mathcal{D} := \{y \in \mathcal{Y} : y \cdot \nabla G(y_0) \text{ exists}\}$$

is a dense linear subspace of $\mathcal{Y}$ and the map

$$\mathcal{D} \ni y \mapsto y \cdot \nabla G(y_0) \in \mathbb{C}$$

is a bounded linear functional.

(2) *Assume that U is open. We say that G is* Fréchet differentiable *at y_0 if there exists a bounded linear functional v such that*

$$\lim_{y \to 0} \frac{G(y_0 + y) - G(y_0) - v \cdot y}{\|y\|} = 0.$$

If such a functional exists, it is necessarily unique and is denoted $\nabla G(y_0)$.

Note that if the Fréchet derivative exists, then so does the Gâteaux derivative, and they are equal.

For example, consider the function $\operatorname{Dom} H^{\frac{1}{2}} \ni y \mapsto G(y) = (y|Hy)$, where H is a positive self-adjoint operator. The set $\operatorname{Dom} H^{\frac{1}{2}}$ is obviously finitely open. G is Gâteaux differentiable at y_0 iff $y_0 \in \operatorname{Dom} H$. It is Fréchet differentiable iff H is bounded.

2.3 Functional calculus

2.3.1 Holomorphic functional calculus

Let $\mathcal{H}$ be a Banach space and $A \in B(\mathcal{H})$. The basic construction of the *holomorphic functional calculus* is described in the following definition:

Definition 2.51 *Let f be a function on* $\operatorname{spec} A$ *that extends to a function holomorphic on an open neighborhood of* $\operatorname{spec} A$. *Let γ be a closed curve encircling* $\operatorname{spec} A$ *counterclockwise and contained in the domain of f. We set*

$$f(A) := \frac{1}{2\pi i} \oint_{\gamma} f(z)(z\mathbb{1} - A)^{-1} dz. \tag{2.5}$$

It is easy to see that (2.5) does not depend on the choice of the curve γ.

Let Θ be a subset of $\operatorname{spec} A$.

Definition 2.52 *The* characteristic function *of the set Θ is defined as*

$$\mathbb{1}_{\Theta}(z) := \begin{cases} 1, & z \in \Theta, \\ 0, & z \in \operatorname{spec} A \backslash \Theta. \end{cases}$$

Suppose that Θ is a relatively open and closed subset of $\operatorname{spec} A$. Then the function $\mathbb{1}_{\Theta}$ satisfies the assumptions of the holomorphic spectral calculus.

Definition 2.53 $\mathbb{1}_{\Theta}(A)$ *is called the* (Riesz) spectral projection *of A onto Θ.*

Clearly, if γ encircles Θ, staying outside of $\operatorname{spec} A \backslash \Theta$, then

$$\mathbb{1}_{\Theta}(A) = \frac{1}{2\pi i} \oint_{\gamma} (z\mathbb{1} - A)^{-1} dz. \tag{2.6}$$

2.3.2 Functional calculus for normal operators

In the case of Hilbert spaces, besides the holomorphic calculus, we have another functional calculus based on the spectral theorem, which applies to normal operators.

Let us be more precise. Let $\mathcal{H}$ be a real or complex Hilbert space.

Definition 2.54 *An operator A on $\mathcal{H}$ is called* normal *if* $\operatorname{Dom} A = \operatorname{Dom} A^*$ *and* $(A\Phi|A\Psi) = (A^*\Phi|A^*\Psi)$, $\Phi, \Psi \in \operatorname{Dom} A$.

Self-adjoint and unitary operators are normal. In the case of normal operators the spectral theorem can be used to extend the functional calculus to a much larger class of functions.

Let A be a normal operator on a complex Hilbert space.

Definition 2.55 *If $f : \operatorname{spec} A \to \mathbb{C}$ is Borel, we define $f(A)$ by the* functional calculus for normal operators.

For normal operators we can extend the definition of spectral projections to a much larger class of sets.

Definition 2.56 *Let Θ be a Borel subset of $\operatorname{spec} A$. The operator $\mathbb{1}_\Theta(A)$ is called the* spectral projection *of A onto Θ.*

Let us now consider the functional calculus on real Hilbert spaces. Let $\mathcal{H}$ be a real Hilbert space and A a normal operator on $\mathcal{H}$. Then we can apply the functional calculus to the operator $A_\mathbb{C}$ on $\mathbb{C}\mathcal{H}$. Note that $\operatorname{spec} A_\mathbb{C}$ satisfies $\overline{\operatorname{spec} A_\mathbb{C}} = \operatorname{spec} A_\mathbb{C}$. If a Borel function f on $\operatorname{spec} A$ satisfies

$$f(\bar{z}) = \overline{f(z)}, \tag{2.7}$$

then $f(A_\mathbb{C})$ preserves $\mathcal{H}$, and the formula $f(A) := f(A_\mathbb{C})\big|_{\mathcal{H}}$ defines an operator on $\mathcal{H}$.

These conditions are satisfied, for instance, if A is a self-adjoint operator on $\mathcal{H}$ and f is a real Borel function. Note that in this case $f(A)$ is a self-adjoint operator on $\mathcal{H}$.

Let us describe another application of functional calculus on real Hilbert spaces that we will need. Let $R \in O(\mathcal{H})$ be such that $\operatorname{Ker}(R + \mathbb{1}) = \{0\}$. Consider the function $f(z) = z^t$ for $t \in \mathbb{R}$, where if $t \notin \mathbb{Z}$ we take the principal branch of z^t, with a cut along the negative semi-axis. Note that z^t is not defined for $z = -1$. However, $\mathbb{1}_{\{-1\}}(R_\mathbb{C}) = 0$; therefore $R_\mathbb{C}^t$ is well defined. Moreover z^t satisfies (2.7), so we can define R^t. Note that $R^t \in O(\mathcal{H})$ and $R^t R^s = R^{t+s}$. For $|t| \leq 1$, we have $\operatorname{Ker}(R^t + \mathbb{1}) = \{0\}$ and $(R^t)^s = R^{ts}$.

2.3.3 Spectrum of the product of operators

It is well known that if $A, B \in B(\mathcal{H})$, then

$$\operatorname{spec}(AB)\backslash\{0\} = \operatorname{spec}(BA)\backslash\{0\}.$$

This is also true if AB and BA are closed with $\mathrm{spec}\,(AB)$, $\mathrm{spec}\,(BA) \neq \mathbb{C}$; see Hardt–Konstantinov–Mennicken (2000). We will need the following related facts:

Proposition 2.57 (1) *Let A, B be two linear operators on a Hilbert space $\mathcal{H}$ such that AB and BA are closed. Let $z \in \mathbb{C}$ such that $z \notin \mathrm{spec}\,(AB) \cup \mathrm{spec}\,(BA)$. Then*

$$A(z\mathbb{1} - BA)^{-1} = (z\mathbb{1} - AB)^{-1}A.$$

Moreover, if $A, B \in B(\mathcal{H})$ and f is holomorphic near $\mathrm{spec}\,(AB) \cup \mathrm{spec}\,(BA)$, then

$$Af(BA) = f(AB)A.$$

(2) *If $A \in Cl(\mathcal{H})$ and f is a bounded Borel function, then*

$$Af(A^*A) = f(AA^*)A.$$

Proof Let $\Phi \in \mathrm{Dom}\,A$ and $(z\mathbb{1} - BA)\Psi = \Phi$. Then $BA\Psi = z\Psi - \Phi \in \mathrm{Dom}\,A$ and $ABA\Psi = zA\Psi - A\Phi$ hence $A\Psi \in \mathrm{Dom}\,AB$ and $(z\mathbb{1} - AB)A\Psi = A\Phi$. This proves (1).

To prove (2) we note that A^*A and AA^* are self-adjoint, so the identity $A(z\mathbb{1} - A^*A)^{-1} = (z\mathbb{1} - AA^*)^{-1}A$ for $z \in \mathbb{C}\backslash\mathbb{R}$ is true by (1). It extends by the usual argument to all bounded Borel functions. □

2.3.4 Scale of Hilbert spaces associated with a positive operator

Let $\mathcal{H}$ be a real or complex Hilbert space.

Definition 2.58 *For an operator B on $\mathcal{H}$ we will write $B \geq 0$ if it is positive self-adjoint. If in addition 0 is not an eigenvalue of B, then we will write $B > 0$.*

Let $B > 0$. Let us introduce the scale of Hilbert spaces associated with B. The Hilbert space $\mathcal{H}$ will play the role of a "pivot" space.

If $\mathcal{H}$ is real, we will identify $\mathcal{H}^{\#}$ with $\mathcal{H}$, and if $\mathcal{H}$ is complex, we identify $\mathcal{H}^*$ with $\mathcal{H}$, using the scalar product.

Definition 2.59 *We equip $\mathrm{Dom}\,B^{-s}$ with the scalar product $(\Phi|\Psi)_{-s} := (B^{-s}\Phi|B^{-s}\Psi)$ and the norm $\|B^{-s}\Phi\|$. We set*

$$B^s\mathcal{H} := \left(\mathrm{Dom}\,B^{-s}\right)^{\mathrm{cpl}}.$$

Proposition 2.60 (1) $B^{-s}\mathcal{H} = \mathrm{Dom}\,B^s$ *if $s \geq 0$ and $0 \notin \mathrm{spec}\,B$.*
(2) $B^t : \mathrm{Dom}\,B^{-s} \cap \mathrm{Dom}\,B^t \to \mathrm{Dom}\,B^{-s-t}$ *extends continuously to a unitary map from $B^s\mathcal{H}$ to $B^{s+t}\mathcal{H}$.*
(3) $(B^t)^s\mathcal{H} = B^{st}\mathcal{H}$.

(4) *If $\mathcal{H}$ is complex, the sesquilinear product $(\Psi|\Phi)$ on $\mathrm{Dom}\, B^s \times \mathrm{Dom}\, B^{-s}$ extends continuously to $B^{-s}\mathcal{H} \times B^s\mathcal{H}$ and one can unitarily identify $(B^s\mathcal{H})^*$ with $B^{-s}\mathcal{H}$.*

(5) *If $\mathcal{H}$ is real, the bilinear product $\langle\Psi|\Phi\rangle$ on $\mathrm{Dom}\, B^s \times \mathrm{Dom}\, B^{-s}$ extends continuously to $B^{-s}\mathcal{H} \times B^s\mathcal{H}$ and one can isometrically identify $(B^s\mathcal{H})^{\#}$ with $B^{-s}\mathcal{H}$.*

Definition 2.61 *If B_1, B_2 are two positive self-adjoint operators, we write $B_1 \leq B_2$ if $\mathrm{Dom}\, B_2^{\frac{1}{2}} \subset \mathrm{Dom}\, B_1^{\frac{1}{2}}$ and*

$$\|B_1^{\frac{1}{2}}\Phi\|^2 \leq \|B_2^{\frac{1}{2}}\Phi\|^2, \ \Phi \in \mathrm{Dom}\, B_2^{\frac{1}{2}}.$$

If $0 \leq B_1 \leq B_2$, then the Kato–Heinz theorem says that $0 \leq B_1^\alpha \leq B_2^\alpha$ for $\alpha \in [0,1]$. If $0 < B_1 \leq B_2$, then also $0 \leq B_2^{-\alpha} \leq B_1^{-\alpha}$, for $\alpha \in [0,1]$. This implies the following fact:

Proposition 2.62 *Let $0 < B_1 \leq B_2$ and $-\frac{1}{2} \leq \alpha \leq \frac{1}{2}$. Then the natural embeddings*

$$I_\alpha : B_1^\alpha\mathcal{H} \to B_2^\alpha\mathcal{H}$$

are contractive and $I_\alpha^* = I_{-\alpha}$.

Note also the following useful fact, which follows from the three lines theorem.

Proposition 2.63 *Let $B > 0$ be a self-adjoint operator. Let $\Psi \in \mathrm{Dom}\, B$. Then*

$$\{z \ : \ 0 \leq \mathrm{Re}\, z \leq 1\} \ni z \mapsto B^z\Psi$$

is a continuous function holomorphic in the interior of the domain and satisfying the bound

$$\|B^z\Psi\| \leq \|\Psi\|^{1-\mathrm{Re}\, z}\|B\Psi\|^{\mathrm{Re}\, z}.$$

2.3.5 C_0-semi-groups

Let $\mathcal{H}$ be a real or complex Hilbert space.

Definition 2.64 *A C_0-semi-group is a one-parameter semi-group $[0,\infty[\ni t \mapsto U(t) \in B(\mathcal{H})$ continuous in the strong topology. Every C_0-semi-group $U(t)$ has the generator A defined by*

$$\mathrm{Dom}\, A := \left\{\Phi \in \mathcal{H} \ : \ \mathrm{s} - \lim_{t \to 0} t^{-1}(U(t)\Phi - \Phi) =: A\Phi \ exists\right\}.$$

In such a case we will write $U(t) =: \mathrm{e}^{tA}$.

The generator of a C_0-semi-group is always closed and densely defined.

The set of generators of C_0-groups in $O(\mathcal{H})$ and $U(\mathcal{H})$ coincides with the set of anti-self-adjoint operators. This fact is known as *Stone's theorem*.

Definition 2.65 *If* $\mathbb{R} \ni t \mapsto U(t)$ *is a unitary* C_0*-group, then the* self-adjoint generator *of* $U(t)$ *is the operator* B *defined as* $U(t) = \mathrm{e}^{\mathrm{i}tB}$.

Definition 2.66 *A is a* maximal dissipative *operator if it is a closed densely defined operator such that* $\mathrm{Re}(\Phi|A\Phi) \leq 0$ *for* $\Phi \in \mathrm{Dom}\,A$ *and* $\mathrm{Ran}(-A + \lambda\mathbb{1}) = \mathcal{H}$ *for some* $\lambda > 0$.

The Hille–Yosida theorem says that the set of generators of C_0-semi-groups of contractions coincides with the set of maximal dissipative operators. A is maximal accretive if $-A$ is maximal dissipative.

2.3.6 Local Hermitian semi-groups

Let $\mathcal{H}$ be a real or complex Hilbert space. Clearly, if $[0,\infty[\ni t \mapsto U(t) \in B(\mathcal{H})$ is a C_0-semi-group of self-adjoint contractions, then $U(t) = \mathrm{e}^{-tA}$ for A positive self-adjoint.

The notion of local Hermitian semi-groups, due to Klein–Landau (1981a) and Fröhlich (1980), allows us to extend this construction to the case of semi-groups of *unbounded* Hermitian operators. It is particularly important in the Euclidean approach to quantum field theory, especially at positive temperatures.

Definition 2.67 *Let* $T > 0$. *A* local Hermitian semi-group $\{P(t), \mathcal{D}_t\}_{t \in [0,T]}$ *is a family of linear operators* $P(t)$ *on* $\mathcal{H}$ *and subspaces* $\mathcal{D}_t$ *of* $\mathcal{H}$ *such that*

(1) $\mathcal{D}_0 = \mathcal{H}$, $\mathcal{D}_t \supset \mathcal{D}_s$ *if* $0 \leq t \leq s \leq T$ *and* $\mathcal{D} = \underset{0 < t \leq T}{\cup} \mathcal{D}_t$ *is dense in* $\mathcal{H}$;

(2) $P(t)$ *is a Hermitian linear operator with* $\mathrm{Dom}\,P(t) = \mathcal{D}_t$ *such that* $P(0) = \mathbb{1}$, $P(s)\mathcal{D}_t \subset \mathcal{D}_{t-s}$ *for* $0 \leq s \leq t \leq T$, *and* $P(t)P(s) = P(t+s)$ *on* $\mathcal{D}_{t+s}$ *for* $t, s, t+s \in [0,T]$;

(3) $t \mapsto P(t)$ *is weakly continuous, i.e. for* $\Phi \in \mathcal{D}_s$ *the map* $[0,s] \ni t \mapsto (\Phi, P(t)\Phi)$ *is continuous.*

Remark 2.68 *In the literature, local Hermitian semi-groups are often called* local symmetric semi-groups.

An example of a local Hermitian semi-group is $P(t) = \mathrm{e}^{-tH}$, $\mathcal{D}_t = \mathrm{Dom}\,\mathrm{e}^{-tH}$, with $T = \infty$, if H is a self-adjoint operator on $\mathcal{H}$. The following theorem shows that all local Hermitian semi-groups are restrictions of groups of unbounded self-adjoint operators of this form.

Theorem 2.69 *Let* $\{P(t), \mathcal{D}_t\}_{t \in [0,T]}$ *be a local Hermitian semi-group on* $\mathcal{H}$. *Then there exists a unique self-adjoint operator* H *on* $\mathcal{H}$ *such that*

(1) $\mathcal{D}_t \subset \mathrm{Dom}\,\mathrm{e}^{-tH}$, $\mathrm{e}^{-tH}\big|_{\mathcal{D}_t} = P(t)$ *for* $0 \leq t \leq T$;

(2) *For any* $0 < T' \leq T$, $\underset{0 < t \leq T'}{\cup}\underset{0 < s < t}{\cup} P(s)\mathcal{D}_t$ *is a core for* H.

(The core of a Hermitian operator is defined in Subsect. 2.3.7). For the proof one needs a definition and a lemma due to Widder (1934).

Definition 2.70 *A continuous function* $r : [T_1, T_2] \to \mathbb{R}$ *is* OS *positive if for any* $n \in \mathbb{N}$ *and* $t_1, \ldots, t_n \in \mathbb{R}$ *such that* $T_1 \leq t_i + t_j \leq T_2$ *the matrix* $[r(t_i + t_j)]_{1 \leq i, j \leq n}$ *is positive.*

Lemma 2.71 *The continuous function* $r : [T_1, T_2] \to \mathbb{R}$ *is* OS *positive iff there exists a positive measure* ν *such that* $\lambda \mapsto e^{-t\lambda}$ *belongs to* $L^1(\mathbb{R}, d\nu)$ *for each* $t \in [T_1, T_2]$ *and*

$$r(t) = \int_{\mathbb{R}} e^{-t\lambda} d\nu(\lambda).$$

Proof of Thm. 2.69. We fix $0 < t < T$ and $\Phi \in \mathcal{D}_t$ and set $r(s) = \|P(s/2)\Phi\|^2$ for $s \in [0, 2t]$. The function r is continuous by the weak continuity of $P(s)$. Using the symmetry and semi-group property we see that r is OS positive on $[0, 2t]$. By Lemma 2.71, there exists a measure ν on $\mathbb{R}$ such that $r(s) = \int_{\mathbb{R}} e^{-s\lambda} d\nu(\lambda)$, $s \in [0, 2t]$. We note that

$$(P(s_1)\Phi | P(s_2)\Phi) = r(s_1 + s_2) = \int_{\mathbb{R}} e^{-s_1\lambda} e^{-s_2\lambda} d\nu(\lambda), \ 0 \leq s_1, s_2 \leq t. \quad (2.8)$$

For $z \in \mathbb{C}$, set $g_z(\lambda) := e^{-z\lambda}$. Since the span of $\{g_s \ : \ 0 \leq s \leq t\}$ is dense in the Hilbert space $L^2(\mathbb{R}, d\nu)$, we see that the map

$$J : L^2(\mathbb{R}, d\nu) \ni g_s \mapsto P(s)\Phi \in \mathcal{H}$$

extends by linearity and density to a unitary map between $L^2(\mathbb{R}, d\nu)$ and the closed span of $\{P(s)\Phi \ : \ s \in [0, t]\}$. The map

$$z \mapsto g_z(\lambda) \in L^2(\mathbb{R}, d\nu)$$

is clearly holomorphic in the strip $\{0 < \text{Re}\, z < t\}$ and continuous up to the boundary. Applying J, we obtain that the map $s \to P(s)\Phi$ is the restriction to $[0, t]$ of a map $z \mapsto \Phi(z)$ with the same properties. We define now

$$U(y)\Phi := \Phi(iy), \quad y \in \mathbb{R}. \quad (2.9)$$

Clearly, $U(y)$ is defined on $\mathcal{D}$. We claim that $U(y)$ extends to $\mathcal{H}$ as a strongly continuous unitary group. To prove that $U(y)$ is isometric, we use the identity

$$(\Phi(z_1) | \Phi(z_2)) = \int_{\mathbb{R}} e^{-(\overline{z_1} - z_2)\lambda} d\nu(\lambda),$$

which follows from (2.8) by analytic continuation. The map $U(y)$ is clearly linear on $\mathcal{D}$, if we note that $U(y)\Phi$ is independent of the space $\mathcal{D}_t$ to which Φ belongs and use that two vectors $\Phi, \Psi \in \mathcal{D}$ always belong to a common space $\mathcal{D}_t$. The strong continuity of $y \mapsto U(y)$ follows from the norm continuity of $\Phi(z)$.

To prove the group property, we pick $\Phi \in \mathcal{D}_t$ and set $\Phi(s_1, s_2) = P(s_1)P(s_2)\Phi = P(s_1 + s_2)\Phi$ for $s_1, s_2, s_1 + s_2 \in [0, t]$. We first analytically continue $\Phi(s_1, s_2)$ in s_1 to $\Phi(iy_1, s_2) = U(y_1)P(s_1)\Phi$ and then in s_2 to $\Phi(iy_1, iy_2) = U(y_1)U(y_2)\Phi$. Since $P(s)\Phi$ analytically continues to $\Phi(z)$, we see that

$P(s_1 + s_2)f$ analytically continues in (s_1, s_2) to $\Phi(iy_1 + iy_2) = U(y_1 + y_2)f$. Therefore, $U(y_1)U(y_2)\Phi = U(y_1 + y_2)\Phi$.

We now uniquely define a self-adjoint operator H by $U(y) =: e^{-iyH}$. We note that if $\Phi \in \mathcal{D}_t$, then

$$(\Phi|U(y)\Phi) = \int_{\mathbb{R}} e^{-iy\lambda} d\nu(\lambda),$$

hence $d\nu(\lambda) = d(\Phi|\mathbb{1}_{]-\infty,\lambda]}(H)\Phi)$, which implies that $\Phi \in \text{Dom} \, e^{-tH}$. The two functions $e^{-iyH}\Phi$ and $\Phi(iy)$ coincide and are the boundary values of the functions $e^{-zH}\Phi$ and $\Phi(z)$, both holomorphic in the strip $\{0 < \text{Re} \, z < t\}$ and continuous up to the boundary. It follows that these two holomorphic functions coincide everywhere and hence in particular

$$\Phi(t) = P(t)\Phi = e^{-tH}\Phi.$$

This shows the existence of a self-adjoint operator H satisfying (1). If H_1, H_2 are two such operators, then the same analytic continuation argument shows that $e^{-iyH_1}\Phi = e^{-iyH_2}\Phi$ for $\Phi \in \mathcal{D}$, which implies that $H_1 = H_2$. We refer to Klein–Landau (1981a) for the proof of (2). $\square$

2.3.7 Essential self-adjointness

Let A be a Hermitian linear operator on a Hilbert space $\mathcal{H}$, i.e. such that $A \subset A^*$.

Definition 2.72 *A is called* essentially self-adjoint *if A^{cl} is self-adjoint. If the domain $\mathcal{D}$ of A needs to be specified, one says that A is* essentially self-adjoint *on $\mathcal{D}$. If a self-adjoint operator A is the closure of $A\big|_{\mathcal{D}}$, one says that $\mathcal{D}$ is a* core *for A.*

Definition 2.73 *If A is any operator, vectors $\Phi \in \bigcap_n \text{Dom} \, A^n$ satisfying for some $t > 0$*

$$\sum_{n=0}^{\infty} \frac{t^n \|A^n \Phi\|}{n!} < \infty$$

are called analytic vectors *of A.*

Let us give three criteria for essential self-adjointness, all due to Nelson.

Theorem 2.74 (1) (Nelson's commutator theorem) *Let A be Hermitian and B self-adjoint positive on $\mathcal{H}$ with $\text{Dom} \, B \subset \text{Dom} \, A$. Assume that*

$$\|A\Phi\|^2 \leq C\|(B + \mathbb{1})\Phi\|^2, \quad |(A\Phi|B\Phi) - (B\Phi|A\Phi)| \leq C(\Phi|(B + \mathbb{1})\Phi),$$
$$\Phi \in \text{Dom} \, B.$$

Then A is essentially self-adjoint on $\text{Dom} \, B$.

(2) (Nelson's invariant domain theorem) *Consider $U_t = e^{itA}$, a strongly continuous unitary group on $\mathcal{H}$. Let $\mathcal{D}$ be a dense subspace of $\mathcal{H}$ such that*

$\mathcal{D} \subset \operatorname{Dom} A$ *and* $\mathcal{D}$ *is invariant under* U_t. *Then* A *is essentially self-adjoint on* $\mathcal{D}$.

(3) (Nelson's analytic vectors theorem) *Let* A *be a Hermitian operator possessing a dense space of analytic vectors. Then it is essentially self-adjoint on this space.*

A useful application of the notion of essential self-adjointness are the following two versions of Trotter's product formula:

Theorem 2.75 (1) *Let* A, B *be two self-adjoint operators on* $\mathcal{H}$ *such that* $A + B$ *with domain* $\operatorname{Dom} A \cap \operatorname{Dom} B$ *is essentially self-adjoint. Then*

$$\mathrm{e}^{\mathrm{i}t(A+B)^{\mathrm{cl}}} = \mathrm{s} - \lim_{n \to \infty} \left(\mathrm{e}^{\mathrm{i}tA/n} \mathrm{e}^{\mathrm{i}tB/n} \right)^n .$$

(2) *Suppose in addition that* A, B *are bounded below. Then*

$$\mathrm{e}^{-t(A+B)^{\mathrm{cl}}} = \mathrm{s} - \lim_{n \to \infty} \left(\mathrm{e}^{-tA/n} \mathrm{e}^{-tB/n} \right)^n , \quad t \geq 0.$$

2.3.8 Commuting self-adjoint operators

Let A_1, A_2 be self-adjoint operators on $\mathcal{H}$.

Definition 2.76 *We say that* A_1 *and* A_2 **commute** *if all their bounded Borel functions commute in the usual sense. (It is enough to demand e.g. that* $\mathrm{e}^{\mathrm{i}t_1 A_1}$ *commutes with* $\mathrm{e}^{\mathrm{i}t_2 A_2}$ *for any* $t_1, t_2 \in \mathbb{R}$.)

If $A_1, \ldots, A_n$ are commuting self-adjoint operators, then for any Borel function F on $\mathbb{R}^n$ we can define $F(A_1, \ldots, A_n)$ by the self-adjoint calculus.

One can generalize this as follows. Let $\mathcal{X}$ be a real vector space.

Definition 2.77 *We will say that*

$$\mathcal{X} \ni x \mapsto \langle x|A\rangle \in Cl_{\mathrm{h}}(\mathcal{H}) \tag{2.10}$$

is an $\mathcal{X}^{\#}$-*vector of commuting self-adjoint operators if there exists a unitary representation* $\mathcal{X} \ni x \mapsto U(x) \in U(\mathcal{H})$ *such that, for all* $x \in \mathcal{X}$, $\mathbb{R} \ni t \mapsto U(tx)$ *is strongly continuous and* $U(tx) = \mathrm{e}^{\mathrm{i}t\langle x|A\rangle}$.

Consider a vector of commuting self-adjoint operators (2.10). Clearly, $\langle x_1|A\rangle$, $\langle x_2|A\rangle$ commute for any $x_1, x_2 \in \mathcal{X}$. If F is a Borel function that depends on a finite-dimensional subspace of $\mathcal{X}^{\#}$, we can define $F(A)$ by the self-adjoint functional calculus.

Definition 2.78 C^{∞} *vectors for* (2.10) *are elements of*

$$\bigcap_{n=1}^{\infty} \bigcap_{x_1, \ldots, x_n \in \mathcal{X}} \operatorname{Dom}\left(\langle x_1|A\rangle \cdots \langle x_n|A\rangle \right).$$

2.3.9 Conjugations adapted to a self-adjoint operator

Let $\mathcal{H}$ be a complex Hilbert space. Recall that τ is a conjugation on $\mathcal{H}$ if it is an anti-unitary involution.

Proposition 2.79 *Let A be a self-adjoint operator on a (complex) Hilbert space $\mathcal{H}$. Then there exists a conjugation τ such that $\tau A \tau = A$. We then say that τ is adapted to A.*

Proof By the spectral theorem, there exists a collection $\{Q_i, \mu_i\}_{i \in I}$ of measure spaces such that $\mathcal{H} = \underset{i \in I}{\oplus} L^2(Q_i, \mu_i)$ and A is unitarily equivalent to the multiplication by a real measurable function. Then we take the standard conjugation on $\underset{i \in I}{\oplus} L^2(Q_i, \mu_i)$. $\qquad\square$

2.4 Polar decomposition

Every operator on a Hilbert space possesses a canonical decomposition into the product of a positive operator and a partial isometry. It is called the *polar decomposition*. In this section we discuss various forms and consequences of the polar decomposition of an operator on a complex or real Hilbert space.

We will mostly consider the polar decomposition for operators that have a trivial kernel and co-kernel. In this case the decomposition into a positive operator and a partial isometry (which in this case is a unitary, resp. orthogonal operator) is unique, and not only canonical.

2.4.1 Polar decomposition

Let $\mathcal{H}, \mathcal{K}$ be real or complex Hilbert spaces and $A \in Cl(\mathcal{H}, \mathcal{K})$.

Theorem 2.80 *There exist a unique positive operator $|A| \in Cl(\mathcal{H})$ and a unique partial isometry $U \in B(\mathcal{H}, \mathcal{K})$ such that $A = U|A|$ and $\mathrm{Ker}\,|A| = (\mathrm{Ran}\,U)^{\perp}$. We have $|A| := (A^*A)^{\frac{1}{2}}$. Moreover one has $A = |A^*|U$ for $|A^*| = (AA^*)^{\frac{1}{2}}$.*

Definition 2.81 *The decomposition $A = |U|A$ described in Thm. 2.80 is called the* polar decomposition *of A.*

We will actually mostly need a special case of the polar decomposition, described in the following proposition:

Proposition 2.82 *Assume that $\mathrm{Ker}\,A = \{0\}$ and $\mathrm{Ran}\,A$ is dense in $\mathcal{K}$. Then there exists a unique positive operator $|A|$ and a unique orthogonal, resp. unitary operator U such that*

$$A = U|A| = |A^*|U. \tag{2.11}$$

2.4.2 *Polar decomposition of self-adjoint and anti-self-adjoint operators*

In the self-adjoint case the polar decomposition has additional properties:

Proposition 2.83 *Let A be a self-adjoint operator on a real or complex Hilbert space. Assume that $\operatorname{Ker} A = \{0\}$. Let $A = U|A|$ be the polar decomposition of A. Then $|A|U = U|A|$ and $U^2 = \mathbb{1}$.*

Next let us consider anti-self-adjoint operators. Only the real case is interesting, because in the complex case the multiplication of anti-self-adjoint operators by the imaginary unit makes them self-adjoint. Therefore, until the end of this subsection $\mathcal{H}$ will be a real Hilbert space.

Proposition 2.84 (1) *Let A be an anti-self-adjoint operator on $\mathcal{H}$ such that $\operatorname{Ker} A = \{0\}$. Let $A = U|A|$ be its polar decomposition. Then $U \in O(\mathcal{H})$, $U^2 = -\mathbb{1}$ (U is a Kähler anti-involution) and $U|A| = |A|U$.*

(2) *Let $R \in O(\mathcal{H})$ such that $\operatorname{Ker}(R^2 - \mathbb{1}) = \{0\}$. Define $C = \frac{1}{2}(R + R^*)$. Then $-\mathbb{1} \leq C \leq \mathbb{1}$. Moreover, we have the polar decomposition $\frac{1}{2}(R - R^*) = V\sqrt{\mathbb{1} - C^2}$, where $V \in O(\mathcal{H})$, $V^2 = -\mathbb{1}$ and $[V, C] = 0$. Finally, we have $R = C + V\sqrt{\mathbb{1} - C^2}$.*

Proof (1) The identity $A = U|A| = |A^*|U$ implies that $U = -U^*$ since $A = -A^*$. Since $U \in O(\mathcal{H})$, we have $U^2 = -\mathbb{1}$.

(2) Since $R \in O(\mathcal{H})$, we get that $-\mathbb{1} \leq C \leq \mathbb{1}$. The operator $\frac{1}{2}(R - R^*)$ is anti-self-adjoint and has a zero kernel since $\operatorname{Ker}(R^2 - \mathbb{1}) = \{0\}$. Moreover,

$$\frac{1}{2}(R - R^*)^* \frac{1}{2}(R - R^*) = \frac{1}{4}(2\mathbb{1} - R^2 - R^{*2}) = \mathbb{1} - C^2. \qquad (2.12)$$

Applying (1), we get that $V^2 = -\mathbb{1}$ and $[V, \sqrt{\mathbb{1} - C^2}] = 0$. Also $\sqrt{\mathbb{1} - C^2}[V, C] = [\sqrt{\mathbb{1} - C^2}V, C] = [R, C] = 0$. Since by (2.12) we know that $\operatorname{Ker}(\mathbb{1} - C^2) = 0$, this implies that $[V, C] = 0$. $\qquad \square$

Let $R \in O(\mathcal{H})$. Set $\mathcal{H}_{\pm} := \operatorname{Ker}(R \mp \mathbb{1})$ and $\mathcal{H}_1 := (\mathcal{H}_- + \mathcal{H}_+)^{\perp}$. Then $\mathcal{H}_1$ is a subspace invariant w.r.t. R and $(R^2 - \mathbb{1})\big|_{\mathcal{H}_1}$ has a trivial kernel. Thus Prop. 2.84 can be applied also in situations when $\operatorname{Ker} A$ and $\operatorname{Ker}(R^2 - \mathbb{1})$ are non-trivial.

Corollary 2.85 (1) *Let A be an anti-self-adjoint compact operator. Then there exists an o.n. basis $\{e_{i\pm}, e_j\}_{i \in I, j \in J}$ and real numbers $\{\lambda_i\}_{i \in I}$ with $\lambda_i > 0$ such that*

$$Ae_{i+} = \lambda_i e_{i-}, \quad Ae_{i-} = -\lambda_i e_{i+}, \quad Ae_j = 0.$$

(2) *Let $R \in O(\mathcal{H}) \cap (\mathbb{1} + B_\infty(\mathcal{H}))$. Then there exist an o.n. basis $\{e_{i\pm}, f_j, g_k\}_{i \in I, j \in J, k \in K}$ and numbers $\{\theta_i\}_{i \in I}$ with $\operatorname{Im} \theta_i > 0$ such that*

$$Re_{i+} = \theta_i e_{i-}, \quad Re_{i-} = \bar{\theta}_i e_{i+}, \quad Rf_j = f_j, \quad Rg_k = -g_k.$$

Proof Since A preserves $(\operatorname{Ker} A)^{\perp}$, we can assume that $\operatorname{Ker} A = \{0\}$. Let $A = V|A|$ the polar decomposition of A. Let $\{\lambda_i\}_{i \in I}$ be the eigenvalues of $|A|$ and $\mathcal{H}_i = \operatorname{Ker}(|A| - \lambda_i)$. Then $\mathcal{H}_i$ is invariant under V, so V is a Kähler anti-involution of $\mathcal{H}_i$. Let $(e_1, \cdots, e_n)$ be an o.n. basis of the complex Hilbert space $\mathbb{C}\mathcal{H}_i$. We set $e_{j+} = e_j$, $e_{j-} = Ve_j$, so that $(e_{1+}, \cdots e_{n+}, e_{1-}, \ldots, e_{n-})$ is an o.n. basis of the real Hilbert space $\mathcal{H}_i$ and $Ae_{j+} = \lambda_i e_{j-}$, $Ae_{j-} = -\lambda_i e_{j+}$. Collecting the above bases of $\mathcal{H}_i$ we obtain the first statement of the corollary. $\qquad\square$

Proposition 2.86 *Let $(\mathcal{Y}, \nu, \omega, \mathrm{j})$ be a complete Kähler space.*

(1) *Let A be a self-adjoint or anti-self-adjoint operator on $(\mathcal{Y}, \nu)$ such that $\operatorname{Ker} A = \{0\}$ and $A\mathrm{j} = \mathrm{j}A$. Let $|A|$, U be as in Prop. 2.83 or Prop. 2.84 (1). Then $\mathrm{j}|A| = |A|\mathrm{j}$, $U\mathrm{j} = \mathrm{j}U$.*

(2) *Let $R \in O(\mathcal{Y})$ such that $\operatorname{Ker}(R^2 - \mathbb{1}) = \{0\}$ and $R\mathrm{j} = \mathrm{j}R$. Let C, V be as in Prop. 2.84 (2). Then $V\mathrm{j} = \mathrm{j}V$ and $\mathrm{j}C = C\mathrm{j}$.*

Proof To prove (1) we use that $\mathrm{j}^* = -\mathrm{j}$, since (ν, j) is Kähler, and hence $[A^*, \mathrm{j}] = 0$. This implies that $[A^*A, \mathrm{j}] = 0$ and hence $[|A|, \mathrm{j}] = 0$, $[V, \mathrm{j}] = 0$. The proof of (2) is similar. $\qquad\square$

2.4.3 Polar decomposition of symmetric and anti-symmetric operators

In this subsection $\mathcal{H}$ is a complex Hilbert space. We use the notation $A^{\#} = \overline{A}^{*}$ defined in Subsect. 2.2.3. Recall that $Cl_{\mathrm{s/a}}(\overline{\mathcal{H}}, \mathcal{H})$ stands for the set of operators A from $\overline{\mathcal{H}}$ to $\mathcal{H}$ satisfying $A = A^{\#}$, resp. $A = -A^{\#}$.

Proposition 2.87 *Let $A \in Cl_{\mathrm{s/a}}(\overline{\mathcal{H}}, \mathcal{H})$ such that $\operatorname{Ker} A = \{0\}$. Consider the polar decomposition $A = U|A|$. Then we have*

$$U \in U(\overline{\mathcal{H}}, \mathcal{H}), \quad U|A| = |\overline{A}|U, \quad \overline{U}U = \pm\mathbb{1}. \tag{2.13}$$

Proof Consider the real Hilbert space $\mathcal{H}_{\mathbb{R}}$, that is, the realification of $\mathcal{H}$. It can be identified with the realification of $\overline{\mathcal{H}}$. Let $A_{\mathbb{R}}$ denote the operator A understood as an operator on $\mathcal{H}_{\mathbb{R}}$. It is easy to see that

$$(A^{\#})_{\mathbb{R}} = (A_{\mathbb{R}})^{\#},$$

where the superscript $^{\#}$ is defined in the complex sense on the left and in the real sense on the right. Therefore, $A_{\mathbb{R}}^{\#} = \pm A_{\mathbb{R}}$. By the real case of Prop. 2.83, resp. Prop. 2.84 (1), we obtain $A_{\mathbb{R}} = U_{\mathbb{R}}|A_{\mathbb{R}}|$ with $U_{\mathbb{R}} \in O(\mathcal{H}_{\mathbb{R}})$:

$$U_{\mathbb{R}} \in O(\mathcal{H}_{\mathbb{R}}), \quad U_{\mathbb{R}}|A_{\mathbb{R}}| = |A_{\mathbb{R}}|U_{\mathbb{R}}, \quad U_{\mathbb{R}}^2 = \pm\mathbb{1}. \tag{2.14}$$

Then we go back from $\mathcal{H}_{\mathbb{R}}$ to $\mathcal{H}$ and $\overline{\mathcal{H}}$, and (2.14) becomes (2.13). $\qquad\square$

Corollary 2.88 (1) *Let $A \in B_\mathrm{s}(\overline{\mathcal{H}}, \mathcal{H})$ be compact. Then there exists an o.n. basis of $(\operatorname{Ker} A)^\perp$, $\{e_i\}_{i \in I}$, and positive numbers $\{\lambda_i\}_{i \in I}$ such that $A\overline{e}_i = \lambda_i e_i$.*

(2) *Let $A \in B_\mathrm{a}(\overline{\mathcal{H}}, \mathcal{H})$ be compact. Then there exists an o.n. basis of $(\operatorname{Ker} A)^\perp$, $\{e_{i+}, e_{i-}\}_{i \in I}$, and positive numbers $\{\lambda_i\}_{i \in I}$ such that $A\overline{e}_{i+} = \lambda_i e_{i-}$, $A\overline{e}_{i-} = -\lambda_i e_{i+}$.*

2.5 Notes

The standard reference for operators on Hilbert spaces is the four-volume monograph by Reed–Simon (1975, 1978a,b, 1980), and also the books by Kato (1976) and by Davies (1980).

The Fredholm and regularized determinants are discussed e.g. in Simon (1979).

Thm. 2.69 about local Hermitian semi-groups is shown in Klein–Landau (1981a) and Fröhlich (1980).

3

Tensor algebras

In this chapter we study various constructions related to the tensor product of vector spaces. In particular, we introduce *symmetric* and *anti-symmetric tensor algebras*, whose Hilbert space versions are called *bosonic* and *fermionic Fock spaces*. Fock spaces are fundamental tools used to describe quantum field theories in terms of particles.

We also discuss the notions of *determinants, volume forms* and *Pfaffians*, which are closely related to anti-symmetric tensors.

3.1 Direct sums and tensor products

There are several non-equivalent versions of the tensor product of two infinite-dimensional vector spaces. We will introduce two of them, which are especially useful: the *algebraic tensor product* and the *tensor product in the sense of Hilbert spaces*. The former will be denoted with $\overset{\mathrm{al}}{\otimes}$ and the latter with $\otimes$.

There is a similar problem with the direct sum of an infinite number of vector spaces, where we will introduce the *algebraic direct sum* $\overset{\mathrm{al}}{\oplus}$ and the *direct sum in the sense of Hilbert spaces* $\oplus$.

3.1.1 Direct sums

Recall that if $\mathcal{Y}_1, \ldots, \mathcal{Y}_n$ is a finite family of vector spaces, then

$$\underset{1 \leq i \leq n}{\oplus} \mathcal{Y}_i$$

stands for the direct sum of the spaces $\mathcal{Y}_i$, $i = 1, \ldots, n$; see Def. 1.2. It is equal to the Cartesian product $\underset{1 \leq i \leq n}{\prod} \mathcal{Y}_i$ with the obvious operations.

The notion of the direct sum can be generalized in several ways to the case of an infinite family of vector spaces. One of the most useful is described below.

Let $\{\mathcal{Y}_i\}_{i \in I}$ be a family of vector spaces.

Definition 3.1 *The* algebraic direct sum *of vector spaces* $\{\mathcal{Y}_i\}_{i \in I}$, *denoted*

$$\underset{i \in I}{\overset{\mathrm{al}}{\oplus}} \mathcal{Y}_i, \tag{3.1}$$

is the subspace of the Cartesian product $\underset{i \in I}{\prod} \mathcal{Y}_i$ *consisting of families with all but a finite number of terms equal to zero.*

Note that for a finite family of spaces the symbols $\oplus$ and $\overset{\text{al}}{\oplus}$ can be used interchangeably.

If $\{\mathcal{Y}_i\}_{i \in I}$ is a family of Hilbert spaces, then $\overset{\text{al}}{\underset{i \in I}{\oplus}} \mathcal{Y}_i$ has a natural scalar product

$$(\{y_i\}_{i \in I} | \{w_i\}_{i \in I}) = \sum_{i \in I} (y_i | w_i),$$

where $\{y_i\}_{i \in I}, \{w_i\}_{i \in I}$ are elements of $\overset{\text{al}}{\underset{i \in I}{\oplus}} \mathcal{Y}_i$.

Definition 3.2 *The* direct sum in the sense of Hilbert spaces *is defined as*

$$\underset{i \in I}{\oplus} \mathcal{Y}_i := \left(\overset{\text{al}}{\underset{i \in I}{\oplus}} \mathcal{Y}_i \right)^{\text{cpl}}.$$

3.1.2 Direct sums of operators

Let $\{\mathcal{Y}_i\}_{i \in I}, \{\mathcal{W}_i\}_{i \in I}$ be families of vector spaces.

Definition 3.3 *If $a_i \in L(\mathcal{Y}_i, \mathcal{W}_i)$, $i \in I$, then their* direct sum *is defined as the unique operator* $\underset{i \in I}{\oplus} a_i$ *in* $L\left(\overset{\text{al}}{\underset{i \in I}{\oplus}} \mathcal{Y}_i, \overset{\text{al}}{\underset{i \in I}{\oplus}} \mathcal{W}_i \right)$ *satisfying*

$$\left(\underset{i \in I}{\oplus} a_i \right) \{y_i\}_{i \in I} := \{a_i y_i\}_{i \in I}.$$

Let $\{\mathcal{Y}_i\}_{i \in I}, \{\mathcal{W}_i\}_{i \in I}$ be families of Hilbert spaces, and a_i, $i \in I$, be closable operators from $\mathcal{Y}_i$ to $\mathcal{W}_i$ with domains $\text{Dom}\, a_i$. Then the operator $\underset{i \in I}{\oplus} a_i$ with the domain $\overset{\text{al}}{\underset{i \in I}{\oplus}} \text{Dom}\, a_i$ is closable since

$$\underset{i \in I}{\oplus} a_i^* \subset \left(\underset{i \in I}{\oplus} a_i \right)^*.$$

Definition 3.4 *The closure of* $\underset{i \in I}{\oplus} a_i \in L\left(\overset{\text{al}}{\underset{i \in I}{\oplus}} \mathcal{Y}_i, \overset{\text{al}}{\underset{i \in I}{\oplus}} \mathcal{W}_i \right)$ *is denoted by the same symbol* $\underset{i \in I}{\oplus} a_i \in Cl\left(\underset{i \in I}{\oplus} \mathcal{Y}_i, \underset{i \in I}{\oplus} \mathcal{W}_i \right)$.

Clearly, $\underset{i \in I}{\oplus} a_i$ is bounded iff a_i are bounded and $\underset{i \in I}{\sup} \|a_i\| < \infty$, and then

$$\| \underset{i \in I}{\oplus} a_i \| = \underset{i \in I}{\sup} \|a_i\|.$$

Similarly, $\underset{i \in I}{\oplus} a_i$ is essentially self-adjoint on $\overset{\text{al}}{\underset{i \in I}{\oplus}} \text{Dom}\, a_i$ iff a_i are essentially self-adjoint.

3.1.3 Algebraic tensor product

Let $\mathcal{Y}, \mathcal{W}$ be vector spaces over $\mathbb{K}$. Let $\mathcal{Z} = c_c(\mathcal{Y} \times \mathcal{W}, \mathbb{K})$, that is, the space of finite linear combinations of $(y, w) \in \mathcal{Y} \times \mathcal{W}$ with coefficients in $\mathbb{K}$ (see Def 2.6). Let $\mathcal{Z}_0$ be the subspace of $\mathcal{Z}$ spanned by elements of the form

$$(y, w_1 + w_2) - (y, w_1) - (y, w_2), \ (y_1 + y_2, w) - (y_1, w) - (y_2, w),$$
$$(\lambda y, w) - \lambda(y, w), \ (y, \lambda w) - \lambda(y, w), \ \lambda \in \mathbb{K}, \ y, y_1, y_2 \in \mathcal{Y}, \ w, w_1, w_2 \in \mathcal{W}.$$

Definition 3.5 *The* algebraic tensor product *of $\mathcal{Y}$ and $\mathcal{W}$ is defined as*

$$\mathcal{Y} \overset{\mathrm{al}}{\otimes} \mathcal{W} := \mathcal{Z}/\mathcal{Z}_0.$$

The formula $y \otimes w := (y, w) + \mathcal{Z}_0$ defines the bilinear map

$$\mathcal{Y} \times \mathcal{W} \ni (y, w) \mapsto y \otimes w \in \mathcal{Y} \overset{\mathrm{al}}{\otimes} \mathcal{W},$$

called the tensor multiplication.

We have natural isomorphisms

$$\mathcal{Y} \simeq \mathbb{K} \overset{\mathrm{al}}{\otimes} \mathcal{Y} \simeq \mathcal{Y} \overset{\mathrm{al}}{\otimes} \mathbb{K}.$$

More generally, let $\mathcal{Y}_1, \ldots, \mathcal{Y}_n$ be a finite family of vector spaces. Let $\mathcal{Z} := c_c(\mathcal{Y}_1 \times \cdots \times \mathcal{Y}_n, \mathbb{K})$, that is, the vector space over $\mathbb{K}$ of finite linear combinations of $(y_1, \ldots, y_n) \in \mathcal{Y}_1 \times \cdots \times \mathcal{Y}_n$. Let $\mathcal{Z}_0$ be the subspace of $\mathcal{Z}$ spanned by elements of the form

$$(\ldots, y_j + y_j', \ldots) - (\ldots, y_j, \ldots) - (\ldots, y_j', \ldots),$$
$$(\ldots, \lambda y_j, \ldots) - \lambda(\ldots, y_j, \ldots), \ \lambda \in \mathbb{K}, \ y_i, y_i' \in \mathcal{Y}_i, \ i = 1, \ldots, n.$$

Definition 3.6 *The* algebraic tensor product *of $\mathcal{Y}_1, \ldots, \mathcal{Y}_n$ is defined as*

$$\mathcal{Y}_1 \overset{\mathrm{al}}{\otimes} \cdots \overset{\mathrm{al}}{\otimes} \mathcal{Y}_n := \mathcal{Z}/\mathcal{Z}_0.$$

The formula $y_1 \otimes \cdots \otimes y_n := (y_1, \ldots, y_n) + \mathcal{Z}_0$ defines the n-linear map

$$\mathcal{Y}_1 \times \cdots \times \mathcal{Y}_n \ni (y_1, \ldots, y_n) \mapsto y_1 \otimes \cdots \otimes y_n \in \mathcal{Y}_1 \overset{\mathrm{al}}{\otimes} \cdots \overset{\mathrm{al}}{\otimes} \mathcal{Y}_n,$$

called the tensor multiplication.

We have a natural identification

$$\mathcal{Y}_1 \overset{\mathrm{al}}{\otimes} (\mathcal{Y}_2 \overset{\mathrm{al}}{\otimes} \mathcal{Y}_3) \simeq (\mathcal{Y}_1 \overset{\mathrm{al}}{\otimes} \mathcal{Y}_2) \overset{\mathrm{al}}{\otimes} \mathcal{Y}_3 \simeq \mathcal{Y}_1 \overset{\mathrm{al}}{\otimes} \mathcal{Y}_2 \overset{\mathrm{al}}{\otimes} \mathcal{Y}_3. \tag{3.2}$$

The tensor multiplication $\otimes$ is associative.

Remark 3.7 *Note that we can replace the set $\{1, \ldots, n\}$, labeling the spaces $\mathcal{Y}_i$ in Def. 3.6, by any finite set I. Then we obtain the definition of $\underset{i \in I}{\overset{\mathrm{al}}{\otimes}} \mathcal{Y}_i$.*

If $\mathcal{Y}, \mathcal{W}$ are real vector spaces, then we have the identification

$$\mathbb{C}(\mathcal{Y} \overset{\mathrm{al}}{\otimes} \mathcal{W}) \simeq \mathbb{C}\mathcal{Y} \overset{\mathrm{al}}{\otimes} \mathbb{C}\mathcal{W}. \tag{3.3}$$

Clearly, if $\mathcal{Y}$ and $\mathcal{W}$ are complex spaces, then $\overline{\mathcal{Y} \overset{\text{al}}{\otimes} \mathcal{W}}$ can be identified with $\overline{\mathcal{Y}} \overset{\text{al}}{\otimes} \overline{\mathcal{W}}$.

If one of the spaces $\mathcal{Y}$ or $\mathcal{W}$ is finite-dimensional then we will often write $\mathcal{Y} \otimes \mathcal{W}$ instead of $\mathcal{Y} \overset{\text{al}}{\otimes} \mathcal{W}$.

If $\mathcal{Y}$ and $\mathcal{W}$ are finite-dimensional, then $(\mathcal{Y} \otimes \mathcal{W})^{\#}$ will be identified with $\mathcal{W}^{\#} \otimes \mathcal{Y}^{\#}$ using the following convention: if $\xi \in \mathcal{Y}^{\#}$, $\theta \in \mathcal{W}^{\#}$ then

$$\langle \theta \otimes \xi | y \otimes w \rangle := \langle \xi | y \rangle \langle \theta | w \rangle. \tag{3.4}$$

(Note the reversal of the order.)

3.1.4 Tensor product in the sense of Hilbert spaces

If $\mathcal{Y}$, $\mathcal{W}$ are Hilbert spaces, then $\mathcal{Y} \overset{\text{al}}{\otimes} \mathcal{W}$ has a unique scalar product such that

$$(y_1 \otimes w_1 | y_2 \otimes w_2) := (y_1 | y_2)(w_1 | w_2), \quad y_1, y_2 \in \mathcal{Y}, \quad w_1, w_2 \in \mathcal{W}.$$

Definition 3.8 *We set*

$$\mathcal{Y} \otimes \mathcal{W} := (\mathcal{Y} \overset{\text{al}}{\otimes} \mathcal{W})^{\text{cpl}}, \tag{3.5}$$

and call it the tensor product of $\mathcal{Y}$ and $\mathcal{W}$ in the sense of Hilbert spaces.

If one of the spaces $\mathcal{Y}$ or $\mathcal{W}$ is finite-dimensional, then (3.5) coincides with $\mathcal{Y} \overset{\text{al}}{\otimes} \mathcal{W}$.

The remaining part of the basic theory of the tensor product in the sense of Hilbert spaces is analogous to that of the algebraic tensor product described in the previous subsection.

3.1.5 Bases of tensor products

Let $\mathcal{Y}, \mathcal{W}$ be finite-dimensional vector spaces. If $\{e_i\}_{i \in I}$ is a basis of $\mathcal{Y}$ and $\{f_j\}_{j \in J}$ is a basis of $\mathcal{W}$, then

$$\{e_i \otimes f_j\}_{(i,j) \in I \times J}$$

is a basis of $\mathcal{Y} \otimes \mathcal{W}$.

If $\{e^i\}_{i \in I}$ is the dual basis in $\mathcal{Y}^{\#}$ and $\{f^j\}_{j \in J}$ is the dual basis in $\mathcal{W}^{\#}$ then

$$\{f^j \otimes e^i\}_{(j,i) \in J \times I}$$

is the dual basis in $(\mathcal{Y} \otimes \mathcal{W})^{\#} \simeq \mathcal{W}^{\#} \otimes \mathcal{Y}^{\#}$.

Suppose now that $\mathcal{Y}$ and $\mathcal{W}$ are Hilbert spaces. If $\{e_i\}_{i \in I}$ is an o.n. basis of $\mathcal{Y}$ and $\{f_j\}_{j \in J}$ is an o.n. basis of $\mathcal{W}$, then $\{e_i \otimes f_j\}_{(i,j) \in I \times J}$ is an o.n. basis of $\mathcal{Y} \otimes \mathcal{W}$.

3.1.6 Operators in tensor products

Let $\mathcal{Y}_1, \mathcal{Y}_2, \mathcal{W}_1, \mathcal{W}_2$ be vector spaces.

Definition 3.9 *If $a_1 \in L(\mathcal{Y}_1, \mathcal{W}_1)$ and $a_2 \in L(\mathcal{Y}_2, \mathcal{W}_2)$, then $a_1 \otimes a_2$ is defined as the unique operator in $L(\mathcal{Y}_1 \overset{al}{\otimes} \mathcal{Y}_2, \mathcal{W}_1 \overset{al}{\otimes} \mathcal{W}_2)$ such that*

$$(a_1 \otimes a_2)(y_1 \otimes y_2) := a_1 y_1 \otimes a_2 y_2.$$

If $\mathcal{Y}_1, \mathcal{Y}_2, \mathcal{W}_1, \mathcal{W}_2$ are Hilbert spaces and a_1, resp. a_2, are closable operators from $\mathcal{Y}_1$ to $\mathcal{W}_1$, resp. from $\mathcal{Y}_2$ to $\mathcal{W}_2$, then $a_1 \otimes a_2$ with the domain $\mathrm{Dom}\, a_1 \overset{al}{\otimes} \mathrm{Dom}\, a_2$ is closable, since

$$a_1^* \otimes a_2^* \subset (a_1 \otimes a_2)^*.$$

Definition 3.10 *The closure of $a_1 \otimes a_2 \in L(\mathcal{Y}_1 \otimes \mathcal{Y}_2, \mathcal{W}_1 \otimes \mathcal{W}_2)$ will be denoted by the same symbol $a_1 \otimes a_2 \in Cl(\mathcal{Y}_1 \otimes \mathcal{Y}_2, \mathcal{W}_1 \otimes \mathcal{W}_2)$.*

If both a_1 and a_2 are non-zero, then $a_1 \otimes a_2$ is bounded iff both a_1 and a_2 are bounded, and then $\|a_1 \otimes a_2\| = \|a_1\|\|a_2\|$.

If both a_1 and a_2 are essentially self-adjoint, then $a_1 \otimes a_2$ is essentially self-adjoint on $\mathrm{Dom}\, a_1 \overset{al}{\otimes} \mathrm{Dom}\, a_2$.

3.1.7 Permutations

Let $\mathcal{Y}_1, \ldots, \mathcal{Y}_n$ be vector spaces.

Definition 3.11 *Let S_n denote the permutation group of n elements and $\sigma \in S_n$. $\Theta(\sigma)$ is defined as the unique operator in $L(\mathcal{Y}_1 \overset{al}{\otimes} \cdots \overset{al}{\otimes} \mathcal{Y}_n, \mathcal{Y}_{\sigma^{-1}(1)} \overset{al}{\otimes} \cdots \overset{al}{\otimes} \mathcal{Y}_{\sigma^{-1}(n)})$ such that*

$$\Theta(\sigma) y_1 \otimes \cdots \otimes y_n = y_{\sigma^{-1}(1)} \otimes \cdots \otimes y_{\sigma^{-1}(n)}.$$

If $\mathcal{Y}_1, \ldots, \mathcal{Y}_n$ are Hilbert spaces, then $\Theta(\sigma)$ is unitary.

3.1.8 Identifications

Let $\mathcal{Y}, \mathcal{W}$ be vector spaces, with $\mathcal{W}$ finite-dimensional. Then there exists a unique linear map $L(\mathcal{W}, \mathcal{Y}) \to \mathcal{Y} \otimes \mathcal{W}^\#$ such that

$$|y\rangle\langle\xi| \mapsto y \otimes \xi.$$

If $\mathcal{Y}, \mathcal{W}$ are Hilbert spaces, then there exists a unique unitary map $B^2(\mathcal{W}, \mathcal{Y}) \to \mathcal{Y} \otimes \overline{\mathcal{W}}$ such that

$$|y)(w| \mapsto y \otimes \overline{w}.$$

Note the identity that uses the above identification, valid for $y \in \mathcal{Y}$, $w \in \mathcal{W}$, $B \in B^2(\mathcal{W}, \mathcal{Y})$:

$$(y|Bw) = (y \otimes \overline{w}|B).$$

3.1.9 Infinite tensor product of grounded Hilbert spaces

It is well known that there are problems with the definition of the tensor product of an infinite family of Hilbert spaces. The most useful definition of such a tensor product depends on the choice of a normalized vector in each of these spaces.

Definition 3.12 *A pair* $(\mathcal{H}, \Omega)$ *consisting of a Hilbert space and a vector* $\Omega \in \mathcal{H}$ *of norm 1 is called a* grounded Hilbert space.

Let $\left\{(\mathcal{H}_i, \Omega_i)\right\}_{i \in I}$ be a family of grounded Hilbert spaces. If $J_1 \subset J_2 \subset I$ are two finite sets, we introduce the isometric identification

$$\underset{i \in J_1}{\otimes} \mathcal{H}_i \ni \Psi \mapsto \Psi \otimes \underset{i \in J_2 \setminus J_1}{\otimes} \Omega_i \in \underset{i \in J_2}{\otimes} \mathcal{H}_i.$$

Definition 3.13 *The* tensor product of grounded Hilbert spaces $\left\{(\mathcal{H}_i, \Omega_i)\right\}_{i \in I}$ *is defined as*

$$\underset{i \in I}{\otimes}(\mathcal{H}_i, \Omega_i) := \left(\bigcup_{J \in 2^I_{\mathrm{fin}}} \underset{i \in J}{\otimes} \mathcal{H}_i\right)^{\mathrm{cpl}}.$$

The image of $\Psi \in \underset{i \in J}{\otimes} \mathcal{H}_i$ *will be denoted by*

$$\Psi \otimes \underset{i \in I \setminus J}{\otimes} \Omega_i.$$

Such vectors are called finite vectors. *Similarly, if* $B \in B(\underset{i \in J}{\otimes} \mathcal{H}_i)$, *we will use the obvious notation*

$$B \otimes \underset{i \in I \setminus J}{\otimes} \mathbb{1}_{\mathcal{H}_i} \in B\left(\underset{i \in I}{\otimes} \mathcal{H}_i\right).$$

Clearly, if I is a finite set, then $\underset{i \in I}{\otimes}(\mathcal{H}_i, \Omega_i) = \underset{i \in I}{\otimes} \mathcal{H}_i$ for any family of normalized vectors Ω_i. Moreover, for $I_1 \cap I_2 = \emptyset$ we have

$$\underset{i \in I_1}{\otimes}(\mathcal{H}_i, \Omega_i) \otimes \underset{i \in I_2}{\otimes}(\mathcal{H}_i, \Omega_i) \simeq \underset{i \in I_1 \cup I_2}{\otimes}(\mathcal{H}_i, \Omega_i).$$

3.1.10 Infinite tensor product of vectors and operators

Theorem 3.14 *Let* $\Phi_i \in \mathcal{H}_i$, $i \in I$, *have norm 1. Set*

$$\Psi_J := \underset{i \in J}{\otimes} \Phi_i \otimes \underset{i \in I \setminus J}{\otimes} \Omega_i \tag{3.6}$$

for $J \in 2^I_{\mathrm{fin}}$. *Then the net* $\{\Psi_J\}_{J \in 2^I_{\mathrm{fin}}}$ *is convergent iff the infinite product* $\prod_{i \in I}(\Omega_i|\Phi_i)$ *is convergent.*

Definition 3.15 *The vector* $\lim_J \Psi_J$ *will be denoted by* $\bigotimes_{i \in I} \Phi_i$.

Proof of Thm. 3.14. Assume first that the net $\{\Psi_J\}_{J \in 2^I_{\text{fin}}}$ is convergent in $\bigotimes_{i \in I} (\mathcal{H}_i, \Omega_i)$. If $I_0 = \{i \in I : (\Phi_i | \Omega_i) = 0\}$ is infinite, then clearly $\lim_J (\Psi_J | \Psi) = 0$ for all finite vectors Ψ. Since finite vectors are dense this is a contradiction, since $\lim_J \|\Psi_J\| = 1$. Therefore, I_0 is finite.

It remains to prove that the net $\left\{ \prod_{i \in J \setminus I_0} (\Phi_i | \Omega_i) \right\}_{J \in 2^I_{\text{fin}}}$ has a non-zero limit. Clearly,

$$\lim_J \Psi_J = \bigotimes_{i \in I_0} \Phi_i \otimes \bigotimes_{i \in I \setminus I_0} \Phi_i. \tag{3.7}$$

If $I_0 \subset J$, then

$$\left(\Psi_J \Big| \bigotimes_{i \in I_0} \Phi_i \otimes \bigotimes_{i \in I \setminus I_0} \Omega_i \right) = \prod_{i \in J \setminus I_0} (\Phi_i | \Omega_i),$$

which proves that the net $\left\{ \prod_{i \in J \setminus I_0} (\Phi_i | \Omega_i) \right\}_{J \in 2^I_{\text{fin}}}$ is convergent in $\mathbb{C}$. If the limit is 0, then, since $(\Phi_i | \Omega_i) \neq 0$ for $i \in I \setminus I_0$, we obtain that the vector $\bigotimes_{i \in I \setminus I_0} \Phi_i$ is orthogonal to all finite vectors in $\bigotimes_{i \in I \setminus I_0} (\mathcal{H}_i, \Omega_i)$, which using (3.7) yields a contradiction, since $\lim_J \|\Psi_J\| = 1$. Therefore, the infinite product $\prod_{i \in I} (\Phi_i | \Omega_i)$ is convergent.

Conversely, assume that the infinite product $\prod_{i \in I} (\Phi_i | \Omega_i)$ is convergent. Then

$$\sum_{i \in I} |1 - (\Phi_i | \Omega_i)| < \infty.$$

Note that if $J_1 \subset J_2$, then

$$\|\Psi_{J_1} - \Psi_{J_2}\|^2 = 2 - 2 \operatorname{Re} \prod_{i \in J_2 \setminus J_1} (\Phi_i | \Omega_i).$$

Therefore, the net $\{\Psi_J\}_{J \in 2^I_{\text{fin}}}$ is Cauchy, and hence converges in $\bigotimes_{i \in I} (\mathcal{H}_i, \Omega_i)$. $\square$

Using Thm. 3.14, we immediately obtain the following theorem.

Theorem 3.16 *Let* $A_i \in B(\mathcal{H}_i)$ *be contractions. Then there exists the strong limit of*

$$B_J := \bigotimes_{i \in J} A_i \otimes \bigotimes_{i \in I \setminus J} \mathbb{1}_{\mathcal{H}_i} \tag{3.8}$$

iff the infinite product $\prod_{i \in I} (\Omega_i | A_i \Omega_i)$ *is convergent.*

Definition 3.17 *The operator* $\lim_J B_J$ *will be denoted by* $\bigotimes_{i \in I} A_i$.

3.2 Tensor algebra

In this section we introduce the *tensor algebra over a vector space*. This concept has two basic versions: we can consider the *algebraic tensor algebra*, or if the vector space has the structure of a Hilbert space, the *complete tensor algebra* (which is also a Hilbert space), called sometimes the *full Fock space*. Full Fock spaces play the central role in the so-called *free probability*. For us, they are mainly intermediate constructions to be used in the discussion of *bosonic* and *fermionic Fock spaces*.

3.2.1 Full Fock space

Let $\mathcal{Y}$ be a vector space.

Definition 3.18 *Let $\overset{al}{\otimes}{}^n\mathcal{Y}$ (or $\mathcal{Y}^{\overset{al}{\otimes}n}$) denote the n-th algebraic tensor power of $\mathcal{Y}$. We will write $\overset{al}{\otimes}{}^0\mathcal{Y} := \mathbb{K}$. The algebraic tensor algebra over $\mathcal{Y}$ is defined as*

$$\overset{al}{\otimes}\mathcal{Y} := \underset{0 \leq n < \infty}{\overset{al}{\bigoplus}} \overset{al}{\otimes}{}^n\mathcal{Y}.$$

The element $1 \in \overset{al}{\otimes}{}^0\mathcal{Y}$ is called the vacuum *and denoted by Ω. If $\mathcal{Y}$ is a finite-dimensional space, we will often write $\otimes^n\mathcal{Y}$ instead of $\overset{al}{\otimes}{}^n\mathcal{Y}$.*

$\overset{al}{\otimes}\mathcal{Y}$ is an associative algebra with the operation $\otimes$ and the identity Ω.

Assume now that $\mathcal{Y}$ is a Hilbert space,

Definition 3.19 *We will write $\otimes^n\mathcal{Y}$ (or $\mathcal{Y}^{\otimes n}$) for the n-th tensor power of $\mathcal{Y}$ in the sense of Hilbert spaces. Clearly, it is equal to $\left(\overset{al}{\otimes}{}^n\mathcal{Y}\right)^{\mathrm{cpl}}$. We set*

$$\otimes\mathcal{Y} := \underset{0 \leq n < \infty}{\bigoplus} \otimes^n\mathcal{Y} = \left(\overset{al}{\otimes}\mathcal{Y}\right)^{\mathrm{cpl}}.$$

$\otimes\mathcal{Y}$ *is called the* complete tensor algebra *or the* full Fock space.

We will also need notation for the finite particle full Fock space

$$\otimes^{\mathrm{fin}}\mathcal{Y} := \underset{0 \leq n < \infty}{\overset{al}{\bigoplus}} \otimes^n\mathcal{Y}.$$

$\otimes\mathcal{Y}$ and $\otimes^{\mathrm{fin}}\mathcal{Y}$ are associative algebras with the operation $\otimes$ and the identity Ω.

3.2.2 Operators $\mathrm{d}\Gamma$ and Γ in full Fock spaces

The definitions of this subsection have obvious algebraic counterparts. For simplicity, we restrict ourselves to the Hilbert space case and assume that $\mathcal{Y}, \mathcal{Y}_1, \mathcal{Y}_2$ are Hilbert spaces.

Definition 3.20 *Let p be a linear operator from $\mathcal{Y}_1$ to $\mathcal{Y}_2$. Then we define $\Gamma^n(p) := p^{\otimes n}$ with domain $\overset{\mathrm{al}}{\otimes}{}^n \mathrm{Dom}\, p$, and the operator $\Gamma(p)$ from $\otimes \mathcal{Y}_1$ to $\otimes \mathcal{Y}_2$*

$$\Gamma(p) := \bigoplus_{n=0}^{\infty} \Gamma^n(p)$$

with domain $\overset{\mathrm{al}}{\otimes} \mathrm{Dom}\, p$.

By Subsects. 3.1.2 and 3.1.6 we see that if p is closable, resp. essentially self-adjoint, then so is $\Gamma(p)$. $\Gamma(p)$ is bounded iff $\|p\| \leq 1$. $\Gamma(p)$ is unitary iff p is.

Definition 3.21 *If h is a linear operator on $\mathcal{Y}$, we set*

$$\mathrm{d}\Gamma^n(h) := \sum_{j=1}^{n} \mathbb{1}_{\mathcal{Y}}^{\otimes j-1} \otimes h \otimes \mathbb{1}_{\mathcal{Y}}^{\otimes(n-j)}$$

with domain $\overset{\mathrm{al}}{\otimes}{}^n \mathrm{Dom}\, h$, and

$$\mathrm{d}\Gamma(h) := \bigoplus_{n=0}^{\infty} \mathrm{d}\Gamma^n(h)$$

with domain $\overset{\mathrm{al}}{\otimes} \mathrm{Dom}\, h$.

Again, if h is closable, resp. essentially self-adjoint, then so is $\mathrm{d}\Gamma(h)$.

Definition 3.22 *The* number operator *and the* parity operator *are defined respectively as*

$$N := \mathrm{d}\Gamma(\mathbb{1}), \tag{3.9}$$
$$I := (-1)^N = \Gamma(-\mathbb{1}). \tag{3.10}$$

Proposition 3.23 (1) *Let $h, h_1, h_2 \in B(\mathcal{Y})$, $p_1 \in B(\mathcal{Y}, \mathcal{Y}_1)$, $p_2 \in B(\mathcal{Y}_1, \mathcal{Y}_2)$, $\|p_1\|, \|p_2\| \leq 1$. We then have*

$$\Gamma(\mathrm{e}^h) = \mathrm{e}^{\mathrm{d}\Gamma(h)},$$
$$\Gamma(p_2)\Gamma(p_1) = \Gamma(p_2 p_1),$$
$$[\mathrm{d}\Gamma(h_1), \mathrm{d}\Gamma(h_2)] = \mathrm{d}\Gamma([h_1, h_2]).$$

(2) *Let $\Phi, \Psi \in \otimes^{\mathrm{fin}} \mathcal{Y}$, $h \in B(\mathcal{Y})$, $p \in B(\mathcal{Y}, \mathcal{Y}_1)$. Then*

$$\Gamma(p)\, \Phi \otimes \Psi = (\Gamma(p)\Phi) \otimes (\Gamma(p)\Psi),$$
$$\mathrm{d}\Gamma(h)\, \Phi \otimes \Psi = (\mathrm{d}\Gamma(h)\Phi) \otimes \Psi + \Phi \otimes (\mathrm{d}\Gamma(h)\Psi).$$

3.3 Symmetric and anti-symmetric tensors

In this section we describe *symmetric*, resp. *anti-symmetric tensor algebras*. Their Hilbert space versions are also called *bosonic*, resp. *fermionic Fock spaces*.

Unfortunately, there seems to be no uniform terminology, and especially notation, in this context in the literature. We try to introduce a coherent notation, which in particular stresses parallel properties of the symmetric and anti-symmetric cases.

3.3.1 Fock spaces

Let $\mathcal{Y}$ be a vector space. Recall that in Subsect. 3.1.7, for $\sigma \in S_n$ we defined the operators $\Theta(\sigma) \in L(\overset{\text{al}}{\otimes}{}^n)$. Clearly,

$$S_n \ni \sigma \mapsto \Theta(\sigma) \in L(\overset{\text{al}}{\otimes}{}^n \mathcal{Y})$$

is a representation of the permutation group.

Definition 3.24 *We define the following operators on* $\overset{\text{al}}{\otimes}{}^n \mathcal{Y}$:

$$\Theta_{\mathrm{s}}^n := \frac{1}{n!} \sum_{\sigma \in S_n} \Theta(\sigma),$$

$$\Theta_{\mathrm{a}}^n := \frac{1}{n!} \sum_{\sigma \in S_n} \mathrm{sgn}(\sigma)\Theta(\sigma).$$

We will write s/a *as a subscript which can mean either* s *or* a.

It is easy to check that $\Theta_{\mathrm{s/a}}^n$ is a projection.

Definition 3.25 *Introduce the following projections acting on* $\overset{\text{al}}{\otimes} \mathcal{Y}$:

$$\Theta_{\mathrm{s/a}} := \bigoplus_{0 \leq n < \infty} \Theta_{\mathrm{s/a}}^n.$$

We set

$$\overset{\text{al}}{\Gamma}{}_{\mathrm{s/a}}^n(\mathcal{Y}) := \Theta_{\mathrm{s/a}}^n \overset{\text{al}}{\otimes}{}^n \mathcal{Y},$$

$$\overset{\text{al}}{\Gamma}{}_{\mathrm{s/a}}(\mathcal{Y}) := \overset{\text{al}}{\underset{0 \leq n < \infty}{\bigoplus}} \overset{\text{al}}{\Gamma}{}_{\mathrm{s/a}}^n(\mathcal{Y}) = \Theta_{\mathrm{s/a}} \overset{\text{al}}{\otimes} \mathcal{Y}.$$

$\overset{\text{al}}{\Gamma}{}_{\mathrm{s/a}}(\mathcal{Y})$ *are called the* algebraic symmetric, resp. anti-symmetric tensor algebras *or* algebraic bosonic, resp. fermionic Fock spaces.

If $\mathcal{Y}$ *is a finite-dimensional space, we can write* $\Gamma_{\mathrm{s/a}}^n(\mathcal{Y})$ *instead of* $\overset{\text{al}}{\Gamma}{}_{\mathrm{s/a}}^n(\mathcal{Y})$.

Elements of $\overset{\text{al}}{\Gamma}{}_{\mathrm{s/a}}^n(\mathcal{Y})$ consist of symmetric, resp. anti-symmetric tensors, as expressed in the following proposition:

Proposition 3.26 *Let* $\Psi \in \overset{\text{al}}{\otimes}{}^n \mathcal{Y}$. *Then*

(1) $\Psi \in \overset{\text{al}}{\Gamma}{}_{\mathrm{s}}^n(\mathcal{Y})$ *iff* $\Theta(\sigma)\Psi = \Psi$, $\sigma \in S_n$;
(2) $\Psi \in \overset{\text{al}}{\Gamma}{}_{\mathrm{a}}^n(\mathcal{Y})$ *iff* $\Theta(\sigma)\Psi = \mathrm{sgn}(\sigma)\Psi$, $\sigma \in S_n$.

Assume now that $\mathcal{Y}$ is a Hilbert space. Then $\Theta_{\mathrm{s/a}}^n$ and $\Theta_{\mathrm{s/a}}$ are orthogonal projections.

Definition 3.27 *We define*

$$\Gamma_{s/a}^n(\mathcal{Y}) := \Theta_{s/a}^n \otimes^n \mathcal{Y} = \left(\overset{\text{al}}{\Gamma}_{s/a}^n(\mathcal{Y})\right)^{\text{cpl}},$$

$$\Gamma_{s/a}(\mathcal{Y}) := \overset{\infty}{\underset{n=0}{\oplus}} \Gamma_{s/a}^n(\mathcal{Y}) = \Theta_{s/a} \otimes \mathcal{Y} = \left(\overset{\text{al}}{\Gamma}_{s/a}(\mathcal{Y})\right)^{\text{cpl}}.$$

$\Gamma_{s/a}(\mathcal{Y})$ *is called the* bosonic, *resp.* fermionic Fock space.

Note that $\Gamma_{s/a}(\mathcal{Y})$ itself is a Hilbert space (as a closed subspace of $\otimes \mathcal{Y}$).

Definition 3.28 *We will need notation for the* finite particle bosonic, *resp.* fermionic Fock space:

$$\Gamma_{s/a}^{\text{fin}}(\mathcal{Y}) := \overset{\text{al}}{\underset{0 \le n < \infty}{\oplus}} \Gamma_{s/a}^n(\mathcal{Y}).$$

3.3.2 Symmetric and anti-symmetric tensor products

Let $\Psi, \Phi \in \overset{\text{al}}{\Gamma}_{s/a}(\mathcal{Y})$.

Definition 3.29 *We define the* symmetric, *resp.* anti-symmetric tensor product of Φ and Ψ:

$$\Psi \otimes_{s/a} \Phi := \Theta_{s/a} \Psi \otimes \Phi.$$

$\overset{\text{al}}{\Gamma}_{s/a}(\mathcal{Y})$ is an associative algebra with the operation $\otimes_{s/a}$ and the identity Ω. Note that the set of vectors of the form

$$\underset{n \text{ times}}{\underbrace{y \otimes \cdots \otimes y}} = \underset{n \text{ times}}{\underbrace{y \otimes_s \cdots \otimes_s y}}, \tag{3.11}$$

for $y \in \mathcal{Y}$, spans $\overset{\text{al}}{\Gamma}_s^n(\mathcal{Y})$.

Definition 3.30 *For brevity we will denote (3.11) by* $y^{\otimes n}$.

The notation $\otimes_a$ that we introduced is not common in the literature. Instead, one usually prefers a different closely related operation:

Definition 3.31 *The* wedge product of vectors Φ and Ψ *is defined as*

$$\Psi \wedge \Phi := \frac{(p+q)!}{p!q!} \Psi \otimes_a \Phi, \quad for \ \Psi \in \overset{\text{al}}{\Gamma}_a^p(\mathcal{Y}), \ \Phi \in \overset{\text{al}}{\Gamma}_a^q(\mathcal{Y}). \tag{3.12}$$

The advantage of the wedge product over $\otimes_a$ is visible if we compare the following identities:

$$y_1 \wedge \cdots \wedge y_n = \sum_{\sigma \in S_n} \text{sgn}(\sigma) \, y_{\sigma(1)} \otimes \cdots \otimes y_{\sigma(n)},$$

$$y_1 \otimes_a \cdots \otimes_a y_n = \frac{1}{n!} \sum_{\sigma \in S_n} \text{sgn}(\sigma) \, y_{\sigma(1)} \otimes \cdots \otimes y_{\sigma(n)}, \quad y_1, \cdots, y_n \in \mathcal{Y}.$$

Note that $\wedge$ is also associative.

Definition 3.32 *One often writes* $\wedge^n \mathcal{Y}$ *and* $\wedge \mathcal{Y}$ *for* $\overset{\text{al}}{\Gamma}{}_{\text{a}}^{n}(\mathcal{Y})$ *and* $\overset{\text{al}}{\Gamma}{}_{\text{a}}(\mathcal{Y})$.

Definition 3.33 *If* $\mathcal{Y}$ *is a Hilbert space, we can define* $\otimes_{\text{s/a}}$ *and* $\wedge$ *in* $\Gamma_{\text{s/a}}(\mathcal{Y})$ *in the same way, with the same properties.*

3.3.3 $d\Gamma$ *and* Γ *operators*

For brevity we restrict ourselves to the case of Hilbert spaces.

Let p be a closable operator from $\mathcal{Y}$ to $\mathcal{W}$. Then $\Gamma^n(p)$ maps $\Gamma_{\text{s/a}}^{n}(\mathcal{Y})$ into $\Gamma_{\text{s/a}}^{n}(\mathcal{W})$. Hence $\Gamma(p)$ maps $\Gamma_{\text{s/a}}(\mathcal{Y})$ into $\Gamma_{\text{s/a}}(\mathcal{W})$.

Definition 3.34 *We will use the same symbols* $\Gamma^n(p)$ *and* $\Gamma(p)$ *to denote the corresponding restricted operators.* $\Gamma(p)$ *is sometimes called the* second quantization *of* p.

Let h be a closable operator on $\mathcal{Y}$. Then $d\Gamma^n(h)$ maps $\Gamma_{\text{s/a}}^{n}(\mathcal{Y})$ into itself. Hence, $d\Gamma(h)$ maps $\Gamma_{\text{s/a}}(\mathcal{Y})$ into itself.

Definition 3.35 *We will use the same symbols* $d\Gamma^n(h)$ *and* $d\Gamma(h)$ *to denote the corresponding restricted operators. Perhaps the correct name of* $d\Gamma(h)$ *should be the* infinitesimal second quantization *of* h.

Note that in the context of bosonic, resp. fermionic Fock spaces the operators $\Gamma(\cdot)$ and $d\Gamma(\cdot)$ still have the properties described in Prop. 3.23 (1). Prop. 3.23 (2) needs to be replaced by the following statement:

Proposition 3.36 *Let* $p \in B(\mathcal{Y}, \mathcal{Y}_1)$, $h \in B(\mathcal{Y})$, $\Psi, \Phi \in \Gamma_{\text{s/a}}^{\text{fin}}(\mathcal{Y})$. *Then*

$$\Gamma(p)\, \Psi \otimes_{\text{s/a}} \Phi = (\Gamma(p)\Psi) \otimes_{\text{s/a}} (\Gamma(p)\Phi),$$
$$d\Gamma(h)\, \Psi \otimes_{\text{s/a}} \Phi = (d\Gamma(h)\Psi) \otimes_{\text{s/a}} \Phi + \Psi \otimes_{\text{s/a}} (d\Gamma(h)\Phi).$$

3.3.4 *Identifications*

Let $\mathcal{Y}$ be a finite-dimensional vector space. Then $\Gamma_{\text{s/a}}^{2}(\mathcal{Y})$ can be identified with $L_{\text{s/a}}(\mathcal{Y}^{\#}, \mathcal{Y})$, which were defined in Defs. 1.18 and 1.29.

Let $\mathcal{Y}$ be a Hilbert space. Recall that $B^2(\mathcal{Y}, \mathcal{W})$ denotes the space of Hilbert–Schmidt operators from $\mathcal{Y}$ to $\mathcal{W}$. We introduce the following symbols for the spaces of symmetric and anti-symmetric Hilbert–Schmidt operators:

$$B_{\text{s}}^{2}(\overline{\mathcal{Y}}, \mathcal{Y}) := \left\{ a \in B^2(\overline{\mathcal{Y}}, \mathcal{Y}) \ : \ a^{\#} = a \right\},$$
$$B_{\text{a}}^{2}(\overline{\mathcal{Y}}, \mathcal{Y}) := \left\{ a \in B^2(\overline{\mathcal{Y}}, \mathcal{Y}) \ : \ a^{\#} = -a \right\},$$

where as usual one identifies $\mathcal{Y}^{\#}$ with $\overline{\mathcal{Y}}$ using the Hilbert structure of $\mathcal{Y}$. Then the unitary map of Subsect. 3.1.8 allows us to unitarily identify $\Gamma_{\text{s/a}}^{2}(\mathcal{Y})$ with $B_{\text{s/a}}^{2}(\overline{\mathcal{Y}}, \mathcal{Y})$.

3.3.5 Bases in bosonic Fock spaces

Let $\mathcal{Y}$ be a finite-dimensional vector space and $\{e_i \ : \ i = 1, \ldots, d\}$ a basis of $\mathcal{Y}$.

Definition 3.37 *For $\vec{k} = (k_1, \ldots, k_d) \in \mathbb{N}^d$, we set*

$$|\vec{k}| := k_1 + \cdots k_d, \quad \vec{k}! := k_1! \ldots k_d!,$$

$$e_{\vec{k}} := e_1^{\otimes k_1} \otimes_{\mathrm{s}} \cdots \otimes_{\mathrm{s}} e_d^{\otimes k_d}, \quad e_{\vec{0}} := \Omega.$$

Then

$$\{e_{\vec{k}} \ : \ \vec{k} \in \mathbb{N}^d, \ |\vec{k}| = n\} \tag{3.13}$$

is a basis of $\Gamma_{\mathrm{s}}^n(\mathcal{Y})$.

The dual of $\Gamma_{\mathrm{s}}^n(\mathcal{Y})$ can be identified with $\Gamma_{\mathrm{s}}^n(\mathcal{Y}^{\#})$. Let $\{e^i \ : \ i = 1, \ldots, d\}$ be the dual basis of $\mathcal{Y}^{\#}$.

Definition 3.38 *We set $e^{\vec{k}} := (e^1)^{\otimes k_d} \otimes_{\mathrm{s}} \cdots \otimes_{\mathrm{s}} (e^d)^{\otimes k_1}$, for $k \in \mathbb{N}^d$.*

Then

$$\left\{ \frac{|\vec{k}|!}{\vec{k}!} e^{\vec{k}} \ : \ \vec{k} \in \mathbb{N}^d, \ |\vec{k}| = n \right\}$$

is the basis of $\Gamma_{\mathrm{s}}^n(\mathcal{Y}^{\#})$ dual to (3.13).

Let $\mathcal{Y}$ be now a Hilbert space with an o.n. basis $\{e_i\}_{i \in I}$.

Definition 3.39 *Recall that $c_{\mathrm{c}}(I, \mathbb{N})$ denotes the set of functions $I \to \mathbb{N}$ with all but a finite number of values equal to zero. If $\vec{k} \in c_{\mathrm{c}}(I, \mathbb{N})$, then the definitions of $|\vec{k}|$, $\vec{k}!$ and $e_{\vec{k}}$ have obvious versions in the present context.*

Then

$$\left\{ \frac{\sqrt{|\vec{k}|!}}{\sqrt{\vec{k}!}} e_{\vec{k}} \ : \ \vec{k} \in c_{\mathrm{c}}(I, \mathbb{N}), \ |\vec{k}| = n \right\}$$

is an o.n. basis of $\Gamma_{\mathrm{s}}^n(\mathcal{Y})$.

3.3.6 Bases in fermionic Fock spaces

Let $\mathcal{Y}$ be a finite-dimensional vector space and $\{e_i \ : \ i = 1, \ldots, d\}$ a basis of $\mathcal{Y}$.

Definition 3.40 *For $J = \{i_1, \cdots, i_n\} \subset \{1, \ldots, d\}$ with $1 \le i_1 < \cdots < i_n \le d$, set*

$$e_J := e_{i_1} \otimes_{\mathrm{a}} \cdots \otimes_{\mathrm{a}} e_{i_n}.$$

Then

$$\{e_J \ : \ J \subset \{1, \ldots, d\}, \ \#J = n\} \tag{3.14}$$

is a basis of $\Gamma_{\mathrm{a}}^n(\mathcal{Y})$.

The dual of $\Gamma_{\mathrm{a}}^n(\mathcal{Y})$ can be identified as above with $\Gamma_{\mathrm{a}}^n(\mathcal{Y}^{\#})$.

Let $\{e^i \ : \ i = 1, \ldots, d\}$ be the dual basis in $\mathcal{Y}^{\#}$.

Definition 3.41 *For* $J = \{i_1, \cdots, i_n\} \subset \{1, \ldots, d\}$ *with* $1 \le i_1 < \cdots < i_n \le d$ *put* $e^J := e^{i_n} \otimes_{\mathrm{a}} \cdots \otimes_{\mathrm{a}} e^{i_1}$.

Then

$$\left\{ \#J! e^J \ : \ J \subset \{1, \ldots, d\}, \ \#J = n \right\}$$

is the basis of $\Gamma_{\mathrm{a}}^n(\mathcal{Y}^{\#})$ dual to (3.14).

Let $\mathcal{Y}$ be now a Hilbert space with an o.n. basis $\{e_i \ : \ i \in I\}$. Let us choose a total order in the set I.

Definition 3.42 *For a finite subset* J *of* I, *we define* e_J *in an obvious way.*

Then

$$\left\{ \sqrt{\#J!} e_J \ : \ J \subset I, \ , \#J = n \right\}$$

is an o.n. basis of $\Gamma_{\mathrm{a}}^n(\mathcal{Y})$.

3.3.7 Exponential law for Fock spaces

For brevity we restrict ourselves again to the case of Hilbert spaces. Let $\mathcal{Y}_1$ and $\mathcal{Y}_2$ be Hilbert spaces and let $j_i : \mathcal{Y}_i \to \mathcal{Y}_1 \oplus \mathcal{Y}_2$ be the canonical embeddings.

We introduce an identification

$$U : \Gamma_{\mathrm{s/a}}^{\mathrm{fin}}(\mathcal{Y}_1) \overset{\mathrm{al}}{\otimes} \Gamma_{\mathrm{s/a}}^{\mathrm{fin}}(\mathcal{Y}_2) \to \Gamma_{\mathrm{s/a}}^{\mathrm{fin}}(\mathcal{Y}_1 \oplus \mathcal{Y}_2)$$

as follows. Let $\Psi_1 \in \Gamma_{\mathrm{s/a}}^{n_1}(\mathcal{Y}_1)$, $\Psi_2 \in \Gamma_{\mathrm{s/a}}^{n_2}(\mathcal{Y}_2)$. Then

$$U \Psi_1 \otimes \Psi_2 := \sqrt{\tfrac{(n_1 + n_2)!}{n_1! n_2!}} \left(\Gamma(j_1) \Psi_1 \right) \otimes_{\mathrm{s/a}} \left(\Gamma(j_2) \Psi_2 \right). \tag{3.15}$$

Theorem 3.43 (1) *U extends to a unitary operator from* $\Gamma_{\mathrm{s/a}}(\mathcal{Y}_1) \otimes \Gamma_{\mathrm{s/a}}(\mathcal{Y}_2)$ *to* $\Gamma_{\mathrm{s/a}}(\mathcal{Y}_1 \oplus \mathcal{Y}_2)$.
(2) *$U \Omega_1 \otimes \Omega_2 = \Omega$.*
(3) *If* $h_i \in B(\mathcal{Y}_i)$, *then*

$$\mathrm{d}\Gamma(h_1 \oplus h_2) U = U \big(\mathrm{d}\Gamma(h_1) \otimes \mathbb{1} + \mathbb{1} \otimes \mathrm{d}\Gamma(h_2) \big). \tag{3.16}$$

(4) *If* $p_i \in B(\mathcal{Y}_i)$, *then*

$$\Gamma(p_1 \oplus p_2) U = U \Gamma(p_1) \otimes \Gamma(p_2). \tag{3.17}$$

Proof Let us prove (1). To simplify the notation let us restrict ourselves to the symmetric case. Let $\Psi_1 \in \Gamma_s^{n_1}(\mathcal{Y}_1)$, $\Psi_2 \in \Gamma_s^{n_2}(\mathcal{Y}_2)$. Then

$$\Gamma(j_1)\Psi_1 \otimes_s \Gamma(j_2)\Psi_2 = \frac{1}{(n_1+n_2)!} \sum_{\sigma \in S_{n_1+n_2}} \Theta(\sigma)\Gamma(j_1)\Psi_1 \otimes \Gamma(j_2)\Psi_2$$

$$= \frac{n_1!n_2!}{(n_1+n_2)!} \sum_{[\sigma] \in S_{n_1+n_2}/S_{n_1} \times S_{n_2}} \Theta(\sigma)\Gamma(j_1)\Psi_1 \otimes \Gamma(j_2)\Psi_2.$$

Now the elements of the sum on the right are mutually orthogonal. Hence

$$\|\Gamma(j_1)\Psi_1 \otimes_s \Gamma(j_2)\Psi_2\|^2 = \left(\frac{n_1!n_2!}{(n_1+n_2)!}\right)^2 \sum_{[\sigma] \in S_{n_1+n_2}/S_{n_1} \times S_{n_2}} \|\Theta(\sigma)\Psi_1 \otimes \Psi_2\|^2$$

$$= \frac{n_1!n_2!}{(n_1+n_2)!} \|\Psi_1 \otimes \Psi_2\|^2. \qquad \Box$$

Using the concept of the tensor product of grounded Hilbert spaces, one can easily generalize the exponential law to the case of an infinite number of Fock spaces. In fact, let $\mathcal{Y}_i$, $i \in I$ be a family of Hilbert spaces and denote by $j_i : \mathcal{Y}_i \to \underset{i \in I}{\oplus} \mathcal{Y}_i$ the canonical embeddings. Let Ω_i denote the vacuum in $\Gamma_{s/a}(\mathcal{Y}_i)$. Then

$$U\Psi_{i_1} \otimes \cdots \otimes \Psi_{i_n} \otimes \underset{i \in I \setminus \{i_1,\dots,i_n\}}{\otimes} \Omega$$

$$:= \frac{\sqrt{(i_1 + \cdots + i_n)!}}{\sqrt{i_1!} \cdots \sqrt{i_n!}} \Gamma(j_{i_1})\Psi_1 \otimes_s \cdots \otimes_s \Gamma(j_{i_n})\Psi_{i_n}$$

extends to a unitary map

$$U : \underset{i \in I}{\otimes} (\Gamma_{s/a}(\mathcal{Y}_i), \Omega_i) \to \Gamma_{s/a}\left(\underset{i \in I}{\oplus} \mathcal{Y}_i\right).$$

3.3.8 Dimension of Fock spaces

Let $\dim \mathcal{Y} = d$. Then it is easy to see that

$$\dim \Gamma_s^n(\mathcal{Y}) = \frac{(d+n-1)!}{(d-1)!n!},$$

$$\dim \Gamma_a^n(\mathcal{Y}) = \frac{n!}{d!(n-d)!}.$$

We have the following generating functions for the above quantities:

$$(1-t)^{-d} = \sum_{n=0}^{\infty} t^n \frac{(d+n-1)!}{(d+1)!n!},$$

$$(1+t)^d = \sum_{n=0}^{d} t^n \frac{n!}{d!(n-d)!}. \tag{3.18}$$

Recall that we have the identifications

$$\Gamma_{s/a}^n(\mathcal{Y}_1 \oplus \mathcal{Y}_2) \simeq \underset{m=0}{\overset{n}{\oplus}} \Gamma_{s/a}^m(\mathcal{Y}_1) \otimes \Gamma_{s/a}^{n-m}(\mathcal{Y}_2). \tag{3.19}$$

Assume that $\dim \mathcal{Y}_1 = d_1$, $\dim \mathcal{Y}_2 = d_2$. Then comparing the dimensions of both sides of (3.19) we obtain the following identities:

$$\frac{(d_1+d_2+n-1)!}{n!(d_1+d_2-1)!} = \sum_{m=0}^{n} \frac{(d_1+m-1)!}{m!(d_1-1)!} \frac{(d_2+n-m-1)!}{(n-m)!(d_2-1)!},$$

$$\frac{(d_1+d_2)!}{n!(d_1+d_2-n)!} = \sum_{m=0}^{n} \frac{d_1!}{m!(d_1-m)!} \frac{d_2!}{(n-m)!(d_2-n+m)!}.$$

These identities can be easily shown using the generating functions (3.18) and the identities

$$(1-t)^{-d_1}(1-t)^{-d_2} = (1-t)^{-(d_1+d_2)},$$

$$(1+t)^{d_1}(1+t)^{d_2} = (1+t)^{(d_1+d_2)}.$$

3.3.9 Super-Fock spaces

Let $(\mathcal{Y}, \epsilon)$ be a super-space (that is, a vector space equipped with an involution; see Subsect. 1.1.15). Then we introduce the action of the permutation group in $L(\overset{\mathrm{al}}{\otimes}{}^{n} \mathcal{Y})$ as follows:

Definition 3.44 *Let $\sigma \in \mathcal{S}_n$. Then $\Theta_\epsilon(\sigma)$ will denote the unique linear operator on $\overset{\mathrm{al}}{\otimes}{}^{n} \mathcal{Y}$ with the following property. Let $y_1, \ldots, y_n \in \mathcal{Y}$ be homogeneous. Then*

$$\Theta_\epsilon(\sigma) y_1 \otimes \cdots \otimes y_n = \mathrm{sgn}_\epsilon(\sigma)\, y_{\sigma^{-1}(1)} \otimes \cdots \otimes y_{\sigma^{-1}(n)},$$

where $\mathrm{sgn}_\epsilon(\sigma)$ is the sign of the permutation σ restricted to the odd elements.

Definition 3.45 *We define*

$$\Theta_\epsilon^n := \frac{1}{n!} \sum_{\sigma \in \mathcal{S}_n} \Theta_\epsilon(\sigma).$$

Clearly, Θ_ϵ^n is a projection on $\overset{\mathrm{al}}{\otimes}{}^{n}\mathcal{Y}$.

Definition 3.46 *We set*

$$\overset{\mathrm{al}}{\Gamma}{}_\epsilon^{n}(\mathcal{Y}) := \Theta_\epsilon^n \overset{\mathrm{al}}{\otimes}{}^{n} \mathcal{Y},$$

$$\overset{\mathrm{al}}{\Gamma}{}_\epsilon(\mathcal{Y}) := \overset{\mathrm{al}}{\underset{0 \le n \le \infty}{\bigoplus}} \overset{\mathrm{al}}{\Gamma}{}_\epsilon^{n}(\mathcal{Y}).$$

If $\mathcal{Y}$ is a finite-dimensional space, we can write $\Gamma_\epsilon^n(\mathcal{Y})$ instead of $\overset{\mathrm{al}}{\Gamma}{}_\epsilon^{n}(\mathcal{Y})$.

If $\mathcal{Y}$ is a Hilbert space, then Θ_ϵ^n are orthogonal projections.

Definition 3.47 *We define the* super-Fock spaces

$$\Gamma_\epsilon^n(\mathcal{Y}) := \Theta_\epsilon^n \otimes^n \mathcal{Y},$$

$$\Gamma_\epsilon(\mathcal{Y}) := \overset{\infty}{\underset{n=0}{\bigoplus}} \Gamma_\epsilon^n(\mathcal{Y}).$$

We extend various definitions from the context of bosonic, resp. fermionic Fock spaces to super-spaces in an obvious way. In particular, we define the operation $\otimes_\epsilon$, creation, resp. annihilation operators (generalizing the definitions of Sect. 3.4 below) and the operators $\Gamma(\cdot)$ and $d\Gamma(\cdot)$.

$\Gamma_\epsilon^n(\mathcal{Y})$ is naturally a super-space with the involution $\Gamma(\epsilon)$.

Super-Fock spaces enjoy the exponential property analogous to that described in Thm. 3.43 for bosonic and fermionic Fock spaces. Thus if $(\mathcal{Y}, \epsilon)$, $(\mathcal{W}, \varepsilon)$ are two super-Hilbert spaces, then

$$\Gamma_{\epsilon \oplus \varepsilon}(\mathcal{Y} \oplus \mathcal{W}) \simeq \Gamma_\epsilon(\mathcal{Y}) \otimes \Gamma_\varepsilon(\mathcal{W}). \tag{3.20}$$

In particular, if $\mathcal{Y} = \mathcal{Y}_0 \oplus \mathcal{Y}_1$ is the decomposition into the even and odd subspace, we then have

$$\Gamma_\epsilon(\mathcal{Y}) \simeq \Gamma_s(\mathcal{Y}_0) \otimes \Gamma_a(\mathcal{Y}_1), \tag{3.21}$$

which can be treated as an alternative definition of a super-Fock space.

We will often drop the index ϵ in (3.21)

Note that if $c \in L(\mathcal{Y})$ is odd, then $d\Gamma(c)^2 = d\Gamma(c^2)$. In the matrix notation:

$$d\Gamma\left(\begin{bmatrix} 0 & c_{01} \\ c_{10} & 0 \end{bmatrix}\right)^2 = d\Gamma\left(\begin{bmatrix} c_{01}c_{10} & 0 \\ 0 & c_{10}c_{01} \end{bmatrix}\right).$$

This identity plays an important role in super-symmetric quantum physics.

3.4 Creation and annihilation operators

Creation and *annihilation operators* belong to the most useful constructions of quantum physics. This section is devoted to their basic properties, in both the bosonic and the fermionic case.

Throughout this section we will use the standard convention for the scalar product in the Fock spaces. Some of the properties of creation and annihilation operators actually look simpler on *modified Fock spaces*, which will be discussed in Subsect. 3.5.7.

Throughout the section, $\mathcal{Z}$, $\mathcal{Z}_1$ and $\mathcal{Z}_2$ are Hilbert spaces.

3.4.1 Creation and annihilation operators: abstract approach

We prepare for the definitions of the creation and annihilation operators with two lemmas in an attract setting. We start with the bosonic case.

Lemma 3.48 *Let $\mathcal{H}$ be a Hilbert space, $\mathcal{D} \subset \mathcal{H}$ a dense subspace, and c, a two linear operators on $\mathcal{D}$ such that*

(1) $c, a : \mathcal{D} \to \mathcal{D}$;
(2) $c \subset a^*$, $a \subset c^*$;

(3) $ac - ca = \mathbb{1}$, *as an operator identity on* $\mathcal{D}$;
(4) ca *is essentially self-adjoint on* $\mathcal{D}$.

Then c, a are closable with $a^{\mathrm{cl}} = c^*$, $c^{\mathrm{cl}} = a^*$. If we write a for a^{cl}, one has

$$aa^* - a^*a = \mathbb{1}, \text{ as a quadratic form identity on } \mathrm{Dom}(a) \cap \mathrm{Dom}(a^*).$$

Proof Since $c \subset a^*$ and $a \subset c^*$, c^* and a^* are densely defined, and hence c and a are closable. Moreover, since $c \subset a^*$ we have $c^{\mathrm{cl}} \subset a^{\mathrm{cl}*}$. From now on we will denote a^{cl}, c^{cl} simply by a, c.

Set $N := (ca)^{\mathrm{cl}}$. Using (4), we see that N is a positive self-adjoint operator and $\mathcal{D}$ is a core for N. Since

$$\|a\Phi\|^2 = (\Phi|ca\Phi), \ \|c\Phi\|^2 = (\Phi|ca\Phi) + (\Phi|\Phi) \text{ for } \Phi \in \mathcal{D},$$

we see that $\mathrm{Dom}\,a = \mathrm{Dom}\,c = \mathrm{Dom}\,N^{\frac{1}{2}}$. This implies that

$$a(N + \mathbb{1})^{-\frac{1}{2}}, \ c(N + \mathbb{1})^{-\frac{1}{2}}, \ (N + \mathbb{1})^{-\frac{1}{2}}c, \ (N + \mathbb{1})^{-\frac{1}{2}}a \in B(\mathcal{H}). \tag{3.22}$$

Next, for $\Phi, \Psi \in \mathcal{D}$, we have

$$|(c\Phi|\Psi)| = |(\Phi|a\Psi)| = |((N + \mathbb{1})^{\frac{1}{2}}\Phi|(N + \mathbb{1})^{-\frac{1}{2}}a\Psi)| \leq C\|(N + \mathbb{1})^{\frac{1}{2}}\Phi\|\|\Psi\|.$$

Since $\mathcal{D}$ is dense in $\mathrm{Dom}\,N^{\frac{1}{2}}$ and in $\mathrm{Dom}\,a$, we obtain that $\mathrm{Dom}\,N^{\frac{1}{2}} \subset \mathrm{Dom}\,a^*$, and $a^*|_{\mathrm{Dom}\,N^{\frac{1}{2}}} = c$.

To prove that $a^* = c$, it remains to prove that $\mathrm{Dom}\,a^* = \mathrm{Dom}\,N^{\frac{1}{2}}$. Note that $\Phi \in \mathrm{Dom}\,N^{\frac{1}{2}}$ iff

$$\|N(\epsilon N + 1)^{-1}\Phi\| \leq C\epsilon^{-\frac{1}{2}}, \ \epsilon > 0. \tag{3.23}$$

From the identity $a(N + \mathbb{1}) = Na$ valid on $\mathcal{D}$, we deduce first that $(\epsilon N + \mathbb{1})^{-1}a(\epsilon N + \mathbb{1} - \epsilon) = a$ on $\mathcal{D}$ and then on $\mathrm{Dom}\,N$, and then that

$$(\epsilon N + \mathbb{1})^{-1}a = a(\epsilon N + \mathbb{1} - \epsilon)^{-1} \text{ on } \mathcal{H}, \tag{3.24}$$

since both operators are bounded by (3.22). For $\Phi \in \mathrm{Dom}\,a^*$ and $\Psi \in \mathcal{D}$, we have

$$\begin{aligned}
|(\Phi|N(\epsilon N + \mathbb{1})^{-1}\Psi)| &= |(\Phi|(\epsilon N + \mathbb{1})^{-1}ca\Psi)| \\
&= |(\Phi|(\epsilon N + \mathbb{1})^{-1}ac\Psi) - (\Phi|(\epsilon N + \mathbb{1})^{-1}\Psi)| \\
&= (\Phi|a(\epsilon N + \mathbb{1} - \epsilon)^{-1}c\Psi) - (\Phi|(\epsilon N + \mathbb{1})^{-1}\Psi)| \\
&\leq C\|(\epsilon N + \mathbb{1} - \epsilon)^{-1}c\Psi\| + C\|\Psi\| \\
&\leq C\|(N + \mathbb{1})^{\frac{1}{2}}(\epsilon N + \mathbb{1} - \epsilon)^{-1}\|\|(N + \mathbb{1})^{-\frac{1}{2}}c\Psi\| + C\|\Psi\| \\
&\leq C\epsilon^{-\frac{1}{2}}\|\Psi\|,
\end{aligned}$$

where we have used (3.24) and the fact that $\Phi \in \mathrm{Dom}\,a^*$. Using (3.23), we obtain that $\Phi \in \mathrm{Dom}\,N^{\frac{1}{2}}$, which completes the proof that $a^* = c$, and hence that $c^* = a$. The quadratic form identity on $\mathrm{Dom}\,a \cap \mathrm{Dom}\,a^*$ follows then by density from the operator identity on $\mathcal{D}$. $\qquad\square$

The following lemma describes properties of fermionic creation and annihilation operators in the abstract setting:

Lemma 3.49 *Let $\mathcal{H}$ be a Hilbert space, $\mathcal{D} \subset \mathcal{H}$ a dense subspace, and c, a two linear operators on $\mathcal{D}$ such that*

(1) $c, a : \mathcal{D} \to \mathcal{D}$;

(2) $c \subset a^*$, $a \subset c^*$;

(3) $a^2 = c^2 = 0$, $ac + ca = \mathbb{1}$ *as operator identities on $\mathcal{D}$.*

Then c, a extend as bounded operators on $\mathcal{H}$, $c = a^*$ and $\|a\| = \|c\| = 1$.

Proof We obtain from (2) and (3) that

$$\|c\Phi\|^2 + \|a\Phi\|^2 = \|\Phi\|^2, \quad \Phi \in \mathcal{D},$$

and hence c and a extend as bounded operators on $\mathcal{H}$ with $a = c^*$, $c = a^*$. Next we use

$$a^* a a^* a = a^* a - (a^*)^2 a^2 = a^* a,$$

and hence $\|a\|^4 = \|a^* a a^* a\| = \|a^* a\| = \|a\|^2$. By $[a, a^*]_+ = \mathbb{1}$, $\|a\|$ cannot be 0. Therefore, $\|a\| = \|a^*\| = \mathbb{1}$. $\qquad\qquad\qquad\qquad\qquad\qquad\qquad\qquad\qquad\qquad\qquad\square$

3.4.2 Creation and annihilation operators on Fock spaces

We consider the bosonic or fermionic Fock space $\Gamma_{s/a}(\mathcal{Z})$.

Definition 3.50 *Let $w \in \mathcal{Z}$. The* creation operator *of w, resp. the* annihilation operator *of w, are defined as operators on $\Gamma_{s/a}^{\text{fin}}(\mathcal{Z})$ by*

$$c(w)\Psi := \sqrt{n+1} w \otimes_{s/a} \Psi,$$
$$a(w)\Psi := \sqrt{n}(w| \otimes \mathbb{1}_{\mathcal{Z}}^{\otimes(n-1)} \Psi, \quad \Psi \in \Gamma_{s/a}^n(\mathcal{Z}).$$

Theorem 3.51 (Bosonic case) *In the bosonic case, the operators $c(w)$ and $a(w)$ are densely defined and closable. We denote their closures by the same symbols. They satisfy $a(w)^* = c(w)$. Therefore, we will write $a^*(w)$ instead of $c(w)$.*

(1) *The following quadratic form identities are valid:*

$$[a^*(w_1), a^*(w_2)] = [a(w_1), a(w_2)] = 0,$$
$$[a(w_1), a^*(w_2)] = (w_1|w_2)\mathbb{1}.$$

(2) *For $\Psi \in \Gamma_s(\mathcal{Z})$, $w \in \mathcal{Z}$,*

$$\|a(w)\Psi\| \le \|w\| \|N^{\frac{1}{2}}\Psi\|, \quad \|a^*(w)\Psi\| \le \|w\| \|(N+\mathbb{1})^{\frac{1}{2}}\Psi\|.$$

Proof We apply Lemma 3.48 to $c(w)$, $a(w)$ with $\mathcal{D} = \Gamma_s^{\text{fin}}(\mathcal{Z})$ (without loss of generality we can assume that $\|w\| = 1$). Then $c(w)a(w) = \mathrm{d}\Gamma(|w)(w|)$, which is essentially self-adjoint on $\mathcal{D}$.

We have

$$a^*(w)a(w) = \mathrm{d}\Gamma(|w)(w|), \quad a(w)a^*(w) = \mathrm{d}\Gamma(|w)(w|) + \|w\|^2\mathbb{1}.$$

Using that $|w)(w| \leq \|w\|^2 \mathbb{1}$ on $\mathcal{Z}$, we get

$$d\Gamma(|w)(w|) \leq \|w\|^2 N,$$

which implies (2).   □

Theorem 3.52 (Fermionic case) *In the fermionic case, the operators $c(w)$ and $a(w)$ are densely defined and bounded. We denote their closures by the same symbols. They satisfy $a(w)^* = c(w)$. Therefore, we will write $a^*(w)$ instead of $c(w)$.*

(1) *The following operator identities are valid:*

$$[a^*(w_1), a^*(w_2)]_+ = [a(w_1), a(w_2)]_+ = 0,$$

$$[a(w_1), a^*(w_2)]_+ = (w_1|w_2)\mathbb{1}.$$

(2) $\|a(w)\| = \|a^*(w)\| = \|w\|.$

Proof We apply Lemma 3.49 to $c(w)$, $a(w)$ with $\mathcal{D} = \Gamma_a^{\mathrm{fin}}(\mathcal{Z})$ (without loss of generality we can assume that $\|w\| = 1$).   □

Proposition 3.53 *If $p \in B(\mathcal{Z}_1, \mathcal{Z}_2)$ and $h \in Cl(\mathcal{Z})$, one has*

(1) $a(w_2)\Gamma(p) = \Gamma(p)a(p^*w_2)$, $\Gamma(p)a^*(w_1) = a^*(pw_1)\Gamma(p)$,
(2) $[d\Gamma(h), a(w)] = -a(h^*w)$, $[d\Gamma(h), a^*(w)] = a^*(hw)$,

the last two identities being quadratic form identities on $\overset{\text{al}}{\Gamma}_{\mathrm{s/a}}(\mathrm{Dom}\, h)$.

For further reference we note the following obvious facts:

$$\{\Psi \in \Gamma_{\mathrm{s/a}}(\mathcal{Z}) \;:\; a(w)\Psi = 0, \; w \in \mathcal{Z}\} = \mathbb{C}\Omega, \qquad (3.25)$$

$$\mathrm{Span}^{\mathrm{cl}}\left\{\prod_{i=0}^{n} a^*(w_i)\Omega, \; w_1, \ldots, w_n \in \mathcal{Z}, \; n = 0, 1, \ldots\right\} = \Gamma_{\mathrm{s/a}}(\mathcal{Z}). \quad (3.26)$$

Remark 3.54 *The notation for creation and annihilation operators introduced in this section is typical for the mathematically oriented literature. In the physical literature it is common to assume that the one-particle space has a distinguished o.n. basis $\{e_j\}_{j \in J}$. One writes a_j^* and a_j instead of $a^*(e_j)$ and $a(e_j)$, $j \in J$.*

Clearly, every vector $w \in \mathcal{Z}$ can then be written as $\sum_{i \in J} w_j e_j$, and we have the following dictionary between "mathematician's" and "physicist's" notations:

$$a^*(w) = \sum_{j \in J} w_j a_j^*,$$

$$a(w) = \sum_{j \in J} \overline{w}_j a_j. \qquad (3.27)$$

Note that the latter notation is heavier and depends on the choice of a basis, but has a useful advantage: it does not hide the anti-linearity of the annihilation operator.

Sometimes, instead of choosing an o.n. basis of $\mathcal{Z}$ it is more natural to assume that $\mathcal{Z} = L^2(\mathcal{Q}, dq)$ for some measure space $(\mathcal{Q}, dq)$. Clearly, $w \in \mathcal{Z}$ can be represented as a function $\mathcal{Q} \ni q \mapsto w(q)$. One introduces "operator-valued distributions" $\mathcal{Q} \ni q \mapsto a_q^*, a_q$, which are then "smeared out" with test functions to obtain creation and annihilation operators:

$$a^*(w) = \int w(q) a_q^* dq,$$

$$a(w) = \int \overline{w(q)} a_q dq. \tag{3.28}$$

(3.28) can be viewed as a generalization of (3.27).

The following operator seems to have no name, but is useful, especially on fermionic Fock spaces:

Definition 3.55 *Set*

$$\Lambda := (-1)^{N(N-1)/2}. \tag{3.29}$$

The following property is valid in both the bosonic and the fermionic case:

$$\Lambda a^*(z)\Lambda = -I a^*(z) = a^*(z)I,$$
$$\Lambda a(z)\Lambda = I a(z) = -a(z)I, \tag{3.30}$$

where I denotes the parity operator. In the fermionic case, (3.30) allows the conversion of the anti-commutation relations into commutation relations:

$$[\Lambda a^*(z_1)\Lambda, a^*(z_2)] = [\Lambda a(z_1)\Lambda, a(z_2)] = 0,$$

$$[\Lambda a^*(z_1)\Lambda, a(z_2)] = I(z_2|z_1).$$

3.4.3 Exponential law for creation and annihilation operators

Let N_i, I_i, Λ_i be the operators on $\Gamma_{s/a}(\mathcal{Z}_i)$ defined as in (3.9), (3.10) and (3.29). Recall that the unitary operator $U : \Gamma_{s/a}(\mathcal{Z}_1) \otimes \Gamma_{s/a}(\mathcal{Z}_2) \to \Gamma_{s/a}(\mathcal{Z}_1 \oplus \mathcal{Z}_2)$ was defined in Thm. 3.43.

The exponential law for creation and annihilation operators is slightly different in the bosonic and fermionic cases:

Proposition 3.56 *Let* $(w_1, w_2) \in \mathcal{Z}_1 \oplus \mathcal{Z}_2$.

(1) *In the bosonic case we have*

$$a^*(w_1, w_2)U = U(a^*(w_1) \otimes \mathbb{1} + \mathbb{1} \otimes a^*(w_2)),$$
$$a(w_1, w_2)U = U(a(w_1) \otimes \mathbb{1} + \mathbb{1} \otimes a(w_2)).$$

(2) *In the fermionic case, we have*

$$a^*(w_1, w_2)U = U(a^*(w_1) \otimes \mathbb{1} + I_1 \otimes a^*(w_2)),$$
$$a(w_1, w_2)U = U(a(w_1) \otimes \mathbb{1} + I_1 \otimes a(w_2)).$$

Proposition 3.57 (1) $IU = UI_1 \otimes I_2$,
(2) $\Lambda U = U(\Lambda_1 \otimes \Lambda_2)(-1)^{N_1 \otimes N_2}$.
(3) *In the fermionic case,*

$$\Lambda a^*(w_1, w_2)\Lambda\ U = U(a^*(w_1)I_1 \otimes I_2 + \mathbb{1} \otimes a^*(w_2)I_2),$$

$$\Lambda a(w_1, w_2)\Lambda\ U = U(-a(w_1)I_1 \otimes I_2 - \mathbb{1} \otimes a(w_2)I_2).$$

Proof We use

$$NU = U(N_1 \otimes \mathbb{1} + \mathbb{1} \otimes N_2),$$

$$\frac{1}{2}N(N - \mathbb{1}) = \frac{1}{2}N_1(N_1 - \mathbb{1}) + \frac{1}{2}N_2(N_2 - \mathbb{1}) + N_1 N_2,$$

$$(-1)^{N_1 \otimes N_2}(a(w) \otimes \mathbb{1})(-1)^{N_1 \otimes N_2} = a(w) \otimes I_2. \qquad \square$$

3.4.4 Multiple creation and annihilation operators

Let $\Phi \in \Gamma_{\mathrm{s/a}}^m(\mathcal{Z})$.

Definition 3.58 *We define the* operator of creation of Φ *with the domain* $\Gamma_{\mathrm{s/a}}^{\mathrm{fin}}(\mathcal{Z})$ *as*

$$a^*(\Phi)\Psi := \sqrt{(n+1)\cdots(n+m)}\,\Phi \otimes_{\mathrm{s/a}} \Psi, \quad \Psi \in \Gamma_{\mathrm{s/a}}^n(\mathcal{Z}).$$

$a^*(\Phi)$ is a densely defined closable operator. We denote its closure by the same symbol.

Definition 3.59 *We set*

$$a(\Phi) := (a^*(\Phi))^*.$$

$a(\Phi)$ *is called the* operator of annihilation of Φ.

For $w_1, \ldots, w_m \in \mathcal{Z}$ we have

$$a^*(w_1 \otimes_{\mathrm{s/a}} \cdots \otimes_{\mathrm{s/a}} w_m) = a^*(w_1)\cdots a^*(w_m),$$

$$a(w_1 \otimes_{\mathrm{s/a}} \cdots \otimes_{\mathrm{s/a}} w_m) = a(w_m)\cdots a(w_1).$$

Note that in the fermionic case we have

$$a(\Lambda w_1 \otimes_{\mathrm{a}} \cdots \otimes_{\mathrm{a}} w_m) = a(w_1)\cdots a(w_m),$$

where Λ was defined in (3.29).

3.5 Multi-linear symmetric and anti-symmetric forms

We continue to discuss symmetric and anti-symmetric tensors. In this section we will look at them mostly as multi-linear functions. This leads to somewhat different notational conventions.

Let $\mathcal{Y}$ be a real or complex vector space.

3.5.1 Polynomials

Let $\Psi \in \overset{\text{al}}{\Gamma}{}^n_{\text{s/a}}(\mathcal{Y})$. Then Ψ determines the function

$$\mathcal{Y}^\# \times \cdots \times \mathcal{Y}^\# \ni (v_1, \ldots, v_n) \mapsto \Psi(v_1, \ldots, v_n)$$

$$:= \langle v_1 \otimes_{\text{s/a}} \cdots \otimes_{\text{s/a}} v_n \,|\, \Psi \rangle \in \mathbb{K}. \tag{3.31}$$

Definition 3.60 *The space* $\overset{\text{al}}{\Gamma}_{\text{s/a}}(\mathcal{Y})$ *will often be denoted by* $\text{Pol}_{\text{s/a}}(\mathcal{Y}^\#)$, *if we want to stress the interpretation of its elements given by (3.31). (Pol stands for "poly-linear" or "a polynomial".) It will be called the symmetric, resp. anti-symmetric tensor algebra written in the* polynomial notation.

More generally, if $\mathcal{Y} = \mathcal{Y}_0 \oplus \mathcal{Y}_1$ *is a super-space, the super-tensor algebra* $\overset{\text{al}}{\Gamma}_\epsilon(\mathcal{Y})$ *will be also sometimes denoted by* $\text{Pol}_\epsilon(\mathcal{Y}^\#)$. *Clearly,*

$$\text{Pol}_\epsilon(\mathcal{Y}^\#) \simeq \text{Pol}_{\text{s}}(\mathcal{Y}^\#_0) \overset{\text{al}}{\otimes} \text{Pol}_{\text{a}}(\mathcal{Y}^\#_1). \tag{3.32}$$

Thus an element of $\text{Pol}_\epsilon(\mathcal{Y}^\#)$ *is a polynomial in commuting variables from* $\mathcal{Y}_0$ *and in anti-commuting variables from* $\mathcal{Y}_1$.

We will often drop the subscript ϵ *in (3.32).*

In the symmetric case we can make yet another identification. Let $\Psi = \sum_{n=0}^{\infty} \Psi_n$ with $\Psi_n \in \text{Pol}^n_{\text{s}}(\mathcal{Y}^\#)$.

Definition 3.61 *We introduce the function called the* polynomial function *associated with* Ψ:

$$\mathcal{Y}^\# \ni v \mapsto \Psi(v) := \sum_{n=0}^{\infty} \langle v^{\otimes n} \,|\, \Psi_n \rangle. \tag{3.33}$$

Note that if we know the function (3.33), we have full knowledge of $\Psi \in \text{Pol}_{\text{s}}(\mathcal{Y}^\#)$.

In the following proposition $\Psi, \Phi \in \text{Pol}_{\text{s}}(\mathcal{Y}^\#)$ are interpreted as polynomial functions and $v \in \mathcal{Y}^\#$:

Proposition 3.62 (1) $\Gamma(p)\Psi(v) = \Psi(p^\# v)$.
(2) $\Psi \otimes_{\text{s}} \Phi(v) = \Psi(v)\Phi(v)$.

Motivated by Prop. 3.62, we will often replace $\Psi \otimes_{\text{s}} \Phi$ with $\Psi \cdot \Phi$. We will often do the same in the anti-symmetric case as well.

In (3.31), $v_1, \ldots, v_n$ are elements of $\mathcal{Y}^\#$. In (3.33) and in Prop. 3.62, v has the same meaning. Sometimes, however, we will write $\Psi(v)$ without having in mind a concrete $v \in \mathcal{Y}^\#$. We will treat the symbol v as the "generic variable in $\mathcal{Y}^\#$"; see Subsect. 2.1.2.

In the anti-symmetric case we do not have an analog of (3.33). Nevertheless, following the common usage of theoretical physics, one often calls elements of $\text{Pol}^n_{\text{a}}(\mathcal{Y}^\#)$ "polynomials in non-commuting variables from $\mathcal{Y}^\#$". This suggests the notation $\Psi(v)$ instead of $\Psi(v_1, \ldots, v_n)$. In this context v is just the name of the generic variable in $\mathcal{Y}^\#$. Similarly, motivated by Prop. 3.62 (1), we will write $\Psi(p^\# v)$ instead of $\Gamma(p)\Psi(v)$. This point of view will be further developed in Chap. 7.

3.5.2 *Multiplication and differentiation operators*

As mentioned above, we will use the letter v as the name of the generic variable in $\mathcal{Y}^\#$. This symbol will appear in the multiplication and derivative operators that we define below.

Definition 3.63 *For $y \in \mathcal{Y}$, the* operator of multiplication by y *is defined by*

$$y(v)\Psi := y \otimes_{\mathrm{s/a}} \Psi, \quad \Psi \in \mathrm{Pol}_{\mathrm{s/a}}(\mathcal{Y}^\#).$$

We will often write $y \cdot v$ instead of $y(v)$.

More generally, if $\Phi \in \mathrm{Pol}_{\mathrm{s/a}}(\mathcal{Y}^\#)$, $\Phi(v)$ will denote the operator of multiplication by Φ:

$$\Phi(v)\Psi := \Phi \otimes_{\mathrm{s/a}} \Psi.$$

Definition 3.64 *For $w \in \mathcal{Y}^\#$, the* derivative in the direction of w *is defined by*

$$w(\nabla_v)\Psi := n\langle w| \otimes \mathbb{1}^{\otimes(n-1)}\Psi, \quad \Psi \in \mathrm{Pol}_{\mathrm{s/a}}^n(\mathcal{Y}^\#). \tag{3.34}$$

We will often write $w \cdot \nabla_v$ instead of $w(\nabla_v)$.

More generally, if $\Phi \in \mathrm{Pol}_{\mathrm{s/a}}(\mathcal{Y})$, we define the derivative $\Phi(\nabla_v)$. *For $\Phi \in \mathrm{Pol}_{\mathrm{s/a}}^m(\mathcal{Y})$, it acts on $\Psi \in \mathrm{Pol}_{\mathrm{s/a}}^n(\mathcal{Y}^\#)$ as*

$$\Phi(\nabla_v)\Psi = n(n-1)\cdots(n-m+1)\langle\Phi| \otimes \mathbb{1}^{\otimes(n-m)}\Psi. \tag{3.35}$$

Then we extend this definition by linearity.

Note that in the symmetric case the differentiation operator defined above is the usual differentiation of polynomials. In particular, $w(\nabla_v)$ in (3.34) coincides with the directional derivative Def. 2.50.

The operators of multiplication and differentiation are essentially equivalent to the creation and annihilation operators. We will discuss this equivalence in Subsect. 3.5.7.

In the following propositions $y, y_1, y_2 \in \mathcal{Y}$, $w, w_1, w_2 \in \mathcal{Y}^\#$ and $\Phi, \Psi \in \mathrm{Pol}_{\mathrm{s/a}}(\mathcal{Y}^\#)$.

Proposition 3.65 (Symmetric case)

(1) $[y_1(v), y_2(v)] = 0$, $[w_1(\nabla_v), w_2(\nabla_v)] = 0$,

(2) $[w(\nabla_v), y(v)] = \langle w|y\rangle \mathbb{1}$,

(3) $w(\nabla_v)\Psi \otimes_{\mathrm{s}} \Phi = (w(\nabla_v)\Psi) \otimes_{\mathrm{s}} \Phi + \Psi \otimes_{\mathrm{s}} (w(\nabla_v)\Phi)$,

(4) $w(\nabla_v) = \sum_{i=1}^n \mathbb{1}^{\otimes(i-1)} \otimes \langle w| \otimes \mathbb{1}^{\otimes(n-j)}$ *on* $\mathrm{Pol}_{\mathrm{s}}^n(\mathcal{Y}^\#)$.

Proposition 3.66 (Anti-symmetric case)

(1) $[y_1(v), y_2(v)]_+ = 0$, $[w_1(\nabla_v), w_2(\nabla_v)]_+ = 0$,

(2) $[w(\nabla_v), y(v)]_+ = \langle w|y \rangle \mathbb{1}$,

(3) $w(\nabla_v)\Psi \otimes_{\mathrm{a}} \Phi = (w(\nabla_v)\Psi) \otimes_{\mathrm{a}} \Phi + (I\Psi) \otimes_{\mathrm{a}} (w\nabla_v)\Phi)$,

(4) $w(\nabla_v) = \sum_{i=1}^{n} (-1)^{i-1} \mathbb{1}^{(i-1)\otimes} \otimes \langle w| \otimes \mathbb{1}^{(n-j)\otimes}$ on $\mathrm{Pol}_{\mathrm{a}}^n(\mathcal{Y}^\#)$.

Proposition 3.67 *Let $p, h \in L(\mathcal{Y})$. In both the symmetric and the anti-symmetric case we have*

(1) $\Gamma(p)y(v) = py(v)\Gamma(p)$, $\quad w(\nabla_v)\Gamma(p) = \Gamma(p)p^\# w(\nabla_v)$,

(2) $[\mathrm{d}\Gamma(h), y(v)] = hy(v)$, $\quad [\mathrm{d}\Gamma(h), w(\nabla_v)] = -h^\# w(\nabla_v)$,

(3) $\Phi(\nabla_v)\Psi(\nabla_v) = (\Phi \otimes_{\mathrm{s/a}} \Psi)(\nabla_v)$.

$\Psi \in \mathrm{Pol}_{\mathrm{s/a}}^n(\mathcal{Y}^\#)$ can be treated as an n-linear function on $(\mathcal{Y}^\#)^n$. Let us denote the generic variable of the j-th $\mathcal{Y}^\#$ by v_j. We can write an identity

$$\nabla_v \Psi = (\nabla_{v_1} + \cdots + \nabla_{v_n})\Psi, \qquad (3.36)$$

where on the left we use the functional notation, and on the right we treat Ψ as a function depending on n separate variables. (3.36) should be compared with (4) of Props. 3.65 and 3.66. Note that in the anti-symmetric case one has to remember that ∇_{v_i} anti-commutes with the operator of multiplication by v_j, hence the alternating sign.

3.5.3 Right derivative

Definition 3.68 *In the anti-symmetric case the derivative defined in Def. 3.64 should actually be called the* left *derivative. One can also introduce another operator with the name of the* right *derivative. For $w \in \mathcal{Y}^\#$, the right derivative in the direction of w acts on $\Psi \in \mathrm{Pol}_{\mathrm{a}}^n(\mathcal{Y}^\#)$ as*

$$w(\overleftarrow{\nabla}_v)\Psi := n\mathbb{1}^{(n-1)\otimes} \otimes \langle w|\Psi.$$

More generally, if $\Phi \in \mathrm{Pol}_{\mathrm{a}}(\mathcal{Y})$, we can define the right *derivative $\Phi(\overleftarrow{\nabla}_v)$. For $\Phi \in \mathrm{Pol}_{\mathrm{a}}^m(\mathcal{Y})$ and $\Psi \in \mathrm{Pol}_{\mathrm{a}}^n(\mathcal{Y}^\#)$, it is given by*

$$\Phi(\overleftarrow{\nabla}_v)\Psi = n(n-1)\cdots(n-m+1)\mathbb{1}^{(n-m)\otimes} \otimes \langle \Phi|\Psi. \qquad (3.37)$$

Note that we need to invert the order (compare with Prop. 3.67 (3)):

$$\Phi_1(\overleftarrow{\nabla}_v)\Phi_2(\overleftarrow{\nabla}_v)\Psi = (\Phi_2 \otimes_{\mathrm{a}} \Phi_1)(\overleftarrow{\nabla}_v)\Psi. \qquad (3.38)$$

Here is the relation between the left and right derivative:

$$\Phi(\nabla_v)\Psi = (-1)^{n(m-n)}\Phi(\overleftarrow{\nabla}_v)\Psi, \quad \Phi \in \mathrm{Pol}_{\mathrm{a}}^m(\mathcal{Y}), \Psi \in \mathrm{Pol}_{\mathrm{a}}^n(\mathcal{Y}^\#).$$

3.5.4 Exponential law in the polynomial notation

The exponential law described in Subsect. 3.3.7 is not the only convention used in the context of the tensor product of symmetric and anti-symmetric tensor algebras. In fact, there exists another convention that avoids the complicated multiplier involving the square roots of factorials. This convention is commonly used in the "algebraic case" (when we are not interested in the Hilbert space structure).

Let $\mathcal{Y}_1$, $\mathcal{Y}_2$ be two vector spaces. Let $j_i : \mathcal{Y}_i \to \mathcal{Y}_1 \oplus \mathcal{Y}_2$, $i = 1, 2$, be the canonical embeddings.

Definition 3.69

$$U^{\mathrm{mod}} : \mathrm{Pol}_{\mathrm{s/a}}(\mathcal{Y}_1^{\#}) \overset{\mathrm{al}}{\otimes} \mathrm{Pol}_{\mathrm{s/a}}(\mathcal{Y}_2^{\#}) \to \mathrm{Pol}_{\mathrm{s/a}}((\mathcal{Y}_1 \oplus \mathcal{Y}_2)^{\#})$$

is defined as the unique linear map such that if $\Psi_1 \in \mathrm{Pol}_{\mathrm{s/a}}^{n_1}(\mathcal{Y}_1^{\#})$, $\Psi_2 \in \mathrm{Pol}_{\mathrm{s/a}}^{n_2}(\mathcal{Y}_2^{\#})$, *then (in the tensor notation)*

$$U^{\mathrm{mod}}\Psi_1 \otimes \Psi_2 := (\Gamma(j_1)\Psi_1) \otimes_{\mathrm{s/a}} (\Gamma(j_2)\Psi_2). \tag{3.39}$$

We will use v_i as the generic variables in $\mathcal{Y}_i^{\#}$, $i = 1, 2$. In the "polynomial notation", $\Psi := U^{\mathrm{mod}}\Psi_1 \otimes \Psi_2$ will be simply written as

$$\Psi(v) = \Psi_1(v_1) \otimes_{\mathrm{s/a}} \Psi_2(v_2), \quad v = (v_1, v_2). \tag{3.40}$$

Often, we will even omit $\otimes_{\mathrm{s/a}}$ between the factors. Note that in the symmetric case, if we use the "polynomial interpretation", the exponential law is just the usual multiplication of polynomials in two separate variables, which is consistent with the notation (3.40).

Clearly, the identities (3.16) and (3.17) hold with U replaced with U^{mod}.

Proposition 3.70 (1) (Symmetric case)

$$(y_1, y_2)(v)U^{\mathrm{mod}} = U^{\mathrm{mod}}(y_1(v_1) \otimes \mathbb{1} + \mathbb{1} \otimes y_2(v_2)),$$
$$(w_1, w_2)(\nabla_v)U^{\mathrm{mod}} = U^{\mathrm{mod}}(w_1(\nabla_{v_1}) \otimes \mathbb{1} + \mathbb{1} \otimes w_2(\nabla_{v_2})).$$

(2) (Anti-symmetric case)

$$(y_1, y_2)(v)U^{\mathrm{mod}} = U^{\mathrm{mod}}(y_1(v_1) \otimes \mathbb{1} + I_1 \otimes y_2(v_2)),$$
$$(w_1, w_2)(\nabla_v)U^{\mathrm{mod}} = U^{\mathrm{mod}}(w_1(\nabla_{v_1}) \otimes \mathbb{1} + I_1 \otimes w_2(\nabla_{v_2})).$$

3.5.5 Holomorphic continuation of polynomials

Let $\mathcal{Y}$ be a real vector space. The identification (3.3) leads to the following isomorphism:

$$\mathbb{C}\Gamma_{\mathrm{s/a}}(\mathcal{Y}) \simeq \Gamma_{\mathrm{s/a}}(\mathbb{C}\mathcal{Y}). \tag{3.41}$$

In the polynomial notation this isomorphism is written as

$$\mathbb{C}\mathrm{Pol}_{s/a}(\mathcal{Y}^{\#}) \simeq \mathrm{Pol}_{s/a}(\mathbb{C}\mathcal{Y}^{\#}).$$

Note that in the polynomial interpretation $\Psi \in \mathbb{C}\mathrm{Pol}^n_{s/a}(\mathcal{Y}^{\#})$ is a complex multi-linear function on $\mathcal{Y}^{\#}$, whereas the corresponding $\Psi_{\mathbb{C}} \in \mathrm{Pol}_{s/a}(\mathbb{C}\mathcal{Y}^{\#})$ is a multi-linear function on $\mathbb{C}\mathcal{Y}^{\#}$, which restricted to $\mathcal{Y}^{\#}$ equals Ψ.

Definition 3.71 *The polynomial $\Psi_{\mathbb{C}}$ will be called the* holomorphic extension *of* Ψ.

(Of course, instead of polynomials one can consider more general holomorphic functions.)

3.5.6 Polynomials on complex spaces

Let $\mathcal{Z}$ be a complex vector space. Recall that $\mathcal{Z}_{\mathbb{R}}$ denotes its realification. We can distinguish four basic families of polynomials related to $\mathcal{Z}$:

Definition 3.72 (1) *Elements of* $\mathrm{Pol}_{s/a}(\mathcal{Z}_{\mathbb{R}})$ *are called* real-valued polynomials.
(2) *Elements of* $\mathbb{C}\mathrm{Pol}_{s/a}(\mathcal{Z}_{\mathbb{R}})$ *are called* complex-valued polynomials.
(3) *Elements of* $\mathrm{Pol}_{s/a}(\mathcal{Z})$ *are called* holomorphic polynomials.
(4) *Elements of* $\mathrm{Pol}_{s/a}(\overline{\mathcal{Z}})$ *are called* anti-holomorphic polynomials.

As sets, $\mathcal{Z}_{\mathbb{R}}$, $\mathcal{Z}$ and $\overline{\mathcal{Z}}$ can be identified. With these identifications, $\mathbb{C}\mathrm{Pol}_{s/a}(\mathcal{Z}_{\mathbb{R}})$ is the largest family – it contains the other three.

Let us use the notation and results from Subsect. 1.3.6. In particular, we recall the space $\mathrm{Re}(\mathcal{Z} \oplus \overline{\mathcal{Z}}) = \{(z, \overline{z}) : z \in \mathcal{Z}\}$, whose complexification can be identified with $\mathcal{Z} \oplus \overline{\mathcal{Z}}$. We have the obvious map (which according to Def. 1.84 is called T_1^{-1})

$$\mathcal{Z}_{\mathbb{R}} \ni z \mapsto (z, \overline{z}) \in \mathrm{Re}(\mathcal{Z} \oplus \overline{\mathcal{Z}}). \tag{3.42}$$

Its complexification is

$$\mathbb{C}\mathcal{Z}_{\mathbb{R}} \ni z_1 + \mathrm{i}z_2 \mapsto (z_1 + \mathrm{i}z_2, \overline{z}_1 + \mathrm{i}\overline{z}_2) \in \mathcal{Z} \oplus \overline{\mathcal{Z}}. \tag{3.43}$$

With these identifications, we have

$$\mathrm{Pol}_{s/a}(\mathcal{Z}_{\mathbb{R}}) \simeq \mathrm{Pol}_{s/a}\left(\mathrm{Re}(\mathcal{Z} \oplus \overline{\mathcal{Z}})\right),$$
$$\mathbb{C}\mathrm{Pol}_{s/a}(\mathcal{Z}_{\mathbb{R}}) \simeq \mathrm{Pol}_{s/a}(\mathbb{C}\mathcal{Z}_{\mathbb{R}}) \simeq \mathrm{Pol}_{s/a}(\mathcal{Z} \oplus \overline{\mathcal{Z}}) \simeq \mathrm{Pol}_{s/a}(\mathcal{Z}) \otimes \mathrm{Pol}_{s/a}(\overline{\mathcal{Z}}).$$

In the last line, we first used (3.41), then (3.43), and finally the exponential law.

In the symmetric case, the polynomial functions corresponding to all four cases of Def. 3.72 can be viewed as functions on the same space $\mathcal{Z}_{\mathbb{R}}$. By allowing convergent series, we can also consider more general functions, in particular *holomorphic* and *anti-holomorphic functions on $\mathcal{Z}$*, with obvious definitions.

3.5.7 Modified Fock spaces

Let $\mathcal{Z}$ be a Hilbert space. Recall that the scalar product in the Fock space $\Gamma_{\mathrm{s/a}}(\mathcal{Z})$ is inherited from the scalar product in the tensor algebra $\otimes \mathcal{Z}$. This choice has some disadvantages. Instead, many authors adopt a different convention, which we will describe in this subsection.

Recall that N denotes the number operator on $\Gamma_{\mathrm{s/a}}(\mathcal{Z})$.

Definition 3.73 *Let us set* $\Gamma_{\mathrm{s/a}}^{\mathrm{mod}}(\mathcal{Z}) := \mathrm{Dom}\sqrt{N!}$ *equipped with the scalar product* $(\Psi|\Phi)_{\mathrm{mod}} := (\Psi|N!\Phi)$. *We introduce also the unitary operator*

$$\Gamma_{\mathrm{s/a}}(\mathcal{Z}) \ni \Psi \mapsto T^{\mathrm{mod}}\Psi := \frac{1}{\sqrt{N!}}\Psi \in \Gamma_{\mathrm{s/a}}^{\mathrm{mod}}(\mathcal{Z}). \qquad (3.44)$$

Sometimes we will write Ψ^{mod} *for* $T^{\mathrm{mod}}\Psi$.

The operators $\mathrm{d}\Gamma(h)$ and $\Gamma(p)$ keep the same form after conjugation by T^{mod}. If $\{e_i\}_{i\in I}$ is an o.n. basis of $\mathcal{Z}$, then

$$\left\{ \frac{1}{\sqrt{\vec{k}!}} e_{\vec{k}} : \vec{k} \in (\mathbb{N}^I)_{\mathrm{fin}} \right\}$$

is an o.n. basis of $\Gamma_{\mathrm{s}}^{\mathrm{mod}}(\mathcal{Z})$, and

$$\{ e_J : J \in 2_{\mathrm{fin}}^I \}$$

is an o.n. basis of $\Gamma_{\mathrm{a}}^{\mathrm{mod}}(\mathcal{Z})$, where $e_{\vec{k}}$ and e_J are defined in Subsects. 3.3.5 and 3.3.6.

Often we will consider the "polynomial notation" for $\Gamma_{\mathrm{s/a}}^{\mathrm{al}}(\mathcal{Z})$, where $\mathcal{Z}$ is a Hilbert space. In this case, it is convenient to use elements of the topological dual of $\mathcal{Z}$, instead of the algebraic dual, as arguments of the polynomial. The topological dual of $\mathcal{Z}$ is identified with $\overline{\mathcal{Z}}$. Thus the polynomial notation for $\Gamma_{\mathrm{s/a}}^{\mathrm{al}}(\mathcal{Z})$ will be $\mathrm{Pol}_{\mathrm{s/a}}(\overline{\mathcal{Z}})$. Clearly, $\mathrm{Pol}_{\mathrm{s/a}}(\overline{\mathcal{Z}})$ is dense in $\Gamma_{\mathrm{s/a}}^{\mathrm{mod}}(\mathcal{Z})$.

The generic variable of $\overline{\mathcal{Z}}$ will be often denoted $\overline{z}$. Thus an element of $\mathrm{Pol}_{\mathrm{s/a}}(\overline{\mathcal{Z}})$ in the polynomial notation will be written as $\Psi(\overline{z})$. If $w \in \mathcal{Z}$, then the corresponding multiplication and differentiation operators are $w(\overline{z})$ and $\overline{w}(\nabla_{\overline{z}})$. They are related to the creation and annihilation operators as

$$T^{\mathrm{mod}}a(w)(T^{\mathrm{mod}})^{-1} = \overline{w}(\nabla_{\overline{z}}), \quad T^{\mathrm{mod}}a^*(w)(T^{\mathrm{mod}})^{-1} = w(\overline{z}). \qquad (3.45)$$

If $\mathcal{Z}_1, \mathcal{Z}_2$ are Hilbert spaces, then the map U^{mod} defined in Def. 3.69 extends to a unitary map from $\Gamma_{\mathrm{s/a}}^{\mathrm{mod}}(\mathcal{Z}_1) \otimes \Gamma_{\mathrm{s/a}}^{\mathrm{mod}}(\mathcal{Z}_2)$ to $\Gamma_{\mathrm{s/a}}^{\mathrm{mod}}(\mathcal{Z}_1 \oplus \mathcal{Z}_2)$. It is related to the map $U : \Gamma_{\mathrm{s/a}}(\mathcal{Z}_1) \otimes \Gamma_{\mathrm{s/a}}(\mathcal{Z}_2) \to \Gamma_{\mathrm{s/a}}(\mathcal{Z}_1 \oplus \mathcal{Z}_2)$ defined in Subsect. 3.3.7 by

$$U^{\mathrm{mod}} = T^{\mathrm{mod}}U(T_1^{\mathrm{mod}} \otimes T_2^{\mathrm{mod}})^{-1},$$

where $T^{\mathrm{mod}}, T_1^{\mathrm{mod}}, T_2^{\mathrm{mod}}$ are the unitary identifications of the corresponding Fock and modified Fock spaces; see (3.44).

3.6 Volume forms, determinant and Pfaffian

In this section we recall some well-known concepts related to anti-symmetric tensors, such as *volume forms*, the *determinant* of a matrix and the *Pfaffian* of an anti-symmetric matrix. They are usually introduced in a coordinate-dependent fashion. In our presentation, we try to stress the coordinate-independent approach based on the anti-symmetric tensor algebra.

3.6.1 Volume forms

Let $\mathcal{X}$ be a (real or complex) d-dimensional space. A special role is played by the space $\wedge^d \mathcal{X}^\#$ of anti-symmetric d-forms on $\mathcal{X}$, which is one-dimensional.

Definition 3.74 *A non-zero element of $\wedge^d \mathcal{X}^\#$ will be called a* volume form *on* $\mathcal{X}$. *If the name of the generic variable in $\mathcal{X}$ is x, then a volume form on $\mathcal{X}$ will be often denoted by* $\mathrm{d}x$.

Suppose that we choose a basis $(e_1, \dots, e_d)$ in $\mathcal{X}$. Let $(e^1, \dots, e^d)$ be the dual basis in $\mathcal{X}^\#$. Then we have a distinguished volume form on $\mathcal{X}$ defined by

$$\Xi = e^d \wedge \cdots \wedge e^1. \tag{3.46}$$

(Note the reverse order and the use of $\wedge$ and not of $\otimes_a$.) We have

$$\langle \Xi | e_1 \otimes_a \cdots \otimes_a e_d \rangle = 1.$$

Definition 3.75 *If $\mathcal{X}$ is a Euclidean, resp. unitary space of dimension d, then we say that a volume form Ξ is* compatible with its Euclidean, *resp.* unitary structure *if there exists an o.n. basis of $\mathcal{X}$ $(e_1, \dots, e_d)$ such that*

$$\Xi = e^d \wedge \cdots \wedge e^1.$$

If $\mathcal{X}_i$, $i = 1, 2$, are vector spaces with volume forms Ξ_i, then on $\mathcal{X}_1 \oplus \mathcal{X}_2$ we take the volume form $\Xi_2 \wedge \Xi_1$. (Note again the reverse order and the use of $\wedge$ and not of $\otimes_a$.)

Definition 3.76 *If we use the notation $\mathrm{d}x_i$ for Ξ_i, we will often write $\mathrm{d}x_2 \mathrm{d}x_1$ for $\mathrm{d}x_2 \wedge \mathrm{d}x_1$.*

Definition 3.77 *If Ξ is a distinguished volume form on $\mathcal{X}$, then we have a distinguished volume form Ξ^{dual} on $\mathcal{X}^\#$ determined by*

$$\langle \Xi | \Xi^{\mathrm{dual}} \rangle = d!.$$

If in coordinates Ξ is given by (3.46), then

$$\Xi^{\mathrm{dual}} = e_1 \wedge \cdots \wedge e_d.$$

Note that $(\Xi^{\mathrm{dual}})^{\mathrm{dual}} = \Xi$. We will often use ξ as the generic variable of $\mathcal{X}^\#$, and then the dual volume form on $\mathcal{X}^\#$ will be denoted by $\mathrm{d}\xi$.

3.6.2 Hodge star operator

Let $d = \dim \mathcal{X}$. Let us fix a volume form $\Xi \in \Gamma_a^d(\mathcal{X}^\#) = \mathrm{Pol}_a^d(\mathcal{X})$ on $\mathcal{X}$.

Definition 3.78 *The* Hodge star operator *is defined as the map*

$$\theta : \mathrm{Pol}_a(\mathcal{X}^\#) \to \mathrm{Pol}_a(\mathcal{X})$$

by

$$\langle \Phi | \theta \Psi \rangle := \frac{1}{(d-n)!} \langle \Psi \otimes_a \Phi | \Xi \rangle, \quad \Psi \in \mathrm{Pol}_a^n(\mathcal{X}^\#), \quad \Phi \in \mathrm{Pol}_a^{d-n}(\mathcal{X}^\#).$$

Note that θ maps $\mathrm{Pol}_a^n(\mathcal{X}^\#)$ onto $\mathrm{Pol}_a^{d-n}(\mathcal{X})$. We will see in Subsect. 7.1.7 that the Hodge star operator can be viewed as an analog of the *Fourier transformation*.

Let us fix a basis $(e_1, \ldots, e_d)$ of $\mathcal{X}$ such that $\Xi = e^d \wedge \cdots \wedge e^1$. Let $\sigma \in S_d$ be a permutation and $0 \leq n \leq d$. Then

$$\theta \, e_{\sigma(1)} \otimes_a \cdots \otimes_a e_{\sigma(n)} = \mathrm{sgn}(\sigma) \, e^{\sigma(d)} \otimes_a \cdots \otimes_a e^{\sigma(n+1)}.$$

3.6.3 Liouville volume forms

Let $(\mathcal{Y}, \omega)$ be a symplectic space of dimension $2d$. Note that $\omega \in L_a(\mathcal{Y}, \mathcal{Y}^\#) \simeq \Gamma_a^2(\mathcal{Y}^\#)$.

Definition 3.79 $\mathcal{Y}$ *possesses a distinguished volume form called the* Liouville form,

$$\Xi^{\mathrm{Liouv}} := \frac{1}{d!} \wedge^d \omega. \tag{3.47}$$

Recall that $\mathcal{Y}^\#$ is equipped with the symplectic form ω^{-1}. Thus it possesses its own Liouville form $\frac{1}{d!} \wedge^d \omega^{-1}$. It is easy to see that it equals the volume form dual to Ξ^{Liouv}.

3.6.4 Liouville volume forms on $\mathcal{X}^\# \oplus \mathcal{X}$

Assume that $\mathcal{X}$ is a vector space of dimension d. Consider $\mathcal{Y} = \mathcal{X}^\# \oplus \mathcal{X}$ with its canonical symplectic form (1.9). If we choose an arbitrary basis $e_1, \ldots, e_d$ of $\mathcal{X}$ and $e^1, \ldots, e^d$ is the dual basis, then one can use the wedge product to write the canonical symplectic form as

$$\omega = \sum_{i=1}^d e_i \wedge e^i. \tag{3.48}$$

Hence the Liouville form on $\mathcal{X}^\# \oplus \mathcal{X}$ is

$$e_1 \wedge e^1 \wedge \cdots e_d \wedge e^d = e_1 \wedge \cdots \wedge e_d \wedge e^d \wedge \cdots \wedge e^1. \tag{3.49}$$

Proposition 3.80 *Choose an arbitrary volume form Ξ on $\mathcal{X}$. Then the Liouville volume form on $\mathcal{Y}$ is equal to $\Xi^{\mathrm{dual}} \wedge \Xi$.*

Proof We choose a basis of $\mathcal{X}$ and the dual basis of $\mathcal{X}^{\#}$ as above. Now for any volume form Ξ on $\mathcal{X}$ there exists $\lambda \neq 0$ such that

$$\Xi = \lambda e^d \wedge \cdots \wedge e^1, \qquad \Xi^{\mathrm{dual}} = \lambda^{-1} e_1 \wedge \cdots \wedge e_d. \qquad \square$$

The symplectic form ω^{-1} on $\mathcal{Y}^{\#} = \mathcal{X} \oplus \mathcal{X}^{\#}$ can be also written as (3.48). Hence the Liouville form on $\mathcal{Y}^{\#}$ can be written as (3.49).

Recall that often the symbols $\mathrm{d}x$ is used for a fixed volume form on $\mathcal{X}$, and its dual form on $\mathcal{X}^{\#}$ is denoted by $\mathrm{d}\xi$. Then the symbol $\mathrm{d}x\mathrm{d}\xi$ denotes the Liouville volume form on $\mathcal{X}^{\#} \oplus \mathcal{X}$ and on $\mathcal{X} \oplus \mathcal{X}^{\#}$.

3.6.5 Densities and Lebesgue measures

Let $\mathcal{X}$ be a real d-dimensional vector space.

Definition 3.81 *An element of $\wedge^d \mathcal{X}^{\#} / \{1, -1\}$ will be called a* density *on $\mathcal{X}$. The density associated with a volume form Ξ will be denoted by $|\Xi|$. Thus $|\Xi| = \{\Xi, -\Xi\}$. If $|\Xi|$ is a density on $\mathcal{X}$, we define the corresponding* dual density *on $\mathcal{X}^{\#}$ by $|\Xi|^{\mathrm{dual}} := |\Xi^{\mathrm{dual}}|$.*

Clearly, the set of volume forms compatible with a Euclidean structure is a density.

Recall from Def. 3.74 that if the generic variable of $\mathcal{X}$ is denoted x, then $\mathrm{d}x$ denotes a fixed volume form on $\mathcal{X}$. Thus, according to Def. 3.81, the corresponding density should be denoted by $|\mathrm{d}x|$.

Definition 3.82 *By a* Lebesgue measure *on $\mathcal{X}$ we mean a non-zero translation invariant Borel measure on $\mathcal{X}$ finite on compact sets.*

If $|\Xi|$ is a density on $\mathcal{X}$, then $|\Xi|$ induces a Lebesgue measure μ on $\mathcal{X}$ by setting

$$\mu\left(V(e_1, \ldots, e_d)\right) := |\langle \Xi | e_1 \otimes_{\mathrm{a}} \cdots \otimes_{\mathrm{a}} e_d \rangle|,$$

where $V(e_1, \ldots, e_d) := \left\{ \sum_{i=1}^d t_i e_i \ : \ t_i \in [0,1] \right\}$ is the parallelepiped with edges $e_1, \ldots, e_d$. Conversely a Lebesgue measure on $\mathcal{X}$ yields a unique density on $\mathcal{X}$. Therefore, we will often identify the concepts of a Lebesgue measure and a density.

The integral w.r.t. a Lebesgue measure is called a *Lebesgue integral*. If F is a function on $\mathcal{X}$, its Lebesgue integral is denoted $\int F(x) \mathrm{d}x$ (although, according to Def. 3.81, the notation $\int F(x) |\mathrm{d}x|$ would be more appropriate, since a Lebesgue integral depends only on the density $|\mathrm{d}x|$, and not on the volume form $\mathrm{d}x$). For further reference let us list elementary properties of a Lebesgue integral:

$$\int \Phi(\nabla_x) f(x) \mathrm{d}x = 0, \qquad \Phi \in \mathbb{C}\mathrm{Pol}_s^{\geq 1}(\mathcal{X}^\#);$$

$$\int f(x+y)\mathrm{d}x = \int f(x)\mathrm{d}x, \quad y \in \mathcal{X};$$

$$\int f(mx)\mathrm{d}x = (\det m)^{-1} \int f(x)\mathrm{d}x, \quad m \in L(\mathcal{X}). \tag{3.50}$$

3.6.6 Determinants

Definition 3.83 *If $a = [a_{ij}]$ is a $d \times d$ matrix, one defines its* determinant *as*

$$\det(a) := \sum_{\sigma \in S_d} \mathrm{sgn}(\sigma) \prod_{i=1}^{d} a_{i\sigma(i)}.$$

It is possible to give a manifestly coordinate-independent definition of the determinant. Let $\mathcal{X}$ be a d-dimensional vector space over $\mathbb{K}$.

Definition 3.84 *For $a \in L(\mathcal{X})$, its* determinant *is defined as the unique number* $\det a$ *satisfying*

$$\Gamma(a)\big|_{\wedge^d \mathcal{X}} =: \det a\, \mathbb{1}.$$

Clearly, this definition is possible, because $\Gamma(a)$ sends $\wedge^d \mathcal{X}$ into itself. If $(e_1, \ldots, e_d)$ is a basis of $\mathcal{X}$ and $(e^1, \ldots, e^d)$ its dual basis, then $\det a$ coincides with the determinant of the matrix $[\langle e^i | a e_j \rangle]$.

Proposition 3.85 (1) *If $\mathcal{X}$ is real and $a \in L(\mathcal{X})$, then $\det a = \det a_{\mathbb{C}}$.*
(2) *If $a, b \in L(\mathcal{X})$, then $\det ab = \det a \det b$.*
(3) *If $a_i \in L(\mathcal{X}_i)$, $i = 1, 2$, then $\det(a_1 \oplus a_2) = \det a_1 \det a_2$.*
(4) *If $a \in L(\mathcal{X})$, then $\det a = \det a^\#$.*

3.6.7 Determinant of a bilinear form

Now let $\mathcal{X}$ be a finite-dimensional vector space equipped with a density $|\Xi|$.

Definition 3.86 *If $\zeta \in L(\mathcal{X}, \mathcal{X}^\#)$, we define the* determinant *of ζ w.r.t. the density $|\Xi|$ as the unique number $\det \zeta$ satisfying*

$$\Gamma(\zeta^\#)\Xi = \det \zeta \; \Xi^{\mathrm{dual}}. \tag{3.51}$$

(Note that the above definition does not depend on the sign of Ξ.)

If $(e_1, \ldots, e_d)$ is a basis of $\mathcal{X}$ such that $\Xi = e^d \wedge \cdots \wedge e^1$, then $\det \zeta$ is equal to the determinant of the matrix $[\langle e_i, \zeta e_j \rangle]$.

If $|\Xi|$ is compatible with a Euclidean scalar product ν, then the determinant of ζ w.r.t. $|\Xi|$ is equal to the determinant of the operator $\nu^{-1}\zeta \in L(\mathcal{X})$.

3.6.8 Orientations of vector spaces

Let $\mathcal{X}$ be a finite-dimensional real vector space.

Definition 3.87 *Two bases of $\mathcal{X}$ are said to be* equivalent *if the determinant of the matrix of the change of basis is positive. An* orientation *of $\mathcal{X}$ is the choice of one of two equivalence classes of bases. Bases in this class are called* compatible *with the orientation. A space equipped with an orientation is called* oriented.

Sometimes it is useful to have the concept of an orientation also on a complex vector space. Its definition is identical to that on the real vector space. The only difference is that on a complex vector space the set of orientations is not made of two elements but is homeomorphic to a circle.

3.6.9 Volume forms on complex spaces

Let $\mathcal{Z}$ be a complex space of dimension d equipped with a complex volume form denoted by Ξ. On $\overline{\mathcal{Z}}$ we have the volume form $\overline{\Xi}$ defined by $\langle \overline{\Xi} | \overline{\Psi} \rangle = \overline{\langle \Xi | \Psi \rangle}$, $\Psi \in \Gamma_a^d(\mathcal{Z})$. We will also need $\overline{\Xi}^{\text{rev}} = (-1)^{\frac{1}{2}d(d-1)}\overline{\Xi}$. We will usually denote Ξ by $\mathrm{d}z$ and $\overline{\Xi}^{\text{rev}}$ by $\mathrm{d}\overline{z}$. If $e^1, \ldots, e^d$ is a basis of $\mathcal{Z}^{\#}$ and $\mathrm{d}z = e^d \wedge \cdots \wedge e^1$ then $\mathrm{d}\overline{z} = \overline{e}^1 \wedge \cdots \wedge \overline{e}^d$.

On $\overline{\mathcal{Z}} \oplus \mathcal{Z}$ we have a distinguished volume form given by

$$\mathrm{i}^{-d}\overline{\Xi}^{\text{rev}} \wedge \Xi = \mathrm{i}^{-d}\mathrm{d}\overline{z} \wedge \mathrm{d}z. \tag{3.52}$$

We claim that the restriction of $\mathrm{i}^{-d}\mathrm{d}\overline{z} \wedge \mathrm{d}z$ to $\mathrm{Re}(\overline{\mathcal{Z}} \oplus \mathcal{Z})$ is a real volume form. This can be seen by noting that the canonical conjugation

$$\overline{\mathcal{Z}} \oplus \mathcal{Z} \ni (\overline{z}_1, z_2) \mapsto \epsilon(\overline{z}_1, z_2) := (\overline{z}_2, z_1) \in \overline{\mathcal{Z}} \oplus \mathcal{Z}$$

fixes $\mathrm{Re}(\overline{\mathcal{Z}} \oplus \mathcal{Z})$ and transforms $\mathrm{i}^{-d}\mathrm{d}\overline{z} \wedge \mathrm{d}z$ into its complex conjugate.

Recall that the realification of $\mathcal{Z}$ is denoted $\mathcal{Z}_{\mathbb{R}}$. It is a real $2d$-dimensional space. $\mathcal{Z}_{\mathbb{R}}$ has a distinguished real volume form obtained by pulling back $\mathrm{i}^{-d}\mathrm{d}\overline{z} \wedge \mathrm{d}z$ from $\mathrm{Re}(\mathcal{Z} \oplus \overline{\mathcal{Z}})$ to $\mathcal{Z}_{\mathbb{R}}$ by the transformation

$$\mathcal{Z}_{\mathbb{R}} \ni z \mapsto (z, \overline{z}) \in \mathrm{Re}(\mathcal{Z} \oplus \overline{\mathcal{Z}}), \tag{3.53}$$

(which we encountered before, e.g. in (3.42)). The Lebesgue measure obtained from this volume form will be adopted as the standard measure on $\mathcal{Z}$. Thus, a typical notation for the Lebesgue integral of a function $\mathcal{Z} \ni z \mapsto F(z)$ will be

$$\int F(z)\mathrm{i}^{-d}\mathrm{d}\overline{z} \wedge \mathrm{d}z. \tag{3.54}$$

Let us give an argument for why $\mathrm{i}^{-d}\mathrm{d}\overline{z} \wedge \mathrm{d}z$ is a natural choice for the distinguished volume form on a complex vector space. Assume that $\mathcal{Z}$ is a unitary space and that $\mathrm{d}z$ is compatible with its structure (see Subsect. 3.6.5). We have seen in Subsect. 1.3.9 that $\mathrm{Re}(\overline{\mathcal{Z}} \oplus \mathcal{Z})$ is equipped with a Euclidean scalar product and with a symplectic form. We claim that $i^{-d}d\overline{z} \wedge dz$ is compatible with these two structures. To see this, note that if $(e^1, \ldots, e^d)$ is an o.n. basis in $\mathcal{Z}^\# \simeq \overline{\mathcal{Z}}$, then

$$i^{-d}d\overline{z}dz = i^{-d}\overline{e}^1 \wedge \cdots \wedge \overline{e}^d \wedge e^d \wedge \cdots \wedge e^1$$

$$= \frac{e^1 + \overline{e}^1}{\sqrt{2}} \wedge \frac{-ie^1 + i\overline{e}^1}{\sqrt{2}} \wedge \cdots \wedge \frac{e^d + \overline{e}^d}{\sqrt{2}} \wedge \frac{-ie^d + i\overline{e}^d}{\sqrt{2}},$$

and $(\frac{e^1 + \overline{e}^1}{\sqrt{2}}, \frac{-ie^1 + i\overline{e}^1}{\sqrt{2}}, \cdots, \frac{e^d + \overline{e}^d}{\sqrt{2}}, \frac{-ie^d + i\overline{e}^d}{\sqrt{2}})$ is both an o.n. and a symplectic basis in $\text{Re}(\overline{\mathcal{Z}} \oplus \mathcal{Z})$.

Remark 3.88 *The following remark may sound academic, but actually it is related to a true computational nuisance – factors of $\sqrt{2}$ in various formulas, which are often difficult to keep track of.*

We saw that the volume form $i^{-d}d\overline{z} \wedge dz$ is compatible with the natural Euclidean structure on $\text{Re}(\mathcal{Z} \oplus \overline{\mathcal{Z}})$. Its pullback to $\mathcal{Z}_\mathbb{R}$, however, is not compatible with the usual Euclidean structure on $\mathcal{Z}_\mathbb{R}$, that is, with the real scalar product $\text{Re}\,\overline{z}_1 \cdot z_2$. To see this, note that the map (3.53) is not orthogonal. (3.53) becomes orthogonal only after we multiply it by $\frac{1}{\sqrt{2}}$. A volume form compatible with the Euclidean structure of $\mathcal{Z}_\mathbb{R}$ is $(2i)^{-d}d\overline{z} \wedge dz$.

One can say that when we consider integrals on $\mathcal{Z}$, we actually view them as integrals on $\text{Re}(\mathcal{Z} \oplus \overline{\mathcal{Z}})$, where the integrand has been pulled back from $\mathcal{Z}$ onto $\text{Re}(\mathcal{Z} \oplus \overline{\mathcal{Z}})$ by (3.53). Therefore, when normalizing the Lebesgue measure in (3.54), we prefer the convention adapted to $\text{Re}(\mathcal{Z} \oplus \overline{\mathcal{Z}})$ rather than to $\mathcal{Z}$.

3.6.10 Pfaffians

Definition 3.89 *Let $d \in \mathbb{N}$. We denote by Pair_{2d} the set of pairings of $\{1, \ldots, 2d\}$, i.e. the set of partitions of $\{1, \ldots, 2d\}$ into pairs.*

A pairing can be uniquely written as

$$((i_1, j_1), (i_2, j_2), \ldots, (i_d, j_d)),$$

where $i_k < j_k$ and $i_1 < i_2 < \cdots < i_d$, and we can identify Pair_{2d} with the subset of permutations

$$\text{Pair}_{2d} = \big\{\sigma \in S_{2d} \,:\, \sigma(2i-1) < \sigma(2i), \ \sigma(2i-1) < \sigma(2i+1), \ 1 \leq i \leq d\big\}.$$

It is easy to see that Pair_{2d} has $\frac{(2d)!}{d!2^d}$ elements.

Definition 3.90 *If $\zeta = [\zeta_{ij}]$ is a $2d \times 2d$ anti-symmetric matrix, one defines its Pfaffian by*

$$\text{Pf}(\zeta) := \sum_{\sigma \in \text{Pair}_{2d}} \text{sgn}(\sigma) \prod_{i=1}^{d} \zeta_{\sigma(2i-1), \sigma(2i)}.$$

It is possible to give a manifestly coordinate-independent definition of the Pfaffian. Now let $\mathcal{Y}$ be a (real or complex) vector space of dimension $2d$, equipped with the volume form Ξ.

Definition 3.91 *For $\zeta \in L_a(\mathcal{Y}^\#, \mathcal{Y}) \simeq \Gamma_a^2(\mathcal{Y})$, its Pfaffian w.r.t. Ξ is defined by*

$$\mathrm{Pf}(\zeta) := \frac{1}{2^d d!} \langle \otimes_a^d \zeta | \Xi \rangle. \tag{3.55}$$

An alternative definition is

$$\wedge^d \zeta =: d! \mathrm{Pf}(\zeta) \Xi^{\mathrm{dual}}. \tag{3.56}$$

If $(e^1, \ldots, e^{2d})$ is a basis of $\mathcal{Y}^\#$ such that $\Xi = e^{2d} \wedge \cdots \wedge e^1$, then $\mathrm{Pf}(\zeta)$ coincides with the Pfaffian of the matrix $[\langle e^i | \zeta e^j \rangle]$.

Proposition 3.92 (1) *If $\zeta \in L_a(\mathcal{Y}^\#, \mathcal{Y})$, $r \in L(\mathcal{Y})$, then*

$$\mathrm{Pf}(r \zeta r^\#) = \mathrm{Pf}(\zeta) \det r.$$

(2) *Let $\zeta_i \in L_a(\mathcal{Y}_i^\#, \mathcal{Y}_i)$, $i = 1, 2$. Then*

$$\mathrm{Pf}\left(\begin{bmatrix} \zeta_1 & 0 \\ 0 & \zeta_2 \end{bmatrix} \right) = \mathrm{Pf}(\zeta_1) \mathrm{Pf}(\zeta_2),$$

where the Pfaffian on the l.h.s. is computed w.r.t. $\Xi_1 \wedge \Xi_2$.

(3) *For $\zeta \in L_a(\mathcal{Y}^\#, \mathcal{Y})$, one has*

$$\mathrm{Pf}(\zeta)^2 = \det \zeta,$$

where $\det a$ is computed w.r.t. the density $|\Xi^{\mathrm{dual}}|$.

(4) *Let $\mathcal{X}$ be a finite-dimensional vector space equipped with a volume form Ξ and let us equip $\mathcal{Y} = \mathcal{X}^\# \oplus \mathcal{X}$ with the volume form $\Xi^{\mathrm{dual}} \wedge \Xi$. Let $a \in L(\mathcal{X})$, so that $\begin{bmatrix} 0 & a^\# \\ -a & 0 \end{bmatrix} \in L_a(\mathcal{Y}^\#, \mathcal{Y})$. Then*

$$\mathrm{Pf}\left(\begin{bmatrix} 0 & a^\# \\ -a & 0 \end{bmatrix} \right) = \det a.$$

3.7 Notes

The tensor product of Hilbert spaces is studied e.g. in the monograph by Reed–Simon (1980). The notions of Fock spaces and second quantization were originally introduced by Fock (1932). Mathematical expressions can be found e.g. in Reed–Simon (1980), Simon (1974), Bratteli–Robinson (1996) and Glimm–Jaffe (1987).

4

Analysis in $L^2(\mathbb{R}^d)$

In this chapter we describe basic properties of operators acting on $L^2(\mathbb{R}^d)$. After a preliminary Sect. 4.1, we will study the *Weyl commutation relations* and prove the famous *Stone–von Neumann uniqueness theorem*. Then we define the so-called x, D-*quantization*, with position to the left and the momentum to the right. We will compare it to the D, x-*quantization*, which uses the reverse order of position and momentum. The *Weyl–Wigner quantization*, in some sense superior to the x, D- and D, x-quantizations, will be introduced in Chap. 8, which can be viewed as a continuation of the present chapter.

4.1 Distributions and the Fourier transformation

Throughout this section, $\mathcal{X}$ is a real vector space of dimension d with a Lebesgue measure $\mathrm{d}x$. As in Subsect. 3.6.5, the dual space $\mathcal{X}^{\#}$ is then equipped with a canonical Lebesgue measure, which we denote $\mathrm{d}\xi$. If additionally $\mathcal{X}$ is equipped with a Euclidean structure, we take $\mathrm{d}x$ to be the unique compatible Lebesgue measure (see Subsect. 3.6.5).

4.1.1 Distributions

Let Ω be an open subset of $\mathcal{X}$.

Definition 4.1 $C_{\mathrm{c}}^{\infty}(\Omega)$ *denotes the space of smooth functions compactly supported in* Ω. *We equip* $C_{\mathrm{c}}^{\infty}(\Omega)$ *with the usual topology and rename it* $\mathcal{D}(\Omega)$. $\mathcal{D}'(\Omega)$ *denotes its topological dual. Elements of* $\mathcal{D}'(\Omega)$ *are called* distributions.

A large class of distributions in $\mathcal{D}'(\Omega)$ is given by functions $f \in L_{\mathrm{loc}}^1(\Omega)$ with the action on $\Phi \in C_{\mathrm{c}}^{\infty}(\Omega)$ given by

$$\langle f|\Phi\rangle := \int f(x)\Phi(x)\mathrm{d}x. \tag{4.1}$$

We will use the integral notation on the r.h.s. of (4.1) also in the case of distributions that do not belong to $L_{\mathrm{loc}}^1(\Omega)$. Here are some examples with $\Omega = \mathcal{X} = \mathbb{R}$:

$$\int \delta(t)\Phi(t)\mathrm{d}t := \Phi(0),$$

$$\int (t \pm \mathrm{i}0)^{\lambda}\Phi(t)\mathrm{d}t := \lim_{\epsilon \searrow 0} \int (t \pm \mathrm{i}\epsilon)^{\lambda}\Phi(t)\mathrm{d}t.$$

4.1.2 Pullback of distributions

Let $\chi : \Omega_1 \to \Omega_2$ be a diffeomorphism between two open sets $\Omega_i \subset \mathbb{R}^d$, $i = 1, 2$.

Definition 4.2 *One defines the* pullback $\chi^\# : \mathcal{D}'(\Omega_2) \to \mathcal{D}'(\Omega_1)$ *by*

$$\int \chi^\# f(x_1) \Phi(x_1) \mathrm{d}x_1 := \int f(x_2) \Phi \circ \chi^{-1}(x_2) |\det \nabla \chi^{-1}(x_2)| \mathrm{d}x_2, \quad \Phi \in \mathcal{D}(\Omega_1).$$

Clearly, if $f \in L^1_{\mathrm{loc}}(\Omega_2)$, then $\chi^\# f(x_1) = f \circ \chi(x_1)$.

The pullback of distributions can be generalized to a large class of transformations between sets of different dimension. Let $\Omega_i \subset \mathbb{R}^{d_i}$, $i = 1, 2$, be two open sets and $\tau : \Omega_1 \to \Omega_2$ a *submersion*, that is, a smooth map whose derivative is everywhere surjective. We can find an open set $\Omega_3 \subset \mathbb{R}^{d_1 - d_2}$ and a diffeomorphism $\chi : \Omega_1 \to \Omega_2 \times \Omega_3$ such that

$$\pi_{\Omega_2} \circ \chi = \tau, \tag{4.2}$$

where π_{Ω_2} is the projection onto Ω_2. We then define the map $\tau^\# : \mathcal{D}'(\Omega_2) \to \mathcal{D}'(\Omega_1)$ as

$$\tau^\# f := \chi^\# (f \otimes 1),$$

where we consider $f \otimes 1$ as an element of $\mathcal{D}'(\Omega_2 \times \Omega_3)$. One can show that $\tau^\#$ is independent on the choice of χ satisfying (4.2).

Definition 4.3 *The map* $\tau^\# : \mathcal{D}'(\Omega_2) \to \mathcal{D}'(\Omega_1)$ *is also called the* pullback of *distributions.*

In particular, if $f \in L^1_{\mathrm{loc}}(\Omega_2)$, then

$$\int \tau^\# f(x_1) \Phi(x_1) \mathrm{d}x_1 = \int f \circ \tau(x_1) \Phi(x_1) \mathrm{d}x_1. \tag{4.3}$$

We will use the notation of the r.h.s. of (4.3) also for the pullback of distributions that do not belong to $L^1_{\mathrm{loc}}(\Omega)$. For instance,

$$\int \delta(\tau(t)) \Phi(t) \mathrm{d}t = \sum_{\tau(s)=0} |\tau'(s)|^{-1} \Phi(s).$$

4.1.3 Schwartz functions and distributions

Definition 4.4 *The space of* Schwartz functions *on* $\mathcal{X}$ *is defined as*

$$\mathcal{S}(\mathcal{X}) := \left\{ \Psi \in C^\infty(\mathcal{X}) \ : \ \int |x^\alpha \nabla_x^\beta \Psi(x)|^2 \mathrm{d}x < \infty, \quad \alpha, \beta \in \mathbb{N}^d \right\}. \tag{4.4}$$

(In the definition we use an identification of $\mathcal{X}$ with $\mathbb{R}^d$. It is clear that $\mathcal{S}(\mathcal{X})$ does not depend on this identification.)

Remark 4.5 *The definition (4.4) is equivalent to*

$$\mathcal{S}(\mathcal{X}) = \left\{ \Psi \in C^\infty(\mathcal{X}) \ : \ |x^\alpha \nabla_x^\beta \Psi(x)| \leq c_{\alpha,\beta}, \quad \alpha, \beta \in \mathbb{N}^d \right\}. \tag{4.5}$$

The definition (4.5) is more common in the literature, even though one can argue that (4.4) is more natural.

Definition 4.6 *On $\mathcal{S}(\mathcal{X})$ we introduce semi-norms*

$$\|\Psi\|_{\alpha,\beta} := \left(\int |x^\alpha \nabla_x^\beta \Psi(x)|^2 \mathrm{d}x \right)^{\frac{1}{2}},$$

which make it into a Fréchet space. $\mathcal{S}'(\mathcal{X})$ denotes its topological dual.

Note that we have continuous inclusions

$$\mathcal{D}(\mathcal{X}) \subset \mathcal{S}(\mathcal{X}) \subset L^2(\mathcal{X}) \subset \mathcal{S}'(\mathcal{X}) \subset \mathcal{D}'(\mathcal{X}).$$

4.1.4 Derivatives

Let f be a complex function on $\mathcal{X}$. Recall from the real case of Def. 2.50 (1) that the *derivative of f at $x_0 \in \mathcal{X}$ in the direction $q \in \mathcal{X}$* is defined by

$$q \cdot \nabla_x f(x_0) := \frac{\mathrm{d}}{\mathrm{d}t} f(x_0 + tq)\big|_{t=0}. \tag{4.6}$$

Proposition 4.7 *The derivative of a C^1 function at a point is a complex linear functional on $\mathcal{X}$, that is, $\nabla_x f(x_0) \in \mathbb{C}\mathcal{X}^\#$.*

Definition 4.8 *If $f \in C^2(\mathcal{X}, \mathbb{R})$, its Hessian at $x_0 \in \mathcal{X}$ is denoted $\nabla_x^{(2)} f(x_0) \in L_\mathrm{s}(\mathcal{X}, \mathcal{X}^\#)$ and defined by*

$$q_2 \cdot \nabla_x^{(2)} f(x_0) q_1 := \frac{\mathrm{d}^2}{\mathrm{d}t_1 \mathrm{d}t_2} f(x_0 + t_1 q_1 + t_2 q_2)\big|_{t_1=t_2=0}, \quad q_1, q_2 \in \mathcal{X}.$$

If $\zeta \in L_\mathrm{s}(\mathcal{X}^\#, \mathcal{X})$, then $\nabla_x \cdot \zeta \nabla_x$ denotes the corresponding differential operator:

$$\nabla_x \cdot \zeta \nabla_x f(x_0) := \mathrm{Tr}\, \zeta \nabla_x^{(2)} f(x).$$

If $\mathcal{X}$ is a Euclidean space with the scalar product denoted by $x_1 \cdot x_2$, then $\nabla_x \cdot \nabla_x = \nabla_x^2 = \Delta_x$ stands for the Laplacian.

4.1.5 Complex derivatives

Let $\mathcal{Z}$ be a complex vector space. Let f be a complex function on $\mathcal{Z}$.

Definition 4.9 *The* holomorphic, *resp.* anti-holomorphic derivative of f at $z_0 \in \mathcal{Z}$ in the direction of $w \in \mathcal{Z}$, resp. $\overline{w} \in \overline{\mathcal{Z}}$ is defined by

$$w \cdot \nabla_z f(z_0) := \frac{1}{2} \frac{\mathrm{d}}{\mathrm{d}t} \left(f(z_0 + tw) - \mathrm{i}f(z_0 + \mathrm{i}tw) \right)\big|_{t=0},$$

$$\overline{w} \cdot \nabla_{\overline{z}} f(z_0) := \frac{1}{2} \frac{\mathrm{d}}{\mathrm{d}t} \left(f(z_0 + tw) + \mathrm{i}f(z_0 + \mathrm{i}tw) \right)\big|_{t=0}.$$

Proposition 4.10 *The holomorphic, resp. anti-holomorphic derivative of a C^1 function at a point is a linear, resp. anti-linear functional on $\mathcal{Z}$, that is, $\nabla_z f(z_0) \in \mathcal{Z}^\#$, resp. $\nabla_{\overline{z}} f(z_0) \in \overline{\mathcal{Z}}^\#$.*

Recall from the complex case of Def. 2.50 (1) that f possesses a (complex) derivative at z_0 in the direction of w if there exists the limit

$$\lim_{u \to 0} \frac{f(z_0 + uw) - f(z_0)}{u}, \tag{4.7}$$

where u is a complex parameter.

Definition 4.11 *Assume that $\mathcal{Z}$ is finite-dimensional and let $U \subset \mathcal{Z}$ be an open set. We say that $f : U \to \mathbb{C}$ is* holomorphic *in U if it possesses a complex derivative at each $z_0 \in U$.*

Proposition 4.12 *A function $f : U \to \mathbb{C}$ is holomorphic iff $f \in L^1_{\mathrm{loc}}(U)$ and $\nabla_{\overline{z}} f = 0$ in U in the distribution sense. Then (4.7) equals $w \cdot \nabla_z f(z_0)$.*

We consider also the realification of $\mathcal{Z}$, denoted $\mathcal{Z}_{\mathbb{R}}$, where the multiplication by i is denoted by j.

Let $\nabla_z^{\mathbb{R}}$ denote the usual (real) derivative on $\mathcal{Z}_{\mathbb{R}}$. We can express the holomorphic and anti-holomorphic derivative in terms of the real derivative:

$$w \cdot \nabla_z = \frac{1}{2} \left(w \cdot \nabla_z^{\mathbb{R}} - \mathrm{i}(\mathrm{j}w) \cdot \nabla_z^{\mathbb{R}} \right),$$

$$\overline{w} \cdot \nabla_{\overline{z}} = \frac{1}{2} \left(w \cdot \nabla_z^{\mathbb{R}} + \mathrm{i}(\mathrm{j}w) \cdot \nabla_z^{\mathbb{R}} \right),$$

$$w \cdot \nabla_z + \overline{w} \cdot \nabla_{\overline{z}} = w \cdot \nabla_z^{\mathbb{R}}. \tag{4.8}$$

(On the left w is treated as an element of $\mathcal{Z}$ and on the right w as a real vector in $\mathcal{Z}_{\mathbb{R}}$.)

Note that if we make the identification $\mathcal{Z}_{\mathbb{R}} \ni w \mapsto (w, \overline{w}) \in \mathcal{Z} \oplus \overline{\mathcal{Z}}$, as in (1.31), then (4.8) can be written as $\nabla_z + \nabla_{\overline{z}} = \nabla_z^{\mathbb{R}}$.

4.1.6 Position and momentum operators

Definition 4.13 *For $\eta \in \mathcal{X}^{\#}$ and $q \in \mathcal{X}$ we set*

$$(\eta \cdot x \Psi)(x) := \eta \cdot x \Psi(x), \quad \mathrm{Dom}\, \eta \cdot x := \left\{ \Psi \in L^2(\mathcal{X}) : \int |\eta \cdot x|^2 |\Psi(x)|^2 \mathrm{d}x < \infty \right\},$$

$$(q \cdot D \Psi)(x) := -\mathrm{i} q \cdot \nabla \Psi(x), \quad \mathrm{Dom}\, q \cdot D := \left\{ \Psi \in L^2(\mathcal{X}) : \int |q \cdot \nabla_x \Psi(x)|^2 \mathrm{d}x < \infty \right\}.$$

$\eta \cdot x$ and $q \cdot D$ are called respectively position *and* momentum operators *and are self-adjoint operators.*

Remark 4.14 *In the formulas above the symbol x is used with as many as three different meanings:*

(1) *as an element of the space $\mathcal{X}$, e.g. in $\Psi(x)$ or in $\eta \cdot x$ on the right of $:=$;*
(2) *as the name of the "generic variable in $\mathcal{X}$"; e.g. in $\mathrm{d}x$ or ∇_x;*
(3) *as a vector of self-adjoint operators, e.g. in $\eta \cdot x$ on the left of $:=$.*

This ambiguous usage of the same symbol, although sometimes confusing, seems to be difficult to avoid and is often employed. Sometimes one tries to differentiate the third meaning by decorating x in some way, e.g. writing $\hat{x}$.

Proposition 4.15 *The Schwartz space $\mathcal{S}(\mathcal{X})$ is the largest subspace of $L^2(\mathcal{X})$ contained in the domain of position and momentum operators and preserved by all the operators $\eta{\cdot}x$ and $q{\cdot}D$.*

The operator $\eta{\cdot}x$ and $q{\cdot}D$, viewed as operators on $\mathcal{S}(\mathcal{X})$, satisfy the so-called *Heisenberg commutation relations*:

$$[\eta_1{\cdot}x, \eta_2{\cdot}x] = [q_1{\cdot}D, q_2{\cdot}D] = 0, \qquad [\eta{\cdot}x, q{\cdot}D] = \mathrm{i}\eta{\cdot}q\mathbb{1}. \tag{4.9}$$

Definition 4.16 *The algebra of differential operators with polynomial coefficients will be denoted $\mathrm{CCR}^{\mathrm{pol}}(\mathcal{X}^{\#} \oplus \mathcal{X})$.*

Elements of $\mathrm{CCR}^{\mathrm{pol}}(\mathcal{X}^{\#} \oplus \mathcal{X})$ act naturally on $\mathcal{S}(\mathcal{X})$. By duality, they also act on $\mathcal{S}'(\mathcal{X})$.

Remark 4.17 *In Subsect. 8.3.1 we will define a more general class of algebras, denoted $\mathrm{CCR}^{\mathrm{pol}}(\mathcal{Y})$, where $\mathcal{Y}$ is a symplectic space.*

Remark 4.18 *The algebra $\mathrm{CCR}^{\mathrm{pol}}(\mathcal{X}^{\#} \oplus \mathcal{X})$ is sometimes called the* Weyl alge-bra. *However, we prefer to use this name for a different class of algebras; see Subsect. 8.3.5.*

4.1.7 Fourier transformation

Definition 4.19 *We denote by $C_{\infty}(\mathcal{X})$ the Banach space of continuous functions on $\mathcal{X}$ tending to 0 at ∞.*

Definition 4.20 *For $f \in L^1(\mathcal{X})$ the Fourier transform of f, denoted either $\mathcal{F}f$ or $\hat{f}$, is given by the formula*

$$\hat{f}(\xi) = \int f(x)\mathrm{e}^{-\mathrm{i}x\cdot\xi}\mathrm{d}x.$$

It is well known that $\mathcal{F}$ extends to a unique bounded operator from $L^2(\mathcal{X}, \mathrm{d}x)$ to $L^2(\mathcal{X}^{\#}, \mathrm{d}\xi)$, where $\mathrm{d}\xi$ is the dual Lebesgue measure on $\mathcal{X}^{\#}$.

The Riemann–Lebesgue lemma says that if $f \in L^1(\mathcal{X})$, then $\hat{f} \in C_{\infty}(\mathcal{X}^{\#})$. $(2\pi)^{-\frac{d}{2}}\mathcal{F}$ is unitary, and we have the Fourier inversion formula

$$f(x) = (2\pi)^{-d} \int \hat{f}(\xi)\mathrm{e}^{\mathrm{i}x\cdot\xi}\mathrm{d}\xi.$$

The space $\mathcal{S}(\mathcal{X})$ is mapped by $\mathcal{F}$ continuously onto $\mathcal{S}(\mathcal{X}^{\#})$. $\mathcal{F}$ can be extended to a unique continuous linear map from $\mathcal{S}'(\mathcal{X})$ onto $\mathcal{S}'(\mathcal{X}^{\#})$.

4.1.8 Gaussian integrals

Let $\nu \in L_s(\mathcal{X}, \mathcal{X}^\#)$ be positive definite. Let $\eta \in \mathbb{C}\mathcal{X}^\#$. Then

$$(2\pi)^{-\frac{d}{2}} \int e^{-\frac{1}{2}x\cdot\nu x + \eta\cdot x} dx = (\det \nu)^{-\frac{1}{2}} e^{\frac{1}{2}\eta\cdot\nu^{-1}\eta}. \qquad (4.10)$$

Note that the determinant $\det \nu$ is defined w.r.t. the Lebesgue measure dx (see Subsect. 3.6.6). In particular, if $f(x) = e^{-\frac{1}{2}x\cdot\nu x}$, then

$$\hat{f}(\xi) = (2\pi)^{\frac{d}{2}} (\det \nu)^{-\frac{d}{2}} e^{-\frac{1}{2}\xi\cdot\nu^{-1}\xi}. \qquad (4.11)$$

If $\nu \in L_s(\mathcal{X}, \mathcal{X}^\#)$ is not necessarily positive definite and $\eta \in \mathcal{X}^\#$, then

$$\lim_{R\to\infty} (2\pi)^{-\frac{d}{2}} \int_{|x|<R} e^{\frac{1}{2}x\cdot\nu x + i\eta\cdot x} dx = |\det \nu|^{-\frac{1}{2}} e^{\frac{1}{4}\pi \, \mathrm{inert}\,\nu} e^{-\frac{1}{2}\eta\cdot\nu^{-1}\eta}. \qquad (4.12)$$

In particular, if $g(x) = e^{\frac{1}{2}x\cdot\nu x}$, then

$$\hat{g}(\xi) = (2\pi)^{\frac{d}{2}} e^{\frac{1}{4}\pi \, \mathrm{inert}\,\nu} (\det \nu)^{-\frac{d}{2}} e^{-\frac{1}{2}\xi\cdot\nu^{-1}\xi}.$$

We will sometimes abuse the notation and write $\det(-i\nu)^{-\frac{1}{2}}$ for $|\det \nu|^{-\frac{1}{2}} e^{\frac{1}{4}\pi \, \mathrm{inert}\,\nu}$.

4.1.9 Gaussian integrals for complex variables

Let $\mathcal{Z}$ be a complex space of dimension d. Recall from Subsect. 3.6.9 that the integral of a function $\mathcal{Z} \ni z \mapsto F(z)$ over $\mathcal{Z}$ is interpreted as the integral of the pullback of F by

$$\mathcal{Z} \ni z \mapsto (z, \overline{z}) \in \mathrm{Re}(\mathcal{Z} \oplus \overline{\mathcal{Z}})$$

on the space $\mathrm{Re}(\mathcal{Z} \oplus \overline{\mathcal{Z}})$, and $i^{-d} d\overline{z} dz$ is used as the standard volume form.

Let us translate formula (4.10) into the context of complex variables. Let $\beta \in L_h(\mathcal{Z}, \mathcal{Z}^*)$ be positive definite, and $w_1, w_2 \in \mathcal{Z}^*$. Then

$$(2\pi i)^{-d} \int e^{-\overline{z}\cdot\beta z + w_1\cdot\overline{z} + \overline{w}_2\cdot z} d\overline{z} dz = (\det \beta)^{-1} e^{w_1\cdot\beta^{-1}\overline{w}_2}, \qquad (4.13)$$

where $\det \beta$ is computed w.r.t. the volume form dz.

Let us explain the proof of (4.13). As mentioned above, the integral in (4.13) is interpreted as an integral on the real vector space $\mathrm{Re}(\overline{\mathcal{Z}} \oplus \mathcal{Z})$. We choose any scalar product on $\mathcal{Z}$ compatible with dz. Note from Subsect. 3.6.9 that the volume form $i^{-d} d\overline{z} dz$ is compatible with the Euclidean scalar product on $\mathrm{Re}(\overline{\mathcal{Z}} \oplus \mathcal{Z})$. We identify β with an element of $L(\mathcal{Z})$ using the unitary structure of $\mathcal{Z}$. Then, setting

$$v = (z, \overline{z}), \quad m = 2d, \quad dv = i^{-d} d\overline{z} dz,$$

$$\nu := \begin{bmatrix} 0 & \overline{\beta} \\ \beta & 0 \end{bmatrix}, \quad \xi = (w_1, \overline{w}_2) \in \mathbb{C}\mathrm{Re}(\mathcal{Z} \oplus \overline{\mathcal{Z}}) \simeq \mathbb{C}\mathrm{Re}(\overline{\mathcal{Z}} \oplus \mathcal{Z})^\#,$$

we see that (4.13) reduces to (4.10). To compute the determinant of ν as an operator on $\mathrm{Re}(\overline{\mathcal{Z}} \oplus \mathcal{Z})$, we use that

$$\det \nu = \det \nu_{\mathbb{C}} = \det \beta \det \overline{\beta} = \det \beta \det \beta^{\#} = \det \beta^2,$$

since $\beta = \beta^*$. Then (4.13) follows from (4.10).

4.1.10 Convolution operators

Definition 4.21 *If $f \in \mathcal{S}'(\mathcal{X})$, $\Psi \in \mathcal{S}(\mathcal{X})$, then their* convolution product $f \star \Psi$ *is defined by*

$$f \star \Psi(x) := \int f(x - x_1)\Psi(x_1)\mathrm{d}x_1.$$

We have

$$\mathcal{F}(f \star \Psi) = (\mathcal{F}f)(\mathcal{F}\Psi).$$

Recall that $D = \frac{1}{i}\nabla_x$ is a vector of commuting self-adjoint operators. Note that $\mathcal{F}D\mathcal{F}^{-1} = \xi$ where ξ is the operator of multiplication by $\xi \in \mathcal{X}^{\#}$ on $L^2(\mathcal{X}^{\#})$.
Note the identities

$$f(D)\Psi(x) = (2\pi)^{-d}\int e^{i(x-y)\cdot\xi}f(\xi)\Psi(y)\mathrm{d}\xi\mathrm{d}y$$

$$= (2\pi)^{-d}\int \hat{f}(y - x)\Psi(y)\mathrm{d}y, \quad f \in \mathcal{S}(\mathcal{X}^{\#}).$$

If $\nu \in L_s(\mathcal{X}^{\#}, \mathcal{X})$, then

$$e^{-\frac{1}{2}D_x\cdot\nu D_x}\Psi(x) = e^{\frac{1}{2}\nabla_x\cdot\nu\nabla_x}\Psi(x) \tag{4.14}$$

$$= (2\pi)^{-\frac{d}{2}}(\det\nu)^{-\frac{1}{2}}\int e^{-\frac{1}{2}(x-x_1)\cdot\nu^{-1}(x-x_1)}\Psi(x_1)\mathrm{d}x_1.$$

As a consequence, we obtain the following identity for $\Psi \in \mathbb{C}\mathrm{Pol}_s(\mathcal{X})$:

$$(2\pi)^{-\frac{d}{2}}\int \Psi(x)e^{-\frac{1}{2}x\cdot\nu x}\mathrm{d}x = |\det\nu|^{-\frac{1}{2}}\left(e^{\frac{1}{2}\nabla_x\cdot\nu^{-1}\nabla_x}\Psi\right)(0). \tag{4.15}$$

As an example of (4.14) let us note

$$e^{-itD_x\cdot D_\xi}\Psi(x,\xi) = (2\pi t)^{-d}\int e^{\frac{i}{t}(x-x_1)\cdot(\xi-\xi_1)}\Psi(x_1,\xi_1)\mathrm{d}x_1\mathrm{d}\xi_1.$$

Let us write the analog of (4.14) on a complex space $\mathcal{Z}$ of dimension d, for $\beta \in L(\mathcal{Z}, \mathcal{Z}^*)$ and $\beta > 0$:

$$e^{\nabla_z\cdot\beta\nabla_{\overline{z}}}\Psi(\overline{z}, z) = (2\pi i)^{-d}(\det\beta)^{-1}\int e^{-(\overline{z}-\overline{z}_1)\cdot\beta^{-1}(z-z_1)}\Psi(\overline{z}_1, z_1)\mathrm{d}\overline{z}_1\mathrm{d}z_1. \tag{4.16}$$

4.1.11 Sesquilinear forms on $S(\mathcal{X})$

Definition 4.22 $A \in B(L^2(\mathcal{X}))$ *is called an* S-type operator *if it is given by an integral kernel in* $S(\mathcal{X} \times \mathcal{X})$, *that is, there exists* $A(\cdot, \cdot) \in S(\mathcal{X} \times \mathcal{X})$ *such that*

$$A\Psi(x) := \int A(x, y)\Psi(y)\mathrm{d}y.$$

The set of S-type operators is denoted $\mathrm{CCR}^S(\mathcal{X}^\# \oplus \mathcal{X})$.

Definition 4.23 *Continuous linear functionals on* $\mathrm{CCR}^S(\mathcal{X}^\# \oplus \mathcal{X})$ *are called* S'-type forms. *Their space is denoted by* $\mathrm{CCR}^{S'}(\mathcal{X}^\# \oplus \mathcal{X})$.

Clearly, elements of $\mathrm{CCR}^{S'}(\mathcal{X}^\# \oplus \mathcal{X})$ are represented by distributions in $S'(\mathcal{X} \oplus \mathcal{X})$. We have the obvious pairing for $B \in \mathrm{CCR}^{S'}(\mathcal{X}^\# \oplus \mathcal{X})$ and $A \in \mathrm{CCR}^S(\mathcal{X}^\# \oplus \mathcal{X})$:

$$B(A) = \int \int B(x, y) A(x, y) \mathrm{d}x \mathrm{d}y.$$

Let

$$\mathrm{CCR}^S(\mathcal{X}^\# \oplus \mathcal{X}) \ni A \mapsto B(A) \in \mathbb{C} \tag{4.17}$$

be an S'-type form. Clearly, for any $\Psi_1, \Psi_2 \in S(\mathcal{X})$, the operator $|\Psi_2)(\Psi_1|$ belongs to $\mathrm{CCR}^S(\mathcal{X}^\# \oplus \mathcal{X})$. Thus we obtain a sesquilinear form

$$S(\mathcal{X}) \times S(\mathcal{X}) \ni (\Psi_1, \Psi_2) \mapsto B(|\Psi_2)(\Psi_1|) \in \mathbb{C}. \tag{4.18}$$

We can interpret (4.18) as the action of $B\Psi_2$ on $\overline{\Psi_1}$, where B is a continuous linear map from $S(\mathcal{X})$ to $S'(\mathcal{X})$. Thus (4.18) can be written as $(\Psi_1|B\Psi_2)$. We call it the "operator notation for (4.18)", and we will use it henceforth.

We can write

$$\mathrm{CCR}^S(\mathcal{X}^\# \oplus \mathcal{X}) \subset B(L^2(\mathcal{X})) \subset \mathrm{CCR}^{S'}(\mathcal{X}^\# \oplus \mathcal{X}).$$

Theorem 4.24 (The Schwartz kernel theorem) B *is a continuous linear transformation from* $S(\mathcal{X})$ *to* $S'(\mathcal{X})$ *iff* B *belongs to* $\mathrm{CCR}^{S'}(\mathcal{X}^\# \oplus \mathcal{X})$, *that is, iff there exists a distribution* $B(\cdot, \cdot) \in S'(\mathcal{X} \oplus \mathcal{X})$ *such that*

$$(\Psi_1|B\Psi_2) = \int \overline{\Psi_1(x_1)} B(x_1, x_2) \Psi_2(x_2) \mathrm{d}x_1 \mathrm{d}x_2, \quad \Psi_1, \Psi_2 \in S(\mathcal{X}).$$

Definition 4.25 *The distribution* $B(\cdot, \cdot) \in S'(\mathcal{X} \oplus \mathcal{X})$ *is called the* distributional *kernel of the transformation* B.

Definition 4.26 *We define the* adjoint form B^* *by* $(\Psi_1|B^*\Psi_2) = \overline{(\Psi_2|B\Psi_1)}$. *If* B_1 *or* B_2^* *are continuous operators on* $S(\mathcal{X})$, *then we can define* $B_2 \circ B_1$ *as an element of* $\mathrm{CCR}^{S'}(\mathcal{X}^\# \oplus \mathcal{X})$ *by*

$$(\Psi_1|B_2 \circ B_1\Psi_2) := (\Psi_1|B_2(B_1\Psi)), \ or \ (\Psi_1|B_2 \circ B_1\Psi_2) := (B_2^*\Psi|B_1\Psi).$$

In particular this is possible if B_1 *or* $B_2 \in \mathrm{CCR}^{\mathrm{pol}}(\mathcal{X}^\# \oplus \mathcal{X})$.

4.1.12 Hilbert–Schmidt and trace-class operators on $L^2(\mathcal{X})$

Note that $B \in B^2(L^2(\mathcal{X}))$ iff the distributional kernel of B belongs to $L^2(\mathcal{X} \oplus \mathcal{X})$. Moreover, if $B_1, B_2 \in B^2(L^2(\mathcal{X}))$, then

$$\mathrm{Tr}\, B_1^* B_2 = \int \overline{B_1(x_2, x_1)} B_2(x_1, x_2) \mathrm{d}x_1 \mathrm{d}x_2.$$

Consider a trace-class operator $B \in B^1(L^2(\mathcal{X}))$. On the formal level we have the formula

$$\mathrm{Tr}\, B = \int B(x, x) \mathrm{d}x.$$

The following theorem gives some of many possible rigorous versions of the above identity:

Theorem 4.27 (1) *If $B \in \mathrm{CCR}^{\mathcal{S}}(\mathcal{X}^{\#} \oplus \mathcal{X})$, then*

$$\mathrm{Tr}\, B = \int B(x, x) \mathrm{d}x.$$

(2) *Fix an arbitrary Euclidean structure on $\mathcal{X}$. If $B \in B^1(L^2(\mathcal{X}))$ then*

$$\mathrm{Tr}\, B = \lim_{\epsilon \searrow 0} (2\pi/\epsilon)^{\frac{d}{2}} \int \mathrm{e}^{-\frac{1}{2\epsilon}(x_1 - x_2)^2} B(x_1, x_2) \mathrm{d}x_1 \mathrm{d}x_2.$$

Proof (1) is left to the reader. To prove (2) we set $P_\epsilon := \mathrm{e}^{-\frac{\epsilon}{2} D^2}$. Note that $0 \leq P_\epsilon \leq \mathbb{1}$ and $\mathrm{w} - \lim_{\epsilon \to 0} P_\epsilon = \mathbb{1}$. By Subsect. 2.2.6, we know that $\mathrm{Tr}\, B = \lim_{\epsilon \to 0} \mathrm{Tr}(P_\epsilon B) = \lim_{\epsilon \to 0} \mathrm{Tr}(P_{\epsilon/2} B P_{\epsilon/2})$. By (4.14), P_ϵ has the kernel

$$(2\pi\epsilon)^{-\frac{d}{2}} \mathrm{e}^{-\frac{1}{2\epsilon}(x_x - x_2)^2},$$

and $P_{\epsilon/2} B P_{\epsilon/2}$ has kernel $B \star T_\epsilon$, where $T_\epsilon(x_1, x_2) = (\pi\epsilon)^{-d} \mathrm{e}^{-\frac{\epsilon}{\epsilon}(x_1^2 + x_2^2)}$. Now $B \star T_\epsilon \in \mathcal{S}(\mathcal{X} \oplus \mathcal{X})$, and by (1) we get

$$\mathrm{Tr}(P_{\epsilon/2} B P_{\epsilon/2}) = (\pi\epsilon)^{-d} \int \mathrm{e}^{-\frac{1}{\epsilon}(x - x_1)^2 - \frac{1}{\epsilon}(x - x_2)^2} B(x_1, x_2) \mathrm{d}x_1 \mathrm{d}x_2 \mathrm{d}x.$$

Next we use (4.14) and the fact that $\mathrm{e}^{-\frac{\epsilon}{4} D^2} \mathrm{e}^{-\frac{\epsilon}{4} D^2} = \mathrm{e}^{-\frac{\epsilon}{2} D^2}$ to perform the integral in x, which yields

$$\mathrm{Tr}(P_{\epsilon/2} B P_{\epsilon/2}) = (2\pi\epsilon)^{-\frac{d}{2}} \int \mathrm{e}^{-\frac{1}{2\epsilon}(x_1 - x_2)^2} B(x_1, x_2) \mathrm{d}x_1 \mathrm{d}x_2. \qquad \square$$

4.2 Weyl operators

As in the previous section, $\mathcal{X}$ is a finite-dimensional real vector space with the Lebesgue measure $\mathrm{d}x$.

The *Heisenberg commutation relations* (4.9) involve two unbounded operators: position and momentum. This makes them problematic as rigorous statements.

In the early period of quantum mechanics Weyl noticed that for many purposes it is preferable to replace the Heisenberg commutation relations by relations involving the unitary groups generated by the position and momentum, since then called the *Weyl commutation relations*. These relations involve only bounded operators, hence their meaning is clear. On the formal level they are equivalent to the Heisenberg relations.

Linear combinations of the position and the momentum are self-adjoint. Their exponentials are often called *Weyl operators*. They are very useful in quantum mechanics.

One of the central results of mathematical foundations of quantum mechanics is the *Stone–von Neumann theorem*, which says that the properties of the position and momentum, up to a unitary equivalence, are essentially determined by the Weyl relations.

4.2.1 Definition of Weyl operators

Let us consider the one-parameter unitary groups on $L^2(X)$

$$\mathcal{X}^\# \ni \eta \mapsto e^{i\eta \cdot x} \in U\big(L^2(\mathcal{X})\big),$$
$$\mathcal{X} \ni q \mapsto e^{iq \cdot D} \in U\big(L^2(\mathcal{X})\big)$$

generated by the position and the momentum operators.

Theorem 4.28 *Let $\eta \in \mathcal{X}^\#$, $q \in \mathcal{X}$. We have the so-called Weyl commutation relations,*

$$e^{i\eta \cdot x} e^{iq \cdot D} = e^{-i\eta \cdot q} e^{iq \cdot D} e^{i\eta \cdot x}. \tag{4.19}$$

The operator $\eta \cdot x + q \cdot D$ is essentially self-adjoint on $\mathcal{S}(\mathcal{X})$. For $\Psi \in L^2(\mathcal{X})$ we have

$$e^{i(\eta \cdot x + q \cdot D)} \Psi(x) = e^{\frac{1}{2}\eta \cdot q + i\eta \cdot x} \Psi(x + q). \tag{4.20}$$

Moreover, the following identities are true:

$$\begin{aligned}
e^{i(\eta \cdot x + q \cdot D)} &= e^{\frac{1}{2}\eta \cdot q} e^{i\eta \cdot x} e^{iq \cdot D} = e^{-\frac{1}{2}\eta \cdot q} e^{iq \cdot D} e^{i\eta \cdot x} \\
&= e^{\frac{1}{2}\eta \cdot x} e^{iq \cdot D} e^{\frac{1}{2}\eta \cdot x} = e^{\frac{1}{2}q \cdot D} e^{i\eta \cdot x} e^{\frac{1}{2}q \cdot D}.
\end{aligned} \tag{4.21}$$

Proof Clearly, we have

$$e^{iq \cdot D} \Psi(x) = \Psi(x + q).$$

This easily implies (4.19).

Define

$$U(t) := e^{\frac{1}{2}t^2 \eta \cdot q} e^{it\eta \cdot x} e^{itq \cdot D},$$

or

$$U(t)\Psi(x) := e^{\frac{1}{2}t^2 \eta \cdot q + it\eta \cdot x} \Psi(x + tq).$$

We compute

$$\partial_t U(t)\Psi = \mathrm{i}(\eta x + qD)U(t)\Psi, \quad \Psi \in \mathcal{S}(\mathcal{X}).$$

Clearly, if $\Psi \in \mathcal{S}(\mathcal{X})$, then $U(t)\Psi \in \mathcal{S}(\mathcal{X})$ for all t. Therefore, by Nelson's invariant domain theorem, Thm. 2.74 (2),

$$U(t) = \mathrm{e}^{\mathrm{i}t(\eta \cdot x + q \cdot D)}$$

and $\mathcal{S}(\mathcal{X})$ is a core of $\eta \cdot x + q \cdot D$. This implies (4.20).

The identities (4.21) follow from (4.20).   □

Theorem 4.29 *If $B \in B\big(L^2(\mathcal{X})\big)$ commutes with all operators in*

$$\{\mathrm{e}^{\mathrm{i}\eta \cdot x}, \ \mathrm{e}^{\mathrm{i}q \cdot D} \ : \ \eta \in \mathcal{X}^{\#}, \ q \in \mathcal{X}\}, \tag{4.22}$$

then B is proportional to identity. In other words, the set (4.22) is irreducible in $B\big(L^2(\mathcal{X})\big)$.

Proof $L^\infty(\mathcal{X})$, identified with multiplication operators in $L^2(\mathcal{X})$, is a maximal Abelian algebra in $B\big(L^2(\mathcal{X})\big)$. By the Fourier transformation, linear combinations of operators of the form $\mathrm{e}^{\mathrm{i}\eta \cdot x}$ are #-weakly dense in $L^\infty(\mathcal{X})$. Hence if B commutes with all operators $\mathrm{e}^{\mathrm{i}\eta \cdot x}$, it has to be of the form $f(x)$ with $f \in L^\infty(\mathcal{X})$.

We have $\mathrm{e}^{\mathrm{i}q \cdot D} f(x)\mathrm{e}^{-\mathrm{i}q \cdot D} = f(x+q)$. Hence if $f(x)$ commutes with $\mathrm{e}^{\mathrm{i}q \cdot D}$, then $f(x+q) = f(x)$. If this is the case for all $q \in \mathcal{X}$, f has to be constant.   □

Theorem 4.30 *Let $\Psi \in L^2(\mathcal{X})$. Then $\Psi \in \mathcal{S}(\mathcal{X})$ iff*

$$\mathcal{X}^{\#} \oplus \mathcal{X} \ni (\eta, q) \mapsto (\Psi | \mathrm{e}^{\mathrm{i}\eta \cdot x} \mathrm{e}^{\mathrm{i}q \cdot D} \Psi) \tag{4.23}$$

belongs to $\mathcal{S}(\mathcal{X}^{\#} \oplus \mathcal{X})$.

Proof (4.23) is a partial Fourier transform of the function $\mathcal{X} \oplus \mathcal{X} \ni (x,q) \mapsto \overline{\Psi}(x)\Psi(x+q)$. Thus (4.23) belongs to $\mathcal{S}(\mathcal{X}^{\#} \oplus \mathcal{X})$ iff $\overline{\Psi}(x)\Psi(x+q)$ belongs to $\mathcal{S}(\mathcal{X} \oplus \mathcal{X})$, which is equivalent to $\Psi \in \mathcal{S}(\mathcal{X})$.   □

4.2.2 Quantum Fourier transform

Operators can be represented as an integral of $\mathrm{e}^{\mathrm{i}\eta \cdot x}\mathrm{e}^{\mathrm{i}q \cdot D}$. This fact resembles the Fourier transformation; therefore we call it the *quantum Fourier transformation*.

The following proposition will be used in our analysis of the x, D and Weyl quantizations:

Proposition 4.31 (1) *Let $w \in L^1(\mathcal{X}^{\#} \oplus \mathcal{X})$. Then the operator*

$$(2\pi)^{-d} \int w(\eta, q)\mathrm{e}^{\mathrm{i}\eta \cdot x}\mathrm{e}^{\mathrm{i}q \cdot D}\, \mathrm{d}\eta \mathrm{d}q \tag{4.24}$$

belongs to $B_\infty(L^2(\mathcal{X}))$ and is bounded by $(2\pi)^{-d}\|w\|_1$.

(2) *Let $B \in B^1(L^2(\mathcal{X}))$. Then the function*

$$w(\eta, q) := \mathrm{Tr}\, B\mathrm{e}^{-\mathrm{i}q \cdot D}\mathrm{e}^{-\mathrm{i}\eta \cdot x} \tag{4.25}$$

belongs to $C_\infty(\mathcal{X}^{\#} \oplus \mathcal{X})$ and is bounded by $\mathrm{Tr}|B|$.

(3) If $B \in B^1(L^2(\mathcal{X}))$ and w is defined by (4.25), then

$$B = (2\pi)^{-d} \int w(\eta, q) \mathrm{e}^{\mathrm{i}\eta \cdot x} \mathrm{e}^{\mathrm{i}q \cdot D} \,\mathrm{d}\eta \mathrm{d}q, \qquad (4.26)$$

as a quadratic form identity on $\mathcal{S}(\mathcal{X})$.

(4) If, moreover, $w \in L^1(\mathcal{X}^\# \oplus \mathcal{X})$, then (4.26) is an operator identity on $L^2(\mathcal{X})$.

Remark 4.32 *Note that (4.26) follows from the following formal identity:*

$$\mathrm{Tr}\, \mathrm{e}^{\mathrm{i}\eta \cdot x} \mathrm{e}^{\mathrm{i}q \cdot D} = (2\pi)^d \delta(\eta)\delta(q).$$

Proof (1) Let $w_n \in \mathcal{S}(\mathcal{X}^\# \oplus \mathcal{X})$ be a sequence such that $w_n \to w$ in $L^1(\mathcal{X}^\# \oplus \mathcal{X})$ and

$$B_n = (2\pi)^{-d} \int w_n(\eta, q) \mathrm{e}^{\mathrm{i}\eta \cdot x} \mathrm{e}^{\mathrm{i}q \cdot D} \,\mathrm{d}\eta \mathrm{d}q.$$

Then the integral kernel of B_n belongs to $\mathcal{S}(\mathcal{X})$, hence B_n is Hilbert–Schmidt. Besides, $B_n \to B$ in $B(L^2(\mathcal{X}))$; therefore B is compact as the norm limit of compact operators.

(2) The map $\mathcal{X}^\# \oplus \mathcal{X} \ni (\eta, q) \mapsto \mathrm{e}^{-\mathrm{i}q \cdot D} \mathrm{e}^{-\mathrm{i}\eta \cdot x} \in B(L^2(\mathcal{X}))$ is continuous for the weak topology and $\mathrm{e}^{-\mathrm{i}q \cdot D} \mathrm{e}^{-\mathrm{i}\eta \cdot x}$ tends weakly to 0 when $(\eta, q) \to \infty$. This easily implies that $w \in C_\infty(\mathcal{X}^\# \oplus \mathcal{X})$.

(3) Let us fix $\Psi \in \mathcal{S}(\mathcal{X})$. It is enough to show that

$$(\Psi|B\Psi) = (2\pi)^{-d} \int w(\eta, q)(\Psi|\mathrm{e}^{\mathrm{i}\eta \cdot x} \mathrm{e}^{\mathrm{i}q \cdot D} \Psi)\mathrm{d}\eta \mathrm{d}q. \qquad (4.27)$$

For B of finite rank, (4.27) follows by a direct computation. Let us extend it to B of trace class.

From (2) we know that the map

$$B^1(L^2(\mathcal{X})) \ni B \mapsto w \in C_\infty(\mathcal{X}^\# \oplus \mathcal{X})$$

is continuous. Clearly, $(\eta, q) \mapsto (\Psi|\mathrm{e}^{\mathrm{i}\eta \cdot x} \mathrm{e}^{\mathrm{i}q \cdot D} \Psi)$ belongs to $\mathcal{S}(\mathcal{X}^\# \oplus \mathcal{X})$. The maps

$$B^1(L^2(\mathcal{X})) \ni B \mapsto (\Psi|B\Psi),$$

$$C_\infty(\mathcal{X}^\# \oplus \mathcal{X}) \ni w \mapsto (2\pi)^{-d} \int w(\eta, q)(\Psi|\mathrm{e}^{\mathrm{i}\eta \cdot x} \mathrm{e}^{\mathrm{i}q \cdot D} \Psi)\mathrm{d}\eta \mathrm{d}q$$

are continuous. Hence we can extend (4.27) to an arbitrary $B \in B^1(L^2(\mathcal{X}))$ by density.

(4) Clearly, if $w \in L^1(\mathcal{X}^\# \oplus \mathcal{X})$, the r.h.s. of (4.26) is a norm convergent integral. $\qquad \square$

Proposition 4.33 *Let us equip $\mathcal{X}$ with a Euclidean structure. Let $P_0 := |\Phi_0)(\Phi_0|$, where $\Phi_0 \in L^2(\mathcal{X})$ is given by*

$$\Phi_0(x) := \pi^{-\frac{d}{4}} \mathrm{e}^{-\frac{1}{2}x^2}.$$

Then

(1) $P_0 = P_0^* = P_0^2$,

(2) $\pi^{\frac{d}{2}} P_0 e^{x^2} f(x) P_0 = P_0 \int f(x) \mathrm{d}x$, *for* $f \in L^1(\mathcal{X})$,

(3) $P_0 = (2\pi)^{-d} \int e^{-\frac{1}{4}\eta^2 - \frac{1}{4}q^2 + \mathrm{i}\frac{1}{2}q \cdot \eta} e^{\mathrm{i}\eta \cdot x} e^{\mathrm{i}q \cdot D} \mathrm{d}\eta \mathrm{d}q$,

(4) $\mathrm{Span}^{\mathrm{cl}} \{ e^{\mathrm{i}\eta \cdot x} e^{\mathrm{i}q \cdot D} \Phi_0,\ \eta \in \mathcal{X}^\#, q \in \mathcal{X} \} = L^2(\mathcal{X})$.

Proof (1) is immediate, since $\|\Phi_0\| = 1$. To prove (3), we note that $e^{-\mathrm{i}\eta \cdot x} P_0 e^{-\mathrm{i}q \cdot D}$ has the kernel $\pi^{-\frac{d}{2}} e^{-\frac{1}{2}x^2} e^{-\mathrm{i}\eta x} e^{-\frac{1}{2}(y+q)^2}$, which belongs to $\mathcal{S}(\mathcal{X} \oplus \mathcal{X})$. Hence, by Thm. 4.27,

$$\mathrm{Tr}(P_0 e^{-\mathrm{i}\eta \cdot x} e^{-\mathrm{i}q \cdot D}) = \pi^{-\frac{d}{2}} \int e^{-\frac{1}{2}x^2} e^{-\mathrm{i}\eta \cdot x} e^{-\frac{1}{2}(x+q)^2} \mathrm{d}x = e^{-\frac{1}{4}\eta^2 - \frac{1}{4}q^2 + \frac{\mathrm{i}}{2}\eta \cdot q}.$$

Then we apply Prop. 4.31. (2) and (4) are left to the reader. □

4.2.3 Stone–von Neumann theorem

Theorem 4.34 (Stone–von Neumann theorem) *Suppose that $\mathcal{X}$ is a finite-dimensional vector space and we are given a pair of strongly continuous unitary representations of the Abelian groups $\mathcal{X}^\#$ and $\mathcal{X}$ on a Hilbert space $\mathcal{H}$,*

$$\mathcal{X}^\# \ni \eta \mapsto V(\eta) \in U(\mathcal{H}),$$
$$\mathcal{X} \ni q \mapsto T(q) \in U(\mathcal{H}),$$

satisfying the Weyl commutation relations

$$V(\eta)T(q) = e^{-\mathrm{i}\eta \cdot q} T(q)V(\eta).$$

Then there exists a Hilbert space $\mathcal{K}$ and a unitary operator

$$U : L^2(\mathcal{X}) \otimes \mathcal{K} \to \mathcal{H}$$

such that

$$V(\eta)U = U e^{\mathrm{i}\eta \cdot x} \otimes \mathbb{1}_\mathcal{K},$$

$$T(q)U = U e^{\mathrm{i}q \cdot D} \otimes \mathbb{1}_\mathcal{K}.$$

Proof *Step 1.* Clearly, the groups $V(\eta)$ and $T(q)$ can be written as

$$V(\eta) = e^{\mathrm{i}\eta \cdot \tilde{x}}, \qquad T(q) = e^{\mathrm{i}q \cdot \tilde{D}},$$

for some vectors of self-adjoint operators on $\mathcal{H}$, $\tilde{x}$ and $\tilde{D}$. We can define

$$P_0 := (2\pi)^{-d} \int e^{-\frac{1}{4}\eta^2 - \frac{1}{4}q^2 + \frac{1}{2}\eta \cdot q} e^{\mathrm{i}\eta \tilde{x}} e^{\mathrm{i}q \tilde{D}} \mathrm{d}\eta \mathrm{d}q$$

$$= (2\pi)^{-d} \int e^{-\frac{1}{4}\eta^2 - \frac{1}{4}q^2 - \frac{1}{2}\eta \cdot q} e^{\mathrm{i}q \cdot \tilde{D}} e^{\mathrm{i}\eta \cdot \tilde{x}} \mathrm{d}\eta \mathrm{d}q, \qquad (4.28)$$

and $\mathcal{K} := \mathrm{Ran}\, P_0$. The definition of P_0 is suggested by Prop. 4.33. The identities of Prop. 4.33 are true for P_0 defined in (4.28), since they only rely on the Weyl commutation relations. Hence we get

$$P_0 = P_0^* = P_0^2,$$

$$\pi^{\frac{d}{2}} P_0 \mathrm{e}^{\tilde{x}^2} f(\tilde{x}) P_0 = P_0 \int f(x)\mathrm{d}x, \quad f \in L^1(\mathcal{X}). \tag{4.29}$$

Step 2. Let

$$U\Phi \otimes \Psi := \pi^{\frac{d}{4}} \mathrm{e}^{\frac{1}{2}\tilde{x}^2} \Phi(\tilde{x})\Psi, \text{ for } \Phi \in \mathcal{S}(\mathcal{X}), \ \Psi \in \mathcal{K}.$$

(Note that, by (4.29), $f \in L^2(\mathcal{X})$ implies $\mathrm{e}^{\frac{1}{2}\tilde{x}^2} f(\tilde{x}) P_0 \in B(\mathcal{H})$.) We have

$$
\begin{aligned}
(U\Phi_1 \otimes \Psi_1 | U\Phi_2 \otimes \Psi_2) &= \pi^{\frac{d}{2}} (\Psi_1 | \mathrm{e}^{\tilde{x}^2} \overline{\Phi_1(\tilde{x})} \Phi_2(\tilde{x})\Psi_2) \\
&= \pi^{\frac{d}{2}} (\Psi_1 | P_0 \mathrm{e}^{\tilde{x}^2} \overline{\Phi_1(\tilde{x})} \Phi_2(\tilde{x}) P_0 \Psi_2) \\
&= (\Psi_1 | \Psi_2) \int \overline{\Phi_1(x)} \Phi_2(x) \mathrm{d}x,
\end{aligned}
$$

by (4.29). Hence U uniquely extends to an isometry from $L^2(\mathcal{X}) \otimes \mathcal{K}$ into $\mathcal{H}$.
Step 3. We prove that U intertwines the Weyl commutation relations. To this end, using (4.29), we first obtain

$$\mathrm{e}^{\mathrm{i}q\tilde{D}} P_0 = \mathrm{e}^{-q\tilde{x} - \frac{1}{2}q^2} P_0. \tag{4.30}$$

Thus, for $\Psi \in \mathcal{K}$,

$$\mathrm{e}^{\mathrm{i}q \cdot \tilde{D}} \Psi = \mathrm{e}^{-q \cdot \tilde{x} - \frac{1}{2}q^2} \Psi.$$

Hence

$$
\begin{aligned}
\mathrm{e}^{\mathrm{i}q \cdot \tilde{D}} U\Phi \otimes \Psi &= \pi^{\frac{d}{4}} \mathrm{e}^{\mathrm{i}q \cdot \tilde{D}} \mathrm{e}^{\frac{1}{2}\tilde{x}^2} \Phi(\tilde{x})\Psi \\
&= \pi^{\frac{d}{4}} \mathrm{e}^{\frac{1}{2}(\tilde{x}+q)^2} \Phi(\tilde{x}+q) \mathrm{e}^{\mathrm{i}q \cdot \tilde{D}} \Psi \\
&= \pi^{\frac{d}{4}} \mathrm{e}^{\frac{1}{2}\tilde{x}^2} \Phi(\tilde{x}+q)\Psi = U \, \mathrm{e}^{\mathrm{i}q \cdot D} \Psi \otimes \Phi.
\end{aligned}
$$

It is easier to check that U intertwines the position operators:

$$\mathrm{e}^{\mathrm{i}\eta \cdot \tilde{x}} U\Phi \otimes \Psi = \pi^{\frac{d}{4}} \mathrm{e}^{\mathrm{i}\eta \cdot \tilde{x}} \mathrm{e}^{\frac{1}{2}\tilde{x}^2} \Psi(\tilde{x})\Phi = U\mathrm{e}^{\mathrm{i}\eta \cdot \tilde{x}} \Phi \otimes \Psi.$$

Step 4. Finally, let us show that U is surjective. Clearly, if $\Psi \in \mathcal{K}$, then $U\Phi_0 \otimes \Psi = \Psi$, where we recall that $\Phi_0 = \pi^{-\frac{d}{4}} \mathrm{e}^{-\frac{1}{2}x^2}$. Hence $\mathcal{K} \subset \mathrm{Ran}\, U$. Thus, using Prop. 4.33 (3) and the intertwining property of U, it is enough to show that the span of

$$\{ \mathrm{e}^{\mathrm{i}\eta \cdot \tilde{x}} \mathrm{e}^{\mathrm{i}q \cdot \tilde{D}} \Psi \ : \ \eta \in \mathcal{X}^\#, \ q \in \mathcal{X}, \ \Psi \in \mathcal{K} \} \tag{4.31}$$

is dense in $\mathcal{H}$.

Let $\Xi \in \mathcal{H}$ and $f(\eta, q) := (\Xi | \mathrm{e}^{\mathrm{i}\eta \cdot \tilde{x}} \mathrm{e}^{\mathrm{i}q \cdot \tilde{D}} \Xi)$. Assume that Ξ is orthogonal to (4.31). Then

$$
\begin{aligned}
0 &= (\Xi | \mathrm{e}^{\mathrm{i}\eta \cdot \tilde{x}} \mathrm{e}^{\mathrm{i}q \cdot \tilde{D}} P_0 \mathrm{e}^{-\mathrm{i}\eta \cdot \tilde{x}} \mathrm{e}^{-\mathrm{i}q \cdot \tilde{D}} \Xi) \\
&= (2\pi)^{-d} \int \mathrm{d}q_1 \mathrm{d}\eta_1 \, f(\eta_1, q_1) \mathrm{e}^{-\frac{1}{4}\eta^2 - \frac{1}{4}q^2 - \frac{1}{2}\eta_1 \cdot q_1 + \mathrm{i}(q \cdot \eta_1 - \eta \cdot q_1) - \mathrm{i}q \cdot \eta}.
\end{aligned}
$$

By the properties of the Fourier transformation, $f(\eta, q) = 0$ a.e. (almost everywhere). But $(\eta, q) \mapsto f(\eta, q)$ is a continuous function and $f(0, 0) = \|\Xi\|^2$. So $\Xi = 0$. $\square$

4.3 x, D-quantization

As in both previous sections, $\mathcal{X}$ is a finite-dimensional real vector space with the Lebesgue measure $\mathrm{d}x$.

Looking at operators on $L^2(\mathcal{X})$ as a *quantization of classical symbols*, that is, of functions on the *classical phase space* $\mathcal{X} \otimes \mathcal{X}^\#$, has a long tradition in quantum physics. In mathematics the usefulness of this point of view seems to have been discovered much later. Apparently, among pure mathematicians this started with a paper of Kohn–Nirenberg (1965). The *calculus of pseudo-differential operators* introduced in that paper proved to be very successful in the study of partial differential equations and originated a branch of mathematics called *microlocal analysis*.

In this section we discuss the two most naive kinds of quantizations, commonly used in the context of partial differential equations – the x, D, and D, x-*quantizations*. Other kinds of quantization, in particular the *Weyl quantization*, will be discussed later in Chap. 8.

We will start with a discussion of quantization of polynomial symbols, where certain properties have elementary algebraic proofs. (Actually, these proofs generalize to the case where the symbols depend polynomially only on, say, momenta.) The definition of the x, D- and D, x-quantizations has a natural generalization to a much larger class of symbols, that of tempered distributions, which we will consider in the following subsection.

4.3.1 Quantization of polynomial symbols

Recall that $\mathrm{CCR}^{\mathrm{pol}}(\mathcal{X}^\# \oplus \mathcal{X})$ denotes the algebra of operators on $\mathcal{S}(\mathcal{X})$ generated by x and D.

Clearly, if $f \in \mathbb{C}\mathrm{Pol}_s(\mathcal{X})$, then $f(x)$ is well defined as an operator on $\mathcal{S}(\mathcal{X})$. Such operators form a commutative sub-algebra in $\mathrm{CCR}^{\mathrm{pol}}(\mathcal{X}^\# \oplus \mathcal{X})$.

Likewise, if $g \in \mathbb{C}\mathrm{Pol}_s(\mathcal{X}^\#)$, then $g(D)$ is well defined as an operator on $\mathcal{S}(\mathcal{X})$. Such operators form another commutative algebra in $\mathrm{CCR}^{\mathrm{pol}}(\mathcal{X}^\# \oplus \mathcal{X})$.

Definition 4.35 *We define the x, D-quantization, resp. the D, x-quantization as the maps*

$$\mathbb{C}\mathrm{Pol}_s(\mathcal{X} \oplus \mathcal{X}^\#) \ni b \mapsto \mathrm{Op}^{x, D}(b) \in \mathrm{CCR}^{\mathrm{pol}}(\mathcal{X}^\# \oplus \mathcal{X}),$$
$$\mathbb{C}\mathrm{Pol}_s(\mathcal{X} \oplus \mathcal{X}^\#) \ni b \mapsto \mathrm{Op}^{D, x}(b) \in \mathrm{CCR}^{\mathrm{pol}}(\mathcal{X}^\# \oplus \mathcal{X}),$$

as follows: if $b(x, \xi) = f(x)g(\xi)$, $f \in \mathbb{C}\mathrm{Pol}_s(\mathcal{X})$, $g \in \mathbb{C}\mathrm{Pol}_s(\mathcal{X}^\#)$, we set

$$\mathrm{Op}^{x, D}(b) := f(x)g(D), \tag{4.32}$$
$$\mathrm{Op}^{D, x}(b) := g(D)f(x). \tag{4.33}$$

We extend the definition to $\mathbb{C}\mathrm{Pol}_s(\mathcal{X} \oplus \mathcal{X}^\#)$ *by linearity.*

We will treat the ordering x, D as the standard one. Instead of $\mathrm{Op}^{x,D}(b)$ one often uses the notation $b(x, D)$.

Remark 4.36 *The* x, D-*quantization is sometimes called the* Kohn–Nirenberg *quantization.*

Definition 4.37 *The maps inverse to (4.32) and (4.33) are denoted*

$$\mathrm{CCR}^{\mathrm{pol}}(\mathcal{X}^\# \oplus \mathcal{X}) \ni B \mapsto s_B^{x,D} \in \mathbb{C}\mathrm{Pol}_s(\mathcal{X} \oplus \mathcal{X}^\#), \qquad (4.34)$$

$$\mathrm{CCR}^{\mathrm{pol}}(\mathcal{X}^\# \oplus \mathcal{X}) \ni B \mapsto s_B^{D,x} \in \mathbb{C}\mathrm{Pol}_s(\mathcal{X} \oplus \mathcal{X}^\#), \qquad (4.35)$$

and the polynomials $s_B^{x,D}$ *and* $s_B^{D,x}$ *are called the* x, D- *and* D, x-*symbols of the operator* B.

Theorem 4.38 (1) *If* $b \in \mathbb{C}\mathrm{Pol}_s(\mathcal{X} \oplus \mathcal{X}^\#)$, *then*

$$\mathrm{Op}^{x,D}(b)^* = \mathrm{Op}^{D,x}(\bar{b}). \qquad (4.36)$$

(2) *If* $b_-, b_+ \in \mathbb{C}\mathrm{Pol}_s(\mathcal{X} \oplus \mathcal{X}^\#)$, *and* $\mathrm{Op}^{D,x}(b_-) = \mathrm{Op}^{x,D}(b_+)$, *then*

$$b_+(x, \xi) = e^{iD_x \cdot D_\xi} b_-(x, \xi)$$

$$= (2\pi)^{-d} \int e^{-i(x-x_1)\cdot(\xi-\xi_1)} b_-(x_1, \xi_1) dx_1 d\xi_1. \qquad (4.37)$$

(3) *If* $b_1, b_2 \in \mathbb{C}\mathrm{Pol}_s(\mathcal{X} \oplus \mathcal{X}^\#)$ *then* $\mathrm{Op}^{x,D}(b_1)\mathrm{Op}^{x,D}(b_2) = \mathrm{Op}^{x,D}(b)$, *for*

$$b(x, \xi) = e^{iD_{\xi_1} \cdot D_{x_2}} b_1(x_1, \xi_1) b_2(x_2, \xi_2) \Big|_{\substack{x_1 = x_2 = x, \\ \xi_1 = \xi_2 = \xi}}$$

$$= (2\pi)^{-d} \int e^{-i(x-x_1)\cdot(\xi-\xi_1)} b_1(x, \xi_1) b_2(x_1, \xi) dx_1 d\xi_1. \qquad (4.38)$$

The operator $e^{iD_x \cdot D_\xi}$ *in (4.37) and the similar operator in (4.38) are understood as the sums of differential operators. In the case of this theorem, the sum is finite, because we deal with polynomial symbols.*

The integral formulas in (4.37) and (4.38) should be understood in the sense of oscillatory integrals.

Proof To prove (4.37) it is sufficient to consider monomials. By a simple combinatorial argument,

$$(\eta_1 \cdot x) \cdots (\eta_n \cdot x)(q_1 \cdot D) \cdots (q_m \cdot D)$$

$$= \sum_{k=0}^{\min(n,m)} \sum_{i_1 < \cdots < i_k} \sum_{\text{distinct } j_1,\ldots,j_k} (\eta_{i_1} \cdot q_{j_1}) \cdots (\eta_{i_k} \cdot q_{j_k})$$

$$\times \prod_{i \in \{1,\ldots,m\} \setminus \{j_1,\ldots,j_k\}} (q_i \cdot D) \prod_{i \in \{1,\ldots,n\} \setminus \{i_1,\ldots,i_k\}} (\eta_i \cdot x)$$

$$= \mathrm{Op}^{D,x} \Big(\sum_{k=0}^{\min(n,m)} \frac{1}{k!} (-\mathrm{i}\nabla_x \cdot \nabla_\xi)^k (q_1 \cdot \xi) \cdots (q_m \cdot \xi)(\eta_1 \cdot x) \cdots (\eta_n \cdot x) \Big).$$

(4.38) follows easily from (4.37). In fact, it is enough to assume that $b_i(x,\xi) = f_i(x)g_i(\xi)$. Set $a(x,\xi) = f_2(x)g_1(\xi)$. Then

$$\mathrm{Op}^{x,D}(b_1)\mathrm{Op}^{x,D}(b_2) = f_1(x)\mathrm{Op}^{D,x}(a)g_2(D)$$

$$= f_1(x)\mathrm{Op}^{x,D}(\tilde{b})g_2(D) = b(x,D),$$

where

$$\tilde{b}(x,\xi) = \mathrm{e}^{-\mathrm{i}\nabla_x \cdot \nabla_\xi} a(x,\xi), \quad b(x,\xi) = f_1(x)\tilde{b}(x,\xi)g_2(\xi).$$

$\square$

Formulas (4.37) and (4.38) follow also (in a much larger generality) from integral formulas considered in the next subsection.

The following formula is a version of *Wick's theorem*. It follows from (4.38). We will see similar theorems later on for other quantizations.

Theorem 4.39 *Let $b_1,\ldots,b_n,b \in \mathbb{C}\mathrm{Pol}_\mathrm{s}(\mathcal{X} \oplus \mathcal{X}^{\#})$ and*

$$b(x,D) = b_1(x,D) \cdots b_n(x,D).$$

Then

$$b(x,\xi) = \exp\Big(\mathrm{i}\sum_{i<j} D_{\xi_i} \cdot D_{x_j}\Big) b_1(x_1,\xi_1) \cdots b_n(x_n,\xi_n)\Big|_{\substack{x = x_1 = \cdots = x_n, \\ \xi = \xi_1 = \cdots = \xi_n.}}$$

4.3.2 Quantization of distributional symbols

Recall that $\mathrm{CCR}^{\mathcal{S}'}(\mathcal{X}^{\#} \oplus \mathcal{X})$ denotes the family of operators (or, actually, quadratic forms on $\mathcal{S}(\mathcal{X})$) whose distributional kernels belong to $\mathcal{S}'(\mathcal{X} \times \mathcal{X})$.

Definition 4.40 *If $b \in \mathcal{S}'(\mathcal{X} \oplus \mathcal{X}^{\#})$, then we define $\mathrm{Op}^{x,D}(b)$ and $\mathrm{Op}^{D,x}(b)$ as the elements of $\mathrm{CCR}^{\mathcal{S}'}(\mathcal{X}^{\#} \oplus \mathcal{X})$ whose distributional kernels are*

$$\mathrm{Op}^{x,D}(b)(x_1, x_2) = (2\pi)^{-d} \int_{\mathcal{X}^{\#}} b(x_1, \xi) \mathrm{e}^{\mathrm{i}(x_1 - x_2) \cdot \xi} \mathrm{d}\xi,$$

$$\mathrm{Op}^{D,x}(b)(x_1, x_2) = (2\pi)^{-d} \int_{\mathcal{X}^{\#}} b(x_2, \xi) \mathrm{e}^{\mathrm{i}(x_1 - x_2) \cdot \xi} \mathrm{d}\xi. \qquad (4.39)$$

Theorem 4.41 (1) *For $b \in \mathbb{C}\mathrm{Pol}_{\mathrm{s}}(\mathcal{X} \oplus \mathcal{X}^{\#}) \subset \mathcal{S}'(\mathcal{X} \oplus \mathcal{X}^{\#})$, the above definition coincides with (4.32) and (4.33).*
(2) *The maps*

$$\mathcal{S}'(\mathcal{X} \oplus \mathcal{X}^{\#}) \ni b \mapsto \mathrm{Op}^{x,D}(b) \in \mathrm{CCR}^{\mathcal{S}'}(\mathcal{X}^{\#} \oplus \mathcal{X}),$$

$$\mathcal{S}'(\mathcal{X} \oplus \mathcal{X}^{\#}) \ni b \mapsto \mathrm{Op}^{D,x}(b) \in \mathrm{CCR}^{\mathcal{S}'}(\mathcal{X}^{\#} \oplus \mathcal{X})$$

are bijective. Denote their inverses (symbols) as in (4.34) and (4.35). Then for $B \in \mathrm{Op}(\mathcal{S}'(\mathcal{X} \oplus \mathcal{X}^{\#}))$ we have

$$s_B^{x,D}(x, \xi) = \int_{\mathcal{X}} B(x, x - y) \mathrm{e}^{-\mathrm{i}\xi \cdot y} \mathrm{d}y,$$

$$s_B^{D,x}(x, \xi) = \int_{\mathcal{X}} B(x + y, x) \mathrm{e}^{-\mathrm{i}\xi \cdot y} \mathrm{d}y. \qquad (4.40)$$

(3) *The formulas (4.36) and (4.37) are true.*
(4) *The formula (4.38) is true, for instance, if either $b_1 \in \mathcal{S}'(\mathcal{X} \oplus \mathcal{X}^{\#})$ and $b_2 \in \mathbb{C}\mathrm{Pol}_{\mathrm{s}}(\mathcal{X} \oplus \mathcal{X}^{\#})$, or the other way around.*
(5) *(4.38) is also true if the Fourier transforms of b_1 and b_2 belong to $L^1(\mathcal{X}^{\#} \oplus \mathcal{X})$.*
(6) *We have $b(x, D) \in B^2(L^2(\mathcal{X}))$ iff $b \in L^2(\mathcal{X} \oplus \mathcal{X}^{\#})$. Moreover,*

$$\mathrm{Tr}\, b(x, D)^* a(x, D) = (2\pi)^{-d} \int_{\mathcal{X} \oplus \mathcal{X}^{\#}} \overline{b(x, \xi)} a(x, \xi) \mathrm{d}x \mathrm{d}\xi, \quad a, b \in L^2(\mathcal{X} \oplus \mathcal{X}^{\#}).$$

Proof (2) follows from (4.39) by the inversion of the Fourier transform. (4.37) follows by combining the first formula of (4.39) with the second formula of (4.40). $\qquad \square$

Example 4.42 *Fix a Euclidean structure in $\mathcal{X}$. Let P_0 be the orthogonal projection onto the normalized vector $\Phi_0 = \pi^{-\frac{d}{4}} \mathrm{e}^{-\frac{1}{2}x^2}$ (as in Prop. 4.33). The integral kernel of P_0 is*

$$P_0(x, y) = \pi^{-\frac{d}{2}} \mathrm{e}^{-\frac{1}{2}x^2 - \frac{1}{2}y^2}.$$

Its x, D- and D, x-symbols are

$$s_{P_0}^{x,D}(x, \xi) = 2^{\frac{d}{2}} \mathrm{e}^{-\frac{1}{2}x^2 - \frac{1}{2}\xi^2 - \mathrm{i}x \cdot \xi},$$

$$s_{P_0}^{D,x}(x, \xi) = 2^{\frac{d}{2}} \mathrm{e}^{-\frac{1}{2}x^2 - \frac{1}{2}\xi^2 + \mathrm{i}x \cdot \xi}.$$

4.4 Notes

An exposition of the theory of distributions can be found e.g. in Schwartz (1966) and Gelfand–Vilenkin (1964).

The Stone–von Neumann theorem was announced by Stone in 1930, but the first published proof was given by von Neumann (1931). Proofs can be found in Emch (1972) and Bratteli–Robinson (1996).

The $x, D-$ and $D, x-$ quantization goes back to a paper by Kohn–Nirenberg (1965).

5

Measures

The first section of this chapter is devoted to a review of basic definitions of measure theory. Among other topics, we recall basic properties of *positivity preserving operators*, which provide tools useful in constructive quantum field theory.

The rest of this chapter is devoted to measures on infinite-dimensional Hilbert spaces. It is well known that there are no Borel translation invariant measures on infinite-dimensional vector spaces. However, one can define useful measures on such spaces which are not translation invariant. In particular, the notion of a *Gaussian measure* has a natural generalization to the infinite-dimensional case.

Measures on an infinite-dimensional Hilbert space $\mathcal{X}$ is quite a subtle topic. A naive approach to this subject leads to the notion of a *weak distribution*, which is a family of measures on finite-dimensional subspaces satisfying a natural compatibility condition. It is natural to ask whether a weak distribution is generated by a measure on $\mathcal{X}$. In general, the answer is negative. In order to obtain such a measure, one has to consider a larger measure space containing $\mathcal{X}$. Many choices of such a larger space are possible. A class of such choices that we describe in detail are Hilbert spaces $B\mathcal{X}$ for a self-adjoint operator B satisfying certain conditions.

Measures on Hilbert spaces play an important role in probability theory and quantum field theory. One of them is the *Wiener measure*, used to describe Brownian motion. There are also natural representations of the Fock space as the L^2 space with respect to a Gaussian measure: the so-called *real-wave* and *complex-wave CCR representations*, which we will consider in Chap. 9.

Note that for most practical purposes many subtleties of measures in infinite dimensions can be ignored. In applications, an important role is played by such concepts as L^p spaces, the integral, the positivity a.e., etc. It is important that there exists an underlying measure space, so that we can use tools of measure theory. However, which measure space we actually take is irrelevant. Therefore, the choice of the operator B mentioned above is usually not important for applications.

5.1 General measure theory

In this section we recall basic concepts and facts of measure and integration theory.

5.1.1 *σ-algebras*

Let Q be a set. Let 2^Q denote the family of its subsets. Let us introduce some useful kinds of subfamilies of 2^Q.

Definition 5.1 *Let* $\mathfrak{R} \subset 2^Q$.

(1) *We say that* $\mathfrak{R}$ *is a* ring *if* $A, B \in \mathfrak{R} \Rightarrow A \backslash B,\ A \cup B \in \mathfrak{R}$.

(2) $\mathfrak{R}$ *is a* σ-ring *if it is a ring and* $A_1, A_2, ... \in \mathfrak{R} \Rightarrow \bigcup_{n=1}^{\infty} A_n \in \mathfrak{R}$.

Definition 5.2 *Let* $\mathfrak{S} \subset 2^Q$.

(1) $\mathfrak{S}$ *is an* algebra *if it is a ring and* $Q \in \mathfrak{S}$.

(2) $\mathfrak{S}$ *is a* σ-algebra *if it is a σ-ring and an algebra.*

Definition 5.3 *If* $\mathfrak{T} \subset 2^Q$, *then there exists the smallest ring, σ-ring, algebra and σ-algebra containing* $\mathfrak{T}$. *It is called the* ring, σ-ring, algebra, *resp.* σ-algebra generated by $\mathfrak{T}$.

Definition 5.4 *If* $(Q_i, \mathfrak{S}_i)$, $i = 1, 2$, *are spaces equipped with σ-algebras, we say that* $F : Q_1 \to Q_2$ *is* measurable *if for any* $A \in \mathfrak{S}_2$, $F^{-1}(A) \in \mathfrak{S}_1$.

5.1.2 *Measures*

Let $(Q, \mathfrak{S})$ be a space equipped with a σ-algebra.

Definition 5.5 *A* finite complex measure *is a function*

$$\mathfrak{S} \ni A \mapsto \mu(A) \in \mathbb{C}$$

such that $\mu(\emptyset) = 0$ *and for any* $A_1, A_2, ... \in \mathfrak{S}$, $A_i \cap A_j = \emptyset$, $i \neq j$,

$$\bigcup_{j=1}^{\infty} A_j = A \Rightarrow \mu(A) = \sum_{j=1}^{\infty} \mu(A_j), \tag{5.1}$$

where the above sum is absolutely convergent. A finite real, *resp.* finite positive measure *on* $(Q, \mathfrak{S})$ *has the same definition, except that we replace* $\mathbb{C}$ *with* $\mathbb{R}$, *resp.* $[0, \infty[$. *In the case of a positive measure we usually drop the word* positive. *(In this case the requirement of the absolute convergence of the series in (5.1) is automatically satisfied, and hence can be dropped from the definition).*

We say that a positive finite measure μ *is a* probability measure *if* $\mu(Q) = 1$.

In the positive case Def. 5.5 has a well-known generalization that allows the measure to take infinite values.

Definition 5.6 *A* (positive) measure, *is a function*

$$\mathfrak{S} \ni A \mapsto \mu(A) \in [0, \infty]$$

such that $\mu(\emptyset) = 0$ and for any $A_1, A_2, \ldots \in \mathfrak{S}$, $A_i \cap A_j = \emptyset$, $i \neq j$,

$$\bigcup_{j=1}^{\infty} A_j = A \implies \mu(A) = \sum_{j=1}^{\infty} \mu(A_j). \tag{5.2}$$

Such a triple $(Q, \mathfrak{S}, \mu)$ is often called a measure space. *If in addition μ is a probability measure, $(Q, \mathfrak{S}, \mu)$ is called a* probability space.

A measure space $(Q, \mathfrak{S}, \mu)$ is *complete* if $B \subset A$ with $A \in \mathfrak{S}$ and $\mu(A) = 0$ implies $B \in \mathfrak{S}$. If $(Q, \mathfrak{S}, \mu)$ is a measure space, one sets

$$\mathfrak{S}^{\mathrm{cpl}} := \{ B \in 2^Q \ : \ \exists A_1, A_2 \in \mathfrak{S} \text{ with } A_1 \subset B \subset A_2, \ \mu(A_2 \setminus A_1) = 0 \},$$

$$\mu^{\mathrm{cpl}}(B) := \mu(A_1).$$

Then $(Q, \mathfrak{S}^{\mathrm{cpl}}, \mu^{\mathrm{cpl}})$ is a complete measure space called the *completion of* $(Q, \mathfrak{S}, \mu)$. It admits more measurable sets and functions and therefore is more convenient for the theory of integration.

5.1.3 Pre-measures

Generalizing Def. 5.6 to the real or complex case poses problems because the series in (5.1) could be divergent. In this case, one of the possible solutions is to use the concept of a pre-measure, which is defined only on a ring, takes finite values and is conditionally σ-additive.

Let $(Q, \mathfrak{R})$ be a space equipped with a ring.

Definition 5.7 *A complex pre-measure on $(Q, \mathfrak{R})$ is a function*

$$\mathfrak{R} \ni A \mapsto \nu(A) \in \mathbb{C}$$

such that $\nu(\emptyset) = 0$ and for any $A_1, A_2, \ldots \in \mathfrak{S}$, $A_i \cap A_j = \emptyset$, $i \neq j$,

$$\bigcup_{j=1}^{\infty} A_j = A \in \mathfrak{R} \implies \mu(A) = \sum_{j=1}^{\infty} \mu(A_j). \tag{5.3}$$

where the above sum is absolutely convergent. A real, resp. *positive pre-measure on $(Q, \mathfrak{R})$ has the same definition, except that we replace $\mathbb{C}$ with $\mathbb{R}$,* resp. *$[0, \infty[$.*

The following well-known theorem allows us to extend in a canonical way a positive pre-measure to a positive measure.

Theorem 5.8 *Suppose that $(Q, \mathfrak{R})$ is a space with a ring and $\nu : \mathfrak{R} \to [0, \infty[$ is a positive pre-measure. Let $\mathfrak{S}$ be a σ-algebra containing $\mathfrak{R}$. Then*

$$\mu(A) := \sup \{ \nu(B) \ : \ B \in \mathfrak{R}, \ B \subset A \}, \quad A \in \mathfrak{S}, \tag{5.4}$$

is a measure on $\mathfrak{S}$ extending ν. If $\mathfrak{S}$ coincides with the σ-algebra generated by $\mathfrak{R}$, then μ is the unique measure on $\mathfrak{S}$ extending ν.

5.1.4 Borel measures and pre-measures

Let Q be a topological space. The following two families of subsets of Q play a distinguished role in measure theory:

Definition 5.9 (1) *The σ-algebra generated by the family of open sets of Q will be called the* Borel σ-algebra *of Q and denoted $\mathfrak{B}(Q)$.*

(2) *The ring that consists of pre-compact Borel sets in Q will be denoted $\mathfrak{K}(Q)$. (We say that a set is* pre-compact *if its closure is compact).*

Definition 5.10 *A complex, real, resp. positive Borel pre-measure on Q is a complex, real resp. positive pre-measure on $(Q, \mathfrak{K}(Q))$. $\mathrm{Meas}(Q)$ will denote the space of complex Borel pre-measures.*

Definition 5.11 *μ is a* positive Borel measure *on Q if it is a measure on $(Q, \mathfrak{B}(Q))$ that is finite on $\mathfrak{K}(Q)$ and*

$$\mu(A) = \sup\{\mu(B) \ : \ B \in \mathfrak{K}(Q), \ B \subset A\}, \quad A \in \mathfrak{B}(Q). \tag{5.5}$$

$\mathrm{Meas}^+(Q)$ *will denote the space of positive Borel measures on Q.*

Note that every positive Borel pre-measure possesses a unique extension to a Borel measure. Conversely, every positive Borel measure restricted to $\mathfrak{K}(Q)$ is a positive Borel pre-measure.

Definition 5.12 *Let μ be a complex Borel pre-measure on Q. The* total variation *of μ is the positive Borel measure $|\mu|$ defined for $A \in \mathfrak{B}(Q)$ by*

$$|\mu|(A) := \sup \sum_{i=1}^{\infty} |\mu(A_i)|,$$

where the supremum is taken over all families $A_1, A_2, \dots \in \mathfrak{K}(Q)$ such that $A_i \cap A_j = \emptyset$, $i \neq j$ and $A_i \subset A$. $\mathrm{Meas}^1(Q)$ will denote the space of finite complex Borel pre-measures on Q equipped with the norm $|\mu|(Q)$, which makes it into a Banach space.

5.1.5 Integral

Let $(Q, \mathfrak{S})$ be a space with a σ-algebra.

Definition 5.13 *Let $\mathcal{M}_+(Q, \mathfrak{S})$, resp. $\mathcal{M}(Q, \mathfrak{S})$ denote the set of $\mathfrak{S}$-measurable functions with values in $[0, \infty[$, resp. $\mathbb{C}$.*

Let $(Q, \mathfrak{S}, \mu)$ be a measure space.

We will often abbreviate $(Q, \mathfrak{S})$ to Q and $(Q, \mathfrak{S}, \mu)$ to (Q, μ).

Definition 5.14 *Let $\mathcal{N}(Q, \mu)$, denote the subset of $\mathcal{M}(Q)$ consisting of functions vanishing outside of a set of measure zero. We set $M_+(Q, \mu) := \mathcal{M}_+(Q)/\mathcal{N}(Q, \mu)$ and $M(Q, \mu) := \mathcal{M}(Q)/\mathcal{N}(Q, \mu)$.*

Definition 5.15 *For $f \in \mathcal{M}_+(Q)$, in a standard way we define its integral, which is an element of $[0, \infty]$ and is denoted*

$$\int f \mathrm{d}\mu. \tag{5.6}$$

Clearly, (5.6) does not change if we add to f a function vanishing outside a set of measure zero, hence it makes sense to write $\int f \mathrm{d}\mu$ also for $f \in M_+(Q, \mu)$.

5.1.6 L^p spaces

Definition 5.16 *For $f \in \mathcal{M}_+(Q)$ we define*

$$\operatorname{ess\,sup} f := \inf \left\{ \sup f|_{Q \setminus N} \ : \ N \in \mathfrak{S}, \ \mu(N) = 0 \right\}. \tag{5.7}$$

Clearly, (5.7) does not change if we add to f a function vanishing outside a set of measure zero, hence it makes sense to write $\operatorname{ess\,sup} f$ also for $f \in M_+(Q, \mu)$.

Definition 5.17 *For $1 \leq p \leq \infty$ and $f \in M(Q, \mu)$, we set*

$$\|f\|_p := \left(\int_Q |f|^p \mathrm{d}\mu \right)^{1/p},$$

$$\|f\|_\infty := \operatorname{ess\,sup} |f|.$$

We also introduce in the standard way the Banach spaces $L^p(Q, \mu) \subset M(Q, \mu)$. For $f \in L^1(Q, \mu)$, we define its integral, denoted by $\int f \mathrm{d}\mu$.

If q is used as the generic variable in Q, then instead of (5.6) one can write $\int f(q) \mathrm{d}\mu(q)$. Often, especially if Q is a finite-dimensional vector space and μ is a Lebesgue measure on Q, we will write $\int f(q) \mathrm{d}q$ for (5.6).

If the measure μ is obvious from the context, we will often drop μ from our notation and we will write $L^p(Q)$, $M(Q)$ etc. for $L^p(Q, \mu)$, $M(Q, \mu)$,

Let $1 \leq p, q \leq \infty$, $p^{-1} + q^{-1} = 1$. If $f, g \in M(Q)$, the *Hölder's inequality* says

$$\|fg\|_1 \leq \|f\|_p \|g\|_q,$$

Definition 5.18 *We will write $L^p_+(Q)$ for $L^p(Q) \cap M_+(Q)$.*

Definition 5.19 *Let $g \in M(Q)$. We say that g is* strictly positive *(w.r.t. μ), and we write $g > 0$, if $g \geq 0$ and $\mu\left(\{q \ : \ g(q) = 0\}\right) = 0$.*

Proposition 5.20 *Let $g \in L^p(Q)$, $1 \leq p, q \leq \infty$, $p^{-1} + q^{-1} = 1$.*

(1) $g \geq 0$ *iff*

$$\int_Q fg \mathrm{d}\mu \geq 0, \quad f \in L^q_+(Q). \tag{5.8}$$

(2) $g > 0$ *iff*

$$\int_Q fg\mathrm{d}\mu > 0, \quad f \in L^q_+(Q), \ f \neq 0.$$

If the measure is finite, then $q \geq p$ implies $L^q(Q) \subset L^p(Q)$.

5.1.7 *Operators on L^p spaces*

In this subsection we recall properties of linear operators on L^p spaces.

Let μ_i be a measure on $(Q_i, \mathfrak{S}_i)$, $i = 1, 2$.

Definition 5.21 $T \in B\big(L^2(Q_1), L^2(Q_2)\big)$ *is called*

(1) positivity preserving *if* $f \geq 0 \ \Rightarrow\ Tf \geq 0$,
(2) positivity improving *if* $f \geq 0$, $f \neq 0 \Rightarrow Tf > 0$.

Note that T is positivity preserving (resp. improving) iff T^* is.

Let us assume in addition that μ_i, $i = 1, 2$, are probability measures.

Definition 5.22 $T \in B\big(L^2(Q_1), L^2(Q_2)\big)$ *is called* hyper-contractive *if T is a contraction and there exists $p > 2$ such that T is bounded from $L^2(Q_1)$ into $L^p(Q_2)$.*

Let μ be a probability measure on $(Q, \mathfrak{S})$. Clearly, the constant function 1 belongs to $L^2(Q)$.

Definition 5.23 $T \in B\big(L^2(Q)\big)$ *is* doubly Markovian *if T is positivity preserving and $T1 = T^*1 = 1$.*

We recall some classic results.

Proposition 5.24 *A doubly Markovian map T extends to a contraction on $L^p(Q)$ for all $1 \leq p \leq \infty$.*

Theorem 5.25 (Perron-Frobenius) *Let H be a bounded below self-adjoint operator on $L^2(Q)$, such that e^{-tH} is positivity preserving for $t \geq 0$ and $E = \inf \mathrm{spec}(H)$ is an eigenvalue. Then the following are equivalent:*

(1) $\inf \mathrm{spec}(H)$ *is a simple eigenvalue with a strictly positive eigenvector.*
(2) e^{-tH} *is positivity improving for all $t > 0$.*

5.1.8 *Conditional expectations*

Let μ be a measure on $(Q, \mathfrak{S})$. Let $\mathfrak{S}_0$ be a sub-σ-algebra of $\mathfrak{S}$. Let μ_0 denote the restriction of the measure μ to $\mathfrak{S}_0$.

For $1 \leq p \leq \infty$, elements of $L^p(Q, \mu)$ that are $\mathfrak{S}_0$-measurable form a closed subspace of $L^p(Q, \mu)$ that can be identified with $L^p(Q, \mu_0)$.

Definition 5.26 *We denote by $E_{\mathfrak{S}_0}$ the orthogonal projection from $L^2(Q, \mu)$ onto the subspace $L^2(Q, \mu_0)$. $E_{\mathfrak{S}_0}$ is called the* conditional expectation *w.r.t. $\mathfrak{S}_0$.*

The following properties are well known.

Proposition 5.27 *Let μ be a probability measure.*

(1) *$E_{\mathfrak{S}_0}$ extends to a contraction on $L^p(Q, \mu)$ for all $1 \leq p \leq \infty$.*
(2) *$E_{\mathfrak{S}_0}$ extends to an operator from $M_+(Q, \mathfrak{S})$ to $M_+(Q, \mathfrak{S}_0)$.*
(3) *If $g \in L^\infty(Q, \mu)$ is $\mathfrak{S}_0$-measurable, then $E_{\mathfrak{S}_0}(gf) = gE_{\mathfrak{S}_0}(f)$ whenever both sides are defined.*
(4) *If $\varphi : \mathbb{R} \to \mathbb{R}$ is convex and positive, then*

$$\varphi(E_{\mathfrak{S}_0} f) \leq E_{\mathfrak{S}_0}(\varphi(f)) \ a.e.$$

(5) *If $\mathfrak{S}_0 \subset \mathfrak{S}_1$ are two sub-σ-algebras of $\mathfrak{S}$, then $E_{\mathfrak{S}_0} \leq E_{\mathfrak{S}_1}$.*
(6) *Let $\{\mathfrak{S}_n\}_{n \in \mathbb{N}}$ be an increasing sequence of sub-σ-algebras of $\mathfrak{S}$ such that $\mathfrak{S}$ is generated by $\bigcup_{n \in \mathbb{N}} \mathfrak{S}_n$. Then*

$$\mathrm{s} - \lim_{n \to \infty} E_{\mathfrak{S}_n} = \mathbb{1}, \ in \ L^p(Q, \mu), \ 1 \leq p < \infty.$$

(7) *Let $F \in L^1(Q, \mu)$ with $F > 0$ a.e. and set $\mathrm{d}\mu_F = \left(\int_Q F \mathrm{d}\mu\right)^{-1} F \mathrm{d}\mu$. Denote by $E_{\mathfrak{S}_0}^F$ the conditional expectation for the measure μ_F. Then*

$$E_{\mathfrak{S}_0}^F(f) = \frac{E_{\mathfrak{S}_0}(Ff)}{E_{\mathfrak{S}_0}(F)}.$$

5.1.9 Convergence in measure

Let (Q, μ) be a probability space. In this subsection we review various notions of convergence for nets of functions on a probability space.

Definition 5.28 *The* topology of *convergence in measure on $M(Q)$ is defined by the following family $V(\epsilon, \delta)$ of neighborhoods of 0:*

$$V(\epsilon, \delta) := \left\{ f \in M(Q) \ : \ \mu(\{q \ : \ |f(q)| > \epsilon\}) < \delta \right\}.$$

It is a metric topology for the distance

$$d(f, g) - \sum_{n=0}^{\infty} 2^{-n} \mu\left(\{q \ : \ |f(q) - g(q)| \geq ?^{-n}\}\right)$$

The following proposition is immediate:

Proposition 5.29 *If $f_n \to f$ a.e. then $f_n \to f$ in measure.*

We also recall the useful notion of the *equi-integrability*.

Definition 5.30 *A family $\{f_i\}_{i \in I}$ in $M(Q)$ is* equi-integrable *if*

$$\lim_{n \to +\infty} \sup_{i \in I} \int_Q |f_i| \mathbb{1}_{[n,\infty[}(f_i) \mathrm{d}\mu = 0.$$

The following two results are well-known:

Proposition 5.31 *Let $\{f_i\}_{i \in I}$ belong to $M(Q)$. Then the following hold:*

(1) *If $f := \sup_{i \in I} |f_i|$ is in $L^1(Q)$, then $\{f_i\}_{i \in I}$ is equi-integrable.*

(2) *If $\sup_{i \in I} \|f_i\|_p < \infty$ for some $p > 1$, then $\{f_i\}_{i \in I}$ is equi-integrable.*

Theorem 5.32 (Lebesgue–Vitali theorem) *Let $1 \le p < \infty$, $(f_n)_{n \in \mathbb{N}}$ belong to $L^p(Q)$ and $f \in M(Q)$. Then the following are equivalent:*

(1) *$f \in L^p(Q)$ and $f_n \to f$ in $L^p(Q)$.*

(2) *$(|f_n|^p)_{n \in \mathbb{N}}$ is equi-integrable and $f_n \to f$ in measure.*

5.1.10 Measure preserving transformations

Let μ be a probability measure on $(Q, \mathfrak{S})$. Clearly, $L^\infty(Q)$ is a commutative W^*-algebra equipped with a faithful normal state, which we also denote by μ, that is,

$$\mu(f) := \int f \mathrm{d}\mu, \quad f \in L^\infty(Q).$$

(See Subsect. 6.2.7 for the terminology on W^*-algebras.) Conversely, every commutative W^*-algebra equipped with a faithful normal state can be represented as $L^\infty(Q)$ for some probability space $(Q, \mathfrak{S}, \mu)$. However, in general there may be many non-isomorphic choices of probability spaces that lead to the same W^*-algebra and state.

Clearly, if r is a measure preserving bijection on Q, then $r_\# f := f \circ r^{-1}$ defines an isometry on $L^p(Q)$ for all $1 \le p \le \infty$. In the case of $p = \infty$, it is in addition a σ-continuous $*$-automorphism of the commutative W^*-algebra $L^\infty(Q)$ preserving the state μ. However, if we are given a σ-continuous $*$-automorphism of $L^\infty(Q)$, we have no guarantee that there exists an underlying bijection of Q. Therefore, in the following proposition we do not insist on the existence of an underlying bijection for $*$-automorphisms of $L^\infty(Q)$.

Proposition 5.33 (1) *A $*$-automorphism of $L^\infty(Q)$ that preserves the state μ extends to an isometry of $L^p(Q)$ for all $1 \le p \le +\infty$.*

(2) *Let $\mathbb{R} \ni t \mapsto U(t)$ be a group of $*$-automorphisms of $L^\infty(Q)$ preserving the state μ. Then the following statements are equivalent:*

    (i) *For some $1 \le p < \infty$ and all $f \in L^p(Q)$, $\mathbb{R} \ni t \mapsto U(t)f \in L^p(Q)$ is norm continuous.*

(ii) *For all $f \in L^\infty(Q)$, $\mathbb{R} \ni t \mapsto U(t)f$ is continuous in measure.*

(iii) *For all $1 \le p < \infty$ and $f \in L^p(Q)$, $\mathbb{R} \ni t \mapsto U(t)f \in L^p(Q)$ is norm continuous.*

(iv) *For all $f \in L^\infty(Q)$, $\mathbb{R} \ni t \mapsto U(t)f$ is σ-weakly continuous.*

Proof Let T be a $*$-automorphism of $\mathcal{M}(Q)$ as in (1). Clearly, T preserves the L^p norm of simple functions for all $1 \le p < \infty$. Therefore, T is an isometry of L^p for $1 \le p < \infty$. Then using that $\|f\|_\infty = \|m(f)\|_{B(L^2(Q))}$ if $m(f)$ is the operator of multiplication by f, we obtain also that T is an isometry of $L^\infty(Q)$.

We now prove (2). Since $\int |f|^p \mathrm{d}\mu \ge \epsilon^p \mu(\{|f| \ge \epsilon\})$, we obtain that (i)$\Rightarrow$(ii). Let us prove that (ii)$\Rightarrow$(iii). Using (1) it suffices by density to show that

$$\lim_{t \to 0} \int |U(t)f - f|^p \mathrm{d}\mu = 0, \text{ for } f \in L^\infty. \tag{5.9}$$

We write

$$\int |U(t)f - f|^p \mathrm{d}\mu \le \mu\left(\{|U(t)f - f| \ge \epsilon\}\right) 2^p \|f\|_\infty^p + \epsilon^p.$$

Choosing first ϵ and then t small enough we obtain (5.9). To complete the proof of the lemma it suffices to prove that (iii) $\Rightarrow$ (iv) $\Rightarrow$ (i). Since $\int_Q fU(t)g\mathrm{d}\mu = \int_Q U(-t)fg\mathrm{d}\mu$ for $g \in L^\infty$, $f \in L^1$, we see that (iii) $\Rightarrow$ (iv). Using that $\|U(t)g - g\|_2^2 = 2\|g\|^2 - 2\mathrm{Re} \int_Q U(t)\overline{g}g\mathrm{d}\mu$ for $g \in L^\infty$, we obtain by a density argument that (iv) $\Rightarrow$ (i). $\qquad \square$

5.1.11 Relative continuity

Let μ be a measure on $(Q, \mathfrak{S})$.

Proposition 5.34 *Let $F \in M_+(Q)$. Then*

$$\mathfrak{S} \ni A \mapsto \nu(A) := \int \mathbb{1}_A F \mathrm{d}\mu \tag{5.10}$$

is a measure.

Definition 5.35 *The measure (5.10) is called the* measure with the density F w.r.t. the measure μ *and is denoted $\nu = F\mu$. We will also write $\frac{\mathrm{d}\nu}{\mathrm{d}\mu} := F$.*

Proposition 5.36 (1) *For F, G measurable functions we have*

$$F = G \ \mu\text{-a.e.} \Rightarrow F\mu = G\mu.$$

(2) *If $F\mu$ is σ-finite, then the converse implication is also true.*

Definition 5.37 *Let ν be a measure on $(Q, \mathfrak{S})$. ν is called* continuous w.r.t. μ *(or μ-continuous), if*

$$\mu(N) = 0 \Rightarrow \nu(N) = 0, \quad N \in \mathcal{F}.$$

Theorem 5.38 (Radon–Nikodym theorem) *Let μ be σ-finite. Let ν be a measure on $(Q, \mathfrak{S})$. Then the following conditions are equivalent:*

(1) *there exists a positive measurable function F such that $\nu = F\mu$.*
(2) *ν is μ-continuous. The function F is called the* Radon–Nikodym *derivative of ν w.r.t. μ and denoted by $\frac{d\nu}{d\mu}$.*

Note that, in the notation of Def. 5.35, the map

$$L^2(Q,\nu) \ni f \mapsto \left(\frac{d\nu}{d\mu}\right)^{\frac{1}{2}} f \in L^2(Q,\mu)$$

is unitary.

5.1.12 Moments of a measure

Let μ be a probability measure on $(Q,\mathfrak{S})$.

Proposition 5.39 *Let $f : Q \to \mathbb{R}$ be a measurable function. Let*

$$C(t) = \int e^{itf}\, d\mu, \quad t \in \mathbb{R}.$$

(1) *$f \in \bigcap_{p \in \mathbb{N}} L^p(Q)$ iff $C(t) \in C^\infty(\mathbb{R})$, and then*

$$\int f^p\, d\mu = (-i)^p \frac{d^p}{dt^p} C(0).$$

(2) *Assume that $C(t)$ extends holomorphically to $\{|\mathrm{Im}\, z| < R_0\}$. Then for all $|\mathrm{Im}\, z| < R_0$, $e^{izf} \in L^1(Q)$ and*

$$C(z) = \int e^{izf}\, d\mu.$$

Proof Let us first prove (1). The $\Rightarrow$ part is immediate by differentiating under the integral sign. It remains to prove $\Leftarrow$. It suffices to prove that $f \in L^{2n}(Q)$ for all $n \in \mathbb{N}$ by induction on n. For $\Phi \in L^2(Q)$, $f\Phi \in L^2(Q)$ iff $\Phi \in \mathrm{Dom}\, m(f)$, where $m(f)$ denotes the operator of multiplication by f on $L^2(Q)$. This is equivalent to $\|(e^{itf} - \mathbb{1})\Phi\|^2 \le Ct^2$ for $|t| \le 1$. If $\Phi = 1$, we get

$$\|(e^{itf} - \mathbb{1})\Phi\|^2 = \int_Q (2 - e^{itf} - e^{-itf}) d\mu$$
$$= 2C(0) - C(t) - C(-t) = O(t^2),$$

since $C(t)$ is C^2, and hence $f \in L^2(Q)$. Assume now that $f \in L^{2n}(Q)$. We then have

$$\frac{d^{2n}}{dt^{2n}} C(t) = i^{2n} \int f^{2n} e^{itf}\, d\mu.$$

Applying the above remark to $\Phi = f^{2n}$, we get

$$\|(e^{itf} - \mathbb{1})f^{2n}\|^2 = \int (2 - e^{itf} - e^{-itf}) f^{2n} d\mu$$

$$= 2C^{(2n)}(0) - C^{(2n)}(t) - C^{(2n)}(-t) = O(t^2),$$

since $C(t)$ is C^{2n+2}. Hence $f \in L^{2n+2}(Q)$.

To prove (2), it clearly suffices to show that $e^{\pm Rf} \in L^1(Q)$ for all $0 < R < R_0$. By Cauchy's inequalities, we get for all $0 < R < R_0$

$$|C^{(n)}(0)| \le C_R R^{-n} n!,$$

and hence

$$\int f^{2n} d\mu \le C_R R^{-2n} (2n)!,$$

$$\int |f|^{2n+1} d\mu \le \left(\int f^{2n} d\mu \right)^{\frac{1}{2}} \left(\int f^{2n+2} d\mu \right)^{\frac{1}{2}} \le C_R R^{-(2n+1)} \sqrt{2n!} \sqrt{(2n+2)!}.$$

Using Stirling's formula, we see that $\sqrt{2n!}\sqrt{(2n+2)!} \sim (2n+1)!$, and hence

$$\int |f|^{2n+1} d\mu \le C_R' R^{-(2n+1)} (2n+1)!.$$

From these bounds, by expanding the exponential, we deduce that $e^{\pm Rf} \in L^1(Q)$ for all $R < R_0$ □

5.2 Finite measures on real Hilbert spaces

In this section we describe the basic theory of probability measures on real Hilbert spaces.

Throughout this section, $\mathcal{X}$ will be a real *separable* Hilbert space. For $x_1, x_2 \in \mathcal{X}$ we denote their scalar product by $x_1 \cdot x_2$.

5.2.1 Cylinder sets and cylinder functions

Let $\mathcal{Y}$ be a closed subspace of $\mathcal{X}$. Recall that $P_{\mathcal{Y}}$ denotes the orthogonal projection on $\mathcal{Y}$. Recall also that $\mathfrak{B}(\mathcal{Y})$ stands for the σ-algebra of Borel sets in $\mathcal{Y}$. We will write $\mathfrak{B}$ for $\mathfrak{B}(\mathcal{X})$.

Definition 5.40 Fin$(\mathcal{X})$ *will denote the family of finite-dimensional subspaces of $\mathcal{X}$. For $\mathcal{Y} \in \mathrm{Fin}(\mathcal{X})$ and $A \subset \mathcal{Y}$, the set*

$$P_{\mathcal{Y}}^{-1}(A) := \{x \in \mathcal{X} \ : \ P_{\mathcal{Y}} x \in A\}$$

is called the cylinder set of base A. *Denote by $\mathfrak{B}^{\mathcal{Y}}$ the σ-algebra of cylinder sets of bases in $\mathfrak{B}(\mathcal{Y})$.*

$$\mathfrak{B}_{\mathrm{cyl}} := \bigcup_{\mathcal{Y} \in \mathrm{Fin}(\mathcal{X})} \mathfrak{B}^{\mathcal{Y}}$$

is the algebra of all cylinder sets.

Clearly, $\mathfrak{B}^{\mathcal{Y}_1} \subset \mathfrak{B}^{\mathcal{Y}_2}$ if $\mathcal{Y}_1 \subset \mathcal{Y}_2$.

Proposition 5.41 $\mathfrak{B}$ *is the σ-algebra generated by $\mathfrak{B}_{\mathrm{cyl}}$.*

Definition 5.42 *We say that $F : \mathcal{X} \to \mathbb{C}$ is* based on *$\mathcal{Y} \in \mathrm{Fin}(\mathcal{X})$ if it is measurable w.r.t. $\mathfrak{B}^{\mathcal{Y}}$. F is called a* cylinder function *if it is based on $\mathcal{Y}$ for some $\mathcal{Y} \in \mathrm{Fin}(\mathcal{X})$.*

Each cylinder function is of the form $F(x) = F_{\mathcal{Y}}(P_{\mathcal{Y}}x)$ for some measurable function $F_{\mathcal{Y}}$ on $\mathcal{Y}$.

5.2.2 Finite-dimensional distributions of a measure

Until the end of this section we fix a probability measure μ on $(\mathcal{X}, \mathfrak{B})$.

Definition 5.43 *If $\mathcal{Y} \in \mathrm{Fin}(\mathcal{X})$, we define the probability measure $\mu_{\mathcal{Y}}$ on $(\mathcal{Y}, \mathfrak{B}(\mathcal{Y}))$ by*

$$\mu_{\mathcal{Y}}(A) := \mu\big(P_{\mathcal{Y}}^{-1}(A)\big), \quad A \in \mathfrak{B}(\mathcal{Y}).$$

The collection $\{\mu_{\mathcal{Y}} : \mathcal{Y} \in \mathrm{Fin}(\mathcal{X})\}$ is called the set of finite-dimensional distributions *of the measure μ.*

Finite-dimensional distributions satisfy the following *compatibility condition*:

$$\mu_{\mathcal{Y}_1}(A) = \mu_{\mathcal{Y}_2}\big(P_{\mathcal{Y}_1}^{-1}(A) \cap \mathcal{Y}_2\big), \quad A \in \mathfrak{B}(\mathcal{Y}_1), \quad \mathcal{Y}_1 \subset \mathcal{Y}_2. \tag{5.11}$$

Proposition 5.44 *The set of finite-dimensional distributions uniquely determines the measure μ on the whole $\mathfrak{B}$.*

Proof Finite-dimensional distributions uniquely determine μ on $\mathfrak{B}_{\mathrm{cyl}}$. But $\mathfrak{B}_{\mathrm{cyl}}$ generates $\mathfrak{B}$. $\qquad\square$

5.2.3 Characteristic functional of a measure

Recall that $\mathcal{X}^{\#}$ denotes the space dual to $\mathcal{X}$. Even though there exists a canonical identification of $\mathcal{X}$ and $\mathcal{X}^{\#}$, it is sometimes convenient to distinguish between $\mathcal{X}$ and $\mathcal{X}^{\#}$.

Definition 5.45 *For $\xi \in \mathcal{X}^{\#}$, we set*

$$\hat{\mu}(\xi) := \int_{\mathcal{X}} \mathrm{e}^{-\mathrm{i}\xi \cdot x} \mathrm{d}\mu(x).$$

The function $\hat{\mu} : \mathcal{X}^{\#} \to \mathbb{C}$ is called the characteristic functional *of μ, or the* Fourier transform *of μ.*

Proposition 5.46 *The characteristic functional of μ satisfies the following three conditions:*

(1) $\hat{\mu}(0) = 1$,

(2) $\sum\limits_{i,j=1}^{N} \hat{\mu}(\xi_i - \xi_j)\overline{z}_i z_j \geq 0$, $\quad \xi_i \in \mathcal{X}^{\#}$, $\quad z_i \in \mathbb{C}$,

(3) $\mathcal{X}^{\#} \ni \xi \mapsto \hat{\mu}(\xi) \in \mathbb{C}$ *is sequentially continuous for the weak topology of $\mathcal{X}^{\#}$.*

The condition (2) above is called *positive definiteness*.

Proposition 5.47 *The characteristic functional $\hat{\mu}$ uniquely determines the measure μ.*

Proof The restriction of $\hat{\mu}$ to $\mathcal{Y}^{\#}$ for $\mathcal{Y} \in \text{Fin}(\mathcal{X})$ is the Fourier transform of $\mu_{\mathcal{Y}}$, so $\hat{\mu}$ determines the finite-dimensional distributions of μ. By Prop. 5.44 this determines μ. □

5.2.4 Moment functions

Proposition 5.48 *Let $p_0 \geq 0$. Assume that for all $\xi \in \mathcal{X}^{\#}$, the function $x \mapsto \xi \cdot x$ belongs to $L^{p_0}(\mathcal{X}, \mathrm{d}\mu)$. Then, for $0 \leq p \leq p_0$, there exists C such that*

$$\gamma_p(\xi) := \int_{\mathcal{X}} |\xi \cdot x|^p \mathrm{d}\mu(x) \leq C\|\xi\|^p. \tag{5.12}$$

Proof For $\epsilon > 0$, set

$$\gamma_{p,\epsilon}(\xi) := \int_{\mathcal{X}} |\xi \cdot x|^p \mathrm{e}^{-\epsilon\|x\|^2} \mathrm{d}\mu(x).$$

For $n \in \mathbb{N}$, set

$$A_n := \{\xi \in \mathcal{X}^{\#} \ : \ \gamma_p(\xi) \leq n\},$$
$$A_{n,\epsilon} := \{\xi \in \mathcal{X}^{\#} \ : \ \gamma_{p,\epsilon}(\xi) \leq n\}.$$

Clearly, $\gamma_{p,\epsilon}(\xi) \nearrow \gamma_p(\xi)$ when $\epsilon \to 0$, hence $A_n = \bigcap_{\epsilon>0} A_{n,\epsilon}$. Since $\xi \mapsto \gamma_{p,\epsilon}(\xi)$ is norm continuous, $A_{n,\epsilon}$ is closed and so is A_n as an intersection of closed sets. Finally $\mathcal{X}^{\#} = \bigcup_{n \in \mathbb{N}} A_n$.

Since $\mathcal{X}^{\#}$ has a non-empty interior, there exists by the Baire property a set A_m with a non-empty interior. Let $\xi_0 \in \mathcal{X}^{\#}$, $\delta > 0$ such that $B(\xi_0, \delta) \subset A_m$. If $\|\xi\| \leq \delta$, we write $\xi = \xi_0 + \xi_1$, $\xi_1 = \xi - \xi_0 \in A_m$. Using that

$$|\xi \cdot x|^p \leq C \sum_{p_1+p_2=p} |\xi_0 \cdot x|^{p_1} |\xi_1 \cdot x|^{p_2}$$

and the Hölder inequality, we obtain that

$$\gamma_p(\xi) \leq C, \quad \|\xi\| \leq \delta,$$

which proves (5.12). □

Definition 5.49 *Assume that the conditions of Prop. 5.48 are satisfied. The moment functions of order $1 \leq p \leq p_0$ of the measure μ are the maps*

$$(\xi_1, \ldots, \xi_p) \mapsto \sigma_p(\xi_1, \ldots, \xi_p) := \int_{\mathcal{X}} (\xi_1 \cdot x) \cdots (\xi_p \cdot x) \mathrm{d}\mu(x).$$

Moment functions are well defined by the Hölder inequality.

The following proposition follows directly from Props. 5.39 and 5.48:

Proposition 5.50 (1) *The moment functions σ_p are multi-linear symmetric functionals on $\mathcal{X}^\#$.*

(2)

$$|\sigma_p(\xi_1,\ldots,\xi_p)| \leq C\|\xi_1\|\cdots\|\xi_p\|. \tag{5.13}$$

(3) *μ admits moments of all orders iff its characteristic functional $\hat{\mu}$ is weakly infinitely differentiable. We then have*

$$\sigma_p(\xi_1,\ldots,\xi_p) = (-\mathrm{i})^p \frac{\partial^p}{\partial t_1 \cdots \partial t_p} \hat{\mu}\Big(\sum_{i=1}^p t_i\xi_i\Big)\Big|_{t_1=\cdots=t_p=0}.$$

By Prop. 5.50 and the Riesz theorem, if the assumptions of Prop. 5.48 hold with $n = 1$, then there exists $q \in \mathcal{X}$ such that

$$\xi \cdot q = \int_{\mathcal{X}} (\xi \cdot x)\mathrm{d}\mu(x), \quad \xi \in \mathcal{X}^\#.$$

Definition 5.51 *The vector q is called the* mean of the measure μ.

Again by Prop. 5.50, if assumptions of Prop. 5.48 hold with $n = 2$ and q is the mean of μ, there exists a bounded positive $A \in B_\mathrm{s}(\mathcal{X})$ such that

$$\xi_1 \cdot A\xi_2 = \int_{\mathcal{X}} \big(\xi_1 \cdot (x - q)\big)\big(\xi_2 \cdot (x - q)\big)\mathrm{d}\mu(x), \quad \xi_1, \xi_2 \in \mathcal{X}^\#.$$

Definition 5.52 *The operator A is called the* covariance of the measure μ.

Proposition 5.53 *Assume that the measure μ has mean zero and*

$$\int_{\mathcal{X}} \|x\|^2 \mathrm{d}\mu(x) < \infty.$$

Then the covariance A of μ is trace-class and

$$\mathrm{Tr}\, A = \int_{\mathcal{X}} \|x\|^2 \mathrm{d}\mu(x).$$

Proof It suffices to let $n \to \infty$ in the equality

$$\sum_{i=1}^n e_i \cdot Ae_i = \int_{\mathcal{X}} \sum_{i=1}^n (x \cdot e_i)^2 \mathrm{d}\mu(x),$$

where $(e_i)_{i\in\mathbb{N}}$ is an o.n. basis of $\mathcal{X}$. $\square$

5.2.5 Density of exponentials

Theorem 5.54 *Let $\mathcal{D}$ be a dense subspace of $\mathcal{X}^{\#}$. Then the space*

$$\mathrm{Span}\{e^{i\xi \cdot x} : \xi \in \mathcal{D}\}$$

is dense in $L^2(\mathcal{X})$.

Proof Let $G \in L^2(\mathcal{X})$ such that

$$\int_{\mathcal{X}} e^{i\xi \cdot x} G(x) \mathrm{d}\mu(x) = 0, \quad \xi \in \mathcal{D}. \tag{5.14}$$

Without loss of generality we can assume that G is real-valued. Let

$$B_1 = \{x \in \mathcal{X} : G(x) \geq 0\}, \quad B_2 = \{x \in \mathcal{X} : G(x) < 0\}.$$

We can define the finite measures

$$\mu_1(A) := \int_A \mathbb{1}_{B_1}(x) G(x) \mathrm{d}\mu(x), \quad \mu_2(A) = -\int_A \mathbb{1}_{B_2}(x) G(x) \mathrm{d}\mu(x),$$

where $A \in \mathfrak{B}$. From (5.14), we deduce that

$$\int_{\mathcal{X}} e^{i\xi \cdot x} \mathrm{d}\mu_1(x) = \int_{\mathcal{X}} e^{i\xi \cdot x} \mathrm{d}\mu_2(x), \quad \xi \in \mathcal{D}. \tag{5.15}$$

$\mathcal{D}$ is a dense subspace of $\mathcal{X}^{\#}$. Hence it is weakly sequentially dense in $\mathcal{X}^{\#}$. Since the characteristic functional of a measure is sequentially continuous for the weak topology, (5.15) extends to all $\xi \in \mathcal{X}^{\#}$. So μ_1 and μ_2 have the same characteristic functionals, and hence are identical, i.e. $\mu_1(A) = \mu_2(A)$ for all $A \in \mathfrak{B}$. But $\mu_i(A) = \mu_i(A \cap B_i)$, $i = 1, 2$, and $B_1 \cap B_2 = \emptyset$. Hence, $\mu_1 = \mu_2 = 0$. This implies that $G(x) = 0$ μ-a.e., and hence $G = 0$. $\square$

5.2.6 Density of continuous polynomials

Let $\mathcal{D}$ be a subspace of $\mathcal{X}^{\#}$.

Definition 5.55 *Functions on $\mathcal{X}$ of the form $(\xi_1 \cdot x) \cdots (\xi_p \cdot x)$, for $\xi_1, \ldots, \xi_n \in \mathcal{D}$, are called* monomials based on $\mathcal{D}$. *Finite linear combinations (with complex coefficients) of monomials based on $\mathcal{D}$ are called* polynomials based on $\mathcal{D}$.

Note that polynomials based on $\mathcal{X}^{\#}$ are continuous functions. Therefore, they are sometimes called *continuous polynomials*.

If the measure μ admits moments of all orders, then all continuous polynomials belong to $L^2(\mathcal{X})$. The following theorem gives a sufficient condition for the density of continuous polynomials in $L^2(\mathcal{X})$.

Theorem 5.56 *Let $\mathcal{D} \subset \mathcal{X}^{\#}$ be a dense subspace of $\mathcal{X}^{\#}$. Assume that for all $\xi \in \mathcal{D}$ there exists $R(\xi) > 0$ such that the function*

$$\mathbb{R} \ni t \mapsto \hat{\mu}(t\xi) \in \mathbb{C}$$

extends holomorphically to $|\operatorname{Im} t| < R(\xi)$. *Then polynomials based on* $\mathcal{D}$ *are dense in* $L^2(\mathcal{X})$.

Proof Let $G \in L^2(\mathcal{X})$ be a vector orthogonal to all polynomials based on $\mathcal{D}$. Without loss of generality we can assume that G is real-valued. We then have

$$\int_{\mathcal{X}} G(x)(\xi \cdot x)^n \mathrm{d}\mu(x) = 0, \quad \xi \in \mathcal{D}, \quad n \in \mathbb{N}.$$

Let us fix $\xi \in \mathcal{D}$ and let $2R < R(\xi)$. Then by Prop. 5.39 we know that $\mathrm{e}^{2R|\xi \cdot x|} \in L^1(Q)$ and

$$\int G(x)\mathrm{e}^{\mathrm{i}R\xi \cdot x} \mathrm{d}\mu(x) = \lim_{n \to \infty} \int G(x) \sum_{k=1}^{n} \frac{(\mathrm{i}R\xi \cdot x)^k}{k!} \mathrm{d}\mu(x).$$

We can exchange sum and integral, since the integrand in the r.h.s. is less than

$$|G(x)|\mathrm{e}^{R|\xi \cdot x|} \leq \frac{1}{2}\left(|G(x)|^2 + \mathrm{e}^{2R|\xi \cdot x|}\right) \in L^1(\mathcal{X}).$$

We obtain hence that

$$\int G(x)\mathrm{e}^{\mathrm{i}R\xi \cdot x} \mathrm{d}\mu(x) = 0,$$

and, by differentiating w.r.t. R,

$$\int G(x)\mathrm{e}^{\mathrm{i}R\xi \cdot x}(\xi \cdot x)^n \mathrm{d}\mu(x) = 0, \quad n \in \mathbb{N}.$$

Arguing as above with $G(x)$ replaced by $G(x)\mathrm{e}^{\mathrm{i}R\xi \cdot x}$, we obtain

$$\int G(x)\mathrm{e}^{\mathrm{i}R\xi \cdot x}\mathrm{e}^{\mathrm{i}R\xi \cdot x} \mathrm{d}\mu(x) = 0.$$

Hence, repeating this argument, we obtain

$$\int G(x)\mathrm{e}^{\mathrm{i}mR\xi \cdot x} \mathrm{d}\mu(x) = 0, \quad m \in \mathbb{N}.$$

If we choose $m \in \mathbb{N}$ and $2R < R(\xi)$ such that $mR = 1$, we finally obtain

$$\int G(x)\mathrm{e}^{\mathrm{i}\xi \cdot x} \mathrm{d}\mu(x) = 0, \quad \xi \in \mathcal{D}.$$

Applying Thm. 5.54, we obtain that $G = 0$. $\qquad\qquad\square$

5.3 Weak distributions and the Minlos–Sazonov theorem

Throughout this section, $\mathcal{X}$ is a separable real Hilbert space.

Suppose that we have a compatible family of measures on finite-dimensional subspaces of $\mathcal{X}$. We can ask whether this family comes from a measure on a certain measure space. Often, there is no such a measure on $\mathcal{X}$ itself. However, if we enlarge $\mathcal{X}$, usually in a non-unique way, then such a measure may exist.

5.3.1 Weak distributions

Definition 5.57 *A collection* $\mu_* = \{\mu_{\mathcal{Y}} : \mathcal{Y} \in \mathrm{Fin}(\mathcal{X})\}$ *is called a* weak distribution *or a* generalized measure *if, for each* $\mathcal{Y} \in \mathrm{Fin}(\mathcal{X})$, $\mu_{\mathcal{Y}}$ *is a Borel probability measure on* $\mathcal{Y}$, *and these measures satisfy the compatibility condition (5.11).*

Note that cylinder functions can be "integrated" w.r.t. a weak distribution μ_*. In fact, we can set

$$\int_{\mathcal{X}} F \mathrm{d}\mu_* := \int_{\mathcal{Y}} F_{\mathcal{Y}} \mathrm{d}\mu_{\mathcal{Y}}, \tag{5.16}$$

where $F(x) = F_{\mathcal{Y}}(P_{\mathcal{Y}} x)$. Because of the compatibility condition (5.11), the r.h.s. of (5.16) is independent of the choice of $\mathcal{Y}$ on which F is based.

For each $\mathcal{Y} \in \mathrm{Fin}(\mathcal{X})$ and $1 \le p < \infty$, we can define the space $L^p(\mathcal{Y}, \mu_{\mathcal{Y}})$. For $\mathcal{Y}_1 \subset \mathcal{Y}_2$, we have natural isometric embeddings

$$L^p(\mathcal{Y}_1, \mu_{\mathcal{Y}_1}) \subset L^p(\mathcal{Y}_2, \mu_{\mathcal{Y}_2}).$$

Definition 5.58 *The* generalized L^p space *associated with a generalized measure μ_* is defined as the inductive limit of the spaces $L^p(\mathcal{Y}, \mu_{\mathcal{Y}})$, that is,*

$$\mathbf{L}^p(\mathcal{X}, \mu_*) := \left(\bigcup_{\mathcal{Y} \in \mathrm{Fin}(\mathcal{X})} L^p(\mathcal{Y}, \mu_{\mathcal{Y}}) \right)^{\mathrm{cpl}}.$$

5.3.2 Weak distributions generated by a measure

Definition 5.59 *Let μ be a measure on $(\mathcal{X}, \mathfrak{B})$. A weak distribution $\mu_* = \{\mu_{\mathcal{Y}} : \mathcal{Y} \in \mathrm{Fin}(\mathcal{X})\}$ is said to be* generated by μ *if it is the set of finite-dimensional distributions of μ.*

The following necessary and sufficient condition for this to happen is given in Skorokhod (1974):

Theorem 5.60 *A weak distribution μ_* is generated by a probability measure iff*

$$\lim_{R \to \infty} \left(\sup_{\mathcal{Y} \in \mathrm{Fin}(\mathcal{X})} \int_{\mathcal{Y}} \mathbb{1}_{[R,\infty[}(\|y\|) \mathrm{d}\mu_{\mathcal{Y}}(y) \right) = 0. \tag{5.17}$$

5.3.3 Characteristic functionals of weak distributions

The following proposition coincides with the famous Bochner theorem if $\mathcal{X}$ is finite-dimensional:

Proposition 5.61 *Let $F : \mathcal{X} \to \mathbb{C}$ be a function satisfying the following conditions:*

(1) $F(0) = 1$,

(2) $\sum_{i,j=1}^{n} F(\xi_i - \xi_j) z_i \overline{z_j} \geq 0$, $\quad \xi_1, \ldots, \xi_n \in \mathcal{X}$, $\quad z_1, \ldots, z_n \in \mathbb{C}$,

(3) $\mathcal{Y} \ni \xi \mapsto F(\xi) \in \mathbb{C}$ *is continuous for all* $\mathcal{Y} \in \mathrm{Fin}(\mathcal{X})$.

Then there exists a weak distribution $\{\mu_{\mathcal{Y}} : \mathcal{Y} \in \mathrm{Fin}(\mathcal{X})\}$ *such that, for any* $\mathcal{Y} \in \mathrm{Fin}(\mathcal{X})$,

$$F(\xi) = \int_{\mathcal{Y}} \mathrm{e}^{-\mathrm{i}\xi \cdot y} \mathrm{d}\mu_{\mathcal{Y}}(y), \quad \xi \in \mathcal{Y}. \tag{5.18}$$

Note that the functions $\mathcal{X} \ni x \mapsto \mathrm{e}^{\mathrm{i}\xi \cdot x}$ are cylinder functions, hence the integral in the r.h.s. of (5.18) is well defined.

Definition 5.62 *A function F satisfying (1), (2) and (3) of Prop. 5.61 will be called a* weak characteristic functional.

Proof of Prop. 5.61.   For any $\mathcal{Y} \in \mathrm{Fin}(\mathcal{X})$, the restriction of F to $\mathcal{Y}$ satisfies the hypotheses of Bochner's theorem (see Reed–Simon (1978b)). Hence there exists a probability measure $\mu_{\mathcal{Y}}$ on $\mathcal{Y}$ such that (5.18) holds. It remains to check the compatibility condition (5.11). To check this, it suffices to show that, if $\mathcal{Y}_1 \subset \mathcal{Y}_2$, for each bounded continuous function G on $\mathcal{Y}_1$ one has

$$\int_{\mathcal{Y}_2} G \circ P_{\mathcal{Y}_1} \mathrm{d}\mu_{\mathcal{Y}_2} = \int_{\mathcal{Y}_1} G \mathrm{d}\mu_{\mathcal{Y}_1}. \tag{5.19}$$

This is clearly satisfied for $G(y) = \mathrm{e}^{\mathrm{i}\xi \cdot y}$ for $\xi \in \mathcal{Y}_1$. Next we can find a bounded sequence (G_n) of finite linear combinations of $\mathrm{e}^{\mathrm{i}\xi \cdot x}$ for $\xi \in \mathcal{Y}_1$ which converges a.e. to G, from which (5.19) follows. $\square$

5.3.4 Minlos–Sazonov theorem

Theorem 5.63 (Minlos–Sazonov theorem) *Let $F : \mathcal{X}^{\#} \to \mathbb{C}$ be a weak characteristic functional. Then the following are equivalent:*

(1) *F is the characteristic functional of a probability measure μ on $(\mathcal{X}, \mathcal{B})$.*

(2) *There exists a positive trace-class operator S on $\mathcal{X}$ such that $\mathcal{X} \ni \xi \mapsto F(\xi) \in \mathbb{C}$ is continuous if we equip $\mathcal{X}$ with the norm $\|\xi\|_S := (\xi|S\xi)^{\frac{1}{2}}$.*

Proof   (1)$\Rightarrow$(2). Assume that F is the characteristic functional of a measure μ. Note that

$$|F(\xi_1) - F(\xi_2)|^2 = 2\mathrm{Re}\big(1 - F(\xi_1 - \xi_2)\big). \tag{5.20}$$

Now, for $R > 0$,

$$\mathrm{Re}(1 - F(\xi)) = \int_{\mathcal{X}} (1 - \cos(\xi \cdot x))\, \mathrm{d}\mu(x)$$

$$\leq \frac{1}{2} \int_{\|x\| \leq R} (\xi \cdot x)^2 \mathrm{d}\mu(x) + 2 \int_{\|x\| \geq R} \mathrm{d}\mu(x),$$

where we used $1 - \cos\theta \leq \inf(\frac{\theta^2}{2}, 2)$. Since $\int_{\|x\| \leq R} \|x\|^2 \mathrm{d}\mu(x) < \infty$, we obtain from Prop. 5.53 that there exists a trace-class operator A_R such that

$$\int_{\|x\| \leq R} (\xi \cdot x)^2 \mathrm{d}\mu(x) = \xi \cdot A_R \xi.$$

This yields

$$\mathrm{Re}(1 - F(\xi)) \leq \xi \cdot A_R \xi + 2\mu(\{\|x\| \geq R\}).$$

Now let $\epsilon > 0$. Fixing $R_\epsilon > 0$ such that $2\mu(\{\|x\| \geq R_\epsilon\}) \leq \frac{1}{2}\epsilon$, and then taking $S_\epsilon = 2\epsilon^{-1} A_{R_\epsilon}$, we prove that for any $\epsilon > 0$ there exists a trace class S_ϵ such that $(\xi | S_\epsilon \xi) \leq 1$ implies

$$\mathrm{Re}(1 - F(\xi)) \leq \epsilon.$$

Now let $\epsilon_k \to 0$. Let S_k be positive trace-class operators such that $\mathrm{Re}(1 - F(\xi)) \leq \epsilon_k$ if $(\xi | S_k \xi) \leq 1$. We pick a sequence $(\lambda_k) > 0$ such that $\sum_k \lambda_k \mathrm{Tr}\, S_k < \infty$. Then $S = \sum_k \lambda_k S_k$ is trace-class. Moreover, if $(\xi | S\xi) \leq \lambda_k$, then $(\xi | S_k \xi) \leq 1$, and hence $\mathrm{Re}(1 - F(\xi)) \leq \epsilon_k$.

$(1) \Leftarrow (2)$. Since F satisfies the conditions of Prop. 5.61, we can construct from F a weak distribution $\{\mu_{\mathcal{Y}} : \mathcal{Y} \in \mathrm{Fin}(\mathcal{X})\}$. To construct a measure from the weak distribution, we will use Thm. 5.60.

Let us fix $\delta > 0$. Let ϵ be such that $(\xi | S\xi) \leq \epsilon$ implies $\mathrm{Re}(1 - F(\xi)) \leq \delta$. Since $\mathrm{Re}(1 - F(\xi)) \leq 2$, we clearly have

$$\mathrm{Re}(1 - F(\xi)) \leq \delta + \frac{2}{\epsilon}(\xi | S\xi).$$

Let $\mathcal{Y} \in \mathrm{Fin}(\mathcal{X})$, $\alpha > 0$, $\dim \mathcal{Y} = d$. By (4.10), for $y \in \mathcal{Y}$ we have

$$\mathrm{e}^{-\frac{1}{2}\alpha \|y\|^2} = (2\pi\alpha)^{-\frac{1}{2}d} \int_{\mathcal{Y}} \mathrm{e}^{\mathrm{i}y \cdot \xi} \mathrm{e}^{-\frac{1}{2\alpha}\|\xi\|^2} \mathrm{d}\xi,$$

and hence

$$\int \left(1 - \mathrm{e}^{-\frac{1}{2}\alpha \|y\|^2}\right) \mathrm{d}\mu_{\mathcal{Y}}(y) = (2\pi\alpha)^{-\frac{1}{2}d} \int_{\mathcal{Y}} \mathrm{e}^{-\frac{1}{2\alpha}\|\xi\|^2} (1 - F(\xi)) \mathrm{d}\xi$$

$$= (2\pi\alpha)^{-\frac{1}{2}d} \int_{\mathcal{Y}} \mathrm{e}^{-\frac{1}{2\alpha}\|\xi\|^2} \mathrm{Re}(1 - F(\xi)) \mathrm{d}\xi$$

$$\leq (2\pi\alpha)^{-\frac{1}{2}d} \int_{\mathcal{Y}} \mathrm{e}^{-\frac{1}{2\alpha}\|\xi\|^2} \left(\delta + \frac{2}{\epsilon}\xi \cdot S\xi\right) \mathrm{d}\xi$$

$$= \delta + 2\frac{\alpha}{\epsilon} \mathrm{Tr}\, P_{\mathcal{Y}} S P_{\mathcal{Y}}$$

$$\leq \delta + 2\frac{\alpha}{\epsilon} \mathrm{Tr}\, S,$$

using (4.15). Next we have

$$1 - e^{-\frac{1}{2}\alpha\|y\|^2} \geq (1 - e^{-\frac{1}{2}\alpha R^2})\mathbb{1}_{[R,\infty[}(\|y\|),$$

which yields

$$\int_{\mathcal{Y}} \mathbb{1}_{[R,\infty[}(\|y\|)\mathrm{d}\mu_{\mathcal{Y}}(y) \leq (1 - e^{-\frac{1}{2}\alpha R^2})^{-1}\int_{\mathcal{Y}}(1 - e^{-\frac{1}{2}\alpha\|y\|^2})\mathrm{d}\mu_{\mathcal{Y}}(y)$$

$$\leq (1 - e^{-\frac{1}{2}\alpha R^2})^{-1}\left(\delta + 2\frac{\alpha}{\epsilon}\mathrm{Tr}\,S\right).$$

Fixing first $\delta > 0$, then $\alpha > 0$, and then letting $R \to \infty$, we see that condition (5.17) is satisfied. This completes the proof of the theorem. $\qquad\square$

5.3.5 Measures on enlarged spaces

Using the Minlos–Sazonov theorem, it is possible to realize many weak characteristic functionals on $\mathcal{X}$ (and even on a dense subspace of $\mathcal{X}$) as characteristic functionals of measures on a *larger* Hilbert space.

In the theorem below the Hilbert space $B^{\frac{1}{2}}\mathcal{X}$ is defined as in Subsect. 2.3.4. We follow the usual convention for scales of real Hilbert spaces: $\mathcal{X}^{\#}$ is identified with $\mathcal{X}$, but $(B^{\frac{1}{2}}\mathcal{X})^{\#}$ is identified with $B^{-\frac{1}{2}}\mathcal{X}$ using the scalar product on $\mathcal{X}$.

Theorem 5.64 *Let $F : \mathcal{X} \to \mathbb{C}$ be a weak characteristic functional continuous for the norm of $\mathcal{X}$. Let $B > 0$ be a self-adjoint operator on $\mathcal{X}$ such that B^{-1} is trace-class. Then there exists a Borel probability measure μ_B on the Hilbert space $B^{\frac{1}{2}}\mathcal{X}$ such that*

$$F(\xi) = \int_{B^{\frac{1}{2}}\mathcal{X}} e^{\mathrm{i}\xi\cdot x}\mathrm{d}\mu_B(x), \quad \xi \in B^{-\frac{1}{2}}\mathcal{X}.$$

Proof Since B^{-1} is trace-class, B is bounded away from zero, and hence $B^{-\frac{1}{2}}\mathcal{X} = \mathrm{Dom}\,B^{\frac{1}{2}} \subset \mathcal{X}$. Let F_B be the restriction of the functional F to $B^{-\frac{1}{2}}\mathcal{X}$. Clearly, F_B is continuous if we equip $B^{-\frac{1}{2}}\mathcal{X}$ with the norm $(\xi|B^{-1}\xi)^{\frac{1}{2}}_{B^{-\frac{1}{2}}\mathcal{X}} = (\xi|\xi)^{\frac{1}{2}}_{\mathcal{X}}$. Hence F_B is a weak characteristic functional on $B^{-\frac{1}{2}}\mathcal{X}$.

B^{-1} can be restricted to $B^{-\frac{1}{2}}\mathcal{X}$. Interpreted in this way, it will be denoted $B^{-1}\big|_{B^{-\frac{1}{2}}\mathcal{X}}$. It is then unitarily equivalent to B^{-1} as an operator on $\mathcal{X}$. Indeed, $B^{-\frac{1}{2}} : \mathcal{X} \to B^{-\frac{1}{2}}\mathcal{X}$, $B^{\frac{1}{2}} : B^{-\frac{1}{2}}\mathcal{X} \to \mathcal{X}$ are unitary and

$$B^{-1}\big|_{B^{-\frac{1}{2}}\mathcal{X}} = B^{-\frac{1}{2}}B^{-1}B^{\frac{1}{2}}.$$

Hence, if B^{-1} is trace-class, then so is $B^{-1}\big|_{B^{-\frac{1}{2}}\mathcal{X}}$. Therefore, we can apply now Thm. 5.63, which implies that F_B is the characteristic functional of a Borel probability measure μ_B on the dual $(B^{-\frac{1}{2}}\mathcal{X})^{\#}$. By Prop. 2.60, $(B^{-\frac{1}{2}}\mathcal{X})^{\#}$ can be identified with $B^{\frac{1}{2}}\mathcal{X}$. This completes the proof of the theorem. $\qquad\square$

Remark 5.65 *Sometimes the functional F is not continuous for the topology of $\mathcal{X}$, but for a certain norm $(\xi|A\xi)^{\frac{1}{2}}$, where $A > 0$ is a self-adjoint operator on $\mathcal{X}$.*

This case can be easily reduced to the case $A = \mathbb{1}$ by replacing $\mathcal{X}$ by $A^{-\frac{1}{2}}\mathcal{X}$. The condition on B becomes that $B^{-\frac{1}{2}}AB^{-\frac{1}{2}}$ is trace-class on $\mathcal{X}$.

Remark 5.66 *Note that we still use the notation x for the generic variable in the enlarged space $B^{\frac{1}{2}}\mathcal{X}$.*

5.3.6 Comparison of enlarged spaces

Proposition 5.67 *Let F be as in Thm. 5.64 and let $B_i > 0$, $i = 1, 2$, be two self-adjoint operators on $\mathcal{X}$. Assume that B_1^{-1} is trace-class and $B_1 \leq B_2$. Then B_2^{-1} is trace-class. Let μ_i be the associated probability measures on $B_i^{\frac{1}{2}}\mathcal{X}$. Then $B_1^{\frac{1}{2}}\mathcal{X}$ is a Borel subset of $B_2^{\frac{1}{2}}\mathcal{X}$ and*

$$\mu_2(C) = \mu_1(C \cap B_1^{\frac{1}{2}}\mathcal{X}), \quad C \in \mathfrak{B}(B_2^{\frac{1}{2}}\mathcal{X}).$$

For the proof we will use the following lemma:

Lemma 5.68 *Let $\mathcal{X}$ be a real Hilbert space and $A \in B(\mathcal{X})$. Then $\operatorname{Ran} A \in \mathfrak{B}(\mathcal{X})$.*

Proof We use the polar decomposition $A = U|A|$ of A, where U is a partial isometry. It is clear that partial isometries map Borel sets onto Borel sets. Therefore, it suffices to show that $\operatorname{Ran}|A|$ is Borel. By the spectral theorem,

$$\operatorname{Ran}|A| = \left\{ x \in \mathcal{X}, \ \sup_{n \in \mathbb{N}} \left\| \left(|A| + n^{-1} \right)^{-1} x \right\|_{\mathcal{X}} < \infty \right\}$$
$$= \bigcup_{m \in \mathbb{N}} \bigcap_{n \in \mathbb{N}} \left\{ x \in \mathcal{X}, \ \left\| \left(|A| + n^{-1} \right)^{-1} x \right\|_{\mathcal{X}} < m \right\}.$$

This proves that $\operatorname{Ran}|A| \in \mathfrak{B}(\mathcal{X})$. $\qquad\square$

Proof of Prop. 5.67. $B_1^{\frac{1}{2}}\mathcal{X}$ equals $AB_2^{\frac{1}{2}}\mathcal{X}$, where $A = B_1^{\frac{1}{2}}B_2^{-\frac{1}{2}} \in B(B_2^{\frac{1}{2}}\mathcal{X})$. Hence, by Lemma 5.68, $B_1^{\frac{1}{2}}\mathcal{X} \in \mathfrak{B}(B_2^{\frac{1}{2}}\mathcal{X})$.

Recall from Subsect. 2.3.4 that we have a natural embedding $I : B_1^{\frac{1}{2}}\mathcal{X} \to B_2^{\frac{1}{2}}\mathcal{X}$. Its adjoint is an embedding $I^{\#} : B_2^{-\frac{1}{2}}\mathcal{X} \to B_1^{-\frac{1}{2}}\mathcal{X}$. Both $B_2^{-\frac{1}{2}}\mathcal{X}$ and $B_1^{-\frac{1}{2}}\mathcal{X}$ are embedded in $\mathcal{X}$. Thus, for $\xi \in B_2^{-\frac{1}{2}}\mathcal{X}$ treated as an element of $\mathcal{X}$, we can write $I^{\#}\xi = \xi$.

Define a measure $\tilde{\mu}_2$ on $\mathfrak{B}(B_2^{\frac{1}{2}}\mathcal{X})$ by

$$\tilde{\mu}_2(C) = \mu_1(I^{-1}C) = \mu_1(C \cap B_1^{\frac{1}{2}}\mathcal{X}), \quad C \in \mathfrak{B}(B_2^{\frac{1}{2}}\mathcal{X}).$$

For $\xi \in B_2^{-\frac{1}{2}}\mathcal{X}$, we have

$$\int_{B_2^{\frac{1}{2}}\mathcal{X}} e^{-i\xi \cdot x_2} \, d\tilde{\mu}_2(x_2) = \int_{B_1^{\frac{1}{2}}\mathcal{X}} e^{-i\xi \cdot Ix_1} \, d\mu_1(x_1) = \int_{B_1^{\frac{1}{2}}\mathcal{X}} e^{-iI^{\#}\xi \cdot x_1} \, d\mu_1(x_1)$$
$$= F(I^{\#}\xi) = F(\xi) = \int_{B_2^{\frac{1}{2}}\mathcal{X}} e^{-i\xi \cdot x_2} \, d\mu_2(x_2).$$

This implies that the characteristic functionals of $\tilde{\mu}_2$ and μ_2 are equal. Hence $\mu_2 = \tilde{\mu}_2$. This completes the proof of the proposition. $\square$

5.4 Gaussian measures on real Hilbert spaces

Let $\mathcal{X}$ be a real Hilbert space. We would like to discuss *Gaussian measures* on real Hilbert spaces and the corresponding $\mathbf{L}^2$ *spaces*. This section has a natural continuation in Sect. 9.3, where we discuss the *real-wave representation of CCR*.

5.4.1 Gaussian measures

Proposition 5.69 *Let A be a positive self-adjoint operator on $\mathcal{X}$ and q be a bounded linear functional on $A^{-\frac{1}{2}}\mathcal{X}$.*

(1) *The function*

$$\operatorname{Dom} A \ni \xi \mapsto F(\xi) = \mathrm{e}^{\mathrm{i} q \cdot \xi - \frac{1}{2} \xi \cdot A \xi} \tag{5.21}$$

is a weak characteristic functional.

(2) *It is the characteristic functional of a probability measure μ on $\mathcal{X}$ iff A is trace-class.*

Proof (1) To prove the conditions of Prop. 5.61 we can assume that $\mathcal{X}$ is finite-dimensional. Setting $\mathcal{X}_1 = \operatorname{Ker} A$, we decompose $\mathcal{X}$ as $\mathcal{X}_1 \oplus \mathcal{X}_2$ and $q = (q_1, q_2)$. Let A_2 be A restricted to $\mathcal{X}_2$. Using (4.10) we see that F is the Fourier transform of the probability measure $\mathrm{d}\mu = \mathrm{d}\mu_1 \otimes \mathrm{d}\mu_2$ for

$$\mathrm{d}\mu_1(x_1) = \delta(x_1 - q_1)\mathrm{d}x_1,$$
$$\mathrm{d}\mu_2(x_2) = (2\pi)^{-\frac{1}{2}\dim \mathcal{Y}_2} \det A_2^{-\frac{1}{2}} \mathrm{e}^{-\frac{1}{2}(x_2 - q_2) \cdot A_2^{-1}(x_2 - q_2)}\mathrm{d}x_2.$$

(2) Let us prove $\Leftarrow$. We have

$$\operatorname{Re}(1 - F(\xi)) = (1 - \mathrm{e}^{-\frac{1}{2}\xi \cdot A \xi}) + \mathrm{e}^{-\frac{1}{2}\xi \cdot A \xi}(1 - \cos(q \cdot \xi))$$

$$\leq \tfrac{1}{2}\xi \cdot A \xi + c|q \cdot \xi|^2.$$

Since q is bounded on $A^{-\frac{1}{2}}\mathcal{X}$ we obtain that $|\operatorname{Re}(1 - F(\xi))| \leq C\xi \cdot A\xi$. By (5.20) this proves the continuity of F for the norm given by A, which is trace-class. So we can apply the Minlos–Sazonov theorem.

Let us now prove $\Rightarrow$. Let us assume that F is the characteristic functional of a measure μ. By translating the measure μ we can assume that $q = 0$. Splitting $\mathcal{X}$ as $\operatorname{Ker} A \oplus \operatorname{Ker} A^\perp$, we may assume that A is non-degenerate. If A is not compact, we can find a sequence $(\xi_n)_{n \in \mathbb{N}}$ such that $\mathrm{w} - \lim_{n \to \infty} \xi_n = 0$ and $\lim_{n \to \infty} \xi_n \cdot A\xi_n = \lambda \neq 0$. This contradicts the weak continuity of F. Hence A is a compact operator.

Now let $(e_j)_{j\in\mathbb{N}}$ be an o.n. basis of eigenvectors of A for the eigenvalues $(\lambda_j)_{j\in\mathbb{N}}$. Let $\mathcal{Y}_n = \mathrm{Span}\{e_1,\ldots,e_n\}$, P_n be the orthogonal projection on $\mathcal{Y}_n$ and $A_n = P_n A P_n$. Let μ_n denote the measure $\mu_{\mathcal{Y}_n}$ on $\mathcal{Y}_n$, y_n the generic variable on $\mathcal{Y}_n$ and dy_n the Lebesgue measure on $\mathcal{Y}_n$. By (4.10), we know that

$$d\mu_n(y_n) = (2\pi)^{-\frac{n}{2}} \det A_n^{-\frac{1}{2}} e^{-\frac{1}{2}y_n \cdot A^{-1}y_n} \, dy_n.$$

Hence, for $\epsilon > 0$,

$$\int_{\mathcal{X}} e^{-\frac{\epsilon}{2}\|P_n x\|^2} \, d\mu(x) = \int_{\mathcal{Y}_n} e^{-\frac{\epsilon}{2}\|y_n\|^2} \, d\mu_n(y_n) = \prod_{j=1}^n (1+\epsilon\lambda_j)^{-\frac{1}{2}}.$$

Now

$$1 = \lim_{\epsilon \searrow 0} \lim_{n\to\infty} \int_{\mathcal{X}} e^{-\frac{\epsilon}{2}\|P_n x\|^2} \, d\mu(x) = \lim_{\epsilon \searrow 0} \prod_{j=1}^\infty (1+\epsilon\lambda_j)^{-\frac{1}{2}}.$$

This implies that $\prod_{j=1}^\infty (1+\epsilon\lambda_j) < \infty$ for small enough $\epsilon > 0$, and hence the series $\sum_{j=1}^\infty \lambda_j$ is convergent and A is trace-class. $\qquad\square$

Definition 5.70 *The measure defined in Prop. 5.69 will be called the* Gaussian measure on $\mathcal{X}$ *of mean q and covariance A and will be denoted by*

$$C\delta(x_1 - a_1)e^{-\frac{1}{2}(x_2-q_2)\cdot A_2^{-1}(x_2-q_2)}dx_1 dx_2, \tag{5.22}$$

or, if $\mathrm{Ker}\, A = 0$, *by*

$$Ce^{-\frac{1}{2}(x-q)\cdot A^{-1}(x-q)}dx. \tag{5.23}$$

Note that C in (5.22) and (5.23) has the meaning of the "normalizing constant" that makes (5.22) a probability measure.

Remark 5.71 *Prop. 5.69 provides an example of a weak distribution on $\mathcal{X}$ which is not generated by a probability measure on $\mathcal{X}$.*

5.4.2 Gaussian measures on enlarged spaces

In this subsection we consider the case of a covariance for which (5.21) is only a weak characteristic functional.

Let A be a positive self-adjoint operator on $\mathcal{X}$. Consider the function

$$\mathcal{X} \ni \xi \mapsto e^{-\frac{1}{2}\xi \cdot A\xi}. \tag{5.24}$$

It is a weak characteristic functional. It is not a characteristic functional of a measure unless A is trace-class.

Definition 5.72 *The generalized measure given by the weak characteristic functional (5.24) will be called the* generalized Gaussian measure on $\mathcal{X}$ *with* covariance A. We will denote by

$$\mathbf{L}^2(\mathcal{X}, \mathrm{e}^{-\frac{1}{2}x \cdot A^{-1}x}\mathrm{d}x)$$

the corresponding $\mathbf{L}^2$ *space. We will call it the* Gaussian $\mathbf{L}^2$ *space over* $\mathcal{X}$ *with covariance* A.

If B is a positive self-adjoint operator B on $\mathcal{X}$ such that $B^{-\frac{1}{2}}AB^{-\frac{1}{2}}$ is trace-class, then $\mathbf{L}^2(\mathcal{X}, \mathrm{e}^{\frac{1}{2}xA^{-1}x}\mathrm{d}x)$ is naturally isomorphic to $L^2(B^{\frac{1}{2}}\mathcal{X}, \mathrm{d}\mu_B)$, where

$$\int_{B^{\frac{1}{2}}\mathcal{X}} \mathrm{e}^{\mathrm{i}\xi \cdot x}\mathrm{d}\mu_B(x) = \mathrm{e}^{-\frac{1}{2}\xi \cdot A\xi}, \quad \xi \in B^{-\frac{1}{2}}\mathcal{X}.$$

Note that there is no canonical choice of the operator B.

Definition 5.73 *Following (5.23), the measure* μ_B *will often be denoted*

$$C\mathrm{e}^{-\frac{1}{2}x \cdot A^{-1}x}\mathrm{d}x.$$

(Note that this notation hides the dependence on B, which plays only an auxiliary technical role.)

Consider in particular the case of covariance $\mathbb{1}$. $\mathbf{L}^2(\mathcal{X}, \mathrm{e}^{-\frac{1}{2}x^2}\mathrm{d}x)$ can be realized as an L^2 space over $\mathcal{X}$ iff $\mathcal{X}$ is finite-dimensional. $\mathbf{L}^2(\mathcal{X}, \mathrm{e}^{-\frac{1}{2}x^2}\mathrm{d}x)$ is then equal to $L^2(\mathcal{X}, (2\pi)^{-\frac{1}{2}d}\mathrm{e}^{-\frac{1}{2}x^2}\mathrm{d}x)$, where $d = \dim \mathcal{X}$ and $\mathrm{d}x$ is the Lebesgue measure on $\mathcal{X}$ compatible with the Euclidean structure.

Remark 5.74 *(5.24) is a weak characteristic functional even if the positive operator* A *has a non-zero kernel. If this is the case, then the corresponding Gaussian* $\mathbf{L}^2$ *space can be identified with* $\mathbf{L}^2(\mathcal{X}_1, \mathrm{e}^{-\frac{1}{2}x_1 \cdot A_1^{-1}x_1}\mathrm{d}x_1)$, *where* $\mathcal{X}_1 := (\mathrm{Ker}\, A)^{\perp}$, A_1 *is the restriction of* A *to* $\mathcal{X}_1$ *and* x_1 *is the generic variable of* $\mathcal{X}_1$.

5.4.3 Exponential law for Gaussian spaces

In this subsection, for simplicity we restrict ourselves to covariance $\mathbb{1}$.

Proposition 5.75 *Let* $\mathcal{X}_1$, $\mathcal{X}_2$ *be two real Hilbert spaces. Set* $\mathcal{X} := \mathcal{X}_1 \oplus \mathcal{X}_2$. *Then the map*

$$U : \mathbb{C}\mathrm{Pol}(\mathcal{X}_1) \otimes \mathbb{C}\mathrm{Pol}(\mathcal{X}_2) \to \mathbb{C}\mathrm{Pol}(\mathcal{X})$$
$$P_1(x_1) \otimes P_2(x_2) \mapsto P(x_1)P(x_2)$$

extends to a unitary map

$$U : \mathbf{L}^2(\mathcal{X}_1, \mathrm{e}^{-\frac{1}{2}x_1^2}\mathrm{d}x_1) \otimes \mathbf{L}^2(\mathcal{X}_2, \mathrm{e}^{-\frac{1}{2}x_2^2}\mathrm{d}x_2) \to \mathbf{L}^2(\mathcal{X}, \mathrm{e}^{-\frac{1}{2}x^2}\mathrm{d}x).$$

Proof Let us choose two operators B_1, B_2 such that B_i^{-1} is trace-class on $\mathcal{X}_i$, and use $L^2(B_i^{\frac{1}{2}}\mathcal{X}_i, \mathrm{d}\mu_{B_i})$ as representatives for $\mathbf{L}^2(\mathcal{X}_i, \mathrm{e}^{-\frac{1}{2}x_i^2}\mathrm{d}x_i)$. Then the map U extends to a unitary map from $L^2(B_1^{\frac{1}{2}}\mathcal{X}_1, \mathrm{d}\mu_{B_1}) \otimes L^2(B_2^{\frac{1}{2}}\mathcal{X}_2, \mathrm{d}\mu_{B_2})$ into

$L^2(B^{\frac{1}{2}}\mathcal{X}, \mathrm{d}\mu_B)$ for $B = B_1 \oplus B_2$. We have

$$\int_{B^{\frac{1}{2}}\mathcal{X}} \mathrm{e}^{\mathrm{i}\xi \cdot x} \mathrm{d}\mu_B(x) = \mathrm{e}^{-\frac{1}{2}\xi_1^2 - \frac{1}{2}\xi_2^2} = \mathrm{e}^{-\frac{1}{2}\xi^2},$$

which shows that $L^2(B^{\frac{1}{2}}\mathcal{X}, \mathrm{d}\mu_B)$ is a representative of $\mathbf{L}^2(\mathcal{X}, \mathrm{e}^{-\frac{1}{2}x^2}\mathrm{d}x)$. $\qquad\square$

5.4.4 Polynomials in Gaussian spaces

Let A, B be positive operators with $B^{-\frac{1}{2}}AB^{-\frac{1}{2}}$ trace-class. We identify $\mathbf{L}^2(\mathcal{X}, \mathrm{e}^{-\frac{1}{2}x \cdot A^{-1}x}\mathrm{d}x)$ with $L^2(B^{\frac{1}{2}}\mathcal{X}, \mathrm{d}\mu_B)$.

Proposition 5.76 *Polynomials based on $B^{-\frac{1}{2}}\mathcal{X}$ are dense in $\mathbf{L}^2(\mathcal{X}, \mathrm{e}^{-\frac{1}{2}x^2}\mathrm{d}x)$.*

Proof Clearly, for $\xi \in B^{-\frac{1}{2}}\mathcal{X}$, the function

$$\mathbb{C} \ni t \mapsto \hat{\mu}_B(t\xi) = \int_{B^{\frac{1}{2}}\mathcal{X}} \mathrm{e}^{-\mathrm{i}t\xi \cdot x} \mathrm{d}\mu_B(x) = \mathrm{e}^{-\frac{t^2}{2}\xi \cdot A\xi}$$

is entire. Hence the statement follows from Thm. 5.56. $\qquad\square$

Clearly, we have the inclusion $B^{-\frac{1}{2}}\mathcal{X} \subset A^{-\frac{1}{2}}\mathcal{X}$. If we regard $B^{\frac{1}{2}}\mathcal{X}$ as the underlying space, then only polynomials based on $B^{-\frac{1}{2}}\mathcal{X}$ are continuous functions. Those based on $A^{-\frac{1}{2}}\mathcal{X}$ do not have to be continuous. However, they are L^p integrable, as the following proposition shows.

Proposition 5.77 *Polynomials based on $A^{-\frac{1}{2}}\mathcal{X}$ belong to $\bigcap_{1 \leq p < \infty} L^p(B^{\frac{1}{2}}\mathcal{X}, \mathrm{d}\mu_B)$ and, for $\xi \in A^{-\frac{1}{2}}\mathcal{X}$, we have*

$$\int_{B^{\frac{1}{2}}\mathcal{X}} (\xi \cdot x)^{2n+1} \mathrm{d}\mu_B(x) = 0,$$
$$\int_{B^{\frac{1}{2}}\mathcal{X}} (\xi \cdot x)^{2n} \mathrm{d}\mu_B(x) = \frac{2n!}{2^n n!}(\xi \cdot A\xi)^n. \tag{5.25}$$

Proof Using Prop. 5.50, we obtain (5.25) for $\xi \in B^{-\frac{1}{2}}\mathcal{X}$.

Using (5.25), we see that if $(\xi_n)_{n \in \mathbb{N}}$ is a sequence in $B^{-\frac{1}{2}}\mathcal{X}$ converging to some $\xi \in \mathcal{X}$, then the sequence of functions $(\xi_n \cdot x)^m$ is Cauchy in $\bigcap_{1 \leq p < \infty} L^p(B^{\frac{1}{2}}\mathcal{X}, \mathrm{d}\mu_B)$. Hence we can define the function

$$(\xi \cdot x)^m := \lim_{n \to \infty} (\xi_n \cdot x)^m,$$

which belongs to $\bigcap_{1 \leq p < \infty} L^p(B^{\frac{1}{2}}\mathcal{X}, \mathrm{d}\mu_B)$. $\qquad\square$

5.4.5 Relative continuity of Gaussian measures

Let A_i, $i = 1, 2$, be two bounded positive operators on $\mathcal{X}$. For simplicity we assume that $A_i > 0$, i.e. $\operatorname{Ker} A_i = \{0\}$. Let B^{-1} be trace-class. Consider the Gaussian measures μ_i with the covariances A_i, $i = 1, 2$, on the space $B^{\frac{1}{2}}\mathcal{X}$.

Theorem 5.78 (Feldmann–Hajek theorem) *The measures μ_1 and μ_2 are absolutely continuous w.r.t. one another iff $A_1^{-\frac{1}{2}} A_2 A_1^{-\frac{1}{2}} - \mathbb{1} \in B^2(\mathcal{X})$.*

Let us now discuss the Radon–Nikodym derivative $\frac{\mathrm{d}\mu_2}{\mathrm{d}\mu_1}(x)$ under the hypotheses of Thm. 5.78. For simplicity we assume that $A_1 = \mathbb{1}$ and denote A_2 by A, μ_1 by μ and μ_2 by $\tilde{\mu}$. It is easy to obtain the corresponding statements in the general case by replacing $\mathcal{X}$ by $A_1^{-\frac{1}{2}}\mathcal{X}$ (see Subsect. 11.4.6).

Proposition 5.79 *Assume that $\mathbb{1} - A \in B^2(\mathcal{X})$. Then the following hold:*

(1) *Let $\{\pi_n\}_{n \in \mathbb{N}}$ be an increasing sequence of finite rank orthogonal projections in $\mathcal{X}$ with $\mathrm{s} - \lim \pi_n = \mathbb{1}$. Set*

$$F_n(x) := (\det \pi_n A \pi_n)^{-\frac{1}{2}} \mathrm{e}^{\frac{1}{2} x \cdot \pi_n (\mathbb{1} - A^{-1}) \pi_n x}, \ n \in \mathbb{N}.$$

Then $\{F_n\}_{n \in \mathbb{N}}$ converges in $L^1(B^{\frac{1}{2}}\mathcal{X}, \mathrm{d}\mu)$ to a positive function F with $\int F \mathrm{d}\mu = 1$.

(2) *If $\mathbb{1} - A \in B^1(\mathcal{X})$, then*

$$F(x) = (\det A)^{-\frac{1}{2}} \mathrm{e}^{\frac{1}{2} x \cdot (\mathbb{1} - A^{-1}) x}.$$

(3) *One has $\frac{\mathrm{d}\tilde{\mu}}{\mathrm{d}\mu}(x) = F(x)$.*

Remark 5.80 *Statement (3) of Prop. 5.79 shows that F is independent on the choice of $\{\pi_n\}$. Note also that $x \mapsto x \cdot (\mathbb{1} - A^{-1}) x$ is continuous on $B^{\frac{1}{2}}\mathcal{X}$, hence $x \mapsto \mathrm{e}^{\frac{1}{2} x \cdot (\mathbb{1} - A^{-1}) x}$ is measurable on $B^{\frac{1}{2}}\mathcal{X}$, although not integrable if $\mathbb{1} - A \notin B^1(\mathcal{X})$. Therefore, a convenient notation for F is*

$$F(x) = C \mathrm{e}^{\frac{1}{2} x \cdot (\mathbb{1} - A^{-1}) x},$$

where C is the "normalizing constant", as in Def. 5.70.

The proof of this theorem will be given later on; see Subsect. 11.4.6.

5.5 Gaussian measures on complex Hilbert spaces

Let $\mathcal{Z}$ be a separable (complex) Hilbert space. We denote by $\overline{z}_1 \cdot z_2$ the scalar product of $z_1, z_2 \in \mathcal{Z}$.

We will discuss *Gaussian L^2 spaces of anti-holomorphic functions on $\mathcal{Z}$*. This section has a natural continuation in Sect. 9.2, where we discuss the *complex-wave representation of CCR*.

5.5.1 Holomorphic and anti-holomorphic functions

Recall from Subsect. 3.5.6 that inside the space of all complex polynomials $\mathbb{C}\mathrm{Pol}(\mathcal{Z}_{\mathbb{R}})$ we have the subspace $\mathrm{Pol}(\mathcal{Z})$, resp. $\mathrm{Pol}(\overline{\mathcal{Z}})$ of holomorphic, resp. anti-holomorphic polynomials spanned by $\prod_{i=1}^{p} \overline{w}_i \cdot z$, resp. $\prod_{i=1}^{p} w_i \cdot \overline{z}$, for $w_i \in \mathcal{Z}$.

The following definition generalizes the notion of a holomorphic function to an arbitrary dimension.

Definition 5.81 *A function $F : \mathcal{Z} \to \mathbb{C}$ is* holomorphic, *resp.* anti-holomorphic *if its restriction to any finite-dimensional complex subspace of $\mathcal{Z}$ is holomorphic, resp. anti-holomorphic.*

5.5.2 Measures on complex Hilbert spaces

Recall from Subsect. 3.6.9 that, in the context of the integration, a complex space $\mathcal{Z}$ is often identified with $\mathrm{Re}(\mathcal{Z} \oplus \overline{\mathcal{Z}})$ by the map

$$\mathcal{Z} \ni z \mapsto (z, \overline{z}) \in \mathrm{Re}(\mathcal{Z} \oplus \overline{\mathcal{Z}}). \tag{5.26}$$

This suggests adoption of the following convention for characteristic functionals on complex spaces:

Definition 5.82 *If μ is a Borel probability measure on $\mathcal{Z}$, its characteristic functional is defined by*

$$\mathcal{Z} \ni w \mapsto \hat{\mu}(w) := \int_{\mathcal{Z}} \mathrm{e}^{-2\mathrm{i}\mathrm{Re}\overline{w} \cdot z} \mathrm{d}\mu(z) = \int_{\mathcal{Z}} \mathrm{e}^{-\mathrm{i}\overline{w} \cdot z - \mathrm{i}w \cdot \overline{z}} \mathrm{d}\mu(z).$$

5.5.3 Gaussian measures on complex spaces

Now let $A > 0$ be a trace-class self-adjoint operator on $\mathcal{Z}$. There exists a unique measure μ on $\mathcal{Z}$ such that

$$\hat{\mu}(w) = \mathrm{e}^{-\overline{w} \cdot Aw}, \quad w \in \mathcal{Z}. \tag{5.27}$$

This follows from Prop. 5.69, if we consider $\mathcal{Z}$ as the real Hilbert space $\mathcal{Z}_{\mathbb{R}}$ equipped with the scalar product $\mathrm{Re}\,\overline{z}_1 \cdot z_2$.

Definition 5.83 *The measure μ defined by (5.27) will be denoted $C\mathrm{e}^{-\overline{z} \cdot A^{-1} z} \mathrm{d}\overline{z}\mathrm{d}z$ and called the* Gaussian measure of covariance A.

Let $\mathcal{Z}$ be finite-dimensional of complex dimension d with a fixed (complex) volume form $\mathrm{d}z$. By Subsect. 4.1.9, we then have

$$C\mathrm{e}^{-\overline{z} \cdot A^{-1} z} \mathrm{d}\overline{z}\mathrm{d}z = \det A^{-1} (2\pi\mathrm{i})^{-d} \mathrm{e}^{-\overline{z} \cdot A^{-1} z} \mathrm{d}\overline{z}\mathrm{d}z. \tag{5.28}$$

(The notation $\mathrm{i}^{-d}\mathrm{d}\overline{z}\mathrm{d}z$ is explained in Subsect. 3.6.9.)

Definition 5.84 *We denote by* $L^2_{\mathbb{C}}(\mathcal{Z}, Ce^{-\bar{z}\cdot A^{-1}z}\mathrm{d}\bar{z}\mathrm{d}z)$, *resp.* $L^2_{\mathbb{C}}(\overline{\mathcal{Z}}, Ce^{-\bar{z}\cdot A^{-1}z}\,\mathrm{d}\bar{z}\mathrm{d}z)$ *the closure in* $L^2(\mathcal{Z}_{\mathbb{R}}, Ce^{-\bar{z}\cdot A^{-1}z}\mathrm{d}\bar{z}\mathrm{d}z)$ *of* $\mathrm{Pol}(\mathcal{Z})$, *resp.* $\mathrm{Pol}(\overline{\mathcal{Z}})$.

Theorem 5.85 *The space* $L^2_{\mathbb{C}}(\mathcal{Z}, Ce^{-\bar{z}\cdot A^{-1}z}\mathrm{d}\bar{z}\mathrm{d}z)$, *resp.* $L^2_{\mathbb{C}}(\overline{\mathcal{Z}}, Ce^{-\bar{z}\cdot A^{-1}z}\mathrm{d}\bar{z}\mathrm{d}z)$ *coincides with the space of holomorphic, resp. anti-holomorphic functions in* $L^2(\mathcal{Z}_{\mathbb{R}}, Ce^{-\bar{z}\cdot A^{-1}z}\mathrm{d}\bar{z}\mathrm{d}z)$.

Proof It suffices to consider the holomorphic case.

Let $\mathcal{Y} \subset \mathcal{Z}$ be a finite-dimensional complex subspace. If G is a function on $\mathcal{Z}$, let $G_{|\mathcal{Y}}$ be its restriction to $\mathcal{Y}$. Let $F \in L^2_{\mathbb{C}}(\mathcal{Z}, Ce^{-\bar{z}\cdot A^{-1}z}\mathrm{d}\bar{z}\mathrm{d}z)$, and (P_n) a sequence in $\mathrm{Pol}(\mathcal{Z})$ converging to F in $L^2(\mathcal{Z}_{\mathbb{R}}, Ce^{-\bar{z}\cdot A^{-1}z}\mathrm{d}\bar{z}\mathrm{d}z)$. If $\mathcal{Y}$ is finite-dimensional then $(P_n)_{|\mathcal{Y}}$ converges to $F_{|\mathcal{Y}}$ in $L^2(\mathcal{Y}_{\mathbb{R}}, Ce^{-\bar{z}\cdot A^{-1}z}\mathrm{d}\bar{z}\mathrm{d}z)$, hence in $\mathcal{D}'(\mathcal{Y}_{\mathbb{R}})$. By Prop. 4.12 it follows that $F_{|\mathcal{Y}}$ is holomorphic.

Conversely, let $F \in L^2(\mathcal{Z}_{\mathbb{R}}, Ce^{-\bar{z}\cdot A^{-1}z}\mathrm{d}\bar{z}\mathrm{d}z)$ be a holomorphic function, and assume that F is orthogonal to all holomorphic polynomials. Let $(e_j)_{j\in\mathbb{N}}$ be an o.n. basis of eigenvectors of A for the eigenvalues $(\lambda_j)_{j\in\mathbb{N}}$. We fix d and restrict F to $\mathrm{Span}\{e_1, \ldots, e_d\}$. If we identify $\mathbb{C}^d$ with $\mathrm{Span}\{e_1, \ldots, e_d\}$ by the map

$$(z_1, \ldots, z_d) \mapsto \sum_{i=1}^{d} \frac{z_j}{\sqrt{\lambda_j}} e_j,$$

we are reduced to considering a holomorphic function G on $\mathbb{C}^d$, which is orthogonal to all holomorphic polynomials for the measure $(2\pi i)^{-d}e^{-\bar{z}\cdot z}\mathrm{d}\bar{z}\mathrm{d}z$.

For $\vec{n} = (n_1, \ldots, n_d) \in \mathbb{N}^d$ we recall that $\vec{n}! := n_1! \ldots n_d!$, $\partial_z^n = \partial_{z_1}^{n_1} \ldots \partial_{z_d}^{n_d}$. From Cauchy's formula, we get

$$\partial_z^n G(0) = \frac{\vec{n}!}{(2\pi)^d} \int_{[0,2\pi]^d} G(r_1 e^{i\theta_1}, \ldots, r_d e^{i\theta_d}) \prod_{j=1}^{d} e^{in_j\theta_j} r_j^{-n_j}\,\mathrm{d}\theta_1 \ldots \mathrm{d}\theta_d.$$

If $C(n) = \prod_{j=1}^{d} \int_0^{+\infty} r^{2n_j+1}e^{-r^2}\,\mathrm{d}r$, we obtain

$$C(n)\partial_z^n G(0) = \vec{n}!2^{-d} \int G(z_1, \ldots, z_d) \prod_{j=1}^{d} \bar{z}_j^{n_j} e^{-\bar{z}\cdot z}(2i\pi)^{-d}\mathrm{d}\bar{z}\mathrm{d}z.$$

Hence, if $G \in L^2(\mathbb{C}^d, (2i\pi)^{-d}e^{-\bar{z}\cdot z}\mathrm{d}\bar{z}\mathrm{d}z)$ is holomorphic and orthogonal to the holomorphic polynomials, we have $\partial_z^{\vec{n}} G(0) = 0$ for all $\vec{n}$ and hence $G(z) \equiv 0$.

This implies that the restriction of F to $\mathrm{Span}\{e_1, \ldots, e_d\}$ is equal to 0 for all d. In particular, F is orthogonal to all real polynomials generated by $\mathrm{Re}(\bar{e}_j \cdot z)$ and $\mathrm{Im}(\bar{e}_j \cdot z)$. Since these polynomials are dense in $L^2(\mathcal{Z}_{\mathbb{R}}, e^{-\bar{z}\cdot A^{-1}z}\mathrm{d}\bar{z}\mathrm{d}z)$, we have $F \equiv 0$. $\qquad\square$

5.5.4 *Generalized Gaussian measures on complex spaces*

We now extend Def. 5.84 to generalized Gaussian measures that cannot be realized as measures on $\mathcal{Z}$. For simplicity, we assume that the covariance of the measure is given by the scalar product of the underlying (complex) Hilbert space.

Definition 5.86 *Denote by* $\mathbf{L}^2_{\mathbb{C}}(\mathcal{Z}, e^{-\overline{z}\cdot z}\,d\overline{z}dz)$, *resp.* $\mathbf{L}^2_{\mathbb{C}}(\overline{\mathcal{Z}}, e^{-\overline{z}\cdot z}\,d\overline{z}dz)$ *the closure in* $\mathbf{L}^2(\mathcal{Z}_{\mathbb{R}}, e^{-\overline{z}\cdot z}\,d\overline{z}dz)$ *of the space of holomorphic, resp. anti-holomorphic polynomials on* $\mathcal{Z}$. *The space* $\mathbf{L}^2_{\mathbb{C}}(\mathcal{Z}, e^{-\overline{z}\cdot z}\,d\overline{z}dz)$, *resp.* $\mathbf{L}^2_{\mathbb{C}}(\overline{\mathcal{Z}}, e^{-\overline{z}\cdot z}\,d\overline{z}dz)$ *will be called the* holomorphic, *resp.* anti-holomorphic Gaussian $\mathbf{L}^2$ *space with covariance* $\mathbb{1}$.

Proposition 5.87 *Let* $B \geq 0$ *be an operator such that* B^{-1} *is trace-class. Identify* $\mathbf{L}^2(\mathcal{Z}_{\mathbb{R}}, e^{-\overline{z}\cdot z}\,d\overline{z}dz)$ *with* $L^2(B^{\frac{1}{2}}\mathcal{Z}_{\mathbb{R}}, Ce^{-\overline{z}\cdot z}\,d\overline{z}dz)$ *in the usual way. Then* $\mathbf{L}^2_{\mathbb{C}}(\mathcal{Z}, e^{-\overline{z}\cdot z}\,d\overline{z}dz)$, *resp.* $\mathbf{L}^2_{\mathbb{C}}(\overline{\mathcal{Z}}, e^{-\overline{z}\cdot z}\,d\overline{z}dz)$ *coincide with* $L^2_{\mathbb{C}}(B^{\frac{1}{2}}\mathcal{Z}, Ce^{-\overline{z}\cdot z}\,d\overline{z}dz)$, *resp.* $L^2_{\mathbb{C}}(\overline{B^{\frac{1}{2}}\mathcal{Z}}, Ce^{-\overline{z}\cdot z}\,d\overline{z}dz)$.

5.5.5 Isomorphism with modified Fock spaces

Recall the modified Fock space $\Gamma^{\mathrm{mod}}_{\mathrm{s}}(\mathcal{Z})$, defined as the completion of $\overset{\mathrm{al}}{\Gamma}_{\mathrm{s}}(\mathcal{Z})$ with the scalar product given by $(\Phi|\Psi)_{\Gamma^{\mathrm{mod}}_{\mathrm{s}}(\mathcal{Z})} := \left(\Phi|\frac{1}{N!}\Psi\right)_{\Gamma_{\mathrm{s}}(\mathcal{Z})}$. Moreover, we recall from Subsect. 3.5.1 that $\overset{\mathrm{al}}{\Gamma}_{\mathrm{s}}(\mathcal{Z})$ can be identified with $\mathrm{Pol}_{\mathrm{s}}(\overline{\mathcal{Z}})$, which is dense in $\mathbf{L}^2_{\mathbb{C}}(\overline{\mathcal{Z}}, e^{-\overline{z}\cdot z}\,d\overline{z}dz)$. It turns out that this identification extends to a unitary map:

Theorem 5.88 *The map*

$$\overset{\mathrm{al}}{\Gamma}_{\mathrm{s}}(\mathcal{Z}) \ni \Phi \mapsto \Phi(\cdot) \in \mathrm{Pol}_{\mathrm{s}}(\overline{\mathcal{Z}})$$

given by

$$\Phi(\overline{z}) := \sum_{n=0}^{\infty}(z^{\otimes n}|\Phi)$$

extends by continuity to a unitary map

$$\Gamma^{\mathrm{mod}}_{\mathrm{s}}(\mathcal{Z}) \ni \Phi \mapsto \Phi(\cdot) \in \mathbf{L}^2_{\mathbb{C}}(\overline{\mathcal{Z}}, e^{-\overline{z}\cdot z}\,d\overline{z}dz). \tag{5.29}$$

The proof of the above theorem for $\dim \mathcal{Z} = 1$ follows immediately from the following simple computation:

Lemma 5.89 *Let* $z \in \mathbb{C}$. *Then*

$$(2\pi\mathrm{i})^{-1}\int_{\mathbb{C}} e^{-\overline{z}\cdot z}z^m\overline{z}^n\,dzd\overline{z} = n!\delta_{n,m}. \tag{5.30}$$

Proof We identify $\mathbb{C}$ with $\mathbb{R}^2$. In the polar coordinates $z = re^{\mathrm{i}\phi}$, the l.h.s. of (5.30) equals

$$\pi\int_0^{2\pi}d\phi\int_0^{\infty}dre^{\mathrm{i}\phi(m-n)}r^{m+n+1}e^{-r^2}. \tag{5.31}$$

For $n \neq m$ the integral w.r.t. ϕ yields zero. For $n = m$ we get

$$\frac{1}{2} \int_0^\infty r^{2m+1} e^{-r^2} \, \mathrm{d}r = \int_0^\infty r^{2m} e^{-r^2} \, \mathrm{d}r^2 = m!.$$

Alternatively, we can rewrite (5.30) as

$$\mathrm{i}^{n+m} \partial_t^n \partial_{\bar t}^m (2\pi \mathrm{i})^{-1} \int e^{-\bar z \cdot z} e^{-\mathrm{i}z\bar t - \mathrm{i}\bar z t} \mathrm{d}z\mathrm{d}\bar z \Big|_{t=0} = \mathrm{i}^{n+m} \partial_t^n \partial_{\bar t}^m e^{-|t|^2} \Big|_{t=0}$$

$$= n! \delta_{nm}. \qquad \square$$

Proof of Thm. 5.88. For notational simplicity assume that $\dim \mathcal{Z} < \infty$. Let $(e_1, \ldots, e_n)$ be an o.n. basis of $\mathcal{Z}$. Recall that $\{e_{\vec k} \ : \ \vec k \in \mathbb{N}^n\}$ is an o.n. basis of $\Gamma_{\mathrm s}^{\mathrm{mod}}(\mathcal{Z})$, where

$$e_{\vec k} := \frac{1}{\sqrt{\vec k!}} e_1^{\otimes k_1} \otimes_{\mathrm s} \cdots \otimes_{\mathrm s} e_n^{\otimes k_n},$$

$e_{\vec 0} = \Omega$ and $\vec k! = k_1! \cdots k_n!$. The vector $e_{\vec k}$ is mapped onto the polynomial

$$e_{\vec k}(\bar z) = \frac{1}{\sqrt{\vec k!}} \prod_{i=1}^n (e_i \cdot \bar z)^{k_i}.$$

Using Lemma 5.89 we see that $\{e_{\vec k}(\cdot) \ : \ \vec k \in \mathbb{N}^n\}$ form an o.n. basis of $L_{\mathbb{C}}^2(\bar{\mathcal{Z}}, Ce^{-\bar z \cdot z} \mathrm{d}\bar z \mathrm{d}z)$. $\qquad \square$

The following proposition is an illustration of the formalism of Gaussian complex spaces.

Proposition 5.90 *Let* $F \in \mathbf{L}_{\mathbb{C}}^2(\bar{\mathcal{Z}}, e^{-\bar z \cdot z} \mathrm{d}\bar z \mathrm{d}z)$. *Then*

$$F(\bar z_0) = \int F(\bar z) e^{\bar z \cdot z_0} C e^{-\bar z \cdot z} \mathrm{d}\bar z \mathrm{d}z, \quad z_0 \in \mathcal{Z}.$$

Proof The integral on the r.h.s. is well defined, since $\bar z \mapsto e^{\bar z \cdot z_0}$ belongs to $\mathbf{L}_{\mathbb{C}}^2(\bar{\mathcal{Z}}, e^{-\bar z \cdot z} \mathrm{d}\bar z \mathrm{d}z)$. By density and linearity it suffices to check the identity for monomials. We have

$$\int_{\mathcal{Z}} \prod_{i=1}^p (e_i \cdot \bar z)^{n_i} e^{\bar z \cdot z_0} C e^{-\bar z \cdot z} \mathrm{d}\bar z \mathrm{d}z = \int_{\mathcal{Z}} \prod_{i=1}^p \partial_{t_i}^{n_i} \exp\left(z \cdot z_0 + \sum_{i=1}^p t_i e_i \cdot \bar z\right)$$

$$\times C e^{-\bar z \cdot z} \mathrm{d}\bar z \mathrm{d}z \Big|_{t=0}$$

$$= \prod_{i=1}^p \partial_{t_i}^{n_i} \exp\left(\sum_{i=1}^p t_i e_i \cdot \bar z_0\right)\Big|_{t=0}$$

$$= \prod_{i=1}^p (e_i \cdot \bar z_0)^{n_i}.$$

This completes the proof of the proposition. $\qquad \square$

5.6 Notes

General measure theory is studied e.g. in the monographs by Halmos (1950) and Bauer (1968).

Properties of positivity preserving maps are discussed e.g. in Reed–Simon (1978b).

The notion of equi-integrability and the Lebesgue–Vitali theorem can be found in Kallenberg (1997). Measures on Hilbert spaces is the subject of a monograph by Skorokhod (1974). The proof of Prop. 5.41 can be found e.g. in Chap. I.1 of Skorokhod (1974).

The Feldman–Hajek theorem about relative continuity of Gaussian measures was proved independently by Feldman (1958) and Hajek (1958).

6

Algebras

In this chapter we recall basic definitions related to algebras, especially C^*- and W^*-*algebras.*

Operator algebras are often used in mathematical formulations of quantum theory to describe observables of quantum systems. This is especially useful if we consider infinitely extended systems. They are also convenient to express the Einstein causality properties of relativistic quantum fields.

It is also common to express canonical commutation and anti-commutation relations in terms of algebras. This is especially natural in the case of the CAR. In fact, we will use algebras to treat the CAR in a representation-independent way in Chap. 14. Algebras are less useful in the case of the CCR. We will discuss various choices of CCR algebras in Sect. 8.3.

The theory of W^*-algebras, including elements of the *modular theory*, will be especially needed in Chap. 17, devoted to quasi-free states.

6.1 Algebras

6.1.1 Associative algebras

Let $\mathfrak{A}$ be a vector space over $\mathbb{K} = \mathbb{C}$ or $\mathbb{R}$.

Definition 6.1 $\mathfrak{A}$ *is called an* algebra over $\mathbb{K}$ *if it is equipped with a multiplication satisfying*

$$A(B + C) = AB + AC, \quad (B + C)A = BA + CA,$$
$$(\alpha\beta)(AB) = (\alpha A)(\beta B), \quad \alpha, \beta \in \mathbb{K}, \quad A, B, C \in \mathfrak{A}.$$

If in addition

$$A(BC) = (AB)C, \quad A, B, C \in \mathfrak{A},$$

then we say that it is an associative algebra.

Unless indicated otherwise, by an *algebra* we will mean an *associative algebra*.

Definition 6.2 *A subspace* $\mathfrak{I}$ *of an algebra* $\mathfrak{A}$ *is called a* (two-sided) ideal *of* $\mathfrak{A}$ *if $A \in \mathfrak{A}$ and $B \in \mathfrak{I}$ implies $AB, BA \in \mathfrak{I}$.*

If $\mathfrak{I}$ is an ideal of $\mathfrak{A}$, then $\mathfrak{A}/\mathfrak{I}$ is naturally an algebra.

Definition 6.3 *An algebra* $\mathfrak{A}$ *is called* simple *if* $\mathfrak{A}$ *has no ideals except for* $\{0\}$ *and itself, and* $\mathfrak{A} \neq \mathbb{K}$ *with the multiplication given by* $AB = 0$ *for all* $A, B \in \mathfrak{A}$.

For every subset $\mathfrak{T}$ of an algebra $\mathfrak{A}$ there exists the smallest ideal containing $\mathfrak{T}$.

Definition 6.4 *This ideal is called the* ideal generated by $\mathfrak{T}$ *and is denoted by* $\mathfrak{J}(\mathfrak{T})$.

Definition 6.5 *If* $\mathfrak{A}$, $\mathfrak{B}$ *are algebras, then a linear map* $\pi : \mathfrak{A} \to \mathfrak{B}$ *satisfying* $\pi(A_1 A_2) = \pi(A_1)\pi(A_2)$ *is called a* homomorphism. *It is called an* anti-homomorphism *if* $\pi(A_1 A_2) = \pi(A_2)\pi(A_1)$. *(In the well-known way, we also define* isomorphisms, automorphisms *etc.)*

6.1.2 *-algebras

Definition 6.6 *We say that an algebra* $\mathfrak{A}$ *is a* $*$-algebra *if it is equipped with an anti-linear involution* $\mathfrak{A} \ni A \mapsto A^* \in \mathfrak{A}$ *such that* $(AB)^* = B^* A^*$.

Let $\mathfrak{A}$ be a $*$-algebra. If $\mathfrak{J}$ is a $*$-invariant ideal of $\mathfrak{A}$, then $\mathfrak{A}/\mathfrak{J}$ is naturally a $*$-algebra.

Definition 6.7 *If* $\mathfrak{A}$, $\mathfrak{B}$ *are* $*$-algebras, then a homomorphism $\pi : \mathfrak{A} \to \mathfrak{B}$ *satisfying* $\pi(A^*) = \pi(A)^*$ *is called a* $*$-homomorphism. *(We also define* $*$-isomorphisms, $*$-automorphisms *etc.)* $\mathrm{Aut}(\mathfrak{A})$ *will denote the group of* $*$-automorphisms *of* $\mathfrak{A}$.

6.1.3 Algebras generated by symbols and relations

Suppose that $\mathcal{A}$ is a set.

Recall that $c_c(\mathcal{A}, \mathbb{K})$ denotes the vector space over $\mathbb{K}$ consisting of finite linear combinations of elements indexed by the set $\mathcal{A}$. We adopt the convention that the element of $c_c(\mathcal{A}, \mathbb{K})$ corresponding to $A \in \mathcal{A}$ is denoted simply by A. Recall also that $\overset{\mathrm{al}}{\otimes}\mathcal{Y}$ denotes the algebraic tensor algebra over the vector space $\mathcal{Y}$.

Definition 6.8 (1) *The* unital universal algebra over $\mathbb{K}$ with generators $\mathcal{A}$ *is defined as*

$$\mathfrak{A}(\mathcal{A}, \mathbb{1}) := \overset{\mathrm{al}}{\otimes} c_c(\mathcal{A}, \mathbb{K}),$$

where we write $A_1 A_2 \cdots A_n$ *instead of* $A_1 \otimes A_2 \otimes \cdots \otimes A_n$, $A_1, \ldots, A_n \in \mathcal{A}$ *and the unit element is denoted by* $\mathbb{1}$.

(2) *The* universal unital $*$-algebra with generators $\mathcal{A}$ *is the* $*$-algebra $\mathfrak{A}(\mathcal{A} \sqcup \mathcal{A}^*, \mathbb{1})$ *equipped with the involution* $*$ *such that* $(A_1 A_2 \cdots A_n)^* = A_n^* \cdots A_2^* A_1^*$, $\mathbb{1} = \mathbb{1}^*$.

Definition 6.9 (1) *Let* $\mathfrak{R} \subset \mathfrak{A}(\mathcal{A}, \mathbb{1})$. *The* unital algebra with generators $\mathcal{A}$ and relations $R = 0$, $R \in \mathfrak{R}$, *is defined as* $\mathfrak{A}(\mathcal{A}, \mathbb{1})/\mathfrak{I}(\mathfrak{R})$.

(2) *Let* $\mathfrak{R} \subset \mathfrak{A}(\mathcal{A} \cup \mathcal{A}^*, \mathbb{1})$ *be* $*$-*invariant. The* unital $*$-algebra with generators $\mathcal{A}$ *and relations* $R = 0$, $R \in \mathfrak{R}$, *is defined as* $\mathfrak{A}(\mathcal{A} \cup \mathcal{A}^*, \mathbb{1})/\mathfrak{I}(\mathfrak{R})$.

6.1.4 Super-algebras

Recall from Subsect. 1.1.15 that $(\mathcal{Y}, \epsilon)$ is a *super-space* if $\mathcal{Y}$ is a vector space and $\epsilon \in L(\mathcal{Y})$ satisfies $\epsilon^2 = \mathbb{1}$. We then have a decomposition $\mathcal{Y} = \mathcal{Y}_0 \oplus \mathcal{Y}_1$ into its even and odd subspace.

Definition 6.10 $(\mathfrak{A}, \alpha)$ *is called a* super-algebra *if* $\mathfrak{A}$ *is an algebra and* α *is an involutive automorphism of* $\mathfrak{A}$.

We then have a decomposition $\mathfrak{A} = \mathfrak{A}_0 \oplus \mathfrak{A}_1$ into even and odd subspace. Clearly, for pure elements $A, B \in \mathfrak{A}$ of parity $|A|$, resp. $|B|$, the parity of AB is $|A| + |B|$.

Note that $\mathfrak{A}_0$ is a sub-algebra of $\mathfrak{A}$.

Definition 6.11 *We say that a super-algebra* $\mathfrak{A}$ *is* super-commutative *iff* $AB = (-1)^{|A||B|} AB$.

Below we give two typical examples of associative super-algebras:

Example 6.12 (1) *Let* $(\mathcal{Y}, \epsilon)$ *be a super-space. Then* $L(\mathcal{Y})$ *equipped with the involution*

$$\alpha(A) = \epsilon A \epsilon \tag{6.1}$$

is a super-algebra. It will be denoted $gl(\mathcal{Y}, \epsilon)$.

(2) $\overset{\text{al}}{\Gamma}_\epsilon(\mathcal{Y})$ *equipped with* $\otimes_\epsilon$ *is a super-commutative super-algebra (see Subsect. 3.3.9).*

6.2 C^*- and W^*-algebras

In this section we recall basic terminology from the theory of C^*- and W^*-algebras.

6.2.1 Banach algebras

Definition 6.13 *An algebra* $\mathfrak{A}$ *is called a* normed algebra *if it is equipped with a norm* $\|\cdot\|$ *satisfying*

$$\|AB\| \leq \|A\| \|B\|, \quad A, B \in \mathfrak{A}.$$

It is called a Banach algebra *if it is complete in this norm.*

6.2.2 C^*-algebras

Definition 6.14 *We say that $\mathfrak{A}$ is a C^*-algebra if it is a complex Banach $*$-algebra satisfying*

$$\|A^*\| = \|A\|, \quad \|A^*A\| = \|A\|^2, \quad A \in \mathfrak{A}. \tag{6.2}$$

Definition 6.15 *Let $\mathfrak{A}$ be a complex normed $*$-algebra (not necessarily complete). We say that its norm is a C^*-norm if it satisfies (6.2).*

Clearly, the completion of an algebra equipped with a C^*-norm is a C^*-algebra.

If $\mathcal{H}$ is a Hilbert space, then $B(\mathcal{H})$ equipped with the Hermitian conjugation and the operator norm is a C^*-algebra.

Definition 6.16 *A norm closed $*$-sub-algebra of $B(\mathcal{H})$ is called a* concrete C^*-algebra.

Clearly, every concrete C^*-algebra is a C^*-algebra. Conversely, every C^*-algebra is $*$-isomorphic to a concrete C^*-algebra.

Any $*$-homomorphism, resp. $*$-isomorphism between two C^*-algebras is a contraction, resp. isometry.

Definition 6.17 *We define the set of* positive *elements of $\mathfrak{A}$ as the set of self-adjoint elements with spectrum in $[0, \infty[$, or equivalently, of elements of the form A^*A. The set of positive elements of $\mathfrak{A}$ is denoted $\mathfrak{A}_+$.*

Definition 6.18 *Let $\mathfrak{A}$ be a C^*-algebra. A* C^*-dynamics *on $\mathfrak{A}$ is a one-parameter group $\mathbb{R} \ni t \mapsto \tau^t \in \mathrm{Aut}(\mathfrak{A})$ such that for each $A \in \mathfrak{A}$ the map $t \mapsto \tau^t(A)$ is continuous. Such a pair $(\mathfrak{A}, \tau)$ is called a* C^*-dynamical system.

6.2.3 Representations of C^*-algebras

Let $\mathcal{H}$ be a Hilbert space and $\mathfrak{A} \subset B(\mathcal{H})$.

Definition 6.19 *The* commutant *of $\mathfrak{A}$ is defined as*

$$\mathfrak{A}' := \{B \in B(\mathcal{H}) : AB = BA, \quad A \in \mathfrak{A}\}.$$

Let $\mathfrak{A} \subset B(\mathcal{H})$ be a $*$-algebra.

Definition 6.20 *$\mathfrak{A}$ is called* irreducible *if the only closed subspaces of $\mathcal{H}$ invariant under $\mathfrak{A}$ are $\{0\}$ and $\mathcal{H}$, or equivalently if $\mathfrak{A}' = \mathbb{C}\mathbb{1}$. $\mathfrak{A}$ is called* non-degenerate *if $\mathfrak{A}\mathcal{H}$ is dense in $\mathcal{H}$.*

Let $\mathfrak{A}$ be a C^*-algebra.

Definition 6.21 *$(\mathcal{H}, \pi)$ is a* representation *of $\mathfrak{A}$ if $\mathcal{H}$ is a Hilbert space and π is a $*$-homomorphism of $\mathfrak{A}$ into $B(\mathcal{H})$. π is called* faithful *if $\mathrm{Ker}\,\pi = \{0\}$.*

(*Faithful* in this context is the synonym of *injective*.) Since $\mathrm{Ker}\,\pi$ is a closed two-sided ideal of $\mathfrak{A}$, any non-trivial representation of a simple C^*-algebra is faithful. Actually, a stronger statement is true: a C^*-algebra is simple iff all its representations are faithful.

Let $(\mathcal{H}, \pi)$ be a representation of a C^*-algebra $\mathfrak{A}$.

Definition 6.22 *A closed subspace $\mathcal{H}_1 \subset \mathcal{H}$ is* invariant *if $\pi(A)\mathcal{H}_1 \subset \mathcal{H}_1$ for all $A \in \mathfrak{A}$. $(\mathcal{H}_1, \pi_1)$ is a* sub-representation *of $(\mathcal{H}, \pi)$ if $\mathcal{H}_1$ is an invariant subspace of $\mathcal{H}$ and $\pi_1 = \pi\big|_{\mathcal{H}_1}$.*

Definition 6.23 *We say that $(\mathcal{H}, \pi)$ is the* direct sum *of $(\mathcal{H}_1, \pi_1)$ and $(\mathcal{H}_2, \pi_2)$ if $\mathcal{H} = \mathcal{H}_1 \oplus \mathcal{H}_2$ and $(\mathcal{H}_i, \pi_i)$ are sub-representations of $(\mathcal{H}, \pi)$.*

Note that if $\mathcal{H}_1$ is invariant, then so is $\mathcal{H}_2 := \mathcal{H}_1^\perp$. $(\mathcal{H}, \pi)$ is then the direct sum of $(\mathcal{H}_1, \pi_1)$, $(\mathcal{H}_2, \pi_2)$, with $\pi_1 := \pi\big|_{\mathcal{H}_1}$, $\pi_2 := \pi\big|_{\mathcal{H}_2}$.

Definition 6.24 *We say that a representation $(\mathcal{H}, \pi)$ of a C^*-algebra is* irreducible *if $\pi(\mathfrak{A})$ is irreducible. Equivalently $\pi(\mathfrak{A})' = \mathbb{C}\mathbb{1}$, or π has no non-trivial sub-representations.*

Definition 6.25 *The representation $(\mathcal{H}, \pi)$ is called* non-degenerate *if $\pi(\mathfrak{A})$ is non-degenerate.*

Definition 6.26 *The representation $(\mathcal{H}, \pi)$ is called* factorial *if $\pi(\mathfrak{A}) \cap \pi(\mathfrak{A})' = \mathbb{C}\mathbb{1}$.*

Let $\mathcal{E} \subset \mathcal{H}$.

Definition 6.27 (1) *$\mathcal{E}$ is called* cyclic *for π if $\{\pi(A)\,\Phi : A \in \mathfrak{A},\ \Phi \in \mathcal{E}\}$ is dense in $\mathcal{H}$.*

(2) *$\mathcal{E}$ is called* separating *for π if*

$$\pi(A)\Phi = 0, \quad \Phi \in \mathcal{E} \quad \Rightarrow \quad A = 0.$$

Clearly, if $(\mathcal{H}, \pi)$ is irreducible, all non-zero vectors in $\mathcal{H}$ are cyclic.

6.2.4 Intertwiners and unitary equivalence

Let $(\mathcal{H}_1, \pi_1)$, $(\mathcal{H}_2, \pi_2)$ be two representations of a C^*-algebra $\mathfrak{A}$.

Definition 6.28 *An operator $B \in B(\mathcal{H}_1, \mathcal{H}_2)$* intertwines *$\pi_1$ and π_2 if*

$$B\pi_1(A) = \pi_2(A)B, \quad A \in \mathfrak{A}.$$

If π_1 and π_2 have an intertwiner in $U(\mathcal{H}_1, \mathcal{H}_2)$, they are called *unitarily equivalent*.

The following theorem can be called *Schur's lemma for C^*-algebras*:

Theorem 6.29 *If $(\mathcal{H}_1, \pi_1)$, $(\mathcal{H}_2, \pi_2)$ are irreducible, then the set of intertwiners equals either $\{0\}$ or $\{\lambda U : \lambda \in \mathbb{C}\}$ for some $U \in U(\mathcal{H}_1, \mathcal{H}_2)$.*

Proof If B intertwines π_1 and π_2, B^* intertwines π_2 and π_1, hence $B^*B \in \pi_1(\mathfrak{A})'$ and $BB^* \in \pi_2(\mathfrak{A})'$. By irreducibility, $B^*B = \lambda_1 \mathbb{1}$, $BB^* = \lambda_2 \mathbb{1}$ for some $\lambda_1, \lambda_2 \in \mathbb{R}$. Now

$$\lambda_1^2 \mathbb{1} = BB^*BB^* = B\lambda_2 B^* = \lambda_2 \lambda_1 \mathbb{1}. \tag{6.3}$$

If $\lambda_1 = 0$, then $B = 0$, and hence $\lambda_2 = 0$. Hence (6.3) implies that $\lambda_1 = \lambda_2$, which means that $B = \lambda U$ for some $U \in U(\mathcal{H}_1, \mathcal{H}_2)$. If B_1 and B_2 are two intertwiners, then a similar argument shows that $B_1 B_2^*$ is proportional to identity. This means that B_1 is proportional to B_2. $\qquad\square$

6.2.5 States

Let $\mathfrak{A}$ be a C^*-algebra.

Definition 6.30 *A linear functional on $\mathfrak{A}$ is called* positive *if it maps positive elements to positive numbers.*

A positive linear functional is automatically continuous.

Definition 6.31 *A positive linear functional is called a* state *if its norm is 1. In the case of a unital C^*-algebra it is equivalent to requiring that $\omega(\mathbb{1}) = 1$.*

Definition 6.32 *A state ω is called* faithful *if $\omega(A) = 0$ and $A \in \mathfrak{A}_+$ implies $A = 0$.*

Definition 6.33 *A state ω is called* tracial *if*

$$\omega(AB) = \omega(BA), \quad A, B \in \mathfrak{A}.$$

6.2.6 GNS representations

Let $(\mathcal{H}, \pi)$ be a $*$-representation of $\mathfrak{A}$, Ω a normalized vector in $\mathcal{H}$. Then

$$\omega(A) = (\Omega | \pi(A)\Omega) \tag{6.4}$$

defines a state on $\mathfrak{A}$.

Definition 6.34 *If (6.4) is true, we say that Ω is a* vector representative *of ω.*

Definition 6.35 *$(\mathcal{H}, \pi, \Omega)$ is called a* cyclic $*$-representation *if $(\mathcal{H}, \pi)$ is a $*$-representation and Ω is a cyclic vector.*

Theorem 6.36 (Gelfand–Najmark–Segal theorem) *Let ω be a state on $\mathfrak{A}$. Then there exists a cyclic $*$-representation $(\mathcal{H}_\omega, \pi_\omega, \Omega_\omega)$ such that Ω_ω is a vector representative of ω. Such a representation is unique up to a unitary equivalence.*

Definition 6.37 *The cyclic $*$-representation described in Thm. 6.36 is called the* GNS *representation (for Gelfand–Najmark–Segal) associated with ω.*

6.2.7 W^*-algebras

Definition 6.38 *We say that $\mathfrak{M}$ is a W^*-algebra if it is a C^*-algebra such that there exists a Banach space whose dual is isomorphic to $\mathfrak{M}$ as a Banach space. This Banach space is unique up to an isometry. It is called the* pre-dual *of $\mathfrak{M}$ and is denoted $\mathfrak{M}_\#$. The topology on $\mathfrak{M}$ given by the functionals from $\mathfrak{M}_\#$ (the $*$-weak topology in the terminology of Banach spaces) is called the σ-weak topology. Functionals in $\mathfrak{M}_\#$ are called* normal functionals.

It follows from the general theory of Banach spaces that $\mathfrak{M}_\#$ coincides with the space of all σ-weakly continuous functionals on $\mathfrak{M}$.

Definition 6.39 *The set*

$$\{B \in \mathfrak{M} \ : \ AB = BA, \ A \in \mathfrak{M}\}$$

is called the center *of $\mathfrak{M}$. A W^*-algebra with a trivial center is called a* factor.

Two-sided σ-weakly closed ideals $\mathfrak{I}$ of a W^*-algebra $\mathfrak{M}$ have a simple form: they are equal to $\mathfrak{I} = \mathfrak{M}E$, for a projection E in the center of $\mathfrak{M}$. Clearly, all two-sided σ-weakly closed ideals of a factor are trivial.

If ω is a σ-weakly continuous state, then the map π_ω given by the GNS representation is σ-weakly continuous.

Definition 6.40 *Let $\mathfrak{M}$ be a W^*-algebra. A W^*-dynamics on $\mathfrak{M}$ is a one-parameter group $\mathbb{R} \ni t \mapsto \tau^t \in \mathrm{Aut}(\mathfrak{M})$ such that for each $A \in \mathfrak{M}$ the map $t \mapsto \tau^t(A)$ is σ-weakly continuous. Such a pair $(\mathfrak{M}, \tau)$ is called a W^*-dynamical system.*

6.2.8 Von Neumann algebras

Let $\mathcal{H}$ be a Hilbert space. Then $B(\mathcal{H})$ is a W^*-algebra, since it is the dual of $B^1(\mathcal{H})$ (the space of trace-class operators on $\mathcal{H}$). Thus $B^1(\mathcal{H})$ is the pre-dual of $B(\mathcal{H})$ and the topology on $B(\mathcal{H})$ given by functionals in $B^1(\mathcal{H})$ is its σ-weak topology.

Definition 6.41 *Every C^*-sub-algebra of $B(\mathcal{H})$ closed w.r.t. the σ-weak topology is called a* concrete W^*-algebra. *If in addition it contains $\mathbb{1}_\mathcal{H}$, then it is called a* von Neumann algebra.

Clearly, all concrete W^*-algebras are W^*-algebras. Conversely, a W^*-algebra is isomorphic to a von Neumann algebra.

Definition 6.42 *Let* $\mathfrak{M}_i \subset B(\mathcal{H}_i)$, $i = 1, 2$. *Let* $\rho : \mathfrak{M}_1 \to \mathfrak{M}_2$ *be an isomorphism. We say that* ρ *is* spatially implementable *if there exists* $U \in U(\mathcal{H}_1, \mathcal{H}_2)$ *such that* $\rho(A) = UAU^*$, $A \in \mathfrak{M}_1$.

If $\mathfrak{A} \subset B(\mathcal{H})$ is $*$-invariant, then $\mathfrak{A}'$ is a von Neumann algebra.

An equivalent characterization of a von Neumann algebra is given by *von Neumann's double commutant theorem*, stating that a $*$-algebra $\mathfrak{M}$ is a von Neumann algebra iff

$$\mathfrak{M} = \mathfrak{M}''.$$

The *von Neumann density theorem* says that if $\mathfrak{A} \subset B(\mathcal{H})$ is a non-degenerate $*$-algebra, then $\mathfrak{A}$ is dense in $\mathfrak{A}''$ in the weak, strong, strong*, σ-weak, σ-strong and σ-strong* topologies.

The *Kaplansky density theorem* says that if $\mathfrak{A} \subset B(\mathcal{H})$ is a $*$-algebra, then the unit ball of $\mathfrak{A}$ is σ-weakly dense in the unit ball of $\mathfrak{A}''$.

Let $\mathfrak{M} \subset B(\mathcal{H})$ be a von Neumann algebra, and A a closed densely defined operator on $\mathcal{H}$. Let $A = U|A|$, where U is a partial isometry, be its polar decomposition.

Definition 6.43 A *is called* affiliated *to* $\mathfrak{M}$ *if the operators* U *and* $\mathbb{1}_\Delta(|A|)$ *belong to* $\mathfrak{M}$ *for all Borel sets* $\Delta \subset \mathbb{R}$.

Clearly, a von Neumann algebra $\mathfrak{M} \subset B(\mathcal{H})$ is a factor iff $\mathfrak{M} \cap \mathfrak{M}' = \mathbb{C}\mathbb{1}_\mathcal{H}$, or equivalently, $(\mathfrak{M} \cup \mathfrak{M}')'' = B(\mathcal{H})$. Below we give a more elaborate criterion for being a factor.

Proposition 6.44 *Let* $\mathfrak{M} \subset B(\mathcal{H})$ *be a von Neumann algebra. Suppose that*

(1) $\Omega \in \mathcal{H}$ *is a cyclic vector for* $(\mathfrak{M} \cup \mathfrak{M}')''$;
(2) *There exists a set* $\mathfrak{L} \subset (\mathfrak{M} \cup \mathfrak{M}')''$ *such that* $\{\Psi \in \mathcal{H} \ : \ A\Psi = 0, \ A \in \mathfrak{L}\} = \mathbb{C}\Omega$.

Then $\mathfrak{M}$ *is a factor.*

Proof Suppose that $\mathfrak{M}$ is not a factor and Ω is cyclic for $(\mathfrak{M} \cup \mathfrak{M}')''$. Then there exists an orthogonal projection $P \in \mathfrak{M} \cap \mathfrak{M}'$ different from 0 and $\mathbb{1}$. If $P\Omega = 0$, then $(\mathbb{1} - P)(\mathfrak{M} \cup \mathfrak{M}')''\Omega = (\mathfrak{M} \cup \mathfrak{M}')''(\mathbb{1} - P)\Omega = (\mathfrak{M} \cup \mathfrak{M}')''\Omega$. Hence Ω is not cyclic for $(\mathfrak{M} \cup \mathfrak{M}')''$. Therefore, $P\Omega \neq 0$. Likewise, we show that $(\mathbb{1} - P)\Omega \neq 0$.

Now let $\mathfrak{L}$ be as in (2). Then since $P \in \mathfrak{M} \cap \mathfrak{M}'$ one has

$$A\left(c_1 P + c_2(\mathbb{1} - P)\right)\Omega = 0, \quad A \in \mathfrak{L}, \quad c_1, c_2 \in \mathbb{C}.$$

But for $c_1 \neq c_2$, the vector $\left(c_1 P + c_2(\mathbb{1} - P)\right)\Omega$ is not proportional to Ω. $\qquad\square$

6.2.9 UHF algebras

In this subsection we describe an example of a C^*-algebra which plays an important role in mathematical physics, and in particular in the theory of CAR.

For any $n = 1, 2, \ldots$, we introduce the identifications

$$B(\otimes^n \mathbb{C}^2) \ni A \mapsto A \otimes \mathbb{1}_{\mathbb{C}^2} \in B(\otimes^{n+1} \mathbb{C}^2).$$

Definition 6.45 *Define*

$$\mathrm{UHF}_0(2^\infty) := \bigcup_{k=1}^{\infty} B(\otimes^n \mathbb{C}^2), \quad \mathrm{UHF}(2^\infty) := \mathrm{UHF}_0(2^\infty)^{\mathrm{cpl}}.$$

$UHF(2^\infty)$ *is called the* uniformly hyper-finite C^*-algebra of type 2^∞.

6.2.10 Hyper-finite type II_1 factor

We continue to consider the C^*-algebra $\mathrm{UHF}(2^\infty)$ introduced in the last subsection. On $B(\otimes^n \mathbb{C}^2)$ we have a tracial state

$$\mathrm{tr}A := 2^{-n} \mathrm{Tr}\, A.$$

This state extends to a state on the whole $\mathrm{UHF}(2^\infty)$. Let $(\pi_{\mathrm{tr}}, \mathcal{H}_{\mathrm{tr}}, \Omega_{\mathrm{tr}})$ be the GNS representation given by the state tr on $\mathrm{UHF}(2^\infty)$.

Definition 6.46 *The W^*-algebra*

$$\mathrm{HF} := \pi_{\mathrm{tr}} \left(\mathrm{UHF}(2^\infty) \right)''. \tag{6.5}$$

is called the hyper-finite type II_1 factor.

Clearly,

$$\mathrm{tr}(A) := (\Omega_{\mathrm{tr}} | A \Omega_{\mathrm{tr}})$$

defines a tracial state on HF.

6.2.11 Conditional expectations

Let $\mathfrak{N}$ be a unital C^*-sub-algebra of a C^*-algebra $\mathfrak{M}$. We assume that the unit of $\mathfrak{M}$ is contained in $\mathfrak{N}$.

Definition 6.47 *We say that $E : \mathfrak{M} \to \mathfrak{N}$ is $\mathfrak{N}$-linear if $A \in \mathfrak{M}$, $B \in \mathfrak{N}$ implies $E(AB) = E(A)B$, $E(BA) = BE(A)$.*

We say that E is a conditional expectation *if*

(1) *$A \geq 0$ implies $E(A) \geq 0$,*
(2) *E is $\mathfrak{N}$-linear,*
(3) *$E(\mathbb{1}) = \mathbb{1}$.*

Proposition 6.48 *Let ω be a normal tracial faithful state on a W^*-algebra $\mathfrak{M}$. Then there exists a unique conditional expectation from $\mathfrak{M}$ with range equal to $\mathfrak{N}$ such that $\omega(A) = \omega(E(A))$.*

6.3 Tensor products of algebras

Let $\mathfrak{A},\mathfrak{B}$ be algebras. Then $\mathfrak{A} \overset{\mathrm{al}}{\otimes} \mathfrak{B}$ is naturally an algebra. If in addition $\mathfrak{A},\mathfrak{B}$ are $*$-algebras, then so is $\mathfrak{A} \overset{\mathrm{al}}{\otimes} \mathfrak{B}$.

One can define natural tensor products also in the category of C^*- and W^*-algebras. The definitions of these constructions are given in this section.

6.3.1 Tensor product of C^*-algebras

Let $\mathfrak{A},\mathfrak{B}$ be C^*-algebras. We choose an arbitrary injective $*$-representation $(\mathcal{H}, \pi)$ of $\mathfrak{A}$ and $(\mathcal{K}, \rho)$ of $\mathfrak{B}$. Then $\mathfrak{A} \overset{\mathrm{al}}{\otimes} \mathfrak{B}$ has an obvious $*$-representation in $B(\mathcal{H} \otimes \mathcal{K})$. It equips $\mathfrak{A} \overset{\mathrm{al}}{\otimes} \mathfrak{B}$ with a C^* norm. It can be shown that this norm does not depend on the representations $(\mathcal{H}, \pi)$ and $(\mathcal{K}, \rho)$.

Definition 6.49 *The C^*-algebra*

$$\mathfrak{A} \otimes \mathfrak{B} := (\mathfrak{A} \overset{\mathrm{al}}{\otimes} \mathfrak{B})^{\mathrm{cpl}},$$

is called the minimal C^*-tensor product of $\mathfrak{A}$ and $\mathfrak{B}$.

6.3.2 Tensor product of W^*-algebras

Let $\mathfrak{M},\mathfrak{N}$ be W^*-algebras. We choose an arbitrary injective σ-continuous $*$-representation $(\mathcal{H}, \pi)$ of $\mathfrak{M}$ and $(\mathcal{K}, \rho)$ of $\mathfrak{N}$. Then $\mathfrak{M} \overset{\mathrm{al}}{\otimes} \mathfrak{N}$ has an obvious $*$-representation in $B(\mathcal{H} \otimes \mathcal{K})$. Let $\mathcal{X}$ denote the Banach space of linear functionals on $\mathfrak{M} \overset{\mathrm{al}}{\otimes} \mathfrak{N}$ given by density matrices in $B^1(\mathcal{H} \otimes \mathcal{K})$. One can show that $\mathcal{X}$ does not depend on the choice of representations $(\mathcal{H}, \pi)$ and $(\mathcal{K}, \rho)$.

Definition 6.50 *We set*

$$\mathfrak{M} \otimes \mathfrak{N} := \mathcal{X}^{\#},$$

and call it the W^*-tensor product of $\mathfrak{M}$ and $\mathfrak{N}$.

Clearly, $\mathfrak{M} \overset{\mathrm{al}}{\otimes} \mathfrak{N}$ is σ-weakly dense in $\mathfrak{M} \otimes \mathfrak{N}$. We extend the multiplication from $\mathfrak{M} \overset{\mathrm{al}}{\otimes} \mathfrak{N}$ to $\mathfrak{M} \otimes \mathfrak{N}$ by the σ-weak continuity. One can check that $\mathfrak{M} \otimes \mathfrak{N}$ is a W^*-algebra.

Remark 6.51 *According to our convention, the meaning of $\otimes$ between two algebras depends on the context. It depends on whether we treat the algebras as C^*- or W^*-algebras.*

6.4 Modular theory

In this section we give a concise resumé of the *modular theory*. The modular theory is one of the most interesting parts of the theory of operator algebras. It sheds light on the structure of general W^*-algebras. It plays an important role in applications of operator algebras to quantum statistical physics. Key concepts of the modular theory include the *modular automorphism* and *conjugation* due to Tomita–Takesaki, *KMS states* and *standard forms* introduced by Araki, Connes and Haagerup.

6.4.1 Standard representations

Let $\mathcal{H}$ be a Hilbert space.

Definition 6.52 *A* self-dual cone $\mathcal{H}^+$ *is a subset of* $\mathcal{H}$ *with the property*

$$\mathcal{H}^+ = \big\{ \Phi \in \mathcal{H} \; : \; (\Phi|\Psi) \geq 0, \; \Psi \in \mathcal{H}^+ \big\}.$$

Let $\mathfrak{M}$ be a W^*-algebra.

Definition 6.53 *A quadruple* $(\mathcal{H}, \pi, J, \mathcal{H}^+)$ *is a standard representation of a W^*-algebra $\mathfrak{M}$ if $\pi : \mathfrak{M} \to B(\mathcal{H})$ is a faithful σ-weakly continuous representation, J is a conjugation on $\mathcal{H}$ and $\mathcal{H}^+$ is a self-dual cone in $\mathcal{H}$ with the following properties:*

(1) $J\pi(\mathfrak{M})J = \pi(\mathfrak{M})'$,
(2) $J\pi(A)J = \pi(A)^*$ *for A in the center of* $\mathfrak{M}$,
(3) $J\Phi = \Phi$ *for* $\Phi \in \mathcal{H}^+$,
(4) $\pi(A)J\pi(A)\mathcal{H}^+ \subset \mathcal{H}^+$ *for* $A \in \mathfrak{M}$.

Every W^*-algebra admits a unique (up to unitary equivalence) standard representation.

The standard representation has several important properties.

Theorem 6.54 (1) *For every σ-weakly continuous state ω on $\mathfrak{M}$ there exists a unique vector $\Omega \in \mathcal{H}^+$ such that $\omega(A) = (\Omega|A\Omega)$.*

(2) *For every $*$-automorphism τ of $\mathfrak{M}$ there exists a unique $U \in U(\mathcal{H})$ such that*

$$\pi(\tau(A)) = U\pi(A)U^*, \quad U\mathcal{H}^+ \subset \mathcal{H}^+.$$

(3) *If $\mathbb{R} \ni t \mapsto \tau^t$ is a W^*-dynamics on $\mathfrak{M}$, there exists a unique self-adjoint operator L on $\mathcal{H}$ such that*

$$\pi(\tau^t(A)) = \mathrm{e}^{\mathrm{i}tL}\pi(A)\mathrm{e}^{-\mathrm{i}tL}, \quad \mathrm{e}^{\mathrm{i}tL}\mathcal{H}^+ \subset \mathcal{H}^+. \tag{6.6}$$

Definition 6.55 *The operator L that appears in (6.6) is called the* standard Liouvillean *of the W^* dynamics $t \mapsto \tau^t$.*

Definition 6.56 *Given a standard representation $(\mathcal{H}, \pi, J, \mathcal{H}^+)$, we also have the* right representation $\pi_\mathrm{r} : \overline{\mathfrak{M}} \to B(\mathcal{H})$ *given by* $\pi_\mathrm{r}(\overline{A}) := J\pi(A)J$. *Note that $\pi_\mathrm{r}(\overline{\mathfrak{M}}) = \pi(\mathfrak{M})'$. We will often write π_l for π and call it the* left *representation.*

6.4.2 Tomita–Takesaki theory

Let $\mathfrak{M}$ be a W^*-algebra, $(\mathcal{H}, \pi)$ a faithful σ-weakly continuous representation of $\mathfrak{M}$ and Ω a cyclic and separating vector for $\pi(\mathfrak{M})$.

Definition 6.57 *Define the operator S_0 with domain $\pi(\mathfrak{M})\Omega$ by*

$$S_0 \pi(A)\Omega := \pi(A^*)\Omega, \ A \in \mathfrak{M}.$$

One can show that S_0 is closable.

Definition 6.58 *S is defined as the closure of S_0.*

For further reference let us note the following proposition, which follows by the von Neumann density theorem:

Proposition 6.59 *If $\mathfrak{A} \subset \mathfrak{M}$ is a $*$-algebra weakly dense in $\mathfrak{M}$, then $\{A\Omega : A \in \mathfrak{A}\}$ is an essential domain for S.*

Definition 6.60 *The modular operator Δ and modular conjugation J are defined by the polar decomposition:*

$$S =: J\Delta^{\frac{1}{2}}.$$

Definition 6.61 *The natural positive cone is defined by*

$$\mathcal{H}^+ := \big\{\pi(A)J\pi(A)\Omega \ : \ A \in \mathfrak{M}\big\}^{\text{cl}}.$$

Theorem 6.62 *$(\mathcal{H}, \pi, J, \mathcal{H}^+)$ is a standard representation of $\mathfrak{M}$. Given $(\mathcal{H}, \pi)$, it is the unique standard representation such that $\Omega \in \mathcal{H}^+$.*

6.4.3 KMS states

Let $(\mathfrak{M}, \tau)$ be a W^*-dynamical system. Consider $\beta > 0$ (having the interpretation of the inverse temperature). Let ω be a normal state on $\mathfrak{M}$.

Definition 6.63 *ω is called a (τ, β)-KMS state if for all $A, B \in \mathfrak{M}$ there exists a function $F_{A,B}(z)$ holomorphic in the strip $I_\beta = \{z \in \mathbb{C} : 0 < \operatorname{Im} z < \beta\}$, bounded and continuous on its closure, such that the KMS boundary condition holds:*

$$F_{A,B}(t) = \omega\big(A\tau^t(B)\big), \quad F_{A,B}(t + \mathrm{i}\beta) = \omega\big(\tau^t(B)A\big), \quad t \in \mathbb{R}. \qquad (6.7)$$

Below we quote a number of properties of KMS states.

Proposition 6.64 (1) *One has $|F_{A,B}(z)| \leq \|A\|\|B\|$, uniformly on I_β^{cl}.*
(2) *A KMS state is τ^t-invariant.*
(3) *Let $\mathfrak{A}$ be a $*$-algebra weakly dense in $\mathfrak{M}$ and τ-invariant. If (6.7) holds for all $A, B \in \mathfrak{A}$, then it holds for all $A, B \in \mathfrak{M}$.*

Proposition 6.65 *A KMS state on a factor is faithful.*

Definition 6.66 *If $\mathfrak{M} \subset B(\mathcal{H})$ and $\Phi \in \mathcal{H}$, we say that Φ is a (τ, β)-KMS vector if $(\Phi| \cdot \Phi)$ is a (τ, β)-KMS state.*

6.4.4 Type I factors: irreducible representation

Definition 6.67 *Algebras isomorphic to $B(\mathcal{H})$, where $\mathcal{H}$ is a Hilbert space, are called* type I factors.

Such algebras are the most elementary W^*-algebras. In this and the next subsection we describe various concepts of the theory of W^*-algebras as applied to type I factors.

The space of σ-weakly continuous functionals on $B(\mathcal{H})$ (the pre-dual of $B(\mathcal{H})$) can be identified with $B^1(\mathcal{H})$ (trace-class operators) by the formula

$$\psi(A) = \operatorname{Tr} \gamma A, \quad \gamma \in B^1(\mathcal{H}), \quad A \in B(\mathcal{H}). \tag{6.8}$$

In particular, σ-weakly continuous states are determined by density matrices. A state given by a density matrix γ is faithful iff $\operatorname{Ker} \gamma = \{0\}$.

Proposition 6.68 (1) *Every $*$-automorphism of $B(\mathcal{H})$ is of the form*

$$\tau(A) = UAU^* \tag{6.9}$$

for some $U \in U(\mathcal{H})$. If $U_1, U_2 \in U(\mathcal{H})$ satisfy (6.9), then there exists $\mu \in \mathbb{C}$ with $|\mu| = 1$ such that $U_1 = \mu U_2$.
(2) *Every W^*-dynamics $\mathbb{R} \ni t \mapsto \tau_t$ on $B(\mathcal{H})$ is of the form*

$$\tau_t(A) = \mathrm{e}^{\mathrm{i}tH} A \mathrm{e}^{-\mathrm{i}tH} \tag{6.10}$$

for some self-adjoint H. If H_1 is another self-adjoint operator satisfying (6.10), then there exists $c \in \mathbb{R}$ such that $H_1 = H + c$.

Definition 6.69 *In the context of (6.9) we say that U implements τ. In the context of (6.10) we say that H is a Hamiltonian of $\{\tau_t\}_{t \in \mathbb{R}}$.*

A state given by (6.8) is invariant w.r.t. the W^*-dynamics (6.10) iff H commutes with γ.

There exists a (β, τ)-KMS state iff $\operatorname{Tr} \mathrm{e}^{-\beta H} < \infty$, and then it has the density matrix $\mathrm{e}^{-\beta H} / \operatorname{Tr} \mathrm{e}^{-\beta H}$.

6.4.5 Type I factors: representation on Hilbert–Schmidt operators

Clearly, the representation of $B(\mathcal{H})$ on $\mathcal{H}$ is not in the standard form. To construct a standard form of $B(\mathcal{H})$, consider the Hilbert space of Hilbert–Schmidt operators on $\mathcal{H}$, denoted $B^2(\mathcal{H})$.

Definition 6.70 *We introduce two injective representations:*

$$B(\mathcal{H}) \ni A \mapsto \pi_{\mathrm{l}}(A) \in B\big(B^2(\mathcal{H})\big), \quad \pi_{\mathrm{l}}(A)B := AB, \quad B \in B^2(\mathcal{H});$$

$$\overline{B(\mathcal{H})} \ni \overline{A} \mapsto \pi_{\mathrm{r}}(\overline{A}) \in B\big(B^2(\mathcal{H})\big), \quad \pi_{\mathrm{r}}(\overline{A})B := BA^*, \quad B \in B^2(\mathcal{H}). \tag{6.11}$$

We set $J_{\mathcal{H}}B := B^*$, $B \in B^2(\mathcal{H})$.

With the above notation, $J_{\mathcal{H}}\pi_{\mathrm{l}}(A)J_{\mathcal{H}} = \pi_{\mathrm{r}}(\overline{A})$ and

$$\big(B^2(\mathcal{H}), \pi_l, J_{\mathcal{H}}, B_+^2(\mathcal{H})\big)$$

is a standard representation of $B(\mathcal{H})$.

If a state on $B(\mathcal{H})$ is given by a density matrix $\gamma \in B_+^1(\mathcal{H})$, then its standard vector representative is $\gamma^{\frac{1}{2}} \in B_+^2(\mathcal{H})$. If $\tau \in \mathrm{Aut}\big(B(\mathcal{H})\big)$ is implemented by $W \in U(\mathcal{H})$, then its standard implementation is $\pi_{\mathrm{l}}(W)\pi_{\mathrm{r}}(\overline{W})$. If the W^*-dynamics $t \mapsto \tau^t$ has a Hamiltonian H, then its standard Liouvillean is $\pi_{\mathrm{l}}(H) - \pi_{\mathrm{r}}(\overline{H})$.

6.5 Non-commutative probability spaces

Throughout the section, $\mathfrak{R}$ is a W^*-algebra and ω a normal faithful tracial state on $\mathfrak{R}$.

The two most important examples of such a pair $(\mathfrak{R}, \omega)$ are as follows:

Example 6.71 (1) *Let* $(Q, \mathfrak{S}, \mu)$ *be a set with a σ-algebra and a probability measure. Then taking* $\mathfrak{R} = L^\infty(Q, \mu)$ *and*

$$\omega(F) = \int_Q F \mathrm{d}\mu, \quad F \in L^\infty(Q, \mu),$$

we obtain an example of a W^*-algebra with a normal tracial state.

(2) *The algebra* HF *with the state* tr, *described in Subsect. 6.2.10, is another example.*

Recall that the triple $(Q, \mathfrak{S}, \mu)$ of Example 6.71 (1) is called a *probability space*. Therefore, some authors call a couple consisting of a W^*-algebra and a normal tracial faithful state a *non-commutative probability space*. In any case, this section is in many ways analogous to parts of Sect. 5.1, where (commutative) probability spaces were considered.

6.5.1 Measurable operators

Let us start with an abstract construction of measurable operators.

Definition 6.72 *The* measure topology *on the W^*-algebra $\mathfrak{R}$ is given by the family $V(\epsilon, \delta)$ of neighborhoods of 0 defined for $\epsilon, \delta > 0$ as*

$$V(\epsilon, \delta) := \{A \in \mathfrak{R} \; : \; \|AP\| < \epsilon, \; \omega(\mathbb{1} - P) < \delta,$$

$$\text{for some orthogonal projection } P \in \mathfrak{R}\}.$$

$\mathcal{M}(\mathfrak{R})$ *denotes the completion of* $\mathfrak{R}$ *for the measure topology. Elements of* $\mathcal{M}(\mathfrak{R})$ *are called* (abstract) *measurable operators.*

Let us now assume that $\mathfrak{R}$ is isometrically embedded in $B(\mathcal{H})$.

Definition 6.73 *A closed densely defined operator on* $\mathcal{H}$ *is called a* (concrete) *measurable operator iff it is affiliated to* $\mathfrak{R}$ *and*

$$\lim_{R \to +\infty} \omega\big(\mathbb{1}_{[R,+\infty[}(|A|)\big) = 0.$$

It can be shown that one can identify $\mathcal{M}(\mathfrak{R})$ with the set of concrete measurable operators on $\mathcal{H}$. Thus $\mathcal{M}(\mathfrak{R})$ becomes a subset of $Cl(\mathcal{H})$.

Proposition 6.74 *Let* $A, B \in \mathcal{M}(\mathfrak{R})$. *Then* $A + B$ *and* AB *are closable.* $(A + B)^{\mathrm{cl}}$ *and* $(AB)^{\mathrm{cl}}$ *belong again to* $\mathcal{M}(\mathfrak{R})$ *and do not depend on the representation of* $\mathfrak{R}$.

Using the above proposition, we endow $\mathcal{M}(\mathfrak{R})$ with the structure of a $*$-algebra. One extends ω to the subset $\mathcal{M}_+(\mathfrak{R})$ of positive operators in $\mathcal{M}(\mathfrak{R})$ by setting

$$\omega(A) := \lim_{\epsilon \to 0^+} \omega\big(A(\mathbb{1} + \epsilon A)^{-1}\big) \in [0, +\infty].$$

6.5.2 Non-commutative L^p spaces

Definition 6.75 *For* $1 \le p < \infty$ *one sets*

$$L^p(\mathfrak{R}, \omega) := \big\{A \in \mathcal{M}(\mathfrak{R}) \ : \ \omega(|A|^p) < \infty\big\},$$

equipped with the norm $\|A\|_p := \omega\big(|A|^p\big)^{1/p}$.
For $p = \infty$ *one sets* $L^\infty(\mathfrak{R}, \omega) := \mathfrak{R}$, *and* $\|A\|_\infty := \|A\|$.

We will often drop ω from $L^p(\mathfrak{R}, \omega)$, where it does not cause confusion. The spaces $L^p(\mathfrak{R})$ are Banach spaces with $\mathfrak{R}$ as a dense subspace.

Note that if $A \in L^1(\mathfrak{R})$, then $\mathfrak{M} \ni B \mapsto \omega(AB) \in \mathbb{C}$ is a normal functional of norm $\|A\|_1 = \omega(|A|)$. This defines an isometric identification between $L^1(\mathfrak{R})$ and $\mathfrak{R}_\#$, the space of normal functionals on $\mathfrak{R}$.

Let $(\mathcal{H}_\omega, \pi_\omega, \Omega_\omega)$ be the GNS representation for the state ω. Then $L^2(\mathfrak{R})$ can be unitarily identified with the space $\mathcal{H}_\omega$, as an extension of the map

$$\mathfrak{R} \ni A \mapsto A\Omega_\omega \in \mathcal{H}_\omega. \tag{6.12}$$

We have $L^q(\mathfrak{R}) \subset L^p(\mathfrak{R})$ if $q \ge p$.

Proposition 6.76 (1) *For* $A \in L^p(\mathfrak{R})$, $1 \le p \le \infty$, *one has* $\|A\|_p = \|A^*\|_p$. *In particular,* $A \mapsto A^*$ *is anti-unitary on* $L^2(\mathfrak{R})$.

(2) *The* non-commutative Hölder's inequality *holds: for all* $1 \leq r, p, q \leq \infty$ *with* $p^{-1} + q^{-1} = r^{-1}$, *if* $A \in L^p(\mathfrak{R})$, $B \in L^q(\mathfrak{R})$, *then* $AB \in L^r(\mathfrak{R})$ *and*

$$\|AB\|_r \leq \|A\|_p \|B\|_q. \tag{6.13}$$

(3) $\|A\|_p = \sup\{\omega(AB) : B \in \mathfrak{R}, \|B\|_q \leq 1\}$, $p^{-1} + q^{-1} = 1$, $p > 1$.

Definition 6.77 *An element* A *of* $L^p(\mathfrak{R})$ *is* positive *if it is positive as an unbounded operator on* $\mathcal{H}$. *We denote by* $L^p_+(\mathfrak{R})$ *the set of positive elements of* $L^p(\mathfrak{R})$.

For all $1 \leq p \leq \infty$, $\mathfrak{R}_+$ is dense in $L^p_+(\mathfrak{R})$ and the sets $L^p_+(\mathfrak{R})$ are closed in $L^p(\mathfrak{R})$.

Lemma 6.78 (1) $A \in \mathfrak{R}_+$ *iff* $\omega(AB) \geq 0$, $B \in \mathfrak{R}_+$.
(2) $A \in L^p_+(\mathfrak{R})$ *iff* $\omega(AB) \geq 0$, $B \in L^q_+(\mathfrak{R})$.

6.5.3 Operators between non-commutative L^p spaces

Let $(\mathfrak{R}_i, \omega_i)$, $i = 1, 2$, be two W^*-algebras with normal tracial faithful states.

Definition 6.79 $T \in B\big(L^2(\mathfrak{R}_1), L^2(\mathfrak{R}_2)\big)$ *is called*

(1) positivity preserving *if* $A \geq 0 \Rightarrow TA \geq 0$,
(2) hyper-contractive *if* T *is a contraction and there exists* $p > 2$ *such that* T *is bounded from* $L^2(\mathfrak{R}_1)$ *to* $L^p(\mathfrak{R}_2)$.

Using Lemma 6.78 we see as in the commutative case that T is positivity preserving iff T^* is.

Let $(\mathfrak{R}, \omega)$ be a W^*-algebra with a normal tracial faithful state.

Definition 6.80 $T \in B\big(L^2(\mathfrak{R})\big)$ *is called* doubly Markovian *if it is positivity preserving and* $T\mathbb{1} = T^*\mathbb{1} = \mathbb{1}$.

Theorem 6.81 *A doubly Markovian map* T *extends to a contraction on* $L^p(\mathfrak{R})$ *for all* $1 \leq p \leq \infty$.

Proof Using that $\pm T \leq \|T\|_\infty \mathbb{1}$ and the fact that T is positivity preserving, we obtain that T is a contraction on $L^\infty(\mathfrak{R})$. Applying Prop. 6.76 (3) and the above result to T^*, we see that T is a contraction on $L^1(\mathfrak{R}, \omega)$. By the non-commutative version of Stein's interpolation theorem (see Prop. 3 of Gross (1972)), this extends to all $1 < p < \infty$. $\qquad\square$

6.5.4 Conditional expectations on non-commutative spaces

Let $\mathfrak{R}_1$ be a W^*-sub-algebra of $\mathfrak{R}$. Let ω_1 be the restriction of ω to $\mathfrak{R}_1$. Clearly, $L^p(\mathfrak{R}_1, \omega_1)$ injects isometrically into $L^p(\mathfrak{R}, \omega)$.

Definition 6.82 *Denote by $E_{\mathfrak{R}_1}$ the orthogonal projection from $L^2(\mathfrak{R}, \omega)$ onto $L^2(\mathfrak{R}_1, \omega_1)$.*

Proposition 6.83 (1) $E_{\mathfrak{R}_1}$ *uniquely extends to a contraction from $L^p(\mathfrak{R})$ into $L^p(\mathfrak{R}_1)$ for all $1 \le p \le \infty$.*

(2) $E_{\mathfrak{R}_1}$ *is doubly Markovian.*

(3) *Let $A \in L^p(\mathfrak{R})$, $B \in L^q(\mathfrak{R}_1)$, $p^{-1} + q^{-1} = 1$. Then*

$$E_{\mathfrak{R}_1}(AB) = E_{\mathfrak{R}_1}(A)B, \quad E_{\mathfrak{R}_1}(BA) = BE_{\mathfrak{R}_1}(A).$$

(4) $E_{\mathfrak{R}_1}$ *considered as an operator on $L^\infty(\mathfrak{R}) = \mathfrak{R}$ is the unique conditional expectation onto $\mathfrak{R}_1$ described in Prop. 6.48, that is, satisfying*

$$\omega(A) = \omega(E(A)), \quad A \in \mathfrak{R}.$$

6.6 Notes

A comprehensive reference to operator algebras is the three-volume monograph of Takesaki. In particular, Takesaki (1979) contains basics and Takesaki (2003) contains the modular theory. Another useful reference, aimed at applications in mathematical physics, is the two-volume monograph of Bratteli–Robinson (1987, 1996). In particular, proofs of the properties of KMS states of Subsect. 6.4.2 can be found in Bratteli–Robinson (1996).

Non-commutative probability spaces are analyzed in Takesaki (2003), following Segal (1953a,b), Kunze (1958) and Wilde (1974).

7

Anti-symmetric calculus

In almost every respect there exists a strong analogy between symmetric and anti-symmetric tensors, between bosons and fermions. It is often convenient to stress this analogy in terminology and notation.

Symmetric tensors over a vector space can be treated as polynomial functions on its dual. Such functions can be multiplied, differentiated and integrated, and we can change their variables.

There exists a similar language in the case of anti-symmetric tensors. It has been developed mostly by Berezin, hence it is sometimes called the *Berezin calculus*. It is often used by physicists, because it allows them to treat bosons and fermions within the same formalism.

Anti-symmetric calculus has a great appeal – it often allows us to express the analogy between the bosonic and fermionic cases in an elegant way. On the other hand, readers who see it for the first time can find it quite confusing and strange. Therefore, we devote this chapter to a presentation of elements of anti-symmetric calculus.

Note that the main goal of this chapter is to present a certain intriguing notation. Essentially no new concepts of independent importance are introduced here. Therefore, a reader in a hurry can probably skip this chapter on the first reading.

This chapter can be viewed as a continuation of Chap. 3, and especially of Sect. 3.6. In particular, we will use the anti-symmetric multiplication, differentiation and the Hodge star introduced already in Chap. 3.

7.1 Basic anti-symmetric calculus

Let $\mathcal{Y}$ be a vector space over $\mathbb{K}$ of dimension m. Let v denote the generic variable in $\mathcal{Y}^\#$ and y the generic variable in $\mathcal{Y}$. We remind the reader that $\Gamma_a^n(\mathcal{Y})$ denotes the n-th anti-symmetric tensor power of $\mathcal{Y}$.

7.1.1 Functional notation

Recall from Subsect. 3.5.1 that $\Psi \in \Gamma_a^n(\mathcal{Y})$ can be considered as a multi-linear anti-symmetric form

$$\mathcal{Y}^\# \times \cdots \times \mathcal{Y}^\# \ni (v_1, \ldots, v_n) \mapsto \Psi(v_1, \ldots, v_n) = \langle \Psi | v_1 \otimes_a \cdots \otimes_a v_n \rangle. \quad (7.1)$$

When we want to stress the meaning of an anti-symmetric tensor as a multi-linear form, we often write $\mathrm{Pol}_a^n(\mathcal{Y}^{\#})$ instead of $\Gamma_a^n(\mathcal{Y})$.

Definition 7.1 *It is convenient to write* $\Psi(v)$ *for (7.1), where* v *stands for the generic name of the variable in* $\mathcal{Y}^{\#}$ *and not for an individual element of* $\mathcal{Y}^{\#}$*. We will call it the* functional notation.

(We mentioned this notation already in Subsect. 3.5.1).

Sometimes we will consider a vector space with a different name, and then we will change the generic name of its dual variable used in the functional notation. For instance, $\Phi \in \mathrm{Pol}_a(\mathcal{Y}_i)$, resp. $\Psi \in \mathrm{Pol}_a(\mathcal{Y}_1 \oplus \mathcal{Y}_2)$, in the functional notation will be written as $\Phi(v_i)$, resp. $\Psi(v_1, v_2)$.

Remark 7.2 *Note that the same symbols have a different meaning in (7.1) and in the functional notation. In (7.1),* v_i *stands for an "individual element of* $\mathcal{Y}^{\#}$*". In the functional notation,* v_i *is the "name of the generic variable".*

7.1.2 Change of variables in anti-symmetric polynomials

Let $\mathcal{Y}_1, \mathcal{Y}_2$ be two finite-dimensional vector spaces. As mentioned above, v_1, v_2 will denote the generic variables in $\mathcal{Y}_1^{\#}$ and $\mathcal{Y}_2^{\#}$.

Consider $r \in L(\mathcal{Y}_1, \mathcal{Y}_2)$ and $\Psi \in \mathrm{Pol}_a^n(\mathcal{Y}_1^{\#})$. Then $\Gamma(r)\Psi$, understood as a multi-linear functional, acts as

$$\mathcal{Y}^{\#} \times \cdots \times \mathcal{Y}^{\#} \ni (v_1, \ldots, v_n) \mapsto \Gamma(r)\Psi(v_1, \ldots, v_n) = \Psi(r^{\#} v_1, \ldots, r^{\#} v_n). \quad (7.2)$$

Definition 7.3 *The functional notation for* $\Gamma(r)\Psi$ *is* $(\Gamma(r)\Psi)(v_2)$ *or, as suggested by (7.2),* $\Psi(r^{\#} v_2)$.

For example, let

$$\begin{aligned} j : \mathcal{Y} &\to \mathcal{Y} \oplus \mathcal{Y} \\ y &\mapsto y \oplus y, \end{aligned} \quad (7.3)$$

so that $j^{\#}(v_1, v_2) = v_1 + v_2$. Then the two possible functional notations for $\Gamma(j)\Psi$ are $(\Gamma(j)\Psi)(v_1, v_2)$ or $\Psi(v_1 + v_2)$.

7.1.3 Multiplication and differentiation operators

Definition 7.4 *If* $\Psi_1, \Psi_2 \in \Gamma_a(\mathcal{Y})$*, then* $\Psi_1 \otimes_a \Psi_2$ *will be denoted simply by* $\Psi_1 \cdot \Psi_2$*, if we consider* Ψ_1, Ψ_2 *as elements of* $\mathrm{Pol}_a(\mathcal{Y}^{\#})$*. The functional notation will be either* $\Psi_1 \cdot \Psi_2(v)$ *or* $\Psi_1(v)\Psi_2(v)$.

Recall that in Subsect. 3.5.2 we defined *multiplication* and *differentiation* operators. For $\Psi \in \mathrm{Pol}_a^n(\mathcal{Y}^{\#})$ they are given by

$$\begin{aligned} y(v)\Psi &:= y \otimes_a \Psi, && y \in \mathcal{Y}, \\ w(\nabla_v)\Psi &:= n\langle w| \otimes \mathbb{1}_{\mathcal{Y}}^{\otimes(n-1)}\Psi, && w \in \mathcal{Y}^{\#}. \end{aligned}$$

Therefore, v can be given the meaning of a $\mathcal{Y}^{\#}$-vector of anti-commuting operators on $\mathrm{Pol}_a(\mathcal{Y}^{\#})$. Similarly, ∇_v is a $\mathcal{Y}$-vector of anti-commuting operators on $\mathrm{Pol}_a(\mathcal{Y}^{\#})$.

Let $(e_1, \ldots, e_m)$ be a basis in $\mathcal{Y}$ and $(e^1, \ldots, e^m)$ be the corresponding dual basis in $\mathcal{Y}^{\#}$. The following operator on $\mathrm{Pol}_a(\mathcal{Y}^{\#} \oplus \mathcal{Y}^{\#})$ is clearly independent of the choice of the basis:

$$v_1 \cdot \nabla_{v_2} := \sum_{i=1}^{m} e_i(v_1) e^i(\nabla_{v_2}).$$

As an exercise in anti-symmetric calculus, it is instructive to check the following analog of *Taylor's formula*:

Proposition 7.5 *Let* $\Psi \in \mathrm{Pol}_a(\mathcal{Y}^{\#})$. *Then*

$$\Psi(v_1 + v_2) = e^{v_1 \cdot \nabla_{v_2}} \Psi(v_2).$$

Note that $(v_1 \cdot \nabla_{v_2})^p = 0$ for $p > \dim \mathcal{Y}$, so the exponential is well defined.

Proof of Prop. 7.5. Let

$$d = \begin{bmatrix} 0 & \mathbb{1} \\ 0 & 0 \end{bmatrix} \in L(\mathcal{Y} \oplus \mathcal{Y}),$$

$$j_2 = \begin{bmatrix} 0 \\ \mathbb{1} \end{bmatrix}, \quad j = \begin{bmatrix} \mathbb{1} \\ \mathbb{1} \end{bmatrix} \in L(\mathcal{Y}, \mathcal{Y} \oplus \mathcal{Y}).$$

Then $e^d j_2 = j$. This implies that $\Gamma(j) = e^{d\Gamma(d)} \Gamma(j_2)$. If we fix a basis $(e_1, \ldots, e_m)$ of $\mathcal{Y}$, then $d = \sum_{i=1}^{m} |e_i \oplus 0\rangle\langle 0 \oplus e^i|$. Hence,

$$d\Gamma(d) = \sum_{i=1}^{m} a^*(e_i \oplus 0) a(0 \oplus e^i)$$

$$= \sum_{i=1}^{m} e_i(v_1) e^i(\nabla_{v_2}) = v_1 \cdot \nabla_{v_2},$$

where we have used the functional notation for creation and annihilation operators:

$$a^*(e_i \oplus 0) = e_i(v_1), \quad a(0 \oplus e^i) = e^i(\nabla_{v_2}).$$

But

$$\Gamma(j)\Psi(v_1, v_2) = \Psi(v_1 + v_2), \quad \Gamma(j_2)\Psi(v_1, v_2) = \Psi(v_2). \qquad \square$$

7.1.4 Berezin integrals

Recall that in Subsects. 3.5.2 and 3.5.3 we defined the left and right differentiation. Even though it sounds a little strange, the right differentiation will be renamed as integration.

Let us be more precise. Let $\mathcal{Y}_1$ be a subspace of $\mathcal{Y}$ of dimension m_1. Its generic variable will be denoted v_1. Fix a volume form on $\mathcal{Y}_1$, that is, let $\Xi_1 \in \mathrm{Pol}_a^{m_1}(\mathcal{Y}_1)$ be a non-zero form.

Definition 7.6 *The partial right Berezin integral over $\mathcal{Y}_1$ of $\Psi \in \mathrm{Pol}_a(\mathcal{Y}^\#)$ is defined as*

$$\int \Psi(v)\mathrm{d}v_1 := \Xi_1(\overleftarrow{\nabla}_v)\Psi(v). \tag{7.4}$$

Note that (7.4) depends only on $(\mathcal{Y}/\mathcal{Y}_1)^\# \simeq \mathcal{Y}_1^{\mathrm{an}}$, where the superscript an stands for the annihilator (see Def. 1.11). Thus the Berezin integral produces an element of $\mathrm{Pol}_a(\mathcal{Y}_1^{\mathrm{an}})$.

In particular, if we take a volume form Ξ on $\mathcal{Y}$, i.e. a non-zero element of $\mathrm{Pol}_a^m(\mathcal{Y})$, then the *right Berezin integral over $\mathcal{Y}$*

$$\int \Psi(v)\mathrm{d}v = \langle \Xi | \Psi \rangle \tag{7.5}$$

yields a number.

Let $\mathcal{Y} = \mathcal{Y}_1 \oplus \mathcal{Y}_2$. The generic variable on $\mathcal{Y}^\# = \mathcal{Y}_1^\# \oplus \mathcal{Y}_2^\#$ is denoted $v = (v_1, v_2)$. Fix volume forms $\Xi_i \in \mathrm{Pol}_a^{m_i}(\mathcal{Y}_i)$. Equip $\mathcal{Y}^\#$ with the volume form $\Xi = \Xi_2 \wedge \Xi_1$. The corresponding Berezin integrals are denoted $\int \cdot\, \mathrm{d}v_i$, and $\int \cdot\, \mathrm{d}v$. Then we have the following version of the Fubini theorem:

$$\int \Psi(v)\mathrm{d}v = \int \left(\int \Psi(v_1, v_2)\mathrm{d}v_1 \right) \mathrm{d}v_2. \tag{7.6}$$

Thus, we can omit the parentheses and denote (7.6) by $\int\int \Psi(v_1, v_2)\mathrm{d}v_1\mathrm{d}v_2$.

Definition 7.7 *Apart from the right Berezin integral one considers the* partial left Berezin integral over $\mathcal{Y}_1$. *For $\Psi \in \mathrm{Pol}_a^n(\mathcal{Y}^\#)$, the left and right integrals are related to one another by*

$$\int \mathrm{d}v_1 \Psi(v) = (-1)^{m_1 n} \int \Psi(v)\mathrm{d}v_1.$$

In particular, we have the *left Berezin integral over $\mathcal{Y}$*:

$$\int \mathrm{d}v\Psi(v) = (-1)^m \int \Psi(v)\mathrm{d}v.$$

The following identities are easy to check for $\Psi \in \mathrm{Pol}_a(\mathcal{Y}^\#)$:

$$\int \Phi(\nabla_v)\Psi(v)\mathrm{d}v = 0, \qquad \Phi \in \mathrm{Pol}_a^{\geq 1}(\mathcal{Y});$$

$$\int \Psi(v + w)\mathrm{d}v = \int \Psi(v)\mathrm{d}v, \quad w \in \mathcal{Y}^\#;$$

$$\int \Psi(mv)\mathrm{d}v = (\det m) \int \Psi(v)\mathrm{d}v, \quad m \in L(\mathcal{Y}^\#). \tag{7.7}$$

Remark 7.8 *The identities of (7.7) are essentially the same as their analogs in the case of the usual integral described in (3.50) except for one important* *difference: the determinant in the formula for the change of variables has the opposite power.*

This is related to another difference between the Berezin and the usual integral. In the Berezin integral, such as (7.5), the natural meaning of the symbol dv *is a fixed volume form on* $\mathcal{Y}$*. In the usual integral, in the analogous situation, its meaning would be a volume form (or actually the corresponding density) on* $\mathcal{Y}^{\#}$*.*

Remark 7.9 *In the definition of the Berezin integral it does not matter whether the space* $\mathcal{Y}$ *is real or complex. However, if we want to have a closer analogy with the usual integral, we should assume that it is real. In this case, we can allow* $\Psi \in \mathbb{C}\mathrm{Pol}_a(\mathcal{Y}^{\#})$ *in (7.4), so that we can integrate complex polynomials.*

7.1.5 Berezin calculus in coordinates

So far, our presentation of anti-symmetric calculus has been coordinate-free. In most of the literature, it is introduced in a different way. One assumes from the very beginning that coordinates have been chosen and all definitions are coordinate-dependent. This approach has its advantages; in particular, it is a convenient way to check various identities. In this subsection we describe the anti-symmetric calculus in coordinates.

Definition 7.10 $v_1, \dots, v_m$ *denote symbols satisfying the relations*

$$v_i v_j = -v_j v_i. \tag{7.8}$$

They are called Grassmann *or* anti-commuting variables. *If* $I = \{i_1, \dots, i_p\}$ *with* $1 \le i_1 < \cdots < i_p \le m$*, we set* $\prod_{i \in I} v_i := v_{i_1} \cdots v_{i_p}$*.*

The space of expressions

$$\sum_{I \subset \{1,\dots,m\}} \alpha_I \prod_{i \in I} v_i, \quad \alpha_I \in \mathbb{K}$$

is an algebra naturally isomorphic to $\mathrm{Pol}_a(\mathbb{K}^m)$.

Remark 7.11 *Recall that in Remark 7.2 we distinguished two meanings of symbols* $v_1, v_2 \dots$. *The same symbols are used in Def. 7.10 with a third meaning. They stand for* anti-commuting variables *in* $\mathbb{K}^m$ *(generators of the algebra* $\mathrm{Pol}_a(\mathbb{K}^m)$*). The first meaning was as* individual vectors *in* $\mathcal{Y}^{\#}$*; see e.g. (7.1) and (7.12). The second was as the* generic variables *in* $\mathcal{Y}_i^{\#}$*; see (7.2).*

Definition 7.12 *Let* $I \subset \{1, \dots, m\}$*. We denote by* $\mathrm{sgn}(I)$ *the signature of* $(i_1, \dots, i_p, i_{p+1}, \dots, i_m)$*, where*

$$I = \{i_1, \dots, i_p\}, \quad I^c := \{1, \dots, m\} \backslash \{i_1, \dots, i_p\} = \{i_{p+1}, \dots, i_m\},$$

with $i_1 < \cdots < i_p$ *and* $i_{p+1} < \cdots < i_m$*.*

Definition 7.13 *The* Hodge star operator *is defined as*

$$\theta v_{i_1} \cdots v_{i_p} := \operatorname{sgn}(I)\, v_{i_m} \cdots v_{i_{m-p+1}}, \tag{7.9}$$

where $\{i_1, \ldots, i_m\}$ *and* $\operatorname{sgn}(I)$ *are as in Def. 7.12.*

Definition 7.14 *For* $i = 1, \ldots, m$, v_i *will denote not only an element of* $\operatorname{Pol_a}(\mathbb{K}^m)$, *but also the operator of left multiplication by* v_i *acting on* $\operatorname{Pol_a}(\mathbb{K}^m)$. *These operators clearly satisfy the relations (7.8). We consider also the partial derivatives* ∇_{v_i} *satisfying the relations*

$$[\nabla_{v_i}, \nabla_{v_i}]_+ = 0, \quad [\nabla_{v_i}, v_j]_+ = \delta_{ij}.$$

The action of the partial derivatives on the variables is given by

$$\nabla_{v_i} 1 = 0, \quad \nabla_{v_i} v_j = \delta_{ij}.$$

We introduce also the Berezin integral *w.r.t. the variable* v_i. *Its notation consists of two symbols:* $\int$ *and* $\mathrm{d}v_i$. *The rules of manipulating with* $\mathrm{d}v_i$ *are*

$$\mathrm{d}v_i \mathrm{d}v_j = -\mathrm{d}v_j \mathrm{d}v_i, \quad \mathrm{d}v_i v_j = -v_j \mathrm{d}v_i.$$

The rules of evaluating the integrals are

$$\int \mathrm{d}v_i = 0, \quad \int v_i \mathrm{d}v_j = \delta_{ij}.$$

For example, if $\sigma \in S_m$, *then*

$$\int v_{\sigma(1)} \cdots v_{\sigma(p)} \mathrm{d}v_m \cdots \mathrm{d}v_1 = \begin{cases} 0, & \text{if } p < m, \\ \operatorname{sgn}(\sigma), & \text{if } p = m. \end{cases}$$

Now let $\mathcal{Y}$ be a vector space of dimension m. If we fix a basis $(e_1, ..., e_m)$ of $\mathcal{Y}$, we can identify $\mathcal{Y}$ and $\mathcal{Y}^\#$ with $\mathbb{K}^m$, and hence $\operatorname{Pol_a}(\mathcal{Y}^\#)$ and $\operatorname{Pol_a}(\mathcal{Y})$ with $\operatorname{Pol_a}(\mathbb{K}^m)$. We see that v_i coincide with $e_i(v)$, ∇_{v_i} with $e^i(\nabla_v)$, and the Hodge star defined in (7.9) coincides with the Hodge star defined in Subsect. 3.6.2. If we use the volume form $e^m \wedge \cdots \wedge e^1$ on $\mathcal{Y}$, then

$$\int \Psi(v)\mathrm{d}v = \int \Psi(v_1, \ldots, v_m)\mathrm{d}v_m \cdots \mathrm{d}v_1,$$

$$\int \mathrm{d}v \Psi(v) = \int \mathrm{d}v_m \cdots \mathrm{d}v_1 \Psi(v_1, \ldots, v_m).$$

7.1.6 Differential operators and convolutions

The Hodge star operator transforms differentiation into convolution:

Theorem 7.15 *Let* $\Psi, \Phi \in \operatorname{Pol_a}(\mathcal{Y}^\#)$. *Then*

$$(\theta\Psi)(\nabla_v)\Phi(v) = (-1)^m \int \mathrm{d}w\, \Psi(w)\Phi(v + w).$$

If dim $\mathcal{Y}$ *is even, then the formula simplifies to*

$$(\theta\Psi)(\nabla_v)\Phi(v) = \int \Psi(w)\Phi(v+w)\mathrm{d}w.$$

Proof We fix a basis $(e_1, \ldots, e_m)$ of $\mathcal{Y}$ and use the anti-symmetric calculus in coordinates. Without loss of generality we can assume that $\Phi(v) = v_1 \cdots v_n$ and $\Psi(v) = \prod_{j \in J} v_j$. Then, using the notation of Def. 7.12,

$$\Psi(w)\Phi(v+w) = \sum_{I \subset \{1,\ldots,n\}} \mathrm{sgn}(I) \prod_{j \in J} w_j \prod_{i \in I} w_i \prod_{k \in \{1,\ldots,n\}\setminus I} v_k. \qquad (7.10)$$

The Berezin integral

$$\int \mathrm{d}w\,\Psi(w)\Phi(v+w) \qquad (7.11)$$

is non-zero only if $J = \{j_1, \ldots, j_p, n+1, \ldots, m\}$. The only term on the r.h.s. of (7.10) giving a non-zero contribution corresponds to $I = \{j_{p+1}, \ldots, j_n\}$. We have

$$\int \mathrm{d}w_m \cdots \mathrm{d}w_1\, \mathrm{sgn}(j_{p+1}, \ldots, j_n, j_1, \ldots, j_p, n+1, \ldots, m)$$
$$\times\, w_{j_1} \cdots w_{j_p} \cdot w_{n+1} \cdots w_m \cdot w_{j_{p+1}} \cdots w_{j_n} v_{j_1} \cdots v_{j_p}$$
$$= (-1)^m\, \mathrm{sgn}(j_{p+1}, \ldots, j_n, j_1, \ldots, j_p, n+1, \ldots, m)$$
$$\times\, \mathrm{sgn}(j_1, \ldots, j_p, n+1, \ldots, m, j_{p+1}, \ldots, j_n) v_{j_1} \cdots v_{j_p}.$$

On the other hand, using that

$$\theta\Psi(y) = \mathrm{sgn}(j_1, \ldots, j_p, n+1, \ldots, m, j_{p+1}, \ldots, j_n) y_{j_n} \cdots y_{j_{p+1}}$$

and

$$\Phi(v) = \mathrm{sgn}(j_{p+1}, \ldots, j_n, j_1, \ldots, j_p, n+1, \ldots, m)\, v_{j_{p+1}} \cdots v_{j_n} \cdot v_{j_1} \cdots v_{j_p},$$

we get

$$(\theta\Psi)(\nabla_v)\Psi(v) = \mathrm{sgn}(j_{p+1}, \ldots, j_n, j_1, \ldots, j_p, n+1, \ldots, m)$$
$$\times\, \mathrm{sgn}(j_1, \ldots j_p, n+1, \ldots, m, j_{p+1}, \ldots, j_n)\, v_{j_1} \cdots v_{j_p}.$$

This proves the first statement of the theorem. If m is even, then the left and right Berezin integrals coincide, which proves the second statement. $\qquad\square$

7.1.7 Anti-symmetric exponential

Definition 7.16 *The* anti-symmetric exponential *of* $\Phi \in \mathrm{Pol}_a(\mathcal{Y}^\#)$ *is defined as*

$$\mathrm{e}^\Phi(v) := \sum_{n=0}^\infty \frac{1}{n!} \Phi^n(v).$$

(Note that the series terminates after a finite number of terms.)

If at least one of the terms Φ_1, Φ_2 is even, then

$$e^{\Phi_1 + \Phi_2}(v) = e^{\Phi_1} e^{\Phi_2}(v).$$

The following propositions justify the analogy between the Hodge star operator and the Fourier transform.

Let $\mathcal{Y}$ be a vector space equipped with the volume form Ξ. Let us equip $\mathcal{Y}^{\#}$ with the volume form Ξ^{dual}.

Proposition 7.17 *Let $\Psi \in \mathrm{Pol}_{\mathrm{a}}(\mathcal{Y}^{\#})$. Then*

$$\theta\Psi(y) = (-1)^m \int \mathrm{d}v\Psi(v) \cdot e^{v \cdot y},$$

$$\Psi(v) = (-1)^m \int \mathrm{d}y\theta\Psi(y) \cdot e^{y \cdot v}.$$

In particular, if m is even, then

$$\theta\Psi(y) = \int \Psi(v) \cdot e^{v \cdot y} \mathrm{d}v,$$

$$\Psi(v) = \int \theta\Psi(y) \cdot e^{y \cdot v} \mathrm{d}y.$$

Proof We use the anti-symmetric calculus in coordinates and assume that $\Psi(v) = v_1 \cdots v_p$. We have

$$e^{v \cdot y} = e^{\sum_{i=1}^m v_i \cdot y_i} = \sum_{K \subset \{1, \ldots, m\}} \prod_{i \in K} v_i \cdot y_i.$$

This yields

$$\int \mathrm{d}v\Psi(v) e^{v \cdot y} = \int \mathrm{d}v_m \cdots \mathrm{d}v_1 v_1 \cdots v_p \cdot v_{p+1} \cdot y_{p+1} \cdots v_m \cdot y_m$$

$$= \int \mathrm{d}v_m \cdots \mathrm{d}v_1 v_1 \cdots v_m y_m \cdots y_{p+1} = (-1)^m y_m \cdots y_{p+1} = \theta\Psi(y).$$

The second identity can be proved similarly, using that $\mathrm{d}y = \mathrm{d}y_1 \cdots \mathrm{d}y_m$. $\square$

7.1.8 Anti-symmetric Gaussians

Let $\zeta \in \mathrm{Pol}_{\mathrm{a}}^2(\mathcal{Y}^{\#}) \simeq L_{\mathrm{a}}(\mathcal{Y}^{\#}, \mathcal{Y})$.

Definition 7.18 *The functional notation for*

$$\mathcal{Y}^{\#} \times \mathcal{Y}^{\#} \ni (v_1, v_2) \mapsto v_1 \cdot \zeta v_2 \tag{7.12}$$

will be either $\zeta(v)$ or, more often, $v \cdot \zeta v$. The functional notation for e^{ζ} will be either $e^{\zeta}(v)$ or $e^{v \cdot \zeta v}$.

The following proposition should be compared to (4.11) and (4.14), the corresponding identities for the usual Gaussians.

Proposition 7.19 *Let $\mathcal{Y}$ be a vector space of even dimension equipped with a volume form. Then*

(1) $\left(\theta e^{\frac{1}{2}\varsigma}\right)(y) = \int e^{y\cdot v}e^{\frac{1}{2}v\cdot\varsigma v}dv = \mathrm{Pf}(\varsigma)e^{\frac{1}{2}y\cdot\varsigma^{-1}y}$.

(2) $e^{\frac{1}{2}\nabla_v\cdot\varsigma^{-1}\nabla_v}\Phi(v) = \mathrm{Pf}(\varsigma)^{-1}\int e^{\frac{1}{2}w\cdot\varsigma w}\Phi(v+w)dw, \quad \Phi \in \mathrm{Pol}_a(\mathcal{Y}^\#)$.

(3) $\int e^{\frac{1}{2}v\cdot\varsigma v}dv = \mathrm{Pf}(\varsigma)$.

Proof Let us consider ς as an element of $L_a(\mathcal{Y}^\#, \mathcal{Y})$. Let us equip $\mathcal{Y}$ with a Euclidean structure ν compatible with the volume form Ξ and note that $\varsigma\nu$ is an anti-self-adjoint operator on $\mathcal{Y}$. Applying Corollary 2.85, we can find a basis $(e_1, \ldots, e_{2m})$ of $\mathcal{Y}$ such that

$$\varsigma = \sum_{i=1}^{m} \mu_i \left(|e_{2i-1}\rangle\langle e_{2i}| - |e_{2i}\rangle\langle e_{2i-1}|\right). \tag{7.13}$$

Note that

$$\mathrm{Pf}(\varsigma) = \prod_{i=1}^{m} \mu_i.$$

We can rewrite (7.13) as

$$\frac{1}{2}\varsigma = \sum_{i=1}^{m} \varsigma_i,$$

where $\varsigma_i = \mu_i e_{2i-1}\cdot e_{2i}$. Since $\varsigma_i^2 = 0$ and $\varsigma_i\varsigma_j = \varsigma_j\varsigma_i$, we have

$$e^{\frac{1}{2}\varsigma} = \sum_{I\subset\{1,\ldots,m\}} \prod_{i\in I} \varsigma_i.$$

Now

$$\theta \prod_{i\in I} \varsigma_i = \left(\prod_{i=1}^{m} \mu_i\right) \prod_{i\in I^c} \mu_i^{-1} e^{2i}\cdot e^{2i-1}.$$

This yields $\theta e^{\frac{1}{2}\varsigma} = \mathrm{Pf}(\varsigma)e^{\frac{1}{2}\varsigma^{-1}}$. By Prop. 7.17, we know that

$$\theta e^{\frac{1}{2}\varsigma}(y) = \int e^{\frac{1}{2}v\cdot\varsigma v}e^{v\cdot y}dv. \tag{7.14}$$

The two exponentials in the integral commute since they are both of even degree, and the function on the l.h.s. is an even function of y, which proves that (7.14) equals $\int e^{y\cdot v}e^{\frac{1}{2}v\cdot\varsigma v}dv$.

(2) follows from (1) and statement (1) of Thm. 7.15 for $\Psi(v) = e^{\frac{1}{2}v\cdot\varsigma v}$.

(3) follows from (2) for $\Phi = 1$. $\qquad\square$

7.2 Operators and anti-symmetric calculus

Throughout the section $\mathcal{X}$ is a vector space with $\dim\mathcal{X} = d$. Anti-symmetric calculus is especially useful in the context of the space $\mathcal{X} \oplus \mathcal{X}^\#$. This space has an even dimension and a natural volume form, which is helpful in the context of anti-symmetric calculus. We will see that the space $\mathrm{Pol_a}(\mathcal{X} \oplus \mathcal{X}^\#)$ is well suited to describe linear operators on $\Gamma_\mathrm{a}(\mathcal{X}^\#) = \mathrm{Pol_a}(\mathcal{X})$.

7.2.1 Berezin integral on $\mathcal{X} \oplus \mathcal{X}^\#$

In Subsect. 1.1.16, and then in Subsect. 3.6.4, we considered symplectic spaces of the form $\mathcal{X}^\# \oplus \mathcal{X}$ and $\mathcal{X} \oplus \mathcal{X}^\#$. They can be viewed as dual to one another. The canonical symplectic form on $\mathcal{X}^\# \oplus \mathcal{X}$ is denoted by ω. Consequently, the canonical symplectic form on $\mathcal{X} \oplus \mathcal{X}^\#$ is denoted by ω^{-1}. The corresponding Liouville forms are defined as $\frac{1}{d!} \wedge^d \omega$, resp. $\frac{1}{d!} \wedge^d \omega^{-1}$. If we choose a volume form Ξ on $\mathcal{X}$ and the volume form Ξ^{dual} on $\mathcal{X}^\#$, then the Liouville volume forms on both $\mathcal{X}^\# \oplus \mathcal{X}$ and $\mathcal{X} \oplus \mathcal{X}^\#$ are $\Xi^{\mathrm{dual}} \wedge \Xi$.

The generic variable of $\mathcal{X}$ will be denoted by x and of $\mathcal{X}^\#$ by ξ. The corresponding Berezin integrals will be denoted by $\int \cdot \, \mathrm{d}x$, resp. $\int \cdot \, \mathrm{d}\xi$. Hence the Berezin integral of $\Phi \in \mathrm{Pol_a}(\mathcal{X} \oplus \mathcal{X}^\#)$ w.r.t. the Liouville volume form will be denoted by

$$\int \Phi(x, \xi) \mathrm{d}\xi \mathrm{d}x.$$

If we fix a basis $(e_1, \ldots, e_d)$ of $\mathcal{X}$ and if $(e^1, \ldots, e^d)$ is the dual basis of $\mathcal{X}^\#$, then the symplectic form ω on $\mathcal{X}^\# \oplus \mathcal{X}$ and ω^{-1} on $\mathcal{X} \oplus \mathcal{X}^\#$ is

$$\sum_{i=1}^{d} e_i \wedge e^i. \tag{7.15}$$

The volume forms on $\mathcal{X}$, resp. $\mathcal{X}^\#$ are $e^d \wedge \cdots \wedge e^1$, resp. $e_1 \wedge \cdots \wedge e_d$, which, inside Berezin, integrals, is written as $\mathrm{d}x^d \cdots \mathrm{d}x^1$, resp. $\mathrm{d}\xi_1 \cdots \mathrm{d}\xi_d$.

Definition 7.20 *We will use the following shorthand functional notation:*

$$x \cdot \xi := \sum x^i \xi_i = \sum_{i=1}^{d} e^i(x) \cdot e_i(\xi)$$

$$= -\frac{1}{2}(x, \xi) \cdot \omega^{-1}(x, \xi),$$

$$\nabla_x \cdot \nabla_\xi := \nabla_{x^i} \cdot \nabla_{\xi_i} = \sum_{i=1}^{d} e_i(\nabla_x) \cdot e^i(\nabla_\xi)$$

$$= \frac{1}{2}(\nabla_x, \nabla_\xi) \cdot \omega(\nabla_x, \nabla_\xi),$$

where we have used various notational conventions to express the same object.

As an application we have the following proposition:

Proposition 7.21

$$\mathrm{e}^{t \nabla_x \cdot \nabla_\xi} \Phi(x, \xi) = t^d \int \mathrm{e}^{t^{-1}(\xi - \xi') \cdot (x - x')} \Phi(x', \xi') \mathrm{d}x' \mathrm{d}\xi'.$$

Proof By (7.15), $\mathrm{Pf}(\omega^{-1}) = 1$. Hence the proposition follows from Prop. 7.19 applied to $\zeta = t\omega^{-1}$. $\qquad\qquad\square$

7.2.2 Operators on the space of anti-symmetric polynomials

Let $B \in L\big(\mathrm{Pol}_a(\mathcal{X})\big)$.

Definition 7.22 *The* Bargmann kernel *of B is an element of $\mathrm{Pol}_a(\mathcal{X} \oplus \mathcal{X}^{\#})$, denoted B^{Bar}, obtained from $\frac{1}{\sqrt{N!}} B \frac{1}{\sqrt{N!}}$ by the following identification:*

$$L\big(\mathrm{Pol}_a(\mathcal{X})\big) \simeq \mathrm{Pol}_a(\mathcal{X}) \otimes \mathrm{Pol}_a(\mathcal{X})^{\#}$$
$$\simeq \mathrm{Pol}_a(\mathcal{X}) \otimes \mathrm{Pol}_a(\mathcal{X}^{\#}) \simeq \mathrm{Pol}_a(\mathcal{X} \oplus \mathcal{X}^{\#}). \qquad (7.16)$$

In the first identification we use the identification of $L(\mathcal{V})$ with $\mathcal{V} \otimes \mathcal{V}^{\#}$ described in Subsect. 3.1.8. The second involves the identification of $\mathrm{Pol}_a(\mathcal{X})^{\#}$ with $\mathrm{Pol}_a(\mathcal{X}^{\#})$; see (3.4). The third is the exponential law for anti-symmetric tensor algebras; see Subsect. 3.5.4.

Note that B^{Bar} is the fermionic analog of the Bargmann kernel of an operator introduced in Def. 9.51.

Let us compute the Bargmann kernel in a basis. Recall that we have the following notation: for $I = \{i_1, \ldots, i_n\} \subset \{1, \ldots, d\}$ with $i_1 < \cdots < i_n$,

$$e_I := e_{i_1} \cdots e_{i_n}, \qquad e^I := e^{i_n} \cdots e^{i_1}.$$

In the functional notation these are written as

$$e_I(\xi) := e_{i_1}(\xi) \cdots e_{i_n}(\xi), \quad e^I(x) := e^{i_n}(x) \cdots e^{i_1}(x).$$

We saw in Subsect. 3.3.6 that $\{e_I : I \subset \{1, \ldots, d\}\}$ is a basis of $\mathrm{Pol}_a(\mathcal{X})^{\#}$, and $\{\#I! e^I : I \subset \{1, \ldots, d\}\}$ is the dual basis of $\mathrm{Pol}_a(\mathcal{X})$. Clearly, $B \in L\big(\mathrm{Pol}_a(\mathcal{X})\big)$ can be written in terms of its matrix elements as

$$B = \sum_{I,J \subset \{1,\ldots,d\}} B_{I,J} \#I! |e^I\rangle\langle e_J|,$$

for

$$B_{I,J} = \#J! \langle e_I | B e^J \rangle.$$

Thus

$$\frac{1}{\sqrt{N!}} B \frac{1}{\sqrt{N!}} = \sum_{I,J \subset \{1,\ldots,d\}} B_{I,J} \sqrt{\#I!} |e^I\rangle\langle e_J| \frac{1}{\sqrt{\#J!}}.$$

Therefore, the identification (7.16) leads to the formula

$$B^{\mathrm{Bar}}(x, \xi) = \sum_{I,J \subset \{1,\ldots,d\}} B_{I,J} \sqrt{\#I!} e^I(x) \cdot e_J(\xi) \frac{1}{\sqrt{\#J!}}. \qquad (7.17)$$

Recall that Θ_{a}^k denotes the projection onto $\mathrm{Pol}_{\mathrm{a}}^k(\mathcal{X} \oplus \mathcal{X}^\#)$ (see Def. 3.24). Recall also that in Subsect. 3.5.7 we introduced the following notation: if $\Phi \in \mathrm{Pol}_{\mathrm{a}}(\mathcal{X}^\#)$, $\Psi \in \mathrm{Pol}_{\mathrm{a}}(\mathcal{X})$, then we write

$$\Psi^{\mathrm{mod}} := \frac{1}{\sqrt{N!}} \Psi, \qquad \Phi^{\mathrm{mod}} := \frac{1}{\sqrt{N!}} \Phi.$$

Theorem 7.23 (1) *Let $B \in L(\mathrm{Pol}_{\mathrm{a}}(\mathcal{X}))$, $0 \leq k \leq d$. Then*

$$\mathrm{Tr}\, B\Theta_{\mathrm{a}}^k = \frac{1}{(d-k)!} \int (x \cdot \xi)^{d-k} B^{\mathrm{Bar}}(x, \xi) \mathrm{d}x \mathrm{d}\xi,$$

$$\mathrm{Tr}\, B = \int \mathrm{e}^{x \cdot \xi} B^{\mathrm{Bar}}(x, \xi) \mathrm{d}x \mathrm{d}\xi.$$

(2) *Let $\Phi \in \mathrm{Pol}_{\mathrm{a}}(\mathcal{X}^\#)$, $\Psi \in \mathrm{Pol}_{\mathrm{a}}(\mathcal{X})$. Then*

$$\langle \Phi | \Theta_{\mathrm{a}}^k \Psi \rangle = \frac{1}{(d-k)!} \int (x \cdot \xi)^{d-k} \Psi^{\mathrm{mod}}(x) \Phi^{\mathrm{mod}}(\xi) \mathrm{d}x \mathrm{d}\xi,$$

$$\langle \Phi | \Psi \rangle = \int \mathrm{e}^{x \cdot \xi} \Psi^{\mathrm{mod}}(x) \Phi^{\mathrm{mod}}(\xi) \mathrm{d}x \mathrm{d}\xi.$$

Proof Using the basis of $\mathcal{X}$ and $\mathcal{X}^\#$, we can write

$$\frac{1}{(d-k)!} (x \cdot \xi)^{d-k} = \sum_{\#K=d-k} \prod_{i \in K} e^i(x) \cdot e_i(\xi).$$

By (7.17),

$$\frac{1}{(d-k)!}(x \cdot \xi)^{d-k} B^{\mathrm{Bar}}(x, \xi)$$

$$= \sum_{\#K=d-k} \prod_{i \in K} e^i(x) \cdot e_i(\xi) \sum_{I,J} B_{I,J} \sqrt{\#I!} e^I(x) \cdot e_J(\xi) \frac{1}{\sqrt{\#J!}}. \qquad (7.18)$$

In the integral of (7.18), only the terms of degree (d, d) contribute. Therefore, we can replace (7.18) by

$$\sum_{\#I=k} \prod_{i \in I^c} e^i(x) \cdot e_i(\xi) B_{I,I} e^I(x) \cdot e_I(\xi).$$

Since $e^I \cdot e_I = \prod_{i \in I} e^i(x) \cdot e_i(\xi)$ and $\prod_{i=1}^d e^i \cdot e_i = e^d \cdots e^1 \cdot e_1 \cdots e_d$, we get

$$\frac{1}{(d-k)!} \int (x \cdot \xi)^{d-k} B^{\mathrm{Bar}}(x, \xi) \mathrm{d}x \mathrm{d}\xi = \sum_{\#I=k} B_{I,I}$$

$$= \mathrm{Tr}(B\Theta_{\mathrm{a}}^k).$$

This proves the first statement of (1). The second follows by taking the sum over $1 \leq k \leq d$.

(2) follows from (1) by noting that if $B = |\Psi\rangle\langle\Phi|$, then $B^{\mathrm{Bar}}(x, \xi) = \Psi^{\mathrm{mod}}(x) \cdot \Phi^{\mathrm{mod}}(\xi)$ and $\mathrm{Tr}|\Psi\rangle\langle\Phi| = \langle\Phi|\Psi\rangle$. $\square$

7.2.3 Integral kernel of an operator

Let $B \in L(\mathrm{Pol}_a(\mathcal{X}))$. It is easy to see that there exists a unique $B(\cdot, \cdot) \in \mathrm{Pol}_a(\mathcal{X} \oplus \mathcal{X})$ such that for $\Psi \in \mathrm{Pol}_a(\mathcal{X})$

$$B\Psi(x) = \int B(x, y)\Psi(y)\mathrm{d}y,$$

where we use y as the generic variable in the second copy of $\mathcal{X}$.

Definition 7.24 *We will call* $B(x, y)$ *the* integral kernel of B *(w.r.t. the volume form Ξ).*

Clearly, if $\mathcal{X}$ is real, the integral kernel introduced in the above definition is the fermionic analog of the usual integral kernel, such as in Thm. 4.24.

7.2.4 x, ∇_x-quantization

Definition 7.25 *We define the* x, ∇_x-*quantization, resp. the* ∇_x, x-*quantization as the maps*

$$\mathrm{Pol}_a(\mathcal{X} \oplus \mathcal{X}^\#) \ni b \mapsto \mathrm{Op}^{x, \nabla_x}(b) \in L(\mathrm{Pol}_a(\mathcal{X})),$$
$$\mathrm{Pol}_a(\mathcal{X} \oplus \mathcal{X}^\#) \ni b \mapsto \mathrm{Op}^{\nabla_x, x}(b) \in L(\mathrm{Pol}_a(\mathcal{X})),$$

defined as follows: Let $b_1 \in \mathrm{Pol}_a(\mathcal{X})$, $b_2 \in \mathrm{Pol}_a(\mathcal{X}^\#)$. *Then for* $b(x, \xi) = b_1(x)b_2(\xi)$ *we set*

$$\mathrm{Op}^{x, \nabla_x}(b) := b_1(x)b_2(\nabla_x),$$

and for $b(x, \xi) = b_2(\xi)b_1(x)$ *we set*

$$\mathrm{Op}^{x, \nabla_x}(b) := b_2(\nabla_x)b_1(x).$$

We extend the definition to $\mathrm{Pol}_a(\mathcal{X} \oplus \mathcal{X}^\#)$ *by linearity.*

If $\mathcal{X}$ is real, the (fermionic) ∇_x, x- and x, ∇_x-quantizations introduced above are parallel to the (bosonic) D, x- and x, D-quantizations discussed in Subsect. 4.3.1. If $\mathcal{X}$ is complex, they essentially coincide with the fermionic Wick and anti-Wick quantizations, which will be discussed in Subsect. 13.3.1.

Theorem 7.26 *Assume that d is even.*

(1) *Let* $b \in \mathrm{Pol}_a(\mathcal{X} \oplus \mathcal{X}^\#)$. *Then the integral kernels of the quantizations of b are*

$$\mathrm{Op}^{x, \nabla_x}(b)(x, y) = \int b(x, \xi)\mathrm{e}^{(x-y)\cdot\xi}\mathrm{d}\xi,$$

$$\mathrm{Op}^{\nabla_x, x}(b)(x, y) = \int b(y, \xi)\mathrm{e}^{(x-y)\cdot\xi}\mathrm{d}\xi.$$

(2) *If $b_+, b_- \in \mathrm{Pol}_a(\mathcal{X} \oplus \mathcal{X}^\#)$ and $\mathrm{Op}^{x,\nabla_x}(b_+) = \mathrm{Op}^{\nabla_x,x}(b_-)$, then*

$$b_+(x,\xi) = \mathrm{e}^{\nabla_x \cdot \nabla_\xi} b_-(x,\xi)$$

$$= \int b_-(x_1,\xi_1) \mathrm{e}^{(\xi-\xi_1)\cdot(x-x_1)} \mathrm{d}x_1 \mathrm{d}\xi_1.$$

(3) *If $b_1, b_2 \in \mathrm{Pol}_a(\mathcal{X} \oplus \mathcal{X}^\#)$ and $\mathrm{Op}^{x,\nabla_x}(b_1)\mathrm{Op}^{x,\nabla_x}(b_2) = \mathrm{Op}^{x,\nabla_x}(b)$, then*

$$b(x,\xi) = \mathrm{e}^{\nabla_{x_2}\cdot\nabla_{\xi_1}} b_1(x_1,\xi_1) b_2(x_2,\xi_2) \Big|_{\substack{x_1 = x_2 = x, \\ \xi_1 = \xi_2 = \xi}}$$

$$= \int \mathrm{e}^{(\xi-\xi_1)\cdot(x-x_1)} b_1(x,\xi_1) b_2(x_1,\xi) \mathrm{d}x_1 \mathrm{d}\xi_1.$$

Proof We will give a proof of (1) for the x, ∇_x-quantization. We can assume that $b(x,\xi) = b_1(x)b_2(\xi)$. Then using Thm. 7.15 and Prop. 7.17, we obtain

$$b_1(x)b_2(\nabla_x)\Psi(x) = \int b_1(x)\theta^{-1} b_2(y)\Psi(x+y)\mathrm{d}y$$

$$= \int b_1(x)b_2(\xi)\mathrm{e}^{\xi\cdot y}\Psi(x+y)\mathrm{d}\xi\mathrm{d}y$$

$$= \int b_1(x)b_2(\xi)\mathrm{e}^{(x-y)\cdot\xi}\Psi(y)\mathrm{d}\xi\mathrm{d}y. \qquad \square$$

7.3 Notes

The material of this chapter is based on the work of Berezin (1966, 1983).

8

Canonical commutation relations

Throughout this chapter $(\mathcal{Y}, \omega)$ is a pre-symplectic space, that is, $\mathcal{Y}$ is a real vector space equipped with an anti-symmetric form ω. From the point of view of classical mechanics, $\mathcal{Y}$ will have the interpretation of the *dual of a phase space*, or, as we will say for brevity, of a *dual phase space*. Note that for quantum mechanics dual phase spaces seem more fundamental that phase spaces.

In this chapter we introduce the concept of a *representation of the canonical commutation relations* (a *CCR representation*). According to a naive definition, a CCR representation is a linear map

$$\mathcal{Y} \ni y \mapsto \phi^\pi(y) \tag{8.1}$$

with values in self-adjoint operators on a certain Hilbert space satisfying

$$[\phi^\pi(y_1), \phi^\pi(y_2)] = \mathrm{i} y_1 \cdot \omega y_2 \, \mathbb{1}. \tag{8.2}$$

We will call (8.2) the *canonical commutation relations in the Heisenberg form*. They are unfortunately problematic, because one needs to supply them with the precise meaning of the commutator of unbounded operators on the left hand side.

Weyl proposed replacing (8.2) with the relations satisfied by the operators $\mathrm{e}^{\mathrm{i} \phi^\pi(y)}$. These operators are bounded, and therefore one does not need to discuss domain questions. In our definition of CCR representations we will use the *canonical commutation relations in the Weyl form* (8.4). Under additional regularity assumptions they imply the CCR in the Heisenberg form.

We will introduce two kinds of CCR representations. The usual definition is appropriate to describe *neutral bosons*. In the case of *charged bosons* a somewhat different formalism is used, which we introduce under the name "*charged CCR representations*". Charged CCR representations can be viewed as special cases of (neutral) CCR representations, where the dual phase space $\mathcal{Y}$ is complex and a somewhat different notation is used.

8.1 CCR representations

8.1.1 Definition of a CCR representation

Let $\mathcal{H}$ be a Hilbert space. Recall that $U(\mathcal{H})$ denotes the set of unitary operators on $\mathcal{H}$.

Definition 8.1 *A representation of the canonical commutation relations or a CCR representation over* $(\mathcal{Y}, \omega)$ *in* $\mathcal{H}$ *is a map*

$$\mathcal{Y} \ni y \mapsto W^{\pi}(y) \in U(\mathcal{H}) \tag{8.3}$$

satisfying

$$W^{\pi}(y_1)W^{\pi}(y_2) = e^{-\frac{1}{2}y_1 \cdot \omega y_2} W^{\pi}(y_1 + y_2). \tag{8.4}$$

$W^{\pi}(y)$ *is then called the* Weyl operator corresponding to $y \in \mathcal{Y}$.

Remark 8.2 *The superscript* π *is an example of the "name" of a given CCR representation. It is attached to* W, *which is the generic symbol for "Weyl operators". Later on the same superscript will be attached to other generic symbols, e.g. field operators* ϕ.

Remark 8.3 *Sometimes we will call (8.3)* neutral *CCR representations, to distinguish them from* charged *CCR representations introduced in Def. 8.35.*

Proposition 8.4 *Consider a CCR representation (8.3). Let* $y, y_1, y_2 \in \mathcal{Y}$, $t_1, t_2 \in \mathbb{R}$. *Then*

$$W^{\pi*}(y) = W^{\pi}(-y), \qquad W^{\pi}(0) = \mathbb{1},$$
$$W^{\pi}(t_1 y)W^{\pi}(t_2 y) = W^{\pi}\big((t_1 + t_2)y\big),$$
$$W^{\pi}(y_1)W^{\pi}(y_2) = e^{-iy_1 \cdot \omega y_2} W^{\pi}(y_2)W^{\pi}(y_1). \tag{8.5}$$

Definition 8.5 *A CCR representation (8.3) is called* regular *if*

$$\mathbb{R} \ni t \mapsto W^{\pi}(ty) \in U(\mathcal{H}) \quad \text{is strongly continuous for any } y \in \mathcal{Y}. \tag{8.6}$$

8.1.2 CCR representations over a direct sum

CCR representations can be easily tensored with one another:

Proposition 8.6 *If*

$$\mathcal{Y}_i \ni y_i \mapsto W^i(y) \in U(\mathcal{H}_i), \quad i = 1, 2, \tag{8.7}$$

are two CCR representations, then

$$\mathcal{Y}_1 \oplus \mathcal{Y}_2 \ni (y_1, y_2) \mapsto W^1(y_1) \otimes W^2(y_2) \in U(\mathcal{H}_1 \otimes \mathcal{H}_2)$$

is also a CCR representation.

8.1.3 Cyclicity and irreducibility

Consider a CCR representation (8.3). The following concepts are parallel to the analogous concepts in the representation theory of groups or C^*-algebras:

Definition 8.7 *We say that a subset $\mathcal{E} \subset \mathcal{H}$ is cyclic for (8.3) if $\mathrm{Span}\{W^{\pi}(y)\Psi :$
$\Psi \in \mathcal{E},\ y \in \mathcal{Y}\}$ is dense in $\mathcal{H}$. We say that $\Psi_0 \in \mathcal{H}$ is cyclic for (8.3) if $\{\Psi_0\}$ is
cyclic for (8.3).*

Definition 8.8 *We say that the CCR representation (8.3) is irreducible if
the only closed subspaces of $\mathcal{H}$ invariant under the $W^{\pi}(y)$ for $y \in \mathcal{Y}$ are $\{0\}$
and $\mathcal{H}$.*

Proposition 8.9 (1) *A CCR representation is irreducible iff $B \in B(\mathcal{H})$ and
$[W^{\pi}(y), B] = 0$ for all $y \in \mathcal{Y}$ implies that B is proportional to identity.*
(2) *In the case of an irreducible representation, all non-zero vectors in $\mathcal{H}$ are
cyclic.*

8.1.4 Characteristic functions of CCR representations

Definition 8.10 *We say that $\mathcal{Y} \ni y \mapsto G(y) \in \mathbb{C}$ is a characteristic function if
for $\alpha_1, \ldots, \alpha_n \in \mathbb{C}$, $y_1, \ldots, y_n \in \mathcal{Y}$ and $n \in \mathbb{N}$ we have*

$$\sum_{i,j=1}^{n} \overline{\alpha}_i \alpha_j G(-y_i + y_j) \mathrm{e}^{\frac{i}{2} y_i \cdot \omega y_j} \geq 0. \tag{8.8}$$

Note that for any CCR representation $y \mapsto W(y) \in U(\mathcal{H})$ and any vector $\Psi \in \mathcal{H}$

$$G(y) := (\Psi | W(y)\Psi) \tag{8.9}$$

is a characteristic function. We will see that every characteristic function comes
from a certain CCR representation and a cyclic vector, as in (8.9).

Until the end of this subsection we assume that $y \mapsto G(y)$ is a characteristic
function. Set $\mathcal{H}_0 = c_c(\mathcal{Y}, \mathbb{C})$, as in Def. 2.6, that is, $\mathcal{H}_0$ is the vector space of
finitely supported functions on $\mathcal{Y}$. Equip it with the sesquilinear form $(\cdot|\cdot)$ defined
by

$$(\delta_{y_1} | \delta_{y_2}) := \mathrm{e}^{\frac{i}{2} y_1 \cdot \omega y_2} G(-y_1 + y_2).$$

It follows from (8.8) that $(\cdot|\cdot)$ is semi-positive definite. Let $\mathcal{N}$ be the space of
vectors in $\xi \in \mathcal{H}_0$ such that $(\xi|\xi) = 0$. Set $\mathcal{H} := (\mathcal{H}_0 / \mathcal{N})^{\mathrm{cpl}}$.

For any $y \in \mathcal{Y}$ we define a linear operator $W_0(y)$ on $\mathcal{H}_0$ by

$$W_0(y)\delta_{y_1} := \mathrm{e}^{\frac{i}{2} y \cdot \omega y_1} \delta_{y_1 + y}.$$

The operator $W_0(y)$ preserves the form $(\cdot|\cdot)$, hence it preserves $\mathcal{N}$. Therefore, it
defines a linear operator $W(y)$ on $\mathcal{H}_0 / \mathcal{N}$ by

$$W(y)\xi := W_0(y)\xi + \mathcal{N}, \quad \xi \in \mathcal{H}_0.$$

$W(y)$ extends to a unitary operator on $\mathcal{H}$. We set $\Psi := \delta_0 + \mathcal{N}$.

Proposition 8.11 *Consider the family of operators*

$$\mathcal{Y} \ni y \mapsto W(y) \in U(\mathcal{H}) \qquad (8.10)$$

constructed above from a characteristic function $y \mapsto G(y)$.

(1) *(8.10) is a CCR representation,* Ψ *is a cyclic vector and* $G(y) = (\Psi|W(y)\Psi)$.
(2) *The following conditions are equivalent:*
  (i) *(8.10) is regular.*
  (ii) $\mathbb{R} \ni t \mapsto G(y_1 + ty_2)$ *is continuous for any* $y_1, y_2 \in \mathcal{Y}$.

8.1.5 Intertwining operators

Let

$$\mathcal{Y} \ni y \mapsto W^1(y) \in U(\mathcal{H}_1), \qquad (8.11)$$

$$\mathcal{Y} \ni y \mapsto W^2(y) \in U(\mathcal{H}_2) \qquad (8.12)$$

be CCR representations over the same pre-symplectic space $\mathcal{Y}$.

Definition 8.12 *We say that an operator* $A \in B(\mathcal{H}_1, \mathcal{H}_2)$ *intertwines (8.11) and (8.12) iff*

$$AW^1(y) = W^2(y)A, \quad y \in \mathcal{Y}.$$

We say that (8.11) and (8.12) are unitarily equivalent *if there exists* $U \in U(\mathcal{H}_1, \mathcal{H}_2)$ *intertwining (8.11) and (8.12).*

The proof of the following proposition is essentially identical to the proof of Thm. 6.29:

Proposition 8.13 *If the representations (8.11) and (8.12) are irreducible, then the set of operators intertwining them is either* $\{0\}$ *or* $\{\lambda U : \lambda \in \mathbb{C}\}$ *for some* $U \in U(\mathcal{H})$.

8.1.6 Schrödinger representation

Let $\mathcal{X}$ be a finite-dimensional real vector space. Equip $\mathcal{X}^\# \oplus \mathcal{X}$ with its canonical symplectic form. It follows from Thms. 4.28 and 4.29 that the map

$$\mathcal{X}^\# \oplus \mathcal{X} \ni (\eta, q) \mapsto e^{i(\eta \cdot x + q \cdot D)} \in U(L^2(\mathcal{X})) \qquad (8.13)$$

is an irreducible regular CCR representation.

Definition 8.14 *(8.13) is called the* Schrödinger representation *over* $\mathcal{X}^\# \oplus \mathcal{X}$.

Conversely let $(\mathcal{Y}, \omega)$ be a finite-dimensional symplectic space and

$$\mathcal{Y} \ni y \mapsto W(y) \in U(\mathcal{H}) \qquad (8.14)$$

be a regular CCR representation. By Thm. 1.47, there exists a space $\mathcal{X}$ such that $\mathcal{Y}$ can be identified with $\mathcal{X}^\# \oplus \mathcal{X}$ as symplectic spaces. Thus we can rewrite (8.14) as

$$\mathcal{X}^\# \oplus \mathcal{X} \ni (\eta, q) \mapsto W(\eta, q)$$

satisfying

$$W(\eta_1, q_1) W(\eta_2, q_2) = e^{-\frac{i}{2}(\eta_1 \cdot q_2 - \eta_2 \cdot q_1)} W(\eta_1 + \eta_2, q_1 + q_2).$$

The maps

$$\mathcal{X}^\# \ni \eta \mapsto W(\eta, 0),$$
$$\mathcal{X} \ni q \mapsto W(0, q)$$

are strongly continuous unitary groups satisfying

$$W(\eta, 0) W(0, q) = e^{-i\eta \cdot q} W(0, q) W(\eta, 0).$$

The following theorem is a corollary to the Stone–von Neumann theorem:

Theorem 8.15 *Under the above stated assumptions, there exists a Hilbert space $\mathcal{K}$ and a unitary operator $U : L^2(\mathcal{X}) \otimes \mathcal{K} \to \mathcal{H}$ such that*

$$W(\eta, q) U = U e^{i(\eta \cdot x + q \cdot D)} \otimes \mathbb{1}_\mathcal{K}.$$

The representation is irreducible iff $\mathcal{K} = \mathbb{C}$.

Proof It suffices to use Thm. 4.34 and the identities

$$W(\eta, q) = e^{-\frac{i}{2}\eta \cdot q} W(\eta, 0) W(0, q), \quad e^{i(\eta \cdot x + q \cdot D)} = e^{-\frac{i}{2}\eta \cdot q} e^{i\eta \cdot x} e^{iq \cdot D}.$$

$\square$

The following corollary follows directly from Thm. 4.29 and Prop. 8.13:

Corollary 8.16 *Suppose that $\mathcal{Y}$ is a finite-dimensional symplectic space. Let $\mathcal{Y} \ni y \mapsto W_i(y) \in U(\mathcal{H})$, $i = 1, 2$, be two regular irreducible CCR representations. Then there exists $U \in U(\mathcal{H}_1, \mathcal{H}_2)$, unique up to a phase factor, such that $UW_1(y) = W_2(y)U$.*

8.1.7 Weighted Schrödinger representations

Suppose that $\mathcal{X}$ is a finite-dimensional vector space with a Lebesgue measure dx. Fix $m \in L^2_{\mathrm{loc}}(\mathcal{X})$ such that $m \neq 0$ a.e.. Define the measure $d\mu(x) = |m|^2(x)dx$. Then

$$L^2(\mathcal{X}, d\mu) \ni \Psi \mapsto U\Psi := m\Psi \in L^2(\mathcal{X}, dx)$$

is a unitary operator. If in addition $m \in L^2(\mathcal{X})$, then $U1 = m$.

The following theorem is obvious:

Theorem 8.17

$$\mathcal{X}^* \oplus \mathcal{X} \ni (\eta, q) \mapsto U^* e^{i\eta \cdot x + iq \cdot D} U$$

$$= e^{i\eta \cdot x + iq \cdot D + m^{-1}(x)q \cdot \nabla m(x)} \in U\big(L^2(\mathcal{X}, d\mu)\big)$$

is a regular irreducible CCR representation.

Remark 8.18 *If $V(x) := \frac{1}{2} m^{-1}(x) \Delta m(x)$ is sufficiently regular, then we can define the Schrödinger operator $H := -\frac{1}{2}\Delta + V(x)$. If $m \in L^2(\mathcal{X})$, then we have $Hm = 0$.*

The operator H in the $L^2(\mathcal{X}, d\mu)$ representation looks like

$$U^* H U = -\frac{1}{2}\Delta - m^{-1}(x)\nabla m(x) \cdot \nabla.$$

It is called the Dirichlet form *corresponding to H. If $m \in L^2(\mathcal{X})$, then 1 is its eigenstate with the eigenvalue 0.*

8.1.8 Examples of non-regular CCR representations

In most applications to quantum physics, CCR representations are regular. However, non-regular representations are also useful. In this subsection we describe a couple of examples of non-regular CCR representations.

Recall that, for a set I, $l^2(I)$ denotes the Hilbert space of square summable families of complex numbers indexed by I.

Example 8.19 *Consider the Hilbert space $l^2(\mathcal{Y})$ and the following operators:*

$$W^{\mathrm{d}}(y)f(x) := e^{-\frac{i}{2} y \cdot \omega x} f(x + y). \tag{8.15}$$

Then

$$\mathcal{Y} \ni y \mapsto W^{\mathrm{d}}(y) \in U\big(l^2(\mathcal{Y})\big) \tag{8.16}$$

is a CCR representation.

Note that $\mathbb{R} \ni t \mapsto W^{\mathrm{d}}(ty)$ is not strongly continuous for non-zero $y \in \mathcal{Y}$. Hence (8.16) is non-regular.

Example 8.20 *Let $\mathcal{X}$ be a real vector space (of any dimension). Recall that $\mathcal{X}^\# \oplus \mathcal{X}$ is naturally a symplectic space. On $l^2(\mathcal{X})$ define the following operators:*

$$V(\eta)f(x) := e^{i\eta \cdot x} f(x), \quad \eta \in \mathcal{X}^\#;$$
$$T(q)f(x) := f(x - q), \quad q \in \mathcal{X}.$$

Then

$$\mathcal{X}^\# \oplus \mathcal{X} \ni (\eta, q) \mapsto V(\eta)T(q)e^{\frac{i}{2}\eta \cdot q} \in U\big(l^2(\mathcal{X})\big) \tag{8.17}$$

is a CCR representation.

Note that $\mathbb{R} \ni t \mapsto T(tq)$ *is not strongly continuous for non-zero* $q \in \mathcal{X}$. *Hence (8.17) is non-regular.*

8.1.9 Bogoliubov transformations

Let

$$\mathcal{Y} \ni y \mapsto W(y) \in U(\mathcal{H}) \tag{8.18}$$

be a CCR representation.

Recall that $\mathcal{Y}^{\#}$ denotes the space of linear functionals on $\mathcal{Y}$, and $Sp(\mathcal{Y})$ the group of symplectic transformations of $\mathcal{Y}$. Let $v \in \mathcal{Y}^{\#}$, $r \in Sp(\mathcal{Y})$. Clearly, the map

$$\mathcal{Y} \ni y \mapsto W^{v,r}(y) := e^{iv \cdot y} W(ry) \in U(\mathcal{H}) \tag{8.19}$$

is a CCR representation.

Definition 8.21 *(8.19) can be called the* Bogoliubov transformation *of (8.18) by (v,r). Alternatively, if $r = \mathbb{1}$, it can be called the* Bogoliubov translation *by v or, if $v = 0$, the* Bogoliubov rotation *by r.*

The pairs (v,r) that appear in (8.19) are naturally interpreted as elements of the group $\mathcal{Y}^{\#} \rtimes Sp(\mathcal{Y})$, the semi-direct product of $\mathcal{Y}^{\#}$ and $Sp(\mathcal{Y})$, with the product given by

$$(v_2, r_2)(v_1, r_1) := (r_1^{\#} v_2 + v_1, r_2 r_1).$$

Note that $\mathcal{Y}^{\#} \rtimes Sp(\mathcal{Y})$ can be viewed as a subgroup of the affine group $\mathcal{Y}^{\#} \rtimes Sp(\mathcal{Y}^{\#}) = ASp(\mathcal{Y}^{\#})$, with the homomorphic embedding

$$\mathcal{Y}^{\#} \rtimes Sp(\mathcal{Y}) \ni (v,r) \mapsto (r^{\#-1}v, r^{\#-1}) \in ASp(\mathcal{Y}^{\#}).$$

Proposition 8.22 (1) *If $(v_1, r_1), (v_2, r_2) \in \mathcal{Y}^{\#} \rtimes Sp(\mathcal{Y})$, then*

$$\left(W^{(v_1, r_1)}\right)^{(v_2, r_2)}(y) = W^{(v_1, r_1)(v_2, r_2)}(y).$$

(2) *The set of $(v,r) \in \mathcal{Y}^{\#} \rtimes Sp(\mathcal{Y})$ such that (8.19) is unitarily equivalent to (8.18) is a subgroup of $\mathcal{Y}^{\#} \rtimes Sp(\mathcal{Y})$ containing $\omega \mathcal{Y} \rtimes \{\mathbb{1}\} \subset \mathcal{Y}^{\#} \rtimes \{\mathbb{1}\}$.*
(3) *(8.19) is regular iff (8.18) is.*
(4) *(8.19) is irreducible iff (8.18) is.*

Proof To see that for $v \subset \omega \mathcal{Y}$ (8.19) and (8.18) are equivalent, we note

$$W^{v,\mathbb{1}}(y) = W(\omega^{-1}v)W(y)W(-\omega^{-1}v). \qquad \square$$

Proposition 8.23 *Let $\mathcal{Y}$ be finite-dimensional and ω symplectic. Then*

(1) *(8.18) and (8.19) are unitarily equivalent for any $(v,r) \in \mathcal{Y}^{\#} \rtimes Sp(\mathcal{Y})$.*

(2) *Let* $\mathrm{Op}(\cdot)$ *and* $\mathrm{Op}^{(v,r)}(\cdot)$ *denote the Weyl quantization w.r.t. (8.18) and (8.19) respectively. (See (8.42) later on for the definition of the Weyl quantization.) For* $b \in \mathcal{S}'(\mathcal{Y}^{\#})$, *set*

$$b^{v,r}(w) = b(r^{\#}w + v), \quad w \in \mathcal{Y}^{\#}.$$

Then $\mathrm{Op}^{(v,r)}(b) = \mathrm{Op}\left(b^{(v,r)}\right)$.

8.2 Field operators

Throughout the section, $(\mathcal{Y}, \omega)$ is a pre-symplectic space and we are given a regular CCR representation

$$\mathcal{Y} \ni y \mapsto W^{\pi}(y) \in U(\mathcal{H}). \tag{8.20}$$

8.2.1 Definition of field operators

By regularity and (8.6), $\mathbb{R} \ni t \mapsto W^{\pi}(ty)$ is a strongly continuous unitary group. By Stone's theorem, for any $y \in \mathcal{Y}$, we can define its self-adjoint generator

$$\phi^{\pi}(y) := -\mathrm{i}\frac{\mathrm{d}}{\mathrm{d}t}W^{\pi}(ty)\big|_{t=0}.$$

In other words, $\mathrm{e}^{\mathrm{i}\phi^{\pi}(y)} = W^{\pi}(y)$.

Definition 8.24 $\phi^{\pi}(y)$ *will be called the* field operator *corresponding to* $y \in \mathcal{Y}$. *(Sometimes the name* Segal field operator *is used.)*

Theorem 8.25 *Let* $y, y_1, y_2 \in \mathcal{Y}$.

(1) $W^{\pi}(y)$ *leaves invariant* $\mathrm{Dom}\,\phi^{\pi}(y_1)$ *and*

$$[\phi^{\pi}(y), W^{\pi}(y_1)] = y_1 \cdot \omega y\, W^{\pi}(y_1). \tag{8.21}$$

(2) $\phi^{\pi}(ty) = t\phi^{\pi}(y)$, $t \in \mathbb{R}$.
(3) *One has* $\mathrm{Dom}\,\phi^{\pi}(y_1) \cap \mathrm{Dom}\,\phi^{\pi}(y_2) \subset \mathrm{Dom}\,\phi^{\pi}(y_1 + y_2)$ *and*

$$\phi^{\pi}(y_1 + y_2) = \phi^{\pi}(y_1) + \phi^{\pi}(y_2), \quad on\ \mathrm{Dom}\,\phi^{\pi}(y_1) \cap \mathrm{Dom}\,\phi^{\pi}(y_2). \tag{8.22}$$

(4) *In the sense of quadratic forms on* $\mathrm{Dom}\,\phi^{\pi}(y_1) \cap \mathrm{Dom}\,\phi^{\pi}(y_2)$, *we have*

$$[\phi^{\pi}(y_1), \phi^{\pi}(y_2)] = \mathrm{i}y_1 \cdot \omega y_2\, \mathbb{1}. \tag{8.23}$$

Proof (8.21) follows immediately from differentiating in t the identity

$$W^{\pi}(ty)W^{\pi}(y_1) = W^{\pi}(y_1)W^{\pi}(ty)\mathrm{e}^{-\mathrm{i}ty \cdot \omega y_1}.$$

To obtain (8.22), we note that, for $\Psi \in \operatorname{Dom} \phi^\pi(y_1) \cap \operatorname{Dom} \phi^\pi(y_2)$,

$$
\begin{aligned}
t^{-1}\big(W^\pi(t(y_1+y_2)) - \mathbb{1}\big)\Psi \;=\;& \mathrm{e}^{-\frac{i}{2}t^2 y_1 \cdot \omega y_2} W^\pi(ty_1) t^{-1}\big(W^\pi(ty_2) - \mathbb{1}\big)\Psi \\
&+ \mathrm{e}^{-\frac{i}{2}t^2 y_1 \cdot \omega y_2} t^{-1}\big(W^\pi(ty_1) - \mathbb{1}\big)\Psi \\
&+ t^{-1}\big(\mathrm{e}^{-\frac{i}{2}t^2 y_1 \cdot \omega y_2} - \mathbb{1}\big)\Psi \\
\underset{t\to 0}{\longrightarrow}\;& \mathrm{i}\phi(y_2)\Psi + \mathrm{i}\phi(y_1)\Psi.
\end{aligned}
$$

By differentiating the identity

$$
\big(W^\pi(t_1 y_1)\Psi_1 \,\big|\, W^\pi(t_2 y_2)\Psi_2\big) = \mathrm{e}^{-\mathrm{i}t_1 t_2 y_1 \cdot \omega y_2}\big(W^\pi(t_2 y_2)\Psi_1 \,\big|\, W^\pi(t_1 y_1)\Psi_2\big)
$$

w.r.t. t_1 and t_2, and setting $t_1 = t_2 = 0$, we obtain (8.23). $\qquad\square$

Sometimes it is convenient to introduce CCR representations with help of field operators, as described in the following proposition. We recall that $Cl_{\mathrm{h}}(\mathcal{H})$ denotes the set of self-adjoint operators on $\mathcal{H}$.

Proposition 8.26 *Let* $\mathcal{Y} \ni y \mapsto \phi^\pi(y) \in Cl_{\mathrm{h}}(\mathcal{H})$ *be a map such that*

(1) $\phi^\pi(ty) = t\phi^\pi(y)$, $t \in \mathbb{R}$;
(2) $\mathrm{e}^{\mathrm{i}\phi^\pi(y_1)}\mathrm{e}^{\mathrm{i}\phi^\pi(y_2)} = \mathrm{e}^{-\frac{i}{2}y_1 \cdot \omega y_2}\mathrm{e}^{\mathrm{i}\phi^\pi(y_1+y_2)}$, $y_1, y_2 \in \mathcal{Y}$.

Then $\mathcal{Y} \ni y \mapsto W^\pi(y) := \mathrm{e}^{\mathrm{i}\phi^\pi(y)}$ *is a regular CCR representation, and* $\phi^\pi(y)$ *are the corresponding Segal field operators.*

Remark 8.27 *Let* $\mathcal{X} \subset \mathcal{Y}$ *be an isotropic subspace. Then the field operators* $\phi^\pi(q)$ *with* $q \in \mathcal{X}$ *commute with one another. Hence*

$$
\phi^\pi(q), \; q \in \mathcal{X},
$$

is an $\mathcal{X}^{\#}$*-vector of commuting self-adjoint operators (see Def. 2.77). If* f *is a cylindrical Borel function on* $\mathcal{X}^{\#}$*, then the operator* $f(\phi^\pi)$ *is well defined by the functional calculus.*

8.2.2 Common domain of field operators

Definition 8.28 *The* Schwartz space *for the CCR representation (8.20) is defined as the intersection of* $\operatorname{Dom} \phi^\pi(y_1) \cdots \phi^\pi(y_n)$ *for* $y_1, \ldots, y_n \in \mathcal{Y}$. *It is denoted* $\mathcal{H}^{\infty,\pi}$ *and has the structure of a topological vector space with semi-norms* $\|\phi^\pi(y_1) \cdots \phi^\pi(y_n)\Psi\|$.

Clearly, polynomials in $\phi^\pi(y)$ act as operators on $\mathcal{H}^{\infty,\pi}$.

Theorem 8.29 *Let* $\mathcal{Y}$ *be finite-dimensional. Then*

(1) $\mathcal{H}^{\infty,\pi}$ *is dense in* $\mathcal{H}$.
(2) *If* $\omega = 0$, *then* $\mathcal{H}^{\infty,\pi}$ *coincides with the space of* C^∞ *vectors for the vector of commuting self-adjoint operators* ϕ^π.
(3) *If* ω *is non-degenerate, then* $\Psi \in \mathcal{H}^{\infty,\pi}$ *iff the function* $\mathcal{Y} \ni y \mapsto (\Psi|W^\pi(y)\Psi)$ *belongs to* $\mathcal{S}(\mathcal{Y})$.

(4) *If* $\mathcal{Y} = \mathcal{X}^{\#} \oplus \mathcal{X}$ *and (8.20) is the Schrödinger representation in* $L^2(\mathcal{X})$, *then* $\mathcal{H}^{\infty,\pi}$ *equals* $\mathcal{S}(\mathcal{X})$.

Proof (2) is obvious. (3) follows from Thms. 8.15 and 4.30. (4) follows from Thm. 4.15.

Let us prove (1). Set $\mathcal{Y}_0 = \operatorname{Ker} \omega$. Let $\mathcal{Y}_1 \subset \mathcal{Y}$ be a complementary space to $\mathcal{Y}_0$. $\mathcal{Y}_1$ is symplectic, hence we can assume that, for some space $\mathcal{X}$, $\mathcal{Y}_1 = \mathcal{X}^{\#} \oplus \mathcal{X}$ with the canonical symplectic form. By Thm. 8.15, there exists a unitary map $U : L^2(\mathcal{X}) \otimes \mathcal{K} \to \mathcal{H}$ such that

$$W^{\pi}(y_1) = UW(y_1) \otimes \mathbb{1}_{\mathcal{K}} U^*, \quad y_1 \in \mathcal{Y}_1,$$

where $W(y)$ denote the Weyl operators in the Schrödinger representation. Now we know from (3) that $U \, \mathcal{S}(\mathcal{X}) \overset{a1}{\otimes} \mathcal{K}$ is contained in the Schwartz space for $\mathcal{Y}_1 \ni y_1 \mapsto W^{\pi}(y_1)$.

Using that $\mathcal{Y}_0$ and $\mathcal{Y}_1$ are orthogonal for ω and Thm. 4.29, we obtain that $U^* W^{\pi}(y_0)U = \mathbb{1} \otimes W^{\pi_0}(y_0)$ for $y_0 \in \mathcal{Y}_0$, where $\mathcal{Y}_0 \ni y_0 \mapsto W^{\pi_0}(y_0) \in U(\mathcal{K})$ is a CCR representation. By (2), the corresponding Schwartz space $\mathcal{K}^{\infty,\pi_0}$ is dense in $\mathcal{K}$. Thus $U \, \mathcal{S}(\mathcal{X}) \overset{a1}{\otimes} \mathcal{K}^{\infty,\pi_0} \subset \mathcal{H}^{\infty,\pi}$ is dense in $\mathcal{H}$. $\qquad\square$

If $\mathcal{Y}$ has an arbitrary dimension, then Thm. 8.29 is still useful, because it can be applied to finite-dimensional subspaces of $\mathcal{Y}$. In particular, Thm. 8.29 implies that for an arbitrary symplectic space $\mathcal{Y}$, the spaces $\operatorname{Dom} \phi^{\pi}(y_1) \cap \operatorname{Dom} \phi^{\pi}(y_2)$ considered in Thm. 8.25 are dense in $\mathcal{H}$.

8.2.3 *Non-self-adjoint fields*

As in Subsect. 1.3.5, we can equip $\mathbb{C}\mathcal{Y}$ with the anti-symmetric form $\omega_{\mathbb{C}}$.

Definition 8.30 *For* $w = y_1 + \mathrm{i}y_2$, $y_1, y_2 \in \mathcal{Y}$, *we define the* field operator

$$\phi^{\pi}(w) := \phi^{\pi}(y_1) + \mathrm{i}\phi^{\pi}(y_2) \ \textit{with domain} \ \operatorname{Dom} \phi^{\pi}(y_1) \cap \operatorname{Dom} \phi^{\pi}(y_2).$$

Proposition 8.31 (1) *For* $w = y_1 + \mathrm{i}y_2$, $y_1, y_2 \in \mathcal{Y}$,

$$\phi^{\pi}(w) \ \textit{is closed on} \ \operatorname{Dom} \phi^{\pi}(y_1) \cap \operatorname{Dom} \phi^{\pi}(y_2).$$

(2) *For* $w_1, w_2 \in \mathbb{C}\mathcal{Y}$, $\lambda_1, \lambda_2 \in \mathbb{C}$,

$$\phi^{\pi}(\lambda_1 w_1 + \lambda_2 w_2) = \lambda_1 \phi^{\pi}(w_1) + \lambda_2 \phi^{\pi}(w_2) \ \textit{on} \ \operatorname{Dom} \phi^{\pi}(w_1) \cap \operatorname{Dom} \phi^{\pi}(w_2).$$

(3) *For* $w_1, w_2 \in \mathbb{C}\mathcal{Y}$,

$$[\phi^{\pi}(w_1), \phi^{\pi}(w_2)] = \mathrm{i}w_1 {\cdot} \omega_{\mathbb{C}} w_2 \, \mathbb{1} \ \textit{as a quadratic form on}$$
$$\operatorname{Dom} \phi^{\pi}(w_1) \cap \operatorname{Dom} \phi^{\pi}(w_2).$$

Proof By Thm. 8.25, we have, for $\Psi \in \operatorname{Dom} \phi^{\pi}(y_1) \cap \operatorname{Dom} \phi^{\pi}(y_2)$,

$$\|\phi^{\pi}(y_1 + \mathrm{i}y_2)\Psi\|^2 = \|\phi^{\pi}(y_1)\Psi\|^2 + \|\phi^{\pi}(y_2)\Psi\|^2 - y_1 {\cdot} \omega y_2 \|\Psi\|^2. \tag{8.24}$$

We know that $\phi^\pi(y_1)$ and $\phi^\pi(y_2)$ are self-adjoint, hence closed. Therefore, $\mathrm{Dom}\,\phi^\pi(y_1)$ and $\mathrm{Dom}\,\phi^\pi(y_2)$ are complete in the graph norms. Hence so is $\mathrm{Dom}\,\phi^\pi(y_1)\cap\mathrm{Dom}\,\phi^\pi(y_2)$ in the intersection norm. This proves (1). (2) is immediate and (3) follows immediately from Thm. 8.25 (4). $\qquad\square$

8.2.4 CCR over a Kähler space

In this subsection we assume that ω is symplectic. We fix a CCR representation (8.20). We use the notation and results of Subsects. 1.3.6, 1.3.8 and 1.3.9.

The following proposition shows that choosing a sufficiently large subspace of commuting field operators that annihilate a certain vector is equivalent to fixing a Kähler structure in $(\mathcal{Y},\omega)$.

Proposition 8.32 *Suppose that $\mathcal{Z}$ is a complex subspace of $\mathbb{C}\mathcal{Y}$ such that*

(1) $\mathbb{C}\mathcal{Y} = \mathcal{Z}\oplus\overline{\mathcal{Z}}$,
(2) $\overline{z}_1,\overline{z}_2\in\overline{\mathcal{Z}}$ *implies* $\phi^\pi(\overline{z}_1)\phi^\pi(\overline{z}_2) = \phi^\pi(\overline{z}_2)\phi^\pi(\overline{z}_1)$ *(or, equivalently, $\overline{\mathcal{Z}}$ is isotropic for $\omega_\mathbb{C}$).*

Then there exists a unique pseudo-Kähler anti-involution j on $(\mathcal{Y},\omega)$ such that

$$\mathcal{Z} = \{y - \mathrm{i}\mathrm{j}y : y\in\mathcal{Y}\}. \tag{8.25}$$

If in addition
(3) *there exists a non-zero $\Omega\in\mathcal{H}$ such that $\Omega\in\mathrm{Dom}\,\phi^\pi(\overline{z})$ and $\phi^\pi(\overline{z})\Omega = 0$, $z\in\mathcal{Z}$, then j is Kähler.*

Proof By (1), each $y\subset\mathcal{Y}$ can be written uniquely as $y = z_y + \overline{z}_y$. Clearly, z_y depends linearly on y. We have $\overline{\mathrm{i}(2z_y - y)} = \mathrm{i}(2z_y - y)$. Hence $\mathrm{j}y := \mathrm{i}(2z_y - y)$ defines $\mathrm{j}\in L(\mathcal{Y})$, and (8.25) is true.

(2) implies

$$0 = (y_1 + \mathrm{i}\mathrm{j}y_1)\cdot\omega_\mathbb{C}(y_2 + \mathrm{i}\mathrm{j}y_2)$$
$$= y_1\cdot\omega y_2 - (\mathrm{j}y_1)\cdot\omega(\mathrm{j}y_2) + \mathrm{i}\big((\mathrm{j}y_1)\cdot\omega y_2 + y_1\cdot\omega\mathrm{j}y_2\big).$$

Hence

$$y_1\cdot\omega y_2 - (\mathrm{j}y_1)\cdot\omega(\mathrm{j}y_2) = (\mathrm{j}y_1)\cdot\omega y_2 + y_1\cdot\omega\mathrm{j}y_2 = 0,$$

which shows that j is symplectic and infinitesimally symplectic, hence pseudo-Kähler.

Then we compute using (3):

$$0 = \|\phi(y + \mathrm{i}\mathrm{j}y)\Omega\|^2$$
$$= (\Omega|\phi^\pi(y)^2\Omega) + (\Omega|\phi^\pi(\mathrm{j}y)^2\Omega) - \mathrm{i}(\Omega|[\phi^\pi(\mathrm{j}y),\phi^\pi(y)]\Omega) \geq -y\cdot\omega\mathrm{j}y. \qquad\square$$

Motivated in part by the above proposition, let us fix j, a pseudo-Kähler anti-involution on $(\mathcal{Y}, \omega)$. Recall that the space $\mathcal{Z}$ given by (8.25) is called the holomorphic subspace of $\mathbb{C}\mathcal{Y}$ (see Subsect. 1.3.6).

Definition 8.33 *We define the* (abstract) creation *and* annihilation *operators associated with* j *by*

$$a^{\pi*}(z) := \phi^{\pi}(z), \quad a^{\pi}(z) := \phi^{\pi}(\overline{z}), \quad z \in \mathcal{Z}.$$

By Prop. 8.31, if $z = y - \mathrm{ij}y \in \mathcal{Z}$, then $a^{\pi}(z) = \phi^{\pi}(y) + \mathrm{i}\phi^{\pi}(\mathrm{j}y)$, $a^{\pi*}(z) = \phi^{\pi}(y) - \mathrm{i}\phi^{\pi}(\mathrm{j}y)$ are closed operators on $\mathrm{Dom}\,\phi^{\pi}(y) \cap \mathrm{Dom}\,\phi^{\pi}(\mathrm{j}y)$.

Proposition 8.34 (1) *One has* $\phi^{\pi}(z, \overline{z}) = a^{\pi*}(z) + a^{\pi}(z)$, $z \in \mathcal{Z}$.
(2)

$$[a^{\pi*}(z_1), a^{\pi*}(z_2)] = 0, \quad [a^{\pi}(z_1), a^{\pi}(z_2)] = 0,$$
$$[a^{\pi}(z_1), a^{\pi*}(z_2)] = \overline{z}_1 \cdot z_2 \mathbb{1}, \quad z_1, z_2 \in \mathcal{Z}.$$

Proof (1) is immediate, since $(z, \overline{z}) = (z, 0) + (0, \overline{z})$. The first line of (2) follows from the fact that $\mathcal{Z}$, $\overline{\mathcal{Z}}$ are isotropic for $\omega_{\mathbb{C}}$ (see Subsect. 1.3.9). To prove the second line we write

$$[a^{\pi}(z_1), a^{\pi*}(z_2)] = [\phi^{\pi}(\overline{z_1}), \phi^{\pi}(z_2)] = \mathrm{i}\overline{z_1} \cdot \omega_{\mathbb{C}} z_2 \mathbb{1}$$
$$= -\mathrm{i}\overline{z_1} \cdot \mathrm{j}_{\mathbb{C}} z_2 \mathbb{1} = \overline{z_1} \cdot z_2 \mathbb{1},$$

using Subsect. 1.3.9 and the fact that $\mathrm{j}_{\mathbb{C}} z_2 = \mathrm{i} z_2$, since $z_2 \in \mathcal{Z}$. $\qquad\square$

Note that in the case of a Fock representation, considered in Chap. 9, the space $\mathcal{Y}$ has a natural Kähler structure. The abstract creation and annihilation operators defined in Def. 8.33 coincide then with the usual creation and annihilation operators.

If the space $\mathcal{Y}$ is equipped with a charge 1 symmetry, then we have a natural pseudo-Kähler structure (see Subsect. 1.3.11). The corresponding creation and annihilation operators are then called *charged field operators*. However, in this case we prefer to use a slightly different formalism, which is described in the next subsection.

8.2.5 Charged CCR representations

CCR representations, as defined in Def. 8.1, are used mainly to describe neutral bosons. Therefore, sometimes we will call them *neutral CCR representations*. In the context of charged bosons one uses another formalism described in the following definition.

Definition 8.35 *Let* $(\mathcal{Y}, \omega)$ *be a charged pre-symplectic space, that is, a complex vector space equipped with an anti-Hermitian form denoted* $(y_1|\omega y_2)$, $y_1, y_2 \in \mathcal{Y}$

(see Subsect. 1.2.11). Let $\mathcal{H}$ be a Hilbert space. We say that a map

$$\mathcal{Y} \ni y \mapsto \psi^\pi(y) \in Cl(\mathcal{H})$$

is a charged CCR representation *if there exists a regular CCR representation of* $(\mathcal{Y}_\mathbb{R}, \mathrm{Re}\,(\cdot|\omega\cdot))$

$$\mathcal{Y} \ni y \mapsto W^\pi(y) = \mathrm{e}^{\mathrm{i}\phi^\pi(y)} \in U(\mathcal{H}) \tag{8.26}$$

such that

$$\psi^\pi(y) = \frac{1}{\sqrt{2}}(\phi^\pi(y) + \mathrm{i}\phi^\pi(\mathrm{i}y)), \quad y \in \mathcal{Y}.$$

Proposition 8.36 *Suppose that $\mathcal{Y} \ni y \mapsto \psi^\pi(y)$ is a charged CCR representation. Let $y, y_1, y_2 \in \mathcal{Y}$. We have:*

(1) $\psi^\pi(\lambda y) = \overline{\lambda}\psi^\pi(y), \lambda \in \mathbb{C}$.
(2) *On* $\mathrm{Dom}\,\psi^\pi(y_1) \cap \mathrm{Dom}\,\psi^\pi(y_2)$ *we have* $\psi^\pi(y_1 + y_2) = \psi^\pi(y_1) + \psi^\pi(y_2)$.
(3) *In the sense of quadratic forms, we have the identities*

$$[\psi^{\pi*}(y_1), \psi^{\pi*}(y_2)] = [\psi^\pi(y_1), \psi^\pi(y_2)] = 0,$$
$$[\psi^\pi(y_1), \psi^{\pi*}(y_2)] = \mathrm{i}(y_1|\omega y_2)\mathbb{1}.$$

By definition, a charged CCR representation determines the neutral CCR representation (8.26) on the symplectic space $(\mathcal{Y}_\mathbb{R}, \mathrm{Re}(\cdot|\omega\cdot))$ with the fields given by

$$\phi^\pi(y) := \frac{1}{\sqrt{2}}\left(\psi^\pi(y) + \psi^{\pi*}(y)\right), \quad y \in \mathcal{Y}. \tag{8.27}$$

In addition, $(\mathcal{Y}_\mathbb{R}, \mathrm{Re}(\cdot|\omega\cdot))$ is equipped with a charge 1 symmetry

$$U(1) \ni \theta \mapsto \mathrm{e}^{\mathrm{i}\theta} \in Sp(\mathcal{Y}_\mathbb{R}).$$

Conversely, charged CCR representations arise when the underlying symplectic space of a (neutral) CCR representation is equipped with a charge 1 symmetry. Let us make this precise. Suppose that $(\mathcal{Y}, \omega)$ is a symplectic space and

$$\mathcal{Y} \ni y \mapsto \mathrm{e}^{\mathrm{i}\phi(y)} \in U(\mathcal{H})$$

is a regular neutral CCR representation. Suppose that

$$U(1) \ni \theta \mapsto u_\theta = \cos\theta\mathbb{1} + \sin\theta\mathrm{j_{ch}} \in Sp(\mathcal{Y})$$

is a charge 1 symmetry. We know from Prop. 1.94 (2) that $\mathrm{j_{ch}}$ is a pseudo-Kähler anti-involution. Set

$$\psi^\pi(y) = \frac{1}{\sqrt{2}}(\phi^\pi(y) + \mathrm{i}\phi^\pi(\mathrm{j_{ch}}y)), \quad \psi^{\pi*}(y) = \frac{1}{\sqrt{2}}(\phi^\pi(y) - \mathrm{i}\phi^\pi(\mathrm{j_{ch}}y)), \quad y \in \mathcal{Y}.$$

Then we obtain a charged CCR representation over $\mathcal{Y}^{\mathbb{C}}$ with the complex structure given by $\mathrm{j_{ch}}$ and the anti-Hermitian form

$$(y_1|\omega y_2) := y_1 \cdot \omega y_2 - \mathrm{i} y_1 \cdot \omega \mathrm{j_{ch}} y_2, \quad y_1, y_2 \in \mathcal{Y}.$$

We can look at this construction as follows. By the standard procedure described in the previous subsection, we introduce the holomorphic subspace for $\mathrm{j_{ch}}$, that is,

$$\mathcal{Z}_{\mathrm{ch}} := \{y - \mathrm{i} \mathrm{j_{ch}} y \ : \ y \in \mathcal{Y}\} \subset \mathbb{C}\mathcal{Y}.$$

Introduce the creation and annihilation operators associated with $\mathrm{j_{ch}}$:

$$a_{\mathrm{ch}}^{\pi}(z) := \phi^{\pi}(\bar{z}), \quad a_{\mathrm{ch}}^{\pi*}(z) := \phi^{\pi*}(z), \quad z \in \mathcal{Z}_{\mathrm{ch}}.$$

We have a natural identification of the space $\mathcal{Z}_{\mathrm{ch}}$ with $\mathcal{Y}$:

$$\mathcal{Y} \ni y \mapsto z = \frac{1}{\sqrt{2}}(\mathbb{1} - \mathrm{i} \mathrm{j_{ch}})y \in \mathcal{Z}_{\mathrm{ch}}. \tag{8.28}$$

Then

$$\psi^{\pi}(y) := a_{\mathrm{ch}}^{\pi}(z), \quad \psi^{\pi*}(y) := a_{\mathrm{ch}}^{\pi*}(z).$$

8.2.6 CCR over a symplectic space with conjugation

Let $\mathcal{X}$ be a real vector space. Let $\mathcal{V}$ be a subspace of $\mathcal{X}^{\#}$. Consider the space $\mathcal{V} \oplus \mathcal{X}$ equipped with its canonical pre-symplectic form ω. Clearly, it is also equipped with a conjugation

$$\tau(\eta, q) = (\eta, -q), \quad (\eta, q) \in \mathcal{V} \oplus \mathcal{X}.$$

Let

$$\mathcal{V} \oplus \mathcal{X} \ni (\eta, q) \mapsto \mathrm{e}^{\mathrm{i}\phi^{\pi}(\eta, q)} \in U(\mathcal{H})$$

be a regular CCR representation.

Definition 8.37 *The* (abstract) *position and momentum operators are* $\mathcal{X}$- *and* $\mathcal{X}^{\#}$-*vectors of commuting self-adjoint operators defined by*

$$\eta \cdot x^{\pi} := \phi^{\pi}(\eta, 0), \quad \eta \in \mathcal{V};$$
$$q \cdot D^{\pi} := \phi^{\pi}(0, q), \quad q \in \mathcal{X}.$$

A natural conjugation on the symplectic space $\mathcal{Y}$ is available in the case of the Schrödinger representation. In this case the operators defined in Def. 8.37 are the usual momentum and position operators.

Recall that for the Schrödinger representation the symplectic space is finite-dimensional. One often considers a conjugation on an infinite-dimensional symplectic space. This is the case for the real-wave representation (see Sect. 9.3), which to some extent can be viewed as a generalization of the Schrödinger representation to infinite dimensions. However, besides a conjugation, the real-wave representation requires an additional structure: $\mathcal{Y}$ needs to be a Kähler space. CCR relations over a Kähler space with a conjugation are discussed in the following subsection.

8.2.7 CCR over a Kähler space with conjugation

Suppose that $\mathcal{X}$ is a real Hilbert space and $c > 0$ is an operator on $\mathcal{X}$. Set

$$\mathcal{Y} := (2c)^{-\frac{1}{2}}\mathcal{X} \oplus (2c)^{\frac{1}{2}}\mathcal{X},$$

which is a symplectic space with a conjugation. (Note that $(2c)^{-\frac{1}{2}}\mathcal{X}$ can be viewed as the space dual to $(2c)^{\frac{1}{2}}\mathcal{X}$). Consider a regular CCR representation

$$(2c)^{-\frac{1}{2}}\mathcal{X} \oplus (2c)^{\frac{1}{2}}\mathcal{X} \ni (\eta, q) \mapsto e^{i\phi^\pi(\eta,q)} \in U(\mathcal{H}).$$

Let x^π and D^π be the position and momentum operators introduced in Def. 8.37. We introduce the following definition:

Definition 8.38 *For* $w \in \mathbb{C}c^{-\frac{1}{2}}\mathcal{X}$ *we define* Schrödinger-type creation *and* annihilation operators

$$a_{\mathrm{sch}}^{\pi*}(w) := \frac{1}{2}w \cdot x^\pi - icw \cdot D^\pi, \quad a_{\mathrm{sch}}^\pi(w) := \frac{1}{2}\overline{w} \cdot x^\pi + ic\overline{w} \cdot D^\pi.$$

By Subsect. 8.2.3, $a_{\mathrm{sch}}^\pi(w)$ and $a_{\mathrm{sch}}^{\pi*}(w)$ are closed and the adjoints of each other on their natural domains.

Proposition 8.39 (1) *For* $\eta \in \mathbb{C}(2c)^{-\frac{1}{2}}\mathcal{X}$, $q \in \mathbb{C}(2c)^{\frac{1}{2}}\mathcal{X}$, *we have*

$$\eta \cdot x^\pi = a_{\mathrm{sch}}^{\pi*}(\eta) + a_{\mathrm{sch}}^\pi(\overline{\eta}), \quad q \cdot D^\pi = \frac{1}{2i}\big(a_{\mathrm{sch}}^\pi(c^{-1}\overline{q}) - a_{\mathrm{sch}}^{\pi*}(c^{-1}q)\big). \quad (8.29)$$

(2) *For* $w_1, w_2 \in \mathbb{C}\mathcal{X}$,

$$\big[a_{\mathrm{sch}}^\pi(w_1), a_{\mathrm{sch}}^{\pi*}(w_2)\big] = \overline{w_1} \cdot cw_2 \,\mathbb{1},$$

$$\big[a_{\mathrm{sch}}^\pi(w_1), a_{\mathrm{sch}}^\pi(w_2)\big] = \big[a_{\mathrm{sch}}^{\pi*}(w_1), a_{\mathrm{sch}}^{\pi*}(w_2)\big] = 0. \quad (8.30)$$

It is easy to interpret Schrödinger-type creation and annihilation operators in terms of an appropriate Kähler structure on $\mathcal{Y}$ with a conjugation, following the terminology of Subsect. 1.3.10. Let us equip $\mathcal{Y} = (2c)^{-\frac{1}{2}}\mathcal{X} \oplus (2c)^{\frac{1}{2}}\mathcal{X}$ with the anti-involution

$$\mathrm{j} := \begin{bmatrix} 0 & -(2c)^{-1} \\ 2c & 0 \end{bmatrix}.$$

Clearly, the pair j, ω is Kähler. The corresponding scalar product of $(\eta_i, q_i) \in \mathcal{Y}$, $i = 1, 2$, is

$$(\eta_1, q_1) \cdot (\eta_2, q_2) = \eta_1 \cdot 2c\eta_2 + q_1 \cdot (2c)^{-1}q_2. \quad (8.31)$$

·

Let us consider the map

$$\mathbb{C}c^{-\frac{1}{2}}\mathcal{X} \ni w \mapsto z := \frac{\mathbb{1}-\mathrm{i}\mathrm{j}}{2}(w,0) = \left(\frac{1}{2}w, -\mathrm{i}cw\right) \in \mathbb{C}\mathcal{Y}. \qquad (8.32)$$

(8.32) is unitary onto $\mathcal{Z}$, the holomorphic subspace of $\mathbb{C}\mathcal{Y}$ associated with j. Then we have

$$a^{\pi}_{\mathrm{sch}}(w) = a^{\pi}(z), \quad a^{\pi*}_{\mathrm{sch}}(w) = a^{\pi*}(z),$$

where $a^{\pi*}(z)$, resp. $a^{\pi}(z)$, are the creation, resp. annihilation operators associated with the anti-involution j, as in Subsect. 8.2.4.

In what follows we drop the superscript π. A standard choice of c is $c = \mathbb{1}$, for which

$$\mathrm{j} = \begin{bmatrix} 0 & -\frac{1}{2}\mathbb{1} \\ 2\mathbb{1} & 0 \end{bmatrix},$$

and leads to the formulas

$$a^*_{\mathrm{sch}}(w) = \frac{1}{2}w \cdot x - \mathrm{i}w \cdot D, \, a_{\mathrm{sch}}(w) = \frac{1}{2}\overline{w} \cdot x + \mathrm{i}\overline{w} \cdot D,$$

$$w \cdot x = a^*_{\mathrm{sch}}(w) + a_{\mathrm{sch}}(\overline{w}), \, w \cdot D = \frac{1}{2\mathrm{i}}(-a^*_{\mathrm{sch}}(w) + a_{\mathrm{sch}}(\overline{w})), \quad w \in \mathbb{C}\mathcal{X}.$$

This choice is the most convenient in the context of the *real-wave representation*, which will be described later.

In another choice, which is often found in the literature, one takes $c = \frac{1}{2}\mathbb{1}$ and multiplies $a_{\mathrm{sch}}(w)$ and $a^*_{\mathrm{sch}}(w)$ by $\sqrt{2}$ to keep the commutation relation $[a_{\mathrm{sch}}(w_1), a^*_{\mathrm{sch}}(w_2)] = \overline{w_1} \cdot w_2$, which leads to the formulas

$$a^*_{\mathrm{sch}}(w) = \frac{1}{\sqrt{2}}w \cdot x - \frac{\mathrm{i}}{\sqrt{2}}w \cdot D, \quad a_{\mathrm{sch}}(w) = \frac{1}{\sqrt{2}}\overline{w} \cdot x + \frac{\mathrm{i}}{\sqrt{2}}\overline{w} \cdot D,$$

$$w \cdot x = \frac{1}{\sqrt{2}}\left(a^*_{\mathrm{sch}}(w) + a_{\mathrm{sch}}(\overline{w})\right), \quad w \cdot D = \frac{1}{\mathrm{i}\sqrt{2}}\left(-a^*_{\mathrm{sch}}(w) + a_{\mathrm{sch}}(\overline{w})\right), \quad w \in \mathbb{C}\mathcal{X}.$$

This choice is more symmetric, but leads to the appearance of ugly square roots of 2; therefore we will not use it.

8.3 CCR algebras

In some approaches to quantum physics the initial step consists in choosing a *-algebra, usually a C^*- or W^*-algebra, which is supposed to describe observables of a system. Only after choosing a state (or a family of states) and making the corresponding GNS construction, we obtain a representation of this *-algebra in a Hilbert space. This philosophy allows us to study a quantum system in a representation-independent fashion.

Many authors try to apply this approach to bosonic systems. This raises the question whether one can associate with a given pre-symplectic space $(\mathcal{Y}, \omega)$

a natural and useful *-algebra describing the canonical commutation relations over $\mathcal{Y}$.

The analogous question has a rather satisfactory answer in the fermionic case. In particular, there exists an obvious choice of a C^*-algebra describing the CAR over a given Euclidean space. It turns out, however, that in the bosonic case the situation is much more complicated, since for a given pre-symplectic space several natural choices of CCR algebra are possible.

This question is discussed in this section. Throughout this section, $(\mathcal{Y}, \omega)$ is a pre-symplectic space and we discuss various *-algebras associated with $\mathcal{Y}$. We will see that each choice has its drawbacks. In the literature, the most popular choice seems to be the Weyl CCR algebra, which we discuss in Subsect. 8.3.5. One can, however, argue that, at least in the case of regular representations, it is more natural to use what we call the regular CCR algebra discussed in Subsect. 8.3.4. Some authors prefer to use the polynomial CCR algebra, discussed in Subsect. 8.3.1, which is purely algebraic and is not a C^*-algebra.

Unfortunately, the C^*-algebraic approach to bosonic systems has some serious problems. Many authors apply it in the case of free dynamics (given by Bogoliubov automorphisms). In the case of physically interesting interacting dynamics, the C^*-algebraic approach is not easy to apply. In fact, in the case of bosonic systems with infinite-dimensional phase spaces it is usually difficult to find a natural C^*-algebra preserved by a non-trivial dynamics. Sometimes, in such a case one can apply W^*-algebras, which we do not discuss here.

In the approach to canonical commutation relations discussed in this book, the central role is played by CCR representations, as defined in Def. 8.1. We view various CCR algebras introduced in this section more as academic curiosities than as basic tools. Therefore, the reader in a hurry may skip this section on the first reading.

8.3.1 Polynomial CCR *-algebras

In this subsection we discuss the polynomial CCR *-algebra over $\mathcal{Y}$. Note that for non-zero ω we cannot represent $\mathrm{CCR}^{\mathrm{pol}}(\mathcal{Y})$ as an algebra of bounded operators on a Hilbert space. The usefulness of this *-algebra for rigorous mathematical physics is rather limited.

Definition 8.40 *The polynomial CCR *-algebra over $\mathcal{Y}$, denoted by $\mathrm{CCR}^{\mathrm{pol}}(\mathcal{Y})$, is defined to be the unital complex *-algebra generated by elements $\phi(y)$, $y \in \mathcal{Y}$, with relations*

$$\phi(\lambda y) = \lambda \phi(y), \ \ \lambda \in \mathbb{R}, \ \ \phi(y_1 + y_2) = \phi(y_1) + \phi(y_2),$$

$$\phi^*(y) = \phi(y), \ \ \phi(y_1)\phi(y_2) - \phi(y_2)\phi(y_1) = \mathrm{i} y_1 {\cdot} \omega y_2 \, \mathbb{1}.$$

Let us describe basic properties of $\mathrm{CCR}^{\mathrm{pol}}(\mathcal{Y})$.

Proposition 8.41 (1) *Let* $r \in ASp(\mathcal{Y})$. *Then there exists a unique* $*$-*isomorphism* $\hat{r} : \mathrm{CCR}^{\mathrm{pol}}(\mathcal{Y}) \to \mathrm{CCR}^{\mathrm{pol}}(\mathcal{Y})$ *such that* $\hat{r}(\phi(y)) = \phi(ry)$, $y \in \mathcal{Y}$.
(2) *Let* $\mathcal{Y}_1$ *be a subspace of* $\mathcal{Y}$. *Then* $\mathrm{CCR}^{\mathrm{pol}}(\mathcal{Y}_1)$ *is naturally embedded in* $\mathrm{CCR}^{\mathrm{pol}}(\mathcal{Y})$, *such that, for* $y \in \mathcal{Y}_1$, $\phi(y)$ *in the sense of* $\mathrm{CCR}^{\mathrm{pol}}(\mathcal{Y}_1)$ *coincide with* $\phi(y)$ *in the sense of* $\mathrm{CCR}^{\mathrm{pol}}(\mathcal{Y})$. *If moreover* $\mathcal{Y}_1 \neq \mathcal{Y}$, *then* $\mathrm{CCR}^{\mathrm{pol}}(\mathcal{Y}_1) \neq \mathrm{CCR}^{\mathrm{pol}}(\mathcal{Y})$.

Definition 8.42 $\hat{r}$ *defined in Prop. 8.41 is called the* Bogoliubov automorphism *of* $\mathrm{CCR}^{\mathrm{pol}}(\mathcal{Y})$ *corresponding to* r.

Proposition 8.43 *Let* $\mathcal{H}$ *be a Hilbert space and let* $\mathcal{Y} \ni y \mapsto \mathrm{e}^{\mathrm{i}\phi^\pi(y)} \in U(\mathcal{H})$ *be a regular CCR representation. Recall that* $\mathcal{H}^{\infty,\pi}$ *denotes the Schwartz space for a given regular CCR representation, and was defined in Def. 8.28. Then there exists a unique* $*$-*representation* $\pi : \mathrm{CCR}^{\mathrm{pol}}(\mathcal{Y}) \to L(\mathcal{H}^{\infty,\pi})$ *such that* $\pi(\phi(y)) = \phi^\pi(y)$.

8.3.2 Stone–von Neumann CCR algebras

In this subsection we always assume that $(\mathcal{Y}, \omega)$ is a finite-dimensional presymplectic space. We set $\mathcal{Y}_0 := \mathrm{Ker}\,\omega \subset \mathcal{Y}$. In this case there exists a natural candidate for a CCR algebra suggested by the Stone–von Neumann theorem (Thm. 8.15), which implies the following proposition:

Proposition 8.44 (1) *Let* $\mathfrak{M}_i \subset B(\mathcal{H}_i)$, $i = 1, 2$, *be von Neumann algebras with distinguished unitary elements* $W_i(y)$ *depending* σ-*weakly continuously on* $y \in \mathcal{Y}$. *Let* $\mathfrak{Z}_i$ *be the centers of* $\mathfrak{M}_i$. *Assume that*
   (i) $W_i(y_1)W_i(y_2) = \mathrm{e}^{-\frac{1}{2}y_1 \cdot \omega y_2} W_i(y_1 + y_2)$, $y_1, y_2 \in \mathcal{Y}$;
   (ii) $\mathrm{Span}\{W_i(y) : y \in \mathcal{Y}\}$ *is* σ-*weakly dense in* $\mathfrak{M}_i$;
   (iii) $\mathfrak{Z}_i$ *are* $*$-*isomorphic to* $L^\infty(\mathcal{Y}_0^\#)$;
   (iv) $\mathfrak{M}_i' = \mathfrak{Z}_i$.
Then there exists a unique σ-*weakly continuous* $*$-*isomorphism* $\rho : \mathfrak{M}_1 \to \mathfrak{M}_2$ *such that*

$$\rho(W_1(y)) = W_2(y), \quad y \in \mathcal{Y}.$$

Moreover, there exists a unitary operator $U : \mathcal{H}_1 \to \mathcal{H}_2$ *such that* $\rho(\cdot) = U \cdot U^*$. *If* U_i, $i = 1, 2$, *are two such operators, then* $U_1^* U_2 \in \mathfrak{Z}_1$ *and* $U_1 U_2^* \in \mathfrak{Z}_2$.
(2) *Identify* $\mathcal{Y}$ *with* $\mathcal{Y}_0 \oplus \mathcal{X}^\# \oplus \mathcal{X}$ *and* ω *with the canonical symplectic form on* $\mathcal{X}^\# \oplus \mathcal{X}$ *extended by zero on* $\mathcal{Y}_0$. *Let* v *denote the generic variable in* $\mathcal{Y}_0^\#$ *and the corresponding multiplication operator. Then the von Neumann algebra*

$$L^\infty(\mathcal{Y}_0^\#) \otimes B(L^2(\mathcal{X})) \subset B(L^2(\mathcal{Y}_0^\# \oplus \mathcal{X})) \tag{8.33}$$

and the family of its elements $W(y) := \mathrm{e}^{\mathrm{i}(y_0 v + \eta x + qD)}$, $y = (y_0, \eta, q) \in \mathcal{Y}_0 \oplus \mathcal{X}^\# \oplus \mathcal{X}$, *satisfy the requirements of (1).* ($\otimes$ *used in (8.33) is the tensor multiplication in the category of* W^*-*algebras; see Subsect. 6.3.2.)*

Prop. 8.44 suggests the following definition:

Definition 8.45 *A Stone–von Neumann CCR algebra over $\mathcal{Y}$ is defined as a von Neumann algebra $\mathfrak{M}$ with distinguished unitary elements $W(y)$, $y \in \mathcal{Y}$, satisfying the conditions of Prop. 8.44. It is denoted $\mathrm{CCR}(\mathcal{Y})$ and the Hilbert space it acts on is denoted $\mathcal{H}_\mathcal{Y}$.*

Prop. 8.44 shows that $\mathrm{CCR}(\mathcal{Y})$ is defined uniquely up to a spatially implementable *-isomorphism. Clearly, if $\omega = 0$, then $\mathrm{CCR}(\mathcal{Y}) \simeq L^\infty(\mathcal{Y}^\#)$. If ω is symplectic, then $\mathrm{CCR}(\mathcal{Y}) = B(\mathcal{H}_\mathcal{Y})$.

Definition 8.46 *Let $y \in \mathcal{Y}$. The corresponding abstract field operator $\phi(y)$ is defined as the self-adjoint operator on $\mathcal{H}_\mathcal{Y}$ such that $W(y) = e^{i\phi(y)}$.*

Note that the operators $\phi(y)$ are affiliated to $\mathrm{CCR}(\mathcal{Y})$.

Note also that the definition of the Stone–von Neumann CCR algebra is simpler if ω is symplectic – we can then drop (iii) and (iv) from Prop. 8.44.

The following proposition is an analog of Prop. 8.41 about polynomial CCR *-algebras. But whereas Prop. 8.41 was a trivial algebraic fact, Prop. 8.47 is somewhat deeper.

Proposition 8.47 (1) *Let $r \in ASp(\mathcal{Y})$. Then there exists a unique spatially implementable *-isomorphism $\hat{r} : \mathrm{CCR}(\mathcal{Y}) \to \mathrm{CCR}(\mathcal{Y})$ such that $\hat{r}(W(y)) = W(ry)$, $y \in \mathcal{Y}$.*
(2) *Let $\mathcal{Y}_1 \subset \mathcal{Y}$. Then there is a unique embedding of $\mathrm{CCR}(\mathcal{Y}_1)$ in $\mathrm{CCR}(\mathcal{Y})$, such that, for $y \in \mathcal{Y}_1$, $W(y)$ in the sense of $\mathrm{CCR}(\mathcal{Y}_1)$ coincide with $W(y)$ in the sense of $\mathrm{CCR}(\mathcal{Y})$. If moreover $\mathcal{Y}_1 \neq \mathcal{Y}$, then $\mathrm{CCR}(\mathcal{Y}_1) \neq \mathrm{CCR}(\mathcal{Y})$.*

Definition 8.48 *$\hat{r}$ defined in Prop. 8.47 is called the* Bogoliubov automorphism *of $\mathrm{CCR}(\mathcal{Y})$ corresponding to r.*

Here is yet another reformulation of the Stone–von Neumann theorem (see Thm. 8.15):

Theorem 8.49 *Let $(\mathcal{Y}, \omega)$ be symplectic. Let $\mathcal{Y} \ni y \mapsto W^\pi(y) \in U(\mathcal{H})$ be a regular CCR representation. Then there exists a unique σ-weakly continuous *-representation $\pi : \mathrm{CCR}(\mathcal{Y}) \to B(\mathcal{H})$ such that $\pi(W(y)) = W^\pi(y)$, $y \in \mathcal{Y}$. Moreover, π is isometric and*

$$\pi(\mathrm{CCR}(\mathcal{Y})) = \{W^\pi(y) \ : \ y \in \mathcal{Y}\}''.$$

If in addition the representation is irreducible, then there also exists a unitary operator $U : \mathcal{H} \to \mathcal{H}_\mathcal{Y}$, unique up to a phase factor, such that $\pi(\cdot) = U \cdot U^*$.

8.3.3 S- and S'-type operators

In this subsection we fix a finite-dimensional symplectic space $(\mathcal{Y}, \omega)$ and consider the von Neumann algebra $\mathrm{CCR}(\mathcal{Y}) = B(\mathcal{H}_\mathcal{Y})$. We will describe an abstract version of the constructions described in Subsect. 4.1.11.

Definition 8.50 $\Psi \in \mathcal{H}_\mathcal{Y}$ *is called an* $\mathcal{S}$*-type vector if the function*

$$\mathcal{Y} \ni y \mapsto (\Psi|W(y)\Psi)$$

belongs to $\mathcal{S}(\mathcal{Y})$*. The abstract Schwartz space for* $\mathcal{Y}$ *is defined as the set of* $\mathcal{S}$*-type vectors. It is denoted* $\mathcal{H}^\infty$*.*

Clearly, $\phi(y)$, $y \in \mathcal{Y}$, leaves $\mathcal{H}^\infty$ invariant. Thus we can define a family of semi-norms

$$\mathcal{H}^\infty \ni \Psi \mapsto \|\phi(y_1)\cdots\phi(y_n)\Psi\|, \quad y_1, \ldots, y_n \in \mathcal{Y},$$

which equip $\mathcal{H}^\infty$ with the structure of a Fréchet space.

Definition 8.51 $\mathcal{H}^{-\infty}$ *is defined as the topological dual to* $\mathcal{H}^\infty$*. It is called the abstract* $\mathcal{S}'$ *space for* $\mathcal{Y}$*.*

Note that $\mathrm{CCR}^{\mathrm{pol}}(\mathcal{Y})$ can be represented as an algebra of linear operators on $\mathcal{H}^\infty$, as well as on $\mathcal{H}^{-\infty}$.

Definition 8.52 $A \in \mathrm{CCR}(\mathcal{Y})$ *is called an* $\mathcal{S}$*-type operator iff it is trace-class and the function*

$$\mathcal{Y} \ni y \mapsto \mathrm{Tr}\, AW(y)$$

belongs to $\mathcal{S}(\mathcal{Y})$*. The set of* $\mathcal{S}$*-type operators is denoted* $\mathrm{CCR}^\mathcal{S}(\mathcal{Y})$*.*

Clearly, $\mathrm{CCR}^\mathcal{S}(\mathcal{Y})$ is a $*$-algebra. It is equipped with a topology by the family of semi-norms

$$\mathrm{CCR}^\mathcal{S}(\mathcal{Y}) \ni A \mapsto |\mathrm{Tr}\, \phi(y_1)\cdots\phi(y_n)A|, \quad y_1, \ldots, y_n \in \mathcal{Y}.$$

Definition 8.53 *Continuous linear functionals on* $\mathrm{CCR}^\mathcal{S}(\mathcal{Y})$ *are called* $\mathcal{S}'$*-type forms over* $\mathcal{Y}$*. Their space is denoted by* $\mathrm{CCR}^{\mathcal{S}'}(\mathcal{Y})$*.*

Let

$$\mathrm{CCR}^\mathcal{S}(\mathcal{Y}) \ni A \mapsto B(A) \in \mathbb{C} \tag{8.34}$$

be an $\mathcal{S}'$-type form. Clearly, for any $\Psi_1, \Psi_2 \in \mathcal{H}^\infty$, the operator $|\Psi_2)(\Psi_1|$ belongs to $\mathrm{CCR}^\mathcal{S}(\mathcal{Y})$. Thus, (8.34) defines a continuous sesquilinear form on $\mathcal{H}^\infty$:

$$\mathcal{H}^\infty \times \mathcal{H}^\infty \ni (\Psi_1, \Psi_2) \mapsto B\left(|\Psi_2)(\Psi_1|\right) \in \mathbb{C}.$$

In what follows we will use the "operator notation", writing $(\Psi_1|B\Psi_2)$ instead of $B\left(|\Psi_2)(\Psi_1|\right)$. Thus bounded operators can be viewed as elements of $\mathrm{CCR}^{\mathcal{S}'}(\mathcal{Y})$, so that we have

$$\mathrm{CCR}^\mathcal{S}(\mathcal{Y}) \subset \mathrm{CCR}(\mathcal{Y}) \subset \mathrm{CCR}^{\mathcal{S}'}(\mathcal{Y}).$$

As in Subsect. 4.1.11, we define the adjoint form B^* by $(\Psi_1|B^*\Psi_2) = \overline{(\Psi_2|B\Psi_1)}$. If B_1 or B_2^* extend as continuous operators on $\mathcal{H}^\infty$, then we can define $B_2 \circ B_1$ as an element of $\mathrm{CCR}^{\mathcal{S}'}(\mathcal{Y})$ by

$$(\Psi_1 | B_2 \circ B_1 \Psi_2) := (\Psi_1 | B_2(B_1 \Psi)), \text{ or } (\Psi_1 | B_2 \circ B_1 \Psi_2) := (B_2^* \Psi | B_1 \Psi).$$

In particular this is possible if B_1 or $B_2 \in \mathrm{CCR}^{\mathrm{pol}}(\mathcal{Y})$.

If $\mathcal{Y} \simeq \mathcal{X}^\# \oplus \mathcal{X}$ and we consider the Schrödinger representation on $L^2(\mathcal{X})$, then $\mathrm{CCR}^{\mathcal{S}}(\mathcal{Y})$ coincides with the set of operators whose integral kernel is in $\mathcal{S}(\mathcal{X} \times \mathcal{X})$. $\mathrm{CCR}^{\mathcal{S}'}(\mathcal{Y})$ consists then of forms whose distributional kernel is in $\mathcal{S}'(\mathcal{X} \times \mathcal{X})$, which were considered already in Subsect. 4.1.11.

8.3.4 Regular CCR algebras

Until the end of this section, $(\mathcal{Y}, \omega)$ is a pre-symplectic space of arbitrary dimension. Recall that $\mathrm{Fin}(\mathcal{Y})$ denotes the set of finite-dimensional subspaces of $\mathcal{Y}$.

In this subsection we introduce the notion of the regular CCR C^*-algebra over $\mathcal{Y}$. In the literature, it is rarely used. Weyl CCR C^*-algebras are more common. Nevertheless, it is a natural construction. Its use was advocated by I. E. Segal.

Let $\mathcal{Y}_1, \mathcal{Y}_2 \in \mathrm{Fin}(\mathcal{Y})$ and $\mathcal{Y}_1 \subset \mathcal{Y}_2$. We can define their Stone–von Neumann CCR algebras, as in Def. 8.45. By Prop. 8.47, we have a natural embedding,

$$\mathrm{CCR}(\mathcal{Y}_1) \subset \mathrm{CCR}(\mathcal{Y}_2).$$

We can define the algebraic regular CCR $*$-algebra as the inductive limit of Stone–von Neumann CCR algebras:

Definition 8.54 *We set*

$$\mathrm{CCR}_{\mathrm{alg}}^{\mathrm{reg}}(\mathcal{Y}) := \bigcup_{\mathcal{Y}_1 \in \mathrm{Fin}(\mathcal{Y})} \mathrm{CCR}(\mathcal{Y}_1). \tag{8.35}$$

Clearly, $\mathrm{CCR}_{\mathrm{alg}}^{\mathrm{reg}}(\mathcal{Y})$ is a $*$-algebra equipped with a C^*-norm.

Definition 8.55 *We define the* regular CCR C^*-algebra *over* $\mathcal{Y}$ *as*

$$\mathrm{CCR}^{\mathrm{reg}}(\mathcal{Y}) := \left(\mathrm{CCR}_{\mathrm{alg}}^{\mathrm{reg}}(\mathcal{Y}) \right)^{\mathrm{cl}}.$$

Clearly, $\mathrm{CCR}^{\mathrm{reg}}(\mathcal{Y})$ is a generalization of the Stone–von Neumann algebra $\mathrm{CCR}(\mathcal{Y})$ from Def. 8.45.

We have an obvious extension of Prop. 8.47:

Proposition 8.56 (1) *Let* $r \in ASp(\mathcal{Y})$. *Then there exists a unique $*$-isomorphism* $\hat{r} : \mathrm{CCR}^{\mathrm{reg}}(\mathcal{Y}) \to \mathrm{CCR}^{\mathrm{reg}}(\mathcal{Y})$ *such that* $\hat{r}(W(y)) = W(ry)$, $y \in \mathcal{Y}$.

(2) *Let* $\mathcal{Y}_1 \subset \mathcal{Y}$. *Then* $\mathrm{CCR}^{\mathrm{reg}}(\mathcal{Y}_1)$ *is naturally embedded in* $\mathrm{CCR}^{\mathrm{reg}}(\mathcal{Y})$. *If moreover* $\mathcal{Y}_1 \neq \mathcal{Y}$, *then* $\mathrm{CCR}^{\mathrm{reg}}(\mathcal{Y}_1) \neq \mathrm{CCR}^{\mathrm{reg}}(\mathcal{Y})$.

Proof Let us give a proof of (2). Working in the Schrödinger representation we see that $\|W(y_1) - W(y_2)\| = 2$ if $y_1 \neq y_2$. Hence, if $y \in \mathcal{Y} \backslash \mathcal{Y}_1$, then $W(y) \notin \mathrm{CCR}^{\mathrm{reg}}(\mathcal{Y}_1)$. $\qquad\square$

Definition 8.57 $\hat{r}$ *defined in Prop. 8.56 is called the* Bogoliubov automorphism of $\mathrm{CCR}^{\mathrm{reg}}(\mathcal{Y})$ *corresponding to* r.

The following proposition is an extension of Thm. 8.49:

Proposition 8.58 *Suppose that* ω *is symplectic. Let* $\mathcal{Y} \ni y \mapsto W^{\pi}(y) \in U(\mathcal{H})$ *be a regular CCR representation. Then there exists a unique $*$-representation* $\pi : \mathrm{CCR}^{\mathrm{reg}}(\mathcal{Y}) \to B(\mathcal{H})$ *such that* $\pi(W(y)) = W^{\pi}(y)$, $y \in \mathcal{Y}$, *and which, for* $\mathcal{Y}_1 \in \mathrm{Fin}(\mathcal{Y})$, *is* σ-*weakly continuous on the sub-algebras* $\mathrm{CCR}(\mathcal{Y}_1) \subset \mathrm{CCR}^{\mathrm{reg}}(\mathcal{Y})$. *Moreover,* π *is isometric.*

Proof We use the fact that if ω is symplectic then we can restrict the union in (8.35) to run over finite-dimensional symplectic subspaces of $\mathcal{Y}$. $\qquad\square$

8.3.5 Weyl CCR algebra

In this subsection we introduce the notion of the Weyl CCR C^*-algebra over $\mathcal{Y}$. This is the C^*-algebra generated by elements satisfying the Weyl CCR relations over $\mathcal{Y}$. Mathematical physicists use Weyl CCR algebras often in their description of bosonic systems.

Note that Weyl CCR algebras can be viewed as non-commutative generalizations of algebras of almost periodic functions. Indeed, $\mathrm{CCR}^{\mathrm{Weyl}}_{\mathrm{alg}}(\mathcal{Y})$ consists of almost periodic functions on $\mathcal{Y}$ if $\omega = 0$.

Let us start with the definition of algebraic Weyl CCR algebras.

Definition 8.59 $\mathrm{CCR}^{\mathrm{Weyl}}_{\mathrm{alg}}(\mathcal{Y})$ *is defined as the $*$-algebra generated by the elements* $W(y)$, $y \in \mathcal{Y}$, *with relations*

$$W(y)^* = W(-y), \quad W(y_1)W(y_2) = e^{-\frac{i}{2} y_1 \cdot \omega y_2} W(y_1 + y_2), \quad y, y_1, y_2 \in \mathcal{Y}.$$

Let $\mathcal{Y} \ni y \mapsto W^{\pi}(y) \in U(\mathcal{H}^{\pi})$ be a CCR representation. Clearly, there exists a unique unital $*$-isomorphism $\pi : \mathrm{CCR}^{\mathrm{Weyl}}_{\mathrm{alg}}(\mathcal{Y}) \to B(\mathcal{H}^{\pi})$ such that $\pi(W(y)) = W^{\pi}(y)$.

Let $\mathcal{R}(\mathcal{Y})$ be the class of CCR representations over $\mathcal{Y}$. $\mathcal{R}(\mathcal{Y})$ is non-empty. In fact, we always have the (non-regular) CCR representation $\mathcal{Y} \ni y \mapsto W^{\mathrm{d}}(y) \in U(l^2(\mathcal{Y}))$ defined in (8.15). It yields a corresponding faithful representation $\pi^{\mathrm{d}} : \mathrm{CCR}^{\mathrm{Weyl}}_{\mathrm{alg}}(\mathcal{Y}) \to B(l^2(\mathcal{Y}))$.

Definition 8.60 *For* $A \in \mathrm{CCR}^{\mathrm{Weyl}}_{\mathrm{alg}}(\mathcal{Y})$ *we set*

$$\|A\| := \sup\{\|\pi(A)\| \; : \; \pi \in \mathcal{R}(\mathcal{Y})\}. \tag{8.36}$$

The Weyl CCR C^*-algebra *is defined as*

$$\mathrm{CCR}^{\mathrm{Weyl}}(\mathcal{Y}) := \left(\mathrm{CCR}^{\mathrm{Weyl}}_{\mathrm{alg}}(\mathcal{Y})\right)^{\mathrm{cpl}}.$$

Clearly, $\| \cdot \|$ defined in (8.36) is a C^*-norm and $\mathrm{CCR}^{\mathrm{Weyl}}(\mathcal{Y})$ is a C^*-algebra.

Proposition 8.61 (1) *Let* $r \in ASp(\mathcal{Y})$. *Then there exists a unique* *-isomorphism* $\hat{r} : \mathrm{CCR}^{\mathrm{Weyl}}(\mathcal{Y}) \to \mathrm{CCR}^{\mathrm{Weyl}}(\mathcal{Y})$ *such that* $\hat{r}(W(y)) = W(ry)$, $y \in \mathcal{Y}$.
(2) *Let* $\mathcal{Y}_1 \subset \mathcal{Y}$. *Then* $\mathrm{CCR}^{\mathrm{Weyl}}(\mathcal{Y}_1)$ *is naturally embedded in* $\mathrm{CCR}^{\mathrm{Weyl}}(\mathcal{Y})$. *If moreover* $\mathcal{Y}_1 \neq \mathcal{Y}$, *then* $\mathrm{CCR}^{\mathrm{Weyl}}(\mathcal{Y}_1) \neq \mathrm{CCR}^{\mathrm{Weyl}}(\mathcal{Y})$.
(3) *If* $\mathcal{Y} \neq \{0\}$, *then* $\mathrm{CCR}^{\mathrm{Weyl}}(\mathcal{Y})$ *is non-separable.*

Proof (1) and (2) are obvious analogs of Prop. 8.41. (3) follows from the fact that $y_1 \neq y_2$ implies $\|W(y_1) - W(y_2)\| = 2$. $\qquad\square$

Definition 8.62 $\hat{r}$ *defined in Prop. 8.61 is called the* Bogoliubov automorphism *of* $\mathrm{CCR}^{\mathrm{Weyl}}(\mathcal{Y})$ *corresponding to* r.

Let us give an analog of Prop. 8.43:

Proposition 8.63 *Let* $\mathcal{Y} \ni y \mapsto W^\pi(y) \in U(\mathcal{H})$ *be a CCR representation. Then there exists a unique *-homomorphism*

$$\pi : \mathrm{CCR}^{\mathrm{Weyl}}(\mathcal{Y}) \to B(\mathcal{H})$$

such that $\pi(W(y)) = W^\pi(y)$.

If ω is symplectic, the algebra $\mathrm{CCR}^{\mathrm{Weyl}}(\mathcal{Y})$ enjoys especially good properties. In particular, there is no need to consider the norms given by all possible representations, since all of them are equal.

Theorem 8.64 *Let* ω *be symplectic. Then*

(1) *For* $A \in \mathrm{CCR}^{\mathrm{Weyl}}(\mathcal{Y})$, $\|A\| = \|\pi^{\mathrm{d}}(A)\|$.
(2) *Every representation* π *described in Thm. 8.63 is isometric.*
(3) $\mathrm{CCR}^{\mathrm{Weyl}}(\mathcal{Y})$ *is simple.*

Proof Let us prove (2).
For $y \in \mathcal{Y}$ we define $R(y) \in U(l^2(\mathcal{Y}))$ by setting

$$(R(y)f)(x) := f(x + y), \quad f \in l^2(\mathcal{V}).$$

Let $\hat{\mathcal{Y}}$ be the Pontryagin dual of $\mathcal{Y}$ (the space of characters on the group $\mathcal{Y}$ with values in $\{z \in \mathbb{C} : |z| = 1\}$). Let $\mathcal{F} : l^2(\mathcal{Y}) \to L^2(\hat{\mathcal{Y}})$ be the (unitary) Fourier transformation. Then $\mathcal{F}R(y)\mathcal{F}^* = \hat{R}(y)$, where $\hat{R}(y) \in U(L^2(\hat{\mathcal{Y}}))$ is defined by

$$(\hat{R}(y)g)(\chi) := \chi(y)g(\chi), \quad g \in L^2(\hat{\mathcal{Y}}), \ \chi \in \hat{\mathcal{Y}}.$$

Consider now a CCR representation

$$\mathcal{Y} \ni y \mapsto W^{\pi}(y) \in U(\mathcal{H}).$$

On the Hilbert space $\mathcal{H} \otimes l^2(\mathcal{Y}) \simeq l^2(\mathcal{Y}, \mathcal{H})$ we introduce the unitary operator U defined by

$$U\Phi(x) := W^{\pi}(x)\Phi(x), \quad \Phi \in l^2(\mathcal{Y}, \mathcal{H}), \quad x \in \mathcal{Y}.$$

Note that

$$U\, W^{\pi}(y) \otimes R(y)\, U^* = \mathbb{1} \otimes W^{\mathrm{d}}(y). \tag{8.37}$$

Now

$$\begin{aligned}
\left\| \sum \lambda_i W^{\mathrm{d}}(y_i) \right\| &= \left\| \sum \lambda_i W^{\pi}(y_i) \otimes R(y_i) \right\| \\
&= \left\| \sum \lambda_i W^{\pi}(y_i) \otimes \hat{R}(y_i) \right\| \\
&= \sup_{\chi \in \hat{\mathcal{Y}}} \left\| \sum \lambda_i W^{\pi}(y_i) \chi(y_i) \right\| \\
&= \sup_{x \in \mathcal{Y}} \left\| \sum \lambda_i W^{\pi}(y_i) \mathrm{e}^{-\mathrm{i}x \cdot \omega y_i} \right\| \\
&= \left\| \sum \lambda_i W^{\pi}(y_i) \right\|.
\end{aligned}$$

First we applied (8.37). Next we used that $\mathcal{F}$ is unitary. Then we used that the set of characters $\chi_x(y) := \mathrm{e}^{-\mathrm{i}x \cdot \omega y}$ for $x \in \mathcal{Y}$ is dense in $\hat{\mathcal{Y}}$, since ω is non-degenerate. Finally we noted that

$$W^{\pi}(y_i)\mathrm{e}^{-\mathrm{i}x \cdot \omega y_i} = W^{\pi}(x)W^{\pi}(y_i)W^{\pi}(-x).$$

(2) immediately implies (1).

By Subsect. 6.2.3, (2) implies (3).   □

8.4 Weyl–Wigner quantization

The Weyl–Wigner quantization has a long and complicated history. It also has many names.

It was first proposed by Weyl in 1927 in his book on group theory in quantum mechanics (Weyl (1931)). Hence it is commonly called the *Weyl quantization*.

Wigner was the first who considered its inverse, at least in the case of an operator of the form $|\Psi)(\Psi|$; see Wigner (1932b). Hence the name *Wigner function* is commonly used to denote the inverse of the Weyl quantization.

Apparently, for some time the link between the Weyl quantization and the Wigner function was not understood. This link seems to have been clarified only in the late 1940s by Moyal (1949). Moyal also found a version of the formula (8.41). The non-commutative operation $*$ defined by $b := b_1 * b_2$ as in (8.41) is often called the *Moyal star*. Moyal also found the identity (8.44).

Our terminology, "the Weyl–Wigner quantization" and "the Weyl–Wigner symbol", is thus a compromise between the names "Weyl quantization" and "Wigner function". In the literature, one can also find the name *Weyl–Wigner–Moyal quantization.*

One can argue that the Weyl–Wigner quantization is the most important kind of quantization. It is certainly the most canonical quantization – its definition depends only on the symplectic structure of the phase space. It is, however, not so useful if the phase space has infinite-dimension.

Historically, Weyl introduced this quantization in the context of the Schrödinger representation, which hides the symplectic invariance of this concept. Therefore, in our presentation we start from manifestly symplectically invariant definitions, which involve a regular CCR representation. The case of the Schrödinger representation is discussed later, in Subsect. 8.4.3.

8.4.1 Quantization of polynomial symbols

In this subsection we will consider the Weyl–Wigner quantization only for polynomial symbols. More general symbols will be considered in the following subsections. (In the subsequent subsections we will, however, restrict ourselves to finite-dimensional symplectic spaces $\mathcal{Y}$).

Suppose that $(\mathcal{Y}, \omega)$ is an arbitrary pre-symplectic space. Let

$$\mathcal{Y} \ni y \mapsto e^{i\phi(y)} \in U(\mathcal{H}) \tag{8.38}$$

be a regular CCR representation. Let $\mathcal{H}^{\infty}$ denote the abstract Schwartz space for this representation introduced in Def. 8.28. Recall that $\mathrm{CCR}^{\mathrm{pol}}(\mathcal{Y})$ denotes the polynomial CCR algebra over $\mathcal{Y}$, which can be treated as an algebra of operators on $\mathcal{H}^{\infty}$.

Definition 8.65 *Let $y_1, \ldots, y_n \in \mathcal{Y}$. We can treat these as polynomials on $\mathcal{Y}^{\#}$ and take their product $y_1 \cdots y_n \in \mathrm{Pol}_{\mathrm{s}}^n(\mathcal{Y}^{\#})$. We define*

$$\mathrm{Op}(y_1 \cdots y_n) := \frac{1}{n!} \sum_{\sigma \in S_n} \phi(y_{\sigma 1}) \cdots \phi(y_{\sigma n}). \tag{8.39}$$

The map extends uniquely to a linear bijective map

$$\mathbb{C}\mathrm{Pol}_{\mathrm{s}}(\mathcal{Y}^{\#}) \ni b \mapsto \mathrm{Op}(b) \in \mathrm{CCR}^{\mathrm{pol}}(\mathcal{Y}). \tag{8.40}$$

Theorem 8.66 (1) $\mathrm{Op}(b)^* = \mathrm{Op}(\bar{b})$, *for $b \in \mathbb{C}\mathrm{Pol}_{\mathrm{s}}(\mathcal{Y}^{\#})$.*
(2) *If $y \in \mathbb{C}\mathcal{Y}$, then*

$$\mathrm{Op}(y) = \phi(y).$$

More generally, let $\mathcal{X}$ be an isotropic subspace in $\mathcal{Y}^{\#}$, so that the operators $\phi(y)$, $y \in \mathcal{X}$ commute with one another. Then, for $f \in \mathbb{C}\mathrm{Pol}_s(\mathcal{X})$, $\mathrm{Op}(f)$ coincides with $f(\phi)$ defined by the functional calculus.

(3) *If $b_1, b_2 \in \mathbb{C}\mathrm{Pol}_s(\mathcal{Y}^{\#})$, then $\mathrm{Op}(b_1)\mathrm{Op}(b_2) = \mathrm{Op}(b)$ for*

$$b(v) = \exp\left(-\frac{\mathrm{i}}{2}D_{v_1} \cdot \omega D_{v_2}\right)b_1(v_1)b_2(v_2)\Big|_{v=v_1=v_2}. \qquad (8.41)$$

(4) *If $b \in \mathbb{C}\mathrm{Pol}_s(\mathcal{Y}^{\#})$ and $y \in \mathbb{C}\mathcal{Y}$, then*

$$\frac{1}{2}\big(\phi(y)\mathrm{Op}(b) + \mathrm{Op}(b)\phi(y)\big) = \mathrm{Op}(yb).$$

Remark 8.67 *We refer to Remark 8.27 for the notation used in (2). The r.h.s. of (3) can be interpreted as a finite sum of differential operators.*

The following theorem is a version of the Wick theorem adapted to the Weyl–Wigner quantization.

Theorem 8.68 *If $b, b_1, \dots, b_n \in \mathbb{C}\mathrm{Pol}_s(\mathcal{Y}^{\#})$ and*

$$\mathrm{Op}(b) = \mathrm{Op}(b_1) \cdots \mathrm{Op}(b_n),$$

then

$$b(v) = \exp\left(\frac{\mathrm{i}}{2}\sum_{i<j}\nabla_{v_i} \cdot \omega \nabla_{v_j}\right)b_1(v_1) \cdots b_n(v_n)\Big|_{v=v_1=\cdots=v_n}.$$

8.4.2 Quantization of distributional symbols

In this subsection we assume that the form ω is symplectic and $\mathcal{Y}$ is finite-dimensional. We set $2d = \dim \mathcal{Y}$. Denote by $\mathrm{d}y$ the Liouville measure on $\mathcal{Y}$ defined in Subsect. 3.6.3. The dual space $\mathcal{Y}^{\#}$ is equipped with the symplectic form ω^{-1} and the dual measure $\mathrm{d}v$.

Consider a regular irreducible CCR representation (8.38). In this subsection we extend the Weyl–Wigner quantization to $\mathcal{S}'(\mathcal{Y}^{\#})$.

Recall that for $b \in \mathcal{S}'(\mathcal{Y}^{\#})$ the Fourier transform of b, denoted $\hat{b} \in \mathcal{S}'(\mathcal{Y})$, satisfies

$$b(v) = (2\pi)^{-2d}\int_{\mathcal{Y}}\hat{b}(y)\mathrm{e}^{\mathrm{i}y \cdot v}\mathrm{d}y, \quad v \in \mathcal{Y}^{\#}.$$

Definition 8.69 *If $b \in \mathcal{S}'(\mathcal{Y}^{\#})$, then $\mathrm{Op}(b) \in \mathrm{CCR}^{\mathcal{S}'}(\mathcal{Y})$ is defined by the formula*

$$\big(\Psi_1|\mathrm{Op}(b)\Psi_2\big) := (2\pi)^{-2d}\int_{\mathcal{Y}}\hat{b}(y)\big(\Psi_1|W(y)\Psi_2\big)\mathrm{d}y \qquad (8.42)$$

$$= (2\pi)^{-2d}\int_{\mathcal{Y}}\int_{\mathcal{Y}^{\#}}b(v)\big(\Psi_1|W(y)\Psi_2\big)\mathrm{e}^{-\mathrm{i}v \cdot y}\mathrm{d}y\mathrm{d}v, \quad \Psi_1, \Psi_2 \in \mathcal{H}^{\infty}.$$

Recall that $\mathcal{H}^\infty$ is the space of $\mathcal{S}$-type vectors for the representation (8.38). We know from Thm. 8.29 that if $\Psi_1, \Psi_2 \in \mathcal{H}^\infty$, then $\mathcal{Y} \ni y \mapsto (\Psi_1 | W(y)\Psi_2)$ is a Schwartz function. Therefore, the integral (8.42) is well defined.

The following theorem extends some of statements of Thm. 8.66 to the case of distributional symbols:

Theorem 8.70 (1) *If $b \in \mathbb{C}\mathrm{Pol}_\mathrm{s}(\mathcal{Y}^\#)$, then the definition (8.39) coincides with (8.42).*

(2) $W(y) = \mathrm{Op}(\mathrm{e}^{\mathrm{i}y(\cdot)})$. *More generally, if $\mathcal{X}$ is an isotropic subspace of $\mathcal{Y}$, and $f \in \mathcal{S}'(\mathcal{X}^\#) \subset \mathcal{S}'(\mathcal{Y}^\#)$ is a measurable function, then $\mathrm{Op}(f)$ coincides with $f(\phi)$ defined by the functional calculus.*

(3) $\mathrm{Op}(b)^* = \mathrm{Op}(\overline{b})$.

(4) *If $b_1 \in \mathbb{C}\mathrm{Pol}_\mathrm{s}(\mathcal{Y}^\#)$, $b_2, b \in \mathcal{S}'(\mathcal{Y}^\#)$ and $\mathrm{Op}(b_1)\mathrm{Op}(b_2) = \mathrm{Op}(b)$, then*

$$b(v) := \exp\left(-\frac{\mathrm{i}}{2}D_{v_1}\cdot\omega D_{v_2}\right)b_1(v_1)b_2(v_2)\Big|_{v=v_1=v_2}$$

$$= \pi^{-2d}\int_{\mathcal{Y}^\#}\int_{\mathcal{Y}^\#} \mathrm{e}^{2\mathrm{i}(v-v_1)\cdot\omega^{-1}(v-v_2)}b_1(v_1)b_2(v_2)\mathrm{d}v_1\,\mathrm{d}v_2.$$

(5) *For $v \in \mathcal{Y}^\#$, $W(-\omega^{-1}v)\mathrm{Op}(b)W(\omega^{-1}v) = \mathrm{Op}(b(\cdot-v))$.*

(6) *The map*

$$\mathcal{S}'(\mathcal{Y}^\#) \ni b \mapsto \mathrm{Op}(b) \in \mathrm{CCR}^{\mathcal{S}'}(\mathcal{Y}) \tag{8.43}$$

is bijective.

(7) $\mathrm{Op}(b) \in B^2(\mathcal{H})$ *iff $b \in L^2(\mathcal{Y}^\#)$, and*

$$\mathrm{Tr}\,\mathrm{Op}(b)^*\mathrm{Op}(a) = (2\pi)^{-d}\int \overline{b(v)}a(v)\mathrm{d}v, \quad a, b \in L^2(\mathcal{Y}^\#). \tag{8.44}$$

Proof To prove (1), it is enough to consider $y_0 \in \mathcal{Y}$ and $b(v) = (y_0\cdot v)^n$, because such polynomials span $\mathbb{C}\mathrm{Pol}_\mathrm{s}(\mathcal{Y}^\#)$. The Fourier transform of b is $\hat{b} = (2\pi)^{2d}\mathrm{i}^n(y_0\cdot\nabla_y)^n\delta_0$. Hence

$$\mathrm{Op}(b) = (-\mathrm{i})^n(y_0\cdot\nabla_y)^n W(y)\big|_{y=0} = \phi(y_0)^n.$$

(2) follows from the spectral theorem and (3) is immediate.

To prove (4), set

$$b_0(v_1, v_2) = \mathrm{e}^{-\frac{\mathrm{i}}{2}D_{v_1}\cdot\omega D_{v_2}}b_1(v_1)b_2(v_2).$$

Clearly,

$$b_0(y_1, y_2) = \mathrm{e}^{-\frac{\mathrm{i}}{2}y_1\cdot\omega y_2}\ddot{b}(y_1)\ddot{b}(y_2).$$

Moreover,

$$b(v) = b_0(v, v) = (2\pi)^{-4d}\int \hat{b}(y_1, y_2)\mathrm{e}^{\mathrm{i}(y_1+y_2)\cdot v}\mathrm{d}y_1\,\mathrm{d}y_2$$

$$= (2\pi)^{-4d}\int \hat{b}_1(y_1)\hat{b}_2(y_2)\mathrm{e}^{-\frac{\mathrm{i}}{2}y_1\cdot\omega y_2}\mathrm{e}^{\mathrm{i}(y_1+y_2)\cdot v}\mathrm{d}y_1\,\mathrm{d}y_2.$$

Hence

$$\mathrm{Op}(b) = (2\pi)^{-4d} \iint \hat{b}_1(y_1)\hat{b}_2(y_2)\mathrm{e}^{-\frac{1}{2}y_1 \cdot \omega y_2} W(y_1 + y_2)\mathrm{d}y_1\,\mathrm{d}y_2$$

$$= (2\pi)^{-4d} \iint \hat{b}_1(y_1)\hat{b}_2(y_2)W(y_1)W(y_2)\mathrm{d}y_1\,\mathrm{d}y_2$$

$$= \mathrm{Op}(b_1)\mathrm{Op}(b_2).$$

To prove the last two items of the theorem it is convenient to use the Schrödinger representation, considered in the next subsection. □

Definition 8.71 *The inverse of (8.43) will be called the Weyl–Wigner symbol. If $B \in \mathrm{CCR}^{\mathcal{S}'}(\mathcal{Y})$, its Weyl–Wigner symbol will be denoted by $s_B \in \mathcal{S}'(\mathcal{Y}^\#)$.*

8.4.3 Weyl–Wigner quantization in the Schrödinger representation

Let $\mathcal{X}$ be a finite-dimensional real vector space. Consider the Schrödinger representation

$$\mathcal{X}^\# \oplus \mathcal{X} \ni (\eta, q) \mapsto \mathrm{e}^{\mathrm{i}(\eta \cdot x + q \cdot D)} \in U\big(L^2(\mathcal{X})\big).$$

Remark 8.72 *In the Schrödinger representation one often writes $b^w(x, D)$ instead of $\mathrm{Op}(b)$.*

Theorem 8.73 (1) *Let $b \in \mathcal{S}'(\mathcal{X} \oplus \mathcal{X}^\#)$. The distributional kernel of $B = \mathrm{Op}(b)$ can be computed as follows:*

$$B(x, y) = (2\pi)^{-d} \int b\Big(\frac{x + y}{2}, \xi\Big)\mathrm{e}^{\mathrm{i}(x-y)\cdot\xi}\mathrm{d}\xi. \tag{8.45}$$

(2) *Let $B \in \mathrm{CCR}^{\mathcal{S}'}(\mathcal{X}^\# \oplus \mathcal{X})$. The symbol of B can be obtained from its distributional kernel by the formula*

$$s_B(x, \xi) = \int B\Big(x + \frac{y}{2}, x - \frac{y}{2}\Big)\mathrm{e}^{-\mathrm{i}\xi \cdot y}\mathrm{d}y.$$

(3) *The relationship between the x, D-symbol and the Weyl–Wigner symbol is as follows: If $\mathrm{Op}^{x,D}(b_+) = \mathrm{Op}(b)$, then*

$$b_+(x, \xi) = \mathrm{e}^{\frac{\mathrm{i}}{2}D_x \cdot D_\xi} b(x, \xi)$$

$$= \pi^{-d} \int \mathrm{e}^{-\mathrm{i}2(x-x_1)\cdot(\xi-\xi_1)} b(x_1, \xi_1)\mathrm{d}x_1\,\mathrm{d}\xi_1.$$

(4) *If* $\mathrm{Op}(b) = \mathrm{Op}(b_1)\mathrm{Op}(b_2)$, *then*

$$b(x,\xi)$$

$$= \exp\frac{\mathrm{i}}{2}\Big(D_{\xi_1}\cdot D_{x_2} - D_{\xi_2}\cdot D_{x_1}\Big)b_1(x_1,\xi_1)b_2(x_2,\xi_2)\Big|_{\substack{x = x_1 = x_2 \\ \xi = \xi_1 = \xi_2}}$$

$$= \pi^{-2d}\int e^{2\mathrm{i}(x-x_1)\cdot(\xi-\xi_2)-(x-x_2)\cdot(\xi-\xi_1)}b_1(x_1,\xi_1)b_2(x_2,\xi_2)\mathrm{d}x_1\mathrm{d}\xi_1\mathrm{d}x_2\mathrm{d}\xi_2.$$

Proof Let us prove (1). It is enough to check (8.45) for $b(x,\xi) := e^{\mathrm{i}(\eta\cdot x+q\cdot\xi)}$. We know that

$$\mathrm{Op}(b) = W(\eta,q) = e^{\mathrm{i}(\eta\cdot x+q\cdot D)} = e^{\frac{\mathrm{i}}{2}\eta\cdot x}e^{\mathrm{i}q\cdot D}e^{\frac{\mathrm{i}}{2}\eta\cdot x},$$

which has the integral kernel

$$B(x,y) = (2\pi)^{-d}\int_{\mathcal{X}^{\#}} e^{\frac{\mathrm{i}}{2}\eta\cdot(x+y)}e^{\mathrm{i}q\cdot\xi+\mathrm{i}\xi\cdot(x-y)}\mathrm{d}\xi.$$

Properties (2), (3), (4) then follow from (1). $\qquad\square$

Example 8.74 *Let P_0 be the operator considered in Example 4.42 (the orthogonal projection onto $\pi^{-\frac{d}{4}}e^{-\frac{1}{2}x^2}$). Then*

$$s_{P_0}(x,\xi) = 2^d e^{-x^2-\xi^2}.$$

8.4.4 Parity operator

Let $(\mathcal{Y},\omega)$ be a symplectic space of dimension $2d$. Consider a regular irreducible CCR representation (8.38). Let δ_v denote the delta function at $v \in \mathcal{Y}^{\#}$.

Definition 8.75 *Define the* parity operator

$$I := \mathrm{Op}(\pi^d\delta_0). \tag{8.46}$$

Theorem 8.76 (1) *I is self-adjoint and $I^2 = \mathbb{1}$.*
(2) *$I\mathrm{Op}(b)I = \mathrm{Op}(b_0)$, where $b_0(v) = b(-v)$.*
(3) *In the Schrödinger representation,*

$$I\Psi(x) = \Psi(-x). \tag{8.47}$$

Proof Let us show (3) first. The distributional kernel of I in the Schrödinger representation is

$$I(x,y) = 2^{-d}\int \delta\Big(\frac{x+y}{2},\xi\Big)e^{\mathrm{i}(x-y)\cdot\xi}\mathrm{d}\xi$$

$$= 2^{-d}\delta\Big(\frac{x+y}{2}\Big) = \delta(x+y).$$

This proves (3).

In the case of the Schrödinger representation, (1) and (2) follow immediately from (3). But every regular irreducible CCR representation is equivalent to the Schrödinger representation. □

Definition 8.77 *Define the* parity operator *centered at v as*

$$I_v := \mathrm{Op}(\pi^d \delta_v) = W(-\omega^{-1}v)IW(\omega^{-1}v), \quad v \in \mathcal{Y}^{\#}.$$

Theorem 8.78 (1) I_v *is self-adjoint and* $I_v^2 = \mathbb{1}$.
(2) $I_v \mathrm{Op}(b) I_v = \mathrm{Op}(b_v)$, *where* $b_v(w) = b(2v - w)$.
(3) *In the Schrödinger representation,*

$$I_{(q,\eta)} \Psi(x) = \mathrm{e}^{2\mathrm{i}\eta \cdot (x-q)} \Psi(2q - x). \tag{8.48}$$

The following theorem is an analog of Prop. 4.31.

Theorem 8.79 (1) *If* $b \in L^1(\mathcal{Y}^{\#})$, *then* $\mathrm{Op}(b)$ *is a compact operator. In terms of an absolutely norm convergent integral, we can write*

$$\mathrm{Op}(b) = \pi^{-d} \int I_v b(v) \mathrm{d}v. \tag{8.49}$$

Hence,

$$\|\mathrm{Op}(b)\| \leq \pi^{-d} \|b\|_1. \tag{8.50}$$

(2) *If* $B \in B^1(\mathcal{H})$, *then* $s_B \in C_\infty(\mathcal{Y}^{\#})$ *and*

$$s_B(v) = 2^d \mathrm{Tr} I_v B. \tag{8.51}$$

Hence

$$|s_B(v)| \leq 2^d \mathrm{Tr}|B|.$$

Proof Clearly, $b = \int b(v) \delta_v \mathrm{d}v$. Therefore, (8.49) follows from $I_v = \mathrm{Op}(\pi^d \delta_v)$. (8.49) implies (8.50).

Let $b \in L^1(\mathcal{Y}^{\#})$. Let $b_n \in L^2(\mathcal{Y}^{\#}) \cap L^1(\mathcal{Y}^{\#})$ such that $b_n \to b$ in $L^1(\mathcal{Y}^{\#})$. By Thm. 8.70 (6), the operators $\mathrm{Op}(b_n)$ are Hilbert–Schmidt and hence compact. By (8.50), we have $\mathrm{Op}(b_n) \to \mathrm{Op}(b)$ in norm. Therefore, b is compact.

Let us prove (2). Let $a \in L^2 \cap L^1$ and let B be trace-class. Then B is also Hilbert–Schmidt. Using first Thm. 8.70 (7), then (8.49), and finally the trace-class property of B, we obtain

$$(2\pi)^{-d} \int \overline{a(v)} s_B(v) \mathrm{d}v = \mathrm{Tr}\, \mathrm{Op}(a)^* B = \pi^{-d} \mathrm{Tr}\left(\int \overline{a(v)} I_v \mathrm{d}v B \right)$$

$$= (2\pi)^{-d} \int \overline{a(v)} 2^d \mathrm{Tr} I_v B \mathrm{d}v.$$

This proves the identity (8.51) for almost all v.

Using the fact that $v \mapsto I_v$ is strongly continuous and B is trace-class we see that $v \mapsto 2^d \mathrm{Tr} I_v B$ is continuous. Using $\mathrm{w} - \lim\limits_{\|v\| \to \infty} I_v = 0$ and Prop. 2.40, we conclude that $\lim\limits_{\|v\| \to \infty} 2^d \mathrm{Tr} I_v B = 0$. $\qquad\qquad\qquad\qquad\qquad\square$

Remark 8.80 *The Weyl–Wigner symbol of a quantum state can be measured. The first such experiment involved the motional degrees of freedom of an ion and was performed by Leibfried* et al. *(1996).*

In the case of a light mode this was first done in a simple and elegant experiment by Wódkiewicz, Radzewicz, Banaszek and Krasiński (described in Banaszek et al. *(1999)). A mode of a laser light was trapped between two mirrors. By applying an external source of light its state was "translated" in the phase space. The parity was measured by counting the number of scattered photons. Then the formula (8.51) was used to compute the Weyl–Wigner symbol of a given quantum state.*

8.5 General coherent vectors

By translating a fixed normalized vector with Weyl operators we obtain a family of vectors parametrized by the phase space. These vectors will be called *coherent vectors*. The family of coherent vectors has properties similar in some respects to those of an o.n. basis.

Coherent vectors can be used to define two kinds of quantizations. These two quantizations go under various names. We use the names proposed by Berezin (1966): the *covariant* and *contravariant quantizations*. These two quantizations are often used in applications.

In the literature the name "coherent vector" (or "coherent state") usually has a narrower meaning, of a Gaussian vector translated in phase space. Up to a phase factor, Gaussian coherent vectors can be also defined as eigenvectors of the annihilation operator. The covariant, resp. contravariant quantization w.r.t. Gaussian coherent vectors are also known as the *Wick*, resp. *anti-Wick quantization*. (Other names are used as well.)

In this section we describe the properties of general coherent vectors. We also discuss the covariant and contravariant quantization related to a given family of coherent vectors.

Gaussian coherent vectors, as well as Wick and anti-Wick quantizations, will be discussed in Chap. 9 about the Fock representation.

Throughout this section $\mathcal{Y}$ is a finite-dimensional symplectic space of dimension $2d$. $\mathcal{Y} \ni y \mapsto W(y) \in U(\mathcal{H})$ is an irreducible regular CCR representation. $\Psi_0 \in \mathcal{H}$ is a fixed normalized vector and $P_0 := |\Psi_0)(\Psi_0|$ is the corresponding orthogonal projection.

8.5.1 Coherent states transformation

Definition 8.81 *The family of* coherent vectors *associated with the vector* Ψ_0 *is defined by*

$$\Psi_v := W(-\omega^{-1}v)\Psi_0, \quad v \in \mathcal{Y}^\#.$$

The orthogonal projection onto Ψ_v, *called the* coherent state, *will be denoted*

$$P_v := W(-\omega^{-1}v)P_0W(\omega^{-1}v), \quad v \in \mathcal{Y}^\#.$$

Remark 8.82 *One often assumes that, for any* $y \in \mathcal{Y}$, $\Psi_0 \in \mathrm{Dom}\ \phi(y)$ *and*

$$\big(\Psi_0|\phi(y)\Psi_0\big) = 0.$$

This assumption implies that $\Psi_v \in \mathrm{Dom}\ \phi(y)$ *and*

$$\big(\Psi_v|\phi(y)\Psi_v\big) = v{\cdot}y, \quad v \in \mathcal{Y}^\#.$$

Thus Ψ_v *is localized in the phase space around* $v \in \mathcal{Y}^\#$. *Note, however, that we will not use the above assumption in this section.*

Definition 8.83 *The* coherent states transform *of* $\Phi \in \mathcal{H}$ *is defined as*

$$\mathcal{Y}^\# \ni v \mapsto T^{\mathrm{FBI}}\Phi(v) := (2\pi)^{-\frac{d}{2}}(\Psi_v|\Phi).$$

The coherent state transform is sometimes also called the FBI transform, *for Fourier, Bros and Iagolnitzer.*

Example 8.84 *Assume for the moment that* $\mathcal{Y} = \mathcal{X}^\# \oplus \mathcal{X}$ *and* $\mathcal{H} = L^2(\mathcal{X})$. *Consider the Schrödinger representation* $\mathcal{X}^\# \oplus \mathcal{X} \ni (\eta, q) \mapsto \mathrm{e}^{\mathrm{i}(\eta{\cdot}x + q{\cdot}D)} \in U(L^2(\mathcal{X}))$. *Fix a normalized vector* $\Psi \in L^2(\mathcal{X})$. *Let* $(q, \eta) \in \mathcal{Y}^\# = \mathcal{X} \oplus \mathcal{X}^\#$. *The coherent vectors and states are then given by*

$$\Psi_{(q,\eta)}(x) = \mathrm{e}^{\mathrm{i}(-q{\cdot}D + \eta{\cdot}x)}\Psi(x) = \mathrm{e}^{\mathrm{i}\eta{\cdot}x - \frac{1}{2}q{\cdot}\eta}\Psi(x - q),$$

$$P_{(q,\eta)}(x_1, x_2) = \Psi(x_1 - q)\overline{\Psi}(x_2 - q)\mathrm{e}^{\mathrm{i}(x_1 - x_2){\cdot}\eta}.$$

Theorem 8.85 (1)

$$(2\pi)^{-d}\int P_v\,\mathrm{d}v = \mathbb{1}, \quad \text{as a weak integral.} \tag{8.52}$$

(2) *If* $\Phi \in \mathcal{H}$, *then* $T^{\mathrm{FBI}}\Phi \in L^2(\mathcal{Y}^\#) \cap C_\infty(\mathcal{Y}^\#)$ *and*

$$\|T^{\mathrm{FBI}}\Phi\|_2 = \|\Phi\|_\mathcal{H}, \quad \|T^{\mathrm{FBI}}\Phi\|_\infty \le (2\pi)^{-\frac{d}{2}}\|\Phi\|_\mathcal{H}. \tag{8.53}$$

In particular, T^{FBI} *is an isometry from* $\mathcal{H}$ *into* $L^2(\mathcal{Y}^\#)$.

(3) *The FBI transformation intertwines the representation* W *with a certain representation of CCR on* $L^2(\mathcal{Y}^\#)$:

$$\mathrm{e}^{\mathrm{i}y{\cdot}(\frac{1}{2}v - \omega D_v)}T^{\mathrm{FBI}} = T^{\mathrm{FBI}}W(y), \quad y \in \mathcal{Y}.$$

Proof To prove (1) we use the Schrödinger representation. Let $\Phi \in L^2(\mathcal{X})$. Then

$$\int\limits_{\mathcal{X} \oplus \mathcal{X}^\#} (\Phi | P_{(q,\eta)} \Phi) \mathrm{d}q \mathrm{d}\eta$$

$$= \int\limits_{\mathcal{X} \oplus \mathcal{X}^\#} \int\limits_{\mathcal{X}} \int\limits_{\mathcal{X}} \overline{\Phi(x_1)} \Psi_0(x_1 - q) \overline{\Psi_0(x_2 - q)} \mathrm{e}^{\mathrm{i}(x_1 - x_2) \cdot \eta} \Phi(x_2) \mathrm{d}x_1 \mathrm{d}x_2 \mathrm{d}q \mathrm{d}\eta$$

$$= (2\pi)^d \int\limits_{\mathcal{X}} \int\limits_{\mathcal{X}} \overline{\Phi(x)} \Psi_0(x - q) \overline{\Psi_0(x - q)} \Phi(x) \mathrm{d}x \mathrm{d}q = (2\pi)^d \|\Phi\|^2 \|\Psi_0\|^2.$$

The first statement from (2) follows immediately from (1), the second from the definition of T^{FBI} and the fact that $W(y)$ tends weakly to 0 when $y \to \infty$.

To prove (3) we compute

$$(T^{\mathrm{FBI}} W(y) \Phi)(v) = (\Psi_0 | W(\omega^{-1} v) W(y) \Phi)$$

$$= \mathrm{e}^{\frac{\mathrm{i}}{2} y \cdot v} (\Psi_0 | W(\omega^{-1}(v + \omega y)) \Phi)$$

$$= \mathrm{e}^{\frac{\mathrm{i}}{2} y \cdot v} \mathrm{e}^{\mathrm{i}(\omega y) \cdot D_v} (\Psi_0 | W(\omega^{-1} v) \Phi) = \left(\mathrm{e}^{\mathrm{i} y \cdot (\frac{1}{2} v - \omega D_v)} T^{\mathrm{FBI}} \Phi \right)(v).$$

$\square$

8.5.2 Contravariant quantization

Recall that $\mathrm{Meas}(\mathcal{Y}^\#)$ denotes the space of complex Borel pre-measures on $\mathcal{Y}^\#$. The subspace of $\mathrm{Meas}(\mathcal{Y}^\#)$ consisting of finite Borel measures is denoted $\mathrm{Meas}^1(\mathcal{Y}^\#)$. If $b \in L^1_{\mathrm{loc}}(\mathcal{Y}^\#)$, then $\mathrm{d}\mu = b\mathrm{d}v$ belongs to $\mathrm{Meas}(\mathcal{Y}^\#)$ and $\|\mu\|_1 = \|b\|_1$. In such a case, μ is absolutely continuous w.r.t. the Lebesgue measure $\mathrm{d}v$ and b is its Radon–Nikodym derivative w.r.t. $\mathrm{d}v$. Thus $L^1_{\mathrm{loc}}(\mathcal{Y}^\#)$, resp. $L^1(\mathcal{Y}^\#)$ can be viewed as subspaces of $\mathrm{Meas}(\mathcal{Y}^\#)$, resp. $\mathrm{Meas}^1(\mathcal{Y}^\#)$. In such a case, we will abuse notation and write simply $b \in \mathrm{Meas}(\mathcal{Y}^\#)$.

Actually, we will abuse the notation even further. We will write $b\mathrm{d}v$ instead of $\mathrm{d}\mu$ even if $\mu \in \mathrm{Meas}(\mathcal{Y}^\#)$ is not absolutely continuous w.r.t. the Lebesgue measure $\mathrm{d}v$. Thus b will denote the "Radon–Nikodym derivative of μ w.r.t. $\mathrm{d}v$", even if strictly speaking such a derivative does not exist.

By smearing out coherent states with a classical symbol we obtain the so-called contravariant quantization. In the following proposition we describe properties of the contravariant quantization. Note in particular that positive symbols correspond to positive operators.

Proposition 8.86 *Let $b \in L^\infty(\mathcal{Y}^\#) + \mathrm{Meas}^1(\mathcal{Y}^\#)$. Then the formula*

$$(\Phi | \mathrm{Op}^{\mathrm{ct}}(b) \Phi) := (2\pi)^{-d} \int (\Phi | P_v \Phi) b(v) \mathrm{d}v \tag{8.54}$$

defines $\mathrm{Op}^{\mathrm{ct}}(b) \in B(\mathcal{H})$. *We have*

$$\|\mathrm{Op}^{\mathrm{ct}}(b)\| \leq (2\pi)^{-d}\|b\|_1, \quad b \in \mathrm{Meas}^1(\mathcal{Y}^\#), \tag{8.55}$$

$$\|\mathrm{Op}^{\mathrm{ct}}(b)\| \leq \|b\|_\infty, \qquad b \in L^\infty(\mathcal{Y}^\#). \tag{8.56}$$

Definition 8.87 $\mathrm{Op}^{\mathrm{ct}}(b) \in B(\mathcal{H})$ *defined in (8.54) is called the* contravariant *quantization of* b.

Proof of Prop. 8.86. If $b \in \mathrm{Meas}^1(\mathcal{Y}^\#)$, then the integral on the r.h.s. of (8.54) is finite and we obtain (8.55).

If $b \in L^\infty(\mathcal{Y}^\#)$, we can write

$$\mathrm{Op}^{\mathrm{ct}}(b) = T^{\mathrm{FBI}*}b(v)T^{\mathrm{FBI}}, \tag{8.57}$$

where on the r.h.s. $b(v)$ has the meaning of a multiplication operator on $L^2(\mathcal{Y}^\#)$, and we obtain (8.56).

In the general case, we can write

$$b = b_0 + b_1, \quad b_0 \in L^\infty(\mathcal{Y}^\#), \quad b_1 \in \mathrm{Meas}^1(\mathcal{Y}^\#), \tag{8.58}$$

and set

$$\mathrm{Op}^{\mathrm{ct}}(b) := \mathrm{Op}^{\mathrm{ct}}(b_0) + \mathrm{Op}^{\mathrm{ct}}(b_1). \tag{8.59}$$

It is easy to see that (8.59) does not depend on the decomposition (8.58). $\square$

Proposition 8.88 (1) $\mathrm{Op}^{\mathrm{ct}}(1) = \mathbb{1}_{\mathcal{H}}$.
(2) $\mathrm{Op}^{\mathrm{ct}}(b)^* = \mathrm{Op}^{\mathrm{ct}}(\overline{b})$.
(3) *If* $v \in \mathcal{Y}^\#$, *then*

$$W(-\omega^{-1}v)\mathrm{Op}^{\mathrm{ct}}(b)W(\omega^{-1}v) = \mathrm{Op}^{\mathrm{ct}}(b(\cdot - v)).$$

(4) *If* $b \in L^\infty(\mathcal{Y}^\#)$ *is real-valued, then*

$$\mathrm{ess\,inf}\, b \leq \mathrm{Op}^{\mathrm{ct}}(b) \leq \mathrm{ess\,sup}\, b. \tag{8.60}$$

(5) *Let* $b \geq 0$. *Then* $\mathrm{Op}^{\mathrm{ct}}(b) \geq 0$. *Moreover,* $b \in \mathrm{Meas}^1(\mathcal{Y}^\#)$ *iff* $\mathrm{Op}^{\mathrm{ct}}(b) \in B^1(\mathcal{H})$, *and*

$$\mathrm{Tr}\,\mathrm{Op}^{\mathrm{ct}}(b) = (2\pi)^{-d}\int b(v)\mathrm{d}v. \tag{8.61}$$

(6) *If* $b \in \mathrm{Meas}^1(\mathcal{Y}^\#)$, *then* $\mathrm{Op}^{\mathrm{ct}}(b) \in B^1(\mathcal{H})$ *and (8.61) is true.*
(7) *Suppose that* $b \in L^\infty_\infty(\mathcal{Y}^\#) + \mathrm{Meas}^1(\mathcal{Y}^\#)$, *where* $L^\infty_\infty(\mathcal{Y}^\#)$ *denotes the set of* $b \in L^\infty(\mathcal{Y}^\#)$ *such that* $\lim_{|v| \to \infty} b(v) = 0$. *Then* $\mathrm{Op}^{\mathrm{ct}}(b)$ *is compact.*

Proof (3) follows from $W(-\omega^{-1}v)P_w W(\omega^{-1}v) = P_{w+v}$. (8.60) follows immediately from (8.57).

We will now prove (5). Let b be positive. Let $\{e_i\}_{i \in I}$ be an o.n. basis of $\mathcal{H}$. By Fubini's theorem, we get

$$\operatorname{Tr} \operatorname{Op}^{\mathrm{ct}}(b) = \sum_{i \in I} (e_i | \operatorname{Op}^{\mathrm{ct}}(b) e_i) = (2\pi)^{-d} \int \sum_{i \in I} b(v)(e_i | P_v e_i) \mathrm{d}v$$

$$= (2\pi)^{-d} \int b(v) \operatorname{Tr} P_v \, \mathrm{d}v = (2\pi)^{-d} \int b(v) \mathrm{d}v,$$

which proves (5).

To show (6) we use (5) and the decomposition $b = b_1 + ib_2 - b_3 - ib_4$, where $b_i \in \mathrm{Meas}^1$ and $b_i \geq 0$. Finally, if $b = b_0 + \mu$ for $b_0 \in L_\infty^\infty(\mathcal{Y}^\#)$, $\mu \in \mathrm{Meas}^1(\mathcal{Y}^\#)$, we write $b_n = \mathbb{1}_{[0,n]}(|v|)b_0 + \mu$, so that $b_n \in \mathrm{Meas}^1(\mathcal{Y}^\#)$, $\operatorname{Op}^{\mathrm{ct}}(b_n) \in B^1(\mathcal{H})$, and $\|\operatorname{Op}^{\mathrm{ct}}(b_n - b)\| \leq \|b_n - b\|_\infty \to 0$ when $n \to \infty$. This proves (7). $\qquad \square$

Definition 8.89 *If the map*

$$L^\infty(\mathcal{Y}^\#) + \mathrm{Meas}^1(\mathcal{Y}^\#) \ni b \mapsto \operatorname{Op}^{\mathrm{ct}}(b) \in B(\mathcal{H})$$

is injective, then its inverse is called the contravariant symbol. *For $B \in B(\mathcal{H})$, its contravariant symbol will be denoted $\mathrm{s}_B^{\mathrm{ct}}$.*

8.5.3 Covariant quantization

In this subsection we describe the covariant quantization, which in a sense is the operation dual to the contravariant quantization. Strictly speaking, the operation that has a natural definition and good properties is not the covariant quantization but the covariant symbol of an operator.

Definition 8.90 *Let $B \in B(\mathcal{H})$. Then we define its* covariant symbol *by*

$$\mathrm{s}_B^{\mathrm{cv}}(v) := \operatorname{Tr} P_v B$$

$$= \left(W(-\omega^{-1} v) \Psi_0 | B W(-\omega^{-1} v) \Psi_0 \right), \quad v \in \mathcal{Y}^\#.$$

Theorem 8.91 (1) $\mathrm{s}_{\mathbb{1}}^{\mathrm{cv}} = 1$, $\mathrm{s}_{B^*}^{\mathrm{cv}} = \overline{\mathrm{s}_B^{\mathrm{cv}}}$.
(2) *If $B \in B(\mathcal{H})$, $B_v := W(-\omega^{-1} v) B W(\omega^{-1} v)$, then*

$$\mathrm{s}_{B_v}^{\mathrm{cv}} = \mathrm{s}_B^{\mathrm{cv}}(\cdot - v).$$

(3) *If $B \in B(\mathcal{H})$, then $\mathrm{s}_B^{\mathrm{cv}} \in C(\mathcal{Y}^\#) \cap L^\infty(\mathcal{Y}^\#)$ and*

$$\|\mathrm{s}_B^{\mathrm{cv}}\|_\infty \leq \|B\|. \tag{8.62}$$

(4) *Let $B \geq 0$. Then $\mathrm{s}_B^{\mathrm{cv}} \geq 0$. Moreover, $B \in B^1(\mathcal{H})$ iff $\mathrm{s}_B^{\mathrm{cv}} \in L^1(\mathcal{Y}^\#)$, and*

$$\operatorname{Tr} B = (2\pi)^{-d} \int \mathrm{s}_B^{\mathrm{cv}}(v) \mathrm{d}v. \tag{8.63}$$

(5) *If $B \in B^1(\mathcal{H})$, then $\mathrm{s}_B^{\mathrm{cv}} \in L^1(\mathcal{Y}^\#)$ and (8.63) is true.*
(6) *If B is compact, then $\mathrm{s}_B^{\mathrm{cv}} \in C_\infty(\mathcal{Y}^\#)$.*

Proof (1) and (2) are immediate. Let us show (3). It is easy to see that the inequality (8.62) is true. Moreover

$$v \mapsto W(\omega^{-1}v)BW(-\omega^{-1}v) \in B(\mathcal{H})$$

is strongly continuous. Hence $v \mapsto s_B(v)$ is continuous. To prove (6), we note that Ψ_v goes weakly to zero as $|v| \to \infty$. Hence, for compact B, $s_B(v) \to 0$ as $|v| \to \infty$.

To show (4), we use (8.52) and apply the trace to the identity

$$B = (2\pi)^{-d} \int_{\mathcal{Y}^\#} B^{\frac{1}{2}} P_v B^{\frac{1}{2}} \mathrm{d}v.$$

The interchange of trace and integral is justified by Fubini's theorem. To prove (5), we note that, if $B \in B^1(\mathcal{H})$, we can decompose it as $B = B_1 + \mathrm{i}B_2 - B_3 - \mathrm{i}B_4$, with $B_i \geq 0$, $B_i \in B^1(\mathcal{H})$. □

Definition 8.92 *If the map*

$$B(\mathcal{H}) \ni B \mapsto s_B^{\mathrm{cv}} \in C(\mathcal{Y}^\#) \cap L^\infty(\mathcal{Y}^\#)$$

is injective, then its inverse will be called the covariant quantization. *If b is a function on $\mathcal{Y}^\#$, its covariant quantization will be denoted* $\mathrm{Op}^{\mathrm{cv}}(b)$.

8.5.4 Connections between various quantizations

In this subsection we show how to pass between the covariant, Weyl–Wigner and contravariant quantizations. Note that there is a preferred direction: from contravariant to Weyl–Wigner, and then from Weyl–Wigner to covariant. Going back is less natural.

Let $w \in \mathcal{Y}^\#$. Let us compute various symbols of P_w:

$$s_{P_w}^{\mathrm{cv}}(v) = |(\Psi_{w-v}|\Psi_0)|^2,$$
$$s_{P_w}(v) = 2^d(\Psi_{w-v}|I\Psi_{w-v}),$$
$$s_{P_w}^{\mathrm{ct}}(v) = (2\pi)^d \delta(v - w).$$

The functions described in the following proposition will be used in formulas connecting various quantizations:

Proposition 8.93 *Set*

$$k_1(v) := (2\pi)^{-d} s_{P_0}(v) = \pi^{-d}(\Psi_0|I_v\Psi_0), \tag{8.64}$$
$$k_2(v) := (2\pi)^{-d} s_{P_0}^{\mathrm{cv}}(v) = (2\pi)^{-d}|(\Psi_v|\Psi_0)|^2. \tag{8.65}$$

Then k_2 is an even function in $C_\infty(\mathcal{Y}^\#)$, $k_1 \in L^1(\mathcal{Y}^\#) \cap C_\infty(\mathcal{Y}^\#)$ and

$$k_2(v) = \int k_1(w - v)k_1(w)\mathrm{d}w. \tag{8.66}$$

Proof Assume first that $\Psi_0 \in \mathcal{H}^\infty$. (Recall that $\mathcal{H}^\infty$ is defined in Def. 8.28.) Then $s_{P_0} \in \mathcal{S}(\mathcal{Y}^\#)$ and using (8.49) we have

$$P_0 = \pi^{-d} \int s_{P_0}(w) I_w \, dw$$

as a norm convergent integral. Next, by (8.51),

$$s_{P_0}(w - v) = 2^d \mathrm{Tr}(I_{w-v} P_0)$$
$$= 2^d \mathrm{Tr}(I_w P_v).$$

Hence,

$$s_{P_0}^{cv}(v) = \mathrm{Tr}(P_0 P_v) = \pi^{-d} \mathrm{Tr} \int s_{P_0}(w) I_w P_v \, dw$$

$$= (2\pi)^{-d} \int s_{P_0}(w) s_{P_0}(w - v) \, dw.$$

If $\Psi_0 \in \mathcal{H}$, we choose a sequence (Ψ_n) of normalized vectors in $\mathcal{H}^\infty$, such that $\Psi_n \to \Psi_0$ when $n \to \infty$. Then $s_{P_n} \to s_{P_0}$ in $L^2(\mathcal{Y}^\#)$, and $s_{P_n}^{cv} \to s_{P_0}^{cv}$ in $C_\infty(\mathcal{Y}^\#)$. (8.66) holds for Ψ_n. By letting $n \to \infty$, it also holds for Ψ_0. $\qquad\square$

Define the integral operator

$$K\Psi(v) := \int k_1(v - w) \Psi(w) \, dw. \qquad (8.67)$$

Then the identity (8.66) means that

$$K^* K \Psi(v) = \int k_2(v - w) \Psi(w) \, dw,$$

where K^* is the adjoint w.r.t. the scalar product of $L^2(\mathcal{Y}^\#)$.

Theorem 8.94 *We have the following identities between various symbols of an operator B, valid for example if $s_B^{ct} \in L^2(\mathcal{Y}^\#)$:*

$$s_B(v) = \int s_B^{ct}(w) k_1(v - w) \, dw, \quad \text{or } s_B = K s_B^{ct},$$

$$s_B^{cv}(v) = \int s_B(w) k_1(w - v) \, dw, \quad \text{or } s_B^{cv} = K^* s_B,$$

$$s_B^{cv}(v) = \int s_B^{ct}(w) k_2(w - v) \, dw, \quad \text{or } s_B^{cv} = K^* K s_B^{ct}.$$

8.5.5 Gaussian coherent vectors

Let us consider the Schrödinger representation on $L^2(\mathcal{X})$ and fix a Euclidean metric on $\mathcal{X}$. Consider the normalized Gaussian vector

$$\Psi_{(0,0)}(x) = \pi^{\frac{d}{4}} e^{-\frac{1}{2}x^2}. \qquad (8.68)$$

The corresponding coherent vectors are

$$\Psi_{(q,\eta)}(x) = \pi^{-\frac{d}{4}} e^{i\eta \cdot x - \frac{1}{2} q \cdot \eta - \frac{1}{2}(x-q)^2}, \quad (q, \eta) \in \mathcal{X} \oplus \mathcal{X}^\#. \qquad (8.69)$$

In the literature, when one speaks about coherent states, one usually has in mind (8.69). They are also called *Gaussian or Glauber's coherent states*. We will say more about them in the next chapter, because they appear naturally in the context of the Fock representation; see Chap. 9.

The covariant, resp. contravariant quantization for Gaussian coherent states coincides with the so-called Wick, resp. anti-Wick quantization, which will be discussed in Sect. 9.4. The corresponding integral kernels k_1, k_2 introduced in (8.64) and (8.65), and the corresponding operators K and K^*K are

$$k_1(x,\xi) = \pi^{-d}e^{-x^2-\xi^2}, \qquad\qquad K = K^* = e^{-\frac{1}{4}(D_x^2+D_\xi^2)},$$

$$k_2(x,\xi) = (2\pi)^{-d}e^{-\frac{1}{2}x^2-\frac{1}{2}\xi^2}, \qquad K^*K = e^{-\frac{1}{2}(D_x^2+D_\xi^2)}.$$

Thus in the Schrödinger representation one can distinguish five most natural quantizations. Their respective relations are nicely described by the following diagram, sometimes called the *Berezin diagram*:

anti-Wick
quantization

$\Big\downarrow e^{-\frac{1}{4}(D_x^2+D_\xi^2)}$

| D,x-
quantization | $e^{\frac{1}{2}D_x\cdot D_\xi}$
$\longrightarrow$ | Weyl–Wigner
quantization | $e^{\frac{1}{2}D_x\cdot D_\xi}$
$\longrightarrow$ | x,D-
quantization |

$\Big\downarrow e^{-\frac{1}{4}(D_x^2+D_\xi^2)}$

Wick
quantization

8.6 Notes

The relations

$$e^{i\eta\cdot x}e^{iq\cdot D} = e^{-iq\cdot\eta}e^{iq\cdot D}e^{i\eta\cdot x}, \quad \eta,q \in \mathbb{R} \tag{8.70}$$

were first stated by Weyl (1931). The proof of the Stone–von Neumann theorem can be found in von Neumann (1931); see also Emch (1972) and Bratteli–Robinson (1996). The canonical commutation relations for systems with many degrees of freedom were used by Dirac (1927) to describe quantized electromagnetic field.

We sketched the early history of the Weyl–Wigner(–Moyal) quantization in the introduction, with basic references Weyl (1931), Wigner (1932b) and Moyal (1949). In pure mathematics it became well known quite late. It was recognized in the so-called microlocal analysis – a powerful approach to the study of partial differential equations; see especially Hörmander (1985). It is also very useful in closely related semi-classical analysis; see e.g. Robert (1987).

The fact that the Weyl–Wigner quantization of the delta function is proportional to the parity operator was discovered only in the 1970s by Grossman (1976).

The Weyl CCR algebra was studied by, among others, Manuceau (1968) and Slawny (1971). Thm. 8.64 comes from Slawny (1971); see also Bratteli–Robinson (1996).

The original and still the most common meaning of the term "coherent state" is what we call a "Gaussian coherent state". These were first studied by Schrödinger (1926). They were extensively applied in quantum optics by Glauber (1963), for which he was awarded the Nobel Prize. Glauber introduced the name "coherent state" and, together with Cahill, studied quantizations based on coherent states in Cahill–Glauber (1969).

Various forms of quantization involving a family of general coherent states, in particular the covariant and contravariant quantizations, were studied by Berezin (1966). For a discussion of quantization see also Berezin–Shubin (1991) and Folland (1989).

The concept of coherent states has been generalized even further to the context of a rather general Lie group with a distinguished subgroup by Perelomov (1972).

The name "FBI transformation" comes from Fourier–Bros–Iagolnitzer. The FBI transformation was used by Iagolnitzer (1975) to study microlocal properties of distributions.

9

CCR on Fock space

This chapter is devoted to the study of the *Fock representation of the canonical commutation relations*. This representation is used as the basic tool in quantum many-body theory and quantum field theory. Unlike the Schrödinger CCR representation, it allows us to consider phase spaces of infinite dimension.

Throughout this chapter, $\mathcal{Z}$ is a Hilbert space. This space will be called the *one-particle space*. The Fock CCR representation will act in the bosonic Fock space $\Gamma_s(\mathcal{Z})$.

As in Sect. 1.3, we introduce the space

$$\mathcal{Y} = \mathrm{Re}(\mathcal{Z} \oplus \overline{\mathcal{Z}}) := \{(z, \overline{z}) \ : \ z \in \mathcal{Z}\},$$

which will serve as the *dual phase space* of our system. It will be equipped with the structure of a Kähler space consisting of the anti-involution j, the Euclidean scalar product · and the symplectic form ω:

$$\mathrm{j}(z, \overline{z}) := (\mathrm{i}z, \overline{\mathrm{i}z}), \tag{9.1}$$

$$(z, \overline{z}) \cdot (w, \overline{w}) := 2\mathrm{Re}(z|w), \tag{9.2}$$

$$(z, \overline{z}) \cdot \omega(w, \overline{w}) := 2\mathrm{Im}(z|w) = -(z, \overline{z}) \cdot \mathrm{j}(w, \overline{w}). \tag{9.3}$$

In principle, we can identify $\mathcal{Z}$ with $\mathcal{Y}$ by

$$\mathcal{Z} \ni z \mapsto \frac{1}{\sqrt{2}}(z + \overline{z}) \in \mathcal{Y}, \tag{9.4}$$

but we choose not to do so.

$\mathbb{C}\mathcal{Y}$ is identified with $\mathcal{Z} \oplus \overline{\mathcal{Z}}$ by the map

$$\mathbb{C}\mathcal{Y} \ni (z_1, \overline{z}_1) + \mathrm{i}(z_2, \overline{z}_2) \mapsto (z_1 + \mathrm{i}z_2, \overline{z_1 - \mathrm{i}z_2}) \in \mathcal{Z} \oplus \overline{\mathcal{Z}}.$$

The complexifications of (9.1), (9.2) and (9.3) are

$$\mathrm{j}_{\mathbb{C}}(z_1, \overline{z}_2) = (\mathrm{i}z_1, -\mathrm{i}\overline{z}_2),$$

$$(z_1, \overline{z}_2) \cdot_{\mathbb{C}} (w_1, \overline{w}_2) = (z_1|w_1) + (w_2|z_2), \tag{9.5}$$

$$(z_1, \overline{z}_2) \cdot \omega_{\mathbb{C}}(w_1, \overline{w}_2) = \frac{1}{\mathrm{i}}((z_1|w_1) - (w_2|z_2)). \tag{9.6}$$

$\mathcal{Y}^\#$, the space dual to $\mathcal{Y}$, is canonically identified with $\mathrm{Re}(\overline{\mathcal{Z}} \oplus \mathcal{Z})$ by using the scalar product (9.2), and $\mathbb{C}\mathcal{Y}^\#$ is identified with $\overline{\mathcal{Z}} \oplus \mathcal{Z}$.

9.1 Fock CCR representation

9.1.1 *Field operators on Fock spaces*

Consider the bosonic Fock space $\Gamma_s(\mathcal{Z})$. Recall that, for $z \in \mathcal{Z}$, $a^*(z)$, resp. $a(z)$ denote the corresponding creation, resp. annihilation operators.

Definition 9.1 *For $w = (z_1, \overline{z}_2) \in \mathcal{Z} \oplus \overline{\mathcal{Z}}$ we define the unbounded operator*

$$\phi(w) := a^*(z_1) + a(z_2) \quad \text{with domain } \Gamma_s^{\mathrm{fin}}(\mathcal{Z}).$$

Proposition 9.2 (1) *For $w \in \mathcal{Z} \oplus \overline{\mathcal{Z}}$, $\Gamma_s^{\mathrm{fin}}(\mathcal{Z})$ is an invariant subspace of entire analytic vectors for $\phi(w)$.*

(2) *The operators $\phi(y)$ for $y \in \mathrm{Re}(\mathcal{Z} \oplus \overline{\mathcal{Z}})$ are essentially self-adjoint. We will still denote by $\phi(y)$ their closures.*

(3) *The operators $\phi(w)$ for $w \in \mathcal{Z} \oplus \overline{\mathcal{Z}}$ are closable. We will still denote by $\phi(w)$ their closures.*

(4) *The map $\mathcal{Z} \oplus \overline{\mathcal{Z}} \ni w \mapsto \phi(w)$ is $\mathbb{C}$-linear on $\Gamma_s^{\mathrm{fin}}(\mathcal{Z})$.*

(5) *For $w_1, w_2 \in \mathbb{C}\mathcal{Y}$, we have*

$$[\phi(w_1), \phi(w_2)] = iw_1 \cdot \omega_{\mathbb{C}} w_2 \, \mathbb{1} \text{ on } \Gamma_s^{\mathrm{fin}}(\mathcal{Z}). \tag{9.7}$$

(6) *If $w = y_1 + iy_2$ with $y_1, y_2 \in \mathcal{Y}$, then $\mathrm{Dom}\,\phi(w) = \mathrm{Dom}\,\phi(y_1) \cap \mathrm{Dom}\,\phi(y_2)$.*

Proof Let $\Psi \in \Gamma_s^{\mathrm{fin}}(\mathcal{Z})$. From Thm. 3.51 we obtain

$$\|\phi(w)\Psi\| \leq \|w\| \|(N+\mathbb{1})^{\frac{1}{2}}\Psi\|.$$

By induction on n we obtain then that

$$\|\phi(w)^n \Psi\| \leq \|w\|^n \|(\tfrac{(N+n)!}{N!})^{\frac{1}{2}}\Psi\|. \tag{9.8}$$

This proves (1).

Now (2) follows from Nelson's commutator theorem; see Thm. 2.74 (1).

To prove (3) note that $\phi(\overline{w}) \subset \phi(w)^*$. So $\phi(w)$ is closable.

(4) and (5) follow by direct computation. (6) follows from (5) by repeating the argument of the proof of Prop. 8.31. $\qquad \square$

Corollary 9.3 *Let $z \in \mathcal{Z}$. Then $a(z)$, $a^*(z)$ are closable. Denoting their closures with the same symbols, for $y = (z, \overline{z})$, we have*

$$a^*(z) = \frac{1}{2}\big(\phi(y) - i\phi(jy)\big), \quad a(z) = \frac{1}{2}\big(\phi(y) + i\phi(jy)\big),$$

$$\mathrm{Dom}\,a^*(z) = \mathrm{Dom}\,a(z) = \mathrm{Dom}\,\phi(y) \cap \mathrm{Dom}\,\phi(jy).$$

Remark 9.4 *We have seen in Subsect. 1.3.9 that the map (9.4) is unitary. Using this identification, one can parametrize field operators by vectors of $\mathcal{Z}$ instead of vectors of $\mathcal{Y} = \mathrm{Re}(\mathcal{Z} \oplus \overline{\mathcal{Z}})$. This leads to the definition*

$$\phi(z) := \frac{1}{\sqrt{2}}\big(a^*(z) + a(z)\big), \quad z \in \mathcal{Z},$$

which is commonly found in the literature. In most of our work we will try to avoid this definition.

9.1.2 Weyl operators on Fock spaces

Theorem 9.5 (1) *If $w_1, w_2 \in \mathbb{C}\mathcal{Y}$ and $\Psi \in \Gamma_{\mathrm{s}}^{\mathrm{fin}}(\mathcal{Z})$, then the relationship*

$$e^{i\phi(w_1)} e^{i\phi(w_2)} \Psi = e^{-\frac{1}{2} w_1 \cdot \omega_{\mathbb{C}} w_2} e^{i\phi(w_1 + w_2)} \Psi \qquad (9.9)$$

holds, where the exponentials are defined in terms of the power series and all the series involved in (9.9) are absolutely convergent.

(2) *Set*

$$W(y) := e^{i\phi(y)}, \quad y \in \mathcal{Y}.$$

Then the map

$$\mathcal{Y} \ni y \mapsto W(y) \in U\big(\Gamma_{\mathrm{s}}(\mathcal{Z})\big) \qquad (9.10)$$

is a regular irreducible CCR representation, if we equip $\mathcal{Y}$ with the symplectic form ω defined in (9.3).

(3) *If $p \in U(\mathcal{Z})$, $(z, \overline{z}) \in \mathcal{Y}$, we have*

$$\Gamma(p) W(z, \overline{z}) = W(pz, \overline{p}z) \Gamma(p).$$

(4) *The map (9.10) is strongly continuous if we equip $\mathcal{Y}$ with the norm topology.*

Definition 9.6 *(9.10) is called the* Fock CCR representation *on $\Gamma_{\mathrm{s}}(\mathcal{Z})$.*

Proof To prove (1), we use the *Baker–Campbell formula*, which says the following: if A, B are operators such that $[A, B]$ commutes with A and B, then

$$e^A e^B = e^{\frac{1}{2}[A,B]} e^{A+B} \qquad (9.11)$$

as an identity between formal power series. We apply this formula to $A = i\phi(w_1)$, $B = i\phi(w_2)$, using (9.7). We use (9.8) to prove the norm convergence of the series appearing in (9.11).

Let us now prove (2). For $y_1, y_2 \in \mathrm{Re}(\mathcal{Z} \oplus \overline{\mathcal{Z}})$, both sides of (9.9) extend to unitary operators, so (9.9) is valid on the whole space $\Gamma_{\mathrm{s}}(\mathcal{Z})$. Therefore, (9.10) is a CCR representation. Since $W(y) = e^{i\phi(y)}$, this representation is regular.

Let us prove that it is irreducible. Let P be an orthogonal projection acting on $\Gamma_{\mathrm{s}}(\mathcal{Z})$ such that $[P, W(y)] = 0$ for all $y \in \mathrm{Re}(\mathcal{Z} \oplus \overline{\mathcal{Z}})$. Then $[P, \phi(y)] = 0$ on $\Gamma_{\mathrm{s}}^{\mathrm{fin}}(\mathcal{Z})$ for all $y \in \mathrm{Re}(\mathcal{Z} \oplus \overline{\mathcal{Z}})$, and hence $[P, a^*(z)] = [P, a(z)] = 0$ for all $z \in \mathcal{Z}$. It follows that $a(z) P\Omega = 0$. Hence, by (3.25), $P\Omega = 0$ or $P\Omega = \Omega$. By (3.26) and the fact that $[P, a^*(z)] = 0$, we obtain that $P = 0$ or $P = \mathbb{1}$.

To prove (4), we first see using the CCR that it suffices to prove the continuity of (9.10) at $y = 0$. Now, for $\Psi \in \Gamma_{\mathrm{s}}^{\mathrm{fin}}(\mathcal{Z})$ we have

$$\|(W(y) - \mathbb{1})\Psi\| \leq \|\phi(y)\Psi\| \leq \|y\| \|(N + \mathbb{1})^{\frac{1}{2}}\Psi\|. \qquad \square$$

Recall that we defined the parity operator as $I := (-1)^N$ in (3.10). If $\mathcal{Y}$ is finite-dimensional, we defined the parity operator as $I := \mathrm{Op}(\pi^d \delta_0)$ in (8.46).

Proposition 9.7 *In the finite-dimensional case, the definitions of the parity operator of (3.10) and of (8.46) coincide.*

9.1.3 Exponentials of creation and annihilation operators

Theorem 9.8 *Let $z \in \mathcal{Z}$.*

(1) *The operators $\mathrm{e}^{\phi(z,\bar{z})}$ are essentially self-adjoint on $\Gamma_\mathrm{s}^\mathrm{fin}(\mathcal{Z})$.*

(2) *$\mathrm{e}^{a^*(z)}$ and $\mathrm{e}^{a(z)}$ are closable operators on $\Gamma_\mathrm{s}^\mathrm{fin}(\mathcal{Z})$ and their closures have the domains*

$$\mathrm{Dom}\,\mathrm{e}^{a^*(z)} = \mathrm{Dom}\,\mathrm{e}^{a(z)} = \mathrm{Dom}\,\mathrm{e}^{\frac{1}{2}\phi(z,\bar{z})}.$$

(3) *In the sense of quadratic forms, we can write*

$$W(-\mathrm{i}z, \mathrm{i}\bar{z}) = \mathrm{e}^{-\frac{1}{2}\bar{z}\cdot z}\mathrm{e}^{a^*(z)}\mathrm{e}^{-a(z)}. \tag{9.12}$$

(4)

$$(\Omega|W(z,\bar{z})\Omega) = \mathrm{e}^{-\frac{1}{2}\bar{z}\cdot z}. \tag{9.13}$$

Proof (1) Using the exponential law in Prop. 3.56, it suffices to consider the case when $\dim\mathcal{Z} = 1$. For $z \in \mathcal{Z}$, we consider the unique conjugation τ such that $\tau z = z$ and introduce the associated real-wave representation defined in Thm. 9.20. This allows us to identify $\Gamma_\mathrm{s}(\mathcal{Z})$ with $L^2(\mathbb{R}, (2\pi)^{-\frac{1}{2}}\mathrm{e}^{-\frac{1}{2}x^2}\mathrm{d}x)$, $\Gamma_\mathrm{s}^\mathrm{fin}(\mathcal{Z})$ with the space of polynomials, and $\phi(z,\bar{z})$ with the operator of multiplication by αx for some $\alpha \in \mathbb{R}$. Then (1) is equivalent to the fact that the space of polynomials is dense in $L^2(\mathbb{R}, \mathrm{d}\mu)$ for $\mathrm{d}\mu = (2\pi)^{-\frac{1}{2}}(1 + \mathrm{e}^{\alpha x})^2\mathrm{e}^{-x^2/2}\mathrm{d}x$, which is well known.

(2) We have

$$\mathrm{e}^{a(z)} \subset (\mathrm{e}^{a^*(z)})^*, \qquad \mathrm{e}^{a^*(z)} \subset (\mathrm{e}^{a(z)})^*.$$

Hence $\mathrm{e}^{a^*(z)}$ and $\mathrm{e}^{a(z)}$ are closable on $\Gamma_\mathrm{s}^\mathrm{fin}(\mathcal{Z})$. Next we use the Baker–Campbell formula (9.11) on $\Gamma_\mathrm{s}^\mathrm{fin}(\mathcal{Z})$ to get

$$\mathrm{e}^{a(z)}\mathrm{e}^{a^*(z)} = \mathrm{e}^{\frac{1}{2}\bar{z}\cdot z}\mathrm{e}^{\phi(z,\bar{z})}, \qquad \mathrm{e}^{a^*(z)}\mathrm{e}^{a(z)} = \mathrm{e}^{-\frac{1}{2}\bar{z}\cdot z}\mathrm{e}^{\phi(z,\bar{z})}.$$

Thus, for $\Psi \in \Gamma_\mathrm{s}^\mathrm{fin}(\mathcal{Z})$,

$$\|\mathrm{e}^{a^*(z)}\Psi\|^2 = \mathrm{e}^{\frac{1}{2}\bar{z}\cdot z}\|\mathrm{e}^{\frac{1}{2}\phi(z,\bar{z})}\Psi\|^2, \qquad \|\mathrm{e}^{a(z)}\Psi\|^2 = \mathrm{e}^{-\frac{1}{2}\bar{z}\cdot z}\|\mathrm{e}^{\frac{1}{2}\phi(z,\bar{z})}\Psi\|^2.$$

Then we apply (1).

(3) follows from (9.11) and implies (4). $\qquad\square$

9.1.4 Gaussian coherent vectors on Fock spaces

Let $z \in \mathcal{Z}$.

Definition 9.9 *We define*

$$\Omega_z := W(-iz, i\overline{z})\Omega = e^{-\frac{1}{2}\overline{z}\cdot z} e^{a^*(z)}\Omega = e^{-\frac{1}{2}\overline{z}\cdot z} \sum_{n=0}^{\infty} \frac{z^{\otimes n}}{\sqrt{n!}}. \tag{9.14}$$

The vectors Ω_z will be called Glauber's *or* Gaussian coherent vectors. *Let P_z be the orthogonal projection onto Ω_z, so that*

$$P_z = W(-iz, i\overline{z})|\Omega)(\Omega|W(iz, -i\overline{z}).$$

Note that $(-iz, i\overline{z}) = -\omega^{-1}(\overline{z}, z)$. Hence, in the notation of Sect. 8.5, Ω_z equals $\Psi_{\overline{z},z}$ for $\Psi_{0,0} = \Omega$. Gaussian coherent vectors are eigenvectors of annihilation operators. Besides, one can say that Ω_z is localized in phase space around $(\overline{z}, z)$. This is expressed in the following proposition:

Proposition 9.10 *Let $w, z \in \mathcal{Z}$. Then $a(w)\Omega_z = (w|z)\Omega_z$. Therefore,*

$$(\Omega_z|a^*(w)\Omega_z) = (z|w),$$
$$(\Omega_z|a(w)\Omega_z) = (w|z),$$
$$(\Omega_z|\phi(w,\overline{w})\Omega_z) = 2\mathrm{Re}(z|w) = (z,\overline{z}) \cdot (w,\overline{w}).$$

9.2 CCR on anti-holomorphic Gaussian L^2 spaces

Let $\mathcal{Z}$ be a separable Hilbert space. We will use z as the generic variable in $\mathcal{Z}$.

Recall that if $\dim \mathcal{Z} < \infty$, then $(2i)^{-d}\mathrm{d}\overline{z}\mathrm{d}z$ is the volume form on $\mathcal{Z}_{\mathbb{R}}$ and $(2\pi i)^{-d}e^{-\overline{z}\cdot z}\mathrm{d}\overline{z}\mathrm{d}z$ defines the Gaussian measure for the covariance $\mathbb{1}$, which is a probability measure on $\mathcal{Z}_{\mathbb{R}}$. We can also define the corresponding Hilbert space of anti-holomorphic functions, denoted $L^2_{\overline{\mathbb{C}}}(\overline{\mathcal{Z}}, (2\pi i)^{-d}e^{-\overline{z}\cdot z}\mathrm{d}\overline{z}\mathrm{d}z)$. Thus if $F, G \in L^2_{\overline{\mathbb{C}}}(\overline{\mathcal{Z}}, (2\pi i)^{-d}e^{-\overline{z}\cdot z}\mathrm{d}\overline{z}\mathrm{d}z)$, then their scalar product is given by

$$(F|G) := (2\pi i)^{-d} \int \overline{F(\overline{z})}G(\overline{z})e^{-\overline{z}\cdot z}\mathrm{d}\overline{z}\mathrm{d}z.$$

Recall from Subsect. 5.5.4 that this Hilbert space has a natural generalization to the case of an arbitrary dimension, denoted $\mathbf{L}^2_{\overline{\mathbb{C}}}(\overline{\mathcal{Z}}, e^{-\overline{z}\cdot z}\mathrm{d}\overline{z}\mathrm{d}z)$ and called the anti-holomorphic Gaussian L^2 space over the space $\mathcal{Z}$.

The bosonic Fock space $\Gamma_s(\mathcal{Z})$ is naturally isomorphic to $\mathbf{L}^2_{\overline{\mathbb{C}}}(\overline{\mathcal{Z}}, e^{-\overline{z}\cdot z}\mathrm{d}\overline{z}\mathrm{d}z)$. This makes it possible to interpret Fock CCR representations in terms of operators acting on anti-holomorphic Gaussian L^2 spaces.

This section can be viewed as a continuation of Sect. 5.5 on Gaussian measures on complex Hilbert spaces.

9.2.1 Bosonic complex-wave representation

Theorem 9.11 (1) *The map* $T^{\mathrm{cw}} : \Gamma_{\mathrm{s}}(\mathcal{Z}) \to \mathbf{L}^2_{\mathbb{C}}(\overline{\mathcal{Z}}, \mathrm{e}^{-\overline{z}\cdot z}\,\mathrm{d}\overline{z}\mathrm{d}z)$ *given by*

$$T^{\mathrm{cw}}\Psi(\overline{z}) := \sum_{n=0}^{\infty} \frac{1}{\sqrt{n!}}(z^{\otimes n}|\Psi),$$

$$= \mathrm{e}^{\frac{1}{2}\overline{z}\cdot z}(\Omega_z|\Psi), \quad \Psi \in \Gamma_{\mathrm{s}}(\mathcal{Z}),$$

is unitary. (In the second line we use Gaussian coherent vectors Ω_z.)
(2) *For* $w \in \mathcal{Z}$ *we have*

$$T^{\mathrm{cw}}\Omega = 1,$$
$$T^{\mathrm{cw}}a^*(w) = w \cdot \overline{z}\, T^{\mathrm{cw}},$$
$$T^{\mathrm{cw}}a(w) = \overline{w} \cdot \nabla_{\overline{z}}\, T^{\mathrm{cw}},$$
$$(T^{\mathrm{cw}}\Gamma(p)\Psi)(\overline{z}) = T^{\mathrm{cw}}\Psi(p^{\#}\overline{z}), \quad p \in B(\mathcal{Z}), \quad \Psi \in \Gamma_{\mathrm{s}}(\mathcal{Z}).$$

(3) *We have a regular irreducible CCR representation*

$$\mathrm{Re}(\mathcal{Z} \oplus \overline{\mathcal{Z}}) \ni (w, \overline{w}) \mapsto \mathrm{e}^{\mathrm{i}(w \cdot \overline{z} + \overline{w} \cdot \nabla_{\overline{z}})} \in U(\mathbf{L}^2_{\mathbb{C}}(\overline{\mathcal{Z}}, \mathrm{e}^{-\overline{z}\cdot z}\,\mathrm{d}\overline{z}\mathrm{d}z)). \qquad (9.15)$$

(4) *The CCR representation (9.15) is equivalent to the Fock representation:*

$$T^{\mathrm{cw}}\mathrm{e}^{\mathrm{i}\phi(w,\overline{w})} = \mathrm{e}^{\mathrm{i}(w \cdot \overline{z} + \overline{w} \cdot \nabla_{\overline{z}})}T^{\mathrm{cw}}, \quad w \in \mathcal{Z}.$$

(5) *(9.15) acts on* $F \in \mathbf{L}^2_{\mathbb{C}}(\overline{\mathcal{Z}}, \mathrm{e}^{-\overline{z}\cdot z}\,\mathrm{d}\overline{z}\mathrm{d}z)$ *as follows:*

$$\mathrm{e}^{\mathrm{i}(w \cdot \overline{z} + \overline{w} \cdot \nabla_{\overline{z}})}F(\overline{z}) = \mathrm{e}^{\mathrm{i}w \cdot \overline{z} - \frac{1}{2}\overline{w} \cdot w}F(\overline{z} + \mathrm{i}\overline{w}), \qquad w \in \mathcal{Z}.$$

Proof (1) follows from Thm. 5.88. (2)–(4) follow immediately from Thm. 5.88 and Subsect. 3.5.2. To prove (5) we use the Baker–Campbell–Hausdorff formula.
□

Definition 9.12 *Following Segal, we will call* $T^{\mathrm{cw}}\Psi$ *the* complex-wave transform *of* Ψ. *(It is also sometimes called the* Bargmann *or* Bargmann–Segal transform *of* $\Psi \in \Gamma_{\mathrm{s}}(\mathcal{Z})$. *Berezin calls it the* generating functional *of* Ψ.*)*

(9.15) will be called the complex-wave CCR *representation. (It is also called the* Bargmann *or* Bargmann–Segal *representation.)*

9.2.2 Coherent vectors in the complex-wave representation

Let $w \in \mathcal{Z}$. The complex-wave transform of the Gaussian coherent vector Ω_w is

$$T^{\mathrm{cw}}\Omega_w(\overline{z}) = \mathrm{e}^{-\frac{1}{2}\overline{w}w}\mathrm{e}^{\overline{z}\cdot w}.$$

As an exercise in the complex-wave representation let us calculate the scalar product of two such vectors:

$$(\Omega_{w_1}|\Omega_{w_2}) = (2\pi\mathrm{i})^{-d} \int \mathrm{e}^{-\frac{1}{2}|w_1|^2 - \frac{1}{2}|w_2|^2 + z \cdot \overline{w}_1 + \overline{z} \cdot w_2 - |z|^2}\,\mathrm{d}\overline{z}\mathrm{d}z$$

$$= \mathrm{e}^{-\frac{1}{2}|w_1|^2 - \frac{1}{2}|w_2|^2 + \overline{w}_1 \cdot w_2}.$$

Definition 9.13 *Let* $\dim_{\mathbb{C}} \mathcal{Z} = d$ *be finite. The* Gaussian FBI transform *is the map* $T^{\mathrm{FBI}} : \Gamma_{\mathrm{s}}(\mathcal{Z}) \to L^2(\mathrm{Re}(\mathcal{Z} \oplus \overline{\mathcal{Z}}))$ *defined by*

$$\mathrm{Re}(\overline{\mathcal{Z}} \oplus \mathcal{Z}) \ni (\overline{z}, z) \mapsto T^{\mathrm{FBI}} \Psi(\overline{z}, z) := (2\pi)^{-\frac{d}{2}} (\Omega_z | \Psi). \qquad (9.16)$$

Clearly, the Gaussian FBI transform is a special case of the FBI transform defined in Subsect. 8.5.1, where we put $\Psi_0 = \Omega$.

By (9.16), in the finite-dimensional case we have the following simple relationship between the Gaussian FBI transformation and the complex-wave transformation:

$$T^{\mathrm{FBI}} \Psi(z, \overline{z}) = (2\pi)^{\frac{d}{2}} \mathrm{e}^{-\frac{1}{2}\overline{z} \cdot z} T^{\mathrm{cw}} \Psi(\overline{z}). \qquad (9.17)$$

This gives the following alternative proof of the unitarity of T^{cw}:

$$(\Psi_1 | \Psi_2) = \mathrm{i}^{-d} \int \overline{T^{\mathrm{FBI}} \Psi_1(\overline{z}, z)} T^{\mathrm{FBI}} \Psi_2(\overline{z}, z) \mathrm{d}\overline{z} \mathrm{d}z \qquad (9.18)$$

$$= (2\pi \mathrm{i})^{-d} \int \mathrm{e}^{-\frac{1}{2}|z|^2} \overline{T^{\mathrm{cw}} \Psi_1(\overline{z})} \mathrm{e}^{-\frac{1}{2}|z|^2} T^{\mathrm{cw}} \Psi_2(\overline{z}) \mathrm{d}\overline{z} \mathrm{d}z \qquad (9.19)$$

$$= (T^{\mathrm{cw}} \Psi_1 | T^{\mathrm{cw}} \Psi_2)_{\mathbf{L}^2_{\mathbb{C}}(\overline{\mathcal{Z}}, \mathrm{e}^{-\overline{z} \cdot z} \mathrm{d}\overline{z} \mathrm{d}z)}.$$

In (9.18) we used that $\mathrm{i}^{-d} \mathrm{d}\overline{z} \mathrm{d}z$ is the canonical measure on the symplectic space $\mathrm{Re}(\overline{\mathcal{Z}} \oplus \mathcal{Z})$ and that T^{FBI} is isometric; see (8.53).

9.3 CCR on real Gaussian $\mathbf{L}^2$ spaces

If the complex dimension of $\mathcal{Z}$ is finite and equals the real dimension of $\mathcal{X}$, then the Fock representation on $\Gamma_{\mathrm{s}}(\mathcal{Z})$ is unitarily equivalent to the Schrödinger representation on $L^2(\mathcal{X})$. In order to describe this equivalence, one needs to fix a conjugation on the Kähler space $\mathrm{Re}(\mathcal{Z} \oplus \overline{\mathcal{Z}})$, which allows us to separate field operators into "momentum" and "position" operators. In addition, one needs to fix a Euclidean structure on $\mathcal{X}$, which allows us to distinguish the Gaussian vector that is mapped to the Fock vacuum.

In the case of an infinite dimension we do not have a Schrödinger representation, since there is no Lebesgue measure on infinite-dimensional vector spaces. However, in this case we have the so-called *real-wave representations*, which can serve as a substitute for Schrödinger representations. Real-wave representations will be the main topic of this section. They are CCR representations acting on real Gaussian $\mathbf{L}^2$ spaces. They are unitarily equivalent to Fock representations.

Throughout this section, $\mathcal{X}$ is a real Hilbert space and $c \in B_{\mathrm{s}}(\mathcal{X})$ is invertible and positive. x will be used as the generic variable in $\mathcal{X}$.

Recall that if $\dim \mathcal{X} < \infty$, then $(2\pi)^{-\frac{d}{2}} (\det c)^{-\frac{1}{2}} \mathrm{e}^{-\frac{1}{2}x \cdot c^{-1}x} \mathrm{d}x$ is a probability measure on $\mathcal{X}$. Thus we can define the corresponding Hilbert space

$$L^2(\mathcal{X}, (2\pi)^{-\frac{d}{2}} (\det c)^{-\frac{1}{2}} \mathrm{e}^{-\frac{1}{2}x \cdot c^{-1}x} \mathrm{d}x).$$

As described in Def. 5.72, this can be generalized to the case of an arbitrary dimension, and then it is called the Gaussian $\mathbf{L}^2$ space for the covariance c and denoted

$$\mathbf{L}^2(\mathcal{X}, e^{-\frac{1}{2}x \cdot c^{-1}x}dx). \tag{9.20}$$

In this section we describe the real-wave representation acting on (9.20).

This section can be viewed as a continuation of Sect. 5.4 on Gaussian measures on real Hilbert spaces.

9.3.1 Real-wave CCR representation

Let $\eta, q \in \mathcal{X}$. We set

$$\eta \cdot x_{\mathrm{rw}} := \eta \cdot x,$$

$$q \cdot D_{\mathrm{rw}} := q \cdot (\frac{1}{i}\nabla_x + \frac{i}{2}c^{-1}x), \quad \text{as operators on } \mathbf{L}^2(\mathcal{X}, e^{-\frac{1}{2}x \cdot c^{-1}x}dx).$$

Theorem 9.14 (1) *The operator* $\eta \cdot x_{\mathrm{rw}} + q \cdot D_{\mathrm{rw}}$ *is essentially self-adjoint on* $\mathbb{C}\mathrm{Pol}_s(\mathcal{X})$.

(2) *The map*

$$\mathcal{X} \oplus \mathcal{X} \ni (\eta, q) \mapsto e^{i(\eta \cdot x_{\mathrm{rw}} + q \cdot D_{\mathrm{rw}})} \in U(\mathbf{L}^2(\mathcal{X}, e^{-\frac{1}{2}x \cdot c^{-1}x}dx)) \tag{9.21}$$

is an irreducible regular CCR representation.

(3) *For* $F \in \mathbf{L}^2(\mathcal{X}, e^{-\frac{1}{2}x \cdot c^{-1}x}dx)$ *one has*

$$e^{i(\eta \cdot x_{\mathrm{rw}} + q \cdot D_{\mathrm{rw}})}F(x) = e^{\frac{1}{2}q \cdot (\eta + \frac{1}{2}c^{-1}q)}e^{ix \cdot (\eta + \frac{1}{2}c^{-1}q)}F(x + q).$$

Proof We consider the one-parameter group

$$U_t F(x) := e^{\frac{1}{2}t^2 q \cdot (\eta + \frac{1}{2}c^{-1}q)}e^{itx \cdot (\eta + \frac{1}{2}c^{-1}q)}F(x + tq), \quad t \in \mathbb{R}.$$

Let $\mathcal{D} := \mathrm{Span}\{e^{w \cdot x}, \ w \in \mathbb{C}\mathcal{X}\}$. From Subsect. 5.2.5, we know that $\mathcal{D}$ is dense in $\mathbf{L}^2(\mathcal{X}, e^{-\frac{1}{2}x \cdot c^{-1}x}dx)$. Clearly, $\mathcal{D}$ is invariant under U_t, and U_t is a strongly continuous group of isometries of $\mathcal{D}$, hence it extends to a strongly continuous unitary group. $\mathcal{D}$ is included in the domain of its generator, which equals $\eta \cdot x_{\mathrm{rw}} + q \cdot D_{\mathrm{rw}}$ on $\mathcal{D}$. By Nelson's invariant domain theorem, Thm. 2.74 (2), we obtain that $\eta \cdot x_{\mathrm{rw}} + q \cdot D_{\mathrm{rw}}$ is essentially self-adjoint on $\mathcal{D}$.

To show the essential self-adjointness on $\mathbb{C}\mathrm{Pol}_s(\mathcal{X})$, we note that $\mathcal{D}$ is in the closure of $\mathbb{C}\mathrm{Pol}_s(\mathcal{X})$ for the graph norm: in fact, for $w \in \mathbb{C}\mathcal{X}$, the series

$$\sum_{n=0}^{+\infty} \frac{(w \cdot x)^n}{n!}$$

converges to $e^{w \cdot x}$ for the graph norm of $\eta \cdot x_{\mathrm{rw}} + q \cdot D_{\mathrm{rw}}$. This proves (1) and (3). (2) follows immediately from (3). $\qquad\square$

Definition 9.15 *The CCR representation (9.21) is called the* real-wave *representation of covariance* c.

Note that the operators x_{rw}, D_{rw} are examples of abstract position and momentum operators considered in Subsect. 8.2.6.

We equip $\mathcal{X} \oplus \mathcal{X}$ with the complex structure

$$\mathrm{j} = \begin{bmatrix} 0 & -(2c)^{-1} \\ 2c & 0 \end{bmatrix}, \tag{9.22}$$

which is Kähler. Thus $\mathcal{X} \oplus \mathcal{X}$ becomes a Kähler space with a conjugation. Therefore, as in Subsect. 8.2.7, for $w \in \mathbb{C}\mathcal{X}$ we can introduce the associated Schrödinger-type creation and annihilation operators:

$$a_{\mathrm{rw}}(w) = \overline{w} \cdot c\nabla_x, \quad a_{\mathrm{rw}}^*(w) = w \cdot x - w \cdot c\nabla_x.$$

Proposition 9.16 *Let* $w, w_1, w_2 \in \mathbb{C}\mathcal{X}$.

(1) *The operators* $a_{\mathrm{rw}}(w)$ *and* $a_{\mathrm{rw}}^*(w)$ *are closable on* $\mathbb{C}\mathrm{Pol}_{\mathrm{s}}(\mathcal{X})$.
(2) *We have*

$$[a_{\mathrm{rw}}(w_1), a_{\mathrm{rw}}^*(w_2)] = (w_1|cw_2)\mathbb{1},$$

$$[a_{\mathrm{rw}}(w_1), a_{\mathrm{rw}}(w_2)] = [a_{\mathrm{rw}}^*(w_1), a_{\mathrm{rw}}^*(w_2)] = 0.$$

(3) $F \in \mathbf{L}^2(\mathcal{X}, \mathrm{e}^{-\frac{1}{2}x \cdot c^{-1}x}\,\mathrm{d}x)$ *satisfies*

$$a_{\mathrm{rw}}(w)F = 0, \quad w \in \mathbb{C}\mathcal{X},$$

iff F *is proportional to* 1.

Proof (1) follows from Prop. 8.31 and (2) is a special case of (8.30).

Let F be such that $a_{\mathrm{rw}}(w)F = 0$ for $w \in \mathbb{C}\mathcal{X}$ and $(F|1) = 0$. In particular, for each $G \in \mathbb{C}\mathrm{Pol}_{\mathrm{s}}(\mathcal{X})$,

$$(a_{\mathrm{rw}}^*(w)G|F) = 0.$$

Clearly, the span of vectors of the form $\prod_{i=1}^{n} a_{\mathrm{rw}}^*(w_i)1$ equals the space of polynomials in $\mathbb{C}\mathrm{Pol}(\mathcal{X})$ of degree greater than 1. So F is orthogonal to $\mathbb{C}\mathrm{Pol}(\mathcal{X})$, and hence $F = 0$, which proves (3). $\qquad \square$

The usual choice is $c = \mathbb{1}$, which leads to the complex structure

$$\mathrm{j} = \begin{bmatrix} 0 & -\frac{1}{2}\mathbb{1} \\ 2\mathbb{1} & 0 \end{bmatrix}.$$

Remark 9.17 *The advantage of the real-wave representation is the fact that we can make an identification*

$$\mathbf{L}^2(\mathcal{X}, \mathrm{e}^{-\frac{1}{2}x \cdot c^{-1}x}\,\mathrm{d}x) \simeq L^2(Q, \mu)$$

for an L^2 *space over some true measure space* $(Q, \mathfrak{S}, \mu)$. *There is no unique choice of the measure space* $(Q, \mathfrak{S}, \mu)$, *especially in the case of an infinite-dimensional* $\mathcal{X}$, *but it essentially does not matter which one we take. A class of possible choices is described in Subsect. 5.4.2: we can set* $Q = B^{\frac{1}{2}}\mathcal{X}$, *where* $B > 0$ *is an operator on* $\mathcal{X}$ *with* B^{-1} *trace-class, but there are many others; see the discussion in Simon (1974). Therefore, the real-wave representation is sometimes called the* Q-*space representation of the bosonic Fock space.*

9.3.2 Real-wave CCR representation in finite dimension

If the dimension of $\mathcal{X}$ is finite, then the real-wave representation is a special case of a weighted Schrödinger representation with

$$m(x) = (2\pi)^{-\frac{d}{4}} (\det c)^{-\frac{1}{4}} e^{-\frac{1}{4}x \cdot c^{-1} x}. \qquad (9.23)$$

(9.23) is the pointwise positive ground state of

$$H = -\Delta + \frac{1}{4}x \cdot c^{-2} x - \frac{1}{2}\operatorname{Tr} c^{-1}.$$

The Dirichlet form for (9.23) in the Hilbert space $L^2\left(\mathcal{X}, (2\pi)^{-\frac{d}{2}}(\det c)^{-\frac{1}{2}} e^{-\frac{1}{2}x \cdot c^{-1} x} dx\right)$ equals

$$-\Delta + x \cdot c^{-1} \nabla_x.$$

The unitary operator

$$L^2\left(\mathcal{X}, (2\pi)^{-\frac{d}{4}}(\det c)^{-\frac{1}{2}} e^{-\frac{1}{2}x \cdot c^{-1} x} dx\right) \ni F \mapsto T^{\mathrm{sch}} F := m(x)F \in L^2(\mathcal{X})$$

intertwines the Schrödinger and the real-wave representations:

$$e^{i(\eta \cdot x + q \cdot D)} T^{\mathrm{sch}} = T^{\mathrm{sch}} e^{i(\eta \cdot x_{\mathrm{rw}} + q \cdot D_{\mathrm{rw}})}.$$

9.3.3 Wick transformation

The real-wave representation on $\mathbf{L}^2(\mathcal{X}, e^{-\frac{1}{2}x \cdot c^{-1} x} dx)$ is unitarily equivalent to the Fock representation on $\Gamma_s(c^{-\frac{1}{2}}\mathbb{C}\mathcal{X})$. This follows by a general argument from Prop. 9.16 and the fact that polynomials are dense; see Subsect. 5.2.6.

In this subsection we will construct an explicit unitary transformation that intertwines the real-wave representation and the Fock representation.

Definition 9.18 *For* $F \in \mathbb{C}\mathrm{Pol}_s(\mathcal{X})$, *we define*

$$:F := a_{\mathrm{rw}}^*(F)1 \in \mathbb{C}\mathrm{Pol}_s(\mathcal{X}).$$

The map $F \mapsto :F:$ *is called the* Wick transformation *w.r.t. the covariance* c.

The following proposition shows how one can compute $: G :$.

Proposition 9.19 (1) *For* $G \in \mathbb{C}\mathrm{Pol}_s(\mathcal{X})$, *one has*

$$:G(x): = e^{-\frac{1}{2}\nabla_x \cdot c \nabla_x} G(x) = e^{\frac{1}{2}x \cdot c^{-1} x} G(-c\nabla_x) e^{-\frac{1}{2}x \cdot c^{-1} x}. \qquad (9.24)$$

(2) *For* $G(x) \in \mathbb{C}\mathrm{Pol}_s(\mathcal{X})$, *one has*

$$G(x) = e^{\frac{1}{2}\nabla_x \cdot c \nabla_x} :G(x): = e^{-\frac{1}{2}x \cdot c^{-1} x} :G:(c\nabla_x) e^{\frac{1}{2}x \cdot c^{-1} x}.$$

Proof Let $w \in \mathbb{C}\mathcal{X}$. The following operator identities are valid on $\mathbb{C}\mathrm{Pol}_s(\mathcal{X})$:

$$a_{\mathrm{rw}}^*(w) = w \cdot x - w \cdot c\nabla_x$$
$$= \mathrm{e}^{-\frac{1}{2}\nabla_x \cdot c\nabla_x}(w \cdot x)\mathrm{e}^{\frac{1}{2}\nabla_x \cdot c\nabla_x} = \mathrm{e}^{\frac{1}{2}x \cdot c^{-1}x}(-w \cdot c\nabla_x)\mathrm{e}^{-\frac{1}{2}x \cdot c^{-1}x}.$$

This yields, for $G \in \mathbb{C}\mathrm{Pol}_s(\mathcal{X})$, the operator identity

$$a_{\mathrm{rw}}^*(G) = \mathrm{e}^{-\frac{1}{2}\nabla_x \cdot c\nabla_x}G(x)\mathrm{e}^{\frac{1}{2}\nabla_x \cdot c\nabla_x} = \mathrm{e}^{\frac{1}{2}x \cdot c^{-1}x}G(-c\nabla_x)\mathrm{e}^{-\frac{1}{2}x \cdot c^{-1}x}.$$

By applying it to the polynomial 1, we obtain

$$:G: = a_{\mathrm{rw}}^*(G)1 = \mathrm{e}^{-\frac{1}{2}\nabla_x \cdot c\nabla_x}G = \mathrm{e}^{\frac{1}{2}x \cdot c^{-1}x}G(-c\nabla_x)\mathrm{e}^{-\frac{1}{2}x \cdot c^{-1}x},$$

which proves (1). Clearly, (2) follows from (1). $\qquad\square$

Note that the space $\mathbb{C}\mathrm{Pol}_s(\mathcal{X})$ can be identified with $\mathrm{Pol}_s(\overline{\mathbb{C}\mathcal{X}})$ (by analytic continuation/restriction; see Subsect. 3.5.6). Let z denote the generic variable in $\mathbb{C}\mathcal{X}$. The following theorem is immediate:

Theorem 9.20 (1) *The map*

$$\mathrm{Pol}_s(\overline{\mathbb{C}\mathcal{X}}) \ni F \mapsto :F: \in \mathbb{C}\mathrm{Pol}_s(\mathcal{X})$$

extends to a unitary map

$$\mathbf{L}_{\mathbb{C}}^2(\overline{\mathbb{C}\mathcal{X}}, \mathrm{e}^{-\bar{z}\cdot c^{-1}z}\,\mathrm{d}\bar{z}\mathrm{d}z) \ni F \mapsto :F: \in \mathbf{L}^2(\mathcal{X}, \mathrm{e}^{-\frac{1}{2}x \cdot c^{-1}x}\,\mathrm{d}x). \tag{9.25}$$

(2) *(9.25) intertwines the complex-wave and real-wave CCR representations:*

$$:\mathrm{e}^{\mathrm{i}(w \cdot \bar{z} + \bar{w} \cdot \nabla_{\bar{z}})}F: = \mathrm{e}^{\mathrm{i}(a_{\mathrm{rw}}^*(w) + a_{\mathrm{rw}}(w))}:F:, \quad F \in \mathbf{L}_{\mathbb{C}}^2(\overline{\mathbb{C}\mathcal{X}}, \mathrm{e}^{-\bar{z}\cdot c^{-1}z}\,\mathrm{d}\bar{z}\mathrm{d}z), \quad w \in \mathbb{C}\mathcal{X}.$$

(3) *For $w \in \mathbb{C}\mathcal{X}$, we have*

$$:\mathrm{e}^{w \cdot x}: = \mathrm{e}^{w \cdot x}\mathrm{e}^{-\frac{1}{2}w \cdot cw}. \tag{9.26}$$

Remark 9.21 *(9.26) is often used as the definition of the Wick transformation.*

Using Subsect. 9.2.1, we can unitarily identify the real-wave representation on $\mathbf{L}^2(\mathcal{X}, \mathrm{e}^{-\frac{1}{2}x \cdot c^{-1}x}\,\mathrm{d}x)$ and the Fock representation on $\Gamma_s(c^{-\frac{1}{2}}\mathbb{C}\mathcal{X})$. This is described in the next theorem.

Theorem 9.22 *Set*

$$\Gamma_s(c^{-\frac{1}{2}}\mathbb{C}\mathcal{X}) \ni \Phi \mapsto T^{\mathrm{rw}}\Phi := :T^{\mathrm{cw}}\Phi: \in \mathbf{L}^2(\mathcal{X}, \mathrm{e}^{-\frac{1}{2}x \cdot c^{-1}x}\,\mathrm{d}x). \tag{9.27}$$

Then

(1) T^{rw} *is unitary.*
(2) T^{rw} *is the unique bounded linear map such that*

$$T^{\mathrm{rw}}\Omega = 1, \quad \text{and} \quad T^{\mathrm{rw}}\mathrm{e}^{\mathrm{i}a^*(\eta) + a(\eta)} = \mathrm{e}^{\mathrm{i}\eta \cdot x_{\mathrm{rw}}}T^{\mathrm{rw}}, \quad \eta \in c^{-\frac{1}{2}}\mathcal{X}.$$

(3) T^{rw} *is the unique bounded linear map such that*

$$T^{\mathrm{rw}}\Omega = 1, \ \ and \ T^{\mathrm{rw}} \prod_{i=1}^{n} a^*(w_i) = \prod_{i=1}^{n} a^*_{\mathrm{rw}}(w_i) T^{\mathrm{rw}}, \ \ w_i \in c^{-\frac{1}{2}}\mathbb{C}\mathcal{X}.$$

Remark 9.23 *In the case of a single variable, that is,* $\mathcal{X} = \mathbb{R}$, *and* $c = 1$, *the Wick transformation for monomials is the same as the Gram–Schmidt orthogonalization procedure with the weight* $\mathrm{e}^{-\frac{1}{2}x^2}$. *The polynomials* $:x^n:$ *are rescaled Hermite polynomials. More precisely, if one adopts the following definition of Hermite polynomials:*

$$\mathrm{e}^{2xt - t^2} =: \sum_{n=0}^{\infty} \frac{t^n}{n!} H_n(x),$$

then

$$:x^n: = \sqrt{2}^n H_n\left(\tfrac{x}{\sqrt{2}}\right).$$

9.3.4 Integrals of polynomials with a Gaussian weight

In this subsection, for simplicity, we assume that $c = 1$.

In physics one often computes integrals of a polynomial times the Gaussian weight. The Wick transformation helps to perform such an integral, as is seen from (9.29):

Theorem 9.24 *Let* $F \in \mathbb{C}\mathrm{Pol}_{\mathrm{s}}(\mathcal{X})$. *Then*

$$\int_{\mathcal{X}} F(x)\mathrm{e}^{-\frac{1}{2}x^2}\,\mathrm{d}x = \left(\mathrm{e}^{\frac{1}{2}\nabla_x^2}F\right)(0), \tag{9.28}$$

$$\int_{\mathcal{X}} :F(x):\mathrm{e}^{-\frac{1}{2}x^2}\,\mathrm{d}x = F(0). \tag{9.29}$$

Proof We can assume that $\mathcal{X}$ is of finite dimension. Recall the identity (4.14):

$$\mathrm{e}^{\frac{1}{2}\nabla_x^2}F(y) = (2\pi)^{-\frac{d}{2}} \int \mathrm{e}^{-\frac{1}{2}(y-x)^2} F(x)\mathrm{d}x. \tag{9.30}$$

In (9.30) we set $y = 0$, which proves (9.28).

To prove (9.29) we use (9.28) and Prop. 9.19. □

Note that the r.h.s. of (9.28) can be expanded in a finite sum and leads to the well-known sum over all possible "pairings". This is the simplest version of what is usually called the *Wick theorem*.

A more complicated version of the Wick theorem is given below. It has a well-known graphical interpretation in terms of diagrams, which we will discuss in Chap. 20.

Theorem 9.25 *Let* $F_1, \ldots, F_n \in \mathbb{C}\mathrm{Pol}_s(\mathcal{X})$. *Then*

$$:F_1(x):\cdots:F_n(x): \tag{9.31}$$

$$= :\exp\left(\sum_{i<j} \nabla_{x_i}\nabla_{x_j}\right) F_1(x_1)\cdots F_n(x_n)\big|_{x=x_1=\cdots=x_n} :,$$

$$(2\pi)^{-\frac{d}{2}} \int :F_1(x):\cdots:F_n(x): e^{-\frac{1}{2}x^2}\,\mathrm{d}x \tag{9.32}$$

$$= \exp\left(\sum_{i<j} \nabla_{x_i}\nabla_{x_j}\right) F_1(x_1)\cdots F_n(x_n)\big|_{0=x_1=\cdots=x_n}.$$

Proof To prove (9.31), we write

$$:F_1(x):\cdots:F_n(x):$$

$$= e^{-\frac{1}{2}\nabla_{x_1}^2} F_1(x_1)\cdots e^{-\frac{1}{2}\nabla_{x_n}^2} F_n(x_n)\big|_{x=x_1=\cdots=x_n}$$

$$= :e^{\frac{1}{2}\nabla_x^2}\left(e^{-\frac{1}{2}\nabla_{x_1}^2} F_1(x_1)\cdots e^{-\frac{1}{2}\nabla_{x_n}^2} F_n(x_n)\big|_{x=x_1=\cdots=x_n}\right):$$

$$= :e^{\frac{1}{2}(\nabla_{x_1}+\cdots+\nabla_{x_n})^2 - \frac{1}{2}\nabla_{x_1}^2 - \cdots - \frac{1}{2}\nabla_{x_n}^2} F_1(x_1)\cdots F_n(x_n)\big|_{x=x_1=\cdots=x_n}:.$$

In the last step we used that

$$\nabla_x f(x,\ldots,x) = (\nabla_{x_1} + \cdots + \nabla_{x_n}) f(x_1,\cdots,x_n)\big|_{x=x_1=\cdots=x_n}.$$

(9.32) follows from (9.31) and (9.29).   □

9.3.5 Operators in the real-wave representation

Definition 9.26 *For an operator a on $\mathcal{X}$, we will write*

$$\Gamma_{\mathrm{rw}}(a) := T^{\mathrm{rw}}\Gamma(a_{\mathbb{C}})T^{\mathrm{rw}*},$$

where we recall that $a_{\mathbb{C}}$ denotes the extension of a to $\mathbb{C}\mathcal{X}$.

Suppose that $c > 0$ is an operator on $\mathcal{X}$. Clearly,

$$\Gamma_{\mathrm{rw}}(c^{-\frac{1}{2}}) : \mathbf{L}^2(\mathcal{X}, e^{-\frac{1}{2}x^2}\,\mathrm{d}x) \to \mathbf{L}^2(c^{-\frac{1}{2}}\mathcal{X}, e^{-\frac{1}{2}x\cdot c^{-1}x}\mathrm{d}x)$$

is a unitary operator. Therefore, in what follows we will stick to the covariance $\mathbb{1}$.

Recall from Remark 9.17 that $\mathbf{L}^2(\mathcal{X}, e^{-\frac{1}{2}x^2}\,\mathrm{d}x)$ can be interpreted as $L^2(Q, \mu)$ for some measure space (Q, μ). Let F be a bounded Borel function on Q. Then one can define $F(x_{\mathrm{rw}})$, which is a bounded operator on $\mathbf{L}^2(\mathcal{X}, e^{-\frac{1}{2}x^2}\,\mathrm{d}x)$. It can be also interpreted as an element of $\mathbf{L}^2(\mathcal{X}, e^{-\frac{1}{2}x^2}\,\mathrm{d}x)$, and then it will simply be written F. Clearly, $F(x_{\mathrm{rw}})1 = F$.

Proposition 9.27 *Let u be an orthogonal operator on $\mathcal{X}$. Then*

$$\Gamma_{\mathrm{rw}}(u)F(x_{\mathrm{rw}})\Gamma_{\mathrm{rw}}(u)^{-1} = (\Gamma_{\mathrm{rw}}(u)F)(x_{\mathrm{rw}}). \tag{9.33}$$

Proof A dense set of vectors in $\mathbf{L}^2(\mathcal{X}, \mathrm{e}^{-\frac{1}{2}x^2}\,\mathrm{d}x)$ is given by $G(x_{\mathrm{rw}})1 = G$ for G bounded Borel functions on $B^{\frac{1}{2}}\mathcal{X}$. We have the commutation property

$$\Gamma_{\mathrm{rw}}(u)F(x_{\mathrm{rw}})\Gamma_{\mathrm{rw}}(u)^{-1}G(x_{\mathrm{rw}}) = G(x_{\mathrm{rw}})\Gamma_{\mathrm{rw}}(u)F(x_{\mathrm{rw}})\Gamma_{\mathrm{rw}}(u)^{-1}. \tag{9.34}$$

Hence, applying (9.34) to the vacuum 1 we obtain

$$\Gamma_{\mathrm{rw}}(u)F(x_{\mathrm{rw}})\Gamma_{\mathrm{rw}}(u)^{-1}G = G(x_{\mathrm{rw}})\Gamma_{\mathrm{rw}}(u)F = (\Gamma_{\mathrm{rw}}(u)F)(x_{\mathrm{rw}})G.$$

$\square$

Proposition 9.28 *Let* $\mathcal{X}_1$ *be a closed subspace of* $\mathcal{X}$. *Let* e_1 *be the orthogonal projection on* $\mathcal{X}_1$. *Let* $\mathfrak{B}_1$ *be the sub-σ-algebra of functions based in* $\mathcal{X}_1$, *and* $E_{\mathfrak{B}_1}$ *the corresponding conditional expectation. Then*

$$E_{\mathfrak{B}_1} = \Gamma_{\mathrm{rw}}(e_1).$$

Proposition 9.29 *Let* $a \in B(\mathcal{X})$. *Then*

(1) *If* $\|a\| \leq 1$, $\Gamma_{\mathrm{rw}}(a)$ *is doubly Markovian, hence it is a contraction on* $L^p(Q, \mathrm{d}\mu)$ *for all* $1 \leq p \leq \infty$.

(2) *If* $\|a\| < 1$, *then* $\Gamma_{\mathrm{rw}}(a)$ *is positivity improving.*

Proof We drop rw from Γ_{rw} and x_{rw}.

We first prove (1). We write a as j^*uj, where

$$\mathcal{X} \ni x \mapsto j(x) := x \oplus 0 \in \mathcal{X} \oplus \mathcal{X}$$

is isometric and

$$u = \begin{bmatrix} a & (\mathbb{1} - aa^*)^{\frac{1}{2}} \\ (\mathbb{1} - a^*a)^{\frac{1}{2}} & a^* \end{bmatrix}$$

is orthogonal. Using Subsect. 5.4.3, we see that if we take $(Q \times Q, \mu \otimes \mu)$ as the Q-space for $\mathcal{X} \oplus \mathcal{X}$, then the map $\Gamma(j)$ is

$$L^2(Q, \mathrm{d}\mu) \ni f \mapsto f \otimes 1 \in L^2(Q, \mathrm{d}\mu) \otimes L^2(Q, \mathrm{d}\mu) \simeq L^2(Q \times Q, \mathrm{d}\mu \otimes \mathrm{d}\mu),$$

which is positivity preserving.

The map $\Gamma(u)$ is clearly positivity preserving. In fact, recall that $F(x)$ is the operator of multiplication by a measurable function F on $L^2(Q, \mu)$. By (9.33) and the unitarity of u, $(\Gamma(u)F)(x) = \Gamma(u)F(x)\Gamma(u)^{-1}$. Since $F \geq 0$ a.e. iff $F(x) \geq 0$, we see that $\Gamma(u)$ is positivity preserving. Finally $\Gamma(j^*) = \Gamma(j)^*$ is also positivity preserving by the remark after Def. 5.21. Hence $\Gamma(a)$ is positivity preserving. Since $\Gamma(a)$ and $\Gamma(a)^*$ preserve 1, $\Gamma(a)$ is doubly Markovian.

Let us now prove (2). We write $\Gamma(a) = \Gamma(\|a\|)\Gamma(b)$, where $a =: \|a\|b$. Then $\|b\| \leq 1$, and thus $\Gamma(b)$ is positivity preserving by (1). If $f \geq 0$ and $f \neq 0$, then $\int_Q \Gamma(b)f\,\mathrm{d}\mu = \int_Q f\,\mathrm{d}\mu > 0$, so $\Gamma(b)$ preserves the set of non-zero positive functions. So it suffices to prove that $\Gamma(\|a\|)$ is positivity improving.

Let $f, g \geq 0$ with $f, g \neq 0$. The function $F(t) = (f|\Gamma(e^{-t})g)$ is positive on $\mathbb{R}^+$ by (1). It tends to $(1|f)(1|g)$ at $+\infty$, since $\Gamma(e^{-t}) = e^{-tN}$, where N is the number operator. Since F extends holomorphically to $\{z : \operatorname{Re} z > 0\}$, it has isolated zeroes in $\mathbb{R}^+$. Let $t > 0$ and $0 < t_0 < t$ such that $F(t_0) > 0$. Set $f_1 = \Gamma(e^{-t_0/2})f$, $g_1 = \Gamma(e^{-t_0/2})g$. Then $f_1, g_1 \geq 0$ and $(f_1|g_1) = F(t_0) > 0$. Therefore, $f_1 g_1 \neq 0$ and $h = \min(f_1, g_1) \neq 0$. This yields

$$
\begin{aligned}
(f|\Gamma(e^{-t})g) &= (f_1|\Gamma(e^{-(t-t_0)})g_1) \\
&\geq (h|\Gamma(e^{-(t-t_0)})g_1) \geq (h|\Gamma(e^{-(t-t_0)})h) \\
&= \|\Gamma(e^{-(t-t_0)/2})h\|^2 > 0,
\end{aligned}
$$

which completes the proof of (2).   □

Below we recall Nelson's famous hyper-contractivity theorem.

Theorem 9.30 *Let $a \in B(\mathcal{X})$ and $1 < p < q < \infty$. If*

$$
\|a\| \leq (p-1)^{\frac{1}{2}}(q-1)^{-\frac{1}{2}},
$$

then $\Gamma_{\mathrm{rw}}(a)$ is a contraction from $L^p(Q, d\mu)$ to $L^q(Q, d\mu)$.

9.4 Wick and anti-Wick bosonic quantization

As elsewhere in this chapter, $\mathcal{Z}$ is a Hilbert space, $\mathcal{Y} = \operatorname{Re}(\mathcal{Z} \oplus \overline{\mathcal{Z}})$, $\mathcal{Y}^\# = \operatorname{Re}(\overline{\mathcal{Z}} \oplus \mathcal{Z})$, $\mathbb{C}\mathcal{Y} = \mathcal{Z} \oplus \overline{\mathcal{Z}}$ and $\mathbb{C}\mathcal{Y}^\# = \overline{\mathcal{Z}} \oplus \mathcal{Z}$. We recall from Subsect. 3.5.6 that $\mathbb{C}\mathrm{Pol}_{\mathrm{s}}(\mathcal{Y}^\#)$ is identified with $\mathrm{Pol}_{\mathrm{s}}(\mathbb{C}\mathcal{Y}^\#)$. We can go from one representation to the other by analytic continuation/restriction. Thus we will freely switch between a polynomial in $\mathbb{C}\mathrm{Pol}_{\mathrm{s}}(\mathcal{Y}^\#)$ and $\mathrm{Pol}_{\mathrm{s}}(\overline{\mathcal{Z}} \oplus \mathcal{Z})$:

$$
\begin{aligned}
\operatorname{Re}(\overline{\mathcal{Z}} \oplus \mathcal{Z}) \ni (\bar{z}, z) &\mapsto b(\bar{z}, z), \\
\overline{\mathcal{Z}} \oplus \mathcal{Z} \ni (\bar{z}_1, z_2) &\mapsto b(\bar{z}_1, z_2).
\end{aligned}
$$

We consider the Fock CCR representation

$$
\mathcal{Y} \ni y \mapsto e^{i\phi(y)} \in U(\Gamma_{\mathrm{s}}(\mathcal{Z})).
$$

Recall that $\mathrm{CCR}^{\mathrm{pol}}(\mathcal{Y})$ is the $*$-algebra generated by $\phi(y)$, $y \in \mathcal{Y}$. It can be faithfully represented by operators on the space $\Gamma_{\mathrm{s}}^{\mathrm{fin}}(\mathcal{Z})$.

We will define and study the bosonic Wick and anti-Wick quantization. The Wick quantization is the most frequently used quantization in quantum field theory and many-body quantum physics.

9.4.1 Wick and anti-Wick ordering

Let $b \in \mathrm{Pol}_{\mathrm{s}}(\overline{\mathcal{Z}})$. Recall that in Subsect. 3.4.4 we defined the multiple creation and annihilation operators $a^*(b)$ and $a(b)$. Note that the possibility of unambiguously defining $a^*(b)$ and $a(b)$ follows from the fact that $\mathcal{Z}$ and $\overline{\mathcal{Z}}$ are isotropic subspaces of $\mathbb{C}\mathcal{Y}$ for $\omega_{\mathbb{C}}$.

Definition 9.31 *For* $b_1, b_2 \in \mathrm{Pol}_s(\overline{\mathcal{Z}})$ *we set*

$$\mathrm{Op}^{a^*,a}(b_1\overline{b_2}) := a^*(b_1)a(b_2),$$

$$\mathrm{Op}^{a,a^*}(\overline{b_2}b_1) := a(b_2)a^*(b_1).$$

These maps extend by linearity to maps

$$\mathbb{C}\mathrm{Pol}_s(\mathcal{Y}^{\#}) \ni b \mapsto \mathrm{Op}^{a^*,a}(b) \in \mathrm{CCR}^{\mathrm{pol}}(\mathcal{Y}),$$

$$\mathbb{C}\mathrm{Pol}_s(\mathcal{Y}^{\#}) \ni b \mapsto \mathrm{Op}^{a,a^*}(b) \in \mathrm{CCR}^{\mathrm{pol}}(\mathcal{Y}), \tag{9.35}$$

called the Wick *and* anti-Wick *bosonic quantizations.*

Definition 9.32 *The inverse maps to (9.35) will be denoted by*

$$\mathrm{CCR}^{\mathrm{pol}}(\mathcal{Y}) \ni B \mapsto s_B^{a^*,a} \in \mathbb{C}\mathrm{Pol}_s(\mathcal{Y}^{\#}),$$

$$\mathrm{CCR}^{\mathrm{pol}}(\mathcal{Y}) \ni B \mapsto s_B^{a,a^*} \in \mathbb{C}\mathrm{Pol}_s(\mathcal{Y}^{\#}).$$

The polynomial $s_B^{a^*,a}$, *resp.* s_B^{a,a^*} *is called the* Wick, *resp.* anti-Wick *symbol of the operator* B.

Remark 9.33 *Suppose that we fix an o.n. basis* $\{e_i \; : \; i \in I\}$ *in* $\mathcal{Z}$. *Every polynomial* $b \in \mathrm{Pol}_s(\overline{\mathcal{Z}} \oplus \mathcal{Z})$ *can be written as*

$$\sum_{\nu,\beta} b_{\nu,\beta} \overline{z}^\nu z^\beta,$$

where ν, β *are multi-indices, that is, elements of* $\{0, 1, 2, \dots\}^I$. *Then*

$$\mathrm{Op}^{a^*,a}(b) = \sum_{\nu,\beta} b_{\nu,\beta} a^{*\nu} a^\beta, \tag{9.36}$$

$$\mathrm{Op}^{a,a^*}(b) = \sum_{\nu,\beta} b_{\nu,\beta} a^\beta a^{*\nu}. \tag{9.37}$$

The r.h.s. of (9.36), resp. (9.37) is probably the most straightforward, even if often somewhat heavy, notation for the Wick, resp. anti-Wick quantization.

More generally, one can assume that $\mathcal{Z} = L^2(\Xi, \mathrm{d}\xi)$, where $(\Xi, \mathrm{d}\xi)$ is a measure space. Then polynomials on $\mathcal{Z}$ can be written as

$$\sum_{n,m} \int \cdots \int b(\xi_1, \dots \xi_n; \xi'_m, \dots, \xi'_1) \overline{z}_{\xi_1} \cdots \overline{z}_{\xi_n} z_{\xi'_m} \cdots z_{\xi'_1},$$

and one writes

$$\sum_{n,m} b(\xi_1,\dots\xi_n;\xi'_m,\dots,\xi'_1)a^*_{\xi_1}\cdots a^*_{\xi_n}\,a_{\xi'_m}\cdots a_{\xi'_1} \quad \text{instead of } \mathrm{Op}^{a^*,a}(b),$$

$$\sum_{n,m} b(\xi_1,\dots\xi_n;\xi'_m,\dots,\xi'_1)a_{\xi'_m}\cdots a_{\xi'_1}\,a^*_{\xi_1}\cdots a^*_{\xi_n} \quad \text{instead of } \mathrm{Op}^{a,a^*}(b).$$

Thus a^*_ξ and a_ξ are treated as "operator-valued measures", which acquire their meaning after being "smeared out" with "test functions".

The following theorem is the analog of Thm. 4.38 devoted to the x, D- and D, x-quantizations.

Theorem 9.34 *Let $b, b_-, b_+, b_1, b_2 \in \mathrm{Pol}_s(\overline{\mathcal{Z}}, \mathcal{Z})$.*

(1) $\mathrm{Op}^{a,a^*}(b)^* = \mathrm{Op}^{a,a^*}(\overline{b})$ *and* $\mathrm{Op}^{a^*,a}(b)^* = \mathrm{Op}^{a^*,a}(\overline{b})$.

(2) *For $w \in \mathcal{Z}$,*

$$\mathrm{Op}^{a^*,a}(wb) = a^*(w)\mathrm{Op}^{a^*,a}(b), \quad \mathrm{Op}^{a^*,a}(\overline{w}b) = \mathrm{Op}^{a^*,a}(b)a(w),$$

$$[\mathrm{Op}^{a^*,a}(b), a^*(w)] = \mathrm{Op}^{a^*,a}(w\nabla_z b), \quad [a(w), \mathrm{Op}^{a^*,a}(b)] = \mathrm{Op}^{a^*,a}(\overline{w}\nabla_{\overline{z}}b).$$

$$(\Omega|\mathrm{Op}^{a^*,a}(b)\Omega) = b(0). \tag{9.38}$$

(3) *If $\mathrm{Op}^{a,a^*}(b_-) = \mathrm{Op}^{a^*,a}(b_+)$, then*

$$b_+(\overline{z}, z) = e^{\nabla_{\overline{z}}\nabla_z} b_-(\overline{z}, z)$$

$$= (2\pi\mathrm{i})^{-d}\int e^{-(\overline{z}-\overline{z_1})(z-z_1)} b_+(\overline{z_1}, z_1)\mathrm{d}\overline{z_1}\mathrm{d}z_1, \quad \text{if } \dim\mathcal{Z} = d.$$

(4) *If $\mathrm{Op}^{a^*,a}(b_1)\mathrm{Op}^{a^*,a}(b_2) = \mathrm{Op}^{a^*,a}(b)$, then*

$$b(\overline{z}, z) = e^{\nabla_{z_1}\nabla_{\overline{z_1}}} b_1(\overline{z}, z_1) b_2(\overline{z_1}, z)\big|_{z_1=z}$$

$$= (2\pi\mathrm{i})^{-d}\int e^{-(\overline{z}-\overline{z_1})(z-z_1)} b_1(\overline{z}, z_1) b_2(\overline{z_1}, z)\mathrm{d}z_1\mathrm{d}\overline{z_1}, \quad \text{if } \dim\mathcal{Z} = d.$$

If $\mathrm{Op}^{a,a^*}(b_1)\mathrm{Op}^{a,a^*}(b_2) = \mathrm{Op}^{a,a^*}(b)$, then

$$b(\overline{z}, z) = e^{-\nabla_{\overline{z_1}}\nabla_{z_1}} b_1(\overline{z_1}, z) b_2(\overline{z}, z_1)\big|_{z_1=z}.$$

Proof If we use the complex-wave representation, we see that the Wick, resp. anti-Wick quantization can be viewed as the $\overline{z}, \nabla_{\overline{z}}$, resp. $\nabla_{\overline{z}}, \overline{z}$ quantization. Therefore, we can apply the same combinatorial arguments as in the proof of Thm. 4.38.   □

Remark 9.35 *The exponentials of differential operators in the above formulas can always be understood as finite sums of differential operators, since we consider polynomial symbols. Note also that in the expression for the anti-Wick symbol of a product of two operators there is no integral formula.*

The theorem that we state below is what is usually meant by *Wick's theorem.* We will discuss its diagrammatic interpretation in Chap. 20. It is an analog of Thm. 4.39.

Theorem 9.36 *Let* $b_1, \ldots, b_n, b \in \mathbb{C}\mathrm{Pol}_s(\mathcal{Y}^\#)$ *and*

$$\mathrm{Op}^{a^*,a}(b) = \mathrm{Op}^{a^*,a}(b_1) \cdots \mathrm{Op}^{a^*,a}(b_n).$$

Then

$$b(\overline{z}, z) \tag{9.39}$$
$$= \exp\left(\sum_{i<j} \nabla_{\overline{z}_i} \cdot \nabla_{z_j}\right) b_1(\overline{z}_1, z_1) \cdots b_n(\overline{z}_n, z_n)\big|_{z=z_1=\cdots=z_n},$$

$$(\Omega|\mathrm{Op}^{a^*,a}(b)\Omega) \tag{9.40}$$
$$= \exp\left(\sum_{i<j} \nabla_{\overline{z}_i} \cdot \nabla_{z_j}\right) b_1(\overline{z}_1, z_1) \cdots b_n(\overline{z}_n, z_n)\big|_{0=z_1=\cdots=z_n}.$$

Proof (9.39) is shown by the same arguments as Thm. 4.39. (9.40) follows from (9.39) and (9.38). □

9.4.2 Relation between Wick, anti-Wick and Weyl–Wigner quantizations

Let us assume that $\dim \mathcal{Z} < \infty$, so that the Weyl–Wigner quantization of a polynomial in $\mathbb{C}\mathrm{Pol}_s(\mathcal{Y}^\#)$ is well defined.

The following theorem gives the connection between the Weyl–Wigner and the Wick and the anti-Wick quantizations. We express these connections using two alternative notations: either we treat them as functions of the complex variables $(\overline{z}_1, z_2) \in \overline{\mathcal{Z}} \oplus \mathcal{Z}$, or we treat the symbols as functions of the real variable $v \in \mathrm{Re}(\overline{\mathcal{Z}} \oplus \mathcal{Z})$.

Theorem 9.37 *Let* $b_-, b, b_+ \in \mathbb{C}\mathrm{Pol}_s(\mathcal{Y}^\#)$. *Let*

$$\mathrm{Op}^{a^*,a}(b_+) = \mathrm{Op}(b) = \mathrm{Op}^{a,a^*}(b_-).$$

(1) *One can express the Wick symbol in terms of the Weyl–Wigner symbol:*

$$b_+(\overline{z}, z) = \mathrm{e}^{\frac{1}{2}\nabla_{\overline{z}} \cdot \nabla_z} b(\overline{z}, z)$$
$$= (\pi\mathrm{i})^{-d} \int \mathrm{e}^{-2(\overline{z}-\overline{z}_1)\cdot(z-z_1)} b(\overline{z}_1, z_1)\mathrm{d}\overline{z}_1\mathrm{d}z_1,$$
$$b_+(v) = \mathrm{e}^{\frac{1}{4}\nabla_v^2} b(v)$$
$$= \pi^{-d} \int \mathrm{e}^{-(v-v_1)^2} b(v_1)\mathrm{d}v_1.$$

(2) *One can express the Weyl–Wigner symbol in terms of the anti-Wick symbol:*

$$b(\bar{z}, z) = e^{\frac{1}{2}\nabla_{\bar{z}} \cdot \nabla_z} b_-(\bar{z}, z)$$

$$= (\pi i)^{-d} \int e^{-2(\bar{z}-\bar{z}_1)\cdot(z-z_1)} b_-(\bar{z}_1, z_1) d\bar{z}_1 dz_1,$$

$$b(v) = e^{\frac{1}{4}\nabla_v^2} b_-(v)$$

$$= \pi^{-d} \int e^{-(v-v_1)^2} b_-(v_1) dv_1.$$

Proof Let $b_1, b_2 \in \mathrm{Pol}_s(\overline{\mathcal{Z}})$, $b_+(\bar{z}, z) = b_1(\bar{z})\bar{b}_2(z)$. We have

$$\mathrm{Op}^{a^*, a}(b_+) = a^*(b_1)a(b_2)$$

$$= \mathrm{Op}(b_1)\mathrm{Op}(\bar{b}_2) = \mathrm{Op}(b).$$

Using the formula for the product of two Weyl–Wigner quantized operators, we obtain

$$b(\bar{z}, z) = e^{\frac{1}{2}(\nabla_{\bar{z}_1}, \nabla_{z_1})\cdot\omega(\nabla_{\bar{z}_2}, \nabla_{z_2})} b_1(\bar{z}_1)\bar{b}_2(z_2)\Big|_{(\bar{z},z)=(\bar{z}_1, z_2)}$$

$$= e^{-\frac{1}{2}(\nabla_{\bar{z}_1}\cdot\nabla_{z_2} - \nabla_{\bar{z}_2}\cdot\nabla_{z_1})} b_1(\bar{z}_1)\bar{b}_2(z_2)\Big|_{(\bar{z},z)=(\bar{z}_1, z_2)}$$

$$= e^{-\frac{1}{2}\nabla_{\bar{z}}\cdot\nabla_z} b_1(\bar{z})\bar{b}_2(z),$$

where in the second line we use the definition (9.6) of the symplectic form ω. This proves the first formula of (1). The second follows from the first, using the identities of Subsect. 4.1.9. (2) follows from (1) and Thm. 9.34 (3). $\qquad\square$

9.4.3 Wick and anti-Wick quantization as covariant and contravariant quantization

For $z \in \mathcal{Z}$, we consider the Gaussian coherent vectors Ω_z and the corresponding projections P_z in $\Gamma_s(\mathcal{Z})$, defined in Def. 9.9. We will show that the Wick, resp. anti-Wick quantizations coincide with the covariant, resp. contravariant quantization for Gaussian coherent vectors.

Theorem 9.38 (1) *Let $B \in \mathrm{CCR}^{\mathrm{pol}}(\mathcal{Y})$. Then for all $z \in \mathcal{Z}$, $\Omega_z \in \mathrm{Dom}\, B$ and*

$$s_B^{a^*, a}(\bar{z}, z) = (\Omega_z | B\Omega_z), \quad z \in \mathcal{Z}. \tag{9.41}$$

(2) *Let $b \in \mathbb{C}\mathrm{Pol}_s(\mathcal{Y}^\#)$. Let the dimension of $\mathcal{Z}$ be finite. Then*

$$\mathrm{Op}^{a, a^*}(b) = (2\pi i)^{-d} \int b(z, \bar{z}) P_z dz d\bar{z}. \tag{9.42}$$

(The integral should be understood in terms of a sesquilinear form on an appropriate domain.)

Proof Let $b_1, b_2 \in \mathrm{Pol}_s(\overline{\mathcal{Z}})$. Set

$$b(\bar{z}, z) := b_1(\bar{z})\overline{b_2(\bar{z})} \in \mathrm{Pol}_s(\overline{\mathcal{Z} \oplus \mathcal{Z}}).$$

Then

$$(\Omega_z | \mathrm{Op}^{a^*,a}(b)\Omega_z) = (\Omega_z | a^*(b_1)a(b_2)\Omega_z)$$
$$= (\Omega | W(\mathrm{i}z, -\mathrm{i}\overline{z})a^*(b_1)a(b_2)W(-\mathrm{i}z, \mathrm{i}\overline{z})\Omega)$$
$$= \left(\Omega | (a^*(b_1) + b_1(\overline{z}))(a(b_2) + \overline{b_2(\overline{z})})\Omega\right)$$
$$= b_1(\overline{z})\overline{b_2(\overline{z})} \quad = \quad b(\overline{z}, z).$$

This proves (9.41). Next, we compute

$$\mathrm{Op}^{a,a^*}(b) = a(b_2)a^*(b_1)$$
$$= (2\pi\mathrm{i})^{-d} \int a(b_2)P_z a^*(b_1)\mathrm{d}\overline{z}\mathrm{d}z$$
$$= (2\pi\mathrm{i})^{-d} \int W(\mathrm{i}z - \mathrm{i}\overline{z})(a(b_2) + \overline{b_2(\overline{z})})P_0(a^*(b_1)$$
$$+ b_1(\overline{z}))W(-\mathrm{i}z + \mathrm{i}\overline{z})\mathrm{d}\overline{z}\mathrm{d}z$$
$$= (2\pi\mathrm{i})^{-d} \int \overline{b_2(\overline{z})}b_1(\overline{z})P_z\mathrm{d}\overline{z}\mathrm{d}z = (2\pi\mathrm{i})^{-d} \int b(\overline{z}, z)P_z\mathrm{d}\overline{z}\mathrm{d}z.$$

This proves (9.42). $\qquad\qquad\qquad\qquad\qquad\qquad\qquad\qquad\qquad\qquad\square$

Remark 9.39 *Thm. 9.38 (1) says that the Wick symbol coincides with the covariant symbol defined with the help of Gaussian coherent states. Thus, using the notation of Sect. 8.5, (9.41) can be denoted $s_B^{\mathrm{cv}}(\overline{z} + z)$. (Strictly speaking, however, operators in $\mathrm{CCR}^{\mathrm{pol}}(\mathcal{Y})$ are usually unbounded, so they do not belong to the class considered in Sect. 8.5.)*

Thm. 9.38 (2) says that the anti-Wick quantization coincides with the contravariant quantization for Gaussian coherent states. Thus, using the notation of Sect. 8.5, (9.42) can be denoted $\mathrm{Op}^{\mathrm{ct}}(b)$. (Strictly speaking, however, functions in $\mathbb{C}\mathrm{Pol}(\mathcal{Y}^{\#})$ usually do not belong to $\mathrm{Meas}^1(\mathcal{Y}^{\#}) + L^{\infty}(\mathcal{Y}^{\#})$, so they do not belong to the class considered in Sect. 8.5.)

9.4.4 Wick symbols on Fock spaces

So far, we have defined the Wick symbol only for operators in $\mathrm{CCR}^{\mathrm{pol}}(\mathcal{Y})$. In this case, it is a polynomial on $\mathrm{Re}(\overline{\mathcal{Z}} \oplus \mathcal{Z})$.

We will now extend the definition of the Wick symbol to a rather large class of quadratic forms on $\Gamma_{\mathrm{s}}(\mathcal{Z})$.

Definition 9.40 *Let B be a quadratic form on $\Gamma_{\mathrm{s}}(\mathcal{Z})$ such that Ω_z belongs to its domain for any $z \in \mathcal{Z}$. We define the Wick symbol of B as*

$$s_B^{a^*,a}(\overline{z}, z) := (\Omega_z | B\Omega_z). \qquad (9.43)$$

By Thm. 9.38 (1), the above definition of the Wick symbol agrees with Def. 9.32 for $B \in \mathrm{CCR}^{\mathrm{pol}}(\mathcal{Y})$. In (9.43), the Wick symbol is viewed as a function on $\mathcal{Y}^{\#} = \mathrm{Re}(\overline{\mathcal{Z}} \oplus \mathcal{Z})$. An alternative point of view on the Wick symbol uses holomorphic functions on $\overline{\mathcal{Z}} \oplus \mathcal{Z}$.

Proposition 9.41 *The holomorphic extension of (9.43) to $\overline{\mathcal{Z}} \oplus \mathcal{Z}$ (see Def. 5.81) is*

$$s_B^{a^*,a}(\overline{z_1}, z_2) = \mathrm{e}^{-\overline{z}_1 \cdot z_2 + \frac{1}{2}\overline{z}_1 \cdot z_1 + \frac{1}{2}\overline{z}_2 \cdot z_2}(\Omega_{z_1} | B\Omega_{z_2}).$$

Proposition 9.42 *Let B be a positive closed quadratic form such that $\Gamma_s^{\mathrm{fin}}(\mathcal{Z}) \subset \mathrm{Dom}\, B$ and for each $z \in \mathcal{Z}$ the series*

$$\sum_{n,m=0}^{\infty} \frac{1}{\sqrt{n!}}(z^{\otimes n}|Bz^{\otimes m})\frac{1}{\sqrt{m!}}$$

is absolutely convergent. Then the Wick symbol of B and its holomorphic extension are

$$s_B^{a^*,a}(\overline{z}, z) = \mathrm{e}^{-\overline{z}\cdot z}\sum_{n,m=0}^{\infty}\frac{1}{\sqrt{n!}}(z^{\otimes n}|Bz^{\otimes m})\frac{1}{\sqrt{m!}}, \tag{9.44}$$

$$s_B^{a^*,a}(\overline{z_1}, z_2) = \mathrm{e}^{-\overline{z_1}\cdot z_2}\sum_{n,m=0}^{\infty}\frac{1}{\sqrt{n!}}(z_1^{\otimes n}|Bz_2^{\otimes m})\frac{1}{\sqrt{m!}}. \tag{9.45}$$

Proof Recalling that

$$\Omega_z = \mathrm{e}^{-\frac{1}{2}\overline{z}\cdot z}\sum_{n=0}^{\infty}\frac{z^{\otimes n}}{\sqrt{n!}},$$

and using that B is closed, we see that $\Omega_z \in \mathrm{Dom}\, B$ and $(\Omega_z | B\Omega_z)$ is given by the convergent series in (9.44). Applying the Cauchy–Schwarz inequality, we obtain that the series in the r.h.s. of (9.45) is absolutely convergent. Then we use Prop. 9.41. $\qquad\square$

In the following proposition we compute the Wick symbol of various operators in the sense of Def. 9.40:

Proposition 9.43 (1) *For $h \in B(\mathcal{Z})$, we have $s_{\mathrm{d}\Gamma(h)}^{a^*,a}(\overline{z}, z) = \overline{z}\cdot hz$.*

(2) *If p is a contraction on $\mathcal{Z}$, we have $s_{\Gamma(p)}^{a^*,a}(\overline{z}, z) = \mathrm{e}^{-\overline{z}\cdot z + \overline{z}\cdot pz}$.*

Example 9.44 *The anti-Wick, Weyl–Wigner and Wick symbols of $P_0 = |\Omega)(\Omega|$ (the projection onto Ω) are given below (compare with Examples 4.42 and 8.74):*

$$s_{P_0}^{a,a^*}(\overline{z}, z) = (2\pi)^d\delta_0,$$
$$s_{P_0}(\overline{z}, z) = 2^d\mathrm{e}^{-2\overline{z}\cdot z},$$
$$s_{P_0}^{a^*,a}(\overline{z}, z) = \mathrm{e}^{-\overline{z}\cdot z}.$$

9.4.5 Wick quantization: the operator formalism

Recall from Subsect. 8.5.3 that in general it is easier to find the covariant symbol of an operator than to compute the covariant quantization of a symbol. This remark applies to the Wick quantization. In this subsection we will describe this more difficult direction.

It is convenient to represent Wick symbols as operators acting on the Fock space. We need, however, to restrict ourselves to a rather small class of such operators.

Recall that N is the number operator and $\mathbb{1}_{\{n\}}(N)$ is the orthogonal projection from $\Gamma_{\mathrm{s}}(\mathcal{Z})$ onto $\Gamma_{\mathrm{s}}^{n}(\mathcal{Z})$.

Definition 9.45 *For $b \in B\big(\Gamma_{\mathrm{s}}(\mathcal{Z})\big)$, set $b_{n,m} := \mathbb{1}_{\{n\}}(N) b \mathbb{1}_{\{m\}}(N)$. Let*

$$B^{\mathrm{fin}}\big(\Gamma_{\mathrm{s}}(\mathcal{Z})\big)$$
$$:= \big\{ b \in B\big(\Gamma_{\mathrm{s}}(\mathcal{Z})\big) \ : \ there \ exists \ n_0 \ such \ that \ b_{n,m} = 0 \ for \ n,m > n_0 \big\}.$$

Definition 9.46 *Let $b \in B^{\mathrm{fin}}\big(\Gamma_{\mathrm{s}}(\mathcal{Z})\big)$. Then we define its* Wick quantization, *denoted by $\mathrm{Op}^{a^*,a}(b)$, as the quadratic form on $\Gamma_{\mathrm{s}}^{\mathrm{fin}}(\mathcal{Z})$ defined for $\Phi, \Psi \in \Gamma_{\mathrm{s}}^{\mathrm{fin}}(\mathcal{Z})$ as*

$$(\Phi|\mathrm{Op}^{a^*,a}(b)\Psi) = \sum_{n,m=0}^{\infty} \sum_{k=0}^{\min(m,n)} \frac{\sqrt{n!m!}}{k!} (\Phi|b_{n-k,m-k} \otimes \mathbb{1}_{\mathcal{Z}}^{\otimes k} \Psi),$$

$$= \sum_{n,m=0}^{\infty} \sum_{k=0}^{\infty} \frac{\sqrt{(n+k)!(m+k)!}}{k!} (\Phi|b_{n,m} \otimes \mathbb{1}_{\mathcal{Z}}^{\otimes k} \Psi).$$

The above definition is essentially an extension of Def. 9.31.

Proposition 9.47 *Let $b \in B^{\mathrm{fin}}\big(\Gamma_{\mathrm{s}}(\mathcal{Z})\big)$. Set $B = \mathrm{Op}^{a^*,a}(b)$, with the Wick quantization defined as in Def. 9.46. Then the Wick symbol of B in the sense of Def. 9.40 and its holomorphic extension are*

$$\mathrm{s}_B^{a^*,a}(\overline{z}, z) = \sum_{n,m=0}^{\infty} (z^{\otimes n}|bz^{\otimes m}), \tag{9.46}$$

$$\mathrm{s}_B^{a^*,a}(z_1, z_2) = \sum_{n,m=0}^{\infty} (z_1^{\otimes n}|bz_2^{\otimes m}). \tag{9.47}$$

Consequently, if $b \in \mathbb{C}\mathrm{Pol}_{\mathrm{s}}(\mathcal{Y}^{\#}) \simeq \mathrm{Pol}(\overline{\mathcal{Z}} \oplus \mathcal{Z})$ is identified with $b \in B^{\mathrm{fin}}\big(\Gamma_{\mathrm{s}}(\mathcal{Z})\big)$ with the help of (9.46) or (9.47), then Def. 9.31 coincides with Def. 9.40.

Proof B clearly satisfies the hypotheses of Def. 9.40, since $b \in B^{\mathrm{fin}}\big(\Gamma_{\mathrm{s}}(\mathcal{Z})\big)$. Using (9.44), we obtain

$$s_B^{a^*,a}(\overline{z},z)\mathrm{e}^{\overline{z}\cdot z} = \sum_{n,m=0}^{\infty} \frac{1}{\sqrt{n!}}(z^{\otimes n}|Bz^{\otimes m})\frac{1}{\sqrt{m!}}$$

$$= \sum_{n,m=0}^{\infty} \sum_{k=0}^{\min(n,m)} \frac{1}{k!}(z^{\otimes n}|b_{n-k,m-k}\otimes \mathbb{1}_{\mathcal{Z}}^{\otimes k}z^{\otimes m})$$

$$= \sum_{n,m=0}^{\infty} \sum_{k=0}^{\min(n,m)} \frac{1}{k!}(\overline{z}\cdot z)^k(z^{\otimes(n-k)}|b_{n-k,m-k}z^{\otimes(m-k)})$$

$$= \sum_{n,m=0}^{\infty} \sum_{k=0}^{\infty} \frac{1}{k!}(\overline{z}\cdot z)^k(z^{\otimes n}|b_{n,m}z^{\otimes m}) = \sum_{n,m=0}^{\infty}(z^{\otimes n}|b_{n,m}z^{\otimes m})\mathrm{e}^{\overline{z}\cdot z}.$$

$\square$

In the following identities it is convenient to use the new, more general definition of the Wick quantization:

Proposition 9.48 *In the following identities* $b \in B^{\mathrm{fin}}\big(\Gamma_{\mathrm{s}}(\mathcal{Z})\big)$, $h \in B(\mathcal{Z}) \subset B^{\mathrm{fin}}\big(\Gamma_{\mathrm{s}}(\mathcal{Z})\big)$, $p \in B(\mathcal{Z}_1, \mathcal{Z}_2)$.

$$\mathrm{Op}^{a^*,a}(h) = \mathrm{d}\Gamma(h);$$
$$[\mathrm{d}\Gamma(h), \mathrm{Op}^{a^*,a}(b)] = \mathrm{Op}^{a^*,a}([\mathrm{d}\Gamma(h), b]);$$
$$\Gamma(p)\mathrm{Op}^{a^*,a}(b\Gamma(p)) = \mathrm{Op}^{a^*,a}(\Gamma(p)b)\Gamma(p);$$
$$\Gamma(p)\mathrm{Op}^{a^*,a}(b) = \mathrm{Op}^{a^*,a}(\Gamma(p)b\Gamma(p^*))\Gamma(p), \quad \textit{if } p \textit{ is isometric;}$$
$$\Gamma(p)\mathrm{Op}^{a^*,a}(b)\Gamma(p^*) = \mathrm{Op}^{a^*,a}(\Gamma(p)b\Gamma(p^*)), \quad \textit{if } p \textit{ is unitary.}$$

The following proposition describes the special class of particle preserving operators:

Proposition 9.49 *If* $b \in B\big(\Gamma_{\mathrm{s}}^m(\mathcal{Z})\big)$, *then*

$$\frac{1}{m!}(\Phi|\mathrm{Op}^{a^*,a}(b)\Psi) = \sum_{k=1}^{\infty} \frac{(m+k)!}{m!k!}(\Phi|b \otimes \mathbb{1}_{\mathcal{Z}}^k \Psi)$$

$$= \sum_{k=1}^{\infty} \sum_{1\leq i_1 < \cdots < i_m \leq m+k} (\Phi|b_{i_1,\ldots,i_m}^{m+k} \Psi).$$

The operators $b_{i_1,\ldots,i_m}^{m+k} \in B\big(\Gamma_{\mathrm{s}}^{m+k}(\mathcal{Z})\big)$ *are defined as follows:*

$$b_{i_1,\ldots,i_m}^{m+k} := \Theta(\sigma)\, b \otimes \mathbb{1}_{\mathcal{Z}}^{\otimes k}\, \Theta(\sigma)^{-1} \in B\big(\Gamma_{\mathrm{s}}^{m+k}(\mathcal{Z})\big),$$

where $\sigma \in S_n$ *is any permutation that transforms* $(1,\ldots,m)$ *onto* $(i_1,\ldots,i_m)$. *Thus* $b_{i_1,\ldots,i_m}^{m+k}$ *is the "m-body interaction" acting on the* i_1-th,.. *through* i_m-th *particles.*

9.4.6 Estimates on Wick polynomials

Let $b \in B\big(\Gamma_{\mathrm{s}}^q(\mathcal{Z}), \Gamma_{\mathrm{s}}^p(\mathcal{Z})\big) \subset B^{\mathrm{fin}}\big(\Gamma_{\mathrm{s}}(\mathcal{Z})\big)$ for $p, q \in \mathbb{N}$. The following estimates are known as N_τ *estimates*.

Proposition 9.50 *Let $m > 0$ be a self-adjoint operator on $\mathcal{Z}$. Then for all Ψ_1, $\Psi_2 \in \Gamma_{\mathrm{s}}(\mathcal{Z})$ one has*

$$\left| \left(\mathrm{d}\Gamma(m)^{-p/2} \Psi_1 \,\big|\, \mathrm{Op}^{a^*,a}(b) \mathrm{d}\Gamma(m)^{-q/2} \Psi_2 \right) \right|$$
$$\leq \| \Gamma(m)^{-\frac{1}{2}} b \Gamma(m)^{-\frac{1}{2}} \| \, \| \Psi_1 \| \, \| \Psi_2 \|. \tag{9.48}$$

In particular, $\mathrm{Op}^{a^*,a}(b)$ *extends to an operator on* $\Gamma_{\mathrm{s}}(\mathcal{Z})$ *with domain* $\mathrm{Dom}\, N^{(p+q)/2}$.

Proof Noting that $N\mathrm{Op}^{a^*,a}(b) = \mathrm{Op}^{a^*,a}(b)(N + p - q)$, we see that the second statement follows from the first for $m = \mathbb{1}$.

To prove the first statement, we will assume for simplicity that $\mathcal{Z}$ is separable (the non-separable case can be treated by the same arguments, replacing sequences by nets). It clearly suffices to prove (9.48) for Ψ_1, $\dot\Psi_2$ such that $\Psi_i = \Gamma(\pi)\Psi_i$, where π is a finite rank projection. Moreover, if (π_n) is an increasing sequence of orthogonal projections with $\mathrm{s} - \lim \pi_n = \mathbb{1}$, and if $b_n = \Gamma(\pi_n) b \Gamma(\pi_n)$, it suffices to prove (9.48) for $\mathrm{Op}^{a^*,a}(b_n)$. Therefore, we may assume that $\mathcal{Z}$ is finite-dimensional. Let $(e_1, \ldots, e_n)$ be an o.n. basis of eigenvectors for m and $m_k = (e_k | m e_k)$. For $\vec{k} = (k_1, \ldots, k_d) \in \mathbb{N}^d$, we define $e_{\vec{k}}$ as in Subsect. 3.3.5. We set

$$f_{\vec{k}} := \frac{\sqrt{|k|!}}{\sqrt{\vec{k}!}} e_{\vec{k}}.$$

Let us consider the operator

$$A : \begin{array}{l} \Gamma_{\mathrm{s}}(\mathcal{Z}) \to \Gamma_{\mathrm{s}}(\mathcal{Z}) \otimes \mathcal{Z}, \\ \Psi \mapsto \sum_{i=1}^n a(e_i) \Psi \otimes e_i, \end{array}$$

and define by induction

$$A_q : \begin{array}{l} \Gamma_{\mathrm{s}}(\mathcal{Z}) \to \Gamma_{\mathrm{s}}(\mathcal{Z}) \otimes \otimes_{\mathrm{s}}^q \mathcal{Z}, \\ A_q := \big(A \otimes \mathbb{1}_{\otimes_{\mathrm{s}}^{q-1} \mathcal{Z}} \big) A_{q-1}. \end{array}$$

It is easy to verify that

$$A_q \Psi = \sum_{|\vec{l}| = q} \frac{|\vec{l}|!}{\vec{l}!} a(e_{\vec{l}}) \Psi \otimes e_{\vec{l}} - \sum_{|\vec{l}| = q} a(f_{\vec{l}}) \Psi \otimes f_{\vec{l}}. \tag{9.49}$$

Since $\{f_{\vec{l}}\}_{|\vec{l}| = q}$ is an o.n. basis of $\otimes_{\mathrm{s}}^q \mathcal{Z}$, we have

$$b = \sum_{|\vec{k}| = p, |\vec{l}| = q} b_{\vec{k}, \vec{l}} | f_{\vec{k}})(f_{\vec{l}} |, \quad b_{\vec{k}, \vec{l}} = (f_{\vec{k}} | b f_{\vec{l}}),$$

and hence

$$\mathrm{Op}^{a^*,a}(b) = \sum_{|\vec{k}|=p,\,|\vec{l}|=q} b_{\vec{k},\vec{l}}\, a^*\big(|f_{\vec{k}}\rangle\big) a\big(\langle f_{\vec{l}}|\big). \tag{9.50}$$

From (9.49) and (9.50), we get that

$$\mathrm{Op}^{a^*,a}(b) = A_p^* \left(\mathbb{1}_{\Gamma_s(\mathcal{Z})} \otimes b\right) A_q. \tag{9.51}$$

Inserting factors of $\Gamma(m)^{\frac{1}{2}}$, we see that (9.48) follows if we prove that

$$\left\| \mathbb{1}_{\Gamma_s(\mathcal{Z})} \otimes \Gamma(m)^{\frac{1}{2}} A_p \mathrm{d}\Gamma(m)^{-q/2} \right\| \le 1. \tag{9.52}$$

To prove (9.52), we note that, first for $\alpha = 1$ and then for any $\alpha \in \mathbb{R}$, one has

$$A\mathrm{d}\Gamma(m)^\alpha = \big(\mathrm{d}\Gamma(m) \otimes \mathbb{1}_{\mathcal{Z}} + \mathbb{1}_{\Gamma_s(\mathcal{Z})} \otimes m\big)^\alpha A. \tag{9.53}$$

Applying (9.53) for $\alpha = -\frac{1}{2}$, we obtain by induction on q that

$$A_q \mathrm{d}\Gamma(m)^{-q/2}$$

$$= \big(A\otimes\mathbb{1}_{\otimes_s^{q-1}\mathcal{Z}}\big) \Big(\mathrm{d}\Gamma(m) \otimes \mathbb{1}_{\otimes_s^{q-1}\mathcal{Z}} + \mathbb{1}_{\Gamma_s(\mathcal{Z})} \otimes \mathrm{d}\Gamma^q(m)\Big)^{-\frac{1}{2}} A_{q-1}\mathrm{d}\Gamma(m)^{-(q-1)/2},$$

and hence

$$\Big(\mathbb{1}_{\Gamma_s(\mathcal{Z})} \otimes \Gamma^q(m)^{\frac{1}{2}}\Big) A_q \mathrm{d}\Gamma(m)^{-q/2}$$

$$= \Big(\big((\mathbb{1}_{\Gamma_s(\mathcal{Z})}\otimes m^{\frac{1}{2}})A\big)\otimes\mathbb{1}_{\otimes_s^{q-1}\mathcal{Z}}\Big) \Big(\mathrm{d}\Gamma(m) \otimes \mathbb{1}_{\otimes_s^{q-1}\mathcal{Z}} + \mathbb{1}_{\Gamma_s(\mathcal{Z})} \otimes \mathrm{d}\Gamma^{q-1}(m)\Big)^{-\frac{1}{2}}$$

$$\times \Big(\mathbb{1}_{\Gamma_s(\mathcal{Z})} \otimes \Gamma^{q-1}(m)^{\frac{1}{2}}\Big) A_{q-1}\mathrm{d}\Gamma(m)^{-(q-1)/2}. \tag{9.54}$$

As a special case of (9.51), we have

$$\mathrm{d}\Gamma(b) = A^* \left(\mathbb{1}_{\Gamma_s(\mathcal{Z})} \otimes m\right) A,$$

which implies that

$$\left\| \big(\mathbb{1}_{\Gamma_s(\mathcal{Z})} \otimes m^{\frac{1}{2}}\big) A\mathrm{d}\Gamma(m)^{-\frac{1}{2}} \right\| \le 1.$$

Clearly, this implies that the first factor in the r.h.s. of (9.54) has norm less than 1, which implies (9.52). □

9.4.7 Bargmann kernel of an operator

Recall that in Def. 9.12 for any $\Psi \in \Gamma_s(\mathcal{Z})$ we defined its complex-wave transform $T^{\mathrm{cw}}\Psi \in \mathbf{L}_{\mathbb{C}}^2(\overline{\mathcal{Z}}, \mathrm{e}^{-\overline{z}\cdot z}\mathrm{d}\overline{z}\mathrm{d}z)$. In the context of the complex-wave transformation one sometimes introduces the so-called Bargmann kernel of an operator, which can be used as an alternative to its distributional kernel, and also to its Wick symbol.

For simplicity, in (2) and (3) of Prop. 9.52 below we assume that the dimension of $\mathcal{Z}$ is finite.

Definition 9.51 *Let $B \in B^{\mathrm{fin}}(\Gamma_s(\mathcal{Z}))$. We define the* Bargmann *or* complex-wave kernel of B as

$$\overline{\mathcal{Z}} \oplus \mathcal{Z} \ni (\overline{z_1}, z_2) \mapsto B^{\mathrm{Bar}}(\overline{z_1}, z_2) := \sum_{n,m=0}^{\infty} (z_1^{\otimes n} \frac{1}{\sqrt{n!}} | B \frac{1}{\sqrt{m!}} z_2^{\otimes m})$$

$$= (e^{a^*(z_1)} \Omega | B e^{a^*(z_2)} \Omega).$$

Proposition 9.52 (1) *The relationship between the Bargmann kernel and the Wick symbol of an operator B on $\Gamma_s(\mathcal{Z})$ is given by the following identity:*

$$B^{\mathrm{Bar}}(\overline{z_1}, z_2) = e^{\overline{z_1} \cdot z_2} s_B^{a^*, a}(\overline{z_1}, z_2) = e^{\frac{1}{2}\overline{z_1} \cdot z_1 + \frac{1}{2}\overline{z_2} \cdot z_2} (\Omega_{z_1} | B \Omega_{z_2}).$$

(2) *Let $B \in B^{\mathrm{fin}}(\Gamma_s(\mathcal{Z}))$, $\Psi \in \Gamma_s^{\mathrm{fin}}(\mathcal{Z})$. Then one has*

$$(T^{\mathrm{cw}} B \Psi)(\overline{z_1}) = (2\pi i)^{-d} \int B^{\mathrm{Bar}}(\overline{z_1}, z_2) T^{\mathrm{cw}} \Psi(z_2) e^{-\overline{z_2} \cdot z_2} d\overline{z_2} dz_2. \quad (9.55)$$

(3) *Let $B_1, B_2 \in B^{\mathrm{fin}}(\Gamma_s(\mathcal{Z}))$. Then*

$$(B_1 B_2)^{\mathrm{Bar}}(\overline{z_1}, z_2) = (2\pi i)^{-d} \int B_1^{\mathrm{Bar}}(\overline{z_1}, z_0) B_2^{\mathrm{Bar}}(\overline{z_0}, z_2) e^{-\overline{z_0} \cdot z_0} d\overline{z_0} dz_0. $$

$$(9.56)$$

Proof (1) is obvious. To prove (2) and (3) we use

$$\mathbb{1} = (2\pi i)^{-d} \int P_z d\overline{z} dz. \quad (9.57)$$

We obtain

$$(\Omega_{z_1} | B \Psi) = (2\pi i)^{-d} \int (\Omega_{z_1} | B \Omega_{z_2})(\Omega_{z_2} | \Psi) dz_2 d\overline{z_2}, \quad (9.58)$$

$$(\Omega_{z_1} | B_1 B_2 \Omega_{z_2}) = (2\pi i)^{-d} \int (\Omega_{z_1} | B_1 \Omega_{z_0})(\Omega_{z_0} | B_2 \Omega_{z_2}) dz_0 d\overline{z_0}. \quad (9.59)$$

Now (9.58) implies (2) and (9.59) implies (3). □

9.4.8 Link between the two Wick operations

In this subsection we use the conventions of Subsect. 9.3.1. In particular, we consider a real Hilbert space $\mathcal{X}$ equipped with a positive operator c. We consider the Kähler space with involution $(2c)^{-\frac{1}{2}}\mathcal{X} \oplus (2c)^{\frac{1}{2}}\mathcal{X}$ equipped with the Kähler anti-involution $\mathrm{j} = \begin{bmatrix} 0 & -(2c)^{-1} \\ 2c & 0 \end{bmatrix}$ (see (9.22)). Recall that the Wick transformation w.r.t. the covariance c is given by

$$:G(x): = e^{-\frac{1}{2}\nabla_x \cdot c \nabla_x} G(x).$$

The following proposition explains the link between the Wick transformation on functions on $\mathcal{X}$ and the Wick ordering of operators.

Proposition 9.53 *Let $F \in \mathbb{C}\mathrm{Pol}_\mathrm{s}(\mathcal{X})$. Then*

$$\mathrm{Op}^{a^*,a}(F) = :F(x_{\mathrm{rw}}):,$$

where on the r.h.s. we use the functional calculus, as explained in Remark 8.27.

Proof From Thm. 9.37, we have $\mathrm{Op}^{a^*,a}(F) = \mathrm{Op}(\mathrm{e}^{-\frac{1}{4}\nabla_v^2}F)$. Setting $\nabla_v = (\nabla_x, \nabla_\xi)$, we have

$$\nabla_v^2 = \nabla_x \cdot 2c\nabla_x + \nabla_\xi \cdot (2c)^{-1}\nabla_\xi.$$

Thus, on a function that depends only on x, we have

$$\mathrm{e}^{-\frac{1}{4}\nabla_v^2}F(x) = \mathrm{e}^{-\frac{1}{2}\nabla_x \cdot c\nabla_x}F(x) = :F(x):.$$

Furthermore, for such functions the Weyl–Wigner quantization coincides with the functional calculus.   □

It is often convenient to use multiplication operators expressed as $:F(x_{\mathrm{rw}}):$, as explained in Prop. 9.53. In particular, let $w \in \mathbb{C}\mathcal{X}$. Recall that

$$w \cdot x_{\mathrm{rw}} = a^*(w) + a(\overline{w}).$$

For later use let us note the identity

$$:(w \cdot x_{\mathrm{rw}})^p: = \sum_{r=0}^{p} \binom{p}{r} a^*(w)^r a(\overline{w})^{p-r}. \tag{9.60}$$

9.5 Notes

The essential self-adjointness of bosonic field operators was established by Cook (1953).

A modern exposition of the mathematical formalism of second quantization can also be found e.g. in Glimm–Jaffe (1987) and Baez–Segal–Zhou (1991).

The complex-wave representation goes back to the work of Bargmann (1961) and Segal (1963). Therefore, it is often called the Bargmann or Bargmann–Segal representation. The name "complex-wave representation" was coined by Segal (1978); see also Baez–Segal–Zhou (1991).

The name "real-wave representation" also comes from Baez–Segal–Zhou (1991). The properties of second quantized operators in the real-wave representation were first established by Nelson (1973). The proof of Prop. 9.29 (1) follows Nelson (1973), and that of Prop. 9.29 (2) follows Simon (1974). Nelson's hyper-contractivity theorem, Thm. 9.30, is proven in Nelson (1973).

The Wick theorem goes back to a paper of Wick (1950) about the evaluation of the S-matrix.

The "N_τ estimates" were used in constructive quantum field theory and are due to Glimm–Jaffe (1985).

Wick quantization in the context of particle preserving Hamiltonians is used, for example, in Dereziński (1998).

10

Symplectic invariance of CCR in finite-dimensions

This is the first chapter devoted to the symplectic invariance of the CCR. In this chapter we restrict ourselves to regular CCR representations over finite-dimensional symplectic spaces.

In an infinite-dimensional symplectic space there is no distinguished topology. This problem is absent in a finite-dimensional space. This motivates a separate discussion of the finite-dimensional case.

The chapter is naturally divided in two parts. In the first three sections we consider symplectic invariance without invoking any conjugation on the symplectic space. We consider an arbitrary irreducible regular CCR representation over a finite-dimensional symplectic space and do not explicitly use the Schrödinger representation.

In the last two subsections we fix a conjugation, so that our symplectic space can be written as $\mathcal{Y} = \mathcal{X}^{\#} \oplus \mathcal{X}$, and we consider the Schrödinger representation on $L^2(\mathcal{X})$.

10.1 Classical quadratic Hamiltonians

Throughout this section $(\mathcal{Y}, \omega)$ is a finite-dimensional symplectic space. Recall that $(\mathcal{Y}^{\#}, \omega^{-1})$ is also a symplectic space. As before we denote by y the generic element of $\mathcal{Y}$ and by v the generic element of $\mathcal{Y}^{\#}$.

Remark 10.1 *It is natural to consider the two symplectic spaces $\mathcal{Y}$ and $\mathcal{Y}^{\#}$ in parallel. It is a little difficult to decide which space should be viewed as the principal one: $\mathcal{Y}^{\#}$ is perhaps more important from the point of view of classical mechanics, since it plays the role of the phase space, whereas the dual phase space $\mathcal{Y}$ is more natural quantum mechanically, since we use it in the CCR relations.*

Recall that $\zeta \in L_s(\mathcal{Y}^{\#}, \mathcal{Y})$ iff $\zeta \in L(\mathcal{Y}^{\#}, \mathcal{Y})$ and $\zeta^{\#} = \zeta$. We write $\zeta \geq 0$ if $v \cdot \zeta v \geq 0$, $v \in \mathcal{Y}^{\#}$. We write $\zeta > 0$ if in addition $\mathrm{Ker}\, \zeta = \{0\}$.

The following section is a preparation for the next two where we consider a regular CCR representation over $\mathcal{Y}$.

10.1.1 Symplectic transformations

Let $r \in L(\mathcal{Y})$. Recall that $r \in Sp(\mathcal{Y})$ iff

$$r^{\#} \omega r = \omega. \tag{10.1}$$

This is equivalent to $r^\# \in Sp(\mathcal{Y}^\#)$, which means

$$r\omega^{-1}r^\# = \omega^{-1}. \qquad (10.2)$$

We have an isomorphism of groups,

$$Sp(\mathcal{Y}) \ni r \mapsto \omega r \omega^{-1} = (r^\#)^{-1} \in Sp(\mathcal{Y}^\#).$$

Let $a \in L(\mathcal{Y})$. Recall that $a \in sp(\mathcal{Y})$ iff $a^\# \omega + \omega a = 0$. This is equivalent to $a^\# \in sp(\mathcal{Y}^\#)$, which means $a\omega^{-1} + \omega^{-1}a^\# = 0$. Note that

$$sp(\mathcal{Y}) \ni a \mapsto \omega a \omega^{-1} = -a^\# \in sp(\mathcal{Y}^\#)$$

is an isomorphism of Lie algebras.

10.1.2 Poisson bracket

Definition 10.2 *For $b_1, b_2 \in C^1(\mathcal{Y}^\#)$ we define the* Poisson bracket

$$\{b_1, b_2\}(v) := \omega \nabla b_1(v) \cdot \nabla b_2(v) = -\nabla b_1(v) \cdot \omega \nabla b_2(v).$$

$C^\infty(\mathcal{Y}^\#)$ equipped with $\{\cdot, \cdot\}$ is a Lie algebra.

Definition 10.3 *By a* quadratic, *resp.* purely quadratic polynomial *we will mean a polynomial of degree ≤ 2, resp. $= 2$.*

Recall that the space of complex quadratic, resp. purely quadratic polynomials on $\mathcal{Y}^\#$ is denoted $\mathbb{C}\mathrm{Pol}_s^{\leq 2}(\mathcal{Y}^\#)$, resp. $\mathbb{C}\mathrm{Pol}_s^2(\mathcal{Y}^\#)$. Both are Lie subalgebras of $C^\infty(\mathcal{Y}^\#)$ w.r.t. the Poisson bracket. More precisely, if $\lambda_i \in \mathbb{C}$, $y_i \in \mathbb{C}\mathcal{Y}$, $\zeta_i \in \mathbb{C}L_s(\mathcal{Y}^\#, \mathcal{Y})$ and $\chi_i(v) := \lambda_i + y_i v + \frac{1}{2}v \cdot \zeta_i v$, then

$$\{\chi_1, \chi_2\}(v) = -y_1 \cdot \omega y_2 + (\zeta_2 \omega y_1 - \zeta_1 \omega y_2) \cdot v$$
$$+ \frac{1}{2}v \cdot (-\zeta_1 \omega \zeta_2 + \zeta_2 \omega \zeta_1)v.$$

If $\chi \in \mathbb{C}\mathrm{Pol}_s^{\leq 2}(\mathcal{Y}^\#)$, so that $\chi(v) = \lambda + y \cdot v + \frac{1}{2}v \cdot \zeta v$ with $\lambda \in \mathbb{C}$, $y \in \mathbb{C}\mathcal{Y}$ and $\zeta \in \mathbb{C}L_s(\mathcal{Y}^\#, \mathcal{Y})$, then

$$\mathbb{C}\mathcal{Y}^\# \ni v \mapsto \omega \nabla \chi(v) = \omega y + \omega \zeta v \in \mathbb{C}\mathcal{Y}^\#$$

is an affine transformation on $\mathbb{C}\mathcal{Y}^\#$. We have surjective homomorphisms of Lie algebras

$$\mathbb{C}\mathrm{Pol}_s^{\leq 2}(\mathcal{Y}^\#) \ni \chi \mapsto \omega \nabla \chi \in asp(\mathbb{C}\mathcal{Y}^\#),$$
$$\mathrm{Pol}_s^{\leq 2}(\mathcal{Y}^\#) \ni \chi \mapsto \omega \nabla \chi \in asp(\mathcal{Y}^\#)$$

(see Def. 1.102 for the definition of $asp(\mathbb{C}\mathcal{Y}^\#)$ and $asp(\mathcal{Y}^\#)$).

Definition 10.4 *If $(w, a^\#) \in asp(\mathcal{Y}^\#)$ and $\omega \nabla \chi(v) = w + a^\# v$, then we say that χ is a* Hamiltonian *of $(w, a^\#)$.*

Clearly, every element of $asp(\mathcal{Y}^\#)$ has a one parameter family of Hamiltonians χ differing by a constant. We will usually demand that $\chi(0) = 0$, which fixes the choice of a Hamiltonian in a canonical way. With this choice,

$$\chi(v) = (\omega^{-1}w)\cdot v + \frac{1}{2}v\cdot\omega^{-1}a^\#v.$$

Let $\chi \in \mathrm{Pol}_s^{\leq 2}(\mathcal{Y}^\#, \mathbb{R})$, and let v_t solve

$$\frac{d}{dt}v_t = \omega\nabla\chi(v_t), \quad v_0 = v.$$

Clearly, $v_t = e^{t\omega\nabla\chi}v$. Moreover, if $b \in C^1(\mathcal{Y}^\#)$ and if we set $b_t(v) = b(v_t)$, then

$$\frac{d}{dt}b_t(v) = \{\chi, b_t\}(v) = \{\chi, b\}(v_t). \tag{10.3}$$

10.1.3 Spectrum of symplectic transformations

Recall that a subspace $\mathcal{Y}_1$ of $\mathcal{Y}$ is called symplectic iff ω restricted to $\mathcal{Y}_1$ is non-degenerate. The following proposition is immediate:

Proposition 10.5 *Let* $\mathcal{Y} = \mathcal{Y}_1 \oplus \cdots \oplus \mathcal{Y}_k$, *and let* $\mathcal{Y}_1, \ldots, \mathcal{Y}_k$ *be mutually ω-orthogonal subspaces. Then all* $\mathcal{Y}_i$, $i = 1, \ldots, k$, *are symplectic.*

Definition 10.6 *An element* $r \in Sp(\mathcal{Y})$ *such that* $\mathrm{Ker}(r + \mathbb{1}) = \{0\}$ *will be called regular.*

Proposition 10.7 *Let* $r \in Sp(\mathcal{Y})$.

(1) $\mathrm{spec}\, r_\mathbb{C}$ *is invariant under* $\mathbb{C} \ni z \mapsto z^{-1} \in \mathbb{C}$.
(2) *For* $\lambda \in \mathrm{spec}\, r_\mathbb{C} \cap \{\mathrm{Im}\, z \geq 0, \ |z| \geq 1\} =: \Lambda_r$ *set* $P_\lambda := \mathbb{1}_{\{\lambda,\lambda^{-1},\bar{\lambda},\bar{\lambda}^{-1}\}}(r_\mathbb{C})$. *Then* P_λ *are real projections, constitute a partition of unity, commute with* r *and* $P_\lambda^\# \omega_\mathbb{C} = \omega_\mathbb{C} P_\lambda$.
(3) *If we set* $\mathcal{Y}_\lambda := P_\lambda\mathcal{Y}$, *then* $\mathcal{Y}_\lambda$ *are symplectic, mutually ω-orthogonal, invariant for r and* $\mathcal{Y} = \underset{\lambda\in\Lambda_r}{\oplus}\mathcal{Y}_\lambda$.
(4) *Set* $\mathcal{Y}_\mathrm{sg} := \mathcal{Y}_{-1}$ *and* $\mathcal{Y}_\mathrm{reg} := \underset{\lambda\in\Lambda_r\setminus\{-1\}}{\oplus}\mathcal{Y}_\lambda$. *Then* $\mathcal{Y} = \mathcal{Y}_\mathrm{sg} \oplus \mathcal{Y}_\mathrm{reg}$. *If we set* $\kappa := (-\mathbb{1}) \oplus \mathbb{1}$, *then*

$$r = \kappa r_0 = r_0\kappa, \tag{10.4}$$

κ *is a symplectic involution and* $r_0 \in Sp(\mathcal{Y})$ *is regular.*

Proof $r^\#\omega = \omega r^{-1}$ implies (1). We also obtain

$$\omega_\mathbb{C}(z\mathbb{1} - r_\mathbb{C}^{-1})^{-1} = (z\mathbb{1} - r_\mathbb{C}^\#)^{-1}\omega_\mathbb{C}.$$

Hence

$$\omega_\mathbb{C}\mathbb{1}_{\{\lambda^{-1}\}}(r_\mathbb{C}) = \omega_\mathbb{C}\mathbb{1}_{\{\lambda\}}(r_\mathbb{C}^{-1}) = \mathbb{1}_{\{\lambda\}}(r_\mathbb{C}^\#)\omega_\mathbb{C}.$$

Therefore,

$$\omega_\mathbb{C}\mathbb{1}_{\{\lambda,\lambda^{-1}\}}(r_\mathbb{C}) = \mathbb{1}_{\{\lambda,\lambda^{-1}\}}(r_\mathbb{C}^\#)\omega_\mathbb{C} = \mathbb{1}_{\{\lambda,\lambda^{-1}\}}(r_\mathbb{C})^\#\omega_\mathbb{C}.$$

If $|\lambda| = 1$, then $P_\lambda = \mathbb{1}_{\{\lambda, \lambda^{-1}\}}(r_\mathbb{C})$. If $|\lambda| \neq 1$, then

$$P_\lambda = \mathbb{1}_{\{\lambda, \lambda^{-1}\}}(r_\mathbb{C}) + \mathbb{1}_{\{\bar\lambda, \bar\lambda^{-1}\}}(r_\mathbb{C}).$$

In both cases, P_λ is real and can be restricted to $\mathcal{Y}$. This proves (2).

(2) implies (3), which yields (4). $\qquad\qquad\qquad\qquad\qquad\qquad\qquad\square$

There exists a classification of quadratic forms in a symplectic case due to Williamson. The following proposition is its special case for a positive semi-definite quadratic form, which is all that we need. Note that it would have a much simpler proof if we assumed that the form is positive definite.

Proposition 10.8 *Let $\zeta \in L_s(\mathcal{Y}^\#, \mathcal{Y})$ and $\zeta \geq 0$. Then we can find $p \leq m \leq d$, $\lambda_1, \ldots, \lambda_p > 0$ and a basis $(e_1, \ldots, e_{2d})$ in $\mathcal{Y}$ so that, if the corresponding dual basis of $\mathcal{Y}^\#$ is $(e^1, \ldots, e^{2d})$, then*

$$\omega e_{2j-1} = -e^{2j}, \quad \omega e_{2j} = e^{2j-1}, \, j = 1, \ldots, d; \qquad (10.5)$$

$$\zeta e^{2j-1} = \lambda_j e_{2j-1}, \, \zeta e^{2j} = \lambda_j e_{2j}, \, j = 1, \ldots, p; \qquad (10.6)$$

$$\zeta e^{2j-1} = e_{2j-1}, \quad \zeta e^{2j} = 0, \qquad j = p+1, \ldots, m; \qquad (10.7)$$

$$\zeta e^{2j-1} = 0, \qquad \zeta e^{2j} = 0, \qquad j = m+1, \ldots, d.$$

Consequently, $\mathrm{spec}\, \omega\zeta \subset i\mathbb{R}$. Besides, $\mathrm{spec}\, (-(\omega\zeta)^2) \subset]0, \infty[$ and $(\omega\zeta)^2$ is diagonalizable. If $\zeta > 0$, then $\omega\zeta$ is diagonalizable as well.

Note that we have two forms on $\mathcal{Y}^\#$: ζ and ω^{-1}. The complements of $\mathcal{V} \subset \mathcal{Y}^\#$ w.r.t. these forms have standard symbols $\mathcal{V}^{\zeta\perp}$ and $\mathcal{V}^{\omega^{-1}\perp}$. For brevity, we will write $\mathcal{V}^\perp$ for the former and $\mathcal{V}^\circ$ for the latter.

For the proof of Prop. 10.8 we need two lemmas. We set

$$\mathcal{V}_1 := \mathrm{Ker}\, \zeta, \, \mathcal{V}_2 := \mathcal{V}_1^\circ, \, \mathcal{V}_3 := (\mathcal{V}_2)^\perp, \, \mathcal{V}_4 := \mathcal{V}_3^\circ.$$

Lemma 10.9 *We have a direct sum decomposition,*

$$\mathcal{Y}^\# = \mathcal{V}_3 \oplus \mathcal{V}_4,$$

which is both ω^{-1}- and ζ-orthogonal, and ζ is non-degenerate on $\mathcal{V}_4$.

Proof We have

$$\mathcal{V}_1 = (\mathcal{Y}^\#)^\perp \subset \mathcal{V}_2^\perp = \mathcal{V}_3, \qquad (10.8)$$

hence

$$\mathcal{V}_4 = \mathcal{V}_3^\circ \subset \mathcal{V}_1^\circ = \mathcal{V}_2. \qquad (10.9)$$

Clearly,

$$\mathcal{V}_2 \subset (\mathcal{V}_2^\perp)^\perp = \mathcal{V}_3^\perp. \qquad (10.10)$$

From (10.9) and (10.10), we get

$$\mathcal{V}_4 \subset \mathcal{V}_3^\perp. \qquad (10.11)$$

Let us show that $\mathcal{V}_3 \cap \mathcal{V}_4 = \{0\}$. Assume that $v \in \mathcal{V}_3 \cap \mathcal{V}_4$ and $v \neq 0$. By (10.11), we have $v \cdot \zeta v = 0$, hence $v \in \mathcal{V}_1$.

By the non-degeneracy of ω^{-1}, there exists v' such that $v' \cdot \omega v \neq 0$. Let us fix a basis $(e^1, \dots, e^q)$ of $\mathcal{V}_2$ such that

$$e^i \cdot \zeta e^j = 0, \quad \text{for } i \neq j,$$
$$e^i \cdot \zeta e^i = 1, \quad \text{for } 1 \leq i \leq p,$$
$$e^i \cdot \zeta e^i = 0, \quad \text{for } p+1 \leq i \leq q.$$

We set $v'' = v' - \sum_{i=1}^p (v' \cdot \zeta e^i) e^i$ so that $v'' \cdot \zeta e^i = 0$ for $1 \leq i \leq q$, and hence $v'' \in \mathcal{V}_2^\perp = \mathcal{V}_3$. Since $v'' - v' \in \mathcal{V}_2 = \mathcal{V}_1^\circ$, we have $v'' \cdot \omega^{-1} v = v' \cdot \omega^{-1} v \neq 0$. Therefore, $v \notin \mathcal{V}_3^\circ = \mathcal{V}_4$, which is a contradiction.

Hence, $\mathcal{V}_3 \cap \mathcal{V}_4 = \{0\}$ and $\mathcal{Y}^\# = \mathcal{V}_3 \oplus \mathcal{V}_4$. The direct sum is clearly ω^{-1}-orthogonal, and also ζ-orthogonal by (10.11). Finally, $\mathcal{V}_4 \cap \mathcal{V}_1 \subset \mathcal{V}_4 \cap \mathcal{V}_3 = \{0\}$ by (10.8). Hence, ζ is non-degenerate on $\mathcal{V}_4$. $\qquad\square$

Lemma 10.10 *There exists a direct sum decomposition*

$$\mathcal{Y}^\# = \mathcal{V}_8 \oplus \mathcal{V}_7 \oplus \mathcal{V}_4$$

which is both ω^{-1}- and ζ-orthogonal, such that ζ is non-degenerate on $\mathcal{V}_4$, $\mathrm{Ker}\,\zeta \cap \mathcal{V}_7$ is Lagrangian in $\mathcal{V}_7$, and $\zeta = 0$ on $\mathcal{V}_8$.

Proof Let $\mathcal{V}_5 \subset \mathcal{V}_3$ be a maximal subspace on which ζ is non-degenerate. By (10.8), $\mathcal{V}_1 \subset \mathcal{V}_3$, so $\mathcal{V}_3 = \mathcal{V}_1 \oplus \mathcal{V}_5$. Set

$$\mathcal{V}_6 := \mathcal{V}_1 \cap \mathcal{V}_2, \quad \mathcal{V}_7 := \mathcal{V}_6 + \mathcal{V}_5, \quad \mathcal{V}_8 := \mathcal{V}_7^\circ \cap \mathcal{V}_3.$$

We claim first that

$$\mathcal{V}_5 \cap (\mathcal{V}_1 + \mathcal{V}_2) = \{0\}, \tag{10.12}$$
$$\mathcal{V}_6 \cap \mathcal{V}_5^\circ = \{0\}. \tag{10.13}$$

In fact

$$\mathcal{V}_5 \cap (\mathcal{V}_1 + \mathcal{V}_2) \subset \mathcal{V}_3 \cap (\mathcal{V}_1 + \mathcal{V}_2) = \mathcal{V}_2^\perp \cap (\mathcal{V}_1 + \mathcal{V}_2) \subset \mathcal{V}_1.$$

Hence, $\mathcal{V}_5 \cap (\mathcal{V}_1 + \mathcal{V}_2) \subset \mathcal{V}_5 \cap \mathcal{V}_1 = \{0\}$.

Similarly,

$$\mathcal{V}_6 \cap \mathcal{V}_5^\circ \subset \mathcal{V}_2 \cap \mathcal{V}_5^\circ = \mathcal{V}_1^\circ \cap \mathcal{V}_5^\circ = (\mathcal{V}_1 + \mathcal{V}_5)^\circ = \mathcal{V}_3^\circ = \mathcal{V}_4.$$

Hence, $\mathcal{V}_6 \cap \mathcal{V}_5^\circ \subset \mathcal{V}_3 \cap \mathcal{V}_4 = \{0\}$.

Recall that if E_1, E_2, F are subspaces of E, then

$$(E_1 + E_2) \cap F = E_1 \cap F + E_2 \cap F, \quad \text{if } E_i \subset F \text{ for } i = 1 \text{ or } 2. \tag{10.14}$$

Let us prove that

$$\mathcal{V}_7 \cap \mathcal{V}_8 = \{0\}. \tag{10.15}$$

In fact,

$$\mathcal{V}_8 = \mathcal{V}_6^\circ \cap \mathcal{V}_5^\circ \cap \mathcal{V}_3 = (\mathcal{V}_1 + \mathcal{V}_2) \cap \mathcal{V}_5^\circ \cap \mathcal{V}_3,$$

hence

$$\mathcal{V}_7 \cap \mathcal{V}_8 = (\mathcal{V}_5 + \mathcal{V}_6) \cap (\mathcal{V}_1 + \mathcal{V}_2) \cap \mathcal{V}_5^\circ.$$

Since $\mathcal{V}_6 \subset \mathcal{V}_1 + \mathcal{V}_2$, we have

$$(\mathcal{V}_5 + \mathcal{V}_6) \cap (\mathcal{V}_1 + \mathcal{V}_2) = \mathcal{V}_5 \cap (\mathcal{V}_1 + \mathcal{V}_2) + \mathcal{V}_6 \cap (\mathcal{V}_1 + \mathcal{V}_2) = \mathcal{V}_6,$$

using (10.12). Next, $\mathcal{V}_6 \cap \mathcal{V}_5^\circ = \{0\}$ by (10.13), which proves (10.15).

It follows that $\mathcal{V}_3 = \mathcal{V}_7 \oplus \mathcal{V}_8$, and that this decomposition is ω^{-1}-orthogonal. Since $\mathcal{V}_8 \subset (\mathcal{V}_1 + \mathcal{V}_2) \cap \mathcal{V}_3 \subset \mathcal{V}_1$, the decomposition is also ζ-orthogonal and $\zeta = 0$ on $\mathcal{V}_8$.

Finally, $\operatorname{Ker}\zeta \cap \mathcal{V}_7 = \mathcal{V}_1 \cap \mathcal{V}_7 = \mathcal{V}_1 \cap \mathcal{V}_2$ and

$$(\mathcal{V}_1 \cap \mathcal{V}_2)^\circ \cap \mathcal{V}_7 = (\mathcal{V}_1 \cap \mathcal{V}_2) \cap \mathcal{V}_7 = (\mathcal{V}_1 + \mathcal{V}_2) \cap (\mathcal{V}_1 \cap \mathcal{V}_2 + \mathcal{V}_5)$$
$$= \mathcal{V}_1 \cap \mathcal{V}_2 + (\mathcal{V}_1 + \mathcal{V}_2) \cap \mathcal{V}_5 = \mathcal{V}_1 \cap \mathcal{V}_2,$$

by (10.12). Hence, $\operatorname{Ker}\zeta$ is Lagrangian in $\mathcal{V}_7$. $\square$

Proof of Prop. 10.8. We first consider separately three cases:

Case 1: ζ is non-degenerate. We can treat $\mathcal{Y}$ as a Euclidean space and apply Corollary 2.85 to the anti-symmetric operator $\omega\zeta$ with a trivial kernel.

Case 2: $\operatorname{Ker}\zeta$ *is a Lagrangian subspace of $\mathcal{Y}^\#$.* Let $\mathcal{V}$ a maximal subspace of $\mathcal{Y}^\#$ on which ζ is non-degenerate. We check that $\mathcal{V}$ is Lagrangian and $\mathcal{Y}^\# = \mathcal{V} \oplus \operatorname{Ker}\zeta$. We choose a ζ-orthogonal basis $(e^1, \dots, e^d)$ of $\mathcal{V}$ and complete it to a symplectic basis of $\mathcal{Y}^\#$ by setting $e^{2j} = \omega\zeta e^j$, for $1 \leq j \leq d$.

Case 3: $\zeta = 0$. We choose any symplectic basis of $\mathcal{Y}^\#$.

In the general case we use Lemma 10.10 and apply Case 1 to $\mathcal{V}_4$, Case 2 to $\mathcal{V}_7$ and Case 3 to $\mathcal{V}_8$. The remaining statements of the proposition are immediate. $\square$

Proposition 10.11 *Let $\zeta \in \mathbb{C}L_s(\mathcal{Y}^\#, \mathcal{Y})$, with $\operatorname{Re}\zeta > 0$. Then $\operatorname{spec}(\omega\zeta) \subset \mathbb{C}\backslash\mathbb{R}$.*

Proof Set $\zeta = \zeta_1 + i\zeta_2$, with ζ_1, ζ_2 real and $\zeta_1 > 0$. Let $w \in \mathbb{C}\mathcal{Y}^\#$ with $\omega\zeta w = \lambda w$ be an eigenvector of $\omega\zeta$ for a real eigenvalue λ. Let $w = w_1 + iw_2$, with $w_1, w_2 \in \mathcal{Y}$. Then,

$$2\lambda i\langle\omega^{-1}w_1|w_2\rangle = \lambda\langle\omega^{-1}\overline{w}|w\rangle$$
$$= \langle\omega^{-1}\overline{w}|\omega\zeta w\rangle = -\langle\overline{w}|\zeta w\rangle$$
$$= -\langle w_1|\zeta w_1\rangle + \langle w_2|\zeta w_2\rangle.$$

Since $\lambda \in \mathbb{R}$, taking the real part yields $\langle w_1|\zeta_1 w_1\rangle + \langle w_2|\zeta_1 w_2\rangle = 0$, and hence $w = 0$. $\square$

10.1.4 Poisson bracket on charged symplectic spaces

Recall that a complex space $\mathcal{Y}$ equipped with a non-degenerate anti-Hermitian form $\overline{y}_1 \cdot \omega y_2$ is called a charged symplectic space. Its realification $\mathcal{Y}_{\mathbb{R}}$ is equipped with an anti-involution given by the imaginary unit denoted by j and the symplectic form

$$y_1 \cdot \omega_{\mathbb{R}} y_2 := \operatorname{Re} \overline{y}_1 \cdot \omega y_2 = \frac{1}{2}(\overline{y}_1 \cdot \omega y_2 + y_1 \cdot \overline{\omega y_2}). \tag{10.16}$$

Recall that v denotes the generic variable of $\mathcal{Y}^{\#}$ (which in this subsection is complex). Recall from Subsect. 4.1.5 that every function F on $\mathcal{Y}^{\#}$ has the usual derivative $\nabla_v^{\mathbb{R}} F(v) \in \mathbb{C}\mathcal{Y}_{\mathbb{R}}$, the holomorphic derivative $\nabla_v F(v) \in \mathcal{Y}$ and the anti-holomorphic derivative $\nabla_{\overline{v}} F(v) \in \overline{\mathcal{Y}}$, related by the identities

$$u \cdot \nabla_v = \frac{1}{2}\left(u \cdot \nabla_v^{\mathbb{R}} - \mathrm{i}(\mathrm{j}u) \cdot \nabla_v^{\mathbb{R}}\right),$$

$$\overline{u} \cdot \nabla_{\overline{v}} = \frac{1}{2}\left(u \cdot \nabla_v^{\mathbb{R}} + \mathrm{i}(\mathrm{j}u) \cdot \nabla_v^{\mathbb{R}}\right),$$

$$u \cdot \nabla_v^{\mathbb{R}} = u \cdot \nabla_v + \overline{u} \cdot \nabla_{\overline{v}}, \qquad u \in \mathcal{Y}^{\#}. \tag{10.17}$$

The symplectic form $\omega_{\mathbb{R}}$ allows us to define a Poisson bracket. Its expression in terms of real derivatives is

$$\{F, G\}(v) = -\nabla_v^{\mathbb{R}} F(v) \cdot \omega_{\mathbb{R}} \nabla_v^{\mathbb{R}} G(v).$$

Proposition 10.12 *The Poisson bracket expressed in terms of the holomorphic and anti-holomorphic derivative is*

$$\{F, G\}(v) = -\frac{1}{2}\nabla_{\overline{v}} F(v) \cdot \omega \nabla_v G(v) - \frac{1}{2}\nabla_v F(v) \cdot \overline{\omega} \nabla_{\overline{v}} G(v).$$

Proof We can write $\mathbb{C}\mathcal{Y}_{\mathbb{R}} \simeq \mathcal{Y} \oplus \overline{\mathcal{Y}}$; see (1.33). By (10.17), $\nabla_v^{\mathbb{R}} F = (\nabla_v F, \nabla_{\overline{v}} F)$, $\nabla_v^{\mathbb{R}} G = (\nabla_v G, \nabla_{\overline{v}} G)$, as elements of $\mathcal{Y} \oplus \overline{\mathcal{Y}}$. Besides, by (10.16), $\omega_{\mathbb{R}}$ written as a matrix $\mathcal{Y} \oplus \overline{\mathcal{Y}} \to \mathcal{Y}^{\#} \oplus \overline{\mathcal{Y}}^{\#}$ is

$$\omega_{\mathbb{R}} = \frac{1}{2}\begin{bmatrix} 0 & \overline{\omega} \\ \omega & 0 \end{bmatrix}. \qquad \square$$

10.2 Quantum quadratic Hamiltonians

As in the previous section, $(\mathcal{Y}, \omega)$ is a finite-dimensional symplectic space. We also fix an irreducible regular CCR representation $\mathcal{Y} \ni y \mapsto W(y) \in U(\mathcal{H})$. Recall that, for $b \in \mathcal{S}'(\mathcal{Y}^{\#})$, $\operatorname{Op}(b)$ denotes the Weyl–Wigner quantization of b.

Recall also that $\mathbb{C}\mathrm{Pol}_{\mathrm{s}}^{\leq?}(\mathcal{Y}^{\#})$ denotes the set of polynomials on $\mathcal{Y}^{\#}$ of degree ≤ 2. $\mathrm{CCR}_{\leq 2}^{\mathrm{pol}}(\mathcal{Y})$ will denote the set of operators on $\mathcal{H}$ obtained as the Weyl–Wigner quantization of elements of $\mathbb{C}\mathrm{Pol}_{\mathrm{s}}^{\leq 2}(\mathcal{Y}^{\#})$. These operators will be called (quantum) quadratic Hamiltonians. (Obviously, in the above definition we can replace the Weyl–Wigner quantization with $x, D-$, $D, x-$, Wick or anti-Wick quantizations.)

This section is devoted to the study of quantum quadratic Hamiltonians. We will see in particular that they behave to a large extent in a classical way.

10.2.1 Commutation properties of quadratic Hamiltonians

Recall that $\nabla^{(2)}b$ denotes the second derivative of $b \in \mathcal{S}'(\mathcal{Y}^\#)$. We treat it as a distribution on $\mathcal{Y}^\#$ with values in $L_s(\mathcal{Y}^\#, \mathcal{Y})$.

The following theorem is one of the most striking expressions of the correspondence principle between classical and quantum mechanics.

Theorem 10.13 (1) *For* $\chi \in \mathbb{C}\mathrm{Pol}_s^{\leq 2}(\mathcal{Y}^\#)$, $b \in \mathcal{S}'(\mathcal{Y}^\#)$ *we have*

$$\mathrm{i}[\mathrm{Op}(\chi), \mathrm{Op}(b)] = \mathrm{Op}(\{\chi, b\}), \tag{10.18}$$

$$\frac{1}{2}\big(\mathrm{Op}(\chi)\mathrm{Op}(b) + \mathrm{Op}(b)\mathrm{Op}(\chi)\big) = \mathrm{Op}\left(\chi b + \frac{1}{8}\mathrm{Tr}\,\omega(\nabla^{(2)}\chi)\omega\nabla^{(2)}b\right).$$

(2) *The map*

$$\mathbb{C}\mathrm{Pol}_s^{\leq 2}(\mathcal{Y}^\#) \ni \chi \mapsto \mathrm{Op}(\chi) \in \mathrm{CCR}_{\leq 2}^{\mathrm{pol}}(\mathcal{Y})$$

is a $*$-isomorphism of Lie algebras, where $\mathbb{C}\mathrm{Pol}_s^{\leq 2}(\mathcal{Y}^\#)$ *is equipped with the Poisson bracket* $\{\cdot, \cdot\}$ *and* $\mathrm{CCR}_{\leq 2}^{\mathrm{pol}}(\mathcal{Y})$ *is equipped with* $\mathrm{i}[\cdot, \cdot]$.

Proof (1) follows from (8.41) by expanding the exponential. (2) follows immediately from (10.18). □

In the following definition one cannot replace the Weyl–Wigner quantization by the other four basic quantizations.

Definition 10.14 *We denote by* $\mathrm{CCR}_2^{\mathrm{pol}}(\mathcal{Y})$ *the set of operators obtained by the Weyl–Wigner quantization of polynomials in* $\mathbb{C}\mathrm{Pol}_s^2(\mathcal{Y}^\#)$. *Elements of this space will be called* purely quadratic (quantum) Hamiltonians.

It will be convenient to introduce the following notation for purely quadratic Hamiltonians:

Definition 10.15 *If* $\zeta \in L_s(\mathbb{C}\mathcal{Y}^\#, \mathbb{C}\mathcal{Y})$, *then* $\mathrm{Op}(\zeta)$ *will denote the Weyl–Wigner quantization of* $\mathcal{Y}^\# \ni v \mapsto v \cdot \zeta v$.

Note that if $\chi(v) = v \cdot \zeta v$, then $\nabla \chi(v) = 2\zeta v$ and $\nabla^{(2)}\chi = 2\zeta$.

Proposition 10.16 (1) *For* $\zeta_1, \zeta_2 \in L_s(\mathbb{C}\mathcal{Y}^\#, \mathbb{C}\mathcal{Y})$,

$$[\mathrm{Op}(\zeta_1), \mathrm{Op}(\zeta_2)] = 2\mathrm{i}\mathrm{Op}(\zeta_2 \cdot \omega\zeta_1 - \zeta_1 \cdot \omega\zeta_2).$$

Hence

$$sp(\mathcal{Y}) \ni a \mapsto \frac{\mathrm{i}}{2}\mathrm{Op}(a\omega^{-1}) \in \mathrm{CCR}_2^{\mathrm{pol}}(\mathcal{Y})$$

is an isomorphism of Lie algebras.

(2) *For $\zeta \in L_s(\mathbb{C}\mathcal{Y}^\#, \mathbb{C}\mathcal{Y})$, $y \in \mathcal{Y}$,*

$$W(y)\mathrm{Op}(\zeta)W(y)^* = \mathrm{Op}(\zeta) + 2\phi(\zeta\omega y) - (y\cdot\omega\zeta\omega y)\mathbb{1}.$$

Proof (1) immediately follows from (10.18). To prove (2), we use
$W(y)\phi(y_1)W(y)^* = \phi(y_1) + y_1 \cdot \omega y \mathbb{1}$. □

10.2.2 Infimum of positive quadratic Hamiltonians

Quantizations of positive quadratic Hamiltonians are positive. One can give a
formula for their infimum, which in quantum physics is responsible for the so-
called *Casimir effect*.

Theorem 10.17 *Let $\zeta \in L_s(\mathcal{Y}^\#, \mathcal{Y})$, $\zeta \geq 0$. Then $\mathrm{Op}(\zeta)$ extends to a bounded
below self-adjoint Hamiltonian and*

$$\inf \mathrm{Op}(\zeta) = \frac{1}{2}\mathrm{Tr}\sqrt{-(\zeta\omega)^2}. \tag{10.19}$$

Remark 10.18 *By Prop. 10.8, $-(\omega\zeta)^2$ is a diagonalizable operator with non-
negative eigenvalues, hence $\sqrt{-(\omega\zeta)^2}$ is well defined.*

Proof of Thm. 10.17. Let $(e_1, \ldots, e_{2d}, e^1, \ldots, e^{2d})$ be as in the proof of Prop.
10.8. Writing ϕ_i for $\phi(e_i)$, we obtain

$$\mathrm{Op}(\zeta) = \sum_{j=1}^{p}\lambda_j(\phi_{2j-1}^2 + \phi_{2j}^2) + \sum_{k=p+1}^{m}\phi_{2k-1}^2.$$

Clearly, $\inf \phi_k^2 = 0$. By the well-known properties of the harmonic oscillator,
$\inf(\phi_{2j+1}^2 + \phi_{2j+2}^2) = 1$. Thus

$$\inf \mathrm{Op}(\zeta) = \sum_{j=1}^{p}\lambda_j.$$

Now,

$$-(\zeta\omega)^2 e_j = \lambda_j^2 e_j, \quad 1 \leq j \leq 2p,$$
$$-(\zeta\omega)^2 e_j = 0, \qquad 2p+1 \leq k \leq 2d.$$

Thus, $\mathrm{Tr}\sqrt{-(\omega\zeta)^2} = 2\sum_{j=1}^{p}\lambda_j$. □

10.2.3 Scale of oscillator spaces

In the Fock space $\Gamma_s(\mathcal{Z})$, a distinguished role is played by the number operator
N. It allows us to define a family of weighted Hilbert spaces $(N+\mathbb{1})^t\Gamma_s(\mathcal{Z})$,
which is often used in applications.

Recall that in this section we consider a regular CCR representation over a finite-dimensional symplectic space $\mathcal{Y} \ni y \mapsto W(y) \in U(\mathcal{H})$. In this framework, in general we do not have a single distinguished operator similar to N. However, a similar role is played by the whole family of positive definite quadratic Hamiltonians. They define a family of equivalent norms, as shown by the following proposition.

Proposition 10.19 *Let* $\zeta, \zeta_1 \in L_s(\mathcal{Y}^\#, \mathcal{Y})$, *where* $\zeta, \zeta_1 > 0$. *Then for any* $t > 0$ *there exist* $0 < C_t$ *such that*

$$C_t^{-1} \| \mathrm{Op}(\zeta_1)^t \Psi \| \leq \| \mathrm{Op}(\zeta)^t \Psi \| \leq C_t \| \mathrm{Op}(\zeta_1)^t \Psi \|, \quad \Psi \in \mathcal{H}. \tag{10.20}$$

Proof Choose a basis, as in Prop. 10.8. Using this basis, we can identify $\mathcal{H}$ with $\Gamma_s(\mathbb{C}^d)$ and $\mathrm{Op}(\zeta)$ with $\mathrm{d}\Gamma(h) + \frac{\mathrm{Tr}h}{2}\mathbb{1}$, where the operator h is diagonal and has positive eigenvalues. Using the natural o.n. basis of $\Gamma_s(\mathbb{C}^d)$ we easily check that for any $n = 1, 2, \ldots$ there exists C_n such that

$$\| \mathrm{Op}(\zeta_1)^n \Psi \|^2 \leq C_n \| \mathrm{Op}(\zeta)^n \Psi \|^2.$$

By interpolation, this implies the first inequality in (10.20). Reversing the role of ζ and ζ_1 we obtain the second inequality. $\qquad\square$

Definition 10.20 *For any* $t \geq 0$, *the* t-*th oscillator space* $\mathcal{H}^t$ *is defined as* $\mathrm{Dom}\,\mathrm{Op}(\zeta)^t$, *where* $\zeta \in L_s(\mathcal{Y}^\#, \mathcal{Y})$, $\zeta > 0$. *By Thm. 10.19,* $\mathcal{H}^t$ *does not depend on the choice of* ζ *and has the structure of a Hilbertizable space. We set* $\mathcal{H}^{-t} := (\mathcal{H}^t)^*$.

Recall that in Def. 8.50 we defined $\mathcal{H}^\infty$ and in Def. 8.51 we defined $\mathcal{H}^{-\infty}$. They are related to spaces $\mathcal{H}^t$ as follows:

$$\mathcal{H}^\infty = \bigcap_{t>0} \mathcal{H}^t, \quad \mathcal{H}^{-\infty} := \bigcup_{t<0} \mathcal{H}^t. \tag{10.21}$$

10.2.4 Quadratic Hamiltonians as closed operators

Prop. 10.19 shows that all $\mathrm{Op}(\zeta)$ with $\zeta > 0$ have the same domain and in particular are essentially self-adjoint on $\mathcal{H}^\infty$. The following theorem describes more general classes of quadratic Hamiltonians.

Theorem 10.21 (1) *Let* $\chi \in \mathrm{Pol}_s^{\leq 2}(\mathcal{Y}^\#)$ *(χ is a real quadratic polynomial). Then* $\mathrm{Op}(\chi)$ *is essentially self-adjoint on* $\mathcal{H}^\infty$.

(2) *Let* $\chi \in \mathbb{C}\mathrm{Pol}_s^{\leq 2}(\mathcal{Y}^\#)$ *(χ is a complex quadratic polynomial). Assume that the purely quadratic part of* χ *is positive definite. (In other words,* $\chi(v) = \mu + y \cdot v + \frac{1}{2} v \cdot \zeta v$ *with* $\mu \in \mathbb{C}$, $y \in \mathbb{C}\mathcal{Y}$, $\zeta \in \mathbb{C}L_s(\mathcal{Y}^\#, \mathcal{Y})$ *and* $\mathrm{Re}\,\zeta > 0$.) *Then* $\mathrm{Op}(\chi)$ *is closed on* $\mathcal{H}^1$ *and maximal accretive.*

Proof Fix $\zeta_0 \in L_s(\mathcal{Y}, \mathcal{Y}^{\#})$ such that $\zeta_0 > 0$, and set $N = \mathrm{Op}(\zeta_0)$. $\mathrm{Op}(\chi)$ is Hermitian on $\mathcal{H}^{\infty}$. By (10.20), we have

$$\|\mathrm{Op}(\chi)\Phi\| \leq C\|N\Phi\|, \quad \Phi \in \mathcal{H}^1.$$

Next we have $[\mathrm{Op}(\chi), iN] = \mathrm{Op}\{\chi, \zeta\}$. Since $\{\chi, \zeta_0\} \in \mathrm{Pol}^{\leq 2}(\mathcal{Y}^{\#})$, we have $\{\chi, \zeta_0\} \leq C\zeta_0$ for some C, and by Thm. 10.17 we get $[\mathrm{Op}(\chi), iN] \leq C(N + \mathbb{1})$. Applying Nelson's commutator theorem, Thm. 2.74 (1), we obtain that $\mathrm{Op}(\chi)$ is essentially self-adjoint on $\mathrm{Dom}\, N$, hence also on $\mathcal{H}^{\infty}$, since N is essentially self-adjoint on $\mathcal{H}^{\infty}$. This proves (1).

To prove (2), we set $\chi_1 = \mathrm{Re}\,\chi$, $\chi_2 = \mathrm{Im}\,\chi$, $B_i = \mathrm{Op}(\chi_i)$, $B = \mathrm{Op}(\chi)$. We note that

$$\pm[B_1, iB_2] = \mathrm{Op}(\{\chi_1, \chi_2\}) \leq C(B_1 + \mathbb{1}), \tag{10.22}$$

by Thm. 10.17. We write

$$\begin{aligned}
B^*B &= B_1^2 + B_2^2 + [B_1, iB_2] \\
&\geq B_1^2 - C_1(B_1 + \mathbb{1}) \geq \tfrac{1}{2}B_1^2 - C_2\mathbb{1},
\end{aligned} \tag{10.23}$$

using (10.22), which shows that B is closed on $\mathrm{Dom}\, B_1$. Next

$$\mathrm{Re}(\Psi|B\Psi) = (\Psi|B_1\Psi) \geq 0, \tag{10.24}$$

by Thm. 10.17. It remains to prove that $B + \lambda$ is invertible for large enough λ. Inequalities (10.24) and (10.23) for B replaced by $B + \lambda\mathbb{1}$ show that $\mathrm{Ker}(B + \lambda\mathbb{1}) = \{0\}$ and that $\mathrm{Ran}(B + \lambda\mathbb{1})$ is closed. Next we have

$$\begin{aligned}
&\tfrac{1}{2}(B + \lambda\mathbb{1})(B_1 + c\mathbb{1})^{-1} + \tfrac{1}{2}(B_1 + c\mathbb{1})^{-1}(B^* + \lambda\mathbb{1}) \\
&= (B_1 + c\mathbb{1})^{-\frac{1}{2}}(B_1 + \lambda\mathbb{1})(B_1 + c\mathbb{1})^{-\frac{1}{2}} + \tfrac{1}{2}(B_1 + c\mathbb{1})^{-1}[B_1, iB_2](B_1 + c\mathbb{1})^{-1} \\
&\geq (\lambda\mathbb{1} - C)(B_1 + c\mathbb{1})^{-1},
\end{aligned}$$

again using (10.22). If $\Psi \in \mathrm{Ran}(B + \lambda\mathbb{1})^{\perp}$, then

$$\mathrm{Re}\left(\Psi|(B + \lambda\mathbb{1})(B_1 + c\mathbb{1})^{-1}\Psi\right) = 0,$$

and hence $\Psi = 0$ if λ is large. This completes the proof of (2). $\square$

10.2.5 One-parameter groups of Bogoliubov $*$-automorphisms

Classical quadratic Hamiltonians generate one-parameter groups of linear symplectic transformations. On the quantum level one can assign two roles to a quadratic Hamiltonian H: $i[H, \cdot]$ generates a one-parameter group of $*$-automorphisms $\mathrm{e}^{itH} \cdot \mathrm{e}^{-itH}$, and iH generates the one-parameter unitary group e^{itH}. The following theorem describes the former group. The latter group, which is somewhat more difficult, is discussed in the following section.

Theorem 10.22 *Let* $\chi \in \mathrm{Pol}_s^{\leq 2}(\mathcal{Y}^\#)$, *i.e.* χ *is a real quadratic polynomial. Let* $b \in \mathcal{S}'(\mathcal{Y}^\#)$ *and* $b_t(v) = b(e^{t\omega\nabla\chi}v)$. *Then*

$$e^{it\mathrm{Op}(\chi)}\mathrm{Op}(b)e^{-it\mathrm{Op}(\chi)} = \mathrm{Op}(b_t). \tag{10.25}$$

In particular, for $y \in \mathcal{Y}$,

$$e^{it\mathrm{Op}(\chi)}W(y)e^{-it\mathrm{Op}(\chi)} = W(e^{-t\nabla\chi\omega}y). \tag{10.26}$$

Proof Let $\Phi, \Psi \in \mathcal{H}^\infty$, $b \in \mathcal{S}(\mathcal{Y}^\#)$. By (10.3),

$$\frac{\mathrm{d}}{\mathrm{d}t}b_t(v) = \{\chi, b_t\}(v).$$

Set

$$\Phi_t := e^{it\mathrm{Op}(\chi)}\Phi, \quad \Psi_t := e^{it\mathrm{Op}(\chi)}\Psi.$$

We know that $\mathrm{Op}(\chi)$ is self-adjoint and $\Phi, \Psi \in \mathrm{Dom}\,\mathrm{Op}(\chi)$. Hence, by Thm. 10.13 (1),

$$\frac{\mathrm{d}}{\mathrm{d}t}(\Phi_t|\mathrm{Op}(b_t)\Psi_t) = -\mathrm{i}(\Phi_t|[\mathrm{Op}(\chi),\mathrm{Op}(b_t)]\Psi_t)$$
$$+(\Phi_t|\mathrm{Op}(\{\chi,b_t\})\Psi_t) = 0.$$

Hence,

$$e^{-it\mathrm{Op}(\chi)}\mathrm{Op}(b_t)e^{it\mathrm{Op}(\chi)} = \mathrm{Op}(b).$$

This proves (10.25) for $b \in \mathcal{S}(\mathcal{Y}^\#)$. We extend (10.25) to $\mathcal{S}'(\mathcal{Y}^\#)$ by duality. $\square$

For further reference, let us restate Thm. 10.22 for purely quadratic Hamiltonians.

Corollary 10.23 *Let* $\zeta \in L_s(\mathcal{Y}^\#, \mathcal{Y})$. *Then for* $b \in \mathcal{S}'(\mathcal{Y})^\#$, $b_t(v) = b(e^{t\omega\nabla\zeta}v)$,

$$e^{\frac{it}{2}\mathrm{Op}(\zeta)}\mathrm{Op}(b)e^{-\frac{it}{2}\mathrm{Op}(\zeta)} = \mathrm{Op}(b_t).$$

In particular, for $y \in \mathcal{Y}$,

$$e^{\frac{it}{2}\mathrm{Op}(\zeta)}W(y)e^{-\frac{it}{2}\mathrm{Op}(\zeta)} = W(e^{-t\zeta\omega}y).$$

10.3 Metaplectic group

In this section, as in the previous one, $(\mathcal{Y}, \omega)$ is a finite-dimensional symplectic space and $\mathcal{Y} \ni y \mapsto W(y) \in U(\mathcal{H})$ is an irreducible regular CCR representation. In this section we study unitary operators of the form $e^{\mathrm{i}H}$, where H is a purely quadratic Hamiltonian. We show that they form a group, called the metaplectic group, isomorphic to the double cover of the symplectic group.

10.3.1 Implementation of Bogoliubov transformations

It follows from the Stone–von Neumann theorem that, for a finite-dimensional symplectic space, all Bogoliubov rotations can be implemented by unitary operators. The set of such unitary implementers forms a group.

Definition 10.24 *We define $Mp^c(\mathcal{Y})$ to be the set of $U \in U(\mathcal{H})$ such that*

$$\{UW(y)U^* \; : \; y \in \mathcal{Y}\} = \{W(y) \; : \; y \in \mathcal{Y}\}.$$

Proposition 10.25 *Let $U \in Mp^c(\mathcal{Y})$. Then there exists a unique $r \in Sp(\mathcal{Y})$ satisfying*

$$UW(y)U^* = W(ry), \quad y \in \mathcal{Y}. \tag{10.27}$$

The map $Mp^c(\mathcal{Y}) \to Sp(\mathcal{Y})$ obtained this way is a group homomorphism.

Definition 10.26 *If (10.27) is satisfied, we say that U implements r.*

Note that (10.27) is equivalent to

$$U\mathrm{Op}(a)U^* = \mathrm{Op}(a \circ r^\#), \quad a \in \mathcal{S}'(\mathcal{Y}^\#). \tag{10.28}$$

There also exists a smaller group that is sufficient to implement all linear symplectic transformations. Its definition is more involved. As a preparation for this definition, with every $r \in Sp(\mathcal{Y})$ we associate a pair of unitaries $\pm U_r$ differing by a sign:

Definition 10.27 (1) *Let $r \in Sp(\mathcal{Y})$ be regular (see Def. 10.6). Let $\gamma \in sp(\mathcal{Y})$ be the Cayley transform of r, that is, $\gamma = \frac{1-r}{1+r}$ (see Subsect. 1.4.6). Set*

$$\pm U_r := \pm\mathrm{Op}(f),$$
$$where \quad f(v) = \det(\mathbb{1} + \gamma)^{\frac{1}{2}} e^{iv \cdot \gamma \omega^{-1} v}. \tag{10.29}$$

(2) *Let $r \in Sp(\mathcal{Y})$ be arbitrary. Let $r = r_0\kappa$ be the canonical decomposition of r into a regular $r_0 \in Sp(\mathcal{Y})$ and an involution $\kappa \in Sp(\mathcal{Y})$ given by (10.4). Let $\mathcal{Y} = \mathcal{Y}_{\mathrm{sg}} \oplus \mathcal{Y}_{\mathrm{reg}}$ be the decomposition of the symplectic space such that $\kappa = (-\mathbb{1}) \oplus \mathbb{1}$. Let $m = \dim \mathcal{Y}_{\mathrm{sg}}$. Then we set*

$$\pm U_r := \pm U_\kappa U_{r_0}$$

for

$$U_\kappa := \pm\mathrm{Op}((i\pi)^{m/2}\delta_{\mathrm{sg}}),$$

where δ_{sg} is the Dirac delta function at zero on $\mathcal{Y}_{\mathrm{sg}}^\#$ times 1 on $\mathcal{Y}_{\mathrm{reg}}^\#$.

Note that under the assumptions of Def. 10.27 (2), our CCR representation can be decomposed as the tensor product of a representation over $\mathcal{Y}_{\mathrm{sg}}$ and over $\mathcal{Y}_{\mathrm{reg}}$, and then

$$\pm U_\kappa = \pm i^{m/2} I_{\mathrm{sg}} \otimes \mathbb{1}_{\mathrm{reg}},$$

where I_{sg} is the parity operator corresponding to $\mathcal{Y}_{\mathrm{sg}}$, defined as in Subsect. 8.4.4.

Definition 10.28 $Mp(\mathcal{Y})$ *is the set of operators of the form* $\pm U_r$ *for some* $r \in Sp(\mathcal{Y})$. *It is called the* metaplectic group *of* $\mathcal{Y}$.

Theorem 10.29 *Let* $r \in Sp(\mathcal{Y})$.

(1) *The set of elements of* $Mp(\mathcal{Y})$ *implementing* r *consists of a pair operators differing by the sign* $\pm U_r = \{U_r, -U_r\}$.
(2) *The set of elements of* $Mp^{\mathrm{c}}(\mathcal{Y})$ *implementing* r *consists of operators of the form* μU_r *with* $|\mu| = 1$.
(3) *If* $r_1, r_2 \in Sp(\mathcal{Y})$, *then* $U_{r_1} U_{r_2} = \pm U_{r_1 r_2}$.

The above statements can be summarized by the following commuting diagram consisting of exact horizontal and vertical sequences:

$$
\begin{array}{ccccc}
1 & & 1 & & 1 \\
\downarrow & & \downarrow & & \downarrow \\
1 \to & \mathbb{Z}_2 & \to & U(1) & \to U(1) \to 1 \\
\downarrow & & \downarrow & & \downarrow \\
1 \to Mp(\mathcal{Y}) & \to & Mp^{\mathrm{c}}(\mathcal{Y}) & \to & U(1) \to 1 \\
\downarrow & & \downarrow & & \downarrow \\
1 \to Sp(\mathcal{Y}) & \to & Sp(\mathcal{Y}) & \to & 1 \\
\downarrow & & \downarrow & & \\
1 & & 1 & &
\end{array}
\tag{10.30}
$$

The meaning of all the arrows in the above diagram should be obvious. In particular, the horizontal arrow $U(1) \to U(1)$ is just $\mu \to \mu^2$.

It remains to prove Thm. 10.29. We start by considering the case of regular symplectic maps. Recall that the formula for $\pm U_r$ is then given in (10.29).

Lemma 10.30 *Let* $r \in Sp(\mathcal{Y})$ *be regular. Then*

(1) U_r *intertwines* r, *i.e.*

$$U_r \phi(y) = \phi(ry) U_r, \quad y \in \mathcal{Y}.$$

(2) U_r *is unitary.*
(3) *If* $r_1, r_2, r \in Sp(\mathcal{Y})$ *are regular and* $r_1 r_2 = r$, *then* $U_{r_1} U_{r_2} = \pm U_r$.

Proof Let $y \in \mathcal{Y}$. Set $b(v) = v \cdot \gamma \omega^{-1} v$. Using Thm. 10.13 (1), we obtain

$$\mathrm{Op}(e^{\mathrm{i} b}) \phi(y) = \mathrm{Op}(e^{\mathrm{i} b}) \mathrm{Op}(y) = \mathrm{Op}\left(e^{\mathrm{i} b} y - \frac{\mathrm{i}}{2} \{y, e^{\mathrm{i} b}\}\right).$$

Now,

$$e^{ib}y - \frac{i}{2}\{y, e^{ib}\} = e^{ib}(\mathbb{1} - \gamma)y$$

$$= e^{ib}(\mathbb{1} + \gamma)ry = e^{ib}ry + \frac{i}{2}\{ry, e^{ib}\}.$$

Hence,

$$\mathrm{Op}(e^{ib})\phi(y) = \mathrm{Op}\left(e^{ib}(ry) + \frac{i}{2}\{ry, e^{ib}\}\right) = \phi(ry)\mathrm{Op}(e^{ib}).$$

Thus $\mathrm{Op}(e^{ib})$ intertwines r, and hence (1) is true.

Let $b_1, b_2, b,\ \gamma_1, \gamma_2, \gamma$ be related to r_1, r_2, r as in (10.29). We know that $\mathrm{Op}(e^{ib_1})\mathrm{Op}(e^{ib_2})$ intertwines r. Likewise, we know that $\mathrm{Op}(e^{ib})$ intertwines r. Hence, for some c,

$$\mathrm{Op}(e^{ib_1})\mathrm{Op}(e^{ib_2}) = c\,\mathrm{Op}(e^{ib}).$$

Next using Thm. 8.70 and formula (4.12), we obtain that $\mathrm{Op}(e^{ib_1})\mathrm{Op}(e^{ib_2})$ has the symbol

$$\int \exp\left(iv_1 \cdot \gamma_1\omega^{-1}v_1 + iv_2 \cdot \gamma_2\omega^{-1}v_2 - 2iv_1 \cdot \omega^{-1}v - 2iv \cdot \omega^{-1}v_2 + 2iv_1 \cdot \omega^{-1}v_2\right)\frac{dv_1\,dv_2}{\pi^{2d}}$$

$$= \pi^{-2d}\int \exp\left(i(v_1, v_2) \cdot \sigma(v_1, v_2) + 2i\theta \cdot (v_1, v_2)\right)\frac{dv_1\,dv_2}{\pi^{2d}},$$

$$(10.31)$$

where

$$\theta := (\omega^{-1}v, -\omega^{-1}v), \quad \sigma := \begin{bmatrix} \gamma_1\omega^{-1} & \omega^{-1} \\ -\omega^{-1} & \gamma_2\omega^{-1} \end{bmatrix}.$$

(10.31) equals

$$\det(-i\sigma)^{-\frac{1}{2}}\exp(-i\theta \cdot \sigma^{-1}\theta).$$

Setting $v = 0$ and using Subsect. 1.1.2, we obtain

$$c = \det(-i\sigma)^{-\frac{1}{2}} = \pm\det(\mathbb{1} + \gamma_1\gamma_2)^{-\frac{1}{2}}.$$

Next, by (1.49),

$$\mathbb{1} + \gamma = (\mathbb{1} + \gamma_2)(\mathbb{1} + \gamma_1\gamma_2)^{-1}(\mathbb{1} + \gamma_1).$$

Hence,

$$\det(\mathbb{1} + \gamma)^{\frac{1}{2}} = \pm\det(\mathbb{1} + \gamma_2)^{\frac{1}{2}}(\mathbb{1} + \gamma_1\gamma_2)^{\frac{1}{2}}(\mathbb{1} + \gamma_1)^{\frac{1}{2}}.$$

Therefore,

$$\det(\mathbb{1} + \gamma)^{\frac{1}{2}}\mathrm{Op}(e^{ib}) = \pm\det(\mathbb{1} + \gamma)^{\frac{1}{2}}\mathrm{Op}(e^{ib_1})(\mathbb{1} + \gamma_1)^{\frac{1}{2}}\mathrm{Op}(e^{ib_2}).$$

This proves (3).

It remains to prove that U_r is unitary. We have $U_r^* = \lambda U_{r^{-1}}$, for

$$\lambda = \pm \det \overline{(\mathbb{1} + \gamma)}^{\frac{1}{2}} \det(\mathbb{1} - \gamma)^{-\frac{1}{2}}.$$

Since by (3) $U_{r^{-1}} U_r = \pm U_{\mathbb{1}} = \pm \mathbb{1}$, it suffices to verify that $|\lambda| = 1$. But using that $\det a = \det a^{\#}$, we get

$$\det(\mathbb{1} + \gamma) = \det(\omega(\mathbb{1} - \gamma)\omega^{-1}) = \det(\mathbb{1} - \gamma).$$

This implies that $|\det(\mathbb{1}+\gamma)^{\frac{1}{2}}| = |\det(\mathbb{1}-\gamma)^{\frac{1}{2}}|$, which completes the proof of (2). $\square$

To treat the general case we will need more lemmas.

Lemma 10.31 *Let r_1, r_2, r_3 be regular. Then we can write r_2 as $r_2 = \tilde{r}_2 \hat{r}_2$ with $\tilde{r}_2, \hat{r}_2, r_1 \tilde{r}_2$ and $\hat{r}_2 r_3$ regular.*

Proof Let $D = \{r \in Sp(\mathcal{Y}) : r \text{ and } r_1 r \text{ are regular}\}$. Clearly, D is an open dense subset of $Sp(\mathcal{Y})$ containing $\mathbb{1}$. Hence, we can write r_2 as $r_2 = \tilde{r}_2 \hat{r}_2$, where $\tilde{r}_2, r_1 \tilde{r}_2$ are regular and $\mathbb{1} - \hat{r}_2$ can be made as small as we wish. Then if $\mathbb{1} - \hat{r}_2$ is sufficiently small, $\hat{r}_2$ and $\hat{r}_2 r_3$ are regular. $\square$

Lemma 10.32 *Let $r_i, \tilde{r}_i \in Sp(\mathcal{Y})$ be regular for $1 \le i \le p$. Assume that $r_1 \cdots r_p = \tilde{r}_1 \cdots \tilde{r}_p$. Then*

$$U_{r_1} \cdots U_{r_p} = \pm U_{\tilde{r}_1} \cdots U_{\tilde{r}_p}.$$

Proof If r is regular, so is r^{-1}, and hence, by Lemma 10.30, $U_{r^{-1}} = \pm U_r^{-1}$. Therefore, we are reduced to proving that

$$r_1 \cdots r_p = \mathbb{1} \implies U_{r_1} \cdots U_{r_p} = \pm \mathbb{1}. \tag{10.32}$$

Using Lemma 10.31, we write $r_1 r_2 r_3$ as $r_1 \tilde{r}_2 \hat{r}_2 r_3$. Then, by Lemma 10.30, we get

$$U_{r_2} = \pm U_{\tilde{r}_2} U_{\hat{r}_2}, \quad U_{r_1} U_{r_2} U_{r_3} = \pm U_{r_1} U_{\tilde{r}_2} U_{\hat{r}_2} U_{r_3} = \pm U_{r_1 \tilde{r}_2} U_{\hat{r}_2 r_3}.$$

Relabeling the r_i, we are reduced to showing (10.32) with p replaced by $p - 1$. Continuing in this way we end up with

$$r_1 r_2 = \mathbb{1} \implies U_{r_1} U_{r_2} = \pm \mathbb{1},$$

which holds since r_1, r_2 and $\mathbb{1}$ are regular. $\square$

Lemma 10.33 *Let κ be a symplectic involution, so that there exists a decomposition $\mathcal{Y} = \mathcal{Y}_{\mathrm{reg}} \oplus \mathcal{Y}_{\mathrm{sg}}$ into mutually ω-orthogonal subspaces and $\kappa = (-\mathbb{1}) \oplus \mathbb{1}$. Decompose $\mathcal{Y}_{\mathrm{sg}}$ further as $\mathcal{Y}_{\mathrm{sg}} = \mathcal{X}_{\mathrm{sg}} \oplus \mathcal{X}_{\mathrm{sg}}$, where $\mathcal{X}_{\mathrm{sg}}$ is a Euclidean space, with the standard symplectic form on $\mathcal{Y}_{\mathrm{sg}}$. Set*

$$u := \begin{bmatrix} 0 & -\mathbb{1} & 0 \\ \mathbb{1} & 0 & 0 \\ 0 & 0 & \mathbb{1} \end{bmatrix} \in Sp(\mathcal{Y}).$$

Then $u \in Sp(\mathcal{Y})$ is regular, $u^2 = \kappa$ and $\pm U_\kappa = \pm U_u^2$.

Proof The lemma follows by the properties of the evolution generated by the harmonic oscillator; see Subsect. 10.5.1. □

Proof of Thm. 10.29. Let us show that $Mp(\mathcal{Y})$ is a group. Let $r_1, r_2, r_3 \in Sp(\mathcal{Y})$, $r_1 r_2 = r_3$. Let $r_i = \kappa_i r_{0i}$ be the decomposition described in (10.4). Applying Lemma 10.33, we can write $\kappa_i = u_i^2$, where $u_i \in Sp(\mathcal{Y})$ are regular. Thus, we have

$$r_1 r_2 = r_{01} u_1^2 r_{02} u_2^2 = r_{03} u_3^2 = r_3.$$

By definition and then Lemma 10.33,

$$\pm U_{r_1} U_{r_2} = \pm U_{r_{01}} U_{u_1}^2 U_{r_{02}} U_{u_2}^2, \tag{10.33}$$

$$\pm U_{r_3} = \pm U_{r_{03}} U_{u_3}^2. \tag{10.34}$$

By Lemma 10.32, (10.33) equals (10.34). This proves also that $Mp(\mathcal{Y}) \to Sp(\mathcal{Y})$ is a homomorphism with the kernel consisting of $\{\mathbb{1}, -\mathbb{1}\} \simeq \mathbb{Z}_2$.

It is obvious that $Mp^c(\mathcal{Y})$ is a group. It clearly contains $Mp(\mathcal{Y})$, and hence the homomorphism $Mp^c(\mathcal{Y}) \to Sp(\mathcal{Y})$ is onto. By the irreducibility of the CCR representation, the kernel of this homomorphism is $U(1)$. □

10.3.2 Semi-groups generated by quadratic Hamiltonians

In the next two theorems, we will compute the Weyl–Wigner symbol of the semi-group generated by a maximal accretive quadratic Hamiltonian and of the unitary group generated by a self-adjoint quadratic Hamiltonian. We start with the case of a maximal accretive Hamiltonian.

Theorem 10.34 *Let $y \in \mathbb{C}\mathcal{Y}$, $\zeta \in \mathbb{C}L_s(\mathcal{Y}^\#, \mathcal{Y})$. Assume that $\mathrm{Re}\,\zeta > 0$. Consider the complex quadratic polynomial*

$$\chi(v) = y \cdot v + v \cdot \zeta v. \tag{10.35}$$

Then for $t \geq 0$, the bounded operator $\mathrm{e}^{-t\mathrm{Op}(\chi)}$ has the Weyl–Wigner symbol

$$f_t(v) = (\det \cos t\omega\zeta)^{-\frac{1}{2}} \exp\left(-v \cdot \omega^{-1} \mathrm{tg}(t\omega\zeta)v \right. \tag{10.36}$$

$$\left. -y \cdot (\omega\zeta)^{-1}\mathrm{tg}(t\omega\zeta)v + \frac{1}{4}y \cdot \left(t\mathbb{1} - (\omega\zeta)^{-1}\mathrm{tg}(t\omega\zeta)\right)\zeta^{-1}y\right).$$

The next theorem describes the case of a quadratic self-adjoint Hamiltonian.

Theorem 10.35 *Let $y \in \mathcal{Y}$, $\zeta \in L_s(\mathcal{Y}^\#, \mathcal{Y})$. Consider the real quadratic polynomial χ defined as in (10.35). For $t \in \mathbb{R}$, let $q_t(v)$ be the Weyl–Wigner symbol of the unitary operator $\mathrm{e}^{-it\mathrm{Op}(\chi)}$.*

(1) *If $\pi\mathbb{Z} \cap \mathrm{spec}\, t\omega\zeta = \emptyset$, then*

$$g_t(v) = (\det \cosh t\omega\zeta)^{-\frac{1}{2}} \exp\left(iv \cdot \omega^{-1}\tanh(t\omega\zeta)v \right. \tag{10.37}$$

$$\left. + iy \cdot (\omega\zeta)^{-1}\tanh(t\omega\zeta)v + \frac{i}{4}y \cdot \left((\omega\zeta)^{-1}\tanh(t\omega\zeta) - t\mathbb{1}\right)\zeta^{-1}y\right).$$

(2) *In the general case, to find the Weyl–Wigner symbol of* $e^{-it\mathrm{Op}(\chi)}$ *we can do as follows. We choose* $\zeta_1 \in L_s(\mathcal{Y}^\#, \mathcal{Y})$ *with* $\zeta_1 > 0$. *We set* $\zeta_\epsilon := \zeta + i\epsilon\zeta_1$, *and let* $g_{\epsilon,t}$ *be defined by* (10.37), *where* ζ *is replaced with* ζ_ϵ. *Then*

$$
g_t(v) = \begin{cases} \lim\limits_{\epsilon \searrow 0} g_{\epsilon,t}(v), & t \geq 0; \\[2mm] \lim\limits_{\epsilon \nearrow 0} g_{\epsilon,t}(v), & t \leq 0. \end{cases}
$$

Proof of Thm. 10.34. We first note that $e^{-t\mathrm{Op}(\chi)}$ is well defined as a strongly continuous semi-group, since $\mathrm{Op}(\chi)$ is maximal accretive. Note also from Lemma 10.11 that $\omega\zeta$ and $\zeta\omega$ have no real eigenvalues, so the operator $\mathrm{tg}(t\omega\zeta)$ is well defined by the holomorphic functional calculus and $\cos(t\omega\zeta)^{\frac{1}{2}} \neq 0$.

Since

$$
\partial_t e^{-t\mathrm{Op}(\chi)} = -\frac{1}{2}\big(\mathrm{Op}(\chi)e^{-t\mathrm{Op}(\chi)} + e^{-t\mathrm{Op}(\chi)}\mathrm{Op}(\chi)\big),
$$

it suffices, using Thm. 10.13, to verify that

$$
\begin{cases} \partial_t f_t(v) = -\chi(v)f_t(v) + \frac{1}{8}\mathrm{Tr}(\nabla^{(2)}\chi)\omega\nabla^{(2)}f_t(v)\omega, \\[2mm] f_0(v) = 1. \end{cases} \tag{10.38}
$$

We have

$$
\begin{aligned}
\partial_t f_t(v) &= f_t(v)\Big(-v\cdot\zeta\cos^{-2}(t\omega\zeta)v - y\cdot\cos^{-2}(t\omega\zeta)v \\
&\quad + \frac{1}{4}y\cdot\big(\mathbb{1} - \cos^{-2}(t\omega\zeta)\big)\zeta^{-1}y - \frac{1}{2}\partial_t\log\det\cos(t\omega\zeta)\Big) \\
&= f_t(v)\Big(-v\cdot\zeta v - y\cdot v - v\cdot\zeta\,\mathrm{tg}^2(t\omega\zeta)v - y\cdot\mathrm{tg}^2(t\omega\zeta)v \\
&\quad + \frac{1}{4}y\cdot\mathrm{tg}^2(t\omega\zeta)\zeta^{-1}y + \frac{1}{2}\mathrm{Tr}\,\omega\zeta\,\mathrm{tg}(t\omega\zeta)\Big).
\end{aligned} \tag{10.39}
$$

Now,

$$
\nabla^{(2)}\chi(v) = 2\zeta,
$$

$$
\nabla^{(2)}f_t(v) = -f_t(v)\Big(2\omega^{-1}\mathrm{tg}(t\omega\zeta)
$$
$$
- \big|\mathrm{tg}(t\omega\zeta)(2\omega^{-1}v + (\omega\zeta)^{-1}y)\big\rangle\big\langle\mathrm{tg}(t\omega\zeta)(2\omega^{-1}v + (\omega\zeta)^{-1})y\big|\Big).
$$

Using that $\mathrm{Tr}|y_1\rangle\langle y_2| = \langle y_2|y_1\rangle$, we get

$$
\mathrm{Tr}(\nabla^{(2)}\chi)\omega\nabla^{(2)}f_t(v)\omega = f_t(v)\Big(4\mathrm{Tr}\,\omega\zeta\,\mathrm{tg}(t\omega\zeta) \tag{10.40}
$$
$$
+ 8v\cdot\omega\zeta\,\mathrm{tg}^2(t\omega\zeta)v + 8v\cdot\mathrm{tg}^2(t\omega\zeta)y + 2y\cdot\zeta^{-1}y\Big).
$$

Comparing (10.39) and (10.40) we see that (10.38) is true. $\qquad\square$

Proof of Thm. 10.35. We may assume that $t \geq 0$. For $\epsilon > 0$, set $\chi_\epsilon(v) = \chi(v) - i\epsilon v \cdot \zeta_1 v$. Then we have

$$e^{-it\mathrm{Op}(\chi)} = \mathrm{s} - \lim_{\epsilon \searrow 0} e^{-it\mathrm{Op}(\chi_\epsilon)}.$$

This implies that $e^{-it\mathrm{Op}(\chi_\epsilon)}$ converges to $e^{-it\mathrm{Op}(\chi)}$ in $\mathrm{CCR}^{\mathcal{S}'}(\mathcal{Y})$. This implies that the Weyl–Wigner symbol of $e^{-it\mathrm{Op}(\chi_\epsilon)}$ converges to the Weyl–Wigner symbol of $e^{-it\mathrm{Op}(\chi)}$ in $\mathcal{S}'(\mathcal{Y}^{\#})$. Hence Thm. 10.35 follows from Thm. 10.34. $\square$

The following theorem provides an alternative definition of the metaplectic group:

Theorem 10.36 (1) *Let $\zeta \in L_\mathrm{s}(\mathcal{Y}^{\#}, \mathcal{Y})$. Then $e^{i\mathrm{Op}(\zeta)} \in Mp(\mathcal{Y})$.*
(2) *Conversely, $Mp(\mathcal{Y})$ is generated by operators of the form $e^{i\mathrm{Op}(\zeta)}$ with $\zeta \in L_\mathrm{s}(\mathcal{Y}^{\#}, \mathcal{Y})$.*

Proof By Thm. 10.35, $e^{i\mathrm{Op}(\zeta)} = \mathrm{Op}(g)$, where

$$g(v) = (\det \cosh t\omega\zeta)^{-\frac{1}{2}} \exp\left(iv \cdot \omega^{-1} \tanh(t\omega\zeta)v\right).$$

Set $r = e^{-\zeta\omega}$. Then

$$\gamma = \frac{e^{\zeta\omega} - e^{-\zeta\omega}}{e^{\zeta\omega} + e^{-\zeta\omega}} = \tanh \zeta\omega,$$

$$\mathbb{1} + \gamma = 2(\mathbb{1} + r)^{-1} = \frac{e^{\zeta\omega}}{\cosh \zeta\omega}.$$

Taking into account that $\det e^{\zeta\omega} = 1$, we obtain that

$$g(v) = \det(\mathbb{1} + \gamma\omega)^{\frac{1}{2}} e^{iv \cdot \gamma v}.$$

This proves (1).

All elements of $Sp(\mathcal{Y})$ in a neighborhood of $\mathbb{1}$ are of the form $r = e^a$ for $a \in sp(\mathcal{Y})$. By (1), the corresponding $\pm U_r$ are of the form $e^{i\mathrm{Op}(\zeta)}$ for $\zeta \in L_\mathrm{s}(\mathcal{Y}^{\#}, \mathcal{Y})$. But the whole group $Sp(\mathcal{Y})$ is generated by a neighborhood of $\mathbb{1}$. This proves (2). $\square$

10.3.3 $Mp(\mathcal{Y})$ as the two-fold covering of $Sp(\mathcal{Y})$

Definition 10.37 *Let G be a path-connected topological group. A covering group of G is a path-connected topological group $\tilde{G}$ with a surjective homomorphism $\pi : \tilde{G} \to G$. If for each $g \in G$ the set $\pi^{-1}(g)$ has n elements, then $\tilde{G}$ is called an n fold covering of G.*

Introducing an arbitrary Kähler structure on $\mathcal{Y}$ and considering the polar decomposition, we easily see that $Sp(\mathcal{Y})$ is path-connected. The same argument shows that its fundamental group, that is, $\pi_1(Sp(\mathcal{Y}))$, equals $\mathbb{Z}$. Hence, for any $n \in \{1, 2, \ldots, \aleph_0\}$ the n-fold covering of $Sp(\mathcal{Y})$ exists and is unique up to an isomorphism.

The group $Mp(\mathcal{Y})$ is clearly path-connected, since $e^{it\mathrm{Op}(\zeta)}$, $t \in [0,1]$, is a continuous path joining $\mathbb{1}$ and $e^{i\mathrm{Op}(\zeta)}$. For $U \in Mp(\mathcal{Y})$, let $\pi(U) \in Sp(\mathcal{Y})$ denote the symplectic transformation r implemented by U. By Thm. 10.29, $\pi^{-1}(r) = \{U_r, -U_r\}$. Hence, $Mp(\mathcal{Y})$ is the double covering of $Sp(\mathcal{Y})$.

10.4 Symplectic group on a space with conjugation

Throughout this section we fix a finite-dimensional space $\mathcal{X}$ and consider the space $\mathcal{X}^{\#} \oplus \mathcal{X}$ equipped with the symplectic form ω and the conjugation τ given by

$$\omega = \begin{bmatrix} 0 & \mathbb{1} \\ -\mathbb{1} & 0 \end{bmatrix}, \quad \tau = \begin{bmatrix} \mathbb{1} & 0 \\ 0 & -\mathbb{1} \end{bmatrix}.$$

Recall that its dual is isomorphic to $\mathcal{X} \oplus \mathcal{X}^{\#}$ with the symplectic form ω^{-1} and conjugation $\tau^{\#}$:

$$\omega^{-1} = \begin{bmatrix} 0 & -\mathbb{1} \\ \mathbb{1} & 0 \end{bmatrix}, \quad \tau^{\#} = \begin{bmatrix} \mathbb{1} & 0 \\ 0 & -\mathbb{1} \end{bmatrix}.$$

The Poisson bracket on $\mathcal{X} \oplus \mathcal{X}^{\#}$ takes the familiar form

$$\{b_1, b_2\} = \nabla_\xi b_1 \cdot \nabla_x b_2 - \nabla_x b_1 \cdot \nabla_\xi b_2, \quad b_1, b_2 \in C^1(\mathcal{X} \oplus \mathcal{X}^{\#}).$$

Recall from Thm. 1.47 that every finite-dimensional symplectic space can be equipped with a conjugation and is isomorphic to $\mathcal{X}^{\#} \oplus \mathcal{X}$. This section is devoted to a discussion of symplectic and infinitesimally symplectic transformations in a symplectic space with conjugation. It is a preparation for the next section, where we consider the Schrödinger CCR representation on $L^2(\mathcal{X})$.

As already discussed in Remark 10.1, we actually have two symplectic spaces with conjugation at our disposal: $\mathcal{Y} = \mathcal{X}^{\#} \oplus \mathcal{X}$ and $\mathcal{Y}^{\#} = \mathcal{X} \oplus \mathcal{X}^{\#}$. They are dual to one another and, as we know, both are relevant, as seen e.g. from the relations (10.27) and (10.28). We will explicitly describe $Sp(\mathcal{Y}^{\#})$ and $sp(\mathcal{Y}^{\#})$, since they appear more naturally in the quantization of classical symbols (but, obviously, it is easy to pass to $Sp(\mathcal{Y})$ and $sp(\mathcal{Y})$, to which they are naturally isomorphic).

10.4.1 Symplectic transformations on a space with conjugation

Let $a^{\#} \in L(\mathcal{X} \oplus \mathcal{X}^{\#})$. $a^{\#}$ belongs to $sp(\mathcal{X} \oplus \mathcal{X}^{\#})$ iff

$$a^{\#} = \begin{bmatrix} c & \beta \\ -\alpha & -c^{\#} \end{bmatrix},$$

where $\alpha \in L_{\mathrm{s}}(\mathcal{X}, \mathcal{X}^{\#})$, $c \in L(\mathcal{X})$, $\beta \in L_{\mathrm{s}}(\mathcal{X}^{\#}, \mathcal{X})$.

Let $(q, \eta) \in \mathcal{X} \oplus \mathcal{X}^{\#}$ and $a^{\#} \in sp(\mathcal{X} \oplus \mathcal{X}^{\#})$. Clearly, $\big((q, \eta), a^{\#}\big) \in asp(\mathcal{X} \oplus \mathcal{X}^{\#})$. Its Hamiltonian is

$$\mathcal{X} \oplus \mathcal{X}^{\#} \ni (x, \xi) \mapsto \chi(x, \xi) = -\eta \cdot x + q \cdot \xi + \frac{1}{2} x \cdot \alpha x + \xi \cdot c x + \frac{1}{2} \xi \cdot \beta \xi.$$

Let $r^{\#} \in L(\mathcal{X} \oplus \mathcal{X}^{\#})$. Write $r^{\#}$ as

$$r^{\#} = \begin{bmatrix} a & b \\ c & d \end{bmatrix}. \tag{10.41}$$

$r^{\#} \in Sp(\mathcal{X} \oplus \mathcal{X}^{\#})$ iff

$$a^{\#} d - c^{\#} b = \mathbb{1}, \quad c^{\#} a = a^{\#} c, \quad d^{\#} b = b^{\#} d, \tag{10.42}$$

or, equivalently,

$$a d^{\#} - b c^{\#} = \mathbb{1}, \quad a b^{\#} = b a^{\#}, \quad c d^{\#} = d c^{\#}. \tag{10.43}$$

In fact, (10.42) is equivalent to (10.1) and (10.43) is equivalent to (10.2). We have

$$r^{\#-1} = \begin{bmatrix} d^{\#} & -b^{\#} \\ -c^{\#} & a^{\#} \end{bmatrix}.$$

10.4.2 Generating function of a symplectic transformation

In the next theorem we prove a factorization result for symplectic transformations, similar to the one discussed in Subsect. 1.1.2. It will be used to define its generating function.

Theorem 10.38 *Let $r^{\#} \in Sp(\mathcal{X} \oplus \mathcal{X}^{\#})$ be as in (10.41) with b invertible.*

(1) *We have the factorization*

$$r^{\#} = \begin{bmatrix} a & b \\ c & d \end{bmatrix} = \begin{bmatrix} \mathbb{1} & 0 \\ e & \mathbb{1} \end{bmatrix} \begin{bmatrix} 0 & b \\ -b^{\#-1} & 0 \end{bmatrix} \begin{bmatrix} \mathbb{1} & 0 \\ f & \mathbb{1} \end{bmatrix}, \tag{10.44}$$

where

$$e = db^{-1} = b^{\#-1} d^{\#} \in L_{\mathrm{s}}(\mathcal{X}, \mathcal{X}^{\#}),$$
$$f = b^{-1} a = a^{\#} b^{\#-1} \in L_{\mathrm{s}}(\mathcal{X}, \mathcal{X}^{\#}).$$

(2) *Define $S \in \mathbb{C}\mathrm{Pol}_{\mathrm{s}}^{\leq 2}(\mathcal{X} \oplus \mathcal{X})$ by setting*

$$\mathcal{X} \times \mathcal{X} \ni (x_1, x_2) \mapsto S(x_1, x_2) := (b^{-1} q) \cdot x_1 + (-eq + \eta) \cdot x_2$$
$$+ \frac{1}{2} x_1 \cdot f x_1 - x_1 \cdot b^{-1} x_2 + \frac{1}{2} x_2 \cdot e x_2.$$

Then

$$\begin{bmatrix} q \\ \eta \end{bmatrix} + \begin{bmatrix} a & b \\ c & d \end{bmatrix} \begin{bmatrix} x_1 \\ \xi_1 \end{bmatrix} = \begin{bmatrix} x_2 \\ \xi_2 \end{bmatrix} \tag{10.45}$$

iff

$$\nabla_{x_1} S(x_1, x_2) = -\xi_1, \quad \nabla_{x_2} S(x_1, x_2) = \xi_2. \tag{10.46}$$

Proof The proofs are direct computations, using (10.42) and (10.43). $\quad\square$

Definition 10.39 *The function $S(x_1, x_2)$ is called a* generating function of the affine symplectic transformation *(10.45).*

10.4.3 Point transformations

Definition 10.40 *Elements of $sp(\mathcal{X} \oplus \mathcal{X}^\#)$ that commute with the conjugation $\tau^\#$ are called* infinitesimal point transformations.

Their set is the image of the following injective homomorphism:

$$gl(\mathcal{X}) \ni c \mapsto \begin{bmatrix} c & 0 \\ 0 & -c^\# \end{bmatrix} \in sp(\mathcal{X} \oplus \mathcal{X}^\#). \tag{10.47}$$

$\mathcal{X} \oplus \mathcal{X}^\# \ni (x, \xi) \mapsto \xi \cdot cx = x \cdot c^\# \xi$ is the Hamiltonian of (10.47).

Definition 10.41 *Elements of $Sp(\mathcal{X} \oplus \mathcal{X}^\#)$ that commute with the conjugation $\tau^\#$ are called* point transformations.

Their set is the image of the following injective homomorphism:

$$GL(\mathcal{X}) \ni m \mapsto \begin{bmatrix} m & 0 \\ 0 & m^{\#-1} \end{bmatrix} \in Sp(\mathcal{X} \oplus \mathcal{X}^\#). \tag{10.48}$$

We have

$$\exp \begin{bmatrix} c & 0 \\ 0 & -c^\# \end{bmatrix} = \begin{bmatrix} e^c & 0 \\ 0 & (e^c)^{\#-1} \end{bmatrix}.$$

10.4.4 Transformations fixing $\mathcal{X}^\#$

The set of elements of $sp(\mathcal{X} \oplus \mathcal{X}^\#)$ that send $\mathcal{X}^\#$ to zero is the image of the following injective homomorphism of Lie algebras with the trivial bracket:

$$L_s(\mathcal{X}, \mathcal{X}^\#) \ni \alpha \mapsto \begin{bmatrix} 0 & 0 \\ \alpha & 0 \end{bmatrix} \in sp(\mathcal{X} \oplus \mathcal{X}^\#). \tag{10.49}$$

The Hamiltonian of (10.49) is $-\frac{1}{2} x \cdot \alpha x$.

The set of elements of $Sp(\mathcal{X} \oplus \mathcal{X}^\#)$ that fix elements of $\mathcal{X}^\#$ is the image of the following injective homomorphism of groups (where $L_s(\mathcal{X}, \mathcal{X}^\#)$ is equipped with the addition):

$$L_s(\mathcal{X}, \mathcal{X}^\#) \ni \alpha \mapsto \begin{bmatrix} \mathbb{1} & 0 \\ \alpha & \mathbb{1} \end{bmatrix} \in Sp(\mathcal{X} \oplus \mathcal{X}^\#). \tag{10.50}$$

We have

$$\exp \begin{bmatrix} 0 & 0 \\ \alpha & 0 \end{bmatrix} = \begin{bmatrix} \mathbb{1} & 0 \\ \alpha & \mathbb{1} \end{bmatrix}.$$

10.4.5 Transformations fixing $\mathcal{X}$

The set of elements of $sp(\mathcal{X} \oplus \mathcal{X}^{\#})$ that send $\mathcal{X}$ to zero is the image of the following injective homomorphism of Lie algebras with the trivial bracket:

$$L_{\mathrm{s}}(\mathcal{X}^{\#}, \mathcal{X}) \ni \beta \mapsto \begin{bmatrix} 0 & \beta \\ 0 & 0 \end{bmatrix} \in sp(\mathcal{X} \oplus \mathcal{X}^{\#}). \tag{10.51}$$

The Hamiltonian of (10.51) is $-\frac{1}{2}\xi \cdot \beta \xi$.

The set of elements of $Sp(\mathcal{X} \oplus \mathcal{X}^{\#})$ that fix elements of $\mathcal{X}$ is the image of the following injective homomorphism of groups (where $L_{\mathrm{s}}(\mathcal{X}^{\#}, \mathcal{X})$ is equipped with the addition):

$$L_{\mathrm{s}}(\mathcal{X}^{\#}, \mathcal{X}) \ni \beta \mapsto \begin{bmatrix} \mathbb{1} & \beta \\ 0 & \mathbb{1} \end{bmatrix} \in Sp(\mathcal{X} \oplus \mathcal{X}^{\#}). \tag{10.52}$$

We have

$$\exp \begin{bmatrix} 0 & \beta \\ 0 & 0 \end{bmatrix} = \begin{bmatrix} \mathbb{1} & \beta \\ 0 & \mathbb{1} \end{bmatrix}.$$

The generating function for the transformation (10.52) is

$$S(x_1, x_2) = -\frac{1}{2}(x_1 - x_2) \cdot \beta^{-1}(x_1 - x_2).$$

10.4.6 Harmonic oscillator

We choose a scalar product on $\mathcal{X}$ and use it to identify $\mathcal{X}^{\#}$ with $\mathcal{X}$.

Consider the Hamiltonian $\chi(x, \xi) = \frac{1}{2}x^2 + \frac{1}{2}\xi^2$. It generates the flow

$$e^{t\omega \nabla \chi} \begin{bmatrix} x_0 \\ \xi_0 \end{bmatrix} = \begin{bmatrix} \cos t & \sin t \\ -\sin t & \cos t \end{bmatrix} \begin{bmatrix} x_0 \\ \xi_0 \end{bmatrix} = \begin{bmatrix} x_t \\ \xi_t \end{bmatrix}.$$

Its generating function is $S(x_0, x_t) = \frac{(x_0^2 + x_t^2) \cos t - 2x_0 \cdot x_t}{2 \sin t}$.

10.4.7 Transformations swapping $\mathcal{X}$ and $\mathcal{X}^{\#}$

Let $b \in L(\mathcal{X}^{\#}, \mathcal{X})$. Then the following transformation is symplectic:

$$\begin{bmatrix} 0 & b \\ -b^{\#-1} & 0 \end{bmatrix}.$$

Its generating function is $S(x_1, x_2) = -x_1 \cdot b^{-1} x_2$.

10.5 Metaplectic group in the Schrödinger representation

As in the previous section, $\mathcal{X}$ is a finite-dimensional real vector space. In this section we describe the metaplectic group $Mp(\mathcal{X}^\# \oplus \mathcal{X})$ in the Schrödinger CCR representation on $L^2(\mathcal{X})$.

10.5.1 Metaplectic group in $L^2(\mathbb{R})$

We start with the one-dimensional case. Let us consider the Schrödinger representation in $L^2(\mathbb{R})$ over $\mathbb{R} \oplus \mathbb{R}$. We will describe some examples of subgroups of the metaplectic group $Mp(\mathbb{R} \oplus \mathbb{R}) \subset U(L^2(\mathbb{R}))$.

Example 10.42 *Let $\chi(x,\xi) = x{\cdot}\xi$. Then $\mathrm{Op}(\chi) = \frac{1}{2}(x{\cdot}D + D{\cdot}x)$ and $\mathrm{e}^{-it\mathrm{Op}(\chi)}$ belongs to the metaplectic group. We have*

$$\mathrm{e}^{-it\mathrm{Op}(\chi)}\Psi(x) = \mathrm{e}^{-\frac{1}{2}t}\Psi(\mathrm{e}^{-t}x), \quad \Psi \in L^2(\mathcal{X}).$$

Example 10.43 *The multiplication operator $\mathrm{e}^{-\frac{1}{2}tx^2}$ belongs to the metaplectic group.*

Example 10.44 *The operator $\mathrm{e}^{-\frac{1}{2}tD^2}$ belongs to the metaplectic group. Its integral kernel equals*

$$(2\pi it)^{-\frac{1}{2}}\mathrm{e}^{\frac{i}{2}\frac{(x-y)^2}{t}}.$$

10.5.2 Harmonic oscillator

We still consider the one-dimensional case. Let $\chi(x,\xi) := \frac{1}{2}\xi^2 + \frac{1}{2}x^2$. Then $\mathrm{Op}(\chi) = \frac{1}{2}D^2 + \frac{1}{2}x^2$. The Weyl–Wigner symbol of $\mathrm{e}^{-t\mathrm{Op}(\chi)}$ is

$$w(t,x,\xi) = (\mathrm{ch}\tfrac{t}{2})^{-1}\exp(-(x^2 + \xi^2)\mathrm{th}\tfrac{t}{2}). \qquad (10.53)$$

Its integral kernel is given by the so-called *Mehler's formula*

$$W(t,x,y) = \pi^{-\frac{1}{2}}(\mathrm{sh}t)^{-\frac{1}{2}}\exp\left(\frac{-(x^2 + y^2)\mathrm{ch}t + 2xy}{2\mathrm{sh}t}\right).$$

$\mathrm{e}^{-it\mathrm{Op}(\chi)}$ has the Weyl–Wigner symbol

$$w(it,x,\xi) = (\cos\tfrac{t}{2})^{-1}\exp\left(-i\,(x^2 + \xi^2)\mathrm{tg}\tfrac{t}{2}\right) \qquad (10.54)$$

and the integral kernel

$$W(it,x,y) = \pi^{-\frac{1}{2}}|\sin t|^{-\frac{1}{2}}\mathrm{e}^{-\frac{i\pi}{4}}\mathrm{e}^{-\frac{i\pi}{2}[\frac{t}{\pi}]}\exp\left(\frac{-(x^2 + y^2)\cos t + 2xy}{2i\sin t}\right).$$

Above, $[c]$ denotes the integral part of c.

It is easy to see that (10.53), resp. (10.54) are special cases of (10.36), resp. (10.37).

We have $W(\mathrm{i}t + 2\mathrm{i}\pi, x, y) = -W(\mathrm{i}t, x, y)$. Note the special cases

$$W(0, x, y) = \delta(x - y),$$

$$W(\tfrac{\mathrm{i}\pi}{2}, x, y) = (2\pi)^{-\frac{1}{2}} \mathrm{e}^{-\frac{\mathrm{i}\pi}{4}} \mathrm{e}^{-\mathrm{i}xy},$$

$$W(\mathrm{i}\pi, x, y) = \mathrm{e}^{-\frac{\mathrm{i}\pi}{2}} \delta(x + y),$$

$$W(\tfrac{\mathrm{i}3\pi}{2}, x, y) = (2\pi)^{-\frac{1}{2}} \mathrm{e}^{-\frac{\mathrm{i}3\pi}{4}} \mathrm{e}^{\mathrm{i}xy}.$$

Corollary 10.45 (1) *The operator with kernel* $\pm(2\pi\mathrm{i})^{-\frac{1}{2}} \mathrm{e}^{-\mathrm{i}xy}$ *belongs to the metaplectic group and implements* $\begin{bmatrix} 0 & -1 \\ 1 & 0 \end{bmatrix}$.

(2) *The operator with kernel* $\pm\mathrm{i}\delta(x + y)$ *belongs to the metaplectic group and implements* $\begin{bmatrix} -1 & 0 \\ 0 & -1 \end{bmatrix}$.

10.5.3 Quadratic Hamiltonians in the Schrödinger representation

Until the end of the section we consider $\mathcal{X}$ of any finite dimension. Any $\chi \in \mathbb{C}\mathrm{Pol}_s^{\leq 2}(\mathcal{Y}^{\#})$ is of the form

$$\mathcal{X} \oplus \mathcal{X}^{\#} \ni (x, \xi) \mapsto \chi(x, \xi) = \alpha(x) + \xi{\cdot}cx + \beta(\xi),$$

where $\alpha \in \mathbb{C}\mathrm{Pol}_s^{\leq 2}(\mathcal{X})$, $\beta \in \mathbb{C}\mathrm{Pol}_s^{\leq 2}(\mathcal{X}^{\#})$ and $c \in L(\mathcal{X})$. We have

$$\mathrm{Op}^{x,D}(\chi) = \alpha(x) + x{\cdot}c^{\#}D + \beta(D),$$

$$\mathrm{Op}(\chi) = \mathrm{Op}^{x,D}(\chi) + \tfrac{1}{2}\mathrm{Tr}\, c.$$

10.5.4 Integral kernel of elements of the metaplectic group

First we describe various examples of elements of the metaplectic group.

Proposition 10.46 *If* $m \in GL(\mathcal{X})$ *with* $\det m \neq 0$, *then the operator*

$$\pm T_m \Psi(x) := \pm(\det m)^{\frac{1}{2}} \Psi(mx) \tag{10.55}$$

belongs to $Mp(\mathcal{X}^{\#} \oplus \mathcal{X})$ *and implements* $\begin{bmatrix} m^{\#} & 0 \\ 0 & m^{-1} \end{bmatrix} \in Sp(\mathcal{X}^{\#} \oplus \mathcal{X})$.

Proof Assume first that $\det m > 0$. Let $c \in gl(\mathcal{X})$ such that $m = \mathrm{e}^c$. Recall that if $\chi(x, \xi) = x{\cdot}c^{\#}\xi$, then

$$\mathrm{Op}(\chi) = x{\cdot}c^{\#}D + \frac{\mathrm{i}}{2}\mathrm{Tr}\, c, \quad \mathrm{Op}^{x,D}(\chi) = x{\cdot}c^{\#}D.$$

But $\mathrm{e}^{\frac{1}{2}\mathrm{Tr}\, c} = (\det \mathrm{e}^c)^{\frac{1}{2}} = (\det m)^{\frac{1}{2}}$.

Suppose now that $\det m < 0$. Fix an arbitrary Euclidean structure in $\mathcal{X}$. We can write m as $m_1 m_2$ where $\det m_1 > 0$ and $m_2 = \mathbb{1} - 2|e\rangle\langle e|$, where $e \in \mathcal{X}$,

$\|e\| = 1$. We have $\begin{bmatrix} m^\# & 0 \\ 0 & m^{-1} \end{bmatrix} = \begin{bmatrix} m_1^\# & 0 \\ 0 & m_1^{-1} \end{bmatrix} \begin{bmatrix} \mathbb{1} - 2|e\rangle\langle e| & 0 \\ 0 & \mathbb{1} - 2|e\rangle\langle e| \end{bmatrix}$. The first term we implement as above, the second by the exponential of a one dimensional harmonic oscillator; see Corollary 10.45 (2). □

Proposition 10.47 *Let* $\alpha \in L_s(\mathcal{X}, \mathcal{X}^\#)$. *Then* $e^{-\frac{i}{2}x \cdot \alpha x} \in Mp(\mathcal{X}^\# \oplus \mathcal{X})$ *and implements* $\begin{bmatrix} \mathbb{1} & \alpha \\ 0 & \mathbb{1} \end{bmatrix}$.

Proposition 10.48 *Let* $b \in L(\mathcal{X}, \mathcal{X}^\#)$. *Then the operator with the kernel*

$$\pm(2\pi i)^{-\frac{d}{2}} (\det b)^{-\frac{1}{2}} e^{ix_1 \cdot b^{-1} x_2} \tag{10.56}$$

belongs to $Mp(\mathcal{X}^\# \oplus \mathcal{X})$ *and implements* $\begin{bmatrix} 0 & -b^{-1} \\ b^\# & 0 \end{bmatrix}$.

Proof Equip $\mathcal{X}$ with a scalar product. We can identify $\mathcal{X}$ with $\mathcal{X}^\#$ and write

$$\begin{bmatrix} 0 & -b^{-1} \\ b^\# & 0 \end{bmatrix} = \begin{bmatrix} 0 & -\mathbb{1} \\ \mathbb{1} & 0 \end{bmatrix} \begin{bmatrix} b^\# & 0 \\ 0 & b^{-1} \end{bmatrix}.$$

By Corollary 10.45, the operator with integral kernel $\pm(2\pi i)^{-\frac{d}{2}} e^{-ix_1 \cdot x_2}$ belongs to $Mp(\mathcal{X} \oplus \mathcal{X})$ and implements $\begin{bmatrix} 0 & -\mathbb{1} \\ \mathbb{1} & 0 \end{bmatrix}$. By Prop. 10.46, $\begin{bmatrix} b^\# & 0 \\ 0 & b^{-1} \end{bmatrix}$ is implemented by (10.55). Then we use the chain rule. □

Let us now describe the case of an (almost) arbitrary $r \in Sp(\mathcal{X}^\# \oplus \mathcal{X})$. We can write $r^\# = \begin{bmatrix} a & b \\ c & d \end{bmatrix}$. Recall from Thm. 10.38 that, if b is invertible, we can factorize $r^\#$ as

$$r^\# = \begin{bmatrix} \mathbb{1} & 0 \\ e & \mathbb{1} \end{bmatrix} \begin{bmatrix} 0 & b \\ -b^{\#-1} & 0 \end{bmatrix} \begin{bmatrix} \mathbb{1} & 0 \\ f & \mathbb{1} \end{bmatrix},$$

and introduce the generating function of $r^\#$:

$$\mathcal{X} \times \mathcal{X} \ni (x_1, x_2) \mapsto S(x_1, x_2) := \frac{1}{2} x_1 \cdot f x_1 - x_1 \cdot b^{-1} x_2 + \frac{1}{2} x_2 \cdot e x_2.$$

The following theorem is one of the most beautiful expressions of the correspondence between classical and quantum mechanics, since the distributional kernel of the (quantum) unitary operator U_r is expressed purely in terms of the generating function for the symplectic transformation $r^\#$.

Theorem 10.49 *Let* $r \in Sp(\mathcal{X}^\# \oplus \mathcal{X})$ *be such that* b *is invertible. Then the operators* $\pm U_r \in Mp(\mathcal{X}^\# \oplus \mathcal{X})$ *implementing* r *have their integral kernels equal to*

$$\pm U_r(x_1, x_2) = \pm(2\pi i)^{-\frac{d}{2}} \sqrt{-\det \nabla_{x_1} \nabla_{x_2} S} \, e^{-iS(x_1, x_2)}.$$

Proof We can write

$$r = \begin{bmatrix} 1 & f \\ 0 & 1 \end{bmatrix} \begin{bmatrix} 0 & -b^{-1} \\ b^{\#} & 0 \end{bmatrix} \begin{bmatrix} 1 & e \\ 0 & 1 \end{bmatrix} = r_f r_b r_e.$$

r_f and r_e are implemented in $Mp(\mathcal{X}^{\#} \oplus \mathcal{X})$ by $U_e = \mathrm{e}^{-\frac{1}{2}x \cdot ex}$ and $U_f = \mathrm{e}^{-\frac{1}{2}x \cdot fx}$. r_b is implemented in $Mp(\mathcal{X}^{\#} \oplus \mathcal{X})$ by U_b, which has the integral kernel (10.56). Hence $r = r_f r_b r_e$ is implemented by $U_f U_b U_e$, which has the integral kernel

$$\pm (2\pi\mathrm{i})^{-\frac{d}{2}} (\det b)^{-\frac{1}{2}} \mathrm{e}^{-\frac{1}{2}x_1 \cdot f x_1} \mathrm{e}^{\mathrm{i} x_1 \cdot b^{-1} x_2} \mathrm{e}^{-\frac{1}{2}x_2 \cdot e x_2}. \qquad \square$$

10.6 Notes

Normal forms of quadratic Hamiltonians were first established by Williamson (1936). Thus, Prop. 10.8 is a special case of Williamson's theorem.

The fact that Bogoliubov rotations are implemented by a projective unitary representation of the symplectic group was noted by Segal (1959). Its implementation by a representation of the two-fold covering of the symplectic group, the so-called metaplectic representation, is attributed to Weil (1964) and Shale (1962). The metaplectic group plays an important role in the concept of the Maslov index, the semi-classical approximation and microlocal analysis; see Maslov (1972), Leray (1978), Guillemin–Sternberg (1977) and Hörmander (1985). The semi-classical approximation and microlocal analysis are asymptotic theories (where the small parameter is the Planck constant $\hbar$ or the inverse λ^{-1} of the momentum scale). One can obtain for example extensions of Thm. 10:35 or Thm. 10.49 to non-quadratic Hamiltonians or non-linear symplectic maps. In these extensions the expressions of Weyl–Wigner symbols or distributional kernels are given by asymptotic expansions in terms of the small parameter. In the linear case these expansions have only one term and are exact.

The first famous application of the symplectic invariance of CCR seems to be Bogoliubov's theory of the excitation spectrum of the homogeneous Bose gas (Bogoliubov (1947a); see also Fetter–Walecka (1971) and Cornean–Dereziński–Ziń (2009)).

11

Symplectic invariance of the CCR
on Fock spaces

This chapter is a continuation of Chap. 9, where we studied Fock CCR representations. Our goal is to extend the results of Chap. 10 about the symplectic invariance of canonical commutation relations to the case of Fock CCR representations in any dimension.

11.1 Symplectic group on a Kähler space

The basic framework of this section, as well as most other sections of this chapter, is the same as that of Chap. 9.

In particular, throughout the section, $(\mathcal{Y}, \cdot, \omega, \mathrm{j})$ is a complete Kähler space. We recall that for $r \in B(\mathcal{Y})$, $r^{\#}$ denotes the adjoint of r for the Euclidean scalar product of $\mathcal{Y}$.

Recall that the holomorphic space $\mathcal{Z}$ in $\mathbb{C}\mathcal{Y}$ is defined as $\mathrm{Ran}\, \frac{1}{2}(\mathbb{1} - \mathrm{ij})$, so that $\mathbb{C}\mathcal{Y} = \mathcal{Z} \oplus \overline{\mathcal{Z}}$. $\mathcal{Z}$ is a (complex) Hilbert space.

In this section we study the symplectic group in a complete Kähler space of any dimension. We treat the symplectic form ω as the basic structure of the Kähler space $\mathcal{Y}$. However, the additional structure on $\mathcal{Y}$ plays an important role. In particular, it gives $\mathcal{Y}$ a Hilbertian topology, which is especially useful when we consider the infinite-dimensional case.

11.1.1 Basic properties

Definition 11.1 *The group of linear transformations on $\mathcal{Y}$ that are bounded, symplectic and have a bounded inverse will be denoted by $Sp(\mathcal{Y})$. Similarly, the Lie algebra of bounded infinitesimally symplectic transformations on $\mathcal{Y}$ will be denoted by $sp(\mathcal{Y})$.*

Note that $sp(\mathcal{Y})$ is the set of generators of norm continuous one-parameter groups in $Sp(\mathcal{Y})$.

We can use the anti-involution j instead of the symplectic form ω to describe various properties of symplectic and infinitesimally symplectic transformations.

The following proposition can be compared with Prop. 1.37.

Proposition 11.2 (1) $r \in Sp(\mathcal{Y})$ *iff*

$$a)\ r^{\#}\mathrm{j}r = \mathrm{j}, \ and \ b)\ r\mathrm{j}r^{\#} = \mathrm{j}.$$

(2) $r \in Sp(\mathcal{Y})$ iff $r^{\#} \in Sp(\mathcal{Y})$.

(3) If $r \in Sp(\mathcal{Y})$, then $r^{-1} = -\mathrm{j}r^{\#}\mathrm{j}$.

Proposition 11.3 (1) $a \in sp(\mathcal{Y})$ iff $a^{\#}\mathrm{j} + \mathrm{j}a = 0$.

(2) $a \in sp(\mathcal{Y})$ iff $a^{\#} \in sp(\mathcal{Y})$.

11.1.2 Unitary group on a Kähler space

Recall that a complete Kähler space $\mathcal{Y}$ can be viewed as a complex Hilbert space. It is then denoted by $\mathcal{Y}^{\mathbb{C}}$, with the imaginary unit given by j and the scalar product given by $(y_1|y_2) := y_1 \cdot y_2 + \mathrm{i}y_1 \cdot \omega y_2$ (see (1.37)).

Proposition 11.4 *We have the following characterizations of the unitary group and Lie algebra on a Kähler space:*

$$U(\mathcal{Y}^{\mathbb{C}}) = O(\mathcal{Y}) \cap Sp(\mathcal{Y}) = O(\mathcal{Y}) \cap GL(\mathcal{Y}^{\mathbb{C}}) = Sp(\mathcal{Y}) \cap GL(\mathcal{Y}^{\mathbb{C}}),$$
$$u(\mathcal{Y}^{\mathbb{C}}) = o(\mathcal{Y}) \cap sp(\mathcal{Y}) = o(\mathcal{Y}) \cap gl(\mathcal{Y}^{\mathbb{C}}) = sp(\mathcal{Y}) \cap gl(\mathcal{Y}^{\mathbb{C}}).$$

It is easy to characterize elements of $U(\mathcal{Y}^{\mathbb{C}})$ and $u(\mathcal{Y}^{\mathbb{C}})$ by their extensions to $\mathbb{C}\mathcal{Y} = \mathcal{Z} \oplus \overline{\mathcal{Z}}$.

Proposition 11.5 (1) $r \in U(\mathcal{Y}^{\mathbb{C}})$ iff

$$r_{\mathbb{C}} = \begin{bmatrix} p & 0 \\ 0 & \overline{p} \end{bmatrix},$$

with $p \in U(\mathcal{Z})$.

(2) $a \in u(\mathcal{Y}^{\mathbb{C}})$ iff

$$a_{\mathbb{C}} = \mathrm{i} \begin{bmatrix} -h & 0 \\ 0 & \overline{h} \end{bmatrix},$$

with $h = h^*$.

11.1.3 Symplectic transformations on Kähler spaces

We recall that if $a \in B(\mathcal{Z}_1, \mathcal{Z}_2)$, then $a^{\#} := \overline{a}^* \in B(\overline{\mathcal{Z}}_2, \overline{\mathcal{Z}}_1)$. We recall also that $B_{\mathrm{s}}(\overline{\mathcal{Z}}, \mathcal{Z})$ denotes the set of $g \in B(\overline{\mathcal{Z}}, \mathcal{Z})$ such that $g^{\#} = g$, and $B_{\mathrm{h}}(\mathcal{Z})$ denotes the set of $h \in B(\mathcal{Z})$ such that $h^* = h$ (see Subsect. 2.2.3).

Proposition 11.6 $r \in Sp(\mathcal{Y})$ iff its extension to $\mathbb{C}\mathcal{Y}$ equals

$$r_{\mathbb{C}} = \begin{bmatrix} p & q \\ \overline{q} & \overline{p} \end{bmatrix}, \tag{11.1}$$

with $p \in B(\mathcal{Z})$, $q \in B(\overline{\mathcal{Z}}, \mathcal{Z})$, and the following conditions hold:

conditions implied by Prop. 11.2 (1a): $\quad p^*p - q^{\#}\overline{q} = \mathbb{1}, \quad p^*q - q^{\#}\overline{p} = 0,$

conditions implied by Prop. 11.2 (1b): $\quad pp^* - qq^* = \mathbb{1}, \quad pq^{\#} - qp^{\#} = 0.$

Proposition 11.7 $a \in sp(\mathcal{Y})$ *iff its extension to* $\mathbb{C}\mathcal{Y}$ *equals*

$$a_{\mathbb{C}} = \mathrm{i} \begin{bmatrix} -h & g \\ -\bar{g} & h \end{bmatrix}, \tag{11.2}$$

with $h \in B_{\mathrm{h}}(\mathcal{Z})$ *and* $g \in B_{\mathrm{s}}(\bar{\mathcal{Z}}, \mathcal{Z})$.

We describe now a convenient factorization of a symplectic map. Let $r \in Sp(\mathcal{Y})$, and let p, q be defined as in (11.1). Note that

$$pp^* \geq \mathbb{1}, \quad p^*p \geq \mathbb{1}.$$

Hence p^{-1} and p^{*-1} are bounded operators, and we can set

$$c := q^{\#}(p^{\#})^{-1}, \tag{11.3}$$

$$d := q\bar{p}^{-1}. \tag{11.4}$$

Recall that, for $a, b \in B_{\mathrm{h}}(\mathcal{Z})$, $a < b$ means $a \leq b$ and $\mathrm{Ker}(b - a) = \{0\}$.

Proposition 11.8 (1) *We have* $c, d \in B_{\mathrm{s}}(\bar{\mathcal{Z}}, \mathcal{Z})$ *and*

$$c^*c < \mathbb{1}, \quad d^*d < \mathbb{1}. \tag{11.5}$$

(2) *The following equivalent characterizations of* c, d *hold:*

$$c = p^{-1}q, \tag{11.6}$$

$$d = (p^*)^{-1}q^{\#}. \tag{11.7}$$

(3) *One has the following factorization:*

$$r_{\mathbb{C}} = \begin{bmatrix} \mathbb{1} & d \\ 0 & \mathbb{1} \end{bmatrix} \begin{bmatrix} (p^*)^{-1} & 0 \\ 0 & \bar{p} \end{bmatrix} \begin{bmatrix} \mathbb{1} & 0 \\ \bar{c} & \mathbb{1} \end{bmatrix}. \tag{11.8}$$

(4) *We have*

$$(r_{\mathbb{C}}^*)^{-1} = \begin{bmatrix} p & -q \\ -\bar{q} & \bar{p} \end{bmatrix},$$

$$(r_{\mathbb{C}}r_{\mathbb{C}}^* - \mathbb{1})(r_{\mathbb{C}}r_{\mathbb{C}}^* + \mathbb{1})^{-1} = \begin{bmatrix} 0 & d \\ \bar{d} & 0 \end{bmatrix},$$

$$(r_{\mathbb{C}}^*r_{\mathbb{C}} - \mathbb{1})(r_{\mathbb{C}}^*r_{\mathbb{C}} + \mathbb{1})^{-1} = \begin{bmatrix} 0 & c \\ \bar{c} & 0 \end{bmatrix}. \tag{11.9}$$

(5) *We have the identities*

$$\mathbb{1} - cc^* = (p^*p)^{-1}, \quad \mathbb{1} - d^*d = (\bar{p}\,\bar{p}^*)^{-1}. \tag{11.10}$$

Proof For example, the first inequality of (11.5) follows from the fact that $pp^* > qq^*$, which implies that $(p^*)^{-1}p^{-1} = (pp^*)^{-1} < (q^*)^{-1}q^{-1}$. Now $c^*c = q^*(p^*)^{-1}p^{-1}q < \mathbb{1}$. □

11.1.4 Positive symplectic transformations

Symplectic transformations that are at the same time positive enjoy special properties. We devote this subsection to a discussion of their basic properties.

Let $r \in Sp(\mathcal{Y})$ such that $r = r^\#$ and $r > 0$ as an operator on $(\mathcal{Y}, \cdot)$. Recall that the unitary structure on $\mathbb{C}\mathcal{Y}$ is obtained from the Euclidean structure of $\mathcal{Y}$ as in Subsect. 1.3.4. Hence, $r_\mathbb{C} = r_\mathbb{C}^*$ and $r_\mathbb{C} > 0$. We have

$$r_\mathbb{C} = \begin{bmatrix} p & q \\ \bar{q} & \bar{p} \end{bmatrix},$$

where $p = p^* > 0$ and $q = q^\#$. The conditions in Prop. 11.6 simplify to

$$p^2 - q\bar{q} = \mathbb{1}, \quad pq - q\bar{p} = 0.$$

We have

$$r_\mathbb{C}^{-1} = \begin{bmatrix} p & -q \\ -\bar{q} & \bar{p} \end{bmatrix}.$$

In the case of positive symplectic transformations some of the identities of Prop. 11.8 simplify:

Proposition 11.9 *Let $r \in Sp(\mathcal{Y})$ such that $r = r^\#$ and $r > 0$. Let $c \in B_\mathrm{s}(\overline{\mathcal{Z}}, \mathcal{Z})$ be defined as in (11.3). Then $c^*c < \mathbb{1}$,*

$$r_\mathbb{C} = \begin{bmatrix} (\mathbb{1} - cc^*)^{-\frac{1}{2}} & (\mathbb{1} - cc^*)^{-\frac{1}{2}}c \\ c^*(\mathbb{1} - cc^*)^{-\frac{1}{2}} & (\mathbb{1} - c^*c)^{-\frac{1}{2}} \end{bmatrix} \tag{11.11}$$

$$= \begin{bmatrix} \mathbb{1} & c \\ 0 & \mathbb{1} \end{bmatrix} \begin{bmatrix} (\mathbb{1} - cc^*)^{\frac{1}{2}} & 0 \\ 0 & (\mathbb{1} - c^*c)^{-\frac{1}{2}} \end{bmatrix} \begin{bmatrix} \mathbb{1} & 0 \\ c^* & \mathbb{1} \end{bmatrix},$$

$$(r_\mathbb{C}^2 - \mathbb{1})(r_\mathbb{C}^2 + \mathbb{1})^{-1} = \begin{bmatrix} 0 & c \\ \bar{c} & 0 \end{bmatrix}. \tag{11.12}$$

Conversely, let $c \in B_\mathrm{s}(\overline{\mathcal{Z}}, \mathcal{Z})$ satisfy $c^*c < \mathbb{1}$, and let r be defined by (11.11). Then $r \in Sp(\mathcal{Y})$, $r = r^\#$, $r > 0$.

Proof The properties of c follow directly from the properties of p, q given above.

Next let $c \in B_\mathrm{s}(\overline{\mathcal{Z}}, \mathcal{Z})$ with $c^*c < \mathbb{1}$. Clearly, $cc^* = \overline{c^*c} < \mathbb{1}$, and hence the operators $\mathbb{1} - cc^*$ and $\mathbb{1} - c^*c$ are invertible. We check that the operator r defined by (11.11) is a positive symplectic transformation. $\square$

Positive symplectic transformations can be obtained as exponentials of self-adjoint infinitesimally symplectic transformations:

Proposition 11.10 *Let $a \in sp(\mathcal{Y})$ such that $a = a^\#$. Then $a_\mathbb{C} = a_\mathbb{C}^*$, and hence there exists $g \in B_\mathrm{s}(\overline{\mathcal{Z}}, \mathcal{Z})$ such that*

$$a_\mathbb{C} = \mathrm{i} \begin{bmatrix} 0 & g \\ -\bar{g} & 0 \end{bmatrix}. \tag{11.13}$$

Moreover, $r = e^a$ belongs to $Sp(\mathcal{Y})$ and satisfies $r = r^\#$, $r > 0$ and

$$r_{\mathbb{C}} = \begin{bmatrix} \cosh\sqrt{gg^*} & \mathrm{i}\frac{\sinh\sqrt{gg^*}}{\sqrt{gg^*}}g \\ -\mathrm{i}g^*\frac{\sinh\sqrt{gg^*}}{\sqrt{gg^*}} & \cosh\sqrt{g^*g} \end{bmatrix}, \tag{11.14}$$

$$c = \mathrm{i}\frac{\tanh\sqrt{gg^*}}{\sqrt{gg^*}}g. \tag{11.15}$$

11.1.5 Polar decomposition of symplectic maps

Proposition 11.11 (1) *Let $r \in Sp(\mathcal{Y})$ such that $r > 0$. Then, for each $\epsilon \in \mathbb{R}$, $r^\epsilon \in Sp(\mathcal{Y})$.*

(2) *Let $r \in Sp(\mathcal{Y})$. Then there exist unique $k \in Sp(\mathcal{Y})$, $u \in U(\mathcal{Y}^{\mathbb{C}})$ such that $k = k^\#$, $k > 0$ and $r = ku$. The operators u, k are given by the polar decomposition of r as an operator on the real Hilbert space $(\mathcal{Y}, \cdot)$.*

Proof Let $r \in Sp(\mathcal{Y})$ such that $r = r^\#$ and $r > 0$. Then $r\mathrm{j} = \mathrm{j}r^{-1}$, since $r \in Sp(\mathcal{Y})$. This implies that $(z-r)^{-1}\mathrm{j} = \mathrm{j}(z-r^{-1})^{-1}$ for $z \in \mathbb{C}\backslash\mathbb{R}$, and hence

$$f(r)\mathrm{j} = \mathrm{j}f(r^{-1}), \text{ for any measurable function } f.$$

In particular, for $\epsilon \in \mathbb{R}$ we have $r^\epsilon\mathrm{j} = \mathrm{j}r^{-\epsilon}$, and hence $r^\epsilon \in Sp(\mathcal{Y})$. This proves (1).

Now let $r \in Sp(\mathcal{Y})$. Set $k = (rr^\#)^{\frac{1}{2}}$. By (1), $k \in Sp(\mathcal{Y})$. Set $u = k^{-1}r$. Clearly, $u \in Sp(\mathcal{Y})$. By the properties of the polar decomposition in $(\mathcal{Y}, \cdot)$ we have $u \in O(\mathcal{Y})$. Hence $u \in U(\mathcal{Y}^{\mathbb{C}})$, which proves (2). $\square$

11.1.6 Restricted symplectic group

In this subsection we introduce a subgroup of the symplectic group on the Kähler space that plays an important role in Shale's theorem, a basic result of the theory of CCR representations on Fock spaces.

Proposition 11.12 *Let $r \in Sp(\mathcal{Y})$. Consider p, q, c, d defined by (11.1), (11.3) and (11.4). The following conditions are equivalent:*

(1) $\mathrm{j} - r^{-1}\mathrm{j}r \in B^2(\mathcal{Y})$.

(2) $r\mathrm{j} - \mathrm{j}r \in B^2(\mathcal{Y})$.

(3) $\mathrm{Tr}q^*q < \infty$.

(4) $\mathrm{Tr}(p^*p - \mathbb{1}) < \infty$.

(5) $\mathrm{Tr}(pp^* - \mathbb{1}) < \infty$.

(6) $d \in B^2(\overline{\mathcal{Z}}, \mathcal{Z})$.

(7) $c \in B^2(\overline{\mathcal{Z}}, \mathcal{Z})$.

Proof Clearly, (1)⇔ (2).

We have

$$(rj - jr)_{\mathbb{C}} = \begin{bmatrix} 0 & -2iq \\ 2i\overline{q} & 0 \end{bmatrix}, \qquad (rj - jr)_{\mathbb{C}}^*(rj - jr)_{\mathbb{C}} = \begin{bmatrix} 4q^{\#}\overline{q} & 0 \\ 0 & 4q^*q \end{bmatrix}.$$

Hence

$$\mathrm{Tr}(rj - jr)_{\mathbb{C}}^*(rj - jr)_{\mathbb{C}} = 4(\mathrm{Tr}\, q^{\#}\overline{q} + \mathrm{Tr}\, q^*q) = 8\mathrm{Tr}\, q^*q = 8\mathrm{Tr}\,(p^*p - \mathbb{1}),$$

using that $q^{\#}\overline{q} = p^*p - \mathbb{1}$. This implies that (2)⇔(3)⇔(4). If $v \in U(\mathcal{Z})$ and $p = v|p| = |p^*|v$ is the polar decomposition of p, we have $pp^* = vp^*pv^*$. So (4)⇔(5).

The identities $c = p^{-1}q$ and $d = q\overline{p}^{-1}$ and the fact that p is invertible show that (3)⇒(6) and (3)⇒(7). The identities $\mathbb{1} - cc^* = (p^*p)^{-1}$ and $\mathbb{1} - d^*d = (\overline{p}p^{\#})^{-1}$ show that (7)⇒(4) and (6)⇒(5). $\qquad\square$

Definition 11.13 *Let $Sp_j(\mathcal{Y})$ be the set of $r \in Sp(\mathcal{Y})$ satisfying the conditions of Prop. 11.12. The set $Sp_j(\mathcal{Y})$ is called the* restricted symplectic group*. We equip it with the metric*

$$d_j(r_1, r_2) := \|p_1 - p_2\| + \|q_1 - q_2\|_2. \tag{11.16}$$

Equivalent metrics are $\|[j, r_1 - r_2]_+\| + \|[j, r_1 - r_2]\|_2$ *and* $\|r_1 - r_2\| + \|[j, r_1 - r_2]\|_2$.

We say that $a \in sp_j(\mathcal{Y})$ if $a \in sp(\mathcal{Y})$ and $aj - ja \in B^2(\mathcal{Y})$, or equivalently $g \in B_s^2(\overline{\mathcal{Z}}, \mathcal{Z})$, where we use the decomposition (11.2).

Proposition 11.14 (1) *$Sp_j(\mathcal{Y})$ is a topological group.*
(2) *$sp_j(\mathcal{Y})$ is a Lie algebra.*
(3) *If $a \in sp_j(\mathcal{Y})$ then $e^a \in Sp_j(\mathcal{Y})$.*

Proof The fact that $Sp_j(\mathcal{Y})$ is a topological group is clear, since $[r_1r_2, j] = r_1[r_2, j] + [r_1, j]r_2$. To prove (3), we write

$$e^a j - je^a = \sum_{n=0}^{\infty} \frac{1}{n!}[a^n, j],$$

and use that $\|[a^n, j]\|_2 \le n\|a\|^{n-1}\|[a, j]\|_2$, which yields

$$\|e^a j - je^a\|_2 \le e^{\|a\|}\|aj - ja\|_2. \qquad\square$$

11.1.7 Anomaly-free symplectic group

In this subsection we introduce another, much smaller subgroup of the symplectic group on the Kähler space. Its name is suggested by the well-known terminology used in quantum field theory.

Definition 11.15 *Let $Sp_{j,af}(\mathcal{Y})$ be the set of $r \in Sp_j(\mathcal{Y})$ such that $2j - (jr + rj) \in B^1(\mathcal{Y})$, or equivalently $p - \mathbb{1}_{\mathcal{Z}} \in B^1(\mathcal{Z})$, where we use the* *decomposition (11.1). $Sp_{j,af}(\mathcal{Y})$ will be called the anomaly-free symplectic group and will be equipped with the metric*

$$d_{j,af}(r_1, r_2) := \|p_1 - p_2\|_1 + \|q_1 - q_2\|_2.$$

An equivalent metric is $\|[j, r_1 - r_2]_+\|_1 + \|[j, r_1 - r_2]\|_2$.

We also define $sp_{j,af}(\mathcal{Y})$ to be the set of $a \in sp_j(\mathcal{Y})$ such that $aj + ja \in B^1(\mathcal{Y})$, or equivalently $h \in B^1(\mathcal{Y})$, where we use the decomposition (11.2).

Proposition 11.16 (1) $Sp_{j,af}(\mathcal{Y})$ *is a topological group.*
(2) $sp_{j,af}(\mathcal{Y})$ *is a Lie algebra.*
(3) *If $a \in sp_{j,af}(\mathcal{Y})$ then $e^a \in Sp_{j,af}(\mathcal{Y})$.*

Proof If $r \in Sp_{j,af}(\mathcal{Y})$, then $\mathbb{1} - r \in B^2(\mathcal{Y})$. It follows that if $r_1, r_2 \in Sp_{j,af}(\mathcal{Y})$, then $r_1 r_2 - (r_1 + r_2) + \mathbb{1} \in B^1(\mathcal{Y})$, which easily implies that $r_1 r_2 \in Sp_{j,af}(\mathcal{Y})$ and proves (1). To prove (2), note that $sp_{j,af}(\mathcal{Y}) \subset B^2(\mathcal{Y})$. To prove (3), we use that if $a \in sp_{j,af}(\mathcal{Y})$, then $e^a - (\mathbb{1} + a) \in B^1(\mathcal{Y})$. □

Proposition 11.17 (1) *Let $r \in Sp(\mathcal{Y})$ be positive. Then $r \in Sp_j(\mathcal{Y})$ iff $r \in Sp_{j,af}(\mathcal{Y})$.*
(2) *Let $a \in sp(\mathcal{Y})$ be self-adjoint. Then $a \in sp_j(\mathcal{Y})$ iff $a \in sp_{j,af}(\mathcal{Y})$.*

Proof (1) We know that $r \in Sp_j(\mathcal{Y})$ iff $c \in B^2(\overline{\mathcal{Z}}, \mathcal{Z})$. But (11.11) then implies that $r \in Sp_{j,af}(\mathcal{Y})$.

(2) By the decomposition (11.2), $a \in sp_j(\mathcal{Y})$ is self-adjoint iff $h = 0$ and $g \in B_s^2(\overline{\mathcal{Z}}, \mathcal{Z})$. □

Proposition 11.18 *Let $r \in B(\mathcal{Y})$ and let $r = r_0 u$ be its polar decomposition. Then*

(1) $r \in Sp(\mathcal{Y})$ *iff $r_0 \in Sp(\mathcal{Y})$.*
(2) $r \in Sp_j(\mathcal{Y})$ *iff $r_0 \in Sp_j(\mathcal{Y})$.*
(3) $r \in Sp_{j,af}(\mathcal{Y})$ *iff $r_0 \in Sp_{j,af}(\mathcal{Y})$ and $u \in Sp_{j,af}(\mathcal{Y})$.*

11.1.8 Pairs of Kähler structures on symplectic spaces

In this subsection we study the relationship between two Kähler anti-involutions on a given symplectic space. One of them is denoted j and is treated as the basic one. The other is denoted j_1.

In the first proposition, j_1 is obtained by conjugating j with an arbitrary symplectic map.

Proposition 11.19 *Let $r \in Sp(\mathcal{Y})$. Set $j_1 = r^{-1}jr$.*

(1) j_1 *is a Kähler anti-involution.*
(2) $r \in U(\mathcal{Y}^{\mathbb{C}})$ *iff $j_1 = j$.*
(3) *If $r = u|r|$ is the polar decomposition of r, then $j_1 = |r|^{-1}j|r|$.*
(4) $j_1 = jr^\# r$.

Proof Since $r \in Sp(\mathcal{Y})$, we have $y \cdot \omega j_1 y = y \cdot \omega r^{-1} j r y = (ry) \cdot \omega j r y$, which shows that (ω, j_1) is Kähler and proves (1). Clearly, $r \in U(\mathcal{Y}^{\mathbb{C}})$ iff $[r, j] = 0$, which is equivalent to $j_1 = j$. This proves (2). If $r = u|r|$, then $[u, j] = 0$, hence $j_1 = |r|^{-1} j|r|$, which proves (3). (4) follows from Prop. 11.2 (1a). $\qquad\square$

The next proposition is a partial converse of the previous one. In particular, we compute a positive symplectic map that transforms j into j_1.

Theorem 11.20 (1) *Let* j_1 *be an anti-involution such that* (ω, j_1) *is Kähler. Then* $k := -jj_1$ *is a positive symplectic transformation.*
(2) *Let* $k \in Sp(\mathcal{Y})$ *be positive. Then* $j_1 := jk$ *is a Kähler anti-involution.*
(3) *Let* k, j_1 *be as in (2). Then* $r = k^{\frac{1}{2}}$ *defined in Subsect. 2.3.2 satisfies*

$$r \in Sp(\mathcal{Y}), \quad r = r^{\#}, \quad r > 0, \quad r^{-1} j r = j_1. \tag{11.17}$$

(r is positive symplectic and intertwines j and j_1.)
(4) *There exists* $c \in B_s(\overline{\mathcal{Z}}, \mathcal{Z})$ *such that*

$$\left(\frac{k - \mathbb{1}}{k + \mathbb{1}}\right)_{\mathbb{C}} = \begin{bmatrix} 0 & c \\ \overline{c} & 0 \end{bmatrix}. \tag{11.18}$$

(5) *We have*

$$r_{\mathbb{C}} = \begin{bmatrix} (\mathbb{1} - cc^*)^{-\frac{1}{2}} & (\mathbb{1} - cc^*)^{-\frac{1}{2}} c \\ c^*(\mathbb{1} - cc^*)^{-\frac{1}{2}} & (\mathbb{1} - c^*c)^{-\frac{1}{2}} \end{bmatrix}, \tag{11.19}$$

$$k_{\mathbb{C}} = \begin{bmatrix} (\mathbb{1} + cc^*)(\mathbb{1} - cc^*)^{-1} & 2(\mathbb{1} - cc^*)^{-1} c \\ 2c^*(\mathbb{1} - cc^*)^{-1} & (\mathbb{1} + c^*c)(\mathbb{1} - c^*c)^{-1} \end{bmatrix}, \tag{11.20}$$

$$j_{1\mathbb{C}} = i \begin{bmatrix} (\mathbb{1} + cc^*)(\mathbb{1} - cc^*)^{-1} & 2(\mathbb{1} - cc^*)^{-1} c \\ -2c^*(\mathbb{1} - cc^*)^{-1} & -(\mathbb{1} + c^*c)(\mathbb{1} - c^*c)^{-1} \end{bmatrix}. \tag{11.21}$$

Proof Since $j, j_1 \in Sp(\mathcal{Y})$, $k = -jj_1 \in Sp(\mathcal{Y})$. Since (ω, j_1) is Kähler, we have

$$0 = (j_1 y_1) \cdot \omega y_2 + y_1 \cdot \omega j_1 y_2 = -(j_1 y_1) \cdot j y_2 - y_1 \cdot j j_1 y_2,$$

i.e. $j_1^{\#} j = -jj_1$. Hence

$$(jj_1)^{\#} = j_1^{\#} j^{\#} = -j_1^{\#} j = jj_1,$$

i.e. $k = k^{\#}$. Again using that (ω, j_1) is Kähler, we get $-y \cdot jj_1 y = y \cdot \omega j_1 y > 0$, i.e. $k > 0$. This proves (1).

Let us prove (3). The fact that $r \in Sp(\mathcal{Y})$ follows from Prop. 11.11 (1). Using that $r = r^{\#}$ and $r \in Sp(\mathcal{Y})$, we obtain that $j_1 = jr^2 = r^{-1} jr$.

Set $b := \frac{k - \mathbb{1}}{k + \mathbb{1}}$. We check that $jb = -bj$. This implies (4). Then using (11.12) we see that $r_{\mathbb{C}}$ equals (11.11), which is repeated as (11.19). Then we use $k = r^2$ and $j_1 = jk$ to obtain (11.20) and (11.21). $\qquad\square$

Proposition 11.21 *Let $\mathcal{Z}$ and $\mathcal{Z}_1$ be the holomorphic subspaces of $\mathbb{C}\mathcal{Y} \simeq \mathcal{Z} \oplus \overline{\mathcal{Z}}$ for the anti-involutions* j *and* j_1. *Let* c *be as above. Then*

$$\mathcal{Z}_1 = \{(z, -\overline{c}z) \ : \ z \in \mathcal{Z}\},$$

$$\overline{\mathcal{Z}}_1 = \{(-c\overline{z}, \overline{z}) \ : \ z \in \mathcal{Z}\}.$$

Proof Every vector of $\mathcal{Z}_1$ is of the form $(\mathbb{1} - \mathrm{i}j_1)y_1$ for some $y_1 \in \mathcal{Y}$. Since $k > 0$, every vector of $\mathcal{Y}$ is of the form $y_1 = (\mathbb{1} + k)^{-1}y$ for some $y \in \mathcal{Y}$. Now

$$(\mathbb{1} - \mathrm{i}j_1)(\mathbb{1} + k)^{-1}y = (\mathbb{1} - \mathrm{i}jk)(\mathbb{1} + k)^{-1}y$$

$$= \mathbb{1}_{\mathcal{Z}}y - \mathbb{1}_{\overline{\mathcal{Z}}}(\tfrac{k-\mathbb{1}}{k+\mathbb{1}})y$$

$$= z - \overline{c}z,$$

where $z = \mathbb{1}_{\mathcal{Z}}y \in \mathcal{Z}$. Hence every vector of $\mathcal{Z}_1$ is of the form $z - \overline{c}z$ for $z \in \mathcal{Z}$. Applying the canonical conjugation on $\mathbb{C}\mathcal{Y}$ we obtain the corresponding result for $\overline{\mathcal{Z}}_1$. $\qquad\square$

The following proposition will be used to describe the unitary equivalence of two Fock CCR representations (one of the versions of Shale's theorem).

Proposition 11.22 *Let* $\mathrm{j}, \mathrm{j}_1, k, c$ *be as above. The following conditions are equivalent:*

(1) $\mathrm{j} - \mathrm{j}_1 \in B^2(\mathcal{Y})$.
(2) $\mathbb{1} - k \in B^2(\mathcal{Y})$.
(3) $c \in B^2(\overline{\mathcal{Z}}, \mathcal{Z})$.
(4)   *There exists a positive* $r \in Sp_{\mathrm{j,af}}(\mathcal{Y})$ *such that* $\mathrm{j}_1 = r\mathrm{j}r^{-1}$.
(5)   *There exists* $r \in Sp_{\mathrm{j}}(\mathcal{Y})$ *such that* $\mathrm{j}_1 = r\mathrm{j}r^{-1}$.

Proof The identity $-\mathrm{j}(\mathrm{j} - \mathrm{j}_1) = \mathbb{1} - k$ and $\mathrm{j} \in Sp(\mathcal{Y})$ imply the equivalence of (1) and (2).

(11.18) and the boundedness of $(\mathbb{1} + k)^{-1}$ show that (2) implies (3).

Since $c^*c < \mathbb{1}_{\overline{\mathcal{Z}}}$ and $c = c^\#$, we have $cc^* < \mathbb{1}_{\overline{\mathcal{Z}}}$, and hence $(\mathbb{1}_{\overline{\mathcal{Z}}} - c^*c)^{-1}$ and $(\mathbb{1}_{\overline{\mathcal{Z}}} - cc^*)^{-1}$ are bounded. From (11.20) we obtain that (3) implies (2).

(4) $\Rightarrow$ (5) is obvious. (5) $\Rightarrow$ (1) is obvious. (3)$\Rightarrow$(4) follows from (11.19). $\quad\square$

Remark 11.23 *The Hilbert–Schmidt property in conditions (1) and (2) uses the real scalar product on* $\mathcal{Y}$ *that belongs to the Kähler structure* $(\cdot, \omega, \mathrm{j})$. *Therefore, conditions (1) and (2) may not seem symmetric w.r.t. the anti-involutions* j *and* j_1. *Nevertheless, if they are satisfied, then the scalar products* $\cdot$ *and* $\cdot k$ *are related with the operator* k, *which is bounded with a bounded inverse, hence the set of Hilbert–Schmidt operators w.r.t. the scalar products* $\cdot$ *and* $\cdot k$ *coincide.*

11.1.9 Conjugation adapted to a pair of Kähler involutions

Generically, a pair of Kähler anti-involutions determines a conjugation for both Kähler structures, as expressed in the following proposition:

Proposition 11.24 *Suppose that* j_1 *is an anti-involution such that* (ω, j_1) *is Kähler. Then the following is true:*

(1) *There exists* $\tau \in B(\mathcal{Y})$ *such that*

$$\tau^2 = \mathbb{1}, \quad \tau j \tau = -j, \quad \tau j_1 \tau = -j_1.$$

(τ is a conjugation for both j *and* j_1.*)*
(2) *Let* $k = -jj_1$. *Then* $\tau k \tau = k$.
(3) *If* $\mathrm{Ker}(j - j_1) = \{0\}$, *or equivalently* $\mathrm{Ker}(\mathbb{1} - k) = \{0\}$, *then we can take*

$$\tau := \mathbb{1}_{]1,+\infty[}(k) - \mathbb{1}_{]0,1[}(k).$$

Proof Recall that $k^\# = k$, $k > 0$.

Assume first that $\mathrm{Ker}(j - j_1) = \{0\}$, and hence $\mathrm{Ker}(\mathbb{1} - k) = \{0\}$. Set $\tau = \mathbb{1}_{]1,+\infty[}(k) - \mathbb{1}_{]0,1[}(k)$. We have $\tau^2 = \mathbb{1}$ and $\tau \in O(\mathcal{Y})$. Using that $kj = jk^{-1}$, we see that $\tau j = -j\tau$. Since $\tau k = k\tau$, we have also $\tau j_1 = -j_1 \tau$.

If $\mathrm{Ker}(j - j_1) \neq \{0\}$, we set $\mathcal{Y}_1 := \mathbb{1}_{\{1\}}(k)\mathcal{Y}$. The spaces $\mathcal{Y}_1$ and $\mathcal{Y}_0 := \mathcal{Y}_1^\perp$ are invariant under j since $kj = jk^{-1}$. We first construct the conjugation τ_0 on $\mathcal{Y}_0$ as above. On $\mathcal{Y}_1$ we have $j = j_1$. We can choose an arbitrary anti-unitary involution τ_1 of $\mathcal{Y}_1^{\mathbb{C}}$. Then we set $\tau = \tau_1 \oplus \tau_0$ on $\mathcal{Y} = \mathcal{Y}_1 \oplus \mathcal{Y}_0$. $\qquad\square$

Recall that $\mathcal{Y}^{\pm\tau} := \{y \in \mathcal{Y} : \tau y = \pm y\}$. Set $\mathcal{X} = \mathcal{Y}^{-\tau}$. As in Subsect. 1.3.10, we can identify the Kähler space with conjugation $\mathcal{Y}$ with $\mathcal{X} \oplus \mathcal{X}$ by the map

$$\mathcal{Y} \ni y \mapsto \left(j\frac{1}{2\sqrt{2}}(\mathbb{1} + \tau)y, \frac{1}{\sqrt{2}}(\mathbb{1} - \tau)y \right) \in \mathcal{X} \oplus \mathcal{X},$$

which corresponds to the choice $c = \mathbb{1}$ in (1.38). We set $m := k^{-1}\big|_{\mathcal{X}}$, which is a positive self-adjoint operator on the real Hilbert space $\mathcal{X}$. The symplectic form on $\mathcal{Y} \simeq \mathcal{X} \oplus \mathcal{X}$ is

$$(x_1^+, x_1^-) \cdot \omega (x_2^+, x_2^-) = x_1^+ \cdot x_2^- - x_1^- \cdot x_2^+.$$

Proposition 11.25 *We have*

$$\tau = \begin{bmatrix} \mathbb{1} & 0 \\ 0 & -\mathbb{1} \end{bmatrix}, \quad j = \begin{bmatrix} 0 & -\frac{1}{2}\mathbb{1} \\ 2\mathbb{1} & 0 \end{bmatrix}, \quad j_1 = \begin{bmatrix} 0 & -(2m)^{-1} \\ 2m & 0 \end{bmatrix}, \quad k = \begin{bmatrix} m & 0 \\ 0 & m^{-1} \end{bmatrix}.$$

Proof The matrix representation of τ and j on $\mathcal{X} \oplus \mathcal{X}$ was shown in Subsect. 1.3.10. To compute k, we note that if $y \in \mathcal{Y}$ is identified with $(jx_1, x_2) \in \mathcal{X} \oplus \mathcal{X}$, then ky is identified with $(kjx_1, kx_2) = (jmx_1, m^{-1}x_2)$ since $jk = k^{-1}j$. The formula for j_1 follows from $j_1 = jk$. $\qquad\square$

11.2　Bosonic quadratic Hamiltonians on Fock spaces

The basic framework of this section is the same as that of the previous one. Recall, in particular, that $\mathcal{Z}$ is a Hilbert space and $\mathcal{Y} := \mathrm{Re}(\mathcal{Z} \oplus \overline{\mathcal{Z}})$ is the corresponding Kähler space. We consider the Fock CCR representation over $\mathcal{Y}$ in $\Gamma_\mathrm{s}(\mathcal{Z})$.

11.2.1　Wick and anti-Wick quantizations of quadratic polynomials

Let us consider various kinds of complex quadratic polynomials on $\mathcal{Y}$ and their quantizations. We recall that $B^\mathrm{fd}(\mathcal{Z})$ denotes the set of finite-dimensional operators on $\mathcal{Z}$.

Let $h \in B^\mathrm{fd}(\mathcal{Z})$. Consider the polynomial

$$\mathcal{Y}^\# \ni (\overline{z}, z) \mapsto \overline{z}{\cdot}hz. \tag{11.22}$$

The Wick, Weyl–Wigner and anti-Wick quantizations of (11.22) are, respectively,

$$\mathrm{d}\Gamma(h), \qquad \mathrm{d}\Gamma(h) + \frac{\mathrm{Tr}\,h}{2}\mathbb{1}, \qquad \mathrm{d}\Gamma(h) + (\mathrm{Tr}\,h)\mathbb{1}.$$

Note that the anti-Wick and Weyl–Wigner quantizations of (11.22) can be extended to the case $h \in B^1(\mathcal{Z})$, that is, to trace class h. The Wick quantization of (11.22) can be defined, e.g. for $h \in B(\mathcal{Z})$, or even for much more general h.

Suppose that $g \in B^\mathrm{fd}_\mathrm{s}(\overline{\mathcal{Z}}, \mathcal{Z}) \simeq \overset{2}{\overset{\mathrm{s}}{\Gamma}}(\mathcal{Z})$. Consider the polynomial

$$\mathcal{Y}^\# \ni (\overline{z}, z) \mapsto \overline{z}{\cdot}g\overline{z}. \tag{11.23}$$

The Wick, anti-Wick and Weyl–Wigner quantizations of (11.23) coincide with the two-particle creation operator $a^*(g)$, according to the notation of Subsect. 3.4.4. Following the notation of Def. 9.46, this can be written as $\mathrm{Op}^{a^*,a}(|g))$. It can be defined as a closable operator also if $g \in B^2_\mathrm{s}(\overline{\mathcal{Z}}, \mathcal{Z}) \simeq \Gamma^2_\mathrm{s}(\mathcal{Z})$. It will act on $\Psi_n \in \Gamma^n_\mathrm{s}(\mathcal{Z})$ as

$$a^*(g)\Psi_n := \sqrt{(n+2)(n+1)}\, g \otimes_\mathrm{s} \Psi_n. \tag{11.24}$$

(Note that on the right of (11.24) g is treated as an element of $\Gamma^2_\mathrm{s}(\mathcal{Z})$.)

The complex conjugate of (11.23) equals

$$\mathcal{Y}^\# \ni (z, \overline{z}) \mapsto z{\cdot}g^*z = z{\cdot}\overline{g}z. \tag{11.25}$$

Its Wick, anti-Wick and Weyl–Wigner quantizations coincide with the two-particle annihilation operator $a(g)$; see again Subsect. 3.4.4. Following the notation of Def. 9.46, this can be written as $\mathrm{Op}^{a^*,a}((g|)$. It is clear that $a(g)$ extends to a closable operator iff $g \in B^2_\mathrm{s}(\overline{\mathcal{Z}}, \mathcal{Z}) \simeq \Gamma_\mathrm{s}(\mathcal{Z})$, and $a(g)^* = a^*(g)$.

A general element of $\mathbb{C}\mathrm{Pol}^2_\mathrm{s}(\mathcal{Y}^\#)$ is

$$\mathcal{Y}^\# \ni (\overline{z}, z) \mapsto 2\overline{z}{\cdot}hz + \overline{z}{\cdot}g_1\overline{z} + z{\cdot}\overline{g}_2 z, \tag{11.26}$$

where $h \in B^{\mathrm{fd}}(\mathcal{Z})$, $g_1, g_2 \in B^{\mathrm{fd}}_{\mathrm{s}}(\overline{\mathcal{Z}}, \mathcal{Z})$. We can write (11.26) as

$$(\overline{z}, z) \cdot \zeta(\overline{z}, z), \quad \text{where} \quad \zeta_{\mathbb{C}} = \begin{bmatrix} g_1 & h \\ h^{\#} & \overline{g}_2 \end{bmatrix} \in B_{\mathrm{s}}(\overline{\mathcal{Z}} \oplus \mathcal{Z}, \mathcal{Z} \oplus \overline{\mathcal{Z}}). \qquad (11.27)$$

(Recall that $\mathbb{C}\mathcal{Y} \simeq \mathcal{Z} \oplus \overline{\mathcal{Z}}$, $\mathbb{C}\mathcal{Y}^{\#} \simeq \overline{\mathcal{Z}} \oplus \mathcal{Z}$ and we use elements of $B^{\mathrm{fd}}_{\mathrm{s}}(\mathcal{Y}^{\#}, \mathcal{Y})$ for symbols of bosonic quadratic Hamiltonians, as in Def. 10.15.)

The quantizations of ζ are

$$\mathrm{Op}^{a^*,a}(\zeta) = 2\mathrm{d}\Gamma(h) + a^*(g_1) + a(g_2), \qquad (11.28)$$

$$\mathrm{Op}(\zeta) = 2\mathrm{d}\Gamma(h) + (\mathrm{Tr}\, h)\mathbb{1} + a^*(g_1) + a(g_2), \qquad (11.29)$$

$$\mathrm{Op}^{a,a^*}(\zeta) = 2\mathrm{d}\Gamma(h) + (2\mathrm{Tr}\, h)\mathbb{1} + a^*(g_1) + a(g_2).$$

Clearly,

$$\mathrm{Op}(\zeta) = \frac{1}{2}\left(\mathrm{Op}^{a^*,a}(\zeta) + \mathrm{Op}^{a,a^*}(\zeta)\right).$$

We can extend the definition of $\mathrm{Op}(\zeta)$ and $\mathrm{Op}^{a,a^*}(\zeta)$ to the case of $g_1, g_2 \in B^2_{\mathrm{s}}(\overline{\mathcal{Z}}, \mathcal{Z})$ and $h \in B^1(\mathcal{Z})$. $\mathrm{Op}^{a^*,a}(\zeta)$ is defined under much more general conditions. All these quantizations are Hermitian operators iff h is Hermitian and $g_1 = g_2$.

11.2.2 Bosonic Schwinger term

For simplicity, in this subsection we assume that $\mathcal{Z}$ is finite-dimensional. Recall from Thm. 10.13 that the Weyl–Wigner quantization restricted to quadratic symbols yields an isomorphism of the Lie algebra $sp(\mathcal{Y})$ into quadratic Hamiltonians in $\mathrm{CCR}^{\mathrm{pol}}(\mathcal{Y})$. This is no longer true in the case of the Wick quantization, where the so-called Schwinger term appears. This is described in the following proposition:

Proposition 11.26 *Let* $\zeta, \zeta_i \in B_{\mathrm{s}}(\mathcal{Y}^{\#}, \mathcal{Y})$, $i = 1, 2$. *Then*

$$\mathrm{Op}(\zeta) = \mathrm{Op}^{a^*,a}(\zeta) - \frac{1}{2}(\mathrm{Tr}\, \zeta\omega\mathrm{j})\,\mathbb{1}, \qquad (11.30)$$

$$\left[\mathrm{Op}^{a^*,a}(\zeta_1), \mathrm{Op}^{a^*a}(\zeta_2)\right] = 2\mathrm{i}\mathrm{Op}^{a^*,a}(\zeta_2\omega\zeta_1 + \zeta_1\omega\zeta_2) - \mathrm{i}(\mathrm{Tr}\,[\zeta_2\omega, \zeta_1\omega]\mathrm{j})\,\mathbb{1}.$$

Proof Let ζ be as in (11.27). We have $\omega_{\mathbb{C}} = \mathrm{i}\begin{bmatrix} 0 & \mathbb{1} \\ -\mathbb{1} & 0 \end{bmatrix} \in B(\mathcal{Z} \oplus \overline{\mathcal{Z}}, \overline{\mathcal{Z}} \oplus \mathcal{Z})$.
Therefore,

$$\zeta_{\mathbb{C}}\omega_{\mathbb{C}} = \mathrm{i}\begin{bmatrix} -h & g_1 \\ -\overline{g}_2 & h^{\#} \end{bmatrix}, \quad \zeta_{\mathbb{C}}\omega_{\mathbb{C}}\mathrm{j}_{\mathbb{C}} = \begin{bmatrix} h & g_1 \\ \overline{g}_2 & h^{\#} \end{bmatrix}.$$

Therefore, (11.30) follows from (11.29).

Now to compute the Schwinger term we apply Prop. 10.16 (1) and (11.30). □

11.2.3 Infimum of bosonic quadratic Hamiltonians

For simplicity, in this subsection we again assume that $\mathcal{Z}$ is finite-dimensional. The Wick quantization of a positive quadratic symbol is then bounded from below. In the following theorem we compute the infimum of the spectrum of such Hamiltonians.

Theorem 11.27 *Let* $h \in B_{\mathrm{h}}(\mathcal{Z})$, $g \in B_{\mathrm{s}}^2(\overline{\mathcal{Z}}, \mathcal{Z})$. *Suppose that for* $z \in \mathcal{Z}$

$$(\overline{z}, z) \cdot \zeta(\overline{z}, z) = 2\overline{z} \cdot hz + z \cdot g^* z + \overline{z} \cdot g\overline{z} \geq 0. \tag{11.31}$$

Set

$$\zeta_{\mathbb{C}} = \begin{bmatrix} g & h \\ h^{\#} & g^* \end{bmatrix} \in B_{\mathrm{s}}(\overline{\mathcal{Z}} \oplus \mathcal{Z}, \mathcal{Z} \oplus \overline{\mathcal{Z}}). \tag{11.32}$$

Then

$$\inf \mathrm{Op}^{a^*,a}(\zeta) = \frac{1}{2} \mathrm{Tr} \left(\begin{bmatrix} h^2 - gg^* & -hg + gh^{\#} \\ g^*h - h^{\#}g^* & h^{\#2} - g^*g \end{bmatrix}^{\frac{1}{2}} - \begin{bmatrix} h & 0 \\ 0 & h^{\#} \end{bmatrix} \right).$$

Proof We have $(\zeta_{\mathbb{C}}\omega_{\mathbb{C}})^2 = -\begin{bmatrix} h^2 - gg^* & -hg + gh^{\#} \\ g^*h - h^{\#}g^* & h^{\#2} - g^*g \end{bmatrix}$. Thus, by Thm. 10.17,

$\inf \mathrm{Op}^{a^*,a}(\zeta) + \mathrm{Tr}\, h = \inf \mathrm{Op}(\zeta)$

$$= \frac{1}{2} \mathrm{Tr}\, |\zeta\omega| = \frac{1}{2} \mathrm{Tr} \begin{bmatrix} h^2 - gg^* & -hg + gh^{\#} \\ g^*h - h^{\#}g^* & h^{\#2} - g^*g \end{bmatrix}^{\frac{1}{2}}.$$

$\square$

11.2.4 Gaussian vectors

Let $c \in \Gamma_{\mathrm{s}}^2(\mathcal{Z})$. Recall that we can define the two-particle creation operator $a^*(c)$ acting on $\Gamma_{\mathrm{s}}^{\mathrm{fin}}(\mathcal{Z})$ as in Subsect. 3.4.4, and that we can identify c with an operator $c \in B_{\mathrm{s}}^2(\overline{\mathcal{Z}}, \mathcal{Z})$ (see Subsect. 3.3.4). Since $c \in B^2(\overline{\mathcal{Z}}, \mathcal{Z})$, c^*c is trace-class, so $\det(\mathbb{1} - c^*c)$ is well defined. If we assume also that $c^*c < \mathbb{1}$, then $\det(\mathbb{1} - c^*c) > 0$. So we can define

$$\Omega_c := \det(\mathbb{1} - c^*c)^{\frac{1}{4}} \mathrm{e}^{\frac{1}{2}a^*(c)}\Omega. \tag{11.33}$$

Theorem 11.28 (1) *If* $c \in B_{\mathrm{s}}^2(\overline{\mathcal{Z}}, \mathcal{Z})$ *and* $c^*c < \mathbb{1}$, *then* Ω_c *is a normalized vector in* $\Gamma_{\mathrm{s}}(\mathcal{Z})$.

(2) *Let* k *be a positive number with* $k^2|c| < \mathbb{1}$. *Then* Ω_c *belongs to* $\mathrm{Dom}\, k^N$ *and* $k^N\Omega_c = \Omega_{k^2c}$, *where* N *is the number operator.*

(3) *Suppose that* c *is a densely defined operator from* $\overline{\mathcal{Z}}$ *to* $\mathcal{Z}$ *such that* $(z_1|c\overline{z}_2) = (z_2|c\overline{z}_1)$, *i.e.* $c \subset c^{\#}$. *Suppose that there exists* $\Psi \in \Gamma_{\mathrm{s}}(\mathcal{Z})$ *satisfying*

$$\big(a(z) - a^*(c\overline{z})\big)\Psi = 0, \quad \overline{z} \in \mathrm{Dom}\, c,$$

in the weak sense. Then $c \in B_s^2(\bar{\mathcal{Z}}, \mathcal{Z})$ and $c^*c < \mathbb{1}$. Moreover, Ψ is proportional to Ω_c.

(4) *Let $c_1, c_2 \in B_s^2(\bar{\mathcal{Z}}, \mathcal{Z})$ and $c_i^* c_i < \mathbb{1}$. Then*

$$(\Omega_{c_1} | \Omega_{c_2}) = \det(\mathbb{1} - c_1^* c_1)^{\frac{1}{4}} \det(\mathbb{1} - c_2^* c_2)^{\frac{1}{4}} \det(\mathbb{1} - c_1^* c_2)^{-\frac{1}{2}}.$$

Proof First let Ψ be as in (3), and let $z_1, z_2, \ldots$ be a sequence of vectors in Dom $\bar{c}$. For $I \subset \{1, 2, \ldots\}$ finite, set

$$M(I) = \Big(\prod_{i \in I} a^*(z_i) \Omega \big| \Psi \Big).$$

From

$$0 = \Big(\prod_{i \in \{1, \cdots n\}} a^*(z_i) \Omega \big| (a^*(z_{n+1}) - a(c\bar{z}_{n+1})) \Psi \Big)$$

we obtain

$$M(\{1, \cdots, n+1\}) = \sum_{i=1}^{n} (z_i | c\bar{z}_{n+1}) M(\{1, \cdots, n\} \setminus \{i\}).$$

This yields

$$(a^*(z_1) \cdots a^*(z_{2m+1}) \Omega | \Psi) = 0,$$

$$(a^*(z_1) \cdots a^*(z_{2m}) \Omega | \Psi) = \lambda \sum_{\sigma \in \mathrm{Pair}_{2m}} \prod_{i=0}^{m-1} (z_{\sigma(2i+1)} | c\bar{z}_{\sigma(2i+2)}),$$

where $\lambda := (\Omega | \Psi)$ and Pair_{2m} is the set of pairings in $\{1, \ldots, 2m\}$ (see Subsect. 3.6.10).

In particular, for $\bar{z}_1, \bar{z}_2 \in \mathrm{Dom}\, c$ this gives the following formula for the two-particle component of Ψ:

$$\sqrt{2}(z_1 \otimes_s z_2 | \Psi) = \lambda(z_1 | c\bar{z}_2). \qquad (11.34)$$

Since $\Psi \in \Gamma_s(\mathcal{Z})$, the l.h.s. of (11.34) can be extended to a bounded functional on $\Gamma_s^2(\mathcal{Z})$. This implies that either $\lambda = 0$ or $c \in \Gamma_s^2(\mathcal{Z})$, and then the l.h.s. of (11.34) equals $\lambda(z_1 \otimes_s z_2 | c)$.

We have

$$(z_1 \otimes_s \cdots \otimes_s z_{2m} | c^{\otimes_s m}) = \frac{1}{2m!} \sum_{\sigma \in S_{2m}} \prod_{i=0}^{m-1} (z_{\sigma(2i+1)} | c\bar{z}_{\sigma(2i+2)})$$

$$= \frac{m! 2^m}{2m!} \sum_{\sigma \in \mathrm{Pair}_{2m}} \prod_{i=0}^{m-1} (z_{\sigma(2i+1)} | c\bar{z}_{\sigma(2i+2)}),$$

which implies that

$$\Psi_{2m} = \lambda \frac{\sqrt{(2m)!}}{2^m m!} c^{\otimes_s m} = \lambda \frac{1}{2^m m!} (a^*(c))^m \Omega, \qquad \Psi_{2m+1} = 0,$$

i.e. $\Psi = \lambda e^{\frac{1}{2} a^*(c)} \Omega$.

Let us compute $\|e^{\frac{1}{2}a^*(c)}\Omega\|^2$. Since c^*c is trace-class, we can by Corollary 2.88 find an o.n. basis $\{e_i, : i \in I\}$ of $\mathcal{Z}$ such that $c\overline{e_i} = \lambda_i e_i$, $\lambda_i \geq 0$. Using this basis, we can unitarily identify $\mathcal{Z}$ with $\underset{i \in I}{\oplus} \mathbb{C}$. By the exponential law of Subsect. 3.5.4, we unitarily identify $\Gamma_s(\mathcal{Z})$ with $\underset{i \in I}{\otimes} (\Gamma_s(\mathbb{C}), \Omega)$. Under this identification,

$$e^{\frac{1}{2}a^*(c)}\Omega \simeq \underset{i \in I}{\otimes} e^{\frac{1}{2}\lambda_i a^{*2}}\Omega,$$

$$\left\|e^{\frac{1}{2}a^*(c)}\Omega\right\|^2_{\Gamma_s(\mathcal{Z})} = \prod_{i \in I} \left\|e^{\frac{1}{2}\lambda_i a^{*2}}\Omega\right\|^2_{\Gamma_s(\mathbb{C})}$$

$$= \prod_{i \in I} \sum_{m=0}^{\infty} \frac{(2m)!\lambda_i^{2m}}{2^{2m}(m!)^2} \tag{11.35}$$

$$= \prod_{i \in I} \left(1 - \lambda_i^2\right)^{-\frac{1}{2}} = \det\left(\mathbb{1} - c^*c\right)^{-\frac{1}{2}}.$$

This shows that the vector Ω_c in (1) is normalized. Moreover, if $\lambda_i \geq 1$ for some $i \in I$, then one of the series on the r.h.s. of (11.35) is divergent, which contradicts the fact that the vector Ψ is normalizable. This shows the necessity of the condition $c^*c < \mathbb{1}$ and completes the proof of (3).

Let us now prove (2). Since $e^{itN}a^*(c)e^{-itN} = e^{2it}a^*(c)$, we obtain that

$$e^{itN}\Omega_c = \det(\mathbb{1} - c^*c)^{\frac{1}{4}}e^{\frac{1}{2}a^*(c_t)}\Omega,$$

for $c_t = e^{2it}c$. It follows that if $k^4c^*c < \mathbb{1}$, $e^{itN}\Omega_c$ extends holomorphically in $\{z \in \mathbb{C} : \text{Im}\, z > -\log k\}$ and is uniformly bounded on this set. Therefore, $\Omega_c \in \text{Dom}\, k^N$ and $k^N\Omega_c = \det(\mathbb{1} - c^*c)^{1/4}e^{\frac{1}{2}a^*(k^2c)}\Omega$, which proves (2).

It remains to prove (4). Let us first assume that $\mathcal{Z}$ is finite-dimensional. In the complex-wave representation, $e^{\frac{1}{2}a^*(c)}\Omega$ equals $e^{\frac{1}{2}\overline{z}\cdot c\overline{z}}$ and

$$\left(e^{\frac{1}{2}a^*(c_1)}\Omega|e^{\frac{1}{2}a^*(c_2)}\Omega\right) = (2\pi i)^{-d}\int_{\text{Re}(\overline{\mathcal{Z}}\oplus\mathcal{Z})} e^{-|z|^2}e^{\frac{1}{2}z\cdot\overline{c_1}z}e^{\frac{1}{2}\overline{z}\cdot c_2\overline{z}}dzd\overline{z}. \tag{11.36}$$

To compute this integral, we use the arguments in Subsect. 4.1.9. We are led to compute $\det \nu$, where ν is an operator on $\text{Re}(\overline{\mathcal{Z}} \oplus \mathcal{Z})$ given by

$$\nu_{\mathbb{C}} = \det \begin{bmatrix} \mathbb{1} & -\overline{c}_1 \\ -c_2 & \mathbb{1} \end{bmatrix}.$$

But $\det \nu = \det \nu_{\mathbb{C}} = \det(\mathbb{1} - \overline{c}_1 c_2)$. From (4.10), we obtain that (11.36) equals

$$\left(e^{\frac{1}{2}a^*(c_1)}\Omega|e^{\frac{1}{2}a^*(c_2)}\Omega\right) = \det(\mathbb{1} - \overline{c}_1 c_2)^{-\frac{1}{2}}.$$

Let us now prove (4) in the general case. For simplicity, we will assume that $\mathcal{Z}$ is separable (the non-separable case can be treated by the same argument, replacing sequences by nets). Let us fix an increasing sequence of finite rank projections $\pi_1, \pi_2, \ldots$ such that $s - \lim_{n\to\infty} \pi_n = \mathbb{1}$. For $c \in B_s^2(\overline{\mathcal{Z}}, \mathcal{Z})$ we set $c_n = \pi_n c\overline{\pi}_n$, so that $c_n \to c$ in the Hilbert–Schmidt norm. We claim that

$$\lim_{n \to \infty} \Omega_{c_n} = \Omega_c. \tag{11.37}$$

By approximating c_1, c_2 by $c_{1,n}, c_{2,n}$, this implies (4) in the general case.

It remains to prove (11.37). Using (4) in the finite-dimensional case, we get that

$$\|\Omega_{c_n} - \Omega_{c_m}\|^2 = 2 - 2\mathrm{Re}(\Omega_{c_n}|\Omega_{c_m}) \to 0 \text{ when } n, m \to \infty,$$

hence the sequence $\Omega_{c_1}, \Omega_{c_2}, \ldots$ converges to a normalized vector Ψ. There exists $k > 1$ and $k_0 < 1$ such that, for all n, $k^4 c_n^* c_n < k_0^4 \mathbb{1}$. Using (2), we obtain that $\Psi \in \mathrm{Dom}\, N$ and that Ω_{c_n} converges to Ψ in $\mathrm{Dom}\, N$. Therefore, we can let $n \to \infty$ in the identity

$$\big(a(z) - a^*(c_n \bar{z})\big)\Omega_{c_n} = 0$$

to get

$$\big(a(z) - a^*(c\bar{z})\big)\Psi = 0.$$

Since $(\Psi|\Omega) = \lim_{n \to \infty}(\Omega_{c_n}|\Omega) \geq 0$, (3) implies that $\Psi = \Omega_c$. $\qquad\square$

11.2.5 *Gaussian vectors in the real-wave representation*

Let $c \in B_s^2(\overline{\mathcal{Z}}, \mathcal{Z})$ such that $c^* c < \mathbb{1}$. Let k be the positive symplectic transformation defined in terms of c by formula (11.20), that is,

$$k_{\mathbb{C}} = \begin{bmatrix} (\mathbb{1} + cc^*)(\mathbb{1} - cc^*)^{-1} & 2(\mathbb{1} - cc^*)^{-1}c \\ 2c^*(\mathbb{1} - cc^*)^{-1} & (\mathbb{1} + c^*c)(\mathbb{1} - c^*c)^{-1} \end{bmatrix}.$$

As discussed in Subsect. 11.1.9, we can identify $\mathcal{Y}$ with $\mathcal{X} \oplus \mathcal{X}$, where $\mathcal{X}$ is a real Hilbert space, the symplectic form on $\mathcal{Y}$ has the standard form and

$$\mathrm{j} = \begin{bmatrix} 0 & -\frac{1}{2}\mathbb{1} \\ 2\mathbb{1} & 0 \end{bmatrix}, \quad k = \begin{bmatrix} m & 0 \\ 0 & m^{-1} \end{bmatrix},$$

where $m \in B_s(\mathcal{X})$, $m > 0$ and $\mathbb{1} - m \in B^2(\mathcal{X})$.

In Sect. 9.3 we described the unitary map T^{rw} between the Fock space $\Gamma_s(\mathcal{Z})$ and the Gaussian $\mathbf{L}^2$ space $\mathbf{L}^2(\mathcal{X}, \mathrm{e}^{-\frac{1}{2}x^2}\,\mathrm{d}x)$ intertwining the Fock and the real-wave representations such that $T^{\mathrm{rw}}\Omega = 1$.

Proposition 11.20 *In the real wave representation, we have*

$$T^{\mathrm{rw}}\Omega_c(x) = C\mathrm{e}^{\frac{1}{4}x \cdot (\mathbb{1} - m^{-1})x},$$

where C is a "normalizing constant" (see Prop. 5.79). If $\mathbb{1} - m \in B^1(\mathcal{X})$ then

$$T^{\mathrm{rw}}\Omega_c(x) = (\det m)^{-\frac{1}{4}}\mathrm{e}^{\frac{1}{4}x \cdot (\mathbb{1} - m^{-1})x}.$$

Proof Assume first that $\mathcal{Y}$ is finite-dimensional. The proposition can then be proved by a direct computation, using that

$$\mathbf{L}^2(\mathcal{X}, e^{-\frac{1}{2}x^2} dx) = L^2(\mathcal{X}, (2\pi)^{-\frac{d}{2}} e^{-\frac{1}{2}x^2} dx).$$

In the general case, we use the same approximation argument as in the proofs of Thm. 11.28 and of Thm. 5.78, given in Subsect. 11.4.6. □

11.2.6 Two-particle creation and annihilation operators

In this subsection we discuss certain properties of two-particle creation and annihilation operators.

Proposition 11.30 *Let* $c \in \Gamma_s^2(\mathcal{Z}) \simeq B_s^2(\overline{\mathcal{Z}}, \mathcal{Z})$. *Then*

(1) $a^*(c)$ *and* $a(c)$ *with domain* $\Gamma_s^{fin}(\mathcal{Z})$ *are densely defined closable operators.*
(2) $a^*(c) + a(c)$ *is essentially self-adjoint on* $\Gamma_s^{fin}(\mathcal{Z})$.
(3) $e^{\frac{1}{2}a^*(c)}$ *and* $e^{\frac{1}{2}a(c)}$ *are closable on* $\Gamma_s^{fin}(\mathcal{Z})$ *iff* $c^*c < \mathbb{1}$, *and we have*

$$e^{-\frac{1}{2}a^*(c)}a(z) = \left(a(z) + a^*(c\bar{z})\right)e^{-\frac{1}{2}a^*(c)}, \qquad z \in \mathcal{Z}; \qquad (11.38)$$

$$e^{\frac{1}{2}a(c)}a^*(z) = \left(a^*(z) + a(c\bar{z})\right)e^{\frac{1}{2}a(c)}, \qquad z \in \mathcal{Z}. \qquad (11.39)$$

Proof We have $a(c) \subset a^*(c)^*$, $a^*(c) \subset a(c)^*$, which proves (1).

To prove (2) we will use Nelson's commutator theorem, Thm. 2.74 (1), with the comparison operator $N + \mathbb{1}$. By Prop. 9.50 we get that $(N + \mathbb{1})^{-1}a^*(c)$ and $a(c)(N + \mathbb{1})^{-1}$ are bounded. Since

$$Na^*(c) = a^*(c)(N + 2\mathbb{1}), \quad Na(c) = a(c)(N - 2\mathbb{1}),$$

we obtain that $a^*(c)(N + \mathbb{1})^{-1}$ is bounded, so $a^*(c) + a(c)$ is bounded on Dom N. Next, since

$$[N, a^*(c) + a(c)] = 2(a^*(c) - a(c)),$$

we get that $\pm[N, a^*(c) + a(c)] \leq C(N + \mathbb{1})$. Applying Nelson's commutator theorem, we see that $a^*(c) + a(c)$ is essentially self-adjoint on Dom N, hence on $\Gamma_s^{fin}(\mathcal{Z})$.

It remains to prove (3). Clearly, $e^{\frac{1}{2}a(c)}$ is defined on $\Gamma_s^{fin}(\mathcal{Z})$. We note also that $e^{\frac{1}{2}a^*(c)}\Omega \in \Gamma_s(\mathcal{Z})$ iff $c^*c < \mathbb{1}$, by Thm. 11.28. It remains to prove that if $c^*c < \mathbb{1}$, then $e^{\frac{1}{2}a^*(c)}$ is defined on $\Gamma_s^{fin}(\mathcal{Z})$. This is equivalent to showing that $e^{\frac{1}{2}a^*(c)}\Omega \in$ Dom $\prod_{i=1}^{n} a^*(z_i)$ for all $z_1, \ldots, z_n \in \mathcal{Z}$. But this follows from Thm. 11.28 (2). Since $e^{\frac{1}{2}a(c)} \subset (e^{\frac{1}{2}a^*(c)})^*$, $e^{\frac{1}{2}a^*(c)} \subset (e^{\frac{1}{2}a(c)})^*$, we see that $e^{\frac{1}{2}a(c)}$ and $e^{\frac{1}{2}a^*(c)}$ are closable. Identities (11.38) and (11.39) are direct computations. □

11.3 Bosonic Bogoliubov transformations on Fock spaces

We use the same framework as in the previous section. $(\mathcal{Y}, \cdot, \omega, \mathrm{j})$ is a complete Kähler space. $\mathcal{Z}$ is the holomorphic subspace of $\mathbb{C}\mathcal{Y}$, so that we can identify $\mathcal{Y}$ with $\mathrm{Re}(\mathcal{Z} \oplus \overline{\mathcal{Z}})$.

$$\mathcal{Y} \ni y \mapsto W(y) \in U\big(\Gamma_\mathrm{s}(\mathcal{Z})\big) \tag{11.40}$$

is the Fock CCR representation.

The central result of this section is the version of Shale's theorem which says that a symplectic transformation is implementable in the Fock CCR representation iff it belongs to the *restricted symplectic group*. The corresponding automorphism of the algebra of operators will be called the *Bogoliubov automorphism*. Unitary operators implementing Bogoliubov automorphisms (the so-called *Bogoliubov implementers*) form a group $Mp_\mathrm{j}^\mathrm{c}(\mathcal{Y})$, which can be viewed as the natural generalization of the group $Mp^\mathrm{c}(\mathcal{Y})$ to the case of an infinite dimension.

We will also describe the group $Mp_{\mathrm{j},\mathrm{af}}(\mathcal{Y})$, which is a generalization of the group $Mp(\mathcal{Y})$ to infinite dimensions. Note that both $Mp_\mathrm{j}^\mathrm{c}(\mathcal{Y})$ and $Mp_{\mathrm{j},\mathrm{af}}(\mathcal{Y})$ depend on the Kähler structure on $\mathcal{Y}$ (which is expressed by putting j as a subscript).

11.3.1 Symplectic transformations in the Fock representation

Definition 11.31 *We define $Mp_\mathrm{j}^\mathrm{c}(\mathcal{Y})$ to be the set of $U \in U\big(\Gamma_\mathrm{s}(\mathcal{Z})\big)$ such that*

$$\{UW(y)U^* \; : \; y \in \mathcal{Y}\} = \{W(y) \; : \; y \in \mathcal{Y}\}.$$

We equip $Mp_\mathrm{j}^\mathrm{c}(\mathcal{Y})$ with the strong operator topology.

Definition 11.32 *Let $r \in Sp(\mathcal{Y})$.*

(1) *We say that $U \in B\big(\Gamma_\mathrm{s}(\mathcal{Z})\big)$ intertwines r if*

$$UW(y)U^* = W(ry), \quad y \in \mathcal{Y}. \tag{11.41}$$

(2) *If in addition U is unitary, then we say that U implements r.*
(3) *If there exists $U \in U\big(\Gamma_\mathrm{s}(\mathcal{Z})\big)$ that implements r, then we say that r is* implementable on $\Gamma_\mathrm{s}(\mathcal{Z})$.

We will prove:

Theorem 11.33 (Shale's theorem about Bogoliubov transformations) (1) *Let $r \in Sp(\mathcal{Y})$. Then r is implementable iff $r \in Sp_\mathrm{j}(\mathcal{Y})$.*

(2) *Let $U \in Mp_\mathrm{j}^\mathrm{c}(\mathcal{Y})$. Then there exists a unique $r \in Sp_\mathrm{j}(\mathcal{Y})$ such that r is implemented by U. $Mp_\mathrm{j}^\mathrm{c}(\mathcal{Y})$ is a group and the map $Mp_\mathrm{j}^\mathrm{c}(\mathcal{Y}) \to Sp_\mathrm{j}(\mathcal{Y})$ obtained this way is a group homomorphism.*

(3) *Let $r \in Sp_j(\mathcal{Y})$. Let p, d, c be defined as in Subsect. 11.1.3. Define*

$$U_r^j = |\det pp^*|^{-\frac{1}{4}} e^{-\frac{1}{2}a^*(d)} \Gamma\big((p^*)^{-1}\big) e^{\frac{1}{2}a(c)}. \tag{11.42}$$

Then U_r^j is the unique unitary operator implementing r in the Fock representation such that

$$(\Omega|U_r^j\Omega) > 0. \tag{11.43}$$

All operators implementing r in the Fock representation are of the form μU_r, where $|\mu| = 1$.

(4) *We have an exact sequence*

$$1 \to U(1) \to Mp_j^c(\mathcal{Y}) \to Sp_j(\mathcal{Y}) \to 1. \tag{11.44}$$

Proof of Thm. 11.33. Let us prove (3). By Prop. 3.53, we have

$$\Gamma(p^{*-1})a^*(z) = a^*(p^{*-1}z)\Gamma(p^{*-1}), \tag{11.45}$$
$$\Gamma(p^{*-1})a(z) = a(pz)\Gamma(p^{*-1}). \tag{11.46}$$

Set $V := e^{-\frac{1}{2}a^*(d)}\Gamma\big((p^*)^{-1}\big)e^{\frac{1}{2}a(c)}$. Using (11.45), (11.46), (11.38) and (11.39), we see that

$$Va^*(z) = \big(a^*(p^{*-1}z + d\overline{pc}z) + a(pc\overline{z})\big)V = \big(a^*(pz) + a(q\overline{z})\big)V,$$
$$Va(z) = \quad\big(a(pz) + a^*(d\overline{pz})\big)V \quad\quad = \big(a^*(q\overline{z}) + a(pz)\big)V.$$

Therefore,

$$V\phi(y) = \phi(ry)V, \quad y \in \mathcal{Y}.$$

Thus V intertwines the representations (11.40) and (11.41). These representations are irreducible. Hence, by Prop. 8.13, V is proportional to a unitary operator. Clearly,

$$V\Omega = e^{-\frac{1}{2}a^*(d)}\Omega.$$

By Thm. 11.28,

$$\|V\Omega\|^2 = \det(\mathbb{1} - d^*d)^{-\frac{1}{2}} = (\det pp^*)^{\frac{1}{2}}.$$

Hence, $U_r^j = (\det pp^*)^{-\frac{1}{4}}V$ is unitary. $(\Omega|U_r^j\Omega) = (\det pp^*)^{-\frac{1}{4}} > 0$, hence U_r^j also satisfies the condition (11.43). The uniqueness is obvious.

Let $r \in Sp(\mathcal{Y})$. Suppose that $UW(y) = W(ry)U$, $y \in \mathcal{Y}$. Define the operator c as in (11.3). Then the vector $U\Omega$ satisfies the conditions of Thm. 11.28 (3). Hence, $c \in B_s^2(\overline{\mathcal{Z}}, \mathcal{Z})$. By Prop. 11.12, this is equivalent to $r \in Sp_j(\mathcal{Y})$. $\quad\square$

11.3.2 One-parameter groups of Bogoliubov transformations

Let $h \in B_h(\mathcal{Z})$, $g \in B_s^2(\overline{\mathcal{Z}}, \mathcal{Z})$. Let $\zeta \in B_s(\mathcal{Y}^\#, \mathcal{Y})$ be defined by $\zeta_{\mathbb{C}} = \begin{bmatrix} g & h \\ \overline{h} & \overline{g} \end{bmatrix}$. Recall that

$$\mathrm{Op}^{a^*,a}(\zeta) = 2\mathrm{d}\Gamma(h) + a^*(g) + a(g)$$

is a self-adjoint operator. If in addition $h \in B_h^1(\mathcal{Z})$, then we can use the Weyl–Wigner quantization to quantize ζ, obtaining

$$\mathrm{Op}(\zeta) = 2\mathrm{d}\Gamma(h) + a^*(g) + a(g) + (\mathrm{Tr}\, h)\mathbb{1}. \tag{11.47}$$

Let $a \in sp(\mathcal{Y})$ be given by

$$a_{\mathbb{C}} = \zeta_{\mathbb{C}}\omega_{\mathbb{C}} = \mathrm{i} \begin{bmatrix} -h & g \\ -\overline{g} & \overline{h} \end{bmatrix}$$

(see (11.2)). Let $r_t = \mathrm{e}^{ta}$ and

$$r_{t\mathbb{C}} = \begin{bmatrix} p_t & q_t \\ \overline{q}_t & \overline{p}_t \end{bmatrix}.$$

For $t \in \mathbb{R}$ we set

$$d_t := q_t\overline{p}_t^{-1}, \quad c_t := q_t^\#\,(p_t^\#)^{-1}.$$

The following theorem gives the unitary group generated by the Wick and Weyl–Wigner quantizations of ζ:

Theorem 11.34 (1) *For any* $t \in \mathbb{R}$, $p_t\mathrm{e}^{-\mathrm{i}th} - \mathbb{1} \in B^1(\mathcal{Z})$, $d_t, c_t \in B^2(\overline{\mathcal{Z}}, \mathcal{Z})$ *and*

$$\mathrm{e}^{\mathrm{i}t\mathrm{Op}^{a^*,a}(\zeta)} = \left(\det p_t\mathrm{e}^{-\mathrm{i}th}\right)^{-\frac{1}{2}} \mathrm{e}^{-\frac{1}{2}a^*(d_t)}\Gamma(p_t^{*-1})\mathrm{e}^{\frac{1}{2}a(c_t)}. \tag{11.48}$$

Besides, (11.48) implements r_t.

(2) *If in addition* $h \in B_h^1(\mathcal{Z})$, *then* $p_t - \mathbb{1} \in B^1(\mathcal{Z})$ *and*

$$\mathrm{e}^{\mathrm{i}t\mathrm{Op}(\zeta)} = (\det p_t)^{-\frac{1}{2}}\mathrm{e}^{-\frac{1}{2}a^*(d_t)}\Gamma(p_t^{*-1})\mathrm{e}^{\frac{1}{2}a(c_t)}. \tag{11.49}$$

(In both (11.48) and (11.49) the branch of the square root is determined by continuity.)

11.3.3 Implementation of positive symplectic transformations

Let us consider a positive $r \in Sp_j(\mathcal{Y})$. From formula (11.11) we know that there exists $c \in B_s^2(\overline{\mathcal{Z}}, \mathcal{Z})$ such that

$$r_{\mathbb{C}} = \begin{bmatrix} \mathbb{1} & c \\ 0 & \mathbb{1} \end{bmatrix} \begin{bmatrix} (\mathbb{1} - cc^*)^{\frac{1}{2}} & 0 \\ 0 & (\mathbb{1} - c^*c)^{-\frac{1}{2}} \end{bmatrix} \begin{bmatrix} \mathbb{1} & 0 \\ c^* & \mathbb{1} \end{bmatrix}.$$

By Thm. 11.33, r is then implemented by

$$U_r^j = \det(\mathbb{1} - cc^*)^{\frac{1}{4}} e^{-\frac{1}{2}a^*(c)} \Gamma(\mathbb{1} - cc^*)^{\frac{1}{2}} e^{\frac{1}{2}a(c)}. \tag{11.50}$$

We recall also that $a \in sp_j(\mathcal{Y})$ is self-adjoint iff

$$a_{\mathbb{C}} = i \begin{bmatrix} 0 & -g \\ g^* & 0 \end{bmatrix}, \tag{11.51}$$

for $g \in B_s^2(\overline{\mathcal{Z}}, \mathcal{Z})$.

The following formula is essentially a special case of (11.48) for $r_t = e^{ta}$:

$$U_{r_t}^j = e^{\frac{it}{2}\left(a^*(g) + a(g)\right)} \tag{11.52}$$

$$= (\det \cosh(t\sqrt{gg^*}))^{-\frac{1}{2}} e^{\frac{it}{2}a^*\left(\frac{\tanh\sqrt{gg^*}}{\sqrt{gg^*}}g\right)} \Gamma(\cosh(t\sqrt{gg^*}))^{-1} e^{-\frac{it}{2}a\left(\frac{\tanh\sqrt{gg^*}}{\sqrt{gg^*}}g\right)}.$$

Suppose now that τ is a conjugation as in Thm. 11.24. By Prop. 11.25 we can identify $\mathcal{Y}$ with $\mathcal{X} \oplus \mathcal{X}$, so that for $m \in L_s(\mathcal{X})$, $m > 0$,

$$r = \begin{bmatrix} m & 0 \\ 0 & m^{-1} \end{bmatrix}.$$

Recall also that we defined the unitary map T^{rw} between the Fock space $\Gamma_s(\mathcal{Z})$ and Gaussian $\mathbf{L}^2$ space $\mathbf{L}^2(\mathcal{X}, e^{-\frac{1}{2}x^2}\,dx)$ intertwining the Fock and real-wave representations such that $T^{\text{rw}}\Omega = 1$.

Proposition 11.35 *In the real-wave representation on* $\mathbf{L}^2(\mathcal{X}, e^{-\frac{1}{2}x^2}\,dx)$ *the operator* U_r^j *takes the form*

$$T^{\text{rw}} U_r^j T^{\text{rw}*} F(x) = (\det m)^{\frac{1}{2}} e^{\frac{1}{4}x\cdot(\mathbb{1}-m^2)x} F(mx).$$

11.3.4 Metaplectic group in the Fock representation

$Sp_j(\mathcal{Y}) \ni r \mapsto U_r^j$ is not a representation – it is only a projective representation. To construct a true representation we need to restrict ourselves to $Sp_{j,\text{af}}(\mathcal{Y})$. Thus we will obtain a generalization of the metaplectic representation to infinite dimensions.

Definition 11.36 *For* $r \in Sp_{j,\text{af}}(\mathcal{Y})$ *we define*

$$\pm U_r := \pm (\det p^*)^{-\frac{1}{2}} e^{-\frac{1}{2}a^*(d)} \Gamma((p^*)^{-1}) e^{\frac{1}{2}a(c)}. \tag{11.53}$$

(We take both signs of the square root, thus $\pm U_r$ *denotes a pair of operators differing by the sign.)*

Definition 11.37 *We denote by* $Mp_{j,\text{af}}(\mathcal{Y})$ *the set of operators of the form* $\pm U_r$ *for some* $r \in Sp_{j,\text{af}}(\mathcal{Y})$. *We equip it with the strong operator topology.*

Theorem 11.38 $Mp_{j,\text{af}}(\mathcal{Y})$ *is a topological group. We have the exact sequence*

$$1 \to \mathbb{Z}_2 \to Mp_{j,\text{af}}(\mathcal{Y}) \to Sp_{j,\text{af}}(\mathcal{Y}) \to 1. \tag{11.54}$$

If $\mathcal{Y}$ is finite-dimensional, then $Mp_{j,af}(\mathcal{Y})$ coincides with $Mp(\mathcal{Y})$ introduced in Def. 10.28.

The proof of this theorem is based on the following lemma:

Lemma 11.39 (1) $\pm U_r$ *are unitary.*
(2) $U_r W(y) U_r^* = W(ry)$.
(3) $U_{r_1} U_{r_2} = \pm U_{r_1 r_2}$.
(4) *If $\mathcal{Y}$ is finite-dimensional, then $\pm U_r$ coincides with $\pm U_r$ introduced in Def. 10.27.*

Proof The operators U_r differ by a phase factor from U_r^j from Thm. 11.33. Therefore, they are unitary and implement r. This proves (1) and (2).

Let us prove (3). We know that

$$(\Omega|U_{r_1 r_2}\Omega) = \pm(\det p^*)^{-\frac{1}{2}} = \pm\big(\det(p_1 p_2 + q_1 \overline{q}_2)^*\big)^{-\frac{1}{2}}. \tag{11.55}$$

Moreover,

$$(\Omega|U_{r_1} U_{r_2}\Omega) = \pm\big(e^{\frac{1}{2}a^*(c_1)}\Omega|e^{-\frac{1}{2}a^*(d_2)}\Omega\big)(\det p_1^*)^{-\frac{1}{2}}(\det p_2^*)^{-\frac{1}{2}}$$

$$= \pm\det(\mathbb{1} + d_2 c_1^*)^{-\frac{1}{2}}(\det p_1^*)^{-\frac{1}{2}}(\det p_2^*)^{-\frac{1}{2}},$$

using Thm. 11.28 (4) and the fact that c_1, d_2 are symmetric. Using the formulas in Subsect. 11.1.3, we see that

$$(p_1 p_2 + q_1 \overline{q}_2)^* = p_2^*(\mathbb{1} + d_2 c_1^*)p_1^*,$$

which implies that

$$(\Omega|U_{r_1} U_{r_2}\Omega) = \pm(\Omega|U_{r_1 r_2}\Omega). \tag{11.56}$$

We know that $U_{r_1 r_2}$ and $U_{r_1} U_{r_2}$ differ by a phase factor. This phase factor must be equal to ± 1 by (11.56), which completes the proof of (3). $\qquad\square$

The following theorem gives an alternative definition of the group $Mp_{j,af}(\mathcal{Y})$:

Theorem 11.40 $Mp_{j,af}(\mathcal{Y})$ *is the subgroup of $U\big(\Gamma_s(\mathcal{Z})\big)$ generated by $e^{\mathrm{iOp}(\zeta)}$, where $\mathrm{Op}(\zeta)$ are defined as in (11.47) with $g \in B_s^2(\overline{\mathcal{Z}}, \mathcal{Z})$ and $h \in B_h^1(\mathcal{Z})$.*

11.4 Fock sector of a CCR representation

The main result of this section is a necessary and sufficient criterion for two Fock CCR representations to be unitarily equivalent. This result goes under the name *Shale's theorem* and is closely related to Thm. 11.33 about the implementability of bosonic Bogoliubov transformations, which we also call Shale's theorem.

Another, closely related, subject of this chapter can be described as follows. Consider a symplectic space $\mathcal{Y}$ and a CCR representation in a Hilbert space $\mathcal{H}$. Suppose that we are given a Kähler anti-involution j. We will describe how to find a subspace of $\mathcal{H}$ on which this representation is unitarily equivalent to a multiple of the Fock representation associated with j.

The basic framework of this section is slightly different from that of the preceding sections of this chapter. As previously, in this section $(\mathcal{Y}, \omega)$ is a symplectic space and $j \in L(\mathcal{Y})$ is a Kähler anti-involution. We do not, however, assume that $\mathcal{Y}$ is complete.

The notation and terminology of this section is based on Subsects. 1.3.6 and 1.3.8, 1.3.9, and was also recalled at the beginning of Chap. 9. Recall, in particular, that $y_1 \cdot y_2 := -y_1 \cdot \omega j y_2$ is the corresponding symmetric form, so that $(\mathcal{Y}, \cdot, \omega, j)$ is a Kähler space (not necessarily complete). $\mathcal{Z} := \frac{1 - ijc}{2} \mathbb{C}\mathcal{Y}$ is the corresponding holomorphic space. We have identifications $\mathbb{C}\mathcal{Y} \simeq \mathcal{Z} \oplus \overline{\mathcal{Z}}$ and

$$\mathcal{Y} \ni y \mapsto \left(\frac{1}{2}(y - ijy), \frac{1}{2}(y + ijy) \right) \in \mathrm{Re}(\mathcal{Z} \oplus \overline{\mathcal{Z}}). \tag{11.57}$$

We recall also that $\mathcal{Z}$ inherits a unitary structure. If $(z_i, \overline{z_i}) = y_i$, then

$$y_1 \cdot y_2 = 2\mathrm{Re}(z_1 | z_2), \tag{11.58}$$
$$y_1 \cdot \omega y_2 = 2\mathrm{Im}(z_1 | z_2).$$

Recall that the completion of $\mathcal{Y}$ is denoted $\mathcal{Y}^{\mathrm{cpl}}$. Clearly, $\mathcal{Z}^{\mathrm{cpl}}$ is the holomorphic subspace of the complete Kähler space $\mathcal{Y}^{\mathrm{cpl}}$.

11.4.1 Vacua of CCR representations

Suppose that

$$\mathcal{Y} \ni y \mapsto W^\pi(y) = e^{i\phi^\pi(y)} \in U\big(\Gamma_{\mathrm{s}}(\mathcal{Z})\big) \tag{11.59}$$

is a regular CCR representation. As in Subsect. 8.2.4, we introduce the creation, resp. annihilation operators $a^{\pi*}(z)$, resp. $a^\pi(z)$ by

$$a^{\pi*}(z) := \phi^\pi(z), \quad a^\pi(z) := \phi^\pi(\overline{z}), \quad z \in \mathcal{Z}. \tag{11.60}$$

Note that these operators depend not only on the representation π, but also on the Kähler anti-involution j, so in some situations it is natural to call them j-*creation*, resp. j-*annihilation operators*.

Definition 11.41 *The space of j-vacua is defined as*

$$\mathcal{K}^\pi := \big\{ \Psi \in \mathcal{H} \ : \ a^\pi(z)\Psi = 0, \ z \in \mathcal{Z} \big\}.$$

Proposition 11.42 (1) $\mathcal{K}^\pi$ *is a closed subspace of* $\mathcal{H}$.
(2) *Let* $\Psi \in \mathcal{H}$. *Then*

$$\Psi \notin \mathcal{K}^\pi \quad \Leftrightarrow \quad \big(\Psi | W^\pi(y)\Psi\big) = \|\Psi\|^2 e^{-\frac{1}{4}\|y\|^2}, \ y \in \mathcal{Y}.$$

(3) *Elements of $\mathcal{K}^\pi$ are analytic vectors of $\phi^\pi(y)$, $y \in \mathcal{Y}$.*

(4) *If $\Phi, \Psi \in \mathcal{K}^\pi$, then*

$$(\Phi|W^\pi(y)\Psi) = (\Phi|\Psi)e^{-\frac{1}{4}\|y\|^2}, \qquad y \in \mathcal{Y};$$

$$(\Phi|\phi^\pi(y_1)\phi^\pi(y_2)\Psi) = \frac{1}{2}(\Phi|\Psi)(y_1 \cdot y_2 + iy_1 \cdot \omega y_2), \quad y_1, y_2 \in \mathcal{Y}.$$

Proof Let us suppress the superscript π to simplify notation.

(1). The space of vacua $\mathcal{K}$ is closed as an intersection of kernels of closed operators.

Let us prove (2) $\Leftarrow$. Let $\Psi \in \mathcal{H}$ such that $(\Psi|W(y)\Psi) = \|\Psi\|^2 e^{-\frac{1}{4}\|y\|^2}$. Without loss of generality we can assume that $\|\Psi\| = 1$. Taking the first two terms of the Taylor expansion of

$$\mathbb{R} \ni t \mapsto (\Psi|W(ty)\Psi) = \|\Psi\|^2 e^{-\frac{1}{4}t^2\|y\|^2},$$

we obtain

$$(\Psi|\phi(y)\Psi) = 0, \qquad (\Psi|\phi(y)^2\Psi) = \frac{1}{2}\|y\|^2. \tag{11.61}$$

In particular, $\Psi \in \mathrm{Dom}\,\phi(y)$, $y \in \mathcal{Y}$. Let $z = y - \mathrm{i}jy \in \mathcal{Z}$. Then

$$a^*(z)a(z) = \phi(z)\phi(\bar{z})$$
$$= (\phi(y) - \mathrm{i}\phi(y))(\phi(y) + \mathrm{i}\phi(jy)) = \phi(y)^2 + \phi(jy)^2 - \|y\|^2.$$

Hence,

$$\|a(z)\Psi\|^2 = (\Psi|\phi(y)^2\Psi) + (\Psi|\phi(jy)^2\Psi) - \|y\|^2 = 0.$$

To prove (2) $\Rightarrow$, note that if $\Psi \in \mathcal{K}$, then $\Psi \in \mathrm{Dom}\,\phi(y)$, $y \in \mathcal{Y}$. In particular, the function

$$\mathbb{R} \ni t \mapsto F(t) = (\Psi|W(ty)\Psi)$$

is C^1. Let $y \in \mathcal{Y}$, $z = \frac{1}{2}(y - \mathrm{i}jy) \in \mathcal{Z}$. Using Thm. 8.25 (1), we get

$$\phi(y)W(ty) = a^*(z)W(ty) + W(ty)a(z) + \frac{\mathrm{i}t}{2}\|y\|^2 W(ty). \tag{11.62}$$

This yields

$$\frac{\mathrm{d}}{\mathrm{d}t}F(t) = \mathrm{i}(\Psi|\phi(y)W(ty)\Psi)$$

$$= \frac{\mathrm{i}}{2}(a(z)\Psi|W(ty)\Psi) + \frac{\mathrm{i}}{2}(\Psi|W(ty)a(z)\Psi) - \frac{t}{2}\|y\|^2(\Psi, W(ty)\Psi)$$

$$= -\frac{t}{2}\|y\|^2 F(t).$$

Since $F(0) = \|\Psi\|^2$, we get that $F(t) = \|\Psi\|^2 e^{-\frac{1}{4}\|y\|^2}$.

From (2) we know that $F(t)$ is analytic, hence by the spectral theorem $\Psi \in \mathcal{K}$ is an analytic vector for $\phi(y)$, $y \in \mathcal{Y}$, which proves (3). Finally, (4) follows from (11.61) by polarization. $\qquad\square$

11.4.2 Fock CCR representations

As in Sect. 3.4, for $z \in \mathcal{Z}^{\mathrm{cpl}}$ we introduce creation, resp. annihilation operators $a^*(z)$, resp. $a(z)$ acting on the bosonic Fock space $\Gamma_{\mathrm{s}}(\mathcal{Z}^{\mathrm{cpl}})$.

Definition 11.43 *The regular CCR representation*

$$\mathcal{Y} \ni y \mapsto W(y) = \mathrm{e}^{\mathrm{i} a^*(z) + \mathrm{i} a(z)} \in U\big(\Gamma_{\mathrm{s}}(\mathcal{Z}^{\mathrm{cpl}})\big), \ y = (z, \overline{z}), \tag{11.63}$$

is called the Fock representation over the Kähler space $\mathcal{Y}$.

This is a slight generalization of the definition used in Subsect. 9.1.1, since we allow for a non-complete space $\mathcal{Y}$. Clearly, the representation (11.63) can be extended to a representation over $\mathcal{Y}^{\mathrm{cpl}}$ in an obvious way.

Note that j-creation, resp. j-annihilation operators defined for the CCR representation (11.63) coincide with the usual creation, resp. annihilation operators $a^*(z)$, resp. $a(z)$. Likewise, a vector $\Psi \in \Gamma_{\mathrm{s}}(\mathcal{Z}^{\mathrm{cpl}})$ is a j-vacuum for (11.63) iff it is proportional to Ω.

We can also consider another Kähler anti-involution j_1, not necessarily equal to j. The following theorem describes the vacua inside $\Gamma_{\mathrm{s}}(\mathcal{Z}^{\mathrm{cpl}})$ corresponding to j_1. It is essentially a restatement of parts of Thm. 11.28.

Theorem 11.44 (1) *Let* $c \in B^2_{\mathrm{s}}(\overline{\mathcal{Z}}^{\mathrm{cpl}}, \mathcal{Z}^{\mathrm{cpl}})$, $cc^* < \mathbb{1}$, *and let* j_1 *be the Kähler anti-involution determined by* c *as in Subsect. 11.1.8. Then* Ω_c *is the unique vector in* $\Gamma_{\mathrm{s}}(\mathcal{Z}^{\mathrm{cpl}})$ *satisfying the following conditions:*

(i) $\|\Omega_c\| = 1$,

(ii) $(\Omega_c | \Omega) > 0$,

(iii) Ω_c *is a* j_1-*vacuum.*

(2) *The statement (1)(iii) is equivalent to*

$$\big(a(z) - a^*(c\overline{z})\big)\Omega_c = 0, \quad z \in \mathcal{Z}. \tag{11.64}$$

11.4.3 Unitary equivalence of Fock CCR representations

Suppose that we are given a symplectic space $(\mathcal{Y}, \omega)$ endowed with two Kähler structures, defined e.g. by two Kähler anti-involutions. Each Kähler structure determines a Fock representation. In this subsection we will prove a necessary and sufficient condition for the equivalence of these two representations.

Theorem 11.45 (Shale's theorem about Fock representations) *Let* $\mathcal{Z}$, $\mathcal{Z}_1$ *be the holomorphic subspaces of* $\mathbb{C}\mathcal{Y}$ *corresponding to Kähler anti-involutions* j *and* j_1. *Let*

$$\mathcal{Y} \ni y \mapsto \mathrm{e}^{\mathrm{i}\phi(y)} \in U\big(\Gamma_{\mathrm{s}}(\mathcal{Z})\big), \tag{11.65}$$
$$\mathcal{Y} \ni y \mapsto \mathrm{e}^{\mathrm{i}\phi_1(y)} \in U\big(\Gamma_{\mathrm{s}}(\mathcal{Z}_1)\big) \tag{11.66}$$

be the corresponding Fock CCR representations. Then the following statements are equivalent:

(1) *There exists a unitary operator* $W : \Gamma_s(\mathcal{Z}) \to \Gamma_s(\mathcal{Z}_1)$ *such that*

$$W\phi(y) = \phi_1(y)W. \tag{11.67}$$

(2) $j - j_1$ *is Hilbert–Schmidt (or any of the equivalent conditions of Prop. 11.22 hold).*

Proof Let a_1^*, a_1, Ω_1 denote the creation and annihilation operators and the vacuum for the representation (11.66).

$(2) \Rightarrow (1)$. Assume that $j - j_1 \in B^2(\mathcal{Y})$. We know by Prop. 11.22 that there exists $r \in Sp_j(\mathcal{Y})$ such that $j_1 = rjr^\#$. Thus, by Thm. 11.33 there exists $U_r \in U(\Gamma_s(\mathcal{Z}))$ such that $U_r\phi(y)U_r^* = \phi(ry)$.

Note that $r_{\mathbb{C}}$ is a unitary operator on $\mathbb{C}\mathcal{Y}$ and $r_{\mathbb{C}}\mathcal{Z} = \mathcal{Z}_1$. Set $u := r_{\mathbb{C}}\big|_{\mathcal{Z}}$. Then $u \in U(\mathcal{Z}, \mathcal{Z}_1)$, hence $\Gamma(u) \in U(\Gamma_s(\mathcal{Z}), \Gamma_s(\mathcal{Z}_1))$ and

$$\Gamma(u)a^*(z)\Gamma(u)^* = a_1^*(uz), \quad \Gamma(u)a(z)\Gamma(u)^* = a_1(uz), \quad z \in \mathcal{Z}.$$

Consequently, $\Gamma(u)\phi(y)\Gamma(u)^* = \phi_1(ry)$. Therefore, $W := \Gamma(u)U_r^*$ satisfies (11.67).

$(1) \Rightarrow (2)$. Suppose that the representations (11.65) and (11.66) are equivalent with the help of the operator $W \in U(\Gamma_s(\mathcal{Z}_1), \Gamma_s(\mathcal{Z}))$. Then $\Psi := W\Omega_1$ satisfies

$$\big(a(z) - a^*(c\bar{z})\big)\Psi = 0, \quad z \in \mathcal{Z}.$$

By Thm. 11.28, this implies that $c \in B^2(\bar{\mathcal{Z}}, \mathcal{Z})$. Hence, $j - j_1 \in B^2(\mathcal{Y})$. $\qquad\square$

11.4.4 Fock sector of CCR representations

Let us go back to an arbitrary regular CCR representation (11.59) over a Kähler space $\mathcal{Y}$. We will describe how to determine the largest sub-representation of (11.59) unitarily equivalent to a multiple of the j-Fock representation over $\mathcal{Y}$ in $\Gamma_s(\mathcal{Z}^{cpl})$.

Theorem 11.46 *Set*

$$\mathcal{H}^\pi := \text{Span}^{cl}\{W^\pi(y)\Psi : \Psi \in \mathcal{K}^\pi, \ y \in \mathcal{Y}\}. \tag{11.68}$$

(1) $\mathcal{H}^\pi$ *is invariant under* $W^\pi(y)$, $y \in \mathcal{Y}$.
(2) *There exists a unique unitary operator*

$$U^\pi : \mathcal{K}^\pi \otimes \Gamma_s(\mathcal{Z}^{cpl}) \to \mathcal{H}^\pi$$

satisfying

$$U^\pi \ \Psi \otimes W(y)\Omega = W^\pi(y)\Psi, \quad \Psi \in \mathcal{K}^\pi, \ y \in \mathcal{Y},$$

where $W(y)$ *denote Weyl operators in the Fock representation on* $\Gamma_s(\mathcal{Z}^{cpl})$.

(3)
$$U^{\pi}\ \mathbb{1}\otimes W(y) = W^{\pi}(y)U^{\pi}, \quad y \in \mathcal{Y}. \tag{11.69}$$

(4) *If there exists an operator* $U : \Gamma_{\mathrm{s}}(\mathcal{Z}^{\mathrm{cpl}}) \to \mathcal{H}$ *such that* $UW(y) = W^{\pi}(y)U$, $y \in \mathcal{Y}$, *then* $\mathrm{Ran}\,U \subset \mathcal{H}^{\pi}$.

(5) $\mathcal{H}^{\pi}$ *depends on* j *only through the equivalence class of* j *w.r.t. the relation*
$$\mathrm{j}_1 \sim \mathrm{j}_2 \ \Leftrightarrow \ \mathrm{j}_1 - \mathrm{j}_2 \in B^2(\mathcal{Y}). \tag{11.70}$$

Definition 11.47 *Introduce the equivalence relation (11.70) in the set of Kähler anti-involutions on* $\mathcal{Y}$. *Let* $[\mathrm{j}]$ *denote the equivalence class w.r.t. this relation. Then the subspace* $\mathcal{H}^{\pi}$ *defined in (11.68) is called the* $[\mathrm{j}]$-*Fock sector of a CCR representation* W^{π}.

Proof of Thm. 11.46. Clearly, $\mathcal{H}^{\pi}$ is invariant under $W^{\pi}(y)$, $y \in \mathcal{Y}$. Now let $\Psi_i \in \mathcal{K}^{\pi}$, $y_i \in \mathcal{Y}$ for $i = 1, 2$. We have
$$\left(W^{\pi}(y_1)\Psi_1 | W^{\pi}(y_2)\Psi_2\right) = \mathrm{e}^{\frac{\mathrm{i}}{2}y_1\cdot\omega y_2}\left(\Psi_1 | W^{\pi}(y_2 - y_1)\Psi_2\right)$$
$$= (\Psi_1|\Psi_2)\mathrm{e}^{\frac{\mathrm{i}}{2}y_1\cdot\omega y_2}\mathrm{e}^{-\frac{1}{4}(y_2-y_1)\cdot(y_2-y_1)}.$$

Similarly, if $(z_i, \overline{z_i}) = y_i$, we have
$$\left(\Psi_1 \otimes W(y_1)\Omega | \Psi_2 \otimes W(y_2)\Omega\right) = (\Psi_1|\Psi_2)\mathrm{e}^{\frac{1}{2}(z_1,\overline{z_1})\cdot\omega(z_2,\overline{z_2})}\mathrm{e}^{-\frac{1}{2}\overline{(z_2-z_1)}\cdot(z_2-z_1)},$$

using (9.13). Using (11.58), we obtain that
$$\left(W^{\pi}(y_1)\Psi_1 | W^{\pi}(y_2)\Psi_2\right) = \left(\Psi_1 \otimes W(y_1)\Omega | \Psi_2 \otimes W(y_2)\Omega\right). \tag{11.71}$$

Let us set
$$U^{\pi}\Psi \otimes W(y)\Omega := W^{\pi}(y)\Psi$$

and extend U^{π} to $\mathcal{K}^{\pi} \otimes \overset{\mathrm{al}}{\Gamma}_{\mathrm{s}}(\mathcal{Z}^{\mathrm{cpl}})$ by linearity. By the identity (11.71), we see that U^{π} is well defined and isometric. It extends as a unitary operator between $\mathcal{K}^{\pi} \otimes \Gamma_{\mathrm{s}}(\mathcal{Z}^{\mathrm{cpl}})$ and $\mathcal{H}^{\pi}$. Property (3) follows from the definition of U^{π}.

Finally, let $U : \Gamma_{\mathrm{s}}(\mathcal{Z}^{\mathrm{cpl}}) \to \mathcal{H}$ be an operator such that $UW(y) = W^{\pi}(y)U$. Then, by the argument leading to (11.71), we see that $U\mathbb{C}\Omega \subset \mathcal{K}^{\pi}$. Therefore, $\mathrm{Ran}\,U \subset \mathrm{Ran}\,U^{\pi}$.

Let us prove (5). For $i = 1, 2$ denote by $\mathcal{H}_i$, $\mathcal{K}_i$ the spaces $\mathcal{H}^{\pi_i}$, $\mathcal{K}^{\pi_i}$ corresponding to the anti-involution j_i. It suffices to prove that $\mathcal{H}_2 \subset \mathcal{H}_1$. We first claim that
$$\mathcal{K}_2 \cap \mathcal{H}_1 \neq \{0\}. \tag{11.72}$$

In fact, by Thm. 11.46 (1) we know that $\mathcal{H}_1$ is invariant under $W(y)$, $y \in \mathcal{Y}$. Besides, $y \mapsto W(y)\big|_{\mathcal{H}_1}$ is unitarily equivalent to a multiple of the Fock representation on $\Gamma_{\mathrm{s}}(\mathcal{Z}_1)$. Since $\mathrm{j}_1 - \mathrm{j}_2 \in B^2(\mathcal{Y})$, it follows from Thm. 11.45 that $y \mapsto W(y)\big|_{\mathcal{H}_1}$ contains vacua for j_2, hence (11.72) holds.

We claim now that $\mathcal{K}_2 \subset \mathcal{H}_1$, which implies that $\mathcal{H}_2 \subset \mathcal{H}_1$. If $\mathcal{K}_2 \not\subset \mathcal{H}_1$, then $\mathcal{K}_2 \cap \mathcal{H}_1^\perp$ is the non-trivial space of j_2-vacua for $y \mapsto W(y)|_{\mathcal{H}_1^\perp}$.

Applying the analog of (11.72) to $\mathcal{H}_1^\perp$, we see that $\mathcal{H}_1^\perp$ should contain vacua for j_1, which is absurd. Hence, $\mathcal{K}_2 \subset \mathcal{H}_1$, which completes the proof. $\qquad\square$

Proposition 11.48 *If the CCR representation (11.59) is irreducible and $\mathcal{K}^\pi \neq \{0\}$, then it is unitarily equivalent to the [j]-Fock representation.*

Proof Since (11.59) is irreducible, we have $\mathcal{H}^\pi = \{0\}$ or $\mathcal{H}^\pi = \mathcal{H}$, which proves the proposition. $\qquad\square$

11.4.5 Number operator of regular CCR representations

In this subsection we consider an arbitrary regular CCR representation (11.59) over a Kähler space $\mathcal{Y}$ with a Kähler anti-involution j. We now discuss the notion of the *number operator* N^π associated with (11.59). The number operator N^π allows for a direct description of the Fock sector $\mathcal{H}^\pi$. In some cases this description can be used to show that $\mathcal{H}^\pi = \mathcal{H}$. This is in particular the case for a finite-dimensional $\mathcal{Y}$, when Thm. 11.52 gives an alternative proof of the Stone–von Neumann theorem (Thm. 8.49).

Here is the first definition of N^π.

Definition 11.49 *Let N be the usual number operator on the bosonic Fock space. Let U^π be defined as in Thm. 11.46. Define* $\operatorname{Dom} N^\pi := U^\pi \mathcal{K}^\pi \otimes \operatorname{Dom} N$, *which is a dense subspace of $\mathcal{H}^\pi$. The number operator in the representation π is the operator on $\mathcal{H}$ with the domain $\operatorname{Dom} N^\pi$ defined by*

$$N^\pi := U^\pi(\mathbb{1} \otimes N)U^{\pi *}.$$

(Note that N^π need not be densely defined.)

Before we give an alternative definition of N^π, let us recall some facts about quadratic forms. We will assume that a positive quadratic form is defined on the whole space $\mathcal{H}$ and takes values in $[0, \infty]$. The domain of a positive quadratic form b is defined as

$$\operatorname{Dom} b := \{\Phi \in \mathcal{H} \ : \ b(\Phi) < \infty\}.$$

If the form b is closed, then there exists a unique positive self-adjoint operator B such that

$$\operatorname{Dom} b = \operatorname{Dom} B^{\frac{1}{2}}, \quad b(\Phi) = (\Phi | B\Phi).$$

If A is a closed operator, then $\|A\Phi\|^2$ is a closed form. The sum of closed forms is a closed form, and the supremum of a family of closed forms is a closed form.

Definition 11.50 *For each finite-dimensional subspace $\mathcal{V} \subset \mathcal{Z}$, set*

$$n_{\mathcal{V}}^{\pi}(\Phi) := \sum_{i=1}^{\dim \mathcal{V}} \| a^{\pi}(v_i)\Phi \|^2,$$

where $\{v_i\}_{i=1}^{\dim \mathcal{V}}$ is an o.n. basis of $\mathcal{V}$. (If $\Phi \notin \mathrm{Dom}\, a^{\pi}(v_i)$ for some i, set $n_{\mathcal{V}}^{\pi}(\Phi) = \infty$.)

The quadratic form $n_{\mathcal{V}}^{\pi}$ does not depend on the choice of the basis $\{v_i\}_{i=1}^{\dim \mathcal{V}}$ of $\mathcal{V}$. Moreover, by Thm. 8.29, $n_{\mathcal{V}}^{\pi}$ is densely defined.

Definition 11.51 *The number quadratic form n^{π} is defined by*

$$n^{\pi}(\Phi) := \sup_{\mathcal{V}} n_{\mathcal{V}}^{\pi}(\Phi), \quad \Phi \in \mathcal{H}.$$

The following theorem says that the number quadratic form of Def. 11.51 gives the number operator introduced in Def. 11.49.

Theorem 11.52 *Let n^{π} be the number quadratic form associated with W^{π} and j. Then $\mathrm{Dom}\, n^{\pi} = \mathrm{Dom}(N^{\pi})^{\frac{1}{2}}$ and*

$$n^{\pi}(\Phi) = (\Phi | N^{\pi}\Phi), \quad \Phi \in \mathrm{Dom}\, N^{\pi}.$$

In particular, $\mathcal{H}^{\pi} = (\mathrm{Dom}\, n^{\pi})^{\mathrm{cl}}$.

To prepare for the proof of the above theorem, note that n^{π} defines a positive operator (with a possibly non-dense domain), which we temporarily denote $\tilde{N}^{\pi}$, such that $\mathrm{Dom}\, n^{\pi} = \mathrm{Dom}(\tilde{N}^{\pi})^{\frac{1}{2}}$ and

$$n^{\pi}(\Phi) = (\Phi | \tilde{N}^{\pi}\Phi), \quad \Phi \in \mathrm{Dom}\, \tilde{N}^{\pi}. \tag{11.73}$$

Our aim is to show that $\tilde{N}^{\pi} = N^{\pi}$.

Note also that

$$\mathrm{Dom}\, n^{\pi} \subset \mathrm{Dom}\, \phi^{\pi}(y), \quad y \in \mathcal{Y}. \tag{11.74}$$

Lemma 11.53 *If $\Phi \in \mathrm{Dom}(\tilde{N}^{\pi})^{\frac{1}{2}}$ and F is a Borel function, then*

$$a^{\pi}(z)F(\tilde{N}^{\pi} - \mathbb{1})\Phi = F(\tilde{N}^{\pi})a^{\pi}(z)\Phi. \tag{11.75}$$

Proof Let us suppress the superscript π to simplify notation. First we note that $W(y)$ maps $\mathrm{Dom}\, \tilde{N}^{\frac{1}{2}}$ into itself and have

$$n(W(y)\Phi) = n(\Phi) + \left(\Phi | \phi(\mathrm{j}y)\Phi\right) + \frac{1}{2}\|y\|^2 \|\Phi\|^2. \tag{11.76}$$

In fact, using (8.21) we see that (11.76) is true if we replace n with $n_{\mathcal{V}}$, where $\mathcal{V}$ is a finite-dimensional subspace of $\mathcal{Y}$ containing y. Then (11.76) follows immediately.

By the polarization identity, (11.76) has the following consequence for $\Phi, \Psi \in$ Dom $\tilde{N}^{\frac{1}{2}}$:

$$\left(\tilde{N}^{\frac{1}{2}} W(y)\Phi | \tilde{N}^{\frac{1}{2}} W(y)\Psi\right) \tag{11.77}$$

$$= \left(\tilde{N}^{\frac{1}{2}}\Phi | \tilde{N}^{\frac{1}{2}}\Psi\right) + \left(\Phi | \phi(\mathrm{j}y)\Psi\right) + \frac{1}{2}\|y\|^2 (\Phi|\Psi).$$

Replacing Φ by $W(y)^*\Phi$ and using the invariance of Dom $\tilde{N}^{\frac{1}{2}}$ under $W(y)$, we can rewrite (11.77) as follows:

$$\left(\tilde{N}^{\frac{1}{2}}\Phi | \tilde{N}^{\frac{1}{2}} W(y)\Psi\right) \tag{11.78}$$

$$= \left(\tilde{N}^{\frac{1}{2}} W(y)^*\Phi | \tilde{N}^{\frac{1}{2}}\Psi\right) + \left(W(y)^*\Phi | \phi(\mathrm{j}y)\Psi\right) + \frac{1}{2}\|y\|^2 \left(W(y)^*\Phi|\Psi\right).$$

Next assume in addition that $\Phi, \Psi \in$ Dom $\tilde{N}$. Then we can rewrite (11.78) as

$$\left(\tilde{N}\Phi | W(y)\Psi\right) \tag{11.79}$$

$$= \left(W(y)^*\Phi | \tilde{N}\Psi\right) + \left(W(y)^*\Phi | \phi(\mathrm{j}y)\Psi\right) + \frac{1}{2}\|y\|^2 \left(W(y)^*\Phi|\Psi\right).$$

We replace y by ty and differentiate (11.79) w.r.t. t. (Differentiating is allowed by (11.74).) We obtain

$$\left(\tilde{N}\Phi | \phi(y)\Psi\right) = \left(\phi(y)\Phi | \tilde{N}\Psi\right) - \mathrm{i}\left(\Phi | \phi(\mathrm{j}y)\Psi\right). \tag{11.80}$$

Substituting $\mathrm{j}y$ for y in (11.80), we obtain

$$\left(\tilde{N}\Phi | \phi(\mathrm{j}y)\Psi\right) = -\left(\phi(\mathrm{j}y)\Phi | \tilde{N}\Psi\right) + \mathrm{i}\left(\Phi | \phi(y)\Psi\right). \tag{11.81}$$

Adding up (11.80) and (11.81), we get

$$\left(\tilde{N}\Phi | a(z)\Psi\right) = \left(a^*(z)\Phi | \tilde{N}\Psi\right) - \left(\Phi | a(z)\Psi\right). \tag{11.82}$$

Next let us assume that $\Phi \in$ Dom $\tilde{N}^{\frac{3}{2}}$. Then $\tilde{N}\Phi \in$ Dom $\tilde{N}^{\frac{1}{2}} \subset$ Dom $a(z)$. Hence, (11.82) implies

$$\left(\tilde{N}\Phi | a(z)\Psi\right) = \left(\Phi | a(z)(\tilde{N} - \mathbb{1})\Psi\right). \tag{11.83}$$

Therefore, $a(z)\Psi \in$ Dom $\tilde{N}$, and we have

$$\tilde{N}a(z)\Psi = a(z)(\tilde{N} - \mathbb{1})\Psi, \tag{11.84}$$

or equivalently

$$(\tilde{N} + \lambda\mathbb{1})a(z)\Psi = a(z)(\tilde{N} + \lambda\mathbb{1} - \mathbb{1})\Psi. \tag{11.85}$$

Now let $\Phi \in$ Dom $\tilde{N}^{\frac{1}{2}}$ and $\lambda > 1$. Then $(\tilde{N} + \lambda\mathbb{1} - \mathbb{1})^{-1}\Phi \in$ Dom $\tilde{N}^{\frac{3}{2}}$. Therefore, by (11.85),

$$(\tilde{N} + \lambda\mathbb{1})a(z)(\tilde{N} + \lambda\mathbb{1} - \mathbb{1})^{-1}v = a(z)\Phi. \tag{11.86}$$

Multiplying this by $(\tilde{N} + \lambda\mathbb{1})^{-1}$, we obtain

$$a(z)(\tilde{N} + \lambda\mathbb{1} - \mathbb{1})^{-1}\Phi = (\tilde{N} + \lambda\mathbb{1})^{-1}a(z)\Phi. \tag{11.87}$$

Since linear combinations of functions $(\tilde{N} + \lambda \mathbb{1})^{-1}$ with $\lambda > 0$ are strongly dense in the von Neumann algebra of bounded Borel functions of $\tilde{N}$, and $a(z)$ is closed, (11.87) implies

$$a(z)F(\tilde{N} - \mathbb{1})\Phi = F(\tilde{N})a(z)\Phi, \ \Phi \in \mathrm{Dom}\,\tilde{N}^{\frac{1}{2}},$$

for any bounded Borel function F. $\qquad\qquad\square$

Lemma 11.54 $\mathcal{K}^\pi = \{0\}$ *implies* $\mathrm{Dom}\,\tilde{N}^\pi = \{0\}$.

Proof Again we suppress the superscript π to simplify notation. Suppose that $\mathrm{Dom}\,\tilde{N} \neq \{0\}$. We know that $\tilde{N} \geq 0$. Therefore, $\mathrm{spec}\,\tilde{N}$ is non-degenerate and bounded from below. Hence, $\lambda_0 := \inf \mathrm{spec}\,\tilde{N}$ is a finite number, and

$$\mathrm{Ran}\,\mathbb{1}_{[\lambda_0,\lambda_0+1[}(\tilde{N}) \neq \{0\}.$$

By Lemma 11.53, for any $z \in \mathcal{Z}$

$$a^\pi(z)\mathbb{1}_{[\lambda_0,\lambda_0+1[}(\tilde{N}) = \mathbb{1}_{[\lambda_0-1,\lambda_0[}(\tilde{N})a^\pi(z). \qquad (11.88)$$

But

$$\mathbb{1}_{[\lambda_0-1,\lambda_0[}(\tilde{N}) = 0.$$

Therefore, (11.88) is zero and

$$\mathrm{Ran}\,\mathbb{1}_{[\lambda_0,\lambda_0+1[}(\tilde{N}) \subset \mathcal{K} = \{0\},$$

which is a contradiction. $\qquad\qquad\square$

The following lemma is immediate:

Lemma 11.55 *Suppose that* $\mathcal{H} = \mathcal{H}^0 \oplus \mathcal{H}^1$. *Suppose that*

$$\mathcal{Y} \ni y \mapsto W^\pi(y) \in \mathcal{H}$$

is a CCR representation and $W^\pi(y)$, $y \in \mathcal{Y}$, *leaves* $\mathcal{H}^0$ *invariant. Then* $W^\pi(y)$ *also leaves* $\mathcal{H}^1$ *invariant. Thus we have two CCR representations,*

$$\mathcal{Y} \ni y \mapsto W^\pi(y)\big|_{\mathcal{H}^0}, \quad \mathcal{Y} \ni y \mapsto W^\pi(y)\big|_{\mathcal{H}^1}.$$

Let $\mathcal{K}^i$, $\tilde{N}^i$ *denote the corresponding spaces of vacua and the operators defined by (11.73) for the representations* $i = 0, 1$. *Then*

$$\mathcal{K} = \mathcal{K}^0 \oplus \mathcal{K}^1, \quad \tilde{N} = \tilde{N}^0 \oplus \tilde{N}^1. \qquad (11.89)$$

Proof of Thm. 11.52. We are in the situation of Lemma 11.55: we have two CCR representations, in $\mathcal{H}^0 = \mathcal{H}^\pi$ and in $\mathcal{H}^1 = (\mathcal{H}^0)^\perp$.

By the definition of N^π, we have

$$N^\pi = N^0 \oplus N^1,$$

where $\mathrm{Dom}\, N^1 = \{0\}$. From $U^\pi \mathbb{1} \otimes W(y) = W^\pi(y)U^\pi$, $y \in \mathcal{Y}$, we get $U^\pi \mathbb{1} \otimes a(z) = a^\pi(z)U^\pi$, $z \in \mathcal{Z}$, and hence $\tilde{N}^0 = N^0$.

We know that $\mathcal{K} \subset \mathcal{H}^0$, hence $\mathcal{K}^1 = \{0\}$. By Lemma 11.54, this implies $\mathrm{Dom}\,\tilde{N}^1 = \{0\}$. Therefore, $\tilde{N}^\pi = N^\pi$. $\qquad\square$

11.4.6 Relative continuity of Gaussian measures

In this subsection we prove Thm. 5.78, the Feldman–Hajek theorem, which says that the Gaussian measures with covariances A_1, A_2 are relatively continuous iff $A_1^{-\frac{1}{2}} A_2 A_1^{-\frac{1}{2}} - \mathbb{1}$ is Hilbert–Schmidt.

Proof of Thm. 5.78. To conform with the notation used in this chapter we denote the covariances A_1, A_2 of Thm. 5.78 by a_1, a_2.

Without loss of generality we can assume that $a_1 = \mathbb{1}$. In fact, note that

$$(\xi|a_1\xi)_{\mathcal{X}} = (\xi|\xi)_{a_1^{-\frac{1}{2}}\mathcal{X}}, \quad (\xi|a_2\xi)_{\mathcal{X}} = (\xi|b\xi)_{a_1^{-\frac{1}{2}}\mathcal{X}},$$

for $b = a_1^{-1}a_2$. Since $a_1^{\frac{1}{2}}$ is unitary from $a_1^{-\frac{1}{2}}\mathcal{X}$ to $\mathcal{X}$, we see that $\mathbb{1} - a_1^{-1}a_2 \in B^2(a_1^{-\frac{1}{2}}\mathcal{X})$ iff $\mathbb{1} - a_1^{-\frac{1}{2}}a_2 a_1^{-\frac{1}{2}} \in B^2(\mathcal{X})$. Hence, replacing $\mathcal{X}$ by $a_1^{-\frac{1}{2}}\mathcal{X}$ and a_2 by $a_1^{-1}a_2$, we may assume that $a_1 = \mathbb{1}$ and denote a_2 simply by a.

Let us consider the real-wave representations for the covariances $\mathbb{1}$ and a, as in Sect. 9.3. From Prop. 9.16 we know that they are unitarily equivalent to the Fock representations for the symplectic space $(\mathcal{X} \oplus \mathcal{X}, \omega)$ with the Kähler anti-involutions

$$\mathrm{j} = \begin{bmatrix} 0 & -\frac{1}{2} \\ 2 & 0 \end{bmatrix}, \quad \mathrm{j}_1 = \begin{bmatrix} 0 & -(2a)^{-1} \\ 2a & 0 \end{bmatrix}.$$

We set

$$k = -\mathrm{j}\mathrm{j}_1 = \begin{bmatrix} a & 0 \\ 0 & a^{-1} \end{bmatrix}.$$

It is easy to see that $\mathbb{1} - k$ is Hilbert–Schmidt for the real scalar product on $\mathcal{X} \oplus \mathcal{X}$ coming from the Kähler structure (ω, j) iff $\mathbb{1} - a$ is Hilbert–Schmidt on $\mathcal{X}$.

Proof of $\Rightarrow$. Assume that $\mathbb{1} - a \in B^2(\mathcal{X})$. For simplicity let us denote the two Gaussian $\mathbf{L}^2$ spaces by $L^2(\mathcal{X}, \mathrm{d}\mu)$ and $L^2(\mathcal{X}, \mathrm{d}\tilde{\mu})$.

By Thm. 11.45, we know that the two real-wave representations above are unitarily equivalent. In particular, there exists a unitary operator U intertwining these representations. By restriction to the position operators, we deduce that

$$U : L^2(\mathcal{X}, \mathrm{d}\mu) \to L^2(\mathcal{X}, \mathrm{d}\tilde{\mu}), \quad UF(x)U^{-1} = F(x),$$

for all cylinder continuous functions F on $\mathcal{X}$. By monotone convergence, this identity extends first to all bounded $\mathfrak{B}_{\mathrm{cyl}}$-measurable functions, and then to all measurable functions (see Subsect. 5.2.1). We note then that if $A \in \mathfrak{B}$, then $\mu(A) = 0$, resp. $\mu_1(A) = 0$ iff $\mathbb{1}_A(x) = 0$ as a multiplication operator on $L^2(\mathcal{X}, \mathrm{d}\mu)$, resp. $L^2(\mathcal{X}, \mathrm{d}\tilde{\mu})$, which shows that μ and μ_1 are mutually absolutely continuous.

Proof of $\Leftarrow$. Assume now that μ and $\tilde{\mu}$ are mutually absolutely continuous. Then we have

$$\mathrm{d}\tilde{\mu} = F\mathrm{d}\mu, \quad \text{for } F \geq 0 \text{ a.e., and} \quad \int_{\mathcal{X}} F\mathrm{d}\mu = 1.$$

Clearly, $\Psi := F^{\frac{1}{2}}$ is a unit vector in $L^2(\mathcal{X}, \mathrm{d}\mu)$. We will show that

$$a_{\mathrm{rw}}(w)\Psi - a_{\mathrm{rw}}^*(c\overline{w})\Psi = 0, \quad w \in \mathbb{C}\mathcal{X}, \tag{11.90}$$

in the weak sense, where $c = \frac{a-\mathbb{1}}{a+\mathbb{1}}$ and $a_{\mathrm{rw}}^*(\cdot)$, $a_{\mathrm{rw}}(\cdot)$ are the creation and annihilation operators in the real-wave representation on $L^2(\mathcal{X}, \mathrm{d}\mu)$ defined in Subsect. 9.3.1.

We claim that (11.90) implies that $\mathbb{1} - a \in B^2(\mathcal{X})$. In fact, by Prop. 9.16, the real-wave representation on $L^2(\mathcal{X}, \mathrm{d}\mu)$ with the anti-involution j above is unitarily equivalent to the Fock representation on $\Gamma_{\mathrm{s}}(\mathbb{C}\mathcal{X})$. Applying Thm. 11.28, we deduce from (11.90) that $c \in B^2(\mathbb{C}\mathcal{X})$, i.e. $\mathbb{1} - a \in B^2(\mathcal{X})$.

Note that if $\mathcal{X}$ is finite-dimensional, then

$$F(x) = (\det a)^{-\frac{1}{2}} \mathrm{e}^{-\frac{1}{2}(x|a^{-1}x) + \frac{1}{2}(x|x)},$$
$$\Psi(x) = F^{\frac{1}{2}}(x) = (\det a)^{-\frac{1}{4}} \mathrm{e}^{-\frac{1}{4}(x|a^{-1}x) + \frac{1}{4}(x|x)}. \tag{11.91}$$

Hence,

$$w \cdot \nabla_x \Psi(x) = \frac{1}{2}(\mathbb{1} - a^{-1})w \cdot x\Psi(x),$$

which is equivalent to (11.90). In the general case we will approximate $\mathcal{X}$ by an increasing family of finite-dimensional subspaces $\mathcal{X}_n$ on which (11.91) is valid, and pass to the limit $n \to +\infty$.

We choose an increasing sequence $(\pi_n)_{n \in \mathbb{N}}$ of rank n orthogonal projections in $\mathcal{X}$ such that $\mathrm{s} - \lim_{n \to \infty} \pi_n = \mathbb{1}$. We set

$$\mathcal{X}_n := \pi_n \mathcal{X}, \quad a_n := \pi_n a \pi_n, \quad \mathfrak{B}_n := \mathfrak{B}^{\mathcal{X}_n},$$

where we recall from Sect. 5.2.1 that $\mathfrak{B}^{\mathcal{Y}}$ is the σ-algebra of cylinder sets based on $\mathcal{Y}$.

We note that $\mathfrak{B}$ is generated by $\bigcup_{n \in \mathbb{N}} \mathfrak{B}_n$. This follows from the fact that the polynomials based on $\bigcup_{n \in \mathbb{N}} \mathcal{X}_n$ are dense in $L^2(\mathcal{X}, \mathrm{d}\mu)$, which is a consequence of Thm. 5.56.

We denote by μ_n, resp. $\tilde{\mu}_n$ the Gaussian measures on $\mathcal{X}_n$ with covariances $\mathbb{1}$, resp. a_n. For $\xi \in \mathcal{X}_n$ we have

$$\int_{\mathcal{X}_n} \mathrm{e}^{\mathrm{i}\xi \cdot x} \mathrm{d}\mu_n = \int_{\mathcal{X}} \mathrm{e}^{\mathrm{i}\xi \cdot x} \mathrm{d}\mu, \quad \int_{\mathcal{X}_n} \mathrm{e}^{\mathrm{i}\xi \cdot x} \mathrm{d}\tilde{\mu}_n = \int_{\mathcal{X}} \mathrm{e}^{\mathrm{i}\xi \cdot x} \mathrm{d}\tilde{\mu},$$

which by a density argument implies that

$$\int_{\mathcal{X}_n} u(x)\mathrm{d}\mu_n = \int_{\mathcal{X}} u(x)\mathrm{d}\mu, \quad \int_{\mathcal{X}_n} u(x)\mathrm{d}\tilde{\mu}_n = \int_{\mathcal{X}} u(x)\mathrm{d}\tilde{\mu}, \tag{11.92}$$

if u is a $\mathfrak{B}_n$-measurable integrable function.

We denote by $F_n := E_{\mathfrak{B}_n}(F)$ the *conditional expectation* of F w.r.t. $\mathfrak{B}_n$. We recall that if $(Q, \mathfrak{B}, \mu)$ is a probability space and $\mathfrak{B}_0 \subset \mathfrak{B}$ is a σ-algebra, then $E_{\mathfrak{B}_0}$ is defined on $L^2(Q, \mathrm{d}\mu)$ as the orthogonal projection on the subspace of $\mathfrak{B}_0$-measurable L^2 functions. $E_{\mathfrak{B}_0}$ extends to a contraction on $L^1(Q, \mathrm{d}\mu)$ with $\int_Q E_{\mathfrak{B}_0}(u)\mathrm{d}\mu = \int_Q u\mathrm{d}\mu$. In our case, since $\mathfrak{B}$ is generated by $\bigcup_{n\in\mathbb{N}} \mathfrak{B}_n$, we know that

$$F_n \to F \quad \mu \text{ a.e. and in } L^1(\mathcal{X}, \mathrm{d}\mu). \tag{11.93}$$

If Φ is $\mathfrak{B}_n$-measurable, we have, using (11.92),

$$\int_{\mathcal{X}_n} \Phi(x)\mathrm{d}\tilde{\mu}_n = \int_{\mathcal{X}} \Phi(x)\mathrm{d}\tilde{\mu} = \int_{\mathcal{X}} \Phi(x)F\mathrm{d}\mu = \int_{\mathcal{X}} \Phi(x)F_n\mathrm{d}\mu = \int_{\mathcal{X}_n} \Phi(x)F_n\mathrm{d}\mu_n,$$

which shows that

$$\mathrm{d}\tilde{\mu}_n = F_n\mathrm{d}\mu_n, \quad n \in \mathbb{N}. \tag{11.94}$$

For $P \in \mathbb{C}\mathrm{Pol}_s(\mathcal{X}_n)$ and $w \in \mathbb{C}\mathcal{X}_n$, we have

$$\int_{\mathcal{X}_n} \overline{w \cdot \nabla_x P(x)} F(x)\mathrm{d}\mu_n = \int_{\mathcal{X}_n} \overline{w \cdot \nabla_x P(x)}\mathrm{d}\tilde{\mu}_n,$$

which we can rewrite as

$$(w \cdot \nabla_x P | F)_{L^2(\mathcal{X}_n, \mathrm{d}\mu_n)} = (w \cdot \nabla_x P | 1)_{L^2(\mathcal{X}_n, \mathrm{d}\tilde{\mu}_n)}.$$

On $L^2(\mathcal{X}, \mathrm{d}\mu_n)$, we have

$$(w \cdot \nabla_x)^* = -\overline{w} \cdot \nabla_x + \overline{w} \cdot x, \tag{11.95}$$

and on $L^2(\mathcal{X}, \mathrm{d}\tilde{\mu}_n)$, we have

$$(w \cdot \nabla_x)^* = -\overline{w} \cdot \nabla_x + a_n^{-1}\overline{w} \cdot x. \tag{11.96}$$

This yields

$$\int_{\mathcal{X}_n} \overline{w \cdot \nabla_x P} F_n\mathrm{d}\mu_n = \int_{\mathcal{X}_n} a_n^{-1}\overline{w \cdot x}P F_n\mathrm{d}\mu_n. \tag{11.97}$$

Since $\mathcal{X}_n$ is n-dimensional, μ_n and $\tilde{\mu}_n$ can be realized on $\mathcal{X}_n$ with

$$\mathrm{d}\mu_n = (2\pi)^{-n/2}\mathrm{e}^{-\frac{1}{2}(x|x)}\mathrm{d}x, \quad \mathrm{d}\tilde{\mu}_n = (2\pi)^{-n/2}(\det a_n)^{-\frac{1}{2}}\mathrm{e}^{-\frac{1}{2}(x|a_n^{-1}x)}\mathrm{d}x,$$

so that

$$F_n(x) = (\det a_n)^{-\frac{1}{2}}\mathrm{e}^{-\frac{1}{2}(x|a_n^{-1}x)+\frac{1}{2}(x|x)}.$$

It follows from (11.97) that F_n satisfies in the usual sense the identity

$$w \cdot \nabla_x F_n(x) = (\mathbb{1} - a_n^{-1}) w \cdot x F_n(x).$$

Let us set

$$\Psi_n := F_n^{\frac{1}{2}},$$

so that $\Psi_n \in L^2(\mathcal{X}_n, \mathrm{d}\mu_n)$, $\|\Psi_n\| = 1$. From $\nabla_x \Psi_n = \frac{1}{2} F_n^{-\frac{1}{2}} \nabla_x F_n$, we get that

$$w \cdot \nabla_x \Psi_n(x) = \frac{1}{2}(\mathbb{1} - a_n^{-1}) w \cdot x \Psi_n(x), \quad w \in \mathbb{C}\mathcal{X}_n.$$

Considering now Ψ_n as a function on $\mathcal{X}$, using (11.95) we can rewrite this identity as

$$\int_{\mathcal{X}} \left(-\overline{w} \cdot \nabla_x + \frac{1}{2}(\mathbb{1} + a_n^{-1}) \overline{w} \cdot x \right) \overline{P} \Psi_n \mathrm{d}\mu = 0, \quad P \in \mathbb{C}\mathrm{Pol}_s(\mathcal{X}_n), \quad w \in \mathbb{C}\mathcal{X}_n,$$

or equivalently

$$\int_{\mathcal{X}} \left(-a_n \overline{w} \cdot \nabla_x + \frac{1}{2}(\mathbb{1} + a_n) \overline{w} \cdot x \right) \overline{P} \Psi_n \mathrm{d}\mu = 0, \quad P \in \mathbb{C}\mathrm{Pol}_s(\mathcal{X}_n), \quad w \in \mathbb{C}\mathcal{X}_n. \tag{11.98}$$

We note now that if $w \in \mathcal{X}$ and $w_n := \pi_n w$, then for all $P \in \mathbb{C}\mathrm{Pol}_s(\mathcal{X})$

$$\lim_{n \to \infty} (a_n w_n - a w) \cdot \nabla_x P = 0,$$

$$\lim_{n \to \infty} \left((a_n + \mathbb{1}) w_n \cdot x - (a + \mathbb{1}) w \cdot x \right) P = 0 \quad \text{in } L^2(X, \mu).$$

Since Ψ_n is uniformly bounded in $L^2(\mathcal{X}, \mathrm{d}\mu)$, we deduce from (11.98) that

$$\lim_{n \to \infty} \int_{\mathcal{X}} \left(-a \overline{w} \cdot \nabla_x + \frac{1}{2}(\mathbb{1} + a) \overline{w} \cdot x \right) P \Psi_n \mathrm{d}\mu = 0, \quad P \in \mathbb{C}\mathrm{Pol}_s(\mathcal{X}_m), \; w \in \mathbb{C}\mathcal{X}$$

for some m. $\tag{11.99}$

We claim now that

$$\mathrm{w} - \lim_{n \to \infty} \Psi_n = \Psi. \tag{11.100}$$

In fact, since Ψ_n is uniformly bounded in L^2, it suffices to show that, for all $G \in L^\infty(\mathcal{X}, \mathrm{d}\mu)$,

$$\lim_{n \to \infty} \int_{\mathcal{X}} (\Psi_n - \Psi) G \mathrm{d}\mu = 0. \tag{11.101}$$

Let $\Phi_n = (\Psi_n - \Psi) G \in L^2 \subset L^1$. We know from (11.93) that $\Phi_n \to 0$ μ a.e.. Moreover the sequence (Φ_n) is bounded in L^2, since $G \in L^\infty$. It follows from Subsect. 5.1.9 that the sequence (Φ_n) is equi-integrable, which using the fact that $\Phi_n \to 0$ μ a.e. implies (11.101).

Passing to the limit in (11.99) and using (11.100), we finally get

$$\int_{\mathcal{X}} \left(-a\overline{w} \cdot \nabla_x + \frac{1}{2}(\mathbb{1}+a)\overline{w} \cdot x \right) P\Psi d\mu = 0, \quad w \in \mathcal{X}, \tag{11.102}$$

first for $P \in \mathbb{C}\mathrm{Pol}_s(\mathcal{X}_m)$ for some m, and then by density for all $P \in \mathbb{C}\mathrm{Pol}_s(\mathcal{X})$.

Using the definition of $a_{\mathrm{rw}}(w)$, $a_{\mathrm{rw}}^*(w)$ on $L^2(\mathcal{X}, d\mu)$, we see that (11.102) implies (11.90), which completes the proof of the $\Leftarrow$ part of the theorem. $\square$

Proof of Prop. 5.79. We use the notation of the proof of Thm. 5.78. We have seen that Ψ, resp. Ψ_n are the bosonic Gaussian vectors for $c = (a - \mathbb{1})(a + \mathbb{1})^{-1}$, resp. $c_n = \pi_n c \pi_n$. Since $c_n \to c$ in $B^2(\mathcal{X})$, it follows from Thm. 11.28 that $\Psi_n \to \Psi$ in L^2, hence $F_n \to F$ in L^1. This proves (1) and (3). (2) is left to the reader. $\square$

11.5 Coherent sector of CCR representations

This section is devoted to coherent representations, that is, translations of Fock CCR representations. It is to a large extent parallel to the previous section about Fock CCR representations.

We keep the same notation as in the previous section. In particular, $(\mathcal{Y}, \omega)$ is a symplectic space,

$$\mathcal{Y} \ni y \mapsto W^{\pi}(y) \in U(\mathcal{H})$$

is a regular representation of CCR and $\mathrm{j} \in L(\mathcal{Y})$ is a Kähler anti-involution, so that we obtain a Kähler space $(\mathcal{Y}, \cdot, \omega, \mathrm{j})$.

$\mathcal{Z}$ is the holomorphic subspace of $\mathbb{C}\mathcal{Y}$. As usual, we identify $\mathbb{C}\mathcal{Y}$ with $\mathcal{Z} \oplus \overline{\mathcal{Z}}$. We do not assume that $\mathcal{Y}$, or equivalently $\mathcal{Z}$, is complete.

$\mathcal{Y}^{\#}$ denotes the algebraic dual of $\mathcal{Y}$. Similarly, $\mathcal{Z}^*$ denotes the algebraic anti-dual of $\mathcal{Z}$. We have the identification $\mathcal{Y}^{\#} = \mathrm{Re}(\mathcal{Z}^* \oplus \overline{\mathcal{Z}}^*)$.

11.5.1 Coherent vectors in a CCR representation

Let $f \in \mathcal{Z}^*$, that is, f is an anti-linear functional on $\mathcal{Z}$, possibly unbounded. We also introduce a symbol for the corresponding (possibly unbounded) linear functional on $\mathcal{Y}$, $v = (f, \overline{f})|_{\mathcal{Y}} \in \mathcal{Y}^{\#}$. Clearly,

$$v \cdot (z, \overline{z}) = (f|z) + (z|f).$$

Definition 11.56 *We define the* space of j, f-coherent vectors

$$\mathcal{K}_f^{\pi} := \left\{ \Psi \in \mathcal{H} \ : \ \Psi \in \mathrm{Dom}\, a^{\pi}(z), \quad a^{\pi}(z)\Psi = (z|f)\Psi, \ z \in \mathcal{Z} \right\},$$

where the j-*creation, resp.* j-*annihilation operators* $a^{\pi *}(z)$, *resp.* $a^{\pi}(z)$ *are defined in Subsect. 8.2.4.*

Proposition 11.57 (1) $\mathcal{K}_f^\pi$ *is a closed subspace of* $\mathcal{H}$.

(2) $\Psi \in \mathcal{K}_f^\pi$ *iff* $(\Psi | W^\pi(y)\Psi) = e^{-\frac{1}{4}y^2 + iv\cdot y}$, $y \in \mathcal{Y}$.

(3) *Elements of* $\mathcal{K}_f^\pi$ *are analytic vectors for* $\phi^\pi(y)$, $y \in \mathcal{Y}$.

(4) *If* $\Phi, \Psi \in \mathcal{K}_f^\pi$, *then*

$$(\Phi | W^\pi(y)\Psi) = (\Phi | \Psi)e^{-\frac{1}{4}y^2 + iv\cdot y},$$

$$(\Phi | \phi^\pi(y)\Psi) = (\Phi | \Psi)v \cdot y,$$

$$(\Phi | \phi^\pi(y_1)\phi^\pi(y_2)\Psi) = \frac{1}{2}\left(y_1 \cdot y_2 + iy_1 \cdot \omega y_2\right)(\Phi | \Psi) + (v\cdot y_1)(v\cdot y_2)(\Phi | \Psi).$$

Proof We will suppress the superscript π to simplify notation.

(1) $\mathcal{K}_f$ is closed as an intersection of kernels of closed operators.

Let us prove (2) $\Leftarrow$. Let $\Psi \in \mathcal{H}$ such that $(\Psi | W(y)\Psi) = \|\Psi\|^2 e^{-\frac{1}{4}y^2 + iv\cdot y}$. Without loss of generality we can assume that $\|\Psi\| = 1$. Taking the first two terms in the Taylor expansion of

$$t \mapsto (\Psi | W(ty)\Psi) = e^{-\frac{1}{4}t^2 y^2 + itv\cdot y},$$

we obtain

$$(\Psi | \phi(y)\Psi) = v\cdot y, \quad (\Psi | \phi(y)^2 \Psi) = \frac{1}{2}y^2 + (v\cdot y)^2. \tag{11.103}$$

If $z = \frac{1}{2}(y - ijy) \in \mathcal{Z}$, we have

$$\left(a^*(z) - (f|z)\mathbb{1}\right)\left(a(z) - (z|f)\mathbb{1}\right)$$

$$= \tfrac{1}{4}\phi(y)^2 + \tfrac{1}{4}\phi(jy)^2 + \tfrac{i}{4}[\phi(y), \phi(jy)]$$

$$-\tfrac{1}{2}(z|f)\phi(y) + \tfrac{i}{2}(z|f)\phi(jy) - \tfrac{1}{2}(f|z)\phi(y) - \tfrac{i}{2}(f|z)\phi(jy) + (f|z)(z|f)\mathbb{1}.$$

Using (11.103), we obtain that

$$\left\|\left(a(z) - (z|f)\mathbb{1}\right)\Psi\right\|^2 = 0.$$

Let us prove (2) $\Rightarrow$. Consider $y \in \mathcal{Y}$. Note that if $\Psi \in \mathcal{K}_f$, then $\Psi \in \text{Dom }\phi(y)$. We consider the function

$$\mathbb{R} \ni t \mapsto (\Psi | W(ty)\Psi),$$

as in the proof of Prop. 11.42. For $z = \frac{1}{2}(y - ijy) \in \mathcal{Z}$, we get

$$\tfrac{d}{dt}F(t) = i\left(\Psi | \phi(y)W(ty)\Psi\right)$$

$$= \tfrac{1}{2}\left(a(z)\Psi | W(ty)\Psi\right) + \tfrac{1}{2}\left(\Psi | W(ty)a(z)\Psi\right) - \tfrac{t}{2}y^2\left(\Psi | W(ty)\Psi\right)$$

$$= i\left((f|z) + (z|f)\right)F(t) - \tfrac{t}{2}y^2 F(t),$$

which yields $F(t) = \|\Psi\|^2 e^{-\frac{1}{4}t^2 y^2 + itv\cdot y}$.

(2) immediately implies (3).

(4) follows from (2) and (11.103) by polarization. $\square$

11.5.2 Coherent vectors in Fock spaces

We consider the bosonic Fock space $\Gamma_s(\mathcal{Z}^{cpl})$ and the usual creation and annihilation operators $a^*(z)$, $a(z)$.

Theorem 11.58 *Let $\Psi \in \Gamma_s(\mathcal{Z}^{cpl})$ be an f-coherent vector for the Fock representation over $\mathcal{Y}$, that is, for any $z \in \mathcal{Z}$, $\Psi \in \mathrm{Dom}\, a(z)$ and*

$$a(z)\Psi = (z|f)\Psi.$$

Then the following is true:

(1) *If f is continuous, i.e. $f \in \mathcal{Z}^{cpl}$, then Ψ is proportional to $W(-\mathrm{i}f, \mathrm{i}\overline{f})\Omega$.*
(2) *If f is not continuous, then $\Psi = 0$.*

Proof By induction we show that, for $z_1, \dots, z_n \in \mathcal{Z}$,

$$a(z_{n-1}) \cdots a(z_1)\Psi \in \mathrm{Dom}\, a(z_n), \quad a(z_n) \cdots a(z_1)\Psi = (z_1|f) \cdots (z_n|f)\Psi.$$

This implies

$$(a^*(z_1) \cdots a^*(z_n)\Omega|\Psi) = (z_1|f) \cdots (z_n|f)(\Omega|\Psi). \tag{11.104}$$

In particular,

$$(z|\Psi) = (a^*(z)\Omega|\Psi) = (z|f)(\Omega|\Psi), \quad z \in \mathcal{Z}.$$

Using the fact that $\mathcal{Z}$ is dense in $\mathcal{Z}^{cpl}$, we see that $(\Omega|\Psi)f$ is a bounded functional on $\mathcal{Z}$, hence it belongs to $\mathcal{Z}^{cpl}$. Thus either $f \in \mathcal{Z}^{cpl}$ or $(\Omega|\Psi) = 0$. In the latter case, (11.104) implies that $\Psi = 0$. $\square$

11.5.3 Coherent representations

Consider the usual Weyl operators $W(y)$, $y \in \mathcal{Y}$, on the bosonic Fock space $\Gamma_s(\mathcal{Z}^{cpl})$. Set

$$\mathcal{Y} \ni y \mapsto W_f(y) := W(y)\mathrm{e}^{2\mathrm{i}\mathrm{Re}(f|z)} \in U\big(\Gamma_s(\mathcal{Z}^{cpl})\big). \tag{11.105}$$

Clearly, (11.105) is a regular CCR representation,

Definition 11.59 *(11.105) is called the* j, f*-coherent CCR representation.*

Theorem 11.60 (1) *(11.105) is the translation of the Fock representation by the vector $(f, \overline{f}) \in \mathcal{Y}^\#$ (see Def. 8.21)*
(2) *If $f \in \mathcal{Z}^{cpl}$ (equivalently, $v \in \mathcal{Y}^{cpl}$), then*

$$W_f(y) = W(\mathrm{i}f, -\mathrm{i}\overline{f})W(y)W(-\mathrm{i}f, \mathrm{i}\overline{f}).$$

(3) *If $f \notin \mathcal{Z}^{cpl}$, then (11.105) is not unitarily equivalent to the Fock representation $\mathcal{Y} \ni y \mapsto W(y)$.*

Proof (3) Let $U \in U(\mathcal{H})$ intertwining $W_f(\cdot)$ and $W(\cdot)$. Since $U\Omega$ satisfies the assumptions of Thm. 11.58 we obtain that $U\Omega = 0$, which is a contradiction. $\square$

11.5.4 Coherent sector

Let us go back to an arbitrary CCR representation (11.59) over a Kähler space $\mathcal{Y}$. We will describe how to determine the largest sub-representation of (11.59) unitarily equivalent to a multiple of a coherent representation over $\mathcal{Y}$ in $\Gamma_s(\mathcal{Z}^{\mathrm{cpl}})$.

Definition 11.61 *Introduce the equivalence relation on the set of Kähler anti-involutions on $\mathcal{Y}$ and anti-linear functionals on the corresponding holomorphic space $\mathcal{Z}$:*

$$(j_1, f_1) \sim (j_2, f_2) \quad \Leftrightarrow \quad j_1 - j_2 \in B^2(\mathcal{Y}), \quad f_1 - f_2 \in \mathcal{Z}^{\mathrm{cpl}}.$$

Let $[j, f]$ denote the equivalence class of (j, f) w.r.t. this relation.

The $[j, f]$-coherent sector of the representation W^π is the subspace of $\mathcal{H}$ defined as

$$\mathcal{H}^\pi_{[f]} := \mathrm{Span}^{\mathrm{cl}}\{W^\pi(y)\Psi \; : \; \Psi \in \mathcal{K}^\pi_f, \; y \in \mathcal{Y}\}.$$

The CCR representation W^π is $[j, f]$-coherent if $\mathcal{H}^\pi_{[f]} = \mathcal{H}$.

Theorem 11.62 (1) $\mathcal{H}^\pi_{[f]}$ *is invariant under $W^\pi(y)$, $y \in \mathcal{Y}$.*
(2) *There exists a unique unitary operator*

$$U^\pi_f : \mathcal{K}^\pi_f \otimes \Gamma_s(\mathcal{Z}^{\mathrm{cpl}}) \to \mathcal{H}^\pi_{[f]}$$

satisfying

$$U^\pi_f \; \Psi \otimes W_f(y)\Omega = W^\pi(y)\Psi, \quad \Psi \in \mathcal{K}^\pi_f, \; y \in \mathcal{Y}.$$

(3)

$$U^\pi_f \; \mathbb{1} \otimes W_f(y) = W^\pi(y)U^\pi_f, \quad y \in \mathcal{Y}. \tag{11.106}$$

(4) *If there exists an operator $U : \Gamma_s(\mathcal{Z}^{\mathrm{cpl}}) \to \mathcal{H}$ such that $UW_f(y) = W^\pi(y)U$ for $y \in \mathcal{Y}$, then $\mathrm{Ran}\, U \subset \mathcal{H}^\pi_{[f]}$.*

11.6 van Hove Hamiltonians

In this section we will study self-adjoint operators on bosonic Fock spaces of the form

$$H = \int h(\xi)a^*(\xi)a(\xi)d\xi + \int \overline{w}(\xi)a(\xi)d\xi + \int w(\xi)a^*(\xi)d\xi + c$$
$$= d\Gamma(h) + a(w) + a^*(w) + c\mathbb{1} \tag{11.107}$$

(first written in the "physicist's notation" and then in the "mathematician's notation"). Note that this expression may have only a formal meaning. In some cases, the constant c is actually infinite. Following Schweber (1962), we call (11.107) *van Hove Hamiltonians*. We will see that in the case of an infinite number of degrees of freedom these Hamiltonians have a surprisingly rich theory. We will discuss both classical and quantum van Hove Hamiltonians. Their theories are parallel to one another.

Throughout this section, $\mathcal{Z}$ is a Hilbert space and h is a positive operator on $\mathcal{Z}$ with $\operatorname{Ker} h = \{0\}$. (It is, however, easy to generalize the theory of van Hove Hamiltonians to non-positive h.)

In addition to h, van Hove Hamiltonians depend on the choice of w. The choice $w \in \mathcal{Z}$ turns out to be too narrow.

In order to explain the nature of w, it will be convenient to use the notation introduced in Subsect. 2.3.4. In particular, we will consider the spaces $(h^\alpha + h^\beta)\mathcal{Z}$ for $0 \leq \alpha \leq \beta$. Note that

$$w \in (h^\alpha + h^\beta)\mathcal{Z} \iff \mathbb{1}_{]0,1]}(h)w \in h^\alpha \mathcal{Z}, \quad \mathbb{1}_{[1,+\infty[}(h)w \in h^\beta \mathcal{Z}. \quad (11.108)$$

For $w \in (h^\alpha + h^\beta)\mathcal{Z}$, the behavior of w near $h = 0$ (resp. $h = +\infty$), i.e. at low (resp. high) energies, is encoded by the exponent α (resp. β) and connected with the so-called *infrared* (resp. *ultraviolet*) *problem*. We will always assume that

$$w \in (\mathbb{1} + h)\mathcal{Z}. \quad (11.109)$$

Note that if $w \in (\mathbb{1} + h)\mathcal{Z}$, then $(\mathrm{e}^{\mathrm{i}th} - \mathbb{1})h^{-1}w \in \mathcal{Z}$ for any $t \in \mathbb{R}$.

11.6.1 Classical van Hove dynamics

Definition 11.63 *The classical van Hove dynamics is defined for $t \in \mathbb{R}$ as*

$$\alpha^t(z) := \mathrm{e}^{\mathrm{i}th} z + (\mathrm{e}^{\mathrm{i}th} - \mathbb{1})h^{-1}w, \quad z \in \mathcal{Z}, \quad t \in \mathbb{R}.$$

It is easy to see that $\mathbb{R} \ni t \mapsto \alpha^t$ is a one-parameter group of affine transformations of $\mathcal{Z}$ preserving the scalar product.

Let us note the following property of the dynamics α. Let p_1, p_2 be two complementary orthogonal projections commuting with h. For $i = 1, 2$, let $\mathcal{Z}_i := \operatorname{Ran} p_i$. Thus we have a direct sum decomposition $\mathcal{Z} = \mathcal{Z}_1 \oplus \mathcal{Z}_2$.

Set $h_i := p_i h$, treated as a self-adjoint operator on $\mathcal{Z}_i$. Let α_i be the dynamics on $\mathcal{Z}_i$ defined by h_i, w_i. Then the dynamics α splits as

$$\alpha^t(z_1, z_2) = (\alpha_1^t(z_1), \alpha_2^t(z_2)).$$

In particular, we can take

$$p_1 := \mathbb{1}_{[0,1]}(h), \quad p_2 := \mathbb{1}_{]1,\infty[}(h). \quad (11.110)$$

Then h_1 is bounded and $h_2 \geq \mathbb{1}$. In the case of h_1 the ultraviolet problem is absent, but the infrared problem can show up. In the case of h_2 we have the opposite situation: the infrared problem is absent, but we can face the ultraviolet problem.

Assume for a moment that $w \in h\mathcal{Z}$. Then the van Hove dynamics is equivalent to the free van Hove dynamics:

$$\alpha^t = \tau^{-1} \circ \alpha_0^t \circ \tau, \tag{11.111}$$

where

$$\tau(z) := z + h^{-1}w, \quad \alpha_0^t(z) := e^{ith}z. \tag{11.112}$$

11.6.2 Classical van Hove Hamiltonians

$\mathcal{Z}$ can be interpreted as a charged symplectic space with the form $\bar{z}_1 \cdot \omega z_2 := i(z_1|z_2)$. In the case of finite dimensions we know that for every charged symplectic dynamics $t \mapsto \alpha_t$ there exists a real function H on $\mathcal{Z}$ satisfying

$$\frac{\mathrm{d}}{\mathrm{d}t}\alpha^t(z) = i\nabla_{\bar{z}}H(\alpha^t(z)). \tag{11.113}$$

This function is called a Hamiltonian of α. It is unique up to an additive constant.

In the case of an infinite number of degrees of freedom the situation is more complicated. It is even unclear how to give a general definition of a Hamiltonian of an arbitrary charged symplectic dynamics. It may, for instance, turn out that natural candidates for a Hamiltonian are defined only on a subset of $\mathcal{Z}$, and differentiable on a smaller subset.

The classical van Hove dynamics is an example of a charged symplectic dynamics. If the dimension is finite, it is easy to see that its Hamiltonian is

$$H(z) = (z|hz) + (z|w) + (w|z) + c,$$

where c is an arbitrary real constant.

In infinite dimensions we will see that the van Hove dynamics possesses natural Hamiltonians. Clearly, these Hamiltonians will be defined only up to an arbitrary additive constant. One can ask whether it is possible to fix this constant in a natural way. We will argue that there are two ways to do so, both under some additional assumptions on w in addition to (11.109).

Definition 11.64 *Assume*

$$w \in (\mathbb{1} + h^{\frac{1}{2}})\mathcal{Z}. \tag{11.114}$$

Set $\mathcal{D}_{\mathrm{I}} := \mathrm{Dom}\, h^{\frac{1}{2}}$, *and*

$$H_{\mathrm{I}}(z) := (z|hz) + (z|w) + (w|z), \quad z \in \mathcal{D}_{\mathrm{I}}.$$

We will say that H_{I} *is the* classical van Hove Hamiltonian of the first kind.

Definition 11.65 *Assume*

$$w \in h^{1/2}\mathcal{Z} + h\mathcal{Z}. \tag{11.115}$$

Set $\mathcal{D}_{\mathrm{II}} := \{z \in \mathcal{Z} \,:\, h^{1/2}z + h^{-1/2}w \in \mathcal{Z}\}$, *and*

$$H_{\mathrm{II}}(z) := (h^{1/2}z + h^{-1/2}w | h^{1/2}z + h^{-1/2}w), \quad z \in \mathcal{D}_{\mathrm{II}}.$$

We will say that H_{II} *is the* classical van Hove Hamiltonian of the second kind.

Clearly, both H_{I} and H_{II} are well defined iff

$$w \in h^{1/2}\mathcal{Z} = (\mathbb{1} + h^{\frac{1}{2}})\mathcal{Z} \cap (h^{\frac{1}{2}} + h)\mathcal{Z},$$

and then

$$H_{\mathrm{II}} = H_{\mathrm{I}} + (w | h^{-1}w).$$

Definition 11.66 *Let* w *be a functional satisfying (11.109). We split the dynamics* α *into* $\alpha_1 \oplus \alpha_2$, *the functional* w *into* $w_1 \oplus w_2$ *as explained in Subsect. 11.6.1, with the splitting given by (11.110). Then, by (11.108),* $w_1 \in \mathcal{Z}_1 = (\mathbb{1} + h_1^{\frac{1}{2}})\mathcal{Z}_1$ *and* $w_2 \in h_2\mathcal{Z}_2 = (h_2^{\frac{1}{2}} + h_2)\mathcal{Z}_2$. *So we can define the Hamiltonian* $H_{1,\mathrm{I}}$ *for the dynamics* α_1 *on the domain* $\mathcal{D}_{1,\mathrm{I}}$, *and the Hamiltonian* $H_{2,\mathrm{II}}$ *for the dynamics* α_2 *on the domain* $\mathcal{D}_{2,\mathrm{II}}$.

Set $\mathcal{D} := \mathcal{D}_{1,\mathrm{I}} \oplus \mathcal{D}_{2,\mathrm{II}}$. *A function* H *on* $\mathcal{D}$ *will be called a* classical van Hove Hamiltonian *if it is of the form*

$$H(z_1, z_2) := H_{1,\mathrm{I}}(z_1) + H_{2,\mathrm{II}}(z_2) + c, \quad (z_1, z_2) \in \mathcal{D},$$

where $c \in \mathbb{R}$ *is arbitrary.*

Note that, in general, there exist $w \in (\mathbb{1} + h)\mathcal{Z}$ that do not belong to $(\mathbb{1} + h^{1/2})\mathcal{Z} \cup (h^{1/2} + h)\mathcal{Z}$. For such w, the dynamics α is well defined but neither H_{I} nor H_{II} are well defined.

The following proposition says that van Hove Hamiltonians are in a certain sense Hamiltonians of the van Hove dynamics. Recall that the Gâteaux differentiability was defined in Def. 2.50.

Proposition 11.67 *Let* $w \in (\mathbb{1} + h)\mathcal{Z}$. *Let* H *be the corresponding van Hove Hamiltonian with the domain* $\mathcal{D}$. *Then*

(1) *The function* H *is Gâteaux differentiable at* $z \in \mathcal{Z}$ *iff* $hz + w$ *belongs to* $\mathcal{Z}$, *and then*

$$\nabla_{\bar{z}}H(z) = hz + w.$$

(2) *The dynamics* $t \mapsto \alpha^t(z)$ *is differentiable w.r.t.* t *iff* $h\alpha^t(z) + w \in \mathcal{Z}$, *and then*

$$\frac{\mathrm{d}}{\mathrm{d}t}\alpha^t(z) = \mathrm{i}(h\alpha^t(z) + w),$$

which can be written in the form (11.113).

(3) α^t *leaves* $\mathcal{D}$ *invariant and* H *is constant along the trajectories.*

The following theorem discusses various special features of van Hove Hamiltonians.

Theorem 11.68 *Let H be a van Hove Hamiltonian.*

(1) *0 belongs to $\mathcal{D}$ iff $w \in (\mathbb{1} + h^{1/2})\mathcal{Z}$. If this is the case, then $H = H_{\mathrm{I}} + H(0)$.*
(2) *H is bounded from below iff $w \in (h^{1/2} + h)\mathcal{Z}$. If this is the case, then $H = H_{\mathrm{II}} + \inf H$.*
(3) *H has a minimum iff $w \in h\mathcal{Z}$. This minimum is at $-h^{-1}w$, and then*

$$H_{\mathrm{II}}(z) = (\tau(z)|h\tau(z)),$$

where τ was defined in (11.112).

Proof We split the dynamics, the functional and the vectors in $\mathcal{Z}$. Then the proofs are immediate. □

Formally,

$$H_{\mathrm{I}}(z) = (z|hz) + (w|z) + (z|w),$$
$$H_{\mathrm{II}}(z) = (z|hz) + (w|z) + (z|w) + (w|h^{-1}w).$$

11.6.3 Quantum van Hove dynamics

We again assume $w \in (\mathbb{1} + h)\mathcal{Z}$. Many quantum objects are analogous to their classical counterparts. Typically, in such cases we will use the same symbols in the classical and quantum case, which should not lead to any confusion.

Definition 11.69 *For $B \in B\big(\Gamma_{\mathrm{s}}(\mathcal{Z})\big)$, we set*

$$\alpha^t(B) := V(t)BV(t)^*,$$

where $V(t)$ is a family of unitary operators on $\Gamma_{\mathrm{s}}(\mathcal{Z})$

$$V(t) := \Gamma(\mathrm{e}^{\mathrm{i}th}) \exp\big(a^*((\mathbb{1} - \mathrm{e}^{-\mathrm{i}th})h^{-1}w) - a((\mathbb{1} - \mathrm{e}^{-\mathrm{i}th})h^{-1}w)\big).$$

$t \mapsto \alpha^t$ will be called a quantum van Hove dynamics.

It is easy to check that $V(t)$ is strongly continuous and, for any $t_1, t_2 \in \mathbb{R}$,

$$V(t_1)V(t_2) = c(t_1, t_2)V(t_1 + t_2),$$

for some $c(t_1, t_2) \in \mathbb{C}$, $|c(t_1, t_2)| = 1$. Hence, α is a one-parameter group of $*$-automorphisms of $B\big(\Gamma_{\mathrm{s}}(\mathcal{Z})\big)$, continuous in the strong operator topology.

In order to make the relationship with the classical dynamics clearer, one can note that

$$\alpha^t(a^*(z)) = ((\mathrm{e}^{\mathrm{i}th} - \mathbb{1})h^{-1}w|z) + a^*(\mathrm{e}^{\mathrm{i}th}z), \quad z \in \mathcal{Z}.$$

Let $\mathcal{Z} = \mathcal{Z}_1 \oplus \mathcal{Z}_2$, as in Subsect. 11.6.1. Then we have the identification $\Gamma_s(\mathcal{Z}) = \Gamma_s(\mathcal{Z}_1) \otimes \Gamma_s(\mathcal{Z}_2)$. The dynamics α factorizes as

$$\alpha^t(B_1 \otimes B_2) = \alpha_1^t(B_1) \otimes \alpha_2^t(B_2), \quad B_i \in B(\Gamma_s(\mathcal{Z}_i)), \ i = 1, 2. \tag{11.116}$$

11.6.4 Quantum van Hove Hamiltonians

Definition 11.70 *We say that a self-adjoint operator H is* a quantum van Hove Hamiltonian for the dynamics $t \mapsto \alpha^t$ if

$$\alpha^t(B) = \mathrm{e}^{\mathrm{i}tH} B \mathrm{e}^{-\mathrm{i}tH}. \tag{11.117}$$

By Prop. 6.68, such a Hamiltonian always exists and is unique up to an additive real constant.

Assume for a moment that $w \in h\mathcal{Z}$. Then, up to a constant, van Hove Hamiltonians are unitarily equivalent to the free van Hove Hamiltonian:

$$H := U \mathrm{d}\Gamma(h) U^* + c \mathbb{1},$$

where U is the "dressing operator"

$$U := \exp\big(-a^*(h^{-1}w) + a(h^{-1}w)\big). \tag{11.118}$$

In the general case, (11.118) can be ill defined, and the construction of van Hove Hamiltonians is more complicated.

Definition 11.71 *Let $w \in (\mathbb{1} + h^{1/2})\mathcal{Z}$. Define*

$$U_{\mathrm{I}}(t) := \mathrm{e}^{\mathrm{i}\mathrm{Im}(h^{-1}w|\mathrm{e}^{\mathrm{i}th} h^{-1}w) - \mathrm{i}t(w|h^{-1}w)} V(t).$$

We easily check that $U_{\mathrm{I}}(t)$ is a one-parameter strongly continuous unitary group. Therefore, by the Stone theorem there exists a unique self-adjoint operator H_{I} such that

$$U_{\mathrm{I}}(t) = \mathrm{e}^{\mathrm{i}tH_{\mathrm{I}}}.$$

We will say that H_{I} is the quantum van Hove Hamiltonian of the first kind.

Definition 11.72 *Let $w \in (h^{1/2} + h)\mathcal{Z}$. Define*

$$U_{\mathrm{II}}(t) := \mathrm{e}^{\mathrm{i}\mathrm{Im}(h^{-1}w|\mathrm{e}^{\mathrm{i}th} h^{-1}w)} V(t).$$

We easily check that $U_{\mathrm{II}}(t)$ is a one-parameter strongly continuous unitary group. Therefore, by the Stone theorem there exists a unique self-adjoint operator H_{II} such that

$$U_{\mathrm{II}}(t) = \mathrm{e}^{\mathrm{i}tH_{\mathrm{II}}}.$$

We will say that H_{II} is the quantum van Hove Hamiltonian of the second kind.

Clearly, both H_I and H_II are well defined iff $w \in h^{1/2}\mathcal{Z}$, and then

$$H_\mathrm{II} = H_\mathrm{I} + (w|h^{-1}w).$$

Theorem 11.73 *Let H be a quantum van Hove Hamiltonian of a dynamics $t \mapsto \alpha^t$. Then the following statements are true:*

(1) *Ω belongs to $\mathrm{Dom}\,|H|^{1/2}$ (the form domain of H) iff $w \in (\mathbb{1} + h^{1/2})\mathcal{Z}$. Under this condition $H = H_\mathrm{I} + (\Omega|H\Omega)$.*
(2) *The operator H is bounded from below iff $w \in (h^{1/2} + h)\mathcal{Z}$. Under this condition $H = H_\mathrm{II} + \inf H$, where $\inf H$ denotes the infimum of the spectrum of H.*
(3) *The operator H has a ground state (inf H is an eigenvalue of H) iff $w \in h\mathcal{Z}$. Then, using the dressing operator defined in (11.118), we can write*

$$H_\mathrm{II} = U\mathrm{d}\Gamma(h)U^*. \tag{11.119}$$

Proof We write α as $\alpha_1 \otimes \alpha_2$, w as $w_1 \oplus w_2$, with $w_1 \in \mathcal{Z}_1$, $w_2 \in h_2\mathcal{Z}_2$; see (11.116).

The operator $\mathrm{d}\Gamma(h_1) + a^*(w_1) + a(w_1)$ is essentially self-adjoint on $\mathrm{Dom}\,N_1$, by Nelson's commutator theorem with the comparison operator N_1; see Thm. 2.74 (1). We set

$$H_{1,\mathrm{I}} := \big(\mathrm{d}\Gamma(h_1) + a^*(w_1) + a(w_1)\big)^{\mathrm{cl}}.$$

Clearly, $H_{1,\mathrm{I}}$ is a Hamiltonian of α_1.

Next we set

$$H_{2,\mathrm{II}} := U_2\mathrm{d}\Gamma(h_2)U_2^*, \quad U_2 := \exp\big(-a^*(h_2^{-1}w_2) + a(h_2^{-1}w_2)\big),$$

which is a Hamiltonian of α_2^t. Hence, any Hamiltonian of α is of the form

$$H = H_{1,\mathrm{I}} \otimes \mathbb{1} + \mathbb{1} \otimes H_{2,\mathrm{II}} + c\mathbb{1}.$$

We drop the subscripts I, II in the rest of the proof. Since $\Omega_1 \in \mathrm{Dom}\,H_1$ we see that

$$\Omega = \Omega_1 \otimes \Omega_2 \in \mathrm{Dom}\,|H|^{\frac{1}{2}} \Leftrightarrow \Omega_2 \in \mathrm{Dom}\,H_2^{\frac{1}{2}}$$
$$\Leftrightarrow U_2^*\Omega \in \mathrm{Dom}\,\mathrm{d}\Gamma(h_2)^{\frac{1}{2}} \Leftrightarrow w_2 \in h_2^{\frac{1}{2}}\mathcal{Z}_2.$$

This proves (1).

Let us now prove (2). Since $H_2 \geq 0$, H is bounded below iff H_1 is bounded below. Since $\mathrm{Dom}\,N_1$ is a core for H_1, we have

$$\inf \mathrm{spec}\,H_1 = \inf_{\Psi_1 \in \mathrm{Dom}\,N_1, \, \|\Psi_1\|=1} (\Psi_1|H_1\Psi_1).$$

Set $w_1^\epsilon = \mathbb{1}_{[\epsilon,1]}(h)w_1$, $U_\epsilon = \exp(-a^*(h_1^{-1}w_1^\epsilon) + a(h_1^{-1}w_1^\epsilon))$. Then

$$(\Psi_1|H_1\Psi_1) = \lim_{\epsilon \to 0}\left(\Psi_1|(d\Gamma(h_1) + a^*(w_1^\epsilon) + a(w_1^\epsilon))\Psi_1\right)$$
$$= \lim_{\epsilon \to 0}\left((U_\epsilon^*\Psi_1|d\Gamma(h_1)U_\epsilon^*\Psi_1) - (w_1^\epsilon|h_1^{-1}w_1^\epsilon)\|\Psi_1\|^2\right).$$

It follows that if $w_1 \in h_1^{\frac{1}{2}}\mathcal{Z}_1$, then $H_1 \geq -(w_1|h_1^{-1}w_1)$.

Conversely, assume that H_1 is bounded below. Then, for

$$\Omega_{z_1} = \exp(a^*(z_1) - a(z_1))\Omega_1,$$

we have, by Subsect. 9.1.4,

$$(\Omega_{z_1}|H_1\Omega_{z_1}) = (z_1|h_1 z_1) + (w_1|z_1) + (z_1|w_1).$$

By Thm. 11.68 (2), this implies that $w_1 \in h_1^{\frac{1}{2}}\mathcal{Z}_1$. This completes the proof of (2).

To prove (3), we note that H_2 has the ground state $U_2\Omega_2$. Hence, H has a ground state iff H_1 has one. If $w_1 \in h_1\mathcal{Z}_1$, then $H_1 = U_1 d\Gamma(h_1)U_1^*$ for $U_1 = \exp\left(-a^*(h_1^{-1}w_1) + a(h_1^{-1})w_1\right)$, hence it has a ground state.

Assume now that H_1 has a ground state Ψ. We again split $\mathcal{Z}_1$ into $\mathcal{Z}_1^\epsilon \oplus \mathcal{Z}_1^{\epsilon\perp}$, for $\mathcal{Z}_1^\epsilon = \mathbb{1}_{[0,\epsilon]}(h)\mathcal{Z}_1$. Then H_1 splits into $H_1^\epsilon \otimes \mathbb{1} + \mathbb{1} \otimes H_1^{\epsilon\perp}$, w_1 into $w_1^\epsilon \oplus w_1^{\epsilon\perp}$ and Ψ into $\Psi^\epsilon \otimes \Psi^{\epsilon\perp}$. Since $w_1^{\epsilon\perp} \in h_1^{-1}\mathcal{Z}_1^{\epsilon\perp}$, we have

$$\Psi^{\epsilon\perp} = \exp\left(-a^*(h_1^{-1}w_1^{\epsilon\perp}) - a(h_1^{-1}w_1^{\epsilon\perp})\right)\Omega,$$

and therefore

$$a(z)\Psi = (z|h^{-1}w_1)\Psi, \quad z \in \mathcal{Z}_1^{\epsilon,\perp}.$$

We apply Thm. 11.58 with $\mathcal{Z} = \bigcup_{\epsilon>0} \mathcal{Z}_1^{\epsilon\perp}$, so that $\mathcal{Z}^{\mathrm{cpl}} = \mathcal{Z}_1$, and we obtain that $w_1 \in h_1\mathcal{Z}_1$. $\qquad\square$

Formally,

$$H_{\mathrm{I}} = d\Gamma(h) + a^*(w) + a(w),$$
$$H_{\mathrm{II}} = d\Gamma(h) + a^*(w) + a(w) + (w|h^{-1}w)\mathbb{1}.$$

11.6.5 Nine classes of van Hove Hamiltonians

We can sum up the theory of van Hove Hamiltonians by dividing them into three classes based on the infrared behavior and three classes based on the ultraviolet behavior. Altogether we obtain $3 \times 3 = 9$ classes.

Infrared regularity

1. $(w|\mathbb{1}_{[0,1]}(h)h^{-2}w) < \infty$.

In the classical case, H has a minimum; and in the quantum case, H has a ground state.

2. $(w|\mathbb{1}_{[0,1]}(h)h^{-1}w) < \infty$, $(w|\mathbb{1}_{[0,1]}(h)h^{-2}w) = \infty$.

H is bounded from below and H_{II} is well defined, but in the classical case H has no minimum, and in the quantum case H has no ground state.

3. $(w|\mathbb{1}_{[0,1]}(h)w) < \infty$, $(w|\mathbb{1}_{[0,1]}(h)h^{-1}w) = \infty$.

H is unbounded from below; H_{II} is ill defined.

Ultraviolet regularity

1. $(w|\mathbb{1}_{[1,\infty[}(h)w) < \infty$.

In the classical case the perturbation is bounded; in the quantum case the perturbation is a closable operator.

2. $(w|\mathbb{1}_{[1,\infty[}(h)h^{-1}w) < \infty$, $(w|\mathbb{1}_{[1,\infty[}(h)w) = \infty$.

H_I is well defined, but in the classical case the perturbation is not bounded, and in the quantum case the perturbation is not a closable operator.

3. $(w|\mathbb{1}_{[1,\infty]}(h)h^{-2}w) < \infty$, $(w|\mathbb{1}_{[1,\infty]}(h)h^{-1}w) = \infty$.

The constant c in (11.107) is infinite; H_I is ill defined.

11.7 Notes

The existence of many inequivalent representations of CCR was noticed in the 1950s, e.g. by Segal (1963) and Gårding–Wightman (1954). Shale's theorem was first proven in Shale (1962). Among early works describing implementations of symplectic transformations on Fock spaces let us mention the books by Friedrichs (1953) and by Berezin (1966). They give concrete formulas for the implementation of Bogoliubov transformations in bosonic Fock spaces. Related problems were discussed, often independently, by other researchers, such as Ruijsenaars (1976, 1978) and Segal (1959, 1963).

Infinite-dimensional analogs of the metaplectic representation seem to have been first noted by Lundberg (1976).

The book by Neretin (1996) and the review article by Varilly–Gracia-Bondia (1992) describe the infinite-dimensional metaplectic group.

The Fock sector of a CCR representation is discussed e.g. in Bratteli–Robinson (1996). It is, in particular, useful in the context of scattering theory; see Chap. 22 and Dereziński–Gérard (1999, 2000, 2004).

Coherent representations appeared already in the book by Friedrichs (1953), and were used by Roepstorff (1970). Our presentation follows Dereziński–Gérard (2004).

The ultraviolet problem of van Hove Hamiltonians is discussed e.g. in the books of Berezin (1966), Sect. III.7.4, and of Schweber (1962), following earlier papers by van Hove (1952), Edwards–Peierls (1954) and Tomonaga (1946). The name "van Hove model" is used in Schweber (1962).

The understanding of the infrared problem of van Hove Hamiltonians can be traced back to the papers by Bloch–Nordsieck (1937) and by Kibble (1968).

Our presentation of the theory of van Hove Hamiltonians follows Dereziński (2003).

12

Canonical anti-commutation relations

Throughout this chapter, $(\mathcal{Y}, \nu)$ is a Euclidean space, that is, a real vector space $\mathcal{Y}$ equipped with a positive definite form ν.

In this chapter we introduce the concept of *representations of the canonical anti-commutation relations* (CAR representations). The definition that we use is very similar to the definition of a *representation of the Clifford relations*, which will be discussed in Chap. 15. In the case of CAR representations we assume in addition that operators satisfying the Clifford relations act on a Hilbert space and are self-adjoint, whereas in the standard definition of Clifford relations the self-adjointness is not required.

CAR representations are used in quantum physics to describe fermions. Actually, CAR representations, as introduced in Def. 12.1, are appropriate for the so-called neutral fermions. Most fermions in physics are charged, and for them a slightly different formalism is used, which we introduce under the name *charged CAR representations*. Charged CAR representations can be viewed as a special case of (neutral) CAR representations, where the dual phase space $\mathcal{Y}$ is complex and a somewhat different notation is used.

CAR representations appear in quantum physics in at least two contexts. First, they describe fermionic systems. This is to us the primary meaning of the CAR, and most of our motivation and terminology is derived from it. Second, they describe spinors, that is, representations of the *Spin* and *Pin groups*. In most applications the second meaning is restricted to the finite-dimensional case. We will also discuss the second meaning (including the Spin and Pin groups over infinite-dimensional spaces).

12.1 CAR representations

12.1.1 Definition of a CAR representation

Let $\mathcal{H}$ be a Hilbert space. Recall that $B_{\mathrm{h}}(\mathcal{H})$ denotes the set of bounded self-adjoint operators on $\mathcal{H}$ and $[A, B]_+ := AB + BA$ is the anti-commutator of A and B.

Definition 12.1 *A representation of the canonical anti-commutation relations or a CAR representation over $\mathcal{Y}$ in $\mathcal{H}$ is a linear map*

$$\mathcal{Y} \ni y \mapsto \phi^\pi(y) \in B_{\mathrm{h}}(\mathcal{H}) \tag{12.1}$$

satisfying

$$[\phi^\pi(y_1), \phi^\pi(y_2)]_+ = 2y_1 \cdot \nu y_2 \, \mathbb{1}, \quad y_1, y_2 \in \mathcal{Y}. \tag{12.2}$$

The operators $\phi^\pi(y)$ are called (fermionic) *field operators.*

Remark 12.2 *The superscript π is an example of a "name" of a given CAR representation.*

Remark 12.3 *Unfortunately, the analogy between the CAR (12.2) and the CCR (8.23) is somewhat violated by the number 2 on the r.h.s. of (12.2). The reason for this convention is the identity $\phi^\pi(y)^2 = (y \cdot \nu y)\mathbb{1}$.*

Remark 12.4 *Later on we will sometimes call (12.1)* neutral CAR representations, *to distinguish them from* charged CAR representations *introduced in Def. 12.17.*

In what follows we assume that we are given a CAR representation (12.1). By complex linearity we can extend the definition of $\phi^\pi(y)$ to $\mathbb{C}\mathcal{Y}$:

$$\phi^\pi(y_1 + iy_2) := \phi^\pi(y_1) + i\phi^\pi(y_2), \quad y_1, y_2 \in \mathcal{Y}.$$

Definition 12.5 *The operators $\phi^\pi(w)$ for $w \in \mathbb{C}\mathcal{Y}$ are also called* field operators.

We have

$$[\phi^\pi(w_1), \phi^\pi(w_2)]_+ = 2w_1 \cdot \nu_\mathbb{C} w_2 \, \mathbb{1}, \quad w_1, w_2 \in \mathbb{C}\mathcal{Y},$$

where $\nu_\mathbb{C}$ is the complexification of ν.

We will sometimes use a different terminology. Let I be a set. We will say that $\{\phi_i^\pi : i \in I\} \subset B_h(\mathcal{H})$ is a CAR representation iff

$$[\phi_i^\pi, \phi_j^\pi]_+ = 2\delta_{ij}. \tag{12.3}$$

Clearly, if $\mathcal{Y} \ni y \mapsto \phi^\pi(y)$ is a CAR representation and we choose an o.n. basis $\{e_i : i \in I\}$, then $\phi_i^\pi := \phi^\pi(e_i)$ is a CAR representation in the second meaning.

Theorem 12.6 *Introduce the notation $|y|_\nu := (y \cdot \nu y)^{\frac{1}{2}}$. Let $y \in \mathcal{Y}$, $\dim \mathcal{Y} > 1$.*

(1) $\mathrm{spec}\ \phi^\pi(y) = \{-|y|_\nu, |y|_\nu\}$, $\|\phi^\pi(y)\| = |y|_\nu$.
(2) *Let $t \in \mathbb{C}$, $y \in \mathcal{Y}$. Then $\|t\mathbb{1} + \phi^\pi(y)\| = \max\{|t + |y|_\nu|, |t - |y|_\nu|\}$.*
(3) $e^{i\phi^\pi(y)} = \cos|y|_\nu \, \mathbb{1} + i\frac{\sin|y|_\nu}{|y|_\nu}\phi^\pi(y)$.
(4) *Let $\mathcal{Y}^{\mathrm{cpl}}$ be the completion of $\mathcal{Y}$. Then there exists a unique extension of (12.1) to a continuous map*

$$\mathcal{Y}^{\mathrm{cpl}} \ni y \mapsto \phi^{\pi^{\mathrm{cpl}}}(y) \in B(\mathcal{H}). \tag{12.4}$$

(12.4) is a representation of CAR.

Proof Since $(\phi^\pi)^2(y) = |y|_\nu^2 \, \mathbb{1}$, we have spec $\phi^\pi(y) \subset \{-|y|_\nu, |y|_\nu\}$. If there exists $y_0 \in \mathcal{Y}$ with $y_0 \neq 0$ and $\phi^\pi(y_0) = \lambda \mathbb{1}_\mathcal{H}$, then $\dim \mathcal{Y} = 1$. But we assumed that $\dim \mathcal{Y} > 1$. Therefore, the spectrum of $\phi^\pi(y)$ cannot consist of only one element, which proves (1). Statements (2) and (3) follow from (12.2), and statement (4) follows from (1).   □

Motivated by Thm. 12.6 (4), henceforth we will assume that $\mathcal{Y}$ is a real Hilbert space.

12.1.2 CAR representations over a direct sum

Constructing a CAR representation over a direct sum of two spaces is not as simple as the analogous construction for CCR relations (compare with Prop. 8.6).

Proposition 12.7 *Let* $\mathcal{Y}_1$, $\mathcal{Y}_2$ *be two real Hilbert spaces. Suppose that* $I_1 \in B(\mathcal{H}_1)$ *is such that*

$$\mathcal{Y}_1 \oplus \mathbb{R} \ni (y_1, t) \mapsto \phi^1(y_1) + tI_1 \in B(\mathcal{H}_1),$$
$$\mathcal{Y}_2 \ni y_2 \mapsto \phi^2(y_2) \in B(\mathcal{H}_2)$$

are CAR representations. Then

$$\mathcal{Y}_1 \oplus \mathcal{Y}_2 \ni (y_1, y_2) \mapsto \phi^1(y_1) \otimes \mathbb{1} + I_1 \otimes \phi^2(y_2) \in B(\mathcal{H}_1 \otimes \mathcal{H}_2)$$

is a CAR representation.

12.1.3 Cyclicity and irreducibility

The following concepts are essentially the same as in the case of CCR representations.

Definition 12.8 *We say that a subset* $\mathcal{U} \subset \mathcal{H}$ *is cyclic for (12.1) if*

$$\mathrm{Span}\{\phi^\pi(y) \cdots \phi^\pi(y_n)\Psi \ : \ \Psi \in \mathcal{U}, \ y_1, \ldots, y_n \in \mathcal{Y}\}$$

is dense in $\mathcal{H}$. *We say that* $\Psi_0 \in \mathcal{H}$ *is cyclic for (12.1) if* $\{\Psi_0\}$ *is cyclic for (12.1).*

Definition 12.9 *We say that the representation (12.1) is irreducible if the only closed subspaces of* $\mathcal{H}$ *invariant under the* $\phi^\pi(y)$ *for* $y \in \mathcal{Y}$ *are* $\{0\}$ *and* $\mathcal{H}$.

Proposition 12.10 (1) *A CAR representation is irreducible iff* $B \in B(\mathcal{H})$ *and* $[\phi^\pi(y), B] = 0$ *for all* $y \in \mathcal{Y}$ *implies that* B *is proportional to identity.*
(2) *In the case of an irreducible CAR representation, all non-zero vectors in* $\mathcal{H}$ *are cyclic.*

12.1.4 Intertwining operators

Let

$$\mathcal{Y} \ni y \mapsto \phi^1(y) \in B_{\mathrm{h}}(\mathcal{H}_1), \tag{12.5}$$
$$\mathcal{Y} \ni y \mapsto \phi^2(y) \in B_{\mathrm{h}}(\mathcal{H}_2) \tag{12.6}$$

be CAR representations over the same Euclidean space $\mathcal{Y}$.

Definition 12.11 *We say that an operator $A \in B(\mathcal{H}_1, \mathcal{H}_2)$ intertwines (12.5) and (12.6) iff*

$$A\phi^1(y) = \phi^2(y)A, \quad y \in \mathcal{Y}.$$

We say that it anti-intertwines (12.5) *and* (12.6) *iff*

$$A\phi^1(y) = -\phi^2(y)A, \quad y \in \mathcal{Y}.$$

We say that (12.5) and (12.6) are unitarily equivalent, *resp.* anti-equivalent *if there exists a unitary $U \in U(\mathcal{H}_1, \mathcal{H}_2)$ intertwining, resp. anti-intertwining (12.5) and (12.6).*

Proposition 12.12 *If the representations (8.11) and (8.12) are irreducible, then the set of operators (anti-)intertwining them equals either $\{0\}$ or $\{\lambda U : \lambda \in \mathbb{C}\}$ for some $U \in U(\mathcal{H})$.*

Proof The proof is an obvious modification of the proof of the analogous fact about CCR representations and about C^*-algebras; see Thm. 8.13. □

12.1.5 Volume element

Consider a CAR representation (12.1). Let $\mathcal{X}$ be a finite-dimensional oriented subspace of $\mathcal{Y}$. Let $(e_1, \dots, e_n)$ be an o.n. basis of $\mathcal{X}$ compatible with the orientation.

Definition 12.13 *The* volume element of the subspace $\mathcal{X}$ in the representation (12.1) *is defined by*

$$Q_{\mathcal{X}}^\pi := \phi^\pi(e_1) \cdots \phi^\pi(e_n).$$

In what follows we drop the superscript π. Note that $Q_{\mathcal{X}}$ does not depend on the choice of an oriented o.n. basis. Changing the orientation amounts to changing $Q_{\mathcal{X}}$ into $-Q_{\mathcal{X}}$. We have

$$Q_{\mathcal{X}}^2 = (-1)^{n(n-1)/2} \mathbb{1}, \quad Q_{\mathcal{X}}^* = Q_{\mathcal{X}}^{-1} = (-1)^{n(n-1)/2} Q_{\mathcal{X}}.$$

Thus $Q_{\mathcal{X}}$ is self-adjoint iff $n = 0, 1 \pmod 4$; otherwise it is anti-self-adjoint.
Define $u_{\mathcal{X}} \in O(\mathcal{Y})$ by

$$u_{\mathcal{X}} = (-\mathbb{1}) \oplus \mathbb{1},$$

where we use the decomposition $\mathcal{Y} = \mathcal{X} \oplus \mathcal{X}^\perp$. Clearly,

$$Q_{\mathcal{X}}\phi(y)Q_{\mathcal{X}}^{-1} = (-1)^n \phi(u_{\mathcal{X}}y), \quad y \in \mathcal{Y}.$$

12.1.6 CAR over Kähler spaces

In this subsection we fix a CAR representation (12.1). We use the notation and results of Subsects. 1.3.6, 1.3.8 and 1.3.9.

The following proposition shows that choosing a sufficiently large subspace of anti-commuting field operators is equivalent to fixing a Kähler structure in $(\mathcal{Y}, \nu)$.

Proposition 12.14 *Suppose that* $\mathcal{Z} \subset \mathbb{C}\mathcal{Y}$ *is a subspace such that*

(1) $\mathbb{C}\mathcal{Y} = \mathcal{Z} \oplus \overline{\mathcal{Z}}$;
(2) $z_1, z_2 \in \mathcal{Z}$ *implies* $[\phi^\pi(z_1), \phi^\pi(z_2)]_+ = 0$ *(or equivalently, $\mathcal{Z}$ is isotropic for $\nu_{\mathbb{C}}$).*

Then there exists a unique Kähler anti-involution j *on* $(\mathcal{Y}, \nu)$ *such that*

$$\mathcal{Z} = \{y - \mathrm{i}\mathrm{j}y \ : \ y \in \mathcal{Y}\}. \tag{12.7}$$

Proof (1) implies that there exists a linear map $\mathrm{j} \in L(\mathcal{Y})$ such that $\mathcal{Z}$ is given by (12.7). (2) implies

$$\begin{aligned}
0 &= (y_1 + \mathrm{i}\mathrm{j}y_1) \cdot \nu_{\mathbb{C}}(y_2 + \mathrm{i}\mathrm{j}y_2) \\
&= y_1 \cdot \nu y_2 - (\mathrm{j}y_1) \cdot \nu(\mathrm{j}y_2) + \mathrm{i}\left((\mathrm{j}y_1) \cdot \nu y_2 + y_1 \cdot \nu \mathrm{j}y_2\right).
\end{aligned}$$

Hence,

$$y_1 \cdot \nu y_2 - (\mathrm{j}y_1) \cdot \nu(\mathrm{j}y_2) = 0, \quad (\mathrm{j}y_1) \cdot \nu y_2 + y_1 \cdot \nu \mathrm{j}y_2 = 0,$$

which shows that j is orthogonal and anti-symmetric, hence is a Kähler anti-involution. $\square$

Motivated in part by the above proposition, let us fix j, a Kähler anti-involution on $(\mathcal{Y}, \nu)$. Recall that the space $\mathcal{Z}$ given by (12.7) is called the holomorphic subspace of $\mathbb{C}\mathcal{Y}$.

Definition 12.15 *We define the* j-creation *and* j-annihilation *operators:*

$$a^{\pi*}(z) := \phi^\pi(z), \quad a^\pi(z) := \phi^\pi(\overline{z}), \quad z \in \mathcal{Z}.$$

They are bounded operators, adjoint to one another.

Proposition 12.16 *One has* $\phi^\pi(z, \overline{z}) = a^{\pi*}(z) + a^\pi(z), z \in \mathcal{Z}$,

$$[a^{\pi*}(z_1), a^{\pi*}(z_2)]_+ = 0, \quad [a^\pi(z_1), a^\pi(z_2)]_+ = 0,$$

$$[a^\pi(z_1), a^{\pi*}(z_2)]_+ = (z_1|z_2)\mathbb{1}, \quad z_1, z_2 \in \mathcal{Z}.$$

The Kähler structure appears naturally in the context of Fock representations. It also arises when the Euclidean space $(\mathcal{Y}, \nu)$ is equipped with a charge 1 $U(1)$ symmetry, as in Subsect. 1.3.11. We discuss the latter application in the following subsection.

12.1.7 Charged CAR representations

CAR representations, as defined in Def. 12.1, provide a natural framework for the description of neutral fermions. Therefore, sometimes we will call them *neutral CAR representations*. In the context of charged fermions (much more common than neutral fermions) physicists prefer to use another formalism described in the following definition.

Definition 12.17 *Suppose that* $(\mathcal{Y}, (\cdot|\cdot))$ *is a unitary space and* $\mathcal{H}$ *a Hilbert space. We say that an anti-linear map*

$$\mathcal{Y} \ni y \mapsto \psi^\pi(y) \in B(\mathcal{H})$$

is a charged CAR representation *if*

$$[\psi^{\pi*}(y_1), \psi^{\pi*}(y_2)]_+ = [\psi^\pi(y_1), \psi^\pi(y_2)]_+ = 0,$$
$$[\psi^\pi(y_1), \psi^{\pi*}(y_2)]_+ = (y_1|y_2)\mathbb{1}, \qquad y_1, y_2 \in \mathcal{Y}.$$

Suppose that $y \mapsto \psi^\pi(y)$ is a charged CAR representation. Set

$$\phi^\pi(y) := \big(\psi^\pi(y) + \psi^{\pi*}(y)\big),$$
$$y_1 \cdot_\nu y_2 := \mathrm{Re}(y_1|y_2).$$

Then $\mathcal{Y} \ni y \mapsto \phi^\pi(y) \in B_\mathrm{h}(\mathcal{H})$ is a neutral CAR representation over the Euclidean space $(\mathcal{Y}, \nu)$. In addition, $\mathcal{Y}$ is equipped with a charge 1 symmetry $U(1) \ni \theta \mapsto \mathrm{e}^{\mathrm{i}\theta} \in O(\mathcal{Y})$.

Conversely, charged CAR representations arise when we have a (neutral) CAR representation and the underlying Euclidean space is equipped with a charge 1 $U(1)$ symmetry. Let us make this precise. Suppose that $(\mathcal{Y}, \nu)$ is a Euclidean space and

$$\mathcal{Y} \ni y \mapsto \phi^\pi(y) \in B_\mathrm{h}(\mathcal{H})$$

is a neutral CAR representation. Suppose that

$$U(1) \ni \theta \mapsto u_\theta = \cos\theta \mathbb{1} + \sin\theta \mathrm{j}_\mathrm{ch} \in O(\mathcal{Y})$$

is a charge 1 symmetry. We know that j_ch is a Kähler anti-involution. Following the standard procedure described in the previous subsection, we introduce the holomorphic subspace for j_ch, that is,

$$\mathcal{Z}_\mathrm{ch} := \{y - \mathrm{i}\mathrm{j}_\mathrm{ch}y \ : \ y \in \mathcal{Y}\} \subset \mathbb{C}\mathcal{Y}.$$

We define creation and annihilation operators associated with j_ch. We have a natural identification of the space $\mathcal{Z}_\mathrm{ch}$ with $\mathcal{Y}$:

$$\mathcal{Y} \ni y \mapsto z = \frac{1}{2}(\mathbb{1} - \mathrm{i}\mathrm{j}_\mathrm{ch})y \in \mathcal{Z}_\mathrm{ch}. \tag{12.8}$$

We use the identification (12.8) to introduce charged fields parametrized by elements of $\mathcal{Y}$:

$$\psi^{\pi*}(y) := \phi^{\pi}(z), \qquad \psi^{\pi}(y) := \phi^{\pi}(\bar{z}).$$

Then we obtain a charged CAR representation over $\mathcal{Y}^{\mathbb{C}}$ with the complex structure given by j_{ch} and the scalar product

$$(y_1|y_2) := y_1 \cdot \nu y_2 - iy_1 \cdot \nu j_{\mathrm{ch}} y_2, \quad y_1, y_2 \in \mathcal{Y}. \tag{12.9}$$

12.1.8 Bogoliubov rotations

Consider a CAR representation (12.1). To simplify notation, we drop π, that is, we consider a CAR representation

$$\mathcal{Y} \ni y \mapsto \phi(y) \in B_{\mathrm{h}}(\mathcal{H}). \tag{12.10}$$

Let $r \in O(\mathcal{Y})$. Clearly,

$$\mathcal{Y} \ni y \mapsto \phi^r(y) := \phi(ry) \in U(\mathcal{H}) \tag{12.11}$$

is also a CAR representation.

Definition 12.18 *We say that the representation (12.11) is* the Bogoliubov rotation *or* transformation of the representation (12.10) by $r^{\#}$.

Proposition 12.19 (1) *If $r_1, r_2 \in O(\mathcal{Y})$, then $(\phi^{r_1})^{r_2}(y) = \phi^{r_2 r_1}(y)$.*
(2) *The set of $r \in O(\mathcal{Y})$ such that (12.11) is unitarily equivalent to (12.10) is a subgroup of $O(\mathcal{Y})$.*
(3) *(12.11) is irreducible iff (12.10) is.*

12.2 CAR representations in finite dimensions

Throughout the section we assume that $(\mathcal{Y}, \nu)$ is a finite-dimensional Euclidean space.

In this section we discuss CAR representations in the finite-dimensional case. In the literature the material of this section is usually described as a part of the theory of spinors and Clifford algebras.

12.2.1 Volume element

Suppose that $\mathcal{Y}$ is oriented and we are given a CAR representation (12.1). Let $(e_1, \ldots, e_n)$ be an o.n. basis of $\mathcal{Y}$ compatible with the orientation. The following definition is a special case of Def. 12.13.

Definition 12.20 *The operator*

$$Q^{\pi} := \phi^{\pi}(e_1) \cdots \phi^{\pi}(e_n)$$

will be called the volume element *in the representation (12.1)*

In what follows we drop the superscript π. Let us summarize the properties of Q, which follow from Subsect. 12.1.5:

Theorem 12.21 (1) *Q depends only on the orientation of $\mathcal{Y}$ and changes sign under the change of the orientation.*

(2) *Q is unitary. It is self-adjoint for $n \equiv 0, 1 \,(\mathrm{mod}\,4)$, otherwise anti-self-adjoint. Moreover, $Q^2 = (-1)^{n(n-1)/2}$.*

(3) *$Q\phi(y) = (-1)^{n-1}\phi(y)Q$, $y \in \mathcal{Y}$.*

(4) *If $n = 2m$, then $Q^2 = (-1)^m$, Q anti-commutes with $\phi(y)$, $y \in \mathcal{Y}$, and*

$$\mathbb{R}^{2m+1} \ni (y, t) \mapsto \phi(y) \pm t\mathrm{i}^m Q$$

are two representations of the CAR.

(5) *If $n = 2m + 1$, then $Q^2 = (-1)^m$, Q commutes with $\phi(y)$, $y \in \mathcal{Y}$, and*

$$\mathcal{H} = \mathrm{Ker}(Q - \mathrm{i}^m \mathbb{1}) \oplus \mathrm{Ker}(Q + \mathrm{i}^m \mathbb{1})$$

gives a decomposition of $\mathcal{H}$ into a direct sum of subspaces invariant for the CAR representation.

Definition 12.22 *Let* $\dim \mathcal{Y} = 2m + 1$. *We will say that a CAR representation is compatible with the orientation if $Q = \mathrm{i}^m \mathbb{1}$.*

12.2.2 Pauli matrices

Consider the space $\mathbb{C}^2$. Occasionally we will need its canonical basis, whose elements will be denoted $|\uparrow\rangle, |\downarrow\rangle$.

Definition 12.23 *Pauli matrices are defined as*

$$\sigma_1 = \begin{bmatrix} 0 & 1 \\ 1 & 0 \end{bmatrix}, \quad \sigma_2 = \begin{bmatrix} 0 & -\mathrm{i} \\ \mathrm{i} & 0 \end{bmatrix}, \quad \sigma_3 = \begin{bmatrix} 1 & 0 \\ 0 & -1 \end{bmatrix}.$$

Note that $\sigma_i^2 = 1$, $\sigma_i^* = \sigma_i$, $i = 1, 2, 3$, and

$$\sigma_1\sigma_2 = -\sigma_2\sigma_1 = \mathrm{i}\sigma_3,$$

$$\sigma_2\sigma_3 = -\sigma_3\sigma_2 = \mathrm{i}\sigma_1,$$

$$\sigma_3\sigma_1 = -\sigma_1\sigma_3 = \mathrm{i}\sigma_2.$$

Moreover, $B(\mathbb{C}^2)$ is generated by $\{\sigma_1, \sigma_2\}$. Clearly, $\{\sigma_1, \sigma_2, \sigma_3\}$ is a CAR representation over $\mathbb{R}^3$.

Lemma 12.24 *Let $\{\phi_1, \phi_2\}$ be a CAR representation over $\mathbb{R}^2$ in a Hilbert space $\mathcal{H}$. Then there exists a Hilbert space $\mathcal{K}$ and a unitary operator $U : \mathbb{C}^2 \otimes \mathcal{K} \to \mathcal{H}$ such that*

$$\sigma_1 \otimes \mathbb{1}_\mathcal{K}\, U = U\phi_1, \quad \sigma_2 \otimes \mathbb{1}_\mathcal{K}\, U = U\phi_2.$$

Proof Set $I := i\phi_1\phi_2$. Clearly, $I = I^*$ and $I^2 = \mathbb{1}$. Note that $I \neq \mathbb{1}$, since $I = \mathbb{1}$ would contradict the fact that ϕ_2 is self-adjoint. Hence, $\operatorname{spec} I = \{1, -1\}$. Let $\mathcal{K} := \operatorname{Ker}(I - \mathbb{1})$. We unitarily identify $\mathcal{H}$ with $\mathcal{K} \oplus \mathcal{K}$ by the map

$$\Psi \mapsto U\Psi := \left(\frac{1}{2}(\phi_1 - \phi_1 I)\Psi, \frac{1}{2}(\mathbb{1} + I)\Psi\right).$$

Then $U\phi_1 = \sigma_1 \otimes \mathbb{1}_{\mathcal{K}} U$, $U\phi_2 = \sigma_2 \otimes \mathbb{1}_{\mathcal{K}} U$. □

12.2.3 Jordan–Wigner representation

In this subsection we introduce certain basic CAR representations over a finite-dimensional space. We start with the case of an even dimension, which is simpler.

In the algebra $B(\otimes^m \mathbb{C}^2)$ we introduce the operators

$$\sigma_i^{(j)} := \mathbb{1}^{\otimes(j-1)} \otimes \sigma_i \otimes \mathbb{1}^{\otimes(m-j)}, \quad i = 1, 2, 3, \quad j = 1, \dots, m.$$

Note that $\sigma_3^{(j)} = i\sigma_1^{(j)}\sigma_2^{(j)}$. Moreover, $B(\otimes^m \mathbb{C}^2)$ is generated by

$$\{\sigma_i^{(j)} : j = 1, \dots, m, \ i = 1, 2\}.$$

We also set $I_0 := \mathbb{1}$, $I_j := \sigma_3^{(1)} \cdots \sigma_3^{(j)}$ for $j = 1, \dots, m$. If we set

$$\phi_{2j-1}^{\mathrm{JW}} := I_{j-1}\sigma_1^{(j)}, \quad \phi_{2j}^{\mathrm{JW}} := I_{j-1}\sigma_2^{(j)}, \ j = 1, \dots, m, \tag{12.12}$$

it is easy to see that

$$(\phi_1^{\mathrm{JW}}, \dots, \phi_{2m}^{\mathrm{JW}}) \tag{12.13}$$

is an irreducible CAR representation over $\mathbb{R}^{2m}$ in the Hilbert space $\otimes^m \mathbb{C}^2$. Note that $Q = I_m$.

Definition 12.25 *The CAR representation (12.13) will be called the* Jordan–Wigner representation over $\mathbb{R}^{2m}$.

In the odd-dimensional case with $n = 2m + 1$, there exist two inequivalent irreducible CAR representations, both in $\otimes^m \mathbb{C}^2$. They are obtained by adding the operator $\pm I_m$ to (12.12). In other words,

$$(\phi_1^{\mathrm{JW}}, \dots, \phi_{2m}^{\mathrm{JW}}, I_m), \tag{12.14}$$

$$(\phi_1^{\mathrm{JW}}, \dots, \phi_{2m}^{\mathrm{JW}}, -I_m) \tag{12.15}$$

are irreducible CAR representations over $\mathbb{R}^{2m+1}$. We have $Q = i^m \mathbb{1}$ in the case of (12.14) and $Q = -i^m \mathbb{1}$ in the case of (12.15). Thus the representation (12.14) is compatible with the natural orientation of $\mathbb{R}^{2m+1}$.

Another useful, but reducible, CAR representation over $\mathbb{R}^{2m+1}$ acts on the space $\otimes^{m+1} \mathbb{C}^2$. It is given by

$$(\phi_1^{\mathrm{JW}}, \dots, \phi_{2m}^{\mathrm{JW}}, \phi_{2m+1}^{\mathrm{JW}}). \tag{12.16}$$

It decomposes into the sum of two irreducible sub-representations, one equivalent to (12.14) and the other equivalent to (12.15). We have $Q = i^m \, \mathbb{1}^{\otimes m} \otimes \sigma_1$. The commutant of (12.16) is spanned by $\mathbb{1}^{\otimes (m+1)}$ and Q.

Definition 12.26 *The CAR representation (12.16) will be called the Jordan–Wigner representation over* $\mathbb{R}^{2m+1}$.

12.2.4 Unitary equivalence of the CAR in finite dimensions

The following two theorems can be viewed as the fermionic analog of the Stone–von Neumann theorem. Again, we first deal with the even-dimensional case.

Theorem 12.27 *Let* $(\phi_1, \phi_2, \ldots, \phi_{2m})$ *be a CAR representation over* $\mathbb{R}^{2m}$ *in a Hilbert space* $\mathcal{H}$. *Then there exists a Hilbert space* $\mathcal{K}$ *and a unitary operator* $U : \otimes^m \mathbb{C}^2 \otimes \mathcal{K} \to \mathcal{H}$ *such that*

$$U \phi_j^{\mathrm{JW}} \otimes \mathbb{1}_{\mathcal{K}} = \phi_j U, \quad j = 1, \ldots, 2m.$$

The representation is irreducible iff $\mathcal{K} = \mathbb{C}$.

In particular, every irreducible CAR representation over an even-dimensional space is unitarily equivalent to the corresponding Jordan–Wigner representation.

Proof Set

$$\tilde{I}_0 := \mathbb{1}, \qquad \tilde{I}_j := i^j \phi_1 \cdots \phi_j, \quad j = 1, \ldots, n,$$
$$\tilde{\sigma}_1^{(j)} := \tilde{I}_{j-1} \phi_{2j-1}, \quad \tilde{\sigma}_2^{(j)} := \tilde{I}_{j-1} \phi_{2j}, \quad j = 1, \ldots, m.$$

From the CAR we get

$$\tilde{I}_j^* = \tilde{I}_j, \quad \tilde{I}_j^2 = \mathbb{1},$$
$$\phi_k \tilde{I}_j = -\tilde{I}_j \phi_k, \quad k \le 2j, \quad \phi_k \tilde{I}_j = \tilde{I}_j \phi_k, \quad k > 2j. \tag{12.17}$$

This implies that

$$\tilde{\sigma}_1^{(j)} \tilde{\sigma}_2^{(j)} = \phi_{2j-1} \phi_{2j}, \quad \tilde{I}_j = i^j \tilde{\sigma}_1^{(1)} \tilde{\sigma}_2^{(1)} \cdots \tilde{\sigma}_1^{(j)} \tilde{\sigma}_2^{(j)},$$
$$\phi_{2j-1} = \tilde{I}_{j-1} \tilde{\sigma}_1^j, \quad \phi_{2j} = \tilde{I}_{j-1} \tilde{\sigma}_2^{(j)}. \tag{12.18}$$

We observe that the pairs $\{\tilde{\sigma}_1^{(j)}, \tilde{\sigma}_2^{(j)}\}$ satisfy the CAR over $\mathbb{R}^2$ and commute with each other. Applying Lemma 12.24 inductively we see that there exists a Hilbert space $\mathcal{K}$ and a unitary map $U : \otimes^n \mathbb{C}^2 \otimes \mathcal{K} \to \mathcal{H}$ such that

$$U \sigma_1^{(j)} \otimes \mathbb{1}_{\mathcal{K}} = \tilde{\sigma}_1^{(j)} U, \quad U \sigma_2^{(j)} \otimes \mathbb{1}_{\mathcal{K}} = \tilde{\sigma}_2^{(j)} U, \quad j = 1, \cdots, m.$$

From (12.18) we get that $U I_j \otimes \mathbb{1}_{\mathcal{K}} = \tilde{I}_j U$, and hence

$$U I_{j-1} \sigma_1^{(j)} \otimes \mathbb{1}_{\mathcal{K}} = \phi_{2j-1} U, \quad U I_{j-1} \sigma_2^{(j)} \otimes \mathbb{1}_{\mathcal{K}} = \phi_{2j} U, \quad j = 1, \cdots, m.$$

This completes the proof of the theorem. $\qquad \square$

Theorem 12.28 *Let $(\phi_1, \phi_2, \ldots, \phi_{2m+1})$ be a representation of CAR over $\mathbb{R}^{2m+1}$ in a Hilbert space $\mathcal{H}$. Then there exist Hilbert spaces $\mathcal{K}_-$ and $\mathcal{K}_+$ and a unitary operator $U : \otimes^m \mathbb{C}^2 \otimes (\mathcal{K}_+ \oplus \mathcal{K}_-) \to \mathcal{H}$ such that*

$$U \phi_j^{\mathrm{JW}} \otimes \mathbb{1}_{\mathcal{K}_+ \oplus \mathcal{K}_-} = \phi_j U, \quad j = 1, \ldots, 2m;$$
$$U I_m \otimes (\mathbb{1}_{\mathcal{K}_+} \oplus -\mathbb{1}_{\mathcal{K}_-}) = \phi_{2m+1} U.$$

The representation is irreducible iff $\mathcal{K}_+ \oplus \mathcal{K}_- = \mathbb{C}$.

In particular, every irreducible CAR representation over an odd-dimensional oriented space compatible with its orientation is unitarily equivalent to (12.14).

Proof Let $\tilde{I}_j$, $\tilde{\sigma}_1^{(j)}$, $\tilde{\sigma}_2^{(j)}$ be as in the proof of Thm. 12.27. Let $U_1 : \otimes^m \mathbb{C}^2 \otimes \mathcal{K} \to \mathcal{H}$ be a unitary operator as in Thm. 12.27 for the CAR representation $(\phi_1, \ldots, \phi_{2m})$ over $\mathbb{R}^{2m}$. From (12.17) and the CAR we get

$$[\phi_j, \tilde{I}_m \phi_{2m+1}] = 0, \quad j = 1, \ldots, 2m. \tag{12.19}$$

Since $B(\otimes^m \mathbb{C}^2)$ is generated by $\{\sigma_i^{(j)}, j = 1, \ldots, m, \ i = 1, 2\}$, we see that

$$U_1^* \tilde{I}_m \phi_{2m+1} U_1 = \mathbb{1} \otimes A, \ A \in B(\mathcal{K}).$$

Again using the CAR, we get $(\tilde{I}_m \phi_{2m+1})^2 = \mathbb{1}$. Hence, $A^2 = \mathbb{1}$.

If $A = \pm\mathbb{1}_{\mathcal{K}}$, we get $\tilde{I}_m \phi_{2m+1} = \pm\mathbb{1}_{\mathcal{H}}$. Hence, $\phi_{2m+1} = \pm\tilde{I}_m$. In this case the CAR representation is one of the two constructed in Subsect. 12.2.3. In the general case we have $\mathcal{K} = \mathcal{K}_+ \oplus \mathcal{K}_-$, $A = \mathbb{1}_{\mathcal{K}_+} \oplus -\mathbb{1}_{\mathcal{K}_-}$, and hence

$$U I_m \otimes (\mathbb{1}_{\mathcal{K}_+} \oplus -\mathbb{1}_{\mathcal{K}_-}) = \phi_{2m+1} U.$$

The other identities follow from Thm. 12.27. $\qquad\square$

Corollary 12.29 (1) *Suppose that $\mathcal{Y}$ is an even-dimensional Euclidean space. Let $\mathcal{Y} \ni y \mapsto \phi^1(y) \in B_{\mathrm{h}}(\mathcal{H})$ and $\mathcal{Y} \ni y \mapsto \phi^2(y) \in B_{\mathrm{h}}(\mathcal{H})$ be two irreducible representations of the CAR. Then they are unitarily equivalent.*

(2) *The same is true if $\mathcal{Y}$ is odd-dimensional and oriented, and both representations are compatible with its orientation.*

12.3 CAR algebras: finite dimensions

As in the previous section, we assume that $(\mathcal{Y}, \nu)$ is a finite-dimensional Euclidean space.

In this section we discuss $*$-algebras generated by the CAR in finite dimension. As pure $*$-algebras they are not very interesting – they are full matrix algebras over a 2^m-dimensional space in the case of even dimension, and the direct sum of two such algebras in the case of odd dimension. They become interesting when we consider them together with the linear subspace of distinguished elements $\phi(y)$, $y \in \mathcal{Y}$.

12.3.1 CAR algebra

Let $(\mathcal{Y}, \nu)$ be a finite-dimensional Euclidean space.

Definition 12.30 $\mathrm{CAR}(\mathcal{Y})$ *is the complex unital $*$-algebra generated by elements* $\phi(y)$, $y \in \mathcal{Y}$, *with relations*

$$\phi(\lambda y) = \lambda \phi(y), \ \lambda \in \mathbb{R}, \qquad \phi(y_1 + y_2) = \phi(y_1) + \phi(y_2),$$
$$\phi^*(y) = \phi(y), \qquad \phi(y_1)\phi(y_2) + \phi(y_2)\phi(y_1) = 2 y_1 \cdot \nu y_2 \, \mathbb{1}.$$

The following theorem is a simple algebraic fact:

Proposition 12.31 *If*

$$\mathcal{Y} \ni y \mapsto \phi^\pi(y) \in B(\mathcal{H})$$

is a CAR representation, then there exists a unique $*$-homomorphism

$$\pi : \mathrm{CAR}(\mathcal{Y}) \to B(\mathcal{H})$$

such that $\pi(\mathbb{1}) = \mathbb{1}_{\mathcal{H}}$ and $\pi(\phi(y)) = \phi^\pi(y)$, $y \in \mathcal{Y}$.

Definition 12.32 *Applying Prop. 12.31 to the Jordan–Wigner representations (12.13) or (12.16) we obtain $*$-homomorphisms*

$$\pi^{\mathrm{JW}} : \mathrm{CAR}(\mathbb{R}^{2m}) \to B(\otimes^m \mathbb{C}^2),$$
$$\pi^{\mathrm{JW}} : \mathrm{CAR}(\mathbb{R}^{2m+1}) \to B(\otimes^{m+1} \mathbb{C}^2).$$

Proposition 12.33 (1) *The $*$-homomorphisms π^{JW} for $n = 2m$ are bijective and $\mathrm{CAR}(\mathbb{R}^{2m})$ is $*$-isomorphic to $B(\otimes^m \mathbb{C}^2)$.*
(2) *The $*$-homomorphisms π^{JW} for $n = 2m + 1$ are injective and $\mathrm{CAR}(\mathbb{R}^{2m+1})$ is $*$-isomorphic to $B(\otimes^m \mathbb{C}^2) \oplus B(\otimes^m \mathbb{C}^2)$.*

Proof Choose an o.n. basis $(e_1, \ldots, e_n)$ in $\mathcal{Y}$. For an ordered subset $\{i_1, \ldots, i_k\}$ of $\{1, \ldots, n\}$, set $\phi_{i_1, \ldots, i_k} := \mathrm{i}^{k(k-1)/2} \phi_{i_1} \cdots \phi_{i_k}$. It is easy to prove that the elements $\phi_{i_1, \ldots, i_k}$ are self-adjoint and are a basis of $\mathrm{CAR}(\mathbb{R}^n)$. Their commutation relations are determined by the CAR.

Following the construction of the Jordan–Wigner representations we see that $B(\otimes^m \mathbb{C}^2)$, if $n = 2m$, and $B(\otimes^m \mathbb{C}^2) \oplus B(\otimes^m \mathbb{C}^2)$, if $n = 2m + 1$, have self-adjoint bases satisfying the same relations. $\qquad \square$

The Jordan–Wigner representation determines a unique C^*-norm on $\mathrm{CAR}(\mathcal{Y})$. Henceforth we will treat $\mathrm{CAR}(\mathcal{Y})$ as a C^*-algebra.

If $\mathcal{Y}_1 \subset \mathcal{Y}_2$ are two finite-dimensional spaces, then $\mathrm{CAR}(\mathcal{Y}_1)$ is isometrically embedded in $\mathrm{CAR}(\mathcal{Y}_2)$.

12.3.2 Parity

CAR algebras have a natural $\mathbb{Z}_2$-grading. Therefore, they are examples of super-algebras. Consistently with the terminology of super-algebras, we introduce the following definition:

Definition 12.34 *The map* $\alpha(\phi(y)) := -\phi(y)$ *extends to a unique* *-isomorphism* α *of* $\mathrm{CAR}(\mathcal{Y})$. *For* $j = 0, 1$, *we set*

$$\mathrm{CAR}_j(\mathcal{Y}) := \{B \in \mathrm{CAR}(\mathcal{Y}) \ : \ \alpha(B) = (-1)^j B\}.$$

Elements of $\mathrm{CAR}_0(\mathcal{Y})$, *resp.* $\mathrm{CAR}_1(\mathcal{Y})$ *are called* even, *resp.* odd.

Suppose that $\mathcal{Y}$ is oriented. The following definition is closely related to Def. 12.13.

Definition 12.35 *The* volume element *of the algebra* $\mathrm{CAR}(\mathcal{Y})$ *is defined by*

$$Q := \phi(e_1) \cdots \phi(e_n), \tag{12.20}$$

where $(e_1, \ldots, e_n)$ *is any o.n. basis of* $\mathcal{Y}$ *compatible with its orientation.*

Note that Q is never proportional to $\mathbb{1}$ as an element of $\mathrm{CAR}(\mathcal{Y})$.

Proposition 12.36 (1) *Let* $A \in \mathrm{CAR}_0(\mathcal{Y})$ *commute with* $\phi(y)$, $y \in \mathcal{Y}$ *(and hence with all* $\mathrm{CAR}(\mathcal{Y})$). *Then* A *is proportional to* $\mathbb{1}$.
(2) *Let a non-zero* $A \in \mathrm{CAR}_1(\mathcal{Y})$ *commute with* $\phi(y)$, $y \in \mathcal{Y}$ *(and hence with all* $\mathrm{CAR}(\mathcal{Y})$). *Then* $\dim \mathcal{Y}$ *is odd, and* A *is proportional to* Q.
(3) *Let a non-zero* $A \in \mathrm{CAR}(\mathcal{Y})$ *anti-commute with* $\phi(y)$, $y \in \mathcal{Y}$ *(and hence with all* $\mathrm{CAR}_1(\mathcal{Y})$). *Then* $\dim \mathcal{Y}$ *is even, and* A *is proportional to* Q.

12.3.3 Complex conjugation and transposition

Definition 12.37 *The map* $\mathrm{c}(\phi(w)) := \phi(\overline{w})$, $w \in \mathbb{C}\mathcal{Y}$, *extends to a unique anti-linear* *-isomorphism* c *of* $\mathrm{CAR}(\mathcal{Y})$. *We introduce the* Clifford algebra *over* $(\mathcal{Y}, \nu)$ *as the real sub-algebra*

$$\mathrm{Cliff}(\mathcal{Y}) := \{B \in \mathrm{CAR}(\mathcal{Y}) \ : \ \mathrm{c}(B) = B\}. \tag{12.21}$$

We also introduce the transposition $A^{\#} := \mathrm{c}(A^*)$, *which is a linear anti-automorphism.*

$\mathrm{Cliff}(\mathcal{Y})$ is a real *-algebra with a basis

$$\phi_{i_1} \cdots \phi_{i_k}, \quad \{i_1, \ldots, i_k\} \subset \{1, \ldots, n\}. \tag{12.22}$$

In Chap. 15 we will introduce a more general notion of Clifford algebras, defined for an arbitrary symmetric form on a vector space. The algebra $\mathrm{Cliff}(\mathcal{Y})$ defined in Def. 12.37 corresponds to the special case of a real space equipped with a Euclidean scalar product.

12.3.4 Bogoliubov automorphisms

Proposition 12.38 *If $r \in O(\mathcal{Y})$, then the map $\hat{r}(\phi(y)) := \phi(ry)$ extends to a unique $*$-automorphism $\hat{r}$ of $\mathrm{CAR}(\mathcal{Y})$. We have $\widehat{r_1 r_2} = \hat{r}_1 \hat{r}_2$.*

Definition 12.39 $\hat{r}$ *is called the* Bogoliubov automorphism *associated with r.*

12.4 Anti-symmetric quantization and real-wave CAR representation

In this section we introduce a natural parametrization of operators in a CAR algebra by anti-symmetric polynomials. This parametrization, which we call the *anti-symmetric quantization*, can be viewed as the fermionic analog of the Weyl–Wigner quantization.

We also define a representation given by the GNS construction from the tracial state. This representation has some analogy to the real-wave CCR representation considered in Sect. 9.3; therefore we will call it the *real-wave CAR representation*. In this section we describe the real-wave CAR representation only in the case of a finite number of degrees of freedom. We will extend it to the case of an infinite dimension in Subsect. 12.5.3, and then we will continue its study using the formalism of Fock spaces in Subsect. 13.2.1.

In this section, $(\mathcal{Y}, \nu)$ is a finite-dimensional Euclidean space.

12.4.1 Anti-symmetric quantization

Definition 12.40 *Let $y_1, \ldots, y_n \in \mathbb{C}\mathcal{Y}$. We can treat these as elements of $\mathbb{C}\mathrm{Pol}_a^1(\mathcal{Y}^\#)$ and take their product $y_1 \cdots y_n \in \mathbb{C}\mathrm{Pol}_a(\mathcal{Y}^\#)$. We define*

$$\mathrm{Op}(y_1 \cdots y_n) := \frac{1}{n!} \sum_{\sigma \in S_n} \mathrm{sgn}(\sigma) \, \phi(y_{\sigma(1)}) \cdots \phi(y_{\sigma(n)}) \in \mathrm{CAR}(\mathcal{Y}). \tag{12.23}$$

The map extends uniquely to a linear bijective map

$$\mathbb{C}\mathrm{Pol}_a(\mathcal{Y}^\#) \ni b \mapsto \mathrm{Op}(b) \in \mathrm{CAR}(\mathcal{Y}), \tag{12.24}$$

called the anti-symmetric quantization.

The above definition should be compared with Def. 8.65, where the Weyl–Wigner quantization was introduced.

Definition 12.41 *The inverse of (12.24) will be called the* anti-symmetric symbol. *The anti-symmetric symbol of an operator $B \in \mathrm{CAR}^{\mathrm{alg}}(\mathcal{Y})$ will be denoted $s_B \in \mathbb{C}\mathrm{Pol}_a(\mathcal{Y}^\#)$.*

As usual, N denotes the number operator, which in the context of $\mathbb{C}\mathrm{Pol}_a(\mathcal{Y}^\#)$ is perhaps better called the *degree operator*. Recall that in Chap. 3 we introduced

$$\Lambda := (-1)^{N(N-1)/2}, \quad I := (-1)^N;$$

see (3.9) and (3.29). We will use the functional notation for elements of $\mathbb{C}\mathrm{Pol}_a(\mathcal{Y}^\#)$; see Subsect. 7.1.1. The generic variable in $\mathcal{Y}^\#$ will be denoted v. We equip $\mathcal{Y}$ with a volume form compatible with the scalar product ν. We will use the corresponding Berezin integral on $\mathbb{C}\mathrm{Pol}_a(\mathcal{Y}^\#)$, defined in Subsect. 7.1.4.

Proposition 12.42 (1) $\mathrm{Op}(b)^* = \mathrm{Op}(\Lambda\bar{b})$.

(2) *Let $\mathcal{Z}$ be an isotropic subspace of $\mathbb{C}\mathcal{Y}$ for $\nu_\mathbb{C}$. Let $f_1, \ldots, f_n \in \mathrm{Pol}_a(\mathcal{Z}^\#) \subset \mathbb{C}\mathrm{Pol}_a(\mathcal{Y}^\#)$. Then*

$$\mathrm{Op}(f_1) \cdots \mathrm{Op}(f_n) = \mathrm{Op}(f_1 \cdots f_n).$$

(3) *If $b, b_1, b_2 \in \mathbb{C}\mathrm{Pol}_a(\mathcal{Y}^\#)$ and $\mathrm{Op}(b) = \mathrm{Op}(b_1)\mathrm{Op}(b_2)$, then*

$$b(v) = \exp\left(\nabla_{v_2}\cdot\nu\nabla_{v_1}\right) b_1(v_1)b_2(v_2)\big|_{v_1=v_2=v} \tag{12.25}$$

$$= \int_{\mathcal{Y}^\#} \int_{\mathcal{Y}^\#} e^{(v-v_1)\cdot\nu^{-1}(v-v_2)} b_1(v_1)b_2(v_2)\mathrm{d}v_2\mathrm{d}v_1. \tag{12.26}$$

(4) *If $b \in \mathbb{C}\mathrm{Pol}_a(\mathcal{Y}^\#)$, $y \in \mathbb{C}\mathcal{Y}$, then*

$$\frac{1}{2}\left(\phi(y)\mathrm{Op}(b) + \mathrm{Op}(Ib)\phi(y)\right) = \mathrm{Op}(y \cdot b), \tag{12.27}$$

$$\frac{1}{2}\left(\phi(y)\mathrm{Op}(b) - \mathrm{Op}(Ib)\phi(y)\right) = \mathrm{Op}((\nu y)\cdot\nabla_v b). \tag{12.28}$$

Proof Statements (1) and (2) are immediate. Let us prove (12.25). Let us fix an o.n. basis $(e_1, \ldots, e_d)$ of $\mathcal{Y}$ such that $\Xi = e^n \wedge \cdots \wedge e^1$. We use the Berezin calculus introduced in Subsect. 7.1.5. We rename the variable v_1 as v and the variable v_2 as w. We will write $v_i = e_i \cdot v$, $w_i = e_i \cdot w$. Without loss of generality, we can assume that $b_1(v) = \prod_{i \in I} v_i$, $b_2(w) = \prod_{i \in J} w_i$ for $I, J \subset \{1, \ldots, d\}$. We have

$$\exp\left(\sum_{i=1}^d \nabla_{w_i} \cdot \nabla_{v_i}\right) = \sum_{K \subset \{i,\ldots,d\}} \prod_{i \in K} \nabla_{w_i} \cdot \nabla_{v_i}. \tag{12.29}$$

The only term in (12.29) giving a non-zero contribution to

$$\exp\left(\sum_{i=1}^d \nabla_{w_i} \cdot \nabla_{v_i}\right) b_1(v) \cdot b_2(w)\big|_{w=v}.$$

is $K = I \cap J$. Without loss of generality we can further assume that $1 \le p \le n \le m$ and

$$b_1(v) = v_1 \cdots v_p \cdot v_{p+1} \cdots v_n, \quad b_2(w) = w_n \cdots w_{p+1} \cdot w_{n+1} \cdots w_m.$$

Then

$$\prod_{i=p+1}^{n} \nabla_{w_i} \cdot \nabla_{v_i} b_1(v) \cdot b_2(w)\big|_{w=v} = v_1 \cdots v_p \cdot v_{n+1} \cdots v_m =: b(v).$$

We have

$$Op(b_1) = \phi(e_1) \cdots \phi(e_p) \cdot \phi(e_{p+1}) \cdots \phi(e_n),$$
$$Op(b_2) = \phi(e_n) \cdots \phi(e_{p+1}) \cdot \phi(e_{n+1}) \cdots \phi(e_m),$$

and

$$Op(b_1)Op(b_2) = \phi(e_1) \cdots \phi(e_p) \cdot \phi(e_{n+1}) \cdots \phi(e_m) = Op(b),$$

using the CAR. This proves (12.25).

To obtain (12.26), we apply Prop. 7.19 to the even-dimensional space $\mathcal{X} = \mathcal{Y} \oplus \mathcal{Y}$. Let $x = (y_1, y_2)$, $\xi = (v_1, v_2)$ be the generic variables in $\mathcal{X}$ and $\mathcal{X}^{\#}$. Let $\zeta = \begin{bmatrix} 0 & \nu^{-1} \\ -\nu^{-1} & 0 \end{bmatrix} \in L_a(\mathcal{X}^{\#}, \mathcal{X})$, so that $\frac{1}{2} x \cdot \zeta^{-1} x = y_2 \cdot \nu y_1$ and $\frac{1}{2} \xi \cdot \zeta \xi = v_1 \cdot \nu^{-1} v_2$. The Pfaffian of ζ w.r.t. $dv_2 \wedge dv_1$ is equal to 1, which by Prop. 7.19 proves the second identity of (3).

To prove (4), we can assume without loss of generality that $b(v) = v_{i_1} \cdots v_{i_p}$, and $\langle y|v \rangle = v_j$. Then $Op(b) = \phi(e_{i_1}) \cdots \phi(e_{i_p})$, $\phi(y) = \phi(e_j)$. Using the CAR we get

$$\frac{1}{2}\left(\phi(e_j)\phi(e_{i_1}) \cdots \phi(e_{i_p}) + (-1)^p \phi(e_{i_1}) \cdots \phi(e_{i_p})\phi(e_j)\right)$$
$$= \begin{cases} 0 & \text{if } j \in \{i_1, \ldots, i_p\}, \\ \phi(e_j)\phi(e_{i_1}) \cdots \phi(e_{i_p}) & \text{if } j \notin \{i_1, \ldots, i_p\}, \end{cases}$$

which proves the first statement of (4). The second can be proved similarly. $\square$

Theorem 12.43 *If $b, b_1, \ldots, b_n \in \mathbb{C}Pol_a(\mathcal{Y}^{\#})$ and*

$$Op(b) = Op(b_1) \cdots Op(b_n),$$

then the following version of the Wick theorem for the anti-symmetric quantization is true:

$$b(v) = \exp\left(\sum_{i>j} \nabla_{v_i} \cdot \nu \nabla_{v_j}\right) b_1(v_1) \cdots b_n(v_n)\big|_{v=v_1=\cdots=v_n}.$$

12.4.2 Real-wave CAR representation

Definition 12.44 *For $A \in CAR(\mathcal{Y})$, we define*

$$\text{tr}A = 2^{-m} \operatorname{Tr} \pi^{JW}(A), \quad \dim \mathcal{Y} = 2m,$$
$$\text{tr}A = 2^{-m-1} \operatorname{Tr} \pi^{JW}(A), \quad \dim \mathcal{Y} = 2m+1,$$

where π^{JW} is the Jordan–Wigner representation. tr is called the canonical tracial state on $\mathrm{CAR}(\mathcal{Y})$.

Theorem 12.45 (1) tr *is a tracial state on* $\mathrm{CCR}(\mathcal{Y})$, *which means*

$$\mathrm{tr}(AB) = \mathrm{tr}(BA), \quad A, B \in \mathrm{CAR}(\mathcal{Y}).$$

(2) *It satisfies*

$$\mathrm{tr}(A) = \mathrm{tr}(\mathrm{c}(A)) = \mathrm{tr}(\alpha(A)) = \mathrm{tr}(\hat{r}(A)), \quad A \in \mathrm{CAR}(\mathcal{Y}), \ r \in O(\mathcal{Y}).$$

(3) *If* $b, c \in \mathbb{C}\mathrm{Pol}_{\mathrm{a}}(\mathcal{Y}^{\#}) \simeq \Gamma_{\mathrm{a}}(\mathbb{C}\mathcal{Y})$, *then*

$$\mathrm{tr}(\mathrm{Op}(b)^{*}\mathrm{Op}(c)) = (b|c). \tag{12.30}$$

(4) *For* $y, y_1, y_2 \in \mathcal{X}$, *we have the expectation values*

$$\mathrm{tr}\left(\phi(y)\right) = 0,$$
$$\mathrm{tr}\left(\phi(y_1)\phi(y_2)\right) = y_1 \cdot \nu y_2.$$

More generally,

$$\mathrm{tr}\left(\phi(y_1) \cdots \phi(y_{2m-1})\right) = 0,$$
$$\mathrm{tr}\left(\phi(y_1) \cdots \phi(y_{2m})\right) = \sum_{\sigma \in \mathrm{Pair}_{2m}} \mathrm{sgn}(\sigma) \prod_{j=1}^{m} y_{\sigma(2j-1)} \cdot \nu y_{\sigma(2j)}.$$

Definition 12.46 *Let* $(\pi_{\mathrm{tr}}, \mathcal{H}_{\mathrm{tr}}, \Omega_{\mathrm{tr}})$ *denote the GNS representation of* $\mathrm{CAR}(\mathcal{Y})$ *w.r.t. the state* tr. *The CAR representation*

$$\mathcal{Y} \ni y \mapsto \phi_{\mathrm{tr}}(y) := \pi_{\mathrm{tr}}(\phi(y)) \in B_{\mathrm{h}}(\mathcal{H}_{\mathrm{tr}})$$

will be called the real-wave *or* tracial CAR *representation.*

12.4.3 Real-wave CAR representation in coordinates

Let n be an integer. We are going to describe the real-wave representation over $\mathbb{R}^n$ more explicitly.

Clearly, $\otimes^n \mathbb{C}^2$ has a natural conjugation, denoted as usual $\otimes^n \mathbb{C}^2 \ni \Psi \mapsto \overline{\Psi} \in \otimes^n \mathbb{C}^2$. For typographical reasons, it will sometimes be denoted by χ. The corresponding real subspace of $\otimes^n \mathbb{C}^2$ obviously equals $\otimes^n \mathbb{R}^2$. Linear operators preserving $\otimes^n \mathbb{R}^2$ are called real.

The conjugation of $A \in B(\otimes^n \mathbb{C}^2)$ is denoted by $\overline{A}$ or $\chi A \chi$.

Define the "vacuum vector" $\Omega := |\downarrow\rangle \otimes \cdots \otimes |\downarrow\rangle$. Clearly, $\Omega = \overline{\Omega}$.

Introduce the following operators on $\otimes^n \mathbb{C}^2$:

$$N := \sum_{j=1}^{n} \mathbb{1}^{\otimes j} \otimes \frac{\sigma_3 + \mathbb{1}}{2} \otimes \mathbb{1}^{\otimes(n-j)}, \quad \Lambda := (-1)^{N(N-\mathbb{1})/2}.$$

The role of these operators will become clear in Chap. 13, where we will identify $\otimes^n \mathbb{C}^2$ with the fermionic Fock space $\Gamma_a(\mathbb{C}^n)$ and they will coincide with the operators defined in (3.9) and (3.29). Therefore, in particular, N is called the number operator. For further reference let us note the following identities involving Λ:

$$\Lambda \mathbb{1}^{\otimes(j-1)} \otimes \sigma_1 \otimes \mathbb{1}^{\otimes(n-j)} \Lambda = (-\sigma_3)^{\otimes(j-1)} \otimes \sigma_1 \otimes (-\sigma_3)^{\otimes(n-j)},$$
$$\Lambda \mathbb{1}^{\otimes(j-1)} \otimes \sigma_2 \otimes \mathbb{1}^{\otimes(n-j)} \Lambda = -(-\sigma_3)^{\otimes(j-1)} \otimes \sigma_2 \otimes (-\sigma_3)^{\otimes(n-j)},$$
$$\Lambda \mathbb{1}^{\otimes(j-1)} \otimes \sigma_3 \otimes \mathbb{1}^{\otimes(n-j)} \Lambda = \mathbb{1}^{\otimes(j-1)} \otimes \sigma_3 \otimes \mathbb{1}^{\otimes(n-j)}.$$

In order to describe the real-wave CAR representation over $\mathbb{R}^n$, introduce the following operators

$$\phi_j^l = \sigma_3^{\otimes(j-1)} \otimes \sigma_1 \otimes \mathbb{1}^{\otimes(n-j)},$$
$$\phi_j^r = \mathbb{1}^{\otimes(n-j)} \otimes \sigma_1 \otimes \sigma_3^{\otimes(j-1)} = \Lambda \phi_j^l \Lambda.$$

Theorem 12.47 (1) *We have two mutually commuting CAR representations:*

$$\phi_1^l, \dots, \phi_n^l, \tag{12.31}$$
$$\phi_1^r, \dots, \phi_n^r. \tag{12.32}$$

That means

$$[\phi_i^l, \phi_j^l]_+ = 2\delta_{i,j}, \quad [\phi_i^r, \phi_j^r]_+ = 2\delta_{i,j}, \quad [\phi_i^l, \phi_j^r] = 0, \quad i,j = 1, \dots, n.$$

(2) *Let* $\pi^l : \mathrm{CAR}(\mathbb{R}^n) \to B(\otimes^n \mathbb{C}^2)$ *be the* $*$-*homomorphism obtained by Prop. 12.31 from the CAR representation* (12.31). *Then*

$$\mathrm{tr}(A) = (\Omega | \pi^l(A)\Omega), \quad A \in \mathrm{CAR}(\mathbb{R}^n),$$

and $\pi^l(\mathrm{CAR}(\mathbb{R}^n))\Omega = \otimes^n \mathbb{C}^2$. *Thus* Ω *is a cyclic vector representative for the state* tr, *and hence* π^l *is the GNS representation of* $\mathrm{CAR}(\mathbb{R}^n)$ *for the state* tr.

(3) *Let* J *be the modular conjugation for the state* tr. *Then* $J = \Lambda \chi$ *(where* χ *denotes the complex conjugation). We have*

$$J\phi_j^l J = \phi_j^r, \quad j = 1, \dots, n.$$

(4) *We have*

$$\overline{\phi_i^l} = \phi_i^l, \quad i = 1, \dots, n.$$

Therefore,

$$\pi_{\mathrm{tr}}(c(A)) = \overline{\pi_{\mathrm{tr}}(A)}, \quad A \in \mathrm{CAR}(\mathbb{R}^n).$$

Consequently, $\pi_{\mathrm{tr}}(\mathrm{Cliff}(\mathbb{R}^n))$ *consists of real elements of* $\pi_{\mathrm{tr}}(\mathrm{CAR}(\mathbb{R}^n))$.

(5) *Let Q be the operator defined in (12.20). We have*

$$\pi^{\mathrm{l}}(Q) = J\pi^{\mathrm{l}}(Q)J = \phi_1^{\mathrm{l}} \cdots \phi_n^{\mathrm{l}} = \phi_n^{\mathrm{r}} \cdots \phi_1^{\mathrm{r}}$$
$$= \begin{cases} (-1)^m \sigma_2 \otimes \sigma_1 \otimes \cdots \otimes \sigma_2 \otimes \sigma_1, & n = 2m; \\ (-1)^m \sigma_1 \otimes \sigma_2 \otimes \cdots \otimes \sigma_2 \otimes \sigma_1, & n = 2m+1. \end{cases}$$

By Thm. 12.47 (2), the representation ϕ^{l} can be identified with the real-wave CAR representation defined in Def. 12.46.

The analysis of the real-wave CAR representation, in the case of an arbitrary dimension, will be continued in Subsect. 13.2.1, where we will use the formalism of Fock spaces.

12.5 CAR algebras: infinite dimensions

Throughout this section $(\mathcal{Y}, \nu)$ is a Euclidean space, possibly infinite-dimensional.

One aspect of the theory of CAR algebras simplifies in infinite dimensions: it is not necessary to distinguish between the even and odd cases. On the other hand, the topological aspects become more subtle. In particular, it is natural to define (at least) three different kinds of CAR algebras: the algebraic, the C^*- and the W^*-CAR algebra. (The situation is, however, simpler than in the case of CCR algebras.)

12.5.1 Algebraic CAR algebra

The definition of the *algebraic* CAR algebra is the same as that of the CAR algebra in finite dimension:

Definition 12.48 *The* algebraic CAR algebra over $\mathcal{Y}$, *denoted* $\mathrm{CAR}^{\mathrm{alg}}(\mathcal{Y})$, *is the complex unital $*$-algebra generated by elements $\phi(y)$, $y \in \mathcal{Y}$, with relations*

$$\phi(\lambda y) = \lambda \phi(y), \ \lambda \in \mathbb{R}, \qquad \phi(y_1 + y_2) = \phi(y_1) + \phi(y_2),$$
$$\phi^*(y) = \phi(y), \qquad \phi(y_1)\phi(y_2) + \phi(y_2)\phi(y_1) = 2y_1 \cdot \nu y_2 \, \mathbb{1}.$$

Clearly, Prop. 12.31 extends to infinite dimension, with $\mathrm{CAR}^{\mathrm{alg}}(\mathcal{Y})$ replacing $\mathrm{CAR}(\mathcal{Y})$. The parity α, the complex conjugation c and the transposition $^{\#}$ naturally extend to $\mathrm{CAR}^{\mathrm{alg}}(\mathcal{Y})$. If $r \in O(\mathcal{Y})$, we can introduce a unique $*$-automorphism $\hat{r}$ of $\mathrm{CAR}^{\mathrm{alg}}(\mathcal{Y})$, called the *Bogoliubov automorphism*, satisfying $\hat{r}(\phi(y)) = \phi(ry)$ as in Def. 12.39.

Definition 12.49 *Let* $j = 0, 1$. *We introduce*

$$\mathrm{CAR}_j^{\mathrm{alg}}(\mathcal{Y}) := \{ B \in \mathrm{CAR}^{\mathrm{alg}}(\mathcal{Y}) : \alpha(B) = (-1)^j B \},$$
$$\mathrm{Cliff}^{\mathrm{alg}}(\mathcal{Y}) := \{ B \in \mathrm{CAR}^{\mathrm{alg}}(\mathcal{Y}) : c(B) = B \}.$$

Note that if $\mathcal{Y}$ is infinite-dimensional, $\mathrm{CAR}^{\mathrm{alg}}(\mathcal{Y})$ does not contain an operator analogous to the operator Q defined in (12.20). (The same remark applies to $\mathrm{CAR}^{C^*}(\mathcal{Y})$ and $\mathrm{CAR}^{W^*}(\mathcal{Y})$ defined later on.)

12.5.2 C^*-CAR algebra

Proposition 12.50 *There exists a unique C^*-norm on $\mathrm{CAR}^{\mathrm{alg}}(\mathcal{Y})$.*

Proof We already know that this is true if $\mathcal{Y}$ is finite-dimensional.

If $\mathcal{Y}$ has an infinite dimension, then $\mathrm{CAR}^{\mathrm{alg}}(\mathcal{Y})$ is the union of $\mathrm{CAR}(\mathcal{Y}_1)$ for finite-dimensional subspaces of $\mathcal{Y}$. So $\mathrm{CAR}^{\mathrm{alg}}(\mathcal{Y})$ is equipped with a unique C^*-norm. $\square$

Definition 12.51 *The CAR C^*-algebra over $\mathcal{Y}$ is defined as*

$$\mathrm{CAR}^{C^*}(\mathcal{Y}) := \Big(\mathrm{CAR}^{\mathrm{alg}}(\mathcal{Y})\Big)^{\mathrm{cpl}},$$

where the completion is w.r.t. the C^*-norm defined above. $\mathrm{CAR}^{C^*}(\mathcal{Y})$ is a C^*-algebra.

The relationship between CAR representations and the algebra $\mathrm{CAR}^{C^*}(\mathcal{Y})$ is given by the following theorem:

Proposition 12.52 *If*

$$\mathcal{Y} \ni y \mapsto \phi^\pi(y) \in B_\mathrm{h}(\mathcal{H})$$

is a CAR representation, then there exists a unique $*$-homomorphism of C^*-algebras

$$\pi : \mathrm{CAR}^{C^*}(\mathcal{Y}) \to B(\mathcal{H})$$

such that $\pi(\phi(y)) = \phi^\pi(y)$, $y \in \mathcal{Y}$.

Proof We already know that this is true if we replace $\mathrm{CAR}^{C^*}(\mathcal{Y})$ with $\mathrm{CAR}^{\mathrm{alg}}(\mathcal{Y})$. π extends to $\mathrm{CAR}^{C^*}(\mathcal{Y})$ by continuity.

To see the uniqueness, note that every $*$-homomorphism between C^*-algebras is continuous. $\square$

Clearly, $\mathrm{CAR}^{C^*}(\mathcal{Y})$ coincides with $\mathrm{CAR}^{C^*}(\mathcal{Y}^{\mathrm{cpl}})$. Hence, it is enough to restrict to complete $\mathcal{Y}$.

Proposition 12.53 *The parity α, the complex conjugation c and the transposition $^\#$ are isometric, and hence extend by continuity from $\mathrm{CAR}^{\mathrm{alg}}(\mathcal{Y})$ to $\mathrm{CAR}^{C^*}(\mathcal{Y})$. For $r \in O(\mathcal{Y})$, the same is true concerning the Bogoliubov automorphism $\hat{r}$.*

Proof Let $A \in \mathrm{CAR}^{\mathrm{alg}}(\mathcal{Y})$. Then

$$\mathrm{spec}\, A = \mathrm{spec}\, \alpha(A) = \overline{\mathrm{spec}\, c(A)} = \overline{\mathrm{spec}\, A^{\#}} = \mathrm{spec}\, \hat{r}(A).$$

Therefore, α, c, $^\#$ and $\hat{r}$ do not change the spectral radius of A. Hence, they are isometric. $\square$

Definition 12.54 *We define* $\mathrm{CAR}_i^{C^*}(\mathcal{Y})$, $i = 0, 1$, *and* $\mathrm{Cliff}^{C^*}(\mathcal{Y})$ *as in Def. 12.49.*

Theorem 12.55 *If $\mathcal{Y}$ is an infinite-dimensional real Hilbert space, then* $\mathrm{CAR}^{C^*}(\mathcal{Y})$ *is simple. If in addition $\mathcal{Y}$ is separable, then* $\mathrm{CAR}^{C^*}(\mathcal{Y})$ *is isomorphic to* $\mathrm{UHF}(2^\infty)$ *defined in Subsect. 6.2.9.*

Proof Choose an o.n. basis $(e_{1+}, e_{1-}, e_{2+}, e_{2-}, \dots)$ of $\mathcal{Y}$. Let $\mathcal{Y}_n$ be the space spanned by the first n vectors of this basis. We have a commuting diagram,

$$\begin{array}{ccc} \mathrm{CAR}(\mathcal{Y}_{2m}) & \subset & \mathrm{CAR}(\mathcal{Y}_{2m+2}) \\ \downarrow & & \downarrow \\ B(\otimes^m \mathbb{C}^2) & \to & B(\otimes^{m+1}\mathbb{C}^2), \end{array}$$

where the vertical arrows are $*$-isomorphisms and the lower horizontal arrow is $A \mapsto A \otimes \mathbb{1}_{\mathbb{C}^2}$. Clearly, $\bigcup_{m=1}^\infty \mathcal{Y}_{2m}$ is dense in $\mathcal{Y}$. Hence,

$$\mathrm{CAR}^{C^*}(\mathcal{Y}) = \left(\bigcup_{m=1}^\infty \mathrm{CAR}(\mathcal{Y}_{2m}) \right)^{\mathrm{cpl}}$$

$$\simeq \left(\bigcup_{m=1}^\infty B(\otimes^m \mathbb{C}^2) \right)^{\mathrm{cpl}} = \mathrm{UHF}(2^\infty). \qquad \square$$

12.5.3 W^*-CAR algebra

In Thm. 12.47 we defined the state tr on $\mathrm{CAR}(\mathcal{Y})$ for any finite-dimensional $\mathcal{Y}$. For an arbitrary $\mathcal{Y}$ this gives rise to a state on $\mathrm{CAR}^{\mathrm{alg}}(\mathcal{Y})$, and hence on $\mathrm{CAR}^{C^*}(\mathcal{Y})$, also denoted tr. We can perform the GNS representation using the state tr and obtain the triple $(\mathcal{H}_{\mathrm{tr}}, \pi_{\mathrm{tr}}, \Omega_{\mathrm{tr}})$, where

$$\pi_{\mathrm{tr}} : \mathrm{CAR}^{C^*}(\mathcal{Y}) \to B(\mathcal{H}_{\mathrm{tr}})$$

is a faithful $*$-representation, $\Omega_{\mathrm{tr}} \in \mathcal{H}_{\mathrm{tr}}$ is a vector cyclic for $\pi_{\mathrm{tr}}(\mathcal{H}_{\mathrm{tr}})$ and

$$\mathrm{tr}(A) = (\Omega_{\mathrm{tr}} | \pi_{\mathrm{tr}}(A)\Omega_{\mathrm{tr}}).$$

Definition 12.56 *We define the W^*-CAR algebra of $\mathcal{Y}$ as*

$$\mathrm{CAR}^{W^*}(\mathcal{Y}) := \left(\pi_{\mathrm{tr}}(\mathrm{CAR}^{C^*}(\mathcal{Y})) \right)''.$$

Since π_{tr} is faithful, it defines an isomorphism of $\text{CAR}^{C^*}(\mathcal{Y})$ onto $\pi_{\text{tr}}(\text{CAR}^{C^*}(\mathcal{Y}))$. Therefore, in what follows these two algebras will be identified. Thus $\text{CAR}^{C^*}(\mathcal{Y})$ is a σ-weakly dense sub-algebra of $\text{CAR}^{W^*}(\mathcal{Y})$.

We have a normal state

$$(\Omega_{\text{tr}}|A\Omega_{\text{tr}}), \quad A \in \text{CAR}^{W^*}(\mathcal{Y}). \tag{12.33}$$

On $\text{CAR}^{C^*}(\mathcal{Y})$ it coincides with tr. In what follows, we will write $\text{tr}A$ also for (12.33).

Thm. 12.45 extends with obvious adjustments:

Theorem 12.57 (1) tr *is a tracial state on* $\text{CAR}^{W^*}(\mathcal{Y})$.
(2) *The conjugation* c *and the parity* α *extend to* σ-*weakly continuous involutions on* $\text{CAR}^{W^*}(\mathcal{Y})$ *preserving* tr. *For* $r \in O(\mathcal{Y})$, *the same is true for the Bogoliubov automorphism* $\hat{r}$.
(3) *The identities of Thm. 12.45 (3) and (4) are true.*

Definition 12.58 *We define* $\text{CAR}_i^{W^*}(\mathcal{Y})$, $i = 0, 1$, *and* $\text{Cliff}^{W^*}(\mathcal{Y})$ *as in Def. 12.49.*

Theorem 12.59 *If* $\mathcal{Y}$ *is an infinite-dimensional separable Hilbert space, then* $\text{CAR}^{W^*}(\mathcal{Y})$ *is isomorphic to* HF *(the unique hyper-finite type* II_1 *factor described in Subsect. 6.2.10).*

Recall from Subsect. 6.5.2 that, for any $1 \leq p \leq \infty$, we can define the space $L^p(\text{CAR}^{W^*}(\mathcal{Y}), \text{tr})$. For $p = 1$, it coincides with the space of normal functionals on $\text{CAR}^{W^*}(\mathcal{Y})$. For $p = 2$, it coincides with the GNS Hilbert space for the state tr, denoted also $\mathcal{H}_{\text{tr}}$. Finally, for $p = \infty$, it coincides with $\text{CAR}^{W^*}(\mathcal{Y})$ itself.

For $1 \leq p < \infty$, $\text{CAR}^{\text{alg}}(\mathcal{Y})$ is dense in $L^p(\text{CAR}^{W^*}(\mathcal{Y}), \text{tr})$, so that $L^p(\text{CAR}^{W^*}(\mathcal{Y}), \text{tr})$, can be understood as the completion of $\text{CAR}^{\text{alg}}(\mathcal{Y})$ in the norm $\|A\|_p := (\text{tr}|A|^p)^{1/p}$.

Definition 12.60 *Similarly to the case of a finite dimension, in the general case we define the* tracial *or* real-wave *CAR representation over* $\mathcal{Y}$ *by*

$$\mathcal{Y} \ni y \mapsto \phi_{\text{tr}}(y) := \pi_{\text{tr}}(\phi(y)) \in B_{\text{h}}(\mathcal{H}_{\text{tr}}).$$

12.5.4 Conditional expectations between CAR algebras

Consider a closed subspace $\mathcal{Y}_1$ of $\mathcal{Y}$. Clearly, $\text{CAR}^{W^*}(\mathcal{Y}_1)$ can be viewed as a W^*-sub-algebra of $\text{CAR}^{W^*}(\mathcal{Y})$. Besides, $\text{CAR}^{W^*}(\mathcal{Y})$ is equipped with a tracial state tr. Therefore, by Subsect. 6.5.4, there exists a unique conditional expectation

$$E_{\mathcal{Y}_1} : \text{CAR}^{W^*}(\mathcal{Y}) \to \text{CAR}^{W_1}(\mathcal{Y}_1)$$

such that

$$\text{tr}A = \text{tr}E_{\mathcal{Y}_1}(A), \quad A \in \text{CAR}^{W^*}(\mathcal{Y}).$$

It commutes with the parity:

$$\alpha \circ E_{\mathcal{Y}_1} = E_{\mathcal{Y}_1} \circ \alpha.$$

It restricts to a conditional expectation between the corresponding CAR C^*-algebras:

$$E_{\mathcal{Y}_1} : \mathrm{CAR}^{C^*}(\mathcal{Y}) \to \mathrm{CAR}^{C^*}(\mathcal{Y}_1).$$

If $\{\mathcal{Y}_i\}_{i \in I}$ is an increasing net of closed subspaces of $\mathcal{Y}$ with $\underset{i \in I}{\cup} \mathcal{Y}_i$ dense in $\mathcal{Y}$, we have the norm convergence

$$\lim_i E_{\mathcal{Y}_i}(A) = A, \quad A \in \mathrm{CAR}^{C^*}(\mathcal{Y}), \tag{12.34}$$

and the σ-weak convergence

$$\sigma\mathrm{-w} - \lim_i E_{\mathcal{Y}_i}(A) = A, \quad A \in \mathrm{CAR}^{W^*}(\mathcal{Y}).$$

12.5.5 Irreducibility of infinite-dimensional CAR algebras

The following proposition extends Prop. 12.36 to the infinite-dimensional case.

Proposition 12.61 (1) *Let $A \in \mathrm{CAR}_0^{C^*}(\mathcal{Y})$ commute with $\phi(y)$, $y \in \mathcal{Y}$ (and hence with all $\mathrm{CAR}^{C^*}(\mathcal{Y})$). Then A is proportional to $\mathbb{1}$.*

(2) *Let a non-zero $A \in \mathrm{CAR}_1^{C^*}(\mathcal{Y})$ commute with $\phi(y)$, $y \in \mathcal{Y}$ (and hence with all $\mathrm{CAR}^{C^*}(\mathcal{Y})$). Then $\dim \mathcal{Y}$ is finite and odd, and A is proportional to Q.*

(3) *Let a non-zero $A \in \mathrm{CAR}^{C^*}(\mathcal{Y})$ anti-commute with $\phi(y)$, $y \in \mathcal{Y}$ (and hence with all $\mathrm{CAR}_1^{C^*}(\mathcal{Y})$). Then $\dim \mathcal{Y}$ is finite and even, and A is proportional to Q.*

Proof We pick an increasing net $\mathcal{Y}_i$, $i \in I$ of finite-dimensional subspaces of $\mathcal{Y}$ with $(\bigcup_{i \in I} \mathcal{Y}_i)^{\mathrm{cl}} = \mathcal{Y}$. Let E_i be the conditional expectation onto $\mathrm{CAR}(\mathcal{Y}_i)$. Let $A \in \mathrm{CAR}_0^{C^*}(\mathcal{Y})$ such that $A\phi(y) = \phi(y)A$ for $y \in \mathcal{Y}$. Let $A_i := E_i(A)$. Since $E_i\phi(y) = \phi(y)$, $y \in \mathcal{Y}_i$, we obtain from Prop. 6.83 that $A_i\phi(y) = \phi(y)A_i$, $y \in \mathcal{Y}_i$. By Prop. 12.36, this implies that, for all i, $A_i = \lambda_i \mathbb{1}$. We know that $\lim_i A_i = A$ by (12.34). Hence, $\lim_i \lambda_i =: \lambda$ exists and $A = \lambda \mathbb{1}$. This proves (1).

Let us now prove (2). Let us assume that there exists A with the stated properties, and that $\dim \mathcal{Y}$ is infinite. We pick an increasing net of finite-dimensional subspaces $\mathcal{Y}_i$ of odd dimensions as above, equip them with orientations and denote by Q_i the associated volume elements as in (12.20). Considering the net $A_i := E_i A$, we know by Prop. 12.36 that, for all i, $A_i = \lambda_i Q_i$. Clearly, if $i \leq j$, then $E_i(Q_j) = Q_i$, which implies that λ_i coincide and equal a certain non-zero number λ. Since $A := \lim_i A_i \neq 0$ exists, $\lim_i Q_i \neq 0$. Using now the CAR we see that if $\mathcal{Y}_i$, $\mathcal{Y}_j$ are two finite-dimensional spaces with $\mathcal{Y}_i \subsetneq \mathcal{Y}_j$, then

$\|Q_i - Q_j\| = 1$, which is a contradiction, since $\mathcal{Y}$ is infinite-dimensional. The proof of (3) is similar. $\qquad\square$

The following proposition is the W^*-analog of Prop. 12.61.

Proposition 12.62 (1) *Let* $A \in \mathrm{CAR}_0^{W^*}(\mathcal{Y})$ *commute with* $\phi(y)$, $y \in \mathcal{Y}$ *(and hence with all* $\mathrm{CAR}^{W^*}(\mathcal{Y})$*). Then* A *is proportional to* $\mathbb{1}$.

(2) *Let a non-zero* $A \in \mathrm{CAR}_1^{W^*}(\mathcal{Y})$ *commute with* $\phi(y)$, $y \in \mathcal{Y}$ *(and hence with all* $\mathrm{CAR}^{W^*}(\mathcal{Y})$*). Then* $\dim \mathcal{Y}$ *is finite and odd, and* A *is proportional to* Q.

(3) *Let a non-zero* $A \in \mathrm{CAR}^{W^*}(\mathcal{Y})$ *anti-commute with* $\phi(y)$, $y \in \mathcal{Y}$ *(and hence with all* $\mathrm{CAR}_1^{W^*}(\mathcal{Y})$*). Then* $\dim \mathcal{Y}$ *is finite and even, and* A *is proportional to* Q.

Proof The proof of (1) is completely analogous to Prop. 12.61. Let us explain the modifications for the proof of (2). By the same arguments as in Prop. 12.61, we obtain that $\lim_i Q_i$ exists in the σ-weak topology. Working in the GNS representation for the tracial state, we see that $\lim_i Q_i \Omega$ does not exist. The proof of (3) is similar. $\qquad\square$

12.6 Notes

Clifford relations and Clifford algebras appeared in mathematics before quantum theory, in Clifford (1878). They will be further discussed in Chap. 15.

Canonical anti-commutation relations were introduced in the description of fermions by Jordan–Wigner (1928).

Pauli matrices were introduced by Pauli (1927) to describe spin $\frac{1}{2}$ particles.

Mathematical properties of CAR algebras were extensively studied; see e.g. the review paper by Araki (1987) and the book by Plymen–Robinson (1994).

13

CAR on Fock spaces

This chapter is devoted to the study of *Fock representations of the canonical anti-commutation relations,* a basic tool of quantum many-body theory and quantum field theory. It is parallel to Chap. 9, where Fock CCR representations were studied.

The basic framework is almost the same as in Chap. 9. Throughout this chapter $\mathcal{Z}$ is a Hilbert space, called the *one-particle space.* We will consider the Fock CAR representation acting on the fermionic Fock space $\Gamma_a(\mathcal{Z})$.

As in Sect. 1.3, we introduce the space

$$\mathcal{Y} = \operatorname{Re}(\mathcal{Z} \oplus \overline{\mathcal{Z}}) := \{(z, \overline{z}) \ : \ z \in \mathcal{Z}\},$$

which will serve as the *dual phase space* of our system. Recall that in the bosonic case we equipped $\mathcal{Y}$ with the structure of the *Kähler space* consisting of the anti-involution j, the Euclidean scalar product $\cdot$ and the symplectic form ω:

$$j(z, \overline{z}) := (iz, \overline{iz}), \tag{13.1}$$

$$(z, \overline{z}) \cdot (w, \overline{w}) := 2\operatorname{Re}(z|w), \tag{13.2}$$

$$(z, \overline{z}) \cdot \omega(w, \overline{w}) := 2\operatorname{Im}(z|w) = -(z, \overline{z}) \cdot j(w, \overline{w}). \tag{13.3}$$

In our presentation of the fermionic case, we will need the *Kähler anti-involution* j. The symplectic form ω will not be used. Instead of the scalar product (13.2) we will use another scalar product,

$$(z, \overline{z}) \cdot \nu(w, \overline{w}) := \operatorname{Re}(z|w) = \frac{1}{2}(z, \overline{z}) \cdot (w, \overline{w}).$$

Again we will avoid identifing $\mathcal{Z}$ with $\mathcal{Y}$.

$\mathbb{C}\mathcal{Y}$ is identified with $\mathcal{Z} \oplus \overline{\mathcal{Z}}$ by the map

$$\mathbb{C}\mathcal{Y} \ni (z_1 + \overline{z}_1) + i(z_2 + \overline{z}_2) \mapsto (z_1 + iz_2, \overline{z_1 - iz_2}) \in \mathcal{Z} \oplus \overline{\mathcal{Z}}.$$

$\mathcal{Y}^{\#}$, the space dual to $\mathcal{Y}$, is canonically identified with $\operatorname{Re}(\overline{\mathcal{Z}} \oplus \mathcal{Z})$ by using the scalar product (9.2), and $\mathbb{C}\mathcal{Y}^{\#}$ is identified with $\overline{\mathcal{Z}} \oplus \mathcal{Z}$.

13.1 Fock CAR representation

Consider the fermionic Fock space $\Gamma_a(\mathcal{Z})$. Recall that, for $z \in \mathcal{Z}$, $a^*(z)$, resp. $a(z)$ denote the corresponding creation, resp. annihilation operators defined in Sect. 3.4.

13.1.1 Field operators

Definition 13.1 *For* $y = (z_1, \bar{z}_2) \in \mathbb{C}\mathcal{Y}$, *the corresponding* field operator *acts on* $\Gamma_a(\mathcal{Z})$ *and is defined as*

$$\phi(z_1, \bar{z}_2) := a^*(z_1) + a(z_2). \tag{13.4}$$

Recall that $I := (-1)^N$.

Theorem 13.2 (1) *Operators* $\phi(z_1, \bar{z}_2)$ *are bounded and*

$$\phi(z_1, \bar{z}_2)^* = \phi(z_2, \bar{z}_1).$$

In particular, $\phi(z, \bar{z})$ *are self-adjoint.*

(2) $[\phi(w_1, \bar{w}_2), \phi(z_1, \bar{z}_2)]_+ = (w_2|z_1) + (z_2|w_1).$ *Hence, in particular,*

$$[\phi(w, \bar{w}), \phi(z, \bar{z})]_+ = 2(w, \bar{w}) \cdot \nu(z, \bar{z})\mathbb{1}.$$

(3) $[\phi(z_1, z_2), I]_+ = 0.$

(4) *If* $p \in B(\mathcal{Z})$, *we have*

$$\Gamma(p)\phi(z_1, \bar{z}_2) = \phi(pz_1, \overline{p^{*-1}z_2})\Gamma(p).$$

(5) *If* $h \in B(\mathcal{Z})$, *we have*

$$[\mathrm{d}\Gamma(h), \phi(z_1, \bar{z}_2)] = \phi(hz_1, -\overline{h^*z_2}).$$

(6) *We have an irreducible CAR representation,*

$$\mathcal{Y} \ni (z, \bar{z}) \mapsto \phi(z, \bar{z}) \in B_h(\Gamma_a(\mathcal{Z})). \tag{13.5}$$

For further reference let us record:

Proposition 13.3 *Let* $A \in B(\Gamma_a(\mathcal{Z}))$ *anti-commute with* $\phi(y)$, $y \in \mathcal{Y}$. *Then* A *is proportional to* I.

Definition 13.4 *(13.5) is called the* Fock CAR *representation over* $\mathcal{Y}$ *in* $\Gamma_a(\mathcal{Z})$.

Remark 13.5 *Suppose that* $\mathcal{Z} = \mathbb{C}^m$ *and* $(e_1, \ldots, e_m)$ *is the canonical basis of* $\mathbb{C}^m$. *Clearly,* $\Gamma_a(\mathbb{C})$ *can be identified with* $\mathbb{C}^2$. *Therefore, we have the identification*

$$\otimes^m \mathbb{C}^2 \simeq \otimes^m \Gamma_a(\mathbb{C}) \simeq \Gamma_a(\mathbb{C}^m).$$

Under this identification, ϕ_{2j-1}^{JW}, *resp.* ϕ_{2j}^{JW} *acting on* $\otimes^m \mathbb{C}^2$ *defined in (12.12) coincides with* $\phi(e_j, \bar{e}_j)$, *resp.* $\phi(\mathrm{i}e_j, -\mathrm{i}\bar{e}_j)$ *acting on* $\Gamma_a(\mathbb{C}^m)$. *Note that*

$$(e_1, \bar{e}_1), (\mathrm{i}e_1, -\mathrm{i}\bar{e}_1), \ldots, (e_m, \bar{e}_m), (\mathrm{i}e_m, -\mathrm{i}\bar{e}_m)$$

is an o.n. basis of $(\mathcal{Y}, \nu)$. *Thus, the Jordan–Wigner representation over* $\mathbb{R}^{2m}$ *in* $\otimes^m \mathbb{C}^2$ *coincides with the Fock representation over* $\mathbb{R}^{2m}$ *on* $\Gamma_a(\mathbb{C}^m)$.

13.1.2 Extended Fock representation

Note that I implements the parity transformation, since

$$I\phi(z,\overline{z})I^{-1} = -\phi(z,\overline{z}).$$

Let us extend the scalar product ν to the space $\mathcal{Y} \oplus \mathbb{R}$ by

$$(z_1,\overline{z_1})\cdot\nu(z_2,\overline{z_2}) + t_1 t_2 := \mathrm{Re}(z_1|z_2) + t_1 t_2.$$

Clearly,

$$\mathcal{Y} \oplus \mathbb{R} \ni (z,\overline{z},t) \mapsto \phi(z,\overline{z}) + tI \in B\big(\Gamma_{\mathrm{a}}(\mathcal{Z})\big) \tag{13.6}$$

is also an irreducible representation of the CAR.

Definition 13.6 *(13.6) is called the* extended Fock CAR *representation over* $\mathcal{Y} \oplus \mathbb{R}$ *in* $\Gamma_{\mathrm{a}}(\mathcal{Z})$.

Remark 13.7 *Extending Remark 13.5 in an obvious way, we note that the representation (12.14) over* $\mathbb{R}^{2m+1}$ *in* $\otimes^m \mathbb{C}^2$ *can be identified with the extended Fock representation over* $\mathbb{R}^{2m+1}$ *in* $\Gamma_{\mathrm{a}}(\mathbb{C}^m)$.

13.1.3 Slater determinants

Let $\mathcal{W}$ be a finite-dimensional *oriented* subspace of $\mathcal{Z}$. (For the definition of an oriented complex space see Subsect. 3.6.8.) Let $(w_1,\ldots,w_n)$ be an o.n. basis of $\mathcal{W}$ compatible with the orientation. Then

$$a^*(w_1)\cdots a^*(w_n)\Omega = \sqrt{n!}\; w_1 \otimes_{\mathrm{a}} \cdots \otimes_{\mathrm{a}} w_n \tag{13.7}$$

is a normalized vector.

Definition 13.8 *Vectors of the form (13.7) are called* Slater determinants. *If* $\mathcal{W} = \mathcal{Z}$, *then (13.7) is called a* ceiling vector.

If $u \in U(\mathcal{W})$, then

$$a^*(uw_1)\cdots a^*(uw_n)\Omega = (\det u)a^*(w_1)\cdots a^*(w_n)\Omega.$$

Thus a Slater determinant depends only on the oriented subspace $\mathcal{W}$.

13.2 Real-wave and complex-wave CAR representation on Fock spaces

In Subsects. 12.4.2 and 12.5.3 we introduced the concept of a real-wave CAR representation by using the GNS representation for the canonical tracial state. There exists a convenient alternative description of this representation that uses the Fock CAR representation, which we will discuss in this section.

We will also introduce the complex-wave CAR representation – an analog of the complex-wave CCR representation, which we discussed in Subsect. 9.2.1.

13.2.1 Real-wave CAR representation on Fock spaces

Let $\mathcal{Y}$ be a real Hilbert space. Clearly, $\mathbb{C}\mathcal{Y}$ is a complex Hilbert space possessing a natural conjugation. For typographical reasons, this conjugation will be sometimes denoted χ.

In this subsection we continue to discuss the real-wave representation in an arbitrary dimension.

We will consider the Fock space $\Gamma_a(\mathbb{C}\mathcal{Y})$ equipped with the corresponding conjugation. $\Gamma_a(\mathcal{Y})$ is its real subspace of elements fixed by the conjugation $\Gamma(\chi)$. Linear operators that preserve $\Gamma_a(\mathcal{Y})$ are called real.

Introduce the following operators on the fermionic Fock space $\Gamma_a(\mathbb{C}\mathcal{Y})$:

$$\phi^l(y) := a^*(y) + a(y),$$
$$\phi^r(y) := \Lambda\big(a^*(y) + a(y)\big)\Lambda, \quad y \in \mathcal{Y},$$

where we recall that $\Lambda = (-1)^{N(N-1)/2}$.

Theorem 13.9 (1) *We have two mutually commuting CAR representations:*

$$\mathcal{Y} \ni y \mapsto \phi^l(y) \in B_h\big(\Gamma_a(\mathbb{C}\mathcal{Y})\big), \tag{13.8}$$
$$\mathcal{Y} \ni y \mapsto \phi^r(y) \in B_h\big(\Gamma_a(\mathbb{C}\mathcal{Y})\big). \tag{13.9}$$

That means, for $y_1, y_2 \in \mathcal{Y}$,

$$[\phi^l(y_1), \phi^l(y_2)]_+ = [\phi^r(y_1), \phi^r(y_2)]_+ = 2y_1 \cdot \nu y_2 \mathbb{1}, \quad [\phi^l(y_1), \phi^r(y_2)] = 0.$$

(2) *We have*

$$\phi^l(w) = a^*(w) + a(\chi w), \quad \phi^l(w)^* = \phi(\chi w), \quad w \in \mathbb{C}\mathcal{Y}. \tag{13.10}$$

(3) *Let $\pi^l : \mathrm{CAR}^{C^*}(\mathcal{Y}) \to B\big(\Gamma_a(\mathbb{C}\mathcal{Y})\big)$ be the $*$-homomorphism obtained by Prop. 12.31 from the CAR representations (13.8). Then Ω is a cyclic vector representative for the state tr and the representation π^l. Therefore, π^l is the GNS representation of $\mathrm{CAR}^{C^*}(\mathcal{Y})$ for the state tr and it extends to a $*$-isomorphism of $\mathrm{CAR}^{W^*}(\mathcal{Y})$ onto $\pi^l\big(\mathrm{CAR}^{C^*}(\mathcal{Y})\big)''$.*

(4) *Let J be the modular conjugation for the state tr. Then $J = \Lambda\Gamma(\chi)$. We have*

$$J\phi^l(y)J = \phi^r(y), \quad y \in \mathcal{Y}.$$

(5) *We have*

$$\pi^l(c(A)) = \Gamma(\chi)\pi^l(A)\Gamma(\chi), \quad A \in \mathrm{CAR}(\mathbb{R}^n).$$

Consequently, $\pi^l(\mathrm{Cliff}(\mathcal{Y}))$ consists of real elements of $\pi^l(\mathrm{CAR}(\mathcal{Y}))$.

Proof Statements (1) and (2) are simple computations.

Consider the GNS representation of $CAR^{C^*}(\mathcal{Y})$ w.r.t. the state tr, denoted $(\mathcal{H}_{tr}, \pi_{tr}, \Omega_{tr})$. The Hilbert space $\mathcal{H}_{tr}$ contains $CAR^{C^*}(\mathcal{Y})$ as a dense subspace, equipped with the scalar product

$$\mathrm{tr} A^* B, \quad A, B \in CAR^{C^*}(\mathcal{Y}).$$

Let us define a linear operator

$$\Gamma_a(\mathbb{C}\mathcal{Y}) \ni a \mapsto Ua := \mathrm{Op}(a) \in CAR^{C^*}(\mathcal{Y}) \subset \mathcal{H}_{tr}.$$

The identity

$$\mathrm{tr} \mathrm{Op}(b)^* \mathrm{Op}(c) = (b|c) \tag{13.11}$$

implies that U extends to a unitary operator

$$U : \Gamma_a(\mathbb{C}\mathcal{Y}) \to \mathcal{H}_{tr}.$$

U maps $\Omega \in \Gamma_a(\mathbb{C}\mathcal{Y})$ onto $\mathrm{Op}(1) = \mathbb{1} = \Omega_{tr}$.

We have

$$\pi_{tr}(A) B = AB, \quad A \in CAR^{C^*}(\mathcal{Y}), \quad B \in CAR^{C^*}(\mathcal{Y}) \subset \mathcal{H}_{tr}.$$

In particular, consider $A = \mathrm{Op}(y) = \phi(y)$, $y \in \mathcal{Y}$, and $B = \mathrm{Op}(b)$. Adding up (12.27) and (12.28) we obtain

$$\pi_{tr}(\phi(y)) \mathrm{Op}(b) = \mathrm{Op}(y \cdot b + (\nu y) \cdot \nabla_v b).$$

Therefore,

$$U^* \pi_{tr}(\phi(y)) U = y \cdot v + (\nu y) \cdot \nabla_v$$
$$= a^*(y) + a(y) = \pi^l(y).$$

This proves (3).   □

By the above theorem, we can identify the representation π^l with the real-wave representation π_{tr} considered in Subsect. 12.5.3.

13.2.2 Operators in the real-wave CAR representation

This subsection is parallel to Subsect. 9.3.5, where we studied operators in the real-wave CCR representation. For brevity, we will write $\mathfrak{R}$ for $CAR^{W^*}(\mathcal{Y})$.

Let us use the terminology of non-commutative probability spaces, introduced in Sect. 6.5. By Thm. 13.9, we have a canonical unitary identification

$$L^2(\mathfrak{R}, \mathrm{tr}) \simeq \Gamma_a(\mathbb{C}\mathcal{Y}). \tag{13.12}$$

Thus the real-wave representation acts on a "non-commutative L^2 space", which we view as a justification for the name "real-wave representation of CAR" for the constructions described above.

Definition 13.10 *Let a be a contraction on $\mathcal{Y}$. We define an operator $\Gamma_{\mathrm{rw}}(a)$ on $L^2(\mathfrak{R}, \mathrm{tr}) = \Gamma_{\mathrm{a}}(\mathbb{C}\mathcal{Y})$ by*

$$\Gamma_{\mathrm{rw}}(a) = \Gamma(a_{\mathbb{C}}).$$

Proposition 13.11 (1) *Let a be a contraction on $\mathcal{Y}$, $b, c \in \Gamma_{\mathrm{a}}(\mathbb{C}\mathcal{Y})$ and $c = \Gamma_{\mathrm{rw}}(a)b$. Then, if we use the identification (13.12), we have*

$$\mathrm{Op}(c) = \mathrm{Op}(\Gamma(a)b).$$

(2) *Let $\mathcal{Y}_1$ be a closed subspace of $\mathcal{Y}$ and e_1 the orthogonal projection onto $\mathcal{Y}_1$. Then*

$$\Gamma_{\mathrm{rw}}(e_1) = E_{\mathcal{Y}_1},$$

where $E_{\mathcal{Y}_1}$ is the conditional expectation introduced in Subsect. 12.5.4.

(3) *Let $r \in O(\mathcal{Y})$. Then $\hat{r}$, defined originally as an automorphism of $\mathfrak{R}$, can be extended to a unitary operator on $L^2(\mathfrak{R}, \mathrm{tr})$. If we denote this extension also by $\hat{r}$, we have*

$$\Gamma_{\mathrm{rw}}(r) = \hat{r}.$$

The following fermionic analog of Prop. 9.29 is due to Gross (1972):

Proposition 13.12 *Let $a \in B(\mathcal{Y})$. Then if $\|a\| \leq 1$, $\Gamma_{\mathrm{rw}}(a)$ is positivity preserving. It follows that $\Gamma_{\mathrm{rw}}(a)$ extends to a contraction on $L^p(\mathfrak{R}, \mathrm{tr})$ for all $1 \leq p \leq \infty$.*

Proof We follow the proof in Prop. 9.29, writing a as $j^* u j$. The map $\Gamma_{\mathrm{rw}}(j)$ becomes

$$L^2(\mathfrak{R}, \mathrm{tr}) \ni A \mapsto A \otimes \mathbb{1} \in L^2(\mathfrak{R}, \mathrm{tr}) \otimes L^2(\mathfrak{R}, \mathrm{tr}),$$

which is positivity preserving, as well as $\Gamma_{\mathrm{rw}}(j^*) = \Gamma_{\mathrm{rw}}(j)^*$. If $A \in \mathfrak{R} \otimes \mathfrak{R}$, then $\Gamma_{\mathrm{rw}}(u)A$ as an operator on $\Gamma_{\mathrm{a}}(\mathbb{C}\mathcal{Y} \oplus \mathbb{C}\mathcal{Y})$ equals $\Gamma(u_{\mathbb{C}})A\Gamma(u_{\mathbb{C}})^{-1}$, which belongs to $\mathfrak{R} \otimes \mathfrak{R}$ and is positive if A is. Hence, $\Gamma(u_{\mathbb{C}})$ is positivity preserving. The second statement then follows from Thm. 6.81. $\qquad\square$

The following fermionic version of Nelson's hyper-contractivity theorem is due to Gross (1972) and Carlen–Lieb (1993):

Theorem 13.13 *Let $a \in B(\mathcal{Y})$, $1 < p \leq q < \infty$ and*

$$\|a\| \leq (p \doteq 1)^{\frac{1}{2}}(q-1)^{-\frac{1}{2}}.$$

Then $\Gamma_{\mathrm{rw}}(a)$ is a contraction from $L^p(\mathfrak{R}, \mathrm{tr})$ to $L^q(\mathfrak{R}, \mathrm{tr})$.

13.2.3 Complex-wave CAR representation in finite dimensions

One can reformulate the Fock CAR representation so that it becomes analogous to the complex-wave CCR representation considered in Subsect. 9.2.1. For

simplicity, at first we restrict ourselves to finite-dimensional spaces $\mathcal{Z}$. We identify $\mathcal{Z}^{\#}$ with $\overline{\mathcal{Z}}$ using the scalar product.

Recall that an alternative notation for $\Gamma_{\mathrm{a}}(\mathcal{Z})$ is $\mathrm{Pol}_{\mathrm{a}}(\overline{\mathcal{Z}})$. Elements of $\mathrm{Pol}_{\mathrm{a}}(\overline{\mathcal{Z}})$
are treated as sequences whose n-th element is an anti-symmetric n-linear form
on $\overline{\mathcal{Z}}$. Thus to define $F \in \mathrm{Pol}_{\mathrm{a}}(\overline{\mathcal{Z}})$ we need to specify

$$F(\overline{z}_1, \ldots, \overline{z}_n), \qquad z_1, \ldots, z_n \in \mathcal{Z}, \quad n = 0, 1, 2, \ldots. \tag{13.13}$$

In algebraic formulas we write $F(\overline{z})$ instead of (13.13), treating $\overline{z}$ as the "generic
variable" in $\overline{\mathcal{Z}}$, as discussed in Subsect. 3.5.1.

Likewise, an alternative notation for $\Gamma_{\mathrm{a}}(\overline{\mathcal{Z}})$ is $\mathrm{Pol}_{\mathrm{a}}(\mathcal{Z})$. Applying the complex
conjugation to $F \in \mathrm{Pol}_{\mathrm{a}}(\overline{\mathcal{Z}})$, we obtain $\overline{F} \in \mathrm{Pol}_{\mathrm{a}}(\mathcal{Z})$ such that

$$\overline{F}(z_1, \ldots, z_n) = \overline{F(\overline{z}_1, \ldots, \overline{z}_n)}, \qquad z_1, \ldots, z_n \in \mathcal{Z}, \quad n = 0, 1, 2, \ldots. \tag{13.14}$$

We will commonly write $\overline{F}(z)$ or $\overline{F(\overline{z})}$ instead of (13.14), treating z as the
"generic variable" in $\mathcal{Z}$.

Let us fix a (complex) volume form $\mathrm{d}\overline{z}$ on $\mathcal{Z}$ compatible with the scalar product
of $\mathcal{Z}$, and let $\mathrm{d}z$ be the dual volume form on $\overline{\mathcal{Z}}$. As in Subsect. 7.2.1, if $A \in$
$\mathbb{C}\mathrm{Pol}_{\mathrm{a}}(\mathcal{Z} \oplus \overline{\mathcal{Z}})$, we can define its Berezin integral,

$$\int A(z, \overline{z}) \mathrm{d}z \mathrm{d}\overline{z}.$$

Equip $\mathrm{Pol}_{\mathrm{a}}(\overline{\mathcal{Z}})$ with the scalar product

$$(F|G) := \int \overline{F(\overline{z})} G(\overline{z}) \mathrm{e}^{z \cdot \overline{z}} \mathrm{d}z \mathrm{d}\overline{z}.$$

We define the map $T^{\mathrm{cw}} : \Gamma_{\mathrm{a}}(\mathcal{Z}) \to \mathrm{Pol}_{\mathrm{a}}(\overline{\mathcal{Z}})$ by

$$T^{\mathrm{cw}} \Psi(\overline{z}_1, \ldots, \overline{z}_n) := \frac{1}{\sqrt{n!}} (z_1 \otimes_{\mathrm{a}} \cdots \otimes_{\mathrm{a}} z_n | \Psi), \quad \Psi \in \Gamma_{\mathrm{a}}(\mathcal{Z}), \; z_1, \ldots, z_n \in \mathcal{Z}.$$

Applying Thm. 7.23 (2) to $\mathcal{Y} = \mathcal{Z}$, $\mathcal{Y}^{\#} = \overline{\mathcal{Z}}$, we obtain the following theorem:

Theorem 13.14 (1) *The operator T^{cw} is unitary, that is, for $\Phi, \Psi \in \Gamma_{\mathrm{a}}(\mathcal{Z})$,*

$$(\Phi|\Psi) = \int \overline{T^{\mathrm{cw}} \Phi(\overline{z})} T^{\mathrm{cw}} \Psi(\overline{z}) \mathrm{e}^{z \cdot \overline{z}} \mathrm{d}z \mathrm{d}\overline{z}.$$

(2) *For $w \in \mathcal{Z}$ we have*

$$T^{\mathrm{cw}} \Omega = 1,$$
$$T^{\mathrm{cw}} a^*(w) = w \cdot \overline{z} \, T^{\mathrm{cw}},$$
$$T^{\mathrm{cw}} a(w) = \overline{w} \cdot \nabla_{\overline{z}} \, T^{\mathrm{cw}},$$
$$(T^{\mathrm{cw}} \Gamma(p) \Psi)(\overline{z}) = (T^{\mathrm{cw}} \Psi)(p^{\#} \overline{z}), \quad p \in B(\mathcal{Z}), \; \Psi \in \Gamma_{\mathrm{a}}(\mathcal{Z}).$$

Proposition 13.15 *For $w \in \mathcal{Z}$, define an operator on $\mathrm{Pol}_{\mathrm{a}}(\overline{\mathcal{Z}})$ by*

$$\phi^{\mathrm{cw}}(w, \overline{w}) := w \cdot \overline{z} + \overline{w} \cdot \nabla_{\overline{z}}.$$

The map

$$\mathrm{Re}(\mathcal{Z} \oplus \overline{\mathcal{Z}}) \ni (w, \overline{w}) \mapsto \phi^{\mathrm{cw}}(w, \overline{w}) \in B_{\mathrm{h}}\big(\mathrm{Pol}_{\mathrm{a}}(\overline{\mathcal{Z}})\big)$$

is a CAR representation unitarily equivalent to the Fock representation:

$$\phi^{\mathrm{cw}}(w, \overline{w}) = T^{\mathrm{cw}}\big(a^*(w) + a(w)\big)T^{\mathrm{cw}*}. \tag{13.15}$$

Definition 13.16 *(13.15) is called the* basic form *of the complex-wave CAR representation.*

13.2.4 Complex-wave CAR representation: the general case

If $\mathcal{Z}$ is infinite-dimensional, the Berezin integral does not exist anymore. Therefore, strictly speaking, the definition of the complex-wave CAR representation has to be modified. We will need to use the formalism of non-commutative probability spaces.

Let us start by defining an appropriate real-wave CAR representation with a tracial state that will replace the Berezin integral. Consider the space $\mathcal{Z} \oplus \overline{\mathcal{Z}}$ equipped with a natural conjugation

$$\chi(z_1, \overline{z}_2) := (z_2, \overline{z}_1), \quad (z_1, \overline{z}_2) \in \mathcal{Z} \oplus \overline{\mathcal{Z}},$$

whose real subspace is $\mathrm{Re}(\mathcal{Z} \oplus \overline{\mathcal{Z}})$. $\mathrm{Re}(\mathcal{Z} \oplus \overline{\mathcal{Z}})$ is equipped with the symmetric form

$$(z, \overline{z}) \cdot \nu(z', \overline{z}') := 2\mathrm{Re}(z|z'), \quad (z, \overline{z}), (z', \overline{z}') \in \mathrm{Re}(\mathcal{Z} \oplus \overline{\mathcal{Z}}).$$

Following Thm. 13.9, consider the real-wave CAR representation

$$\mathrm{Re}(\mathcal{Z} \oplus \overline{\mathcal{Z}}) \ni (z, \overline{z}) \mapsto \phi^{\mathrm{rw}}(z, \overline{z}) := a^*(z, \overline{z}) + a(z, \overline{z}) \in B_{\mathrm{h}}\big(\Gamma_{\mathrm{a}}(\mathcal{Z} \oplus \overline{\mathcal{Z}})\big).$$

The fields $\phi^{\mathrm{rw}}(z, \overline{z})$, $z \in \mathcal{Z}$, generate a von Neumann algebra $\mathfrak{R}$ isomorphic to $\mathrm{CAR}^{W^*}\big(\mathrm{Re}(\mathcal{Z} \oplus \overline{\mathcal{Z}})\big)$. As in (13.10), we extend these fields from $\mathrm{Re}(\mathcal{Z} \oplus \overline{\mathcal{Z}})$ to $\mathcal{Z} \oplus \overline{\mathcal{Z}}$ by complex linearity, setting for $(z_1, \overline{z}_2) \in \mathcal{Z} \oplus \overline{\mathcal{Z}}$

$$\phi^{\mathrm{rw}}(z_1, \overline{z}_2) := a^*(z_1, \overline{z}_2) + a(z_2, \overline{z}_1) \in \mathfrak{R}.$$

We have

$$\phi^{\mathrm{rw}}(z_1, \overline{z}_2)^* = \phi^{\mathrm{rw}}(z_2, \overline{z}_1),$$
$$[\phi^{\mathrm{rw}}(z_1, \overline{z}_2), \phi^{\mathrm{rw}}(z_1', \overline{z}_2')]_+ = 2(z_1|z_2') + 2(z_2|z_1'). \tag{13.16}$$

Denote by $\mathfrak{R}_{\overline{\mathrm{cw}}}$ the σ-weakly closed (but non-self-adjoint) sub-algebra of $\mathfrak{R}$ generated by $\phi^{\mathrm{rw}}(z, 0)$, $z \in \mathcal{Z}$.

Theorem 13.17 *There exists a unique bounded linear map*

$$T^{\widetilde{\mathrm{cw}}} : \Gamma_{\mathrm{a}}(\mathcal{Z}) \to L^2(\mathfrak{R}, \mathrm{tr})$$

such that

$$T^{\widetilde{cw}}\Omega := \mathbb{1},$$
$$T^{\widetilde{cw}}a^*(z) = \phi^{\mathrm{rw}}(z,0)T^{\widetilde{cw}}, \quad z \in \mathcal{Z}.$$

The map is isometric, i.e.

$$(\Phi|\Psi) = \mathrm{tr}\big(T^{\widetilde{cw}}\Phi\big)^* T^{\widetilde{cw}}\Psi, \quad \Phi, \Psi \in \Gamma_{\mathrm{a}}(\mathcal{Z}). \tag{13.17}$$

It satisfies

$$T^{\widetilde{cw}}a(z) = \phi^{\mathrm{rw}}(0,\overline{z})T^{\widetilde{cw}}, \quad z \in \mathcal{Z},$$
$$T^{\widetilde{cw}}\Gamma(p) = \Gamma_{\mathrm{rw}}(p \oplus \overline{p})T^{\widetilde{cw}}, \quad p \in B(\mathcal{Z}).$$

The image of $T^{\widetilde{cw}}$ is a commutative sub-algebra of $L^2(\mathfrak{R},\mathrm{tr})$.

Definition 13.18 *The image of $T^{\widetilde{cw}}$ is denoted by $L^2(\mathfrak{R}_{\widetilde{cw}},\mathrm{tr})$.*

Proposition 13.19 *For $z \in \mathcal{Z}$, the multiplication by $\phi^{\mathrm{rw}}(z,\overline{z})$ preserves $L^2(\mathfrak{R}_{\widetilde{cw}},\mathrm{tr})$. Therefore,*

$$\phi^{\widetilde{cw}}(z,\overline{z})A := \phi^{\mathrm{rw}}(z,\overline{z})A, \quad A \in L^2(\mathfrak{R}_{\widetilde{cw}},\mathrm{tr}),$$

defines an operator on $L^2(\mathfrak{R}_{\widetilde{cw}},\mathrm{tr})$. The map

$$\mathrm{Re}(\mathcal{Z} \oplus \overline{\mathcal{Z}}) \ni (z,\overline{z}) \mapsto \phi^{\widetilde{cw}}(z,\overline{z}) \in B_{\mathrm{h}}\big(L^2(\mathfrak{R}_{\widetilde{cw}},\mathrm{tr})\big)$$

is a CAR representation unitarily equivalent to the Fock representation:

$$\phi^{\widetilde{cw}}(z,\overline{z}) = T^{\widetilde{cw}}\big(a^*(z) + a(z)\big)T^{\widetilde{cw}*}. \tag{13.18}$$

Definition 13.20 *(13.18) is called the* alternate form of the complex-wave CAR representation.

13.3 Wick and anti-Wick fermionic quantization

This section is parallel to Sect. 9.4, where the bosonic Wick and anti-Wick quantizations were considered.

The framework of this section is the same as that of the whole chapter. Recall that $\mathcal{Z}$ is a Hilbert space, $\mathcal{Y} = \mathrm{Re}(\mathcal{Z} \oplus \overline{\mathcal{Z}})$ and we identify $\mathcal{Y}^{\#} \simeq \mathrm{Re}(\overline{\mathcal{Z}} \oplus \mathcal{Z})$. Recall from Subsect. 3.5.6 that $\mathbb{C}\mathrm{Pol}_{\mathrm{a}}(\mathcal{Y}^{\#})$ is identified with $\mathrm{Pol}_{\mathrm{a}}(\overline{\mathcal{Z}} \oplus \mathcal{Z})$.

We consider the Fock CAR representation

$$\mathcal{Y} \ni y \mapsto \phi(y) \in B_{\mathrm{h}}\big(\Gamma_{\mathrm{a}}(\mathcal{Z})\big).$$

Recall that $\mathrm{CAR}^{\mathrm{alg}}(\mathcal{Y})$ is the $*$-algebra generated by $\phi(y)$, $y \in \mathcal{Y}$. It can be represented by operators on the space $\Gamma_{\mathrm{a}}(\mathcal{Z})$. Recall that $\Lambda = (-1)^{N(N-1)/2}$.

We will define and study the fermionic Wick and anti-Wick quantizations.

13.3.1 Wick and anti-Wick ordering

Recall that, with $b \in \mathrm{Pol}_a(\overline{\mathcal{Z}})$, in Subsect. 3.4.4 we defined the multiple creation and annihilation operators. We obtain two homomorphisms,

$$\mathrm{Pol}_a(\overline{\mathcal{Z}}) \ni b \mapsto a^*(b) \in \mathrm{CAR}^{\mathrm{alg}}(\mathcal{Y}),$$
$$\mathrm{Pol}_a(\mathcal{Z}) \ni \overline{b} \mapsto a(\Lambda b) \in \mathrm{CAR}^{\mathrm{alg}}(\mathcal{Y}).$$

Note that the possibility of unambiguously defining $a^*(b)$ and $a(b)$ follows from the fact that $\mathcal{Z}$ and $\overline{\mathcal{Z}}$ are isotropic subspaces of $\mathbb{C}\mathcal{Y}$ for the bilinear symmetric form $\nu_{\mathbb{C}}$.

Definition 13.21 *For b_1, $b_2 \in \mathrm{Pol}_a(\overline{\mathcal{Z}})$ we set*

$$\mathrm{Op}^{a^*,a}(b_1 \overline{b}_2) := a^*(b_1)a(\Lambda b_2),$$

$$\mathrm{Op}^{a,a^*}(\overline{b}_2 b_1) := a(\Lambda b_2)a^*(b_1).$$

These maps extend by linearity to maps

$$\mathbb{C}\mathrm{Pol}_a(\mathcal{Y}^\#) \ni b \mapsto \mathrm{Op}^{a^*,a}(b) \in \mathrm{CAR}^{\mathrm{alg}}(\mathcal{Y}),$$
$$\mathbb{C}\mathrm{Pol}_a(\mathcal{Y}^\#) \ni b \mapsto \mathrm{Op}^{a,a^*}(b) \in \mathrm{CAR}^{\mathrm{alg}}(\mathcal{Y}), \tag{13.19}$$

called the Wick *and* anti-Wick fermionic quantizations.

Definition 13.22 *The inverse maps to (13.19) will be denoted by*

$$\mathrm{CAR}^{\mathrm{alg}}(\mathcal{Y}) \ni B \mapsto s_B^{a^*,a} \in \mathbb{C}\mathrm{Pol}_a(\mathcal{Y}^\#),$$

$$\mathrm{CAR}^{\mathrm{alg}}(\mathcal{Y}) \ni B \mapsto s_B^{a,a^*} \in \mathbb{C}\mathrm{Pol}_a(\mathcal{Y}^\#).$$

The anti-symmetric polynomial $s_B^{a^*,a}$, resp. s_B^{a,a^*} is called the Wick, *resp. anti-Wick symbol of the operator B.*

Remark 13.23 *If we fix an o.n. basis $(e_i \ : \ i \in I)$ of $\mathcal{Z}$ parametrized by a totally ordered set I, and write*

$$b = \sum_{\{i_1,\dots,i_m\},\{i'_n,\dots,i'_1\}\subset I} b_{i_1,\dots,i_m;i'_n,\dots,i_1} \overline{z}_{i_1} \cdots \overline{z}_{i_m} z_{i'_n} \cdots z_{i'_1},$$

$$c = \sum_{\{i_1,\dots,i_m\},\{i'_n,\dots,i'_1\}\subset I} c_{i_1,\dots,i_m;i'_n,\dots,i_1} z_{i_1} \cdots z_{i_m} \overline{z}_{i'_n} \cdots \overline{z}_{i'_1},$$

then we have explicit formulas

$$\mathrm{Op}^{a^*,a}(b) = \sum_{\{i_1,\dots,i_m\},\{i'_n,\dots,i'_1\}\subset I} b_{i_1,\dots,i_m;i'_n,\dots,i_1} a_{i_1}^* \cdots a_{i_m}^* a_{i'_n} \cdots a_{i'_1},$$

$$\mathrm{Op}^{a,a^*}(c) = \sum_{\{i_1,\dots,i_m\},\{i'_n,\dots,i'_1\}\subset I} c_{i_1,\dots,i_m;i'_n,\dots,i_1} a_{i_1} \cdots a_{i_m} a_{i'_n}^* \cdots a_{i'_1}^*.$$

Proposition 13.24 (1) $\mathrm{Op}^{a^*,a}(b)^* = \mathrm{Op}^{a^*,a}(\Lambda\bar{b})$ *and* $\mathrm{Op}^{a,a^*}(b)^* = \mathrm{Op}^{a,a^*}(\Lambda\bar{b})$.
(2) *Let* $w \in \mathcal{Z}$, $b \in \mathbb{C}\mathrm{Pol}_a(\mathcal{Y}^\#)$. *Then*

$$\mathrm{Op}^{a^*,a}(w\cdot b) = a^*(w)\mathrm{Op}^{a^*,a}(b), \quad \mathrm{Op}^{a^*,a}(b\cdot\overline{w}) = \mathrm{Op}^{a^*,a}(b)a(w),$$

$$a^*(w)\mathrm{Op}^{a^*,a}(b) - \mathrm{Op}^{a^*,a}(Ib)a^*(w) = \mathrm{Op}^{a^*,a}(w\cdot\nabla_z b),$$

$$a(w)\mathrm{Op}^{a^*,a}(b) - \mathrm{Op}^{a^*,a}(Ib)a(w) = \mathrm{Op}^{a^*,a}(\overline{w}\cdot\nabla_{\overline{z}}b).$$

(3) *If* $\mathrm{Op}^{a,a^*}(b_-) = \mathrm{Op}^{a^*,a}(b_+)$, *then*

$$b_+(\overline{z},z) = e^{\nabla_{\overline{z}}\cdot\nabla_z}b_-(\overline{z},z)$$

$$= \int e^{(z-z_1)\cdot(\overline{z}-\overline{z}_1)}b_-(\overline{z_1},z_1)\mathrm{d}\overline{z}_1\mathrm{d}z_1.$$

(4) *If* $\mathrm{Op}^{a^*,a}(b_1)\mathrm{Op}^{a^*,a}(b_2) = \mathrm{Op}^{a^*,a}(b)$, *then*

$$b(\overline{z},z) = e^{\nabla_{\overline{z}_1}\cdot\nabla_{z_1}}b_1(\overline{z},z_1)b_2(\overline{z}_1,z)\big|_{z_1=z}$$

$$= \int e^{(z-z_1)\cdot(\overline{z}-\overline{z}_1)}b_1(\overline{z},z_1)b_2(\overline{z}_1,z)\mathrm{d}z_1\mathrm{d}\overline{z_1}.$$

(5) *The Wick quantization satisfies*

$$(\Omega|\mathrm{Op}^{a^*,a}(b)\Omega) = b(0), \quad b \in \mathbb{C}\mathrm{Pol}_a(\mathcal{Y}^\#). \tag{13.20}$$

Proof It suffices to prove (1) and (2) when b is a monomial, which is an easy computation.

To prove (3) and (4), we use the complex-wave representation. We see that the Wick, resp. anti-Wick, quantization can be seen as the $\overline{z}, \nabla_{\overline{z}}$ resp. $\nabla_{\overline{z}}, \overline{z}$ quantization. (3) and (4) follow then from Thm. 7.26. $\qquad\square$

The following formula is the fermionic version of what is usually called Wick's theorem. We will give its diagrammatic interpretation in Chap. 20.

Theorem 13.25 *Let* $b_1, \ldots, b_n \in \mathbb{C}\mathrm{Pol}_a(\mathcal{Y}^\#)$. *Let* $b \in \mathbb{C}\mathrm{Pol}_a(\mathcal{Y}^\#)$ *and*

$$\mathrm{Op}^{a^*,a}(b) = \mathrm{Op}^{a^*,a}(b_1)\cdots\mathrm{Op}^{a^*,a}(b_n).$$

Then

$$b(\overline{z},z)$$
$$= \exp\Big(\sum_{i>j}\nabla_{\overline{z}_i}\cdot\nabla_{z_j}\Big)b_1(\overline{z}_1,z_1)\cdots b_n(\overline{z}_n,z_n)\big|_{z=z_1=\cdots=z_n},$$
$$(\Omega|\mathrm{Op}^{a^*,a}(b_1)\cdots\mathrm{Op}^{a^*,a}(b_n)\Omega)$$
$$= \exp\Big(\sum_{i>j}\nabla_{\overline{z}_i}\cdot\nabla_{z_j}\Big)b_1(\overline{z}_1,z_1)\cdots b_n(\overline{z}_n,z_n)\big|_{0=z_1=\cdots=z_n}.$$

13.3.2 Relation between Wick, anti-Wick and anti-symmetric quantizations

We can introduce the anti-symmetric quantization

$$\mathrm{Pol}_a(\overline{\mathcal{Z}} \oplus \mathcal{Z}) \simeq \mathbb{C}\mathrm{Pol}_a(\mathcal{Y}^\#) \ni b \mapsto \mathrm{Op}(b) \in B(\Gamma_a(\mathcal{Z}))$$

as in Sect. 12.4.

Proposition 13.26 *Let $b, b_+, b_- \in \mathbb{C}\mathrm{Pol}_a(\mathcal{Y}^\#)$. Let*

$$\mathrm{Op}^{a^*,a}(b_+) = \mathrm{Op}(b) = \mathrm{Op}^{a,a^*}(b_-).$$

(1) *The anti-symmetric symbol is given in terms of the Wick symbol by*

$$b(\overline{z}, z) = \mathrm{e}^{\frac{1}{2}\nabla_z \cdot \nabla_{\overline{z}}} b_+(\overline{z}, z)$$

$$= 2^d \int \mathrm{e}^{2(\overline{z}-\overline{z}_1)\cdot(z-z_1)} b_+(\overline{z}_1, z_1) \mathrm{d}\overline{z}_1 \mathrm{d}z_1.$$

(2) *The anti-symmetric symbol is given in terms of the anti-Wick symbol by*

$$b(\overline{z}, z) = \mathrm{e}^{-\frac{1}{2}\nabla_z \cdot \nabla_{\overline{z}}} b_-(\overline{z}, z)$$

$$= 2^d \int \mathrm{e}^{-2(\overline{z}-\overline{z}_1)\cdot(z-z_1)} b_-(\overline{z}_1, z_1) \mathrm{d}\overline{z}_1 \mathrm{d}z_1.$$

Proof To prove (1) we can assume that $b_+(\overline{z}, z) = b_1(\overline{z})\overline{b}_2(z)$, so that

$$\mathrm{Op}^{a^*,a}(b_+) = a^*(b_1)a(b_2) = \mathrm{Op}(b_1)\mathrm{Op}(\overline{b}_2) = \mathrm{Op}(b),$$

using that $\mathcal{Z}, \overline{\mathcal{Z}}$ are isotropic for the scalar product ν. Using Prop. 12.42 we get that

$$b(\overline{z}, z) = \mathrm{e}^{(\nabla_{\overline{z}_1}, \nabla_{z_1})\cdot\nu(\nabla_{\overline{z}_2}, \nabla_{z_2})} b_1(\overline{z}_1)\overline{b}_2(z_2)\big|_{z_1=z_2=z}$$

$$= \mathrm{e}^{\frac{1}{2}\nabla_z \cdot \nabla_{\overline{z}}} b_1(\overline{z})\overline{b}_2(z),$$

which proves (1). Statement (2) follows then from Prop. 13.24. $\square$

13.3.3 Wick quantization: the operator formalism

This subsection is parallel to Subsect. 9.4.5 about the bosonic case.

We will now treat Wick symbols as operators acting on the Fock space. We will restrict ourselves to a rather small class of such operators.

Recall that if N is the number operator, then $\mathbb{1}_{\{n\}}(N)$ is the projection from $\Gamma_s(\mathcal{Z})$ onto $\Gamma_s^n(\mathcal{Z})$. Similarly to the bosonic case, for $b \in B(\Gamma_a(\mathcal{Z}))$ we set $b_{n,m} := \mathbb{1}_{\{n\}}(N)b\mathbb{1}_{\{m\}}(N)$ and

$$B^{\mathrm{fin}}(\Gamma_a(\mathcal{Z}))$$

$$= \{b \in B(\Gamma_a(\mathcal{Z})) : \text{ there exists } n_0 \text{ such that } b_{n,m} = 0 \text{ for } n, m > n_0\}.$$

Definition 13.27 *Let $b \in B^{\text{fin}}(\Gamma_a(\mathcal{Z}))$. The* Wick quantization *of b is defined as the quadratic form on $\Gamma_a^{\text{fin}}(\mathcal{Z})$ such that for $\Phi, \Psi \in \Gamma_a^{\text{fin}}(\mathcal{Z})$,*

$$(\Phi | \text{Op}^{a^*,a}(b) \Psi)$$
$$= \sum_{n,m=0}^{\infty} \sum_{k=0}^{\infty} \frac{\sqrt{(n+k)!(m+k)!}}{k!} (\Phi | b_{n,m} \otimes \mathbb{1}_{\mathcal{Z}}^{\otimes k} \Psi). \tag{13.21}$$

The above definition can be viewed as a generalization of Def. 13.21.

Proposition 13.28 *Let $b \in \mathbb{C}\text{Pol}_a(\mathcal{Y}^{\#}) \simeq \text{Pol}_a(\overline{\mathcal{Z}} \oplus \mathcal{Z})$ be identified with $b \in B^{\text{fin}}(\Gamma_a(\mathcal{Z}))$ by*

$$(z_n \otimes_a \cdots \otimes_a z_1 | b_{n,m} z_m' \otimes_a \cdots \otimes_a z_1') \tag{13.22}$$
$$= \frac{(n+m)!}{n!m!} b_{n,m}(\overline{z}_n \otimes_a \cdots \otimes_a \overline{z}_1 \otimes_a z_1' \otimes_a \cdots z_m'), \quad z_1, \ldots, z_n, z_1', \ldots, z_m' \in \mathcal{Z}.$$

Then $\text{Op}^{a^*,a}(b)$ in the sense of Def. 13.21, which involves b in the first meaning, coincides with $\text{Op}^{a^*,a}(b)$ in the sense of Def. 13.27, involving b in the second meaning.

Proof Choose a totally ordered o.n. basis in $\mathcal{Z}$. Let

$$b = e_{i_1} \otimes_a \cdots \otimes_a e_{i_n} \otimes_a \overline{e}_{j_m} \otimes_a \cdots \otimes_a \overline{e}_{j_1}.$$

Then $\text{Op}^{a^*,a}(b)$ in the sense of Def. 13.21 equals

$$a_{i_1}^* \cdots a_{i_n}^* a_{j_m} \cdots a_{j_1}. \tag{13.23}$$

(13.22) identifies b with the operator

$$|e_{i_1} \otimes_a \cdots \otimes_a e_{i_n})(e_{j_1} \otimes_a \cdots \otimes_a e_{j_m}|.$$

$\text{Op}^{a^*,a}(b)$ in the sense of Def. 13.27 is the quadratic form on $\Gamma_a^{\text{fin}}(\mathcal{Z})$ equal to

$$\sum_{k=0}^{\infty} \frac{\sqrt{(n+k)!(m+k)!}}{k!} |e_{i_1} \otimes_a \cdots \otimes_a e_{i_n})(e_{j_1} \otimes_a \cdots \otimes_a e_{j_m}| \otimes \mathbb{1}_{\mathcal{Z}}^{\otimes k}. \tag{13.24}$$

It is easy to see that (13.23) and (13.24) are equal. $\qquad\square$

Proposition 13.29 *Let $b \in B^{\text{fin}}(\Gamma_a(\mathcal{Z}))$, $h \in B(\mathcal{Z}) \subset B^{\text{fin}}(\Gamma_a(\mathcal{Z}))$, $p \in B(\mathcal{Z}, \mathcal{Z})$. Then*

$$\text{Op}^{a^*,a}(b)^* = \text{Op}^{a^*,a}(b^*),$$
$$\text{Op}^{a^*,a}(h) = d\Gamma(h),$$
$$[d\Gamma(h), \text{Op}^{a^*,a}(b)] = \text{Op}^{a^*,a}(hb - bh^*),$$
$$\Gamma(p)\text{Op}^{a^*,a}(b\Gamma(p)) = \text{Op}^{a^*,a}(\Gamma(p)b)\Gamma(p),$$
$$\Gamma(p)\text{Op}^{a^*,a}(b) = \text{Op}^{a^*,a}(\Gamma(p)b\Gamma(p^*))\Gamma(p) \quad \text{if } p \text{ is isometric,}$$
$$\Gamma(p)\text{Op}^{a^*,a}(b)\Gamma(p^*) = \text{Op}^{a^*,a}(\Gamma(p)b\Gamma(p^*)) \quad \text{if } p \text{ is unitary.}$$

The following proposition describes the special class of particle preserving operators. Recall that the operator $\Theta(\sigma)$ is defined in Def. 3.11.

Theorem 13.30 *If $b \in B\big(\Gamma_a^m(\mathcal{Z})\big)$, then*

$$(\Phi|\mathrm{Op}^{a^*,a}(b)\Psi) = \sum_{k=0}^{\infty} \frac{(m+k)!}{k!}(\Phi|b \otimes \mathbb{1}_{\mathcal{Z}}^k \Psi).$$

Thus

$$\frac{1}{m!}\mathrm{Op}^{a^*,a}(b)\big|_{\Gamma_a^{m+k}(\mathcal{Z})} = \left(\sum_{1 \le i_1 < \cdots < i_m \le m+k} b_{i_1,\dots,i_m}^{m+k} \right)\bigg|_{\Gamma_a^{m+k}(\mathcal{Z})},$$

where the operators $b_{i_1,\dots,i_m}^{m+k} \in B\big(\Gamma_a^{m+k}(\mathcal{Z})\big)$ are defined as follows:

$$b_{i_1,\dots,i_m}^{m+k} := \Theta(\sigma)\, b \otimes \mathbb{1}_{\mathcal{Z}}^{\otimes k}\, \Theta(\sigma)^{-1}\big|_{\Gamma_a^{m+k}(\mathcal{Z})},$$

where $\sigma \in S_{m+k}$ is any permutation that transforms $(1,\dots,m)$ onto $(i_1,\dots,i_m)$.

13.3.4 Estimates on Wick polynomials

Fermionic Wick monomials tend to be bounded more often than bosonic ones. Here is an example of this phenomenon:

Proposition 13.31 *Let $h \in B^1(\mathcal{Z})$ be positive. Then $\|\mathrm{d}\Gamma(h)\| = \mathrm{Tr}\,h$.*

We also have a fermionic analog of bosonic N_τ estimates described in Prop. 9.50. The proof in the fermionic case is fully analogous to that in the bosonic case.

Proposition 13.32 *Let $b \in B\big(\Gamma_a^q(\mathcal{Z}), \Gamma_a^p(\mathcal{Z})\big) \subset B^{\mathrm{fin}}\big(\Gamma_a(\mathcal{Z})\big)$ for $p, q \in \mathbb{N}$. Let $m > 0$ be a self-adjoint operator on $\mathcal{Z}$. Then for all $\Psi_1, \Psi_2 \in \Gamma_a(\mathcal{Z})$ one has*

$$\left|\big(\mathrm{d}\Gamma(m)^{-p/2}\Psi_1|\mathrm{Op}^{a^*,a}(b)\mathrm{d}\Gamma(m)^{-q/2}\Psi_2\big)\right|$$
$$\le \|\Gamma(m)^{-\frac{1}{2}}b\Gamma(m)^{-\frac{1}{2}}\|\|\Psi_1\|\|\Psi_2\|.$$

13.4 Notes

The Wick theorem goes back to Wick (1950).

The fermionic real-wave representation is due to Segal (1956). Second quantized operators in the fermionic real-wave representation were studied by Gross (1972) and Carlen–Lieb (1993).

The fermionic complex-wave representation was developed by Shale–Stinespring (1964).

14

Orthogonal invariance of CAR algebras

Let $(\mathcal{Y}, \nu)$ be a real Hilbert space.

Recall that $O(\mathcal{Y})$ denotes the group of orthogonal operators on $\mathcal{Y}$ and that $o(\mathcal{Y})$ denotes the Lie algebra of bounded anti-self-adjoint operators.

The main aim of this chapter is to discuss the invariance of CAR algebras (mostly C^*- but also W^*-CAR algebras) with respect to the orthogonal group. We will restrict ourselves to the results that are independent of a representation. In particular, they will not involve any Fock representation nor a distinguished Kähler structure. Orthogonal invariance of CAR algebras on Fock spaces will be studied separately in Chap. 16.

To some extent, this chapter can be viewed as an analog of Chap. 10 about the symplectic invariance of CCR in finite dimensions. However, in this chapter we consider the case of an arbitrary dimension, since for the CAR this does not introduce any serious additional difficulties, unlike for the CCR.

14.1 Orthogonal groups

14.1.1 Group $O_1(\mathcal{Y})$

Recall that if $\dim \mathcal{Y}$ is finite, besides the group $O(\mathcal{Y})$ and the Lie algebra $o(\mathcal{Y})$ we have the group $SO(\mathcal{Y}) := \{r \in O(\mathcal{Y}) : \det r = 1\}$. If $\dim \mathcal{Y}$ is arbitrary, we still have $O(\mathcal{Y})$ and $o(\mathcal{Y})$, but there seems to be no analog of $SO(\mathcal{Y})$. However, there exists a natural extension of the triple $(O(\mathcal{Y}), SO(\mathcal{Y}), o(\mathcal{Y}))$ to infinite dimensions described in the following definition:

Definition 14.1 *Set*

$$O_1(\mathcal{Y}) := \{r \in O(\mathcal{Y}) : r - \mathbb{1} \in B^1(\mathcal{Y})\},$$
$$SO_1(\mathcal{Y}) := \{r \in O_1(\mathcal{Y}) : \det r = 1\},$$
$$o_1(\mathcal{Y}) := o(\mathcal{Y}) \cap B^1(\mathcal{Y}).$$

We equip all of them with the metric given by the trace-class norm.

Proposition 14.2 (1) $O_1(\mathcal{Y})$ *is a group and* $SO_1(\mathcal{Y})$ *is its subgroup.*
(2) *We have an exact sequence of groups*

$$1 \to SO_1(\mathcal{Y}) \to O_1(\mathcal{Y}) \to \mathbb{Z}_2 \to 1. \tag{14.1}$$

(3) $o_1(\mathcal{Y})$ *is a Lie algebra and if* $a \in o_1(\mathcal{Y})$, *then* $\mathrm{e}^a \in SO_1(\mathcal{Y})$.

Proof We use the fact that the determinant is a homomorphism of $O_1(\mathcal{Y})$ onto $\{1, -1\}$.   □

We have the following characterization of elements of $SO_1(\mathcal{Y})$:

Theorem 14.3 *Let $r \in O_1(\mathcal{Y})$. Then $\dim \mathrm{Ker}(\mathbb{1} + r)$ is finite. Besides, the following conditions are equivalent:*

(1) $r \in SO_1(\mathcal{Y})$.
(2) $\dim \mathrm{Ker}(\mathbb{1} + r)$ *is even.*
(3) *There exists $a \in o_1(\mathcal{Y})$ such that $\mathrm{e}^a = r$.*

Proof Let us prove (2)⇒(3).

Case 1. Assume $\mathrm{Ker}(\mathbb{1} + r) = \{0\}$. We complexify $\mathcal{Y}$ and consider $r_\mathbb{C} \in U(\mathbb{C}\mathcal{Y})$. Since $\mathrm{Ker}(\mathbb{1} + r_\mathbb{C}) = \{0\}$ and $r_\mathbb{C} - \mathbb{1}$ is compact, we see that

$$\mathrm{spec}\, r_\mathbb{C} \subset \{\mathrm{e}^{\mathrm{i}\phi} \; : \; \phi \in]-\pi, \pi[\}.$$

Take e.g. the principal branch of the logarithm (which maps $\mathbb{C} \backslash]-\infty, 0]$ onto $\{-\pi < \mathrm{Im}\, z < \pi\}$) and define $b := \log r_\mathbb{C}$. b is an anti-self-adjoint operator, $b \in B^1(\mathbb{C}\mathcal{Y})$ and $r_\mathbb{C} = \mathrm{e}^b$. It is real, so there exists $a \in o_1(\mathcal{Y})$ such that $b = a_\mathbb{C}$.

Case 2. Assume that $r = -\mathbb{1}$ and $\dim \mathcal{Y}$ is finite and even. We choose an o.n. basis $(e_1, \ldots, e_{2n})$, and set $ce_i := e_{n+i}$, $ce_{n+i} := -e_i$. Then $c^2 = -\mathbb{1}$, $c \in o_1(\mathcal{Y})$ and $\mathrm{e}^{tc} = \mathbb{1}\cos t + c\sin t$. Thus $\mathrm{e}^{\pi c} = -\mathbb{1}$.

In the general case we set $\mathcal{Y}_{\mathrm{sg}} := \mathrm{Ker}(\mathbb{1} + r)$ and $\mathcal{Y}_{\mathrm{reg}} := \mathcal{Y}_{\mathrm{sg}}^\perp$. These are invariant subspaces of r, so that we can apply case 1 and case 2 to them respectively.

Using that the determinant is continuous in the trace norm topology, we see that $t \mapsto \det \mathrm{e}^{ta}$ is continuous for $a \in o_1(\mathcal{Y})$, which proves (3)⇒(1).

Let us prove (1)⇒(2). Assume that $\dim \mathrm{Ker}(\mathbb{1} + r)$ is odd. Let $y_0 \in \mathrm{Ker}(\mathbb{1} + r)$ be a unit vector and $r_0 := \mathbb{1} - 2|y_0\rangle\langle y_0|$. Then $rr_0 \in O_1(\mathcal{Y})$ and $\mathrm{Ker}(\mathbb{1} + rr_0) = \mathrm{Ker}(\mathbb{1} + r) \ominus \mathbb{R}y_0$. Hence, $\dim \mathrm{Ker}(\mathbb{1} + rr_0)$ is even. Therefore, $\det rr_0 = 1$. Noting that $\det r_0 = -1$, this implies $\det r = -1$.   □

14.1.2 Group $O_p(\mathcal{Y})$

There exist other useful extensions of the triple $(O(\mathcal{Y}), SO(\mathcal{Y}), o(\mathcal{Y}))$ to infinite dimensions, which we consider in this subsection.

Throughout this subsection, $1 \le p \le \infty$. Recall that $B^p(\mathcal{Y})$ denotes the p-th trace ideal, $B_\infty(\mathcal{Y})$ the ideal of compact operators on $\mathcal{Y}$.

Definition 14.4 *Set*

$$O_p(\mathcal{Y}) := \begin{cases} \{r \in O(\mathcal{Y}) \; : \; r - \mathbb{1} \in B^p(\mathcal{Y})\}, & 1 \le p < \infty; \\ \{r \in O(\mathcal{Y}) \; : \; r - \mathbb{1} \in B_\infty(\mathcal{Y})\}, & p = \infty; \end{cases}$$

$$o_p(\mathcal{Y}) := \begin{cases} o(\mathcal{Y}) \cap B^p(\mathcal{Y}), & 1 \le p < \infty; \\ o(\mathcal{Y}) \cap B_\infty(\mathcal{Y}), & p = \infty. \end{cases}$$

We equip all of them with the topology of $B^p(\mathcal{Y})$, resp. $B_\infty(\mathcal{Y})$.

Clearly, $O_p(\mathcal{Y}) \subset O_q(\mathcal{Y})$ and $o_p(\mathcal{Y}) \subset o_q(\mathcal{Y})$ for $p \leq q$.

The determinant is not defined on the whole of $O_p(\mathcal{Y})$ for $p > 1$, which makes the definition of $SO_p(\mathcal{Y})$ harder than that of $SO_1(\mathcal{Y})$. Nevertheless, the following analog of Thm. 14.3 can be shown:

Theorem 14.5 *Let* $r \in O_p(\mathcal{Y})$. *Set* $C(\epsilon) := \{z \in \mathbb{C} : |z| = 1, |z - 1| > \epsilon\}$. *Then for any* $\epsilon > 0$, $\dim \mathbb{1}_{C(\epsilon)}(r)$ *is finite. Besides, the following conditions are equivalent:*

(1) *For* $\epsilon > 0$, $\dim \mathbb{1}_{C(\epsilon)}(r)$ *is even.*
(2) $\dim \operatorname{Ker}(\mathbb{1} + r)$ *is even.*
(3) *There exists* $a \in o_p(\mathcal{Y})$ *such that* $\mathrm{e}^a = r$.

Proof $r - \mathbb{1}$ is compact, hence $\dim \mathbb{1}_{C(\epsilon)}(r)$ is finite for $\epsilon > 0$.

(1)$\Rightarrow$(2) is obvious. To prove (1)$\Leftarrow$(2) we note that for any $\lambda \in \operatorname{spec} r$ we have $\dim \mathbb{1}_{\{\lambda\}}(r) = \dim \mathbb{1}_{\{\bar{\lambda}\}}(r)$.

To show (2)$\Leftrightarrow$(3) we repeat verbatim arguments of the proof of Thm. 14.3 (2)$\Leftrightarrow$(3). $\quad\square$

Definition 14.6 *The set of* $r \in O_p(\mathcal{Y})$ *satisfying the conditions of Thm. 14.5 is denoted by* $SO_p(\mathcal{Y})$. *We will write* $\det r = 1$ *for* $r \in SO_p(\mathcal{Y})$ *and* $\det r = -1$ *for* $r \in O_p(\mathcal{Y}) \backslash SO_p(\mathcal{Y})$, *even though, strictly speaking, the determinant is not defined on* $SO_p(\mathcal{Y})$.

Proposition 14.7 (1) $O_p(\mathcal{Y})$ *is a group and* $SO_p(\mathcal{Y})$ *is its subgroup.*
(2) *We have an exact sequence of groups*

$$1 \to SO_p(\mathcal{Y}) \to O_p(\mathcal{Y}) \to \mathbb{Z}_2 \to 1. \tag{14.2}$$

(3) $o_p(\mathcal{Y})$ *is a Lie algebra, and if* $a \in o_p(\mathcal{Y})$, *then* $\mathrm{e}^a \in SO_p(\mathcal{Y})$.

Proof Clearly, $SO_1(\mathcal{Y})$ sits inside $SO_p(\mathcal{Y})$. Let us show that $SO_1(\mathcal{Y})^{\mathrm{cl}} = SO_p(\mathcal{Y})$.

First note that the condition (1) of Thm. 14.5 implies that $SO_p(\mathcal{Y})$ is closed inside $O_p(\mathcal{Y})$.

Let $r \in SO_p(\mathcal{Y})$. Using Thm. 14.5 (3), we can write $r = \mathrm{e}^a$ with $a \in o_p(\mathcal{Y})$. Using the spectral decomposition of a, we can approximate it with $a_n \in o_1(\mathcal{Y})$, so that $a_n \to a$. Hence, $\mathrm{e}^{a_n} \to r$ with $\mathrm{e}^{a_n} \in SO_1(\mathcal{Y})$. Hence, the closure of $SO_1(\mathcal{Y})$ contains $SO_p(\mathcal{Y})$.

Similarly, we show that $(O_1 \backslash SO_1(\mathcal{Y}))^{\mathrm{cl}} = O_p(\mathcal{Y}) \backslash SO_p(\mathcal{Y})$. In fact, every $r \in O_p(\mathcal{Y}) \backslash SO_p(\mathcal{Y})$ can be written as $r = \kappa r_0$ with $\kappa = \mathbb{1} - 2|e\rangle\langle e|$ and $r_0 \in SO_p(\mathcal{Y})$. We approximate r_0 with elements of $SO_1(\mathcal{Y})$ as above.

(2) follows then from the corresponding statement in Prop. 14.2. $\quad\square$

14.2 Quadratic fermionic Hamiltonians

Recall that $\mathrm{Op}(b)$ denotes the anti-symmetric quantization of an anti-symmetric polynomial $b \in \mathbb{C}\mathrm{Pol}_\mathrm{a}(\mathcal{Y}^{\#})$. In this section we study *quadratic fermionic Hamiltonians*, that is, quantizations of elements from $\mathbb{C}\mathrm{Pol}_\mathrm{a}^2(\mathcal{Y}^{\#})$. We will also describe some situations where a quadratic fermionic Hamiltonian can be well defined even though its symbol is not of finite rank.

14.2.1 Fermionic harmonic oscillator

Let $e_1, e_2 \in \mathcal{Y}$ be an orthonormal pair of vectors in $\mathcal{Y}$. Consider the following operator in $\mathrm{CAR}^{C^*}(\mathcal{Y})$, which can be viewed as a fermionic analog of the harmonic oscillator:

$$H := \phi(e_1)\phi(e_2).$$

Clearly, $H = \mathrm{Op}(\zeta)$, where $\zeta = e_1 \otimes_\mathrm{a} e_2$. If we consider ζ as an element of $L_\mathrm{a}(\mathcal{Y}^{\#}, \mathcal{Y})$, then $\zeta = \frac{1}{2}(|e_1\rangle\langle e_2| - |e_2\rangle\langle e_1|)$. Straightforward computations yield the following properties of the fermionic harmonic oscillator:

Proposition 14.8 (1) $H^2 = -\mathbb{1}$, $H = -H^*$, $\mathrm{spec}\,(\mathrm{i}H) = \{-1, 1\}$;
(2) $\mathrm{e}^{tH} = \cos t\mathbb{1} + (\sin t)H$, in particular, $\mathrm{e}^{\pm\frac{\pi}{2}H} = \pm H$;
(3) $\mathrm{e}^{tH}\phi(y)\mathrm{e}^{-tH} = \phi(\mathrm{e}^{4t\zeta\nu^{-1}}y)$, $y \in \mathcal{Y}$, in particular,

$$H\phi(y)H^{-1} = \phi\left(y - 2e_1\langle e_1|y\rangle - 2e_2\langle e_2|y\rangle\right).$$

Let $y_1, y_2 \in \mathcal{Y}$ be a pair of normalized vectors with $\langle y_1|y_2\rangle = \cos\theta$. Let e_1, e_2 be any o.n. basis of $\mathrm{Span}(y_1, y_2)$ with the same orientation as that of y_1, y_2. Then

$$\phi(y_1)\phi(y_2) = \cos\theta\mathbb{1} + \sin\theta\phi(e_1)\phi(e_2)$$
$$= \mathrm{e}^{\theta\phi(e_1)\phi(e_2)} = \mathrm{Op}(\cos\theta + \sin\theta e_1\cdot e_2).$$

14.2.2 Commutation properties of quadratic fermionic Hamiltonians

The following theorem can be viewed as the fermionic analog of Thm. 10.13 (1).

Theorem 14.9 Let $\chi \in \mathbb{C}\mathrm{Pol}_\mathrm{a}^2(\mathcal{Y}^{\#})$ and $b \in \mathbb{C}\mathrm{Pol}_\mathrm{a}(\mathcal{Y}^{\#})$. Then

$$[\mathrm{Op}(\chi), \mathrm{Op}(b)] = 2\mathrm{Op}((\nabla\chi) \cdot \nu\nabla b); \tag{14.3}$$

$$\frac{1}{2}\left(\mathrm{Op}(\chi)\mathrm{Op}(b) + \mathrm{Op}(b)\mathrm{Op}(\chi)\right) = \mathrm{Op}(\chi \cdot b + \nabla_v\cdot\nu(\nabla^{(2)}\chi)\nu\nabla_v b). \tag{14.4}$$

(In the above expression, $\nabla^{(2)}\chi$ is considered as an element of $L_\mathrm{s}(\mathcal{Y}^{\#}, \mathcal{Y})$ and $\nabla_v\cdot\nu(\nabla^{(2)}\chi)\nu\nabla_v$ is a differential operator acting on the anti-symmetric polynomial b.)

Proof Let us use the "functional notation" for anti-symmetric polynomials. Thus v_1, v_2 are "generic variables" in $\mathcal{Y}$.

If $\deg b_1$ or $\deg b_2$ is equal to 2, then

$$e^{\nabla_{v_2} \cdot \nu \nabla_{v_1}} b_1(v_1) b_2(v_2) = b_1(v_1) b_2(v_2) + \nabla_{v_2} \cdot \nu \nabla_{v_1} b_1(v_1) b_2(v_2)$$
$$+ \frac{1}{2} (\nabla_{v_2} \cdot \nu \nabla_{v_1})^2 b_1(v_1) b_2(v_2).$$

We insert $v_1 = v_2 = v$, switch to the coordinate notation as in the proof of Prop. 12.42, and use the summation convention. We obtain that the symbol of $Op(b_1) Op(b_2)$ equals

$$b_1 b_2 + (-1)^{\deg b_1 - 1} \nu_{ij} (\nabla_{v^j} b_1) \nabla_{v^i} b_2$$
$$- \frac{1}{2} \nu_{i,i'} \nu_{j,j'} (\nabla_{v^{i'}} \nabla_{v^{j'}} b_1) \nabla_{v^i} \nabla_{v^j} b_2. \tag{14.5}$$

The formula for $Op(b_2) Op(b_1)$ coincides with (14.5), except that the second term changes sign. Then we replace b_1, b_2 with b, χ. $\qquad\square$

In what follows it will be convenient to change slightly the parametrization of quadratic fermionic Hamiltonians.

Definition 14.10 $B_a^{fd}(\mathcal{Y}^{\#}, \mathcal{Y})$ *will denote the space of finite rank anti-symmetric operators, that is,* $B_a(\mathcal{Y}^{\#}, \mathcal{Y}) \cap B^{fd}(\mathcal{Y}^{\#}, \mathcal{Y})$.

Every $\chi \in \mathbb{C}Pol_a^2(\mathcal{Y}^{\#})$ (a complex homogeneous anti-symmetric quadratic polynomial on $\mathcal{Y}^{\#}$) can be represented as

$$\mathcal{Y}^{\#} \times \mathcal{Y}^{\#} \ni (v, w) \mapsto \chi(v, w) = v \cdot \zeta w, \tag{14.6}$$

for $\zeta \in \mathbb{C}B_a^{fd}(\mathcal{Y}^{\#}, \mathcal{Y})$. Therefore, we have an identification $\mathbb{C}Pol_a^2(\mathcal{Y}^{\#}) \simeq \mathbb{C}B_a^{fd}(\mathcal{Y}^{\#}, \mathcal{Y})$. Note that $\nabla\chi(v) = 2\zeta v$ and $\nabla^{(2)}\chi = 2\zeta$.

Definition 14.11 *We will write* $Op(\zeta)$ *for the anti-symmetric quantization of* (14.6).

Clearly, if we choose orthonormal coordinates in $\mathcal{Y}^{\#}$, then (14.6) equals

$$v \cdot \zeta w = \sum_{1 \le i, j \le m} \zeta_{ij} v_i w_j,$$

where $[\zeta_{ij}]$ is an anti-symmetric matrix and its quantization equals

$$Op(\zeta) = \sum_{i,j=1}^{m} \phi_i \zeta_{ij} \phi_j. \tag{14.7}$$

14.2.3 Quadratic Hamiltonians in C^*-CAR algebras

It is natural to extend the definition of quadratic fermionic Hamiltonians to symbols that are not finite rank. In this subsection, we will consider quadratic Hamiltonians inside the algebra $\mathrm{CAR}^{C^*}(\mathcal{Y})$.

Definition 14.12 $B_{\mathrm{a}}^1(\mathcal{Y}^\#, \mathcal{Y})$ *will denote the space of trace-class anti-symmetric operators, that is,* $B_{\mathrm{a}}(\mathcal{Y}^\#, \mathcal{Y}) \cap B^1(\mathcal{Y}^\#, \mathcal{Y})$.

Theorem 14.13 (1) *The* map $\mathbb{C}B_{\mathrm{a}}^{\mathrm{fd}}(\mathcal{Y}^\#, \mathcal{Y}) \ni \zeta \mapsto \mathrm{Op}(\zeta) \in \mathrm{CAR}^{C^*}(\mathcal{Y})$
extends by continuity to $\zeta \in \mathbb{C}B_{\mathrm{a}}^1(\mathcal{Y}^\#, \mathcal{Y})$.
(2) *Let* $\zeta \in B_{\mathrm{a}}^1(\mathcal{Y}^\#, \mathcal{Y})$. *Then* $\mathrm{Op}(\zeta)$ *is self-adjoint,*

$$\|\mathrm{Op}(\zeta)\| = \mathrm{Tr}|\zeta\nu|, \quad \inf \mathrm{Op}(\zeta) = -\mathrm{Tr}|\zeta\nu|, \quad \sup \mathrm{Op}(\zeta) = \mathrm{Tr}|\zeta\nu|. \tag{14.8}$$

(3) *If* $\zeta_1, \zeta_2 \in \mathbb{C}B_{\mathrm{a}}^1(\mathcal{Y}^\#, \mathcal{Y})$, *then*

$$[\mathrm{Op}(\zeta_1), \mathrm{Op}(\zeta_2)] = 4\mathrm{Op}(\zeta_1\nu\zeta_2 - \zeta_2\nu\zeta_1). \tag{14.9}$$

Thus

$$o_1(\mathcal{Y}) \ni a \mapsto \frac{1}{4}\mathrm{Op}(a\nu^{-1}) \in \mathrm{CAR}^{C^*}(\mathcal{Y})$$

is a homomorphism of Lie algebras, where $\mathrm{CAR}^{C^*}(\mathcal{Y})$ *is equipped with the commutator.*

Proof Assume first that $\mathcal{Y}$ is of finite dimension. Let $\zeta \in B_{\mathrm{a}}(\mathcal{Y}^\#, \mathcal{Y})$. By Corollary 2.85, we can find an orthonormal system $\{e_{i,\pm}\}_{i \in I}$ and positive real numbers $\{\lambda_i\}_{i \in I}$ such that

$$\zeta\nu = \sum_{i=1}^m \lambda_i \big(|e_{i,-}\rangle\langle e_{i,+}| - |e_{i,+}\rangle\langle e_{i,-}|\big). \tag{14.10}$$

Then

$$\mathrm{Op}(\zeta) := \sum_{i \in I} 2\lambda_i \phi(e_{i,-})\phi(e_{i,+}). \tag{14.11}$$

Using the Jordan–Wigner representation adapted to the above o.n. basis, we see that

$$\mathrm{spec}\,\mathrm{Op}(\zeta) = \Big\{\sum_{i \in I} \lambda_i \epsilon_i, \quad \epsilon_i = \pm 1, \quad i \in I\Big\}.$$

Note that

$$|\zeta\nu| = \sum_{i \in I} \lambda_i \big(|e_{i,-}\rangle\langle e_{i,-}| + |e_{i,+}\rangle\langle e_{i,+}|\big).$$

Therefore, $\mathrm{Tr}|\zeta\nu| = \sum_{i \in I} \lambda_i$. This implies (14.8) in the finite-dimensional case.

In the case of $\zeta \in B^1(\mathcal{Y}^\#, \mathcal{Y})$ in arbitrary dimension, we can still use Corollary 2.85 to find an orthonormal system $\{e_{i,\pm}\}_{i \in I}$ and positive real numbers $\{\lambda_i\}_{i \in I}$

such that (14.10) is true. Note that the sum in (14.11) is convergent, which allows us to define $\mathrm{Op}(\zeta)$. An obvious approximation argument extends (14.8) to the infinite-dimensional case.

Let us prove (14.9). By (14.3) applied to $\chi_i \in \mathrm{Pol}_a(\mathcal{Y}^\#)$, $i = 1, 2$,

$$[\mathrm{Op}(\chi_1), \mathrm{Op}(\chi_2)] = 2\mathrm{Op}((\nabla\chi_1) \cdot \nu\nabla\chi_2). \tag{14.12}$$

Let us compute the symbol on the r.h.s. of (14.12):

$$
\begin{aligned}
&((\nabla\chi_1) \cdot \nu\nabla\chi_2)(v, w) \\
&= \frac{1}{2}\big(\nabla\chi_1(v) \cdot \nu\nabla\chi_2(w) - \nabla\chi_2(v) \cdot \nu\nabla\chi_1(w)\big), \quad v, w \in \mathcal{Y}^\#.
\end{aligned}
$$

Then we use $\nabla\chi_i(v) = 2\zeta_i v$, obtaining (14.9). $\square$

14.2.4 Quadratic Hamiltonians in W^*-CAR algebras

Let us now consider quadratic fermionic Hamiltonians in the setting given by the algebra $\mathrm{CAR}^{W^*}(\mathcal{Y})$.

Definition 14.14 Let $B_a^2(\mathcal{Y}^\#, \mathcal{Y})$ denote the set of Hilbert–Schmidt anti-symmetric operators from $\mathcal{Y}^\#$ to $\mathcal{Y}$, that is, $B_a^2(\mathcal{Y}^\#, \mathcal{Y}) := B_a(\mathcal{Y}^\#, \mathcal{Y}) \cap B^2(\mathcal{Y}^\#, \mathcal{Y})$.

For simplicity, let us assume that $\mathcal{Y}$ is infinite-dimensional separable. Let $\zeta \in B_a^2(\mathcal{Y}^\#, \mathcal{Y})$. By diagonalizing $\zeta\nu$, we can bring it to a diagonal form:

$$\zeta\nu = \sum_{i=1}^{\infty} \lambda_i \big(|e_{i,-}\rangle\langle e_{i,+}| - |e_{i,+}\rangle\langle e_{i,-}|\big). \tag{14.13}$$

Set

$$H_n := \sum_{i=1}^{n} 2\lambda_i \phi(e_{i,-})\phi(e_{i,+}).$$

Proposition 14.15 For any $t \in \mathbb{R}$, there exists the strong limit

$$\mathrm{s} - \lim_{n\to\infty} e^{itH_n}. \tag{14.14}$$

The limit (14.14) defines a one-parameter strongly continuous unitary group. It can be written as e^{itH}, where H is a certain self-adjoint operator, possibly unbounded. We denote H by $\mathrm{Op}(\zeta)$.

If $\zeta \in B_a^1(\mathcal{Y}, \mathcal{Y}^\#)$, then the above defined $\mathrm{Op}(\zeta)$ coincides with that defined in Thm. 14.13. Furthermore, the definition does not depend on the choice of an ordered o.n. basis diagonalizing ζ. Moreover, $\mathrm{Op}(\zeta)$ is affiliated to $\mathrm{CAR}^{W^*}(\mathcal{Y})$.

Proof It is enough to suppose that $(e_{i,-}, e_{i,+} : i = 1, 2, \dots)$ is an o.n. basis. We use the inductive limit of the representation described in Subsect. 12.4.3. Thus $\mathrm{CAR}^{W^*}(\mathcal{Y})$ is represented on the infinite tensor product of grounded Hilbert spaces

$$\overset{\infty}{\underset{i=1}{\otimes}}\left(B^2(\mathbb{C}^2),\ \frac{1}{\sqrt{2}}\mathbb{1}\right).$$

The operator e^{itH_n} acts in this representation as the multiplication from the right by

$$\overset{n}{\underset{i=1}{\otimes}}\begin{bmatrix} e^{it\lambda_i} & 0 \\ 0 & e^{it\lambda_i} \end{bmatrix}\otimes\overset{\infty}{\underset{i=n+1}{\otimes}}\begin{bmatrix} 1 & 0 \\ 0 & 1 \end{bmatrix}.$$

Set

$$\Omega := \overset{\infty}{\underset{i=1}{\otimes}}\frac{1}{\sqrt{2}}\begin{bmatrix} 1 & 0 \\ 0 & 1 \end{bmatrix}.$$

Clearly,

$$(\Omega|e^{itH_n}\Omega) = \prod_{i=1}^{n}\cos t\lambda_i. \tag{14.15}$$

(14.15) converges as $n \to \infty$ iff $\sum_{i=1}^{\infty}\lambda_i^2 < \infty$. But by Thm. 3.16, the convergence of (14.15) is equivalent to the *-strong convergence of e^{itH_n}. $\qquad\square$

14.3 *Pin^c* and *Pin* groups

We keep the same notation as in the rest of the chapter. In particular, $(\mathcal{Y},\nu)$ is a real Hilbert space.

The groups $Pin^c(\mathcal{Y})$ and $Pin(\mathcal{Y})$ are well known in finite dimensions. It is convenient to consider them as subgroups of the *-algebra $CAR(\mathcal{Y})$, resp. its real sub-algebra $Cliff(\mathcal{Y})$.

Recall that these algebras are equipped with the parity automorphism α. As usual, the set of even, resp. odd elements of $CAR(\mathcal{Y})$ is denoted $CAR_0(\mathcal{Y})$, resp. $CAR_1(\mathcal{Y})$.

In this section we concentrate on generalizing the groups $Pin^c(\mathcal{Y})$ and $Pin(\mathcal{Y})$ to infinite dimensions. The first generalization will involve subgroups of the algebra $CAR^{C^*}(\mathcal{Y})$, and the second those of $CAR^{W^*}(\mathcal{Y})$.

14.3.1 *Pin^c* and *Pin* groups in finite dimensions

In this subsection we assume in addition that the dimension of $\mathcal{Y}$ is finite. Let us describe the well-known results about Pin^c and Pin groups in finite dimensions. We do not give their proofs, which are well known. Besides, they will follow from the more general results about the infinite-dimensional case to be described later on.

Definition 14.16 *We define $Pin^c(\mathcal{Y})$ as the set of all unitary elements U in* $CAR(\mathcal{Y})$ *such that*

$$\{U\phi(y)U^* \ : \ y \in \mathcal{Y}\} = \{\phi(y) \ : \ y \in \mathcal{Y}\}.$$

We set

$$Spin^c(\mathcal{Y}) := Pin^c(\mathcal{Y}) \cap \mathrm{CAR}_0(\mathcal{Y}),$$
$$Pin(\mathcal{Y}) := Pin^c(\mathcal{Y}) \cap \mathrm{Cliff}(\mathcal{Y}),$$
$$Spin(\mathcal{Y}) := Spin^c(\mathcal{Y}) \cap \mathrm{Cliff}(\mathcal{Y}).$$

The following theorem is immediate:

Theorem 14.17 *Let $U \in Pin^c(\mathcal{Y})$. Then there exists a unique $r \in O(\mathcal{Y})$ such that*

$$U\phi(y)U^* = \det(r)\phi(ry), \quad y \in \mathcal{Y}. \tag{14.16}$$

The map $Pin^c(\mathcal{Y}) \to O(\mathcal{Y})$ obtained this way is a homomorphism of groups.

Definition 14.18 *If (14.16) is satisfied, we say that U det-implements r.*

Note that in the context of CAR and Clifford algebras, the concept of det-implementation turns out to be more natural than that of implementation.

Theorem 14.19 *Let $r \in O(\mathcal{Y})$.*

(1) *The set of elements of $Pin(\mathcal{Y})$ det-implementing r consists of a pair of operators differing by sign, $\pm U_r = \{U_r, -U_r\}$.*
(2) *The set of elements of $Pin^c(\mathcal{Y})$ det-implementing r consists of operators of the form μU_r with $|\mu| = 1$.*
(3) *$r \in SO(\mathcal{Y})$ iff U_r is even; $r \in O(\mathcal{Y})\backslash SO(\mathcal{Y})$ iff U_r is odd.*
(4) *If $r_1, r_2 \in O(\mathcal{Y})$, then $U_{r_1} U_{r_2} = \pm U_{r_1 r_2}$.*

The above statements can be summarized by the following commuting diagrams of Lie groups and their continuous homomorphisms, where all vertical and horizontal sequences are exact:

$$
\begin{array}{ccccccccc}
& & 1 & & 1 & & & & \\
& & \downarrow & & \downarrow & & & & \\
1 & \to & U(1) & \to & U(1) & \to & 1 & & \\
& & \downarrow & & \downarrow & & \downarrow & & \\
1 & \to & Spin^c(\mathcal{Y}) & \to & Pin^c(\mathcal{Y}) & \to & \mathbb{Z}_2 & \to & 1 \\
& & \downarrow & & \downarrow & & \downarrow & & \\
1 & \to & SO(\mathcal{Y}) & \to & O(\mathcal{Y}) & \to & \mathbb{Z}_2 & \to & 1 \\
& & \downarrow & & \downarrow & & \downarrow & & \\
& & 1 & & 1 & & 1 & &
\end{array}
\tag{14.17}
$$

$$
\begin{array}{ccccccccc}
& & 1 & & 1 & & & & \\
& & \downarrow & & \downarrow & & & & \\
1 & \to & \mathbb{Z}_2 & \to & \mathbb{Z}_2 & \to & 1 & & \\
& & \downarrow & & \downarrow & & \downarrow & & \\
1 & \to & Spin(\mathcal{Y}) & \to & Pin(\mathcal{Y}) & \to & \mathbb{Z}_2 & \to & 1 \\
& & \downarrow & & \downarrow & & \downarrow & & \\
1 & \to & SO(\mathcal{Y}) & \to & O(\mathcal{Y}) & \to & \mathbb{Z}_2 & \to & 1 \\
& & \downarrow & & \downarrow & & \downarrow & & \\
& & 1 & & 1 & & 1 & &
\end{array}
\tag{14.18}
$$

$$
\begin{array}{ccccccccc}
& & 1 & & 1 & & 1 & & \\
& & \downarrow & & \downarrow & & \downarrow & & \\
1 & \to & \mathbb{Z}_2 & \to & U(1) & \to & U(1) & \to & 1 \\
& & \downarrow & & \downarrow & & \downarrow & & \\
1 & \to & Pin(\mathcal{Y}) & \to & Pin^c(\mathcal{Y}) & \to & U(1) & \to & 1 \\
& & \downarrow & & \downarrow & & \downarrow & & \\
1 & \to & O(\mathcal{Y}) & \to & O(\mathcal{Y}) & \to & 1 & & \\
& & \downarrow & & \downarrow & & & & \\
& & 1 & & 1 & & & &
\end{array}
\tag{14.19}
$$

It is well known that $SO(\mathcal{Y})$ is connected and its fundamental group $\pi_1(SO(\mathcal{Y}))$ equals $\mathbb{Z}$ if $\dim \mathcal{Y} = 2$ and $\mathbb{Z}_2$ if $\dim \mathcal{Y} > 2$. Thus $SO(\mathcal{Y})$ possesses a unique two-fold covering group, equal to its universal covering if $\dim \mathcal{Y} > 2$. This two-fold covering is isomorphic to $Spin(\mathcal{Y})$.

14.3.2 Pin_1^c and Pin_1 groups

In this subsection we allow $\dim \mathcal{Y}$ to be arbitrary.

Definition 14.20 *Define* $Pin_1^c(\mathcal{Y})$ *as the set of unitary operators* U *in* $\mathrm{CAR}^{C^*}(\mathcal{Y})$ *such that*

$$
\{U\phi(y)U^* : y \in \mathcal{Y}\} = \{\phi(y) : y \in \mathcal{Y}\}.
$$

Set

$$
\begin{aligned}
Spin_1^c(\mathcal{Y}) &:= Pin_1^c(\mathcal{Y}) \cap \mathrm{CAR}_0^{C^*}(\mathcal{Y}), \\
Pin_1(\mathcal{Y}) &:= Pin_1^c(\mathcal{Y}) \cap \mathrm{Cliff}^{C^*}(\mathcal{Y}), \\
Spin_1(\mathcal{Y}) &:= Pin_1(\mathcal{Y}) \cap \mathrm{Cliff}_0^{C^*}(\mathcal{Y}).
\end{aligned}
$$

We equip all these groups with the metric given by the operator norm.

The concept of implementability has an obvious definition:

Definition 14.21 *Let* $U \in \mathrm{CAR}^{C^*}(\mathcal{Y})$ *and* $r \in O(\mathcal{Y})$.

(1) *We say that* U *intertwines* r *if*

$$
U\phi(y) = \phi(ry)U, \quad y \in \mathcal{Y}.
\tag{14.20}
$$

(2) *If in addition U is unitary, then we also say that U implements r.*

(3) *If there exists $U \in \mathrm{CAR}^{C^*}(\mathcal{Y})$ that implements r, then we say that r is implementable in $\mathrm{CAR}^{C^*}(\mathcal{Y})$.*

A more useful concept is given in the definition below.

Definition 14.22 *Let $r \in O(\mathcal{Y})$.*

(1) *We say that $A \in \mathrm{CAR}^{C^*}(\mathcal{Y})$ α-intertwines $r \in O(\mathcal{Y})$ if*

$$\alpha(A)\phi(y) = \phi(ry)A,$$

$$\text{or, equivalently,} \quad A\phi(y) = \phi(ry)\alpha(A), \quad y \in \mathcal{Y}. \tag{14.21}$$

(2) *If in addition A is unitary, then we also say that A α-implements r.*

(3) *If there exists $U \in \mathrm{CAR}^{C^*}(\mathcal{Y})$ that α-implements r, then we say that r is α-implementable in $\mathrm{CAR}^{C^*}(\mathcal{Y})$.*

We will see later that if there exists an invertible A α-intertwining r, then necessarily $r \in O_\infty(\mathcal{Y})$ (actually, $r \in O_1(\mathcal{Y})$). Therefore, $\det r$ is well defined by Def. 14.6, and we can introduce the following definition, essentially equivalent to α-implementability.

Definition 14.23 *Let $r \in O_\infty(\mathcal{Y})$.*

(1) *We say that $A \in \mathrm{CAR}^{C^*}(\mathcal{Y})$ det-intertwines r if*

$$A\phi(y) = \det r\, \phi(ry)A, \quad y \in \mathcal{Y}. \tag{14.22}$$

(2) *If in addition A is unitary then we also say that A det-implements r.*

(3) *If there exists $U \in \mathrm{CAR}^{C^*}(\mathcal{Y})$ that det-implements r, then we say that r is det-implementable in $\mathrm{CAR}^{C^*}(\mathcal{Y})$.*

The following two theorems are the main results of this subsection.

Theorem 14.24 (1) *Let $r \in O(\mathcal{Y})$. Then r is det-implementable in $\mathrm{CAR}^{C^*}(\mathcal{Y})$ iff r is α-implementable in $\mathrm{CAR}^{C^*}(\mathcal{Y})$ iff $r \in O_1(\mathcal{Y})$.*

(2) *Let $U \in Pin_1^c(\mathcal{Y})$. Then there exists a unique $r \in O_1(\mathcal{Y})$ such that r is det-implemented and α-implemented by U in $\mathrm{CAR}^{C^*}(\mathcal{Y})$. The map $Pin_1^c(\mathcal{Y}) \to O_1(\mathcal{Y})$ obtained this way is a homomorphism of groups.*

Theorem 14.25 *All the statements of Thm. 14.19 are true if we replace $O(\mathcal{Y})$, $SO(\mathcal{Y})$, $Pin^c(\mathcal{Y})$, $Spin^c(\mathcal{Y})$, $Pin(\mathcal{Y})$, $Spin(\mathcal{Y})$ with $O_1(\mathcal{Y})$, $SO_1(\mathcal{Y})$, $Pin_1^c(\mathcal{Y})$, $Spin_1^c(\mathcal{Y})$, $Pin_1(\mathcal{Y})$, $Spin_1(\mathcal{Y})$.*

Before we prove Thms. 14.24 and 14.25, let us show the following lemma.

Lemma 14.26 *Let $r \in O(\mathcal{Y})$. Then the following is true:*

(1) *If A α-intertwines r, then A is either even or odd.*

(2) *If there exists an invertible A α-intertwining r, then $r \in O_\infty(\mathcal{Y})$.*
(3) *If $A \in \mathrm{CAR}^{C^*}(\mathcal{Y})$ α-intertwines r, then A det-intertwines r.*

Proof Let $U = U_0 + U_1 \in \mathrm{CAR}^{C^*}(\mathcal{Y})$ α-intertwine r with U_0 even and U_1 odd. Then

$$(U_0 + U_1)\phi(y) = \phi(ry)(U_0 - U_1), \quad y \in \mathcal{Y}. \tag{14.23}$$

Comparing even and odd terms in (14.23), we obtain

$$U_0\phi(y) = \phi(ry)U_0, \quad U_1\phi(y) = -\phi(ry)U_1, \quad y \in \mathcal{Y}. \tag{14.24}$$

Hence, $U_0^*U_0$ and $U_1^*U_1$ commute with $\phi(y)$, $y \in \mathcal{Y}$. Clearly, they are even. Hence, by Prop. 12.61, they are proportional to identity. Hence, the operators U_i are proportional to a unitary operator.

(14.24) implies also that $U_1^*U_0$ anti-commutes with $\phi(y)$, $y \in \mathcal{Y}$. By Prop. 12.61 this implies that $U_1^*U_0$ is even. But $U_1^*U_0$ is odd. Hence, $U_1^*U_0 = 0$. Thus one of the U_i is zero. This proves (1).

Let us now prove (2). Let an invertible $U \in \mathrm{CAR}^{C^*}(\mathcal{Y})$ α-intertwine r. Assume that $r \notin O_\infty(\mathcal{Y})$. Then there exists a sequence $y_n \in \mathcal{Y}$ with $\mathrm{w} - \lim y_n = 0$ and $y_n - ry_n \nrightarrow 0$. It follows that $U\phi(y_n) \mp \phi(y_n)U \to 0$ in norm, if U is even, resp. odd. Hence, $\phi(ry_n - y_n)U$, and consequently $\phi(ry_n - y_n)$ tend to 0 in norm, which is a contradiction.

Now set $\mathcal{Y}_{\mathrm{sg}} = \mathrm{Ker}(\mathbb{1} + r)$. Let E_{sg} be the associated conditional expectation. Then for $y \in \mathcal{Y}_{\mathrm{sg}}$ we have

$$U_{\mathrm{sg}}\phi(y) = \mp\phi(y)U_{\mathrm{sg}},$$

if U is even, resp. odd and $U_{\mathrm{sg}} = E_{\mathrm{sg}}(U) \in \mathrm{CAR}(\mathcal{Y}_{\mathrm{sg}})$. By Prop. 12.36, this implies that $\dim \mathrm{Ker}(\mathbb{1} + r)$ is even, resp. odd, i.e. $\det r = \pm 1$. Therefore, U also det-intertwines r. $\qquad\square$

The following proposition gives another possible equivalent definition of the Spin group. It follows easily from the commutation properties of quadratic Hamiltonians.

Proposition 14.27 *$\mathrm{Spin}_1(\mathcal{Y})$ consists of operators of the form $\mathrm{e}^{\mathrm{Op}(\zeta)}$ where $\zeta \in B^1(\mathcal{Y}^\#, \mathcal{Y})$. More precisely, let $r \in SO_1(\mathcal{Y})$. By Thm. 14.3, there exists $a \in o_1(\mathcal{Y})$ such that $r = \mathrm{e}^a$. Then*

$$\pm U_r = \pm\mathrm{e}^{\frac{1}{4}\mathrm{Op}(a\nu^{-1})} \in \mathrm{Cliff}_0^{C^*}(\mathcal{Y}) \tag{14.25}$$

intertwines r.

Proof of Thm. 14.24. Let $r \in SO_1(\mathcal{Y})$. Then r is det-implementable by Prop. 14.27. Since U_r in (14.25) is even, r is also α-implementable. Thus $\mathrm{Spin}_1(\mathcal{Y}) \to SO_1(\mathcal{Y})$ is onto.

If $r \in O_1(\mathcal{Y}) \backslash SO_1(\mathcal{Y})$, choose any $e \in \mathcal{Y}$ of norm 1. Set $\kappa_e := \mathbb{1} - 2|e\rangle\langle e|$. Clearly, $\phi(e)$ implements $-\kappa_e$. Hence, $\kappa_e r \in SO_1(\mathcal{Y})$ and r is det-implemented by

$\pm\phi(e)U_{\kappa_e r}$. Since $\phi(e)U_{\kappa_e r}$ is odd, r is also α-implementable. Thus $Pin_1(\mathcal{Y}) \rightarrow O_1(\mathcal{Y})$ is onto.

Now let $r \in O(\mathcal{Y})$ be α-implementable. By Prop. 14.26, $r \in O_\infty(\mathcal{Y})$ and r is also det-implementable. It remains to prove that $r \in O_1(\mathcal{Y})$. Without loss of generality we may assume that $\mathcal{Y}$ is separable.

Assume first that $r \in SO_\infty(\mathcal{Y})$. By Thm. 14.5, there exists $a \in o_\infty(\mathcal{Y})$ such that $r = e^a$. By Corollary 2.85, there exists an o.n. basis $(e_{i\pm})_{i\in\mathbb{N}}$ and real numbers $\lambda_i \geq 0$ such that

$$a = \sum_{i\in\mathbb{N}} \lambda_i\big(|e_{i-}\rangle\langle e_{i+}| - |e_{i+}\rangle\langle e_{i-}|\big).$$

We set $\mathcal{Y}_n = \text{Span}\{e_{i\pm}, \ 1 \leq i \leq n\}$. Let E_n be the conditional expectation associated with $\mathcal{Y}_n$. Set $U_n = E_n(U)$, where $U \in \text{CAR}_0^{C^*}(\mathcal{Y})$ implements r. Also set

$$V_n = \exp\left(\sum_{i=1}^n \frac{\lambda_i}{2}\phi(e_{i+})\phi(e_{i-})\right).$$

Applying Prop. 14.8 and Prop. 6.83, we obtain

$$V_n\phi(y) = \phi(ry)V_n, \ \ U_n\phi(y) = \phi(ry)U_n, \ y \in \mathcal{Y}_n.$$

Hence, by Prop. 12.61, $U_n = \lambda_n V_n$, $\lambda_n \in \mathbb{C}$. Clearly, $E_{n-1}(U_n) = U_{n-1}$, and computing in the real-wave representation we see that $E_{n-1}(V_n) = V_{n-1}$, hence λ_n does not depend on n.

Since by (12.34) $U_n \rightarrow U$ in norm, it follows that V_n converges in norm.

Now set $A_i = \phi(e_{i+})\phi(e_{i-})$, so that, by Prop. 14.8,

$$V_n = \prod_{i=1}^n e^{\frac{\lambda_i}{2}A_i} = \prod_{i=1}^n \Big(\cos(\lambda_i/2)\mathbb{1} + \sin(\lambda_i/2)A_i\Big).$$

Computing in the real-wave representation, we check that

$$(\Omega|V_n\Omega) = \prod_{i=1}^n \cos(\lambda_i/2). \tag{14.26}$$

Therefore, the infinite product $\prod_{i\in\mathbb{N}} \cos(\lambda_i/2)$ converges, and hence $\prod_{i\in\mathbb{N}} (\mathbb{1} + \tan(\lambda_i/2)A_i)$ converges in norm. Since the A_i commute, this implies that the product $\prod_{i\in\mathbb{N}} \|\mathbb{1} + \tan(\lambda_i/2)A_j\|$ converges. Since $\|\mathbb{1} + \tan(\lambda_i/2)A_i\| = 1 + \tan(\lambda_i/2)$ for i large enough, this implies that the series $\sum_{i\in\mathbb{N}} \lambda_i$ is convergent. Hence, $a \in o_1(\mathcal{Y})$ and $r \in SO_1(\mathcal{Y})$.

Assume now that $r \in O_\infty(\mathcal{Y})\backslash SO_\infty(\mathcal{Y})$. Let $U \in \text{CAR}_1^{C^*}(\mathcal{Y})$ α-intertwine r. Then, as above, $\phi(e)U \in \text{CAR}_0^{C^*}(\mathcal{Y})$ implements $r\kappa_e \in SO_\infty(\mathcal{Y})$. Hence, $r\kappa_e \in SO_1(\mathcal{Y})$ and $r \in O_1(\mathcal{Y})$. $\qquad\square$

Proof of Thm. 14.25. We deduce the theorem from Thm. 14.19, reducing ourselves to the finite-dimensional case by the same argument as in Prop. 12.61. $\qquad\square$

The implementability of Bogoliubov rotations can be easily deduced from the results about the det-implementability.

Corollary 14.28 $r \in O(\mathcal{Y})$ *is implementable in* $CAR^{C^*}(\mathcal{Y})$ *iff*

(1) $r \in SO_1(\mathcal{Y})$

or

(2) $-r \in O_1(\mathcal{Y}) \backslash SO_1(\mathcal{Y})$.

14.3.3 Pin_2^c and Pin_2 groups

In this subsection we again allow $\dim \mathcal{Y}$ to be arbitrary.

Definition 14.29 *Define* $Pin_2^c(\mathcal{Y})$ *as the set of unitary operators U in* $CAR^{W^*}(\mathcal{Y})$ *such that*

$$\{U\phi(y)U^* \ : \ y \in \mathcal{Y}\} = \{\phi(y) \ : \ y \in \mathcal{Y}\}.$$

Set

$$Spin_2^c(\mathcal{Y}) := Pin_2^c(\mathcal{Y}) \cap CAR_0^{W^*}(\mathcal{Y}),$$
$$Pin_2(\mathcal{Y}) := Pin_2^c(\mathcal{Y}) \cap Cliff^{W^*}(\mathcal{Y}),$$
$$Spin_2(\mathcal{Y}) := Pin_2(\mathcal{Y}) \cap Cliff_0^{W^*}(\mathcal{Y}).$$

We equip all these groups with the σ-weak topology.

We also have the obvious analogs of Defs. 14.21, 14.22 and 14.23, with $CAR^{C^*}(\mathcal{Y})$ replaced with $CAR^{W^*}(\mathcal{Y})$.

Theorem 14.30 (1) *Let $r \in O(\mathcal{Y})$. Then r is det-implementable in* $CAR^{W^*}(\mathcal{Y})$ *iff r is α-implementable in* $CAR^{W^*}(\mathcal{Y})$ *iff $r \in O_2(\mathcal{Y})$.*

(2) *Let $U \in Pin_2^c(\mathcal{Y})$. Then there exists a unique $r \in O_2(\mathcal{Y})$ such that r is det-implemented by U. The map $Pin_2^c(\mathcal{Y}) \to O_2(\mathcal{Y})$ obtained this way is a homomorphism of groups.*

Proof We can follow closely the proofs in Subsect. 14.3.2, with some modifications. Instead of Prop. 12.61 we use Prop. 12.62.

First we show that if $r \in O_2(\mathcal{Y})$, then r is α-implementable and det-implementable, following the proof of the C^* case, using Prop. 14.15 instead of Thm. 14.13.

It remains to prove that if r is α-implementable, then $r \in O_2(\mathcal{Y})$. $r \in O_\infty(\mathcal{Y})$ is proved as in the proof of Lemma 14.26, replacing the norm convergence by the σ-weak convergence.

We then follow the proof of Thm. 14.24, and obtain that V_n converges in the σ-weak topology. We are left to prove that $\sum \lambda_i^2$ is convergent. But this follows from (14.26). $\qquad\square$

Theorem 14.31 *All the statements of Thm. 14.19 are true if we replace* $O(\mathcal{Y})$, $SO(\mathcal{Y})$, $Pin^c(\mathcal{Y})$, $Spin^c(\mathcal{Y})$, $Pin(\mathcal{Y})$, $Spin(\mathcal{Y})$ *with* $O_2(\mathcal{Y})$, $SO_2(\mathcal{Y})$, $Pin_2^c(\mathcal{Y})$, $Spin_2^c(\mathcal{Y})$, $Pin_2(\mathcal{Y})$, $Spin_2(\mathcal{Y})$.

Again, the implementability of Bogoliubov rotations can easily be deduced from the results about the det-implementability.

Corollary 14.32 $r \in O(\mathcal{Y})$ *is implementable in* $\mathrm{CAR}^{W^*}(\mathcal{Y})$ *iff*

(1) $r \in SO_2(\mathcal{Y})$

　　or

(2) $-r \in O_2(\mathcal{Y})\backslash SO_2(\mathcal{Y})$.

14.3.4 Symbol of elements of $Spin(\mathcal{Y})$

We again assume that $\mathcal{Y}$ is finite-dimensional, although the results of this subsection are easily generalized to an arbitrary dimension. In this subsection we study the anti-symmetric symbol of elements of the Spin group.

Proposition 14.33 *Let* $a \in o(\mathcal{Y})$. *Then*

$$\mathrm{e}^{\frac{1}{4}\mathrm{Op}(a\nu^{-1})} = \mathrm{Op}\left((\det\cosh(2a))^{\frac{1}{2}}\mathrm{e}^{\frac{1}{2}\tanh(2a)\nu^{-1}}\right). \tag{14.27}$$

Proof By Corollary 2.85, there exists an orthonormal system $(e_{i,\pm})_{i=1,\dots,n}$ and positive numbers $(\lambda_i)_{i=1,\dots,n}$ such that

$$a = \sum_{i=1}^{n} a_i, \quad a_i = \frac{\lambda_i}{2}\left(|e_{i,-}\rangle\langle e_{i,+}| - \lambda_i|e_{i,+}\rangle\langle e_{i,-}|\right).$$

Note that $[a_i, a_j] = 0$ and $\left[\mathrm{Op}(a_i\nu^{-1}), \mathrm{Op}(a_j\nu^{-1})\right] = 0$.

Therefore,

$$\mathrm{e}^a = \prod_{i=1}^{n}\mathrm{e}^{a_i}, \quad \mathrm{e}^{\frac{1}{4}\mathrm{Op}(a\nu^{-1})} = \prod_{i=1}^{n}\mathrm{e}^{\frac{1}{4}\mathrm{Op}(a_i\nu^{-1})},$$

and we can assume without loss of generality that

$$\dim\mathcal{Y} = 2, \quad a = \frac{\lambda}{2}\left(|e_1\rangle\langle e_2| - |e_2\rangle\langle e_1|\right), \quad \mathrm{Op}(a\nu^{-1}) = \lambda\phi(e_1)\phi(e_2).$$

By Prop. 14.8, we know that

$$\mathrm{e}^{\mathrm{Op}(a\nu^{-1})} = \cos\lambda + (\sin\lambda)\phi(e_1)\phi(e_2)$$
$$= \mathrm{Op}\left(\cos\lambda\left(\mathbb{1} + \lambda^{-1}(\tan\lambda)a\nu^{-1})\right)\right).$$

Thus the anti-symmetric symbol of $\mathrm{e}^{\frac{1}{4}\mathrm{Op}(a\nu^{-1})}$ equals

$$\cos\lambda\left(\mathbb{1} + \lambda^{-1}(\tan\lambda)a\nu^{-1}\right) = (\cos^2\lambda)^{\frac{1}{2}}\mathrm{e}^{\lambda^{-1}(\tan\lambda)a\nu^{-1}}$$
$$= \left(\det\cosh(2a)\right)^{\frac{1}{2}}\mathrm{e}^{\frac{1}{2}\tanh(2a)\nu^{-1}},$$

where we have used $\cos\lambda\mathbb{1} = \cosh(2a)$ and $\lambda^{-1}(\tan\lambda)a = \frac{1}{2}\tanh(2a)$. $\qquad\square$

Definition 14.34 *We say that $r \in O(\mathcal{Y})$ is* regular *if* $\mathrm{Ker}(r + \mathbb{1}) = \{0\}$.

Proposition 14.35 *Let $r \in O(\mathcal{Y})$ be regular. Let $\gamma \in o(\mathcal{Y})$ be its Cayley transform, that is, $\gamma = \frac{\mathbb{1}-r}{\mathbb{1}+r}$ (see Subsect. 1.4.6).*
Then

$$U_r = \pm \mathrm{Op}\big(\det(\mathbb{1} - \gamma)^{-\frac{1}{2}} e^{\frac{1}{2}\gamma\nu^{-1}}\big). \tag{14.28}$$

Proof We can assume that $r = e^a$ with $a \in o(\mathcal{Y})$. Moreover, by Prop. 14.8 we have

$$e^{\frac{1}{4}\mathrm{Op}(a\nu^{-1})} \phi(y) e^{-\frac{1}{4}\mathrm{Op}(a\nu^{-1})} = \phi(e^a y).$$

Next we note that $\tanh(\frac{1}{2}a) = \frac{r-\mathbb{1}}{r+\mathbb{1}} = \gamma$, $\cosh(\frac{1}{2}a) = e^{-\frac{1}{2}a}(\mathbb{1} - \gamma)^{-1}$. Since $\det e^{\frac{1}{2}a} = 1$, this proves (14.28). $\qquad\square$

Let $r_1, r_2 \in SO(\mathcal{Y})$, $r = r_1 r_2$. We know that

$$U_{r_1} U_{r_2} = \pm U_r. \tag{14.29}$$

It is instructive to prove this fact for regular r_1, r_2, r by a direct calculation involving the Berezin calculus.

Let $\gamma_1, \gamma_2, \gamma$ be the Cayley transforms of r_1, r_2, r. By Prop. 12.42, $U_{r_1} U_{r_2}$ has the anti-symmetric symbol

$$\det(\mathbb{1} - \gamma_1)^{-\frac{1}{2}} \det(\mathbb{1} - \gamma_2)^{-\frac{1}{2}}$$

$$\times \iint e^{(v-v_1)\cdot\nu^{-1}(v-v_2)} e^{\frac{1}{2}v_1\cdot\gamma_1\nu^{-1}v_1} e^{\frac{1}{2}v_2\cdot\gamma_2\nu^{-1}v_2} \, dv_2 dv_1$$

$$= \det(\mathbb{1} - \gamma_1)^{-\frac{1}{2}} \det(\mathbb{1} - \gamma_2)^{-\frac{1}{2}}$$

$$\times \iint e^{\theta\cdot(v_1,v_2)+(v_1,v_2)\cdot\sigma(v_1,v_2)} \, dv_2 dv_1, \tag{14.30}$$

where

$$\theta := (-\nu^{-1}v, \nu^{-1}v), \quad \sigma := \begin{bmatrix} \gamma_1\nu^{-1} & \nu^{-1} \\ -\nu^{-1} & \gamma_2\nu^{-1} \end{bmatrix}.$$

By Prop. 7.19, (14.30) equals

$$\det(\mathbb{1} - \gamma_1)^{-\frac{1}{2}} \det(\mathbb{1} - \gamma_2)^{-\frac{1}{2}} \mathrm{Pf}(\sigma) \exp(\frac{1}{2}\theta\cdot\sigma^{-1}\theta). \tag{14.31}$$

Next $\mathrm{Pf}(\sigma) = \pm \det(\sigma)^{\frac{1}{2}}$. Since the Pfaffian and the determinant above are computed w.r.t. a volume form compatible with the Euclidean structure ν, we have

$$\det(\sigma) = \det \begin{bmatrix} \gamma_1 & \mathbb{1} \\ -\mathbb{1} & \gamma_2 \end{bmatrix} = \det(\mathbb{1} + \gamma_1\gamma_2),$$

using (1.4). By (1.49), we know that

$$(\mathbb{1} - \gamma) = (\mathbb{1} - \gamma_2)(\mathbb{1} + \gamma_1\gamma_2)^{-1}(\mathbb{1} - \gamma_1). \tag{14.32}$$

This implies that the first line of (14.31) equals $\det(\mathbb{1} - \gamma)^{-\frac{1}{2}}$.

By (1.3),

$$\sigma^{-1} = \begin{bmatrix} \nu\gamma_2(\gamma_1\gamma_2 + \mathbb{1})^{-1} & -\nu(\gamma_2\gamma_1 + \mathbb{1})^{-1} \\ \nu(\gamma_1\gamma_2 + \mathbb{1})^{-1} & \nu\gamma_1(\gamma_2\gamma_1 + \mathbb{1})^{-1} \end{bmatrix}.$$

Therefore,

$$\theta \cdot \sigma^{-1} \theta$$
$$= v \cdot \left(\gamma_2(\gamma_1\gamma_2 + \mathbb{1})^{-1} + \gamma_1(\gamma_2\gamma_1 + \mathbb{1})^{-1} + (\gamma_2\gamma_1 + \mathbb{1})^{-1} - (\gamma_1\gamma_2 + \mathbb{1})^{-1} \right) \nu^{-1} v$$
$$= v \cdot \left(\mathbb{1} - (\mathbb{1} - \gamma_2)(\mathbb{1} + \gamma_1\gamma_2)^{-1}(\mathbb{1} - \gamma_1) \right) \nu^{-1} v$$
$$= v \cdot \gamma \nu^{-1} v.$$

14.4 Notes

The so-called spinor representations of orthogonal groups were studied by Cartan (1938) and Brauer–Weyl (1935).

The first famous non-trivial application of the orthogonal invariance to quantum physics seems to be the version of the BCS theory due to Bogoliubov, described e.g. in Fetter–Walecka (1971).

A very comprehensive article devoted to CAR C^*-algebras was written by Araki (1987). More literature references to the subject of this chapter can be found in the notes to Chap. 16.

15

Clifford relations

Clifford algebras and *Clifford relations* were studied by mathematicians long before canonical anti-commutation relations were considered by physicists. Actually, the "(neutral) CAR representations" that we introduced in Def. 12.1 could be called "representations of self-adjoint Clifford relations".

We will use the name "Clifford relations" for anti-commutation relations identical to those of Def. 12.1, but without assuming that the underlying vector space is real, the corresponding operators are self-adjoint or that they even act on a Hilbert space.

In our short presentation we will restrict ourselves mostly to Clifford relations over finite-dimensional pseudo-Euclidean spaces. Our main motivation is to describe spinor representations of the *Lorentz group* (in any dimension). Nevertheless, we will consider the case of a general signature as well.

Some real Clifford algebras are closely related to the *quaternion algebra*, denoted by $\mathbb{H}$. Therefore, we devote Sect. 15.2 to a brief summary of its properties.

We will use the shorthand $\mathbb{K}(n) := L(\mathbb{K}^n)$, where $\mathbb{K} = \mathbb{R}, \mathbb{C}, \mathbb{H}$. We will write $[x]$ for the integer part of $x \in \mathbb{R}$.

15.1 Clifford algebras

15.1.1 Representations of Clifford relations

Let $\mathbb{K}$ be an arbitrary field and $\mathcal{Y}$ a vector space over $\mathbb{K}$. We assume that $\mathcal{Y}$ is equipped with a symmetric bilinear form ν.

Let $\mathcal{V}$ be another vector space (possibly over a bigger field).

Definition 15.1 *We will say that a linear map*

$$\mathcal{Y} \ni y \mapsto \gamma^{\pi}(y) \in L(\mathcal{V}) \tag{15.1}$$

is a representation of Clifford relations *or, for brevity,* a Clifford representation over $\mathcal{Y}$ in $\mathcal{V}$ *if*

$$\left[\gamma^{\pi}(y_1), \gamma^{\pi}(y_2)\right]_+ = 2y_1 \cdot \nu y_2 \mathbb{1}, \quad y_1, y_2 \in \mathcal{Y}. \tag{15.2}$$

15.1.2 Clifford algebras

Definition 15.2 *The* Clifford algebra $\mathrm{Cliff}(\mathcal{Y})$ *is the unital algebra over* $\mathbb{K}$ *generated by elements* $\gamma(y)$, $y \in \mathcal{Y}$, *with relations*

$$\gamma(\lambda y) = \lambda \gamma(y), \ \lambda \in \mathbb{K}, \qquad \gamma(y_1 + y_2) = \gamma(y_1) + \gamma(y_2),$$
$$\gamma(y_1)\gamma(y_2) + \gamma(y_2)\gamma(y_1) = 2y_1 \cdot \nu y_2 \, \mathbb{1}.$$

We have the following analog of Prop. 12.31:

Proposition 15.3 *If*

$$\mathcal{Y} \ni y \mapsto \gamma^\pi(y) \in L(\mathcal{V})$$

is a representation of Clifford relations, then there exists a unique homomorphism

$$\pi : \mathrm{Cliff}(\mathcal{Y}) \to L(\mathcal{V})$$

such that $\pi(\mathbb{1}) = \mathbb{1}_\mathcal{V}$ *and* $\pi(\gamma(y)) = \gamma^\pi(y)$, $y \in \mathcal{Y}$.

Many concepts and facts described in the context of the CAR apply almost verbatim to Clifford relations and algebras. For instance, $\alpha(\phi(y)) = -\phi(y)$, $y \in \mathcal{Y}$, extends to a unique involutive automorphism α of $\mathrm{Cliff}(\mathcal{Y})$. Clifford algebras split into their even and odd parts: $\mathrm{Cliff}(\mathcal{Y}) = \mathrm{Cliff}_0(\mathcal{Y}) \oplus \mathrm{Cliff}_1(\mathcal{Y})$. $\mathrm{Cliff}_0(\mathcal{Y})$ is a sub-algebra of $\mathrm{Cliff}(\mathcal{Y})$, which differs from $\mathrm{Cliff}(\mathcal{Y})$ if the field $\mathbb{K}$ has a characteristic different from 2 (which is the case for $\mathbb{K} = \mathbb{R}, \mathbb{C}$).

There also exists a unique anti-automorphism $A \to A^\dagger$, called the *transposition*, which on products of $\gamma(y)$ equals

$$(\gamma(y_1) \cdots \gamma(y_k))^\dagger = \gamma(y_k) \cdots \gamma(y_1).$$

15.1.3 Complex Clifford algebras

Let us consider an n-dimensional space $\mathcal{Y}$ over $\mathbb{C}$ equipped with a non-degenerate form ν. All such forms are isomorphic to one another, so it is enough to assume that $\mathcal{Y} = \mathbb{C}^n$ and $z \cdot \nu z = \sum_{j=1}^{n} (z_j)^2$ for $z = (z_1, \ldots, z_n) \in \mathbb{C}^n$. It is easy to see that in this case

$$\mathrm{Cliff}(\mathbb{C}^{2m}) = \mathbb{C}(2^m),$$
$$\mathrm{Cliff}(\mathbb{C}^{2m+1}) = \mathbb{C}(2^m) \oplus \mathbb{C}(2^m).$$

Thus, as an algebra, $\mathrm{Cliff}(\mathbb{C}^n)$ coincides with $\mathrm{CAR}(\mathbb{R}^n)$ defined in Def. 12.30, where the transposition † coincides with $^\#$. However, we forget about the Hermitian conjugation $*$, the complex conjugation c and the norm $\| \cdot \|$. ($\mathrm{CAR}(\mathbb{R}^n)$ is a C^*-algebra, whereas $\mathrm{Cliff}(\mathbb{C}^n)$ is not.)

Suppose now that the space $\mathcal{Y}$ is oriented (see Subsect. 3.6.8 for the definition of an orientation of a complex space). Let $(e_1, \ldots, e_n)$ be an o.n. basis of $\mathcal{Y}$ compatible with its orientation, and write γ_j for $\gamma(e_j)$.

Definition 15.4 *The* volume element *is defined as*

$$\omega := \gamma_1 \cdots \gamma_n.$$

Note that ω depends on the o.n. basis $(e_1, \ldots, e_n)$ only through its orientation. Set $m := [n/2]$. The following table summarizes the form of the algebras $\mathrm{Cliff}(\mathbb{C}^n)$:

Table 15.1 *Form of* $\mathrm{Cliff}(\mathbb{C}^n)$

$n \ (\mathrm{mod}\, 4)$	ω^2	$\mathrm{Cliff}_0(\mathbb{C}^n)$	$\mathrm{Cliff}(\mathbb{C}^n)$
0	$\mathbb{1}$	$\mathbb{C}(2^{m-1}) \oplus \mathbb{C}(2^{m-1})$	$\mathbb{C}(2^m)$
1	$\mathbb{1}$	$\mathbb{C}(2^m)$	$\mathbb{C}(2^m) \oplus \mathbb{C}(2^m)$
2	$-\mathbb{1}$	$\mathbb{C}(2^{m-1}) \oplus \mathbb{C}(2^{m-1})$	$\mathbb{C}(2^m)$
3	$-\mathbb{1}$	$\mathbb{C}(2^m)$	$\mathbb{C}(2^m) \oplus \mathbb{C}(2^m)$

15.2 Quaternions

In this section we briefly recall the properties of quaternions.

15.2.1 Basic definitions

Definition 15.5 *The real algebra* $\mathbb{H}$ *with basis* $1, \mathrm{i}, \mathrm{j}, \mathrm{k}$ *satisfying the relations*

$$\mathrm{i}^2 = \mathrm{j}^2 = \mathrm{k}^2 = -1, \quad \mathrm{ij} = \mathrm{k}, \quad \mathrm{jk} = \mathrm{i}, \quad \mathrm{ki} = \mathrm{j}$$

is called the algebra of quaternions. *It is equipped with an involution* $*$ *acting as*

$$1^* = 1, \quad \mathrm{i}^* = -\mathrm{i}, \quad \mathrm{j}^* = -\mathrm{j}, \quad \mathrm{k}^* = -\mathrm{k}.$$

For $x \in \mathbb{H}$, *we set*

$$\operatorname{Re} x := \frac{1}{2}(x + x^*), \quad |x| := \sqrt{x^* x}.$$

(Note that $x^* x$ *is always real positive.)*

If $x = x_1 + x_\mathrm{i}\mathrm{i} + x_\mathrm{j}\mathrm{j} + x_\mathrm{k}\mathrm{k}$ with $x_1, x_\mathrm{i}, x_\mathrm{j}, x_\mathrm{k} \in \mathbb{R}$, then

$$\operatorname{Re} x = x_1, \quad |x| = \sqrt{x_1^2 + x_\mathrm{i}^2 + x_\mathrm{j}^2 + x_\mathrm{k}^2}.$$

Note that $|\cdot|$ is a norm on the algebra $\mathbb{H}$. If $x, y \in \mathbb{H}$, then $|xy| = |x||y|$. $\mathbb{H}$ is an example of a real C^*-algebra.

$\mathbb{H}$ is a real Hilbert space with the scalar product

$$\langle x|y \rangle := \operatorname{Re} x^* y = x_1 y_1 + x_i y_i + x_j y_j + x_k y_k, \quad x, y \in \mathbb{H}.$$

Definition 15.6 *An algebra all of whose non-zero elements are invertible is called a* division algebra.

Clearly, $\mathbb{H}$ is a division algebra.

15.2.2 Quaternionic vector spaces

Quaternionic vector spaces and *finite-dimensional quaternionic vector spaces* have obvious definitions. Every finite-dimensional quaternionic vector space is isomorphic to $\mathbb{H}^n$ for some n. Note the identifications

$$\mathbb{R}^n \otimes \mathbb{C} = \mathbb{C}^n, \quad \mathbb{R}^n \otimes \mathbb{H} = \mathbb{H}^n.$$

$\mathbb{H}$-*linear transformations* on a quaternionic vector space have an obvious definition. Note the identifications

$$\mathbb{R}(n) \otimes \mathbb{C} = \mathbb{C}(n), \quad \mathbb{R}(n) \otimes \mathbb{H} = \mathbb{H}(n).$$

Definition 15.7 *Suppose that $\mathcal{X}$ is a quaternionic vector space, equipped (as a real space) with a scalar product $\langle x|y \rangle \in \mathbb{R}$, $x, y \in \mathcal{X}$. We say that this scalar product is* compatible with the quaternionic structure *if*

$$\langle \lambda x | \lambda y \rangle = |\lambda|^2 \langle x|y \rangle, \quad \lambda \in \mathbb{H}, \ x, y \in \mathcal{X}.$$

A quaternionic space with a compatible scalar product complete in the corresponding norm is called a quaternionic Hilbert space.

Every finite-dimensional quaternionic Hilbert space is isomorphic to $\mathbb{H}^n$ with the scalar product

$$\langle x|y \rangle := \sum \operatorname{Re} x_i^* y_i, \quad x, y \in \mathbb{H}^n.$$

15.2.3 Embedding complex numbers in quaternions

Clearly, there exists exactly one continuous injective homomorphism $\mathbb{R} \to \mathbb{H}$. However, there exist many continuous injective homomorphisms $\mathbb{C} \to \mathbb{H}$. Such a homomorphism is determined uniquely if we fix the image of $i \in \mathbb{C}$ inside $\mathbb{H}$. It is natural to denote it also by i.

Let us fix such a homomorphism $\mathbb{C} \to \mathbb{H}$. Now $\mathbb{H}$ becomes a two-dimensional vector space over the field $\mathbb{C}$. The map

$$\mathbb{H} \ni x \mapsto \frac{1}{2}(x - ixi) \in \mathbb{C} \tag{15.3}$$

is a projection. $\mathbb{H}$ is equipped with a sesquilinear scalar product

$$(x|y) := \frac{1}{2}(yx^* - iyx^*i). \tag{15.4}$$

In fact, by (15.3), the values of this scalar product are in $\mathbb{C}$. The computation

$$(x|zy) = \frac{1}{2}(zyx^* - izyx^*i) = z(x|y),$$

$$(zx|y) = \frac{1}{2}(yx^*\bar{z} - iyx^*\bar{z}i) = (x|y)\bar{z}, \quad z \in \mathbb{C},$$

shows that (15.4) is sesquilinear.

Note that the real scalar product is compatible with the complex scalar product: $\langle x|y\rangle = \mathrm{Re}(x|y)$.

$(1, \mathrm{j})$ is an example of an o.n. basis of $\mathbb{H}$ w.r.t. (15.4).

If we fix an embedding (15.3), then quaternionic vector spaces can be reinterpreted as complex vector spaces, and quaternionic Hilbert spaces as complex Hilbert spaces.

Definition 15.8 *If $\mathcal{X}$ is a quaternionic vector space, then $\mathcal{X}_{\mathbb{C}}$ will denote the same $\mathcal{X}$ understood as a complex space. It will be called the* complex form *of $\mathcal{X}$.*

15.2.4 Matrix representation of quaternions

Quaternions can be represented by the Pauli matrices multiplied by i:

$$\pi(1) = \begin{bmatrix} 1 & 0 \\ 0 & 1 \end{bmatrix}, \quad \pi(\mathrm{i}) = \begin{bmatrix} \mathrm{i} & 0 \\ 0 & -\mathrm{i} \end{bmatrix}, \quad \pi(\mathrm{j}) = \begin{bmatrix} 0 & 1 \\ -1 & 0 \end{bmatrix}, \quad \pi(\mathrm{k}) = \begin{bmatrix} 0 & \mathrm{i} \\ \mathrm{i} & 0 \end{bmatrix}.$$

Thus we obtain a representation of quaternions on the Hilbert space $\mathbb{C}^2$:

$$\pi : \mathbb{H} \to B(\mathbb{C}^2). \tag{15.5}$$

In this representation,

$$\pi(x^*) = \pi(x)^*, \quad |x| = \sqrt{\det \pi(x)}. \tag{15.6}$$

We have

$$\pi(\mathbb{H}) = \{\lambda U \ : \ U \in SU(2), \ \lambda \in [0, \infty[\}.$$

Another useful relation, which depends on the representation chosen above, is

$$\pi(\mathbb{H}) = \{A \in B(\mathbb{C}^2) \ : \ A = R\bar{A}R^{-1}\}, \tag{15.7}$$

where $\bar{A}$ is the usual complex conjugation of the matrix A and $R = \pi(\mathrm{j})$. Note that $R\bar{R} = -\mathbb{1}$.

If we replace (15.5) by $W\pi(\cdot)W^*$ for some unitary W, then R is replaced by $R_W := WR\bar{W}^*$. Note that $R_W\bar{R}_W = -\mathbb{1}$ as well.

15.2.5 Real simple algebras

It is well known that one can classify all simple finite-dimensional algebras over $\mathbb{C}$ and $\mathbb{R}$. The complex case is particularly simple.

Theorem 15.9 *Let $\mathfrak{A}$ be a complex finite-dimensional simple algebra. Then there exists a positive integer n such that $\mathfrak{A}$ is isomorphic to $\mathbb{C}(n)$.*

The corresponding classification in the real case is more complicated.

Theorem 15.10 *Let $\mathfrak{A}$ be a real finite-dimensional simple algebra. Then there exists a positive integer n such that $\mathfrak{A}$ is isomorphic to $\mathbb{C}(n)$, $\mathbb{R}(n)$ or $\mathbb{H}(n)$.*

Moreover, suppose that $\pi : \mathfrak{A} \to L(\mathcal{V})$ is a representation of $\mathfrak{A}$ in a complex space $\mathcal{V}$. (Such a representation always exists.) Define the complex conjugate representation $\overline{\pi} : \mathfrak{A} \to L(\overline{\mathcal{V}})$ by $\overline{\pi}(A) := \overline{\pi(A)}$, $A \in \mathfrak{A}$. Then the following are true:

(1) $\mathfrak{A} \simeq \mathbb{C}(n)$ *iff there exists no $R : \overline{\mathcal{V}} \to \mathcal{V}$ linear invertible such that $\pi(A)R = R\overline{\pi}(A)$.*

(2) $\mathfrak{A} \simeq \mathbb{R}(n)$ *iff there exists $R : \overline{\mathcal{V}} \to \mathcal{V}$ linear invertible such that $\pi(A)R = R\overline{\pi}(A)$ and $R\overline{R} = \mathbb{1}$.*

(3) $\mathfrak{A} \simeq \mathbb{H}(n)$ *iff there exists $R : \overline{\mathcal{V}} \to \mathcal{V}$ linear invertible such that $\pi(A)R = R\overline{\pi}(A)$ and $R\overline{R} = -\mathbb{1}$.*

If π is irreducible, then R in (2) and (3) is defined uniquely up to a phase factor.

Remark 15.11 *Note that we have the following equivalent versions of (1), (2) and (3) of the above theorem:*

(1) *There exists no anti-linear invertible χ on $\mathcal{V}$ such that $\pi(A)\chi = \chi\pi(A)$.*

(2) *There exists an anti-linear invertible χ on $\mathcal{V}$ such that $\pi(A)\chi = \chi\pi(A)$ and $\chi^2 = \mathbb{1}$.*

(3) *There exists an anti-linear invertible χ on $\mathcal{V}$ such that $\pi(A)\chi = \chi\pi(A)$ and $\chi^2 = -\mathbb{1}$.*

We can pass from χ to R by $\chi v = R\overline{v}$.

In particular, $\mathbb{R}(n)$ can be embedded in $\mathbb{C}(n)$, and then $R = \mathbb{1}$. $\mathbb{H}(n)$ can be embedded in $\mathbb{C}(2) \otimes \mathbb{C}(n)$, so that $R = \pi(\mathrm{j}) \otimes \mathbb{1}$.

15.3 Clifford relations over $\mathbb{R}^{q,p}$

Let us consider an n-dimensional vector space over $\mathbb{R}$ equipped with a non-degenerate symmetric form ν. All such forms are determined by their signature, that is, a pair of non-negative integers q, p with $n = q + p$, so that by an appropriate choice of a basis the form ν can be written as

$$y \cdot \nu y = -\sum_{j=1}^{q}(y_j)^2 + \sum_{j=q+1}^{n}(y_j)^2. \tag{15.8}$$

Definition 15.12 *The vector space $\mathbb{R}^n$ equipped with form (15.8) will be denoted $\mathbb{R}^{q,p}$.*

In this section we will study representations of Clifford relations over $\mathbb{R}^{q,p}$.

Definition 15.13 *A representation of Clifford relations will then be called a* real, complex, *resp.* quaternionic representation, *if it acts on a real, complex, resp. quaternionic space $\mathcal{V}$. Elements of $\mathcal{V}$ will be called* real, complex, *resp.* quaternionic spinors.

Of course, the complex case is the most important.

15.3.1 Basic facts

Let

$$\mathbb{R}^{q,p} \ni y \mapsto \gamma^{\pi}(y) \in L(\mathcal{V}) \tag{15.9}$$

be a Clifford representation.

Definition 15.14 *We set $\gamma_i^{\pi} := \gamma^{\pi}(e_i)$, where e_i is the canonical basis of $\mathbb{R}^{q,p}$, and the* volume element *of the representation γ^{π} is defined as*

$$\omega^{\pi} = \gamma_1^{\pi} \cdots \gamma_n^{\pi}. \tag{15.10}$$

Proposition 15.15 *Consider the Clifford representation (15.9). Then*

$$\mathbb{R}^{q,p} \ni y \mapsto -\gamma^{\pi}(y) \in L(\mathcal{V}) \tag{15.11}$$

is also a Clifford representation. If n is even, then

$$\omega^{\pi}\gamma^{\pi}(y)(\omega^{\pi})^{-1} = -\gamma^{\pi}(y),$$

so ω^{π} implements the equivalence between (15.9) and (15.11).

The following proposition is proven by mimicking the arguments of Thms. 12.27 and 12.28. Recall that $q + p = n$.

Proposition 15.16 (1) *Let n be even. Then all complex irreducible Clifford representations over $\mathbb{R}^{q,p}$ are equivalent and act on $\mathbb{C}^{n/2}$.*

(2) *Let n be odd. Then there exist exactly two inequivalent complex irreducible Clifford representations over $\mathbb{R}^{q,p}$. Moreover, if (15.9) is irreducible, then so*

is (15.11), and they are inequivalent. They act on $\mathbb{C}^{(n-1)/2}$ and satisfy

$$\omega^\pi = \pm i^{(n-1)/2+q} \mathbb{1}. \tag{15.12}$$

(3) If γ^π is an irreducible complex Clifford representation, then the complex algebra generated by $\gamma^\pi(y)$, $y \in \mathcal{Y}$, is isomorphic to $\mathbb{C}(2^{[n]/2})$.

The following proposition shows that it is easy to pass from the signature q, p to p, q.

Proposition 15.17 *Suppose that $\mathcal{V}$ is complex. Let the linear map $\epsilon : \mathbb{R}^{p,q} \to \mathbb{R}^{q,p}$ be defined by $\epsilon e_j = e_{q+j}$ for $1 \leq j \leq p$, $\epsilon e_{p+j} = e_j$ for $1 \leq j \leq q$, where $e_1, \ldots, e_n$ is the canonical basis. Then*

$$\mathbb{R}^{p,q} \ni y \mapsto i\gamma^\pi(\epsilon y) \in L(\mathcal{V}) \tag{15.13}$$

is a representation of Clifford relations.

15.3.2 Charge reversal

In this section we consider a representation (15.9) of Clifford relations in a complex space $\mathcal{V}$. For simplicity, we drop the superscript π.

Definition 15.18 *Suppose that χ_+ and χ_- are anti-linear operators on $\mathcal{V}$.*

(1) χ_+ *is called a* real charge reversal *if*

$$\chi_+\gamma(y)\chi_+^{-1} = \gamma(y), \quad \chi_+^2 = \mathbb{1}.$$

(2) χ_+ *is called a* quaternionic charge reversal *if*

$$\chi_+\gamma(y)\chi_+^{-1} = \gamma(y), \quad \chi_+^2 = -\mathbb{1}.$$

(3) χ_- *is called a* pseudo-real charge reversal *if*

$$\chi_-\gamma(y)\chi_-^{-1} = -\gamma(y), \quad \chi_-^2 = \mathbb{1}.$$

(4) χ_- *is called a* pseudo-quaternionic charge reversal *if*

$$\chi_-\gamma(y)\chi_-^{-1} = -\gamma(y), \quad \chi_-^2 = -\mathbb{1}.$$

In the case of an irreducible representation, the operators $\chi_\pm$ are determined uniquely up to a phase factor.

Theorem 15.19 *A complex irreducible representation of Clifford relations over $\mathbb{R}^{q,p}$ possesses a charge reversal of the following types:*

$p - q \pmod 8$		
0	real	pseudo-real
1	real	
2	real	pseudo-quaternionic
3		pseudo-quaternionic
4	quaternionic	pseudo-quaternionic
5	quaternionic	
6	quaternionic	pseudo-real
7		pseudo-real

If both χ_- and χ_+ exist (which is the case for all even n), then $\chi_+\chi_-$ is proportional to ω (see Def. 15.14).

Proof Prop. 15.17 shows that it is enough to prove the real and quaternionic parts of Thm. 15.19. In fact, (15.9) is irreducible iff (15.13) is. Moreover, (15.9) possesses a real, resp. quaternionic charge reversal iff (15.13) possesses a pseudo-real, resp. pseudo-quaternionic charge reversal.

For the proof of Thm. 15.19, it is convenient to use *real Pauli matrices*, that is,

$$\theta_1 := \sigma_1 = \begin{bmatrix} 0 & 1 \\ 1 & 0 \end{bmatrix}, \quad \theta_2 := \frac{1}{i}\sigma_2 = \begin{bmatrix} 0 & -1 \\ 1 & 0 \end{bmatrix}, \quad \theta_3 := \sigma_3 = \begin{bmatrix} 1 & 0 \\ 0 & -1 \end{bmatrix}.$$

Note that $\theta_1^2 = -\theta_2^2 = \theta_3^2 = \mathbb{1}$, and

$$\theta_1\theta_2 = -\theta_2\theta_1 = \theta_3,$$
$$\theta_2\theta_3 = -\theta_3\theta_2 = \theta_1,$$
$$\theta_3\theta_1 = -\theta_1\theta_3 = \theta_2.$$

Moreover, $\mathbb{R}(2)$ is generated by θ_1, θ_2.

Let us now start the main part of the proof. Recall that $n = q + p$. For any (q, p) with $m = [(q + p)/2]$, we will construct a family of matrices in $\mathbb{R}(2^m)$,

$$\gamma_1^{q,p}, \ldots, \gamma_{q+p}^{q,p},$$

such that

$$[\gamma_i^{q,p}, \gamma_j^{q,p}]_+ = 0, \quad 0 \le i < j \le n,$$

$(\gamma_j^{q,p})^2 = -\mathbb{1}$ for q distinct j and $(\gamma_j^{q,p})^2 = \mathbb{1}$ for p distinct j. If possible, we will also construct a real matrix $R_+^{q,p}$ such that $R_+^{q,p}\gamma_j^{q,p}(R_+^{q,p})^{-1} = \gamma_j^{q,p}$ and $(R_+^{q,p})^2 = \pm\mathbb{1}$.

First assume that $q+p$ is even. The case $q=p$ is particularly easy. We set

$$\gamma^{q,q}_{2j-1} := \theta_3^{\otimes(j-1)} \otimes \theta_1,$$
$$\gamma^{q,q}_{2j} := \theta_3^{\otimes(j-1)} \otimes \theta_2, \quad j = 1, \dots, q,$$
$$R^{q,q}_+ := \mathbb{1}^{\otimes q}. \tag{15.14}$$

For $q < p$, we set

$$\gamma^{q,p}_{2j} := i\gamma^{\frac{q+p}{2}, \frac{q+p}{2}}_{2j}, \quad j = 1, \dots, \frac{p-q}{2};$$

$$\gamma^{q,p}_k := \gamma^{\frac{q+p}{2}, \frac{q+p}{2}}_k, \quad \text{for remaining } k;$$

$$R^{q,p}_+ := (\theta_1 \otimes \theta_2)^{\frac{p-q}{4}}, \quad \text{for even } \frac{p-q}{2}, \text{ then } (R^{q,p}_+)^2 = (-\mathbb{1})^{\frac{p-q}{4}};$$

$$R^{q,p}_+ := (\theta_1 \otimes \theta_2)^{\frac{p-q-2}{4}} \otimes \theta_1 \otimes \theta_3^p, \quad \text{for odd } \frac{p-q}{2}, \text{ then } (R^{q,p}_+)^2 = (-\mathbb{1})^{\frac{p-q-2}{4}}.$$

For $q > p$, we define

$$\gamma^{q,p}_{2j-1} := i\gamma^{\frac{q+p}{2}, \frac{q+p}{2}}_{2j-1}, \quad j = 1, \dots, \frac{q-p}{2};$$

$$\gamma^{q,p}_k := \gamma^{\frac{q+p}{2}, \frac{q+p}{2}}_k, \quad \text{for remaining } k;$$

$$R^{q,p}_+ := (\theta_2 \otimes \theta_1)^{\frac{q-p}{4}}, \quad \text{for even } \frac{p-q}{2}, \text{ then } (R^{q,p}_+)^2 = (-\mathbb{1})^{\frac{p-q}{4}};$$

$$R^{q,p}_+ := (\theta_2 \otimes \theta_1)^{\frac{p-q+2}{4}} \otimes \theta_2 \otimes \theta_3^p, \quad \text{for odd } \frac{p-q}{2}, \text{ then } (R^{q,p}_+)^2 = (-\mathbb{1})^{\frac{p-q-2}{4}}.$$

This ends the proof of the real and quaternionic cases for $q+p$ even.

Next assume that $q+p$ is odd. This time, the case $q+1=p$ is particularly easy. We set

$$\gamma^{q,q+1}_{2j-1} := \theta_3^{\otimes(j-1)} \otimes \theta_1,$$
$$\gamma^{q,q+1}_{2j} := \theta_3^{\otimes(j-1)} \otimes \theta_2, \quad j = 1, \dots, q,$$
$$\gamma^{q,q+1}_{2q+1} := \theta_3^{\otimes q},$$
$$R^{q,q+1}_+ := \mathbb{1}^{\otimes q}. \tag{15.15}$$

For $q < p-1$, we set

$$\gamma^{q,p}_{2j} := i\gamma^{\frac{q+p-1}{2}, \frac{q+p+1}{2}}_{2j}, \quad j = 1, \dots, \frac{p-q-1}{2};$$

$$\gamma^{q,p}_k := \gamma^{\frac{q+p-1}{2}, \frac{q+p+1}{2}}_k, \quad \text{for remaining } k;$$

$$R^{q,p}_+ := (\theta_1 \otimes \theta_2)^{\frac{p-q-1}{4}}, \quad \text{for even } \frac{p-q-1}{2}, \text{ then } (R^{q,p}_+)^2 = (-\mathbb{1})^{\frac{p-q-1}{4}};$$

$$R^{q,p}_+ \quad \text{does not exist for odd } \frac{p-q-1}{2}.$$

For $q > p - 1$, we define

$$\gamma_{2j-1}^{q,p} := i\gamma_{2j-1}^{\frac{q+p-1}{2},\frac{q+p+1}{2}}, \quad j = 1, \ldots, \frac{q-p+1}{2};$$

$$\gamma_k^{q,p} := \gamma_k^{\frac{q+p-1}{2},\frac{q+p+1}{2}}, \quad \text{for remaining } k;$$

$$R_+^{q,p} := (\theta_2 \otimes \theta_1)^{\frac{q-p+1}{4}}, \quad \text{for even } \frac{p-q-1}{2}, \text{ then } (R_+^{q,p})^2 = (-1)^{\frac{p-q-1}{4}};$$

$$R_+^{q,p} \quad \text{does not exist for odd } \frac{p-q-1}{2}.$$

This ends the proof of the real and quaternionic cases for $q + p$ odd. $\qquad\square$

15.3.3 Real spinors

In this subsection we consider real representations of Clifford relations.

Note that if we have a Clifford representation on a real space, then by replacing this space with its complexification we obtain a complex Clifford representation.

Conversely, if we have a Clifford representation on a complex space $\mathcal{V}$ equipped with a charge reversal χ_+ of real type, then we can decompose $\mathcal{V}$ into a direct sum of real subspaces, $\mathcal{V} = \mathcal{V}^{\chi_+} \oplus \mathcal{V}^{-\chi_+}$, where

$$\mathcal{V}^{\chi_+} := \{v \in \mathcal{V} : \chi_+ v = v\}, \quad \mathcal{V}^{-\chi_+} := \{v \in \mathcal{V} : \chi_+ v = -v\}.$$

Clearly, we can restrict the representation of Clifford relations to real spaces $\mathcal{V}^{\chi_+}$ and $\mathcal{V}^{-\chi_+}$.

Suppose that $p - q$ equals 0, 1 or 2 modulo 8. Recall that in this case irreducible complex Clifford representations are equipped with a real type charge conjugation. Therefore, there exists a real representation of Clifford relations over $\mathbb{R}^{q,p}$ in $\mathbb{R}^{2^{[n/2]}}$. If γ^π is such a representation, then the real algebra generated by $\gamma^\pi(y)$, $y \in \mathcal{Y}$, equals $\mathbb{R}(2^{[n/2]})$.

Clifford representations possessing a real type charge reversal that appeared in the proof of Thm. 15.19 used complex matrices. It is possible to redefine those representations so that they involve purely real matrices. Such representations are often more complicated than those appearing in the proof of Thm. 15.19. In what follows we will construct such Clifford representations for all real cases of (q,p). They will be generalizations of the *Majorana representation*, well known in physics in the case $(1,3)$.

First recall that for $q = p$ the representation described in (15.14) involved only real matrices. Then we describe real representations with one of q, p equal to zero and the other ≤ 8. First we consider the Euclidean case:

$$\gamma_1^{0,1} := \mathbb{1}, \quad \gamma_1^{0,2} := \theta_1, \quad \gamma_1^{0,8} := \theta_1 \otimes \mathbb{1} \otimes \mathbb{1} \otimes \mathbb{1},$$
$$\gamma_2^{0,2} := \theta_3, \quad \gamma_2^{0,8} := \theta_3 \otimes \mathbb{1} \otimes \mathbb{1} \otimes \mathbb{1},$$
$$\gamma_3^{0,8} := \theta_2 \otimes \theta_2 \otimes \theta_1 \otimes \mathbb{1},$$
$$\gamma_4^{0,8} := \theta_2 \otimes \theta_2 \otimes \theta_3 \otimes \mathbb{1},$$
$$\gamma_5^{0,8} := \theta_2 \otimes \mathbb{1} \otimes \theta_2 \otimes \theta_1,$$
$$\gamma_6^{0,8} := \theta_2 \otimes \mathbb{1} \otimes \theta_2 \otimes \theta_3,$$
$$\gamma_7^{0,8} := \theta_2 \otimes \theta_1 \otimes \mathbb{1} \otimes \theta_2,$$
$$\gamma_8^{0,8} := \theta_2 \otimes \theta_3 \otimes \mathbb{1} \otimes \theta_2,$$

$$\omega^{0,1} := \mathbb{1}, \quad \omega^{0,2} := \theta_2, \quad \omega^{0,8} := \theta_2 \otimes \theta_2 \otimes \theta_2 \otimes \theta_2.$$

Next we consider the anti-Euclidean case:

$$\gamma_1^{6,0} := \theta_2 \otimes \mathbb{1} \otimes \mathbb{1}, \quad \gamma_1^{7,0} := \theta_2 \otimes \mathbb{1} \otimes \mathbb{1}, \quad \gamma_1^{8,0} := \theta_2 \otimes \mathbb{1} \otimes \mathbb{1} \otimes \mathbb{1},$$
$$\gamma_2^{6,0} := \theta_1 \otimes \theta_2 \otimes \mathbb{1}, \quad \gamma_2^{7,0} := \theta_1 \otimes \theta_2 \otimes \mathbb{1}, \quad \gamma_2^{8,0} := \theta_3 \otimes \theta_2 \otimes \theta_2 \otimes \theta_2,$$
$$\gamma_3^{6,0} := \theta_1 \otimes \theta_1 \otimes \theta_2, \quad \gamma_3^{7,0} := \theta_1 \otimes \theta_1 \otimes \theta_2, \quad \gamma_3^{8,0} := \theta_3 \otimes \theta_2 \otimes \theta_1 \otimes \mathbb{1},$$
$$\gamma_4^{6,0} := \theta_1 \otimes \theta_3 \otimes \theta_2, \quad \gamma_4^{7,0} := \theta_1 \otimes \theta_3 \otimes \theta_2, \quad \gamma_4^{8,0} := \theta_3 \otimes \theta_2 \otimes \theta_3 \otimes \mathbb{1},$$
$$\gamma_5^{6,0} := \theta_3 \otimes \mathbb{1} \otimes \theta_2, \quad \gamma_5^{7,0} := \theta_3 \otimes \mathbb{1} \otimes \theta_2, \quad \gamma_5^{8,0} := \theta_3 \otimes \mathbb{1} \otimes \theta_2 \otimes \theta_1,$$
$$\gamma_6^{6,0} := \theta_3 \otimes \theta_2 \otimes \theta_1, \quad \gamma_6^{7,0} := \theta_3 \otimes \theta_2 \otimes \theta_1, \quad \gamma_6^{8,0} := \theta_3 \otimes \mathbb{1} \otimes \theta_2 \otimes \theta_3,$$
$$\gamma_7^{7,0} := \theta_3 \otimes \theta_2 \otimes \theta_3, \quad \gamma_7^{8,0} := \theta_3 \otimes \theta_1 \otimes \mathbb{1} \otimes \theta_2,$$
$$\gamma_8^{8,0} := \theta_3 \otimes \theta_3 \otimes \mathbb{1} \otimes \theta_2,$$

$$\omega^{6,0} := \theta_2 \otimes \mathbb{1} \otimes \mathbb{1}, \quad \omega^{7,0} := \mathbb{1} \otimes \mathbb{1} \otimes \mathbb{1}, \quad \omega^{8,0} := \theta_1 \otimes \mathbb{1} \otimes \mathbb{1} \otimes \mathbb{1}.$$

Now let us consider a pair $q < p$. Let $p = q + 8r + u$, $0 \leq u < 8$. Clearly, $u = 0, 1$ or 2. Then we set (where we drop the factors of $\mathbb{1}$ tensor multiplied on the right)

$$\gamma_k^{q,p} := \gamma_k^{q,q}, \qquad\qquad k = 1, \ldots, 2q;$$
$$\gamma_{2q+8i+j}^{q,p} := \omega^{q,q} \otimes (\omega^{0,8})^{\otimes i} \otimes \gamma_j^{0,8}, \quad i = 0, \ldots, r-1, \quad j = 1, \ldots, 8;$$
$$\gamma_{2q+8r+j}^{q,p} := \omega^{q,q} \otimes (\omega^{0,8})^{\otimes r} \otimes \gamma_j^{0,u}, \quad j = 1, \ldots, u;$$
$$R_+^{q,p} := \mathbb{1}^{\otimes 4q} \otimes (R_+^{0,8})^{\otimes r} \otimes R_+^{0,u}. \tag{15.16}$$

Similarly, for a pair $q > p$, we write $q = p + 8r + u$, $0 \leq u < 8$. We have $u = 0, 6$ or 7. We set

$$\gamma_k^{q,p} := \gamma_k^{p,p}, \qquad\qquad k = 1, \ldots, 2p;$$
$$\gamma_{2p+8i+j}^{q,p} := \omega^{p,p} \otimes (\omega^{8,0})^{\otimes i} \otimes \gamma_j^{8,0}, \quad i = 0, \ldots, r-1, \quad j = 1, \ldots, 8;$$
$$\gamma_{2q+8r+j}^{q,p} := \omega^{p,p} \otimes (\omega^{8,0})^{\otimes r} \otimes \gamma_j^{u,0}, \quad j = 1, \ldots, u;$$
$$R_+^{q,p} := \mathbb{1}^{\otimes 4p} \otimes (R_+^{8,0})^{\otimes r} \otimes R_+^{u,0}. \tag{15.17}$$

$\square$

15.3.4 *Quaternionic spinors*

In this subsection we consider quaternionic representations of Clifford relations.

Recall that a quaternionic vector space, after embedding $\mathbb{C}$ in $\mathbb{H}$, can be interpreted as a complex vector space. Therefore, every Clifford representation on a quaternionic vector space $\mathcal{V}$ can be interpreted as a complex Clifford representation on $\mathcal{V}_{\mathbb{C}}$.

Conversely, if we have a complex Clifford representation with a quaternionic charge reversal χ_+, then setting $\mathrm{j} := \chi_+$ we can consider $\mathcal{V}$ as a vector space over $\mathbb{H}$. The Clifford representation then becomes $\mathbb{H}$-linear.

Suppose that $p - q$ equals 4, 5 or 6 modulo 8. Recall that in this case irreducible complex representations possess a charge conjugation of quaternionic type. Therefore, there exists a quaternionic representation of Clifford relations over $\mathbb{R}^{q,p}$ in $\mathbb{H}^{2^{[n/2]-1}}$. If γ^π is such a representation, then the real algebra generated by $\gamma^\pi(y)$, $y \in \mathcal{Y}$, equals $\mathbb{H}(2^{[n/2]-1})$.

It is instructive to construct representations of Clifford relations for all quaternionic cases of (q, p) by matrices in $\mathbb{H}(2^{[n/2]})$.

Note that the matrices $\mathbb{1}, i\theta_1, \theta_2, i\theta_3$ can be viewed as the generators of quaternions. Moreover, a real matrix tensored with a quaternion is a quaternionic matrix.

Let us first describe quaternionic representations with one of q, p equal to zero and the other ≤ 8. First we consider the Euclidean case:

$$
\begin{aligned}
\gamma_1^{0,4} &:= \theta_1 \otimes \mathbb{1}, & \gamma_1^{0,5} &:= \theta_1 \otimes \mathbb{1}, & \gamma_1^{0,6} &:= \theta_1 \otimes \mathbb{1} \otimes \mathbb{1}, \\
\gamma_2^{0,4} &:= \theta_3 \otimes \mathbb{1}, & \gamma_2^{0,5} &:= \theta_3 \otimes \mathbb{1}, & \gamma_2^{0,6} &:= \theta_3 \otimes \mathbb{1} \otimes \mathbb{1}, \\
\gamma_3^{0,4} &:= \theta_2 \otimes i\theta_1, & \gamma_3^{0,5} &:= \theta_2 \otimes i\theta_1, & \gamma_3^{0,6} &:= \theta_2 \otimes \theta_1 \otimes \theta_2, \\
\gamma_4^{0,4} &:= \theta_2 \otimes i\theta_3, & \gamma_4^{0,5} &:= \theta_2 \otimes i\theta_3, & \gamma_4^{0,6} &:= \theta_2 \otimes \theta_3 \otimes \theta_2, \\
& & \gamma_5^{0,5} &:= \theta_2 \otimes \theta_2, & \gamma_5^{0,6} &:= \theta_2 \otimes \mathbb{1} \otimes i\theta_1, \\
& & & & \gamma_6^{0,6} &:= \theta_2 \otimes \mathbb{1} \otimes i\theta_3,
\end{aligned}
$$

$$
R_+^{0,4} := \mathbb{1} \otimes \theta_2, \quad R_+^{0,5} := \mathbb{1} \otimes \theta_2, \quad R_+^{0,6} := \mathbb{1} \otimes \mathbb{1} \otimes \theta_2.
$$

Next we consider the anti-Euclidean case:

$$
\begin{aligned}
\gamma_1^{2,0} &:= \theta_2, & \gamma_1^{3,0} &:= \theta_2, & \gamma_1^{4,0} &:= \theta_2 \otimes \mathbb{1}, \\
\gamma_2^{2,0} &:= i\theta_1, & \gamma_2^{3,0} &:= i\theta_1, & \gamma_2^{4,0} &:= \theta_3 \otimes \theta_2, \\
& & \gamma_3^{3,0} &:= i\theta_3, & \gamma_3^{4,0} &:= \theta_3 \otimes i\theta_1, \\
& & & & \gamma_4^{4,0} &:= \theta_3 \otimes i\theta_3,
\end{aligned}
$$

$$
R_+^{2,0} := \theta_2, \quad R_+^{3,0} := \theta_2, \quad R_+^{4,0} := \mathbb{1} \otimes \theta_2.
$$

The case of arbitrary q, p is dealt with as in the case of real spinors; see (15.16) and (15.17). $\qquad\square$

15.3.5 Representations of Clifford relations
on pseudo-unitary spaces

Let $\mathcal{V}$ be a finite-dimensional complex vector space and

$$\mathbb{R}^{q,p} \ni y \mapsto \gamma(y) \in L(\mathcal{V}) \tag{15.18}$$

be a Clifford representation. Recall that $\mathcal{V}^*$ denotes the space of anti-linear functionals on $\mathcal{V}$. Clearly,

$$\mathbb{R}^{q,p} \ni y \mapsto \pm\gamma(y)^* \in L(\mathcal{V}^*) \tag{15.19}$$

are also Clifford representations. It is natural to ask when (15.18) and (15.19) are equivalent. The following proposition answers this question for irreducible representations.

Proposition 15.20 *Let (15.18) be irreducible.*

(1) *There exists an invertible* $\lambda_+ \in L_h(\mathcal{V}, \mathcal{V}^*)$ *such that*

$$\gamma(y)^* = \lambda_+ \gamma(y) \lambda_+^{-1}.$$

iff p *is odd or* q *is even.*

(2) *There exists an invertible* $\lambda_- \in L_h(\mathcal{V}, \mathcal{V}^*)$ *such that*

$$-\gamma(y)^* = \lambda_- \gamma(y) \lambda_-^{-1}$$

iff q *is odd or* p *is even.*

Proof Let $\gamma_1, \ldots, \gamma_n$ be an irreducible Clifford representation in the canonical basis of $\mathbb{R}^{q,p}$. Then writing $\gamma_j = \mathrm{i}\phi_j$, $j = 1, \ldots, q$ and $\gamma_j = \phi_j$, $j = q+1, \ldots, n$, we obtain an irreducible Clifford representation over $\mathbb{R}^n$, $\phi_1, \ldots, \phi_n$. On the space $\mathcal{V}$ we can fix a scalar product such that $\phi_i = \phi_i^*$, so that we obtain a CAR representation. This scalar product allows us to identify the space $\mathcal{V}$ with $\mathcal{V}^*$.

Obviously, $\gamma_j^* = -\gamma_j$, $j = 1, \ldots, q$, and $\gamma_j^* = \gamma_j$, $j = q+1, \ldots, n$.

Now set

$$\begin{aligned}
\lambda_+ &:= \pm\mathrm{i}^{q/2}\gamma_1 \cdots \gamma_q, && \text{even } q; \\
\lambda_- &:= \pm\mathrm{i}^{(q+1)/2}\gamma_1 \cdots \gamma_q, && \text{odd } q; \\
\lambda_- &:= \pm\mathrm{i}^{p/2}\gamma_{q+1} \cdots \gamma_n, && \text{even } p; \\
\lambda_+ &:= \pm\mathrm{i}^{(p-1)/2}\gamma_{q+1} \cdots \gamma_n, && \text{odd } p.
\end{aligned}$$

We check that $\lambda_\pm^* = \lambda_\pm$, $\lambda_\pm^2 = \mathbb{1}$ and $\lambda_\pm\gamma_i = \pm\gamma_i^*\lambda_\pm$.

Note that if n is odd, then we obtain two distinct formulas for λ_+ or λ_-. Using (15.12), we easily see that they define the same operator. $\square$

If the assumptions of Prop. 15.20 (1) are satisfied, so that λ_+ exists, we endow the space $\mathcal{V}$ with a non-degenerate Hermitian form

$$\mathcal{V} \times \mathcal{V} \ni (v_1, v_2) \mapsto \bar{v}_1 \cdot \lambda_+ v_2.$$

Definition 15.21 *For every $A \in L(\mathcal{V})$, we define its λ_+-adjoint, denoted $A^\dagger$, by*

$$\bar{v}_1 \cdot \lambda_+ A v_2 = \overline{A^\dagger v_1} \cdot \lambda_+ v_2.$$

We have

$$\gamma(y)^\dagger = \gamma(y), \quad y \in \mathcal{Y}. \tag{15.20}$$

If $\pi : \mathrm{Cliff}(\mathbb{R}^{q,p}) \to L(\mathcal{V})$ is a representation, then

$$\pi(A)^\dagger = \pi(A^\dagger), \quad A \in \mathrm{Cliff}(\mathbb{R}^{q,p}). \tag{15.21}$$

If we replace λ_+ with λ_-, then instead of (15.20) we have

$$\gamma(y)^\dagger = -\gamma(y), \quad y \in \mathcal{Y}.$$

Instead of (15.21), we have:

$$\pi(A)^\dagger = \pi(A^\dagger), \quad A \in \mathrm{Cliff}_0(\mathbb{R}^{q,p}). \tag{15.22}$$

15.4 Clifford algebras over $\mathbb{R}^{q,p}$

In this section we continue to study Clifford relations over $\mathbb{R}^{q,p}$. We adopt the representation-independent point of view: we concentrate on the Clifford algebra $\mathrm{Cliff}(\mathbb{R}^{q,p})$.

For $n = 0, 1, 2$, $\mathrm{Cliff}(\mathbb{R}^{n,0})$ are division algebras. In fact, $\mathrm{Cliff}(\mathbb{R}^{0,0}) = \mathbb{R}$, $\mathrm{Cliff}(\mathbb{R}^{1,0}) = \mathbb{C}$ and $\mathrm{Cliff}(\mathbb{R}^{2,0}) = \mathbb{H}$.

15.4.1 Form of Clifford algebras for a general signature

Let q, p be arbitrary non-negative integers, $n = q + p$ and $m := [(q + p)/2]$. Let us consider the real algebra $\mathrm{Cliff}(\mathbb{R}^{q,p})$.

We have the following counterpart of Def. 15.14:

Definition 15.22 *We will write $\gamma_i := \gamma(e_i)$, where e_i is the canonical basis of $\mathbb{R}^{q,p}$. The volume element of $\mathrm{Cliff}(\mathbb{R}^{q,p})$ will be denoted by*

$$\omega = \gamma_1 \cdots \gamma_n. \tag{15.23}$$

Remark 15.23 *In the case $n = 4$ with the Lorentz signature, particle physicists often denote the operator ω by γ_5. This notation is so popular that it is sometimes used in the case of a dimension different from 4.*

It is possible to describe $\mathrm{Cliff}(\mathbb{R}^{q,p})$ for an arbitrary q, p. Table 15.2, a well-known table of real Clifford algebras, should be compared with the analogous table for the complex case (see Table 15.1, Subsect. 15.1.3).

In the case of n odd all the algebras $\mathrm{Cliff}(\mathbb{R}^{q,p})$ have a non-trivial center spanned by $\mathbb{1}, \omega$.

If $\omega^2 = \mathbb{1}$, which corresponds to cases 1 and 5, $\mathrm{Cliff}(\mathbb{R}^{q,p})$ splits into a direct sum and $\omega \simeq \mathbb{1} \oplus (-\mathbb{1})$.

Table 15.2 *Form of* $\mathrm{Cliff}(\mathbb{R}^{q,p})$

$p - q \pmod 8$	ω^2	$\mathrm{Cliff}_0(\mathbb{R}^{q,p})$	$\mathrm{Cliff}(\mathbb{R}^{q,p})$
0	$\mathbb{1}$	$\mathbb{C}(2^{m-1})$	$\mathbb{R}(2^m)$
1	$\mathbb{1}$	$\mathbb{R}(2^m)$	$\mathbb{R}(2^m) \oplus \mathbb{R}(2^m)$
2	$-\mathbb{1}$	$\mathbb{R}(2^{m-1}) \oplus \mathbb{R}(2^{m-1})$	$\mathbb{R}(2^m)$
3	$-\mathbb{1}$	$\mathbb{R}(2^m)$	$\mathbb{C}(2^m)$
4	$\mathbb{1}$	$\mathbb{C}(2^{m-1})$	$\mathbb{H}(2^{m-1})$
5	$\mathbb{1}$	$\mathbb{H}(2^{m-1})$	$\mathbb{H}(2^{m-1}) \oplus \mathbb{H}(2^{m-1})$
6	$-\mathbb{1}$	$\mathbb{H}(2^{m-2}) \oplus \mathbb{H}(2^{m-2})$	$\mathbb{H}(2^m)$
7	$-\mathbb{1}$	$\mathbb{H}(2^{m-1})$	$\mathbb{C}(2^m)$

If $\omega^2 = -\mathbb{1}$, which corresponds to cases 3 and 7, the algebras are complex and $\omega = i\mathbb{1}$.

In the case $p - q \equiv 0, 1, 2 \pmod 8$, $\mathrm{Cliff}(\mathbb{R}^{q,p})$ can be represented as real matrices, which will correspond to the real type in Thm. 15.19. In the case $p - q \equiv 4, 5, 6 \pmod 8$, $\mathrm{Cliff}(\mathbb{R}^{q,p})$ can be represented as quaternionic matrices, which corresponds to the quaternionic type in Thm. 15.19.

$\mathbb{C} \otimes \mathrm{Cliff}(\mathbb{R}^{q,p})$ coincides with the algebra $\mathrm{Cliff}(\mathbb{C}^n)$. In addition, it is equipped with a unique complex conjugation such that $\mathrm{Cliff}(\mathbb{R}^{q,p})$ consists of elements in $\mathbb{C} \otimes \mathrm{Cliff}(\mathbb{R}^{q,p})$ fixed by this conjugation.

There exists a unique isomorphism of complex algebras $\rho : \mathbb{C} \otimes \mathrm{Cliff}(\mathbb{R}^{q,p}) \to \mathbb{C} \otimes \mathrm{Cliff}(\mathbb{R}^{p,q})$ satisfying

$$\rho(\gamma(y)) = i\gamma(y), \quad y \in \mathcal{Y}. \tag{15.24}$$

(Note that on the left $\gamma(y)$ is an element of $\mathbb{C} \otimes \mathrm{Cliff}(\mathbb{R}^{q,p})$, and on the right of $\mathbb{C} \otimes \mathrm{Cliff}(\mathbb{R}^{p,q})$.) Under this isomorphism we have

$$\rho\big(\mathrm{Cliff}_0(\mathbb{R}^{q,p})\big) = \mathrm{Cliff}_0(\mathbb{R}^{p,q}),$$
$$\rho\big(\mathrm{Cliff}_1(\mathbb{R}^{q,p})\big) = i\mathrm{Cliff}_1(\mathbb{R}^{p,q}).$$

15.4.2 Pseudo-Euclidean group

Recall that we can define the group $O(\mathbb{R}^{q,p})$ of linear transformations that preserve the form (15.8). Obviously, we have a natural isomorphism $O(\mathbb{R}^{q,p}) \simeq O(\mathbb{R}^{p,q})$. The determinant defines a homomorphism of $O(\mathbb{R}^{q,p})$ into $\{1, -1\}$. Elements of $O(\mathbb{R}^{q,p})$ with the determinant 1 form a subgroup $SO(\mathbb{R}^{q,p}) \simeq SO(\mathbb{R}^{p,q})$. We have the exact sequence

$$1 \to SO(\mathbb{R}^{q,p}) \to O(\mathbb{R}^{q,p}) \to \mathbb{Z}_2 \to 1. \tag{15.25}$$

Definition 15.24 *For any $r \in O(\mathbb{R}^{q,p})$,*

$$\hat{r}(\gamma(y)) = \gamma(ry), \quad y \in \mathcal{Y},$$

defines a unique automorphism $\hat{r}$ of $\mathrm{Cliff}(\mathbb{R}^{q,p})$.

We have a homomorphism

$$O(\mathbb{R}^{q,p}) \ni r \mapsto \hat{r} \in \mathrm{Aut}\big(\mathrm{Cliff}(\mathbb{R}^{q,p})\big).$$

15.4.3 Pin group for a general signature

Definition 15.25 *We define $Pin(\mathbb{R}^{q,p})$ as the set of all $U \in \mathrm{Cliff}(\mathbb{R}^{q,p})$ such that $UU^{\dagger} = \mathbb{1}$ or $UU^{\dagger} = -\mathbb{1}$, and*

$$\{U\gamma(y)U^{-1} \ : \ y \in \mathcal{Y}\} = \{\gamma(y) \ : \ y \in \mathcal{Y}\}.$$

We set

$$Spin(\mathbb{R}^{q,p}) := Pin(\mathbb{R}^{q,p}) \cap \mathrm{Cliff}_0(\mathbb{R}^{q,p}).$$

Proposition 15.26 *Let $U \in Pin(\mathbb{R}^{q,p})$. Then there exists a unique $r \in O(\mathbb{R}^{q,p})$ such that*

$$U\gamma(y)U^{-1} = \det(r)\gamma(ry), \quad y \in \mathcal{Y}. \tag{15.26}$$

The map $Pin(\mathbb{R}^{q,p}) \to O(\mathbb{R}^{q,p})$ obtained this way is a surjective homomorphism of groups.

Definition 15.27 *If (15.26) is satisfied, we say that U det-implements r.*

Theorem 15.28 *Let $r \in O(\mathbb{R}^{q,p})$.*

(1) *The set of elements of $\mathrm{Cliff}(\mathbb{R}^{q,p})$ det-implementing r consists of a pair of operators differing by sign, $\pm U_r = \{U_r, -U_r\}$.*
(2) *$r \in SO(\mathbb{R}^{q,p})$ iff U_r is even; $r \in O(\mathbb{R}^{q,p}) \backslash SO(\mathbb{R}^{q,p})$ iff U_r is odd.*
(3) *If $r_1, r_2 \in O(\mathbb{R}^{q,p})$, then $U_{r_1} U_{r_2} = \pm U_{r_1 r_2}$.*

The above statements can be summarized by the following commuting diagram of Lie groups and their continuous homomorphisms, where all vertical and horizontal sequences are exact:

$$
\begin{array}{ccccccccc}
& & 1 & & 1 & & & & \\
& & \downarrow & & \downarrow & & & & \\
1 \to & & \mathbb{Z}_2 & \to & \mathbb{Z}_2 & \to & 1 & & \\
& & \downarrow & & \downarrow & & \downarrow & & \\
1 \to & Spin(\mathbb{R}^{q,p}) & \to & Pin(\mathbb{R}^{q,p}) & \to & \mathbb{Z}_2 & \to & 1 & \\
& \downarrow & & \downarrow & & \downarrow & & & \\
1 \to & SO(\mathbb{R}^{q,p}) & \to & O(\mathbb{R}^{q,p}) & \to & \mathbb{Z}_2 & \to & 1 & \\
& \downarrow & & \downarrow & & \downarrow & & & \\
& 1 & & 1 & & 1 & & &
\end{array}
\tag{15.27}
$$

Moreover, $Spin(\mathbb{R}^{q,p})$ coincides with $Spin(\mathbb{R}^{p,q})$ in the sense that if $U_r^{q,p} \in$ $\mathrm{Cliff}_0(\mathbb{R}^{q,p})$ and $U_r^{p,q} \in \mathrm{Cliff}_0(\mathbb{R}^{p,q})$ both implement $r \in O(\mathbb{R}^{q,p}) = O(\mathbb{R}^{p,q})$, then $U_r^{q,p} = \pm U_r^{p,q}$, where we use the isomorphism described at the end of Subsect. 15.4.1.

15.5 Notes

The so-called spinor representations of orthogonal groups were studied by Cartan (1938) and Brauer–Weyl (1935).

In quantum physics, Clifford relations and spinor representations appear in the description of spin $\frac{1}{2}$ particles. In the non-relativistic case, where the group $Spin(3) \simeq SU(2)$ replaces the group of rotations $SO(3)$, this is due to Pauli (1927). In the relativistic case, where the group $Spin^{\uparrow}(1,3) \simeq SL(2, \mathbb{C})$ replaces the Lorentz group, this is due to Dirac (1928).

Introductions to Clifford algebras can be found in Lawson–Michelson (1989) and Trautman (2006).

16

Orthogonal invariance of the CAR on Fock spaces

In this chapter we continue our study of the orthogonal invariance of the CAR. This invariance was already investigated in Chap. 14. However, whereas in Chap. 14 we used the representation-independent framework of CAR algebras, in this chapter we will consider Fock CAR representations in any dimensions. Therefore, to some extent, this chapter can be viewed as a continuation of Chap. 13 about Fock CAR representations.

Note also that this chapter is parallel to Chap. 11 about the symplectic invariance of the CCR on a Fock space.

16.1 Orthogonal group on a Kähler space

The framework of this section, as well as of most other sections of this chapter, is the same as that of Chap. 13 about the Fock representation of the CAR.

In particular, we assume that $(\mathcal{Y}, \nu)$ is a real Hilbert space with a Kähler anti-involution j. If r is a densely defined operator on $L(\mathcal{Y})$, then $r^{\#}$ denotes its adjoint for the scalar product ν. We also use the holomorphic space $\mathcal{Z} := \frac{\mathbb{1}-\mathrm{i}\mathrm{j}}{2}\mathbb{C}\mathcal{Y}$ and the identification $\mathbb{C}\mathcal{Y} = \mathcal{Z} \oplus \overline{\mathcal{Z}}$.

In this section we study the orthogonal group and Lie algebra on a real Hilbert space equipped with a Kähler structure.

This section is parallel to Sect. 11.1 about the symplectic group on a Kähler space.

16.1.1 Basic properties

Recall that $O(\mathcal{Y})$ denotes the group of orthogonal transformations on $\mathcal{Y}$. Elements of $O(\mathcal{Y})$ are automatically bounded with a bounded inverse. Clearly, $r \in O(\mathcal{Y})$ iff

$$(a)\ r^{\#}r = \mathbb{1}, \qquad (b)\ rr^{\#} = \mathbb{1}.$$

In the context of real Hilbert spaces we adopt the following definition for the corresponding Lie algebra:

Definition 16.1 $o(\mathcal{Y})$ *denotes the Lie algebra of* $a \in B(\mathcal{Y})$ *satisfying* $a^{\#} + a = 0$, *that is,* $o(\mathcal{Y}) = B_{\mathrm{a}}(\mathcal{Y})$.

Recall that every $r \in B(\mathcal{Y})$ extended to $\mathbb{C}\mathcal{Y} = \mathcal{Z} \oplus \overline{\mathcal{Z}}$ can be written as

$$r_{\mathbb{C}} = \begin{bmatrix} p & q \\ \overline{q} & \overline{p} \end{bmatrix}. \tag{16.1}$$

Proposition 16.2 $r \in O(\mathcal{Y})$ iff $p \in B(\mathcal{Z})$, $q \in B(\overline{\mathcal{Z}}, \mathcal{Z})$ and the following conditions hold:

> Conditions implied by (a): $\quad p^* p + q^{\#} \overline{q} = \mathbb{1}, \quad p^* q + q^{\#} \overline{p} = 0;$
>
> Conditions implied by (b): $\quad pp^* + qq^* = \mathbb{1}, \quad pq^{\#} + qp^{\#} = 0.$

Proposition 16.3 $a \in o(\mathcal{Y})$ iff its extension to $\mathbb{C}\mathcal{Y}$ equals

$$a_{\mathbb{C}} = \mathrm{i} \begin{bmatrix} h & g \\ -\overline{g} & -\overline{h} \end{bmatrix}, \tag{16.2}$$

with $h \in B_{\mathrm{h}}(\mathcal{Z})$, and $g \in B_{\mathrm{a}}(\overline{\mathcal{Z}}, \mathcal{Z})$ (h is self-adjoint and g is anti-symmetric).

16.1.2 j-*non-degenerate orthogonal maps*

The theory of orthogonal operators on a Kähler space is more complicated than that of symplectic operators on a Kähler space. For a symplectic transformation r, the operator p was automatically invertible, which greatly simplified the analysis. The analogous statement is not always true for a general orthogonal operator. Nevertheless, a large class of orthogonal transformations can be analyzed in a way parallel to symplectic transformations. These transformations, which we will call j-non-degenerate, will be studied in this subsection.

Proposition 16.4 *Let* $r \in O(\mathcal{Y})$. *Then the following conditions are equivalent:*

(1) $\mathrm{Ker}(rj + jr) = \{0\}$.
(2) $\mathrm{Ker}(r^{\#}j + jr^{\#}) = \{0\}$.
(3) $\mathrm{Ker}\, p = \{0\}$.
(4) $\mathrm{Ker}\, p^* = \{0\}$.

Proof (1)⇔(2), because

$$r^{\#}j + jr^{\#} = r^{\#}(jr + rj)r^{\#}.$$

(1)⇔(3), because

$$r_{\mathbb{C}}j_{\mathbb{C}} + j_{\mathbb{C}}r_{\mathbb{C}} = \begin{bmatrix} p & q \\ \overline{q} & \overline{p} \end{bmatrix} \begin{bmatrix} \mathrm{i} & 0 \\ 0 & -\mathrm{i} \end{bmatrix} + \begin{bmatrix} \mathrm{i} & 0 \\ 0 & -\mathrm{i} \end{bmatrix} \begin{bmatrix} p & q \\ \overline{q} & \overline{p} \end{bmatrix} = 2\mathrm{i} \begin{bmatrix} p & 0 \\ 0 & -\overline{p} \end{bmatrix}.$$

Similarly we see that (2)⇔(4). $\qquad \square$

Definition 16.5 $r \in O(\mathcal{Y})$ *is said to be* j-non-degenerate *if the equivalent conditions of Prop. 16.4 are satisfied.*

Recall that if h is a possibly unbounded operator with $\operatorname{Ker} h = \{0\}$, then we can define h^{-1} with domain $\operatorname{Dom} h^{-1} := \operatorname{Ran} h$. The operator h^{-1} is closed iff h is.

Recall also that $Cl_{\mathrm{a}}(\overline{\mathcal{Z}}, \mathcal{Z})$ denotes the set of closed densely defined operators c from $\overline{\mathcal{Z}}$ to $\mathcal{Z}$ satisfying $c^{\#} = -c$.

Let us now describe a convenient factorization of a j-non-degenerate orthogonal map. Note that if r is j-non-degenerate, then $(\operatorname{Ran} p)^{\mathrm{cl}} = (\operatorname{Ker} p^*)^{\perp} = \mathcal{Z}$. Therefore, the following operators are densely defined:

$$d := q\overline{p}^{-1}, \qquad \operatorname{Dom} d := \operatorname{Ran} \overline{p}; \tag{16.3}$$

$$c := -q^{\#}(p^{\#})^{-1}, \quad \operatorname{Dom} c := \operatorname{Ran} p^{\#}. \tag{16.4}$$

Proposition 16.6 (1) *c and d are closable. Let us denote their closures by the same symbols. Then $c, d \in Cl_{\mathrm{a}}(\overline{\mathcal{Z}}, \mathcal{Z})$.*

(2) *We have the following equivalent characterizations of c, d:*

$$d = -p^{*-1}q^{\#}, \quad \operatorname{Dom} d = \{\overline{z} \in \overline{\mathcal{Z}} \ : \ q^{\#}\overline{z} \in \operatorname{Ran} p^*\}; \tag{16.5}$$

$$c = p^{-1}q, \qquad \operatorname{Dom} c = \{\overline{z} \in \overline{\mathcal{Z}} \ : \ q\overline{z} \in \operatorname{Ran} p\}. \tag{16.6}$$

(3) *We have the following factorization, which holds as an operator identity:*

$$r_{\mathbb{C}} = \begin{bmatrix} \mathbb{1} & d \\ 0 & \mathbb{1} \end{bmatrix} \begin{bmatrix} (p^*)^{-1} & 0 \\ 0 & \overline{p} \end{bmatrix} \begin{bmatrix} \mathbb{1} & 0 \\ \overline{c} & \mathbb{1} \end{bmatrix}. \tag{16.7}$$

(4) *The following operator identities are true:*

$$(r_{\mathbb{C}}jcr_{\mathbb{C}}^* - jc)(r_{\mathbb{C}}jcr_{\mathbb{C}}^* + jc)^{-1} = \begin{bmatrix} 0 & d \\ \overline{d} & 0 \end{bmatrix},$$
$$(jc - r_{\mathbb{C}}^*jcr_{\mathbb{C}})(r_{\mathbb{C}}^*jcr_{\mathbb{C}} + jc)^{-1} = \begin{bmatrix} 0 & c \\ \overline{c} & 0 \end{bmatrix}. \tag{16.8}$$

(Note that if r is j-non-degenerate, then $rjr^* + j$ and $r^*jr + j$ are injective with a dense range. Hence, in the identities (16.8) the meaning of the l.h.s. is described in (2.2) and (2.3).)

(5) *The following quadratic form identities are true:*

$$\mathbb{1} + c^{\#}\overline{c} = p^{*-1}p^{-1}, \quad \mathbb{1} + d^*d = \overline{p}^{*-1}\overline{p}^{-1}.$$

Proof Consider $d = q\overline{p}^{-1}$. We have the identity

$$q^{\#}\overline{p} = -p^*q. \tag{16.9}$$

Therefore, $\operatorname{Ran} \overline{p}$ is contained in

$$\{z \in \mathcal{Z} \ : \ q^{\#} z \in \operatorname{Dom} p^{*-1} = \operatorname{Ran} p^*\}. \tag{16.10}$$

But $\operatorname{Ran} \overline{p}$ is dense. Thus (16.10) is dense. By Prop. 2.35 applied to the bounded operator q and the closed operator $\overline{p}^{-1}$,

$$(q\overline{p}^{-1})^{\#} = p^{*-1}q^{\#}. \tag{16.11}$$

But the identity (16.9) implies

$$p^{*-1}q^{\#}\overline{p} = -q,$$

and hence, on $\operatorname{Ran}\overline{p}$,

$$p^{*-1}q^{\#} = -q\overline{p}^{-1} = (q\overline{p}^{-1})^{\#},$$

by (16.11). Therefore, $d \subset -d^{\#}$, and hence d are closable. This easily implies (1) and (2).

We have

$$r_{\mathrm{C}}\mathrm{j}_{\mathrm{C}} - \mathrm{j}_{\mathrm{C}}r_{\mathrm{C}} = 2\mathrm{i}\begin{bmatrix} 0 & -q \\ -\overline{q} & 0 \end{bmatrix}, \quad r_{\mathrm{C}}\mathrm{j}_{\mathrm{C}} + \mathrm{j}_{\mathrm{C}}r_{\mathrm{C}} = 2\mathrm{i}\begin{bmatrix} p & 0 \\ 0 & -\overline{p} \end{bmatrix}.$$

Hence,

$$\begin{bmatrix} 0 & d \\ \overline{d} & 0 \end{bmatrix} = (r_{\mathrm{C}}\mathrm{j}_{\mathrm{C}} - \mathrm{j}_{\mathrm{C}}r_{\mathrm{C}})(r_{\mathrm{C}}\mathrm{j}_{\mathrm{C}} + \mathrm{j}_{\mathrm{C}}r_{\mathrm{C}})^{-1} = (r_{\mathrm{C}}\mathrm{j}_{\mathrm{C}}r_{\mathrm{C}}^{*} - \mathrm{j}_{\mathrm{C}})(r_{\mathrm{C}}\mathrm{j}_{\mathrm{C}}r_{\mathrm{C}}^{*} + \mathrm{j}_{\mathrm{C}})^{-1}.$$

This proves the first identity of (4). $\square$

Next we give a criterion for the j-non-degeneracy.

Lemma 16.7 *Assume that* $\|r - \mathbb{1}\| < 1$. *Then* r *is* j-*non-degenerate.*

Proof Let $y \in \mathcal{Y}$ such that $y \neq 0$ and $(r\mathrm{j} + \mathrm{j}r)y = 0$. Then,

$$2\mathrm{j}y = (\mathbb{1} - r)\mathrm{j}y + \mathrm{j}(\mathbb{1} - r)y.$$

Hence,

$$2\|y\| \leq 2\|\mathbb{1} - r\|\|y\|.$$

Therefore, $1 \leq \|\mathbb{1} - r\|$. $\square$

16.1.3 j-*self-adjoint maps*

To some extent, this subsection can be viewed as parallel to Subsect. 11.1.4 about positive symplectic transformations.

Definition 16.8 *An operator* $r \in Cl(\mathcal{Y})$ *satisfying* $\mathrm{j}r = r^{\#}\mathrm{j}$ *is called* j-*self-adjoint. We say that it is* j-*positive if, in addition,* $\mathrm{j}r\mathrm{j}^{-1} + r^{\#} \geq 0$.

If the extension of r to $\mathbb{C}\mathcal{Y}$ is given by (16.1), then $r \in B(\mathcal{Y})$ is j-self-adjoint iff $q^{\#} = -q$, $p = p^{*}$. It is j-positive iff in addition $p \geq 0$.

Let $r \in B(\mathcal{Y})$ be j-self-adjoint. It belongs to $O(\mathcal{Y})$ iff

$$p^{2} - q\overline{q} = \mathbb{1}, \quad pq - q\overline{p} = 0.$$

We now examine the form of the decomposition (16.7). Let $r \in O(\mathcal{Y})$ be j-non-degenerate. It is j-self-adjoint iff $c = d$, where $c, d \in Cl_{\mathrm{a}}(\overline{\mathcal{Z}}, \mathcal{Z})$ were defined in (16.3) and (16.4). j-non-degenerate j-positive elements of $O(\mathcal{Y})$ can be fully characterized by c:

Proposition 16.9 *Let* $r \in O(\mathcal{Y})$ *be j-non-degenerate and j-positive. Let* $c \in Cl_a(\overline{\mathcal{Z}}, \mathcal{Z})$ *be defined as in (16.4). Then one has*

$$r_{\mathbb{C}} = \begin{bmatrix} (\mathbb{1} + cc^*)^{-\frac{1}{2}} & (\mathbb{1} + cc^*)^{-\frac{1}{2}}c \\ -c^*(\mathbb{1} + cc^*)^{-\frac{1}{2}} & (\mathbb{1} + c^*c)^{-\frac{1}{2}} \end{bmatrix} \tag{16.12}$$

$$= \begin{bmatrix} \mathbb{1} & c \\ 0 & \mathbb{1} \end{bmatrix} \begin{bmatrix} (\mathbb{1} + cc^*)^{\frac{1}{2}} & 0 \\ 0 & (\mathbb{1} + c^*c)^{-\frac{1}{2}} \end{bmatrix} \begin{bmatrix} \mathbb{1} & 0 \\ -c^* & \mathbb{1} \end{bmatrix},$$

$$(r_{\mathbb{C}}^2 - \mathbb{1}_{\mathbb{C}})(r_{\mathbb{C}}^2 + \mathbb{1}_{\mathbb{C}})^{-1} = \begin{bmatrix} 0 & c \\ \overline{c} & 0 \end{bmatrix}. \tag{16.13}$$

Conversely let $c \in Cl_a(\overline{\mathcal{Z}}, \mathcal{Z})$. *Then* r, *defined by (16.12), belongs to* $O(\mathcal{Y})$, *is j-non-degenerate and is j-positive.*

Proof Let $r \in O(\mathcal{Y})$ be j-non-degenerate and j-self-adjoint. We have $d = c = q\overline{p}^{-1}$ and $\mathbb{1} + c^{\#}\overline{c} = p^{-2}$. Using the positivity of p we obtain

$$p = (\mathbb{1} + c^{\#}\overline{c})^{-\frac{1}{2}}.$$

Now

$$q = c\overline{p} = c(\mathbb{1} + c^*c)^{-\frac{1}{2}} = (\mathbb{1} + cc^*)^{-\frac{1}{2}}c.$$

We then apply the decomposition (16.7) and formula (16.1) to get the first statement of the proposition. □

Proposition 16.10 *Let* $a \in Cl(\mathcal{Y})$ *be anti-self-adjoint and j-self-adjoint. Then there exists* $g \in Cl_a(\overline{\mathcal{Z}}, \mathcal{Z})$ *such that*

$$a_{\mathbb{C}} = i \begin{bmatrix} 0 & g \\ g^* & 0 \end{bmatrix}.$$

Moreover, e^a *belongs to* $O(\mathcal{Y})$, *is j-self-adjoint and*

$$e^{a_{\mathbb{C}}} = \begin{bmatrix} \cos\sqrt{gg^*} & i\frac{\sin\sqrt{gg^*}}{\sqrt{gg^*}}g \\ ig^*\frac{\sin\sqrt{gg^*}}{\sqrt{gg^*}} & \cos\sqrt{g^*g} \end{bmatrix}, \tag{16.14}$$

$$c = i\frac{\tan\sqrt{gg^*}}{\sqrt{gg^*}}g.$$

We have a complete description of j-non-degenerate j-self-adjoint elements of $O(\mathcal{Y})$:

Theorem 16.11 *Let* $r \in O(\mathcal{Y})$ *be j-non-degenerate and j-self-adjoint. Then* $r = mr_0m^*$, *where*

$$m_{\mathbb{C}} := \frac{1}{\sqrt{2}} \begin{bmatrix} \mathbb{1} & i(\mathbb{1} - p^2)^{-\frac{1}{2}}q \\ -i(\mathbb{1} - \overline{p}^2)^{-\frac{1}{2}}\overline{q} & \mathbb{1} \end{bmatrix},$$

$$r_{0\mathbb{C}} := \begin{bmatrix} p + i(\mathbb{1} - p^2)^{\frac{1}{2}} & 0 \\ 0 & \overline{p} - i(\mathbb{1} - \overline{p}^2)^{\frac{1}{2}} \end{bmatrix}.$$

The transformation $m_{\mathbb{C}}$ is unitary, hence $r_{\mathbb{C}}$ is unitarily equivalent to the diagonal operator $r_{0\mathbb{C}}$. Consequently

$$\operatorname{spec} r = \operatorname{spec}(p + \mathrm{i}(\mathbb{1} - p^2)^{\frac{1}{2}}) \cup \overline{\operatorname{spec}(p + \mathrm{i}(\mathbb{1} - p^2)^{\frac{1}{2}})}.$$

In particular, r is j-positive iff

$$\operatorname{spec} r \subset \{e^{\mathrm{i}\phi} \ : \ \phi \in [-\pi/2, \pi/2]\}. \tag{16.15}$$

Proposition 16.12 *Let $k \in O(\mathcal{Y})$ be j-self-adjoint and such that $\operatorname{Ker}(k + \mathbb{1}) = \{0\}$. Let $k^t \in O(\mathcal{Y})$ be defined as in Subsect. 2.3.2 for $t \in \mathbb{R}$. Then k^t is also j-self-adjoint and $(k^t)^{\#} = k^{-t}$. Moreover, if $|t| \le \frac{1}{2}$, then k^t is j-non-degenerate and j-positive.*

Proof We work on $\mathbb{C}\mathcal{Y}$ equipped with its unitary structure and consider $k_{\mathbb{C}}$. Clearly, $k_{\mathbb{C}} \mathrm{j}_{\mathbb{C}} = \mathrm{j}_{\mathbb{C}} k_{\mathbb{C}}^*$, so the identity

$$F(k_{\mathbb{C}}) \mathrm{j}_{\mathbb{C}} = \mathrm{j}_{\mathbb{C}} F(k_{\mathbb{C}}^*) \tag{16.16}$$

holds for polynomials and extends by the usual argument to bounded Borel functions on $\operatorname{spec} k_{\mathbb{C}}$. Taking $F(z) = z^t$, we obtain by restriction to $\mathcal{Y}$ that $k^t \mathrm{j} = \mathrm{j}(k^t)^*$, so that k^t is j-self-adjoint.

Clearly, k^t is j-non-degenerate, and also j-positive for $|t| \le \frac{1}{2}$, by criterion (16.15). □

16.1.4 j-polar decomposition

The following theorem gives a canonical decomposition of every j-non-degenerate orthogonal operator into a product of a unitary operator and a j-positive j-non-degenerate operator. This can be treated as a fermionic analog of the polar decomposition of symplectic transformations discussed in Subsect. 11.1.5.

Theorem 16.13 *Let $r \in O(\mathcal{Y})$ be j-non-degenerate. Set $k := -\mathrm{j}r^{\#}\mathrm{j}r$. Then*

(1) *$k \in O(\mathcal{Y})$ is j-self-adjoint;*
(2) *$\operatorname{Ker}(k + \mathbb{1}) = \{0\}$;*
(3) *$k^{\frac{1}{2}}$ is j-positive and j-non-degenerate;*
(4) *For $w := rk^{-\frac{1}{2}} \in U(\mathcal{Y}^{\mathbb{C}})$ we have*

$$r = wk^{\frac{1}{2}}; \tag{16.17}$$

(5) *If in addition r is j-self-adjoint, then $w = w^*$, $w^2 = \mathbb{1}$ and $r = k^{\frac{1}{2}}w = wk^{\frac{1}{2}}$.*

Proof (1) follows from

$$\mathrm{j}k = r^{\#}\mathrm{j}r = k^{\#}\mathrm{j}.$$

(2) is a consequence of

$$\operatorname{Ker}(k + \mathbb{1}) = -\mathrm{j}r^{\#}\operatorname{Ker}(r\mathrm{j} + \mathrm{j}r) = \{0\}.$$

(3) follows from Prop. 16.12.

Let us prove (4). Clearly, $w \in O(\mathcal{Y})$. Moreover,

$$jw = jrk^{-\frac{1}{2}} = jrk^{-1}k^{\frac{1}{2}}$$
$$= rjk^{\frac{1}{2}} = rk^{-\frac{1}{2}}j = wj.$$

So $w \in U(\mathcal{Y}^{\mathbb{C}})$. $\square$

Definition 16.14 *We call (16.17) the* j*-polar decomposition of* r.

16.1.5 Conjugations on Kähler spaces

Conjugations on a unitary space were defined in Subsect. 1.2.10. We recall that they are anti-unitary involutions.

Conjugations on a Kähler space were defined in Subsect. 1.3.10. We recall that κ is a conjugation of the Kähler space $\mathcal{Y}$ if $\kappa \in O(\mathcal{Y})$, $\kappa^2 = \mathbb{1}$ and $\kappa j = -j\kappa$. Note that κ is self-adjoint, as well as anti-symplectic and infinitesimally symplectic.

Clearly, κ is a conjugation on a Kähler space $\mathcal{Y}$ iff it is a conjugation on the corresponding unitary space $\mathcal{Y}^{\mathbb{C}}$. It can be written as

$$\kappa_{\mathbb{C}} = \begin{bmatrix} 0 & t \\ \bar{t} & 0 \end{bmatrix},$$

where $t \in L(\overline{\mathcal{Z}}, \mathcal{Z})$, $t\bar{t} = \mathbb{1}_{\mathcal{Z}}$ and $t^{\#} = t$. If we set $uz := t\bar{z}$, then u is a conjugation of the Hilbert space $\mathcal{Z}$, which means an anti-unitary operator satisfying $u^2 = \mathbb{1}$.

Conversely, any conjugation on $\mathcal{Z}$ determines a conjugation on $\mathcal{Y}$.

Note also that if j is a Kähler anti-involution, then so is $-$j. If $\kappa \in O(\mathcal{Y})$ is a conjugation, then we have $\kappa j \kappa^{\#} = -j$.

16.1.6 Partial conjugations on Kähler spaces

Definition 16.15 *If* $\mathcal{W}$ *is a unitary space, we will say that* $\kappa \in L(\mathcal{W}_{\mathbb{R}})$ *is a partial conjugation if there exists a decomposition of* $\mathcal{W}$ *into an orthogonal direct sum of (complex) subspaces* $\mathcal{W} = \mathcal{W}_{\mathrm{reg}} \oplus \mathcal{W}_{\mathrm{sg}}$ *such that* κ *preserves this decomposition, is the identity on* $\mathcal{W}_{\mathrm{reg}}$ *and is a conjugation on* $\mathcal{W}_{\mathrm{sg}}$.

Definition 16.16 *If* $(\mathcal{Y}, \nu, \mathrm{j})$ *is a Kähler space, we say that* $\kappa \in L(\mathcal{Y})$ *is a partial conjugation if there exists an orthogonal decomposition* $\mathcal{Y} = \mathcal{Y}_{\mathrm{reg}} \oplus \mathcal{Y}_{\mathrm{sg}}$ *such that* κ *and* j *preserve this decomposition,* κ *is the identity on* $\mathcal{Y}_{\mathrm{reg}}$ *and a conjugation on* $\mathcal{Y}_{\mathrm{sg}}$.

Clearly, κ is a partial conjugation on a Kähler space $\mathcal{Y}$ iff it is a partial conjugation of the unitary space $\mathcal{Y}^{\mathbb{C}}$.

Let κ be a partial conjugation of $\mathcal{Y}$. If $\mathbb{1}_{\mathrm{reg}}$ and $\mathbb{1}_{\mathrm{sg}}$ are the orthogonal projections onto $\mathcal{Y}_{\mathrm{reg}}$ and $\mathcal{Y}_{\mathrm{sg}}$, then

$$\kappa j \kappa^{\#} = j\mathbb{1}_{\mathrm{reg}} - j\mathbb{1}_{\mathrm{sg}}.$$

Writing $\mathcal{Z} = \mathcal{Z}_{\mathrm{reg}} \oplus \mathcal{Z}_{\mathrm{sg}}$, we have

$$\kappa_{\mathbb{C}} = \begin{bmatrix} \mathbb{1} & 0 & 0 & 0 \\ 0 & \mathbb{1} & 0 & 0 \\ 0 & 0 & 0 & t \\ 0 & 0 & \bar{t} & 0 \end{bmatrix} \tag{16.18}$$

for $t \in L(\overline{\mathcal{Z}}_{\mathrm{sg}}, \mathcal{Z}_{\mathrm{sg}})$, $\bar{t}t = \mathbb{1}_{\overline{\mathcal{Z}}_{\mathrm{sg}}}$, $t^{\#} = t$.

16.1.7 Decomposition of orthogonal operators

Definition 16.17 *Let $r \in O(\mathcal{Y})$. We define the* regular *and* singular initial *and* final *subspaces for r by*

$$\mathcal{Y}_{-\mathrm{sg}} := \mathrm{Ker}(rj + jr), \quad \mathcal{Y}_{-\mathrm{reg}} := \mathcal{Y}_{-\mathrm{sg}}^{\perp},$$
$$\mathcal{Y}_{+\mathrm{sg}} := \mathrm{Ker}(r^{\#}j + jr^{\#}), \quad \mathcal{Y}_{+\mathrm{reg}} := \mathcal{Y}_{+\mathrm{sg}}^{\perp}.$$

We also introduce the corresponding holomorphic subspaces

$$\mathcal{Z}_{\pm\mathrm{sg}} := \mathbb{C}\mathcal{Y}_{\pm\mathrm{sg}} \cap \mathcal{Z}, \quad \mathcal{Z}_{\pm\mathrm{reg}} := \mathbb{C}\mathcal{Y}_{\pm\mathrm{reg}} \cap \mathcal{Z}.$$

We easily check that r maps $\mathcal{Y}_{-\mathrm{sg}}$ onto $\mathcal{Y}_{+\mathrm{sg}}$ and $\mathcal{Y}_{-\mathrm{reg}}$ onto $\mathcal{Y}_{+\mathrm{reg}}$. j preserves $\mathcal{Y}_{\pm\mathrm{sg}}$ and $\mathcal{Y}_{\pm\mathrm{reg}}$, and hence we have the decompositions

$$\mathbb{C}\mathcal{Y} = \mathcal{Z}_{-\mathrm{reg}} \oplus \overline{\mathcal{Z}}_{-\mathrm{reg}} \oplus \mathcal{Z}_{-\mathrm{sg}} \oplus \overline{\mathcal{Z}}_{-\mathrm{sg}}, \tag{16.19}$$

$$\mathbb{C}\mathcal{Y} = \mathcal{Z}_{+\mathrm{reg}} \oplus \overline{\mathcal{Z}}_{+\mathrm{reg}} \oplus \mathcal{Z}_{+\mathrm{sg}} \oplus \overline{\mathcal{Z}}_{+\mathrm{sg}}. \tag{16.20}$$

Note that $\mathrm{Ker}\, p = \mathcal{Z}_{-\mathrm{sg}}$ and $\mathrm{Ker}\, p^{*} = \mathcal{Z}_{+\mathrm{sg}}$. We can write $r_{\mathbb{C}}$ as a matrix from (16.19) to (16.20) as follows:

$$r_{\mathbb{C}} = \begin{bmatrix} p_{\mathrm{reg}} & q_{\mathrm{reg}} & 0 & 0 \\ \bar{q}_{\mathrm{reg}} & \bar{p}_{\mathrm{reg}} & 0 & 0 \\ 0 & 0 & 0 & q_{\mathrm{sg}} \\ 0 & 0 & \bar{q}_{\mathrm{sg}} & 0 \end{bmatrix}.$$

Clearly,

$$\mathrm{Ker}\, p_{\mathrm{reg}}^{*} = \mathrm{Ker}\, p_{\mathrm{reg}} = \{0\}, \quad q_{\mathrm{sg}}^{*} q_{\mathrm{sg}} = \mathbb{1}_{\overline{\mathcal{Z}}_{-\mathrm{sg}}}. \tag{16.21}$$

Proposition 16.18 *Let $r \in O(\mathcal{Y})$. Then there exists a decomposition $r = \kappa r_{0}$ such that $r_{0} \in O(\mathcal{Y})$ is* j-non-degenerate *and κ is a partial conjugation.*

Proof Let κ be any partial conjugation such that $\kappa j \kappa^{\#} = j\mathbb{1}_{-\mathrm{reg}} - j\mathbb{1}_{-\mathrm{sg}}$, so that in the matrix notation using (16.20)

$$\kappa_{\mathbb{C}} = \begin{bmatrix} 1 & 0 & 0 & 0 \\ 0 & 1 & 0 & 0 \\ 0 & 0 & 0 & t \\ 0 & 0 & \bar{t} & 0 \end{bmatrix}.$$

We can set $r_0 = \kappa r$, which as a matrix from (16.19) to (16.20) has the form

$$
r_{0\mathbb{C}} = \begin{bmatrix} p_{\mathrm{reg}} & q_{\mathrm{reg}} & 0 & 0 \\ \overline{q}_{\mathrm{reg}} & \overline{p}_{\mathrm{reg}} & 0 & 0 \\ 0 & 0 & t\overline{q}_{\mathrm{sg}} & 0 \\ 0 & 0 & 0 & \overline{t}q_{\mathrm{sg}} \end{bmatrix}. \tag{16.22}
$$

Clearly, $r_0 \in O(\mathcal{Y})$, since $r, \kappa \in O(\mathcal{Y})$. (16.21) implies that r_0 is j-non-degenerate. $\square$

Proposition 16.19 *Let $r \in O(\mathcal{Y})$ be j-self-adjoint. Then $\mathcal{Y}_{+\mathrm{reg}} = \mathcal{Y}_{-\mathrm{reg}} =: \mathcal{Y}_{\mathrm{reg}}$ and $\mathcal{Y}_{+\mathrm{sg}} = \mathcal{Y}_{-\mathrm{sg}} =: \mathcal{Y}_{\mathrm{sg}}$. We have the orthogonal decomposition $\mathcal{Y} = \mathcal{Y}_{\mathrm{reg}} \oplus \mathcal{Y}_{\mathrm{sg}}$ preserved by j and r. Let $\mathrm{j} = \mathrm{j}_{\mathrm{reg}} \oplus \mathrm{j}_{\mathrm{sg}}$ and $r = r_{\mathrm{reg}} \oplus r_{\mathrm{sg}}$. Then r_{reg} is $\mathrm{j}_{\mathrm{reg}}$-non-degenerate, and $\mathrm{j}_{\mathrm{reg}}$-self-adjoint on $\mathcal{Y}_{\mathrm{reg}}$ and $r_{\mathrm{sg}}\mathrm{j}_{\mathrm{sg}}$ is a conjugation on $\mathcal{Y}_{\mathrm{sg}}$.*

16.1.8 Restricted orthogonal group

The following subsection is parallel to Subsect. 11.1.6 about the restricted symplectic group. Recall that $B^2(\mathcal{Y})$ denotes the set of Hilbert–Schmidt operators on $\mathcal{Y}$.

Proposition 16.20 *Let $r \in O(\mathcal{Y})$. Let $p, q, Z_{\pm\mathrm{sg}}, Z_{\pm\mathrm{reg}}$ be defined as above. The following conditions are equivalent:*

(1) $\mathrm{j} - r^{-1}\mathrm{j}r \in B^2(\mathcal{Y})$,   (2) $r\mathrm{j} - \mathrm{j}r \in B^2(\mathcal{Y})$.

(3) $\mathrm{Tr}(q^*q) < \infty$,   (4) $\mathrm{Tr}(p^*p - \mathbb{1}) < \infty$,   (5) $\mathrm{Tr}(pp^* - \mathbb{1}) < \infty$.

(6) $\dim Z_{+\mathrm{sg}} < \infty$ and $d \in B^2(\overline{Z}_{+\mathrm{reg}}, Z_{+\mathrm{reg}})$.

(7) $\dim Z_{-\mathrm{sg}} < \infty$ and $c \in B^2(\overline{Z}_{-\mathrm{reg}}, Z_{-\mathrm{reg}})$.

If the above conditions are true, then $\dim \mathcal{Y}_{-\mathrm{sg}} = \dim \mathcal{Y}_{+\mathrm{sg}} < \infty$.

Proof The proof of the equivalence of the first five conditions is identical to the proof in Prop. 11.12. Assume now that condition (3) (and hence (4), (5)) holds. Since $Z_{-\mathrm{sg}} = \mathrm{Ker}\, p$, $Z_{+\mathrm{sg}} = \mathrm{Ker}\, p^*$, these spaces are finite-dimensional, and $p : Z_{-\mathrm{reg}} \to Z_{+\mathrm{reg}}$, $p^* : Z_{+\mathrm{reg}} \to Z_{-\mathrm{reg}}$ are invertible with bounded inverses. It follows then from (3) that $d = q\overline{p}^{-1} \in B^2(\overline{Z}_{+\mathrm{reg}}, Z_{+\mathrm{reg}})$ and $c = q^\#(p^\#)^{-1} \in B^2(\overline{Z}_{-\mathrm{reg}}, Z_{-\mathrm{reg}})$, so (3) $\Rightarrow$ (6), (7). To prove that (6), (7) $\Rightarrow$ (3), we argue similarly, using the identities $\mathbb{1} + c^\#\overline{c} = (pp^*)^{-1}$, $\mathbb{1} + d^*d = (\overline{pp^*})^{-1}$. $\square$

Definition 16.21 *Let $O_\mathrm{j}(\mathcal{Y})$ be the set of $r \in O(\mathcal{Y})$ satisfying the conditions of Prop. 16.20. $O_\mathrm{j}(\mathcal{Y})$ is called the* restricted orthogonal group *and is equipped with the metric*

$$
d_\mathrm{j}(r_1, r_2) := \|p_1 - p_2\| + \|q_1 - q_2\|_2.
$$

Equivalent metrics are $\|[\mathrm{j}, r_1 - r_2]_+\| + \|[\mathrm{j}, r_1 - r_2]\|_2$ *and* $\|r_1 - r_2\| + \|[\mathrm{j}, r_1 - r_2]\|_2$.

Noting that since $\mathcal{Y}_{-\mathrm{sg}}$ *is* j-*invariant its dimension as a real vector space is even, we define*

$$SO_\mathrm{j}(\mathcal{Y}) := \left\{ r \in O_\mathrm{j}(\mathcal{Y}) \ : \ \frac{1}{2}\dim \mathcal{Y}_{-\mathrm{sg}} \text{ is even}\right\}.$$

We set $\det r = 1$ *if* $r \in SO_\mathrm{j}(\mathcal{Y})$ *and* $\det r = -1$ *if* $r \in O_\mathrm{j}(\mathcal{Y})\backslash SO_\mathrm{j}(\mathcal{Y})$.

We say that $a \in o_\mathrm{j}(\mathcal{Y})$ *if* $a \in o(\mathcal{Y})$ *and* $[a,\mathrm{j}] \in B^2(\mathcal{Y})$, *or equivalently if* $g \in B^2(\overline{\mathcal{Z}},\mathcal{Z})$, *where we use the decomposition* (16.2).

Recall that the groups $O_2(\mathcal{Y})$ and $SO_2(\mathcal{Y})$ and the Lie algebra $o_2(\mathcal{Y})$ were defined in Subsect. 14.1.2.

Proposition 16.22 (1) $O_\mathrm{j}(\mathcal{Y})$ *and* $SO_\mathrm{j}(\mathcal{Y})$ *are topological groups containing* $O_2(\mathcal{Y})$ *and* $SO_2(\mathcal{Y})$, *and we have an exact sequence*

$$1 \to SO_\mathrm{j}(\mathcal{Y}) \to O_\mathrm{j}(\mathcal{Y}) \to \mathbb{Z}_2 \to 1.$$

(2) $o_\mathrm{j}(\mathcal{Y})$ *is a Lie algebra containing* $o_2(\mathcal{Y})$.
(3) *If* $a \in o_\mathrm{j}(\mathcal{Y})$, *then* $\mathrm{e}^a \in SO_\mathrm{j}(\mathcal{Y})$.

In the following lemma, which we will prove before we prove the above proposition, we use the concept of the regularized determinant, defined for $b \in B^2(\mathcal{Y})$ as $\det_2(\mathbb{1} + b) := \det\big((\mathbb{1}+b)\mathrm{e}^{-b}\big)$; see (2.4).

Lemma 16.23 *Let* $r \in O_\mathrm{j}(\mathcal{Y})$. *Then* r *is* j-*non-degenerate iff* $\det_2\big(\mathbb{1}+b(r)\big) \neq 0$ *for* $b(r) := \frac{1}{2}\mathrm{j}r^{\#}\,[\mathrm{j},r]$.

Proof We have

$$r\mathrm{j} + \mathrm{j}r = 2r\mathrm{j}\big(\mathbb{1} + b(r)\big).$$

This implies that $\mathrm{Ker}(r\mathrm{j} + \mathrm{j}r) = \mathrm{Ker}\big(\mathbb{1} + b(r)\big)$ Then we use Prop. 2.49. $\qquad\square$

Proof of Prop. 16.22. The fact that $O_\mathrm{j}(\mathcal{Y})$ is a topological group, and $o_\mathrm{j}(\mathcal{Y})$ is a Lie algebra, follows by the same arguments as in Prop. 11.14. To show the remaining facts, we will use Thm. 16.43, to be proven later on.

Since $SO_\mathrm{j}(\mathcal{Y})$ is also the set of $r \in O_\mathrm{j}(\mathcal{Y})$ implementable by even unitaries, we see that $SO_\mathrm{j}(\mathcal{Y})$ is a subgroup of $O_\mathrm{j}(\mathcal{Y})$, using Thm. 16.43 (2)(ii) and (2)(iv).

Let us now prove that $SO_\mathrm{j}(\mathcal{Y})$ is closed. Let $r_n \in SO_\mathrm{j}(\mathcal{Y})$ converge to r, and let U_{r_n} be the corresponding Bogoliubov implementers. By Thm. 16.43 (2)(v), there exist μ_n, $|\mu_n| = 1$ such that $\mu_n U_{r_n} \to U_r$. U_{r_n} are even, and so is U_r. Hence $r \in SO_\mathrm{j}(\mathcal{Y})$.

Let us now prove (3). Set $f(t) = \det_2\big(\mathbb{1} + b(\mathrm{e}^{ta})\big)$, where $b(\mathrm{e}^{ta})$ is given by Lemma 16.23. The map $t \mapsto f(t)$ is real analytic, and $f(0) \neq 0$, hence $f(t)$ is not identically zero. So we can find a sequence $t_n \underset{n\to\infty}{\to} 1$ such that $f(t_n) \neq 0$. By Lemma 16.23, $\mathrm{e}^{t_n a}$ are j-non-degenerate, hence they belong to $SO_\mathrm{j}(\mathcal{Y})$.

But $o_j(\mathcal{Y}) \ni a \mapsto e^a \in O_j(\mathcal{Y})$ is continuous and $SO_j(\mathcal{Y})$ is closed. Hence, $e^a = \lim_{n \to \infty} e^{t_n a}$ belongs to $SO_j(\mathcal{Y})$. □

16.1.9 Anomaly-free orthogonal group

This subsection is parallel to Subsect. 11.1.7 about the anomaly-free symplectic group. Recall that the groups $O_1(\mathcal{Y})$ and $SO_1(\mathcal{Y})$ and the Lie algebra $o_1(\mathcal{Y})$ were defined in Subsect. 14.1.2. Recall also that $B^1(\mathcal{Y})$ denotes the set of trace-class operators on $\mathcal{Y}$.

Definition 16.24 *Let* $O_{j,af}(\mathcal{Y})$ *be the set of* $r \in O_j(\mathcal{Y})$ *such that* $2j - (jr + rj) \in B^1(\mathcal{Y})$, *or equivalently* $p - \mathbb{1}_{\mathcal{Z}} \in B^1(\mathcal{Z})$. $O_{j,af}(\mathcal{Y})$ *will be called the anomaly-free orthogonal group and will be equipped with the metric*

$$d_{j,af}(r_1, r_2) := \|p_1 - p_2\|_1 + \|q_1 - q_2\|_2.$$

An equivalent metric is $\|[j, r_1 - r_2]_+\|_1 + \|[j, r_1 - r_2]\|_2$.

We set $SO_{j,af}(\mathcal{Y}) := O_{j,af}(\mathcal{Y}) \cap SO_j(\mathcal{Y})$.

We say that $a \in o_{j,af}(\mathcal{Y})$ *if* $a \in o_j(\mathcal{Y})$ *and* $aj + ja \in B^1(\mathcal{Y})$, *or equivalently* $h \in B^1(\mathcal{Y})$, *where we use the decomposition (16.2).*

Proposition 16.25 (1) $O_{j,af}(\mathcal{Y})$ *and* $SO_{j,af}(\mathcal{Y})$ *are topological groups containing* $O_1(\mathcal{Y})$ *and* $SO_1(\mathcal{Y})$ *respectively, and we have an exact sequence*

$$1 \to SO_{j,af}(\mathcal{Y}) \to O_{j,af}(\mathcal{Y}) \to \mathbb{Z}_2 \to 1.$$

(2) $o_{j,af}(\mathcal{Y})$ *is a Lie algebra containing* $o_1(\mathcal{Y})$.
(3) *If* $a \in o_{j,af}(\mathcal{Y})$, *then* $e^a \in SO_{j,af}(\mathcal{Y})$.

Proof The proof is completely analogous to that of Prop. 16.22. □

Proposition 16.26 (1) *Let* $r \in O(\mathcal{Y})$ *be* j-*positive. Then* $r \in O_j(\mathcal{Y})$ *iff* $r \in O_{j,af}(\mathcal{Y})$.
(2) *Let* $a \in o(\mathcal{Y})$ *be* j-*self-adjoint. Then* $a \in o_j(\mathcal{Y})$ *iff* $a \in o_{j,af}(\mathcal{Y})$.

Proof (1) We know that $r \in O_j(\mathcal{Y})$ iff $c \in B^2(\overline{\mathcal{Z}}_{reg}, \mathcal{Z}_{reg})$ and $\dim \mathcal{Y}_{sg} < \infty$. But then (16.12) implies $r \in O_{j,af}(\mathcal{Y})$.

(2) By the decomposition (16.2), $a \in o_j(\mathcal{Y})$ iff $h = 0$ and $g \in B^2_a(\overline{\mathcal{Z}}, \mathcal{Z})$. □

We will also need the following lemma:

Lemma 16.27 *Let* $r \in O_{j,af}(\mathcal{Y})$, $\epsilon > 0$. *There exists a decomposition* $r = ts$ *such that* $\mathbb{1} - s$ *is finite rank and* $t \in O_{j,af}(\mathcal{Y})$, $\|\mathbb{1} - t\| \le \epsilon$.

Proof $\mathbb{1} - r$ is compact. Hence, there exists an o.n. basis $(e_1, e_2, \dots)$ in $\mathbb{C}\mathcal{Y}$ such that

$$r_{\mathbb{C}} = \sum_j \lambda_j |e_j)(e_j|$$

with $|\lambda_j| = 1$ and $\lambda_j \to 1$. Then we set

$$t_{\mathbb{C}} := \sum_{|\lambda_j - 1| \leq \epsilon} \lambda_j |e_j)(e_j| + \sum_{|\lambda_j - 1| > \epsilon} |e_j)(e_j|, \tag{16.23}$$

$$s_{\mathbb{C}} := \sum_{|\lambda_j - 1| \leq \epsilon} |e_j)(e_j| + \sum_{|\lambda_j - 1| > \epsilon} \lambda_j |e_j)(e_j|. \tag{16.24}$$

(We easily see that the r.h.s. of (16.23) and (16.24) restrict to operators on $\mathcal{Y}$.) $\qquad\square$

16.1.10 Pairs of Kähler structures on real Hilbert spaces

This subsection is parallel to Subsect. 11.1.8 about pairs of Kähler structures in a symplectic space.

Recall that, as usual in this chapter, $(\mathcal{Y}, \nu)$ is a real Hilbert space. Let us first describe the action of the orthogonal group on Kähler anti-involutions.

Proposition 16.28 *Let $r \in O(\mathcal{Y})$ and let j be a Kähler anti-involution. Then*

(1) $j_1 = r^{-1} j r$ *is a Kähler anti-involution;*
(2) $r \in U(\mathcal{Y}^{\mathbb{C}})$ *iff $j_1 = j$;*
(3) r *is j-non-degenerate iff $\mathrm{Ker}(j + j_1) = \{0\}$.*

In the following theorem, for two Kähler anti-involutions j and j_1 we try to construct $r \in O(\mathcal{Y})$ such that

$$r^{-1} j r = j_1. \tag{16.25}$$

Note that this problem is more complicated for $O(\mathcal{Y})$ than for $Sp(\mathcal{Y})$ (see Subsect. 11.1.8).

Theorem 16.29 (1) *Let j, j_1 be Kähler anti-involutions on a real Hilbert space $\mathcal{Y}$. Then $k := -j j_1$ is a j-self-adjoint orthogonal transformation.*
(2) *Let $k \in O(\mathcal{Y})$ be j-self-adjoint for a Kähler anti-involution j. Then $j_1 := jk$ is a Kähler anti-involution.*
(3) *In what follows we assume that j, j_1, k are as above. Then $\mathrm{Ker}(j + j_1) = \mathrm{Ker}(k + 1\!\!1)$ is invariant under j and j_1, and so is its orthogonal complement.*
(4) *There exists $r \in O(\mathcal{Y})$ satisfying (16.25) iff $\mathrm{Ker}(j + j_1)$ is even- or infinite-dimensional.*
(5) *If there exists a j-positive $r \in O(\mathcal{Y})$ satisfying (16.25), then $\mathrm{Ker}(j + j_1) = \{0\}$.*
(6) *Assume that $\mathrm{Ker}(j + j_1) = \{0\}$. Then $r := k^{\frac{1}{2}}$ defined in Thm. 16.19 is the unique j-positive element of $O(\mathcal{Y})$ satisfying (16.25).*
(7) *There exists $c \in B_a(\overline{\mathcal{Z}}, \mathcal{Z})$ such that*

$$\left(\frac{k - 1\!\!1}{k + 1\!\!1} \right)_{\mathbb{C}} = \begin{bmatrix} 0 & c \\ \bar{c} & 0 \end{bmatrix}. \tag{16.26}$$

(8) *We have*

$$r_{\mathbb{C}} = \begin{bmatrix} (\mathbb{1} + cc^*)^{-\frac{1}{2}} & (\mathbb{1} + cc^*)^{-\frac{1}{2}}c \\ -c^*(\mathbb{1} + cc^*)^{-\frac{1}{2}} & (\mathbb{1} + c^*c)^{-\frac{1}{2}} \end{bmatrix}, \tag{16.27}$$

$$k_{\mathbb{C}} = = \begin{bmatrix} (\mathbb{1} - cc^*)(\mathbb{1} + cc^*)^{-1} & 2(\mathbb{1} + cc^*)^{-1}c \\ -2c^*(\mathbb{1} + cc^*)^{-1} & (\mathbb{1} - c^*c)(\mathbb{1} + c^*c)^{-1} \end{bmatrix}, \tag{16.28}$$

$$j_{1\mathbb{C}} = i \begin{bmatrix} (\mathbb{1} - cc^*)(\mathbb{1} + cc^*)^{-1} & 2(\mathbb{1} + cc^*)^{-1}c \\ 2c^*(\mathbb{1} + cc^*)^{-1} & (c^*c - \mathbb{1})(\mathbb{1} + c^*c)^{-1} \end{bmatrix}. \tag{16.29}$$

Proof (1)–(3) are straightforward.

Set $b = \frac{k-\mathbb{1}}{k+\mathbb{1}}$. We check that $jb = -bj$. This implies (16.26). Then using (16.13), we see that $r_{\mathbb{C}}$ equals (16.12), which is repeated as (16.27).

By the properties of the Cayley transform we have $k = \frac{\mathbb{1}+b}{\mathbb{1}-b}$, which yields (16.28). Alternatively, we can use $k = r^2$. (16.29) follows from $j_1 = jk$. □

Theorem 16.30 *Let $\mathcal{Z}$ and $\mathcal{Z}_1$ be the holomorphic subspaces of $\mathbb{C}\mathcal{Y}$ for the Kähler anti-involutions j and j_1. Suppose that $\mathrm{Ker}(j + j_1) = \{0\}$. Then*

$$\{(z, -\bar{c}z) \; : \; z \in \mathrm{Dom}\,\bar{c}\} \text{ is dense in } \mathcal{Z}_1,$$

$$\{(-c\bar{z}, \bar{z}) \; : \; \bar{z} \in \mathrm{Dom}\,c\} \text{ is dense in } \overline{\mathcal{Z}_1}.$$

Proof Every vector of $\mathcal{Z}_1$ is of the form $(\mathbb{1} - ij_1)y_1$ for $y_1 \in \mathcal{Y}$. Since $\mathrm{Ker}(j + j_1) = \{0\}$, $\mathrm{Ran}\,(k + \mathbb{1})$ is dense in $\mathcal{Y}$, hence the vectors of the form $(\mathbb{1} - ij_1)(\mathbb{1} + k)^{-1}y$, for $y \in \mathrm{Ran}\,(k + \mathbb{1})$ are dense in $\mathcal{Z}_1$. As in the proof of Prop. 11.21, we get that

$$(\mathbb{1} - ij_1)(\mathbb{1} + k)^{-1}y = z - \bar{c}z,$$

for $z = \mathbb{1}_{\mathcal{Z}}y \in \mathrm{Dom}\,\bar{c}$. □

Proposition 16.31 *Let j, j_1, k be as in Thm. 16.29. Set $\mathcal{Y}_{\mathrm{sg}} := \mathrm{Ker}(j + j_1)$ and $\mathcal{Y}_{\mathrm{reg}} := \mathcal{Y}_{\mathrm{sg}}^{\perp}$. Note that $\mathcal{Y}_{\mathrm{reg}}$ and $\mathcal{Y}_{\mathrm{sg}}$ are preserved by j and j_1. Let $\mathcal{Z}_{\mathrm{reg}}$ and $\mathcal{Z}_{\mathrm{sg}}$ be the corresponding holomorphic spaces. Recall also that one can define*

$$c := (k_{\mathbb{C}} - \mathbb{1})(\mathbb{1} + k_{\mathbb{C}})^{-1}\big|_{\overline{\mathcal{Z}}_{\mathrm{reg}}} \in Cl_{\mathrm{a}}(\overline{\mathcal{Z}}_{\mathrm{reg}}, \mathcal{Z}_{\mathrm{reg}}). \tag{16.30}$$

Then the following conditions are equivalent:

(1) $j - j_1 \in B^2(\mathcal{Y})$.
(2) $\mathbb{1} - k \in B^2(\mathcal{Y})$.
(3) $c \in B^2(\overline{\mathcal{Z}}_{\mathrm{reg}}, \mathcal{Z}_{\mathrm{reg}})$ and $\dim \mathcal{Z}_{\mathrm{sg}}$ is finite.
(4) *There exists a j-positive $r \in O_{j,\mathrm{af}}(\mathcal{Y})$ such that $j_1 = rjr^{\#}$.*
(5) *There exists $r \in O_j(\mathcal{Y})$ such that $j_1 = rjr^{\#}$.*

Proof The identity $-j(j - j_1) = \mathbb{1} - k$ and $j \in O(\mathcal{Y})$ imply the equivalence of (1) and (2).

$(2) \Rightarrow (3)$. Since $\mathcal{Y}_{sg} = \text{Ker}(\mathbb{1} + k)$ and $\mathbb{1} - k$ is compact, $\mathcal{Y}_{sg}$ and hence $\mathcal{Z}_{sg}$ are finite-dimensional. Moreover k preserves $\mathcal{Y}_{reg}$ and $(\mathbb{1} + k)^{-1}\big|_{\mathcal{Y}_{reg}}$ is bounded. Using (16.26), we obtain that $c \in B^2(\overline{\mathcal{Z}}_{reg}, \mathcal{Z}_{reg})$.

$(3) \Rightarrow (2)$. By (16.26), we see that $(\mathbb{1} - k)\big|_{\mathcal{Y}_{reg}} \in B^2(\mathcal{Y}_{reg})$. $\mathcal{Y}_{sg} = \text{Ker}(\mathbb{1} + k)$ is finite-dimensional, hence we get that $\mathbb{1} - k \in B^2(\mathcal{Y})$.

$(4) \Rightarrow (5) \Rightarrow (1)$ is obvious. $(3) \Rightarrow (4)$ follows by setting $r := r_{reg} \oplus r_{sg}$, where $r_{reg} \in O(\mathcal{Y}_{reg})$ is defined as in Thm. 16.29 (5), and r_{sg} is any conjugation on $\mathcal{Y}_{sg}$. $\qquad\square$

16.2 Fermionic quadratic Hamiltonians on Fock spaces

As elsewhere in this chapter, $\mathcal{Z}$ is a Hilbert space and $\mathcal{Y} = \text{Re}(\mathcal{Z} \oplus \overline{\mathcal{Z}})$ is the corresponding Kähler space with the dual $\mathcal{Y}^{\#} = \text{Re}(\overline{\mathcal{Z}} \oplus \mathcal{Z})$. We consider the Fock representation over $\mathcal{Y}$ in $\Gamma_a(\mathcal{Z})$.

We study quadratic Hamiltonians on a fermionic Fock space. This section is parallel to Sect. 11.2 about quadratic Hamiltonians on a bosonic Fock space. It is also a continuation of Sect. 14.2, where quadratic fermionic Hamiltonians were studied in an algebraic setting.

16.2.1 Quadratic anti-commuting polynomials and their quantization

Let $h \in B^{fd}(\mathcal{Z})$ (h is finite rank). It corresponds to the anti-symmetric polynomial

$$\mathcal{Y}^{\#} \times \mathcal{Y}^{\#} \ni ((\overline{z}_1, z_1), (\overline{z}_2, z_2)) \mapsto \frac{1}{2}(\overline{z}_1 \cdot hz_2 - z_1 \cdot h^{\#}\overline{z}_2). \tag{16.31}$$

Its Wick, anti-symmetric and anti-Wick quantizations are

$$d\Gamma(h), \quad d\Gamma(h) - \frac{\text{Tr}\, h}{2}\mathbb{1}, \quad d\Gamma(h) - (\text{Tr}\, h)\mathbb{1}.$$

Note that the anti-Wick and anti-symmetric quantizations can be extended to the case $h \in B^1(\mathcal{Z})$ (h is trace-class). The Wick quantization of (16.31) is well defined for much more general h.

Suppose that $g \in \Gamma_a^2(\mathcal{Z}) \simeq B_a^{fd}(\overline{\mathcal{Z}}, \mathcal{Z})$ (g is anti-symmetric finite rank). Consider the polynomial

$$\mathcal{Y}^{\#} \times \mathcal{Y}^{\#} \ni ((\overline{z}_1, z_1), (\overline{z}_2, z_2)) \mapsto (z_1 \otimes_a z_2 | g) = \overline{z}_1 \cdot g\overline{z}_2. \tag{16.32}$$

The Wick, anti-symmetric and anti-Wick quantizations of (16.32) are the "two-particle creation operator" $a^*(g)$ defined in Subsect. 3.4.4. According to the notation of Def. 13.27, this can be written as $\text{Op}^{a^*, a}(|g))$. It can be defined as a bounded operator also if $g \in \Gamma_a^2(\mathcal{Z}) \simeq B_a^2(\overline{\mathcal{Z}}, \mathcal{Z})$. It will act on $\Psi_n \in \Gamma_a^n(\mathcal{Z})$ as

$$a^*(g)\Psi_n := \sqrt{(n+2)(n+1)}\, g \otimes_a \Psi_n. \tag{16.33}$$

(On the right of (16.33) we interpret g as an element of $\Gamma_a^2(\mathcal{Z})$.)

The polynomial complex conjugate to (16.32) times -1 is

$$\mathcal{Y}^\# \times \mathcal{Y}^\# \ni \left((\bar{z}_1, z_1), (\bar{z}_2, z_2)\right) \mapsto (g|z_2 \otimes_a z_1) = z_1 \cdot g^* z_2. \tag{16.34}$$

(-1 comes from the operator Λ; see (3.29).) The Wick, anti-symmetric and anti-Wick quantizations of (16.34) are the "two particle annihilation operator" $a^*(g)^* = a(g)$ defined in Subsect. 3.4.4. According to the notation of Def. 13.27, this can be written as $\mathrm{Op}^{a^*,a}(|g))$.

A general element of $\mathbb{C}\mathrm{Pol}_a^2(\mathcal{Y}^\#)$ is

$$\left((\bar{z}_1, z_1), (\bar{z}_2, z_2)\right) \mapsto \bar{z}_1 \cdot h z_2 - z_1 \cdot h^\# \bar{z}_2 + \bar{z}_1 \cdot g_1 \bar{z}_2 - z_1 \cdot \bar{g}_2 z_2, \tag{16.35}$$

where $h \in B^{\mathrm{fd}}(\mathcal{Z})$, $g_1, g_2 \in B_a^{\mathrm{fd}}(\overline{\mathcal{Z}}, \mathcal{Z})$. We can write (16.35) as

$$(\bar{z}_1, z_1) \cdot \zeta(\bar{z}_2, z_2), \quad \zeta_\mathbb{C} = \begin{bmatrix} g_1 & h \\ -h^\# & -\bar{g}_2 \end{bmatrix}.$$

(Recall that we use elements of $L_a(\mathcal{Y}^\#, \mathcal{Y})$ for symbols of fermionic quadratic Hamiltonians, as in Subsect. 14.2.3.)

The quantizations of ζ are

$$\mathrm{Op}^{a^*,a}(\zeta) = 2\mathrm{d}\Gamma(h) + a^*(g_1) + a(g_2), \tag{16.36}$$

$$\mathrm{Op}(\zeta) = 2\mathrm{d}\Gamma(h) - (\mathrm{Tr}\, h)\mathbb{1} + a^*(g_1) + a(g_2), \tag{16.37}$$

$$\mathrm{Op}^{a,a^*}(\zeta) = 2\mathrm{d}\Gamma(h) - (2\mathrm{Tr}\, h)\mathbb{1} + a^*(g_1) + a(g_2).$$

Note that

$$\mathrm{Op}(\zeta) = \frac{1}{2}\left(\mathrm{Op}^{a^*,a}(\zeta) + \mathrm{Op}^{a,a^*}(\zeta)\right).$$

In particular, we can extend the definition of $\mathrm{Op}(\zeta)$ and $\mathrm{Op}^{a,a^*}(\zeta)$ to the case when $g_1, g_2 \in B^2(\overline{\mathcal{Z}}, \mathcal{Z})$ and $h \in B^1(\mathcal{Z})$. $\mathrm{Op}^{a^*,a}(\zeta)$ is defined under much more general conditions. All these quantizations are self-adjoint iff $h = h^*$ and $g_1 = g_2$.

16.2.2 Fermionic Schwinger term

Recall from Thm. 14.13 that the anti-symmetric quantization restricted to quadratic symbols yields an isomorphism of Lie algebra $o_1(\mathcal{Y})$ into quadratic Hamiltonians in $\mathrm{CAR}^{C^*}(\mathcal{Y})$. This is no longer true in the case of the Wick quantization, where the so-called Schwinger term appears. This is described in the following proposition:

Proposition 16.32 *Let* $\zeta, \zeta_i \in B(\mathcal{Y}^\#, \mathcal{Y})$, $i = 1, 2$. *Then,*

$$\mathrm{Op}(\zeta) = \mathrm{Op}^{a^*,a}(\zeta) + \frac{\mathrm{i}}{2}(\mathrm{Tr}\,\zeta\nu\mathrm{j})\,\mathbb{1}, \tag{16.38}$$

$$[\mathrm{Op}^{a^*,a}(\zeta_1), \mathrm{Op}^{a^*,a}(\zeta_2)] = 4\mathrm{Op}^{a^*,a}(\zeta_1\nu\zeta_2 - \zeta_2\nu\zeta_1) + \mathrm{i}2(\mathrm{Tr}[\zeta_1\nu, \zeta_2\nu]\mathrm{j})\,\mathbb{1}.$$

Proof We have $\nu_{\mathbb{C}} = \dfrac{1}{2}\begin{bmatrix} 0 & \mathbb{1} \\ \mathbb{1} & 0 \end{bmatrix} \in B(\mathcal{Z} \oplus \overline{\mathcal{Z}}, \overline{\mathcal{Z}} \oplus \mathcal{Z})$. Therefore,

$$\zeta_{\mathbb{C}}\nu_{\mathbb{C}} = \frac{1}{2}\begin{bmatrix} h & g \\ -\overline{g} & -h^{\#} \end{bmatrix}, \quad \zeta_{\mathbb{C}}\nu_{\mathbb{C}}\mathrm{j}_{\mathbb{C}} = \frac{1}{2}\begin{bmatrix} ih & -ig \\ -i\overline{g} & ih^{\#} \end{bmatrix}. \tag{16.39}$$

Hence, $-\operatorname{Tr} h = \frac{1}{2}\operatorname{Tr}\zeta\nu\mathrm{j}$, which implies (16.38).

Now, to compute the Schwinger term we note that by (14.9)

$$[\mathrm{Op}^{a^{*},a}(\zeta_{1}), \mathrm{Op}^{a^{*}a}(\zeta_{2})] = 4\mathrm{Op}(\zeta_{1}\nu\zeta_{2} - \zeta_{2}\nu\zeta_{1}).$$

Then we apply (16.38). $\qquad\square$

16.2.3 Infimum of quadratic fermionic Hamiltonians

For simplicity, in this subsection we assume that $\mathcal{Z}$ is a finite-dimensional Hilbert space.

Theorem 16.33 *Let $h \in B_{\mathrm{h}}(\mathcal{Z})$, $g \in B_{\mathrm{a}}^{2}(\overline{\mathcal{Z}}, \mathcal{Z})$. Let*

$$\zeta_{\mathbb{C}} = \begin{bmatrix} g & h \\ -h^{\#} & -\overline{g} \end{bmatrix}. \tag{16.40}$$

Then,

$$\inf \mathrm{Op}^{a^{*},a}(\zeta) = \frac{1}{2}\operatorname{Tr}\left(-\begin{bmatrix} h^{2} + gg^{*} & hg - gh^{\#} \\ g^{*}h - h^{\#}g^{*} & \overline{h}^{2} + g^{*}g \end{bmatrix}^{\frac{1}{2}} + \begin{bmatrix} h & 0 \\ 0 & h^{\#} \end{bmatrix}\right).$$

Proof Clearly, $\zeta\nu$ is self-adjoint and

$$(\zeta_{\mathbb{C}}\nu_{\mathbb{C}})^{2} = \frac{1}{4}\begin{bmatrix} h^{2} + gg^{*} & hg - gh^{\#} \\ g^{*}h - h^{\#}g^{*} & h^{\#2} + g^{*}g \end{bmatrix}.$$

Thus, by Thm. 14.13,

$$\inf \mathrm{Op}^{a^{*},a}(\zeta) - \operatorname{Tr} h = \inf \mathrm{Op}(\zeta)$$

$$= -\operatorname{Tr}|\zeta\nu| = -\frac{1}{2}\operatorname{Tr}\begin{bmatrix} h^{2} + gg^{*} & hg - gh^{\#} \\ g^{*}h - h^{\#}g^{*} & h^{\#2} + g^{*}g \end{bmatrix}^{\frac{1}{2}}. \qquad\square$$

16.2.4 Two-particle creation and annihilation operators

In this subsection we allow the dimension of $\mathcal{Z}$ to be infinite. We study two-particle creation and annihilation operators. Recall that they are defined for $c \in \Gamma_{\mathrm{a}}^{2}(\mathcal{Z}) \simeq B_{\mathrm{a}}^{2}(\overline{\mathcal{Z}}, \mathcal{Z})$.

Proposition 16.34 *Let $c \in \Gamma_{\mathrm{a}}^{2}(\mathcal{Z})$. Then $a(c)$, $a^{*}(c)$ are bounded operators with*

$$\|a(c)\| = \|a^{*}(c)\| = \|c\|_{2}; \tag{16.41}$$

$$\mathrm{e}^{-\frac{1}{2}a^{*}(c)}a(z)\mathrm{e}^{\frac{1}{2}a^{*}(c)} = a(z) - a^{*}(c\overline{z}), \quad z \in \mathcal{Z}; \tag{16.42}$$

$$\mathrm{e}^{\frac{1}{2}a(c)}a^{*}(z)\mathrm{e}^{-\frac{1}{2}a(c)} = a^{*}(z) - a(c\overline{z}), \quad z \in \mathcal{Z}. \tag{16.43}$$

Proof   Since c is anti-symmetric Hilbert–Schmidt, by Corollary 2.88 there exists an o.n. family

$$(w_{1,+}, w_{1,-}, w_{2,+}, w_{2,-} \dots) \tag{16.44}$$

and positive numbers $(\lambda_1, \lambda_2, \dots)$ such that

$$c = \sum_{j=1}^{\infty} \frac{\lambda_j}{2} \big(|w_{j,+}\rangle\langle w_{j,-}| - |w_{j,-}\rangle\langle w_{j,+}|\big).$$

Then,

$$a^*(c) = \sum_{j=1}^{\infty} \lambda_j a^*(w_{j,+}) a^*(w_{j,-}).$$

Using the Jordan–Wigner representation compatible with the o.n. family (16.44), we easily obtain

$$\|a^*(c)\|^2 = \sum_{j=1}^{\infty} \lambda_j^2. \qquad \square$$

16.2.5 Fermionic Gaussian vectors

Let $c \in \Gamma_{\mathrm{a}}^2(\mathcal{Z}) \simeq B_{\mathrm{a}}^2(\overline{\mathcal{Z}}, \mathcal{Z})$. Then c^*c is trace-class, so $\det(\mathbb{1} + c^*c)$ is well defined.

Definition 16.35 *The* fermionic Gaussian vector *associated with c is defined as*

$$\Omega_c := \det(\mathbb{1} + c^*c)^{-\frac{1}{4}} e^{-\frac{1}{2} a^*(c)} \Omega.$$

Theorem 16.36 (1) *If $c \in B_{\mathrm{a}}^2(\overline{\mathcal{Z}}, \mathcal{Z})$, then Ω_c is a normalized vector in $\Gamma_{\mathrm{a}}(\mathcal{Z})$ satisfying*

$$\big(a(z) - a^*(c\overline{z})\big)\Psi = 0, \quad z \in \mathcal{Z}, \quad (\Omega_c|\Omega_c) > 0.$$

(2) *Let $c \in Cl_{\mathrm{a}}(\overline{\mathcal{Z}}, \mathcal{Z})$. Assume that there exists a non-zero $\Psi \in \Gamma_{\mathrm{a}}(\mathcal{Z})$ satisfying*

$$\big(a(z) - a^*(c\overline{z})\big)\Psi = 0, \quad \overline{z} \in \mathrm{Dom}\, c.$$

Then $c \in B_{\mathrm{a}}^2(\overline{\mathcal{Z}}, \mathcal{Z})$. Moreover Ψ is proportional to Ω_c.
(3) *Let $c_1, c_2 \in B_{\mathrm{a}}^2(\overline{\mathcal{Z}}, \mathcal{Z})$. Then*

$$(\Omega_{c_1}|\Omega_{c_2}) = \det(\mathbb{1} + c_1^*c_1)^{-\frac{1}{4}} \det(\mathbb{1} + c_2^*c_2)^{-\frac{1}{4}} \mathrm{Pf} \begin{bmatrix} \overline{c}_1 & -\mathbb{1} \\ \mathbb{1} & c_2 \end{bmatrix}.$$

To make the above theorem complete we need to define the Pfaffian of certain infinite-dimensional operators, which is provided by the following proposition:

Proposition 16.37 *Let* $c_i \in B_a^2(\overline{\mathcal{Z}}, \mathcal{Z})$, $i = 1, 2$. *Set* $\zeta \in B_a(\mathcal{Z} \oplus \overline{\mathcal{Z}}, \overline{\mathcal{Z}} \oplus \mathcal{Z})$ *equal to*

$$\zeta = \begin{bmatrix} \overline{c}_1 & -\mathbb{1}_{\overline{\mathcal{Z}}} \\ \mathbb{1}_{\mathcal{Z}} & c_2 \end{bmatrix}.$$

Let π_n *be an increasing family of finite rank projections on* $\mathcal{Z}$ *with* $\mathrm{s} - \lim\limits_{n\to\infty} \pi_n = \mathbb{1}$. *Set* $\mathcal{Z}_n = \pi_n \mathcal{Z}$, *and* $\zeta_n = (\overline{\pi}_n \oplus \pi_n)\zeta(\pi_n \oplus \overline{\pi}_n)$. *Then*

$$\lim_{n\to\infty} \mathrm{Pf}\zeta_n =: \mathrm{Pf}\zeta$$

exists, where, for each n $\mathrm{Pf}\zeta_n$, *is computed w.r.t. the Liouville form on* $\overline{\mathcal{Z}}_n \oplus \mathcal{Z}_n$. *Moreover*

$$(\mathrm{Pf}\zeta)^2 = \det(\mathbb{1} - \overline{c}_1 c_2).$$

Proof of Thm. 16.36 and Prop. 16.37. Let Ψ be as in (2). Arguing as in the proof of Thm. 11.28, we obtain, for $\overline{z}_i \in \mathrm{Dom}\,c$ and $\lambda := (\Omega|\Psi)$,

$$\left(a^*(z_{2m+1}) \cdots a^*(z_1)\Omega|\Psi\right) = 0,$$

$$\left(a^*(z_{2m}) \cdots a^*(z_1)\Omega|\Psi\right) = \lambda \sum_{\sigma \in \mathrm{Pair}_{2m}} \mathrm{sgn}(\sigma) \prod_{j=1}^{m} \left(z_{\sigma(2j)}|c\overline{z}_{\sigma(2j+1)}\right).$$

Therefore, $\lambda = 0$ implies $\Psi = 0$. Hence, $\lambda \neq 0$. In particular, for $\overline{z}_1, \overline{z}_2 \in \mathrm{Dom}\,c$ this gives the following formula for the two-particle component of Ψ:

$$\sqrt{2}(z_2 \otimes_a z_1|\Psi_2) = \lambda(z_1|c\overline{z}_2). \tag{16.45}$$

As in Thm. 11.28, this implies that $c \in B_a^2(\overline{\mathcal{Z}}, \mathcal{Z})$ and $\Psi_2 = -\frac{\lambda}{\sqrt{2}}c$.

We have

$$(z_1 \otimes_a \cdots \otimes_a z_{2m}|c^{\otimes_a m}) = \frac{m!2^m}{2m!} \sum_{\sigma \in \mathrm{Pair}_m} \prod_{i=0}^{m-1} \mathrm{sgn}(\sigma) \left(z_{\sigma(2i+1)}|c\overline{z}_{\sigma(2i+2)}\right),$$

which implies that

$$\Psi_{2m} = \lambda(-1)^m \frac{\sqrt{(2m)!}}{2^m m!} c^{\otimes_a m} = \lambda(-1)^m \frac{1}{2^m m!} \left(a^*(c)\right)^m \Omega,$$

$$\Psi_{2m+1} = 0,$$

i.e.

$$\Psi = \lambda e^{-\frac{1}{2}a^*(c)}\Omega.$$

Let us now compute $\|\Psi\|^2$. Without loss of generality we can assume $\lambda = 1$. Since c is compact, we can by Corollary 2.88 find an o.n. basis $\{z_{i,+}, z_{i,-}\}_{i \in I}$ of $\overline{(\mathrm{Ran}\,c)^\perp}$, such that $c\overline{z_{i,-}} = \lambda_i z_{i,+}$, $c\overline{z_{i,+}} = -\lambda_i z_{i,+}$. Thus $c^*c\overline{z}_{i,\pm} = \lambda_i^2 \overline{z}_{i,\pm}$. Using the corresponding basis in $\Gamma_a(\mathcal{Z})$, we obtain

$$\|\Psi\|^2 = \prod_{i \in I}(1 + \lambda_i^2) = \det(\mathbb{1} + c^*c)^{\frac{1}{2}}. \tag{16.46}$$

This shows that the vector Ω_c is normalized.

It remains to show (3). Let us first assume that $\mathcal{Z}$ is finite-dimensional. In the fermionic complex-wave representation, $e^{-\frac{1}{2}a^*(c)}\Omega$ equals $e^{-\frac{1}{2}\bar{z}\cdot c\bar{z}}$.

$$\left(e^{-\frac{1}{2}a^*(c_1)}\Omega\Big|e^{-\frac{1}{2}a^*(c_2)}\Omega\right) = \int e^{z\cdot\bar{z}}e^{-\frac{1}{2}z\cdot\overline{c_1}z}e^{-\frac{1}{2}\bar{z}\cdot c_2\bar{z}}dzd\bar{z}$$

$$= \int \exp\frac{1}{2}[\bar{z},z]\cdot\begin{bmatrix}-\overline{c}_1 & -\mathbb{1} \\ \mathbb{1} & -c_2\end{bmatrix}\begin{bmatrix}z \\ \bar{z}\end{bmatrix}dzd\bar{z}$$

$$= \text{Pf}\begin{bmatrix}-\overline{c}_1 & -\mathbb{1} \\ \mathbb{1} & -c_2\end{bmatrix} = \det(\mathbb{1}-\overline{c}_1c_2)^{\frac{1}{2}},$$

using the formulas in Subsect. 1.1.2.

Let us now consider the general case. We first claim that the map

$$B_{\text{a}}^2(\overline{\mathcal{Z}},\mathcal{Z}) \ni c \mapsto \Omega_c \in \Gamma_{\text{a}}(\mathcal{Z}) \tag{16.47}$$

is continuous for the Hilbert–Schmidt norm. Recall from Prop. 16.34 that

$$\|a^*(c)\| = \|c\|_2. \tag{16.48}$$

Note now that if a_1, a_2 are two bounded operators then

$$\|e^{a_1} - e^{a_2}\| \leq \|a_1 - a_2\|\frac{e^{\|a_1\|} - e^{\|a_2\|}}{\|a_1\| - \|a_2\|}.$$

Using (16.48), for $a_i = a^*(c_i)$ with $\|c_i\|_2 \leq C$, this yields,

$$\|e^{-\frac{1}{2}a^*(c_1)} - e^{-\frac{1}{2}a^*(c_2)}\| \leq C'\|c_1 - c_2\|_2.$$

Since $c \mapsto \det(\mathbb{1} + c^*c)^{-\frac{1}{4}}$ is continuous for the Hilbert–Schmidt norm, this proves (16.47).

We can now complete the proof of (3) in the general case. Let us choose an increasing sequence of finite rank projections π_n and set $c_{i,n} = \pi_n c_i \overline{\pi}_n$, $i = 1, 2$. We have $c_{i,n} \to c_i$ in the Hilbert–Schmidt norm. Hence, by (16.47), $\Omega_{c_{i,n}} \to \Omega_{c_i}$, and thus

$$(\Omega_{c_1}|\Omega_{c_2}) = \lim_{n\to\infty}(\Omega_{c_{1,n}}|\Omega_{c_{2,n}}),$$

which proves (3) in the general case. □

16.3 Fermionic Bogoliubov transformations on Fock spaces

We keep the same framework and notation as in the rest of the chapter. That is, $\mathcal{Z}$ is a Hilbert space, $\mathcal{Y} := \text{Re}(\mathcal{Z} \oplus \overline{\mathcal{Z}})$ is the corresponding complete Kähler space, equipped with ν, j. We also consider the Fock CAR representation

$$\mathcal{Y} \ni y \mapsto \phi(y) \in B_{\text{h}}(\Gamma_{\text{a}}(\mathcal{Z})).$$

We are going to study the implementation of orthogonal transformations on a fermionic Fock space. The central result of the section is the Shale–Stinespring theorem, which says that an orthogonal transformation is implementable iff it belongs to the restricted orthogonal group. The unitary operators implementing the corresponding Bogoliubov automorphisms form a group, denoted $Pin_j^c(\mathcal{Y})$, which is one of the generalizations of the Pin^c group from the finite-dimensional case and contains the group $Pin_2^c(\mathcal{Y})$, which is a subgroup of the unitary part of $CAR^{W^*}(\mathcal{Y})$.

We will also describe the group $Pin_{j,af}(\mathcal{Y})$, which is one of the generalizations of the Pin group from the finite-dimensional case. It contains the group $Pin_2(\mathcal{Y})$ as a proper subgroup.

Clearly, both $Pin_j^c(\mathcal{Y})$ and $Pin_{j,af}(\mathcal{Y})$ depend on the Kähler structure of $\mathcal{Y}$.

This section is parallel to Sect. 11.3, where Bogoliubov transformations on bosonic Fock spaces were studied. It can be viewed as a continuation of Sect. 14.3, which described the implementability of Bogoliubov transformations in the C^*- and W^*-CAR algebras.

16.3.1 Extending parity and complex conjugation

Clearly, we can isometrically embed $CAR^{C^*}(\mathcal{Y})$ in $B(\Gamma_a(\mathcal{Y}))$. The parity automorphism α defined on $CAR^{C^*}(\mathcal{Y})$ extends to a weakly continuous involution on the whole $B(\Gamma_a(\mathcal{Z}))$ by setting

$$\alpha(A) := IAI. \tag{16.49}$$

Thus we can speak about even and odd operators on $B(\Gamma_a(\mathcal{Z}))$.

Unfortunately, there seems to be no analog of (16.49) for the complex conjugation $A \mapsto c(A)$ on the Fock space, as seen from the following proposition:

Proposition 16.38 Let $\mathcal{Y}$ be infinite-dimensional. Then $\mathrm{Cliff}^{C^*}(\mathcal{Y})$ is weakly dense in $B(\Gamma_a(\mathcal{Z}))$. Hence, the anti-linear automorphism $A \to c(A)$ cannot be extended from $CAR^{C^*}(\mathcal{Y})$ to a strongly continuous automorphism of $B(\Gamma_a(\mathcal{Z}))$.

Proof It is sufficient to assume that $\mathcal{Z}$ has an o.n. basis $(e_1, e_2, \dots)$. Let $\theta \in \mathbb{R}$. Let $u_n \in U(\mathcal{Z})$ be defined by

$$u_n e_j := \begin{cases} e^{\frac{i2}{n}\theta} e_j, & j = 1, \dots, n; \\ 0, & j = n+1, \dots. \end{cases}$$

One finds that if $e \in \mathcal{Z}$ is a normalized vector, then

$$2ia^*(e)a(e) - \phi(ie, -i\overline{e})\phi(c, \overline{c}) + i.$$

Hence,

$$\Gamma(u_n) = \exp\left(\sum_{j=1}^n \frac{\theta}{n} 2ia^*(e_j)a(e_j)\right) = e^{i\theta} \exp\left(\sum_{j=1}^n \frac{\theta}{n}\phi(ie_j, -i\overline{e_j})\phi(e_j, \overline{e_j})\right).$$

Therefore, $U_n := \Gamma(u_n)e^{-i\theta} \in \text{Cliff}^{C^*}(\mathcal{Y})$. Clearly,

$$s - \lim_{n\to\infty} U_n = e^{i\theta}\mathbb{1}.$$

Consequently, $e^{i\theta}\mathbb{1}$ belongs to the strong closure of $\text{Cliff}^{C^*}(\mathcal{Y})$. Hence, the strong closure of $\text{Cliff}^{C^*}(\mathcal{Y})$ contains $\text{CAR}^{C^*}(\mathcal{Y})$. But $\text{CAR}^{C^*}(\mathcal{Y})$ is strongly dense in $B(\Gamma_a(\mathcal{Z}))$. $\qquad\qquad\square$

16.3.2 Group $Pin_j^c(\mathcal{Y})$

Definition 16.39 *We define $Pin_j^c(\mathcal{Y})$ to be the set of $U \in U(\Gamma_a(\mathcal{Z}))$ such that*

$$\{U\phi(y)U^* : y \in \mathcal{Y}\} = \{\phi(y) : y \in \mathcal{Y}\}.$$

We set

$$Spin_j^c(\mathcal{Y}) := \{U \in Pin_j^c(\mathcal{Y}) : \alpha(U) = U\}.$$

We equip $Pin_j^c(\mathcal{Y})$ with the strong operator topology.

It is obvious that $Pin_j^c(\mathcal{Y})$ is a topological group and $Spin_j^c(\mathcal{Y})$ is its closed subgroup.

The following definitions are parallel to definitions of Sect. 14.3.

Definition 16.40 *Let $A \in B(\Gamma_a(\mathcal{Z}))$ and $r \in O(\mathcal{Y})$.*

(1) *We say that A intertwines r if*

$$A\phi(y) = \phi(ry)A, \quad y \in \mathcal{Y}. \tag{16.50}$$

(2) *If in addition A is unitary then we also say that A implements r.*
(3) *If there exists $U \in U(\Gamma_a(\mathcal{Z}))$ that implements r, then we say that r is implementable in the Fock representation.*

It is clear that the map $Pin_j^c(\mathcal{Y}) \to O(\mathcal{Y})$ defined by (16.50) is a group homomorphism. However, one prefers to use a different homomorphism, arising from the following definition:

Definition 16.41 *Let $r \in O(\mathcal{Y})$.*

(1) *We say that $A \in B(\Gamma_a(\mathcal{Z}))$ α-intertwines $r \in O(\mathcal{Y})$ if*

$$\alpha(A)\phi(y) = \phi(ry)A, \quad y \in \mathcal{Y}.$$

(2) *If in addition A is unitary then we also say that A α-implements r.*
(3) *If there exists $U \in U(\Gamma_a(\mathcal{Z}))$ that α-implements r, then we say that r is α-implementable in the Fock representation.*

We will see in Thm. 16.43 that if r is α-implementable in the Fock representation, then necessarily $r \in O_j(\mathcal{Y})$. Therefore, $\det r$ is well defined by Def. 16.21

and we can introduce the notion of det-implementation, essentially equivalent to α-implementability.

Definition 16.42 *Let $r \in O_j(\mathcal{Y})$.*

(1) *We say that $A \in B(\Gamma_a(\mathcal{Z}))$ det-intertwines r if*

$$A\phi(y) = \det r \, \phi(ry)A, \quad y \in \mathcal{Y}. \qquad (16.51)$$

(2) *If in addition A is unitary, then we also say that A det-implements r.*
(3) *If there exists $U \in U(\Gamma_a(\mathcal{Z}))$ that det-implements r, then we say that r is det-implementable in the Fock representation.*

We will prove

Theorem 16.43 (The Shale–Stinespring theorem about Bogoliubov transformations)

(1) *Let $r \in O(\mathcal{Y})$. The following statements are equivalent:*
  (i) *r is α-implementable in the Fock representation.*
  (ii) *r is det-implementable.*
  (iii) *r is implementable in the Fock representation.*
  (iv) *$r \in O_j(\mathcal{Y})$.*
(2) *Suppose now that $r \in O_j(\mathcal{Y})$. Then the following is true:*
  (i) *There exists $U_r \in \mathrm{Pin}_j^c(\mathcal{Y})$ such that the set of elements of $U(\Gamma_a(\mathcal{Z}))$ α-implementing r consists of operators of the form μU_r with $|\mu| = 1$.*
  (ii) *U_r is even iff $r \in SO_j(\mathcal{Y})$. Otherwise, it is odd. Hence, U_r α-implements r iff it det-implements r.*
  (iii) *The set of elements of $U(\Gamma_a(\mathcal{Z}))$ implementing r consists of operators of the form μU_r with $|\mu| = 1$ if $\det r = 1$ and μU_{-r} with $|\mu| = 1$ if $\det r = -1$.*
  (iv) *If $r_1, r_2 \in O_j(\mathcal{Y})$, then $U_{r_1} U_{r_2} = \mu U_{r_1 r_2}$ for some μ such that $|\mu| = 1$.*
  (v) *If $r_n \to r$ in $O_j(\mathcal{Y})$, then there exist μ_n, $|\mu_n| = 1$, such that $\mu_n U_{r_n} \to U_r$ strongly.*
(3) *Most of the above statements can be summarized by the following commuting diagram of Lie groups and their continuous homomorphisms, where all vertical and horizontal sequences are exact:*

$$
\begin{array}{ccccccccc}
 & & 1 & & 1 & & & & \\
 & & \downarrow & & \downarrow & & & & \\
1 & \to & U(1) & \to & U(1) & \to & 1 & & \\
 & & \downarrow & & \downarrow & & \downarrow & & \\
1 & \to & \mathrm{Spin}_j^c(\mathcal{Y}) & \to & \mathrm{Pin}_j^c(\mathcal{Y}) & \to & \mathbb{Z}_2 & \to & 1 \\
 & & \downarrow & & \downarrow & & \downarrow & & \\
1 & \to & SO_j(\mathcal{Y}) & \to & O_j(\mathcal{Y}) & \to & \mathbb{Z}_2 & \to & 1 \\
 & & \downarrow & & \downarrow & & \downarrow & & \\
 & & 1 & & 1 & & 1 & &
\end{array}
\qquad (16.52)
$$

As a preparation for the proof of the above theorem we will first show the following lemma:

Lemma 16.44 *Let* $r \in O(\mathcal{Y})$. *Then the set of elements of* $A \in B(\Gamma_a(\mathcal{Z}))$ α-*intertwining* r *is either empty or of the form* $\{\mu U : \mu \in \mathbb{C}\}$, *where* U *is unitary. Besides,* U *is even or odd.*

Proof Let $A = A_0 + A_1$ with A_0 even and A_1 odd. Then

$$(A_0 + A_1)\phi(y) = \phi(ry)(A_0 - A_1), \quad y \in \mathcal{Y}. \tag{16.53}$$

Comparing even and odd terms in (16.53), we obtain

$$A_0\phi(y) = \phi(ry)A_0, \quad A_1\phi(y) = -\phi(ry)A_1, \quad y \in \mathcal{Y}. \tag{16.54}$$

Hence, $A_0^* A_0$ and $A_1^* A_1$ commute with $\phi(y), y \in \mathcal{Y}$. Clearly, they are even. Hence, by the irreducibility of the Fock CAR representation, they are proportional to identity. Hence, the operators A_i are proportional to a unitary operator.

(16.54) implies also that $A_1^* A_0$ anti-commutes with $\phi(y), y \in \mathcal{Y}$. By Prop. 13.3, this implies that $A_1^* A_0$ is even. But $A_1^* A_0$ is odd. Hence, $A_1^* A_0 = 0$. Thus one A_i is zero. □

16.3.3 Implementation of partial conjugations

Let $\kappa \in O(\mathcal{Y})$ be an involution with $\mathrm{Ker}(\kappa + 1\!\!1)$ finite. Clearly, $\kappa \in O_1(\mathcal{Y})$. Hence, κ is det-implementable in $\mathrm{Cliff}^{\mathrm{alg}}(\mathcal{Y})$. In fact, if $(e_1, \ldots, e_n)$ is an o.n. basis of $\mathrm{Ker}(\kappa + 1\!\!1)$, then

$$U_\kappa = \phi(e_1) \cdots \phi(e_n) \in \mathrm{Cliff}^{\mathrm{alg}}(\mathcal{Y})$$

det-implements κ.

Recall that $\mathrm{Cliff}^{\mathrm{alg}}(\mathcal{Y})$ can be treated as a sub-algebra of $B(\Gamma_a(\mathcal{Z}))$. Hence, U_κ det-implements κ in the Fock representation.

In the case of the Fock CAR representation one can distinguish a class of orthogonal involutions with special properties – the so-called partial conjugations; see Def. 16.16. Assume now that κ is not only an orthogonal involution, but also a partial conjugation on the Kähler space $\mathcal{Y}$. Let $\mathcal{W} := \frac{1\!\!1 - \mathrm{i} j}{2}\mathrm{Ker}(\kappa + 1\!\!1)$ be the holomorphic subspace associated with $\mathrm{Ker}(\kappa + 1\!\!1)$. It is easy to see that, setting

$$w_j := \frac{1}{2}(e_j - \mathrm{i} j e_j), \ j = 1, \ldots, n,$$

we obtain an o.n. basis of $\mathcal{W}$, and

$$\kappa_{\mathbb{C}} w_j = \overline{w}_j, \quad \kappa_{\mathbb{C}} \overline{w}_j = w_j.$$

Note that in this case U_κ transforms the vacuum into the Slater determinant associated with the subspace $\mathcal{W}$:

$$U_\kappa \Omega = a^*(w_1) \cdots a^*(w_n)\Omega.$$

Moreover, we can easily put U_κ in the Wick-ordered form:

$$U_\kappa = (a^*(w_1) + a(w_1)) \cdots (a^*(w_n) + a(w_n))$$
$$= \sum \operatorname{sgn}(i_1, \ldots, i_k)\, a^*(w_{i_1}) \cdots a^*(w_{i_k}) a(w_{j_1}) \cdots a(w_{j_{n-k}}),$$

where we sum over all $1 \leq i_1 < \cdots < i_k \leq n$, $1 \leq j_1 < \cdots < j_{n-k} \leq n$ with $\{i_1, \ldots, i_k\} \cup \{j_1, \ldots, j_{n-k}\} = \{1, \ldots, n\}$, and $\operatorname{sgn}(i_1, \ldots, i_k)$ is the sign of the permutation $(i_1, \ldots, i_k, j_1, \ldots, j_{n-k})$.

16.3.4 Implementation of j-non-degenerate transformations

For j-non-degenerate orthogonal transformations we can write down a formula for its Bogoliubov implementer that is parallel to that of the bosonic case (11.42).

Theorem 16.45 *Let $r \in O_j(\mathcal{Y})$ be j-non-degenerate. Let p, c, d be defined as in Subsect. 16.1.1. Set*

$$U_r^j = |\det pp^*|^{\frac{1}{4}} e^{\frac{1}{2}a^*(d)} \Gamma((p^*)^{-1}) e^{\frac{1}{2}a(c)}. \tag{16.55}$$

Then U_r^j is the unique unitary operator implementing r such that

$$(\Omega|U_r^j\Omega) > 0. \tag{16.56}$$

We have $\alpha(U_r^j) = U_r^j$. Thus $U_r^j \in Spin_j^c(\mathcal{Y})$.

Proof Let $z \in \mathcal{Z}$. Recall that

$$\Gamma(p)a^*(z)\Gamma(p^{-1}) = a^*(pz), \quad \Gamma(p)a(z)\Gamma(p^{-1}) = a\left(p^{*-1}z\right). \tag{16.57}$$

Using (16.57), (16.42) and (16.43), we obtain

$$U_r^j a^*(z) = \left(a^*(p^{*-1}z - dpc\bar{z}) - a(pc\bar{z})\right)U_r^j$$
$$= \left(a^*(pz) + a(q\bar{z})\right)U_r^j,$$
$$U_r^j a(z) = \left(a(pz) - a^*(d\overline{pz})\right)U_r^j$$
$$= \left(a(pz) + a^*(q\bar{z})\right)U_r^j.$$

Thus U_r^j implements r.

By Lemma 16.44, U_r^j is proportional to a unitary operator. By Thm. 16.36 (1),

$$U_r^j\Omega = \Omega_{-d}$$

is of norm 1. Hence, U_r^j is unitary. Finally, $(\Omega|U_r^j\Omega) = |\det pp^*|^{\frac{1}{4}} > 0$. $\qquad \square$

16.3.5 End of proof of the Shale–Stinespring theorem

In this subsection we finish the proof of the implementability of the restricted orthogonal group.

Lemma 16.46 *Let* $\mathcal{W}$ *be an infinite-dimensional subspace of* $\mathcal{Z}$, *and* $\Psi \in \Gamma_{\mathrm{a}}(\mathcal{Z})$ *such that*

$$a^*(w)\Psi = 0, \quad w \in \mathcal{W}.$$

Then $\Psi = 0$.

Proof We have

$$a(w)a^*(w) = -a^*(w)a(w) + \|w\|^2 \mathbb{1}, \quad w \in \mathcal{Z}.$$

Hence, for any projection π_n of dimension n with range contained in $\mathcal{W}$, we have

$$e^{\mathrm{itd}\Gamma(\pi_n)}\Psi = e^{\mathrm{it}n}\Psi. \tag{16.58}$$

Now suppose that $\dim \mathcal{W} = \infty$. Then we can find a sequence of projections $\pi_n \leq \mathbb{1}_{\mathcal{W}}$ going strongly to an infinite-dimensional projection π. Then the l.h.s. of (16.58) converges to $e^{\mathrm{itd}\Gamma(\pi)}\Psi$ and the r.h.s. has no limit if $\Psi \neq 0$, which is a contradiction. $\qquad\square$

Proof of Thm. 16.43. Let $r \in O_{\mathrm{j}}(\mathcal{Y})$. By Prop. 16.20, r has a finite-dimensional singular space. By Prop. 16.18, it can be represented as a product of a j-non-degenerate transformation and a partial conjugation with a finite dimensional singular space. The former is implementable in $B(\Gamma_{\mathrm{a}}(\mathcal{Z}))$ by Thm. 16.45, and the latter by Subsect. 16.3.3. This proves that r is implementable in $B(\Gamma_{\mathrm{a}}(\mathcal{Z}))$.

Suppose that $r \in O(\mathcal{Y})$ is implemented by $U \in U(\Gamma_{\mathrm{a}}(\mathcal{Z}))$. Let $\Psi := U\Omega$. Note that, for any $z \in \mathcal{Z}$, $a(z)\Omega = 0$ and

$$Ua(z)U^* = a(pz) + a^*(q\bar{z}).$$

Therefore,

$$\bigl(a(pz) + a^*(q\bar{z})\bigr)\Psi = 0, \quad z \in \mathcal{Z}. \tag{16.59}$$

Assume first that r is j-non-degenerate. Then (16.59) implies

$$\bigl(a(z) + a^*(d\bar{z})\bigr)\Psi = 0, \quad z \in \operatorname{Ran} p.$$

Using the fact that $\operatorname{Ran}\bar{p}$ is dense and Thm. 16.36 (2), we obtain that $d \in B^2(\bar{\mathcal{Z}}, \mathcal{Z})$, and hence $r \in O_{\mathrm{j}}(\mathcal{Y})$.

Suppose now that $r \in O(\mathcal{Y})$ is arbitrary. (16.59) yields

$$a^*(z)\Psi = 0, \quad z \in q\operatorname{Ker}\bar{p}. \tag{16.60}$$

By Lemma 16.46, this implies that $q\operatorname{Ker}\bar{p}$ is finite-dimensional. But, for $z \in \operatorname{Ker}\bar{p}$, $\|qz\| = \|z\|$. Hence, $\dim q\operatorname{Ker}\bar{p} = \dim\operatorname{Ker}\bar{p}$. So $\operatorname{Ker} p$ is finite-dimensional.

By Prop. 16.18, we can find a partial conjugation κ such that $r\kappa$ is j-non-degenerate. The dimension of the singular space of κ is $\dim \operatorname{Ker} p$. By Subsect. 16.3.3, κ is implementable. Hence, $r\kappa$ is implementable. By what we have just proven, $r\kappa \in O_j(\mathcal{Y})$. Clearly, $\kappa \in O_j(\mathcal{Y})$. Hence, $r \in O_j(\mathcal{Y})$.

We write r as κr_0, $U_r = U_\kappa U_{r_0}$, where κ is a partial conjugation and r_0 is j-non-degenerate. Then the Gaussian vector $U_{r_0}\Omega$ is an even vector, as seen in the proof of Thm. 16.36. From the form of U_κ given in Subsect. 16.3.3, we see that $U_r\Omega$ is even (resp. odd) if r is such. Hence, $\alpha(U_r) = U_r$ if $\det r = 1$, and $\alpha(U_r) = -U_r$ if $\det r = -1$.

Finally, let us prove (2)(v). Let $r_n \in O_j(\mathcal{Y})$ such that $r_n \to r$. From Thm. 16.43, we know that $U_{r_n r^{-1}} = \pm U_{r_n} U_r^{-1}$. For n large enough, $r_n r^{-1}$ is close to $\mathbb{1}$ in the topology of $O_j(\mathcal{Y})$, hence is j-non-degenerate and belongs to $SO_j(\mathcal{Y})$. From the explicit form of U_r for j-non-degenerate r given in Thm. 16.45, we see that $U_{r_n r^{-1}} \to \mathbb{1}$ strongly, hence $U_{r_n} \to \pm U_r$ strongly. $\qquad\square$

16.3.6 One-parameter groups of Bogoliubov transformations

Let $h \in B_h(\mathcal{Z})$, $g \in B_a^2(\overline{\mathcal{Z}}, \mathcal{Z})$. Let $\zeta = \begin{bmatrix} g & h \\ -h^\# & -\bar{g} \end{bmatrix}$. Recall that

$$\mathrm{Op}^{a^*,a}(\zeta) := 2\mathrm{d}\Gamma(h) + a^*(g) + a(g)$$

is a self-adjoint operator. If in addition $h \in B^1(\mathcal{Z})$, then we can use the anti-symmetric quantization to quantize ζ obtaining

$$\mathrm{Op}(\zeta) := 2\mathrm{d}\Gamma(h) + a^*(g) + a(g) - (\operatorname{Tr} h)\mathbb{1}.$$

Let $a \in o(\mathcal{Y})$ be given by

$$a_\mathbb{C} = \mathrm{i}\zeta_\mathbb{C}\nu_\mathbb{C} = \frac{\mathrm{i}}{2}\begin{bmatrix} h & g \\ -\bar{g} & -h^\# \end{bmatrix};$$

see (16.39). Let $r_t = e^{ta}$ and

$$r_{t\mathbb{C}} = \begin{bmatrix} p_t & q_t \\ \bar{q}_t & \bar{p}_t \end{bmatrix}.$$

For $t \in \mathbb{R}$ such that r_t is non-degenerate, we set

$$d_t := q_t \bar{p}_t^{-1}, \quad c_t := -q_t^\#(p_t^\#)^{-1}.$$

The following formula gives the unitary group generated by $\mathrm{Op}^{a^*,a}(\zeta)$:

Theorem 16.47 (1) *Let $t \in \mathbb{R}$ be such that r_t is non-degenerate. Then $p_t e^{-\mathrm{i}th} - \mathbb{1} \in B^1(\mathcal{Z})$, $d_t, c_t \in B^2(\overline{\mathcal{Z}}, \mathcal{Z})$, and*

$$e^{\mathrm{i}t\mathrm{Op}^{a^*,a}(\zeta)} = \det\left(p_t e^{-\mathrm{i}th}\right)^{\frac{1}{2}} e^{\frac{1}{2}a^*(d_t)}\Gamma(p_t^{*-1})e^{\frac{1}{2}a(c_t)}. \tag{16.61}$$

Besides, (16.61) implements r_t.

(2) *If in addition* $h \in B^1(\mathcal{Z})$, *then*

$$e^{it\mathrm{Op}(\zeta)} = \det p_t^{\frac{1}{2}} e^{\frac{1}{2}a^*(d_t)} \Gamma(p_t^{*-1}) e^{\frac{1}{2}a(c_t)}. \tag{16.62}$$

(In both (16.61) and (16.62) the branch of the square root is determined by continuity.)

16.3.7 Implementation of j-*non-degenerate* j-*positive transformations*

In this subsection we consider a j-non-degenerate j-positive orthogonal transformation r, considered in Subsect. 16.1.3. From formula (16.12) we see that there exists $c \in B_a^2(\overline{\mathcal{Z}}, \mathcal{Z})$ and

$$r_{\mathbb{C}} = \begin{bmatrix} \mathbb{1} & c \\ 0 & \mathbb{1} \end{bmatrix} \begin{bmatrix} (\mathbb{1} + cc^*)^{\frac{1}{2}} & 0 \\ 0 & (\mathbb{1} + c^*c)^{-\frac{1}{2}} \end{bmatrix} \begin{bmatrix} \mathbb{1} & 0 \\ -c^* & \mathbb{1} \end{bmatrix}.$$

By Thm. 16.45, r is then implemented by

$$U_r^{\mathrm{j}} = \det(\mathbb{1} + cc^*)^{-1/4} e^{\frac{1}{2}a^*(c)} \Gamma(\mathbb{1} + cc^*)^{\frac{1}{2}} e^{\frac{1}{2}a(c)}. \tag{16.63}$$

We recall also that $a \in o(\mathcal{Y})$ is j-self-adjoint iff

$$a_{\mathbb{C}} = i \begin{bmatrix} 0 & g \\ g^* & 0 \end{bmatrix}, \tag{16.64}$$

for $g \in Cl_a(\overline{\mathcal{Z}}, \mathcal{Z})$. Clearly, $r = e^a \in O_{\mathrm{j}}(\mathcal{Y})$ iff $a \in B^2(\mathcal{Y})$, i.e. $g \in B_a^2(\overline{\mathcal{Z}}, \mathcal{Z})$. For such a, we obtain an implementable one-parameter group of j-non-degenerate j-positive orthogonal transformations $\mathbb{R} \ni t \mapsto e^{ta} = r_t$. On the quantum level this corresponds to

$$U_{r_t}^{\mathrm{j}} = e^{\frac{1}{2}\left(a^*(g) + a(g)\right)} \tag{16.65}$$

$$= \left(\det \cos(t\sqrt{gg^*})\right)^{\frac{1}{2}} e^{\frac{it}{2}a^*\left(\frac{\tan\sqrt{gg^*}}{\sqrt{gg^*}}g\right)} \Gamma\left(\cos(t\sqrt{gg^*})\right)^{-1} e^{-\frac{it}{2}a\left(\frac{\tan\sqrt{gg^*}}{\sqrt{gg^*}}g\right)}.$$

Clearly, (16.65) is essentially a special case of (16.61).

16.3.8 Pin *group in the Fock representation*

Recall that in Subsect. 14.3.2 for an arbitrary Euclidean space $\mathcal{Y}$ we defined the group $Pin_1(\mathcal{Y})$ satisfying the exact sequence

$$1 \to Pin_1(\mathcal{Y}) \to O_1(\mathcal{Y}) \to \mathbb{Z}_2 \to 1. \tag{16.66}$$

We also defined the group $Pin_1^c(\mathcal{Y})$, which satisfied

$$1 \to Pin_1^c(\mathcal{Y}) \to O_1(\mathcal{Y}) \to U(1) \to 1. \tag{16.67}$$

We had the property

$$1 \to Pin_1(\mathcal{Y}) \to Pin_1^c(\mathcal{Y}) \to U(1) \to 1. \tag{16.68}$$

Recall that both $\mathrm{CAR}^{C^*}(\mathcal{Y})$ and $\mathrm{Cliff}^{C^*}(\mathcal{Y})$ can be embedded in $B(\Gamma_{\mathrm{a}}(\mathcal{Z}))$ (where $\mathcal{Y} = \mathrm{Re}(\mathcal{Z} \oplus \overline{\mathcal{Z}})$). Hence,, we can embed $Pin_1^{\mathrm{c}}(\mathcal{Y})$ and $Pin_1(\mathcal{Y})$ in $U(\Gamma_{\mathrm{a}}(\mathcal{Z}))$. It is natural to ask whether both these groups have natural extensions in the Fock representation.

The group $Pin_{\mathrm{j}}^{\mathrm{c}}(\mathcal{Y})$ defined in Def. 16.39 is, in some sense, the maximal extension of $Pin_1^{\mathrm{c}}(\mathcal{Y})$. In this subsection we will construct the group $Pin_{\mathrm{j,af}}(\mathcal{Y})$, which can be viewed as the maximal extension of $Pin_1(\mathcal{Y})$,

The analog of (16.68) will not however be true for $Pin_{\mathrm{j}}^{\mathrm{c}}(\mathcal{Y})$ and $Pin_{\mathrm{j,af}}(\mathcal{Y})$ if $\mathcal{Y}$ is infinite-dimensional. In fact, in this case the factor group $Pin_{\mathrm{j}}^{\mathrm{c}}(\mathcal{Y})/Pin_{\mathrm{j,af}}(\mathcal{Y})$ is much larger than $U(1)$. In quantum field theory this is responsible for the so-called *anomalies* – symmetries of the classical system that cannot be lifted to the quantum level.

The definition of the group $Pin_{\mathrm{j,af}}(\mathcal{Y})$ is somewhat complicated. We first define its j-non-degenerate elements. Then we use the representation-independent construction, which we discussed in the definition of $Pin_1(\mathcal{Y})$ in Subsect. 14.3.2, to handle j-degenerate elements.

Definition 16.48 *Let $r \in O_{\mathrm{j,af}}(\mathcal{Y})$ be j-non-degenerate and p, c, d be given by Sect. 16.1. We define the pair of operators*

$$U_r = \pm (\det p^*)^{\frac{1}{2}} \mathrm{e}^{\frac{1}{2} a^*(d)} \Gamma(p^{*-1}) \mathrm{e}^{\frac{1}{2} a(c)}. \tag{16.69}$$

$Pin_{\mathrm{j,af}}(\mathcal{Y})$ *is defined as the set of operators $\pm U_t U_s$ in $U(\Gamma_{\mathrm{a}}(\mathcal{Z}))$, where $t \in O_{\mathrm{j,af}}(\mathcal{Y})$ is j-non-degenerate, $s \in O_1(\mathcal{Y})$, $\pm U_t$ is defined as in (16.69) and U_s is defined as in Subsect. 14.3.2. We set $Spin_{\mathrm{j,af}}(\mathcal{Y}) := Pin_{\mathrm{j,af}}(\mathcal{Y}) \cap Spin_{\mathrm{j}}(\mathcal{Y})$.*

Theorem 16.49 $Pin_{\mathrm{j,af}}(\mathcal{Y})$ *is a subgroup of $Pin_{\mathrm{j}}^{\mathrm{c}}(\mathcal{Y})$. $Spin_{\mathrm{j,af}}(\mathcal{Y})$ is a subgroup of $Spin_{\mathrm{j}}^{\mathrm{c}}(\mathcal{Y})$. $Pin_{\mathrm{j}}^{\mathrm{c}}(\mathcal{Y}) \to O_{\mathrm{j}}(\mathcal{Y})$ restricts to a surjective homomorphism $Pin_{\mathrm{j,af}}(\mathcal{Y}) \to O_{\mathrm{j,af}}(\mathcal{Y})$. The pre-image of each $r \subset O_{\mathrm{j,af}}(\mathcal{Y})$ consists of precisely two elements of $Pin_{\mathrm{j,af}}(\mathcal{Y})$ differing by the sign, which will be denoted by $\pm U_r$.*

The above statements can be summarized by the following commuting diagram of groups and their continuous homomorphisms, where all vertical and horizontal sequences are exact:

$$
\begin{array}{ccccccccc}
& & 1 & & 1 & & & & \\
& & \downarrow & & \downarrow & & & & \\
1 & \to & \mathbb{Z}_2 & \to & \mathbb{Z}_2 & \to & 1 & & \\
& & \downarrow & & \downarrow & & \downarrow & & \\
1 & \to & Spin_{\mathrm{j,af}}(\mathcal{Y}) & \to & Pin_{\mathrm{j,af}}(\mathcal{Y}) & \to & \mathbb{Z}_2 & \to & 1 \\
& & \downarrow & & \downarrow & & \downarrow & & \\
1 & \to & SO_{\mathrm{j,af}}(\mathcal{Y}) & \to & O_{\mathrm{j,af}}(\mathcal{Y}) & \to & \mathbb{Z}_2 & \to & 1 \\
& & \downarrow & & \downarrow & & \downarrow & & \\
& & 1 & & 1 & & 1 & &
\end{array} \tag{16.70}
$$

Furthermore, if $r \in O_1(\mathcal{Y})$, then $\pm U_r$ defined in Subsect. 14.3.2 coincides with $\pm U_r$ defined in Thm. 16.49.

The proof of the above theorem is divided into a sequence of steps.

Lemma 16.50 *Suppose that $r \in O_{j,af}$ is j-non-degenerate. Then U_r defined in Subsect. 14.3.2 coincides with $\pm U_r$ defined in (16.69). In particular, we have the following cases:*

(1) *If $w \in U_1(\mathcal{Y}^{\mathbb{C}})$, so that we can write $w_{\mathbb{C}} = \begin{bmatrix} u & 0 \\ 0 & \bar{u} \end{bmatrix}$ for $u \in U_1(\mathcal{Z})$, then*

$$U_w = \pm(\det u)^{\frac{1}{2}} \Gamma(u). \tag{16.71}$$

(2) *If r is j-non-degenerate and j-positive, so that we can write*

$$r_{\mathbb{C}} = \begin{bmatrix} \mathbb{1} & c \\ 0 & \mathbb{1} \end{bmatrix} \begin{bmatrix} p^{-1} & 0 \\ 0 & \bar{p} \end{bmatrix} \begin{bmatrix} \mathbb{1} & 0 \\ -\bar{c}^* & \mathbb{1} \end{bmatrix} \text{ for } c \in B_a^2(\bar{\mathcal{Z}}, \mathcal{Z}), \ p = (\mathbb{1} + cc^*)^{-\frac{1}{2}}, \text{ then}$$

$$\pm U_r = \pm(\det p)^{\frac{1}{2}} e^{\frac{1}{2} a^*(c)} \Gamma(p^{-1}) e^{\frac{1}{2} a(c)} = \pm U_r^j.$$

Proof Consider first (1). We can find $h \in B_h^1(\mathcal{Z})$ such that $u = e^{ih}$. Then $w = e^a$ with $a_{\mathbb{C}} = i \begin{bmatrix} h & 0 \\ 0 & -h^\# \end{bmatrix}$. By Prop. 14.27, $U_w = \pm e^{\frac{1}{4} \mathrm{Op}(a\nu^{-1})}$. But

$$\frac{1}{4}(a\nu^{-1})_{\mathbb{C}} = \frac{i}{2} \begin{bmatrix} 0 & h \\ -h^\# & 0 \end{bmatrix}, \quad \mathrm{Op}(\frac{1}{4} a\nu^{-1}) = i d\Gamma(h) - \frac{i}{2} \mathrm{Tr}\, h.$$

Thus

$$\pm U_w = \pm e^{i d\Gamma(h) - \frac{i}{2} \mathrm{Tr}\, h} = \pm(\det e^{-ih})^{\frac{1}{2}} \Gamma(e^{ih}) = \pm(\det u^*)^{\frac{1}{2}} \Gamma(u).$$

Let us prove (2). Let $r \in O_1(\mathcal{Y})$ be j-non-degenerate and j-positive. We can find $g \in B_a^1(\bar{\mathcal{Z}}, \mathcal{Z})$ such that $a_{\mathbb{C}} = i \begin{bmatrix} 0 & g \\ g^* & 0 \end{bmatrix}$ and $r = e^a$. We have

$$\frac{1}{4}(a\nu^{-1})_{\mathbb{C}} = \frac{i}{2} \begin{bmatrix} g & 0 \\ 0 & g^* \end{bmatrix}, \quad \mathrm{Op}(\frac{1}{4} a\nu^{-1}) = \frac{i}{2}(a^*(g) + a(g)).$$

By Prop. 14.27, $\pm U_r^j = \pm e^{\mathrm{Op}(\frac{1}{4} a\nu^{-1})}$, and by (16.65),

$$\pm e^{\mathrm{Op}(\frac{1}{4} a\nu^{-1})} = \pm(\det p)^{\frac{1}{2}} e^{\frac{1}{2} a^*(c)} \Gamma(p^{-1}) e^{\frac{1}{2} a(c)} = \pm U_r^j.$$

If r is an arbitrary j-non-degenerate element of $O_1(\mathcal{Y})$, by Thm. 16.13, we can write $r = w r_0$ with w unitary and r_0 j-non-degenerate and j-positive. By the proof of Thm. 16.13, $r_0, w \in O_1(\mathcal{Y})$. Then with

$$w_{\mathbb{C}} = \begin{bmatrix} u & 0 \\ 0 & \bar{u} \end{bmatrix}, \quad (r_0)_{\mathbb{C}} = \begin{bmatrix} p_0 & q_0 \\ \bar{q}_0 & \bar{p}_0 \end{bmatrix},$$

we have

$$r_{\mathbb{C}} = (w r_0)_{\mathbb{C}} = \begin{bmatrix} u p_0 & u q_0 \\ \bar{u} \bar{q}_0 & \bar{u} \bar{p}_0 \end{bmatrix}.$$

Using (1) and (2), we obtain

$$\pm U_r = \pm U_w U_{r_0} = \pm (\det u^*)^{\frac{1}{2}} \Gamma(u)(\det p_0)^{\frac{1}{2}} e^{\frac{1}{2}a^*(c_0)} \Gamma(p_0^{-1}) e^{\frac{1}{2}a(c_0)}$$

$$= \pm (\det(up_0)^*)^{\frac{1}{2}} e^{\frac{1}{2}a^*(uc_0)} \Gamma((up_0)^{*-1}) e^{\frac{1}{2}a(c_0)}. \quad (16.72)$$

But $up_0 = p$, $uc_0 = d$ and $c_0 = c$, hence (16.72) coincides with (16.69). □

Lemma 16.51 *Let $r \in O_{j,af}(\mathcal{Y})$ be j-non-degenerate. Then (16.69) defines a pair of even unitary operators differing by a sign implementing r. $\pm U_r$ depends continuously on $r \in O_{j,af}(\mathcal{Y})$, where $O_{j,af}(\mathcal{Y})$ is equipped with its usual topology.*

Proof To see that $\pm U_r$ implements r, it is enough to note that it is proportional to U_r^j defined in (16.55). □

Lemma 16.52 *Let $t, ts \in O_{j,af}(\mathcal{Y})$ be j-non-degenerate and $s \in O_1(\mathcal{Y})$. Then*

$$\pm U_t U_s = \pm U_{ts}, \quad (16.73)$$

where $\pm U_t$, $\pm U_{ts}$ are defined by (16.69) and $\pm U_s$ was defined in Subsect. 14.3.2.

Proof Let $t_n \in O_{j,af}(\mathcal{Y})$ be a sequence convergent in the metric of $O_{j,af}(\mathcal{Y})$ to t. Then, for any $s \in O_{j,af}(\mathcal{Y})$, $t_n s \to ts$ in the same metric.

The set of j-non-degenerate elements is open in $O_{j,af}(\mathcal{Y})$. Therefore, for sufficiently large indices, t_n and $t_n s_n$ are j-non-degenerate. Hence,

$$\pm U_{t_n} \to \pm U_t, \quad \pm U_{t_n s} \to \pm U_{ts}.$$

Therefore, $\pm U_{t_n} U_s \to \pm U_t U_s$.

Suppose in addition that $s \in O_1(\mathcal{Y})$. Since $O_1(\mathcal{Y})$ is dense in $O_{j,af}(\mathcal{Y})$, we can demand that $t_n \in O_1(\mathcal{Y})$ Therefore, $\pm U_{t_n} U_s = \pm U_{t_n s}$. □

Lemma 16.53 *Let $t_1 s_2 = t_2 s_2$, where $t_i \in O_{j,af}(\mathcal{Y})$ are j-non-degenerate and $s_i \in O_1(\mathcal{Y})$. Then*

$$\pm U_{t_1} U_{s_1} = \pm U_{t_2} U_{s_2}, \quad (16.74)$$

where $\pm U_{t_i}$ are defined by (16.69) and $\pm U_{s_i}$ were defined in Subsect. 14.3.2.

Proof We have $t_1(s_1 s_2^\#) = t_2$, and hence, by Lemma 16.52,

$$\pm U_{t_1} U_{s_1 s_2^\#} = \pm U_{t_2}.$$

But $\pm U_{s_1 s_2^\#} = \pm U_{s_1} U_{s_2}^*$, because $s_1, s_2 \in O_1(\mathcal{Y})$. □

Proof of Thm. 16.49. We know by Lemmas 16.27 and 16.7 that every $r \in O_{j,af}(\mathcal{Y})$ can be written as $r = ts$, where $t \in O_{j,af}(\mathcal{Y})$ is j-non-degenerate and $s \in O_1(\mathcal{Y})$. By Lemma 16.53,

$$\pm U_r := \pm U_t U_s,$$

where U_t is defined as in (16.69) and $\pm U_s$ were defined in Subsect. 14.3.3, does not depend on the decomposition.

For $s \in O_{j,af}(\mathcal{Y})$, set

$$\mathcal{U}_s := \{r \in O_{j,af}(\mathcal{Y}) \: : \: rs \text{ is j-non-degenerate}\}.$$

Clearly, $\mathcal{U}_s$ are open in $O_{j,af}(\mathcal{Y})$. Besides, by Lemma 16.51,

$$\mathcal{U}_{\mathbb{1}} \ni r \mapsto \pm U_r \in Pin_{j,af}(\mathcal{Y})/\{\mathbb{1}, -\mathbb{1}\} \tag{16.75}$$

is continuous. Using Def. 16.48 and the continuity of multiplication in $O_{j,af}(\mathcal{Y})$, we see that, for $s \in O_1(\mathcal{Y})$,

$$\mathcal{U}_s \ni r \mapsto \pm U_r \in Pin_{j,af}(\mathcal{Y})/\{\mathbb{1}, -\mathbb{1}\} \tag{16.76}$$

is also continuous. But $\mathcal{U}_s$ with $s \in O_1(\mathcal{Y})$ cover $O_{j,af}(\mathcal{Y})$. Hence,

$$O_{j,af} \ni r \mapsto \pm U_r \in Pin_{j,af}(\mathcal{Y})/\{\mathbb{1}, -\mathbb{1}\} \tag{16.77}$$

is continuous.

We know that

$$\pm U_{r_1} U_{r_2} = \pm U_{r_1 r_2} \tag{16.78}$$

is true for $r_1, r_2 \in O_1(\mathcal{Y})$. But $O_1(\mathcal{Y})$ is dense in $O_{j,af}(\mathcal{Y})$. Hence, (16.78) holds for $r_1, r_2 \in O_{j,af}(\mathcal{Y})$. This proves that $Pin_{j,af}(\mathcal{Y}) \to O_{j,af}(\mathcal{Y})$ is a homomorphism. $\qquad\square$

As an exercise, we give an alternative proof of the group property of $O_{j,af}(\mathcal{Y})$ restricted to j-non-degenerate elements.

Lemma 16.54 *Let* $r = r_1 r_2$ *with* $r_1, r_2 \in O_{j,af}(\mathcal{Y})$. *Assume that* r, r_1, r_2 *are j-non-degenerate. Then*

$$U_{r_1} U_{r_2} = \pm U_{r_1 r_2}. \tag{16.79}$$

Proof We know that

$$(\Omega | U_{r_1 r_2} \Omega) = \pm (\det p^*)^{\frac{1}{2}} = \pm (\det(p_1 p_2 + q_1 \bar{q}_2)^*)^{\frac{1}{2}}. \tag{16.80}$$

Moreover,

$$\begin{aligned}(\Omega | U_{r_1} U_{r_2} \Omega) &= \pm \left(e^{-\frac{1}{2}a^*(c_1)} \Omega | e^{\frac{1}{2}a^*(d_2)} \Omega\right)(\det p_1^*)^{\frac{1}{2}} (\det p_2^*)^{\frac{1}{2}} \\ &= \pm \det(\mathbb{1} + d_2 c_1^*)^{\frac{1}{2}} (\det p_1^*)^{\frac{1}{2}} (\det p_2^*)^{\frac{1}{2}} \\ &= \pm \left(\det(p_1 p_2 + q_1 \bar{q}_2)^*\right)^{\frac{1}{2}}.\end{aligned}$$

Hence,

$$(\Omega | U_{r_1} U_{r_2} \Omega) = \pm (\Omega | U_{r_1 r_2} \Omega).$$

We know that $U_{r_1} U_{r_2}$ and $U_{r_1 r_2}$ implement $r_1 r_2$ and the representation is irreducible. Hence, (16.79) is true. $\qquad\square$

16.4 Fock sector of a CAR representation

The main result of this section is a necessary and sufficient criterion for two Fock CAR representations to be unitarily equivalent. This result goes under the name *Shale–Stinespring theorem*, and is closely related to Thm. 16.43 about the implementability of fermionic Bogoliubov transformations, which can be viewed as another version of the Shale–Stinespring theorem.

Another, closely related, subject of this chapter can be described as follows. We consider a Euclidean space $\mathcal{Y}$ and a CAR representation in a Hilbert space $\mathcal{H}$. We suppose that we are given a Kähler anti-involution j. We will describe how to find a subspace of $\mathcal{H}$ on which this representation is unitarily equivalent to the Fock CAR representation associated with j.

Throughout the section, $(\mathcal{Y}, \nu)$ is a real Hilbert space and j is a Kähler anti-involution on $\mathcal{Y}$.

We use the notation and results of Subsects. 1.3.6, 1.3.8 and 1.3.9. As usual, $\mathcal{Z}, \overline{\mathcal{Z}}$ are the holomorphic and anti-holomorphic subspaces of $\mathbb{C}\mathcal{Y}$. Recall that $\mathcal{Y}$ is identified with $\mathrm{Re}(\mathcal{Z} \oplus \overline{\mathcal{Z}})$ by

$$\mathcal{Y} \ni y \mapsto \left(\frac{1}{2}(y - \mathrm{i}\mathrm{j}y), \frac{1}{2}(y + \mathrm{i}\mathrm{j}y)\right) \in \mathrm{Re}(\mathcal{Z} \oplus \overline{\mathcal{Z}}).$$

We equip $\mathcal{Z}$ with the unitary structure associated with 2ν and j.

16.4.1 Vacua of CAR representations

Let

$$\mathcal{Y} \ni y \mapsto \phi^{\pi}(y) \in B(\mathcal{H})$$

be a representation of CAR over $(\mathcal{Y}, \nu)$. Recall that by complex linearity we extend the definition of $\phi^{\pi}(y)$ to arguments in $\mathbb{C}\mathcal{Y} = \mathcal{Z} \oplus \overline{\mathcal{Z}}$. As in Subsect. 12.1.6, we introduce the creation, resp. annihilation operators $a^{\pi*}(z)$, resp. $a^{\pi}(z)$ by

$$a^{\pi*}(z) := \phi^{\pi}(z), \quad a^{\pi}(z) := \phi^{\pi}(\overline{z}), \quad z \in \mathcal{Z}. \tag{16.81}$$

As in the bosonic case, sometimes we will call them j-*creation*, resp. j-*annihilation operators*.

Definition 16.55 *We define the space of* j-*vacua as*

$$\mathcal{K}^{\pi} := \{\Psi \in \mathcal{H} \ : \ a^{\pi}(z)\Psi = 0, \quad z \in \mathcal{Z}\}.$$

Theorem 16.56 (1) $\mathcal{K}^{\pi}$ *is a closed subspace of* $\mathcal{H}$.
(2) *If* $\Phi, \Psi \in \mathcal{K}^{\pi}$, *then*

$$(\Phi|\phi^{\pi}(y_1)\phi^{\pi}(y_2)\Psi) = (\Phi|\Psi)(y_1 \cdot \nu y_2 - \mathrm{i}y_1 \cdot \nu \mathrm{j}y_2), \quad y_1, y_2 \in \mathcal{Y}.$$

Proof We will suppress the superscript π to simplify notation.

$\mathcal{K}$ is closed as the intersection of kernels of bounded operators. To prove (2), we set $(z_i, \bar{z}_i) = y_i$, so that $\phi(y_i) = a^*(z_i) + a(z_i)$. Using the CAR, we obtain

$$(\Phi|\phi(y_1)\phi(y_2)\Psi) = (z_1|z_2)(\Phi|\Psi).$$

Since $(z_1|z_2) = y_1 \cdot \nu y_2 - iy_1 \cdot \nu j y_2$, we obtain (2).   □

16.4.2 Fock CAR representations

Recall that in Sect. 3.4, for $z \in \mathcal{Z}$, we introduced creation, resp. annihilation operators $a^*(z)$, resp. $a(z)$ acting on the fermionic Fock space $\Gamma_a(\mathcal{Z})$. We have a CAR representation

$$\mathcal{Y} \ni y \mapsto \phi(y) = a^*(z) + a(z) \in B_h(\Gamma_a(\mathcal{Z})), \quad y = (z, \bar{z}). \tag{16.82}$$

As in Def. 13.4, we call (16.82) the *Fock CAR representation*.

Note that j-creation, resp. j-annihilation operators defined for the CAR representation (16.82) coincide with the usual creation, resp. annihilation operators $a^*(z)$, resp. $a(z)$. Likewise, a vector $\Psi \in \Gamma_a(\mathcal{Z})$ is a j-vacuum for (16.82) iff it is proportional to Ω.

We can also consider another Kähler anti-involution j_1, not necessarily equal to j. The following theorem describes the vacua in $\Gamma_a(\mathcal{Z})$ corresponding to j_1. It is essentially a restatement of parts of Thm. 16.36.

Theorem 16.57 (1) *Let $c \in B_a^2(\bar{\mathcal{Z}}, \mathcal{Z})$. Let j_1 be the Kähler anti-involution determined by c, as in Subsect. 16.1.10. Then Ω_c is the unique vector satisfying the following conditions:*

  (i) *$\|\Omega_c\| = 1$,*
  (ii) *$(\Omega_c|\Omega) > 0$,*
  (iii) *Ω_c is a vacuum for j_1.*

(2) *The statement (1)(iii) is equivalent to*

$$(a(z) - a^*(c\bar{z}))\Omega_c = 0, \quad z \in \mathcal{Z}. \tag{16.83}$$

16.4.3 Unitary equivalence of Fock CAR representations

Suppose that we are given a real Hilbert space $(\mathcal{Y}, \nu)$ endowed with two Kähler structures, defined e.g. by two Kähler anti-involutions. Each Kähler structure determines the corresponding Fock CAR representation. In this subsection we will prove a necessary and sufficient condition for the equivalence of these two representations.

Theorem 16.58 (The Shale–Stinespring theorem about Fock representations) *Let $\mathcal{Z}$, $\mathcal{Z}_1$ be the holomorphic subspaces of $\mathbb{C}\mathcal{Y}$ corresponding to the Kähler* *anti-involutions* j *and* j_1. *Let*

$$\mathcal{Y} \ni y \mapsto \phi(y) \in B_{\mathrm{h}}(\Gamma_{\mathrm{a}}(\mathcal{Z})), \tag{16.84}$$

$$\mathcal{Y} \ni y \mapsto \phi_1(y) \in B_{\mathrm{h}}(\Gamma_{\mathrm{a}}(\mathcal{Z}_1)) \tag{16.85}$$

be the corresponding Fock representations of CAR. Then the following statements are equivalent:

(1) *There exists a unitary operator* $W : \Gamma_{\mathrm{a}}(\mathcal{Z}) \to \Gamma_{\mathrm{a}}(\mathcal{Z}_1)$ *such that*

$$W\phi(y) = \phi_1(y)W, \quad y \in \mathcal{Y}. \tag{16.86}$$

(2) $j - j_1$ *is Hilbert–Schmidt (or any of the equivalent conditions of Prop. 16.31 hold).*

Proof Let a_1^*, a_1, Ω_1 denote the creation and annihilation operators and the vacuum for the representation (16.85).

(2)$\Rightarrow$(1). Assume that $j - j_1 \in B^2(\mathcal{Y})$. We know, by Prop. 16.31 (4), that there exists $r \in O_j(\mathcal{Y})$ such that $j_1 = rjr^\#$. Thus, by Thm. 16.43, there exists $U_r \in U(\Gamma_{\mathrm{a}}(\mathcal{Z}))$ such that $U_r\phi(y)U_r^* = \phi(ry)$.

Note that $r_{\mathbb{C}} \in U(\mathbb{C}\mathcal{Y})$ and $r_{\mathbb{C}}\mathcal{Z} = \mathcal{Z}_1$. Set $u := r_{\mathbb{C}}\big|_{\mathcal{Z}}$. Then $u \in U(\mathcal{Z}, \mathcal{Z}_1)$. Note that $\Gamma(u) \in U(\Gamma_{\mathrm{a}}(\mathcal{Z}), \Gamma_{\mathrm{a}}(\mathcal{Z}_1))$, and

$$\Gamma(u)a^*(z)\Gamma(u)^* = a_1^*(uz), \quad \Gamma(u)a(z)\Gamma(u)^* = a_1(uz), \quad z \in \mathcal{Z}.$$

Hence, $\Gamma(u)\phi(y)\Gamma(u)^* = \phi_1(ry)$. Therefore, $W := \Gamma(u)U_r^*$ satisfies (16.86).

(1)$\Rightarrow$(2). Suppose that the representations (16.84) and (16.85) are equivalent with the help of the operator $W \in U(\Gamma_{\mathrm{a}}(\mathcal{Z}_1), \Gamma_{\mathrm{a}}(\mathcal{Z}))$. Let $\mathcal{Y}_{\mathrm{sg}} := \mathrm{Ker}(j + j_1)$ and $\mathcal{Y}_{\mathrm{reg}} := \mathcal{Y}_{\mathrm{sg}}^\perp$. Let $\mathcal{Z}_{\mathrm{sg}} := \frac{\mathbb{1}-\mathrm{ijc}}{2}\mathcal{Y}_{\mathrm{sg}}$, $\mathcal{Z}_{\mathrm{reg}} := \frac{\mathbb{1}-\mathrm{ijc}}{2}\mathcal{Y}_{\mathrm{reg}}$.

Clearly,

$$a_1(w)\Omega_1 = 0, \quad w \in \mathcal{Z}_1,$$

and

$$Wa_1(w)W^* = W\phi_1(\overline{w})W^* = \phi(\overline{w}), \quad w \in \mathcal{Z}_1.$$

Hence,

$$\phi(\overline{w})W\Omega_1 = 0, \quad w \in \mathcal{Z}_1.$$

Hence, in particular,

$$a^*(z)W\Omega_1 = 0, \quad z \in \mathcal{Z}_{\mathrm{sg}}.$$

Lemma 16.46 implies that $\mathcal{Z}_{\mathrm{sg}}$ is finite-dimensional. Let $(w_1, \dots, w_n)$ be an o.n. basis of $\mathcal{Z}_{\mathrm{sg}}$. Set $\Psi := a^*(w_1) \cdots a^*(w_n)W\Omega_1$. Let $c \in Cl_{\mathrm{a}}(\overline{\mathcal{Z}}, \mathcal{Z})$ be defined as in (16.30). Then

$$\big(a(z) - a^*(c\overline{z})\big)\Psi = 0, \quad z \in \mathcal{Z}_{\mathrm{reg}},$$

$$a(z)\Psi = 0, \quad z \in \mathcal{Z}_{\mathrm{sg}}.$$

By Thm. 16.36 (2), this implies that $c \in B^2(\overline{\mathcal{Z}}, \mathcal{Z})$. Hence, $j - j_1 \in B^2(\mathcal{Y})$. $\quad\square$

16.4.4 Fock sector of a CAR representation

Theorem 16.59 *Set*

$$\mathcal{H}^\pi := \mathrm{Span}^{\mathrm{cl}}\Big\{ \prod_{i=1}^{n} a^{\pi*}(z_i)\Psi \ : \ \Psi \in \mathcal{K}^\pi, \ z_1,\ldots,z_n \in \mathcal{Z}, \ n \in \mathbb{N}\Big\}. \quad (16.87)$$

(1) $\mathcal{H}^\pi$ *is invariant under* $\phi^\pi(y)$, $y \in \mathcal{Y}$.

(2) *There exists a unique unitary operator*

$$U^\pi : \mathcal{K}^\pi \otimes \Gamma_{\mathrm{a}}(\mathcal{Z}) \to \mathcal{H}^\pi$$

satisfying

$$U^\pi \Psi \otimes a^*(z_1)\cdots a^*(z_n)\Omega = a^{\pi*}(z_1)\cdots a^{\pi*}(z_n)\Psi, \ \ \Psi \in \mathcal{K}^\pi, \ \ z_1,\ldots,z_n \in \mathcal{Z}.$$

(3)

$$U^\pi \mathbb{1} \otimes \phi(y) = \phi^\pi(y)U^\pi, \ \ y \in \mathcal{Y}.$$

(4) *If there exists an isometry* $U : \Gamma_{\mathrm{a}}(\mathcal{Z}) \to \mathcal{H}$ *such that* $U\phi(y) = \phi^\pi(y)U$, $y \in \mathcal{Y}$, *then* $\mathrm{Ran}\, U \subset \mathcal{H}^\pi$.

(5) $\mathcal{H}^\pi$ *depends on* j *only through its equivalence class w.r.t. the relation*

$$\mathrm{j}_1 \sim \mathrm{j}_2 \ \Leftrightarrow \ \mathrm{j}_1 - \mathrm{j}_2 \in B^2(\mathcal{Y}). \quad (16.88)$$

Definition 16.60 *Introduce the equivalence relation (16.88) in the set of Kähler anti-involutions on* $\mathcal{Y}$. *Let* $[\mathrm{j}]$ *denote the equivalence class w.r.t. this relation. Then the space* $\mathcal{H}^\pi$ *defined in (16.87) is called the* $[\mathrm{j}]$*-Fock sector of the representation* ϕ^π.

Proof of Thm. 16.59. Clearly, $\mathcal{H}^\pi$ is invariant under $\phi^\pi(y)$, $y \in \mathcal{Y}$. We define $U^\pi : \mathcal{K}^\pi \otimes \Gamma_{\mathrm{a}}(\mathcal{Z}) \to \mathcal{H}$ such that the identity in (2) holds. Clearly, U^π is isometric and extends to a unitary map from $\mathcal{K}^\pi \otimes \Gamma_{\mathrm{a}}(\mathcal{Z})$ to $\mathcal{H}^\pi$ satisfying (3). If U is as in (4), then $U\mathbb{C}\Omega \subset \mathcal{K}^\pi$, which shows that $\mathrm{Ran}\, U \subset \mathcal{H}^\pi$. The proof of (5) is identical to the bosonic case. $\square$

As in Subsect. 11.4.4, we have the following proposition:

Proposition 16.61 *Let* j *be a Kähler anti-involution on* $\mathcal{Y}$. *If the CAR representation* ϕ^π *is irreducible and* $\mathcal{K}^\pi \neq \{0\}$, *then* ϕ^π *is unitarily equivalent to the* $[\mathrm{j}]$*-Fock CAR representation.*

16.4.5 Number operator of a CAR representation

As in Subsect. 11.4.5, we discuss the notion of the number operator associated with a CAR representation and a Kähler anti-involution.

Definition 16.62 *We define the* number operator N^π *associated with the CAR representation* ϕ^π *and the Kähler anti-involution* j *by*

$$N^\pi := U^\pi(\mathbb{1} \otimes N)U^{\pi*}, \ \ \mathrm{Dom}\, N := U^\pi \mathcal{K}^\pi \otimes \mathrm{Dom}\, N.$$

As in Subsect. 11.4.5, it is convenient to give an alternative definition of N^π using the number quadratic form.

Definition 16.63 *We define the* number quadratic form n^π *by*

$$n^\pi(\Phi) := \sup_{\mathcal{V}} n^\pi_{\mathcal{V}}(\Phi), \quad \Phi \in \mathcal{H},$$

where $\mathcal{V}$ runs over finite-dimensional subspaces of $\mathcal{Z}$,

$$n^\pi_{\mathcal{V}}(\Phi) := \sum_{i=1}^{\dim \mathcal{V}} \|a^\pi(v_i)\Phi\|^2,$$

$(v_1, \ldots, v_{\dim \mathcal{V}})$ *being an o.n. basis of $\mathcal{V}$.*

Theorem 16.64 *Let n^π be the number quadratic form associated with W^π, j. Then $\operatorname{Dom} n^\pi = \operatorname{Dom}(N^\pi)^{\frac{1}{2}}$ and*

$$n^\pi(\Phi) = (\Phi | N^\pi \Phi), \quad \Phi \in \operatorname{Dom} N^\pi.$$

In particular, $\mathcal{H}^\pi = (\operatorname{Dom} n^\pi)^{\mathrm{cl}}$.

The proof of Thm. 16.64 is completely analogous to Thm. 11.52. Lemmas 11.54 and 11.55 extend to the fermionic case if we replace Lemma 11.53 by the simpler Lemma 16.65 below.

We denote by $\tilde{N}^\pi$ the self-adjoint operator (with a possibly non-dense domain) associated with the quadratic form n^π.

Lemma 16.65 *The operators $a^\pi(z)$ preserve $(\operatorname{Dom} \tilde{N}^\pi)^{\mathrm{cl}}$, and if F is a bounded Borel function, one has*

$$a^\pi(z)F(\tilde{N}^\pi - \mathbb{1}) = F(\tilde{N}^\pi)a^\pi(z), \quad z \in \mathcal{Z}.$$

Proof Let us suppress the superscript π to simplify notation. Considering first the quadratic forms $n_{\mathcal{V}}$ for $\mathcal{V}$ finite-dimensional, we easily obtain

$$n(a(z)\Phi) + n(a^*(z)\Phi) = \|z\|^2 n(\Phi) - 2\|a(z)\Phi\|^2 + \|z\|^2\|\Phi\|^2, \quad \Phi \in \operatorname{Dom} n,$$

which implies that $a(z)$, $a^*(z)$ preserve $\operatorname{Dom}(\tilde{N}^{\frac{1}{2}})$. Similarly, we obtain

$$\left(\Phi \mid \tilde{N}a(z)\Psi\right) = \left(\Phi \mid a(z)\tilde{N}\Psi\right) - \left(\Phi \mid a(z)\Psi\right), \quad \Phi, \Psi \in \operatorname{Dom} \tilde{N}^{\frac{1}{2}}.$$

This implies that $a(z) : \operatorname{Dom} \tilde{N} \to \operatorname{Dom} \tilde{N}$ and

$$a(z)(\tilde{N} - \mathbb{1}) = \tilde{N}a(z). \tag{16.89}$$

From (16.89), we get that $a(z)(\tilde{N} - \lambda\mathbb{1})^{-1} = (\tilde{N} + \mathbb{1} - \lambda\mathbb{1})^{-1}a(z)$, which completes the proof of the lemma. $\qquad\square$

16.5 Notes

The Shale–Stinespring theorem comes from Shale–Stinespring (1964).

Infinite-dimensional analogs of the Pin representation seem to have been first noted by Lundberg (1976).

Among early works describing implementations of orthogonal transformations on Fock spaces let us mention the books by Berezin (1966) and by Friedrichs (1953). They give concrete formulas for the implementation of Bogoliubov transformations in fermionic Fock spaces. Related problems were discussed, often independently, by other researchers, such as Ruijsenaars (1976, 1978).

A comprehensive monograph about the CAR is the book by Plymen–Robinson (1994).

The book by Neretin (1996) and a review article by Varilly–Gracia-Bondia (1994) describe the infinite-dimensional Pin group.

17

Quasi-free states

Suppose that we have a state ψ on the polynomial algebra generated by the fields $\phi(y)$ satisfying the CCR or CAR relations. For simplicity, assume that ψ is *even*, that is, vanishes on odd polynomials. Clearly, this state determines a bilinear form on $\mathcal{Y}$ given by the "2-point function"

$$\mathcal{Y} \times \mathcal{Y} \ni (y_1, y_2) \mapsto \psi\big(\phi(y_1)\phi(y_2)\big). \tag{17.1}$$

We say that a state ψ is *quasi-free* if all expectation values

$$\psi\big(\phi(y_1) \cdots \phi(y_m)\big), \quad y_1, \ldots, y_m \in \mathcal{Y}, \tag{17.2}$$

can be expressed in terms of (17.1) by the sum over all pairings.

This chapter is devoted to a study of (even) quasi-free states, both bosonic and fermionic. This is an important class of states, often used in physical applications. Fock vacuum states belong to this class. It also includes Gibbs states of quadratic Hamiltonians.

Representations obtained by the GNS construction from quasi-free states will be called *quasi-free representations*. They are usually reducible. Many interesting concepts from the theory of von Neumann algebras can be nicely illustrated in terms of quasi-free representations.

Quasi-free states can be easily realized on Fock spaces, using the so-called *Araki–Woods*, resp. *Araki–Wyss representations* in the bosonic, resp. fermionic case. Under some additional assumptions, in particular in the case of a finite number of degrees of freedom, these representations can be obtained as follows. First we consider a Fock space equipped with a quadratic Hamiltonian. Then we perform the GNS construction with respect to the corresponding Gibbs state. Finally, we apply an appropriate Bogoliubov rotation.

The last section of this chapter is devoted to a lattice of von Neumann algebras generated by fields based on real subspaces of the one-particle space. The most interesting result of this section gives a description of the commutant of such an algebra. The proof of this result uses Araki–Woods, resp. Araki–Wyss representations together with the modular theory of von Neumann algebras.

We will extensively use the terminology of the theory of operator algebras, in particular the modular theory of W^*-algebras; see Chap. 6.

17.1 Bosonic quasi-free states

In this section we discuss bosonic quasi-free states. They can be introduced in two different ways: by demanding that n-point functions can be expressed by the 2-point function, or by demanding that their value on Weyl operators is given by a Gaussian function. We choose the latter approach as the basic definition, since it does not involve unbounded operators.

In the literature, in the bosonic case, the name "quasi-free states" is often used to designate a wider class of states, which do not need to be even. For such states the "1-point function"

$$\mathcal{Y} \ni y \mapsto \psi(\phi(y)), \quad y \in \mathcal{Y}, \tag{17.3}$$

may be non-zero and fixes a linear functional on $\mathcal{Y}$. Quasi-free states are then determined by both (17.1) and (17.3). Gaussian coherent states considered in Subsects. 9.1.4 and 11.5.1 are examples of non-even quasi-free states. It is easy to see that an appropriate translation of the fields (see Subsect. 8.1.9) reduces a non-even quasi-free state to an even quasi-free state. Therefore, we will not consider non-even quasi-free states.

17.1.1 Definitions of bosonic quasi-free states

Let $(\mathcal{Y}, \omega)$ be a pre-symplectic space, that is, a real vector space $\mathcal{Y}$ equipped with an anti-symmetric form ω. Recall that $\mathrm{CCR}^{\mathrm{Weyl}}(\mathcal{Y})$ denotes the Weyl CCR algebra, that is, the C^*-algebra generated by operators $W(y)$ satisfying the (Weyl) CCR commutation relations; see Subsect. 8.3.5.

Definition 17.1 (1) *A state ψ on* $\mathrm{CCR}^{\mathrm{Weyl}}(\mathcal{Y})$ *is a* quasi-free state *if there exists $\eta \in L_{\mathrm{s}}(\mathcal{Y}, \mathcal{Y}^{\#})$ (a symmetric form on $\mathcal{Y}$) such that*

$$\psi(W(y)) = \mathrm{e}^{-\frac{1}{2} y \cdot \eta y}, \quad y \in \mathcal{Y}. \tag{17.4}$$

(2) *If $\mathcal{Y} \ni y \mapsto W^{\pi}(y) \in U(\mathcal{H})$ is a CCR representation, a normalized vector $\Psi \in \mathcal{H}$ is called a* quasi-free vector *if*

$$\psi(W(y)) := (\Psi | W^{\pi}(y) \Psi), \quad y \in \mathcal{Y},$$

defines a quasi-free state on $\mathrm{CCR}^{\mathrm{Weyl}}(\mathcal{Y})$.

(3) *A representation $\mathcal{Y} \ni y \mapsto W^{\pi}(y) \in U(\mathcal{H})$ is* quasi-free *if there exists a cyclic quasi-free vector in $\mathcal{H}$.*

(4) *The form η is called the* covariance *of the quasi-free state ψ, and of the quasi-free vector Ψ.*

For a quasi-free state ψ on $\mathrm{CCR}^{\mathrm{Weyl}}(\mathcal{Y})$, let $(\mathcal{H}_{\psi}, \pi_{\psi}, \Omega_{\psi})$ be the corresponding GNS representation. Then, clearly, $\Omega_{\psi} \in \mathcal{H}_{\psi}$ is a quasi-free vector for the CCR representation $\mathcal{Y} \ni y \mapsto \pi_{\psi}(W(y)) \in U(\mathcal{H}_{\psi})$.

The covariance defines the representation uniquely:

Proposition 17.2 *Let $\mathcal{Y} \ni y \mapsto W^i(y) \in U(\mathcal{H}_i)$, $i = 1, 2$, be quasi-free CCR representations with cyclic quasi-free vectors $\Psi_i \in \mathcal{H}_i$, both of covariance η. Then there exists a unique $U \in U(\mathcal{H}_1, \mathcal{H}_2)$ intertwining W^1 with W^2 and satisfying $U\Psi_1 = \Psi_2$.*

Let us note the following important special subclasses of quasi-free representations:

(1) If the pair $(2\eta, \omega)$ is Kähler, the corresponding quasi-free representation is Fock; see Thm. 17.13.
(2) Let $\omega = 0$. Then η can be an arbitrary positive definite form (see Prop. 17.5 below). Without loss of generality we can assume that $\mathcal{Y}$ is complete w.r.t. the scalar product given by η. Let $\mathcal{V} := \mathcal{Y}^\#$, so that $\mathcal{V}$ is a real Hilbert space with the scalar product η^{-1} and the generic variable v. Then the Hilbert space $\mathcal{H}$ can be identified with the Gaussian $\mathbf{L}^2$ space $\mathbf{L}^2(\mathcal{V}, e^{-\frac{1}{2}v \cdot \eta^{-1}v} dv)$, $W(y)$ are the operators of multiplication by $e^{iy \cdot v}$, and the function 1 is the corresponding quasi-free vector.

The following proposition follows from Prop. 8.11:

Proposition 17.3 *Every quasi-free representation is regular.*

We recall that the space $\mathcal{H}^{\infty,\pi}$ associated with a CCR representation W^π is defined in Subsect. 8.2.2. (It is the intersection of domains of products of field operators.)

Proposition 17.4 *A quasi-free vector Ψ for a CCR representation W^π belongs to the subspace $\mathcal{H}^{\infty,\pi}$. Moreover,*

$$\left(\Psi | \phi^\pi(y_1)\phi^\pi(y_2)\Psi\right) = y_1 \cdot \eta y_2 + \frac{i}{2} y_1 \cdot \omega y_2, \quad y_1, y_2 \in \mathcal{Y}. \tag{17.5}$$

Proof We remove the superscript π to simplify notation. For any $y \in \mathcal{Y}$,

$$\left(\Psi | e^{it\phi(y)}\Psi\right) = e^{-\frac{t^2}{2}y \cdot \eta y}. \tag{17.6}$$

Hence, Ψ is an analytic vector for $\phi(y)$. It follows that, for any n, $\Psi \in \text{Dom } \phi(y)^n$, hence $\Psi \in \mathcal{H}^\infty$.

To prove the second statement, we differentiate (17.6) w.r.t. t to get

$$\left(\Psi | \phi(y)^2 \Psi\right) = y \cdot \eta y,$$

which, using linearity and the CCR, implies (17.5). $\qquad\square$

Proposition 17.5 *Let $\eta \in L_s(\mathcal{Y}, \mathcal{Y}^\#)$. Then the following are equivalent:*

(1) $\mathcal{Y} \ni y \mapsto e^{-\frac{1}{2}y \cdot \eta y}$ *is a characteristic function in the sense of Def. 8.10, and hence there exists a quasi-free state satisfying (17.4).*
(2) $\eta_\mathbb{C} + \frac{1}{2}\omega_\mathbb{C} \geq 0$ *on $\mathbb{C}\mathcal{Y}$, where $\eta_\mathbb{C}, \omega_\mathbb{C} \in L(\mathbb{C}\mathcal{Y}, (\mathbb{C}\mathcal{Y})^*)$ are the canonical sesquilinear extensions of η, ω.*

(3) $|y_1 \cdot \omega y_2| \leq 2(y_1 \cdot \eta y_1)^{\frac{1}{2}}(y_2 \cdot \eta y_2)^{\frac{1}{2}}$, $y_1, y_2 \in \mathcal{Y}$.

For the proof we will need the following fact:

Proposition 17.6 *Let* $A = [a_{jk}]$, $B = [b_{jk}] \in B(\mathbb{C}^n)$, *with* A, $B \geq 0$. *Then* $[a_{jk} b_{jk}] =: C \geq 0$.

Proof Writing A and B as sums of positive rank one matrices, it suffices to prove the lemma if A and B are positive of rank one. In this case C is also positive of rank one. □

Corollary 17.7 *Let* $B = [b_{jk}] \in B(\mathbb{C}^n)$ *with* $B \geq 0$. *Then* $[e^{b_{jk}}] \geq 0$.

Proof of Prop. 17.5. We work in the GNS representation and denote by Ψ the corresponding quasi-free vector.

(1) $\Rightarrow$ (2). Using linearity, we deduce from (17.5) that

$$\left(\Psi|\phi(w)^*\phi(w)\Psi\right) = \overline{w}\cdot\eta_\mathbb{C} w + \frac{i}{2}\overline{w}\cdot\omega_\mathbb{C} w, \quad w \in \mathbb{C}\mathcal{Y}. \tag{17.7}$$

It follows that the Hermitian form $\eta_\mathbb{C} + \frac{i}{2}\omega_\mathbb{C}$ is positive semi-definite on $\mathbb{C}\mathcal{Y}$, which proves (2).

Conversely, let $y_1, \ldots, y_n \in \mathcal{Y}$. Set

$$b_{jk} = y_j \cdot \eta y_k + \frac{i}{2}y_j \cdot \omega y_k, \quad j, k = 1, \ldots, n.$$

Then, for $\lambda_1, \ldots, \lambda_n \in \mathbb{C}$,

$$\sum_{1 \leq j,k \leq n} \overline{\lambda_j} b_{jk} \lambda_k = \overline{w}\cdot\eta_\mathbb{C} w + \frac{1}{2}\overline{w}\cdot\omega_\mathbb{C} w, \quad w = \sum_{j=1}^n \lambda_j y_j \in \mathbb{C}\mathcal{Y}.$$

By (2), the matrix $[b_{jk}]$ is positive. By Corollary 17.7, the matrix $[e^{b_{jk}}]$ is positive, and hence the matrix $[e^{-\frac{1}{2}y_j \cdot \eta y_j} b_{jk} e^{-\frac{1}{2}y_k \cdot \eta y_k}]$ is positive. Thus

$$\sum_{j,k=1}^n e^{-\frac{1}{2}(y_k - y_j)\cdot\eta(y_k - y_j)} e^{\frac{i}{2}y_j \cdot \omega y_k} \overline{\lambda_j} \lambda_k$$

$$= \sum_{j,k=1}^n e^{-\frac{1}{2}y_j \cdot \eta y_j} e^{b_{jk}} e^{-\frac{1}{2}y_j \cdot \eta y_j} \overline{\lambda_j} \lambda_k \geq 0.$$

Hence, by Def. 8.10, $\mathcal{Y} \ni y \mapsto e^{-\frac{1}{2}y\cdot\eta y}$ is a characteristic function.

(2) $\Leftrightarrow$ (3). We note that taking complex conjugates (2) implies that

$$\pm\frac{i}{2}\omega_\mathbb{C} \leq \eta_\mathbb{C}, \text{ on } \mathbb{C}\mathcal{Y},$$

or equivalently

$$|\overline{w}_1\cdot\omega_\mathbb{C} w_2| \leq 2(\overline{w}_1\cdot\eta_\mathbb{C} w_1)^{\frac{1}{2}}(\overline{w}_2\cdot\eta_\mathbb{C} w_2)^{\frac{1}{2}}, \quad w_1, w_2 \in \mathbb{C}\mathcal{Y}.$$

For $w_i = y_i \in \mathcal{Y}$, this implies (3).

Conversely, if (3) holds, then

$$2y_1 \cdot \omega y_2 \leq y_1 \cdot \eta y_1 + y_2 \cdot \eta y_2,$$

which, setting $w = y_1 + iy_2$, implies that $\overline{w} \cdot \eta_{\mathbb{C}} w + \frac{i}{2} \overline{w} \cdot \omega_{\mathbb{C}} w \geq 0$. $\qquad\square$

Let ψ be a quasi-free state on $\mathrm{CCR}^{\mathrm{Weyl}}(\mathcal{Y})$, η its covariance and $\mathcal{Y}^{\mathrm{cpl}}$ be the completion of $\mathcal{Y}$ w.r.t. η. Clearly, we can uniquely extend the pre-symplectic form ω to $\mathcal{Y}^{\mathrm{cpl}}$ so that it still satisfies the condition of Prop. 17.5 (3). We can also extend the state ψ uniquely to a quasi-free state on $\mathrm{CCR}^{\mathrm{Weyl}}(\mathcal{Y}^{\mathrm{cpl}})$. Similarly, if W^{π} is a quasi-free CCR representation over $\mathcal{Y}$ satisfying (17.4), we can extend it uniquely to a quasi-free CCR representation over $\mathcal{Y}^{\mathrm{cpl}}$. Therefore, it will not restrict the generality to consider only quasi-free states and representations over $\mathcal{Y}$ complete w.r.t. η. Note, however, that ω may be degenerate on $\mathcal{Y}^{\mathrm{cpl}}$, even if it is non-degenerate on $\mathcal{Y}$.

Proposition 17.8 *Let $\mathcal{Y} \ni y \mapsto W^{\pi}(y) \in U(\mathcal{H})$ be a CCR representation. Let $\Psi \in \mathcal{H}$ be a unit vector. Then the following are equivalent:*

(1) Ψ *is a cyclic quasi-free vector.*
(2) W^{π} *is regular, $\Psi \in \mathcal{H}^{\infty, \pi}$ and, for $y_1, y_2, \ldots \in \mathcal{Y}$,*

$$\big(\Psi | \phi^{\pi}(y_1) \cdots \phi^{\pi}(y_{2m-1}) \Psi\big) = 0,$$

$$\big(\Psi | \phi^{\pi}(y_1) \cdots \phi^{\pi}(y_{2m}) \Psi\big) = \sum_{\sigma \in \mathrm{Pair}_{2m}} \prod_{j=1}^{m} \big(\Psi | \phi^{\pi}(y_{\sigma(2j-1)}) \phi^{\pi}(y_{\sigma(2j)}) \Psi\big).$$

Proof (2) $\Rightarrow$ (1). Let $y \in \mathcal{Y}$. Since the number of elements of Pair_{2m} equals $\frac{1}{2^m} \frac{2m!}{m!}$, we have

$$\big(\Psi | \phi(y)^{2m+1} \Psi\big) = 0, \quad \big(\Psi | \phi(y)^{2m} \Psi\big) = \frac{1}{2^m} \frac{2m!}{m!} (y \cdot \eta y)^m,$$

for

$$y \cdot \eta y = \big(\Psi | \phi^2(y) \Psi\big). \tag{17.8}$$

Using the CCR, we see that the symmetric form η satisfies condition (2) of Prop. 17.5. Moreover, Ψ is an entire vector for $\phi(y)$, and

$$\big(\Psi | e^{i\phi(y)} \Psi\big) = \sum_{m=0}^{\infty} \frac{(-1)^m}{2^m} \frac{1}{m!} (y \cdot \eta y)^m = e^{-\frac{1}{2} y \cdot \eta y}.$$

Hence, Ψ is a quasi-free vector.

(1) $\Rightarrow$ (2). Let Ψ be a quasi-free vector. For $y_1, \ldots, y_n \in \mathcal{Y}$, $t_1, \ldots, t_n \in \mathbb{R}$, we have, using the CCR,

$$\prod_{j=1}^{n} e^{it_j \phi(y_j)} = \exp\Big(-\frac{i}{2} \sum_{1 \leq j < k \leq n} t_j t_k y_j \cdot \omega y_k \Big) \exp\Big(i \sum_{j=1}^{n} t_j \phi(y_j) \Big).$$

Hence,

$$\left(\Psi\Big|\prod_{j=1}^{n} e^{it_j\,\phi(y_j)}\Psi\right) = \exp\left(-\frac{i}{2}\sum_{1\le j<k\le n} t_j t_k\, y_j\cdot\omega y_k\right)\exp\left(-\frac{1}{2}\sum_{1\le j,k\le n} t_j t_k\, y_j\cdot\eta y_k\right)$$

$$= \exp\left(-\sum_{1\le j<k\le n} t_j t_k\left(y_j\cdot\eta y_k+\frac{i}{2}y_j\cdot\omega y_k\right)\right)\exp\left(-\frac{1}{2}\sum_{j=1}^{n} t_j^2\, y_j\cdot\eta y_j\right). \qquad (17.9)$$

From (17.7), we have

$$(\Psi|\phi(y_j)\phi(y_k)\Psi) = y_j\cdot\eta y_k+\frac{i}{2}y_j\cdot\omega y_k =: r_{jk}.$$

Expanding the r.h.s. of (17.9), it follows that $i^n\left(\Psi\Big|\prod_{j=1}^{n}\phi(y_j)\Psi\right)$ is the coefficient of $t_1\cdots t_n$ in the product of the two formal power series

$$\sum_{p\in\mathbb{N}}\frac{1}{p!}\frac{(-1)^p}{2^p}\left(\sum_{j<k}t_j t_k r_{jk}\right)^p\times\sum_{p\in\mathbb{N}}\frac{1}{p!}\frac{(-1)^p}{2^p}\left(\sum_{j=1}^{n}t_j^2\, y_j\cdot\eta y_j\right)^p,$$

or equivalently in the formal power series

$$\sum_{p\in\mathbb{N}}\frac{1}{p!}\frac{(-1)^p}{2^p}\left(\sum_{j<k}t_j t_k r_{jk}\right)^p.$$

If n is odd, this coefficient vanishes. If $n=2m$, the only contributing term is

$$\frac{1}{m!}\frac{(-1)^m}{2^m}\left(\sum_{j<k}t_j t_k r_{jk}\right)^m,$$

which yields the coefficient

$$(-1)^m\sum_{\sigma\in\mathrm{Pair}_{2m}}\prod_{j=1}^{m} r_{\sigma(2j-1)\sigma(2j)},$$

as claimed. $\square$

One could alternatively use the polynomial CCR algebra to describe bosonic quasi-free states. If we want to do this, there is a minor conceptual problem: these algebras are not C^*-algebras, hence strictly speaking the standard definition of a state is no longer valid. Fortunately, it is easy to extend the notion of a state to an arbitrary $*$-algebra by introducing the definition given below.

Definition 17.9 *Let $\mathfrak{A}$ be a unital $*$-algebra. A linear map $\psi:\mathfrak{A}\to\mathbb{C}$ is called a state if for any $A\in\mathfrak{A}$ we have $\psi(A^*A)\ge 0$ and $\psi(\mathbb{1})=1$.*

Note that, given a state on an arbitrary $*$-algebra, the GNS construction can be repeated verbatim from the C^*-algebraic theory.

The following definition is parallel to Def. 17.1 (1):

Definition 17.10 *A state ψ on* $\mathrm{CCR}^{\mathrm{pol}}(\mathcal{Y})$ *is* quasi-free *if*

$$\psi\big(\phi(y_1)\cdots\phi(y_{2m-1})\big) = 0,$$

$$\psi\big(\phi(y_1)\cdots\phi(y_{2m})\big) = \sum_{\sigma\in\mathrm{Pair}_{2m}}\prod_{j=1}^{m}\psi\big(\phi(y_{\sigma(2j-1)})\phi(y_{\sigma(2j)})\big).$$

Clearly, there is an obvious one-to-one correspondence between quasi-free states on $\mathrm{CCR}^{\mathrm{pol}}(\mathcal{Y})$ and quasi-free states on $\mathrm{CCR}^{\mathrm{Weyl}}(\mathcal{Y})$.

17.1.2 Gauge-invariant bosonic quasi-free states

Let $(\mathcal{Y}, \omega)$ be a symplectic space equipped with a pseudo-Kähler anti-involution j. The algebra $\mathrm{CCR}^{\mathrm{Weyl}}(\mathcal{Y})$ is then equipped with the one-parameter group of charge automorphisms, denoted $U(1) \ni \theta \mapsto \widehat{u_\theta}$, defined by

$$\widehat{u_\theta}(W(y)) = W(\mathrm{e}^{\mathrm{j}\theta}y).$$

Definition 17.11 *A state ψ on* $\mathrm{CCR}^{\mathrm{Weyl}}(\mathcal{Y})$ *is called* gauge-invariant *if it is invariant w.r.t. $\widehat{u_\theta}$, that is,*

$$\psi\big(W(y)\big) = \psi\big(W(\mathrm{e}^{\mathrm{j}\theta}y)\big), \qquad y \in \mathcal{Y}, \quad \theta \in U(1). \tag{17.10}$$

In what follows we consider a gauge-invariant quasi-free state ψ with covariance η. Clearly, its gauge-invariance is equivalent to (η, j) being Kähler. (See Prop. 1.95 for a similar statement).

Let us stress that the fact that the two pairs (ω, j) and (η, j) are pseudo-Kähler does not imply that the triple $(\omega, \eta, \mathrm{j})$ is pseudo-Kähler.

Let us introduce the holomorphic space $\mathcal{Z}$ associated with the anti-involution j. Recall that $\mathbb{C}\mathcal{Y} = \mathcal{Z} \oplus \overline{\mathcal{Z}}$. The sesquilinear forms $\omega_{\mathbb{C}}$ and $\eta_{\mathbb{C}}$ can be reduced w.r.t. the direct sum $\mathcal{Z} \oplus \overline{\mathcal{Z}}$. Thus we can write

$$\omega_{\mathbb{C}} = \begin{bmatrix} \omega_{\mathcal{Z}} & 0 \\ 0 & \overline{\omega}_{\mathcal{Z}} \end{bmatrix}, \quad \eta_{\mathbb{C}} = \begin{bmatrix} \eta_{\mathcal{Z}} & 0 \\ 0 & \overline{\eta}_{\mathcal{Z}} \end{bmatrix}, \tag{17.11}$$

where $\eta_{\mathcal{Z}}$ is Hermitian and $\omega_{\mathcal{Z}}$ anti-Hermitian. Note that the condition $\eta_{\mathbb{C}} + \frac{\mathrm{i}}{2}\omega_{\mathbb{C}} \geq 0$, which by Prop. 17.5 is necessary and sufficient for η to be the covariance of a quasi-free state, is equivalent to

$$\eta_{\mathcal{Z}} \pm \frac{\mathrm{i}}{2}\omega_{\mathcal{Z}} \geq 0. \tag{17.12}$$

If the pair (ω, j) is Kähler or, equivalently, $\mathrm{i}\omega_{\mathcal{Z}} \geq 0$, then (17.12) is equivalent to $\eta_{\mathcal{Z}} > \frac{\mathrm{i}}{2}\omega_{\mathcal{Z}}$.

Until the end of the subsection we assume that $(\mathcal{Y}, \omega)$ is a pre-symplectic space and ψ is a quasi-free state on $\mathrm{CCR}^{\mathrm{Weyl}}(\mathcal{Y})$ with covariance η. As explained in Subsect. 17.1.1, without loss of generality we can suppose that $\mathcal{Y}$ is complete for the metric given by η. We will see that under very general conditions there exists a Kähler anti-involution on $\mathcal{Y}$ that makes ψ gauge-invariant.

Theorem 17.12 (1) *Assume that* $\dim \mathrm{Ker}\,\omega$ *is even or infinite. Then there exists an anti-involution* j *such that* ψ *is gauge-invariant for the charge symmetry given by* j.

(2) *If* ω *is symplectic, then the anti-involution* j *described in (1) is unique if we demand in addition that it is Kähler on the symplectic space* $(\mathcal{Y}, \omega)$.

Proof By Prop. 17.5, we see that ω is a bilinear form on the real Hilbert space $(\mathcal{Y}, \eta)$ with norm less than 2. Hence, there exists $b \in B_{\mathrm{a}}(\mathcal{Y})$ (a bounded anti-symmetric operator on $\mathcal{Y}$) with $\|b\| \leq 1$ such that

$$y_1 \cdot \omega y_2 = 2 y_1 \cdot \eta b y_2. \tag{17.13}$$

Set $\mathcal{Y}_{\mathrm{sg}} := \mathrm{Ker}\,b$ and $\mathcal{Y}_{\mathrm{reg}} := \mathcal{Y}_{\mathrm{sg}}^{\perp}$. Since $b = -b^{\#}$, b preserves $\mathcal{Y}_{\mathrm{reg}}$ and we can set $b_{\mathrm{reg}} := b\big|_{\mathcal{Y}_{\mathrm{reg}}}$. From (17.13) we see that $\mathcal{Y}_{\mathrm{reg}}$ and $\mathcal{Y}_{\mathrm{sg}}$ are orthogonal for ω, and that $(\mathcal{Y}_{\mathrm{reg}}, \omega)$ is symplectic. Consider the polar decomposition $b_{\mathrm{reg}} =: -j_{\mathrm{reg}}|b_{\mathrm{reg}}|$ of b_{reg}. Then both $(\eta\big|_{\mathcal{Y}_{\mathrm{reg}}}, j_{\mathrm{reg}})$ and $(\omega\big|_{\mathcal{Y}_{\mathrm{reg}}}, j_{\mathrm{reg}})$ are Kähler. Since $\dim \mathcal{Y}_{\mathrm{sg}}$ is even or infinite, there exists an orthogonal anti-involution j_{sg} on $\mathcal{Y}_{\mathrm{sg}}$. We now set $j := j_{\mathrm{reg}} \oplus j_{\mathrm{sg}}$, which has the required properties. $\qquad\square$

In the proof of the following theorem we will use the material developed in a later part of this section.

Theorem 17.13 *The GNS representation associated with* ψ *is*

(1) *factorial iff* ω *is non-degenerate on* $\mathcal{Y}$,

(2) *irreducible iff* $(2\eta, \omega)$ *is Kähler.*

Proof Set $\mathfrak{M} = \pi_{\psi}(\mathrm{CCR}^{\mathrm{Weyl}}(\mathcal{Y}))''$. We easily see that $\pi_{\psi}(W(y))$ is not proportional to the identity for $y \in \mathcal{Y}\backslash\{0\}$. If $y \in \mathrm{Ker}\,\omega$, then $\pi_{\psi}(W(y)) \in \mathfrak{M} \cap \mathfrak{M}'$. Therefore, if $\mathfrak{M}$ is a factor, then ω is non-degenerate. This proves (1) $\Rightarrow$.

Let us now discuss the GNS representation π_{ψ} when ω is non-degenerate. Let b and j be the operators constructed in the proof of Thm. 17.12. Recall that $b := (2\eta)^{-1}\omega \in B_{\mathrm{a}}(\mathcal{Y})$ and $b = -j|b|$. Let $\mathcal{Z}$ be the corresponding holomorphic subspace. We have

$$\eta z - \frac{\mathrm{i}}{2}\omega z = \eta z - \frac{1}{2}\omega z j z = \eta z (\mathbb{1} - |bz|). \tag{17.14}$$

If we treat our CCR representation as a charged representation in the terminology of the next subsection, then (17.14) can be interpreted as the density ρ; see Def. 17.15.

We split $\mathcal{Y}$ as $\mathcal{Y}_1 \oplus \mathcal{Y}_2$, where

$$\mathcal{Y}_1 := \mathbb{1}_{\{1\}}(|b|)\mathcal{Y}, \quad \mathcal{Y}_2 := \mathbb{1}_{\mathbb{R}\backslash\{1\}}(|b|)\mathcal{Y},$$

and note that $\mathcal{Y}_1$ and $\mathcal{Y}_2$ are orthogonal for η and ω. For $i = 1, 2$ we set $\omega_i = \omega\big|_{\mathcal{Y}_i}$, $\eta_i = \eta\big|_{\mathcal{Y}_i}$, $j_i = j\big|_{\mathcal{Y}_i}$. We denote by ψ_i the quasi-free state on $\mathrm{CCR}^{\mathrm{Weyl}}(\mathcal{Y}_i)$ with covariance η_i, and by $\mathcal{Z}_i \subset \mathcal{Z}$ the holomorphic subspace associated with j_i. We set $\rho_i := \rho\big|_{\mathcal{Z}_i}$.

Note that ω_i are non-degenerate, and that the state ψ on $\mathrm{CCR}^{\mathrm{Weyl}}(\mathcal{Y})$ can be identified to $\psi_1 \otimes \psi_2$ on $\mathrm{CCR}^{\mathrm{Weyl}}(\mathcal{Y}_1) \otimes \mathrm{CCR}^{\mathrm{Weyl}}(\mathcal{Y}_2)$. Therefore, the GNS representation associated with ψ is unitarily equivalent to the tensor product of the GNS representations associated with ψ_1 and ψ_2.

We have $\rho_1 = 0$. Hence, $(2\eta_1, \omega_1)$ is Kähler and the GNS representation associated with ψ_1 is the Fock representation associated with j_1.

Consider the Araki–Woods representation associated with ρ_2 (see Subsect. 17.1.5). By Thm. 17.24 (4), the vacuum Ω is a vector representative for ψ_2. By (17.14), $\mathrm{Ker}\,\rho_2 = \{0\}$, hence Ω is cyclic by Thm. 17.24 (6). Thus the Araki–Woods representation is the GNS representation for ψ_2.

We have $\mathfrak{M} = B\big(\Gamma_{\mathrm{s}}(\mathcal{Z}_1)\big) \otimes \mathrm{CCR}_{\gamma_2,\mathrm{l}}$ (see Def. 17.23). Since by Thm. 17.24 (7), $\mathbb{1} \otimes \mathrm{CCR}_{\gamma_2,\mathrm{r}} \subset \mathfrak{M}'$, we obtain that

$$B\big(\Gamma_{\mathrm{s}}(\mathcal{Z}_1)\big) \otimes B\big(\Gamma_{\mathrm{s}}(\mathcal{Z}_2 \oplus \overline{\mathcal{Z}}_2)\big) \subset B\big(\Gamma_{\mathrm{s}}(\mathcal{Z}_1)\big) \otimes (\mathrm{CCR}_{\gamma_2,\mathrm{l}} \cup \mathrm{CCR}_{\gamma_2,\mathrm{r}})''$$
$$\subset (\mathfrak{M} \cup \mathfrak{M}')'',$$

hence $\mathfrak{M}$ is a factor. This proves (1) $\Leftarrow$.

Now note that the Kähler property implies that ω is non-degenerate. On the other hand, the irreducibility implies the factoriality, which by (1) implies that ω is non-degenerate. Therefore, to prove (2) we can assume the non-degeneracy of ω.

By the discussion above, the GNS representation associated with ψ is equal to the tensor product of the Fock representation associated with $(\mathcal{Y}_1, \omega_1, \mathrm{j}_1)$ and of the Araki–Woods representation associated with $(\mathcal{Z}_2, \rho_2)$, where $\rho_2 > 0$. Every Fock representation is irreducible, while an Araki–Woods representation for a non-zero particle density is not (see Thm. 17.24 (7)). Therefore, the GNS representation associated with ψ is irreducible iff $\mathcal{Y}_2 = \{0\}$ ie. $(2\eta, \omega)$ is Kähler. This proves (2). $\qquad\square$

17.1.3 Quasi-free charged representations

The following subsection is essentially a translation of the previous subsection from the terminology of neutral CCR representation to that of charged CCR representations, which seems more convenient in the context of gauge invariance.

Let $(\mathcal{Y}, \omega)$ be a charged symplectic space. That means the symbols $\mathcal{Y}$ and ω slightly change their meanings compared with the previous subsection: $\mathcal{Y}$ is now a *complex* space and ω is a *charged* symplectic form. To go back to the framework of the previous subsection we need to take the space $\mathcal{Y}_{\mathbb{R}}$, the realification of $\mathcal{Y}$, and equip it with the symplectic form $y_1 \cdot \omega_{\mathbb{R}}\, y_2 := \mathrm{Re}\, y_1 \cdot \omega y_2$, the real part of the charged symplectic form.

Clearly, $\mathcal{Y}$ is equipped with a pseudo-Kähler anti-involution – the imaginary unit. Therefore, all the definitions of the previous subsections make sense. We will write $\mathrm{CCR}^{\mathrm{Weyl}}(\mathcal{Y})$, resp. $\mathrm{CCR}^{\mathrm{pol}}(\mathcal{Y})$ to denote the algebra $\mathrm{CCR}^{\mathrm{Weyl}}(\mathcal{Y}_{\mathbb{R}})$, resp. $\mathrm{CCR}^{\mathrm{pol}}(\mathcal{Y}_{\mathbb{R}})$ equipped with the charge symmetry induced by $U(1) \ni \theta \mapsto \mathrm{e}^{\mathrm{i}\theta}$.

Note that we have a minor notational problem. Throughout our work, we consistently used the letter ψ to denote charged fields. In this chapter this letter is taken (and denotes a state). Therefore, we will use a different letter to denote charged fields – they will be denoted by the letter a, as annihilation operators. In particular, the algebra $\mathrm{CCR}^{\mathrm{pol}}(\mathcal{Y})$ is generated by the operators $a(y)$, $a^*(y)$, $y \in \mathcal{Y}$. Clearly, we can define the concepts of a gauge-invariant state and of a quasi-free state on $\mathrm{CCR}^{\mathrm{pol}}(\mathcal{Y})$.

We also have the corresponding notions on the algebra $\mathrm{CCR}^{\mathrm{Weyl}}(\mathcal{Y})$, generated as usual by $W(y)$, $y \in \mathcal{Y}$. There is a one-to-one correspondence between gauge-invariant quasi-free states on $\mathrm{CCR}^{\mathrm{Weyl}}(\mathcal{Y})$ and $\mathrm{CCR}^{\mathrm{pol}}(\mathcal{Y})$ that can be derived from the formal relation

$$W(y) = \exp\left((\mathrm{i}/\sqrt{2})\big(a^*(y) + a(y)\big)\right).$$

However, when discussing charged CCR relations we prefer to use the polynomial algebra.

Proposition 17.14 (1) *A state ψ on* $\mathrm{CCR}^{\mathrm{pol}}(\mathcal{Y})$ *is gauge-invariant if*

$$\psi\big(a^*(y_1)\cdots a^*(y_n)a(w_m)\cdots a(w_1)\big) = 0, \quad n \neq m, \quad y_1\ldots,y_n,w_m,\ldots,w_1 \in \mathcal{Y}.$$

(2) *It is quasi-free if in addition, for any $y_1\ldots,y_n,w_n,\ldots,w_1 \in \mathcal{Z}$,*

$$\psi\big(a^*(y_1)\cdots a^*(y_n)a(w_n)\cdots a(w_1)\big) = \sum_{\sigma \in S_n} \prod_{j=1}^{n} \psi\big(a^*(y_j)a(w_{\sigma(j)})\big).$$

Definition 17.15 *If ψ is a gauge-invariant quasi-free state on* $\mathrm{CCR}^{\mathrm{pol}}(\mathcal{Y})$, *the positive semi-definite Hermitian form ρ on $\mathcal{Y}$ defined by*

$$(y_2|\rho y_1) := \psi\big(a^*(y_1)a(y_2)\big), \quad y_1,y_2 \in \mathcal{Y},$$

is called the density *associated with ψ. If $\mathrm{i}\omega$ is positive definite, we will also use the alternative name* one-particle density.

Recall that in the framework of neutral CCR relations one introduces the holomorphic space $\mathcal{Z}$. Charged CCR relations amount to identifying the space $\mathcal{Y}$ with $\mathcal{Z}$, as explained e.g. in Subsect. 8.2.5. Under this identification, the Hermitian form $\mathrm{i}\omega$ is transformed into $\mathrm{i}\omega_{\bar{\mathcal{Z}}}$, and the density ρ into $\eta_{\bar{\mathcal{Z}}} - \frac{1}{2}\omega_{\bar{\mathcal{Z}}}$ (see (17.11)). Therefore, (17.12) implies the following proposition.

Proposition 17.16 *A Hermitian form $\rho \in L_{\mathrm{h}}(\mathcal{Y},\mathcal{Y}^*)$ is the density of a gauge-invariant quasi-free state iff*

$$\rho \geq 0, \quad \rho + \mathrm{i}\omega \geq 0.$$

Assume that

$$\mathcal{Y} \ni y \mapsto a^{\pi*}(y) \in Cl(\mathcal{H}) \tag{17.15}$$

is a charged CCR representation. We have the obvious analogs of Def. 17.1 (2) and (3):

Definition 17.17 (1) $\Psi \in \mathcal{H}$ *is called a* gauge-invariant quasi-free vector *if* $\Psi \in \mathcal{H}^{\infty,\pi}$ *and*

$$\psi\big(a^*(y_1)\cdots a^*(y_n)a(w_1)\cdots a(w_m)\big)$$
$$:= \big(\Psi|a^{\pi*}(y_1)\cdots a^{\pi*}(y_n)a^\pi(w_m)\cdots a^\pi(w_1)\Psi\big), \quad y_1,\ldots,y_n,w_m,\ldots,w_1 \in \mathcal{Y},$$

defines a gauge-invariant quasi-free state on $\mathrm{CCR}^{\mathrm{pol}}(\mathcal{Y})$.

(2) *A charged CCR representation (17.15) is called* gauge-invariant quasi-free *if there exists a cyclic gauge-invariant quasi-free vector in* $\mathcal{H}$.

Recall that with every charged CCR representation (17.15) we can associate a unique regular neutral CCR representation

$$\mathcal{Y}_{\mathbb{R}} \ni y \mapsto W^\pi(y) \in U(\mathcal{H}) \tag{17.16}$$

such that

$$W^\pi(y) = \exp\Big((\mathrm{i}/\sqrt{2})\big(a^{\pi*}(y) + a^\pi(y)\big)\Big).$$

It is clear that a vector Ψ is gauge-invariant quasi-free w.r.t. W^π iff it is such w.r.t. $a^{\pi*}$. Likewise, the representation W^π is gauge-invariant quasi-free iff $a^{\pi*}$ is.

17.1.4 Gibbs states of bosonic quadratic Hamiltonians
Density matrix

Let $0 \leq \gamma \leq \mathbb{1}$ be a self-adjoint operator on a Hilbert space $\mathcal{Z}$ with $\mathrm{Ker}(\mathbb{1} - \gamma) = \{0\}$. We associate with γ the self-adjoint operator ρ, called the *one-particle density*, defined by

$$\rho := \gamma(\mathbb{1} - \gamma)^{-1}, \quad \gamma = \rho(\rho + \mathbb{1})^{-1}. \tag{17.17}$$

We assume in addition that γ is trace-class. This is equivalent to assuming that ρ is trace-class. Note the following identity:

$$\mathrm{Tr}\,\Gamma(\gamma) = \det(\mathbb{1} - \gamma)^{-1} = \det(\mathbb{1} + \rho).$$

Thus $\Gamma(\gamma)\det(\mathbb{1} - \gamma)$ is a density matrix (see Def. 2.41).

Definition 17.18 *The state* ψ_γ *on* $B\big(\Gamma_{\mathrm{s}}(\mathcal{Z})\big)$ *is defined by*

$$\psi_\gamma(A) := \mathrm{Tr}\,A\Gamma(\gamma)\det(\mathbb{1} - \gamma), \quad A \subset B\big(\Gamma_{\mathrm{s}}(\mathcal{Z})\big).$$

We identify $\mathcal{Z}$ with $\mathrm{Re}(\mathcal{Z} \oplus \overline{\mathcal{Z}})$ using the usual map $z \mapsto \frac{1}{\sqrt{2}}(z + \overline{z})$. We can faithfully represent the Weyl CCR algebra $\mathrm{CCR}^{\mathrm{Weyl}}(\mathcal{Z})$ in $B\big(\Gamma_{\mathrm{s}}(\mathcal{Z})\big)$. Note that we have a natural charge symmetry on $B\big(\Gamma_{\mathrm{s}}(\mathcal{Z})\big)$ leaving invariant $\mathrm{CCR}^{\mathrm{Weyl}}(\mathcal{Z})$, implemented by $U(1) \ni \theta \mapsto \mathrm{e}^{\mathrm{i}\theta N}$.

Proposition 17.19 *The state ψ_γ restricted to $\mathrm{CCR}^{\mathrm{Weyl}}(\mathcal{Z})$ is gauge-invariant quasi-free. We have*

$$\psi_\gamma\big(W(z)\big) = \exp\Big(-\frac{1}{4}(z|z) - \frac{1}{2}(z|\rho z)\Big) = \exp\Big(-\frac{1}{4}\Big(z\big|\frac{1+\gamma}{1-\gamma}z\Big)\Big), \quad z \in \mathcal{Z}.$$

The "2-point functions" are

$$\psi_\gamma\big(a^*(z_1)a(z_2)\big) = (z_2|\rho z_1),$$
$$\psi_\gamma\big(a(z_1)a^*(z_2)\big) = (z_1|z_2) + (z_1|\rho z_2),$$
$$\psi_\gamma\big(a(z_1)a(z_2)\big) = \psi_\gamma\big(a^*(z_1)a^*(z_2)\big) = 0, \quad z_1, z_2 \in \mathcal{Z}.$$

Proof We can find an o.n. basis $(e_1, e_2, \dots)$ diagonalizing the trace-class operator γ. Using the identification

$$\Gamma_{\mathrm{s}}(\mathcal{Z}) \simeq \overset{\infty}{\underset{i=1}{\otimes}} \big(\Gamma_{\mathrm{s}}(\mathbb{C}e_i), \Omega\big), \tag{17.18}$$

we can confine ourselves to the case of one degree of freedom, which is a well-known computation involving summing up a geometric series. $\square$

Suppose now that γ is non-degenerate. In this case, the state ψ_γ is faithful. If in addition we fix $\beta > 0$ and $\gamma = \mathrm{e}^{-\beta h}$ for some operator h bounded from below, then

$$\Gamma(\gamma)\det(\mathbb{1}-\gamma) = \mathrm{e}^{-\beta \mathrm{d}\Gamma(h)}/\mathrm{Tr}\,\mathrm{e}^{-\beta \mathrm{d}\Gamma(h)}.$$

Thus, in this case, ψ_γ is the Gibbs state at the inverse temperature β for the dynamics generated by the Hamiltonian $\mathrm{d}\Gamma(h)$.

Standard representations on Hilbert–Schmidt operators

Consider the Hilbert space $B^2\big(\Gamma_{\mathrm{s}}(\mathcal{Z})\big)$. It will be convenient to introduce an alternative notation for the Hermitian conjugation: $JB := B^*$.

Recall the representations of $B\big(\Gamma_{\mathrm{s}}(\mathcal{Z})\big)$ and $\overline{B\big(\Gamma_{\mathrm{s}}(\mathcal{Z})\big)}$ on $B^2\big(\Gamma_{\mathrm{s}}(\mathcal{Z})\big)$ introduced in Subsect. 6.4.5:

$$\pi_{\mathrm{l}}(A)B = AB, \quad \pi_{\mathrm{r}}(\overline{A})B := BA^*, \quad B \in B^2\big(\Gamma_{\mathrm{s}}(\mathcal{Z})\big), \quad A \in B\big(\Gamma_{\mathrm{s}}(\mathcal{Z})\big).$$

Clearly, $J\pi_{\mathrm{l}}(A)J^* = \pi_{\mathrm{r}}(A)$.

Thus we can introduce two commuting charged CCR representations,

$$\mathcal{Z} \ni z \mapsto \pi_{\mathrm{l}}\big(a^*(z)\big) \in Cl\big(B^2\big(\Gamma_{\mathrm{s}}(\mathcal{Z})\big)\big), \tag{17.19}$$
$$\overline{\mathcal{Z}} \ni \overline{z} \mapsto \pi_{\mathrm{r}}\big(\overline{a^*(z)}\big) \in Cl\big(B^2\big(\Gamma_{\mathrm{s}}(\mathcal{Z})\big)\big). \tag{17.20}$$

They are interchanged by the operator J:

$$J\pi_{\mathrm{l}}\big(a^*(z)\big)J^* = \pi_{\mathrm{r}}\big(\overline{a^*(z)}\big).$$

The vector

$$\Psi_\gamma := \det(\mathbb{1}-\gamma)^{\frac{1}{2}}\Gamma(\gamma^{\frac{1}{2}})$$

is gauge-invariant quasi-free for the representations (17.19) and (17.20) and the one-particle density ρ. Both (17.19) and (17.20) are gauge-invariant quasi-free charged CCR representations.

Standard representations on the double Fock spaces

Note the following chain of identifications:

$$B^2\big(\Gamma_s(\mathcal{Z})\big) \simeq \Gamma_s(\mathcal{Z}) \otimes \overline{\Gamma_s(\mathcal{Z})}$$
$$\simeq \Gamma_s(\mathcal{Z}) \otimes \Gamma_s(\overline{\mathcal{Z}}) \simeq \Gamma_s(\mathcal{Z} \oplus \overline{\mathcal{Z}}). \tag{17.21}$$

We denote by $T_s : B^2\big(\Gamma_s(\mathcal{Z})\big) \to \Gamma_s(\mathcal{Z} \oplus \overline{\mathcal{Z}})$ the unitary map given by (17.21). Introduce the anti-unitary map

$$\mathcal{Z} \oplus \overline{\mathcal{Z}} \ni (z_1, \overline{z}_2) \mapsto \epsilon(z_1, \overline{z}_2) := (z_2, \overline{z_1}) \in \mathcal{Z} \oplus \overline{\mathcal{Z}}. \tag{17.22}$$

Proposition 17.20 $T_s J T_s^* = \Gamma(\epsilon)$.

By applying T_s to (17.19) and (17.20), we obtain two new commuting charged CCR representations

$$\mathcal{Z} \ni z \mapsto T_s \pi_l\big(a^*(z)\big) T_s^* = a^*(z, 0) \in Cl\big(\Gamma_s(\mathcal{Z} \oplus \overline{\mathcal{Z}})\big), \tag{17.23}$$
$$\overline{\mathcal{Z}} \ni \overline{z} \mapsto T_s \pi_r\big(a^*(z)\big) T_s^* = a^*(0, \overline{z}) \in Cl\big(\Gamma_s(\mathcal{Z} \oplus \overline{\mathcal{Z}})\big). \tag{17.24}$$

They are interchanged by the operator $\Gamma(\epsilon)$:

$$\Gamma(\epsilon) a^*(z, 0) \Gamma(\epsilon)^* = a^*(0, \overline{z}).$$

Again using a basis diagonalizing γ, as in (17.18), the double Fock space on the right of (17.21) can be written as an infinite tensor product,

$$\overset{\infty}{\underset{i-1}{\otimes}} \big(\Gamma_s(\mathbb{C}e_i \oplus \mathbb{C}\overline{e}_i), \Omega\big). \tag{17.25}$$

We have $T_s \Psi_\gamma = \Omega_\gamma$, where

$$\Omega_\gamma := \overset{\infty}{\underset{i=1}{\otimes}} (1 - \gamma_i)^{\frac{1}{2}} e^{\gamma_i^{\frac{1}{2}} a^*(e_i) a^*(\overline{e}_i)} \Omega$$

is a gauge-invariant quasi-free vector for the representations (17.23) and (17.24), and the one-particle density ρ. Clearly, both (17.23) and (17.24) are gauge-invariant quasi-free CCR representations.

Note that if we set

$$c = \begin{bmatrix} 0 & \gamma^{\frac{1}{2}} \\ \overline{\gamma}^{\frac{1}{2}} & 0 \end{bmatrix} \in B_s^2(\overline{\mathcal{Z}} \oplus \mathcal{Z}, \mathcal{Z} \oplus \overline{\mathcal{Z}}), \tag{17.26}$$

then

$$\Omega_\gamma = \det(\mathbb{1} - cc^*)^{\frac{1}{4}} e^{\frac{1}{2} a^*(c)} \Omega,$$

so this is an example of a bosonic Gaussian vector introduced in (11.33), where it was denoted Ω_c.

Araki–Woods form of standard representation

Using the infinite tensor product decomposition (17.25), we define the following transformation on $\Gamma_{\mathrm{s}}(\mathcal{Z} \oplus \overline{\mathcal{Z}})$:

$$R_\gamma := \overset{\infty}{\underset{i=1}{\otimes}} (1 - \gamma_i)^{\frac{1}{2}} e^{-\gamma_i^{\frac{1}{2}} a^*(e_i) a^*(\overline{e}_i)} \Gamma\big((1 - \gamma_i)^{\frac{1}{2}} \mathbb{1}\big) e^{\gamma_i^{\frac{1}{2}} a(e_i) a(\overline{e}_i)}. \quad (17.27)$$

Theorem 17.21 R_γ *is a unitary operator satisfying*

$$R_\gamma \phi(z_1, \overline{z}_2) R_\gamma^* = \phi\big((\rho + \mathbb{1})^{\frac{1}{2}} z_1 + \rho^{\frac{1}{2}} z_2, \overline{\rho}^{\frac{1}{2}} \overline{z}_1 + (\overline{\rho} + \mathbb{1})^{\frac{1}{2}} \overline{z}_2\big),$$

$$R_\gamma a(z_1, \overline{z}_2) R_\gamma^* = a\big((\rho + \mathbb{1})^{\frac{1}{2}} z_1 + \rho^{\frac{1}{2}} z_2, 0\big)$$
$$+ a^*\big(0, \overline{\rho}^{\frac{1}{2}} \overline{z}_1 + (\overline{\rho} + \mathbb{1})^{\frac{1}{2}} \overline{z}_2\big),$$

$$R_\gamma a^*(z_1, \overline{z}_2) R_\gamma^* = a^*\big((\rho + \mathbb{1})^{\frac{1}{2}} z_1 + \rho^{\frac{1}{2}} z_2, 0\big)$$
$$+ a\big(0, \overline{\rho}^{\frac{1}{2}} \overline{z}_1 + (\overline{\rho} + \mathbb{1})^{\frac{1}{2}} \overline{z}_2\big), \quad (z_1, \overline{z}_2) \in \mathcal{Z} \oplus \overline{\mathcal{Z}},$$

$$R_\gamma \Gamma(\epsilon) R_\gamma^* = \Gamma(\epsilon),$$

$$R_\gamma \Omega_\gamma = \Omega,$$

$$R_\gamma \mathrm{d}\Gamma(h, -\overline{h}) R_\gamma^* = \mathrm{d}\Gamma(h, -\overline{h}).$$

Proof Let c be defined as in (17.26). Using

$$\Gamma(\mathbb{1} - cc^*) = \Gamma\big((\mathbb{1} - \gamma) \oplus (\mathbb{1} - \overline{\gamma})\big),$$

we see that

$$R_\gamma := \det(\mathbb{1} - cc^*)^{\frac{1}{4}} e^{-\frac{1}{2} a^*(c)} \Gamma(\mathbb{1} - cc^*)^{\frac{1}{2}} e^{\frac{1}{2} a(c)}.$$

Thus R_γ is an example of an operator whose properties we studied in detail in Sect. 11.3. Thus all the identities that we need to show follow from the fact that R_γ is a unitary operator implementing a positive symplectic map given in (11.50). $\square$

By applying R_γ to (17.23) and (17.24), we obtain two new commuting charged CCR representations

$$\mathcal{Z} \ni z \mapsto a_{\gamma,\mathrm{l}}^*(z) := R_\gamma a^*(z, 0) R_\gamma^*$$
$$= a^*\big((\rho + \mathbb{1})^{\frac{1}{2}} z, 0\big) + a\big(0, \overline{\rho}^{\frac{1}{2}} \overline{z}\big) \in Cl\big(\Gamma_{\mathrm{s}}(\mathcal{Z} \oplus \overline{\mathcal{Z}})\big),$$

$$\overline{\mathcal{Z}} \ni \overline{z} \mapsto a_{\gamma,\mathrm{r}}^*(\overline{z}) := R_\gamma a^*(0, \overline{z}) R_\gamma^*$$
$$= a\big(\rho^{\frac{1}{2}} z, 0\big) + a^*\big(0, (\overline{\rho} + \mathbb{1})^{\frac{1}{2}} \overline{z}\big) \in Cl\big(\Gamma_{\mathrm{s}}(\mathcal{Z} \oplus \overline{\mathcal{Z}})\big).$$

They are interchanged by the operator $\Gamma(\epsilon)$:

$$\Gamma(\epsilon) a_{\gamma,\mathrm{l}}^*(z) \Gamma(\epsilon)^* = a_{\gamma,\mathrm{r}}^*(\overline{z}).$$

We have $R_\gamma \Omega_\gamma = \Omega$, hence the Fock vacuum Ω is a gauge-invariant quasi-free vector for the representations $a_{\gamma,\mathrm{l}}^*$ and $a_{\gamma,\mathrm{r}}^*$, and the one-particle density ρ. Both $a_{\gamma,\mathrm{l}}^*$ and $a_{\gamma,\mathrm{r}}^*$ are gauge-invariant quasi-free CCR representations. They are special cases of *Araki–Woods CCR representations*, which we will consider in the next subsection.

17.1.5 Araki–Woods representations

In this subsection we will see that $a^*_{\gamma,\mathrm{l}}$ and $a^*_{\gamma,\mathrm{r}}$ can be defined more generally, as compared with the framework of the previous subsection.

Let $\mathcal{Z}$ be a Hilbert space. We introduce the operators γ and ρ as at the beginning of Subsect. 17.1.4, except that we do not assume that they are trace-class.

Definition 17.22 *For $z \in \mathrm{Dom}\,\rho^{\frac{1}{2}}$ we define the following closed operators on $\Gamma_{\mathrm{s}}(\mathcal{Z} \oplus \overline{\mathcal{Z}})$, called the* Araki–Woods creation operators:

$$a^*_{\gamma,\mathrm{l}}(z) := a^*\left((\rho+\mathbb{1})^{\frac{1}{2}}z,0\right) + a\left(0,\overline{\rho}^{\frac{1}{2}}\overline{z}\right),$$

$$a^*_{\gamma,\mathrm{r}}(\overline{z}) := a^*\left(\rho^{\frac{1}{2}}z,0\right) + a\left(0,(\overline{\rho}+\mathbb{1})^{\frac{1}{2}}\overline{z}\right).$$

For completeness, let us write down the adjoints of $a^*_{\gamma,\mathrm{l/r}}(z)$, called the *Araki–Woods annihilation operators:*

$$a_{\gamma,\mathrm{l}}(z) := a\left((\rho+\mathbb{1})^{\frac{1}{2}}z,0\right) + a^*\left(0,\overline{\rho}^{\frac{1}{2}}\overline{z}\right),$$

$$a_{\gamma,\mathrm{r}}(\overline{z}) := a\left(\rho^{\frac{1}{2}}z,0\right) + a^*\left(0,(\overline{\rho}+\mathbb{1})^{\frac{1}{2}}\overline{z}\right).$$

We also have the *Araki–Woods Weyl operators:*

$$W_{\gamma,\mathrm{l}}(z) := \exp\left((\mathrm{i}/\sqrt{2})\left(a^*_{\gamma,\mathrm{l}}(z) + a_{\gamma,\mathrm{l}}(z)\right)\right) = W\left((\mathbb{1}+\rho)^{\frac{1}{2}}z,\overline{\rho}^{\frac{1}{2}}\overline{z}\right),$$

$$W_{\gamma,\mathrm{r}}(\overline{z}) := \exp\left((\mathrm{i}/\sqrt{2})\left(a^*_{\gamma,\mathrm{r}}(z) + a_{\gamma,\mathrm{r}}(z)\right)\right) = W\left(\rho^{\frac{1}{2}}z,(\mathbb{1}+\overline{\rho})^{\frac{1}{2}}\overline{z}\right).$$

Definition 17.23 *The von Neumann algebra generated by*

$$\left\{W_{\gamma,\mathrm{l}}(z) \;:\; z \in \mathrm{Dom}\,\rho^{\frac{1}{2}}\right\} \quad resp. \quad \left\{W_{\gamma,\mathrm{r}}(\overline{z}) \;:\; z \in \mathrm{Dom}\,\rho^{\frac{1}{2}}\right\}$$

will be denoted by $\mathrm{CCR}_{\gamma,\mathrm{l}}$, *resp.* $\mathrm{CCR}_{\gamma,\mathrm{r}}$, *and called the* left, *resp.* right *Araki–Woods CCR algebra.*

Theorem 17.24 (1) *The map*

$$\mathcal{Z} \ni z \mapsto a^*_{\gamma,\mathrm{l}}(z) \in B\left(\Gamma_{\mathrm{s}}(\mathcal{Z}\oplus\overline{\mathcal{Z}})\right)$$

is a charged CCR representation. In particular,

$$[a_{\gamma,\mathrm{l}}(z_1), a^*_{\gamma,\mathrm{l}}(z_2)] = (z_1|z_2)\mathbb{1},$$
$$[a^*_{\gamma,\mathrm{l}}(z_1), a^*_{\gamma,\mathrm{l}}(z_2)] = [a_{\gamma,\mathrm{l}}(z_1), a_{\gamma,\mathrm{l}}(z_2)] = 0.$$

It will be called the left Araki–Woods (charged CCR) representation.
(2) *The map*

$$\overline{\mathcal{Z}} \ni \overline{z} \mapsto a^*_{\gamma,\mathrm{r}}(\overline{z}) \in B\left(\Gamma_{\mathrm{s}}(\mathcal{Z}\oplus\overline{\mathcal{Z}})\right)$$

is a charged CCR representation. In particular,

$$[a_{\gamma,\mathrm{r}}(\overline{z}_1), a^*_{\gamma,\mathrm{r}}(\overline{z}_2)] = \overline{(z_1|z_2)}\mathbb{1},$$
$$[a^*_{\gamma,\mathrm{r}}(\overline{z}_1), a^*_{\gamma,\mathrm{r}}(\overline{z}_2)] = [a_{\gamma,\mathrm{r}}(\overline{z}_1), a_{\gamma,\mathrm{r}}(\overline{z}_2)] = 0.$$

It will be called the right Araki–Woods (charged CCR) representation.
(3) *Set*

$$J_s := \Gamma(\epsilon). \tag{17.28}$$

Then

$$J_s a_{\gamma,l}^*(z) J_s = a_{\gamma,r}^*(\overline{z}),$$
$$J_s a_{\gamma,l}(z) J_s = a_{\gamma,r}(\overline{z}).$$

(4) *The vacuum Ω is a bosonic quasi-free vector for $a_{\gamma,l}^*$ with the 2-point functions equal to*

$$(\Omega | a_{\gamma,l}(z_1) a_{\gamma,l}^*(z_2) \Omega) = (z_1|(\rho + \mathbb{1})z_2) = (z_1|(\mathbb{1} - \gamma)^{-1} z_2),$$
$$(\Omega | a_{\gamma,l}^*(z_1) a_{\gamma,l}(z_2) \Omega) = (z_2|\rho z_1) = (z_2|\gamma(\mathbb{1} - \gamma)^{-1} z_1),$$
$$(\Omega | a_{\gamma,l}(z_1) a_{\gamma,l}(z_2) \Omega) = (\Omega | a_{\gamma,l}^*(z_1) a_{\gamma,l}^*(z_2) \Omega) = 0.$$

(5) $CCR_{\gamma,l}$ *is a factor.*
(6) $\mathrm{Ker}\,\gamma = \{0\}$ *iff Ω is separating for $CCR_{\gamma,l}$ iff Ω is cyclic for $CCR_{\gamma,l}$. If this is the case, the modular conjugation for Ω is equal to J_s and the modular operator for Ω is $\Delta = \Gamma(\gamma \oplus \overline{\gamma}^{-1})$.*
(7) *We have*

$$CCR'_{\gamma,l} = CCR_{\gamma,r}.$$

Proof (1)–(4) follow by straightforward computations. Let us prove (5).

We check that $[W_{\gamma,l}(z_1), W_{\gamma,r}(\overline{z}_2)] = 0$ for $z_1, z_2 \in \mathrm{Dom}\,\rho^{\frac{1}{2}}$, which implies that $CCR_{\gamma,r} \subset CCR'_{\gamma,l}$.

Clearly, $(CCR_{\gamma,l} \cup CCR_{\gamma,r})''$ is equal to $\{W(w),\ w \in \mathcal{E}\}''$, for

$$\mathcal{E} = \mathrm{Span}\left\{\left((\rho + \mathbb{1})^{\frac{1}{2}} z_1 + \rho^{\frac{1}{2}} z_2, \overline{\rho}^{\frac{1}{2}} \overline{z_1} + (\overline{\rho} + \mathbb{1})^{\frac{1}{2}} \overline{z_2}\right),\ z_1, z_2 \in \mathrm{Dom}\,\rho^{\frac{1}{2}}\right\}.$$

Clearly, $\mathcal{E}$ is dense in $\mathcal{Z} \oplus \overline{\mathcal{Z}}$. Recall that, by Thm. 9.5, Weyl operators on a Fock space depend strongly continuously on their arguments. Therefore, $\{W(w),\ w \in \mathcal{E}\}'' = \{W(w),\ w \in \mathcal{Z} \oplus \overline{\mathcal{Z}}\}'' = B(\Gamma_s(\mathcal{Z} \oplus \overline{\mathcal{Z}}))$. Thus

$$(CCR_{\gamma,l} \cup CCR'_{\gamma,l})'' \supset (CCR_{\gamma,l} \cup CCR_{\gamma,r})'' = B(\Gamma_s(\mathcal{Z} \oplus \overline{\mathcal{Z}})),$$

which implies that $CCR_{\gamma,l}$ is a factor.

Let us prove the $\Rightarrow$ part of (6). Assume that $\mathrm{Ker}\,\gamma = \{0\}$ and set $\tau^t(A) = \Gamma(\gamma, \overline{\gamma}^{-1})^{it} A \Gamma(\gamma, \overline{\gamma}^{-1})^{-it}$. We have

$$\tau^t(W_{\gamma,l}(z)) = W_{\gamma,l}(\gamma^{it} z),$$

hence τ^t is a W^*-dynamics on $CCR_{\gamma,l}$. We claim that Ω is a $(\tau, 1)$-KMS vector on $CCR_{\gamma,l}$. In fact, we have

$$(\Omega | W_{\gamma,l}(z_1) W_{\gamma,l}(\gamma^{it} z_2) \Omega) = e^{-\frac{1}{4} F(t, z_1, z_2)},$$

for

$$F(t, z_1, z_2) = \left(z_1 \Big| \frac{\gamma + \mathbb{1}}{\mathbb{1} - \gamma} z_1\right) + \left(z_2 \Big| \frac{\gamma + \mathbb{1}}{\mathbb{1} - \gamma} z_2\right) + \left(z_1 \Big| \gamma^{it} \frac{2\gamma}{\mathbb{1} - \gamma} z_2\right)$$
$$+ \left(z_2 \Big| \gamma^{-it} \frac{2}{\mathbb{1} - \gamma} z_1\right),$$

which proves the $(\tau, 1)$-KMS condition for the Weyl operators. By linearity, it holds also for the $*$-algebra of finite linear combinations of $W_{\gamma,1}(z)$ and, by Prop. 6.64, for $CCR_{\gamma,1}$.

Applying then Prop. 6.65 to the factor $CCR_{\gamma,1}$, we obtain that Ω is separating for $CCR_{\gamma,1}$.

We denote by $\mathcal{H}$ the closure of $CCR_{\gamma,1}\Omega$. We would like to show that $\mathcal{H} = \Gamma_s(\mathcal{Z} \oplus \overline{\mathcal{Z}})$, which means that Ω is cyclic for $CCR_{\gamma,1}$. As a byproduct of this proof, we will develop the modular theory of $CCR_{\gamma,1}$.

Clearly, $\mathcal{H}$ is invariant under $CCR_{\gamma,1}$, Ω is cyclic and separating for $CCR_{\gamma,1}$ restricted to $\mathcal{H}$. Let us compute the operators S, Δ and J of the modular theory for Ω and $CCR_{\gamma,1}$ restricted to $\mathcal{H}$.

Let us set

$$\mathcal{H}_1 := \mathrm{Span}\{W_{\gamma,1}(z)\Omega \; : \; z \in \mathrm{Dom}\, \rho^{\frac{1}{2}}\} = \mathrm{Span}\{\Psi_z \; : \; z \in \mathrm{Dom}\, \rho^{\frac{1}{2}}\},$$

for

$$\Psi_z = e^{ia^*((\rho + \mathbb{1})^{\frac{1}{2}} z, \overline{\rho}^{\frac{1}{2}} \overline{z})}\Omega.$$

We have

$$\Gamma(\gamma, \overline{\gamma}^{-1})^{it}\Psi_z = \Psi_{\gamma^{it} z}, \tag{17.29}$$

which implies that the self-adjoint operator $\Gamma(\gamma, \overline{\gamma}^{-1})$ preserves $\mathcal{H}$. Moreover, the r.h.s. of (17.29) extends analytically in t to $t = -i/2$. This shows that $\Psi_z \in \mathrm{Dom}\, \Gamma(\gamma, \overline{\gamma}^{-1})^{\frac{1}{2}}$ and

$$\Gamma(\gamma, \overline{\gamma}^{-1})^{\frac{1}{2}}\Psi_z = e^{ia^*\left(\rho^{\frac{1}{2}} z, (\overline{\rho} + \mathbb{1})^{\frac{1}{2}} \overline{z}\right)}\Omega.$$

Moreover,

$$S\Psi_z = e^{-ia^*\left((\rho + \mathbb{1})^{\frac{1}{2}} z, \overline{\rho}^{\frac{1}{2}} \overline{z}\right)}\Omega \tag{17.30}$$
$$= J_s \Gamma(\gamma, \overline{\gamma}^{-1})^{\frac{1}{2}}\Psi_z.$$

Clearly, $\mathcal{H}_1$ is dense in $\mathcal{H}$.

$$\mathfrak{A} := \mathrm{Span}\{W_{\gamma,1}(z) \; : \; z \in \mathrm{Dom}\, \rho^{\frac{1}{2}}\}$$

is a $*$-algebra $*$-strongly dense in $CCR_{\gamma,1}$ and $\mathcal{H}_1 = \mathfrak{A}\Omega$; therefore, by Subsect. 6.4.2, $\mathcal{H}_1$ is an essential domain of S. Therefore, we can extend (17.30) by density to the whole $\mathcal{H}$, using that J_s is isometric. We obtain

$$S = J_s \Gamma(\gamma, \overline{\gamma}^{-1})^{\frac{1}{2}}\big|_{\mathcal{H}}. \tag{17.31}$$

Since $\operatorname{Ker}\gamma = \{0\}$, the range of $\Gamma(\gamma,\overline{\gamma}^{-1})^{\frac{1}{2}}\big|_{\mathcal{H}}$ is dense in $\mathcal{H}$. Using (17.31), this implies that J_{s} preserves $\mathcal{H}$. Thus

$$S = J_{\mathrm{s}}\big|_{\mathcal{H}}\Gamma(\gamma,\overline{\gamma}^{-1})^{\frac{1}{2}}\big|_{\mathcal{H}}$$

is the polar decomposition of S, defining the modular operator and modular conjugation. Next we see that

$$W_{\gamma,\mathrm{l}}(z_1)J_{\mathrm{s}}W_{\gamma,\mathrm{l}}(z_2)\Omega = W\left((\rho+1)^{\frac{1}{2}}z_1 + \rho^{\frac{1}{2}}z_2, \overline{\rho}^{\frac{1}{2}}\overline{z_1} + (\overline{\rho}+1)^{\frac{1}{2}}\overline{z_2}\right)\Omega.$$

Therefore, $\mathrm{CCR}_{\gamma,\mathrm{l}}J_{\mathrm{s}}\mathrm{CCR}_{\gamma,\mathrm{l}}\Omega$ is dense in $\Gamma_{\mathrm{s}}(\mathcal{Z}\oplus\overline{\mathcal{Z}})$. Since $\mathrm{CCR}_{\gamma,\mathrm{l}}J_{\mathrm{s}}\mathrm{CCR}_{\gamma,\mathrm{l}}\Omega \subset \mathcal{H}$, this proves that $\mathcal{H} = \Gamma_{\mathrm{s}}(\mathcal{Z}\oplus\overline{\mathcal{Z}})$, and hence Ω is cyclic for $\mathrm{CCR}_{\gamma,\mathrm{l}}$. This proves the $\Rightarrow$ part of (6). The proof of the $\Leftarrow$ part of (6) will be given after the proof of (7).

Let us now prove (7). Assume first that $\operatorname{Ker}\gamma = \{0\}$. By the $\Rightarrow$ part of (6), we can apply the Tomita–Takesaki theory to $(\mathrm{CCR}_{\gamma,\mathrm{l}}, \Omega)$ and obtain that $\mathrm{CCR}'_{\gamma,\mathrm{l}} = J_{\mathrm{s}}\mathrm{CCR}_{\gamma,\mathrm{l}}J_{\mathrm{s}}$. By (3), we have $J_{\mathrm{s}}\mathrm{CCR}_{\gamma,\mathrm{l}}J_{\mathrm{s}} = \mathrm{CCR}_{\gamma,\mathrm{r}}$.

For a general γ, we write $\mathcal{Z} = \mathcal{Z}_0 \oplus \mathcal{Z}_1$, for $\mathcal{Z}_0 = \operatorname{Ker}\gamma$. Then $\gamma_1 = \gamma\big|_{\mathcal{Z}_1}$ is injective. Using the exponential law of Subsect. 3.3.7, we have

$$\Gamma_{\mathrm{s}}(\mathcal{Z}\oplus\overline{\mathcal{Z}}) \simeq \Gamma_{\mathrm{s}}(\mathcal{Z}_0) \otimes \Gamma_{\mathrm{s}}(\mathcal{Z}_1 \oplus \overline{\mathcal{Z}_1}) \otimes \Gamma_{\mathrm{s}}(\overline{\mathcal{Z}_0}),$$
$$W_{\gamma,\mathrm{l}}(z) \simeq W(z_0) \otimes W_{\gamma_1,\mathrm{l}}(z_1) \otimes 1,$$
$$W_{\gamma,\mathrm{r}}(\overline{z}) \simeq 1 \otimes W_{\gamma_1,\mathrm{r}}(\overline{z_1}) \otimes W(\overline{z_0}),$$

and hence

$$\mathrm{CCR}_{\gamma,\mathrm{l}} \simeq B(\Gamma_{\mathrm{s}}(\mathcal{Z}_0)) \otimes \mathrm{CCR}_{\gamma_1,\mathrm{l}} \otimes \mathbb{C}1,$$
$$\mathrm{CCR}_{\gamma,\mathrm{r}} \simeq \mathbb{C}1 \otimes \mathrm{CCR}_{\gamma_1,\mathrm{r}} \otimes B(\Gamma_{\mathrm{s}}(\overline{\mathcal{Z}_0})), \tag{17.32}$$

which shows that $\mathrm{CCR}'_{\gamma,\mathrm{l}} = \mathrm{CCR}_{\gamma,\mathrm{r}}$ and completes the proof of (7).

From (17.32), we see that if $\operatorname{Ker}\gamma \neq \{0\}$, Ω is neither cyclic nor separating. This proves the $\Leftarrow$ part of (6). $\square$

17.1.6 Quasi-free CCR representations as Araki–Woods representations

Recall that in Thm. 17.12 we showed that every neutral quasi-free CCR representation over a symplectic space can be reinterpreted as a charged quasi-free CCR representation over a charged symplectic space with $i\omega$ positive definite. Under minor technical assumptions, such representations are unitarily equivalent to Araki–Woods representations. This is described in the theorem below.

Theorem 17.25 *Let* $(\mathcal{Y}, \omega)$ *be a charged symplectic space such that* $i\omega$ *is positive definite on* $\mathcal{Y}$. *Let*

$$\mathcal{Y} \ni y \mapsto a^{\pi*}(y) \in Cl(\mathcal{H})$$

be a charged CCR representation with a cyclic gauge-invariant quasi-free vector Ψ. Let ρ be given by

$$\overline{y}_1 \cdot \rho y_2 = \big(\Psi | a^{\pi *}(y_2) a^{\pi}(y_1)\Psi\big), \quad y_1, y_2 \in \mathcal{Y}.$$

Assume that $\mathcal{Y}$ is complete for the scalar product $2\rho + \mathrm{i}\omega$. Let $\mathcal{Z}$ be the completion of $\mathcal{Y}$ w.r.t. $\mathrm{i}\omega$. Note that ρ can be interpreted as a positive self-adjoint operator on $\mathcal{Z}$ such that $\mathcal{Y} = \mathrm{Dom}\,\rho^{\frac12}$. Set $\gamma = \rho(\mathbb{1} + \rho)^{-1}$. Then there exists a unique isometry $U : \mathcal{H} \to \Gamma_{\mathrm{s}}(\mathcal{Z} \oplus \overline{\mathcal{Z}})$ such that

$$U\Psi = \Omega,$$
$$U a^{\pi *}(y) = a_{\gamma,\mathrm{l}}^*(y) U, \quad y \in \mathcal{Y}. \tag{17.33}$$

17.1.7 Free Bose gas at positive temperatures

In this subsection we would like to describe in general terms how quasi-free bosonic states usually arise in quantum physics. We will also discuss various mathematical formalisms used in this context.

Let h be a positive self-adjoint operator on a Hilbert space $\mathcal{Z}$. Consider a quantum system described by the Hamiltonian $H := \mathrm{d}\Gamma(h)$ acting on the Hilbert space $\Gamma_{\mathrm{s}}(\mathcal{Z})$. Note that $(\Omega| \cdot \Omega)$ describes the ground state of the system. On the algebra $B\big(\Gamma_{\mathrm{s}}(\mathcal{Z})\big)$ we have the dynamics

$$\tau^t(A) := \mathrm{e}^{\mathrm{i}tH} A \mathrm{e}^{-\mathrm{i}tH}, \quad A \in B\big(\Gamma_{\mathrm{s}}(\mathcal{Z})\big), \quad t \in \mathbb{R}.$$

We also have a natural charged CCR representation $\mathcal{Z} \ni z \mapsto a^*(z) \in Cl\big(\Gamma_{\mathrm{s}}(\mathcal{Z})\big)$ and the corresponding neutral CCR representation $\mathcal{Z} \ni z \mapsto W(z) = \exp\big(\mathrm{i}\frac{a^*(z)+a(z)}{\sqrt{2}}\big) \in U\big(\Gamma_{\mathrm{s}}(\mathcal{Z})\big)$. They satisfy

$$\tau^t\big(a^*(z)\big) = a^*(\mathrm{e}^{\mathrm{i}th} z), \quad \tau^t\big(W(z)\big) = W(\mathrm{e}^{\mathrm{i}th} z), \quad z \in \mathcal{Z}.$$

Suppose that we consider the above quantum system at a positive temperature. Let $\beta \geq 0$ denote the inverse temperature. If

$$\mathrm{Tr}\,\mathrm{e}^{-\beta h} < \infty, \tag{17.34}$$

we can consider the Gibbs state given by the density matrix

$$\mathrm{e}^{-\beta\mathrm{d}\Gamma(h)}\big/\mathrm{Tr}\,\mathrm{e}^{-\beta\mathrm{d}\Gamma(h)}. \tag{17.35}$$

Positive-temperature systems are especially interesting for infinitely extended physical systems. For such systems $\mathrm{e}^{-\beta h}$ is rarely trace-class – in fact, typically, h has a continuous spectrum, which rules out (17.34). Therefore, the formalism based on the Gibbs state with the density matrix (17.35) breaks down.

As a typical example of such a system we can consider the (non-relativistic) free Bose gas. Its one-particle Hilbert space and the one-particle Hamiltonian are

$$\mathcal{Z} := L^2(\mathbb{R}^d), \quad h := -\Delta. \tag{17.36}$$

Clearly, in this case (17.34) is not satisfied. Therefore, we need a different formalism to describe positive-temperature systems in this situation.

In the literature one can distinguish three approaches to positive temperatures for infinitely extended systems:

(1) the thermodynamic limit,
(2) the W^* approach,
(3) the C^* approach.

Thermodynamic limit

The thermodynamic limit consists in approximating our system by a sequence of systems in finite volume. Thus we have a sequence of one-particle Hilbert spaces $\mathcal{Z}_L$ with one-particle Hamiltonians h_L. We also need to identify $\mathcal{Z}_{L_1}$ as a subspace of $\mathcal{Z}_{L_2}$ for $L_1 < L_2$, which allows us to embed the corresponding observable algebras $B(\Gamma(\mathcal{Z}_{L_1})) \subset B(\Gamma(\mathcal{Z}_{L_2}))$. Typically, for finite L, the condition

$$\mathrm{Tr}\, e^{-\beta h_L} < \infty \tag{17.37}$$

is satisfied, and so we can use the corresponding Gibbs state. Then we expect that for a fixed L_0 and a large class of observables $A \in B(\Gamma(\mathcal{Z}_{L_0}))$, the expectation value

$$\mathrm{Tr}\, A e^{-\beta \mathrm{d}\Gamma(h_L)} / \mathrm{Tr}\, e^{-\beta \mathrm{d}\Gamma(h_L)}$$

converges to a limit as $L \to \infty$.

In the case of (17.36), we typically take $\mathcal{Z}_L := L^2([-L, L]^d)$, and h_L is the Laplacian with some conditions on the boundary of the box $[-L, L]^d$. For many purposes the choice of boundary conditions should not matter. The Dirichlet or Neumann boundary conditions seem more relevant physically, whereas the periodic boundary conditions might be more convenient mathematically.

Note that this approach involves a significant amount of arbitrariness. One needs to introduce a lot of additional structure, which in the end is irrelevant.

W^* approach

We can describe temperature states by using the Araki–Woods representations. In fact, consider the space $\Gamma_s(\mathcal{Z} \oplus \overline{\mathcal{Z}})$. For $z \in \mathrm{Dom}(e^{\beta h/2})$, define

$$a_\beta^*(z) := a^*\big((\mathbb{1} - e^{-\beta h})^{-\frac{1}{2}} z, 0\big) + a\big(0, (e^{\beta \overline{h}} - \mathbb{1})^{-\frac{1}{2}} \overline{z}\big), \tag{17.38}$$

that is, the Araki–Woods representation for the *Planck density* $(e^{\beta h} - \mathbb{1})^{-1}$. Then

$$\mathrm{Dom}(e^{\beta h/2}) \ni z \mapsto a_\beta^*(z) \in Cl\big(\Gamma_s(\mathcal{Z} \oplus \overline{\mathcal{Z}})\big)$$

is a charged CCR representation. The von Neumann algebra generated by (17.38) will be denoted by CCR_β. Set

$$L := \mathrm{d}\Gamma\left(h \oplus (-\overline{h})\right).$$

Then

$$\tau_\beta^t(A) := \mathrm{e}^{\mathrm{i}tL}A\mathrm{e}^{-\mathrm{i}tL}, \quad A \in \mathrm{CCR}_\beta, \quad t \in \mathbb{R},$$

is a W^*-dynamics on CCR_β and L is its Liouvillean. The state

$$\omega_\beta := (\Omega|A\Omega), \quad A \in \mathrm{CCR}_\beta,$$

is a β-KMS state for the W^*-dynamics τ_β.

Thus we obtain a family of W^*-dynamical systems $(\mathfrak{M}_\beta, \tau_\beta)$ equipped with the state ω_β. One can argue that all of them describe the same physical system and differ only by the temperature. In concrete situations, one can derive the family $(\mathfrak{M}_\beta, \tau_\beta, \omega_\beta)$ using the thermodynamic limit.

Note that the W^* approach does not involve any additional structure (unlike the thermodynamic limit). It is often used in the mathematical physics literature. Implicitly, it is also widely used in the theoretical physics literature.

C^* approach

Consider the C^*-algebra $\mathrm{CCR}^{\mathrm{Weyl}}(\mathcal{Z})$, where $\mathcal{Z}$ is equipped with the symplectic structure $\mathrm{Im}(\cdot|\cdot)$, as well as the charge symmetry $z \mapsto \mathrm{e}^{\mathrm{i}\theta}z$, $\theta \in [0, 2\pi[$. Define the dynamics on $\mathrm{CCR}^{\mathrm{Weyl}}(\mathcal{Z})$ by setting

$$\tau^t\left(W(z)\right) := W(\mathrm{e}^{\mathrm{i}th}z), \quad z \in \mathcal{Z}, \quad t \in \mathbb{R}.$$

It is easy to see that, for any $\beta \in \,]0, \infty]$, there exists on $\mathrm{CCR}^{\mathrm{Weyl}}(\mathcal{Z})$ a unique state β-KMS for the dynamics τ. It is given by

$$\omega_\beta\left(W(z)\right) = \exp\left(-\frac{1}{4}\left(z\Big|\frac{1 + \exp(-\beta h)}{1 - \exp(-\beta h)}z\right)\right), \quad z \in \mathcal{Z}.$$

We can then pass to the GNS representation $(\mathcal{H}_\beta, \pi_\beta, \Omega_\beta)$ and construct the Liouvillean L_β.

In the case of $\beta = \infty$ (the zero temperature), we obtain, up to unitary equivalence, $\mathcal{H}_\infty = \Gamma_\mathrm{s}(\mathcal{Z})$, $\pi_\infty\left(W(z)\right) = W(z)$, $\Omega_\infty = \Omega$ and $L_\infty = H$. This is the quantum system that we started with at the beginning of the subsection.

In the case $\beta < \infty$ (positive temperatures), we obtain the Araki–Woods representation for $\gamma = \mathrm{e}^{-\beta h}$ described in (17.38).

The main advantage of this approach is its conceptual and mathematical elegance. Its starting point is a *single* system, and various temperature states arise naturally by the application of a general principle.

This approach has also a serious disadvantage. The choice of the algebra of observables $\mathrm{CCR}^{\mathrm{Weyl}}(\mathcal{Z})$ is rather arbitrary. In principle, one could replace it by another $*$-algebra related to the CCR over $\mathcal{Z}$, e.g. one of those described in Sect. 8.3. The choice of the CCR algebra does not have much relevance as long as the dynamics is free (that is, as long as it is described by Bogoliubov transformations). The problem becomes more serious when we try to consider a system with a non-trivial interaction. Then, in concrete situations, it is usually not easy to find a C^*-algebra preserved by a given dynamics, and the C^* approach is difficult to apply.

17.2 Fermionic quasi-free states

In this section we describe the theory of fermionic quasi-free states. It is in many ways parallel to that of bosonic quasi-free states. Therefore, each subsection about fermionic quasi-free states has its counterpart in the previous bosonic section.

17.2.1 Definition of fermionic quasi-free states

Let $(\mathcal{Y}, \nu)$ be a real Hilbert space. Recall that $\mathrm{CAR}^{C^*}(\mathcal{Y})$ denotes the CAR C^*-algebra over $\mathcal{Y}$, that is, the C^*-algebra generated by $\phi(y)$, $y \in \mathcal{Y}$, satisfying the CAR relations; see Subsect. 12.5.2.

Definition 17.26 (1) *A state ψ on* $\mathrm{CAR}^{C^*}(\mathcal{Y})$ *is called* quasi-free *if*

$$\psi(\phi(y_1) \cdots \phi(y_{2m-1})) = 0,$$

$$\psi(\phi(y_1) \cdots \phi(y_{2m})) = \sum_{\sigma \in \mathrm{Pair}_{2m}} \mathrm{sgn}(\sigma) \prod_{j=1}^{m} \psi(\phi(y_{\sigma(2j-1)})\phi(y_{\sigma(2j)})),$$

for all $y_1, y_2, \cdots \in \mathcal{Y}$, $m \in \mathbb{N}$.

(2) *If $\mathcal{Y} \ni y \mapsto \phi^\pi(y) \in B_{\mathrm{h}}(\mathcal{H})$ is a CAR representation, $\Psi \in \mathcal{H}$ is called a* quasi-free vector *if*

$$\psi(\phi(y_1) \cdots \phi(y_n)) := (\Psi | \phi^\pi(y_1) \cdots \phi^\pi(y_n)\Psi), \quad y_1, \ldots, y_n \in \mathcal{Y}, \quad n \in \mathbb{N},$$

defines a quasi-free state on $\mathrm{CAR}^{C^*}(\mathcal{Y})$.

(3) *A CAR representation ϕ^π on a Hilbert space $\mathcal{H}$ is* quasi-free *if there exists a cyclic quasi-free vector in $\mathcal{H}$.*

(4) *The anti-symmetric form $\beta \in L_{\mathrm{a}}(\mathcal{Y}, \mathcal{Y}^\#)$ given by*

$$y_1 \cdot \beta y_2 := \frac{1}{\mathrm{i}}\psi([\phi(y_1), \phi(y_2)]), \quad y_1, y_2 \in \mathcal{Y}. \tag{17.39}$$

is called the covariance *of the quasi-free state ψ, and of the quasi-free vector Ψ.*

For a quasi-free state ψ on $\mathrm{CAR}^{C^*}(\mathcal{Y})$, let $(\mathcal{H}_\psi, \pi_\psi, \Omega_\psi)$ be the corresponding GNS representation. Then clearly $\Omega_\psi \in \mathcal{H}_\psi$ is a quasi-free vector for the CAR representation $\mathcal{Y} \ni y \mapsto \pi_\psi(\phi(y)) \in B_{\mathrm{h}}(\mathcal{H}_\psi)$.

The covariance defines the representation uniquely:

Proposition 17.27 *Let $\mathcal{Y} \ni y \mapsto \phi^i(y) \in B_h(\mathcal{H}_i)$, $i = 1, 2$, be quasi-free CAR representations with cyclic quasi-free vectors $\Psi_i \in \mathcal{H}_1$, both of covariance β. Then there exists a unique $U \in U(\mathcal{H}_1, \mathcal{H}_2)$ intertwining ϕ^1 with ϕ^2 satisfying $U\Psi_1 = \Psi_2$.*

Let us note the following important special subclasses of quasi-free representations:

(1) If the pair ν and $\frac{1}{2}\beta$ is Kähler, the corresponding representation is Fock; see Thm. 17.31.
(2) If $\beta = 0$, the corresponding representation is unitarily equivalent to the real-wave (or tracial) representation, discussed already in Subsects. 12.4.2 and 13.2.1.

From the CAR it follows that

$$\psi\big(\phi(y_1)\phi(y_2)\big) = y_1 \cdot \nu y_2 + \frac{\mathrm{i}}{2} y_1 \cdot \beta y_2, \quad y_1, y_2 \in \mathcal{Y}. \tag{17.40}$$

(17.40) implies the following proposition:

Proposition 17.28 *Let $\beta \in L_a(\mathcal{Y}, \mathcal{Y}^\#)$. Then the following are equivalent:*

(1) *There exists a quasi-free state ψ such that (17.40) holds.*
(2) $\nu_{\mathbb{C}} + \frac{\mathrm{i}}{2}\beta_{\mathbb{C}} \geq 0$ *on $\mathbb{C}\mathcal{Y}$*
(3) $|y_1 \cdot \beta y_2| \leq 2(y_1 \cdot \nu y_1)^{\frac{1}{2}}(y_2 \cdot \nu y_2)^{\frac{1}{2}}$, $y_1, y_2 \in \mathcal{Y}$.

Proof The equivalence of (2) and (3) is shown as in Prop. 17.8. To prove (1) $\Rightarrow$ (2) we compute

$$\psi\big(\phi^*(w)\phi(w)\big) = \overline{w} \cdot \nu_{\mathbb{C}} w + \frac{\mathrm{i}}{2} \overline{w} \cdot \beta_{\mathbb{C}} w \geq 0, \quad w \in \mathbb{C}\mathcal{Y}.$$

Let us prove (3) $\Rightarrow$ (1). We fix $\beta \in L_a(\mathcal{Y}, \mathcal{Y}^\#)$ satisfying (3). From Def. 17.26, we obtain a linear functional ψ on the $*$-algebra generated by the $\phi(y)$, $y \in \mathcal{Y}$. It clearly suffices to show that ψ is positive. To check this we may assume that $\mathcal{Y}$ is finite-dimensional.

Using Corollary 2.85 we can find an o.n. basis $(e_1, \ldots, e_{2m}, f_1, \ldots, f_d)$ of $\mathcal{Y}$ such that

$$\beta e_{2j-1} = \lambda_j e_{2j}, \quad \beta e_{2j} = -\lambda_j e_{2j-1}, \quad \beta f_j = 0,$$

for $\lambda_1, \ldots, \lambda_m > 0$. Condition (3) for β is equivalent to $|\lambda_1|, \ldots, |\lambda_m| \leq 2$.

Assume first that $\dim \mathcal{Y} = 2n$. Then, allowing some λ_j to be equal to 0, we can assume that $m = n$. We set $\phi_j = \phi(e_j)$ and use the Jordan–Wigner representation of $\mathrm{CAR}(\mathbb{R}^{2n})$ on $\otimes^n \mathbb{C}^2$ defined in Subsect. 12.2.3. We note that if $|\lambda| \leq 2$, then

$$\rho(\lambda) = \frac{1}{2} \begin{bmatrix} 1 - \lambda/2 & 0 \\ 0 & 1 + \lambda/2 \end{bmatrix}$$

satisfies

$$\rho(\lambda) \geq 0, \quad \mathrm{Tr}\,\rho(\lambda) = 1,$$
$$\mathrm{Tr}(\rho(\lambda)\sigma_1) = \mathrm{Tr}(\rho(\lambda)\sigma_2) = 0, \quad \mathrm{Tr}(\rho(\lambda)\sigma_3) = -\lambda/2. \tag{17.41}$$

We set

$$\rho = \rho(\lambda_1) \otimes \cdots \otimes \rho(\lambda_n).$$

We will prove that

$$\psi(A) = \mathrm{Tr}(\rho A), \quad A \in \mathrm{CAR}(\mathbb{R}^{2n}), \tag{17.42}$$

which implies that ψ is positive.

We first see that

$$\mathrm{Tr}(\rho\phi_{2j-1}\phi_{2j}) = -\mathrm{i}\lambda_j/2, \quad \mathrm{Tr}(\rho\phi_j\phi_k) = 0 \ \text{ if } \ |j-k| \geq 2, \tag{17.43}$$

hence

$$\psi(\phi(y_1)\phi(y_2)) = \mathrm{Tr}(\rho\phi(y_1)\phi(y_2)), \quad y_1, y_2 \in \mathcal{Y}.$$

We claim now that

$$\mathrm{Tr}(\rho\phi_{i_1}\cdots\phi_{i_k}) = 0, \ \text{ if } \ k \ \text{ is odd.} \tag{17.44}$$

We can assume, using the CAR, that $i_1 < \cdots < i_k$. Let i_l be one of the indices. If $l = 2j - 1$, then the j factor of $\phi_{i_1}\cdots\phi_{i_k}$ is equal to $-\mathrm{i}\sigma_2$, except if $i_{l+1} = i_l + 1$, and if $l = 2j$, the j factor of $\phi_{i_1}\cdots\phi_{i_k}$ is equal to $\mathrm{i}\sigma_1$, except if $i_{l-1} = i_l - 1$. It follows from (17.41) that $\mathrm{Tr}(\rho\phi_{i_1}\cdots\phi_{i_k}) = 0$, except when for each $1 \leq l \leq k$ one has $i_{l+1} = i_l + 1$ or $i_{l-1} = i_l - 1$. This condition is not satisfied if k is odd, which proves (17.44). We claim that

$$\mathrm{Tr}(\rho\phi_{i_1}\cdots\phi_{i_{2m}}) = \sum_{\sigma \in \mathrm{Pair}_{2m}} \mathrm{sgn}(\sigma) \prod_{j=1}^{m} \mathrm{Tr}(\rho\phi_{i_{\sigma(2j-1)}}\phi_{i_{\sigma(2j)}}), \tag{17.45}$$

which combined with (17.44) implies (17.42).

The same argument as above shows that the l.h.s. of (17.45) is zero if $(i_1, \ldots, i_{2m})$ is not a collection of pairs $(2j - 1, 2j)$. The same holds for the r.h.s., since in this case, for all $\sigma \in \mathrm{Pair}_{2m}$, at least one of the factors vanishes. It remains to consider the case when

$$(i_1, \ldots, i_{2m}) = (2j_1 - 1, 2j_1, \ldots, 2j_m - 1, 2j_m),$$

for $j_1 < \cdots < j_m$. In this case

$$\phi_{i_1}\cdots\phi_{i_{2m}} = (\mathrm{i}\sigma_3)^{(j_1)}\cdots(\mathrm{i}\sigma_3)^{j_m},$$

and hence the l.h.s. of (17.45) equals

$$\mathrm{Tr}(\rho\phi_{i_1}\cdots\phi_{i_{2m}}) = \prod_{k=1}^{m}(-\mathrm{i}\lambda_{j_k})/2.$$

Since the only pairing contributing to the r.h.s. is $(2j_1 - 1, 2j_1), \ldots,$ $(2j_m - 1, 2j_m)$, we see using (17.43) that (17.45) holds.

Assume now that $\dim \mathcal{Y}$ is odd. Then we set $\mathcal{Y}_1 = \mathcal{Y} \oplus \mathbb{R}$, and consider the (reducible) representation of $\mathrm{CAR}(\mathbb{R}^{2n+1})$ in $\otimes^{n+1}\mathbb{C}^2$ obtained from the Jordan–Wigner representation of $\mathrm{CAR}(\mathbb{R}^{2(n+1)})$. We are then reduced to the previous case. This completes the proof of the proposition. $\qquad\square$

17.2.2 Gauge-invariant fermionic quasi-free states

Suppose that the real Hilbert space $(\mathcal{Y}, \nu)$ is equipped with a Kähler anti-involution j. As in Subsect. 1.3.11, $\mathrm{CAR}^{C^*}(\mathcal{Y})$ is equipped with the one-parameter group of charge automorphisms, denoted $U(1) \ni \theta \mapsto \widehat{u_\theta}$, and defined by

$$\widehat{u_\theta}\big(\phi(y)\big) = \phi(e^{\mathrm{j}\theta} y).$$

Definition 17.29 *A state ψ on $\mathrm{CAR}^{C^*}(\mathcal{Y})$ is called* gauge-invariant *if it is invariant w.r.t. $\widehat{u_\theta}$.*

Consider a fermionic gauge-invariant quasi-free state with covariance β.

Let us introduce the holomorphic space $\mathcal{Z}$ associated with the anti-involution j, so that $\mathbb{C}\mathcal{Y} = \mathcal{Z} \oplus \overline{\mathcal{Z}}$. The sesquilinear forms $\nu_\mathbb{C}$ and $\beta_\mathbb{C}$ can be reduced w.r.t. the direct sum $\mathcal{Z} \oplus \overline{\mathcal{Z}}$. Thus we can write

$$\nu_\mathbb{C} = \begin{bmatrix} \nu_{\mathcal{Z}} & 0 \\ 0 & \overline{\nu}_{\mathcal{Z}} \end{bmatrix}, \quad \beta_\mathbb{C} = \begin{bmatrix} \beta_{\mathcal{Z}} & 0 \\ 0 & \overline{\beta}_{\mathcal{Z}} \end{bmatrix}, \tag{17.46}$$

where $\nu_{\mathcal{Z}}$ is Hermitian and $\beta_{\mathcal{Z}}$ anti-Hermitian. Note that the condition $\nu_\mathbb{C} + \frac{1}{2}\beta_\mathbb{C} > 0$, which by Prop. 17.28 is necessary and sufficient for β to be a covariance of a quasi-free state, is equivalent to

$$\nu_{\mathcal{Z}} \pm \frac{\mathrm{i}}{2}\beta_{\mathcal{Z}} \geq 0. \tag{17.47}$$

Until the end of this subsection we assume that $(\mathcal{Y}, \nu)$ is a real Hilbert space and ψ a quasi-free state on $\mathrm{CAR}^{C^*}(\mathcal{Y})$ with covariance $\beta \in L_\mathrm{a}(\mathcal{Y}, \mathcal{Y}^\#)$.

Theorem 17.30 (1) *Assume that $\mathrm{Ker}\,\beta$ is even- or infinite-dimensional. Then there exists an anti-involution j such that ψ is gauge-invariant for the complex structure given by j.*
(2) *If $\mathrm{Ker}\,\beta = \{0\}$, then the anti-involution j given by (1) is unique if we demand in addition that (β, j) is Kähler.*

Proof By Prop. 17.28, there exists an anti-symmetric operator b such that $\|b\| \leq 1$ and

$$y_1 \cdot \beta y_2 = 2y_1 \cdot \nu b y_2, \quad y_1, y_2 \in \mathcal{Y}.$$

Let $\mathcal{Y}_{\mathrm{sg}} := \mathrm{Ker}\, b$ and $\mathcal{Y}_{\mathrm{reg}} = \mathcal{Y}_{\mathrm{sg}}^{\perp}$. On $\mathcal{Y}_{\mathrm{reg}}$ we use the polar decomposition

$$b_{\mathrm{reg}} = -|b_{\mathrm{reg}}|\mathrm{j}_{\mathrm{reg}} = -\mathrm{j}_{\mathrm{reg}}|b_{\mathrm{reg}}|,$$

so that $\mathrm{j}_{\mathrm{reg}}$ is a Kähler anti-involution on $\mathcal{Y}_{\mathrm{reg}}$ both for $\nu\big|_{\mathcal{Y}_{\mathrm{reg}}}$ and $\beta\big|_{\mathcal{Y}_{\mathrm{reg}}}$. If $\dim \mathcal{Y}_{\mathrm{reg}}$ is even or infinite, we can extend $\mathrm{j}_{\mathrm{reg}}$ to a Kähler anti-involution on $(\mathcal{Y}, \nu)$. $\qquad\square$

The following theorem is the fermionic analog of Thm. 17.13 (2).

Theorem 17.31 *The GNS representation associated with ψ is irreducible iff $(\nu, \frac{1}{2}\beta)$ is Kähler.*

17.2.3 Charged quasi-free CAR representations

The following subsection is essentially a translation of the previous subsection from the terminology of neutral CAR representation to that of charged CAR representations, which seems more convenient in the context of gauge invariance.

Let $(\mathcal{Y}, (\cdot|\cdot))$ be a complex Hilbert space. On $\mathcal{Y}_{\mathbb{R}}$, that is, on the realification of $\mathcal{Y}$, we introduce the real scalar product $\nu := \frac{1}{2}\mathrm{Re}(\cdot|\cdot)$.

Clearly, $\mathcal{Y}$ is equipped with a Kähler anti-involution – the imaginary unit. Therefore, all the definitions off the previous subsections make sense. In particular, the CAR algebra $\mathrm{CAR}^{C^*}(\mathcal{Y}_{\mathbb{R}})$ is equipped with a charge symmetry and we can define the notion of a gauge-invariant state. We will write $\mathrm{CAR}^{C^*}(\mathcal{Y})$ to denote the algebra $\mathrm{CAR}^{C^*}(\mathcal{Y}_{\mathbb{R}})$ equipped with this charge symmetry.

As in the bosonic case, we will denote charged fields using the letter a, and not the usual ψ. Clearly, $\mathrm{CAR}^{C^*}(\mathcal{Y})$ is generated as a $*$-algebra by $a(y) = \frac{1}{2}\big(\phi(y) - \mathrm{i}\phi(\mathrm{i}y)\big)$, $y \in \mathcal{Y}$.

Proposition 17.32 (1) *A state ψ on $\mathrm{CAR}^{C^*}(\mathcal{Y})$ is gauge-invariant if*

$$\psi\big(a^*(y_1)\cdots a^*(y_n)a(w_m)\cdots a(w_1)\big) = 0, \ n \neq m, \ y_1\ldots,y_n, w_m,\ldots, w_1 \in \mathcal{Y}.$$

(2) *It is quasi-free if in addition, for any $y_1\ldots,y_n, w_n,\ldots, w_1 \in \mathcal{Y}$,*

$$\psi\big(a^*(y_1)\cdots a^*(y_n)a(w_n)\cdots a(w_1)\big) = \sum_{\sigma\in S_n} \mathrm{sgn}(\sigma) \prod_{j=1}^{n} \psi\big(a^*(y_j)a(w_{\sigma(j)})\big).$$

Definition 17.33 *If ψ is a gauge-invariant quasi-free state on $\mathrm{CAR}^{C^*}(\mathcal{Y})$, the positive Hermitian operator χ on $\mathcal{Y}$ defined by*

$$(y_2|\chi y_1) := \psi\big(a^*(y_1)a(y_2)\big), \ y_1, y_2 \in \mathcal{Y},$$

is called the one-particle density *of ψ.*

Recall that in the framework of neutral CAR relations one introduces the holomorphic space $\mathcal{Z}$. Charged CAR relations amount to identifying the space $\mathcal{Y}$ with $\mathcal{Z}$, as explained e.g. in Subsect. 12.1.7. Under this identification, the scalar product on $\mathcal{Y}$ is transformed into $2\nu_{\mathcal{Z}}$ and the Hermitian form defined by the one-particle density χ is transformed into $\nu_{\mathcal{Z}} - \frac{1}{2}\beta_{\mathcal{Z}}$. Therefore, (17.47) implies the following proposition.

Proposition 17.34 *A Hermitian operator $\chi \in B_{\mathrm{h}}(\mathcal{Y})$ is the one-particle density of a gauge-invariant quasi-free state iff*

$$0 \leq \chi \leq \mathbb{1}.$$

Suppose now that

$$\mathcal{Y} \ni y \mapsto a^{\pi*}(y) \in B(\mathcal{H}) \tag{17.48}$$

is a charged CAR representation.

Definition 17.35 (1) *$\Psi \in \mathcal{H}$ is called a* gauge-invariant quasi-free vector *if*

$$\psi\big(a^*(y_1)\cdots a^*(y_n)a(w_m)\cdots a(w_1)\big)$$
$$:= \big(\Psi|a^{\pi*}(y_1)\cdots a^{\pi*}(y_n)a^{\pi}(w_m)\cdots a^{\pi}(w_1)\Psi\big), \quad y_1,\ldots,y_n,w_1,\ldots,w_m \in \mathcal{Y},$$

defines a gauge-invariant quasi-free state on $\mathrm{CAR}^{C^*}(\mathcal{Y})$.
(2) *A charged CAR representation (17.48) is* gauge-invariant quasi-free *if there exists a cyclic gauge-invariant quasi-free vector in $\mathcal{H}$.*

Recall that with every neutral CAR representation over a unitary space we associate a charged CAR representation $\mathcal{Y} \ni y \mapsto a^{\pi*}(y) \in B(\mathcal{H})$, such that

$$\phi^{\pi}(y) = a^{\pi*}(y) + a^{\pi}(y).$$

It is clear that a vector Ψ is gauge-invariant quasi-free w.r.t. ϕ^{π} iff it is such w.r.t. $a^{\pi*}$. Likewise, the representation ϕ^{π} is gauge-invariant quasi-free iff $a^{\pi*}$ is.

17.2.4 Gibbs states of fermionic quadratic Hamiltonians
Density matrix

Let $0 \leq \gamma$ be a self-adjoint operator on a Hilbert space $\mathcal{Z}$. We associate with γ the self-adjoint operator $0 \leq \chi < \mathbb{1}$, called the *one-particle density*, defined by

$$\chi := \gamma(\mathbb{1} + \gamma)^{-1}, \quad \gamma = \chi(\mathbb{1} - \chi)^{-1}. \tag{17.49}$$

Note in passing that replacing γ with γ^{-1} is equivalent to replacing χ with $\mathbb{1} - \chi$.

We assume in addition that γ is trace-class. This is equivalent to assuming that χ is trace-class. Note the following identity:

$$\mathrm{Tr}\Gamma(\gamma) = \det(\mathbb{1} + \gamma) = \det(\mathbb{1} - \chi)^{-1}.$$

Thus $\Gamma(\gamma)\det(\mathbb{1} + \gamma)^{-1}$ is a density matrix.

Definition 17.36 *We define the state ψ_γ on $B(\Gamma_{\mathrm{a}}(\mathcal{Z}))$ by*

$$\psi_\gamma(A) := \operatorname{Tr} A\Gamma(\gamma)\det(\mathbb{1}+\gamma)^{-1}, \quad A \in \mathrm{CAR}^{C^*}(\mathcal{Z}).$$

We identify $\mathcal{Z}$ with $\operatorname{Re}(\mathcal{Z} \oplus \overline{\mathcal{Z}})$ using the usual map $z \mapsto (z, \overline{z})$. We can faithfully represent the algebra $\mathrm{CAR}^{C^*}(\mathcal{Z})$ in $B(\Gamma_{\mathrm{a}}(\mathcal{Z}))$.

Proposition 17.37 *The state ψ_γ restricted to $\mathrm{CAR}^{C^*}(\mathcal{Z})$ is gauge-invariant quasi-free. We have*

$$\psi_\gamma\big(a^{\pi*}(z_1)a^\pi(z_2)\big) = (z_2|\chi z_1),$$
$$\psi_\gamma\big(a^\pi(z_1)a^{\pi*}(z_2)\big) = (z_1|z_2) - (z_1|\chi z_2),$$
$$\psi_\gamma\big(a^\pi(z_1)a^\pi(z_2)\big) = \psi_\gamma\big(a^{\pi*}(z_1)a^{\pi*}(z_2)\big) = 0, \quad z_1, z_2 \in \mathcal{Z}.$$

Proof We can find an o.n. basis $(e_1, e_2, \dots)$ diagonalizing the trace-class operator γ. Using the identification

$$\Gamma_{\mathrm{a}}(\mathcal{Z}) \simeq \overset{\infty}{\underset{i=1}{\otimes}} \big(\Gamma_{\mathrm{a}}(\mathbb{C}e_i), \Omega\big), \tag{17.50}$$

we can confine ourselves to the case of one degree of freedom. $\square$

Suppose now that γ is non-degenerate. In this case, the state ψ_γ is faithful. If in addition we fix $\beta \in \mathbb{R}$, we can write $\gamma = \mathrm{e}^{-\beta h}$ for some self-adjoint operator h. Then

$$\Gamma(\gamma)\det(\mathbb{1}+\gamma)^{-1} = \mathrm{e}^{-\beta\mathrm{d}\Gamma(h)}/\operatorname{Tr}\mathrm{e}^{-\beta\mathrm{d}\Gamma(h)}.$$

Thus in this case ψ_γ is the Gibbs state for the dynamics $\mathrm{d}\Gamma(h)$ at the inverse temperature β.

Standard representations on Hilbert–Schmidt operators

Consider the Hilbert space $B^2(\Gamma_{\mathrm{a}}(\mathcal{Z}))$. As in the bosonic case, we will use an alternative notation for the Hermitian conjugation: $JB := B^*$.

We will use the representations of $B(\Gamma_{\mathrm{a}}(\mathcal{Z}))$ and $\overline{B(\Gamma_{\mathrm{a}}(\mathcal{Z}))}$ on $B^2(\Gamma_{\mathrm{a}}(\mathcal{Z}))$ introduced in Subsect. 6.4.5:

$$\pi_{\mathrm{l}}(A)B = AB, \quad \pi_{\mathrm{r}}(\overline{A})B := BA^*, \quad B \in B^2(\Gamma_{\mathrm{a}}(\mathcal{Z})), \quad A \in B(\Gamma_{\mathrm{a}}(\mathcal{Z})).$$

Again, $J\pi_{\mathrm{l}}(A)J^* = \pi_{\mathrm{r}}(A)$.

Thus we can introduce two commuting charged CAR representations

$$\mathcal{Z} \ni z \mapsto \pi_{\mathrm{l}}\big(a^*(z)\big) \in B\big(B^2(\Gamma_{\mathrm{a}}(\mathcal{Z}))\big), \tag{17.51}$$
$$\overline{\mathcal{Z}} \ni \overline{z} \mapsto \pi_{\mathrm{r}}\big(\overline{a^*(z)}\big) \in B\big(B^2(\Gamma_{\mathrm{a}}(\mathcal{Z}))\big). \tag{17.52}$$

They are interchanged by the operator J:

$$J\pi_{\mathrm{l}}\big(a^*(z)\big)J^* = \pi_{\mathrm{r}}\big(\overline{a^*(z)}\big).$$

The vector

$$\Psi_\gamma := \det(\mathbb{1} + \gamma)^{-\frac{1}{2}} \Gamma(\gamma^{\frac{1}{2}})$$

is gauge-invariant quasi-free for the representations (17.51) and (17.52), and the one-particle density χ. If γ is non-degenerate, then both (17.51) and (17.52) are gauge-invariant quasi-free CAR representations.

Standard representations on double Fock spaces

We need to identify the complex conjugate of the Fock space $\overline{\Gamma_a(\mathcal{Z})}$ with the Fock space over the complex conjugate $\Gamma_a(\overline{\mathcal{Z}})$. Recall that in the bosonic case this is straightforward. In the fermionic case, however, we will not use the naive identification, but the identification that "reverses the order of particles", consistent with the convention adopted in (3.4). More precisely, if $z_1, \ldots, z_n \in \mathcal{Z}$, then the identification looks as follows:

$$\overline{\Gamma_a(\mathcal{Z})} \ni \overline{z_1 \otimes_a \cdots \otimes_a z_n} \mapsto \overline{z}_n \otimes_a \cdots \otimes_a \overline{z}_1 \in \Gamma_a(\overline{\mathcal{Z}}). \tag{17.53}$$

(Thus this identification equals Λ times the naive, "non-reversing" identification.)

Note the following chain of identifications:

$$\begin{aligned} B^2\big(\Gamma_a(\mathcal{Z})\big) &\simeq \Gamma_a(\mathcal{Z}) \otimes \overline{\Gamma_a(\mathcal{Z})} \\ &\simeq \Gamma_a(\mathcal{Z}) \otimes \Gamma_a(\overline{\mathcal{Z}}) \simeq \Gamma_a(\mathcal{Z} \oplus \overline{\mathcal{Z}}). \end{aligned} \tag{17.54}$$

We denote by $T_a : B^2\big(\Gamma_a(\mathcal{Z})\big) \to \Gamma_a(\mathcal{Z} \oplus \overline{\mathcal{Z}})$ the unitary map given by (17.54).

Proposition 17.38 $T_a J T_a^* = \Lambda\Gamma(\epsilon)$.

Proof Consider $z_1, \ldots, z_n, w_1, \ldots, w_m \in \mathcal{Z}$ and

$$B = |z_1 \otimes_a \cdots \otimes_a z_n)(w_1 \otimes_a \cdots \otimes_a w_m| \in B^2\big(\Gamma_a(\mathcal{Z})\big).$$

This corresponds to

$$\sqrt{(n+m)!} z_1 \otimes_a \cdots \otimes_a z_n \otimes_a \overline{w_1 \otimes_a \cdots \otimes_a w_m}$$
$$= \sqrt{(n+m)!} z_1 \otimes_a \cdots \otimes_a z_n \otimes_a \overline{w}_m \otimes_a \cdots \otimes_a \overline{w}_1 \in \Gamma_a(\mathcal{Z} \oplus \overline{\mathcal{Z}}).$$

On the other hand,

$$B^* = |w_1 \otimes_a \cdots \otimes_a w_m)(z_1 \otimes_a \cdots \otimes_a z_n|$$

corresponds to

$$\sqrt{(n+m)!} w_1 \otimes_a \cdots \otimes_a w_m \otimes_a \overline{z_1 \otimes_a \cdots \otimes_a z_n}$$
$$= \sqrt{(n+m)!} w_1 \otimes_a \cdots \otimes_a w_m \otimes_a \overline{z}_n \otimes_a \cdots \otimes_a \overline{z}_1$$
$$= (-1)^{\frac{n(n-1)}{2} + \frac{m(m-1)}{2} + nm} \sqrt{(n+m)!} \overline{z}_1 \otimes_a \cdots \otimes_a \overline{z}_n \otimes_a w_m \otimes_a \cdots \otimes_a w_1$$
$$= \Lambda\Gamma(\epsilon) \sqrt{(n+m)!} z_1 \otimes_a \cdots \otimes_a z_n \otimes_a \overline{w}_m \otimes_a \cdots \otimes_a \overline{w}_1,$$

where at the last step we used $\Gamma(\epsilon)z_i = \overline{z}_i$, $\Gamma(\epsilon)\overline{w}_i = w_i$ and

$$\frac{n(n-1)}{2} + \frac{m(m-1)}{2} + nm = \frac{(n+m)(n+m-1)}{2}.$$

□

By applying T_a to (17.51) and (17.52), we obtain two new commuting charged CAR representations

$$\mathcal{Z} \ni z \mapsto T_\mathrm{a}\pi_\mathrm{l}\big(a^*(z)\big)T_\mathrm{a}^* = a^*(z,0) \in B\left(\Gamma_\mathrm{a}(\mathcal{Z} \oplus \overline{\mathcal{Z}})\right), \qquad (17.55)$$

$$\overline{\mathcal{Z}} \ni \overline{z} \mapsto T_\mathrm{a}\pi_\mathrm{r}\big(\overline{a^*(z)}\big)T_\mathrm{a}^* = \Lambda a^*(0,\overline{z})\Lambda \in B\big(\Gamma_\mathrm{a}(\mathcal{Z} \oplus \overline{\mathcal{Z}})\big). \qquad (17.56)$$

They are interchanged by the operator $\Lambda\Gamma(\epsilon)$:

$$\Lambda\Gamma(\epsilon)a^*(z,0)\Gamma(\epsilon)^*\Lambda^* = \Lambda a^*(0,\overline{z})\Lambda, \quad z \in \mathcal{Z}.$$

Again using a basis diagonalizing γ, as in (17.50), the double Fock space on the right of (17.54) can be written as an infinite tensor product

$$\overset{\infty}{\underset{i=1}{\otimes}}\left(\Gamma_\mathrm{a}(\mathbb{C}e_i \oplus \mathbb{C}\overline{e}_i), \Omega\right). \qquad (17.57)$$

We have $T_\mathrm{a}\Psi_\gamma = \Omega_\gamma$, where

$$\Omega_\gamma := \overset{\infty}{\underset{i=1}{\otimes}}(1+\gamma_i)^{-\frac{1}{2}}\mathrm{e}^{\gamma_i^{\frac{1}{2}}a^*(e_i)a^*(\overline{e}_i)}\Omega$$

is gauge-invariant quasi-free for the representations (17.55) and (17.56), and the one-particle density χ. Clearly, both (17.55) and (17.56) are gauge-invariant quasi-free CAR representation.

Note that if we set

$$c = \begin{bmatrix} 0 & \gamma^{\frac{1}{2}} \\ -\overline{\gamma}^{\frac{1}{2}} & 0 \end{bmatrix} \in B_\mathrm{a}^2(\overline{\mathcal{Z}} \oplus \mathcal{Z}, \mathcal{Z} \oplus \overline{\mathcal{Z}}), \qquad (17.58)$$

then

$$\Omega_\gamma = \det(\mathbb{1} + c^*c)^{-\frac{1}{4}}\mathrm{e}^{\frac{1}{2}a^*(c)}\Omega,$$

so this is an example of a fermionic Gaussian vector introduced in Def. 16.35, where it was denoted Ω_c.

Araki–Wyss form of standard representation

Using the infinite tensor product decomposition (17.57), we define the following transformation on $\Gamma_\mathrm{a}(\mathcal{Z} \oplus \overline{\mathcal{Z}})$:

$$R_\gamma := \overset{\infty}{\underset{i=1}{\otimes}}(1+\gamma_i)^{-\frac{1}{2}}\mathrm{e}^{\gamma_i^{\frac{1}{2}}a^*(e_i)a^*(\overline{e}_i)}\Gamma\big((1+\gamma_i)^{\frac{1}{2}}\mathbb{1}\big)\mathrm{e}^{-\gamma_i^{\frac{1}{2}}a(e_i)a(\overline{e}_i)}. \qquad (17.59)$$

Theorem 17.39 R_γ *is a unitary operator satisfying*

$$R_\gamma \phi(z_1, \overline{z_2}) R_\gamma^* = \phi\big((\mathbb{1} - \chi)^{\frac{1}{2}} z_1 + \chi^{\frac{1}{2}} z_2, \overline{\chi}^{\frac{1}{2}} \overline{z_1} + (\mathbb{1} - \overline{\chi})^{\frac{1}{2}} \overline{z_2}\big),$$

$$R_\gamma a(z_1, \overline{z_2}) R_\gamma^* = a\big((\mathbb{1} - \chi)^{\frac{1}{2}} z_1 + \chi^{\frac{1}{2}} z_2, 0\big)$$
$$+ a^*\big(0, \overline{\chi}^{\frac{1}{2}} \overline{z_1} + (\mathbb{1} - \overline{\chi})^{\frac{1}{2}} \overline{z_2}\big),$$

$$R_\gamma a^*(z_1, \overline{z_2}) R_\gamma^* = a^*\big((\mathbb{1} - \chi)^{\frac{1}{2}} z_1 + \chi^{\frac{1}{2}} z_2, 0\big)$$
$$+ a\big(0, \overline{\chi}^{\frac{1}{2}} \overline{z_1} + (\mathbb{1} - \overline{\chi})^{\frac{1}{2}} \overline{z_2}\big), \quad (z_1, \overline{z_2}) \in \mathcal{Z} \oplus \overline{\mathcal{Z}},$$

$$R_\gamma \Gamma(\epsilon) R_\gamma^* = \Gamma(\epsilon),$$

$$R_\gamma \Omega_\gamma = \Omega,$$

$$R_\gamma \mathrm{d}\Gamma(h, -\overline{h}) R_\gamma^* = \mathrm{d}\Gamma(h, -\overline{h}).$$

Proof Let c be defined as in (17.58). Using

$$\Gamma(\mathbb{1} + cc^*) = \Gamma\big((\mathbb{1} + \gamma) \oplus (\mathbb{1} + \overline{\gamma})\big),$$

we see that

$$R_\gamma := \det(\mathbb{1} + cc^*)^{-\frac{1}{4}} e^{\frac{1}{2} a^*(c)} \Gamma(\mathbb{1} + cc^*)^{-\frac{1}{2}} e^{-\frac{1}{2} a(c)}.$$

Thus R_γ belongs to a class of operators that we know very well and we can easily show the properties mentioned in the theorem: it is the unitary operator implementing a j-positive orthogonal transformation given in (16.63). $\square$

By applying R_γ to (17.55) and (17.56), we obtain two new commuting charged CAR representations

$$\mathcal{Z} \ni z \mapsto a_{\gamma,\mathrm{l}}^*(z) := R_\gamma a^*(z, 0) R_\gamma^*$$
$$= a^*\big((\mathbb{1} - \chi)^{\frac{1}{2}} z, 0\big) + a\big(0, \overline{\chi}^{\frac{1}{2}} \overline{z}\big) \in B\left(\Gamma_{\mathrm{a}}(\mathcal{Z} \oplus \overline{\mathcal{Z}})\right),$$

$$\overline{\mathcal{Z}} \ni \overline{z} \mapsto a_{\gamma,\mathrm{r}}^*(\overline{z}) := R_\gamma \Lambda a^*(0, \overline{z}) \Lambda R_\gamma^*$$
$$= \Lambda\big(a(\chi^{\frac{1}{2}} z, 0) + a^*(0, (\mathbb{1} - \chi)^{\frac{1}{2}} \overline{z})\big) \Lambda \in B\left(\Gamma_{\mathrm{a}}(\mathcal{Z} \oplus \overline{\mathcal{Z}})\right).$$

They are interchanged by the operator $\Lambda\Gamma(\epsilon)$:

$$\Lambda\Gamma(\epsilon) a_{\gamma,\mathrm{l}}^*(z) \Gamma(\epsilon)^* \Lambda^* = a_{\gamma,\mathrm{r}}^*(\overline{z}), \quad z \in \mathcal{Z}.$$

We have $R_\gamma \Omega_\gamma = \Omega$, hence the Fock vacuum Ω is a quasi-free vector for the representations $a_{\gamma,\mathrm{l}}^*$ and $a_{\gamma,\mathrm{r}}^*$, and the one-particle density χ. Thus, if γ is non-degenerate, then both are gauge-invariant quasi-free CAR representations. They are special cases of Araki–Wyss charged CAR representations, which we consider more generally in the next subsection.

17.2.5 Araki–Wyss representations

In this subsection we will see that Araki–Wyss representations can be defined more generally, as compared with the framework of the previous subsection.

Let $\mathcal{Z}$ be a Hilbert space. We assume that we are given the operators γ and χ linked by the relation (17.49). This time we drop the condition that γ is trace-class. We assume only that γ is positive, possibly with a non-dense domain, and $0 \leq \chi \leq \mathbb{1}$.

Note that $\mathrm{Dom}\,\gamma = \mathrm{Ran}(\mathbb{1} - \chi) = \mathrm{Ker}(\mathbb{1} - \chi)^{\perp}$. We have $\mathrm{Ker}\,\gamma = \mathrm{Ker}\,\chi$, and set $\mathrm{Ker}\,\gamma^{-1} := \mathrm{Ker}(\mathbb{1} - \chi)$, which amounts to setting $(z|\gamma z) = +\infty$ for $z \notin \mathrm{Dom}\,\gamma$.

Definition 17.40 *For $z \in \mathcal{Z}$, we define the* Araki–Wyss creation operators *on* $\Gamma_{\mathrm{a}}(\mathcal{Z} \oplus \overline{\mathcal{Z}})$:

$$a^*_{\gamma,\mathrm{l}}(z) := a^*\left((\mathbb{1} - \chi)^{\frac{1}{2}} z, 0\right) + a\left(0, \overline{\chi}^{\frac{1}{2}}\overline{z}\right),$$

$$a^*_{\gamma,\mathrm{r}}(\overline{z}) := \left(-a\left(\chi^{\frac{1}{2}} z, 0\right) + a^*\left(0, (\mathbb{1} - \overline{\chi})^{\frac{1}{2}}\overline{z}\right)\right) I$$

$$= \Lambda\left(a\left(\chi^{\frac{1}{2}} z, 0\right) + a^*\left(0, (\mathbb{1} - \overline{\chi})^{\frac{1}{2}}\overline{z}\right)\right)\Lambda = \Lambda a_{\gamma^{-1},\mathrm{l}}(z)\Lambda.$$

For completeness let us write down the adjoints of Araki–Wyss creation operators, called *Araki–Wyss annihilation operators*:

$$a_{\gamma,\mathrm{l}}(z) := a\left((\mathbb{1} - \chi)^{\frac{1}{2}} z, 0\right) + a^*\left(0, \overline{\chi}^{\frac{1}{2}}\overline{z}\right),$$

$$a_{\gamma,\mathrm{r}}(\overline{z}) := \left(a^*\left(\chi^{\frac{1}{2}} z, 0\right) - a\left(0, (\mathbb{1} - \overline{\chi})^{\frac{1}{2}}\overline{z}\right)\right) I$$

$$= \Lambda\left(a^*\left(\chi^{\frac{1}{2}} z, 0\right) + a\left(0, (\mathbb{1} - \overline{\chi})^{\frac{1}{2}}\overline{z}\right)\right)\Lambda = \Lambda a^*_{\gamma^{-1},\mathrm{l}}(z)\Lambda.$$

We also have *Araki–Wyss field operators*:

$$\phi_{\gamma,\mathrm{l}}(z) := a^*_{\gamma,\mathrm{l}}(z) + a_{\gamma,\mathrm{l}}(z) = \phi\left((\mathbb{1} - \chi)^{\frac{1}{2}} z, \overline{\chi}^{\frac{1}{2}}\overline{z}\right),$$

$$\phi_{\gamma,\mathrm{r}}(\overline{z}) := a^*_{\gamma,\mathrm{r}}(\overline{z}) + a_{\gamma,\mathrm{r}}(\overline{z}) = -\mathrm{i}\phi\left(\mathrm{i}\chi^{\frac{1}{2}} z, \mathrm{i}(\mathbb{1} - \overline{\chi})^{\frac{1}{2}}\overline{z}\right) I$$

$$= \Lambda\phi\left(\chi^{\frac{1}{2}} z, (\mathbb{1} - \overline{\chi})^{\frac{1}{2}}\overline{z}\right)\Lambda = \Lambda\phi_{\gamma^{-1},\mathrm{l}}(z)\Lambda.$$

(See (3.30) for identities concerning Λ.)

Definition 17.41 *The von Neumann algebras generated by* $\{a^*_{\gamma,\mathrm{l}}(z) : z \in \mathcal{Z}\}$, *resp.* $\{a^*_{\gamma,\mathrm{r}}(\overline{z}) : z \in \mathcal{Z}\}$ *will be denoted by* $\mathrm{CAR}_{\gamma,\mathrm{l}}$, *resp.* $\mathrm{CAR}_{\gamma,\mathrm{r}}$ *and called the* left, *resp.* right Araki–Wyss algebras.

Clearly,

$$\mathrm{CAR}_{\gamma,\mathrm{r}} = \Lambda\,\mathrm{CAR}_{\gamma,\mathrm{l}}\,\Lambda.$$

Theorem 17.42 (1) *The map*

$$\mathcal{Z} \ni z \mapsto a^*_{\gamma,\mathrm{l}}(z) \in B\left(\Gamma_{\mathrm{a}}(\mathcal{Z} \oplus \overline{\mathcal{Z}})\right)$$

is a charged CAR representation. In particular

$$[a_{\gamma,\mathrm{l}}(z_1), a^*_{\gamma,\mathrm{l}}(z_2)]_+ = (z_1|z_2),$$

$$[a^*_{\gamma,\mathrm{l}}(z_1), a^*_{\gamma,\mathrm{l}}(z_2)]_+ = [a_{\gamma,\mathrm{l}}(z_1), a_{\gamma,\mathrm{l}}(z_2)]_+ = 0.$$

It will be called the left Araki–Wyss (charged CAR) representation.

(2) *The map*

$$\overline{\mathcal{Z}} \ni \overline{z} \mapsto a^*_{\gamma,\mathrm{r}}(\overline{z}) \in B\big(\Gamma_{\mathrm{a}}(\mathcal{Z} \oplus \overline{\mathcal{Z}})\big)$$

is a charged CAR representation. In particular

$$[a_{\gamma,\mathrm{r}}(z_1), a^*_{\gamma,\mathrm{r}}(z_2)]_+ = \overline{(z_1|z_2)},$$
$$[a^*_{\gamma,\mathrm{r}}(z_1), a^*_{\gamma,\mathrm{r}}(z_2)]_+ = [a_{\gamma,\mathrm{r}}(z_1), a_{\gamma,\mathrm{r}}(z_2)]_+ = 0.$$

It will be called the right Araki–Wyss (charged CAR) *representation.*

(3) *Set*

$$J_{\mathrm{a}} := \Lambda\Gamma(\epsilon). \tag{17.60}$$

Then

$$J_{\mathrm{a}} a^*_{\gamma,\mathrm{l}}(z) J_{\mathrm{a}} = a^*_{\gamma,\mathrm{r}}(\overline{z}),$$

$$J_{\mathrm{a}} a_{\gamma,\mathrm{l}}(z) J_{\mathrm{a}} = a_{\gamma,\mathrm{r}}(\overline{z}).$$

(4) *The vacuum Ω is a fermionic quasi-free vector for $a^*_{\gamma,\mathrm{l}}$ with the 2-point functions*

$$(\Omega|a_{\gamma,\mathrm{l}}(z_1)a^*_{\gamma,\mathrm{l}}(z_2)\Omega) = (z_1|(\mathbb{1} - \chi)z_2) = (z_1|(\mathbb{1} + \gamma)^{-1}z_2),$$
$$(\Omega|a^*_{\gamma,\mathrm{l}}(z_1)a_{\gamma,\mathrm{l}}(z_2)\Omega) = (z_2|\chi z_1) = (z_2|\gamma(\mathbb{1} + \gamma)^{-1}z_1),$$
$$(\Omega|a_{\gamma,\mathrm{l}}(z_1)a_{\gamma,\mathrm{l}}(z_2)\Omega) = (\Omega|a^*_{\gamma,\mathrm{l}}(z_1)a^*_{\gamma,\mathrm{l}}(z_2)\Omega) = 0.$$

(5) $\mathrm{CAR}_{\gamma,\mathrm{l}}$ *is a factor.*

(6) $\mathrm{Ker}\,\gamma = \mathrm{Ker}\,\gamma^{-1} = \{0\}$ *(equivalently, $\mathrm{Ker}\,\chi = \mathrm{Ker}(\mathbb{1} - \chi) = \{0\}$) iff Ω is separating for $\mathrm{CAR}_{\gamma,\mathrm{l}}$ iff Ω is cyclic for $\mathrm{CAR}_{\gamma,\mathrm{l}}$. If this is the case, then J_{a} and $\Delta = \Gamma(\gamma \oplus \overline{\gamma}^{-1})$ are the modular conjugation and modular operator for $(\mathrm{CAR}_{\gamma,\mathrm{l}}, \Omega)$.*

(7) *We have*

$$\mathrm{CAR}'_{\gamma,\mathrm{l}} = \mathrm{CAR}_{\gamma,\mathrm{r}}. \tag{17.61}$$

(8) *If $\chi = \frac{1}{2}\mathbb{1}$ (or, equivalently, $\gamma = \mathbb{1}$), then the Araki–Wyss representation coincides with the real-wave representation and $\mathrm{CAR}_{\gamma,\mathrm{l}}$ coincides with $\mathrm{CAR}^{W^*}(\mathcal{Z})$.*

Proof Items (1) to (4) follow by straightforward computations.

The proof of (5) uses Prop. 6.44. First note that

$$[\phi_{\gamma,\mathrm{l}}(z_1), \phi_{\gamma,\mathrm{r}}(\overline{z}_2)] = 0.$$

Consequently $\mathrm{CAR}_{\gamma,\mathrm{l}}$ and $\mathrm{CAR}_{\gamma,\mathrm{r}}$ commute with one another. Therefore,

$$(\mathrm{CAR}_{\gamma,\mathrm{l}} \cup \mathrm{CAR}'_{\gamma,\mathrm{l}})'' \supset \mathrm{CAR}_{\gamma,\mathrm{l}} \cup \mathrm{CAR}_{\gamma,\mathrm{r}}.$$

It is easy to see that Ω is cyclic for $\mathrm{CAR}_{\gamma,\mathrm{l}} \cup \mathrm{CAR}_{\gamma,\mathrm{r}}$, which means that Condition (1) of Prop. 6.44 is satisfied for the vector Ω.

Set

$$b(z) = a_{\gamma,\mathrm{l}}\big((\mathbb{1} - \chi)^{\frac{1}{2}} z\big) + a^*_{\gamma,\mathrm{r}}\big(\overline{\chi^{\frac{1}{2}} z}\big)$$

$$= a(z, 0) + \Big(-a(\chi z) + a^*\big(0, (\mathbb{1} - \overline{\chi})^{\frac{1}{2}} \overline{\chi^{\frac{1}{2}} z}\big)\Big)(\mathbb{1} - I),$$

$$b(\overline{z}) = a_{\gamma,\mathrm{l}}\big(\chi^{\frac{1}{2}} z\big) - a^*_{\gamma,\mathrm{r}}\big((\mathbb{1} - \overline{\chi})^{\frac{1}{2}} \overline{z}\big)$$

$$= a(0, \overline{z}) + \Big(a^*\big((\mathbb{1} - \chi)^{\frac{1}{2}} \chi^{\frac{1}{2}} z, 0\big) - a\big(0, (\mathbb{1} - \overline{\chi})\overline{z}\big)\Big)(\mathbb{1} - I).$$

For Condition (2) of Prop. 6.44, the set $\mathcal{L}$ is defined as

$$\mathcal{L} := \big\{b(z) \ : \ z \in \mathcal{Z}\big\} \cup \big\{b(\overline{z}) \ : \ z \in \mathcal{Z}\big\}.$$

Suppose that Ψ is annihilated by all elements of $\mathcal{L}$. All of them anti-commute with I; therefore they separately annihilate the even and odd parts of Ψ, i.e. $\Psi_\pm := \frac{1}{2}(\mathbb{1} + I)\Psi$. We have

$$b(z)\Psi_+ = a(z, 0)\Psi_+ = 0,$$
$$b(\overline{z})\Psi_+ = a(0, \overline{z})\Psi_+ = 0.$$

Therefore, Ψ_+ is proportional to Ω, the Fock vacuum. Moreover,

$$b(z)\Psi_- = \Big(a\big((\mathbb{1} - 2\chi)z, 0\big) + a^*\big(0, (\mathbb{1} - \overline{\chi})^{\frac{1}{2}} \overline{\chi^{\frac{1}{2}} z}\big)\Big)\Psi_- = 0,$$

$$b(\overline{z})\Psi_- = \Big(a^*\big((\mathbb{1} - \chi)^{\frac{1}{2}} \chi^{\frac{1}{2}} z, 0\big) - a\big(0, (\mathbb{1} - 2\overline{\chi})\overline{z}\big)\Big)\Psi_- = 0. \qquad (17.62)$$

Define $\mathcal{Z}_0 := \mathrm{Ker}(\chi - \frac{1}{2}\mathbb{1})$, and $\mathcal{Z}_1 := \mathcal{Z}_0^\perp$, so that $\mathcal{Z} = \mathcal{Z}_0 \oplus \mathcal{Z}_1$. We can rewrite (17.62) as

$$\Big(a(w_1, 0) + a^*\big(0, (\mathbb{1} - \overline{\chi})^{\frac{1}{2}} \overline{\chi}^{\frac{1}{2}} (\mathbb{1} - 2\chi)^{-1} \overline{w}_1\big)\Big)\Psi_- = 0, \quad w_1 \in \mathcal{Z}_1,$$

$$\Big(a^*\big((\mathbb{1} - \chi)^{\frac{1}{2}} \chi^{\frac{1}{2}} (-\mathbb{1} + 2\chi)^{-1} w_1, 0\big) + a(0, \overline{w}_1)\Big)\Psi_- = 0, \quad w_1 \in \mathcal{Z}_1,$$

$$a^*(0, \overline{w}_0)\Psi_- = a^*(w_0, 0)\Psi_- = 0, \quad w_0 \in \mathcal{Z}_0. \qquad (17.63)$$

By Lemma 16.46, Ψ_- can be non-zero only if $\dim \mathcal{Z}_0$ is finite. If this is the case, by Thm. 16.36 and arguments of Subsect. 16.3.5, Ψ_- is proportional to a fermionic Gaussian vector tensored with an even ceiling vector. In any case, this means that Ψ_- is even. But we know that Ψ_- is odd. Hence, $\Psi_- = 0$.

Therefore, Ψ is proportional to Ω. Hence, Condition (2) of Prop. 6.44 is satisfied. This proves that $\mathrm{CAR}_{\gamma,\mathrm{l}}$ is a factor and ends the proof of (5).

Let us now prove (6). Assume that $\mathrm{Ker}\,\gamma = \mathrm{Ker}\,\gamma^{-1} = \{0\}$. Set

$$\tau^t(A) := \Gamma(\gamma \oplus \overline{\gamma}^{-1})^{\mathrm{it}}\, A\, \Gamma(\gamma \oplus \overline{\gamma}^{-1})^{-\mathrm{it}}.$$

We first see that $\tau^t(\phi_{\gamma,\mathrm{l}}(z)) = \phi_{\gamma,\mathrm{l}}(\gamma^{\mathrm{it}} z)$, hence τ^t preserves $\mathrm{CAR}_{\gamma,\mathrm{l}}$ and is a W^*-dynamics on $\mathrm{CAR}_{\gamma,\mathrm{l}}$. Next we check that Ω is a $(\tau, -1)$-KMS vector. This is straightforward for the field operators $\phi_{\gamma,\mathrm{l}}(z)$. For products of field operators we use the identities of Prop. 17.32. By Prop. 6.64 we extend the KMS condition to

$CAR_{\gamma,1}$. Applying Prop. 6.65 to the factor $CAR_{\gamma,1}$, we obtain that Ω is separating for $CAR_{\gamma,1}$.

We denote by $\mathcal{H}$ the closure of $CAR_{\gamma,1}\Omega$. The vector Ω is cyclic and separating for $CAR_{\gamma,1}$ restricted to $\mathcal{H}$. Therefore, we can compute the operators that belong to the modular theory for Ω, as operators on $\mathcal{H}$.

We fix an o.n. basis $\{f_j\}_{j \in J}$ of $\mathcal{Z}$. Since $\operatorname{Ker}\gamma = \operatorname{Ker}\gamma^{-1} = \{0\}$, we can moreover assume that $f_j \in \operatorname{Dom}\gamma^{\frac{1}{2}} \cap \operatorname{Dom}\gamma^{-\frac{1}{2}}$. Clearly, the family $\{e_i\}_{i \in I} = \{f_j, if_j\}_{j \in J}$ is an o.n. basis of $\mathcal{Z}$ for the Euclidean scalar product $\operatorname{Re}(\cdot | \cdot)$. Set

$$\mathcal{H}_1 := \operatorname{Span}\Big\{ \prod_{i \in I_1} \phi_{\gamma,1}(e_i)\Omega,\ I_1 \subset I \text{ finite}\Big\}.$$

Clearly, $\mathcal{H}_1$ is a dense subspace of $\mathcal{H}$. We will prove that

$$S = J_a \Gamma(\gamma \oplus \overline{\gamma}^{-1})^{\frac{1}{2}} \text{ on } \mathcal{H}_1. \tag{17.64}$$

Let $(e_1, \dots, e_n)$ be a finite family in $\{e_i\}_{i \in I}$ and

$$\Phi := \prod_{i=1}^{n} \phi_{\gamma,1}(e_i)\Omega.$$

We have

$$S\Phi = \prod_{i=n}^{1} \phi_{\gamma,1}(e_i)\Omega = (-1)^{n(n-1)/2} \prod_{i=1}^{n} \phi_{\gamma,1}(e_i)\Omega.$$

Note that

$$\Phi = \prod_{i=1}^{n} \big(a^*(u_i) + a(u_i)\big)\Omega,$$

for $u_i = \big((\mathbb{1} - \chi)^{\frac{1}{2}}e_i, \overline{\chi}^{\frac{1}{2}}\overline{e_i}\big)$. To compute $\Gamma(\gamma \oplus \overline{\gamma}^{-1})^{\frac{1}{2}}\Phi$, we apply Prop. 3.53 (1). We obtain

$$\Gamma(\gamma \oplus \overline{\gamma}^{-1})^{\frac{1}{2}}\Phi$$
$$= \prod_{i=1}^{n} \Big(a^*\big(\chi^{\frac{1}{2}}e_i, (\mathbb{1} - \overline{\chi})^{\frac{1}{2}}\overline{e_i}\big) + a\big(\chi^{-\frac{1}{2}}(\mathbb{1} - \chi)e_i, \overline{\chi}(\mathbb{1} - \overline{\chi})^{-\frac{1}{2}}\overline{e_i}\big)\Big)\Omega,$$

and hence

$$\Gamma(\epsilon)\Gamma(\gamma \oplus \overline{\gamma}^{-1})^{\frac{1}{2}}\Phi$$
$$= \prod_{i=1}^{n} \Big(a^*\big((\mathbb{1} - \chi)^{\frac{1}{2}}e_i, \overline{\chi}^{\frac{1}{2}}\overline{e_i}\big) + a\big(\chi(\mathbb{1} - \chi)^{-\frac{1}{2}}e_i, \overline{\chi}^{-\frac{1}{2}}(\mathbb{1} - \overline{\chi})\overline{e_i}\big)\Big)\Omega.$$

Using (3.30), we finally get that

$$\Lambda\Gamma(\epsilon)\Gamma(\gamma \oplus \overline{\gamma}^{-1})^{\frac{1}{2}}\Phi$$
$$= (-1)^{n(n-1)/2} \prod_{i=1}^{n} \Big(a^*\big((\mathbb{1} - \chi)^{\frac{1}{2}}e_i, \overline{\chi}^{\frac{1}{2}}\overline{e_i}\big) - a\big(\chi(\mathbb{1} - \chi)^{-\frac{1}{2}}e_i, \overline{\chi}^{-\frac{1}{2}}(\mathbb{1} - \overline{\chi})\overline{e_i}\big)\Big)\Omega.$$

Hence, to prove that $S\Phi = J_a\Gamma(\gamma \oplus \overline{\gamma}^{-1})^{\frac{1}{2}}\Phi$, it remains to check that

$$\prod_{i=1}^{n}\left(a^*((\mathbb{1}-\chi)^{\frac{1}{2}}e_i, \overline{\chi}^{\frac{1}{2}}\overline{e_i}) + a((\mathbb{1}-\chi)^{\frac{1}{2}}e_i, \overline{\chi}^{\frac{1}{2}}\overline{e_i})\right)\Omega$$
$$= \prod_{i=1}^{n}\left(a^*((\mathbb{1}-\chi)^{\frac{1}{2}}e_i, \overline{\chi}^{\frac{1}{2}}\overline{e_i}) - a(\chi(\mathbb{1}-\chi)^{-\frac{1}{2}}e_i, \overline{\chi}^{-\frac{1}{2}}(\mathbb{1}-\overline{\chi})\overline{e_i})\right)\Omega.$$

(17.65)

We can Wick-order both sides of (17.65) by moving annihilation operators to the right until they act on Ω. For the l.h.s., we pick terms coming from the anti-commutation relations that are products of

$$L(i,k) := \left((\mathbb{1}-\chi)^{\frac{1}{2}}e_i|(\mathbb{1}-\chi)^{\frac{1}{2}}e_k\right)_{\mathcal{Z}} + \left(\overline{\chi}^{\frac{1}{2}}\overline{e_i}|\overline{\chi}^{\frac{1}{2}}\overline{e_k}\right)_{\overline{\mathcal{Z}}}$$
$$= (e_i|e_k)_{\mathcal{Z}} - (e_i|\chi e_k)_{\mathcal{Z}} + (e_k|\chi e_i)_{\mathcal{Z}}.$$

For the r.h.s., we obtain identical terms with $L(i,k)$ replaced with

$$R(i,k) := -\left(\chi(\mathbb{1}-\chi)^{-\frac{1}{2}}e_i|(\mathbb{1}-\chi)^{\frac{1}{2}}e_k\right)_{\mathcal{Z}} - \left(\overline{\chi}^{-\frac{1}{2}}(\mathbb{1}-\overline{\chi})\overline{e_i}|\overline{\chi}^{\frac{1}{2}}\overline{e_k}\right)_{\overline{\mathcal{Z}}}$$
$$= -(e_k|e_i)_{\mathcal{Z}} - (e_i|\chi e_k)_{\mathcal{Z}} + (e_k|\chi e_i)_{\mathcal{Z}}.$$

Therefore,

$$L(i,k) - R(i,k) = 2\text{Re}(e_i|e_k) = 2\delta_{i,k}.$$

This ends the proof of (17.64).

By Prop. 6.59, we know that the closure of $S|_{\mathcal{H}_1}$ equals S. Moreover, we easily see that $\Gamma(\gamma \oplus \overline{\gamma}^{-1})^{\frac{1}{2}}$ preserves $\mathcal{H}$ and is essentially self-adjoint on $\mathcal{H}_1$. Since J_a is isometric, (17.64) implies that $S = J_a\Gamma(\gamma \oplus \overline{\gamma}^{-1})^{\frac{1}{2}}$, as an identity between closed operators on $\mathcal{H}$. It also proves that the modular conjugation is given by $J_a|_{\mathcal{H}}$ and the modular operator is given by $\Gamma(\gamma, \overline{\gamma}^{-1})|_{\mathcal{H}}$.

Now,

$$\phi_{\gamma,l}(z_1)\cdots\phi_{\gamma,l}(z_n)J_a\phi_{\gamma,l}(w_m)\cdots\phi_{\gamma,l}(w_1)\Omega$$
$$= \phi_{\gamma,l}(z_1)\cdots\phi_{\gamma,l}(z_n)\phi_{\gamma,r}(\overline{w}_m)\cdots\phi_{\gamma,r}(\overline{w}_1)\Omega.$$

This easily implies that $\text{CAR}_{\gamma,l}J_a\text{CAR}_{\gamma,l}\Omega$ is dense in $\Gamma_a(\mathcal{Z}\oplus\overline{\mathcal{Z}})$. But

$$\text{CAR}_{\gamma,l}J_a\text{CAR}_{\gamma,l}\Omega \subset \mathcal{H},$$

hence $\mathcal{H}$ is dense in $\Gamma_a(\mathcal{Z}\oplus\overline{\mathcal{Z}})$, which proves that Ω is cyclic. This ends the proof of the $\Rightarrow$ part of (6), as well as giving the formulas for the modular conjugation and the modular operator.

We first prove (7) under the assumption that $\text{Ker}\,\gamma = \text{Ker}\,\gamma^{-1} = \{0\}$. By the $\Rightarrow$ part of (6), we know that Ω is cyclic and separating for $\text{CAR}_{\gamma,l}$, and J_a is the modular conjugation for Ω. Applying the modular theory, we have $\text{CAR}'_{\gamma,l} = J_a\text{CAR}_{\gamma,l}J_a = \text{CAR}_{\gamma,r}$ by (3).

To prove the general case, we will invoke some of the results to be proven only in the next section. Set $\mathcal{Z}_0 = \text{Ker}\,\chi$, $\mathcal{Z}_2 = \text{Ker}(\mathbb{1}-\chi)$, and write

$$\mathcal{Z} = \mathcal{Z}_0 \oplus \mathcal{Z}_1 \oplus \mathcal{Z}_2.$$

(17.66)

We set

$$\mathcal{V} := \left\{ \left((\mathbb{1} - \chi)^{\frac{1}{2}} z, \overline{\chi}^{\frac{1}{2}} \overline{z} \right) \ : \ z \in \mathcal{Z} \right\},$$

which is a closed real subspace of $\mathcal{W} = \mathcal{Z} \oplus \overline{\mathcal{Z}}$. From (17.66) we obtain

$$\mathcal{V} = \mathcal{V}_0 \oplus \mathcal{V}_1 \oplus \mathcal{V}_2,$$

for

$$\mathcal{V}_0 = \{(z_0, 0) \ : \ z_0 \in \mathcal{Z}_0\}, \qquad \mathcal{V}_2 = \{(0, \overline{z}_2) \ : \ z_2 \in \mathcal{Z}_2\},$$

$$\mathcal{V}_1 = \left\{ \left((\mathbb{1} - \chi_1)^{\frac{1}{2}} z_1, \overline{\chi_1}^{\frac{1}{2}} \overline{z}_1 \right) \ : \ z_1 \in \mathcal{Z}_1 \right\}.$$

We have, with the notation in Subsect. 17.3.1,

$$i\mathcal{V}_0^{\mathrm{perp}} = \{(0, \overline{z}_0) \ : \ z_0 \in \mathcal{Z}_0\}, \qquad i\mathcal{V}_2^{\mathrm{perp}} = \{(z_2, 0) \ : \ z_2 \in \mathcal{Z}_2\},$$

$$i\mathcal{V}_1^{\mathrm{perp}} = \left\{ \left(\chi_1^{\frac{1}{2}} z_1, (\mathbb{1} - \overline{\chi_1})^{\frac{1}{2}} \overline{z}_1 \right) \ : \ z_1 \in \mathcal{Z}_1 \right\}.$$

With the notation of Subsect. 17.3.5, $\mathrm{CAR}_{\gamma,1}$ is identified with $\mathrm{CAR}(\mathcal{V})$, hence (7) follows from Thm. 17.61.

It remains to prove the $\Leftarrow$ part of (6). If $\mathrm{Ker}\,\chi \neq \{0\}$, then

$$\Gamma_a(\{0\} \oplus \overline{\mathcal{Z}_0}) \perp \mathrm{CAR}_{\gamma,1}\Omega$$

and $a_{\gamma,1}(z_0)\Omega = 0$ for $z_0 \in \mathcal{Z}_0$, hence Ω is neither cyclic not separating for $\mathrm{CAR}_{\gamma,1}$.

Similarly, if $\mathrm{Ker}(\mathbb{1} - \chi) \neq \{0\}$, then $\Gamma_a(\mathcal{Z}_2 \oplus \{0\}) \perp \mathrm{CAR}_{\gamma,1}\Omega$ and $a_{\gamma,1}^*(z_2)\Omega = 0$ for $z_2 \in \mathcal{Z}_2$. This completes the proof of (6). $\qquad \square$

17.2.6 Quasi-free CAR representations as Araki–Wyss representations

Every quasi-free charged CAR representation can be realized as an Araki–Wyss representation.

Theorem 17.43 *Let $\mathcal{Z}$ be a Hilbert space. Let*

$$\mathcal{Z} \ni z \mapsto a^{\pi*}(z) \in B(\mathcal{H})$$

be a charged CAR representation with a gauge-invariant cyclic quasi-free vector Ψ. Let χ be defined by

$$\overline{z}_1 \cdot \chi z_2 = \left(\Psi | a^{\pi*}(z_2) a^{\pi}(z_1) \Psi \right), \quad z_1, z_2 \in \mathcal{Z}.$$

Then, for $\gamma := \chi(\mathbb{1} - \chi)^{-1}$, there exists an isometry $U : \mathcal{H} \to \Gamma_a(\mathcal{Z} \oplus \overline{\mathcal{Z}})$ such that

$$U\Psi = \Omega,$$

$$Ua^{\pi*}(z) = a_{\gamma,1}^*(z)U, \quad z \in \mathcal{Z}.$$

17.2.7 Free Fermi gas at positive temperatures

This subsection is parallel to Subsect. 17.1.7 about the free Bose gas. We start with h, a positive self-adjoint operator on a Hilbert space $\mathcal{Z}$. Consider a quantum system described by the Hamiltonian $H := \mathrm{d}\Gamma(h)$ on the Hilbert space $\Gamma_{\mathrm{a}}(\mathcal{Z})$. Clearly, $(\Omega| \cdot \Omega)$ describes the ground state of the system. On the algebra $B(\Gamma_{\mathrm{a}}(\mathcal{Z}))$ we have the dynamics

$$\tau^t(A) := \mathrm{e}^{\mathrm{i}tH} A \mathrm{e}^{-\mathrm{i}tH}, \quad A \in B(\Gamma_{\mathrm{a}}(\mathcal{Z})), \quad t \in \mathbb{R}.$$

We also have a natural charged CAR representation $\mathcal{Z} \ni z \mapsto a^*(z) \in B(\Gamma_{\mathrm{a}}(\mathcal{Z}))$ and the corresponding neutral CAR representation $\mathcal{Z} \ni z \mapsto \phi(z) = a^*(z) + a(z) \in B_{\mathrm{h}}(\Gamma_{\mathrm{a}}(\mathcal{Z}))$. They satisfy

$$\tau^t(a^*(z)) = a^*(\mathrm{e}^{\mathrm{i}th} z), \quad \tau^t(\phi(z)) = \phi(\mathrm{e}^{\mathrm{i}th} z), \quad z \in \mathcal{Z}.$$

Suppose that we consider the above quantum system at a positive temperature. Let $\beta \geq 0$ denote the inverse temperature. If

$$\mathrm{Tr}\, \mathrm{e}^{-\beta h} < \infty, \tag{17.67}$$

we can consider the Gibbs state given by the density matrix

$$\mathrm{e}^{-\beta \mathrm{d}\Gamma(h)} / \mathrm{Tr}\, \mathrm{e}^{-\beta \mathrm{d}\Gamma(h)}. \tag{17.68}$$

Again, the formalism based on the Gibbs state with the density matrix (17.68) breaks down at infinite volume, for instance in the case of (17.36).

As in the case of the Bose gas, we distinguish three possible formalisms for infinitely extended systems:

(1) the thermodynamic limit,
(2) the W^* approach,
(3) the C^* approach.

The general framework of the thermodynamic limit in the Fermi case is analogous to that in the Bose case. Therefore, we do not describe it separately.

W^* approach

The W^*-approach to free Fermi systems is also analogous to that for Bose systems. We just replace Araki–Woods representations with Araki–Wyss representations. Let us, however, describe this in detail, apologizing to the reader for almost verbatim repetitions from the bosonic case.

Consider the space $\Gamma_{\mathrm{a}}(\mathcal{Z} \oplus \overline{\mathcal{Z}})$. For $z \in \mathcal{Z}$, define

$$a_\beta^*(z) := a^*((\mathbb{1} + \mathrm{e}^{-\beta h})^{-\frac{1}{2}} z, 0) + a(0, (\mathbb{1} + \mathrm{e}^{\beta \overline{h}})^{-\frac{1}{2}} \overline{z}). \tag{17.69}$$

Then

$$\mathcal{Z} \ni z \mapsto a_\beta^*(z) \in B(\Gamma_{\mathrm{a}}(\mathcal{Z} \oplus \overline{\mathcal{Z}}))$$

is a charged CAR representation. In fact, it is the Araki–Wyss representation for the *Fermi–Dirac density* $(\mathbb{1} + e^{\beta h})^{-1}$. The von Neumann algebra generated by (17.69) will be denoted by CAR_β. Set

$$L := \mathrm{d}\Gamma\left(h \oplus (-\overline{h})\right).$$

Then

$$\tau_\beta^t(A) := e^{\mathrm{i}tL} A e^{-\mathrm{i}tL}, \quad A \in \mathrm{CAR}_\beta,$$

is a W^*-dynamics on CAR_β. The state

$$\omega_\beta(A) := (\Omega|A\Omega), \quad A \in \mathrm{CAR}_\beta,$$

is a β-KMS state for the W^*-dynamics τ_β.

C^* approach

Again, the C^* approach for fermions follows the same lines as the C^* approach for bosons. There is, however, a difference: there exists a natural choice of a C^*-algebra, which seemed not to be the case for bosons.

Consider the C^*-algebra $\mathrm{CAR}^{C^*}(\mathcal{Z})$, where $\mathcal{Z}$ is equipped with the Euclidean structure $\frac{1}{2}\mathrm{Re}(\cdot|\cdot)$, as well as the usual charge symmetry. Define the dynamics on $\mathrm{CAR}^{C^*}(\mathcal{Z})$ by setting

$$\tau^t\left(a^*(z)\right) := a^*(e^{\mathrm{i}th} z), \quad z \in \mathcal{Z}.$$

It is easy to see that, for any $\beta \in [-\infty, \infty]$, there exists on $\mathrm{CAR}^{C^*}(\mathcal{Z})$ a unique state β-KMS for the dynamics τ. It is the gauge-invariant quasi-free state given by

$$\omega_\beta\left(a(z_1)a^*(z_2)\right) = \left(z_1|(\mathbb{1} + e^{-\beta h})^{-1} z_2\right), \quad z_1, z_2 \in \mathcal{Z}.$$

We can then pass to the GNS representation, obtaining $(\mathcal{H}_\beta, \pi_\beta, \Omega_\beta)$, and the Liouvillean L_β.

In the case of $\beta = \infty$ (the zero temperature), we obtain, up to unitary equivalence, $\mathcal{H}_\infty = \Gamma_\mathrm{s}(\mathcal{Z})$, $\pi_\infty\left(W(z)\right) = W(z)$, $\Omega_\infty = \Omega$ and $L_\infty = H$. This is the quantum system that we started with at the beginning of the subsection.

In the case $-\infty < \beta < \infty$ (positive temperatures), we obtain the Araki–Wyss representation for $\gamma = e^{-\beta h}$ described in (17.69).

Note that in the fermionic case the C^*-algebraic approach is better justified than in the bosonic case. The algebra $\mathrm{CAR}^{C^*}(\mathcal{Z})$ can be viewed as a natural algebra to describe observables of a fermionic system. Because of the boundedness of fermionic fields, it is more likely that we will be able to define a dynamics on this algebra, even in the presence of non-trivial interactions.

17.3 Lattices of von Neumann algebras on a Fock space

Let $\mathcal{W}$ be a complex Hilbert space. With every real closed subspace $\mathcal{V}$ of $\mathcal{W}$ we can naturally associate the von Neumann sub-algebra $\mathfrak{M}(\mathcal{V})$ of $B(\Gamma_{s/a}(\mathcal{W}))$ generated by fields based on $\mathcal{V}$. These von Neumann sub-algebras form a complete lattice. Properties of this lattice are studied in this section. They have important applications in quantum field theory.

The material of this section is closely related to the Araki–Woods and Araki–Wyss representations. In fact, the algebras $\mathrm{CCR}_{\gamma,1}$ and $\mathrm{CAR}_{\gamma,1}$ coincide with the algebras $\mathfrak{M}(\mathcal{V})$ for appropriate real subspaces $\mathcal{V}$ inside $\mathcal{Z} \oplus \overline{\mathcal{Z}}$.

17.3.1 Pair of subspaces in a Hilbert space

In this subsection we consider one of the classic problems of the theory of Hilbert spaces: how to describe a relative position of two closed subspaces.

Suppose that $\mathcal{Y}$ is a real or complex Hilbert space and $\mathcal{P}$, $\mathcal{Q}$ are closed subspaces in $\mathcal{Y}$. Let p, resp. q be the orthogonal projections onto $\mathcal{P}$, resp. $\mathcal{Q}$.

Proposition 17.44 $(\mathcal{P} \cap \mathcal{Q}) + (\mathcal{P}^\perp \cap \mathcal{Q}^\perp) = \mathrm{Ker}(p - q)$.

Proof The $\subset$ part is obvious.

Let $y \in \mathcal{Y}$. If $(p - q)y = 0$, then $y = py + (\mathbb{1} - q)y$, where $py = qy \in \mathcal{P} \cap \mathcal{Q}$ and $(\mathbb{1} - p)y = (\mathbb{1} - q)y \in \mathcal{P}^\perp \cap \mathcal{Q}^\perp$. This shows the $\supset$ part. $\qquad \square$

Proposition 17.45 *The following conditions are equivalent:*

(1) $\mathrm{Ker}(p - q) = \{0\}$.
(2) $\mathcal{P} \cap \mathcal{Q} = \mathcal{P}^\perp \cap \mathcal{Q}^\perp = \{0\}$.
(3) $\mathcal{P} \cap \mathcal{Q} = \{0\}$ *and* $\mathcal{P} + \mathcal{Q}$ *is dense in* $\mathcal{Y}$.

Proof The equivalence of (1) and (2) follows by Prop. 17.44.

The equivalence of (2) and (3) follows by

$$\{0\} = (\mathcal{P} + \mathcal{Q})^\perp = \mathcal{P}^\perp \cap \mathcal{Q}^\perp.$$

$\qquad \square$

Definition 17.46 *We say that a pair* $(\mathcal{P}, \mathcal{Q})$ *is in* generic position *if*

$$\mathcal{P} \cap \mathcal{Q} = \mathcal{P}^\perp \cap \mathcal{Q}^\perp = \mathcal{P}^\perp \cap \mathcal{Q} = \mathcal{P}^\perp \cap \mathcal{Q} = \{0\}.$$

Set $m := p + q - \mathbb{1}$, $n := p - q$, which are bounded self-adjoint operators. The following relations are immediate:

$$\begin{aligned}
n^2 &= \mathbb{1} - m^2 &= p + q - pq - qp, \\
nm &= -mn &= qp - pq, \\
\end{aligned}$$
$$-\mathbb{1} \leq m \leq \mathbb{1}, \ -\mathbb{1} \leq n \leq \mathbb{1}. \tag{17.70}$$

Proposition 17.47 $(\mathcal{P}, \mathcal{Q})$ *are in generic position iff*

$$\operatorname{Ker} m = \operatorname{Ker} n = \{0\}. \tag{17.71}$$

If this is the case, we also have

$$\operatorname{Ker}(m \pm \mathbb{1}) = \{0\}, \quad \operatorname{Ker}(n \pm \mathbb{1}) = \{0\}. \tag{17.72}$$

Proof The following identities follow from Prop. 17.44:

$$\operatorname{Ker} n = \operatorname{Ker} p \cap \operatorname{Ker} q + \operatorname{Ker}(\mathbb{1} - p) \cap \operatorname{Ker}(\mathbb{1} - q),$$
$$\operatorname{Ker} m = \operatorname{Ker} p \cap \operatorname{Ker}(\mathbb{1} - q) + \operatorname{Ker}(\mathbb{1} - p) \cap \operatorname{Ker} q.$$

This yields (17.71). We also obviously have

$$\operatorname{Ker}(n - \mathbb{1}) = \operatorname{Ker}(\mathbb{1} - p) \cap \operatorname{Ker} q, \qquad \operatorname{Ker}(n + \mathbb{1}) = \operatorname{Ker} p \cap \operatorname{Ker}(\mathbb{1} - q),$$
$$\operatorname{Ker}(m - \mathbb{1}) = \operatorname{Ker}(\mathbb{1} - p) \cap \operatorname{Ker}(\mathbb{1} - q), \qquad \operatorname{Ker}(m + \mathbb{1}) = \operatorname{Ker} p \cap \operatorname{Ker} q,$$

which proves (17.72). □

The following result is immediate:

Proposition 17.48 *Set*

$$\mathcal{Y}_0 := \left(\mathcal{P} \cap \mathcal{Q} + \mathcal{P}^\perp \cap \mathcal{Q}^\perp + \mathcal{P}^\perp \cap \mathcal{Q} + \mathcal{P} \cap \mathcal{Q}^\perp\right)^\perp,$$
$$\mathcal{P}_0 := \mathcal{P} \cap \mathcal{Y}_0, \quad \mathcal{Q}_0 := \mathcal{Q} \cap \mathcal{Y}_0.$$

Then the following direct sum decomposition holds:

$$\mathcal{Y} = \mathcal{P} \cap \mathcal{Q} \oplus \mathcal{P}^\perp \cap \mathcal{Q}^\perp \oplus \mathcal{P}^\perp \cap \mathcal{Q} \oplus \mathcal{P} \cap \mathcal{Q}^\perp \oplus \mathcal{Y}_0,$$
$$\mathcal{P} = \mathcal{P} \cap \mathcal{Q} \oplus \quad \{0\} \quad \oplus \quad \{0\} \quad \oplus \mathcal{P} \cap \mathcal{Q}^\perp \oplus \mathcal{P}_0,$$
$$\mathcal{Q} = \mathcal{P} \cap \mathcal{Q} \oplus \quad \{0\} \quad \oplus \mathcal{P}^\perp \cap \mathcal{Q} \oplus \quad \{0\} \quad \oplus \mathcal{Q}_0.$$

Moreover, the pair $(\mathcal{P}_0, \mathcal{Q}_0)$ *is in generic position in* $\mathcal{Y}_0$.

Theorem 17.49 *Let* $(\mathcal{P}, \mathcal{Q})$ *be a pair of subspaces in generic position. Then the following is true:*

(1) *There exists a unitary (orthogonal in the real case) involution* ϵ, *a subspace* $\mathcal{Z}$ *of* $\mathcal{Y}$ *such that* $\mathcal{Z}^\perp = \epsilon \mathcal{Z}$, *and a self-adjoint operator* χ *on* $\mathcal{Z}$ *satisfying*

$$0 < \chi < \tfrac{1}{2}\mathbb{1},$$
$$\left\{ ((\mathbb{1} - \chi)^{\frac{1}{2}} z, \epsilon \chi^{\frac{1}{2}} z) \ : \ z \in \mathcal{Z} \right\} = \mathcal{P},$$
$$\left\{ (\chi^{\frac{1}{2}} z, \epsilon (\mathbb{1} - \chi)^{\frac{1}{2}} z) \ : \ z \in \mathcal{Z} \right\} = \mathcal{Q}.$$

(2) *Set* $\rho := \chi(\mathbb{1} - 2\chi)^{-1}$. *Then*

$$\rho > 0,$$
$$\left\{ ((\mathbb{1} + \rho)^{\frac{1}{2}} z, \epsilon \rho^{\frac{1}{2}} z) \ : \ z \in \operatorname{Dom} \rho^{\frac{1}{2}} \right\} = \mathcal{P},$$
$$\left\{ (\rho^{\frac{1}{2}} z, \epsilon (\mathbb{1} + \rho)^{\frac{1}{2}} z) \ : \ z \in \operatorname{Dom} \rho^{\frac{1}{2}} \right\} = \mathcal{Q}.$$

Proof We introduce the polar decompositions of n and m:

$$n = |n|\epsilon = \epsilon|n|, \quad m = \kappa|m| = |m|\kappa.$$

Clearly, ϵ, κ are unitary/orthogonal operators satisfying $\epsilon^2 = \kappa^2 = \mathbb{1}$. Moreover, using (17.70) we obtain

$$\kappa\epsilon = -\epsilon\kappa, \quad \epsilon m = -m\epsilon, \quad \kappa n = -n\kappa. \tag{17.73}$$

Set

$$\mathcal{Z} := \mathrm{Ker}(\kappa - \mathbb{1}) = \mathrm{Ran}\,\mathbb{1}_{]0,1[}(m).$$

We have

$$\epsilon\mathcal{Z} = \mathrm{Ker}(\kappa + \mathbb{1}) = \mathrm{Ran}\,\mathbb{1}_{]-1,0[}(m),$$

hence $\epsilon\mathcal{Z} = \mathcal{Z}^{\perp}$.

Let $\mathbb{1}_{\mathcal{Z}}$ be the orthogonal projection from $\mathcal{Y}$ onto $\mathcal{Z}$. Clearly,

$$\mathbb{1}_{\mathcal{Z}} = \mathbb{1}_{]0,1[}(m), \quad \epsilon\mathbb{1}_{\mathcal{Z}}\epsilon = \mathbb{1} - \mathbb{1}_{\mathcal{Z}} = \mathbb{1}_{]-1,0[}(m).$$

We claim that $\mathcal{P}$ is the closure of $p\mathcal{Z}$. Indeed, $\mathcal{P}$ is closed and contains $p\mathcal{Z}$. Let $y \in \mathcal{P} \cap (p\mathcal{Z})^{\perp}$. Then

$$0 = (y|p\mathbb{1}_{\mathcal{Z}}y) = (y|\mathbb{1}_{\mathcal{Z}}y) = \|\mathbb{1}_{\mathcal{Z}}y\|^2,$$

hence $y \in \mathcal{Z}^{\perp}$. Therefore, using $q = m + \mathbb{1} - p$, we obtain

$$(y|qy) = (y|my) \leq 0.$$

Hence, $qy = 0$, and so $y \in \mathcal{Q}^{\perp}$. Remember that $y \in \mathcal{P}$, hence, by the generic position, $y = 0$.

Set $\chi := \frac{1}{2}\mathbb{1}_{\mathcal{Z}}(\mathbb{1} - m)$. Clearly, $0 < \chi < \frac{1}{2}\mathbb{1}$. Using $p = \frac{m+n+\mathbb{1}}{2}$, we obtain

$$p\mathbb{1}_{\mathcal{Z}} = \frac{m + \mathbb{1}}{2}\mathbb{1}_{\mathcal{Z}} + \frac{\epsilon|n|}{2}\mathbb{1}_{\mathcal{Z}}$$

$$= \frac{m + \mathbb{1}}{2}\mathbb{1}_{\mathcal{Z}} + \frac{\epsilon(\mathbb{1} - m^2)^{\frac{1}{2}}}{2}\mathbb{1}_{\mathcal{Z}}$$

$$= \left((\mathbb{1} - \chi) + \epsilon\chi^{\frac{1}{2}}(\mathbb{1} - \chi)^{\frac{1}{2}}\right)\mathbb{1}_{\mathcal{Z}}.$$

Thus

$$p\mathcal{Z} = \left((\mathbb{1} - \chi)^{\frac{1}{2}} + \epsilon\chi^{\frac{1}{2}}\right)(\mathbb{1} - \chi)^{\frac{1}{2}}\mathcal{Z}. \tag{17.74}$$

The operator

$$(\mathbb{1} - \chi)^{\frac{1}{2}}\mathbb{1}_{\mathcal{Z}} + \epsilon\chi^{\frac{1}{2}}\mathbb{1}_{\mathcal{Z}}$$

is an isometry from $\mathcal{Z}$ into $\mathcal{Y}$. Therefore,

$$\left((\mathbb{1} - \chi)^{\frac{1}{2}} + \epsilon \chi^{\frac{1}{2}}\right) \mathcal{Z} \tag{17.75}$$

is closed. $(\mathbb{1} - \chi)^{\frac{1}{2}} \mathcal{Z}$ is dense in $\mathcal{Z}$. Therefore, (17.75) is the closure of (17.74).

We proved that $\mathcal{P}$ is the closure of $p\mathcal{Z}$. Hence, (17.75) equals $\mathcal{P}$. This completes the proof of the first identity of (1).

To prove (2), we note that every $z \in \mathcal{Z}$ can be written as

$$z = (\mathbb{1} + 2\rho)^{\frac{1}{2}} z_1, \quad z_1 \in \mathrm{Dom}\, \rho^{\frac{1}{2}}.$$

We then have

$$(\mathbb{1} - \chi)^{\frac{1}{2}} z + \epsilon \chi^{\frac{1}{2}} z = (\mathbb{1} + \rho)^{\frac{1}{2}} z_1 + \epsilon \rho^{\frac{1}{2}} z_1,$$
$$\chi^{\frac{1}{2}} z + \epsilon (\mathbb{1} - \chi)^{\frac{1}{2}} z = \rho^{\frac{1}{2}} z_1 + \epsilon (\mathbb{1} + \rho)^{\frac{1}{2}} z_1,$$

which immediately implies (2). $\qquad \square$

17.3.2 Real subspaces in a Hilbert space

This subsection is devoted to another classic problem, closely related to the previous subsection: how to describe the position of a closed real subspace in a complex Hilbert space. This analysis will then be used in both the bosonic and the fermionic case.

Let $(\mathcal{W}, (\cdot|\cdot))$ be a complex Hilbert space. Then $(\mathcal{W}_{\mathbb{R}}, \mathrm{Re}(\cdot|\cdot))$ is a real Hilbert space and $(\mathcal{W}_{\mathbb{R}}, \mathrm{Im}(\cdot|\cdot))$ is a symplectic space. Clearly, if $\mathcal{V} \subset \mathcal{W}$ is a real vector space, $\mathcal{V} \cap i\mathcal{V}$ and $\mathcal{V} + i\mathcal{V}$ are complex vector spaces.

Definition 17.50 *If* $\mathcal{U} \subset \mathcal{W}$, *then we have three kinds of complements of* $\mathcal{U}$:

$$\mathcal{U}^{\perp} := \{w \in \mathcal{W} \ : \ (v|w) = 0, \ v \in \mathcal{U}\},$$
$$\mathcal{U}^{\mathrm{perp}} := \{w \in \mathcal{W} \ : \ \mathrm{Re}(v|w) = 0, \ v \in \mathcal{U}\},$$
$$i\mathcal{U}^{\mathrm{perp}} = \{w \in \mathcal{W} \ : \ \mathrm{Im}(v|w) = 0, \ v \in \mathcal{U}\} = (i\mathcal{U})^{\mathrm{perp}}.$$

$\mathcal{U}^{\perp}, \mathcal{U}^{\mathrm{perp}}$, *resp.* $i\mathcal{U}^{\mathrm{perp}}$ *will be called the* complex orthogonal, *the* real orthogonal, *resp. the* symplectic complement *of* $\mathcal{U}$.

Clearly, $\mathcal{U}^{\mathrm{perp}}$ and $i\mathcal{U}^{\mathrm{perp}}$ are closed real vector subspaces of $\mathcal{W}$. If $\mathcal{V}$ is a complex vector subspace, then $\mathcal{V}^{\mathrm{perp}} = i\mathcal{V}^{\mathrm{perp}} = \mathcal{V}^{\perp}$.

Let $\mathcal{V}$ be a closed real subspace of $\mathcal{W}$. Let us remark that $(i\mathcal{V})^{\mathrm{perp}} = i(\mathcal{V})^{\mathrm{perp}}$. Moreover, $i(i\mathcal{V}^{\mathrm{perp}})^{\mathrm{perp}} = \mathcal{V}$.

Definition 17.51 *We will say that* $\mathcal{V} \subset \mathcal{W}$ *is in* generic position *if*

$$\mathcal{V} \cap i\mathcal{V} = \mathcal{V} \cap i\mathcal{V}^{\mathrm{perp}} = \{0\}.$$

Proposition 17.52 *The following conditions are equivalent:*

(1) $\mathcal{V}$ *is in generic position.*
(2) $(\mathcal{V}, i\mathcal{V})$ *is in generic position in* $\mathcal{W}_{\mathbb{R}}$.
(3) $(\mathcal{V}, i\mathcal{V}^{\text{perp}})$ *is in generic position in* $\mathcal{W}_{\mathbb{R}}$.

The following result is an analog of Prop. 17.48:

Proposition 17.53 *Let* $\mathcal{V}$ *be a closed real subspace of* $\mathcal{W}$. *Set*

$$\mathcal{W}_1 := \mathcal{V} \cap i\mathcal{V}^{\text{perp}} + (i\mathcal{V} \cap \mathcal{V}^{\text{perp}}), \quad \mathcal{V}_1 := \mathcal{V} \cap i\mathcal{V}^{\text{perp}},$$

$$\mathcal{W}_+ := \mathcal{V} \cap i\mathcal{V}, \qquad\qquad \mathcal{W}_- := \mathcal{V}^{\text{perp}} \cap i\mathcal{V}^{\text{perp}},$$

$$\mathcal{W}_0 := (\mathcal{W}_+ + \mathcal{W}_- + \mathcal{W}_1)^{\perp}, \qquad \mathcal{V}_0 := \mathcal{V} \cap \mathcal{W}_0.$$

Then the following is true:

(1) $\mathcal{W}_-, \mathcal{W}_+, \mathcal{W}_0, \mathcal{W}_1$ *are closed complex subspaces of* $\mathcal{W}$.
(2) *The following direct sum decompositions hold:*

$$\mathcal{W} = \mathcal{W}_1 \oplus \mathcal{W}_+ \oplus \mathcal{W}_- \oplus \mathcal{W}_0,$$

$$\mathcal{V} = \mathcal{V}_1 \oplus \mathcal{W}_+ \oplus \{0\} \oplus \mathcal{V}_0,$$

$$i\mathcal{V}^{\text{perp}} = \mathcal{V}_1 \oplus \{0\} \oplus \mathcal{W}_- \oplus i\mathcal{V}_0^{\text{perp}},$$

where $\mathcal{V}_0^{\text{perp}}$ *is the real orthogonal of* $\mathcal{V}_0$ *inside* $\mathcal{W}_0$.
(3) $\mathcal{W}_1 \cap \mathcal{V} = \mathcal{W}_1 \cap i\mathcal{V}^{\text{perp}} = \mathcal{V}_1 = i\mathcal{V}_1^{\text{perp}}$, *where* $\mathcal{V}_1^{\text{perp}}$ *is the real orthogonal complement of* $\mathcal{V}_1$ *inside* $\mathcal{W}_1$.
(4) $\mathcal{W}_+ \cap \mathcal{V} = \mathcal{W}_+, \quad \mathcal{W}_+ \cap i\mathcal{V}^{\text{perp}} = \{0\}$.
(5) $\mathcal{W}_- \cap \mathcal{V} = \{0\}, \quad \mathcal{W}_- \cap i\mathcal{V}^{\text{perp}} = \mathcal{W}_-$.
(6) $\mathcal{W}_0 \cap \mathcal{V} = \mathcal{V}_0, \quad \mathcal{W}_0 \cap i\mathcal{V}^{\text{perp}} = i\mathcal{V}_0^{\text{perp}}$. *Moreover,* $\mathcal{V}_0$ *is in generic position in* $\mathcal{W}_0$.

In other words, given a closed real subspace $\mathcal{V} \subset \mathcal{W}$, one can decompose $\mathcal{W}$ into four complex subspaces such that $\mathcal{V}$ decomposes into subspaces which are respectively complex, in generic position, Lagrangian and zero.

We can define the operators m, n for the pair of subspaces $\mathcal{V}$, $\mathcal{V}^{\text{perp}}$, as in the previous subsection. They are self-adjoint in the sense of the real Hilbert space $\mathcal{W}_{\mathbb{R}}$. m is linear, whereas n is anti-linear on $\mathcal{W}$. Therefore, κ is unitary and ϵ is anti-unitary. We can use $\epsilon\big|_{\mathcal{Z}}$ as the (external) conjugation and identify $\epsilon\mathcal{Z}$ with $\overline{\mathcal{Z}}$. This gives a unitary identification

$$\mathcal{W} \simeq \mathcal{Z} \oplus \overline{\mathcal{Z}}.$$

Note that

$$\epsilon(z_1, \overline{z}_2) := (z_2, \overline{z}_1)$$

and $\epsilon\mathcal{V} = i\mathcal{V}^{\text{perp}}$.

Now Thm. 17.49 (1) can be reformulated in the following way, which is adapted to the Araki–Wyss representation:

$$\left\{\left((\mathbb{1}-\chi)^{\frac{1}{2}}z, \overline{\chi}^{\frac{1}{2}}\overline{z}\right) \ : \ z \in \mathcal{Z}\right\} = \mathcal{V},$$

$$\left\{\left(\chi^{\frac{1}{2}}z, (\mathbb{1}-\overline{\chi})^{\frac{1}{2}}\overline{z}\right) \ : \ z \in \mathcal{Z}\right\} = \mathrm{i}\mathcal{V}^{\mathrm{perp}}. \tag{17.76}$$

Thm. 17.49 (2) can be reformulated as follows, which is adapted to the Araki–Woods representation:

$$\left\{\left((\mathbb{1}+\rho)^{\frac{1}{2}}z, \overline{\rho}^{\frac{1}{2}}\overline{z}\right) \ : \ z \in \mathrm{Dom}\,\rho^{\frac{1}{2}}\right\} = \mathcal{V},$$

$$\left\{\left(\rho^{\frac{1}{2}}z, (\mathbb{1}+\overline{\rho})^{\frac{1}{2}}\overline{z}\right) \ : \ z \in \mathrm{Dom}\,\rho^{\frac{1}{2}}\right\} = \mathrm{i}\mathcal{V}^{\mathrm{perp}}. \tag{17.77}$$

In the following proposition, which follows immediately from (17.76) and (17.77), for typographical reasons we will write τz for $\overline{z}$, where $z \in \mathcal{Z}$. We consider $\mathcal{W}$ as a Kähler space with the Euclidean, resp. symplectic form given by $\mathrm{Re}(\cdot|\cdot)$, resp. $\mathrm{Im}(\cdot|\cdot)$. It is equipped with an anti-involution and conjugation

$$\mathrm{j} = \begin{bmatrix} \mathrm{i}\mathbb{1} & 0 \\ 0 & -\mathrm{i}\mathbb{1} \end{bmatrix}, \quad \epsilon = \begin{bmatrix} 0 & \tau^{-1} \\ \tau & 0 \end{bmatrix}.$$

We also have the operators χ and ρ on $\mathcal{Z}$. Recall that the notion of a j-positive orthogonal transformation was defined in Def. 16.8.

Proposition 17.54 *Let $\mathcal{V}$ be a closed real vector subspace of a complex Hilbert space $\mathcal{W}$ in generic position.*

(1) *Define the operator r_a on $\mathcal{W}$ by*

$$r_\mathrm{a} = \begin{bmatrix} (\mathbb{1}-\chi)^{\frac{1}{2}} & \chi^{\frac{1}{2}}\tau^{-1} \\ -\overline{\chi}^{\frac{1}{2}}\tau & (\mathbb{1}-\overline{\chi})^{\frac{1}{2}} \end{bmatrix}.$$

Then r_a is a j-positive orthogonal transformation on $\mathcal{W}$ commuting with ϵ, and $r_\mathrm{a}\mathcal{Z} = \mathcal{V}$.

(2) *Define the operator $r_\mathrm{s} : \mathrm{Dom}(\rho^{\frac{1}{2}}) \oplus \overline{\mathrm{Dom}(\rho^{\frac{1}{2}})} \to \mathcal{W}$ by*

$$r_\mathrm{s} = \begin{bmatrix} (\mathbb{1}+\rho)^{\frac{1}{2}} & \rho^{\frac{1}{2}}\tau^{-1} \\ \overline{\rho}^{\frac{1}{2}}\tau & (\mathbb{1}+\overline{\rho})^{\frac{1}{2}} \end{bmatrix}.$$

Then r_s is a positive symplectic transformation on $\mathcal{W}$ commuting with ϵ, and $r_\mathrm{s}\mathrm{Dom}(\rho^{\frac{1}{2}}) = \mathcal{V}$.

Note that the transformations r_s, resp. r_a yield the Bogoliubov rotations implemented by the operators (17.27) and (17.59), which were used in Subsect. 17.1.4, resp. 17.2.4 to introduce the Araki–Woods, resp. Araki–Wyss representations.

17.3.3 Complete lattices

In this subsection we recall some definitions about abstract lattices. They provide a convenient language that can be used to express some properties of a class of von Neumann algebras acting on a Fock space.

Definition 17.55 *Let $(X, \leq)$ be an ordered set. Let $\{x_i \ : \ i \in I\}$ be a non-empty subset of X. One says that $u \in X$ is a* largest minorant *of $\{x_i \ : \ i \in I\}$ if*

(1) $i \in I$ *implies* $u \leq x_i$,
(2) $v \leq x_i$ *for all* $i \in I$ *implies* $v \leq u$.

If $\{x_i \ : \ i \in I\}$ has a largest minorant, then it is unique. The largest minorant of a set $\{x_i \ : \ i \in I\}$ is usually denoted by $\bigwedge_{i \in I} x_i$.

We define similarly the smallest majorant of $\{x_i \ : \ i \in I\}$, which is usually denoted by $\bigvee_{i \in I} x_i$.

One says that $(X, \leq)$ is a complete lattice *if every non-empty subset of X has the largest minorant and the smallest majorant. It is then equipped with the operations $\wedge$ and $\vee$.*

Definition 17.56 *One says that the complete lattice $(X, \leq)$ is* complemented *if it is equipped with a map $X \ni x \mapsto \sim x \in X$ such that*

(1) $\sim (\sim x) = x$,
(2) $x_1 \leq x_2$ *implies* $\sim x_2 \leq \sim x_1$,
(3) $\sim \bigwedge_{i \in I} x_i = \bigvee_{i \in I} \sim x_i$.

The operation $\sim$ will be called a complementation.

Let us give some examples of complemented lattices that will be useful in the sequel.

Example 17.57 (1) *Let $\mathcal{W}$ be a topological vector space. Then the set $\mathrm{Subsp}(\mathcal{W})$ of closed vector subspaces of $\mathcal{W}$ equipped with the order $\subset$ is a complete lattice with*

$$\bigwedge_{i \in I} \mathcal{V}_i = \bigcap_{i \in I} \mathcal{V}_i, \qquad \bigvee_{i \in I} \mathcal{V}_i = \Big(\sum_{i \in I} \mathcal{V}_i \Big)^{\mathrm{cl}}.$$

(2) *If $\mathcal{W}$ is a (real or complex) Hilbert space, then the map $\mathcal{V} \mapsto \mathcal{V}^\perp$ is a complementation on $(\mathrm{Subsp}(\mathcal{W}), \subset)$.*
(3) *If $\mathcal{W}$ is a complex Hilbert space, then $(\mathrm{Subsp}(\mathcal{W}_{\mathbb{R}}), \subset)$ denotes the lattice of closed real subspaces of $\mathcal{W}$. Then $\mathcal{V} \mapsto \mathcal{V}^{\mathrm{perp}}$ and $\mathcal{V} \mapsto i\mathcal{V}^{\mathrm{perp}}$ are complementations on this lattice.*
(4) *Now let $\mathcal{H}$ be a Hilbert space and $\mathrm{vN}(\mathcal{H})$ be the set of von Neumann algebras in $B(\mathcal{H})$ equipped with the order $\subset$. Then $(\mathrm{vN}(\mathcal{H}), \subset)$ is also a complete*

lattice with

$$\bigwedge_{i \in I} \mathfrak{M}_i = \bigcap_{i \in I} \mathfrak{M}_i, \quad \bigvee_{i \in I} \mathfrak{M}_i = \left(\bigcup_{i \in I} \mathfrak{M}_i \right)''.$$

The map $\mathfrak{M} \mapsto \mathfrak{M}'$ *is a complementation on* $(\mathrm{vN}(\mathcal{H}), \subset)$. ·

17.3.4 Lattice of von Neumann algebras on a bosonic Fock space

Let $\mathcal{W}$ be a complex Hilbert space. We identify $\mathcal{W}$ with $\mathrm{Re}(\mathcal{W} \oplus \overline{\mathcal{W}})$ using $w \mapsto \frac{1}{\sqrt{2}}(w, \overline{w})$; see (1.29). Consider the Hilbert space $\Gamma_s(\mathcal{W})$ and the corresponding Fock representation $\mathcal{W} \ni w \mapsto W(w) \in U(\Gamma_s(\mathcal{W}))$.

Definition 17.58 *For a real subspace* $\mathcal{V} \subset \mathcal{W}$, *we define the von Neumann algebra*

$$\mathfrak{M}_s(\mathcal{V}) := \{W(w) \ : \ w \in \mathcal{V}\}'' \subset B(\Gamma_s(\mathcal{W})).$$

Using von Neumann's density theorem and the fact that $\mathcal{W} \in w \mapsto W(w)$ is strongly continuous (see Thm. 9.5), we see that $\mathfrak{M}_s(\mathcal{V}) = \mathfrak{M}_s(\mathcal{V}^{\mathrm{cl}})$. Therefore, in the sequel it suffices to consider closed real subspaces of $\mathcal{W}$.

Theorem 17.59 (1) $\mathfrak{M}_s(\mathcal{V}_1) = \mathfrak{M}_s(\mathcal{V}_2)$ *iff* $\mathcal{V}_1 = \mathcal{V}_2$;
(2) $\mathcal{V}_1 \subset \mathcal{V}_2$ *implies* $\mathfrak{M}_s(\mathcal{V}_1) \subset \mathfrak{M}_s(\mathcal{V}_2)$;
(3) $\mathfrak{M}_s(\mathcal{W}) = B(\Gamma_s(\mathcal{W}))$ *and* $\mathfrak{M}_s(\{0\}) = \mathbb{C}\mathbb{1}$;
(4) $\mathfrak{M}_s(\vee_{i \in I} \mathcal{V}_i) = \vee_{i \in I} \mathfrak{M}_s(\mathcal{V}_i)$;
(5) $\mathfrak{M}_s(\cap_{i \in I} \mathcal{V}_i) = \cap_{i \in I} \mathfrak{M}_s(\mathcal{V}_i)$;
(6) $\mathfrak{M}_s(\mathcal{V})' = \mathfrak{M}_s(i\mathcal{V}^{\mathrm{perp}})$;
(7) $\mathfrak{M}_s(\mathcal{V})$ *is a factor iff* $\mathcal{V} \cap i\mathcal{V}^{\mathrm{perp}} = \{0\}$.

Proof To prove (1), let $\mathcal{V}_1, \mathcal{V}_2$ be two distinct closed subspaces. We may assume that $\mathcal{V}_2 \not\subset \mathcal{V}_1$, and hence $i\mathcal{V}_1^{\mathrm{perp}} \not\subset i\mathcal{V}_2^{\mathrm{perp}}$. For $w \in i\mathcal{V}_1^{\mathrm{perp}} \backslash i\mathcal{V}_2^{\mathrm{perp}}$, we have $W(w) \in \mathfrak{M}_s(\mathcal{V}_1)' \backslash \mathfrak{M}_s(\mathcal{V}_2)'$. This implies that $\mathfrak{M}_s(\mathcal{V}_1)' \neq \mathfrak{M}_s(\mathcal{V}_2)'$, which proves (1).

(2) and (3) are immediate, as are the $\supset$ part of (4) and the $\subset$ part of (5). The $\subset$ part of (4) follows again from the strong continuity of $w \mapsto W(w)$. If we know (6), then the $\supset$ part of (5) follows from the $\subset$ part of (4). (7) follows from (1), (5) and (6).

Thus it remains to prove (6). Assume first that $\mathcal{V}$ is in generic position in $\mathcal{W}$. Then, using Thm. 17.49 and identifying $\epsilon \mathcal{Z}$ with $\overline{\mathcal{Z}}$, we obtain a decomposition $\mathcal{W} = \mathcal{Z} \oplus \overline{\mathcal{Z}}$ and a positive operator ρ on $\mathcal{Z}$ such that

$$\left\{ \left((\mathbb{1} + \rho)^{\frac{1}{2}} z, \overline{\rho}^{\frac{1}{2}} \overline{z} \right) \ : \ z \in \mathcal{Z} \right\} = \mathcal{V}.$$

This implies that $\mathfrak{M}_s(\mathcal{V})$ is the left Araki–Woods algebra $\mathrm{CCR}_{\gamma,\mathrm{l}}$. By Thm. 17.24, we know that the commutant of $\mathrm{CCR}_{\gamma,\mathrm{l}}$ is $\mathrm{CCR}_{\gamma,\mathrm{r}}$. But, again by Thm. 17.49,

$$\left\{ \left(\rho^{\frac{1}{2}} z + (\mathbb{1} + \overline{\rho})^{\frac{1}{2}} \overline{z} \right) \ : \ z \in \mathcal{Z} \right\} = i\mathcal{V}^{\mathrm{perp}}.$$

Therefore, $\mathrm{CCR}_{\gamma,\mathrm{r}}$ coincides with $\mathfrak{M}_{\mathrm{s}}(i\mathcal{V}^{\mathrm{perp}})$. This ends the proof of (6), if $\mathcal{V}$ is in generic position.

For an arbitrary real subspace $\mathcal{V}$, we write as in Prop. 17.53:

$$\mathcal{W} = \mathcal{W}_+ \oplus \mathcal{W}_0 \oplus \mathcal{W}_1 \oplus \mathcal{W}_-,$$

$$\mathcal{V} = \mathcal{W}_+ \oplus \mathcal{V}_0 \oplus \mathcal{V}_1 \oplus \{0\},$$

$$i\mathcal{V}^{\mathrm{perp}} = \{0\} \oplus i\mathcal{V}_0^{\mathrm{perp}} \oplus \mathcal{V}_1 \oplus \mathcal{W}_-,$$

where $\mathcal{V}_0$ is in generic position and $\mathcal{W}_1 = \mathbb{C}\mathcal{V}_1$. Using the exponential law, we have the unitary identifications

$$B\big(\Gamma_{\mathrm{s}}(\mathcal{W})\big) \simeq B\big(\Gamma_{\mathrm{s}}(\mathcal{W}_+)\big) \otimes B\big(\Gamma_{\mathrm{s}}(\mathcal{W}_0)\big) \otimes B\big(\Gamma_{\mathrm{s}}(\mathcal{W}_1)\big) \otimes B\big(\Gamma_{\mathrm{s}}(\mathcal{W}_-)\big),$$

$$\mathfrak{M}_{\mathrm{s}}(\mathcal{V}) \simeq B\big(\Gamma_{\mathrm{s}}(\mathcal{W}_+)\big) \otimes \mathfrak{M}_{\mathrm{s}}(\mathcal{V}_0) \otimes \mathfrak{M}_{\mathrm{s}}(\mathcal{V}_1) \otimes \mathbb{1},$$

$$\mathfrak{M}_{\mathrm{s}}(i\mathcal{V}^{\mathrm{perp}}) \simeq \mathbb{1} \otimes \mathfrak{M}_{\mathrm{s}}(i\mathcal{V}_0^{\mathrm{perp}}) \otimes \mathfrak{M}_{\mathrm{s}}(\mathcal{V}_1) \otimes B\big(\Gamma_{\mathrm{s}}(\mathcal{W}_-)\big).$$

Since $\mathcal{V}_0$ is in generic position in $\mathcal{W}_0$, $\mathfrak{M}_{\mathrm{s}}(\mathcal{V}_0)' = \mathfrak{M}_{\mathrm{s}}(i\mathcal{V}_0^{\mathrm{perp}})$. Since $\mathcal{W}_1 = \mathbb{C}\mathcal{V}_1$, using the real-wave representation of Sect. 9.3, we see that $\mathfrak{M}_{\mathrm{s}}(\mathcal{V}_1)' = \mathfrak{M}_{\mathrm{s}}(\mathcal{V}_1)$. Therefore, $\mathfrak{M}_{\mathrm{s}}(\mathcal{V})' = \mathfrak{M}_{\mathrm{s}}(i\mathcal{V}^{\mathrm{perp}})$, which completes the proof (6). $\qquad\square$

We can interpret Thm. 17.59 as the fact that the map $\mathcal{V} \mapsto \mathfrak{M}_{\mathrm{s}}(\mathcal{V})$ is an order preserving isomorphism between the complete lattice of closed real subspaces of $\mathcal{W}$ and the complete lattice of von Neumann algebras $\mathfrak{M}_{\mathrm{s}}(\mathcal{V}) \subset B\big(\Gamma_{\mathrm{s}}(\mathcal{W})\big)$, preserving the operations $\wedge$, $\vee$, and the complementations given respectively by the symplectic complement and the commutant.

17.3.5 Lattice of von Neumann algebras on a fermionic Fock space

In this subsection we consider the fermionic analog of Thm. 17.59. Again let $\mathcal{W}$ be a complex Hilbert space, and let us identify $\mathcal{W}$ with $\mathrm{Re}(\mathcal{W} \oplus \overline{\mathcal{W}})$ using $w \mapsto (w, \overline{w})$; see (1.29). Consider the Hilbert space $\Gamma_{\mathrm{a}}(\mathcal{W})$ and the corresponding Fock representation $\mathcal{W} \ni w \mapsto \phi(w) \in B_{\mathrm{h}}\big(\Gamma_{\mathrm{a}}(\mathcal{W})\big)$.

Definition 17.60 *For a real subspace $\mathcal{V} \subset \mathcal{W}$, we define the von Neumann algebra*

$$\mathfrak{M}_{\mathrm{a}}(\mathcal{V}) := \{\phi(w) \,:\, w \in \mathcal{V}\}'' \subset B\big(\Gamma_{\mathrm{a}}(\mathcal{W})\big).$$

As usual, set $\Lambda = (-1)^{N(N-1)/2}$.

Note first that, by the norm continuity of $\mathcal{W} \ni w \mapsto \phi(w)$, we have $\mathfrak{M}_{\mathrm{a}}(\mathcal{V}) = \mathfrak{M}_{\mathrm{a}}(\mathcal{V}^{\mathrm{cl}})$. Therefore, in the sequel it suffices to consider closed real subspaces of $\mathcal{W}$.

Theorem 17.61 (1) $\mathfrak{M}_{\mathrm{a}}(\mathcal{V}_1) = \mathfrak{M}_{\mathrm{a}}(\mathcal{V}_2)$ *iff* $\mathcal{V}_1 = \mathcal{V}_2$,
(2) $\mathcal{V}_1 \subset \mathcal{V}_2$ *implies* $\mathfrak{M}_{\mathrm{a}}(\mathcal{V}_1) \subset \mathfrak{M}_{\mathrm{a}}(\mathcal{V}_2)$,
(3) $\mathfrak{M}_{\mathrm{a}}(\mathcal{W}) = B\big(\Gamma_{\mathrm{s}}(\mathcal{W})\big)$ *and* $\mathfrak{M}_{\mathrm{a}}(\{0\}) = \mathbb{C}\mathbb{1}$,
(4) $\mathfrak{M}_{\mathrm{a}}(\vee_{i \in I}\mathcal{V}_i) = \vee_{i \in I}\mathfrak{M}_{\mathrm{a}}(\mathcal{V}_i)$,

(5) $\mathfrak{M}_a(\cap_{i\in I}\mathcal{V}_i) = \cap_{i\in I}\mathfrak{M}_a(\mathcal{V}_i)$,

(6) $\mathfrak{M}_a(\mathcal{V})' = \Lambda\mathfrak{M}_a(i\mathcal{V}^{\mathrm{perp}})\Lambda$.

The proof of Thm. 17.61 is very similar to the proof of Thm. 17.59. The main additional difficulty is the behavior of the fermionic fields under the tensor product, which is studied in the following theorem.

Theorem 17.62 *Let $\mathcal{W}_i$, $i = 1, 2$, be two Hilbert spaces and $\mathcal{W} = \mathcal{W}_1 \oplus \mathcal{W}_2$. Let us unitarily identify $\Gamma_a(\mathcal{W})$ with $\Gamma_a(\mathcal{W}_1) \otimes \Gamma_a(\mathcal{W}_2)$ by the exponential law (see Subsect. 3.3.7). Let $\mathcal{V}_i \subset \mathcal{W}_i$ be closed real subspaces. Then*

$$\mathfrak{M}_a(\mathcal{V}_1 \oplus \mathcal{V}_2) \simeq \left(\mathfrak{M}_a(\mathcal{V}_1)\otimes\mathbb{1}+(-1)^{N_1\otimes N_2}\mathbb{1}\otimes\mathfrak{M}_a(\mathcal{V}_2)(-1)^{N_1\otimes N_2}\right)''; \quad (17.78)$$

$$\mathfrak{M}_a(\mathcal{V}_1 \oplus \{0\}) \simeq \mathfrak{M}_a(\mathcal{V}_1) \otimes \mathbb{1}; \quad (17.79)$$

$$\mathfrak{M}_a(\mathcal{W}_1 \oplus \mathcal{V}_2) \simeq B(\Gamma_a(\mathcal{W}_1)) \otimes \mathfrak{M}_a(\mathcal{V}_2); \quad (17.80)$$

$$\Lambda\mathfrak{M}_a(\mathcal{V}_1 \oplus \mathcal{W}_2)\Lambda \simeq \Lambda_1\mathfrak{M}_a(\mathcal{V}_1)\Lambda_1 \otimes B(\Gamma_a(\mathcal{W}_2)); \quad (17.81)$$

$$\Lambda\mathfrak{M}_a(\{0\} \oplus \mathcal{V}_2)\Lambda \simeq \mathbb{1} \otimes \Lambda_2\mathfrak{M}_a(\mathcal{V}_2)\Lambda_2. \quad (17.82)$$

Proof Clearly, for $v_1 \in \mathcal{V}_1$,

$$\phi(v_1, 0) \simeq \phi(v_1) \otimes \mathbb{1}.$$

Therefore, (17.79) holds. By Thm. 3.56, for $v_2 \in \mathcal{V}_2$, we have

$$\phi(0, v_2) \simeq (-1)^{N_1} \otimes \phi(v_2) = (-1)^{N_1\otimes N_2}\mathbb{1} \otimes \phi(v_2)(-1)^{N_1\otimes N_2}.$$

Therefore,

$$\mathfrak{M}_a(\{0\} \oplus \mathcal{V}_2) \simeq (-1)^{N_1\otimes N_2}\mathbb{1} \otimes \mathfrak{M}_a(\mathcal{V}_2)(-1)^{N_1\otimes N_2}. \quad (17.83)$$

Now (17.79) and (17.83) imply (17.78), which implies

$$\mathfrak{M}_a(\mathcal{V}_1 \otimes \mathcal{W}_2) \simeq (-1)^{N_1\otimes N_2}\mathfrak{M}_a(\mathcal{V}_1) \otimes B(\Gamma_a(\mathcal{V}_2))(-1)^{N_1\otimes N_2}. \quad (17.84)$$

Noting that $\Lambda \simeq (-1)^{N_1\otimes N_2}\Lambda_1 \otimes \Lambda_2$, (17.83) implies (17.82), and (17.84) implies (17.81). $\qquad\qquad\square$

Proof of Thm. 17.61. To prove (1), let $\mathcal{V}_1$, $\mathcal{V}_2$ be two distinct closed subspaces. We may assume that $\mathcal{V}_2 \not\subset \mathcal{V}_1$, and hence $i\mathcal{V}_1^{\mathrm{perp}} \not\subset i\mathcal{V}_2^{\mathrm{perp}}$. For $w \in i\mathcal{V}_1^{\mathrm{perp}}\backslash i\mathcal{V}_2^{\mathrm{perp}}$, using (3.30), we have $\Lambda\phi(w)\Lambda \in \mathfrak{M}_a(\mathcal{V}_1)'\backslash\mathfrak{M}_a(\mathcal{V}_2)'$. This implies that $\mathfrak{M}_a(\mathcal{V}_1)' \neq \mathfrak{M}_a(\mathcal{V}_2)'$, which implies (1). (2) and (3) are immediate. The proof of (4), (5) are similar to the bosonic case, given (6).

It remains to prove (6). Assume first that $\mathcal{V}$ is in generic position in $\mathcal{W}$. By Prop. 17.53, we can write

$$\mathcal{W} = \mathcal{W}_0 \oplus \mathcal{W}_1,$$
$$\mathcal{V} = \mathcal{V}_0 \oplus \mathcal{V}_1,$$
$$i\mathcal{V}^{\mathrm{perp}} = i\mathcal{V}_0^{\mathrm{perp}} \oplus \mathcal{V}_1,$$

where $\mathcal{V}_0$ is in generic position in $\mathcal{W}_0$ and $i\mathcal{V}_1^{\mathrm{perp}} = \mathbb{C}\mathcal{V}_1$, where the orthogonal complement is taken inside $\mathcal{W}$. Again using Thm. 17.49, we obtain a decomposition $\mathcal{W}_0 = \mathcal{Z} \oplus \overline{\mathcal{Z}}$ together with a self-adjoint operator $0 \leq \chi \leq \frac{1}{2}\mathbb{1}$ such that $\mathrm{Ker}\, \chi = \mathrm{Ker}(\chi - \frac{1}{2}\mathbb{1}) = \{0\}$ and

$$\{(\mathbb{1} - \chi)^{\frac{1}{2}}z + \overline{\chi}^{\frac{1}{2}}\overline{z} \, : \, z \in \mathcal{Z}\} \oplus \mathcal{V}_1 = \mathcal{V},$$
$$\{\chi^{\frac{1}{2}}z + (\mathbb{1} - \overline{\chi})^{\frac{1}{2}}\overline{z} \, : \, z \in \mathcal{Z}\} \oplus \mathcal{V}_1 = i\mathcal{V}^{\mathrm{perp}}.$$

Then we are in the framework of Thm. 17.42, which implies that $\mathfrak{M}_a(\mathcal{V})' = \Lambda\mathfrak{M}_a(i\mathcal{V}^{\mathrm{perp}})\Lambda$.

For an arbitrary $\mathcal{V}$, we write

$$\mathcal{W} = \mathcal{W}_+ \oplus \mathcal{W}_0 \oplus \mathcal{W}_1 \oplus \mathcal{W}_-,$$

$$\mathcal{V} = \mathcal{W}_+ \oplus \mathcal{V}_0 \oplus \mathcal{V}_1 \oplus \{0\},$$

$$i\mathcal{V}^{\mathrm{perp}} = \{0\} \oplus i(\mathcal{V}_0 \oplus \mathcal{V}_1)^{\mathrm{perp}} \oplus \mathcal{W}_-,$$

where $\mathcal{V}_0, \mathcal{V}_1$ are as above. Using Thm. 17.62, we have the unitary identifications

$$B(\Gamma_a(\mathcal{W})) \simeq B(\Gamma_a(\mathcal{W}_+)) \otimes B(\Gamma_a(\mathcal{W}_0 \oplus \mathbb{C}\mathcal{V}_1)) \otimes B(\Gamma_a(\mathcal{W}_-)),$$

$$\mathfrak{M}_a(\mathcal{V}) \simeq B(\Gamma_a(\mathcal{W}_+)) \otimes \mathfrak{M}_a(\mathcal{V}_0 \oplus \mathcal{V}_1) \otimes \mathbb{1}.$$

Let N_{01}, resp. N_{01-} be the number operator on $\Gamma_a(\mathcal{W}_0 \oplus \mathcal{W}_1)$, resp $\Gamma_a(\mathcal{W}_0 \oplus \mathcal{W}_1 \oplus \mathcal{W}_-)$. We define Λ_{01}, resp. Λ_{01-} in the obvious way. The commutant of $\mathfrak{M}_a(\mathcal{V})$ is

$$\mathfrak{M}_a(\mathcal{V})' \simeq \mathbb{1} \otimes \mathfrak{M}_a(\mathcal{V}_0 \oplus \mathcal{V}_1)' \otimes B(\Gamma_a(\mathcal{W}_-))$$
$$= \mathbb{1} \otimes \Lambda_{01}\mathfrak{M}_a(i(\mathcal{V}_0 \oplus \mathcal{V}_1)^{\mathrm{perp}})\Lambda_{01} \otimes B(\Gamma_a(\mathcal{W}_-))$$
$$\simeq \mathbb{1} \otimes \Lambda_{01-}\mathfrak{M}_a(i(\mathcal{V}_0 \oplus \mathcal{V}_1 \oplus \{0\})^{\mathrm{perp}})\Lambda_{01-}$$
$$\simeq \Lambda\mathfrak{M}_a(i\mathcal{V}^{\mathrm{perp}})\Lambda,$$

again using Thm. 17.62. □

17.3.6 Even fermionic von Neumann algebras

We continue within the framework of the previous subsection.

Definition 17.63 *For a real subspace $\mathcal{V}$ of $\mathcal{W}$, we introduce the even part of the fermionic von Neumann algebra $\mathfrak{M}_a(\mathcal{V})$:*

$$\mathfrak{M}_{a,0}(\mathcal{V}) := \{A \in \mathfrak{M}_a(\mathcal{V}) \, : \, IAI = A\} = \mathfrak{M}_a(\mathcal{V}) \cap \{I\}'.$$

Recall that we described the commutant of $\mathfrak{M}_a(\mathcal{V})$ in terms of the symplectic complement: $\mathfrak{M}_a(\mathcal{V})' = \Lambda\mathfrak{M}_a(i\mathcal{V}^{\mathrm{perp}})\Lambda$. If we are interested just in the even part of the commutant, the role of the symplectic complement can be to some extent taken by the real orthogonal complement:

Proposition 17.64 *We have $\mathfrak{M}_a(\mathcal{V})' \cap \{I\}' = \mathfrak{M}_{a,0}(\mathcal{V}^{\mathrm{perp}})$.*

Proof Write

$$\mathfrak{M}_a(\mathcal{V})' \cap \{I\}' = \Lambda\mathfrak{M}_a(i\mathcal{V}^{\mathrm{perp}})\Lambda \cap \{I\}'$$
$$= \Lambda\big(\mathfrak{M}_a(i\mathcal{V}^{\mathrm{perp}}) \cap \{I\}'\big)\Lambda = \Lambda\mathfrak{M}_{a,0}(i\mathcal{V}^{\mathrm{perp}})\Lambda.$$

Every element of $\mathfrak{M}_{a,0}(i\mathcal{V}^{\mathrm{perp}})$ is the strong limit of even polynomials in $\phi(v)$, where $v \in i\mathcal{V}^{\mathrm{perp}}$. Since

$$\Lambda\phi(iv_1)\phi(iv_2)\Lambda = \phi(v_1)\phi(v_2), \quad v_1, v_2 \in \mathcal{V},$$

we have

$$\Lambda\mathfrak{M}_{a,0}(i\mathcal{V}^{\mathrm{perp}})\Lambda = \mathfrak{M}_{a,0}(\mathcal{V}^{\mathrm{perp}}).$$

$\square$

17.4 Notes

In the physics literature, quasi-free states go back to the early days of quantum theory. The Planck law and the Fermi–Dirac distribution belong to the oldest formulas of quantum physics – in the terminology of this chapter they describe the density of a thermal state for the free Bose, resp. Fermi gas.

In the mathematical literature, quasi-free states were first identified by Robinson (1965) and Shale–Stinespring (1964). Quasi-free representations were extensively studied, especially by Araki (1964, 1970, 1971), Araki–Shiraishi (1971), Araki–Yamagami (1982), Powers–Stoermer (1970) and van Daele (1971). Applications of quasi-free states to quantum field theory on curved space-times were studied in Kay–Wald (1991), where a result essentially equivalent to Thm. 17.12 was proven.

Araki–Woods representations first appeared in Araki–Woods (1963). Araki–Wyss representations go back to Araki–Wyss (1964).

It is instructive to use the Araki–Woods and Araki–Wyss representations as illustrations for the Tomita–Takesaki theory and for the so-called standard form of a W^*-algebra as in Haagerup (1975); see also Araki (1970), Connes (1974), Bratteli–Robinson (1987), Stratila (1981) and Dereziński–Jakšić–Pillet (2003). They are quite often used in recent works on quantum statistical physics; see e.g. Jakšić–Pillet (2002) and Dereziński–Jakšić (2003).

The relative position of two subspaces in a Hilbert space was first investigated by Dixmier (1948) and Halmos (1969). The study of a position of a real subspace in a complex Hilbert space is an important ingredient of the version of the Tomita–Takesaki theory presented by Rieffel–van Daele (1977).

The theorem about the lattice of real subspaces of a Hilbert space and the corresponding von Neumann algebras on a bosonic Fock space were first proven by Araki (1963); see also Eckmann–Osterwalder (1973). The analogous theorem about von Neumann algebras on a fermionic Fock space was apparently first given in a review article by Dereziński (2006).

Most of the chapter closely follows Dereziński (2006). The proof of the factoriality of algebras $CAR_{\gamma,1}$ is due to Araki (1970).

The use of Araki–Woods and Araki–Wyss representations in the description of quantum systems at positive temperatures was advocated in papers of Jakšić–Pillet (1996, 2002); see also Dereziński–Jakšić (2001).

18

Dynamics of quantum fields

In this chapter we describe how to quantize linear classical dynamics. The starting point will be a dual phase space $\mathcal{Y}$ equipped with a dynamics – a one-parameter group of linear transformations $\{r_t\}_{t \in \mathbb{R}}$ preserving the structure of $\mathcal{Y}$.

The most typical examples of $\mathcal{Y}$ are the space of solutions of the *Klein–Gordon equation* and of the *Dirac equation*, possibly on a curved space-time and in the presence of external potentials. We can also consider other systems, not necessarily relativistic, e.g. motivated by condensed-matter physics.

We describe how to quantize $(\mathcal{Y}, \{r_t\}_{t \in \mathbb{R}})$ obtaining a model of non-interacting quantum field theory. We demand that quantum fields are represented on a Hilbert space and that the dynamics is implemented by a unitary group generated by a *positive Hamiltonian*. In all the cases we consider, the first step of quantization is the construction of the so-called *one-particle space* $\mathcal{Z}$, equipped with a dynamics generated by a positive *one-particle Hamiltonian* h. Then we apply the usual procedure of the *second quantization* to obtain the *Fock space over* $\mathcal{Z}$ equipped with the dynamics given by the *second quantization of* $e^{\mathrm{i}th}$.

The positivity of the Hamiltonian of the quantum system means that we are at the *zero temperature*. We will also consider briefly the case of *positive temperatures*, which involves the construction of a state satisfying the *KMS condition*.

The abstract procedure outlined above is used in concrete situations in quantum field theory to construct *free* (i.e. *non-interacting*) quantum fields and many-body quantum systems. In this chapter we will not discuss the construction of *interacting quantum fields*, which is much more difficult. In the physical literature, one usually tries to construct interacting fields by perturbing free ones, which is one of the reasons for the importance of free fields. We will describe the diagrammatic aspects of the formal perturbation theory in Chap. 20. Some mathematical tools involved in the rigorous construction of interacting fields are described in Chap. 21 and will be applied to bosonic models in two space-time dimensions in Chap. 22.

The space $\mathcal{Y}$ will always have an additional structure preserved by the dynamics. We distinguish four kinds of such structures leading to four kinds of quantization formalisms:

(1) **Neutral bosonic systems.** The space $\mathcal{Y}$ is symplectic. This formalism is used e.g. for real solutions of the Klein–Gordon equation.

(2) **Neutral fermionic systems.** The space $\mathcal{Y}$ is Euclidean. This formalism can be used e.g. for Majorana spinors satisfying the Dirac equation.

(3) **Charged bosonic systems.** The space $\mathcal{Y}$ is charged symplectic (equipped with a non-degenerate anti-Hermitian form). This formalism is used e.g. for complex solutions of the Klein–Gordon equation.

(4) **Charged fermionic systems.** The space $\mathcal{Y}$ is unitary. This formalism is used e.g. for Dirac spinors satisfying the Dirac equation.

Remark 18.1 *In the most common physics applications we encounter the neutral bosonic formalism (e.g. for photons) and the charged fermionic formalism (e.g. for electrons). Charged bosons are also quite common, e.g. charged pions or gluons in the standard model. On the other hand, until recently, the neutral fermionic formalism had mostly theoretical interest. However, in the modern version of the standard model involving massive neutrinos, Majorana spinors can be useful.*

Remark 18.2 *To avoid possible confusion, let us discuss the distinction between the notion of a "phase space" and of a "dual phase space".*

Possible states of a classical system are described by elements (points) of a set $\mathcal{V}$, called a phase space. *$\mathcal{V}$ is typically a manifold, often equipped with an additional structure, e.g. it is a symplectic manifold. The time evolution of a classical system is given by a one-parameter group $\{r_t\}_{t\in\mathbb{R}}$ of isomorphisms of $\mathcal{V}$. Classical observables are described by (real- or complex-valued) functions on $\mathcal{V}$.*

If $\mathcal{V}$ is in addition a vector space, we have in particular linear (i.e. "coordinate") functions $\mathcal{V} \ni v \mapsto v \cdot y \in \mathbb{R}$ labeled $y \in \mathcal{V}^{\#} =: \mathcal{Y}$. We will say that $\mathcal{Y}$ is the dual phase space. *After the bosonic resp. fermionic quantization, we obtain a family of quantum observables $\phi(y)$, $y \in \mathcal{Y}$, which are operators satisfying the CCR, resp. the CAR and are called the* bosonic, *resp.* fermionic fields. *They are labeled by elements of the dual phase space.*

As we see from this discussion, in the quantum case it is the dual $\mathcal{Y}$ of the phase space that has a more fundamental role than the phase space $\mathcal{V}$ itself. Therefore, in our work the starting point is typically $\mathcal{Y}$.

The distinction between the phase space and its dual is rather academic in the fermionic case, where they can be naturally identified using the scalar product. In the bosonic case, if the space $\mathcal{V}$ is symplectic (the form ω is non-degenerate), one can identify the phase space and its dual with help of this form.

The Hamiltonian, which generates a symplectic dynamics, is traditionally defined as a function on the phase space. If the phase space is symplectic we can transport the Hamiltonian by ω from $\mathcal{V}$ to $\mathcal{Y}$, so that it becomes a function on $\mathcal{Y}$. In this chapter, in the bosonic case the phase space will be always symplectic and we will treat Hamiltonians as functions on the dual phase space, as explained above.

One can distinguish three stages of quantization.

(1) **Classical system.** We consider one of the four kinds of the *dual phase space* $\mathcal{Y}$, together with a one-parameter group of its automorphisms, $\mathbb{R} \ni t \mapsto r_t$, which we view as a *classical dynamics*.

(2) **Algebraic quantization.** We choose an appropriate $*$-algebra $\mathfrak{A}$, together with a one-parameter group of $*$-automorphisms $\mathbb{R} \ni t \mapsto \hat{r}_t$. The algebra $\mathfrak{A}$ is sometimes called the *field algebra* of the quantum system. The commutation, resp. anti-commutation relations satisfied by the appropriate distinguished elements of $\mathfrak{A}$ are governed by the (charged) symplectic form, resp. the scalar product on the dual phase space. $\{\hat{r}_t\}_{t \in \mathbb{R}}$ describes the *quantum dynamics in the Heisenberg picture*. The algebra $\mathfrak{A}$ contains operators that are useful in the theoretical description of the system. However, we do not assume that all of its elements are physically observable, even in principle. Therefore, we also distinguish the *algebra of observables*. It is a certain $*$-sub-algebra of $\mathfrak{A}$, invariant with respect to the dynamics, which consists of operators whose measurement is theoretically possible.

(3) **Hilbert space quantization.** We represent the algebra $\mathfrak{A}$ on a certain Hilbert space $\mathcal{H}$. Typically, the representation of the algebra $\mathfrak{A}$ is faithful, so that we can write $\mathfrak{A} \subset B(\mathcal{H})$. We demand that the dynamics is implemented by a one-parameter unitary group on $\mathcal{H}$. In the case of a zero temperature, we want this unitary group to be generated by a positive operator H, called the *Hamiltonian*, so that

$$\hat{r}_t(A) = \mathrm{e}^{\mathrm{i}tH} A \mathrm{e}^{-\mathrm{i}tH}. \tag{18.1}$$

In the case of a positive temperature, the space $\mathcal{H}$ should contain a cyclic vector satisfying the KMS condition with respect to the dynamics. Its generator is called the *Liouvillean* and denoted L.

Note that, among the three stages of quantization described above, the most important are the first and the third. The second stage – the algebraic quantization – can be skipped altogether. In the usual presentation, typical for physics textbooks, it is limited to a formal level – one says that "commuting classical observables" are replaced by "non-commuting quantum observables" satisfying the appropriate commutation, resp. anti-commutation relations. In our presentation we have tried to interpret this statement in terms of well-defined C^*-algebras. This is quite easy in the case of fermions. Unfortunately, in the case of bosons it leads to certain technical difficulties related to the unboundedness of bosonic fields, and involves a considerable amount of arbitrariness in the choice of a C^*-algebra describing bosonic observables. To some extent, the algebraic quantization is merely an exercise of academic interest. Nevertheless, in some situations it sheds light on some conceptual aspects of quantum theory.

One of the confusing conceptual points that we believe our abstract approach can explain is the difference between the dual phase space and the one-particle space. Throughout our work, the former is typically denoted by $\mathcal{Y}$ and the latter by $\mathcal{Z}$. These two spaces are often identified. They have, however, different physical meanings and are equipped with different algebraic structures.

We also discuss abstract properties of two commonly used discrete symmetries of quantum systems: the time reversal and the charge reversals. Their properties can be quite confusing. We believe that the precise language of linear algebra is particularly adapted to explain their properties. Note, for instance, that the charge reversal is anti-linear with respect to the complex structure on the phase space and linear with respect to the complex structure on the one-particle space. On the other hand, the (Wigner) time reversal is anti-linear with respect to both.

We will always assume that the time and charge reversals are involutions on the observables. Only in the neutral bosonic case do they need to be involutive on the fields as well. In the other three cases observables are even in fields; therefore the time and charge reversals can be anti-involutive.

The first two sections of this chapter present the quantization in an abstract way. In the next two sections, we specify it a little more, considering what we call *abstract Klein–Gordon* and *abstract Dirac dynamics*. This presentation allows us to isolate the main features of various constructions used in quantum field theory and many-body quantum physics.

Throughout the chapter, t is the generic name of a real variable denoting the time.

18.1 Neutral systems

This section is devoted to the neutral bosonic and fermionic formalism of quantization.

In the neutral formalism the vector space $\mathcal{Y}$ is real and is equipped with a symplectic form ω in the bosonic case, resp. with a positive definite scalar product ν in the fermionic case. The dynamics describing the *time evolution* is a one-parameter group $\{r_t\}_{t\in\mathbb{R}}$ with values in $Sp(\mathcal{Y})$, resp. $O(\mathcal{Y})$. The problem addressed in this section is to find a CCR, resp. CAR representation $\mathcal{Y} \ni y \mapsto \phi(y)$ on a Hilbert space $\mathcal{H}$ and a self-adjoint operator H on $\mathcal{H}$ such that e^{itH} implements r_t. In the case of a zero temperature, usually one demands that the Hamiltonian H is *positive*.

We will do this by finding a Kähler anti-involution that commutes with the dynamics, and thus leads to a Fock representation in which the dynamics is implementable.

It turns out that this is easy in the fermionic case. The bosonic case is more technical. In particular, one needs to assume that the dynamics is *stable*, which roughly means that the classical Hamiltonian is positive.

One often assumes that the dynamics $\{r_t\}_{t\in\mathbb{R}}$ is a part of a larger group of symmetries G. In other words, our starting point is a homomorphism of a group G into $Sp(\mathcal{Y})$, resp. $O(\mathcal{Y})$. One often asks whether the action of G can be implemented in the Hilbert space $\mathcal{H}$ by unitary operators.

A different kind of a symmetry is the *time reversal*. After quantization, the time reversal is implemented by an anti-unitary operator.

Recall that if a is an operator on a real space $\mathcal{Y}$, then $a_{\mathbb{C}}$, resp. $a_{\overline{\mathbb{C}}}$ denotes its linear, resp. anti-linear extension to $\mathbb{C}\mathcal{Y}$.

18.1.1 Neutral bosonic systems

Let $(\mathcal{Y}, \omega)$ be a symplectic space. Let $\mathbb{R} \ni t \mapsto r_t \in Sp(\mathcal{Y})$ be a one-parameter group.

Algebraic quantization of a symplectic dynamics

It is easy to describe the quantum counterpart of the above classical dynamical system. We take one of the CCR algebras over $(\mathcal{Y}, \omega)$, say $CCR^{Weyl}(\mathcal{Y})$, and equip it with the group of Bogoliubov automorphisms $\{\hat{r}_t\}_{t \in \mathbb{R}}$, defined by

$$\hat{r}_t(W(y)) = W(r_t y), \quad y \in \mathcal{Y}.$$

Stable symplectic dynamics

Typical symplectic dynamics that appear in physics have *positive Hamiltonians*. We will call such dynamics *stable*. We will see that (under some technical conditions) a stable dynamics leads to a uniquely defined Fock representation.

It is easy to make the concept of stability precise if $\dim \mathcal{Y} < \infty$. In this case $\mathcal{Y}$ has a natural topology. Of course, we assume that the dynamics $t \mapsto r_t$ is continuous. Let a be its generator, so that $r_t = e^{ta}$. Clearly, the form β defined by

$$y_1 \cdot \beta y_2 := y_1 \cdot \omega a y_2, \quad y_1, y_2 \in \mathcal{Y}, \tag{18.2}$$

is symmetric.

Definition 18.3 *We say that the group $t \mapsto r_t \in Sp(\mathcal{Y})$ is stable if β is positive definite.*

The definition of a stable dynamics in the case of infinite dimensions is more complicated, because we need to equip $(\mathcal{Y}, \omega)$ with a topology. There are various possibilities for doing this; let us consider the simplest one.

Definition 18.4 *We say that $(\mathcal{Y}, \omega, \beta, \{r_t\}_{t \in \mathbb{R}})$ is a weakly stable dynamics if the following conditions are true:*

(1) *β is a positive definite symmetric form. We equip $\mathcal{Y}$ with the norm $\|y\|_{en} := (y \cdot \beta y)^{\frac{1}{2}}$. We denote by $\mathcal{Y}_{en}$ the completion of $\mathcal{Y}$ w.r.t. this norm.*
(2) *$\mathbb{R} \ni t \mapsto r_t \in Sp(\mathcal{Y})$ is a strongly continuous group of bounded operators. Thus, we can extend r_t to a strongly continuous group on $\mathcal{Y}_{en}$ and define its generator a, so that $r_t = e^{ta}$.*
(3) *$\operatorname{Ker} a = \{0\}$, or equivalently, $\bigcap_{t \in \mathbb{R}} \operatorname{Ker}(r_t - \mathbb{1}) = \{0\}$.*
(4) *$\mathcal{Y} \subset \operatorname{Dom} a$ and $y_1 \cdot \beta y_2 = y_1 \cdot \omega a y_2$, $y_1, y_2 \in \mathcal{Y}$.*

If, in addition, ω is bounded for the topology given by β, so that it can be extended to the whole $\mathcal{Y}_{\mathrm{en}}$, we will say that the dynamics is strongly stable. *In this case $(\mathcal{Y}_{\mathrm{en}}, \omega)$ is a symplectic space.*

Note that β has two roles: it endows $\mathcal{Y}$ with a topology and it is the Hamiltonian for r_t.

Theorem 18.5 *Let $(\mathcal{Y}, \omega, \beta, \{r_t\}_{t \in \mathbb{R}})$ be a weakly stable dynamics. Then*

(1) *r_t are orthogonal transformations on the real Hilbert space $\mathcal{Y}_{\mathrm{en}}$.*
(2) *a is anti-self-adjoint and $\mathrm{Ker}\, a = \{0\}$.*
(3) *The polar decomposition*

$$a =: |a|\mathrm{j} = \mathrm{j}|a|$$

  defines a Kähler anti-involution j and a self-adjoint operator $|a| > 0$ on $\mathcal{Y}_{\mathrm{en}}$.
(4) *The dynamics is strongly stable iff $|a| \geq C$ for some $C > 0$.*

Recall that, given an operator $|a| > 0$ on $\mathcal{Y}_{\mathrm{en}}$, we can define a scale of Hilbert spaces $|a|^s \mathcal{Y}_{\mathrm{en}}$ (see Subsect. 2.3.4). Then r_t and j are bounded on $\mathcal{Y}_{\mathrm{en}} \cap |a|^s \mathcal{Y}_{\mathrm{en}}$ for the norm of $|a|^s \mathcal{Y}_{\mathrm{en}}$. Let $r_{s,t}$ and j_s denote their extensions. Similarly, a and $|a|$ are closable on $\mathcal{Y}_{\mathrm{en}} \cap |a|^s \mathcal{Y}_{\mathrm{en}}$ for the norm $|a|^s \mathcal{Y}_{\mathrm{en}}$. Let a_s, $|a|_s$ denote their closures. Clearly, for any s, $a_s = |a|_s \mathrm{j}_s = \mathrm{j}_s |a|_s$ is the polar decomposition, j_s is an orthogonal anti-involution and $r_{s,t} = \mathrm{e}^{t a_s}$ is an orthogonal one-parameter group.

Let $\cdot_s$ denote the natural scalar product on $|a|^s \mathcal{Y}_{\mathrm{en}}$. Let us express the scalar product and the symplectic form in terms of β:

$$
\begin{aligned}
y_1 \cdot_s y_2 &= y_1 \beta |a|^{-2s} y_2 = \left(|a|^{-2s} y_1\right) \cdot \beta y_2, && (18.3) \\
y_1 \cdot \omega y_2 &= y_1 \cdot \beta a^{-1} y_2 = (a^{-1} y_1) \cdot \beta y_2.
\end{aligned}
$$

Note that the symplectic form does not need to be defined everywhere.

Of particular interest for us is the case $s = \frac{1}{2}$, for which we introduce the notation $\mathcal{Y}_{\mathrm{dyn}} := |a|^{\frac{1}{2}} \mathcal{Y}_{\mathrm{en}}$. In what follows we drop the subscript $s = \frac{1}{2}$ from $r_{s,t}$, j_s, $\cdot_s$, a_s and $|a|_s$.

Proposition 18.6 *$\mathcal{Y}_{\mathrm{dyn}}$ equipped with $(\cdot, \omega, \mathrm{j})$ is a complete Kähler space.*

Proof Setting $s = \frac{1}{2}$ in (18.3), we obtain

$$y_1 \cdot y_2 = y_1 \cdot \beta |a|^{-1} y_2 = y_1 \cdot \omega a |a|^{-1} y_2 = y_1 \cdot \omega \mathrm{j} y_2. \qquad \square$$

Fock quantization of symplectic dynamics

Until the end of this subsection we drop the subscript $_{\mathrm{dyn}}$ from $\mathcal{Y}_{\mathrm{dyn}}$. Let $\mathcal{Z}$ be the holomorphic subspace of $\mathbb{C}\mathcal{Y}$ for the Kähler anti-involution j constructed in Thm. 18.5.

Clearly, $|a|$ commutes with j, hence its complexification $|a|_{\mathbb{C}}$ preserves $\mathcal{Z}$. We set $h := |a|_{\mathbb{C}}\big|_{\mathcal{Z}}$. Note that $h > 0$ and $a_{\mathbb{C}} = \mathrm{i} \begin{bmatrix} h & 0 \\ 0 & -\overline{h} \end{bmatrix}$, if we use the identification

$\mathbb{C}\mathcal{Y} = \mathcal{Z} \oplus \overline{\mathcal{Z}}$. Likewise, $(r_t)_{\mathbb{C}} = (e^{tj|a|})_{\mathbb{C}}$ preserves $\mathcal{Z}$, and we have

$$(r_t)_{\mathbb{C}}\big|_{\mathcal{Z}} = e^{ith}.$$

For $y \in \mathcal{Y}$, define the field operators

$$\phi(y) := a^* \left(\frac{\mathbb{1} - \mathrm{ij}}{2} y \right) + a \left(\frac{\mathbb{1} - \mathrm{ij}}{2} y \right).$$

Then,

$$\mathcal{Y} \ni y \mapsto e^{i\phi(y)} \in U\big(\Gamma_{\mathrm{s}}(\mathcal{Z})\big) \tag{18.4}$$

is a Fock CCR representation. Introduce the positive operator $H := \mathrm{d}\Gamma(h)$ on $\Gamma_{\mathrm{s}}(\mathcal{Z})$. We have

$$e^{itH}\phi(y)e^{-itH} = \phi(r_t y). \tag{18.5}$$

Definition 18.7 *(18.4) is called the* positive energy Fock quantization *of the weakly stable dynamics* $\{r_t\}_{t\in\mathbb{R}}$. *For any* $y \in \mathcal{Y}$, *the corresponding* time t phase space field *is defined as*

$$\phi_t(y) := \phi(r_{-t} y).$$

Quantizing symplectic dynamics with the (classical) Hamiltonian that is not bounded below is in general more difficult. Even if it is possible, the corresponding quantum Hamiltonian will not be bounded from below. There are some situations in physics when non-positive Hamiltonians arise. An example of such situations is the Klein–Gordon field in the space-time describing a rotating black hole, where the phenomenon of *super-radiance* appears; see Gibbons (1975).

Criterion for a weakly stable symplectic dynamics

In practice, our starting point for quantization of a symplectic dynamics can be somewhat different from that described in Def. 18.4. In this subsection we describe a more general framework that leads to a stable dynamics.

Suppose that the symplectic space $\mathcal{Y}$ is equipped with a Hilbertian topology given by a norm $\|\cdot\|$ such that the symplectic form ω is bounded. Let $\{r_t\}_{t\in\mathbb{R}}$ be a strongly continuous symplectic dynamics. Again, we denote its generator by a, so that $r_t = e^{ta}$. It is easy to see that

$$y_1 \cdot \omega a y_2 = -(a y_1) \cdot \omega y_2, \quad y_1, y_2 \in \mathrm{Dom}\, a.$$

Hence,

$$y_1 \cdot \beta y_2 := y_1 \cdot \omega a y_2, \quad y_1, y_2 \subset \mathrm{Dom}\, a.$$

defines a symmetric quadratic form. Let us assume that there exists $c > 0$ such that

$$y \cdot \beta y \geq c\|y\|^2, \quad y \in \mathrm{Dom}\, a. \tag{18.6}$$

Lemma 18.8 *Consider the Hilbert space* $\mathcal{Y}_{en}$ *obtained by completing* $\mathrm{Dom}\, a$ *in the norm* $\|y\|_{en} = (y \cdot \beta y)^{\frac{1}{2}}$. *Then* $\mathcal{Y}_{en}$ *can be viewed as a dense subspace of* $\mathcal{Y}$. *Moreover,* r_t *preserves* $\mathcal{Y}_{en}$ *and is a strongly continuous isometric group on* $\mathcal{Y}_{en}$.

Proof (18.6) guarantees that $\mathcal{Y}_{en}$ can be considered as a subspace of $\mathcal{Y}$.

Let $y \in \mathrm{Dom}\, a$. Then,

$$y \cdot \beta y = y \cdot \omega a y = (r_t y) \cdot \omega r_t a y$$
$$= (r_t y) \cdot \omega a r_t y = (r_t y) \cdot \beta r_t y.$$

Thus r_t is isometric in $\| \cdot \|_{en}$ on $\mathrm{Dom}\, a$ (and hence on $\mathcal{Y}_{en}$). Moreover,

$$(r_t y - y) \cdot \beta (r_t y - y) = (r_t y - y) \cdot \omega (r_t a y - a y) \to 0.$$

Thus r_t is strongly continuous in $\| \cdot \|_{en}$ on $\mathrm{Dom}\, a$ (and hence on $\mathcal{Y}_{en}$). $\square$

Let a_{en} denote the generator of the dynamics $\{r_t\}_{t \in \mathbb{R}}$ restricted to $\mathcal{Y}_{en}$. Clearly, $a_{en} \subset a$.

The following theorem is easy:

Theorem 18.9 *Under the assumptions of this subsection, the space* $\mathrm{Dom}\, a_{en}$ *equipped with* ω, β *and* $\{r_t\}_{t \in \mathbb{R}}$ *restricted to* $\mathrm{Dom}\, a_{en}$ *satisfy the conditions of a weakly stable dynamics of Def. 18.4.*

18.1.2 Neutral fermionic systems

Let $(\mathcal{Y}, \nu)$ be a real Hilbert space. We think of it as the dual phase space of a fermionic system. A strongly continuous one-parameter group $\mathbb{R} \ni t \mapsto r_t \in O(\mathcal{Y})$ will be called an *orthogonal dynamics*. We view it as a classical dynamical system.

Algebraic quantization of an orthogonal dynamics

We choose $\mathrm{CAR}^{C^*}(\mathcal{Y})$ as the field algebra of our system. It is equipped with the one-parameter group of Bogoliubov automorphisms $\{\hat{r}_t\}_{t \in \mathbb{R}}$, defined by

$$\hat{r}_t\big(\phi(y)\big) = \phi(r_t y), \quad y \in \mathcal{Y}.$$

In quantum physics only even fermionic operators are observable. Therefore, it is natural to use the even sub-algebra $\mathrm{CAR}_0^{C^*}(\mathcal{Y})$ as the observable algebra.

Kähler structure for a non-degenerate orthogonal dynamics

Let a be the generator of r_t, so that $r_t = \mathrm{e}^{ta}$ and $a = -a^\#$.

Definition 18.10 *We say that the dynamics* $t \mapsto r_t \in O(\mathcal{Y})$ *is* non-degenerate *if*

$$\mathrm{Ker}\, a = \{0\}, \quad \text{or equivalently} \quad \bigcap_{t \in \mathbb{R}} \mathrm{Ker}(r_t - \mathbb{1}) = \{0\}. \tag{18.7}$$

Theorem 18.11 *The polar decomposition*

$$a =: |a|\mathrm{j} = \mathrm{j}|a|$$

defines an operator $|a| > 0$ *and a Kähler anti-involution* j *on* $\mathcal{Y}$.

Fock quantization of orthogonal dynamics

Let $\mathcal{Z}$ be the holomorphic subspace of $\mathbb{C}\mathcal{Y}$ for the Kähler anti-involution j.

The operator $|a|$ commutes with j. Hence, its complexification $|a|_\mathbb{C}$ preserves $\mathcal{Z}$. We set $h := |a|_\mathbb{C}\big|_{\mathcal{Z}}$. Note that $h > 0$ and $a_\mathbb{C} = \mathrm{i} \begin{bmatrix} h & 0 \\ 0 & -\bar{h} \end{bmatrix}$. Likewise, $(r_t)_\mathbb{C} = (\mathrm{e}^{t\mathrm{j}|a|})_\mathbb{C}$ preserves $\mathcal{Z}$, and we have

$$(r_t)_\mathbb{C}\big|_{\mathcal{Z}} = \mathrm{e}^{\mathrm{i}th}.$$

Consider the Fock representation associated with the Kähler anti-involution j

$$\mathcal{Y} \ni y \mapsto \phi(y) := a^* \left(\frac{\mathbb{1} - \mathrm{ij}}{2} y \right) + a \left(\frac{\mathbb{1} - \mathrm{ij}}{2} y \right) \in B_\mathrm{h}\left(\Gamma_\mathrm{a}(\mathcal{Z}) \right), \qquad (18.8)$$

and the positive operator $H := \mathrm{d}\Gamma(h)$ on $\Gamma_\mathrm{a}(\mathcal{Z})$. We have

$$\mathrm{e}^{\mathrm{i}tH} \phi(y) \mathrm{e}^{-\mathrm{i}tH} = \phi(r_t y). \qquad (18.9)$$

Definition 18.12 *(18.8) is called the* positive energy Fock quantization *of the dynamics* $\{r_t\}_{t \in \mathbb{R}}$. *For any* $y \in \mathcal{Y}$, *the corresponding* time t field *is defined as*

$$\phi_t(y) := \phi(r_{-t} y).$$

18.1.3 Time reversal in neutral systems

Let $(\mathcal{Y}, \omega)$ be a symplectic space in the bosonic case, or let $(\mathcal{Y}, \nu)$ be a real Hilbert space in the fermionic case.

Time reversal and its algebraic quantization

Definition 18.13 *A map* $\tau \in L(\mathcal{Y})$ *is a* time reversal *if*

(1) τ *is anti-symplectic and* $\tau^2 = \mathbb{1}$ *in the bosonic case*,
(2) τ *is orthogonal and* $\tau^2 = \mathbb{1}$ *or* $\tau^2 = -\mathbb{1}$ *in the fermionic case*.

Let us fix a time reversal τ. Let us quantize τ on the algebraic level.

Proposition 18.14 (1) *In the bosonic case, there exists a unique anti-linear* $*$-*homomorphism* $\hat{\tau}$ *of the algebra* $\mathrm{CCR}^{\mathrm{Weyl}}(\mathcal{Y})$ *such that* $\hat{\tau}(W(y)) := W(\tau y)$. $\hat{\tau}^2$ *is the identity.*
(2) *In the fermionic case, there exists a unique anti-linear* $*$-*homomorphism* $\hat{\tau}$ *of the algebra* $\mathrm{CAR}^{C^*}(\mathcal{Y})$ *such that* $\hat{\tau}(\phi(y)) := \phi(\tau y)$. $\hat{\tau}^2$ *is the identity on* $\mathrm{CAR}_0^{C^*}(\mathcal{Y})$ *(the even algebra).*

Definition 18.15 $\hat{\tau}$ *defined in Prop. 18.14 is called the* algebraic time reversal.

Suppose that $\{r_t\}_{t\in\mathbb{R}}$ is a dynamics, where $r_t \in Sp(\mathcal{Y})$ in the bosonic case and $r_t \in O(\mathcal{Y})$ in the fermionic case.

Definition 18.16 *We say that the dynamics $\{r_t\}_{t\in\mathbb{R}}$ is time reversal invariant if*

$$\tau r_t = r_{-t}\tau. \tag{18.10}$$

Clearly, on the algebraic level (18.10) implies $\hat{\tau}\hat{r}_t = \hat{r}_{-t}\hat{\tau}$.

Fock quantization of time reversal

Let $\{r_t\}_{t\in\mathbb{R}}$ be a time reversal invariant dynamics. In the bosonic case we assume that the dynamics is weakly stable; in the fermionic case we assume that it is non-degenerate. In both cases we can introduce a, j, h. We have

$$\tau a = -a\tau, \quad \tau j = -j\tau, \quad \tau|a| = |a|\tau.$$

Note that the anti-linear extension of τ, denoted $\tau_{\overline{\mathbb{C}}}$, preserves $\mathcal{Z}$.

Definition 18.17 *We write $\tau_{\mathcal{Z}} := \tau_{\overline{\mathbb{C}}}\big|_{\mathcal{Z}}$.*

Clearly, $\tau_{\mathcal{Z}}$ is anti-unitary and $\tau_{\mathcal{Z}}h = h\tau_{\mathcal{Z}}$. Moreover,

(1) $\tau_{\mathcal{Z}}^2 = \mathbb{1}$ in the bosonic case,
(2) $\tau_{\mathcal{Z}}^2 = \mathbb{1}$ or $\tau_{\mathcal{Z}}^2 = -\mathbb{1}$ in the fermionic case.

Consider the positive energy quantization of the dynamics on the Fock space $\Gamma_{\mathrm{s/a}}(\mathcal{Z})$.

Definition 18.18 *The Fock quantization of time reversal is defined as the anti-unitary map $T := \Gamma(\tau_{\mathcal{Z}})$.*

We have $THT^{-1} = H$, $Te^{itH}T^{-1} = e^{-itH}$. T implements $\hat{\tau}$ and

$$T\phi(y)T^{-1} = \phi(\tau y), \quad y \in \mathcal{Y}.$$

Recall that I denotes the parity operator defined in (3.10). We have

(1) $T^2 = \mathbb{1}$ in the bosonic case,
(2) $T^2 = \mathbb{1}$ or $T^2 = I$ in the fermionic case.

18.2 Charged systems

In the charged formalism, the classical system is described by a complex vector space $\mathcal{Y}$.

In the bosonic case, it is equipped with an anti-Hermitian form $(\cdot|\omega\cdot)$ – we say that it is a *charged symplectic space*. The dynamics $\{r_t\}_{t\in\mathbb{R}}$ describing the time evolution is assumed to preserve $(\cdot|\omega\cdot)$ – we say that r_t is *charged symplectic*.

In the fermionic case it is equipped with a positive scalar product $(\cdot|\cdot)$. Without decreasing the generality we can assume that it is complete – $\mathcal{Y}$ is a complex Hilbert space. The dynamics $\{r_t\}_{t\in\mathbb{R}}$ preserves $(\cdot|\cdot)$ – it is unitary.

By a positive energy quantization of a charged classical system we mean a charged CCR or CAR representation $\mathcal{Y} \ni y \mapsto \psi^*(y)$ on a Hilbert space $\mathcal{H}$ and a positive self-adjoint operator H on $\mathcal{H}$ such that $\mathrm{e}^{\mathrm{i}tH}$ implements r_t.

The complex structure of $\mathcal{Y}$ is responsible for the action of a $U(1)$ symmetry $\{\mathrm{e}^{\mathrm{i}\theta}\}_{\theta \in [0,2\pi]}$. On the level of the Fock representation it is implemented by $\mathrm{e}^{\mathrm{i}\theta Q}$, where Q is called the *charge operator*.

Recall that charged systems can be viewed as special cases of neutral systems equipped in addition with a certain symmetry. As discussed in Subsect. 1.3.11, a homomorphism $U(1) \ni \theta \mapsto u_\theta \in L(\mathcal{Y})$ on a real space $\mathcal{Y}$ is called a $U(1)$ *symmetry of charge 1* if there exists an anti-involution j_{ch} such that $u_\theta = \cos\theta \mathbb{1} + \sin\theta \mathrm{j}_{\mathrm{ch}}$. Assume that it preserves the symplectic, resp. Euclidean form ω, resp. ν, which is equivalent to saying that j_{ch} is pseudo-Kähler, resp. Kähler. Assume also that the dynamics $\{r_t\}_{t \in \mathbb{R}}$ commutes with this symmetry, which is equivalent to saying that j_{ch} commutes with r_t. If we equip the space with the complex structure given by j_{ch}, then the symmetry u_θ becomes just the multiplication by $\mathrm{e}^{\mathrm{i}\theta}$. It is then natural to replace the real bilinear forms ω, resp. ν by the closely related sesquilinear forms $(\cdot|\omega\cdot)$, resp. $(\cdot|\cdot)$. The invariance of the dynamics w.r.t. the charge symmetry is now expressed by the fact that the dynamics is complex linear. See Subsects. 8.2.5 and 12.1.7 for further details.

In this section we describe in abstract terms the charged formalism. At the end of the section, we discuss the *charge reversal* and the *time reversal* for charged systems.

We will use θ as the generic variable in $U(1) = \mathbb{R}/2\pi\mathbb{Z}$.

18.2.1 Charged bosonic systems

Let $(\mathcal{Y}, (\cdot|\omega\cdot))$ be a charged symplectic space. Let $\mathbb{R} \supset t \mapsto r_t \in ChSp(\mathcal{Y})$ be a charged symplectic dynamics.

Algebraic quantization of a charged symplectic dynamics

By taking $\mathrm{Re}(y_1|\omega y_2)$ we can view $\mathcal{Y}_\mathbb{R}$ as a real symplectic space. We choose $\mathrm{CCR}^{\mathrm{reg}}(\mathcal{Y}_\mathbb{R})$ as the field algebra of our system. This algebra is generated (in the sense described in Subsect. 8.3.4) by the Weyl elements $\mathrm{e}^{\mathrm{i}\psi(y)+\mathrm{i}\psi^*(y)}$, $y \in \mathcal{Y}$, satisfying the relations

$$\mathrm{e}^{\mathrm{i}\psi(y_1)+\mathrm{i}\psi^*(y_1)}\mathrm{e}^{\mathrm{i}\psi(y_2)+\mathrm{i}\psi^*(y_2)} = \mathrm{e}^{-\mathrm{i}\mathrm{Re}(y_1|\omega y_2)}\mathrm{e}^{\mathrm{i}\psi(y_1+y_2)+\mathrm{i}\psi^*(y_1+y_2)}.$$

We equip $\mathrm{CCR}^{\mathrm{reg}}(\mathcal{Y}_\mathbb{R})$ with the automorphism groups $\{\widehat{\mathrm{e}^{\mathrm{i}\theta}}\}_{\theta \in U(1)}$ and $\{\hat{r}_t\}_{t \in \mathbb{R}}$ defined by

$$\widehat{\mathrm{e}^{\mathrm{i}\theta}}\left(\mathrm{e}^{\mathrm{i}\psi(y)+\mathrm{i}\psi^*(y)}\right) = \mathrm{e}^{\mathrm{i}\psi(\mathrm{e}^{\mathrm{i}\theta}y)+\mathrm{i}\psi^*(\mathrm{e}^{\mathrm{i}\theta}y)},$$
$$\hat{r}_t\left(\mathrm{e}^{\mathrm{i}\psi(y)+\mathrm{i}\psi^*(y)}\right) = \mathrm{e}^{\mathrm{i}\psi(r_t y)+\mathrm{i}\psi^*(r_t y)}, \qquad y \in \mathcal{Y}.$$

For the observable algebra it is natural to choose the so-called *gauge-invariant regular CCR algebra* $\mathrm{CCR}_{\mathrm{gi}}^{\mathrm{reg}}(\mathcal{Y})$, which is defined as the set of elements of $\mathrm{CCR}^{\mathrm{reg}}(\mathcal{Y}_{\mathbb{R}})$ fixed by $\widehat{\mathrm{e}^{\mathrm{i}\theta}}$. Note that $\mathrm{CCR}_{\mathrm{gi}}^{\mathrm{reg}}(\mathcal{Y})$ is contained in the even algebra $\mathrm{CCR}_0^{\mathrm{reg}}(\mathcal{Y}_{\mathbb{R}})$ and is preserved by the dynamics $\hat{r}_t$.

Remark 18.19 *In this subsection, for the field algebra of our system we have preferred to choose* $\mathrm{CCR}^{\mathrm{reg}}(\mathcal{Y}_{\mathbb{R}})$ *instead of* $\mathrm{CCR}^{\mathrm{Weyl}}(\mathcal{Y}_{\mathbb{R}})$. *This is motivated by the fact that the only element left invariant by the gauge symmetry* $\widehat{\mathrm{e}^{\mathrm{i}\theta}}$ *in* $\mathrm{CCR}^{\mathrm{Weyl}}(\mathcal{Y}_{\mathbb{R}})$ *is* $\mathbb{1}$, *whereas in the case of* $\mathrm{CCR}^{\mathrm{reg}}(\mathcal{Y}_{\mathbb{R}})$ *we obtain a large gauge-invariant algebra.*

Fock quantization of a charged symplectic dynamics

The concept of stability of dynamics in the charged case is analogous to the neutral case.

Definition 18.20 *We say that* $\big(\mathcal{Y}, (\cdot|\omega\cdot), (\cdot|\beta\cdot), \{r_t\}_{t\in\mathbb{R}}\big)$ *is a weakly stable dynamics if the following conditions are true:*

(1) $(\cdot|\beta\cdot)$ *is a positive definite sesquilinear form. We equip* $\mathcal{Y}$ *with the norm* $\|y\|_{\mathrm{en}} := (y|\beta y)^{\frac{1}{2}}$. *We denote by* $\mathcal{Y}_{\mathrm{en}}$ *the completion of* $\mathcal{Y}$ *w.r.t. this norm.*

(2) *We assume that* $\{r_t\}_{t\in\mathbb{R}}$ *is a strongly continuous group of bounded operators on* $\mathcal{Y}$. *Thus we can extend* r_t *to a strongly continuous group on* $\mathcal{Y}_{\mathrm{en}}$ *and define its generator* $\mathrm{i}b$, *so that* $r_t = \mathrm{e}^{\mathrm{i}tb}$.

(3) $\mathrm{Ker}\, b = \{0\}$, *or equivalently*, $\bigcap_{t\in\mathbb{R}} \mathrm{Ker}(r_t - \mathbb{1}) = \{0\}$.

(4) *We assume that* $\mathcal{Y} \subset \mathrm{Dom}\, b$ *and*

$$(y_1|\beta y_2) := \mathrm{i}(y_1|\omega b y_2), \quad y_1, y_2 \in \mathcal{Y}. \tag{18.11}$$

If in addition ω *is bounded for the topology given by* β, *so that* $(\cdot|\omega\cdot)$ *can be extended to the whole* $\mathcal{Y}_{\mathrm{en}}$, *we will say that the dynamics is* strongly stable.

Theorem 18.21 *Let* $\big(\mathcal{Y}, (\cdot|\omega\cdot), (\cdot|\beta\cdot), \{r_t\}_{t\in\mathbb{R}}\big)$ *be a weakly stable dynamics. Then*

(1) r_t *are unitary transformations on the Hilbert space* $\mathcal{Y}_{\mathrm{en}}$,
(2) b *is self-adjoint and* $\mathrm{Ker}\, b = \{0\}$.

Set $q := \mathrm{sgn}(b)$ and $\mathrm{j} := \mathrm{i}\,\mathrm{sgn}(b)$. Clearly, $|b|$ is positive and $r_t = \mathrm{e}^{t\mathrm{j}|b|}$.

Set $\mathcal{Y}_{\mathrm{dyn}} := |b|^{\frac{1}{2}}\mathcal{Y}_{\mathrm{en}}$. As in Subsect. 18.1.1, we can view r_t, j, b and $|b|$ as defined on $\mathcal{Y}_{\mathrm{dyn}}$. In what follows we drop the subscript $_{\mathrm{dyn}}$ from $\mathcal{Y}_{\mathrm{dyn}}$.

Let $\mathbb{1}_{\pm} := \mathbb{1}_{]0,\infty[}(\pm b) = \mathbb{1}_{\{\pm 1\}}(q)$, $\mathcal{Y}_{\pm} := \mathrm{Ran}\, \mathbb{1}_{\pm}$. Let $\mathcal{Z}$ denote the space $\mathcal{Y}$ equipped with the complex structure given by j. (In other words, $\mathcal{Z} := \mathcal{Y}_+ \oplus \overline{\mathcal{Y}}_-$.)

The operators $|b|$, q and b preserve $\mathcal{Y}_{\pm}$. Hence, they can be viewed as complex linear operators on $\mathcal{Z}$ as well, in which case they will be denoted h, $q_{\mathcal{Z}}$ and $b_{\mathcal{Z}}$.

Consider the space $\Gamma_s(\mathcal{Z})$. For $y \in \mathcal{Y}$, let us introduce the *charged fields* on $\mathcal{Y}$, which are closed operators on $\Gamma_s(\mathcal{Z})$ defined by

$$\psi^*(y) = a^* \left(\mathbb{1}_+ y\right) + a \left(\overline{\mathbb{1}_- y}\right),$$
$$\psi(y) = a \left(\mathbb{1}_+ y\right) + a^* \left(\overline{\mathbb{1}_- y}\right). \tag{18.12}$$

We obtain a charged CCR representation

$$\mathcal{Y} \ni y \mapsto \psi^*(y) \in Cl(\Gamma_s(\mathcal{Z})). \tag{18.13}$$

Define the self-adjoint operators on $\Gamma_s(\mathcal{Z})$

$$H := \mathrm{d}\Gamma(h), \quad Q := \mathrm{d}\Gamma(q_{\mathcal{Z}}).$$

Clearly,

$$\mathrm{e}^{\mathrm{i}tH} \psi(y)\mathrm{e}^{-\mathrm{i}tH} = \psi(\mathrm{e}^{\mathrm{i}tb}y), \quad \mathrm{e}^{\mathrm{i}\theta Q}\psi(y)\mathrm{e}^{-\mathrm{i}\theta Q} = \psi(\mathrm{e}^{\mathrm{i}\theta}y), \quad y \in \mathcal{Y}.$$

Definition 18.22 *(18.13) is called the* positive energy Fock quantization *for the dynamics $\{r_t\}_{t\in\mathbb{R}}$. For any $y \in \mathcal{Y}$, the corresponding* time t field *is defined as*

$$\psi_t(y) := \psi(r_{-t}y).$$

18.2.2 Charged fermionic systems

Let $(\mathcal{Y}, (\cdot|\cdot))$ be a complex Hilbert space describing a charged fermionic system. A strongly continuous one-parameter group $\mathbb{R} \ni t \mapsto r_t \in U(\mathcal{Y})$ will be called a *unitary dynamics*.

Algebraic quantization of a unitary dynamics

Clearly, by taking the real scalar product $y_1 \cdot \nu y_2 := \frac{1}{2}\mathrm{Re}(y_1|y_2)$ we can view $\mathcal{Y}_\mathbb{R}$ as a real Hilbert space. We can associate with our system the field algebra $\mathrm{CAR}^{C^*}(\mathcal{Y}_\mathbb{R})$ with distinguished elements $\psi(y)$, $y \in \mathcal{Y}$. We equip it with the automorphism group $\{\widehat{\mathrm{e}^{\mathrm{i}\theta}}\}_{\theta\in U(1)}$ and $\{\hat{r}_t\}_{t\in\mathbb{R}}$ defined by

$$\widehat{\mathrm{e}^{\mathrm{i}\theta}}\left(\psi(y)\right) = \psi(\mathrm{e}^{\mathrm{i}\theta}y),$$
$$\hat{r}_t\left(\psi(y)\right) = \psi(r_t y), \quad y \in \mathcal{Y}.$$

Similarly to the bosonic case, for the observable algebra we choose the so-called *gauge-invariant CAR algebra* $\mathrm{CAR}_{\mathrm{gi}}^{C^*}(\mathcal{Y})$, which is defined as the set of elements of $\mathrm{CAR}^{C^*}(\mathcal{Y}_\mathbb{R})$ fixed by $\widehat{\mathrm{e}^{\mathrm{i}\theta}}$. Note that $\mathrm{CAR}_{\mathrm{gi}}^{C^*}(\mathcal{Y})$ is contained in the even algebra $\mathrm{CAR}_0^{C^*}(\mathcal{Y}_\mathbb{R})$ and is preserved by the dynamics $\hat{r}_t$.

Fock quantization of a unitary dynamics

Let b be the self-adjoint generator of $\{r_t\}_{t\in\mathbb{R}}$, so that $r_t = \mathrm{e}^{\mathrm{i}tb}$.

Definition 18.23 *We say that the dynamics $t \mapsto r_t \in U(\mathcal{Y})$ is* non-degenerate *if*

$$\mathrm{Ker}\, b = \{0\}, \text{ or equivalently } \bigcap_{t\in\mathbb{R}} \mathrm{Ker}(r_t - \mathbb{1}) = \{0\}. \tag{18.14}$$

Set $q := \mathrm{sgn}(b)$ and $\mathrm{j} := \mathrm{i}\,\mathrm{sgn}(b)$. Clearly, $|b|$ is positive, and $r_t = \mathrm{e}^{t\mathrm{j}|b|}$. Let $\mathbb{1}_\pm := \mathbb{1}_{]0,\infty[}(\pm b) = \mathbb{1}_{\{\pm 1\}}(q)$, $\mathcal{Y}_\pm := \mathrm{Ran}\,\mathbb{1}_\pm$. Let $\mathcal{Z}$ denote the space $\mathcal{Y}$ equipped with the complex structure given by j. (In other words, $\mathcal{Z} := \mathcal{Y}_+ \oplus \overline{\mathcal{Y}_-}$.)

The operators $|b|$, q and b preserve $\mathcal{Y}_\pm$. Hence, they can also be viewed as complex linear operators on $\mathcal{Z}$ as well, in which case they will be denoted h, $q_\mathcal{Z}$ and $b_\mathcal{Z}$.

Consider the space $\Gamma_{\mathrm{a}}(\mathcal{Z})$. For $y \in \mathcal{Y}$, let us introduce the *charged fields* on $\mathcal{Y}$, which are closed operators on $\Gamma_{\mathrm{a}}(\mathcal{Z})$ defined by

$$\psi^*(y) = a^*\left(\mathbb{1}_+ y\right) + a\left(\overline{\mathbb{1}_- y}\right), \tag{18.15}$$

$$\psi(y) = a\left(\mathbb{1}_+ y\right) + a^*\left(\overline{\mathbb{1}_- y}\right). \tag{18.16}$$

We obtain a charged CAR representation

$$\mathcal{Y} \ni y \mapsto \psi^*(y) \in B\big(\Gamma_{\mathrm{a}}(\mathcal{Z})\big). \tag{18.17}$$

Define the self-adjoint operators on $\Gamma_{\mathrm{a}}(\mathcal{Z})$

$$H := \mathrm{d}\Gamma(h), \quad Q := \mathrm{d}\Gamma(q_\mathcal{Z}).$$

Clearly,

$$\mathrm{e}^{\mathrm{i}tH}\,\psi(y)\mathrm{e}^{-\mathrm{i}tH} = \psi(\mathrm{e}^{\mathrm{i}tb}y), \quad \mathrm{e}^{\mathrm{i}\theta Q}\psi(y)\mathrm{e}^{-\mathrm{i}\theta Q} = \psi(\mathrm{e}^{\mathrm{i}\theta}y), \quad y \in \mathcal{Y}.$$

Definition 18.24 *(18.17) is called the* positive energy Fock quantization *of the dynamics $\{r_t\}_{t\in\mathbb{R}}$. For any $y \in \mathcal{Y}$, the corresponding* time t phase space field *is defined as*

$$\psi_t(y) := \psi(r_{-t}y).$$

18.2.3 Charge reversal

Let $(\mathcal{Y}, (\cdot|\omega\cdot))$ be a charged symplectic space in the bosonic case, or let $(\mathcal{Y}, (\cdot|\cdot))$ be a complex Hilbert space in the fermionic case.

Charge reversal and its algebraic quantization

Definition 18.25 $\chi \in L(\mathcal{Y}_\mathbb{R})$ *is a* charge reversal *if $\chi^2 = \mathbb{1}$ or $\chi^2 = -\mathbb{1}$, and*

(1) $(\chi y_1|\omega\chi y_2) = \overline{(y_1|\omega y_2)}$ *(χ is anti-charged symplectic) in the bosonic case;*
(2) $(\chi y_1|\chi y_2) = \overline{(y_1|y_2)}$ *(χ is anti-unitary) in the fermionic case.*

Let us fix a charge reversal χ. Consider now its algebraic quantization.

Proposition 18.26 (1) *In the bosonic case, there exists a unique $*$-automorphism $\hat{\chi}$ of $\mathrm{CCR}^{\mathrm{reg}}(\mathcal{Y}_\mathbb{R})$ such that*

$$\hat{\chi}\big(\mathrm{e}^{\mathrm{i}\psi(y)+\mathrm{i}\psi^*(y)}\big) = \mathrm{e}^{\mathrm{i}\psi(\chi y)+\mathrm{i}\psi^*(\chi y)}.$$

(2) *In the fermionic case, there exists a unique ∗-automorphism $\hat{\chi}$ of* $\mathrm{CAR}^{C^*}(\mathcal{Y}_{\mathbb{R}})$ *such that*

$$\hat{\chi}(\psi^*(y)) = \psi(\chi y).$$

In both the bosonic and the fermionic case, $\hat{\chi}$ leaves invariant the gauge-invariant algebra and is involutive on it.

Definition 18.27 $\hat{\chi}$ *defined in Prop. 18.26 is called the* algebraic charge reversal.

Let us remark that whereas χ is anti-linear, $\hat{\chi}$ is linear.

Suppose that $\{r_t\}_{t\in\mathbb{R}}$ is a charged symplectic or unitary dynamics.

Definition 18.28 *We say that the dynamics is* invariant under the charge reversal χ *if*

$$\chi r_t = r_t \chi. \tag{18.18}$$

Similarly, if we have a group of symmetries $\{r_g\}_{g\in G}$ we say that it is invariant under the charge reversal χ *if $r_g \chi = \chi r_g$, $g \in G$.*

Clearly, on the algebraic level (18.18) implies $\hat{\chi}\hat{r}_t = \hat{r}_t\hat{\chi}$.

Fock quantization of charge reversal

Let $\{r_t\}_{t\in\mathbb{R}}$ be a charge reversal invariant dynamics. In the bosonic case assume that the dynamics is weakly stable. In the fermionic case assume it is non-degenerate. Let b, h, q etc. be constructed as before. In both the bosonic and the fermionic case it follows that

$$\chi|b| = |b|\chi, \quad \chi b = -b\chi, \quad \chi q = -q\chi, \quad \chi \mathrm{j} = \mathrm{j}\chi.$$

Definition 18.29 *We denote $\chi_{\mathcal{Z}}$ the map χ considered on $\mathcal{Z}$.*

Note that $\chi_{\mathcal{Z}}$ is unitary, unlike χ.

Definition 18.30 *The* Fock quantization of the charge reversal *is the unitary* $C := \Gamma(\chi_{\mathcal{Z}})$.

We have $CHC^{-1} = H$, $CQC^{-1} = -Q$. C implements $\hat{\chi}$ and

$$C\psi^*(y)C^{-1} = \psi(\chi y).$$

Note that $C^2 = \mathbb{1}$ or $C^2 = I$, where we recall that I is the parity operator.

Neutral subspace

Assume that $\chi^2 = \mathbb{1}$. Recall that we can define the spaces

$$\mathcal{Y}^{\pm\chi} := \{y \in \mathcal{Y} : y = \pm\chi y\}.$$

The dynamics and the symmetry group restrict to $\mathcal{Y}^{\chi}$ and $\mathcal{Y}^{-\chi}$.

Definition 18.31 *We will call $\mathcal{Y}^{\chi}$ the* neutral subspace *of $\mathcal{Y}$. (In the fermionic case, we will also call it the* Majorana subspace.*)*

Note that $\mathcal{Y} = \mathcal{Y}^\chi \oplus i\mathcal{Y}^\chi$, hence the system can be viewed as a couple of neutral systems.

Let us describe the converse construction. Suppose that we have a neutral system $(\mathcal{Y}, \omega)$ or $(\mathcal{Y}, \nu)$ equipped with the dynamics $\{r_t\}_{t\in\mathbb{R}}$. We can extend it to a charged system as follows. We consider the complexified space $\mathbb{C}\mathcal{Y}$ equipped with the natural conjugation denoted by the "bar". We equip it with the anti-Hermitian form, resp. scalar product

$$(w_1|\omega w_2) := \overline{w}_1 \cdot \omega w_2,$$

$$\text{or} \quad (w_1|w_2) := 2\overline{w}_1 \cdot \nu w_2, \quad w_1, w_2 \in \mathbb{C}\mathcal{Y}.$$

We extend the dynamics r_t to $(r_t)_{\mathbb{C}}$ on $\mathbb{C}\mathcal{Y}$. Clearly, $(r_t)_{\mathbb{C}}$ is a charged symplectic, resp. unitary dynamics with the charge reversal given by $\chi w := \overline{w}$, $w \in \mathbb{C}\mathcal{Y}$. It satisfies $\chi^2 = \mathbb{1}$. One gets back the original system by the restriction to the neutral subspace.

18.2.4 Time reversal in charged systems

Let $(\mathcal{Y}, (\cdot|\omega\cdot))$ be a charged symplectic space in the bosonic case, or let $(\mathcal{Y}, (\cdot|\cdot))$ be a complex Hilbert space in the fermionic case.

In the case of charged systems it is natural to consider two kinds of time reversal. The standard choice is an anti-linear symmetry considered by Wigner. The so-called Racah time reversal is actually historically older than the Wigner time reversal. It is linear and from a purely mathematical point of view may seem more natural.

Wigner time reversal and its algebraic quantization

Definition 18.32 $\tau \in L(\mathcal{Y}_{\mathbb{R}})$ *is a* Wigner time reversal *if* $\tau^2 = \mathbb{1}$ *or* $\tau^2 = -\mathbb{1}$, *and*

(1) $(\tau y_1 | \omega \tau y_2) = -\overline{(y_1|\omega y_2)}$ *(τ is anti-charged anti-symplectic) in the bosonic case;*

(2) $(\tau y_1 | \tau y_2) = \overline{(y_1|y_2)}$ *(τ is anti-unitary) in the fermionic case.*

Let us fix a Wigner time reversal τ.

Proposition 18.33 (1) *There exists a unique anti-linear $*$-automorphism $\hat{\tau}$ on the algebra* $\mathrm{CCR}^{\mathrm{reg}}(\mathcal{Y}_{\mathbb{R}})$ *such that*

$$\hat{\tau}\big(e^{i\psi(y)+i\psi^*(y)}\big) = e^{-i\psi(\tau y)-i\psi^*(\tau y)}.$$

(2) *There exists a unique anti-linear $*$-automorphism $\hat{\tau}$ of the algebra* $\mathrm{CAR}^{C^*}(\mathcal{Y}_{\mathbb{R}})$ *such that*

$$\hat{\tau}\big(\psi(y)\big) = \psi(\tau y).$$

In both the bosonic and the fermionic case, $\hat{\tau}$ leaves invariant the gauge-invariant algebra and is involutive on it.

Definition 18.34 $\hat{\tau}$ *defined in Prop. 18.33 is called the* algebraic Wigner time reversal.

Note that both τ and $\hat{\tau}$ are anti-linear.

Suppose that $\{r_t\}_{t\in\mathbb{R}}$ is a charged symplectic or unitary dynamics.

Definition 18.35 *We say that the dynamics is* invariant under the Wigner time reversal τ *if*

$$\tau r_t = r_{-t}\tau. \tag{18.19}$$

Clearly, on the algebraic level (18.19) implies $\hat{\tau}\hat{r}_t = \hat{r}_{-t}\hat{\tau}$.

Fock quantization of Wigner time reversal

Let $\{r_t\}_{t\in\mathbb{R}}$ be a Wigner time reversal dynamics. In the bosonic case assume that the dynamics is weakly stable. In the fermionic case assume it is non-degenerate. Let b, h, q etc. be constructed as before. In both the bosonic and the fermionic case it follows that

$$\tau|b| = |b|\tau, \quad \tau b = b\tau, \quad \tau q = q\tau, \quad \tau\mathrm{j} = -\mathrm{j}\tau.$$

Thus $\tau\mathcal{Y}_+ = \mathcal{Y}_+, \tau\mathcal{Y}_- = \mathcal{Y}_-$.

Definition 18.36 *Let $\tau_{\mathcal{Z}}$ denote τ considered on $\mathcal{Z}$.*

Note that $\tau_{\mathcal{Z}}$ is anti-unitary.

Definition 18.37 *The* Fock quantization of the Wigner time reversal *is given by the anti-unitary $T := \Gamma(\tau_{\mathcal{Z}})$.*

We have $THT^{-1} = H, Te^{\mathrm{i}tH}T^{-1} = \mathrm{e}^{-\mathrm{i}tH}, TQT^{-1} = Q, Te^{\mathrm{i}\theta Q}T^{-1} = \mathrm{e}^{-\mathrm{i}\theta Q}$. T implements $\hat{\tau}$ and

$$T\psi(y)T^{-1} = \psi(\tau y), \quad T\psi^*(y)T^{-1} = \psi^*(\tau y), \quad y \in \mathcal{Y}.$$

Moreover, $T^2 = \mathbb{1}$ or $T^2 = I$.

Racah time reversal

Definition 18.38 $\kappa \in L(\mathcal{Y})$ *is a* Racah time reversal *if $\kappa^2 = \mathbb{1}$ or $\kappa^2 = -\mathbb{1}$, and*

(1) $(\kappa y_1|\omega\kappa y_2) = -(y_1|\omega y_2)$ *(κ is charged anti-symplectic) in the bosonic case;*
(2) $(\kappa y_1|\kappa y_2) = (y_1|y_2)$ *(κ is unitary) in the fermionic case.*

Let us stress that the Racah time reversal is linear.

Let $\{r_t\}_{t\in\mathbb{R}}$ be a charged symplectic or unitary dynamics.

Definition 18.39 $\{r_t\}_{t\in\mathbb{R}}$ *is invariant under the Racah time reversal if $\kappa r_t = r_{-t}\kappa$.*

Suppose that χ is charge reversal and τ is a Wigner time reversal satisfying

$$\tau\chi = \chi\tau \text{ or } \tau\chi = -\chi\tau.$$

Then it is easy to see that $\kappa := \tau\chi$ is a Racah time reversal. In particular, $\kappa^2 = \mathbb{1}$ or $\kappa^2 = -\mathbb{1}$.

Note that we are free to multiply either χ or τ by i. Therefore, possibly after a redefinition of χ or τ, we can always assume that

$$\tau\chi = \chi\tau. \tag{18.20}$$

Thus we have three commuting symmetries: χ, τ and κ.

Consider in addition $\{r_t\}_{t\in\mathbb{R}}$, a charged dynamics invariant under Wigner's time reversal τ and a charge reversal χ. Let us recall the various commutation properties:

$$\tau|b| = |b|\tau, \quad \tau\mathrm{j} = -\mathrm{j}\tau, \quad \tau\mathrm{i} = -\mathrm{i}\tau, \quad \tau q = q\tau,$$
$$\chi|b| = |b|\chi, \quad \chi\mathrm{j} = \mathrm{j}\chi, \quad \chi\mathrm{i} = -\mathrm{i}\chi, \quad \chi q = -q\chi.$$

τ, χ, κ and $q\kappa$ are all either involutions or anti-involutions. The following list describes various possible behaviors of these four symmetries:

τ^2	χ^2	κ^2	$(q\kappa)^2$
$\mathbb{1}$	$\mathbb{1}$	$\mathbb{1}$	$-\mathbb{1}$
$\mathbb{1}$	$-\mathbb{1}$	$-\mathbb{1}$	$\mathbb{1}$
$-\mathbb{1}$	$-\mathbb{1}$	$\mathbb{1}$	$-\mathbb{1}$
$-\mathbb{1}$	$\mathbb{1}$	$-\mathbb{1}$	$\mathbb{1}$

Note that both κ and $q\kappa$ satisfy the conditions of the Racah time reversal. If $\tau^2 = \chi^2 = \pm\mathbb{1}$, we have $\kappa^2 = \mathbb{1}$, whereas if $\tau^2 = -\chi^2 = \pm\mathbb{1}$, we have $(q\kappa)^2 = \mathbb{1}$. Therefore, one of the operators κ or $q\kappa$ is always an involution.

18.3 Abstract Klein–Gordon equation and its quantization

In Subsects. 18.1.1, resp. 18.2.1 we described how to quantize a symplectic, resp. charged symplectic dynamics. The most important symplectic or charged symplectic dynamics used in quantum field theory is associated with the *wave equation*

$$(\partial_t^2 - \Delta)\zeta = 0$$

or, more generally, to the closely related *Klein–Gordon equation*

$$(\partial_t^2 - \Delta + m^2)\zeta = 0.$$

One of the characteristic features of the wave and Klein–Gordon equation is the second order of the time derivative. In this section we study an abstract version of the wave or Klein–Gordon equation. We forget about the spatial structure of the system, but we keep the second-order temporal derivative. We describe the corresponding symplectic dynamics and its quantization.

In the next chapter we will consider the true wave and Klein–Gordon equation on the space-time and its quantization. We find it instructive and amusing, however, that many of the constructions used in this context can be described in a rather abstract fashion.

18.3.1 Splitting into configuration and momentum space

Suppose that $\mathcal{Y}$ is a symplectic space equipped with a time reversal τ. Recall that it satisfies $\tau^2 = \mathbb{1}$. Thus τ is a conjugation on a real symplectic space. As discussed in Subsect. 1.1.16, we can split the dual phase space into the direct sum of Lagrangian subspaces $\mathcal{Y} = \mathcal{Y}^\tau \oplus \mathcal{Y}^{-\tau}$, where $\mathcal{Y}^{\pm\tau} := \{y \in \mathcal{Y} \ : \ y = \pm\tau y\}$.

$\mathcal{Y}^\tau$ has the interpretation of the *dual of the configuration space*, whereas $\mathcal{Y}^{-\tau}$ has the interpretation of the *dual of the momentum space*.

Recall from Subsect. 1.1.16 that $(\mathcal{Y}^\tau, \mathcal{Y}^{-\tau})$ can be interpreted as a dual pair so that the symplectic form can be written as

$$(\vartheta_1, \varsigma_1) \cdot \omega(\vartheta_2, \varsigma_2) = \vartheta_1 \cdot \varsigma_2 - \varsigma_1 \cdot \vartheta_2 \quad (\vartheta_i, \varsigma_i) \in \mathcal{Y}^\tau \oplus \mathcal{Y}^{-\tau}, \ i = 1, 2. \quad (18.21)$$

The time reversal acts as

$$\tau(\vartheta, \varsigma) = (\vartheta, -\varsigma), \quad (\vartheta, \varsigma) \in \mathcal{Y}^\tau \oplus \mathcal{Y}^{-\tau}. \quad (18.22)$$

Let $\{r_t\}_{t \in \mathbb{R}}$ be a time reversal invariant dynamics. For $(\vartheta, \varsigma) \in \mathcal{Y}^\tau \oplus \mathcal{Y}^{-\tau}$ write $r_t(\vartheta, \varsigma) = (\vartheta(t), \varsigma(t))$. Then there exist $f \in L(\mathcal{Y}^\tau, \mathcal{Y}^{-\tau})$, $g \in L(\mathcal{Y}^{-\tau}, \mathcal{Y}^\tau)$ such that $f = f^\#$, $g = g^\#$ and

$$\partial_t \varsigma(t) = f\vartheta(t), \quad \partial_t \vartheta(t) = -g\varsigma(t).$$

The Hamiltonian of the dynamics is

$$\frac{1}{2}\vartheta \cdot g\vartheta + \frac{1}{2}\varsigma \cdot f\varsigma. \quad (18.23)$$

18.3.2 Neutral Klein–Gordon equation

Let $\mathcal{X}$ be a real Hilbert space. Let $\epsilon > 0$ be a strictly positive self-adjoint operator on $\mathcal{X}$. (Recall that $\epsilon > 0$ means that $\epsilon \geq 0$ and $\mathrm{Ker}\,\epsilon = \{0\}$.)

Definition 18.40 *The equation*

$$\partial_t^2 \zeta(t) + \epsilon^2 \zeta(t) = 0, \quad (18.24)$$

where $\zeta(t)$ is a function from $\mathbb{R}$ to $\mathcal{X}$, will be called an abstract neutral Klein–Gordon equation.

Clearly, if $\zeta(t)$ is a solution, $\zeta(-t)$ is also a solution, so (18.24) is invariant under time reversal.

Examples of (18.24) are the wave or Klein–Gordon equations on *static space-times*; see Chap. 19.

Let us reinterpret (18.24) as a first-order equation. To this end, we consider the space of Cauchy data $\mathcal{Y} = \mathcal{X} \oplus \mathcal{X}$, whose elements are denoted (ϑ, ς) or $\begin{bmatrix} \vartheta \\ \varsigma \end{bmatrix}$. We equip it with the symplectic form

$$(\vartheta_1, \varsigma_1) \cdot \omega (\vartheta_2, \varsigma_2) = \vartheta_1 \cdot \varsigma_2 - \vartheta_2 \cdot \varsigma_1, \qquad (\vartheta_i, \varsigma_i) \in \mathcal{X} \oplus \mathcal{X}, \quad i = 1, 2.$$

Setting

$$\varsigma(t) := \zeta(t), \quad \vartheta(t) := \partial_t \zeta(t), \quad a := \begin{bmatrix} 0 & -\epsilon^2 \\ \mathbb{1} & 0 \end{bmatrix},$$

we rewrite (18.24) as

$$\partial_t \begin{bmatrix} \vartheta(t) \\ \varsigma(t) \end{bmatrix} = a \begin{bmatrix} \vartheta(t) \\ \varsigma(t) \end{bmatrix}, \qquad \begin{bmatrix} \vartheta(0) \\ \varsigma(0) \end{bmatrix} = \begin{bmatrix} \vartheta \\ \varsigma \end{bmatrix}. \tag{18.25}$$

(Note that we put the time derivative first, since we are considering the dual phase space.) We see that (18.25) is solved by

$$\begin{bmatrix} \vartheta(t) \\ \varsigma(t) \end{bmatrix} = \begin{bmatrix} \cos(\epsilon t) & -\epsilon \sin(\epsilon t) \\ \epsilon^{-1} \sin(\epsilon t) & \cos(\epsilon t) \end{bmatrix} \begin{bmatrix} \varsigma \\ \vartheta \end{bmatrix} = e^{ta} \begin{bmatrix} \vartheta \\ \varsigma \end{bmatrix}. \tag{18.26}$$

For bounded ϵ, e^{ta} is a symplectic dynamics on $\mathcal{X} \oplus \mathcal{X}$ with the Hamiltonian

$$\frac{1}{2} \vartheta \cdot \vartheta + \frac{1}{2} \varsigma \cdot \epsilon^2 \varsigma. \tag{18.27}$$

For unbounded ϵ, there is a problem, since $\mathcal{X} \oplus \mathcal{X}$ is not preserved by e^{ta}. In this case, one can replace $\mathcal{X} \oplus \mathcal{X}$ with $\mathcal{Y} = \mathcal{X} \oplus \mathrm{Dom}\,\epsilon$, which is a symplectic space preserved by the dynamics. The dynamics is weakly stable. If in addition $\epsilon \geq m > 0$, then it is stable. The energy space $\mathcal{Y}_{\mathrm{en}}$ is equal to $\mathcal{X} \oplus \epsilon^{-1}\mathcal{X}$.

The Kähler anti-involution of Thm. 18.5 takes the form

$$\mathrm{j} = \begin{bmatrix} 0 & -\epsilon \\ \epsilon^{-1} & 0 \end{bmatrix}. \tag{18.28}$$

The associated Hermitian product is

$$((\vartheta_1, \varsigma_1) | (\vartheta_2, \varsigma_2)) = \vartheta_1 \cdot \epsilon^{-1} \vartheta_2 + \varsigma_1 \cdot \epsilon \varsigma_2 + \mathrm{i}(\vartheta_1 \cdot \varsigma_2 - \vartheta_2 \cdot \varsigma_1). \tag{18.29}$$

The completion of the Kähler space $\mathcal{Y}_{\mathrm{en}}$ for (18.29) is

$$\mathcal{Y}_{\mathrm{dyn}} := \epsilon^{\frac{1}{2}} \mathcal{X} \oplus \epsilon^{-\frac{1}{2}} \mathcal{X}. \tag{18.30}$$

In the standard way we introduce the space $\mathcal{Z} := \frac{1}{2}(\mathbb{1} - \mathrm{ij})\mathbb{C}\mathcal{Y}_{\mathrm{dyn}}$, which will serve as the one-particle space for quantization.

Note that the dual phase space and dynamics of an abstract Klein–Gordon equation belong to the class described in Subsect. 18.3.1. In particular, the time reversal is given by (18.22).

It is natural to introduce the following identification:

$$\mathbb{C}(2\epsilon)^{\frac{1}{2}}\mathcal{X} \ni \vartheta \mapsto U\vartheta := \frac{1\!\!1 - \mathrm{i}\mathrm{j}}{2}(\vartheta, 0) = \left(\frac{1}{2}\vartheta, \frac{\mathrm{i}}{2\epsilon}\vartheta\right) \in \mathcal{Z} \subset \mathbb{C}\mathcal{Y}_{\mathrm{dyn}}. \qquad (18.31)$$

Note that U is unitary.

Recall that the dynamics can be lifted to the space $\mathcal{Z}$ by $e_{\mathcal{Z}}^{ta} := e_{\mathbb{C}}^{ta}\big|_{\mathcal{Z}}$. We have $U^* e_{\mathcal{Z}}^{ta} U = e^{\mathrm{i}t\epsilon}$.

Likewise, the time reversal can be lifted to $\mathcal{Z}$ by $\tau_{\mathcal{Z}} := \tau_{\overline{\mathbb{C}}}\big|_{\mathcal{Z}}$. Now $U^* \tau_{\mathcal{Z}} U$ coincides with the usual canonical conjugation on $\mathbb{C}(2\epsilon)^{\frac{1}{2}}\mathcal{X}$.

Note that $\mathcal{Y}_{\mathrm{dyn}}$ is a complete Kähler space with a conjugation τ. Recall that we considered the CCR over such spaces in Subsect. 8.2.7. The operator $(2c)^{-1}$ of Subsect. 8.2.7 can be identified with ϵ of this subsection. The map U is the same map as (8.32).

Remark 18.41 *An abstract neutral Klein–Gordon equation describes the most general stable dynamics invariant w.r.t. a time reversal. In fact, recall the Hamiltonian (18.23), discussed in Subsect. 18.1.3 about the time-reversal invariance, and assume that it is strictly positive. Then it is easy to see that (18.23) can be brought to the form (18.27).*

18.3.3 Neutral Klein–Gordon equation in an external potential

We consider now the following modification of (18.24):

$$(\partial_t + d)^2\zeta(t) + \epsilon^2\zeta(t) = 0, \qquad (18.32)$$
$$\text{or} \quad \partial_t^2\zeta(t) + 2d\partial_t\zeta(t) + (\epsilon^2 + d^2)\zeta(t) = 0,$$

where $d = -d^*$ is anti-self-adjoint on $\mathcal{X}$. Note that this equation is no longer invariant under time-reversal. Examples of (18.32) are wave or Klein–Gordon equations on *stationary space-times* (see Example 19.43). Setting

$$\varsigma(t) := \zeta(t), \quad \vartheta(t) := \partial_t\zeta(t) + d\zeta(t), \quad a := \begin{bmatrix} -d & -\epsilon^2 \\ 1\!\!1 & -d \end{bmatrix},$$

we can rewrite (18.32) as a first-order equation,

$$\partial_t \begin{bmatrix} \vartheta(t) \\ \varsigma(t) \end{bmatrix} = a \begin{bmatrix} \vartheta(t) \\ \varsigma(t) \end{bmatrix},$$

with a Hamiltonian

$$\frac{1}{2}(\vartheta - d\varsigma)\cdot(\vartheta - d\varsigma) - \frac{1}{2}(d\varsigma)\cdot d\varsigma + \frac{1}{2}(\epsilon\varsigma)\cdot\epsilon\varsigma.$$

If $\epsilon^2 + d^2 > 0$, then the dynamics is weakly stable.

Note that the associated complex structure j does not have a simple expression anymore.

18.3.4 Splitting into complex configuration and momentum spaces

This subsection is the charged version of Subsect. 18.3.1. Suppose that $(\mathcal{Y}, \omega)$ is a charged symplectic space equipped with a Racah time reversal satisfying $\kappa^2 = \mathbb{1}$. Again, we can split the dual phase space into the direct sum of Lagrangian subspaces $\mathcal{Y} = \mathcal{Y}^\kappa \oplus \mathcal{Y}^{-\kappa}$, where $\mathcal{Y}^{\pm\kappa} := \{y \in \mathcal{Y} : y = \pm\kappa y\}$. (Note that spaces $\mathcal{Y}^\kappa$ and $\mathcal{Y}^{-\kappa}$ are both complex.)

$\mathcal{Y}^\kappa$ has the interpretation of the *dual configuration space* and $\mathcal{Y}^{-\kappa}$ of the *dual momentum space*.

$(\mathcal{Y}^\kappa, \mathcal{Y}^{-\kappa})$ can be interpreted as an anti-dual pair, so that the charged symplectic form can be written

$$\overline{(\vartheta_1, \varsigma_1)} \cdot \omega(\vartheta_2, \varsigma_2) = \bar{\vartheta}_1 \cdot \varsigma_2 - \bar{\varsigma}_1 \cdot \vartheta_2, \qquad (\vartheta_i, \varsigma_i) \in \mathcal{Y}^\kappa \oplus \mathcal{Y}^{-\kappa}, \quad i = 1, 2.$$

The Racah time reversal acts as

$$\kappa(\vartheta, \varsigma) = (\vartheta, -\varsigma), \quad (\vartheta, \varsigma) \in \mathcal{Y}^\kappa \oplus \mathcal{Y}^{-\kappa}. \tag{18.33}$$

Let $\{r_t\}_{t\in\mathbb{R}}$ be a dynamics invariant w.r.t the Racah time reversal. For $(\vartheta, \varsigma) \in \mathcal{Y}^\kappa \oplus \mathcal{Y}^{-\kappa}$ write $r_t(\vartheta, \varsigma) = (\vartheta(t), \varsigma(t))$. Then there exist $f \in L(\mathcal{Y}^\kappa, \mathcal{Y}^{-\kappa})$, $g \in L(\mathcal{Y}^{-\kappa}, \mathcal{Y}^\kappa)$ such that $g = g^*$, $f = f^*$ and

$$\partial_t \varsigma(t) = f\vartheta(t), \quad \partial_t \vartheta(t) = -g\varsigma(t).$$

The Hamiltonian of the dynamics is

$$\bar{\vartheta} \cdot g\vartheta + \bar{\varsigma} \cdot f\varsigma.$$

18.3.5 Charged Klein–Gordon equation

Now we describe the charged version of Subsect. 18.3.2. Let $\mathcal{X}$ be a complex Hilbert space. For $\zeta_1, \zeta_2 \in \mathcal{X}$, the scalar product will be denoted by $\bar{\zeta}_1 \cdot \zeta_2$. Consider again a strictly positive self-adjoint operator ϵ on $\mathcal{X}$ and the equation (18.24).

Definition 18.42 *If the space $\mathcal{X}$ is complex, the equation (18.24) will be called an* abstract charged Klein–Gordon equation.

Thus the only difference between the charged and neutral Klein–Gordon equations is the presence of the $U(1)$ symmetry given by the multiplication by $e^{i\theta}$, $\theta \in [0, 2\pi]$.

The Racah time reversal consists in replacing $t \mapsto \zeta(t)$ with $t \mapsto \zeta(-t)$. The charged Klein–Gordon equation is always invariant w.r.t. the Racah time reversal.

Let us fix a complex conjugation on $\mathcal{X}$, denoted by $\zeta \mapsto \bar{\zeta}$, which defines the charge reversal. The Wigner time reversal involves replacing a function $t \mapsto \zeta(t)$ with $t \mapsto \overline{\zeta(-t)}$. If $\bar{\epsilon} = \epsilon$, then (18.24) is also invariant w.r.t. the charge and Wigner time reversal.

Consider the Cauchy problem (18.25). We introduce the space $\mathcal{X} \oplus \mathcal{X}$, equipped with the charged symplectic form

$$\overline{(\vartheta_1, \varsigma_1)} \cdot \omega(\vartheta_2, \varsigma_2) = \overline{\vartheta}_1 \cdot \varsigma_2 - \overline{\varsigma}_1 \cdot \vartheta_2, \quad (\vartheta_i, \varsigma_i) \in \mathcal{X} \oplus \mathcal{X}, \quad i = 1, 2.$$

The Hamiltonian is

$$\overline{\vartheta} \cdot \vartheta + \overline{\epsilon \varsigma} \cdot \epsilon \varsigma.$$

$\mathcal{Y}_{en}$, $\mathcal{Y}_{dyn}$, j, a are given by the same expressions as in Subsect. 18.3.2. In particular, it is natural to replace the original dual phase space $\mathcal{X} \oplus \mathcal{X}$ by $\mathcal{Y}_{dyn} = \epsilon^{\frac{1}{2}} \mathcal{X} \oplus \epsilon^{-\frac{1}{2}} \mathcal{X}$.

In terms of the Cauchy data, the Racah time reversal is given by (18.33). The charge reversal and the Wigner time reversal are given by

$$\chi(\vartheta, \varsigma) = (\overline{\vartheta}, \overline{\varsigma}), \quad \tau(\vartheta, \varsigma) = (\overline{\vartheta}, -\overline{\varsigma}).$$

We can "diagonalize" the dynamics by introducing the map

$$W : \mathcal{Y}_{dyn} = \epsilon^{\frac{1}{2}} \mathcal{X} \oplus \epsilon^{-\frac{1}{2}} \mathcal{X} \ni (\vartheta, \varsigma) \mapsto (\vartheta + i\epsilon\varsigma, \overline{\vartheta} + i\epsilon\overline{\varsigma}) \in (2\epsilon)^{\frac{1}{2}} \mathcal{X} \oplus (2\epsilon)^{\frac{1}{2}} \mathcal{X}.$$

W is a unitary operator satisfying

$$W e^{ta} W^{-1} = e^{it(\epsilon \oplus \epsilon)}, \quad W i W^{-1} = i \mathbb{1} \oplus (-i\mathbb{1}),$$
$$W j W^{-1} = i \mathbb{1} \oplus i \mathbb{1}, \quad W q W^{-1} = \mathbb{1} \oplus (-\mathbb{1}).$$

Thus if we interpret W as an operator on $\mathcal{Z}$ (which differs from $\mathcal{Y}_{dyn}$ only by treating j as the basic complex structure), then $W : \mathcal{Z} \to (2\epsilon)^{\frac{1}{2}} \mathcal{X} \oplus (2\epsilon)^{\frac{1}{2}} \mathcal{X}$ is unitary.

After conjugation by W, the charge and Wigner time reversal become

$$\chi(h_1, \overline{h}_2) = (\overline{h}_2, h_1), \quad \tau(h_1, \overline{h}_2) = (-\overline{h}_1, h_2).$$

Remark 18.43 *This remark is analogous to Remark 18.41 from the neutral case. A charged abstract Klein–Gordon equation describes the most general stable dynamics invariant w.r.t. the Racah time reversal.*

18.3.6 Charged Klein–Gordon equation in an external potential

Again we can consider the complex analog of (18.32). It is more natural to write it as

$$(\partial_t + iV)^2 \zeta(t) + \epsilon^2 \zeta(t) = 0, \tag{18.34}$$
$$\text{or} \quad \partial_t^2 \zeta(t) + 2iV \partial_t \zeta(t) + (\epsilon^2 - V^2)\zeta(t) = 0,$$

where $V = V^*$. An example is obtained by minimally coupling (18.24) to an external electric field.

Setting

$$\varsigma(t) := \zeta(t), \quad \vartheta(t) := \partial_t \zeta(t) + iV\zeta(t),$$

we can rewrite (18.34) as

$$\partial_t \begin{bmatrix} \vartheta(t) \\ \varsigma(t) \end{bmatrix} = \begin{bmatrix} -iV & -\epsilon^2 \\ \mathbb{1} & -iV \end{bmatrix} \begin{bmatrix} \varsigma(t) \\ \vartheta(t) \end{bmatrix}$$

with the Hamiltonian

$$\overline{\vartheta} \cdot \vartheta + \overline{\epsilon\varsigma} \cdot \epsilon\varsigma - i\overline{\vartheta} \cdot V\varsigma + i\overline{\varsigma} \cdot V\vartheta$$
$$= \overline{(\vartheta - iV\varsigma)} \cdot (\vartheta - iV\varsigma) - \overline{V\varsigma} \cdot V\varsigma + \overline{\epsilon\varsigma} \cdot \epsilon\varsigma.$$

If $\epsilon^2 - V^2 > 0$, then the dynamics is weakly stable.

Note that if $\mathcal{X}$ is equipped with a conjugation such that V and ϵ are real, then (18.34) is invariant under the Wigner time reversal.

18.3.7 Quantization of the Klein–Gordon equation

Until the end of this section, we would like to treat the neutral and charged cases together. We do this by embedding the neutral case in the charged case.

More precisely, until the end of the section $\mathcal{X}$ is always a complex Hilbert space with a positive self-adjoint operator ϵ. We consider the abstract charged Klein–Gordon equation for $t \mapsto \zeta(t) \in \mathcal{X}$:

$$\partial_t^2 \zeta(t) + \epsilon^2 \zeta(t) = 0. \tag{18.35}$$

We consider the charged symplectic space of solutions of (18.35), denoted $\mathcal{Y}$. Recall that every such solution can be parametrized by its Cauchy data (ϑ, ς). The space $\mathcal{Y}$ is equipped with a charged symplectic dynamics r_t.

If we want to consider the neutral case, we assume that there exists a conjugation χ on $\mathcal{X}$ that commutes with ϵ. Thus we can restrict the abstract Klein–Gordon equation to $\mathcal{X}^\chi = \{\zeta \in \mathcal{X} : \chi\zeta = \zeta\}$, obtaining the symplectic space of solutions $\mathcal{Y}^\chi$. The space $\mathcal{Y}^\chi$ is equipped with a symplectic dynamics $r_t\big|_{\mathcal{Y}^\chi}$, which satisfies the abstract neutral Klein–Gordon equation.

We apply the positive energy quantization described in Subsects. 18.1.1, resp. 18.2.1, obtaining operator-valued functions

$$\mathcal{Y}^\chi \ni (\vartheta, \varsigma) \mapsto \phi(\vartheta, \varsigma), \quad \text{in the neutral case,}$$
$$\mathcal{Y} \ni (\vartheta, \varsigma) \mapsto \psi(\vartheta, \varsigma), \quad \text{in the charged case,}$$

and the Hamiltonian H such that

$$e^{itH} \phi(\vartheta, \varsigma) e^{-itH} = \phi\big(r_t(\vartheta, \varsigma)\big), \quad (\vartheta, \varsigma) \in \mathcal{Y}^\chi,$$
$$e^{itH} \psi(\vartheta, \varsigma) e^{-itH} = \psi\big(r_t(\vartheta, \varsigma)\big), \quad (\vartheta, \varsigma) \in \mathcal{Y}.$$

Definition 18.44 *The* time t configuration space field *is defined as*

$$\phi_t(\vartheta) := \phi\big(r_{-t}(\vartheta, 0)\big), \quad \vartheta \in \epsilon^{\frac{1}{2}} \mathcal{X}^\chi,$$
$$\psi_t(\vartheta) := \psi\big(r_{-t}(\vartheta, 0)\big), \quad \vartheta \in \epsilon^{\frac{1}{2}} \mathcal{X}.$$

18.3.8 Two-point functions for the Klein–Gordon equation

The remaining part of this section is devoted to various functions related to the Klein–Gordon equation, which are often used in its quantization.

Consider the Klein–Gordon equation (18.35), where $\zeta(t) \in \mathcal{X}$ is replaced with an operator $G(t) \in B(\mathcal{X})$:

$$\partial_t^2 G(t) + \epsilon^2 G(t) = 0. \tag{18.36}$$

The functions $\frac{\sin \epsilon t}{\epsilon}$ and $\frac{e^{it\epsilon}}{2\epsilon}$ solve (18.36) and appear naturally in the quantized theory:

$$[\varphi_{t_1}(\vartheta_1), \varphi_{t_2}(\vartheta_2)] = i\vartheta_1 \cdot \frac{\sin \epsilon(t_1 - t_2)}{\epsilon} \vartheta_2 \, \mathbb{1},$$

$$(\Omega|\varphi_{t_1}(\vartheta_1)\varphi_{t_2}(\vartheta_2)\Omega) = \vartheta_1 \cdot \frac{1}{2\epsilon} e^{i\epsilon(t_1 - t_2)} \vartheta_2, \quad \vartheta_1, \vartheta_2 \in \mathcal{X}^{\chi},$$

$$[\psi_{t_1}(\vartheta_1), \psi_{t_2}^*(\vartheta_2)] = i\overline{\vartheta}_1 \cdot \frac{\sin \epsilon(t_1 - t_2)}{\epsilon} \vartheta_2 \, \mathbb{1},$$

$$(\Omega|\psi_{t_1}(\vartheta_1)\psi_{t_2}^*(\vartheta_2)\Omega) = \overline{\vartheta}_1 \cdot \frac{1}{2\epsilon} e^{i\epsilon(t_1 - t_2)} \vartheta_2, \quad \vartheta_1, \vartheta_2 \in \mathcal{X}.$$

Definition 18.45 $\frac{\sin \epsilon t}{\epsilon}$ *is called the* Pauli–Jordan *or* commutator function.

18.3.9 Green's functions of the abstract Klein–Gordon equation

Let us now consider an inhomogeneous version of Eq. (18.36).

In what follows we will often use the Heaviside function $\theta(t) := \mathbb{1}_{[0,+\infty[}(t)$.

Definition 18.46 $\mathbb{R} \ni t \mapsto G(t) \in B(\mathcal{X})$ *is a* Green's function *or a* fundamental solution *of Eq. (18.35) if it solves*

$$\partial_t^2 G(t) + \epsilon^2 G(t) = \delta(t)\mathbb{1}. \tag{18.37}$$

In particular, we introduce the following Green's functions:

retarded $$G^+(t) := \theta(t)\frac{\sin \epsilon t}{\epsilon},$$

advanced $$G^-(t) := -\theta(-t)\frac{\sin \epsilon t}{\epsilon},$$

Feynman *or* causal $$G_{\mathrm{F}}(t) := \frac{1}{2i\epsilon}\left(e^{it\epsilon}\theta(t) + e^{-it\epsilon}\theta(-t)\right),$$

anti-Feynman *or* anti-causal $$G_{\overline{\mathrm{F}}}(t) := -\frac{1}{2i\epsilon}\left(e^{-it\epsilon}\theta(t) + e^{it\epsilon}\theta(-t)\right),$$

principal value *or* Dirac $$G_{\mathrm{Pv}}(t) := \frac{\mathrm{sgn}(t)}{2}\frac{\sin \epsilon t}{\epsilon}.$$

Note the identities

$$\frac{\sin \epsilon t}{\epsilon} = G^+(t) - G^-(t),$$

$$G_{\mathrm{Pv}}(t) = \frac{1}{2}\left(G^+(t) + G^-(t)\right),$$

$$G^+(t) + G^-(t) = G_{\mathrm{F}}(t) + G_{\overline{\mathrm{F}}}(t).$$

The importance of the Feynman Green's functions in the quantum theory comes from the identities

$$(\Omega | T \left(\varphi_{t_2}(\vartheta_2)\varphi_{t_1}(\vartheta_1) \right) \Omega) = i\vartheta_2 \cdot G_F(t_2 - t_1)\vartheta_1, \quad \vartheta_1, \vartheta_2 \in \mathcal{X}^{\times},$$
$$(\Omega | T \left(\psi_{t_2}(\vartheta_2)\psi_{t_1}^*(\vartheta_1) \right) \Omega) = i\overline{\vartheta}_2 \cdot G_F(t_2 - t_1)\vartheta_1, \quad \vartheta_1, \vartheta_2 \in \mathcal{X},$$

where we have used the time-ordering operation

$$T\left(A_{t_2} A_{t_1}\right) := \theta(t_2 - t_1)A_{t_2} A_{t_1} + \theta(t_1 - t_2)A_{t_1} A_{t_2}.$$

18.3.10 Green's functions of the Klein–Gordon equation as operators

Let $\mathcal{X}$ be as above. For simplicity, we assume that $\mathcal{X}$ is separable. We will need the space

$$L^2(\mathbb{R}) \otimes \mathcal{X} \simeq L^2(\mathbb{R}, \mathcal{X}). \tag{18.38}$$

Note that the unitary identification $\simeq$ in (18.38) is possible thanks to the fact that $\mathcal{X}$ is separable. It means that we can represent elements of $L^2(\mathbb{R}) \otimes \mathcal{X}$ with measurable a.e. defined functions, which e.g. in the temporal representation are written as $\mathbb{R} \ni t \mapsto \zeta(t) \in \mathcal{X}$ and satisfy

$$\int \|\zeta(t)\|^2 \mathrm{d}t < \infty.$$

Clearly, the subspace $(L^1 \cap L^2)(\mathbb{R}, \mathcal{X})$ is dense in $L^2(\mathbb{R}, \mathcal{X})$.

We will distinguish two physical meanings of the variable in $\mathbb{R}$ that appears in (18.38). The first meaning is the time, and the corresponding generic variable in $\mathbb{R}$ will be denoted t. We will then say that we use the *temporal representation* of the extended space. The second meaning will be the energy. Its generic name will be τ and we will speak about the *energy representation*. To pass from one representation to the other we apply the Fourier transformation $\mathcal{F}$, so that $\mathcal{F}^{-1}\tau\mathcal{F} = i^{-1}\partial_t$.

Green's functions of the abstract Klein–Gordon equation can be interpreted as quadratic forms on $(L^1 \cap L^2)(\mathbb{R}, \mathcal{X})$ given (in the temporal representation) by

$$\overline{\zeta}_1 \cdot G\zeta_2 := \int \overline{\zeta_1(t)} \cdot G(t - s)\zeta_2(s)\mathrm{d}t\mathrm{d}s, \tag{18.39}$$

for $\zeta_1, \zeta_2 \in (L^1 \cap L^2)(\mathbb{R}, \mathcal{X})$. In the energy representation they are multiplication operators. Here we list the most important Green's functions in the momentum representation:

$$G^+(\tau) = (\epsilon^2 - (\tau - i0)^2)^{-1},$$
$$G^-(\tau) = (\epsilon^2 - (\tau + i0)^2)^{-1},$$
$$G_F(\tau) = (\epsilon^2 - \tau^2 + i0)^{-1} = \left(\epsilon^2 - (\tau^2 - i0 \operatorname{sgn}(\tau))^2\right)^{-1},$$
$$G_{\overline{F}}(\tau) = (\epsilon^2 - \tau^2 - i0)^{-1} = \left(\epsilon^2 - (\tau^2 + i0 \operatorname{sgn}(\tau))^2\right)^{-1},$$
$$G_{\mathrm{Pv}}(\tau) = \mathrm{Pv}(\epsilon^2 - \tau^2)^{-1}.$$

18.3.11 Euclidean Green's function of the Klein–Gordon equation

Let us introduce the *imaginary time*, that is, let us replace $t \in \mathbb{R}$ with $is \in i\mathbb{R}$. The abstract Klein–Gordon equation is then transformed into

$$-\partial_s^2 \zeta + \epsilon^2 \zeta = 0. \tag{18.40}$$

Definition 18.47 *Equation (18.40) will be called the* abstract Euclidean Klein–Gordon equation.

The use of (18.40) instead of the Klein–Gordon equation is the main feature of the so-called Euclidean approach to quantum field theory.

Definition 18.48 *The* Euclidean Green's function of the abstract Klein–Gordon equation *is defined as*

$$G_{\mathrm{E}}(s) = \frac{1}{2\epsilon}\left(e^{-s\epsilon}\theta(s) + e^{s\epsilon}\theta(-s)\right).$$

Clearly, G_{E} solves

$$-\partial_s^2 G_{\mathrm{E}}(s) + \epsilon^2 G_{\mathrm{E}}(s) = \delta(s)\mathbb{1}. \tag{18.41}$$

The function $G_{\mathrm{E}}(s)$ extends to a continuous function for complex s with $\operatorname{Re} s \geq 0$, holomorphic for $\operatorname{Re} s > 0$. We have

$$\frac{1}{i}G_{\mathrm{E}}(it) = G_{\mathrm{F}}(t), \qquad -\frac{1}{i}G_{\mathrm{E}}(-it) = G_{\overline{\mathrm{F}}}(t).$$

Consider the self-adjoint operator $-\partial_s^2 + \epsilon^2$ on $L^2(\mathbb{R}, \mathcal{X})$. Set

$$G_{\mathrm{E}} := (-\partial_s^2 + \epsilon^2)^{-1}.$$

We then have

$$G_{\mathrm{E}}\zeta(s) = \int_{\mathbb{R}} G_{\mathrm{E}}(s - s_1)\zeta(s_1)\mathrm{d}s_1$$

for $\zeta \in (L^1 \cap L^2)(\mathbb{R}, \mathcal{X})$. In the energy representation it is the operator of multiplication by

$$G_{\mathrm{E}}(\tau) = (\tau^2 + \epsilon^2)^{-1}.$$

Note that if $\epsilon \geq m > 0$, then G_{E} is bounded.

We will use the standard notation for operators on $L^2(\mathbb{R})$. In particular, the operator of multiplication by t in the temporal representation is denoted by t and $D_t := -i\partial_t$. A similar notation will be used for the energy representation, with τ replacing t.

Introduce the following operator on $L^2(\mathbb{R})$ (where we give its form in both the temporal and the energy representation):

$$A := -\frac{1}{2}(tD_t + D_t t) = \frac{1}{2}(\tau D_\tau + D_\tau \tau). \tag{18.42}$$

Clearly,

$$e^{i\theta A}te^{-i\theta A} = e^{-\theta}t, \qquad e^{i\theta A}\tau e^{-i\theta A} = e^{\theta}\tau.$$

Note that

$$\mathbb{R} \ni \theta \mapsto e^{i\theta A} G_{\mathrm{E}} e^{-i\theta A} =: G_{\mathrm{E}}^\theta$$

extends to an analytic function in the strip $-\frac{\pi}{2} < \mathrm{Im}\,\theta < \frac{\pi}{2}$ given in the momentum representation by

$$G_{\mathrm{E}}^\theta(\tau) = (e^{2\theta}\tau^2 + \epsilon^2)^{-1}.$$

Its boundary values coincide with the Feynman and anti-Feynman Green's functions:

$$G_{\mathrm{E}}^{i\frac{\pi}{2}} = G_{\mathrm{F}}, \quad G_{\mathrm{E}}^{-i\frac{\pi}{2}} = G_{\overline{\mathrm{F}}}.$$

This is the famous *Wick rotation*.

18.3.12 Thermal Green's function of the Klein–Gordon equation

Recall that one of the steps of the quantization of symplectic, resp. charged symplectic dynamics is the construction of the one-particle space $\mathcal{Z}$ and the one-particle Hamiltonian h, as described in Subsect. 18.1.1, resp. 18.2.1. For the zero temperature, the main requirement is the positivity of the Hamiltonian, and as the result of the quantization we obtain the Hilbert space $\Gamma_\mathrm{s}(\mathcal{Z})$ and the Hamiltonian $\mathrm{d}\Gamma(h)$.

If we are interested in positive temperatures, we can apply the formalism described in Subsect. 17.1.7, obtaining a β-KMS state ω_β and the corresponding Araki–Woods CCR representation. In particular, we can apply this formalism to the abstract Klein–Gordon equation.

In this subsection we describe the 2-point correlation functions for the abstract Klein–Gordon equation at positive temperatures.

Definition 18.49 *The* thermal Euclidean Green's function at inverse temperature β of the abstract Klein–Gordon equation *is defined for $s \in [0, \beta]$ as*

$$G_{\mathrm{E},\beta}(s) := \frac{e^{-s\epsilon} + e^{(s-\beta)\epsilon}}{2\epsilon(\mathbb{1} - e^{-\beta\epsilon})}.$$

Note that $G_{\mathrm{E},\beta}$ is the unique solution of the problem

$$-\partial_s^2 G_{\mathrm{E},\beta}(s) + \epsilon^2 G_{\mathrm{E},\beta}(s) = 0, \quad s \in\,]0, \beta[,$$
$$G_{\mathrm{E},\beta}(0) = G_{\mathrm{E},\beta}(\beta), \quad \partial_s^- G_{\mathrm{E},\beta}(\beta) - \partial_s^+ G_{\mathrm{E},\beta}(0) = \mathbb{1}_{\mathcal{X}},$$

where $\partial_s^\pm$ denotes the derivative from the right, resp. from the left. In fact, we have

$$G_{\mathrm{E},\beta}(\beta) = G_{\mathrm{E},\beta}(0) = \frac{\mathbb{1} + e^{-\beta\epsilon}}{2\epsilon(\mathbb{1} - e^{-\beta\epsilon})},$$
$$\partial_s^- G_{\mathrm{E},\beta}(\beta) = -\partial_s^+ G_{\mathrm{E},\beta}(0) = \frac{1}{2}\mathbb{1}_{\mathcal{X}}.$$

Let $S_\beta := [0, \beta]$ (with the endpoints identified) be the circle of length β. Clearly, $G_{\mathrm{E},\beta}$ can be interpreted as a function on S_β that solves the equation

$$-\partial_s^2 G_{\mathrm{E},\beta}(s) + \epsilon^2 G_{\mathrm{E},\beta}(s) = \delta(s)\mathbb{1}_\mathcal{X}, \quad \text{on } S_\beta.$$

We denote by $\mathcal{F}_\beta : L^2(S_\beta) \to l^2\left(\frac{2\pi}{\beta}\mathbb{Z}\right)$ the discrete Fourier transform

$$\mathcal{F}_\beta \zeta(\sigma) = \int_0^\beta e^{-is\sigma}\zeta(s)ds, \quad \zeta \in L^2(S_\beta).$$

Its inverse is given by

$$\mathcal{F}_\beta^{-1} v(s) = \beta^{-1} \sum_{\sigma \in \frac{2\pi}{\beta}\mathbb{Z}} e^{is\sigma} v(\sigma), \quad v \in l^2\left(\frac{2\pi}{\beta}\mathbb{Z}\right).$$

If we denote by $\partial_s^{\mathrm{per}}$ the operator ∂_s with periodic boundary conditions, defined by

$$\mathrm{Dom}\, \partial_s^{\mathrm{per}} := \left\{\zeta \in L^2([0,\beta]), \ \partial_s\zeta \in L^2([0,\beta]), \ \zeta(0) = \zeta(\beta)\right\},$$

then

$$\mathcal{F}_\beta \partial_s^{\mathrm{per}} = i\sigma \mathcal{F}_\beta. \tag{18.43}$$

Introduce the space

$$L^2(S_\beta) \otimes \mathcal{X} \simeq L^2(S_\beta, \mathcal{X}). \tag{18.44}$$

Consider the self-adjoint operator $-(\partial_s^{\mathrm{per}})^2 + \epsilon^2$ on $L^2(S_\beta, \mathcal{X})$. Set

$$G_{\mathrm{E},\beta} := \left(-(\partial_s^{\mathrm{per}})^2 + \epsilon^2\right)^{-1}.$$

We then have

$$G_{\mathrm{E},\beta}\zeta(s) = \int_{S_\beta} G_{\mathrm{E},\beta}(s - s_1)\zeta(s_1)ds_1, \quad \text{for } \zeta \in L^2(S_\beta, \mathcal{X}).$$

In the energy representation obtained by applying the discrete Fourier transform $\mathcal{F}_\beta$, $G_{\mathrm{E},\beta}$ becomes the multiplication operator on $l^2\left(\frac{2\pi}{\beta}\mathbb{Z}, \mathcal{X}\right)$ by the Fourier transform of $s \mapsto G_{\mathrm{E},\beta}(s)$, the so-called *Matsubara coefficients*:

$$G_{\mathrm{E},\beta}(\sigma) := (\sigma^2 + \epsilon^2)^{-1}, \quad \sigma \in \frac{2\pi}{\beta}\mathbb{Z}. \tag{18.45}$$

Let us now describe the role of thermal Green's functions in the quantum theory. The function $s \mapsto G_{\mathrm{E},\beta}(s)$ extends to a function continuous in the strip $\mathrm{Re}\, s \in [0, \beta]$ and holomorphic inside this strip. Its boundary values express the 2-point correlation functions for the state ω_β. More precisely, we have the following identities (first in the neutral and then in the charged case):

$$\omega_\beta\left(\varphi_t(\vartheta_1)\varphi_0(\vartheta_2)\right) = \vartheta_1 \cdot G_{\mathrm{E},\beta}(\mathrm{i}t)\vartheta_2,$$
$$\omega_\beta\left(\varphi_0(\vartheta_2)\varphi_t(\vartheta_1)\right) = \vartheta_1 \cdot G_{\mathrm{E},\beta}(\beta + \mathrm{i}t)\vartheta_2, \quad \vartheta_1, \vartheta_2 \in \mathcal{X}^\times,$$

$$\omega_\beta\left(\psi_t(\vartheta_1)\psi_0^*(\vartheta_2)\right) = \bar{\vartheta}_1 \cdot G_{\mathrm{E},\beta}(\mathrm{i}t)\vartheta_2,$$
$$\omega_\beta\left(\psi_0^*(\vartheta_2)\psi_t(\vartheta_1)\right) = \bar{\vartheta}_1 \cdot G_{\mathrm{E},\beta}(\beta + \mathrm{i}t)\vartheta_2, \quad \vartheta_1, \vartheta_2 \in \mathcal{X}.$$

18.4 Abstract Dirac equation and its quantization

In Subsects. 18.1.2, resp. 18.2.2 we described how to quantize orthogonal, resp. unitary dynamics. The most important orthogonal or unitary dynamics used in quantum field theory is given by the *Dirac equation*. We will study this equation and its quantization in the next chapter. In this section we describe various constructions related to the quantization of the Dirac equation in an abstract setting.

An orthogonal dynamics can be defined by an equation of the form

$$(\partial_t - a)\zeta(t) = 0, \tag{18.46}$$

with an anti-self-adjoint a. In the charged case (18.46) can be replaced with

$$(\partial_t - ib)\zeta(t) = 0, \tag{18.47}$$

where b is self-adjoint. Many of the constructions of this section involve only a or b. We choose, however, to use more structure in our presentation. In particular, we multiply (18.46) and (18.47) from the left by an anti-self-adjoint operator γ_0 satisfying $\gamma_0^2 = -\mathbb{1}$. This is used in the relativistic formulation of the Dirac equation to make it covariant.

This section is parallel to Sect. 18.3 about the abstract Klein–Gordon equation. We will see, in particular, that with every abstract Dirac equation we can associate an abstract Klein–Gordon equation. The knowledge of Green's functions for the abstract Klein–Gordon equation can be used to compute Green's functions of the abstract Dirac equation.

18.4.1 Abstract Dirac equation

Let $\mathcal{Y}$ be a real or complex Hilbert space, which will have the meaning of a fermionic dual phase space. Let Γ and γ_0 be anti-self-adjoint operators on $\mathcal{Y}$ such that

$$\gamma_0^2 = -\mathbb{1}, \quad \gamma_0\Gamma + \Gamma\gamma_0 = 0. \tag{18.48}$$

Let $m \geq 0$ be a number called the *mass*.

Definition 18.50 *An equation of the form*

$$(\gamma_0\partial_t + \Gamma - m\mathbb{1})\zeta(t) = 0, \tag{18.49}$$

where $\zeta(t)$ is a function from $\mathbb{R}$ to $\mathcal{Y}$, will be called an abstract Dirac *equation.*

Multiplying (18.49) with $-\gamma_0$, we obtain the equation (18.46) with an anti-self-adjoint operator

$$a := \gamma_0\Gamma - m\gamma_0.$$

Introducing the Cauchy problem

$$\begin{cases} (\gamma_0 \partial_t + \Gamma - m\mathbb{1})\zeta(t) = 0, \\ \zeta(0) = \vartheta, \end{cases} \tag{18.50}$$

we see that (18.49) is solved by $\zeta(t) = e^{ta}\vartheta$.

Using (18.48), we obtain that $a^\# a = \Gamma^\# \Gamma + m^2 \mathbb{1}$, so the dynamics $r_t = e^{ta}$ is non-degenerate if

$$\operatorname{Ker}\Gamma = \{0\} \text{ or } m \neq 0. \tag{18.51}$$

18.4.2 Neutral Dirac equation

If $\mathcal{Y}$ is a real Hilbert space, (18.49) will be called a *abstract neutral Dirac equation*. A time reversal for the neutral Dirac equation is $\tau \in O(\mathcal{Y})$ satisfying $\tau^2 = \pm\mathbb{1}$,

$$\tau\gamma_0 = -\gamma_0\tau, \quad \tau\Gamma = \Gamma\tau, \tag{18.52}$$

or, if $m = 0$,

$$\tau\gamma_0 = \gamma_0\tau, \quad \tau\Gamma = -\Gamma\tau. \tag{18.53}$$

In both cases, $\tau a = -a\tau$, hence if $\zeta(t)$ is a solution of (18.49), then so is $\tau\zeta(-t)$.

18.4.3 Charged Dirac equation

Assume now that $\mathcal{Y}$ is a (complex) Hilbert space. Thanks to the complex structure, we can introduce the self-adjoint operator

$$b := -i\gamma_0\Gamma + im\gamma_0.$$

The Cauchy problem (18.50) is solved by $\zeta(t) = e^{itb}\vartheta$.

A *(Wigner) time reversal* is an anti-unitary τ on $\mathcal{Y}$ satisfying (18.52), or, if $m = 0$, (18.53). In both cases, $\tau b = b\tau$, hence if $\zeta(t)$ is a solution of (18.49), then so is $\tau\zeta(-t)$.

A *charge reversal* is an anti-unitary χ on $\mathcal{Y}$ such that $\chi^2 = \pm\mathbb{1}$ and

$$\chi\gamma_0 = \gamma_0\chi, \quad \chi\Gamma = \Gamma\chi, \tag{18.54}$$

or, if $m = 0$,

$$\chi\gamma_0 = -\gamma_0\chi, \quad \chi\Gamma = -\Gamma\chi. \tag{18.55}$$

In both cases $b\chi = -\chi b$ and the dynamics e^{itb} is invariant under χ.

If $\chi^2 = \mathbb{1}$, then χ is a conjugation on $\mathcal{Y}$ and by restriction to the real subspace $\mathcal{Y}^\chi$ we obtain a neutral Dirac equation, as in Subsect. 18.4.2.

18.4.4 Quantization of the Dirac equation

Until the end of this section we would like to treat the neutral and charged cases together. We will treat the charged case as the basic one and $\mathcal{Y}$ will always be a complex Hilbert space. $\mathcal{Y}$ is identified with the space of solutions of the abstract Dirac equation by considering the initial conditions: $\vartheta = \zeta(0)$. The space $\mathcal{Y}$ is equipped with a unitary dynamics $r_t = \mathrm{e}^{itb}$.

If we want to consider the neutral case, we assume that there exists in $\mathcal{Y}$ a conjugation χ that anti-commutes with b. Thus we can restrict the Dirac equation to $\mathcal{Y}^\chi$ equipped with the Euclidean structure and an orthogonal dynamics $\mathrm{e}^{itb}\big|_{\mathcal{Y}^\chi} = \mathrm{e}^{at}$.

We apply the positive energy quantization described in Subsects. 18.1.2 and 18.2.2, obtaining operator-valued functions

$$\mathcal{Y}^\chi \ni \vartheta \mapsto \phi(\vartheta), \quad \text{in the neutral case,}$$
$$\mathcal{Y} \ni \vartheta \mapsto \psi(\vartheta), \quad \text{in the complex case,}$$

and the Hamiltonian H such that

$$\mathrm{e}^{itH}\phi(\vartheta)\mathrm{e}^{-itH} = \phi(r_t(\vartheta)), \quad \vartheta \in \mathcal{Y}^\chi,$$
$$\mathrm{e}^{itH}\psi(\vartheta)\mathrm{e}^{-itH} = \psi(r_t(\vartheta)), \quad \vartheta \in \mathcal{Y}.$$

The fermionic fields satisfy the anti-commutation relations

$$[\phi(\vartheta_1), \phi(\vartheta_2)]_+ = 2\vartheta_1 \cdot \vartheta_2, \quad \vartheta_1, \vartheta_2 \in \mathcal{Y}^\chi,$$
$$[\psi(\vartheta_1), \psi^*(\vartheta_2)]_+ = \overline{\vartheta}_1 \cdot \vartheta_2, \quad \vartheta_1, \vartheta_2 \in \mathcal{Y}.$$

In the following part of the section, for simplicity we restrict ourselves to the charged case.

18.4.5 Two-point functions for the Dirac equation

Consider the abstract Dirac equation (18.35), where $\zeta(t) \in \mathcal{Y}$ is replaced with an operator $S(t) \in B(\mathcal{Y})$:

$$(\gamma_0 \partial_t + \Gamma - m\mathbb{1})S(t) = 0. \tag{18.56}$$

Recall that θ denotes the Heaviside function. The functions e^{ibt} and $\mathrm{e}^{ibt}\theta(b)$ solve (18.56) and appear naturally in the quantized theory:

$$[\psi_{t_2}(\vartheta_2), \psi_{t_1}^*(\vartheta_1)]_+ = \overline{\vartheta}_2 \cdot \mathrm{e}^{ib(t_2-t_1)}\vartheta_1 \,\mathbb{1},$$
$$(\Omega|\psi_{t_2}(\vartheta_2)\psi_{t_1}^*(\vartheta_1)\Omega) = \overline{\vartheta}_2 \cdot \mathrm{e}^{ib(t_2-t_1)}\theta(b)\vartheta_1, \quad \vartheta_1, \vartheta_2 \in \mathcal{Y}.$$

18.4.6 Green's functions of the Dirac equation

Consider the abstract charged Dirac equation as in Subsect. 18.4.3.

Definition 18.51 *We say that* $\mathbb{R} \ni t \mapsto S(t) \in B(\mathcal{Y})$ *is a* Green's function *or* fundamental solution *of Eq. (18.49) if it solves*

$$(\gamma_0 \partial_t + \Gamma - m\mathbb{1})S(t) = \delta(t) \otimes \mathbb{1}_{\mathcal{Y}}. \tag{18.57}$$

We introduce in particular the following Green's functions:

retarded $\qquad S^+(t) := -\theta(t)e^{itb}\gamma_0,$

advanced $\qquad S^-(t) := -\theta(-t)e^{itb}\gamma_0,$

Feynman $\qquad S_{\mathrm{F}}(t) := -e^{itb}\left(\theta(t)\theta(b) - \theta(-t)\theta(-b)\right)\gamma_0,$

anti-Feynman $\qquad S_{\overline{\mathrm{F}}}(t) := -e^{itb}\left(\theta(t)\theta(-b) - \theta(-t)\theta(b)\right)\gamma_0,$

principal value $\qquad S_{\mathrm{Pv}}(t) := -\dfrac{\mathrm{sgn}(t)}{2}e^{itb}\gamma_0.$

Set

$$\epsilon := |b| = \sqrt{\Gamma^*\Gamma + m^2\mathbb{1}}.$$

We then have

$$(\gamma_0\partial_t + \Gamma - m\mathbb{1})(\gamma_0\partial_t + \Gamma + m\mathbb{1}) = -(\partial_t^2 + \epsilon^2).$$

Thus if $G(t)$ is a Green's function for $\partial_t^2 + \epsilon^2$, then $-(\gamma_0\partial_t + \Gamma + m\mathbb{1})G(t)$ is a Green's function for the Dirac equation. In fact we have the identities

$$S^+(t) = -(\gamma_0\partial_t + \Gamma + m\mathbb{1})G^+(t),$$
$$S^-(t) = -(\gamma_0\partial_t + \Gamma + m\mathbb{1})G^-(t),$$
$$S_{\mathrm{F}}(t) = -(\gamma_0\partial_t + \Gamma + m\mathbb{1})G_{\mathrm{F}}(t),$$
$$S_{\overline{\mathrm{F}}}(t) = -(\gamma_0\partial_t + \Gamma + m\mathbb{1})G_{\overline{\mathrm{F}}}(t),$$
$$S_{\mathrm{Pv}}(t) = -(\gamma_0\partial_t + \Gamma + m\mathbb{1})G_{\mathrm{Pv}}(t),$$

which easily follow from

$$e^{itb} = \cos(\epsilon t) + \mathrm{i}\,\mathrm{sgn}(b)\sin(\epsilon t), \qquad \Gamma + m\mathbb{1} = \mathrm{i}b\gamma_0.$$

The Feynman Green's functions arise in the quantum theory in the following way:

$$\left(\Omega|\mathrm{T}\big(\psi_{t_2}(\vartheta_2)\psi_{t_1}^*(\vartheta_1)\big)\Omega\right) = \overline{\vartheta}_2 \cdot S_{\mathrm{F}}(t_2 - t_1)\vartheta_1, \qquad \vartheta_1, \vartheta_2 \in \mathcal{Y},$$

where we have used the fermionic time-ordering operation: if A_1, A_2 are odd fermionic operators, then

$$\mathrm{T}\left(A_{t_2}A_{t_1}\right) := \theta(t_2 - t_1)A_{t_2}A_{t_1} - \theta(t_1 - t_2)A_{t_1}A_{t_2}. \qquad (18.58)$$

18.4.7 Green's functions of the Dirac equation as operators

Let $\mathcal{Y}$ be as above. For simplicity, we assume that $\mathcal{Y}$ is separable. Similarly to Subsect. 18.3.10, we will use the space

$$L^2(\mathbb{R}) \otimes \mathcal{Y} \simeq L^2(\mathbb{R}, \mathcal{Y}). \qquad (18.59)$$

We will use both the *temporal representation* and the *energy representation* of $L^2(\mathbb{R}, \mathcal{Y})$. Green's functions of the abstract Dirac equation can be interpreted as quadratic forms on $(L^1 \cap L^2)(\mathbb{R}, \mathcal{Y})$, denoted S and given (in the temporal representation) by

$$\bar{\zeta}_1 \cdot S\zeta_2 := \int \overline{\zeta_1(t)} \cdot S(t-s)\zeta_2(s)\mathrm{d}t\mathrm{d}s. \tag{18.60}$$

In the energy representation they are multiplication operators. Here we list the most important Green's functions of the Dirac operator in the energy representation:

$$S^+(\tau) = \left(\mathrm{i}\gamma_0(\tau - \mathrm{i}0) + \Gamma - m\mathbb{1}\right)^{-1},$$
$$S^-(\tau) = \left(\mathrm{i}\gamma_0(\tau + \mathrm{i}0) + \Gamma - m\mathbb{1}\right)^{-1},$$
$$S_{\mathrm{F}}(\tau) = \left(\mathrm{i}\gamma_0(\tau - \mathrm{i}0\,\mathrm{sgn}(\tau)) + \Gamma - m\mathbb{1}\right)^{-1},$$
$$S_{\overline{\mathrm{F}}}(\tau) = \left(\mathrm{i}\gamma_0(\tau + \mathrm{i}0\,\mathrm{sgn}(\tau)) + \Gamma - m\mathbb{1}\right)^{-1},$$
$$S_{\mathrm{Pv}}(\tau) = \mathrm{Pv}\left(\mathrm{i}\gamma_0\tau + \Gamma - m\mathbb{1}\right)^{-1}.$$

18.4.8 Euclidean Green's function of the Dirac equation

Definition 18.52 *The* Euclidean Green's function *for the abstract Dirac equation is defined as*

$$S_{\mathrm{E}}(s) = \left(-\theta(s)\theta(b) + \theta(-s)\theta(-b)\right)\mathrm{e}^{-bs}\mathrm{i}\gamma_0.$$

Note that S_{E} solves

$$(-\mathrm{i}\gamma_0\partial_s + \Gamma - m\mathbb{1})S_{\mathrm{E}}(s) = \delta(s)\mathbb{1}_{\mathcal{Y}}.$$

It is related to the Green's function of the abstract Klein–Gordon equation by

$$S_{\mathrm{E}}(s) = (-\mathrm{i}\gamma_0\partial_s + \Gamma + m\mathbb{1})G_{\mathrm{E}}(s).$$

The function $S_{\mathrm{E}}(s)$ extends to an analytic function for complex s. We have

$$\frac{1}{\mathrm{i}}S_{\mathrm{E}}(\mathrm{i}t) = S_{\mathrm{F}}(t), \quad -\frac{1}{\mathrm{i}}S_{\mathrm{E}}(-\mathrm{i}t) = S_{\overline{\mathrm{F}}}(t).$$

Consider the operator on $L^2(\mathbb{R}, \mathcal{Y})$

$$-\mathrm{i}\gamma_0\partial_s + \Gamma - m\mathbb{1}.$$

In the energy representation, this becomes the operator of multiplication by

$$\gamma_0\tau + \Gamma - m\mathbb{1}.$$

It is closed on its natural domain. Moreover, we have

$$(\gamma_0\tau + \Gamma - m\mathbb{1})^*(\gamma_0\tau + \Gamma - m\mathbb{1}) = (\gamma_0\tau + \Gamma - m\mathbb{1})(\gamma_0\tau + \Gamma - m\mathbb{1})^*$$
$$= \tau^2 + \Gamma^*\Gamma + m^2\mathbb{1} > 0,$$

which implies that $-\mathrm{i}\gamma_0\partial_s + \Gamma - m\mathbb{1}$ is invertible if (18.51) holds. If this is the case, set $S_\mathrm{E} := (-\mathrm{i}\gamma_0\partial_s + \Gamma - m\mathbb{1})^{-1}$. Then,

$$S_\mathrm{E}\zeta(s) = \int_\mathbb{R} S_\mathrm{E}(s - s_1)\zeta(s_1)\mathrm{d}s_1, \quad \zeta \in (L^1 \cap L^2)(\mathbb{R}, \mathcal{Y}).$$

In the energy representation, this is the operator of multiplication by

$$S_\mathrm{E}(\tau) = (\gamma_0\tau + \Gamma - m\mathbb{1})^{-1}.$$

Using the notation in Subsect. 18.3.11, and in particular the generator of dilations A, we see that

$$\mathbb{R} \ni \theta \mapsto e^{\mathrm{i}\theta A} S_\mathrm{E} e^{-\mathrm{i}\theta A} =: S_\mathrm{E}^\theta$$

extends to an analytic function in the strip $-\frac{\pi}{2} < \mathrm{Im}\,\theta < \frac{\pi}{2}$, given in the momentum representation by

$$S_\mathrm{E}^\theta(\tau) = (\gamma_0 e^\theta \tau + \Gamma - m\mathbb{1})^{-1}.$$

Its boundary values coincide with the Feynman and anti-Feynman propagators:

$$S_\mathrm{E}^{\mathrm{i}\frac{\pi}{2}} = S_\mathrm{F}, \quad S_\mathrm{E}^{-\mathrm{i}\frac{\pi}{2}} = S_\mathrm{\bar{F}}.$$

This is the *Wick rotation* in the fermionic case.

18.4.9 Thermal Green's function for the abstract Dirac equation

Clearly, we can apply the positive temperature formalism of Subsect. 17.2.7 to a system described by an abstract Dirac equation, obtaining a β-KMS state ω_β and the corresponding Araki–Wyss CAR representation. In this subsection, parallel to Subsect. 18.3.12, we describe the 2-point functions given by this state.

Definition 18.53 *The* thermal Euclidean Green's function at inverse temperature β of the abstract Dirac equation *is defined for* $s \in [0, \beta]$ *as*

$$S_{\mathrm{E},\beta}(s) := -\mathrm{i}\frac{e^{-sb}}{\mathbb{1} + e^{-\beta b}}\gamma_0.$$

Note that $S_{\mathrm{E},\beta}$ is the unique solution of the problem

$$(-\mathrm{i}\gamma_0\partial_s + \Gamma - m\mathbb{1})S_{\mathrm{E},\beta}(s) = 0, \quad s \in]0, \beta[,$$
$$-\mathrm{i}\gamma_0 S_{\mathrm{E},\beta}(0) = \mathrm{i}\gamma_0 S_{\mathrm{E},\beta}(\beta) + \mathbb{1}_y. \tag{18.61}$$

(18.61) can be interpreted as

$$(-\mathrm{i}\gamma_0\partial_s + \Gamma - m\mathbb{1})S_{\mathrm{E},\beta}(s) - \delta(s)\mathbb{1}_y, \quad \text{on } S_\beta,$$

where we look for functions with anti-periodic boundary conditions at $\beta = 0$. More precisely, we consider functions ζ on S_β such that $\zeta(\beta) = -\zeta(0)$, and the Dirac delta function is defined as $\int_{S_\beta} \delta(s)\zeta(s)\mathrm{d}s = \zeta(0) = -\zeta(\beta)$ (the "right hand side delta function").

$S_\beta \ni s \mapsto S_{E,\beta}(s)$ extends to a function continuous in the strip $\operatorname{Re} s \in [0, \beta]$ and holomorphic inside this strip. If ω_β is the β-KMS state, then

$$\omega_\beta\big(\psi_t(\vartheta_1)\psi_0^*(\vartheta_2)\big) = \overline{\vartheta}_1 \cdot S_{E,\beta}(\mathrm{i}t)(-\mathrm{i}\gamma_0)\vartheta_2,$$
$$\omega_\beta\big(\psi_0^*(\vartheta_2)\psi_t(\vartheta_1)\big) = \overline{\vartheta}_1 \cdot S_{E,\beta}(\beta + \mathrm{i}t)(-\mathrm{i}\gamma_0)\vartheta_2, \quad \vartheta_1, \vartheta_2 \in \mathcal{Y}.$$

Let $\partial_s^{\mathrm{ant}}$ denote the operator ∂_s on the Hilbert space $L^2(S_\beta)$ with the anti-periodic boundary conditions. Its domain is given by

$$\operatorname{Dom} \partial_s^{\mathrm{ant}} := \{\zeta \in L^2(S_\beta),\ \partial_s\zeta \in L^2(S_\beta),\ \zeta(0) = -\zeta(\beta)\}.$$

Note that $\partial_s^{\mathrm{ant}}$ is anti-self-adjoint.

Define the *anti-periodic discrete Fourier transform*

$$\mathcal{F}_\beta^{\mathrm{ant}} : L^2(S_\beta) \to l^2\left(\frac{2\pi}{\beta}\left(\mathbb{Z} + \frac{1}{2}\right)\right)$$

by

$$\mathcal{F}_\beta\zeta(\sigma) = \int_0^\beta \mathrm{e}^{-\mathrm{i}s\sigma}\zeta(s)\mathrm{d}s, \quad \zeta \in L^2(S_\beta).$$

Its inverse is

$$(\mathcal{F}_\beta^{\mathrm{ant}})^{-1}v(s) = \beta^{-1} \sum_{\sigma \in \frac{2\pi}{\beta}(\mathbb{Z}+\frac{1}{2})} \mathrm{e}^{\mathrm{i}s\sigma}v(\sigma), \quad v \in l^2\left(\frac{2\pi}{\beta}\left(\mathbb{Z} + \frac{1}{2}\right)\right).$$

Clearly, $\partial_s^{\mathrm{ant}} = (\mathcal{F}_\beta^{\mathrm{ant}})^{-1}(\mathrm{i}\sigma)\mathcal{F}_\beta^{\mathrm{ant}}$.

On the Hilbert space $L^2(S_\beta) \otimes \mathcal{Y} \simeq L^2(S_\beta, \mathcal{Y})$, we define the closed operator

$$(-\mathrm{i}\gamma_0\partial_s^{\mathrm{ant}} + \Gamma - m\mathbb{1}) = (-\mathrm{i}\gamma_0)(\partial_s^{\mathrm{ant}} + b).$$

Note that $\partial_s^{\mathrm{ant}} + b$ is a normal operator on its natural domain. We set $S_{E,\beta} := (-\mathrm{i}\gamma_0\partial_s^{\mathrm{ant}} + \Gamma - m\mathbb{1})^{-1}$. We then have

$$S_{E,\beta}\zeta(s) = \int_{S_\beta} S_{E,\beta}(s_1)\zeta(s - s_1)\mathrm{d}s_1, \quad \zeta \in L^2(S_\beta, \mathcal{Y}).$$

In the energy representation, obtained by applying $\mathcal{F}_\beta^{\mathrm{ant}}$, this becomes the operator of multiplication by the *fermionic Matsubara coefficients*:

$$S_{E,\beta}(\sigma) = (\gamma_0\sigma + \Gamma - m)^{-1}, \quad \sigma \in \frac{2\pi}{\beta}\left(\mathbb{Z} + \frac{1}{2}\right).$$

Set

$$G_{E,\beta}^{\mathrm{ant}}(s) = \frac{-\mathrm{e}^{-s\epsilon} + \mathrm{e}^{(s-\beta)\epsilon}}{2\epsilon(\mathbb{1} + \mathrm{e}^{-\beta\epsilon})}.$$

Note that $G_{E,\beta}^{\mathrm{ant}}$ is the unique solution of

$$-\partial_s^2 G_{E,\beta}^{\mathrm{ant}}(s) + \epsilon^2 G_{E,\beta}^{\mathrm{ant}}(s) = 0, \quad s \in\ ]0, \beta[,$$
$$G_{E,\beta}^{\mathrm{ant}}(0) = -G_{E,\beta}^{\mathrm{ant}}(\beta), \quad \partial_s^+ G_{E,\beta}^{\mathrm{ant}}(0) + \partial_s^- G_{E,\beta}^{\mathrm{ant}}(\beta) = \mathbb{1}. \tag{18.62}$$

In fact, we have

$$G_{\mathrm{E},\beta}^{\mathrm{ant}}(0) = -G_{\mathrm{E},\beta}^{\mathrm{ant}}(\beta) = \frac{-\mathbb{1} + e^{-\beta\epsilon}}{2\epsilon(\mathbb{1} + e^{-\beta\epsilon})},$$

$$\partial_s^+ G_{\mathrm{E},\beta}^{\mathrm{ant}}(0) = \partial_s^- G_{\mathrm{E},\beta}^{\mathrm{ant}}(\beta) = \frac{1}{2}\mathbb{1}.$$

Thus $G_{\mathrm{E},\beta}^{\mathrm{ant}}$ can be interpreted as the solution of the equation on S_β

$$-\partial_s^2 G_{\mathrm{E},\beta}^{\mathrm{ant}}(s) + \epsilon^2 G_{\mathrm{E},\beta}^{\mathrm{ant}}(s) = \delta(s)\mathbb{1}_\mathcal{X},$$

with anti-periodic boundary conditions at $\beta = 0$ and the right hand side Dirac delta function. Then we can express $S_{\mathrm{E},\beta}$ in terms of $G_{\mathrm{E},\beta}^{\mathrm{ant}}$ as

$$S_{\mathrm{E},\beta}(s) = (-i\gamma_0\partial_s + \Gamma + m\mathbb{1})G_{\mathrm{E},\beta}^{\mathrm{ant}}(s).$$

18.5 Notes

The topics discussed in this chapter in the context of concrete systems, usually based on relativistic equations, can be found in every textbook on quantum field theory, such as Jauch–Röhrlich (1976), Schweber (1962), Weinberg (1995) and Srednicki (2007). The Racah and Wigner time reversals were introduced by Racah (1927) and Wigner (1932a), respectively. Our presentation, in spite of its abstract mathematical language, follows very closely the usual exposition; see in particular Srednicki (2007), Sect. 22, for complex bosons and Srednicki (2007), Sect. 49, for neutral fermions.

The idea of positive quantization of classical linear dynamics goes back to the early days of the quantum field theory. In the fermionic case it was first formulated in terms of the "Dirac sea"; see Dirac (1930). This approach hides the particle–anti-particle symmetry. Its modern formulation is attributed to Fock (1933) and Furry–Oppenheimer (1934).

In the case of bosons, the "Dirac sea" approach is not available. The bosonic positive energy quantization was described by Pauli–Weisskopf (1934).

An interesting outline of the history of quantum field theory, which in particular discusses the topic of the positive energy quantization, is contained in the introduction to the monograph of Weinberg (1995).

The role of complex structures in positive energy quantization was emphasized by Segal (1964) and Weinless (1969).

An abstract formulation of the positive energy quantization was given by Segal, and can be found e.g. in Baez–Segal–Zhou (1991).

Positive temperature Green's functions can be found e.g. in Fetter–Walecka (1971), and from a more mathematical point of view in Birke–Fröhlich (2002).

19

Quantum fields on space-time

In this chapter we describe the most important examples of (non-interacting) relativistic quantum fields. We will use extensively the formalism developed in Chap. 18.

Most textbook presentations of this subject start from the discussion of representations of the *Poincaré group*. They stress that the most fundamental quantum fields are covariant with respect to this group. In our presentation the Poincaré covariance is a secondary property. The property that we emphasize more is the *Einstein causality* of fields. In the mathematical language this is expressed by the fact that observables belonging to *causally separated* subsets of space-time commute with one another. This property can be true even when there is no Poincaré covariance, e.g. due to the presence of an external (vector) potential in a curved space-time.

The chapter is naturally divided in two parts. In the first part we consider the flat *Minkowski space* and in the second an arbitrary *globally hyperbolic manifold*. In both cases we discuss the influence of an external (classical) potential and a variable mass. In the Minkowski case, we discuss separately the Poincaré covariance, which holds if the potential is zero and mass is constant.

The quantization consists of two stages. In the first stage one introduces the CCR or CAR algebra describing the observables of the system. The underlying phase space is the space of solutions of the corresponding equation defined on the space-time. This space is equipped with a bilinear or sesquilinear form, which leads to the appropriate CCR or CAR.

In the second stage one chooses a representation of the algebra of observables on a Hilbert space. In order to determine this representation one usually assumes that the generator of the time evolution of the classical system is time-independent. Then one can apply the formalism described in Chap. 18. In the case of the Klein–Gordon and Dirac equations on Minkowski space this means that the external potential and the mass do not depend on time.

Sects. 19.5, resp. 19.6 are generalizations of Sects. 19.2, resp. 19.3 to a curved space-time. We limit our discussion to the algebraic quantization of Klein–Gordon and Dirac equations on a globally hyperbolic manifold. As a result we obtain a net of CCR, resp. CAR algebras satisfying the Einstein causality.

Our presentation is limited to the most basic elements of the theory of quantum fields on curved space-time. One of the topics that we leave out, which however is easy to figure out mimicking the discussion in Sects. 19.2, resp. 19.3,

is the positive energy quantization of time-independent equations on a stationary space-time; see e.g. Kay (1978). The subject that is more difficult is how
to choose representations of quantum fields in the case of non-stationary curved
space-times, where there is no preferred vacuum state. There has been significant progress in our understanding of this question. It is believed that one should
choose representations generated by states whose correlation functions satisfy a
certain natural microlocal condition, the so-called *Hadamard states*. One can
find more about this subject in the literature; see Brunetti–Fredenhagen–Köhler
(1996) and Brunetti–Fredenhagen–Verch (2003).

19.1 Minkowski space and the Poincaré group

19.1.1 Minkowski space

Consider the Minkowski space $\mathbb{R}^{1,d}$. Recall that it is the vector space $\mathbb{R}^{1+d}$
equipped with a pseudo-Euclidean form of signature $(1, d)$; see Sect. 15.3. In
the coordinates $x = (x^\mu)$, $\mu = 0, 1, \ldots, d$, the pseudo-Euclidean quadratic form
will be denoted

$$\langle x|x \rangle = -(x^0)^2 + \sum_{i=1}^{d}(x^i)^2.$$

Definition 19.1 *A non-zero vector $x \in \mathbb{R}^{1,d}$ is called*

$$
\begin{array}{ll}
\text{time-like} & \text{if } \langle x|x \rangle < 0, \\
\text{causal} & \text{if } \langle x|x \rangle \leq 0, \\
\text{light-like} & \text{if } \langle x|x \rangle = 0, \\
\text{space-like} & \text{if } \langle x|x \rangle > 0.
\end{array}
$$

*The set of causal, resp. time-like vectors is denoted J, resp. I. A causal vector
x is called*

$$
\begin{array}{ll}
\text{future oriented} & \text{if } x^0 > 0, \\
\text{past oriented} & \text{if } x^0 < 0.
\end{array}
$$

*The set of future, resp. past oriented causal vectors is denoted $J^\pm$. The set of
future, resp. past oriented time-like vectors is denoted $I^\pm$.*

Clearly, $I^\pm$ is the interior of $J^\pm$.

Definition 19.2 *For $\mathcal{U} \subset \mathbb{R}^{1,d}$, we set $J(\mathcal{U}) := J + \mathcal{U}$ and $J^\pm(\mathcal{U}) := J^\pm + \mathcal{U}$.
$J(\mathcal{U})$ is called the causal shadow of $\mathcal{U}$ and $J^\pm(\mathcal{U})$ the causal future, resp. past of
$\mathcal{U}$. A function on $\mathbb{R}^{1,d}$ is called space-compact if there exists a compact $\mathcal{U} \subset \mathbb{R}^{1,d}$
such that $\operatorname{supp} f \subset J(\mathcal{U})$. It is called future, resp. past space-compact if there
exists a compact $\mathcal{U} \subset \mathbb{R}^{1,d}$ such that $\operatorname{supp} f \subset J^\pm(\mathcal{U})$.*

The set of space-compact smooth functions will be denoted $C^\infty_{sc}(\mathbb{R}^{1,d})$. The set of future, resp. past space-compact smooth functions will be denoted $C^\infty_{\pm sc}(\mathbb{R}^{1,d})$.

Definition 19.3 *Let $\mathcal{U}_1, \mathcal{U}_2 \subset \mathbb{R}^{1,d}$. We say that $\mathcal{U}_1$ and $\mathcal{U}_2$ are causally separated if $J(\mathcal{U}_1) \cap \mathcal{U}_2 = \emptyset$, or equivalently if $\mathcal{U}_1 \cap J(\mathcal{U}_2) = \emptyset$.*

Definition 19.4 *The operator*

$$\Box := \partial_\mu \partial^\mu = -(\partial^0)^2 + \sum_{i=1}^{d}(\partial^i)^2$$

is called the d'Alembertian.

19.1.2 The Lorentz group

Definition 19.5 *The pseudo-Euclidean group $O(\mathbb{R}^{1,d}) \simeq O(\mathbb{R}^{d,1})$ is called the* Lorentz group *in $1 + d$ dimensions.*

Let $r \in O(\mathbb{R}^{1,d})$. We will say that r is space-time even *if $\det r = 1$ and* space-time odd *if $\det r = -1$.*

Note that $r(J^+) = J^+$ or $r(J^+) = J^-$. In the former case we say that r is orthochronous *and in the latter case we say that r is* anti-orthochronous.

Thus $O(\mathbb{R}^{1,d})$ has four connected components:

(1) the space-time even orthochronous component $O^\uparrow_+(\mathbb{R}^{1,d})$,
(2) the space-time even anti-orthochronous component $O^\downarrow_+(\mathbb{R}^{1,d})$,
(3) the space-time odd orthochronous component $O^\uparrow_-(\mathbb{R}^{1,d})$,
(4) the space-time odd anti-orthochronous component $O^\downarrow_-(\mathbb{R}^{1,d})$.

Clearly, $O^\uparrow_+(\mathbb{R}^{1,d})$, also denoted $SO^\uparrow(\mathbb{R}^{1,d})$, is a normal subgroup and we have an exact sequence

$$1 \to SO^\uparrow(\mathbb{R}^{1,d}) \to O(\mathbb{R}^{1,d}) \to \mathbb{Z}_2 \times \mathbb{Z}_2 \to 1. \qquad (19.1)$$

We have three subgroups of $O(\mathbb{R}^{1,d})$ of index 2:

$$O^\uparrow(\mathbb{R}^{1,d}) = O^\uparrow_+(\mathbb{R}^{1,d}) \cup O^\uparrow_-(\mathbb{R}^{1,d}),$$
$$O_+(\mathbb{R}^{1,d}) = O^\uparrow_+(\mathbb{R}^{1,d}) \cup O^\downarrow_+(\mathbb{R}^{1,d}),$$
$$SO(\mathbb{R}^{1,d}) = O^\uparrow_+(\mathbb{R}^{1,d}) \cup O^\downarrow_-(\mathbb{R}^{1,d}).$$

Definition 19.6 *The* temporal parity *is the homomorphism*

$$O(\mathbb{R}^{1,d}) \ni L \mapsto \rho_L \in \{1, -1\}$$

that equals 1 on an orthochronous and -1 on an anti-orthochronous L.

Definition 19.7 *The* affine extension *of the Lorentz group*

$$AO(\mathbb{R}^{1,d}) = \mathbb{R}^{1+d} \rtimes O(\mathbb{R}^{1,d})$$

is called the Poincaré group *in $1 + d$ dimensions.*

We refer to Def. 1.100 for the definition of the affine extension.

19.1.3 Pin groups for the Lorentzian signature

In the case of the Lorentz group we have the following refinement of Diagram (14.18):

$$
\begin{array}{ccccccc}
& & 1 & & 1 & & \\
& & \downarrow & & \downarrow & & \\
1 & \to & \mathbb{Z}_2 & \to & \mathbb{Z}_2 & \to & 1 \\
& & \downarrow & & \downarrow & & \downarrow \\
1 & \to & Spin^\uparrow(\mathbb{R}^{1,d}) & \to & Pin(\mathbb{R}^{1,d}) & \to & \mathbb{Z}_2 \times \mathbb{Z}_2 & \to & 1 \qquad (19.2) \\
& & \downarrow & & \downarrow & & \downarrow \\
1 & \to & SO^\uparrow(\mathbb{R}^{1,d}) & \to & O(\mathbb{R}^{1,d}) & \to & \mathbb{Z}_2 \times \mathbb{Z}_2 & \to & 1 \\
& & \downarrow & & \downarrow & & \downarrow \\
& & 1 & & 1 & & 1
\end{array}
$$

$Pin(\mathbb{R}^{1,d})$ has four connected components

$$
\begin{aligned}
& Pin_+^\uparrow(\mathbb{R}^{1,d}), \\
& Pin_+^\downarrow(\mathbb{R}^{1,d}), \\
& Pin_-^\uparrow(\mathbb{R}^{1,d}), \\
& Pin_-^\downarrow(\mathbb{R}^{1,d}),
\end{aligned}
$$

which cover the corresponding connected components of $O(\mathbb{R}^{1,d})$ listed in Subsect. 19.1.2.

If we replace $\mathbb{R}^{1,d}$ with $\mathbb{R}^{d,1}$, then all the entries of the above diagram remain the same except for $Pin(\mathbb{R}^{1,d})$ replaced with $Pin(\mathbb{R}^{d,1})$, which are not isomorphic to one another. Both have four connected components, with the obvious notation. Note that we can identify $Spin(\mathbb{R}^{1,d}) \simeq Spin(\mathbb{R}^{d,1})$, hence we can identify $Pin_+^\uparrow(\mathbb{R}^{1,d})$, resp. $Pin_-^\downarrow(\mathbb{R}^{1,d})$, with $Pin_+^\uparrow(\mathbb{R}^{d,1})$, resp. $Pin_-^\downarrow(\mathbb{R}^{d,1})$.

We will also use the affine extensions of the Pin groups

$$
\begin{aligned}
APin(\mathbb{R}^{1,d}) &= \mathbb{R}^{1,d} \rtimes Pin(\mathbb{R}^{1,d}), \\
APin(\mathbb{R}^{d,1}) &= \mathbb{R}^{d,1} \rtimes Pin(\mathbb{R}^{d,1}),
\end{aligned}
$$

which are two-fold coverings of the Poincaré group.

19.1.4 Positive energy representations of Clifford relations

Let $\mathcal{V}$ be a pseudo-unitary vector space equipped with a Hermitian form β. As in Subsect. 15.3.5, we denote by $A^\dagger$ the adjoint of A w.r.t. this form.

Definition 19.8 *We say that a representation of Clifford relations*

$$
\mathbb{R}^{1,d} \ni y \mapsto \gamma(y) \in L(\mathcal{V}) \qquad (19.3)
$$

is a positive energy representation if $\gamma(y) = -\gamma(y)^\dagger$, $y \in \mathcal{Y}$, and

$$i\overline{v} \cdot \beta\gamma(y_0)v > 0, \quad v \in \mathcal{V}, \ v \neq 0,$$

for some time-like future oriented $y_0 \in \mathbb{R}^{1,d}$.

Lemma 19.9 $i\beta\gamma(y_0)$ *is positive definite for some time-like future oriented y_0 iff $i\beta\gamma(y)$ is positive definite for all time-like future oriented y.*

Proof Let y_1, y_2 be two future oriented vectors. We may assume that $\langle y_1|y_1\rangle = \langle y_2|y_2\rangle$. There exists $r \in SO^\uparrow(\mathbb{R}^{1,d})$ such that $y_2 = ry_1$. By Thm. 15.28, there exists $U \in Spin^\uparrow(\mathbb{R}^{1,d})$ implementing r and such that $UU^\dagger = \mathbb{1}$. $Spin^\uparrow$ is connected, hence $UU^\dagger = \mathbb{1}$, i.e. U is pseudo-unitary. Therefore,

$$\overline{v} \cdot \beta\gamma(y_2)v = \overline{v} \cdot \beta U\gamma(y_1)U^\dagger v = \overline{U^\dagger v} \cdot \beta\gamma(y_1)U^\dagger v.$$

Hence, $i\beta\gamma(y_1)$ is positive definite iff $i\beta\gamma(y_2)$ is positive definite. $\qquad\square$

Positive energy representations act on a pseudo-unitary space. After fixing a future oriented time-like vector y_0, their representation space can be equipped with the positive definite scalar product

$$iv_1 \cdot \beta\gamma(y_0)v_2, \quad v_1, v_2 \in \mathcal{V}.$$

19.2 Quantization of the Klein–Gordon equation

The *homogeneous Klein–Gordon equation* has the form

$$(-\Box + m^2)\zeta(x) = 0, \tag{19.4}$$

where $\mathbb{R}^{1,d} \ni x \mapsto \zeta(x)$ is a function on the Minkowski space. This equation is *Poincaré invariant*. That means, elements of the Poincaré group transform solutions of (19.4) into solutions of (19.4).

Besides (19.4), we will consider the Klein–Gordon equation with an external potential and a variable mass,

$$\left(- \left(\partial_\mu + iA_\mu(x)\right)\left(\partial^\mu + iA^\mu(x)\right) + m^2(x)\right)\zeta(x) = 0. \tag{19.5}$$

It is *Poincaré covariant*. That means solutions of (19.5) are transformed by elements of the Poincaré group into solutions of (19.5) with the transformed external potential and mass.

The equation (19.5) has several interesting properties. First of all, its solutions do not propagate faster than the speed of light. In other words, solutions of the Cauchy problem are supported in the causal shadow of the support of its initial conditions. Secondly, the space of real, resp. complex, space-compact solutions (19.5), denoted $\mathcal{Y}$, has a natural symplectic, resp. charged symplectic form given by a local expression. As a consequence, two solutions of (19.5) with the Cauchy data supported in disjoint regions are orthogonal w.r.t. this symplectic, resp. charged symplectic form.

Let us associate with $\mathcal{Y}$ the corresponding CCR algebra. It will satisfy the *Einstein causality* (a property also known by the name *locality*). This means that observables associated with the Cauchy data with disjoint supports will commute. This property is one of the basic postulates of quantum field theory. It is incorporated in the standard sets of axioms of quantum field theory: the *Wightman axioms* (see Streater–Wightman (1964)) and the *Haag–Kastler axioms* (see Haag–Kastler (1964) and Haag (1992)).

The space $\mathcal{Y}$ is equipped with a natural symplectic, resp. charged symplectic dynamics. If the external potential and the mass do not depend on time, this dynamics is generated by a time-independent classical Hamiltonian. If this Hamiltonian is positive, we can apply the positive energy quantization, as described in Subsects. 18.1.1, resp. 18.2.1. We obtain a one-particle Hilbert space $\mathcal{Z}$ and a positive Hamiltonian implementing the dynamics acting on the bosonic Fock space $\Gamma_{\mathrm{s}}(\mathcal{Z})$.

The discussion of the quantization of the Klein–Gordon equation with an external potential and a variable mass gives a good illustration of the difference between the dual phase space $\mathcal{Y}$ and the one-particle space $\mathcal{Z}$. This difference is visible in particular in our discussion of the Poincaré covariance. In particular, we will describe the charge, parity and time reversal covariance on the level of the classical equation, its algebraic quantization and its Hilbert space quantization.

19.2.1 Klein–Gordon operator

Definition 19.10 *Let* $m^2 \in \mathbb{R}$. *The* Klein–Gordon operator with squared mass m^2 *is the operator on* $\mathbb{R}^{1,d}$ *given by*

$$\Box(m^2) := -\partial_\mu \partial^\mu + m^2. \tag{19.6}$$

Definition 19.11 *Let*

$$\mathbb{R}^{1,d} \ni x \mapsto m^2(x) \in \mathbb{R},$$
$$\mathbb{R}^{1,d} \ni x \mapsto A(x) = \big(A^\mu(x)\big) \in \mathbb{R}^{1,d}$$

be smooth functions. The Klein–Gordon operator with squared mass m^2 *and external potential* A *is defined as*

$$\Box(m^2, A) := -\big(\partial_\mu + \mathrm{i}A_\mu(x)\big)\big(\partial^\mu + \mathrm{i}A^\mu(x)\big) + m^2(x). \tag{19.7}$$

Note that in the real case the external potential has to be zero, because of the imaginary unit in front of it. In what follows, for definiteness, we will consider mostly the complex-valued case.

19.2.2 Lagrangian of the Klein–Gordon equation

In our presentation we avoid using the Lagrangian formalism. Nevertheless, it is worth mentioning that (19.5) can be obtained as the Euler–Lagrange equations of a variational problem. The Lagrangian can be taken as

$$L(\zeta, \partial\zeta) = -\frac{1}{2}\partial_\mu\zeta\partial^\mu\zeta - \frac{1}{2}m^2\zeta^2, \quad \text{in the real case,}$$

$$L(\zeta, \overline{\zeta}, \partial\zeta, \partial\overline{\zeta}) = -\overline{(\partial_\mu + iA_\mu)\zeta}(\partial_\mu + iA_\mu)\zeta - m^2\overline{\zeta}\zeta, \quad \text{in the complex case.}$$

In the real case the Euler–Lagrange equations

$$\partial_\zeta L - \partial_\mu \frac{\partial L}{\partial(\partial_\mu\zeta)} = 0 \tag{19.8}$$

yield $\Box(m^2, 0)\zeta = 0$. (Recall that in the real case we do not consider the external potential.)

In the complex case we have two sets of Euler–Lagrange equations. (19.8) yields $\Box(m^2, A)\zeta = 0$. It should be supplemented by

$$\partial_{\overline{\zeta}} L - \partial_\mu \frac{\partial L}{\partial(\partial_\mu\overline{\zeta})} = 0, \tag{19.9}$$

which yields $\Box(m^2, -A)\overline{\zeta}(x) = 0$.

19.2.3 Green's functions

The following theorem describes advanced and retarded Green's functions of the inhomogeneous Klein–Gordon equation.

Theorem 19.12 *Write* $\Box = \Box(m^2, A)$. *For any* $f \in C_c^\infty(\mathbb{R}^{1,d})$ *there exist unique functions* $\zeta^\pm \in C_{\pm\mathrm{sc}}^\infty(\mathbb{R}^{1,d})$, *solutions of*

$$\Box\zeta^\pm = f.$$

Moreover,

$$\zeta^\pm(x) = (G^\pm f)(x) = \int_{\mathbb{R}^{1,d}} G^\pm(x, y)f(y)\mathrm{d}y,$$

where $G^\pm = G^\pm(m^2, A) \in \mathcal{D}'(\mathbb{R}^{1,d} \times \mathbb{R}^{1,d})$ *satisfy*

$$\Box G^\pm = G^\pm\Box = \mathbb{1},$$
$$\mathrm{supp}\, G^\pm \subset \{(x, y) \,:\, x \in J^\pm(y)\},$$
$$\overline{G^\pm(x, y)} = G^\mp(y, x).$$

The proof of Thm. 19.12 can be found e.g. in Bär–Ginoux–Pfäffle (2007).

Definition 19.13 $G^\pm$ *is called the* retarded, *resp.* advanced *Green's function.*

Note that by duality $G^\pm$ can be applied to distributions of compact support.

Definition 19.14 *The* Pauli–Jordan *or* commutator function *is defined as*

$$G(x,y) := G^+(x,y) - G^-(x,y).$$

Note that

$$\Box G = G\Box = 0,$$
$$\operatorname{supp} G \subset \{(x,y) \ : \ x \in J(y)\},$$
$$\overline{G(x,y)} = -G(y,x).$$

19.2.4 Cauchy problem

Let us introduce a non-covariant notation. We will write t for x^0 and x for $(x^1,\ldots,x^d)$. The dot will denote the derivative w.r.t. t. We will also write V for A^0 and A for $(A^1,\ldots,A^d)$. Thus, the free Klein–Gordon operator becomes

$$\Box(m^2) = \partial_t^2 - \Delta_{\mathrm{x}} + m^2, \tag{19.10}$$

and the Klein–Gordon operator with an external potential and a variable squared mass becomes

$$\Box(m^2, V, \mathrm{A}) = \left(\partial_t - \mathrm{i}V(t,\mathrm{x})\right)^2 - \sum_{i=1}^{d}\left(\partial_{\mathrm{x}^i} + \mathrm{i}A_i(t,\mathrm{x})\right)^2 + m^2(t,\mathrm{x}). \tag{19.11}$$

The Pauli–Jordan function G can be used to describe the solution of the Cauchy problem of the Klein–Gordon equation.

Theorem 19.15 *Let $\vartheta, \varsigma \in C_{\mathrm{c}}^{\infty}(\mathbb{R}^d)$. Then there exists a unique $\zeta \in C_{\mathrm{sc}}^{\infty}(\mathbb{R}^{1,d})$ that solves*

$$\Box(m, V, \mathrm{A})\zeta = 0 \tag{19.12}$$

with initial conditions

$$\zeta(0,\mathrm{x}) = \varsigma(\mathrm{x}), \quad \dot{\zeta}(0,\mathrm{x}) = \vartheta(\mathrm{x}) - \mathrm{i}V(0,\mathrm{x})\varsigma(\mathrm{x}).$$

It satisfies $\operatorname{supp}\zeta \subset J(\operatorname{supp}\varsigma \cup \operatorname{supp}\vartheta)$ *and is given by*

$$\zeta(t,\mathrm{x}) = -\int_{\mathbb{R}^d} \left(\partial_s G(t,\mathrm{x};0,\mathrm{y}) - \mathrm{i}G(t,\mathrm{x};0,\mathrm{y})V(0,\mathrm{y})\right)\varsigma(\mathrm{y})\mathrm{d}\mathrm{y}$$
$$+ \int_{\mathbb{R}^d} G(t,\mathrm{x};0,\mathrm{y})\vartheta(\mathrm{y})\mathrm{d}\mathrm{y}.$$

19.2.5 Symplectic form on the space of solutions

Definition 19.16 *Let $\mathcal{Y}(m^2, A)$ (also denoted for brevity $\mathcal{Y}$) be the space of smooth space-compact solutions of the Klein–Gordon equation, that is, $\zeta \in C_{\mathrm{sc}}^{\infty}(\mathbb{R}^{1,d})$ satisfying (19.5).*

Definition 19.17 *Let* $\zeta_1, \zeta_2 \in C^\infty(\mathcal{X})$. *We define*

$$j^\mu(\zeta_1, \zeta_2, x) := \overline{\partial^\mu \zeta_1(x)}\zeta_2(x) - \overline{\zeta_1(x)}\partial^\mu \zeta_2(x) - 2\mathrm{i}A^\mu(x)\overline{\zeta_1(x)}\zeta_2(x).$$

(In the real case the definition is the same except that we do not need the complex conjugation and the term involving the potential is absent.) We easily check that

$$\partial_\mu j^\mu(x) = -\overline{\Box\zeta_1(x)}\zeta_2(x) + \overline{\zeta_1(x)}\Box\zeta_2(x),$$

where $\Box = \Box(m^2, A)$. Therefore, if $\zeta_1, \zeta_2 \in \mathcal{Y}(m^2, A)$, then

$$\partial_\mu j^\mu(x) = 0,$$

and in such a case the flux of j^μ across a space-like subspace $\mathcal{S}$ of co-dimension 1 does not depend on its choice.

If the space-like hyper-subspace $\mathcal{S}$ is given by the parametrization

$$\mathcal{S} = \{(a + \mathrm{b}\cdot \mathrm{x}, \mathrm{x}) \ : \ \mathrm{x} \in \mathbb{R}^d\}, \tag{19.13}$$

for some $a \in \mathbb{R}$ and $\mathrm{b} \in \mathbb{R}^d$ with $|\mathrm{b}| < 1$, then the flux of j^μ across $\mathcal{S}$

$$\overline{\zeta_1}\cdot\omega\zeta_2 = \int_{\mathbb{R}^d}(1 - |\mathrm{b}|^2)\,(j^0 - \mathrm{b}\cdot\mathrm{j})\,(\zeta_1, \zeta_2, a + \mathrm{b}\cdot\mathrm{x}, \mathrm{x})\mathrm{d}\mathrm{x}$$

defines a (charged) symplectic form on $\mathcal{Y}(m^2, A)$.

Note that the (charged) symplectic form ω is defined covariantly under the group $AO^\uparrow(\mathbb{R}^{1,d})$ (it does not depend on the choice of coordinates that preserves the time direction). This is true even if A and m^2 are variable. Under the change of coordinates in $AO^\downarrow(\mathbb{R}^{1,d})$ the (charged) symplectic form changes its sign.

For $\zeta_1, \zeta_2 \in \mathcal{Y}(m^2, V, A)$, the (charged) symplectic form is

$$\overline{\zeta_1}\cdot\omega\zeta_2 = \int_{\mathbb{R}^d}\left(\overline{\dot{\zeta}_1(0, \mathrm{x})}\zeta_2(0, \mathrm{x}) - \overline{\zeta_1(0, \mathrm{x})}\dot{\zeta}_2(0, \mathrm{x}) - 2\mathrm{i}V(0, \mathrm{x})\overline{\zeta_1(0, \mathrm{x})}\zeta_2(0, \mathrm{x})\right)\mathrm{d}\mathrm{x}$$

$$= \int_{\mathbb{R}^d}\left(\overline{\vartheta_1(\mathrm{x})}\varsigma_2(\mathrm{x}) - \overline{\varsigma_1(\mathrm{x})}\vartheta_2(\mathrm{x})\right)\mathrm{d}\mathrm{x}. \tag{19.14}$$

19.2.6 Solutions parametrized by test functions

The Pauli–Jordan function G can be used to construct solutions of the Klein–Gordon equation, which are especially useful in the axiomatic formulation of quantum field theory.

Theorem 19.18 (1) *For any* $f \in C_\mathrm{c}^\infty(\mathbb{R}^{1,d})$, $Gf \in \mathcal{Y}$.
(2) *Every element of* $\mathcal{Y}$ *is of this form.*
(3) $\overline{Gf_1}\cdot\omega Gf_2 = \int \overline{f_1(x)}G(x, y)f_2(y)\mathrm{d}x\mathrm{d}y$.

For an open set $\mathcal{O} \subset \mathbb{R}^{1,d}$ we set

$$\mathcal{Y}(\mathcal{O}) = \mathcal{Y}(\mathcal{O}, m^2, A) := \{Gf \ : \ f \in C_\mathrm{c}^\infty(\mathcal{O})\}.$$

Theorem 19.19 (1) $\mathcal{Y} = \mathcal{Y}(\mathbb{R}^{1,d})$.

(2) $\mathcal{O}_1 \subset \mathcal{O}_2$ *implies* $\mathcal{Y}(\mathcal{O}_1) \subset \mathcal{Y}(\mathcal{O}_2)$.

(3) *If* $\mathcal{O}_1$ *and* $\mathcal{O}_2$ *are causally separated and* $\zeta_i \in \mathcal{Y}(\mathcal{O}_i)$, $i = 1, 2$, *then*

$$\overline{\zeta_1} \cdot \omega \zeta_2 = 0.$$

19.2.7 Algebraic quantization

In the neutral case, starting from symplectic space $(\mathcal{Y}, \omega)$, we define the C^*-algebra

$$\mathfrak{A} = \mathfrak{A}(m^2, A) := \mathrm{CCR}^{\mathrm{Weyl}}(\mathcal{Y}).$$

More generally, we have a similar definition for any open set $\mathcal{O} \subset \mathbb{R}^{1,d}$:

$$\mathfrak{A}(\mathcal{O}) = \mathfrak{A}(\mathcal{O}, m^2, A) := \mathrm{CCR}^{\mathrm{Weyl}}(\mathcal{Y}(\mathcal{O})).$$

In the charged case, we replace the algebra $\mathrm{CCR}^{\mathrm{Weyl}}$ with $\mathrm{CCR}^{\mathrm{reg}}_{\mathrm{gi}}$, as explained in Subsect. 18.2.1.

Theorem 19.20 (1) $\mathfrak{A} = \mathfrak{A}(\mathbb{R}^{1,d})$.

(2) $\mathcal{O}_1 \subset \mathcal{O}_2$ *implies* $\mathfrak{A}(\mathcal{O}_1) \subset \mathfrak{A}(\mathcal{O}_2)$.

(3) *If* $\mathcal{O}_1$ *and* $\mathcal{O}_2$ *are causally separated, and* $B_i \in \mathfrak{A}(\mathcal{O}_i)$, *then*

$$B_1 B_2 = B_2 B_1.$$

19.2.8 Fock quantization

Assume that A and m^2 do not depend on t. For $\zeta \in \mathcal{Y}(m^2, A)$, set

$$r_t \zeta(s, \mathbf{x}) := \zeta(s - t, \mathbf{x}).$$

Clearly,

$$r_t : \mathcal{Y}(m^2, A) \to \mathcal{Y}(m^2, A),$$
$$\hat{r}_t : \mathfrak{A}(m^2, A) \to \mathfrak{A}(m^2, A)$$

are one-parameter groups. (Recall that $\hat{r}_t$ denotes the Bogoliubov automorphism associated with r_t.)

Assume in addition that $m^2(\mathbf{x}) \geq 0$. The Klein–Gordon equation is then a special case of what we called the abstract Klein–Gordon equation in an external potential considered in Subsect. 18.3.6 with

$$\epsilon = \Big(-\sum_{i=1}^{d} (\partial_{\mathbf{x}^i} + \mathrm{i} A_i(\mathbf{x}))^2 + m^2(\mathbf{x}) \Big)^{\frac{1}{2}}.$$

Assuming that $V^2(\mathbf{x}) < \epsilon^2$, we can apply the formalism of the positive energy quantization to the space $\mathcal{Y}(m^2, A)$, described in Subsect. 18.3.6. If $A(\mathbf{x}) \equiv 0$, we can apply Subsect. 18.3.2 in the neutral case or Subsect. 18.3.5 in the charged case.

First we obtain the one-particle Hilbert space $\mathcal{Z}(m^2, A)$ together with a self-adjoint operator h. Then we obtain a (neutral or charged) CCR representation over $\mathcal{Y}(m^2, A)$ in $\Gamma_s(\mathcal{Z}(m^2, A))$. We also obtain a positive Hamiltonian $H = d\Gamma(h)$ that acts on $\Gamma_s(\mathcal{Z}(m^2, A))$, so that

$$\hat{r}_t(B) = e^{\mathrm{i}tH} B e^{-\mathrm{i}tH}, \quad B \in \mathfrak{A}.$$

19.2.9 Charge symmetry and charge reversal

Let us consider the complex case. We then have the action of the $U(1)$ group on each $\mathcal{Y}(\mathcal{O}, m^2, A)$, and hence the corresponding group of automorphisms $\{\widehat{e^{\mathrm{i}\theta}}\}_{\theta \in U(1)}$ on each $\mathfrak{A}(\mathcal{O}, m^2, A)$.

We also have the charge reversal operator χ, defined as $\chi\zeta = \bar{\zeta}$. Clearly, we obtain the isomorphisms

$$\chi : \mathcal{Y}(\mathcal{O}, m^2, A) \to \mathcal{Y}(\mathcal{O}, m^2, -A),$$
$$\hat{\chi} : \mathfrak{A}(\mathcal{O}, m^2, A) \to \mathfrak{A}(\mathcal{O}, m^2, -A).$$

If m^2, A do not depend on time, then we obtain the unitary operator $\chi_{\mathcal{Z}} : \mathcal{Z}(m^2, A) \to \mathcal{Z}(m^2, -A)$ such that

$$\hat{\chi}(B) = \Gamma(\chi_{\mathcal{Z}}) B \Gamma(\chi_{\mathcal{Z}})^{-1}, \quad B \in \mathfrak{A}.$$

19.2.10 Covariance under the Poincaré group

An element $\Lambda = (a, L) \in AO(\mathbb{R}^{1,d})$ acts on $\mathbb{R}^{1,d}$ by $\Lambda x = Lx + a$. It also acts on $\zeta \in C^\infty(\mathbb{R}^{1,d})$ by

$$u_\Lambda \zeta(x) := \zeta(\Lambda^{-1} x),$$

or $u_\Lambda \zeta = \zeta \circ \Lambda^{-1}$, and on $A \in C^\infty(\mathbb{R}^{1,d}, \mathbb{R}^{1,d})$ by

$$u_\Lambda A(x) = LA(\Lambda^{-1} x),$$

or $u_\Lambda A = LA \circ \Lambda^{-1}$. Clearly, u_Λ preserves $C_{\mathrm{sc}}^\infty(\mathbb{R}^{1,d})$. We obtain the isomorphisms

$$u_\Lambda : \mathcal{Y}(\mathcal{O}, m^2, A) \to \mathcal{Y}(\Lambda\mathcal{O}, u_\Lambda m^2, u_\Lambda A). \tag{19.15}$$

(19.15) preserves the (charged) symplectic form ω for $\Lambda \in AO^\uparrow(\mathbb{R}^{1,d})$ and changes its sign for $\Lambda \in AO^\downarrow(\mathbb{R}^{1,d})$.

Actually, the standard choice for the action of the Poincaré group is

$$w_\Lambda := \begin{cases} u_\Lambda, & \text{for orthochronous } \Lambda, \\ \chi u_\Lambda, & \text{for anti-orthochronous } \Lambda. \end{cases}$$

We have

$$w_\Lambda : \mathcal{Y}(\mathcal{O}, m^2, A) \to \mathcal{Y}(\Lambda\mathcal{O}, u_\Lambda m^2, \rho_\Lambda u_\Lambda A), \tag{19.16}$$

$$\hat{w}_\Lambda : \mathfrak{A}(\mathcal{O}, m^2, A) \to \mathfrak{A}(\Lambda\mathcal{O}, u_\Lambda m^2, \rho_\Lambda u_\Lambda A), \tag{19.17}$$

where we recall that ρ_Λ denotes the temporal parity of Λ. Equations (19.16) and (19.17) are linear for Λ orthochronous, and anti-linear otherwise.

Assume now that $m^2(x)$ and $A(x)$ do not depend on time. Consider an element of the Poincaré group $\Lambda = (a, L)$ such that $w_\Lambda m^2$ and $w_\Lambda A$ do not depend on time as well. In particular, this is the case when Λ or $T\Lambda$ belongs to $\mathbb{R}^{1,d} \rtimes O(\mathbb{R}^d)$, where T denotes the time reversal. Under this assumption, we can introduce

$$w_{\Lambda,\mathcal{Z}} : \mathcal{Z}(m^2, V, A) \to \mathcal{Z}(m^2 \circ \Lambda^{-1}, V \circ \Lambda^{-1}, LA \circ \Lambda^{-1}),$$

so that the automorphism $\hat{w}_\Lambda$ is implemented by a unitary or anti-unitary operator:

$$\hat{w}_\Lambda(B) = \Gamma(w_{\Lambda,\mathcal{Z}}) B \Gamma(w_{\Lambda,\mathcal{Z}})^{-1}, \quad B \in \mathfrak{A}.$$

If $V = 0$, $A = 0$, then the whole group $AO^\uparrow(\mathbb{R}^{1,d})$ acts by unitary transformations on the quantum level, and $AO^\downarrow(\mathbb{R}^{1,d})$ acts by anti-unitary transformations. Thus we obtain the action of the whole Poincaré group. In the complex case it should be supplemented by the action of the charge symmetry and the charge reversal.

19.2.11 Parity reversal

An important special element of the Poincaré group is the parity reversal. It is defined as $\mathrm{P} = \mathrm{diag}(1, -1, \ldots, -1)$. For $\pi := w_\mathrm{P}$ we have

$$\pi : \mathcal{Y}(\mathcal{O}, m^2, V, A) \to \mathcal{Y}(\mathrm{P}\mathcal{O}, m^2 \circ \mathrm{P}, V \circ \mathrm{P}, -A \circ \mathrm{P}),$$

$$\hat{\pi} : \mathfrak{A}(\mathcal{O}, m^2, V, A) \to \mathfrak{A}(\mathrm{P}\mathcal{O}, m^2 \circ \mathrm{P}, V \circ \mathrm{P}, -A \circ \mathrm{P}).$$

If in addition m^2, A do not depend on time, then we have the unitary operator

$$\pi_\mathcal{Z} : \mathcal{Z}(m^2, V, A) \to \mathcal{Z}(m^2 \circ \mathrm{P}, V \circ \mathrm{P}, -A \circ \mathrm{P})$$

such that

$$\hat{\pi}(B) = \Gamma(\pi_\mathcal{Z}) B \Gamma(\pi_\mathcal{Z})^{-1}, \quad B \in \mathfrak{A}.$$

19.2.12 Time reversal

Let $\mathrm{T} = \mathrm{diag}(-1, 1, \ldots, 1)$ be the time reversal as an element of the Poincaré group. The standard choice for the time reversal (Wigner's time reversal) is $\tau := w_\mathrm{T} = \chi u_\mathrm{T}$. We have

$$\tau : \mathcal{Y}(\mathcal{O}, m^2, V, A) \to \mathcal{Y}(\mathrm{T}\mathcal{O}, m^2 \circ \mathrm{T}, V \circ \mathrm{T}, -A \circ \mathrm{T}),$$

$$\hat{\tau} : \mathfrak{A}(\mathcal{O}, m^2, V, A) \to \mathfrak{A}(\mathrm{T}\mathcal{O}, m^2 \circ \mathrm{T}, V \circ \mathrm{T}, -A \circ \mathrm{T}).$$

If in addition m^2, A do not depend on time, then we obtain the anti-unitary operator

$$\tau_{\mathcal{Z}} : \mathcal{Z}(m^2, V, A) \to \mathcal{Z}(m^2, V, -A)$$

such that

$$\hat{\tau}(B) = \Gamma(\tau_{\mathcal{Z}}) B \Gamma(\tau_{\mathcal{Z}})^{-1}, \quad B \in \mathfrak{A}.$$

19.2.13 Klein–Gordon equation in the momentum representation

In this subsection we assume that $A = (V, A) = 0$ (the external potential vanishes).

We denote by $\xi = (\tau, k) \in \mathbb{R}^{1,d}$ the variables dual to $(t, x) \in \mathbb{R}^{1,d}$, paired by the Lorentz metric:

$$\langle \xi | x \rangle = -\tau \cdot t + k \cdot x.$$

If $\varsigma \in \mathcal{S}(\mathbb{R}^d)$, $\hat{\varsigma}$ will denote the usual unitary Fourier transform of ς, and if $f \in \mathcal{S}(\mathbb{R}^{1+d})$, we set

$$.\hat{f}(\xi) = (2\pi)^{-\frac{1}{2}(1+d)} \int e^{-i\langle \xi | x \rangle} f(x) dx.$$

Note that

$$\widehat{u_\Lambda \zeta}(\xi) = e^{-i\langle a | \xi \rangle} \hat{\zeta}(L^{-1}\xi), \quad \Lambda = (a, L). \tag{19.18}$$

Definition 19.21 *Define the* mass hyperboloid

$$C_m := \{\xi \in \mathbb{R}^{1,d} : \langle \xi | \xi \rangle + m^2 = 0\},$$

which splits into the two connected components $C_m^\pm = C_m \cap \{\pm \tau > 0\}$.

Note that

$$\delta(\langle \xi | \xi \rangle + m^2) \tag{19.19}$$

is a measure supported on C_m invariant w.r.t. $O(\mathbb{R}^{1,d})$. (We refer to Subsect. 4.1.2 for the notation used in (19.19).)

Set $\epsilon(k) := (k^2 + m^2)^{\frac{1}{2}}$. (19.19) has a decomposition

$$\delta(\langle \xi | \xi \rangle + m^2) = \frac{\delta(\tau - \epsilon(k))}{2\epsilon(k)} + \frac{\delta(\tau + \epsilon(k))}{2\epsilon(k)}$$

into the sum of measures supported on C_m^+ and C_m^-, invariant w.r.t. $O^\uparrow(\mathbb{R}^{1,d})$.

Proposition 19.22 *Let ζ be the solution of (19.4) with initial data (ς, ϑ) at $t = 0$, that is,*

$$\Box \zeta = 0,$$
$$\zeta(0, \mathrm{x}) = \varsigma(\mathrm{x}), \quad \dot{\zeta}(0, \mathrm{x}) = \vartheta(\mathrm{x}).$$

Define a function f on C_m by

$$f(\tau, \mathrm{k}) = \epsilon(\mathrm{k})\hat{\varsigma}(\mathrm{k}) + \mathrm{i}\operatorname{sgn}(\tau)\hat{\vartheta}(\mathrm{k}), \quad (\tau, k) \in C_m.$$

Then,

$$\hat{\zeta}(\xi) = (2\pi)^{\frac{1}{2}} f(\xi)\delta\big(\langle\xi|\xi\rangle + m^2\big), \tag{19.20}$$

$$\overline{\zeta_1}\cdot\omega\zeta_2 = \frac{1}{2\pi}\operatorname{Im}\int \overline{f_2(\xi)}\operatorname{sgn}(\tau)f_1(\xi)\delta\big(\langle\xi|\xi\rangle + m^2\big)\mathrm{d}\xi, \tag{19.21}$$

$$\widehat{\mathrm{j}\zeta}(\xi) = -\mathrm{i}\operatorname{sgn}(\tau)\hat{\zeta}(\xi).$$

Proof Using (18.26), we get that

$$\hat{\zeta}(\tau, \mathrm{k})$$
$$= 2^{-1}(2\pi)^{\frac{1}{2}}\left(\Big(\delta(\tau - \epsilon(\mathrm{k}))\Big)\Big(\hat{\varsigma}(\mathrm{k}) + \mathrm{i}\frac{\hat{\vartheta}(\mathrm{k})}{\epsilon(\mathrm{k})}\Big) + \delta\Big(\tau + \epsilon(\mathrm{k})\Big)\Big(\hat{\varsigma}(\mathrm{k}) - \mathrm{i}\frac{\hat{\vartheta}(\mathrm{k})}{\epsilon(\mathrm{k})}\Big)\right)$$
$$= (2\pi)^{\frac{1}{2}}\Big(\mathrm{i}\operatorname{sgn}(\tau)\hat{\vartheta}(\mathrm{k}) + \epsilon(\mathrm{k})\hat{\varsigma}(\mathrm{k})\Big)\delta\big(\langle\xi|\xi\rangle + m^2\big),$$

which yields (19.20). To see (19.21) we use the expression of ω in terms of the Cauchy data given in Subsect. 18.3.5 and

$$\hat{\varsigma}(\mathrm{k}) = \frac{1}{2\epsilon(\mathrm{k})}\big(f(\epsilon(\mathrm{k}), \mathrm{k}) + f(-\epsilon(\mathrm{k}), \mathrm{k})\big),$$

$$\hat{\vartheta}(\mathrm{k}) = \frac{1}{2\mathrm{i}}\big(f(\epsilon(\mathrm{k}), \mathrm{k}) - f(-\epsilon(\mathrm{k}), \mathrm{k})\big). \qquad \Box$$

Using that $\delta\big(\langle\xi|\xi\rangle + m^2\big)$ is invariant under the Lorentz group, we see that the action of the Poincaré group becomes

$$w_\Lambda f(\xi) = \begin{cases} \mathrm{e}^{-\mathrm{i}\langle a|\xi\rangle} f(L^{-1}\xi), & (a, L) \in AO^\uparrow(\mathbb{R}^{1,d}), \\ \mathrm{e}^{-\mathrm{i}\langle a|\xi\rangle} \overline{f(-L^{-1}\xi)}, & (a, L) \in AO^\downarrow(\mathbb{R}^{1,d}). \end{cases}$$

This is another way to see that w_Λ is symplectic and commutes with j, resp. anti-symplectic and anti-commutes with j, for all $\Lambda \in AO^\uparrow(\mathbb{R}^{1,d})$, resp. $\Lambda \in AO^\downarrow(\mathbb{R}^{1,d})$. Therefore, all elements of $AO^\uparrow(\mathbb{R}^{1,d})$, resp. $AO^\downarrow(\mathbb{R}^{1,d})$ can be implemented by the unitaries, resp. anti-unitaries $\Gamma(w_\Lambda)$ in the Fock representation.

19.3 Quantization of the Dirac equation

The *homogeneous Dirac equation* has the form

$$(\gamma^\mu \partial_\mu - m)\zeta(x) = 0. \tag{19.22}$$

Here, $\mathbb{R}^{1,d} \ni x \mapsto \zeta(x)$ is a function on the Minkowski space with values in Dirac spinors and γ^μ are the Dirac matrices. Equation (19.22) is invariant w.r.t. $APin(\mathbb{R}^{1,d})$, the covering of the Poincaré group.

Besides (19.22) we will consider the Dirac equation with an external potential and a variable mass,

$$\big(\gamma^\mu(\partial_\mu + \mathrm{i}A_\mu(x)) - m(x)\big)\zeta(x) = 0. \tag{19.23}$$

It is covariant w.r.t. the group $APin(\mathbb{R}^{1,d})$.

The Dirac equation has a number of similarities with the Klein–Gordon equation. First of all, solutions of (19.23) do not propagate faster than the speed of light. Secondly, on the space of space-compact solutions of (19.23), denoted by $\mathcal{Y}$, there exists a locally defined sesquilinear form. Recall that in the case of the Klein–Gordon equation this form was symplectic or charged symplectic. In the case of the Dirac equation, this form is a positive definite scalar product given by a local expression. As a consequence, two solutions of (19.23) with the Cauchy data supported in disjoint regions are orthogonal.

In the case of a positive definite scalar product, it is natural to use the fermionic quantization. Thus, let us associate with $\mathcal{Y}$ the corresponding CAR algebra. It will satisfy the fermionic version of the Einstein causality. This means that fields associated with the Cauchy data with disjoint supports will anti-commute. Consequently, even observables associated with data with disjoint supports will commute.

The space $\mathcal{Y}$ has a natural unitary dynamics. If the external potential and the mass do not depend on time, this dynamics has a time invariant generator. If the dynamics is non-degenerate, we can apply the positive energy quantization, as described in Subsect. 18.2.2. We obtain a one-particle Hilbert space $\mathcal{Z}$ and a positive Hamiltonian implementing the dynamics acting on the fermionic Fock space $\Gamma_{\mathrm{a}}(\mathcal{Z})$.

The discussion of the Poincaré invariance and covariance of the Dirac equation, which we give at the end of this section, is more complicated than in the case of the Klein–Gordon equation. In particular, to discuss the charge and time reversal we need some properties of Clifford algebras obtained in Subsect. 15.3.2.

19.3.1 Dirac operator

Let $(\mathcal{V}, \beta)$ be a finite-dimensional pseudo-unitary space. Let

$$\mathbb{R}^{1,d} \ni y \mapsto \gamma(y) \in L(\mathcal{V})$$

be a positive energy representation of $\mathrm{Cliff}(\mathbb{R}^{1,d})$. We fix a future oriented unit vector e in $\mathbb{R}^{1,d}$. Without loss of generality we may assume that $e = e^0$, where $e^0, \ldots, e^d$ is the canonical basis of $\mathbb{R}^{1,d}$. We set $\gamma^\mu = \gamma(e^\mu)$, $\mu = 0, \ldots, d$, and equip $\mathcal{V}$ with the scalar product

$$\overline{v}_1 \cdot v_2 := \mathrm{i}\overline{v}_1 \cdot \beta\gamma^0 v_2.$$

Using this scalar product, we can identify the Hermitian form β with $\mathrm{i}\gamma^0$. Therefore in the rest of this section, β will denote the operator $\mathrm{i}\gamma^0$. The adjoint of A w.r.t. the above scalar product will be denoted as usual by A^*.

As seen in Subsect. 19.1.4, we have

$$(\gamma^0)^2 = -\mathbb{1}, \quad (\gamma^i)^2 = \mathbb{1}, \quad i = 1, \ldots, d;$$
$$\gamma^\mu \gamma^\nu + \gamma^\nu \gamma^\mu = 0, \quad 0 \le \mu < \nu \le d;$$
$$(\gamma^0)^* = -\gamma^0, \quad (\gamma^i)^* = \gamma^i, \quad i = 1, \ldots, d.$$

Definition 19.23 *Let $m \in \mathbb{R}$. The* Dirac operator of mass m *is the operator on* $C^\infty(\mathbb{R}^{1,d}, \mathcal{V})$ *given by*

$$\mathbb{D}(m) := \gamma^\mu \partial_\mu - m.$$

Definition 19.24 *Let*

$$\mathbb{R}^{1,d} \ni x \mapsto m(x) \in \mathbb{R},$$
$$\mathbb{R}^{1,d} \ni x \mapsto A(x) = \big(A^\mu(x)\big) \in \mathbb{R}^{1,d}$$

be smooth functions. The Dirac operator with mass m and external potential A *is defined as*

$$\mathbb{D}(m, A) := \gamma^\mu \big(\partial_\mu + \mathrm{i}A_\mu(x)\big) - m(x).$$

19.3.2 Lagrangian of the Dirac equation

The Dirac equation $\mathbb{D}(m, A)\zeta = 0$ can be obtained as the Euler–Lagrange equation of the following Lagrangian:

$$L_1(\zeta, \overline{\zeta}, \partial\zeta, \partial\overline{\zeta}) := \overline{\zeta} \cdot \beta\big(\gamma^\mu(\partial_\mu + \mathrm{i}A_\mu) - m\big)\zeta. \tag{19.24}$$

It is also the Euler–Lagrange equation of the following more symmetric Lagrangian:

$$L(\zeta, \overline{\zeta}, \partial\zeta, \partial\overline{\zeta})$$
$$:= \frac{1}{2}\big(\overline{\zeta} \cdot \beta\gamma^\mu \partial_\mu \zeta - \overline{\gamma^\mu \partial_\mu \zeta} \cdot \beta\zeta\big) + \mathrm{i}\overline{\zeta} \cdot \beta\gamma^\mu A_\mu \zeta - m\overline{\zeta} \cdot \beta\zeta. \tag{19.25}$$

It is easy to see that (19.24) and (19.25) differ by a full derivative.

Remark 19.25 *In our notation $\overline{\zeta}$ will always denote the complex conjugate of ζ (to be consistent with the usage of $\overline{\zeta}$ elsewhere in our work and in most of the literature). In a large part of the physics literature, in the context of the Dirac equation, $\overline{\zeta}$ has a special meaning: in our notation it means $\overline{\beta\zeta}$.*

19.3.3 Green's functions

Note the identity

$$-\mathbb{D}(-m, A)\mathbb{D}(m, A) = -\big(\partial_\mu + \mathrm{i}A_\mu(x)\big)\big(\partial^\mu + \mathrm{i}A^\mu(x)\big)$$
$$+\gamma^{\mu\nu} F_{\mu\nu}(x) + \gamma^\mu G_\mu(x) + m(x)^2, \tag{19.26}$$

where

$$\gamma^{\mu\nu} := \frac{1}{2\mathrm{i}}[\gamma^\mu, \gamma^\nu],$$
$$F_{\mu\nu}(x) := \partial_\mu A_\nu(x) - \partial_\nu A_\mu(x),$$
$$G_\mu(x) := \partial_\mu m(x).$$

(19.26) is a Klein–Gordon operator with a matrix-valued mass. It has a Green's function, which can be used to express the Green's function of $\mathbb{D}(m, A)$. In fact, let $G^\pm(x, y)$ be the retarded, resp. advanced Green's function of (19.26).

Definition 19.26 $S^\pm(x, y) := -\mathbb{D}(-m, A)G^\pm(x, y)$ *is called the* retarded, resp. advanced *Green's function of the Dirac equation.*

Theorem 19.27 *Write $\mathbb{D}$ for $\mathbb{D}(m, A)$. For any $f \in C_c^\infty(\mathbb{R}^{1,d}, \mathcal{V})$ there exist unique functions $\zeta^\pm \in C_{\pm\mathrm{sc}}^\infty(\mathbb{R}^{1,d}, \mathcal{V})$ that solve*

$$\mathbb{D}\zeta^\pm = f. \tag{19.27}$$

Moreover,

$$\zeta^\pm(x) = (S^\pm f)(x) = \int_{\mathbb{R}^{1,d}} S^\pm(x, y)f(y)\mathrm{d}y,$$

where $S^\pm$ is a distribution in $\mathcal{D}'\big(\mathbb{R}^{1,d} \times \mathbb{R}^{1,d}, L(\mathcal{V})\big)$, which satisfies

$$\mathbb{D}S^\pm = S^\pm\mathbb{D} = \mathbb{1},$$
$$\mathrm{supp}\, S^\pm \subset \{(x, y) : x \in J^\pm(y)\},$$
$$S^\pm(x, y)^* = S^\mp(y, x).$$

Definition 19.28 *We set $S(x, y) = S^+(x, y) - S^-(x, y)$.*

We have

$$\mathbb{D}S = 0,$$
$$\mathrm{supp}\, S \subset \{(x, y) : x \in J(y)\},$$
$$S(x, y)^* = -S(y, x).$$

19.3.4 Cauchy problem

We will use the non-covariant notation introduced in Subsect. 19.2.4. We also set $\alpha^i := -\gamma^0\gamma^i$, $i = 1, \ldots, d$, obtaining a representation of CAR $\beta, \alpha^1, \ldots, \alpha^d$, that is,

$$\beta^2 = \mathbb{1}, \quad (\alpha_i)^2 = \mathbb{1}, \quad i = 1, \ldots, d;$$
$$\beta\alpha_i + \alpha_i\beta = 0, \quad \alpha_i\alpha_j + \alpha_j\alpha_i = 0, \quad 1 \le i < j \le d;$$
$$\beta^* = \beta, \quad \alpha_i^* = \alpha_i, \quad i = 1, \ldots, d.$$

We can rewrite the Dirac equation in the Hamiltonian form

$$\partial_t \vartheta_t = \mathrm{i}b(t)\vartheta_t, \tag{19.28}$$
$$b(t) := -\alpha^i \left(D_i - A_i(t, \mathrm{x}) \right) + V(t, \mathrm{x}) + m(t, \mathrm{x})\beta,$$

where $\vartheta_t(\mathrm{x}) = \zeta(t, \mathrm{x})$.

Theorem 19.29 *Let* $\vartheta \in C_c^\infty(\mathbb{R}^d, \mathcal{V})$. *Then there exists a unique* $\zeta \in C_{\mathrm{sc}}^\infty(\mathbb{R}^{1,d}, \mathcal{V})$ *that solves*

$$\begin{cases} \mathbb{D}(m, A)\zeta(x) = 0, \\ \quad \zeta(0, \mathrm{x}) = \vartheta(\mathrm{x}). \end{cases}$$

It satisfies $\mathrm{supp}\,\zeta \subset J(\mathrm{supp}\,\vartheta)$ *and is given by*

$$\zeta(t, \mathrm{x}) = -\int_{\mathbb{R}^d} S(t, \mathrm{x}; 0, \mathrm{y})\gamma^0 \vartheta(\mathrm{y})\mathrm{d}\mathrm{y}.$$

Moreover, $S(t, \mathrm{x}; 0, \mathrm{y})\gamma^0$ *is the integral kernel of the operator*

$$\mathrm{T}\exp\left(\mathrm{i}\int_0^t b(s)\mathrm{d}s \right).$$

19.3.5 Scalar product in the space of solutions

Definition 19.30 *Let* $\mathcal{Y}(m, A)$ *be the space of smooth space-compact solutions of the Dirac equation, that is,* $\zeta \in C_{\mathrm{sc}}^\infty(\mathbb{R}^{1,d}, \mathcal{V})$ *satisfying (19.23).*

Definition 19.31 *Let* $\zeta_1, \zeta_2 \in C^\infty(\mathbb{R}^{1,d}, \mathcal{V})$. *Set*

$$j^\mu(\zeta_1, \zeta_2, x) := \mathrm{i}\overline{\zeta_1(x)} \cdot \beta\gamma^\mu \zeta_2(x).$$

We easily check that

$$\partial_\mu j^\mu(x) = \mathrm{i}\overline{\mathbb{D}\zeta_1(x)} \cdot \beta\zeta_2(x) + \mathrm{i}\overline{\zeta_1(x)} \cdot \beta\mathbb{D}\zeta_2(x),$$

where $\mathbb{D} = \mathbb{D}(m, A)$. Therefore, if $\zeta_1, \zeta_2 \in \mathcal{Y}(m, A)$, then

$$\partial_\mu j^\mu(x) = 0,$$

and in such a case the flux of j^μ across a space-like hyper-subspace does not depend on its choice. This choice defines a scalar product on $\mathcal{Y}(m, A)$. For instance, if we consider the hyper-subspace (19.13), then we obtain the following expression for this scalar product:

$$\overline{\zeta}_1 \cdot \zeta_2 = \int_{\mathbb{R}^d} (1 - |b|^2) \left(j^0 - \mathrm{b} \cdot \mathrm{j} \right) (\zeta_1, \zeta_2, a + \mathrm{b} \cdot \mathrm{x}, \mathrm{x})\mathrm{d}\mathrm{x}.$$

In terms of the Cauchy data we have

$$\overline{\zeta}_1 \cdot \zeta_2 = \int_{\mathbb{R}^d} \overline{\zeta_1(0, \mathrm{x})} \cdot \zeta_2(0, \mathrm{x})\mathrm{d}\mathrm{x}.$$

19.3.6 Solutions parametrized by test functions

Similarly as in the case of the Klein–Gordon equation, solutions of the Dirac equation can be parametrized by space-time functions:

Theorem 19.32 (1) *For any* $f \in C_c^\infty(\mathbb{R}^{1,d}, \mathcal{V})$, $Sf \in \mathcal{Y}$.
(2) *Every element of* $\mathcal{Y}$ *is of this form.*
(3) $\overline{Sf_1} \cdot Sf_2 = \mathrm{i} \int_{\mathbb{R}^{1,d}} \int_{\mathbb{R}^{1,d}} \overline{f_1(x)} \beta S(x,y) f_2(y) \mathrm{d}x \mathrm{d}y.$

For an open set $\mathcal{O} \subset \mathbb{R}^{1,d}$ we put

$$\mathcal{Y}(\mathcal{O}) = \mathcal{Y}(\mathcal{O}, m, A) := \{Sf \ : \ f \in C_c^\infty(\mathcal{O}, \mathcal{V})\}.$$

Theorem 19.33 (1) $\mathcal{Y} = \mathcal{Y}(\mathbb{R}^{1,d})$.
(2) $\mathcal{O}_1 \subset \mathcal{O}_2$ *implies* $\mathcal{Y}(\mathcal{O}_1) \subset \mathcal{Y}(\mathcal{O}_2)$.
(3) *If* $\mathcal{O}_1$ *and* $\mathcal{O}_2$ *are causally separated and* $\zeta_i \in \mathcal{Y}(\mathcal{O}_i)$, $i = 1, 2$, *then*

$$\overline{\zeta}_1 \cdot \zeta_2 = 0.$$

19.3.7 Algebraic quantization

Let

$$\zeta_1 \cdot \nu \zeta_2 := \frac{1}{2} \mathrm{Re}\, \overline{\zeta}_1 \cdot \zeta_2, \quad \zeta_1, \zeta_2 \in \mathcal{Y}.$$

As explained in Subsect. 18.2.2, the Euclidean space $(\mathcal{Y}_\mathbb{R}, \nu)$ is used to define the field algebra of the fermionic system, the C^*-algebra

$$\mathfrak{A} = \mathfrak{A}(m, A) := \mathrm{CAR}^{C^*}(\mathcal{Y}_\mathbb{R}).$$

More generally, we have a similar definition for any open set $\mathcal{O} \subset \mathbb{R}^{1,d}$:

$$\mathfrak{A}(\mathcal{O}) = \mathfrak{A}(\mathcal{O}, m, A) := \mathrm{CAR}^{C^*}(\mathcal{Y}(\mathcal{O})_\mathbb{R}).$$

As explained in Subsect. 18.2.2, for the observable algebra we take $\mathrm{CAR}_{\mathrm{gi}}^{C^*}(\mathcal{Y})$.

Theorem 19.34 (1) $\mathfrak{A} = \mathfrak{A}(\mathbb{R}^{1,d})$.
(2) $\mathcal{O}_1 \subset \mathcal{O}_2$ *implies* $\mathfrak{A}(\mathcal{O}_1) \subset \mathfrak{A}(\mathcal{O}_2)$.
(3) *If* $\mathcal{O}_1$ *and* $\mathcal{O}_2$ *are causally separated, and* $B_i \in \mathfrak{A}(\mathcal{O}_i)$ *are elements of pure parity, then*

$$B_1 B_2 = (-1)^{|B_1||B_2|} B_2 B_1.$$

19.3.8 Fock quantization

Assume that A and m do not depend on t. For $\zeta \in \mathcal{Y}(m, A)$, set

$$r_t \zeta(s, \mathbf{x}) := \zeta(s - t, \mathbf{x}).$$

Clearly,

$$r_t : \mathcal{Y}(m, A) \to \mathcal{Y}(m, A),$$
$$\hat{r}_t : \mathfrak{A}(m, A) \to \mathfrak{A}(m, A)$$

are one-parameter groups.

Write the Dirac equation in the Hamiltonian form (19.28), where now b does not depend on time. If $\operatorname{Ker} b = \{0\}$, we can apply the formalism of the positive energy quantization to the space $\mathcal{Y}(m, A)$, described in Subsect. 18.1.2 in the neutral case and in Subsect. 18.2.2 in the charged case. This construction leads to the Kähler anti-involution j, the one-particle space $\mathcal{Z}(m, A)$ and the positive one-particle Hamiltonian h on $\mathcal{Z}(m, A)$. We obtain a representation of $\mathfrak{A}(m, A)$ on $\Gamma_a(\mathcal{Z}(m, A))$ such that the dynamics $\hat{r}_t$ is implemented by the Hamiltonian $H := \mathrm{d}\Gamma(h)$:

$$\hat{r}_t(B) = e^{itH} B e^{-itH}, \quad B \in \mathfrak{A}(m, A).$$

19.3.9 Charge symmetry

We have the action of the $U(1)$ group

$$e^{i\theta} : \mathcal{Y}(\mathcal{O}, m, A) \to \mathcal{Y}(\mathcal{O}, m, A),$$
$$\widehat{e^{i\theta}} : \mathfrak{A}(\mathcal{O}, m, A) \to \mathfrak{A}(\mathcal{O}, m, A).$$

If m, A do not depend on time, then we can define the charge operator $Q := \mathrm{d}\Gamma(q_{\mathcal{Z}})$, and we have

$$\widehat{e^{i\theta}}(B) = e^{i\theta Q} B e^{-i\theta Q}, \quad B \in \mathfrak{A}(m, A).$$

19.3.10 Charge reversal

Recall that in Subsect. 15.3.2 we studied the existence of charge reversal in the context of Clifford relations. They are anti-linear operators on $\mathcal{V}$, denoted χ_+ or χ_-, defined by the following conditions:

(1) χ_+ is called a *real charge reversal* if

$$\chi_+ \gamma(y) \chi_+^{-1} = \gamma(y), \quad \chi_+^2 = \mathbb{1};$$

(2) χ_+ is called a *quaternionic charge reversal* if

$$\chi_+ \gamma(y) \chi_+^{-1} = \gamma(y), \quad \chi_+^2 = -\mathbb{1};$$

(3) χ_- is called a *pseudo-real charge reversal* if

$$\chi_- \gamma(y) \chi_-^{-1} = -\gamma(y), \quad \chi_-^2 = \mathbb{1};$$

(4) χ_- is called a *pseudo-quaternionic charge reversal* if

$$\chi_- \gamma(y) \chi_-^{-1} = -\gamma(y), \quad \chi_-^2 = -\mathbb{1}.$$

$\chi_\pm$ can be lifted in an obvious way to the dual phase space $\mathcal{Y}$, and then to the algebra $\mathfrak{A}$. We obtain the operators

$$\chi_\pm : \mathcal{Y}(\mathcal{O}, m, A) \to \mathcal{Y}(\mathcal{O}, \pm m, -A), \tag{19.29}$$

$$\hat{\chi}_\pm : \mathfrak{A}(\mathcal{O}, m, A) \to \mathfrak{A}(\mathcal{O}, \pm m, -A). \tag{19.30}$$

Note that (19.29) is anti-unitary and (19.30) is a (linear) $*$-homomorphism.

Recall from Subsect. 15.3.2 that in the case of an irreducible representation of Clifford relations, the existence and properties of $\chi_\pm$ can be summarized with the following table:

Table 19.1

$d \pmod 8$	χ_+^2	χ_-^2
0	—	$\mathbb{1}$
1	$\mathbb{1}$	$\mathbb{1}$
2	$\mathbb{1}$	—
3	$\mathbb{1}$	$-\mathbb{1}$
4	—	$-\mathbb{1}$
5	$-\mathbb{1}$	$-\mathbb{1}$
6	$-\mathbb{1}$	—
7	$-\mathbb{1}$	$\mathbb{1}$

If m, A do not depend on time, then we can introduce the unitary operator

$$\chi_{\pm, z} : \mathcal{Z}(m, A) \to \mathcal{Z}(\pm m, -A)$$

such that

$$\hat{\chi}_\pm(B) = \Gamma(\chi_{\pm, z}) B \Gamma(\chi_{\pm, z})^{-1}, \quad B \in \mathfrak{A}(m, A).$$

19.3.11 Inversion of the sign in front of the mass

Suppose that η is a unitary operator on $\mathcal{V}$ such that $\eta \gamma^\mu = -\gamma^\mu \eta$. In the case of an irreducible representation, such an operator exists only if d is odd. It is then proportional to ω, whose definition (15.10) we recall:

$$\omega := \gamma^0 \gamma^1 \cdots \gamma^d.$$

If d is even, such operators exist only in reducible representations. Note that

$$\eta : \mathcal{Y}(\mathcal{O}, m, A) \to \mathcal{Y}(\mathcal{O}, -m, A),$$

$$\hat{\eta} : \mathfrak{A}(\mathcal{O}, m, A) \to \mathfrak{A}(\mathcal{O}, -m, A).$$

If m, A do not depend on time, we can introduce the unitary operator

$$\eta_{\mathcal{Z}} : \mathcal{Z}(m, A) \to \mathcal{Z}(-m, A)$$

such that

$$\hat{\eta}(B) = \Gamma(\eta_{\mathcal{Z}}) B \Gamma(\eta_{\mathcal{Z}})^{-1}, \quad B \in \mathfrak{A}(m, A).$$

19.3.12 Covariance under the Poincaré group

Recall that in the case of spinors the role of the Poincaré group is played by its double cover, $APin(\mathbb{R}^{1,d})$. We first describe its representations on spinor-valued functions that may seem the most natural from the mathematical point of view. It is not, however, the standard choice in quantum field theory. Then we describe another representation, which is preferred in standard textbooks.

Recall that in Subsect. 19.2.10 for $\Lambda = (a, L) \in AO(\mathbb{R}^{1,d})$ we defined an operator u_Λ acting on $C^\infty(\mathbb{R}^{1,d})$. Let us define the analog of u_Λ for spinor-valued functions.

The group $Pin(\mathbb{R}^{1,d})$ will be treated as a subgroup of $L(\mathcal{V})$ by the Pin representation. Let $\tilde{\Lambda} = (a, \tilde{L}) \in APin(\mathbb{R}^{1,d})$. Let $\Lambda = (a, L)$ be the corresponding element of the Poincaré group $AO(\mathbb{R}^{1,d})$. $\tilde{\Lambda}$ acts on $\zeta \in C^\infty(\mathbb{R}^{1,d}, \mathcal{V})$ by

$$u_{\tilde{\Lambda}} \zeta(x) := \tilde{L} \zeta(\Lambda^{-1} x),$$

or $u_{\tilde{\Lambda}} \zeta = \tilde{L} \zeta \circ \Lambda^{-1}$. We obtain the unitary map

$$u_{\tilde{\Lambda}} : \mathcal{Y}(\mathcal{O}, m, A) \to \mathcal{Y}(\Lambda \mathcal{O}, \det L \, u_\Lambda m, u_\Lambda A). \tag{19.31}$$

Actually, the standard choice for the action of the Poincaré group is different. It involves additionally the charge reversal operator χ_+ and the operator η that inverts the sign of the mass. We set

$$w_{\tilde{\Lambda}} := \begin{cases} u_{\tilde{\Lambda}}, & \text{for even orthochronous } \Lambda, \\ \eta u_{\tilde{\Lambda}}, & \text{for odd orthochronous } \Lambda, \\ \chi_+ u_{\tilde{\Lambda}}, & \text{for even anti-orthochronous } \Lambda, \\ \eta \chi_+ u_{\tilde{\Lambda}}, & \text{for odd anti-orthochronous } \Lambda. \end{cases}$$

We obtain the transformations

$$w_{\tilde{\Lambda}} : \mathcal{Y}(\mathcal{O}, m, A) \to \mathcal{Y}(\Lambda \mathcal{O}, u_\Lambda m, \rho_\Lambda u_\Lambda A), \tag{19.32}$$

$$\hat{w}_{\tilde{\Lambda}} : \mathfrak{A}(\mathcal{O}, m, A) \to \mathfrak{A}(\Lambda \mathcal{O}, u_\Lambda m, \rho_\Lambda u_\Lambda A). \tag{19.33}$$

(Recall that ρ_Λ denotes the temporal parity of Λ.) (19.32) is unitary and (19.33) is a linear $*$-homomorphism for orthochronous Λ. (19.32) is anti-unitary and (19.33) is an anti-linear $*$-homomorphism for anti-orthochronous Λ. In irreducible representations the operator χ_+ exists only if $d \neq 0 \pmod 4$, and the operator η exists only if d is odd (one can then choose $\eta = \omega$). Thus in irreducible representations the definition of $w_{\tilde\Lambda}$ is possible if $d \equiv 1, 3 \pmod 4$, which includes the physical case $d = 3$.

Assume now that $m(x)$ and $A(x)$ do not depend on time, so that one can consider the Fock quantization. Consider an element $\Lambda = (a, L)$ such that $u_\Lambda m$ and $u_\Lambda A$ do not depend on time as well. As before, this is the case if Λ or $T\Lambda$ belong to $\mathbb{R}^{1,d} \rtimes O(\mathbb{R}^d)$. Note that $O_\pm(\mathbb{R}^d) \subset O_\pm^\uparrow(\mathbb{R}^{1,d})$ and $TO_\pm(\mathbb{R}^d) \subset O_\mp^\downarrow(\mathbb{R}^{1,d})$, where T is the time reversal. From this we deduce the existence of operators

$$w_{\Lambda, z} : \mathcal{Z}(m, V, A) \to \mathcal{Z}(m \circ \Lambda^{-1}, V \circ \Lambda^{-1}, \rho_\Lambda LA \circ \Lambda^{-1})$$

satisfying

$$\hat{w}_\Lambda(A) = \Gamma(w_{\Lambda, z}) A \Gamma(w_{\Lambda, z})^{-1}.$$

The operator $w_{\Lambda, z}$ is unitary for orthochronous Λ; otherwise it is anti-unitary.

19.3.13 Parity reversal

Consider the parity reversal operator $P := \mathrm{diag}(1, -1, \dots, -1)$. This operator is even if d is even and odd if d is odd. Let $\tilde{P}$ be an element of $Pin(\mathbb{R}^{1,d})$ or $Pin(\mathbb{R}^{d,1})$ covering P (which equals $\pm \gamma_1 \cdots \gamma_d$).

The standard operator implementing the parity reversal (which works in any dimension) is $\pi := w_{\tilde P} = \eta u_{\tilde P}$. We have

$$\pi : \mathcal{Y}(\mathcal{O}, m, V, A) \to \mathcal{Y}(P\mathcal{O}, m \circ P, V \circ P, -A \circ P),$$
$$\hat\pi : \mathfrak{A}(\mathcal{O}, m, V, A) \to \mathfrak{A}(P\mathcal{O}, m \circ P, V \circ P, -A \circ P).$$

Therefore, if $V \circ P = V$, $A \circ P = -A$ and $m \circ P = m$, then the system is invariant w.r.t. π. Note that $\pi^2 = -\mathbb{1}$.

If m, A do not depend on time, then we can introduce the unitary operator

$$\pi_z : \mathcal{Z}(m, V, A) \to \mathcal{Z}(m \circ P, V \circ P, -A \circ P)$$

such that

$$\hat\pi(B) = \Gamma(\pi_z) B \Gamma(\pi_z)^{-1}, \quad B \in \mathfrak{A}(m, A).$$

19.3.14 Time reversal

Recall that $T := \mathrm{diag}(-1, 1, \dots, 1)$ denotes the time reversal operator. It is an odd anti-orthochronous operator. Let $\tilde{T}$ be an element of $\tilde{L}$ covering T (which equals $\pm \gamma_0$).

The operator u_T is the so-called *Racah time reversal* (see Subsect. 18.2.4). The *Wigner time reversal*, which is the standard choice, is $\tau := w_T = \eta \chi_+ u_T$. We have

$$\tau : \mathcal{Y}(\mathcal{O}, m, V, A) \to \mathcal{Y}(T\mathcal{O}, m \circ T, V \circ T, -A \circ T), \qquad (19.34)$$

$$\hat{\tau} : \mathfrak{A}(\mathcal{O}, m, V, A) \to \mathfrak{A}(T\mathcal{O}, m \circ T, V \circ T, -A \circ T). \qquad (19.35)$$

(19.34) is anti-unitary and (19.35) is an anti-linear $*$-homomorphism.

If m, A do not depend on time, then we can introduce the anti-unitary operator

$$\tau_{\mathcal{Z}} : \mathcal{Z}(m, V, A) \to \mathcal{Z}(m, V, -A)$$

such that

$$\hat{\tau}(B) = \Gamma(\tau_{\mathcal{Z}}) B \Gamma(\tau_{\mathcal{Z}})^{-1}, \quad B \in \mathfrak{A}(m, V, A).$$

Note the identity

$$\pi \chi_+ \tau = \pm u_{PT}.$$

We thus obtain the unitary operator

$$\pm u_{PT} : \mathcal{Y}(m, A) \to \mathcal{Y}(m \circ PT, -A \circ PT),$$

which is an important ingredient of the famous PCT theorem.

19.3.15 Dirac equation in the momentum representation

Let us assume that m is constant and $A = 0$. The Dirac equation can be written as

$$(\gamma^\mu \partial_\mu - m)\zeta = 0. \qquad (19.36)$$

If ζ is a solution, then

$$(-\Box + m^2)\zeta = 0.$$

Therefore,

$$\widehat{\zeta}(\tau, \mathbf{k}) = (2\pi)^{\frac{1}{2}} \left(f^+(\mathbf{k}) \frac{\delta(\tau - \epsilon(\mathbf{k}))}{2\epsilon(\mathbf{k})} + f^-(\mathbf{k}) \frac{\delta(\tau + \epsilon(\mathbf{k}))}{2\epsilon(\mathbf{k})} \right)$$

defines invariantly the functions $f^\mp$ on the lower and upper hyperboloid of mass m.

If we introduce $\vartheta_t(\mathbf{x}) = \zeta(t, \mathbf{x})$, then (19.36) can be rewritten in the Hamiltonian form

$$\partial_t \vartheta_t = ib\vartheta_t, \quad \text{for} \ \ b := -\alpha \cdot D + \beta m. \qquad (19.37)$$

b is self-adjoint on $L^2(\mathbb{R}^d, \mathcal{V})$.

Proposition 19.35 *Let ζ_i, $i = 1, 2$, be two solutions of (19.36). Then we have*

$$\overline{\zeta_1} \cdot \zeta_2$$
$$= \frac{1}{2m} \int \left(\overline{f_1^+(\mathbf{k})} \cdot \beta f_2^+(\mathbf{k}) \frac{\delta(\tau - \epsilon(\mathbf{k}))}{2\epsilon(\mathbf{k})} - \overline{f_1^-(\mathbf{k})} \cdot \beta f_2^-(\mathbf{k}) \frac{\delta(\tau + \epsilon(\mathbf{k}))}{2\epsilon(\mathbf{k})} \right) d\tau d\mathbf{k},$$

$$\widehat{\mathrm{sgn}(b)}\zeta(\tau, \mathbf{k}) = (2\pi)^{\frac{1}{2}} \left(f^+(\mathbf{k}) \frac{\delta(\tau - \epsilon(\mathbf{k}))}{2\epsilon(\mathbf{k})} - f^-(\mathbf{k}) \frac{\delta(\tau + \epsilon(\mathbf{k}))}{2\epsilon(\mathbf{k})} \right).$$

Proof Set $b(\mathbf{k}) = \gamma^0 \gamma \cdot \mathbf{k} - im\gamma^0$, so that $\widehat{b\vartheta}(\mathbf{k}) = b(\mathbf{k})\hat{\vartheta}(\mathbf{k})$. Note that $b(\mathbf{k})$ is self-adjoint on $(\mathcal{V}, (\cdot|\cdot))$ and $b(\mathbf{k})^2 = \epsilon^2(\mathbf{k})$. Hence,

$$\mathbb{1} = P_+(\mathbf{k}) + P_-(\mathbf{k})$$

for $P_\pm(\mathbf{k}) = \mathbb{1}_{\{\mp\epsilon(\mathbf{k})\}}(b(\mathbf{k}))$.

If ϑ_t is a solution of (19.36) with the initial condition ϑ, then

$$\hat{\vartheta}_t(\mathbf{k}) = e^{itb(\mathbf{k})}\hat{\vartheta}(\mathbf{k}).$$

Taking the Fourier transform we obtain

$$\hat{\zeta}(\tau, \mathbf{k}) = (2\pi)^{\frac{1}{2}} \left(\delta(\tau - \epsilon(\mathbf{k})) P_+(\mathbf{k}) + \delta(\tau + \epsilon(\mathbf{k})) P_-(\mathbf{k}) \right) \hat{\vartheta}(\mathbf{k}).$$

Therefore,

$$f^\pm(\mathbf{k}) = 2\epsilon(\mathbf{k}) P_\pm(\mathbf{k})\hat{\vartheta}(\mathbf{k}). \tag{19.38}$$

We clearly have $\gamma^0 b(\mathbf{k}) + b(\mathbf{k})\gamma^0 = 2im$, which implies that

$$P_\pm(\mathbf{k}) = \pm im^{-1}\epsilon(\mathbf{k}) P_\pm(\mathbf{k})\gamma^0 P_\pm(\mathbf{k}). \tag{19.39}$$

Recall that $\beta = i\gamma^0$. Hence, setting $\vartheta^\pm(\mathbf{k}) := P^\pm(\mathbf{k})\vartheta(\mathbf{k})$,

$$\frac{i}{2m} \int \overline{f_1^+(\mathbf{k})} \cdot \gamma^0 f_2^+(\mathbf{k}) \frac{\delta(\tau - \epsilon(\mathbf{k}))}{2\epsilon(\mathbf{k})} d\tau d\mathbf{k} - \frac{i}{2m} \overline{f_1^-(\mathbf{k})} \cdot \gamma^0 f_2^-(\mathbf{k}) \frac{\delta(\tau + \epsilon(\mathbf{k}))}{2\epsilon(\mathbf{k})} d\tau d\mathbf{k}$$
$$= \frac{i}{2m} \int_{\mathbb{R}^d} 2\epsilon(\mathbf{k})\overline{\hat{\vartheta}_1^+(\mathbf{k})} \cdot \gamma^0 \hat{\vartheta}_2^+(\mathbf{k}) d\mathbf{k} - \frac{i}{2m} \int_{\mathbb{R}^d} 2\epsilon(\mathbf{k})\overline{\hat{\vartheta}_1^-(\mathbf{k})} \cdot \gamma^0 \hat{\vartheta}_2^-(\mathbf{k}) d\mathbf{k}$$
$$= \int_{\mathbb{R}^d} \overline{\hat{\vartheta}_1^+(\mathbf{k})} \cdot \hat{\vartheta}_2^+(\mathbf{k}) d\mathbf{k} + \int_{\mathbb{R}^d} \overline{\hat{\vartheta}_1^-(\mathbf{k})} \cdot \hat{\vartheta}_2^-(\mathbf{k}) d\mathbf{k}$$
$$= \int_{\mathbb{R}^d} \overline{\hat{\vartheta}_1(\mathbf{k})} \cdot \hat{\vartheta}_2(\mathbf{k}) d\mathbf{k} = \int_{\mathbb{R}^d} \overline{\vartheta_1(\mathbf{x})} \cdot \vartheta_2(\mathbf{x}) d\mathbf{x} = \overline{\zeta_1} \cdot \zeta_2.$$

$\square$

19.4 Partial differential equations on manifolds

In this section we introduce basic notation and terminology for the analysis of partial differential equations on manifolds. We will be mostly interested in Lorentzian manifolds, especially the so-called globally hyperbolic manifolds, which serve as models for curved space-times. The material introduced in this section will be needed in Sects. 19.5 and 19.6, where we describe quantization of the algebraic Klein–Gordon and Dirac equations on curved space-times.

19.4.1 Manifolds

Let $\mathcal{X}$ be a manifold of dimension d. We will denote by $C_{\mathrm{c}}^{\infty}(\mathcal{X})$ the space of compactly supported smooth functions on $\mathcal{X}$ and by $\mathcal{D}'(\mathcal{X})$ its dual space – the space of distributions on $\mathcal{X}$.

$T\mathcal{X}$, resp. $T^{\#}\mathcal{X}$ denote the *tangent*, resp. *cotangent bundle over* $\mathcal{X}$ with fibers $T_x\mathcal{X}$, resp. $T_x^{\#}\mathcal{X}$ equal to the *tangent*, resp. *cotangent space to* $\mathcal{X}$ at $x \in \mathcal{X}$. Smooth sections of $T\mathcal{X}$, resp. $T^{\#}\mathcal{X}$ are called *vector fields*, resp. *differential 1-forms on* $\mathcal{X}$.

Suppose that (an open subset of) $\mathcal{X}$ is parametrized by local coordinates $x = (x^1, \dots, x^d)$ from (an open subset of) $\mathbb{R}^d$. Then we have a natural local frame in $T\mathcal{X}$, traditionally denoted $(\partial_{x^1}, \dots, \partial_{x^d})$. Its dual frame is denoted $(\mathrm{d}x^1, \dots, \mathrm{d}x^d)$. We will use the coordinate-dependent notation, tacitly identifying $\mathcal{X}$ (or its open subset) with $\mathbb{R}^d$ (or its open subset). We will use the Einstein summation convention.

By a *(parametrized) curve* in $\mathcal{X}$ we will mean a continuous piecewise C^1 map from an interval in $\mathbb{R}$ into $\mathcal{X}$. A curve is called *inextensible* if none of its piecewise C^1 reparametrizations can be continuously extended beyond its endpoints.

Let $\mathcal{V}$ be a finite-dimensional vector space. In the rest of the section we will discuss differential operators acting on $C^{\infty}(\mathcal{X}, \mathcal{V})$ – the space of smooth functions $\mathcal{X} \to \mathcal{V}$. Of course, our discussion can be easily generalized to differential operators on smooth sections of a *vector bundle* $(E, \mathcal{X})$ with base $\mathcal{X}$ and fibers isomorphic to $\mathcal{V}$. Using local trivializations of E, one can locally reduce the analysis to the trivial bundle $\mathcal{X} \times \mathcal{V}$ considered here. All the objects introduced below have natural definitions covariant under change of coordinates and of local frames, which a reader with a little familiarity with vector bundles can easily guess.

19.4.2 Integration on pseudo-Riemannian manifolds

A manifold $\mathcal{X}$ is called *pseudo-Riemannian* if it is equipped with a smooth pseudo-Euclidean form, called the *metric tensor* $\mathcal{X} \ni x \mapsto g(x) = [g_{\mu\nu}(x)] \in \otimes_{\mathrm{s}}^2 T_x^{\#}\mathcal{X}$. It equips $T_x\mathcal{X}$ with a scalar product, so that $g_{\mu\nu} = (\partial_{x^\mu}|\partial_{x^\nu})$. We will set $|g|(x) := |\det[g_{\mu\nu}(x)]|$.

The inverse of $[g_{\mu\nu}(x)]$ will be denoted by $[g^{\mu\nu}(x)]$. Clearly, it induces the dual scalar product in $T_x^{\#}\mathcal{X}$: $(dx^{\mu}|dx^{\nu}) = g^{\mu\nu}$.

Let dx denote the Lebesgue measure on $\mathbb{R}^d$ transported by a local chart to the manifold $\mathcal{X}$. dv will denote the measure $|g|^{\frac{1}{2}}dx$ on $\mathcal{X}$. It does not depend on the coordinates. Thus if f is a function on $\mathcal{X}$, its integral over $\mathcal{X}$ is denoted $\int f dv$. We equip $C_c(\mathcal{X})$ with the scalar product

$$(f|g) := \int \overline{f} g \, dv. \tag{19.40}$$

Let S be a smooth *hypersurface of* $\mathcal{X}$ (that is, a sub-manifold of co-dimension 1). We can find local coordinates such that

$$S = \{(x^1, \ldots, x^d) \, : \, x^d = 0\}. \tag{19.41}$$

We define a measure on S by $ds := |h|^{\frac{1}{2}}dx^1 \ldots dx^{d-1}$, where $|h| = \det[h_{ij}]$ and $[h_{ij}]$ is obtained from $[g_{ij}]$ by discarding the last column and the last line (so that $i, j = 1, \ldots, d-1$). The measure ds does not depend on the coordinates. If f is a smooth function on S, then we write $\int_S f ds$ for its integral over S.

We say that S has an *external orientation* if for any $x \in S$ a unit normal vector has been chosen, which depends continuously on x. A hypersurface given by (19.41) has a natural external orientation: in the direction of the coordinate x^d. The co-vector $|g^{dd}|^{-\frac{1}{2}}dx^d$ restricted to S is called the *normal co-vector*. It will be denoted n_{μ}. In the coordinates that we use we have

$$n_{\mu} = \begin{cases} 0, & \mu = 1, \ldots, d-1, \\ |g^{dd}|^{-\frac{1}{2}}, & \mu = d. \end{cases}$$

Note that $|g| = |h||g^{dd}|^{-1}$.

If $x \mapsto [f^{\mu}(x)]$ is a vector field, then we define its *flux across* S as

$$\int f^{\mu} n_{\mu} ds. \tag{19.42}$$

Again, (19.42) does not depend on coordinates. To shorten the notation, we will write ds_{μ} instead of $n_{\mu} ds$.

The *Stokes theorem* says that if Ω is an open subset of $\mathcal{X}$ with a sufficiently regular boundary $\partial\Omega$, then

$$\int_{\Omega} |g|^{-\frac{1}{2}} \nabla_{\mu} |g|^{\frac{1}{2}} f^{\mu} dv = \int_{\partial\Omega} f^{\mu} ds_{\mu}, \tag{19.43}$$

where $\nabla_{\mu} = \partial_{x^{\mu}}$ is the μth partial derivative.

19.4.3 Lorentzian manifolds

We will use the terminology for vectors in the Minkowski space introduced in Def. 19.1.

Definition 19.36 *A pseudo-Riemannian manifold* $\mathcal{X}$ *is called* Lorentzian *if the signature of its metric tensor is* $(-1, 1, \ldots, 1)$.

We say that $\mathcal{X}$ is time-orientable *if there exists a global continuous time-like vector field on* $\mathcal{X}$. *If* v *is such a vector field and* $x \in \mathcal{X}$, *a time-like vector* $v' \in T_x\mathcal{X}$ *is future, resp. past oriented if* $\pm v(x) \cdot g(x)v' > 0$. *The manifold* $\mathcal{X}$ *equipped with such a continuous choice of future/past directions is called* time-oriented.

In the remaining part of this subsection, $\mathcal{X}$ is a time-oriented Lorentzian manifold.

Definition 19.37 *A curve in* $\mathcal{X}$ *is called* time-like, causal, *resp.* light-like *if all its tangent vectors are such and all pairs of tangent vectors at break points are in the same causal cone. A curve in* $\mathcal{X}$ *is called* space-like *if all its tangent vectors are such.*

Definition 19.38 *Let* $x \in \mathcal{X}$. *The* causal, *resp.* time-like future, *resp.* past *of* x *is the set of all* $y \in \mathcal{X}$ *that can be reached from* x *by a causal, resp. time-like future-, resp. past-directed curve, and is denoted* $J^{\pm}(x)$, *resp.* $I^{\pm}(x)$. *For* $\mathcal{U} \subset \mathcal{X}$, *its* causal, *resp.* time-like future, *resp.* past *is defined as*

$$J^{\pm}(\mathcal{U}) = \bigcup_{x \in \mathcal{U}} J^{\pm}(x), \quad I^{\pm}(\mathcal{U}) = \bigcup_{x \in \mathcal{U}} I^{\pm}(x).$$

We define also the causal, *resp.* time-like shadow:

$$J(\mathcal{U}) = J^{+}(\mathcal{U}) \cup J^{-}(\mathcal{U}), \quad I(\mathcal{U}) = I^{+}(\mathcal{U}) \cup I^{-}(\mathcal{U}).$$

Definition 19.39 *A* Cauchy hypersurface *is a hypersurface* $S \subset \mathcal{X}$ *such that each inextensible time-like curve intersects* S *at exactly one point.*

If S is a smooth space-like Cauchy hypersurface, it will always be equipped with the external orientation given by the future directed normal vector at each point of S.

Let us quote the following result from the theory of Lorentzian manifolds:

Theorem 19.40 *Let* $\mathcal{X}$ *be a connected Lorentzian manifold. The following are equivalent:*

(1) *The following two conditions hold:*
   (1a) for any $x, y \in \mathcal{X}$, $J^{+}(x) \cap J^{-}(y)$ *is compact,*
   (1b) (causality condition) *there are no closed causal curves.*
(2) *There exists a Cauchy hypersurface.*
(3) $\mathcal{X}$ *is isometric to* $\mathbb{R} \times S$ *with metric* $-\beta dt^2 + g_t$, *where* β *is a smooth positive function,* g_t *is a Riemannian metric on* S *depending smoothly on* $t \in \mathbb{R}$, *and each* $\{t\} \times S$ *is a smooth space-like Cauchy hypersurface in* $\mathcal{X}$.

The above theorem is quoted from Bär–Ginoux–Pfäffle (2007) except for (1b), where in this reference the so-called strong causality condition is given. The fact that the strong causality condition can be replaced by the causality condition is a recent result of Bernal–Sanchez (2007).

Definition 19.41 *A connected Lorentzian manifold satisfying the equivalent conditions of the above theorem is called* globally hyperbolic.

We recall that a Riemannian manifold $(\mathcal{S}, h)$ is *geodesically complete* if all its geodesics can be infinitely extended. By the Hopf–Rinow theorem, this condition is equivalent to the condition that $\mathcal{S}$, equipped with the Riemannian distance, is complete as a metric space; see Sakai (1996).

Example 19.42 *Let $(\mathcal{S}, h)$ be a Riemannian manifold and $I \subset \mathbb{R}$ an open interval. Let $f : I \to]0, \infty[$ be a smooth function. Then $\mathcal{X} = I \times \mathcal{S}$ with the metric $-\mathrm{d}t^2 + f(t)^2 h$ is globally hyperbolic iff $(\mathcal{S}, h)$ is geodesically complete.*

Example 19.43 *Let $(\mathcal{S}, h)$ be a Riemannian manifold and $\tau : \mathcal{S} \to]0, +\infty[$ a smooth function, $v \in T^\# \mathcal{S}$ a smooth 1-form. Then $\mathcal{X} = \mathbb{R} \times \mathcal{S}$, equipped with the Lorentzian metric $-\tau(\mathrm{x})\big(\mathrm{d}t - v_j(\mathrm{x})\mathrm{d}\mathrm{x}^j\big)^2 + h_{jk}(\mathrm{x})\mathrm{d}\mathrm{x}^j\,\mathrm{d}\mathrm{x}^k$, is called a* stationary *space-time. If $v = 0$ it is called* static. *It is called* uniformly static *if there exists $c > 0$ such that $c \leq \tau(\mathrm{x}) \leq c^{-1}$. A uniformly static space-time is globally hyperbolic iff $(\mathcal{S}, h)$ is geodesically complete; see Fulling (1989) .*

It is straightforward to generalize the notion of space-compact functions from the Minkowski space (see Def. 19.2) to a Lorentzian manifold.

Definition 19.44 *A function $f \in C(\mathcal{X})$ is called* space-compact *iff there exists a compact $K \subset \mathcal{X}$ such that $\mathrm{supp}\, f \subset J(K)$. It is called* future, *resp.* past space-compact *iff there exists a compact $K \subset \mathcal{X}$ such that $\mathrm{supp}\, f \subset J^\pm(K)$.*

The set of smooth space-compact functions will be denoted $C^\infty_{\mathrm{sc}}(\mathcal{X})$. The set of smooth future, resp. past space-compact functions will be denoted $C^\infty_{\pm\mathrm{sc}}(\mathcal{X})$.

Finally, let us give the definition of the causal dependence.

Definition 19.45 *Let $\mathcal{X}$ be globally hyperbolic and $\mathcal{O} \subset \mathcal{X}$. We say that $x \in \mathcal{X}$ is* causally dependent on $\mathcal{O}$ *if there exists a neighborhood $\mathcal{U}$ of x and a smooth Cauchy surface $\mathcal{S}$ such that every causal curve starting from $\mathcal{U}$ intersects $\mathcal{S}$ in $\mathcal{O}$.*

If $\mathcal{O}_1, \mathcal{O}_2 \subset \mathcal{X}$ we say that $\mathcal{O}_1$ is causally dependent on $\mathcal{O}_2$, *if every $x \in \mathcal{O}_1$ is causally dependent on $\mathcal{O}_2$.*

19.4.4 *First-order partial differential equations*

We assume that the manifold $\mathcal{X}$ is equipped with a measure, which in local coordinates equals $|g|^{\frac{1}{2}}(x)\mathrm{d}x$. (In this subsection, $|g|^{\frac{1}{2}}$ does not have to come from a metric tensor.)

Let $\mathcal{V}$ be a finite-dimensional vector space over $\mathbb{K} = \mathbb{R}$ or $\mathbb{C}$. To simplify notation we will assume that $\mathbb{K} = \mathbb{C}$. The formulas in the real case are obtained from the complex case by dropping the bars and replacing the Hermitian conjugation $*$ with the transposition $\#$.

In this subsection we study first-order differential equations for functions with values in $\mathcal{V}$. With a large class of such equations we will associate a locally defined vector-valued sesquilinear form J^μ whose divergence vanishes. In physics J^μ is often called a *conserved current*.

This construction has a special importance for hyperbolic equations on hyperbolic manifolds, where it leads to an invariantly defined bilinear or sesquilinear form on the space of solutions. In the case of the Klein–Gordon equation (which, of course, can be reduced to a first-order equation) this form is symplectic for $\mathbb{K} = \mathbb{R}$ and charged symplectic for $\mathbb{K} = \mathbb{C}$. The Klein–Gordon equation on curved space-times will be considered in Sect. 19.5. In the case of the generalized Dirac equation, which on a curved space-time we will consider in Sect. 19.6, this form is a positive definite scalar product. In both cases these forms play a fundamental role in the quantization of the classical equation.

In many textbooks on quantum field theory the conserved current J^μ is derived from the Lagrangian by the Noether theorem using the invariance w.r.t. the charge symmetry $\zeta \mapsto e^{i\theta}\zeta$, $\theta \in U(1)$. In the derivation that we give, complex numbers do not enter at all.

Let us consider a first-order linear equation for $\zeta \in C^\infty(\mathcal{X}, \mathcal{V})$. Every such equation can be written in the form

$$\alpha^\mu(x)\nabla_\mu\zeta(x) + \frac{1}{2}\big(|g|^{-\frac{1}{2}}(x)\nabla_\mu|g|^{\frac{1}{2}}(x)\alpha^\mu(x)\big)\zeta(x) + \theta(x)\zeta(x) = 0, \qquad (19.44)$$

where $\mathcal{X} \ni x \mapsto \alpha^\mu(x), \theta(x) \in L(\mathcal{V}, \mathcal{V}^*)$, and μ enumerates the coordinates of $\mathcal{X}$. The following theorem describes conditions that guarantee the existence of a conserved current for (19.44).

Theorem 19.46 *Suppose that either Condition (19.45) or Condition (19.46) is satisfied:*

$$\alpha^\mu(x)^* = \alpha^\mu(x), \quad \theta(x)^* = -\theta(x), \qquad (19.45)$$
$$or \quad \alpha^\mu(x)^* = -\alpha^\mu(x), \quad \theta(x)^* = \theta(x). \qquad (19.46)$$

Define

$$J^\mu(\zeta_1, \zeta_2, x) := \overline{\zeta_1(x)}\cdot\alpha^\mu(x)\zeta_2(x).$$

Let ζ_1, ζ_2 be solutions of (19.44). Then

$$\nabla_\mu|g|^{\frac{1}{2}}(x)J^\mu(\zeta_1, \zeta_2, x) = 0. \qquad (19.47)$$

Proof

$$|g|^{-\frac{1}{2}}\nabla_\mu|g|^{\frac{1}{2}}J^\mu = \overline{\Big(\alpha^{\mu*}\nabla_\mu + \frac{1}{2}\big(|g|^{-\frac{1}{2}}\nabla_\mu|g|^{\frac{1}{2}}\alpha^\mu\big)^* - \theta^*\Big)\zeta_1} \cdot \zeta_2$$
$$+ \overline{\zeta_1} \cdot \Big(\alpha^\mu\nabla_\mu + \frac{1}{2}\big(|g|^{-\frac{1}{2}}\nabla_\mu|g|^{\frac{1}{2}}\alpha^\mu\big) + \theta\Big)\zeta_2 = 0. \qquad \square$$

Note that in the complex case one can pass from Condition (19.45) to (19.46) by multiplying α^μ and θ by i.

19.5 Generalized Klein–Gordon equation on curved space-time

Throughout this section we assume that $\mathcal{X} \ni x \mapsto g(x) = [g_{\mu\nu}(x)]$ is the metric tensor on a pseudo-Riemannian manifold $\mathcal{X}$. In the first two subsections only we allow $\mathcal{X}$ to be an arbitrary pseudo-Riemannian manifold. Starting with Subsect. 19.5.3, we will assume that $\mathcal{X}$ is globally hyperbolic.

$\mathcal{V}$ will be a real or complex finite-dimensional vector space. For simplicity, most formulas will be given for the complex case. We will consider a vector bundle with a base $\mathcal{X}$ and fiber $\mathcal{V}$. For simplicity, we will always assume that the bundle is trivial and trivialized to $\mathcal{X} \times \mathcal{V}$.

In this section we describe algebraic quantization of a large class of second-order equations on $\mathcal{X}$ with values in $\mathcal{V}$. We will always assume that the principal term of these equations is given by the metric tensor. In the case of Lorentzian manifolds, such equations will be called *generalized Klein–Gordon equations*. Solutions of these equations propagate causally.

Generalized Klein–Gordon equations possess a conserved current, which is symplectic in the real case and charged symplectic in the complex case. Therefore, it is natural to quantize these equations using the CCR. The bosonic algebraic quantization leads to a net of algebras satisfying the Einstein causality.

This section is to a large extent a generalization of Sect. 19.2 to a curved space-time. Unlike in Sect. 19.2, we limit our discussion to the algebraic quantization. We do not discuss the positive energy quantization on a bosonic Fock space, which is possible for the Klein–Gordon equation on a stationary space-time; see Kay (1978).

19.5.1 Klein–Gordon operators

Let $\mathcal{X} \ni x \mapsto \Gamma_\mu(x) \in L(\mathcal{V})$, $\mu = 0, 1, \ldots, d$, be smooth functions.

Definition 19.47

$$\nabla_\mu^\Gamma \zeta := (\nabla_\mu + \Gamma_\mu)\zeta, \quad \zeta \in C^\infty(\mathcal{X}, \mathcal{V}),$$

is called the covariant derivative of ζ with the connection Γ.

Let $\mathcal{X} \ni x \mapsto \rho(x) \in L(\mathcal{V})$ be a smooth function.

Definition 19.48 *The operator*

$$\Box = \Box(\Gamma, \rho) := -|g|^{-\frac{1}{2}} \nabla_\mu^\Gamma |g|^{\frac{1}{2}} g^{\mu\nu} \nabla_\nu^\Gamma + \rho, \tag{19.48}$$

acting on $C^\infty(\mathcal{X}, \mathcal{V})$, *will be called the* generalized Klein–Gordon operator with the connection Γ and the mass-squared term ρ.

We will study the equation

$$\Box\zeta = 0. \tag{19.49}$$

Remark 19.49 *The name* Klein–Gordon equation *is usually associated with the Lorentz signature. Its analog for the Euclidean signature has the traditional name of the* Helmholtz equation.

Let us fix a smooth map $\mathcal{X} \ni x \mapsto \lambda(x) \in L(\mathcal{V}, \mathcal{V}^*)$ with values in positive definite forms on $\mathcal{V}$. One can then introduce a scalar product on $C^\infty(\mathcal{X}, \mathcal{V})$ by

$$\overline{\zeta}_1 \cdot \zeta_2 := \int_{\mathcal{X}} \overline{\zeta_1(x)} \cdot \lambda(x)\zeta_2(x) \mathrm{d}v(x). \tag{19.50}$$

Remark 19.50 *In the literature, a complex vector bundle $\mathcal{X} \times \mathcal{V} \to \mathcal{X}$ equipped with a scalar product $\lambda(x)$ is often called a* Hermitian bundle, *which is not quite correct, since it is equipped with a* positive definite Hermitian form. *A connection satisfying (19.51) is also called* Hermitian.

We will assume the following conditions:

$$\nabla_\mu \lambda = \Gamma_\mu^* \lambda + \lambda \Gamma_\mu, \qquad \lambda\rho = \rho^*\lambda. \tag{19.51}$$

If (19.51) holds, for any smooth open set $\Omega \subset \mathcal{X}$, *Green's formula* is valid:

$$\int_\Omega \left(\overline{\zeta_1(x)} \cdot \lambda(x)\Box\zeta_2(x) - \overline{\Box\zeta_1(x)} \cdot \lambda(x)\zeta_2(x) \right) \mathrm{d}v(x) \tag{19.52}$$

$$= -\int_{\partial\Omega} g^{\mu,\nu}(x) \left(\overline{\zeta_1(x)} \cdot \lambda(x)\nabla_\mu^\Gamma \zeta_2(x) - \overline{\nabla_\mu^\Gamma \zeta_1(x)} \cdot \lambda(x)\zeta_2(x) \right) \mathrm{d}s_\nu(x).$$

This formula follows easily from (19.43).

In the following theorem we describe a conserved current associated with equation (19.49):

Theorem 19.51 *Assume (19.51). Let ζ_1, ζ_2 be solutions of (19.49). Then*

$$J^\mu(\zeta_1, \zeta_2, x)$$
$$:= \overline{\zeta_1(x)} \cdot \lambda(x)g^{\mu\nu}(x)\nabla_\nu^\Gamma \zeta_2(x) - \overline{\nabla_\nu^\Gamma \zeta_1(x)} \cdot \lambda(x)g^{\mu\nu}(x)\zeta_2(x)$$

satisfies

$$\nabla_\mu |g|^{\frac{1}{2}}(x) J^\mu(\zeta_1, \zeta_2, x) = 0. \tag{19.53}$$

Before we prove the above theorem let us remark that we can always assume that $\lambda(x)$ does not depend on x. Then we can drop $\lambda(x)$ from our notation altogether and replace the condition (19.51) with

$$\Gamma_\mu \text{ are anti-self-adjoint and } \rho \text{ is self-adjoint.} \tag{19.54}$$

In fact let us first fix a scalar product $(\cdot|\cdot)$ on $\mathcal{V}$ so that $\lambda(x) > 0$ becomes a self-adjoint operator on $\mathcal{V}$. We have

$$\nabla^\Gamma_\mu \zeta = \lambda^{-\frac{1}{2}} \nabla^{\tilde{\Gamma}}_\mu \lambda^{\frac{1}{2}} \zeta, \quad \text{for } \tilde{\Gamma}_\mu = \lambda^{\frac{1}{2}} \Gamma_\mu \lambda^{-\frac{1}{2}} - \left(\nabla_\mu \lambda^{\frac{1}{2}}\right) \lambda^{-\frac{1}{2}}.$$

From (19.51), we obtain that $\tilde{\Gamma}_\mu$ is anti-self-adjoint for $(\cdot|\cdot)$. In fact, we can rewrite (19.51) as

$$(\nabla_\mu \lambda^{\frac{1}{2}}) \lambda^{\frac{1}{2}} + \lambda^{\frac{1}{2}} \nabla_\mu \lambda^{\frac{1}{2}} = \Gamma^*_\mu \lambda + \lambda \Gamma_\mu,$$

which gives $\tilde{\Gamma}^*_\mu = -\tilde{\Gamma}_\mu$. Moreover, $\tilde{\rho} = \lambda^{\frac{1}{2}} \rho \lambda^{-\frac{1}{2}}$ is self-adjoint for $(\cdot|\cdot)$. Considering the function $\tilde{\zeta} = \lambda^{\frac{1}{2}} \zeta$, we obtain (19.54).

Proof of Thm. 19.51. The theorem follows by direct computation. However, it is perhaps instructive to give a proof that reduces it to a special case of Thm. 19.46.

As remarked above, we can assume (19.54). Let us now introduce $\zeta'_\mu := \nabla^\Gamma_\mu \zeta$. The equation (19.49) yields

$$\left(g^{\mu\nu} \nabla_\mu + \frac{1}{2} |g|^{-\frac{1}{2}} (\nabla_\mu |g|^{\frac{1}{2}} g^{\mu\nu})\right) \zeta'_\nu = \rho\zeta + \left(-|g|^{-\frac{1}{2}} \frac{1}{2} (\nabla_\mu |g|^{\frac{1}{2}} g^{\mu\nu}) - g^{\mu\nu} \Gamma_\mu\right) \zeta'_\nu,$$

$$-\left(g^{\mu\nu} \nabla_\nu + |g|^{-\frac{1}{2}} \frac{1}{2} (\nabla_\nu |g|^{\frac{1}{2}} g^{\mu\nu})\right) \zeta = \left(-|g|^{-\frac{1}{2}} \frac{1}{2} (\nabla_\nu |g|^{\frac{1}{2}} g^{\mu\nu}) + g^{\mu\nu} \Gamma_\nu\right) \zeta - g^{\nu,\mu} \zeta'_\nu.$$

This can be rewritten as

$$\alpha^\mu \nabla_\mu \begin{bmatrix} \zeta \\ \zeta' \end{bmatrix} + \frac{1}{2} |g|^{-\frac{1}{2}} (\nabla_\mu |g|^{\frac{1}{2}} \alpha^\mu) \begin{bmatrix} \zeta \\ \zeta' \end{bmatrix} + \theta \begin{bmatrix} \zeta \\ \zeta' \end{bmatrix} = 0, \qquad (19.55)$$

for

$$\theta = \begin{bmatrix} -\rho & |g|^{-\frac{1}{2}} \frac{1}{2} (\nabla_\mu |g|^{\frac{1}{2}} g^{\mu\nu}) + g^{\mu\nu} \Gamma_\mu \\ |g|^{-\frac{1}{2}} \frac{1}{2} (\nabla_\nu |g|^{\frac{1}{2}} g^{\mu\nu}) - g^{\mu\nu} \Gamma_\nu & g^{\mu\nu} \end{bmatrix},$$

$$\alpha^\mu = \begin{bmatrix} 0 & g^{\mu\nu} \\ -g^{\mu\nu} & 0 \end{bmatrix}.$$

Identifying $\mathcal{V}$ with $\mathcal{V}^*$ using the scalar product $(\cdot|\cdot)$, we obtain an equation of the form (19.44) with $\mathcal{V}$ replaced by $\mathcal{V} \oplus \mathcal{V}^d$. Clearly, $\theta^* = \theta$ and $\alpha^{\mu*} = -\alpha^\mu$. Thus (19.55) satisfies the condition (19.46). Hence,

$$J^\mu = \begin{bmatrix} \zeta_1 \\ \zeta'_1 \end{bmatrix} \alpha^\mu \begin{bmatrix} \zeta_2 \\ \zeta'_2 \end{bmatrix} = \overline{\zeta_1} g^{\mu\nu} \zeta'_{2\nu} - \overline{\zeta'_{1\nu}} g^{\mu\nu} \zeta_2$$

is a conserved current. $\square$

19.5.2 Lagrangian of the Klein–Gordon equation

The equation (19.49) can be obtained as an Euler–Lagrange equation, where the Lagrangian is

$$L(\zeta, \partial\zeta) := -\frac{1}{2}(\partial_\mu + \Gamma_\mu)\zeta \cdot g^{\mu\nu}|g|^{\frac{1}{2}}(\partial_\mu + \Gamma_\mu)\zeta - \frac{1}{2}\zeta \cdot |g|^{\frac{1}{2}}\rho\zeta, \text{in the real case,}$$

$$L(\zeta, \overline{\zeta}, \partial\zeta, \partial\overline{\zeta}) := -\overline{(\partial_\mu + \Gamma_\mu)\zeta} \cdot g^{\mu\nu}|g|^{\frac{1}{2}}(\partial_\mu + \Gamma_\mu)\zeta - \overline{\zeta} \cdot |g|^{\frac{1}{2}}\rho\zeta, \text{in the complex case.}$$

(For simplicity, we have assumed that λ does not depend on x and that (19.54) holds.)

19.5.3 Green's functions of hyperbolic Klein–Gordon equations

From here until the end of the section we assume that $\mathcal{X}$ is globally hyperbolic.

The following theorem is a classic result from the theory of hyperbolic equations. Its proof can be found e.g. in Bär–Ginoux–Pfäffle (2007).

Theorem 19.52 *For any $f \in C_c^\infty(\mathcal{X}, \mathcal{V})$, there exist unique functions $\zeta^\pm \in C_{\pm sc}^\infty(\mathcal{X}, \mathcal{V})$ that solve*

$$\Box\zeta^\pm = f.$$

Moreover,

$$\zeta^\pm(x) = (G^\pm f)(x) := \int G^\pm(x, y)f(y)\mathrm{dv}(y),$$

where $G^\pm \in \mathcal{D}'(\mathcal{X} \times \mathcal{X}, L(\mathcal{V}))$ satisfy

$$\Box G^\pm = G^\pm\Box = \mathbb{1}, \quad \mathrm{supp}\, G^\pm \subset \{(x, y) \ : \ x \in J^\pm(y)\}.$$

If in addition (19.51) holds, then $G^{\pm*} = G^{\mp}$.

Note that by duality $G^\pm$ can be applied to distributions of compact support.

Definition 19.53 G^+, *resp.* G^- *is called the* retarded, *resp.* advanced *Green's function.*

$$G := G^+ - G^-$$

is called the Pauli–Jordan *function.*

In what follows, until the end of the section, we assume (19.51). Note that $G^* = -G$ or, in other words,

$$\overline{v_1} \cdot \lambda(x)G(x, y)v_2 = -\overline{G(y, x)v_1} \cdot \lambda(y)v_2, \quad v_1, v_2 \in \mathcal{V}.$$

19.5.4 Cauchy problem

Theorem 19.54 *Let S be a smooth Cauchy hypersurface. Let $\varsigma, \vartheta \in C_c^\infty(S, \mathcal{V})$. Then there exists a unique $\zeta \in C_{sc}^\infty(\mathcal{X}, \mathcal{V})$ that solves*

$$\Box \zeta = 0 \qquad \qquad (19.56)$$

with initial conditions $\zeta\big|_S = \varsigma$, $n^\mu \nabla_\mu^\Gamma \zeta\big|_S = \vartheta$.

It satisfies $\mathrm{supp}\,\zeta \subset J(\mathrm{supp}\,\varsigma \cup \mathrm{supp}\,\vartheta)$ *and is given by*

$$\zeta(x) = -\int_S \left(\nabla_{y^\mu} G(x,y) - G(x,y)\Gamma_\mu(y)\right)\varsigma(y)g^{\mu\nu}(y)\mathrm{d}s_\nu(y)$$

$$+ \int_S G(x,y)\vartheta(y)\mathrm{d}s(y). \qquad (19.57)$$

Proof The existence and uniqueness of solutions are well known. Let us prove (19.57). We apply Green's formula (19.52) to $\zeta_2 = \zeta$ and to $\zeta_1 = G^{\mp}f$, $f \in C_c^\infty(\mathcal{X}, \mathcal{V})$, $\Omega = J^\pm(S)$. We obtain

$$\int_{J^+(S)} \overline{f} \cdot \lambda\zeta\mathrm{d}v = \int_S g^{\mu\nu}\left(\overline{G^- f} \cdot \lambda\nabla_\mu^\Gamma\zeta - \overline{\nabla_\mu^\Gamma G^- f} \cdot \lambda\zeta\right)\mathrm{d}s_\nu, \qquad (19.58)$$

$$\int_{J^-(S)} \overline{f} \cdot \lambda\zeta\mathrm{d}v = \int_S g^{\mu\nu}\left(-\overline{G^+ f} \cdot \lambda\nabla_\mu^\Gamma\zeta + \overline{\nabla_\mu^\Gamma G^+ f} \cdot \lambda\zeta\right)\mathrm{d}s_\nu. \qquad (19.59)$$

Adding (19.58) and (19.59) we get

$$\int_{\mathcal{X}} \overline{f} \cdot \lambda\zeta\mathrm{d}v = \int_S g^{\mu\nu}\left(-\overline{Gf} \cdot \lambda\nabla_\mu^\Gamma\zeta + \overline{\nabla_\mu^\Gamma Gf} \cdot \lambda\zeta\right)\mathrm{d}s_\nu.$$

This can be rewritten as

$$\int_{\mathcal{X}} \overline{f(x)} \cdot \lambda(x)\zeta(x)\mathrm{d}v(x)$$

$$= -\int_{\mathcal{X}} \mathrm{d}v(x) \int_S \overline{G(y,x)f(x)} \cdot \lambda(y)\nabla_\mu^\Gamma\zeta(y)g^{\mu\nu}(y)\mathrm{d}s_\nu(y)$$

$$+ \int_{\mathcal{X}} \mathrm{d}v(x) \int_S \overline{\nabla_{y^\mu}^\Gamma G(y,x)f(x)} \cdot \lambda(y)\zeta(y)g^{\mu\nu}(y)\mathrm{d}s_\nu(y)$$

$$= \int_{\mathcal{X}} \overline{f(x)} \cdot \lambda(x)\mathrm{d}v(x) \int_S G(x,y)\nabla_\mu^\Gamma\zeta(y)g^{\mu\nu}(y)\mathrm{d}s_\nu(y)$$

$$- \int_{\mathcal{X}} \overline{f(x)} \cdot \lambda(x)\mathrm{d}v(x) \int_S \left(\nabla_{y^\mu} G(x,y) - G(x,y)\Gamma_\mu(y)\right)\zeta(y)g^{\mu\nu}(y)\mathrm{d}s_\nu(y),$$

where in the last line we use $G^* = -G$. Thus (19.57) is true. $\qquad \Box$

19.5.5 Symplectic space of solutions of the Klein–Gordon equation

Let $\mathcal{Y} = \mathcal{Y}(\Gamma, \rho)$ denote the set of solutions of (19.56) in $C_{\mathrm{sc}}^{\infty}(\mathcal{X}, \mathcal{V})$.

Theorem 19.55 *Let $\zeta_1, \zeta_2 \in \mathcal{Y}$. Define $J^{\mu}(\zeta_1, \zeta_2, x)$ as in Thm. 19.51. Then*

$$\overline{\zeta_1} \cdot \omega \zeta_2 := \int_S J^{\mu}(\zeta_1, \zeta_2, x) \mathrm{d}s_{\mu}(x) \tag{19.60}$$

does not depend on the choice of a Cauchy hypersurface S and defines a charged symplectic form on $\mathcal{Y}$.

Proof Let S_1, S_2 be two Cauchy hypersurfaces. We can find a third Cauchy hypersurface S_0 that lies in the future of $\mathrm{supp}\, \zeta_i \cap S_j$, $i = 1, 2$, $j = 1, 2$. Applying the Stokes theorem and (19.53) to the domain between S_0 and S_1, we show that the integrals (19.60) on S_0 and S_1 coincide. By the same argument, the integrals (19.60) on S_0 and S_2 coincide. □

Note that in terms of the Cauchy data we have

$$\overline{\zeta_1} \cdot \omega \zeta_2 = \int_S \left(\overline{\vartheta_1} \cdot \lambda \varsigma_2 - \overline{\varsigma_1} \cdot \lambda \vartheta_2 \right) \mathrm{d}s.$$

19.5.6 Solutions parametrized by test functions

Theorem 19.56 *(1) For any $f \in C_{\mathrm{c}}^{\infty}(\mathcal{X}, \mathcal{V})$, $Gf \in \mathcal{Y}$.*
(2) Every element of $\mathcal{Y}$ is of this form.
(3) $\overline{Gf_1} \cdot \omega Gf_2 = \int \overline{f_1(x)} \cdot \lambda(x) G(x, y) f_2(y) \mathrm{d}v(x) \mathrm{d}v(y)$.

Proof Gf is a solution of (19.56), since G is a solution of (19.56) in its first variable. The fact that Gf is space-compact follows from the support properties of $G^{\pm}$. Hence, (1) is true.

Now let $\zeta \in \mathcal{Y}$. Since ζ is space-compact, we can find cutoff functions $\chi^{\pm} \in C_{\pm\mathrm{sc}}^{\infty}(\mathcal{X})$ such that $\chi^+ + \chi^- = 1$ on $\mathrm{supp}\, \zeta$. Moreover, it follows from Condition (1a) of Thm. 19.40 that $\mathrm{supp}\, \nabla \chi^{\pm} \cap \mathrm{supp}\, \zeta$ is compact. Setting

$$\zeta^{\pm} := \chi^{\pm} \zeta, \quad f := \Box \zeta^+ = -\Box \zeta^-,$$

we see that $f \in C_{\mathrm{c}}^{\infty}(\mathcal{X})$. Hence, by Thm. 19.52, $\zeta^{\pm} = \pm G^{\pm} f$, and $\zeta = Gf$. This proves (2).

Let $f_1, f_2 \in C_{\mathrm{c}}^{\infty}(\mathcal{X}, \mathcal{V})$. In a sufficiently far future we have $Gf_i = G^+ f_i$, $i = 1, 2$. Hence, for a Cauchy surface S in a far future we have

$$\overline{Gf_1} \omega Gf_2 = \int_S g^{\mu\nu} \left(\overline{\nabla_{\mu}^{\Gamma} G^+ f_1} \cdot \lambda G^+ f_2 - \overline{G^+ f_1} \cdot \lambda \nabla_{\mu}^{\Gamma} G^+ f_2 \right) \mathrm{d}s_{\nu}$$

$$= \int_{J^-(S)} \left(\overline{\Box G^+ f_1} \cdot \lambda G^+ f_2 - \overline{G^+ f_1} \cdot \lambda \Box G^+ f_2 \right) \mathrm{d}v$$

$$= \int_{\mathcal{X}} \left(\overline{f_1} \cdot \lambda G^+ f_2 - \overline{G^+ f_1} \cdot \lambda f_2 \right) \mathrm{d}v$$

$$= \int_{\mathcal{X}} \left(\overline{f_1} \cdot \lambda G^+ f_2 - \overline{f_1} \cdot \lambda G^- f_2 \right) \mathrm{d}v.$$

In the first line we use the definition of ω, in the second Green's formula, in the third the fact that f_i are compactly supported, and in the last the fact that $G^{+*} = G^-$. $\qquad\square$

Let $\mathcal{O}$ be an open subset of $\mathcal{X}$. We define

$$\mathcal{Y}(\mathcal{O}) := \{Gf \ : \ f \in C_c^\infty(\mathcal{O}, \mathcal{V})\}.$$

Theorem 19.57 (1) $\mathcal{Y}(\mathcal{X}) = \mathcal{Y}$.
(2) $\mathcal{O}_1 \subset \mathcal{O}_2$ *implies* $\mathcal{Y}(\mathcal{O}_1) \subset \mathcal{Y}(\mathcal{O}_2)$.
(3) *If* $\mathcal{O}_1$ *and* $\mathcal{O}_2$ *are causally separated and* $\zeta_i \in \mathcal{Y}(\mathcal{O}_i)$, $i = 1, 2$, *then*

$$\overline{\zeta}_1 {\cdot} \omega \zeta_2 = 0.$$

(4) *If* $\mathcal{O}_1$ *is causally dependent on* $\mathcal{O}_2$, *then* $\mathcal{Y}(\mathcal{O}_1) \subset \mathcal{Y}(\mathcal{O}_2)$.
(5) ω *is non-degenerate on* $\mathcal{Y}$.

Proof (1) follows from Thm. 19.56 (2). (2) is obvious. (3) follows from the definition of ω and the support properties of $G^\pm$ in Thm. 19.52. To prove (4), it suffices to show that if $\zeta = Gf$ for $f \in C_c^\infty(\mathcal{O}_1)$, then $\zeta = Gg$ for $g \in C_c^\infty(\mathcal{O}_2)$. Using a partition of unity of $\mathcal{O}_1$, we can assume that there exists a Cauchy hypersurface S such that any causal curve starting from $\mathcal{O}_1$ intersects S in $\mathcal{O}_2$. If $\zeta = Gf$ for $f \in C_c^\infty(\mathcal{O}_1)$, we set $\varsigma = \zeta|_S$, $\vartheta = n^\mu \nabla_\mu^\Gamma \zeta|_S$. By Thm. 19.52, the Cauchy data (ϑ, ς) are supported in a compact set $N \subset S$, and by Thm. 19.54 we obtain that $\operatorname{supp} \zeta \subset J^+(N) \cup J^-(N) \subset \mathcal{O}_2^+ \cup \mathcal{O}_2^-$, where $\mathcal{O}_2^\pm = \mathcal{O}_2 \cup S^\pm$ and $S^\pm$ are the future, resp. past of S. Hence, we can find cutoff functions $\chi^\pm$ supported in $\mathcal{O}_2^\pm$ such that $\zeta = \chi^+ \zeta + \chi^- \zeta =: \zeta^+ + \zeta^-$. Setting

$$g := \Box \zeta^+ = -\Box \zeta^-,$$

we obtain that $\operatorname{supp} g \subset \mathcal{O}_2^+ \cap \mathcal{O}_2^- = \mathcal{O}_2$. Moreover, since $\zeta^\pm$ is past, resp. future space-compact, we see by Thm. 19.52 that $\zeta^\pm = \pm G^\pm g$. Hence, $\zeta = Gg$. This completes the proof of the theorem. $\qquad\square$

19.5.7 Algebraic quantization

Let $\mathfrak{A} := \mathrm{CCR}^{\mathrm{Weyl}}(\mathcal{Y})$. More generally, if $\mathcal{O}$ is an open bounded subset in $\mathcal{X}$, let $\mathfrak{A}(\mathcal{O})$ be the sub-algebra of $\mathfrak{A}$ generated by $W(Gf + \overline{Gf})$, where $f \in C_c^\infty(\mathcal{O})$. In other words, $\mathfrak{A}(\mathcal{O}) = \mathrm{CCR}^{\mathrm{Weyl}}(\mathcal{Y}(\mathcal{O}))$.

The family of algebras $\mathfrak{A}(\mathcal{O})$ satisfies the following properties, which express the Einstein causality:

Theorem 19.58 (1) $\mathfrak{A}(\mathcal{X}) = \mathfrak{A}$.
(2) $\mathcal{O}_1 \subset \mathcal{O}_2$ *implies* $\mathfrak{A}(\mathcal{O}_1) \subset \mathfrak{A}(\mathcal{O}_2)$.

(3) *If $\mathcal{O}_1$ and $\mathcal{O}_2$ are space-like separated and $B_i \in \mathfrak{A}(\mathcal{O}_i)$, $i = 1, 2$, then*

$$B_1 B_2 = B_2 B_1.$$

(4) *If $\mathcal{O}_1$ is causally dependent on $\mathcal{O}_2$, then $\mathfrak{A}(\mathcal{O}_1) \subset \mathfrak{A}(\mathcal{O}_2)$.*

19.6 Generalized Dirac equation on curved space-time

The setting of this section is similar to that of Sect. 19.5. Again, we assume that $\mathcal{X}$ is a pseudo-Riemannian manifold. Starting with Subsect. 19.6.3, we assume in addition that $\mathcal{X}$ is globally hyperbolic.

Similarly, as in the previous section we consider also a finite-dimensional space $\mathcal{V}$ and the vector bundle $\mathcal{X} \times \mathcal{V} \to \mathcal{X}$.

The goal of this section is to describe the algebraic quantization of a large class of first-order equations on a curved space-time. We will always assume that the principal term of this equation is of the form $\gamma^\mu(x)\partial_\mu$, where $\gamma^\mu(x)$ satisfy the Clifford relations given by the metric tensor $[g_{\mu\nu}(x)]$. Such an equation will be called a *generalized Dirac equation*. In the case of Lorentzian manifolds, solutions of this equation have causal propagation.

We will see that generalized Dirac equations possess a conserved current. In the case of globally hyperbolic manifolds, this current defines a scalar product on the space of solutions. Therefore, it is natural to quantize this equation using the CAR. Its algebraic quantization leads to a net of algebras satisfying the fermionic version of the Einstein causality.

This section is to a large extent a generalization of Sect. 19.3 to a curved space-time. Unlike in Sect. 19.3, we limit our discussion to the algebraic quantization. We do not discuss the positive energy quantization on a fermionic Fock space, which is possible for the Dirac equation on a stationary space-time.

19.6.1 Dirac operators

Recall that $\mathcal{X}$ is a pseudo-Riemannian manifold. $\mathcal{V}$ can be a real or complex space. For simplicity, we will consider only the complex case.

Definition 19.59 *A map $\mathcal{X} \ni x \mapsto \gamma^\mu(x) \in L(\mathcal{V})$ satisfying*

$$[\gamma^\mu(x), \gamma^\nu(x)]_+ = 2g^{\mu\nu}(x) \tag{19.61}$$

is called a spinor structure *on $\mathcal{X}$.*

Definition 19.60 *Let $\mathcal{X} \ni x \mapsto \theta(x) \in L(\mathcal{V})$. The operator on $C^\infty(\mathcal{X}, \mathcal{V})$*

$$\mathbb{D} = \mathbb{D}(\theta) := \gamma^\mu(x)\nabla_\mu + \theta(x) \tag{19.62}$$

is called a generalized Dirac operator.

We will study the equation

$$\mathbb{D}\zeta = 0. \tag{19.63}$$

Let us fix a smooth map $\mathcal{X} \ni x \mapsto \lambda(x) \in L(\mathcal{V}, \mathcal{V}^*)$ such that $\lambda(x)$ is non-degenerate for all $x \in \mathcal{X}$. We will often assume that

$$\lambda(x)\gamma^{\mu}(x) \quad \text{is self-adjoint,} \tag{19.64}$$

$$\lambda(x)\theta(x) - \frac{1}{2}|g|^{-\frac{1}{2}}(x)\nabla_{\mu}|g|^{\frac{1}{2}}(x)\lambda(x)\gamma^{\mu}(x) \quad \text{is anti-self-adjoint.} \tag{19.65}$$

Using Thm. 19.46 and the Stokes theorem, we obtain the following version of Green's formula:

$$\int_{\Omega} \left(\overline{\zeta_1(x)} \cdot \lambda(x) \mathbb{D}\zeta_2(x) - \overline{\mathbb{D}\zeta_1(x)} \cdot \lambda(x)^* \zeta_2(x) \right) dv(x)$$

$$= \int_{\partial\Omega} \overline{\zeta_1(x)} \cdot \lambda(x) \gamma^{\mu}(x) \zeta_2(x) ds_{\mu}(x). \tag{19.66}$$

For $\zeta_1, \zeta_2 \in C^{\infty}(\mathcal{X}, \mathcal{V})$, we define

$$J^{\mu}(\zeta_1, \zeta_2, x) = \overline{\zeta_1(x)} \cdot \lambda(x) \gamma^{\mu}(x) \zeta_2(x). \tag{19.67}$$

If ζ_1, ζ_2 are solutions of the Dirac equation, by Thm. 19.46, we have

$$\nabla_{\mu}|g|^{\frac{1}{2}}(x) J^{\mu}(\zeta_1, \zeta_2, x) = 0. \tag{19.68}$$

Note also that if we have another Dirac operator $\mathbb{D}_1 := \gamma^{\mu}(x)\nabla_{\mu} + \theta_1(x)$, then

$$\square := \mathbb{D}\mathbb{D}_1$$

is a second-order operator of the form considered in Subsect. 19.5.1.

19.6.2 Lagrangian of the Dirac equation

(19.63) can be obtained as the Euler–Lagrange equation for the following Lagrangians:

$$L_1(\zeta, \overline{\zeta}, \partial\zeta, \partial\overline{\zeta}) := -|g|^{\frac{1}{2}}\overline{\zeta} \cdot \lambda(\gamma^{\mu}\partial_{\mu} + \theta)\zeta, \quad \text{or}$$

$$L(\zeta, \overline{\zeta}, \partial\zeta, \partial\overline{\zeta}) := -\frac{1}{2}|g|^{\frac{1}{2}} \left(\overline{\zeta} \cdot \lambda\gamma^{\mu}\partial_{\mu}\zeta + \overline{\partial_{\mu}\zeta} \cdot \lambda\gamma^{\mu}\zeta \right)$$

$$-\overline{\zeta} \cdot \left(|g|^{\frac{1}{2}}\lambda\theta - \frac{1}{2}(\partial_{\mu}|g|^{\frac{1}{2}}\lambda\gamma^{\mu}) \right)\zeta.$$

19.6.3 Green's functions of hyperbolic Dirac equations

Until the end of the section we assume that $\mathcal{X}$ is globally hyperbolic and $\mathbb{D}$ is a generalized Dirac operator on $\mathcal{X}$.

Theorem 19.61 *For any* $f \in C_c^\infty(\mathcal{X}, \mathcal{V})$, *there exist unique functions* $\zeta^\pm \in C_{\pm\mathrm{sc}}^\infty(\mathcal{X}, \mathcal{V})$ *that solve*

$$\mathbb{D}\zeta^\pm = f. \tag{19.69}$$

Moreover

$$\zeta^\pm(x) = (S^\pm f)(x) := \int S^\pm(x,y) f(y) \mathrm{d}v(y), \tag{19.70}$$

where $S^\pm \in \mathcal{D}'(\mathcal{X} \times \mathcal{X}, L(\mathcal{V}))$ *satisfy*

$$\mathbb{D}S^\pm = S^\pm \mathbb{D} = \mathbb{1}, \quad \mathrm{supp}\, S^\pm \subset \{(x,y) \,:\, x \in J^\pm(y)\}. \tag{19.71}$$

If in addition (19.64) and (19.65) hold, then

$$\lambda(x)^* S^+(x,y) = -S^-(x,y)^* \lambda(y). \tag{19.72}$$

Proof By the remark in Subsect. 19.6.1, $\mathbb{D}^2$ is a generalized Klein–Gordon operator. Let $G^\pm$ be the retarded, resp. advanced Green's functions of $\mathbb{D}^2$. Clearly, $\zeta^\pm = \mathbb{D}G^\pm f$ are solutions of (19.69). To prove the uniqueness, we note that if $\mathbb{D}\zeta^\pm = 0$ and $\mathrm{supp}\,\zeta^\pm \subset J^\pm(K)$ for some compact K, then $\mathbb{D}^2\zeta^\pm = 0$. Hence, $\zeta^\pm = 0$ by Thm. 19.52. We set then $S^\pm = \mathbb{D}G^\pm$. This proves (19.70) and (19.71). Let us prove (19.72). We need to show that

$$\int \overline{f_2} \cdot \lambda^* S^+ f_1 \mathrm{d}v = -\int \overline{f_2} \cdot S^{+*} \lambda f_1 \mathrm{d}v. \tag{19.73}$$

It is enough to set $f_i = \mathbb{D}\zeta_i$ for $\zeta_i = S^+ f_i$. Now, $\lambda \mathbb{D}$ is anti-Hermitian for the scalar product (19.40), hence

$$\int \overline{\mathbb{D}\zeta_2} \cdot \lambda^* S^+ \mathbb{D}\zeta_1 \mathrm{d}v = \int \overline{\lambda \mathbb{D}\zeta_2} \cdot S^+ \mathbb{D}\zeta_1 \mathrm{d}v$$

$$= -\int \overline{\zeta_2} \cdot \lambda \mathbb{D} S^+ \mathbb{D}\zeta_1 \mathrm{d}v = -\int \overline{\zeta_2} \cdot \lambda \mathbb{D}\zeta_1 \mathrm{d}v$$

$$= -\int \overline{S^+ \mathbb{D}\zeta_2} \cdot \lambda \mathbb{D}\zeta_1 \mathrm{d}v = -\int \overline{\mathbb{D}\zeta_2} \cdot S^{+*} \lambda \mathbb{D}\zeta_1 \mathrm{d}v. \qquad \square$$

Note that by duality $S^\pm$ can be applied to distributions of compact support.

Definition 19.62 S^+, *resp.* S^- *is called the* retarded, *resp.* advanced Green's function. *We also set*

$$S := S^+ - S^-.$$

Note that

$$\lambda(x)^* S(x,y) = S(x,y)^* \lambda(x). \tag{19.74}$$

19.6.4 Cauchy problem

Until the end of the section we assume that $\mathcal{X} \ni x \mapsto \lambda(x)$ has been chosen so that (19.64) and (19.65) hold. We also assume that $\lambda(x)$ is non-degenerate for any $x \in \mathcal{X}$.

Theorem 19.63 *Let S be a smooth Cauchy surface. Let $\vartheta \in C_c^\infty(S, V)$. Then there exists a unique $\zeta \in C_{sc}^\infty(\mathcal{X}, V)$ that solves*

$$\mathbb{D}\zeta = 0 \tag{19.75}$$

with initial conditions $\zeta\big|_S = \vartheta$. It satisfies $\operatorname{supp} \zeta \subset J(\operatorname{supp} \vartheta)$ and is given by

$$\zeta(x) = -\int_S S(x, y)\gamma^\mu(y)\vartheta(y)\mathrm{d}s_\mu(y). \tag{19.76}$$

Proof The existence and uniqueness is well known. Let us prove (19.76).

We apply Green's Formula (19.66) to $\zeta_2 = \zeta$ and $\zeta_1 = S^\mp f$, $f \in C_c^\infty(\mathcal{X})$, $\Omega = J^\pm(S)$, obtaining

$$\int_{J^+(S)} \overline{f}\cdot\lambda^*\zeta\mathrm{d}v = \int_S \overline{S^- f}\cdot\lambda\gamma^\mu\zeta\mathrm{d}s_\mu, \tag{19.77}$$

$$\int_{J^-(S)} \overline{f}\cdot\lambda^*\zeta\mathrm{d}v = -\int_S \overline{S^+ f}\cdot\lambda\gamma^\mu\zeta\mathrm{d}s_\mu. \tag{19.78}$$

Adding (19.77) and (19.78), we get

$$\int_\mathcal{X} \overline{f}\cdot\lambda^*\zeta\mathrm{d}v = -\int_S \overline{Sf}\cdot\lambda\gamma^\mu\zeta\mathrm{d}s_\mu.$$

This can be rewritten as

$$\int_\mathcal{X} \overline{f(x)}\cdot\lambda(x)^*\zeta(x)\mathrm{d}v(x) = -\int_\mathcal{X} \mathrm{d}v(x)\int_S \overline{S(y,x)f(x)}\cdot\lambda(y)\gamma^\mu\zeta(y)\mathrm{d}s_\mu(y)$$

$$= -\int_\mathcal{X} \overline{f(x)}\cdot\lambda(x)^*\mathrm{d}v(x)\int_S S(x,y)\gamma^\mu(y)\zeta(y)\mathrm{d}s_\mu(y),$$

where in the last line we use (19.74). $\qquad\square$

19.6.5 Unitary space of solutions of the Dirac equation

Let $\mathcal{Y}$ denote the set of solutions of the Dirac equation in $C_{sc}^\infty(\mathcal{X}, V)$. To equip $\mathcal{Y}$ with a scalar product an additional positivity condition is required. We assume that for all $x \in \mathcal{X}$

$$\lambda(x)\gamma^\mu(x)v_\mu > 0, \text{ if } v \in T_x\mathcal{X} \text{ is time-like and future directed.} \tag{19.79}$$

By Lemma 19.9, it suffices to assume that there exists a time-like future directed vector field v such that

$$\lambda(x)\gamma^\mu(x)v_\mu(x) > 0, \quad x \in \mathcal{X}.$$

Theorem 19.64 *Let $\zeta_1, \zeta_2 \in \mathcal{Y}$. Define $J^\mu(\zeta_1, \zeta_2, x)$ as in (19.67). Then*

$$\overline{\zeta}_1 \cdot \zeta_2 := \int_S J^\mu(\zeta_1, \zeta_2, x)\mathrm{d}s_\mu(x)$$

does not depend on the choice of a Cauchy hypersurface S and defines a positive definite Hermitian form on $\mathcal{Y}$.

Proof To show that $\overline{\zeta}_1 \cdot \zeta_2$ is independent of S we apply the Stokes theorem as in Thm. 19.55, using (19.68). The fact that it is positive definite follows from the positivity condition (19.79). $\qquad\square$

19.6.6 Solutions parametrized by test functions

Theorem 19.65 (1) *For any $f \in C_c^\infty(\mathcal{X}, \mathcal{V})$, $Sf \in \mathcal{Y}$.*
(2) *Every element of $\mathcal{Y}$ is of this form.*
(3) $\overline{Sf_1} \cdot Sf_2 = \int \overline{f_1(x)} \cdot \lambda(x)^* S(x, y) f_2(y) \mathrm{d}v(x) \mathrm{d}v(y).$

Proof (1) follows from the fact that S solves (19.75) in its first coordinate.
(2). Let $\zeta \in \mathcal{Y}$. We can write $\zeta = \zeta^+ + \zeta^-$, where $\zeta^+ \in C_{\pm\mathrm{sc}}^\infty$. Set

$$f := \mathbb{D}\zeta^+ = -\mathbb{D}\zeta^-.$$

Then $f \in C_c^\infty(\mathcal{X}, \mathcal{V})$, $\zeta^\pm = \pm S^\pm f$, and hence $\zeta = Sf$.
(3). Let $f_1, f_2 \in C_c^\infty(\mathcal{X}, \mathcal{V})$. In a sufficiently far future we have $Sf_i = S^+ f_i$, $i = 1, 2$. Hence, for a late Cauchy surface $\mathcal{S}$,

$$\overline{Sf_1} \cdot Sf_2 = \int_{\mathcal{S}} \overline{S^+ f_1} \cdot \lambda \gamma^\mu S^+ f_2 \mathrm{d}s_\mu$$

$$= \int_{J^-(\mathcal{S})} \left(\overline{\mathbb{D}S^+ f_1} \cdot \lambda^* S^+ f_2 - \overline{S^+ f_1} \cdot \lambda \mathbb{D}S^+ f_2 \right) \mathrm{d}v$$

$$= \int_{\mathcal{X}} \left(\overline{f_1} \cdot \lambda^* S^+ f_2 - \overline{S^+ f_1} \cdot \lambda f_2 \right) \mathrm{d}v$$

$$= \int_{\mathcal{X}} \left(\overline{f_1} \cdot \lambda^* S^+ f_2 - \overline{f_1} \cdot \lambda^* S^- f_2 \right) \mathrm{d}v. \qquad\square$$

Let $\mathcal{O}$ be an open subset of $\mathcal{X}$. We define

$$\mathcal{Y}(\mathcal{O}) := \{ Sf \ : \ f \in C_c^\infty(\mathcal{O}, \mathcal{V}) \}.$$

Theorem 19.66 (1) $\mathcal{Y}(\mathcal{X}) = \mathcal{Y}$.
(2) $\mathcal{O}_1 \subset \mathcal{O}_2$ *implies* $\mathcal{Y}(\mathcal{O}_1) \subset \mathcal{Y}(\mathcal{O}_2)$.
(3) *If $\mathcal{O}_1$ and $\mathcal{O}_2$ are space-like separated and $\zeta_i \in \mathcal{Y}(\mathcal{O}_i)$, $i = 1, 2$, then*

$$\overline{\zeta}_1 \cdot \zeta_2 = 0.$$

(4) *If $\mathcal{O}_1$ is causally dependent on $\mathcal{O}_2$, then $\mathcal{Y}(\mathcal{O}_1) \subset \mathcal{Y}(\mathcal{O}_2)$.*

19.6.7 Algebraic quantization

Set $\mathfrak{A} := \mathrm{CAR}^{C^*}(\mathcal{Y})$. Note that it is a graded algebra. More generally, if $\mathcal{O}$ is an open bounded subset in $\mathcal{X}$, let $\mathfrak{A}(\mathcal{O})$ be the C^*-sub-algebra of $\mathfrak{A}$ generated by $\psi(Sf)$, where $f \in C_c^\infty(\mathcal{O})$. In other words, $\mathfrak{A}(\mathcal{O}) = \mathrm{CAR}^{C^*}(\mathcal{Y}(\mathcal{O}))$. The family of algebras $\mathfrak{A}(\mathcal{O})$ satisfies the following properties:

Theorem 19.67 (1) $\mathfrak{A}(\mathcal{X}) = \mathfrak{A}$.

(2) $\mathcal{O}_1 \subset \mathcal{O}_2$ *implies* $\mathfrak{A}(\mathcal{O}_1) \subset \mathfrak{A}(\mathcal{O}_2)$.

(3) *If* $\mathcal{O}_1$ *and* $\mathcal{O}_2$ *are causally separated, and* $B_i \in \mathfrak{A}(\mathcal{O}_i)$ *are elements of parity* $|B_i|$, *then*

$$B_1 B_2 = (-1)^{|B_1||B_2|} B_2 B_1.$$

(4) *If* $\mathcal{O}_1$ *is causally dependent on* $\mathcal{O}_2$, *then* $\mathfrak{A}(\mathcal{O}_1) \subset \mathfrak{A}(\mathcal{O}_2)$.

19.7 Notes

The material of the first three sections is discussed in essentially all textbooks on quantum field theory, such as Jauch–Röhrlich (1976), Schweber (1962), Weinberg (1995) and Srednicki (2007).

Mathematical aspects of quantum field theory on curved space-time were studied by Dimock (1980, 1982). A review of this subject can be found in monographs by Wald (1994) and Fulling (1989).

A short and readable monograph on wave equations on Lorentzian manifolds is Bär–Ginoux–Pfäffle (2007).

20

Diagrammatics

The diagrammatic method is one of the most powerful tools of theoretical physics. It allows us to efficiently organize perturbative computations in statistical physics, quantum many-body theory and quantum field theory. The main feature of this method is a representation of individual terms of a perturbative expansion as *diagrams (graphs)*. Diagrams consist of *vertices* representing terms in the perturbation, *lines* representing pairings between vertices and, possibly, *external legs*.

There exist several kinds of diagrams. We will try to present them in a systematic way.

In Sect. 20.1 we present a diagrammatic formalism whose goal is to organize integration of polynomials with respect to a Gaussian measure. This formalism is used extensively in classical statistical physics. It also plays an important role in quantum physics, especially in the Euclidean approach, since many quantum quantities can be expressed in terms of Gaussian integrals over classical variables.

We use the term "Gaussian integration" in a rather broad sense. Beside commuting "bosonic" variables, we also consider anti-commuting "fermionic" variables, where we use the Berezin integral with respect to a Gaussian weight. Even in the case of commuting variables, the "Gaussian integral" is not necessarily meant in the sense of measure theory. It denotes an algebraic operation performed on polynomials (or formal power series), which in the case of a positive definite covariance coincides with the usual integral with a Gaussian weight. However, we allow the covariance to be complex, or even negative definite, and do not insist that the operation have a measure theoretic meaning.

We distinguish two kinds of spaces on which we perform the integrals: real and complex. As in many other places in our work, we treat these two cases in parallel. Of course, the difference between the real (i.e. neutral), and the complex (i.e. charged) formalism is mainly that of a different notation. In particular, charged lines need to be equipped with an arrow, whereas neutral lines need not.

The terminology that we use is inspired by quantum field theory. Therefore, the variables that enter the integral are associated with "particles"; they are divided into "bosons" and "fermions", each subdivided into "neutral" and "charged" particles.

In the main part of the chapter we describe the diagram formalism used in quantum many-body physics and quantum field theory. As a preparation, we include a brief Sect. 20.2 devoted to the basic terminology of perturbation theory for quantum dynamics. We focus on the concept of the *scattering operator* and of the *energy shift of the ground state*.

We discuss first the situation of a time-dependent perturbation of a fixed free Hamiltonian. In this case, the usual scattering operator is guaranteed to exist, e.g. if the perturbation decays in time sufficiently fast.

If the perturbation is time-independent, one can still try to use the usual definition of the scattering operator. It is well known that this definition works well in quantum mechanics, where the free Hamiltonian is the Laplacian and the perturbation is a short-range potential. However, in quantum field theory the standard definition of the scattering operator is usually inapplicable, even on the level of formal expressions. This is related to the fact that the interacting Hamiltonian has a different ground state than the free Hamiltonian.

There exists a different formalism for scattering theory, which has more applicability and in some situations can be used in quantum field theory. The main idea of this formalism is the so-called *adiabatic switching of the interaction*. More precisely, we multiply the interaction with a time-dependent coupling constant $e^{-\epsilon|t|}$ and introduce the scattering operator depending on the parameter ϵ. Then we take the limit of the scattering operator as $\epsilon \searrow 0$, dividing it by its expectation value with respect to a distinguished vector (typically, the non-interacting vacuum). This procedure is associated with the names of Gell-Mann and Low, and is usually (more or less implicitly) taken as the basic definition of the scattering operator in quantum field theory.

This procedure works, at least on the perturbative level, for sufficiently regular perturbations localized in space. If we assume that the perturbation is translation invariant, which is the usual assumption in quantum field theory, the situation becomes more complicated. In particular, one needs to perform the so-called wave function renormalization. We will not discuss this topic.

Starting with Sect. 20.3, we describe diagrams used in many-body quantum theory and quantum field theory. Our main aim is the computation of the scattering operator and the energy shift of the ground state.

It seems natural to divide diagrams into two categories. The first are the so-called *Friedrichs diagrams* and the second *Feynman diagrams*.

Friedrichs diagrams appear naturally when we want to compute the Wick symbol of a product of Wick-ordered operators. An algorithm for its computation is usually called the *Wick theorem*. It can be given a graphical interpretation, which we describe in Sect. 20.3.

In this formalism, a vertex represents a Wick monomial. It has two kinds of legs, those representing annihilation operators and those representing creation operators. We draw the former on the right of a vertex and the latter on the left.

A typical Hamiltonian in many-body quantum physics and in quantum field theory can be written as the sum of a quadratic term of the form $d\Gamma(h)$ for some one-particle Hamiltonian h and an interaction given by a Wick polynomial. One can use Friedrichs diagrams to compute the scattering operator for such

Hamiltonians, as we describe in Sect. 20.4. A characteristic feature of this formalism is the presence of time labels on all vertices and the fact that diagrams with different time orderings are considered distinct.

Naively, this formalism seems very natural and physically intuitive. In fact, Friedrichs diagrams would provide natural illustrations for typical computations of the early years of quantum field theory (even though diagrams were apparently not used in that time). Weinberg calls a formalism essentially equivalent to that of Friedrichs diagrams the *old-fashioned perturbation theory*.

Since the late 1940s, a different diagram formalism has been developed. Since then it has dominated the calculations of quantum field theory. It originated especially in the work of Feynman, and therefore is called the formalism of *Feynman diagrams*. Again, the main goal is to compute the scattering operator for a Hamiltonian of the form $d\Gamma(h)$ perturbed by a quantization of a Wick polynomial. It is convenient to express this perturbation using the neutral or charged formalism.

In Sect. 20.5. we describe Feynman diagrams used to compute the vacuum expectation value of the scattering operator. They can be essentially interpreted as a special case of the diagrams described in Sect. 20.1 used to compute Gaussian integrals. In this formalism, the order of times associated with individual vertices does not play any role. This allows us to cut down on the number of diagrams, as compared with Friedrichs diagrams. In the case of relativistic theories, each Feynman diagram is manifestly covariant, which is not the case for Friedrichs diagrams. Therefore, Feynman diagrams are usually preferred for practical computations over Friedrichs diagrams.

Feynman diagrams used to compute the Wick symbol of the scattering operator have in addition external legs. These legs are either incoming or outgoing. The former are then paired with creation operators and the latter with annihilation operators. Again, the temporal order of vertices is not relevant.

The main goal of this chapter is a formal description of the diagrammatic method. We will disregard the problems of convergence. We will often treat vector spaces as if they were finite-dimensional, even if in applications they are usually infinite-dimensional.

We try to describe the graphical method using a rigorous formalized language. This is perhaps not always the most natural thing to do. One can argue that an informal account involving a more colloquial language is more convenient in this context. Nevertheless, some readers may appreciate a formalized description.

20.1 Diagrams and Gaussian Integration

In this section we present a diagrammatic formalism used to describe the integration and the Wick transformation w.r.t. a Gaussian measure.

We start with a description of purely graphical and combinatorial elements of the formalism. We will introduce the analytic part later.

We will freely use basic terminology of set theory. In particular, we will always include the possibility that a set can be empty. $\#X$ will denote the number of elements of the set X. Recall also the following definition:

Definition 20.1 *Let* $\{A_j\}_{j\in J}$ *be a family of sets indexed by a set* J. *The* disjoint union *of the sets* A_j, $j \in J$, *is defined as*

$$\bigsqcup_{j\in J} A_j := \big\{(j,a) \ : \ j \in J, \ a \in A_j\big\}.$$

Let us stress that the notion of a disjoint union of sets does not coincide with the notion of the union of disjoint sets.

At some places in this section it will be convenient to totally order sets that we consider. In the case of bosonic particles, the end result does not depend on this ordering. For fermions, however, some quantities may depend on the order, but only modulo even permutations. In order to express this dependence, we introduce the following definition:

Definition 20.2 *Let* A *be a set of* n *elements. Let* p, q *be bijections* $\{1,\ldots,n\} \to A$. *We say that they are equivalent if* $q^{-1} \circ p$ *is an even permutation. There are precisely two equivalence classes of this relation.*

We say that the set A *is* oriented *if one of these equivalence classes is chosen. This equivalence class is then called the* orientation *of* A. *We say that a total order of* A *is* admissible *if it is given by an element of the orientation of* A.

Let $\{A_i\}_{i\in I}$ be a finite family of oriented disjoint sets, each with an even number of elements. Then $\bigcup_{i\in I} A_i$ has a natural orientation.

20.1.1 Vertices

Definition 20.3 *Let* Pr *be a set. Its elements are called* (species of) particles. Pr *is subdivided into disjoint sets* Pr_s *and* Pr_a, *whose elements are called* (species of) bosons, *resp.* fermions. *We assume that the set* Pr_a *is oriented. For* $p \in \mathrm{Pr}_s$ *we set* $\epsilon_p = 1$, *and for* $p \in \mathrm{Pr}_a$ *we set* $\epsilon_p = -1$.

Definition 20.4 Pr_s, *resp.* Pr_a *are subdivided into disjoint sets*

$$\mathrm{Pr}_s = \mathrm{Pr}_s^n \cup \mathrm{Pr}_s^c, \quad \textit{resp.} \quad \mathrm{Pr}_a = \mathrm{Pr}_a^n \cup \mathrm{Pr}_a^c.$$

We set

$$\mathrm{Pr}^n := \mathrm{Pr}_s^n \cup \mathrm{Pr}_a^n, \quad \mathrm{Pr}^c := \mathrm{Pr}_s^c \cup \mathrm{Pr}_a^c.$$

Elements of Pr^n, *resp.* Pr^c *are called* (species of) neutral, *resp.* charged *particles.*

Definition 20.5 *A* multi-degree *is a triple of functions*

$$\mathrm{Pr}^n \ni p \mapsto m_p \in \{0, 1, 2, \dots\},$$
$$\mathrm{Pr}^c \ni q \mapsto m_q^{(+)} \in \{0, 1, 2, \dots\},$$
$$\mathrm{Pr}^c \ni q \mapsto m_q^{(-)} \in \{0, 1, 2, \dots\}.$$

For brevity a multi-degree is typically denoted by a single letter, e.g. m.

We say that m is fermion-even *if*

$$\sum_{p \in \mathrm{Pr}^n_a} m_p + \sum_{q \in \mathrm{Pr}^c_a} \left(m_q^{(+)} + m_q^{(-)}\right) \quad is\ even. \tag{20.1}$$

Definition 20.6 *A* vertex, *denoted F, is a finite set* $\mathrm{Lg}(F)$ *equipped with a map*

$$\mathrm{Lg}(F) \ni l \mapsto \mathrm{pr}(l) \in \mathrm{Pr} \tag{20.2}$$

and a partition into three disjoint subsets

$$\mathrm{Lg}^n(F),\ \mathrm{Lg}^{(+)}(F),\ \mathrm{Lg}^{(-)}(F)$$

such that the image of $\mathrm{Lg}^n(F)$ *under (20.2) is contained in* Pr^n *and the images of* $\mathrm{Lg}^{(+)}(F)$ *and* $\mathrm{Lg}^{(-)}(F)$ *are contained in* Pr^c.

Elements of $\mathrm{Lg}^n(F)$ *are called* neutral legs *of the vertex F. Elements of* $\mathrm{Lg}^{(+)}(F)$, *resp.* $\mathrm{Lg}^{(-)}(F)$ *are called* charge creating, *resp.* charge annihilating legs *of the vertex F.*

We set

$$\mathrm{Lg}_p(F) := \left\{l \in \mathrm{Lg}^n(F)\ :\ \mathrm{pr}(l) = p\right\}, \quad p \in \mathrm{Pr}^n,$$
$$\mathrm{Lg}_q^{(\pm)}(F) := \left\{l \in \mathrm{Lg}^{(\pm)}(F)\ :\ \mathrm{pr}(l) = q\right\}, \quad q \in \mathrm{Pr}^c.$$

We assume that the sets $\mathrm{Lg}_p(F)$, $p \in \mathrm{Pr}^n_a$, *and* $\mathrm{Lg}_q^{(\pm)}(F)$, $q \in \mathrm{Pr}^c_a$, *are oriented.*

The multi-degree *of F is defined by*

$$m_p(F) := \#\mathrm{Lg}_p(F), \quad p \in \mathrm{Pr}^n,$$
$$m_q^{(\pm)}(F) := \#\mathrm{Lg}_q^{(\pm)}(F), \quad q \in \mathrm{Pr}^c.$$

Graphically a leg is depicted by a line segment attached to the vertex at one end. The shape or the decoration of a leg corresponds to the particle type. For example, traditionally, photon legs are represented by wavy lines, while electron legs are represented by straight lines. Moreover lines corresponding to charge creating, resp. charge annihilating legs are decorated with an arrow pointing away, resp. towards the origin of the line. Neutral legs have no arrows at all.

A vertex F is depicted by a dot with the legs of $\mathrm{Lg}(F)$ originating at the dot. Each kind of a vertex is represented by a different dot.

Figure 20.1 A vertex with four neutral legs.

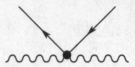

Figure 20.2 A vertex with two electron legs and two photon legs.

20.1.2 Diagrams

Vertices are linked with one another to form *diagrams*. A rigorous definition of a diagram (used in the Gaussian integration) is given below.

Definition 20.7 *Let* $\{F_j\}_{j \in J}$ *be a finite family of vertices. Set*

$$\mathrm{Lg}\Big(\prod_{j \in J} F_j\Big) := \bigsqcup_{j \in J} \mathrm{Lg}(F_j).$$

Elements of $\mathrm{Lg}\big(\prod_{j \in J} F_j\big)$ *are called* legs *of* $\prod_{j \in J} F_j$*. For* $\mathrm{l} = (j, l) \in \mathrm{Lg}\big(\prod_{j \in J} F_j\big)$ *we define* $\mathrm{nr}(\mathrm{l}) := j$ *and* $\mathrm{pr}(\mathrm{l}) := \mathrm{pr}(l)$.

$\mathrm{Lg}\big(\prod_{j \in J} F_j\big)$ *is the union of disjoint sets*

$$\mathrm{Lg}^\pi\Big(\prod_{j \in J} F_j\Big) := \bigsqcup_{j \in J} \mathrm{Lg}^\pi(F_j), \quad \pi = \mathrm{n}, (+), (-).$$

Elements of $\mathrm{Lg}^\mathrm{n}\big(\prod_{j \in J} F_j\big)$ *are called* neutral *legs of* $\prod_{j \in J} F_j$*. Elements of* $\mathrm{Lg}^{(+)}\big(\prod_{j \in J} F_j\big)$*, resp.* $\mathrm{Lg}^{(-)}\big(\prod_{j \in J} F_j\big)$ *are called* charge annihilating*, resp.* charge creating *legs of* $\prod_{j \in J} F_j$*.*

A labeled diagram over $\prod_{j \in J} F_j$ *is a pair* $D = (\mathrm{Lg}(D), \mathrm{Ln}(D))$*, where*

(1) $\mathrm{Lg}(D)$ *is a subset of* $\mathrm{Lg}\big(\prod_{j \in J} F_j\big)$*, whose elements are called* legs *or* external lines *of* D*;*
(2) $\mathrm{Ln}(D)$ *is a partition of* $\mathrm{Lg}\big(\prod_{j \in J} F_j\big) \backslash \mathrm{Lg}(D)$ *into pairs such that, if* $\{\mathrm{l}, \mathrm{l}'\} \in \mathrm{Ln}(D)$*, then*
   (i) $\mathrm{pr}(\mathrm{l}) = \mathrm{pr}(\mathrm{l}')$*;*
   (ii) *if* $\mathrm{l} \in \mathrm{Lg}^{(-)}\big(\prod_{j \in J} F_j\big)$*, then* $\mathrm{l}' \in \mathrm{Lg}^{(+)}\big(\prod_{j \in J} F_j\big)$*.*
Pairs in $\mathrm{Ln}(D)$ *are called* links *or* internal lines *of* D*.*

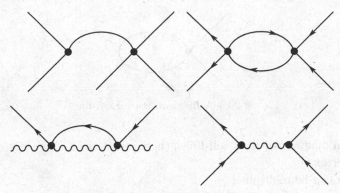

Figure 20.3 Various diagrams.

We set

$$\mathrm{Lg}^{\pi}(D) := \mathrm{Lg}(D) \cap \mathrm{Lg}^{\pi}\Big(\prod_{j \in J} F_j\Big), \quad \pi = \mathrm{n}, (+), (-).$$

We define $\mathrm{Ln}^{\mathrm{n}}(D)$ to be the set of links consisting of neutral particles and $\mathrm{Ln}^{\mathrm{c}}(D)$ to be the set of links consisting of charged particles. Sometimes we will treat charged links as ordered pairs, writing $(\mathrm{l}^{(+)}, \mathrm{l}^{(-)}) \in \mathrm{Ln}^{\mathrm{c}}(D)$ with $\mathrm{l}^{(\pm)} \in \mathrm{Lg}^{(\pm)}\Big(\prod_{j \in J} F_j\Big)$.

We set

$$\mathrm{Lg}_p(D) := \{l \in \mathrm{Lg}^{\mathrm{n}}(D) \ : \ \mathrm{pr}(l) = p\}, \quad p \in \mathrm{Pr}^{\mathrm{n}},$$
$$\mathrm{Lg}_q^{(\pm)}(D) := \{l \in \mathrm{Lg}^{(\pm)}(D) \ : \ \mathrm{pr}(l) = q\}, \quad q \in \mathrm{Pr}^{\mathrm{c}}.$$

The multi-degree of D is defined by

$$m_p(D) := \#\mathrm{Lg}_p(D), \quad p \in \mathrm{Pr}^{\mathrm{n}},$$
$$m_q^{(\pm)}(D) := \#\mathrm{Lg}_q^{(\pm)}(D), \quad q \in \mathrm{Pr}^{\mathrm{c}}.$$

The number of vertices of D is denoted by

$$\mathrm{vert}(D) := \#J.$$

The set of all labeled diagrams over $\Big(\prod_{j \in J} F_j\Big)$ will be denoted $\widetilde{\mathrm{Dg}}\Big(\prod_{j \in J} F_j\Big)$.

Thus to draw a diagram over vertices $\{F_j\}_{j \in J}$, we first draw the vertices themselves, then join some of the legs. We are allowed to join only pairs of legs that belong to the same particle species. In the case of charged particles, we are only allowed to join a charge creating with a charge annihilating leg. Neutral lines have no arrow, whereas charged lines are decorated with an arrow.

Definition 20.8 Let D be a diagram over $\{F_j\}_{j \in J}$. We say that D has no self-lines if $\{1, 1'\} \in \mathrm{Ln}(D)$ implies $\mathrm{nr}(l) \neq \mathrm{nr}(l')$. The set of all labeled diagrams over $\{F_j\}_{j \in J}$ without self-lines will be denoted $\mathrm{Dg}\Big(\prod_{j \in J} F_j\Big)$.

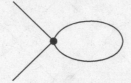

Figure 20.4 A diagram with a self-line.

Thus, in a diagram without self-lines, there are no lines that start and end at the same vertex.

Diagrams can be multiplied:

Definition 20.9 *Consider two diagrams* $D \in \widetilde{\mathrm{Dg}}\left(\prod_{i \in I} F_i\right)$ *and* $D' \in \widetilde{\mathrm{Dg}}\left(\prod_{j \in J} F_j\right)$.

Clearly, $\mathrm{Lg}(D)$ *and* $\mathrm{Lg}(D')$ *can be considered as subsets of* $\mathrm{Lg}\left(\prod_{i \in I} F_i \times \prod_{j \in J} F_j\right)$.

Likewise, $\mathrm{Ln}(D)$ *and* $\mathrm{Ln}(D')$ *can be considered as sets of pairs in* $\mathrm{Lg}\left(\prod_{i \in I} F_i \times \prod_{j \in J} F_j\right)$.

The product of D *and* D', *denoted* $DD' = D'D$, *is defined as the diagram over* $\prod_{i \in I} F_i \times \prod_{j \in J} F_j$ *such that*

$$\mathrm{Lg}(DD') := \mathrm{Lg}(D) \cup \mathrm{Lg}(D'), \quad \mathrm{Ln}(DD') := \mathrm{Ln}(D) \cup \mathrm{Ln}(D').$$

Thus, graphically, multiplication of diagrams consists simply in their juxtaposition. Clearly, a product of diagrams with no self-lines is a diagram with no self-lines.

A vertex is an example of a diagram with no self-lines. The diagram whose set of legs equals $\mathrm{Lg}\left(\prod_{j \in J} F_j\right)$, and whose set of lines is empty equals $\prod_{j \in J} F_j$. This explains the notation used in Def. 20.7.

20.1.3 Connected diagrams

The following concepts have self-explanatory names.

Definition 20.10 *A diagram* $D \in \widetilde{\mathrm{Dg}}\left(\prod_{j \in J} F_j\right)$ *is called* connected *if for all* $j, j' \in J$ *there exist*

$$\{l_n, l'_n\}, \dots, \{l_1, l'_1\} \in \mathrm{Ln}(D)$$

such that $\mathrm{nr}(l_n) = j$, $\mathrm{nr}(l'_k) = \mathrm{nr}(l_{k-1})$, $k = n, \dots, 1$, $\mathrm{nr}(l'_1) = j'$.

For $A \subset \widetilde{\mathrm{Dg}}\left(\prod_{j \in J} F_j\right)$, *we set*

$$A_{\mathrm{con}} := \{D \in A \ : \ D \text{ is connected}\}.$$

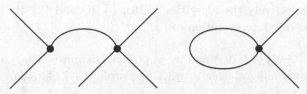

Figure 20.5 A disconnected diagram.

Clearly, each diagram can be decomposed into a product of connected diagrams. This decomposition is unique up to a permutation of factors.

Definition 20.11 *We say that a diagram D has no external legs if* $\mathrm{Lg}(D) = \emptyset$. *If $A \subset \widetilde{\mathrm{Dg}}\left(\prod_{j \in J} F_j\right)$, then we set*

$$A_{\mathrm{nl}} := \{D \in A \; : \; D \text{ has no external legs}\},$$
$$A_{\mathrm{cnl}} := (A_{\mathrm{nl}})_{\mathrm{con}},$$
$$A_{\mathrm{link}} := (A \backslash A_{\mathrm{nl}})_{\mathrm{con}}.$$

20.1.4 Particle spaces and Gaussian integration

Now we introduce the analytical part of the diagram formalism.

Definition 20.12 (1) *For any $p \in \mathrm{Pr}^{\mathrm{n}}$, let $\mathcal{V}_p$ be a real vector space equipped with a form $\sigma_p \in L(\mathcal{V}_p^{\#}, \mathcal{V}_p)$, where σ_p is non-degenerate symmetric if $p \in \mathrm{Pr}_{\mathrm{s}}^{\mathrm{n}}$ and non-degenerate anti-symmetric if $p \in \mathrm{Pr}_{\mathrm{a}}^{\mathrm{n}}$. We set*

$$\mathcal{V}^{\mathrm{n}} := \bigoplus_{p \in \mathrm{Pr}^{\mathrm{n}}} \mathcal{V}_p, \qquad \epsilon^{\mathrm{n}} := \bigoplus_{p \in \mathrm{Pr}^{\mathrm{n}}} \epsilon_p \mathbb{1}_{\mathcal{V}_p}, \qquad \sigma^{\mathrm{n}} := \bigoplus_{p \in \mathrm{Pr}^{\mathrm{n}}} \sigma_p.$$

(2) *For any $q \in \mathrm{Pr}^{\mathrm{c}}$, let $\mathcal{V}_q$ be a complex vector space equipped with a form $\sigma_q \in L(\mathcal{V}_q^{\#}, \overline{\mathcal{V}}_q)$, where σ_q is non-degenerate Hermitian if $q \in \mathrm{Pr}_{\mathrm{s}}^{\mathrm{c}}$ and non-degenerate anti-Hermitian if $q \in \mathrm{Pr}_{\mathrm{a}}^{\mathrm{c}}$. We set*

$$\mathcal{V}^{\mathrm{c}} := \bigoplus_{q \in \mathrm{Pr}^{\mathrm{c}}} \mathcal{V}_q, \qquad \epsilon^{\mathrm{c}} := \bigoplus_{q \in \mathrm{Pr}^{\mathrm{c}}} \epsilon_q \mathbb{1}_{\mathcal{V}_q}, \qquad \sigma^{\mathrm{c}} := \bigoplus_{q \in \mathrm{Pr}^{\mathrm{c}}} \sigma_q.$$

(3) *Set*

$$\mathcal{V} := \mathbb{C}\mathcal{V}^{\mathrm{n}} \oplus \mathcal{V}^{\mathrm{c}} \oplus \overline{\mathcal{V}}^{\mathrm{c}}, \qquad \epsilon = \epsilon_{\mathbb{C}}^{\mathrm{n}} \oplus \epsilon^{\mathrm{c}} \oplus \overline{\epsilon}^{\mathrm{c}}.$$

We will treat $\mathcal{V}^{\mathrm{n}}$, $\mathcal{V}^{\mathrm{c}}$ and $\mathcal{V}$ as super-spaces (see Subsect. 1.1.15). In particular, we can define the set of holomorphic polynomials over $\mathcal{V}$, denoted $\mathrm{Pol}(\mathcal{V})$. As usual, if $G \in \mathrm{Pol}(\mathcal{V})$, then $G(0)$ denotes the zero-th order component of G.

Definition 20.13 *For $G \in \mathrm{Pol}(\mathcal{V})$, we define*

$$\int G := \left(\exp \left(\sum_{p \in \mathrm{Pr}^{\mathrm{n}}} \frac{1}{2} \nabla_{v_p} \sigma_p \nabla_{v_p} + \sum_{q \in \mathrm{Pr}^{\mathrm{c}}} \nabla_{\overline{v}_q} \sigma_q \nabla_{v_q} \right) G \right)(0). \qquad (20.3)$$

Applying respectively the identities (4.15), (4.16) and (2) of Prop. 7.19, we obtain the following interpretation of (20.3):

(1) If $p \in \mathrm{Pr}_\mathrm{s}^\mathrm{n}$, $G \in \mathrm{Pol}_\mathrm{s}(\mathbb{C}\mathcal{V}_p)$ and σ_p is positive definite, then (20.3) coincides with the usual integral over $\mathcal{V}_p$ w.r.t. the probability Gaussian measure with covariance σ_p, that is,

$$\int G = C \int G(v_p) \mathrm{e}^{-\frac{1}{2} v_p \sigma_p^{-1} v_p} \, \mathrm{d}v_p.$$

(2) If $q \in \mathrm{Pr}_\mathrm{s}^\mathrm{c}$, $G \in \mathrm{Pol}_\mathrm{s}(\mathcal{V}_q \oplus \overline{\mathcal{V}}_q)$ and σ_q is positive definite, (20.3) coincides with the usual integral over $\mathcal{V}_{q,\mathbb{R}} \simeq \mathrm{Re}(\mathcal{V}_q \oplus \overline{\mathcal{V}}_q)$ w.r.t. the probability Gaussian measure with covariance σ_q, that is,

$$\int G = C \int G(v_q, \overline{v}_q) \mathrm{e}^{-\overline{v}_q \sigma_q^{-1} v_q} \, \mathrm{d}\overline{v}_q \mathrm{d}v_q.$$

(3) If $p \in \mathrm{Pr}_\mathrm{a}^\mathrm{n}$, $G \in \mathrm{Pol}_\mathrm{a}(\mathbb{C}\mathcal{V}_p)$, then (20.3) coincides with the Berezin integral with the weight $\mathrm{e}^{-\frac{1}{2} v_p \sigma_p^{-1} v_p}$, that is,

$$\int G = C \int G(v_p) \mathrm{e}^{-\frac{1}{2} v_p \sigma_p^{-1} v_p} \, \mathrm{d}v_p.$$

(4) If $q \in \mathrm{Pr}_\mathrm{a}^\mathrm{c}$, $G \in \mathrm{Pol}_\mathrm{a}(\mathcal{V}_q \oplus \overline{\mathcal{V}}_q)$, then (20.3) coincides with the Berezin integral with the weight $\mathrm{e}^{-\overline{v}_q \sigma_q^{-1} v_q}$, that is,

$$\int G = C \int G(v_q, \overline{v}_q) \mathrm{e}^{-\overline{v}_q \sigma_q^{-1} v_q} \, \mathrm{d}\overline{v}_q \mathrm{d}v_q.$$

In all these cases, C is the normalizing constant.

Definition 20.14 *Define the* Wick transform *of $G \in \mathrm{Pol}(\mathcal{V})$ by*

$$: G : := \exp\left(-\sum_{p \in \mathrm{Pr}^\mathrm{n}} \frac{1}{2} \nabla_{v_p} \sigma_p \nabla_{v_p} - \sum_{q \in \mathrm{Pr}^\mathrm{c}} \nabla_{\overline{v}_q} \sigma_q \nabla_{v_q} \right) G. \qquad (20.4)$$

If $G = :G_1:$, then G_1 will be sometimes called the Wick symbol *of G.*

Note that Def. 20.14 generalizes the Wick transform from Def. 9.18, where it was a construction closely related to the Gram–Schmidt orthogonalization. Clearly,

$$G = \exp\left(\sum_{p \in \mathrm{Pr}^\mathrm{n}} \frac{1}{2} \nabla_{v_p} \sigma_p \nabla_{v_p} + \sum_{q \in \mathrm{Pr}^\mathrm{c}} \nabla_{\overline{v}_q} \sigma_q \nabla_{v_q} \right) :G:, \qquad (20.5)$$

$$\int :G: = G(0).$$

20.1.5 Monomials

Let m be a multi-degree.

Definition 20.15 *Set*

$$\otimes^m(\mathcal{V}) := \underset{p\in\mathrm{Pr}^n}{\otimes} \otimes^{m_p}(\mathbb{C}\mathcal{V}_p) \otimes \underset{q\in\mathrm{Pr}^c}{\otimes} \otimes^{m_q^{(+)}}(\overline{\mathcal{V}}_q) \otimes \underset{q\in\mathrm{Pr}^c}{\otimes} \otimes^{m_q^{(-)}}(\mathcal{V}_q), \qquad (20.6)$$

$$\Gamma^m(\mathcal{V}) := \underset{p\in\mathrm{Pr}^n}{\otimes} \Gamma_{\epsilon_p}^{m_p}(\mathbb{C}\mathcal{V}_p) \otimes \underset{q\in\mathrm{Pr}^c}{\otimes} \Gamma_{\epsilon_q}^{m_q^{(+)}}(\overline{\mathcal{V}}_q) \otimes \underset{q\in\mathrm{Pr}^c}{\otimes} \Gamma_{\epsilon_q}^{m_q^{(-)}}(\mathcal{V}_q), \qquad (20.7)$$

$$\mathrm{Pol}^m(\mathcal{V}) := \underset{p\in\mathrm{Pr}^n}{\otimes} \mathrm{Pol}_{\epsilon_p}^{m_p}(\mathbb{C}\mathcal{V}_p) \otimes \underset{q\in\mathrm{Pr}^c}{\otimes} \mathrm{Pol}_{\epsilon_q}^{m_q^{(+)}}(\overline{\mathcal{V}}_q) \otimes \underset{q\in\mathrm{Pr}^c}{\otimes} \mathrm{Pol}_{\epsilon_q}^{m_q^{(-)}}(\mathcal{V}_q). \qquad (20.8)$$

Elements of $\mathrm{Pol}^m(\mathcal{V})$ *are called* complex polynomials of multi-degree m.

Clearly, $\Gamma^m(\mathcal{V}) \subset \otimes^m(\mathcal{V})$ and $\mathrm{Pol}^m(\mathcal{V}) = \Gamma^m(\mathcal{V})^\#$.

Definition 20.16

$$\Theta^m := \underset{p\in\mathrm{Pr}^n}{\otimes} \Theta_{\epsilon_p}^{m_p} \otimes \underset{q\in\mathrm{Pr}^c}{\otimes} \Theta_{\epsilon_q}^{m_q^{(+)}} \otimes \underset{q\in\mathrm{Pr}^c}{\otimes} \Theta_{\epsilon_q}^{m_q^{(-)}}$$

denotes the usual projection of $\otimes^m(\mathcal{V})$ *onto* $\Gamma^m(\mathcal{V})$. *Therefore,* $\Theta^{m\#}$ *is the usual projection of* $\otimes^m(\mathcal{V})^\#$ *onto* $\mathrm{Pol}^m(\mathcal{V})$.

Let F be a vertex with multi-degree m. With every leg of the vertex we associate a space

$$\mathcal{V}_l \simeq \mathcal{V}_p, \; l \in \mathrm{Lg}_p(F), \quad p \in \mathrm{Pr}^n,$$
$$\overline{\mathcal{V}}_k \simeq \overline{\mathcal{V}}_q, k \in \mathrm{Lg}_q^{(+)}(F), q \in \mathrm{Pr}^c,$$
$$\mathcal{V}_k \simeq \mathcal{V}_q, k \in \mathrm{Lg}_q^{(-)}(F), q \in \mathrm{Pr}^c. \qquad (20.9)$$

Within each family

$$\mathrm{Lg}_p(F), p \in \mathrm{Pr}^n,$$
$$\mathrm{Lg}_q^{(+)}(F), q \in \mathrm{Pr}^c,$$
$$\mathrm{Lg}_q^{(-)}(F), q \in \mathrm{Pr}^c,$$

we label the legs by consecutive integers. For fermionic particles we assume that the numbering is admissible. Note that apart from this condition, the numbering is arbitrary and plays only an auxiliary role. Thanks to this numbering, we have a natural bijection between the set of legs of the vertex F and the factors of (20.6). Thus $\otimes^m(\mathcal{V})$ can be identified with

$$\underset{l\in\mathrm{Lg}^n(F)}{\otimes} \mathcal{V}_l \otimes \underset{k\in\mathrm{Lg}^{(+)}(F)}{\otimes} \overline{\mathcal{V}}_k \otimes \underset{k\in\mathrm{Lg}^{(-)}(F)}{\otimes} \mathcal{V}_k, \qquad (20.10)$$

and an element of $\otimes^m(\mathcal{V})^\#$ can be viewed as a multi-linear function on

$$\underset{l\in\mathrm{Lg}^n(F)}{\Pi} \mathcal{V}_l \times \underset{k\in\mathrm{Lg}^{(+)}(F)}{\Pi} \overline{\mathcal{V}}_k \times \underset{k\in\mathrm{Lg}^{(-)}(F)}{\Pi} \mathcal{V}_k. \qquad (20.11)$$

In this way we can associate with a monomial in $\mathrm{Pol}^m(\mathcal{V})$ a vertex F of the same multi-degree. Therefore, we will use the same letter F to denote a monomial and its associated vertex, and we will usually not distinguish between them.

It is convenient to adopt natural names of the corresponding generic variables:

$$v_\mathrm{l} \text{ for the generic variable in } \mathcal{V}_\mathrm{l}, \ \mathrm{l} \in \mathrm{Lg}^\mathrm{n}(F),$$
$$\overline{v}_\mathrm{k} \text{ for the generic variable in } \overline{\mathcal{V}}_\mathrm{k}, \ \mathrm{k} \in \mathrm{Lg}^{(+)}(F),$$
$$v_\mathrm{k} \text{ for the generic variable in } \mathcal{V}_\mathrm{k}, \ \mathrm{k} \in \mathrm{Lg}^{(-)}(F).$$

20.1.6 Evaluation of diagrams

Let $\{F_j\}_{j\in J}$ be a family of fermion-even monomials. Let $D \in \widetilde{\mathrm{Dg}}\left(\prod_{j\in J} F_j\right)$, that is, let D be a diagram over $\prod_{j\in J} F_j$.

Definition 20.17 *The evaluation of D is an element of $\mathrm{Pol}^{m(D)}(\mathcal{V})$ denoted by the same symbol D and given by*

$$D := \Theta^{m(D)\#} \prod_{\ell=\{\mathrm{l},\mathrm{l}'\}\in\mathrm{Ln}^\mathrm{n}(D)} \nabla_{v_\mathrm{l}}\sigma_\ell\nabla_{v_{\mathrm{l}'}}$$

$$\times \prod_{\kappa=(\mathrm{k}^{(+)},\mathrm{k}^{(-)})\in\mathrm{Ln}^\mathrm{c}(D)} \nabla_{\overline{v}_{\mathrm{k}(-)}}\sigma_\kappa\nabla_{v_{\mathrm{k}(+)}} \prod_{j\in J} F_j. \qquad (20.12)$$

Here, if $\ell = \{\mathrm{l},\mathrm{l}'\} \in \mathrm{Lg}^\mathrm{n}(D)$ and $p = \mathrm{pr}(\mathrm{l}) = \mathrm{pr}(\mathrm{l}')$, then σ_ℓ denotes σ_p.
Likewise, if $\kappa = (\mathrm{k}^{(+)},\mathrm{k}^{(-)}) \in \mathrm{Lg}^\mathrm{c}(D)$ and $q = \mathrm{pr}(\mathrm{k}^{(+)}) = \mathrm{pr}(\mathrm{k}^{(-)})$, then σ_κ denotes σ_q.

(20.12) should be interpreted as follows:

(1) We treat $\prod_{j\in J} F_j$ as a multi-linear function depending on the variables

$$v_\mathrm{l} \in \mathcal{V}_\mathrm{l}, \ \mathrm{l} \in \mathrm{Lg}^\mathrm{n}\left(\prod_{j\in J} F_j\right),$$
$$\overline{v}_\mathrm{k} \in \overline{\mathcal{V}}_\mathrm{k}, \ \mathrm{k} \in \mathrm{Lg}^{(+)}\left(\prod_{j\in J} F_j\right),$$
$$v_\mathrm{k} \in \mathcal{V}_\mathrm{k}, \ \mathrm{k} \in \mathrm{Lg}^{(-)}\left(\prod_{j\in J} F_j\right).$$

(2) We perform the differentiation indicated in (20.12), which produces a multi-linear function depending on

$$v_\mathrm{l} \in \mathcal{V}_\mathrm{l}, \ \mathrm{l} \in \mathrm{Lg}^\mathrm{n}(D),$$
$$\overline{v}_\mathrm{k} \in \overline{\mathcal{V}}_\mathrm{k}, \ \mathrm{k} \in \mathrm{Lg}^{(+)}(D),$$
$$v_\mathrm{k} \in \mathcal{V}_\mathrm{k}, \ \mathrm{k} \in \mathrm{Lg}^{(-)}(D).$$

(3) We label the set J by consecutive integers in an arbitrary way. This gives an obvious ordering of the legs in each family

$$\mathrm{Lg}_p\left(\prod_{j\in J} F_j\right), p \in \mathrm{Pr}^{\mathrm{n}},$$

$$\mathrm{Lg}_q^{(+)}\left(\prod_{j\in J} F_j\right), q \in \mathrm{Pr}^{\mathrm{c}},$$

$$\mathrm{Lg}_q^{(-)}\left(\prod_{j\in J} F_j\right), q \in \mathrm{Pr}^{\mathrm{c}}. \qquad (20.13)$$

(4) We order the set of particles. The ordering of fermionic particles should be admissible.
(5) The ordering determined in the previous two points allows us to identify the result of differentiation with an element of $\otimes^{m(D)}(\mathcal{V})^{\#}$.
(6) We symmetrize/anti-symmetrize, obtaining an element of $\mathrm{Pol}^{m(D)}(\mathcal{V})$.

Note that if $D_1, \dots, D_n$ are diagrams and $D = D_1 \cdots D_n$, then the evaluation of D equals the product of the evaluations of D_i, $i = 1, \dots, n$.

Remark that if D has no external legs, then as a monomial it is a number, hence $: D := D$.

Note that the group of permutations of J leaves invariant the monomial associated with diagrams in $\mathrm{Dg}\left(\prod_{j\in J} F_j\right)$ because all monomials F_j are fermion-even.

20.1.7 Gaussian integration of products of monomials

The following theorem shows that diagrams can be efficiently used to compute the Wick symbol and the Gaussian integral of products of monomials. In the bosonic case, for a positive definite σ, (20.15) and (20.17) are graphical interpretations of Thm. 9.25.

Theorem 20.18 *Let* $\{F_1, \dots, F_r\}$ *be fermion-even monomials. Then*

$$F_n \cdots F_1 = \sum_{D\in\widetilde{\mathrm{Dg}}(F_n,\dots,F_1)} : D :, \qquad (20.14)$$

$$: F_n : \cdots : F_1 := \sum_{D\in\mathrm{Dg}(F_n,\dots,F_1)} : D :, \qquad (20.15)$$

$$\int F_n \cdots F_1 = \sum_{D\in\widetilde{\mathrm{Dg}}(F_n,\dots,F_1)_{\mathrm{nl}}} D, \qquad (20.16)$$

$$\int : F_n : \cdots : F_1 := \sum_{D\in\mathrm{Dg}(F_n,\dots,F_1)_{\mathrm{nl}}} D. \qquad (20.17)$$

Proof (20.14) is a restatement of (20.5) applied to $F_n \cdots F_1$, where we repeatedly use the formula (3.36).

(20.15) follows from (20.5) applied to $:F_n:\cdots:F_1:$ and (20.4) applied to F_i. (20.16), resp. (20.17) follow from (20.14), resp. (20.15) by (20.3). $\square$

In Fig. 20.6, we illustrate the above theorem by the diagrams needed to evaluate the identities

$$:\phi^4:\ :\phi^4:\ =\ :\phi^8:+4:\phi^6:+72:\phi^4:+96:\phi^2:+24,$$

$$(2\pi)^{-1/2}\int :\phi^4:\ :\phi^4:e^{-\frac{1}{2}\phi^2}\,d\phi = 24.$$

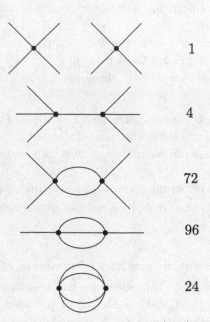

$$
\begin{array}{c}
1\\[3em]
4\\[3em]
72\\[3em]
96\\[3em]
24
\end{array}
$$

Figure 20.6 Diagrams for $:\phi^4:\ :\phi^4:$.

20.1.8 Identical diagrams

Let $\{F_r,\ldots,F_1\}$ be a certain finite set of vertices. For brevity, this set will be denoted by $\mathfrak{V}$.

Recall that S_n denotes the group of permutations of n elements.

Suppose that $n_1,\ldots,n_r \in \{0,1,\ldots\}$. The group $\prod_{i=1}^{r} S_{n_i}$ acts in the obvious way on $\mathrm{Dg}\left(\prod_{i=1}^{r} F_i^{n_i}\right)$.

Definition 20.19 *We set*

$$\mathrm{Dg}\{\mathfrak{V}\} := \bigsqcup_{n_1,\ldots,n_r=1}^{\infty} \mathrm{Dg}\left(\prod_{i=1}^{r} F^{n_i}\right)\Big/\prod_{i=1}^{r} S_{n_i}. \tag{20.18}$$

Elements of (20.18) will be called unlabeled diagrams with vertices in $\mathfrak{V}$.

In other words, $\mathrm{Dg}\{\mathfrak{V}\}$ consists of classes of diagrams made with vertices from $\mathfrak{V}$, which differ just by a permutation. Typically, we will write $[D]$ for an unlabeled diagram, D being a labeled diagram and the square bracket denoting the equivalence class.

20.1.9 Gaussian integration of exponentials

Let $\mathfrak{V} = \{F_1, \ldots, F_r\}$ be a set of fermion-even monomials and λ a coupling constant. Set $G := F_1 + \cdots + F_r$. Our main aim is to compute the Gaussian integral and the Wick symbol of $\exp(\lambda{:}G{:})$.

As indicated before, with each monomial F_i, $i = 1, \ldots, r$, we associate a vertex of the same multi-degree, denoted by the same symbol F_i.

Let $D \in \mathrm{Dg}\left(\prod_{i=1}^{r} F^{n_i}\right)$. The evaluation of the diagram D (see Def. 20.17) does not depend on the action of the group $\prod_{i=1}^{r} S_{n_i}$. Hence, it is well defined for unlabeled diagrams.

Theorem 20.20

$$\exp(\lambda{:}G{:}) = \sum_{[D] \in \mathrm{Dg}\{\mathfrak{V}\}} \lambda^{\mathrm{vert}(D)} {:}D{:}. \tag{20.19}$$

Proof Clearly,

$$\exp(\lambda{:}G{:}) = \sum_{n_1, \ldots, n_r = 0}^{\infty} \frac{1}{n_1! \cdots n_r!} \lambda^{n_1 + \cdots + n_r} {:}F_1{:}^{n_1} \cdots {:}F_r{:}^{n_r}$$

$$= \sum_{n_1, \ldots, n_r = 0}^{\infty} \sum_{D \in \mathrm{Dg}(F_1^{n_1} \cdots F_r^{n_r})} \frac{\lambda^{\mathrm{vert}(D)}}{n_1! \cdots n_r!} {:}D{:}$$

$$= \sum_{n_1, \ldots, n_r = 0}^{\infty} \sum_{[D] \in \mathrm{Dg}(F_1^{n_1} \cdots F_r^{n_r})/\prod_{i=1}^{r} S_{n_i}} \lambda^{\mathrm{vert}(D)} {:}D{:}. \qquad \square$$

Theorem 20.21 (Linked cluster theorem)

$$\exp(\lambda{:}G{:}) = {:} \exp\left(\sum_{[C] \in \mathrm{Dg}\{\mathfrak{V}\}_{\mathrm{con}}} \lambda^{\mathrm{vert}(C)} C\right) {:}, \tag{20.20}$$

$$\log\left(\int \exp(\lambda{:}G{:})\right) = \sum_{[C] \in \mathrm{Dg}\{\mathfrak{V}\}_{\mathrm{cnl}}} \lambda^{\mathrm{vert}(C)} C, \tag{20.21}$$

$$\frac{\exp(\lambda{:}G{:})}{\int \exp(\lambda{:}G{:})} = {:} \exp\left(\sum_{[C] \in \mathrm{Dg}\{\mathfrak{V}\}_{\mathrm{link}}} \lambda^{\mathrm{vert}(C)} C\right) {:}. \tag{20.22}$$

Proof Let us prove (20.20). Let $C_j \in \mathrm{Dg}\left(\prod_{i=1}^{r} F_i^{m_{ji}}\right)$, $j = 1, \ldots, p$, be distinct connected labeled diagrams. Let $k_j \in \{0, 1, \ldots\}$, $j = 1, \ldots, p$. Let $D :=$ $C_1^{k_1} \cdots C_p^{k_p} \in \mathrm{Dg}\left(\prod_{i=1}^{r} F_i^{n_i}\right)$. Clearly, $n_i = \sum_{j=1}^{p} m_{ji}$ and

$$\#[D] = \prod_{i=1}^{r} n_i!, \quad \#[C_j] = \prod_{i=1}^{r} m_{ji}!.$$

An elementary combinatorial argument shows that each diagram in $[D]$ represents $\prod_{i=1}^{r} n_i! \prod_{j=1}^{p} (m_{ji}!)^{-k_j} (k_j!)^{-1}$ times the same diagram in the Cartesian product $\prod_{j=1}^{p} [C_j]^{k_j}$. Therefore,

$$D = \prod_{i=1}^{r} (n_i!)^{-1} \sum_{D' \in [D]} D'$$

$$= \prod_{i=1}^{r} \prod_{j=1}^{p} (m_{ji}!)^{-k_j} (k_j!)^{-1} \prod_{l=1}^{k_j} \sum_{C'_{jl} \in [C_j]} C'_{jl}$$

$$= \prod_{j=1}^{p} (k_j!)^{-1} C_j^{k_j}.$$

Now (20.21) and (20.22) follow from (20.20). $\qquad\square$

20.2 Perturbations of quantum dynamics

In this section we recall the terminology used in quantum physics in the context of a dynamics and its perturbations. We will consider first the case of time-dependent, and then time-independent perturbations.

We recall in particular the basic concepts of scattering theory. Its central notion is the scattering operator. There are several varieties of scattering operators. We recall the standard definition, which is successfully used in the context of Schrödinger operators with short-range potentials. Note, however, that the standard definition usually does not apply to quantum field theory, even on a formal level. We introduce also the adiabatic scattering operator, which is often used to develop the formalism of quantum field theory in standard textbooks.

Our presentation throughout this section will be rather formal. In order to make rigorous some of the formulas we give, one needs to make relatively complicated technical assumptions – we refrain from describing them.

Throughout the section $\mathcal{H}$ is a Hilbert space.

20.2.1 Time-ordered exponentials

Let $\mathbb{R} \ni t \mapsto B_i(t) \in B(\mathcal{H})$, $i = 1, \ldots, n$, be time-dependent families of operators.

Definition 20.22 *Let $t_n, \ldots, t_1 \in \mathbb{R}$ be pairwise distinct. We define Dyson's time-ordered product of $B_n(t_n), \ldots, B_1(t_1)$ by*

$$\mathrm{T}\left(B_n(t_n) \cdots B_1(t_1)\right) := B_{i_n}(t_{i_n}) \cdots B_{i_1}(t_{i_1}),$$

where $(i_n, \ldots, i_1)$ is the permutation of $(n, \ldots, 1)$ such that $t_{i_n} \geq \cdots \geq t_{i_1}$.

Consider now a single family of operators $\mathbb{R} \ni t \mapsto B(t)$.

Definition 20.23 *For $t_+ \geq t_-$, the time-ordered exponential is defined as*

$$\mathrm{Texp}\left(\int_{t_-}^{t_+} B(t)\mathrm{d}t\right) := \sum_{n=0}^{\infty} \int \cdots \int_{t_+ \geq t_n \geq \cdots \geq t_1 \geq t_-} B(t_n) \cdots B(t_1)\mathrm{d}t_n \cdots \mathrm{d}t_1$$

$$= \sum_{n=0}^{\infty} \int_{t_-}^{t_+} \cdots \int_{t_-}^{t_+} \frac{1}{n!} \mathrm{T}\left(B(t_n) \cdots B(t_1)\right) \mathrm{d}t_n \cdots \mathrm{d}t_1.$$

If $t_+ \leq t_-$, then we set

$$\mathrm{Texp}\left(\int_{t_-}^{t_+} B(t)\mathrm{d}t\right) := \left(\mathrm{Texp}\int_{t_+}^{t_-} B(t)\mathrm{d}t\right)^{-1}$$

$$= \sum_{n=0}^{\infty} \int \cdots \int_{t_- \leq t_1 \leq \cdots \leq t_n \leq t_+} (-1)^n B(t_1) \cdots B(t_n)\mathrm{d}t_1 \cdots \mathrm{d}t_n.$$

For brevity, let us write

$$U(t_+, t_-) := \mathrm{Texp}\left(\int_{t_-}^{t_+} B(t)\mathrm{d}t\right).$$

Note that

$$\frac{\mathrm{d}}{\mathrm{d}t_+} U(t_+, t_-) = B(t_+)U(t_+, t_-),$$

$$\frac{\mathrm{d}}{\mathrm{d}t_-} U(t_+, t_-) = -U(t_+, t_-)B(t_-),$$

$$U(t, t) = \mathbb{1},$$

$$U(t_2, t_1)U(t_1, t_0) = U(t_2, t_0).$$

If $B(t) = B$, then $U(t_+, t_-) = e^{(t_+ - t_-)B}$.

20.2.2 Perturbation theory

Let H_0 be a self-adjoint operator. Let $\mathbb{R} \ni t \mapsto V(t)$ be a family of self-adjoint operators. Set $H(t) := H_0 + \lambda V(t)$. Consider the unitary evolution

$$U(t_+, t_-) = \mathrm{Texp}\left(-\mathrm{i}\int_{t_-}^{t_+} H(t)\mathrm{d}t\right).$$

Let $\mathbb{R} \ni t \mapsto A(t)$ be an operator-valued function.

Definition 20.24 *The* operator $A(t)$ in the interaction picture *is defined as*

$$A_I(t) := e^{itH_0} A(t) e^{-itH_0}.$$

The evolution in the interaction picture *is defined as*

$$U_I(t_+, t_-) := e^{it_+ H_0} U(t_+, t_-) e^{-it_- H_0}.$$

Note that

$$U_I(t_+, t_-) = \text{Texp} \left(-i\lambda \int_{t_-}^{t_+} V_I(t) dt \right).$$

In some cases we can take the limit $t_- \to -\infty$ or $t_+ \to \infty$. In particular this is the case if $V(t)$ decays in time sufficiently fast.

Definition 20.25 *The Møller or wave operators, resp. the scattering operator (if they exist) are defined as*

$$S^\pm := s - \lim_{t \to \pm\infty} U_I(0, t),$$

$$S = S^{+*} S^-.$$

Theorem 20.26 (1) *If $S^\pm$ exist, then they are isometric.*
(2) $S := w - \lim_{(t_+, t_-) \to (+\infty, -\infty)} U_I(t_+, t_-).$
(3) *If* $\text{Ran}\, S^+ = \text{Ran}\, S^-$, *then S is unitary.*

Note that the operators $U_I(t_+, t_-)$, S^-, resp. S^{+*} can be viewed as special cases of the scattering operator, if we multiply $V(t)$ by $1\!\!1_{[t_-, t_+]}(t)$, $1\!\!1_{[-\infty, 0[}(t)$, resp. $1\!\!1_{]\infty, 0[}(t)$.

20.2.3 Standard Møller and scattering operators

Until the end of the section, H_0 and V are fixed self-adjoint operators and $H := H_0 + \lambda V$. We have $U(t_+, t_-) = e^{-i(t_+ - t_-)H}$ and $V_I(t) = e^{itH_0} V e^{-itH_0}$.

Clearly, the Møller and scattering operators (if they exist) are

$$S^\pm = s - \lim_{t \to \pm\infty} e^{itH} e^{-itH_0},$$

$$S := w - \lim_{(t_+, t_-) \to (+\infty, -\infty)} e^{it_+ H_0} e^{-i(t_+ - t_-)H} e^{-it_- H_0}.$$

Definition 20.27 *In what follows, we will call $S^\pm$ and S defined as above the standard Møller and scattering operators.*

Theorem 20.28 *Suppose that the standard Møller operators exist.*

(1) *The standard Møller operators satisfy $S^\pm H_0 = H S^\pm$.*
(2) *The standard scattering operator satisfies $H_0 S = S H_0$.*

We have, at least formally,

$$S^+ = \mathrm{Texp}\left(-\int_{+\infty}^0 i\lambda V_{\mathrm{I}}(t)\mathrm{d}t\right),$$

$$S^- = \mathrm{Texp}\left(-\int_{-\infty}^0 i\lambda V_{\mathrm{I}}(t)\mathrm{d}t\right),$$

$$S = \mathrm{Texp}\left(-\int_{-\infty}^{+\infty} i\lambda V_{\mathrm{I}}(t)\mathrm{d}t\right).$$

20.2.4 Stone formula

For $\epsilon > 0$, set

$$\delta_\epsilon(\xi) := \frac{\epsilon}{\pi}(\xi^2 + \epsilon^2)^{-1}.$$

which is a family of approximations of the delta function. For any $a < b$, we have the Stone formula,

$$\mathrm{s} - \lim_{\epsilon \searrow 0} \int_a^b \delta_\epsilon(\xi\mathbb{1} - H)\mathrm{d}\xi = \frac{1}{2}\left(\mathbb{1}_{[a,b]}(H) + \mathbb{1}_{]a,b[}(H)\right).$$

We will formally write

$$\delta(\xi\mathbb{1} - H_0) \quad \text{for} \quad \lim_{\epsilon \searrow 0} \delta_\epsilon(\xi\mathbb{1} - H),$$

$$\left((\xi \pm i0)\mathbb{1} - H_0\right)^{-1} \quad \text{for} \quad \lim_{\epsilon \searrow 0}\left((\xi \pm i\epsilon)\mathbb{1} - H_0\right)^{-1}.$$

These limits do not exist as bounded operators, but can sometimes be given a rigorous meaning as operators between appropriate weighted spaces.

20.2.5 Stationary formulas for Møller and scattering operators

We have the identities (see e.g. Yafaev (1992)):

$$S^\pm = \int\left(\mathbb{1} + \left((\xi \mp i0)\mathbb{1} - H\right)^{-1}\lambda V\right)\delta(\xi\mathbb{1} - H_0)\mathrm{d}\xi$$

$$= \int \sum_{n=0}^\infty \lambda^n\left(\left((\xi \mp i0)\mathbb{1} - H_0\right)^{-1}V\right)^n \delta(\xi\mathbb{1} - H_0)\mathrm{d}\xi.$$

$$S - \mathbb{1} = -2\pi\int \delta(\xi\mathbb{1} - H_0)\left(\lambda V + \lambda^2 V\left((\xi + i0)\mathbb{1} - H\right)^{-1}V\right)\delta(\xi\mathbb{1} - H_0)\mathrm{d}\xi$$

$$= -2\pi\int \delta(\xi\mathbb{1} - H_0)\sum_{n=0}^\infty \lambda^n V\left(\left((\xi + i0)\mathbb{1} - H_0\right)^{-1}V\right)^n \delta(\xi\mathbb{1} - H_0)\mathrm{d}\xi.$$

20.2.6 Problem with eigenvalues

It is easy to show the following fact:

Theorem 20.29 *If the standard Møller operators exist and $H_0\Psi = E\Psi$, then $H\Psi = E\Psi$.*

In practice, the standard formalism of scattering theory is usually applied to Hamiltonians H_0 which only have absolutely continuous spectra. In such a case, Thm. 20.29 is irrelevant.

Thm. 20.29 becomes relevant in models inspired by quantum field theory. Suppose that the starting point of a model is a pair of Hamiltonians H_0 and H (possibly defined only on a formal level). In typical situations both Hamiltonians have a ground state, and these ground states are different. Thm. 20.29 then shows that the standard scattering theory is not applicable. Instead one can sometimes try other approaches, such as the adiabatic approach developed by Gell-Mann and Low, which we describe below.

20.2.7 Adiabatic dynamics

Definition 20.30 *The* adiabatically switched on interaction, *or simply the* adiabatic interaction, *is defined as* $V_\epsilon(t) := e^{-\epsilon|t|}V$, $\epsilon > 0$. *The* adiabatic Hamiltonian *is* $H_\epsilon(t) := H_0 + \lambda V_\epsilon(t)$. *The corresponding dynamics is denoted by* $U_\epsilon(t_+, t_-)$ *and the corresponding dynamics in the interaction picture by* $U_{\epsilon\mathrm{I}}(t_+, t_-)$. *We also define the* adiabatic Møller *and* scattering operators

$$S_\epsilon^\pm := \mathrm{s} - \lim_{t \to \pm\infty} U_{\epsilon\mathrm{I}}(0, t),$$

$$S_\epsilon := S_\epsilon^+ S_\epsilon^{-*}.$$

Note that if the standard Møller operators exist and, if some mild additional assumptions hold, we have

$$S^\pm = \mathrm{s} - \lim_{\epsilon \searrow 0} S_\epsilon^\pm,$$

$$S = \mathrm{w} - \lim_{\epsilon \searrow 0} S_\epsilon.$$

As we argued in Subsect. 20.2.6, in quantum field theory the standard scattering theory usually fails. One needs to use non-standard definition of scattering operators. (Analogs of Møller operators are rarely used in quantum field theory anyway.) One possible modification of the definition of the scattering operator is given below. In this definition, Φ_0 is a distinguished unit vector, typically the ground state of H_0 (e.g. the free vacuum in quantum field theory).

Definition 20.31 *The* Gell-Mann–Low scattering operator *(if it exists) is*

$$S_{\mathrm{GL}} := \mathrm{w} - \lim_{\epsilon \searrow 0} \frac{S_\epsilon}{(\Phi_0 | S_\epsilon \Phi_0)}.$$

20.2.8 Bound state energy

Suppose that Φ_0 and E_0, resp. Φ and E are eigenvectors and eigenvalues of H_0, resp H, so that

$$H_0 \Phi_0 = E_0 \Phi_0, \qquad H\Phi = E\Phi.$$

We assume that Φ, E are small perturbations of Φ_0, E_0 when the coupling constant λ is small enough.

The following heuristic formulas can be sometimes rigorously proven:

$$E - E_0 = \lim_{t \to \pm\infty} \mathrm{i}^{-1} \frac{\mathrm{d}}{\mathrm{d}t} \log(\Phi_0 | \mathrm{e}^{\mathrm{i}tH} \mathrm{e}^{-\mathrm{i}tH_0} \Phi_0), \tag{20.23}$$

$$E - E_0 = \lim_{t \to \pm\infty} (2\mathrm{i})^{-1} \frac{\mathrm{d}}{\mathrm{d}t} \log(\Phi_0 | \mathrm{e}^{-\mathrm{i}tH_0} \mathrm{e}^{\mathrm{i}2tH} \mathrm{e}^{-\mathrm{i}tH_0} \Phi_0). \tag{20.24}$$

To see why we can expect (20.23) to be true, we write

$$(\Phi_0 | \mathrm{e}^{\mathrm{i}tH} \mathrm{e}^{-\mathrm{i}tH_0} \Phi_0) = |(\Phi_0 | \Phi)|^2 \mathrm{e}^{\mathrm{i}t(E-E_0)} + C(t).$$

Then, if we can argue that for large t the term $C(t)$ does not play a role, we obtain (20.23). (20.24) follows by essentially the same argument.

Note that (20.23) involves $\mathrm{e}^{\mathrm{i}tH} \mathrm{e}^{-\mathrm{i}tH_0}$, which can be called an approximate Møller operator, and (20.24) involves $\mathrm{e}^{-\mathrm{i}tH_0} \mathrm{e}^{\mathrm{i}2tH} \mathrm{e}^{-\mathrm{i}tH_0}$, an approximate scattering operator.

Still heuristic, but a little more satisfactory, are the formulas that give the energy shift in terms of the adiabatic Møller and scattering operators:

$$E - E_0 = \lim_{\epsilon \searrow 0} \mathrm{i}\epsilon\lambda\partial_\lambda \log(\Phi_0 | S_\epsilon^+ \Phi_0), \tag{20.25}$$

$$E - E_0 = \lim_{\epsilon \searrow 0} \frac{\mathrm{i}\epsilon\lambda}{2} \partial_\lambda \log(\Phi_0 | S_\epsilon \Phi_0). \tag{20.26}$$

(20.26) is called the *Sucher formula*.

20.3 Friedrichs diagrams and products of Wick monomials

The main aim of the remaining part of this chapter is to describe the perturbation theory for the dynamics of the form $\mathrm{d}\Gamma(h)$ plus the quantization of a (possibly time-dependent) Wick polynomial. We will describe two distinct formalisms for this purpose. In this and the next section we discuss the formalism of *Friedrichs diagrams*. The characteristic feature of these diagrams is the fact that the vertices are ordered in time.

One can argue that the formalism of Friedrichs diagrams was implicitly used in quantum field theory since its birth. Strangely, however, before the late 1940s it was not common to draw pictures to keep track of terms in the perturbation expansion. Apparently, the first to use pictorial representations of the perturbation theory was Stueckelberg and, on a larger scale, Feynman. Their diagrams, however, are different, and will be discussed in Sects. 20.5 and 20.6 under the name *Feynman diagrams*. In Feynman diagrams the order of the time label does not play a role, which usually is a serious advantage. Thus Feynman's invention is not limited to the use of pictorial diagrams. The idea of making pictures when one tries to do computations in perturbation theory is actually quite easy to come by. What was more important and less obvious was to group together various time orderings.

Even though Feynman diagrams dominate, especially in relativistic computations, Friedrichs diagrams are also useful in some situations. In particular, they can be used to compute leading singularities of certain terms of the perturbation expansion.

We divided the discussion of Friedrichs diagrams into two sections. In this section, the goal is to explain how to represent pictorially the symbol of the product of Wick monomials. In the next section we discuss how to compute the scattering operator using Friedrichs diagrams.

20.3.1 Friedrichs vertices

Just as in Sect. 20.1, we start with a description of purely graphical and combinatorial aspects of the formalism. It is quite similar to that of Sect. 20.1, and we will often use the same terms, sometimes with a slightly different meaning.

We assume that we have a set $\mathrm{Pr} = \mathrm{Pr_s} \sqcup \mathrm{Pr_a}$ describing particles, which are bosonic or fermionic, as in Def. 20.3. We assume that the set $\mathrm{Pr_a}$ is oriented.

We adapt the definition of the multi-degree to the context of this section.

Definition 20.32 *A* multi-degree *is a function*

$$\mathrm{Pr} \ni p \mapsto m_p \in \{0, 1, 2, \dots\}.$$

Definition 20.33 *A Friedrichs vertex, denoted W, is a pair of disjoint sets* $(\mathrm{Lg}^+(W), \mathrm{Lg}^-(W))$, *each equipped with a function*

$$\mathrm{Lg}^\pm(W) \ni l \mapsto \mathrm{pr}(l) \in \mathrm{Pr}.$$

Elements of $\mathrm{Lg}^\pm(W)$ are called outgoing, *resp.* incoming legs *of the vertex W. The sets $\mathrm{Lg}_p^\pm(W)$, $p \in \mathrm{Pr_a}$, are oriented.*

The outgoing, *resp.* incoming multi-degree *of W is given by*

$$m_p^\pm(W) := \#\{l \in \mathrm{Lg}^\pm(W) \ : \ \mathrm{pr}(l) = p\}, \quad p \in \mathrm{Pr}.$$

We say that W is fermion-even *if*

$$\sum_{p \in \mathrm{Pr_a}} \left(m_p^+(W) + m_p^-(W)\right) \quad \text{is even.} \tag{20.27}$$

A Friedrichs vertex W is graphically depicted by a dot with legs of $\mathrm{Lg}(W)$ originating at the dot. Incoming legs are on the right and the outgoing legs are on the left. Again, legs for different particle types are depicted by different graphical styles.

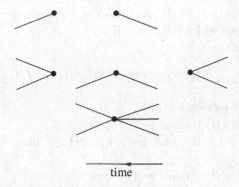

Figure 20.7 Various Friedrichs vertices.

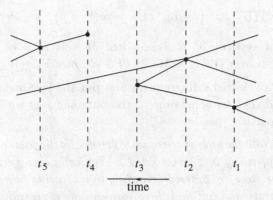

Figure 20.8 A disconnected Friedrichs diagram.

20.3.2 Friedrichs diagrams

Definition 20.34 *Let* $(W_n, \ldots, W_1)$ *be a sequence of Friedrichs vertices. Set*

$$\mathrm{Lg}(W_n, \ldots, W_1) := \bigsqcup_{n \geq j \geq 1} \mathrm{Lg}(W_j),$$

$$\mathrm{Lg}^{\pm}(W_n, \ldots, W_1) := \bigsqcup_{n \geq j \geq 1} \mathrm{Lg}^{\pm}(W_j).$$

Clearly, $\mathrm{Lg}(W_n, \ldots, W_1)$ *is the union of disjoint sets*

$$\mathrm{Lg}^{+}(W_n, \ldots, W_1), \quad \mathrm{Lg}^{-}(W_n, \ldots, W_1).$$

Elements of $\mathrm{Lg}^{\pm}(W_n, \ldots, W_1)$ *are called* incoming, *resp.* outgoing legs *of* $(W_n, \cdots, W_1)$. *For* $l = (j, k) \in \mathrm{Lg}^{\pm}(W_n, \ldots, W_1)$ *we define* $\mathrm{nr}(l) = j$ *and* $\mathrm{pr}(l) = \mathrm{pr}(k)$. *(Note that* $j \in \{n, \ldots, 1\}$ *and* $k \in \mathrm{Lg}(W_j)$*; see Def. 20.1.)*

A Friedrichs diagram B over $(W_n, \cdots, W_1)$ *is a triple* $\left(\mathrm{Lg}^{-}(B), \mathrm{Lg}^{+}(B), \mathrm{Ln}(B)\right)$, *where*

(1) $\mathrm{Lg}^-(B)$ *is a subset of* $\mathrm{Lg}^-(W_n, \dots, W_1)$;
(2) $\mathrm{Lg}^+(B)$ *is a subset of* $\mathrm{Lg}^+(W_n, \dots, W_1)$;
(3) $\mathrm{Ln}(B)$ *is a partition of*

$$\left(\mathrm{Lg}^+(W_n, \dots, W_1) \backslash \mathrm{Lg}^+(B)\right) \cup \left(\mathrm{Lg}^-(W_n; \dots, W_1) \backslash \mathrm{Lg}^-(B)\right)$$

  into two-element sets such that
  (i) $\{1_+, 1_-\} \in \mathrm{Ln}(B)$ *implies* $\mathrm{nr}(1_+) \neq \mathrm{nr}(1_-)$,
  (ii) *if* $\{1_+, 1_-\} \in \mathrm{Ln}(B)$ *and* $\mathrm{nr}(1_+) > \mathrm{nr}(1_-)$, *then* $1_+ \in \mathrm{Lg}^-(B)$, $1_- \in \mathrm{Lg}^+(B)$,
  (iii) $\{1_+, 1_-\} \in \mathrm{Ln}(B)$ *implies* $\mathrm{pr}(1_+) = \mathrm{pr}(1_-)$.

The incoming *and* outgoing multi-degree of B *is defined by*

$$m_p^\pm(B) := \#\{1 \in \mathrm{Lg}^\pm(B) : \mathrm{pr}(1) = p\}, \ p \in \mathrm{Pr}.$$

  The number *of vertices of B is denoted by* $\mathrm{vert}(B) = n$. *The set of all Friedrichs diagrams over* $(W_n, \dots, W_1)$ *will be denoted* $\mathrm{FDg}(W_n, \dots, W_1)$.

Thus, to draw a Friedrichs diagram we first put the Friedrichs vertices in the correct order, and then we join some of the outgoing legs with later incoming legs of the same particle species.

Remark 20.35 *Note that the vertices in a Friedrichs diagram are* ordered, *contrarily to those appearing in Subsect. 20.1.2. Typically each vertex is associated with a time and vertices are ordered according to increasing times. To our knowledge, in the literature one can find three conventions concerning the time arrow in a diagram: time flows to the left, right or upwards. We adopt the convention that time flows to the left, because it agrees with the order of multiplication of operators.*

20.3.3 Connected Friedrichs diagrams

The following definitions are very similar to the analogous definitions of Sect. 20.1.

Definition 20.36 *A Friedrichs diagram B is called* connected *if for all* $j, j' \in \{n, \dots, 1\}$ *there exist*

$$\{1_m, 1'_m\}, \dots, \{1_1, 1'_1\} \in \mathrm{Ln}(B) \tag{20.28}$$

such that $\mathrm{nr}(1_m) = j$, $\mathrm{nr}(1'_k) = \mathrm{nr}(1_{k-1})$, $k = m, \dots, 2$, $\mathrm{nr}(1'_1) = j'$.
  If $A \subset \mathrm{FDg}(W_n, \dots, W_1)$, *then we set*

$$A_{\mathrm{con}} := \{B \in A : B \text{ is connected }\}.$$

Note that the sequence of lines in (20.28) does not have to be ordered in time.

Definition 20.37 *We say that a Friedrichs diagram B has no external legs if* $\mathrm{Lg}^+(B) = \mathrm{Lg}^-(B) = \emptyset$.

If $A \subset \mathrm{FDg}(W_n, \ldots, W_1)$, then we set

$$A_{\mathrm{nl}} := \{B \in A \ : \ B \ \text{has no external legs}\},$$
$$A_{\mathrm{cnl}} := (A_{\mathrm{nl}})_{\mathrm{con}},$$
$$A_{\mathrm{link}} = (A \backslash A_{\mathrm{nl}})_{\mathrm{con}}.$$

20.3.4 One-particle spaces

Definition 20.38 *With each $p \in \mathrm{Pr}$ we associate a complex Hilbert space $\mathcal{Z}_p$. $\mathcal{Z}_p$ is called the* one-particle space *of p. We set*

$$\mathcal{Z} := \bigoplus_{p \in \mathrm{Pr}} \mathcal{Z}_p, \quad \epsilon = \bigoplus_{p \in \mathrm{Pr}} \epsilon_p \mathbb{1}_{\mathcal{Z}_p}.$$

We treat $(\mathcal{Z}, \epsilon)$ as a super-space.

Let $Q \in B^{\mathrm{fin}}(\mathcal{Z})$. Its Wick quantization, denoted as usual $\mathrm{Op}^{a^*, a}(Q)$, is an operator on the Fock space

$$\bigotimes_{p \in \mathrm{Pr}} \Gamma_{\epsilon_p}(\mathcal{Z}_p) \simeq \Gamma(\mathcal{Z}).$$

20.3.5 Incoming and outgoing diagram spaces

Let m be a multi-degree.

Definition 20.39 *We set*

$$\otimes^m \mathcal{Z} := \bigotimes_{p \in \mathrm{Pr}} \otimes^{m_p} \mathcal{Z}_p, \tag{20.29}$$

$$\Gamma^m(\mathcal{Z}) := \bigotimes_{p \in \mathrm{Pr}} \Gamma^{m_p}_{\epsilon_p}(\mathcal{Z}_p), \tag{20.30}$$

$$\mathrm{Pol}^m(\mathcal{Z}) := \bigotimes_{p \in \mathrm{Pr}} \mathrm{Pol}^{m_p}_{\epsilon_p}(\mathcal{Z}_p). \tag{20.31}$$

$\Gamma^m(\mathcal{Z})$ *is called the m-particle space. Let Θ^m denote the usual projection from $\otimes^m \mathcal{Z}$ onto $\Gamma^m(\mathcal{Z})$.*

Let (m^+, m^-) be a pair of multi-degrees. An important role will be played by

$$B\big(\Gamma^{m^-}(\mathcal{Z}), \Gamma^{m^+}(\mathcal{Z})\big). \tag{20.32}$$

(20.32) will sometimes be interpreted as a polynomial in

$$\mathrm{Pol}^{m^-}(\overline{\mathcal{Z}}) \otimes \mathrm{Pol}^{m^+}(\mathcal{Z}).$$

More precisely, with W that belongs to (20.32) we associate

$$W\Big(\{\overline{z}_{i,+,p}\}_{i=1,\ldots,m_p^+, p \in \mathrm{Pr}}, \{z_{j,-,q}\}_{j=1,\ldots,m_q^-, q \in \mathrm{Pr}}\Big)$$

$$= \Big(\bigotimes_{p \in \mathrm{Pr}} \overset{m_p^+}{\underset{i=1}{\otimes}} z_{i,+,p} \Big| W \bigotimes_{q \in \mathrm{Pr}} \overset{m_p^-}{\underset{j=1}{\otimes}} z_{j,-,q}\Big).$$

We will sometimes view (20.32) as a subspace of $B(\otimes^{m^-} \mathcal{Z}, \otimes^{m^+} \mathcal{Z})$: an element of W of (20.32) is extended to an operator on $\otimes^{m^-} \mathcal{Z}$ by setting 0 on the orthogonal complement of $\Gamma^{m^-}(\mathcal{Z})$.

Let W be a vertex of multi-degrees (m^+, m^-). With every leg of the vertex we associate a space

$$\mathcal{Z}_1 \simeq \mathcal{Z}_p, \quad 1 \in \mathrm{Lg}_p^\pm(W), \quad p \in \mathrm{Pr}. \tag{20.33}$$

Within each family $\mathrm{Lg}_p^\pm(W)$ we order the legs by consecutive integers. For fermionic particles we assume that the numbering is admissible. Note that apart from this condition, the numbering is arbitrary and plays only an auxiliary role. Thanks to this numbering, we have a natural bijection between $\mathrm{Lg}^\pm(W)$ and the factors of $\otimes^{m^\pm} \mathcal{Z}$. Thus $\otimes^{m^\pm} \mathcal{Z}$ can be identified with

$$\underset{1 \in \mathrm{Lg}^\pm(W)}{\otimes} \mathcal{Z}_1. \tag{20.34}$$

Consequently, $B(\otimes^{m^-} \mathcal{Z}, \otimes^{m^+} \mathcal{Z})$ can be identified with

$$B\left(\underset{1 \in \mathrm{Lg}^+(W)}{\otimes} \mathcal{Z}_1, \underset{k \in \mathrm{Lg}^-(W)}{\otimes} \mathcal{Z}_k \right). \tag{20.35}$$

Elements of (20.35) can be viewed as multi-linear functions on

$$\underset{1 \in \mathrm{Lg}^+(W)}{\Pi} \overline{\mathcal{Z}}_1 \times \underset{k \in \mathrm{Lg}^-(W)}{\Pi} \mathcal{Z}_k. \tag{20.36}$$

Indeed, consider an element of (20.35), denoted also by W. We associate with it a function

$$W\left(\{\overline{z}_1\}_{1 \in \mathrm{Lg}^+(W)}, \{z_k\}_{k \in \mathrm{Lg}^-(W)} \right) := \left(\underset{1 \in \mathrm{Lg}^+(W)}{\otimes} z_1 \Big| W \underset{k \in \mathrm{Lg}^-(W)}{\otimes} z_k \right), \tag{20.37}$$

where $z_1 \in \mathcal{Z}_{1^+}$, $z_k \in \mathcal{Z}_{1^-}$.

20.3.6 Evaluation of a Friedrichs diagram

Let $W_i \in B\left(\Gamma^{m_i^-}(\mathcal{Z}), \Gamma^{m_i^+}(\mathcal{Z})\right)$, $i = n, \ldots, 1$, be a sequence of Wick monomials. Let B be a Friedrichs diagram over $W_n \cdots W_1$ with $m^\pm = m^\pm(B)$.

Definition 20.40 *The evaluation of the Friedrichs diagram B, usually denoted by the same letter B, is defined by*

$$B := \overline{\Theta^{m^+}}^\# \Theta^{m^-\#} \underset{\{l^+, l^-\} \in \mathrm{Ln}(B)}{\Pi} \nabla_{\overline{z}_{l^+}} \cdot \nabla_{z_{l^-}} \prod_{j=n}^{1} W_j. \tag{20.38}$$

The above definition uses the polynomial interpretation of a Friedrichs vertex. W_j are treated as polynomials, as in (20.37).

(20.38) thus defines a multi-linear function with variables $\overline{z}_l \in \overline{Z}_l$, $l \in$ $Lg^+(W_n, \ldots, W_1)$, $\overline{z}_l \in \overline{Z}_l$, $l \in Lg^-(W_n, \ldots, W_1)$. The differential operator kills some of the variables; only those in $Lg^{\pm}(B)$ survive. Then we apply the symmetrization/anti-symmetrization operators.

It is also possible to give an equivalent definition that uses a purely operator language.

Definition 20.41 *For* $j = 1, \ldots, n$, *we set*

$$L^j(B) = \{l \in Lg^+(B) \ : \ j > nr(l)\}$$
$$\cup \{l \in Lg^-(B) \ : \ nr(l) > j\}$$
$$\cup \{(l_+, l_-) \in Ln(B) \ : \ nr(l_+) > j > nr(l_-)\}.$$

$L^j(B)$ *is called the* set of lines bypassing the vertex W_j.

Note that

$$Lg^-(B) = L^1(B) \cup Lg^-(W_1),$$
$$Lg^+(B) = L^n(B) \cup Lg^+(W_n).$$

Definition 20.42 *For each* $\ell = \{l_+, l_-\} \in Ln(B)$ *with* $nr(l_+) > nr(l_-)$, *let* Z_ℓ *denote the space* $Z_{l_-} \simeq Z_{l_+}$. *Let* $\mathbb{1}_B^j$ *denote the identity on*

$$\bigotimes_{\ell \in L^j(B)} Z_\ell.$$

Let W_j be interpreted as operators in

$$B\left(\bigotimes_{l \in Lg^-(W_i)} Z_l, \bigotimes_{l \in Lg^+(W_i)} Z_l \right).$$

In the operator language, the diagram B can be computed as

$$B = \Theta^{m^+}(W_n \otimes \mathbb{1}_B^n) \cdots (W_1 \otimes \mathbb{1}_B^1) \Theta^{m^-}.$$

20.3.7 Products of operators
Theorem 20.43

$$Op^{a^*, a}(W_n) \cdots Op^{a^*, a}(W_1) = \sum_{B \in FDg(W_n, \ldots, W_1)} Op^{a^*, a}(B),$$

$$(\Omega | Op^{a^*, a}(W_n) \cdots Op^{a^*, a}(W_1) \Omega) = \sum_{B \in FDg(W_n, \ldots, W_1)_{nl}} B.$$

Proof This is essentially a restatement of Thm. 9.36. □

This theorem describes a method of computing the Wick symbol of a product of operators. We first draw the Friedrichs vertices in the appropriate order. Then we draw all possible diagrams by joining the legs. Next we evaluate the diagrams, pairing the external legs with creation and annihilation operators. Finally, we sum up all the contributions.

To compute the vacuum expectation value, we do the same steps, except that we consider only diagrams without external legs.

20.4 Friedrichs diagrams and the scattering operator

In this section we describe how to use the formalism of Friedrichs diagrams to compute two quantities useful for many-body quantum theory and quantum field theory: the scattering operator and the energy shift. From the point of view of diagrams, the new feature is the time label that will appear on each vertex. We will always demand that the order of vertices is consistent with the order of time labels.

Throughout the section we keep the terminology and notation of the previous section.

20.4.1 Multiplication of Friedrichs diagrams

Definition 20.44 *A pair consisting of a Friedrichs vertex and* $t \in \mathbb{R}$ *will be called a* time-labeled Friedrichs vertex. *It will be typically denoted* $W(t)$. *A sequence* $(W_n(t_n), \dots, W_1(t_1))$ *of time-labeled Friedrichs vertices is* time-ordered *if* $t_n > \cdots > t_1$.

We will consider only time-ordered sequences of time-labeled Friedrichs vertices.

Definition 20.45 *Consider two sequences of time-labeled Friedrichs vertices* $(W_n(t_n), \dots, W_1(t_1))$ *and* $(W'_m(t'_m), \dots, W'_1(t'_1))$. *Assume that none of* $t_n, \dots, t_1$ *coincides with* $t'_m, \dots, t'_1$. *Let* $(s_{m+n}, \dots, s_1)$ *be the union of*

$$\{t_n, \dots, t_1\}, \quad \{t'_m, \dots, t'_1\}$$

in decreasing order. Let $(Q_{n+m}(s_{m+n}), \dots, Q_1(s_1))$ *be the time-ordered union of*

$$(W_n(t_n), \dots, W_1(t_1)), \quad (W'_m(t'_m), \dots, W'_1(t'_1)).$$

Note that we obtain an identification of $\mathrm{Lg}^{\pm}(W_n(t_n), \dots, W_1(t_1))$ *and* $\mathrm{Lg}^{\pm}(W'_m(t'_m), \dots, W'_1(t'_1))$ *with complementary subsets of* $\mathrm{Lg}^{\pm}(Q_{n+m}(s_{m+n}), \dots, Q_1(s_1))$.

Consider two Friedrichs diagrams

$$B \in \mathrm{FDg}(W_n(t_n), \dots, W_1(t_1)), \quad B' \in \mathrm{FDg}(W'_m(t'_m), \dots, W'_1(t'_1)).$$

$BB' = B'B$ is   defined   as   the   Friedrichs   diagram   in $\mathrm{FDg}\big(Q_{n+m}(s_{m+n}),\ldots,Q_1(s_1)\big)$ *such that*

$$\mathrm{Lg}^{\pm}(BB') = \mathrm{Lg}^{\pm}(B) \cup \mathrm{Lg}^{\pm}(B'), \quad \mathrm{Ln}(BB') = \mathrm{Ln}(B) \cup \mathrm{Ln}(B').$$

20.4.2 One-particle dynamics

Let h_p be a self-adjoint operator on $\mathcal{Z}_p$. Set $h := \underset{p \in \mathrm{Pr}}{\oplus} h_p$ as an operator on $\mathcal{Z}$. Let

$$H_0 = \mathrm{d}\Gamma(h). \tag{20.39}$$

Note that if W is a Wick monomial, then

$$e^{\mathrm{i}tH_0}\mathrm{Op}^{a^*,a}(W)e^{-\mathrm{i}tH_0} = \mathrm{Op}^{a^*,a}\big(e^{\mathrm{i}tH_0}We^{-\mathrm{i}tH_0}\big), \tag{20.40}$$

where on the right we interpret W as a Wick operator.

Let B be a Friedrichs diagram and $1 \in \mathrm{Lg}^{\pm}(B)$. Then h_1 will denote $h_{\mathrm{pr}(1)}$, understood as an operator on $\mathcal{Z}_1$. Similarly, if $\ell = (1^{(+)},1^{(-)}) \in \mathrm{Ln}(B)$, and $p = \mathrm{pr}(1^{(+)}) = \mathrm{pr}(1^{(-)})$, then h_ℓ denotes h_p understood as an operator from $\mathcal{Z}_{1^{(-)}}$ to $\mathcal{Z}_{1^{(+)}}$.

We will sometimes use the symbol H_0 in a meaning slightly different from (20.39).

Definition 20.46 *Let L be a subset of* $\mathrm{Lg}^+(B) \cup \mathrm{Lg}^-(B) \cup \mathrm{Ln}(B)$. *Then H_0, understood as an operator on the space* $\underset{\ell \in L}{\otimes} \mathcal{Z}_\ell$, *will denote the operator* $\sum_{\ell \in L} h_\ell$.

20.4.3 Time-dependent Wick monomials

Suppose that $\mathbb{R} \ni t \mapsto W_j(t)$, $j = 1,\ldots,r$, are fermion-even Wick monomials depending on time, each with a fixed multi-degree. We represent each W_j with a vertex, independent of the time t but distinct for distinct indices j.

We modify the prescription (20.38).

Definition 20.47 *The evaluation of the diagram B at times $t_n,\ldots,t_1 \in \mathbb{R}$ is*

$$B(t_n,\ldots,t_1) = \prod_{1 \in \mathrm{Lg}_+(B)} e^{\mathrm{i}t_{\mathrm{nr}(1)}h_1} \prod_{1 \in \mathrm{Lg}_-(B)} e^{-\mathrm{i}t_{\mathrm{nr}(1)}\overline{h}_1}$$

$$\times \prod_{\ell - \{1_{+},1_{-}\} \in \mathrm{Ln}(R)} \nabla_{\overline{z}_{1_{+}}} \cdot e^{\mathrm{i}(t_{\mathrm{nr}(1_{+})} - t_{\mathrm{nr}(1_{-})})h_\ell} \nabla_{z_{1_{-}}} \prod_{j=n}^{1} W_j(t_j).$$

In the operator language we have

$$B(t_n,\ldots,t_1) = \Theta_B^+ e^{\mathrm{i}t_n H_0} \big(W_n(t_n) \otimes \mathbb{1}_B^n\big) e^{-\mathrm{i}(t_n - t_{n-1})H_0} \cdots$$
$$\times e^{-\mathrm{i}(t_2 - t_1)H_0} \big(W_1(t_1) \otimes \mathbb{1}_B^1\big) e^{-\mathrm{i}t_1 H_0} \Theta_B^-.$$

The following identity follows immediately from Thm. 20.43 and (20.40).

Theorem 20.48

$$e^{it_n H_0} \mathrm{Op}^{a^*,a}\big(W_n(t_n)\big) e^{-i(t_n - t_{n-1})H_0} \cdots e^{-i(t_2 - t_1)H_0} \mathrm{Op}^{a^*,a}\big(W_1(t_1)\big) e^{-it_1 H_0}$$

$$= \sum_{B \in \mathrm{FDg}(W_n, \ldots, W_1)} \mathrm{Op}^{a^*,a}\big(B(t_n, \ldots, t_1)\big).$$

20.4.4 Diagrams for the scattering operator

Set

$$Q(t) := W_1(t) + \cdots + W_r(t),$$
$$H(t) := H_0 + \lambda \mathrm{Op}^{a^*,a}\big(Q(t)\big),$$

where $W_j(t)$ are self-adjoint Wick monomials. Our main goal is to describe a method of computing the scattering operator S for H_0 and $\{H(t)\}_{t \in \mathbb{R}}$ (see Def. 20.25). Recall that

$$S = \mathrm{Texp}\Big(-i\lambda \int \mathrm{Op}^{a^*,a}\big(Q(t)\big)_{\mathrm{I}} dt\Big), \tag{20.41}$$

where

$$\mathrm{Op}^{a^*,a}\big(Q(t)\big)_{\mathrm{I}} := e^{itH_0} \mathrm{Op}^{a^*,a}\big(Q(t)\big) e^{-itH_0} = \mathrm{Op}^{a^*,a}\big(Q_{\mathrm{I}}(t)\big),$$

with $Q_{\mathrm{I}}(t) = \Gamma(e^{ith})Q(t)\Gamma(e^{-ith})$.

We denote by W_j the Friedrichs vertices in the sense of Def. 20.33 corresponding to the Wick monomials $W_j(t)$. We also often need to use the corresponding time-labeled vertices, which we denote $W_j(t)$.

For brevity, we will denote by $\mathfrak{W}$ the set of vertices $\{W_1, \ldots, W_r\}$.

Definition 20.49 *We introduce the notation*

$$\mathrm{FDg}_n\{\mathfrak{W}\} := \bigcup_{(j_n, \ldots, j_1) \in \{1, \ldots, r\}^n} \mathrm{FDg}(W_{j_n} \cdots W_{j_1}). \tag{20.42}$$

Note that $\mathrm{FDg}(W_{j_n} \cdots W_{j_1})$ are disjoint for distinct sequences $(j_n, \ldots, j_1) \in \{1, \ldots, r\}^n$. Therefore, the union in (20.42) involves disjoint sets.

Note also that when we evaluate a diagram in (20.42) on the monomials $W_i(t_i)$, we obtain a function that depends on $t_n, \ldots, t_1 \in \mathbb{R}$.

The following theorem follows easily from (20.41) and Thm. 20.43:

Theorem 20.50

$$S = \sum_{n=0}^{\infty} \sum_{B \in \mathrm{FDg}_n\{\mathfrak{W}\}} (-i\lambda)^n \int \cdots \int_{t_n > \cdots > t_1} \mathrm{Op}^{a^*,a}\big(B(t_n, \ldots, t_1)\big) dt_n \cdots dt_1.$$

The above theorem is an analog of Thm. 20.20 about the Gaussian integration. Recall that Thm. 20.20 implies Thm. 20.21, the linked cluster theorem for the Gaussian integration. Thm. 20.50 has an analogous consequence:

Theorem 20.51 (Linked cluster theorem for Friedrichs diagrams)

$$S = \mathrm{Op}^{a^*,a}\left(\exp\left(\sum_{n=0}^{\infty}\sum_{B\in\mathrm{FDg}_n\{\mathfrak{W}\}_{\mathrm{con}}}(-\mathrm{i}\lambda)^n\right.\right.$$
$$\left.\left.\times\int\cdots\int_{t_n>\cdots>t_1}B(t_n,\ldots,t_1)\mathrm{d}t_n\cdots\mathrm{d}t_1\right)\right)$$

$$\log\left(\Omega|S\Omega\right) = \sum_{n=0}^{\infty}\sum_{B\in\mathrm{FDg}_n\{\mathfrak{W}\}_{\mathrm{cnl}}}(-\mathrm{i}\lambda)^n\int\cdots\int_{t_n>\cdots>t_1}B(t_n,\ldots,t_1)\mathrm{d}t_n\cdots\mathrm{d}t_1,$$

$$\frac{S}{(\Omega|S\Omega)} = \mathrm{Op}^{a^*,a}\left(\exp\left(\sum_{n=0}^{\infty}\sum_{B\in\mathrm{FDg}_n\{\mathfrak{W}\}_{\mathrm{link}}}(-\mathrm{i}\lambda)^n\right.\right.$$
$$\left.\left.\times\int\cdots\int_{t_n>\cdots>t_1}B(t_n,\ldots,t_1)\mathrm{d}t_n\cdots\mathrm{d}t_1\right)\right).$$

20.4.5 Stationary evaluation of a diagram

Let us now assume that the monomials $W_i(t) = W_i$ do not depend on time and $Q = W_1 + \cdots + W_r$. Let $H := H_0 + \lambda\mathrm{Op}^{a^*,a}(Q)$.

Let $\xi \in \mathbb{C}\backslash\mathrm{spec}(H_0)$.

Definition 20.52 *For a diagram $B \in \mathrm{FDg}(W_n,\ldots,W_1)$ and $\xi \in \mathbb{C}$ we define its stationary evaluation as*

$$B[\xi] := \Theta_B^+\left(W_n \otimes \mathbb{1}_B^n\right)(\xi\mathbb{1} - H_0)^{-1}\cdots(\xi\mathbb{1} - H_0)^{-1}\left(W_1 \otimes \mathbb{1}_B^1\right)\Theta_B^-.$$

20.4.6 Scattering operator for a time-independent Hamiltonian

The Gell-Mann–Low scattering operator

$$S_{\mathrm{GL}} = \lim_{\epsilon\searrow 0}(\Omega|S_\epsilon^+\Omega)^{-1}S_\epsilon$$

is often used as the starting point for computations in quantum field theory. In the following theorem we give two expressions for this operator: a time-dependent one and a stationary one. Note that the division by $(\Omega|S_\epsilon^+\Omega)$ removes diagrams without external legs, which if non-zero would give a divergent contribution.

Theorem 20.53

$$S_{\mathrm{GL}} = \mathrm{Op}^{a^*,a}\left(\exp\left(\sum_{n=0}^{\infty}\sum_{B\in\mathrm{FDg}_n\,\{\mathfrak{W}\}_{\mathrm{link}}}(-i\lambda)^n\int_{t_n>\cdots>t_1}\!\!\cdots\int B(t_n,\ldots,t_1)\mathrm{d}t_n\cdots\mathrm{d}t_1\right)\right)$$

$$= \mathrm{Op}^{a^*,a}\left(\exp\left(-2i\pi\sum_{n=0}^{\infty}\sum_{B\in\mathrm{FDg}_n\,\{\mathfrak{W}\}_{\mathrm{link}}}\lambda^n\right.\right.$$

$$\left.\left.\times\int\delta(H_0-\xi\mathbb{1})B[\xi+i0]\delta(H_0-\xi\mathbb{1})\mathrm{d}\xi\right)\right).$$

Proof The first identity follows from Thm. 20.51. Next we compute the integrand using the operator interpretation of $B(t_n,\ldots,t_1)$:

$$B(t_n,\ldots,t_1) = \Theta_B^+ e^{it_n H_0}\left(W_n\otimes\mathbb{1}_B^n\right)e^{-i(t_n-t_{n-1})H_0}\cdots$$

$$\times e^{-i(t_2-t_1)H_0}\left(W_1\otimes\mathbb{1}_B^1\right)e^{-it_1 H_0}\Theta_B^-$$

$$= \int\delta(H_0-\xi\mathbb{1})\mathrm{d}\xi\Theta_B^+\left(W_n\otimes\mathbb{1}_B^n\right)e^{-iu_n(H_0-\xi\mathbb{1})}\cdots$$

$$\times e^{-iu_2(H_0-\xi\mathbb{1})}\left(W_1\otimes\mathbb{1}_B^1\right)e^{-it_1(H_0-\xi\mathbb{1})}\Theta_B^-,$$

where

$$u_n := t_n-t_{n-1},\ldots,\quad u_2 := t_2-t_1.$$

Now

$$\int_{t_n>\cdots>t_1}\!\!\cdots\int B(t_n,\ldots,t_1)\mathrm{d}t_n\cdots\mathrm{d}t_1$$

$$= \int\mathrm{d}\xi\int_0^{\infty}\mathrm{d}u_n\cdots\int_0^{\infty}\mathrm{d}u_1\int_{-\infty}^{\infty}\mathrm{d}t_1\delta(H_0-\xi\mathbb{1})\Theta_B^+\left(W_n\otimes\mathbb{1}_B^n\right)e^{-iu_n(H_0-\xi\mathbb{1})}\cdots$$

$$\times e^{-iu_2(H_0-\xi\mathbb{1})}\left(W_1\otimes\mathbb{1}_B^1\right)e^{-it_1(H_0-\xi\mathbb{1})}\Theta_B^-$$

$$= 2\pi(-i)^{n-1}\int\mathrm{d}\xi\delta(H_0-\xi\mathbb{1})\Theta_B^+\left(W_n\otimes\mathbb{1}_B^n\right)\left(H_0-(\xi-i0)\mathbb{1}\right)^{-1}\cdots$$

$$\times\left(H_0-(\xi-i0)\mathbb{1}\right)^{-1}\left(W_1\otimes\mathbb{1}_B^1\right)\delta(H_0-\xi\mathbb{1})\Theta_B^-,$$

where we have used the heuristic relations

$$\int_0^{+\infty}e^{iu(H_0-\xi\mathbb{1})}\mathrm{d}u = i\left(H_0-(\xi+i0)\mathbb{1}\right)^{-1},\qquad(20.43)$$

$$\int_{-\infty}^0 e^{iu(H_0-\xi\mathbb{1})}\mathrm{d}u = -i\left(H_0-(\xi-i0)\mathbb{1}\right)^{-1},\qquad(20.44)$$

$$\int e^{it(H_0-\xi\mathbb{1})}\mathrm{d}t = 2\pi\delta(H_0-\xi\mathbb{1}).\qquad(20.45)$$

$\square$

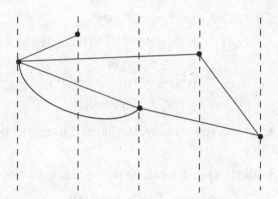

Figure 20.9 Goldstone diagram.

20.4.7 Goldstone theorem

Recall that E denotes the ground-state energy of H, that is, $E := \inf \operatorname{spec} H$. We assume that we can use the heuristic formula for the energy shift

$$E = \lim_{t \to -\infty} \mathrm{i}^{-1} \frac{\mathrm{d}}{\mathrm{d}t} \log(\Omega | \mathrm{e}^{\mathrm{i}tH} \mathrm{e}^{-\mathrm{i}tH_0} \Omega), \qquad (20.46)$$

which follows from (20.23) if we note that $E_0 = 0$. Then we can derive the following diagrammatic expansion for the energy E:

Theorem 20.54 (Goldstone theorem)

$$E = \sum_{n=0}^{\infty} \sum_{B \in \mathrm{FD}g_n \{\mathfrak{W}\}_{\mathrm{cnl}}} \lambda^n B[0].$$

Proof As explained at the end of Subsect. 20.2.2, $\mathrm{e}^{\mathrm{i}tH} \mathrm{e}^{-\mathrm{i}tH_0}$ for $t < 0$ is the scattering operator for the time-dependent perturbation $s \mapsto \lambda \mathbb{1}_{[t,0]}(s) \mathrm{Op}^{a^*,a}(Q)$. Applying Thm. 20.51, we get

$$\log(\Omega | \mathrm{e}^{\mathrm{i}tH} \mathrm{e}^{-\mathrm{i}tH_0} \Omega)$$
$$= \sum_{n=0}^{\infty} \sum_{B \in \mathrm{FD}g_n \{\mathfrak{W}\}_{\mathrm{cnl}}} (-\mathrm{i}\lambda)^n \int \cdots \int_{0 > t_n > \cdots > t_1 > t} B(t_n, \ldots, t_1) \mathrm{d}t_n \cdots \mathrm{d}t_1.$$

So

$$\mathrm{i}^{-1} \frac{\mathrm{d}}{\mathrm{d}t} \log(\Omega | \mathrm{e}^{\mathrm{i}tH} \mathrm{e}^{-\mathrm{i}tH_0} \Omega)$$
$$= \sum_{n=0}^{\infty} \sum_{B \in \mathrm{FD}g_n \{\mathfrak{W}\}_{\mathrm{cnl}}} \mathrm{i}(-\mathrm{i}\lambda)^n \int \cdots \int_{0 > t_n > \cdots > t_2 > t} B(t_n, \ldots, t_2, t) \mathrm{d}t_n \cdots \mathrm{d}t_2.$$

Now introduce

$$u_2 := t - t_2, \ldots, u_n := t_{n-1} - t_n.$$

Then $u_2, \ldots, u_n \leq 0$, $t \leq u_2 + \cdots + u_n \leq 0$, and

$$
\begin{aligned}
B(t_n, \ldots, t_2, t) &= W_n \mathrm{e}^{-\mathrm{i}(t_n - t_{n-1})H_0} \left(W_{n-1} \otimes \mathbb{1}_B^{n-1} \right) \cdots \\
&\quad \times \left(W_2 \otimes \mathbb{1}_B^2 \right) \mathrm{e}^{-\mathrm{i}(t_2 - t)H_0} W_1 \\
&= W_n \mathrm{e}^{\mathrm{i} u_n H_0} \left(W_{n-1} \otimes \mathbb{1}_B^{n-1} \right) \cdots \\
&\quad \times \left(W_2 \otimes \mathbb{1}_B^2 \right) \mathrm{e}^{\mathrm{i} u_2 H_0} W_1 .
\end{aligned}
$$

Then we replace t by $-\infty$ and evaluate the integral using the heuristic relation (20.44). □

Note that an identical expansion can be derived from Sucher's formula,

$$
E = \lim_{\epsilon \searrow 0} \frac{\mathrm{i}\epsilon\lambda}{2} \partial_\lambda \log(\Omega | S_\epsilon \Omega). \tag{20.47}
$$

20.5 Feynman diagrams and vacuum expectation value

We continue to study diagrammatic expansions of many-body quantum physics and quantum field theory. Until the end of this chapter we will, however, use diagrams different from those of the previous two sections: the so-called *Feynman diagrams*. They are closely related to the diagrams discussed in Sect. 20.1. The main topic of that section was integration w.r.t. a Gaussian measure. This includes in particular the Euclidean quantum field theory. We will see that the formalism of Sect. 20.1 can be adapted to compute scattering operators in many-body quantum theory and quantum field theory. In practice, Feynman diagrams are usually preferred over the Friedrichs diagrams of Sects. 20.3 and 20.4. Their main advantages are a smaller number of diagrams and, in the case of relativistic theories, manifest Lorentz covariance of each diagram.

The main idea in passing from Friedrichs diagrams to Feynman diagrams consists in combining the evolution going forwards and backwards in time in a single line. It is done in a different way for neutral and charged particles. The starting point of the formalism is usually a classical system, neutral or charged, described by its dual phase space $\mathcal{Y}_p$. The one-particle space $\mathcal{Z}_p$ is introduced in the standard way, as explained in Chap. 18. In the case of neutral particles, the lines with both time directions are combined into one unoriented line. In the case of charged lines, one combines particles going forwards and anti-particles going backwards in a single line decorated with an arrow pointing forwards. Similarly, one combines particles going backwards and anti-particles going forwards in a single line oriented backwards.

Our discussion of Feynman diagrams in many-body quantum physics and quantum field theory is divided into two sections. In this section we will show how to compute the vacuum expectation value of scattering operators. This method can be interpreted as a special case of the formalism described in Sect. 20.1 on the Gaussian integration. In Sect. 20.6 we will describe how to compute scattering operators. Diagrams needed for scattering operators have some features of Friedrichs diagrams, since their external legs are divided into incoming and outgoing ones.

Throughout this and the next section we will use the terminology and notation of Sect. 20.1. In particular Pr will denote the set of particles, divided into four parts $\mathrm{Pr} = \mathrm{Pr}_\mathrm{s}^\mathrm{n} \cup \mathrm{Pr}_\mathrm{s}^\mathrm{c} \cup \mathrm{Pr}_\mathrm{a}^\mathrm{n} \cup \mathrm{Pr}_\mathrm{a}^\mathrm{c}$, as in Defs. 20.3 and 20.4. We will often write $\Gamma_1(\mathcal{Z})$ instead of $\Gamma_\mathrm{s}(\mathcal{Z})$ and $\Gamma_{-1}(\mathcal{Z})$ instead of $\Gamma_\mathrm{a}(\mathcal{Z})$.

Let us first describe the constructions related to the free dynamics. As usual, it is convenient to describe separately the neutral and charged cases.

20.5.1 Free neutral particles

We assume that for every $p \in \mathrm{Pr}^\mathrm{n}$ we are given a real dual phase space $\mathcal{Y}_p$ equipped with a dynamics $\{r_{p,t}\}_{t \in \mathbb{R}}$. More precisely,

(1) for $p \in \mathrm{Pr}_\mathrm{s}^\mathrm{n}$, $\mathcal{Y}_p$ is a symplectic vector space and $\{r_{p,t}\}_{t \in \mathbb{R}}$ is a stable symplectic dynamics on $\mathcal{Y}_p$;
(2) for $p \in \mathrm{Pr}_\mathrm{a}^\mathrm{n}$, $\mathcal{Y}_p$ is a real Hilbert space and $\{r_{p,t}\}_{t \in \mathbb{R}}$ is a non-degenerate orthogonal dynamics on $\mathcal{Y}_p$.

We use the constructions described in Sect. 18.1. In particular, we write $r_{p,t} = \mathrm{e}^{ta_p}$, and construct the corresponding one-particle spaces $\mathcal{Z}_p$ and the one-particle Hamiltonians $h_p > 0$. Recall that we have a natural decomposition $\mathbb{C}\mathcal{Y}_p = \mathcal{Z}_p \oplus \overline{\mathcal{Z}}_p$, and $a_{p\mathbb{C}} = \mathrm{i}h_p \oplus (-\mathrm{i}\overline{h}_p)$. On the Fock space $\Gamma_{\epsilon_p}(\mathcal{Z}_p)$ we have the Hamiltonian $\mathrm{d}\Gamma(h_p)$ and the fields $\mathcal{Y}_p \ni \zeta \mapsto \phi_p(\zeta)$. We write $\phi_{p,t}(\zeta) := \phi_p(r_{p,-t}\zeta)$.

20.5.2 Free charged particles

We assume that for every $q \in \mathrm{Pr}^\mathrm{c}$ we are given a complex dual phase space $\mathcal{Y}_q$ equipped with a dynamics $\{r_{q,t}\}_{t \in \mathbb{R}}$. More precisely,

(1) for $q \in \mathrm{Pr}_\mathrm{s}^\mathrm{c}$, $\mathcal{Y}_q$ is a charged symplectic vector space and $\{r_{q,t}\}_{t \in \mathbb{R}}$ is a stable charged symplectic dynamics;
(2) for $q \in \mathrm{Pr}_\mathrm{a}^\mathrm{c}$, $\mathcal{Y}_q$ is a complex Hilbert space and $\{r_{q,t}\}_{t \in \mathbb{R}}$ is a non-degenerate unitary dynamics.

Following Sect 18.2, we write $r_{q,t} = \mathrm{e}^{\mathrm{i}tb_q}$, and construct the corresponding one-particle spaces $\mathcal{Z}_q$ and the one-particle Hamiltonians $h_q > 0$. We have a natural decomposition $\mathcal{Y}_q = \mathcal{Y}_q^{(+)} \oplus \mathcal{Y}_q^{(-)}$, with $b_q = b_q^{(+)} \oplus (\ b_q^{(-)})$. Then $\mathcal{Z}_q^{(+)} = \mathcal{Y}_q^{(+)}$, $\mathcal{Z}_q^{(-)} = \overline{\mathcal{Y}}_q^{(-)}$, so that $\mathcal{Z}_q = \mathcal{Z}_q^{(+)} \oplus \mathcal{Z}_q^{(-)}$, $h_q = b_q^{(+)} \oplus \overline{b_q^{(-)}}$. On the Fock space $\Gamma_{\epsilon_q}(\mathcal{Z}_q)$ we have the Hamiltonian $\mathrm{d}\Gamma(h_q)$ and the field $\mathcal{Y}_q \ni \zeta \mapsto \psi_q^*(\zeta)$. We set $\psi_{q,t}^*(\zeta) := \psi_q^*(r_{q,-t}\zeta)$.

20.5.3 Full Hilbert space

Definition 20.55 *Sometimes, for brevity, we will write*

$$\mathcal{Z} = \bigoplus_{p \in \mathrm{Pr}} \mathcal{Z}_p$$

for the total one-particle space.

Clearly, $\mathcal{Z}$ can be treated as a super-space with the grading $\epsilon = \bigoplus_{p \in \mathrm{Pr}} \epsilon_p \mathbb{1}_{\mathcal{Z}_p}$.

The Hilbert space of the system will be

$$\Gamma(\mathcal{Z}) \simeq \bigotimes_{p \in \mathrm{Pr}} \Gamma_{\epsilon_p}(\mathcal{Z}_p).$$

The free Hamiltonian is

$$H_0 = \sum_{p \in \mathrm{Pr}} \mathrm{d}\Gamma(h_p).$$

20.5.4 Wick's time-ordered product

In the presence of fermionic degrees of freedom, it is convenient to modify the definition of the time-ordered product. The so-called Dyson's time-ordered product, defined in Def. 20.22, will be replaced by Wick's time-ordered product, which takes into account the fermionic nature of some operators.

Definition 20.56 *An operator B on $\Gamma(\mathcal{Z})$ is called* bosonic, *resp.* fermionic *if $B = \Gamma(\epsilon)B\Gamma(\epsilon)$, resp. $B = -\Gamma(\epsilon)B\Gamma(\epsilon)$.*

Definition 20.57 *Let $\mathbb{R} \ni t \mapsto B_k(t), \ldots, B_1(t)$ be time-dependent operators, each either bosonic or fermionic. Let $t_n, \ldots, t_1 \in \mathbb{R}$ be pairwise distinct. We define* Wick's time-ordered product *of $B_n(t_n), \ldots, B_1(t_1)$ by*

$$\mathrm{T}\big(B_n(t_n) \cdots B_1(t_1)\big) := \mathrm{sgn}_{\mathrm{a}}(\sigma) B_{\sigma_n}(t_{\sigma_n}) \cdots B_{\sigma_1}(t_{\sigma_1}),$$

where $\sigma_n, \ldots, \sigma_1$ is a permutation of $n, \ldots, 1$ such that $t_{\sigma_n} \geq \cdots \geq t_{\sigma_1}$, and $\mathrm{sgn}_{\mathrm{a}}(\sigma)$ is the sign of the permutation of the fermionic elements among $B_n(t_n), \ldots, B_1(t_1)$.

20.5.5 Feynman 2-point functions: general remarks

An important ingredient of Feynman's diagrammatic approach to quantum field theory is the so-called *Feynman's 2-point functions*. They are given by the vacuum expectation values of time-ordered products of fields. They will be discussed in Subsects. 20.5.6–20.5.9. We will consider separately the neutral and charged cases, which are very similar.

In practice, in the bosonic case one uses two kinds of 2-point functions: the *phase-space* and the *configuration space 2-point functions*. We start with a description of the phase space 2-point functions, since they can be discussed in a parallel way for bosons and fermions. However, in the bosonic case, one usually prefers to use 2-point configuration space functions. They will be discussed separately in Subsects. 20.5.8 and 20.5.9. They are used in particular when the interaction depends only on configuration space, which is often the case.

As usual, we will use $t \in \mathbb{R}$ to denote the time variable. The variable $E \in \mathbb{R}$ will have the meaning of energy.

20.5.6 Feynman's phase space 2-point functions for neutral particles

Let us start with neutral particles, bosonic or fermionic.

Definition 20.58 *For* $p \in \mathrm{Pr}_s^n$, *resp.* $p \in \mathrm{Pr}_a^n$ *the corresponding* Feynman's *phase space 2-point function is the function with values in operators on* $\mathbb{C}\mathcal{Y}_p$ *defined as*

$$S_p(t) := \theta(t)\mathrm{e}^{\mathrm{i}th_p}\, \mathbb{1}_{\mathcal{Z}_p} \pm \theta(-t)\mathrm{e}^{-\mathrm{i}t\overline{h}_p}\, \mathbb{1}_{\overline{\mathcal{Z}}_p}.$$

Note that if $\zeta_1, \zeta_2 \in \mathcal{Y}_p$, then

$$\zeta_1 \cdot S_p(t)\zeta_2 = \big(\Omega | \mathrm{T}\big(\phi_t(\zeta_1)\phi_0(\zeta_2)\big)\Omega\big).$$

The Fourier transform of S_p is

$$\hat{S}_p(E) = (\mathrm{i}h_p - \mathrm{i}E)^{-1}\mathbb{1}_{\mathcal{Z}_p} \mp (-\mathrm{i}\overline{h}_p - \mathrm{i}E)^{-1}\mathbb{1}_{\overline{\mathcal{Z}}_p}.$$

If $p \in \mathrm{Pr}_a^n$, this simplifies to

$$\hat{S}_p(E) = (a_p - \mathrm{i}E)^{-1}.$$

On the space $C_c^\infty(\mathbb{R}, \mathbb{C}\mathcal{Y}_p)$ we obtain a symmetric, resp. anti-symmetric form

$$f_1 \cdot S_p f_2 := \int \int f_1(t_1) \cdot S_p(t_1 - t_2) f_2(t_2) \mathrm{d}t_1 \mathrm{d}t_2.$$

20.5.7 Feynman's phase space 2-point functions for charged particles

Next we consider bosonic and fermionic charged particles.

Definition 20.59 *For* $q \in \mathrm{Pr}_s^c$, *resp.* $q \in \mathrm{Pr}_a^c$ *the corresponding* Feynman's *phase space 2-point function is the function with values in operators on* $\mathcal{Y}_q$ *defined as*

$$S_q(t) := \theta(t)\mathrm{e}^{\mathrm{i}tb_q^{(+)}}\, \mathbb{1}_{\mathcal{Y}_q^{(+)}} \pm \theta(-t)\mathrm{e}^{-\mathrm{i}tb_q^{(-)}}\, \mathbb{1}_{\mathcal{Y}_q^-}.$$

Note that if $\zeta_1, \zeta_2 \in \mathcal{Y}_q$, then

$$\overline{\zeta}_1 \cdot S_q(t)\zeta_2 = \big(\Omega | \mathrm{T}\big(\psi_t(\zeta_1)\psi_0^*(\zeta_2)\big)\Omega\big).$$

The Fourier transform of S_q is

$$\hat{S}_q(E) = \mathrm{i}^{-1}(b_q^{(+)} - E)^{-1}\mathbb{1}_{\mathcal{Y}_q^{(+)}} \mp \mathrm{i}^{-1}(-b_q^{(-)} - E)^{-1}\mathbb{1}_{\mathcal{Y}_q^{(-)}}.$$

If $q \in \mathrm{Pr}_{\mathrm{a}}^{\mathrm{c}}$, this simplifies to

$$\hat{S}_q(E) = \mathrm{i}^{-1}(b_q - E)^{-1}.$$

On the space $C_{\mathrm{c}}^{\infty}(\mathbb{R}, \mathcal{Y}_q)$ we obtain a Hermitian, resp. anti-Hermitian form

$$\overline{f}_1 \cdot S_q f_2 := \int \int \overline{f_1(t_1)} \cdot S_q(t_1 - t_2) f_2(t_2) \mathrm{d}t_1 \mathrm{d}t_2.$$

20.5.8 Feynman's configuration space 2-point functions for neutral bosons

Consider a neutral boson whose phase space is split into a configuration and momentum space.

More precisely, suppose that $p \in \mathrm{Pr}_{\mathrm{s}}^{\mathrm{n}}$ and $\tau_p \in L(\mathcal{Y}_p)$ satisfies

$$(\tau_p y_1) \cdot \omega_p \tau_p y_2 = -y_1 \cdot \omega_p y_2, \quad y_1, y_2 \in \mathcal{Y},$$
$$\tau_p a_p = -\tau_p a_p, \quad \tau_p^2 = \mathbb{1}_{\mathcal{Y}_p}.$$

Set

$$\mathcal{X}_p := \{y \in \mathcal{Y}_p \ : \ \tau_p y = y\},$$
$$\Xi_p := \{y \in \mathcal{Y}_p \ : \ \tau_p y = -y\},$$

and $\mathbb{1}_{\mathcal{X}_p} := \frac{1}{2}(\mathbb{1}_{\mathcal{Y}} + \tau_p)$.

In other words, τ_p is a *time reversal* in the terminology of Def. 18.13, the dynamics is time reversal invariant and $\mathcal{X}_p$, resp. Ξ_p is the corresponding *configuration*, resp. *momentum space* according to Subsect. 18.3.1, and $\mathbb{1}_{\mathcal{X}_p}$ is the projection onto $\mathcal{X}_p$ along Ξ_p. Following our standard notation, $\mathbb{1}_{\mathcal{X}_p, \mathbb{C}}$ denotes the linear extension of $\mathbb{1}_{\mathcal{X}_p}$ to the projection onto $\mathbb{C}\mathcal{X}_p$ along $\mathbb{C}\Xi_p$.

Definition 20.60 *The* configuration space Feynman's 2-point function *is the function with values in operators on* $\mathbb{C}\mathcal{X}_p$ *defined as*

$$D_p(t) := \mathbb{1}_{\mathcal{X}_p, \mathbb{C}} S_p(t) \mathbb{1}_{\mathcal{X}_p, \mathbb{C}},$$

where $S_p(t)$ *was introduced in Def. 20.58.*

Define

$$T_p := \mathbb{1}_{\mathcal{Z}_p} \mathbb{1}_{\mathcal{X}_p, \mathbb{C}}$$

as a map $T_p : \mathbb{C}\mathcal{X}_p \to \mathcal{Z}_p$. Note that $\tau_{p, \mathbb{C}}$ is a unitary map transforming $\mathcal{Z}_p$ onto $\overline{\mathcal{Z}}_p$ and such that $\tau_{p, \mathbb{C}} h_p \tau_{p, \mathbb{C}}^{-1} = \overline{h}_p$. Therefore,

$$D_p(t) = T_p^* \mathrm{e}^{\mathrm{i}|t|h_p} T_p.$$

The Fourier transform of D_p is

$$\hat{D}_p(E) = T_p^* \frac{2h_p}{E^2 - h_p^2} T_p.$$

On the space $C_c^\infty(\mathbb{R}, \mathbb{C}\mathcal{X}_p)$ we obtain a symmetric form,

$$g_1 \cdot D_p g_2 := \int \int g_1(t_1) \cdot D_p(t_1 - t_2) g_2(t_2) dt_1 dt_2.$$

20.5.9 Feynman's configuration space 2-point functions for charged bosons

Consider now a charged boson whose phase space is split into a configuration and momentum space.

More precisely, let $q \in \mathrm{Pr}_s^c$ and suppose that κ_q is a linear map on $\mathcal{Y}_q$ satisfying

$$\overline{\kappa_q y_1} \cdot \omega_q \kappa_q y_2 = -\overline{y}_1 \cdot \omega_q y_2, \quad y_1, y_2 \in \mathcal{Y},$$

$$\kappa_q b_q = -\kappa_q b_q, \quad \kappa_q^2 = \mathbb{1}_{\mathcal{Y}_q}.$$

Set

$$\mathcal{X}_p := \{y \in \mathcal{Y}_p : \kappa_p y = y\},$$
$$\Xi_p := \{y \in \mathcal{Y}_p : \kappa_p y = -y\},$$

and $\mathbb{1}_{\mathcal{X}_p} := \frac{1}{2}(\mathbb{1}_{\mathcal{Y}} + \kappa_p)$. In other words, κ_q is an involutive *Racah time reversal* in the terminology of Def. 18.38, the dynamics is Racah time reversal invariant and $\mathcal{X}_p$, resp. Ξ_p is the *configuration*, resp. *momentum space* in the terminology of Subsect. 18.3.4, and $\mathbb{1}_{\mathcal{X}_p}$ is the projection onto $\mathcal{X}_p$ along Ξ_p.

Definition 20.61 *The configuration space Feynman's 2-point function is the function with values in operators on $\mathcal{X}_q$ defined as*

$$D_q(t) := \mathbb{1}_{\mathcal{X}_q} S_q(t) \mathbb{1}_{\mathcal{X}_q},$$

where $S_q(t)$ was introduced in Def. 20.58.

Define

$$T_q := \mathbb{1}_{\mathcal{Y}_q^{(+)}} \mathbb{1}_{\mathcal{X}_q}$$

as a map $T_q : \mathcal{X}_q \to \mathcal{Y}_q^{(+)}$. Note that κ_q is a unitary map transforming $\mathcal{Y}_q^{(+)}$ onto $\mathcal{Y}_q^{(-)}$ and such that $\kappa_q b_q^{(+)} \kappa_q^{-1} = b_q^{(-)}$. Therefore,

$$D_q(t) = T_q^* e^{\mathrm{i}|t|b_q^{(+)}} T_q.$$

The Fourier transform of D_q is

$$\hat{D}_q(E) = T_q^* \frac{2b_q^{(+)}}{E^2 - (b_q^{(+)})^2} T_q.$$

On the space $C_c^\infty(\mathbb{R}, \mathcal{X}_q')$ we obtain a Hermitian form,

$$\overline{g}_1 \cdot D_q g_2 := \int \int \overline{g_1(t_1)} \cdot D_q(t_1 - t_2) g_2(t_2) dt_1 dt_2.$$

20.5.10 Wick quantization of Feynman polynomials

In the Feynman formalism, perturbations are described by polynomials on the phase space.

Definition 20.62 *We set*

$$\mathcal{Y}^{\mathrm{n}} := \bigoplus_{p\in\mathrm{Pr}^{\mathrm{n}}} \mathcal{Y}_p, \quad \mathcal{Y}^{\mathrm{c}} := \bigoplus_{q\in\mathrm{Pr}^{\mathrm{c}}} \mathcal{Y}_q.$$

A polynomial on $\mathcal{Y} := \mathbb{C}\mathcal{Y}^{\mathrm{n}} \oplus \mathcal{Y}^{\mathrm{c}} \oplus \overline{\mathcal{Y}}^{\mathrm{c}}$ *will be called a* Feynman polynomial.

Definition 20.63 *We set*

$$\mathcal{Z}^{\mathrm{n}} := \bigoplus_{p\in\mathrm{Pr}^{\mathrm{n}}} \mathcal{Z}_p, \quad \mathcal{Z}^{(\pm)} := \bigoplus_{q\in\mathrm{Pr}^{\mathrm{c}}} \mathcal{Z}_q^{(\pm)}.$$

Clearly, $\mathbb{C}\mathcal{Y}^{\mathrm{n}} = \mathcal{Z}^{\mathrm{n}} \oplus \overline{\mathcal{Z}}^{\mathrm{n}}$, $\overline{\mathcal{Y}}^{\mathrm{c}} = \overline{\mathcal{Z}}^{(+)} \oplus \mathcal{Z}^{(-)}$, $\mathcal{Y}^{\mathrm{c}} = \mathcal{Z}^{(+)} \oplus \overline{\mathcal{Z}}^{(-)}$. Therefore, we can identify $\mathcal{Z} \oplus \overline{\mathcal{Z}}$ with $\mathcal{Y}$, where $\mathcal{Z}$ is defined in Def. 20.55. It is convenient to introduce a special notation for this identification.

Definition 20.64 $\rho : \mathcal{Z} \oplus \overline{\mathcal{Z}} \to \mathcal{Y}$ *denotes the map*

$$\rho(z_1^{\mathrm{n}}, z_1^{(+)}, z_1^{(-)}, \overline{z}_2^{\mathrm{n}}, \overline{z}_2^{(+)}, \overline{z}_2^{(-)}) := (z_1^{\mathrm{n}} \oplus \overline{z}_2^{\mathrm{n}}, z_1^{(+)} \oplus \overline{z}_2^{(-)}, \overline{z}_2^{(+)} \oplus z_1^{(-)}).$$

Definition 20.65 *Given a Feynman polynomial* $G \in \mathrm{Pol}(\mathcal{Y})$, *we will write* $G \circ \rho := \Gamma(\rho^{\#})G$, *which is a polynomial in* $\mathrm{Pol}(\overline{\mathcal{Z}} \oplus \mathcal{Z})$. *Its Wick quantization, which is an operator on* $\Gamma(\mathcal{Z})$, *will have a special notation:*

$$:G(\phi, \psi^*, \psi): := \mathrm{Op}^{a^*,a}(G \circ \rho). \tag{20.48}$$

We will use the concept of the multi-degree introduced in Def. 20.5. The following definition is parallel to Def. 20.15:

Definition 20.66 *Given a multi-degree* m, *we define*

$$\mathrm{Pol}^m(\mathcal{Z}) := \Big(\bigotimes_{p\in\mathrm{Pr}^{\mathrm{n}}} \mathrm{Pol}_{\epsilon_p}^{m_p}(\mathcal{Z}_p) \Big)$$

$$\otimes \Big(\bigotimes_{q\in\mathrm{Pr}^{\mathrm{c}}} \mathrm{Pol}_{\epsilon_q}^{m_q^{(+)}}(\mathcal{Z}_q^{(+)}) \Big) \otimes \Big(\bigotimes_{q'\in\mathrm{Pr}^{\mathrm{c}}} \mathrm{Pol}_{\epsilon_{q'}}^{m_{q'}^{(-)}}(\mathcal{Z}_{q'}^{(-)}) \Big),$$

$$\Gamma^m(\mathcal{Z}) := \Big(\bigotimes_{p\in\mathrm{Pr}^{\mathrm{n}}} \Gamma_{\epsilon_p}^{m_p}(\mathcal{Z}_p) \Big)$$

$$\otimes \Big(\bigotimes_{q\in\mathrm{Pr}^{\mathrm{c}}} \Gamma_{\epsilon_q}^{m_q^{(+)}}(\mathcal{Z}_q^{(+)}) \Big) \otimes \Big(\bigotimes_{q'\in\mathrm{Pr}^{\mathrm{c}}} \Gamma_{\epsilon_{q'}}^{m_{q'}^{(-)}}(\mathcal{Z}_{q'}^{(-)}) \Big).$$

We also define

$$H_m := \sum_{p\in\mathrm{Pr}^{\mathrm{n}}} \mathrm{d}\Gamma^{m_p}(h_p) + \sum_{q\in\mathrm{Pr}^{\mathrm{c}}} \mathrm{d}\Gamma^{m_q^{(+)}}(h_q^{(+)}) + \sum_{q'\in\mathrm{Pr}^{\mathrm{c}}} \mathrm{d}\Gamma^{m_{q'}^{(-)}}(h_{q'}^{(-)}),$$

$$\Theta^m := \bigotimes_{p\in\mathrm{Pr}^{\mathrm{n}}} \Theta^{m_p} \otimes \bigotimes_{q\in\mathrm{Pr}^{\mathrm{c}}} \Theta^{m_q^{(+)}} \otimes \bigotimes_{q'\in\mathrm{Pr}^{\mathrm{c}}} \Theta^{m_{q'}^{(-)}}$$

as operators on $\Gamma^m(\mathcal{Z})$.

20.5.11 Evaluation of Feynman diagrams with no external legs

Let $\mathbb{R} \ni t \mapsto F_1(t), \ldots, F_r(t)$ be time-dependent Feynman monomials, each of a constant multi-degree. Set

$$G(t) := F_1(t) + \cdots + F_r(t),$$
$$H(t) := H_0 + \lambda{:}G(t, \phi, \psi^*, \psi){:}.$$

Our aim is to compute the scattering operator

$$S = \text{Texp}\Big(-i\lambda \int e^{itH_0}{:}G(t, \phi, \psi^*, \psi){:}e^{-itH_0}\,dt\Big). \tag{20.49}$$

We use the terminology of Section 20.1. We will denote by $F_1, \ldots, F_r$ distinct vertices of the same multi-degree as the Feynman monomials $F_1(t), \ldots, F_r(t)$. For brevity, we will write $\mathfrak{V}$ for the set $\{F_1, \ldots, F_r\}$.

Let $(F_{j_n}, \ldots, F_{j_1})$ be a sequence of Feynman vertices in $\mathfrak{V}$ and let D be a Feynman diagram over $\prod_{i=n}^{1} F_{j_i}$ with no external legs.

Definition 20.67 *The evaluation of the diagram D at times $t_n, \ldots, t_1 \in \mathbb{R}$ is*

$$D(t_n, \ldots, t_1)$$
$$:= \prod_{\ell=\{l,l'\}\in\text{Ln}_n(D)} \nabla_{y_l} S_\ell(t_{\text{nr}(l)} - t_{\text{nr}(l')}) \nabla_{y_{l'}}$$
$$\times \prod_{\kappa=\{k^{(+)},k^{(-)}\}\in\text{Ln}_c(D)} \nabla_{\overline{y}_{k^{(-)}}} S_\kappa(t_{\text{nr}(k^{(+)})} - t_{\text{nr}(k^{(-)})}) \nabla_{y_{k^{(+)}}} \prod_{i=n}^{1} F_{j_i}(t_i).$$

Remark 20.68 *If for some particle $p \in \text{Pr}_s$ the polynomials F_i depend only on the configuration space, and not on the momentum space, which is often the case for bosons, we can replace the phase space 2-point function S_p by the configuration space 2-point function D_p.*

20.5.12 Vacuum expectation value of the scattering operator

Feynman diagrams with no external legs can be used to compute the vacuum expectation value of scattering operators. The following theorem is closely related to Thm. 20.21.

Theorem 20.69

$$\log(\Omega|S\Omega) = \sum_{n=0}^{\infty} \sum_{D\in\text{Dg}_n\{\mathfrak{V}\}_{\text{cnl}}} (-i\lambda)^n \int \cdots \int D(t_n, \ldots, t_1)\,dt_n \cdots dt_1. \tag{20.50}$$

Proof We first obtain

$$(\Omega|S\Omega) = \sum_{n=0}^{\infty} \sum_{D\in\text{Dg}_n\{\mathfrak{V}\}_{\text{nl}}} (-i\lambda)^n \int \cdots \int D(t_n, \ldots, t_1)\,dt_n \cdots dt_1. \tag{20.51}$$

Following the arguments of Thm. 20.21, (20.51) equals

$$\exp\left(\sum_{n=0}^{\infty} \sum_{D \in \mathrm{Dg}_n \{\mathfrak{V}\}_{\mathrm{cnl}}} (-\mathrm{i}\lambda)^n \int \cdots \int D(t_n, \ldots, t_1) \mathrm{d}t_n \cdots \mathrm{d}t_1\right)$$

$\square$

Here is a reformulation of (20.50) in terms of the Fourier transforms of the diagrams:

$$\log(\Omega|S\Omega) = \sum_{n=0}^{\infty} \sum_{D \in \mathrm{Dg}_n \{\mathfrak{V}\}_{\mathrm{cnl}}} (-\mathrm{i}\lambda)^n \hat{D}(0, \ldots, 0).$$

20.5.13 Energy shift

Assume now that $F_i(t) = F_i$ do not depend on time. Then the Fourier transform $\hat{D}(\tau_n, \ldots, \tau_1)$ is supported in $\tau_n + \cdots + \tau_1 = 0$ and one can write

$$\hat{D}(\tau_n, \ldots, \tau_1) = 2\pi\delta(\tau_n + \cdots + \tau_1)\hat{D}[\tau_n, \ldots, \tau_1],$$

where $\hat{D}[\tau_n, \ldots, \tau_1]$ is defined on $\tau_n + \cdots + \tau_1 = 0$.

Let E be the ground-state energy of H. The (partly heuristic) arguments that gave Thm. 20.54 can be used to give a formula for the energy shift in terms of Feynman diagrams:

Theorem 20.70

$$E = \sum_{n=0}^{\infty} \mathrm{i}(-\mathrm{i}\lambda)^n \sum_{D \in \mathrm{Dg}_n \{\mathfrak{V}\}_{\mathrm{cnl}}} \hat{D}[0, \ldots, 0].$$

Proof The function $D(t_n, \ldots, t_1)$ is translation invariant; therefore it can be written as

$$D(t_n, \ldots, t_1) = d(t_n - t_1, \ldots, t_2 - t_1),$$

for some function d. We compute

$$\hat{D}(\tau_n, \ldots, \tau_1)$$
$$= \int \cdots \int e^{-\mathrm{i}t_n \tau_n - \cdots - \mathrm{i}t_1 \tau_1} D(t_n, \ldots, t_1) \mathrm{d}t_n \cdots \mathrm{d}t_1$$
$$= \int \cdots \int e^{\mathrm{i}s_n \tau_n - \cdots - \mathrm{i}s_2 \tau_2 - \mathrm{i}s_1(\tau_1 + \cdots + \tau_n)} d(s_n, \ldots, s_2) \mathrm{d}s_n \cdots \mathrm{d}s_1$$
$$= 2\pi\delta(\tau_n + \cdots + \tau_1)\hat{d}(\tau_n, \ldots, \tau_2),$$

where we have used the substitution

$$s_j = t_j - t_1, \quad n \geq j \geq 2, \quad s_1 = t_1.$$

Thus, with $\tau_1 = -\tau_n - \cdots - \tau_2$,

$$\hat{D}(\tau_n, \ldots, \tau_1) = \hat{d}(\tau_n, \ldots, \tau_2).$$

Now we would like to use Sucher's formula, (20.26). Terms in the expansion for $\frac{1}{2}i\epsilon\lambda\partial_\lambda \log(\Omega|S_\epsilon\Omega)$ are of the form $i(-i\lambda)^n$ times

$$\frac{1}{2}n\epsilon \int \cdots \int e^{-\epsilon(|t_n|+\cdots|t_1|)} D(t_n,\ldots,t_1)dt_n \cdots dt_1$$

$$= \frac{1}{2}n\epsilon \int e^{-\epsilon(|s_1+s_n|+\cdots+|s_1+s_2|+|s_1|)} d(s_n,\ldots,s_2)ds_n \cdots ds_2.$$

We perform the integral in s_1:

$$\frac{1}{2}n\epsilon \int e^{-\epsilon(|s_1+s_n|+\cdots+|s_1+s_2|+|s_1|)} ds_1$$

$$= \frac{1}{2}n \int e^{-(|u+\epsilon s_n|+\cdots+|u+\epsilon s_2|+|u|)} du$$

$$\to \frac{1}{2}n \int e^{-n|u|} du = 1.$$

Therefore,

$$\lim_{\epsilon\searrow 0} \frac{1}{2}n\epsilon \int e^{-\epsilon(|t_n|+\cdots|t_1|)} D(t_n,\ldots,t_1)dt_n \cdots dt_1$$

$$= \int d(s_n,\ldots,s_2)ds_n \cdots ds_2$$

$$= \hat{d}(0,\ldots,0) = \hat{D}(0,\ldots,0). \qquad \square$$

20.5.14 Polynomials on path spaces

The formalism of Feynman diagrams can be interpreted to some extent as a special case of the formalism described in Sect. 20.1. With this interpretation we say that to obtain the vacuum expectation values of scattering operators we need to integrate over various paths (trajectories).

Paths are functions of time with values in the phase space or the configuration space. We equip path spaces with an appropriate (bilinear or sesquilinear) form defined with the help of the Feynman propagator. In this way we obtain one of the basic ingredients of the formalism of Gaussian integration described in Sect. 20.1: the family of spaces $\mathcal{V}_p$ equipped with a form σ_p.

We can distinguish two kinds of paths: *phase space paths* and *configuration space paths*. Their names are quite awkward; therefore we will abbreviate them: *ph-paths* for the former and *c-paths* for the latter.

Definition 20.71 (1) *Let $p \in \mathrm{Pr}^n$ (p is a neutral particle). For the space of corresponding smooth ph-paths we can take $C_c^\infty(\mathbb{R}, \mathcal{Y}_p)$. It is equipped with the form*

$$f \cdot S_p f' = \int f(t) \cdot S_p(t-t')f'(t')dtdt'.$$

Note that S_p is symmetric, resp. anti-symmetric for $p \in \mathrm{Pr}_s^n$, resp. $p \in \mathrm{Pr}_a^n$.

(2) *Let $q \in \mathrm{Pr}^c$ (q is a charged particle). For the space of corresponding* ph-paths *we can take $C_c^\infty(\mathbb{R}, \mathcal{Y}_q)$. It is equipped with the form*

$$\bar{g} \cdot S_q g' = \int \overline{g(t)} \cdot S_q(t - t') g'(t') dt dt'.$$

Note that S_q is Hermitian, resp. anti-Hermitian for $q \in \mathrm{Pr}_s^c$, resp. $p \in \mathrm{Pr}_a^c$.

In the bosonic case, one often prefers to use paths with values in the configuration space $\mathcal{X}_p$, rather than in the dual phase space $\mathcal{Y}_p$.

Definition 20.72 (1) *Let $p \in \mathrm{Pr}_s^n$ (p is a neutral boson). For the space of corresponding* c-paths *we can take $C_c^\infty(\mathbb{R}, \mathcal{X}_p)$. It is equipped with the form*

$$f \cdot D_p f' = \int f(t) \cdot D_p(t - t') f'(t') dt dt'.$$

Note that D_p is symmetric.

(2) *Let $q \in \mathrm{Pr}_s^c$ (q is a charged boson). For the space of corresponding* c-paths *we can take $C_c^\infty(\mathbb{R}, \mathcal{X}_q)$. It is equipped with the form*

$$\bar{g} \cdot D_q g' = \int \overline{g(t)} \cdot D_q(t - t') g'(t') dt dt'.$$

Note that D_q is Hermitian.

Remark 20.73 *Note that most textbooks start their exposition of the path integration formalism from what we call configuration space paths for neutral bosons.*

For a neutral particle p, the spaces $(C_c^\infty(\mathbb{R}, \mathcal{Y}_p), S_p)$ or $(C_c^\infty(\mathbb{R}, \mathcal{X}_p), D_p)$ can be treated as $(\mathcal{V}_p, p)$ of Def. 20.12 (1). A similar remark applies to charged particles. We introduce the space $\mathcal{V}$ as in Def. 20.12 (3) and note that $\mathcal{V} = C_c^\infty(\mathbb{R}, \mathcal{Y})$, where $\mathcal{Y}$ is defined as in Def. 20.62. We introduce the Wick transform, denoted by double dots, the Gaussian integral, etc.

As discussed in Sect. 20.1, we would like to integrate "monomials of degree m", that is, m-linear symmetric or anti-symmetric functions on $\mathcal{V}^m$. The space of such monomials was denoted by $\mathrm{Pol}^m(\mathcal{V})$. In Sect. 20.1 we assumed that the spaces $\mathcal{V}$ are finite-dimensional, which allowed us to ignore questions about their topology. Path spaces are necessarily infinite-dimensional and difficulties arising from various possible topologies show up. We will keep the notation $\mathrm{Pol}^m(\mathcal{V})$ for monomials of degree m, but we need to make precise what we mean by this. To reduce the complexity of notation, let us assume that we have a single species of particles, which are neutral. They can be bosonic or fermionic. For definiteness, we will use phase space paths.

A reasonable and sufficiently broad definition of $\mathrm{Pol}^m(\mathcal{V})$ is the following. We say that $P \in \mathrm{Pol}^m(\mathcal{V})$ if it is given by a family of distributions

$$P(\cdot, \ldots, \cdot) \in \mathcal{D}'\big(\mathbb{R}^m, (\overset{a^1 \ m}{\otimes} \mathcal{Y})^{\#}\big)$$

with the appropriate symmetry or anti-symmetry properties, and its action on $f_i \in C_c^\infty(\mathbb{R}, \mathcal{Y})$, $i = 1, \ldots, m$, is

$$\langle P|f_1, \ldots, f_m \rangle = \int \cdots \int \langle P(t_m, \ldots, t_1)|f_m(t_m) \cdots f_1(t_1) \rangle \mathrm{d}t_m \cdots \mathrm{d}t_1.$$

We will see that $\mathrm{Pol}^m(\mathcal{V})$ is large enough to contain objects that we need to integrate when computing the scattering operator.

Note that our choice of spaces of the form $C_c^\infty(\mathbb{R}, \mathcal{Y})$ for path spaces is to some extent arbitrary. One could try to replace C_c^∞ by some other class of functions. Nevertheless, one really needs $\mathrm{Pol}^m(\mathcal{V})$ to be quite large, which is made possible with this choice.

When we compute the scattering operator, a special role is played by the time variable. In fact, in this context we often deal with monomials whose associated functions depend on a single time variable, as explained in the following definition.

Definition 20.74 *Let*

$$\mathbb{R} \ni t \mapsto F(t) \in \mathrm{Pol}^m(\mathcal{Y})$$

be a function. We will still denote by F the element of $\mathrm{Pol}^m(\mathcal{V})$ whose associated distribution $F(t_1, \ldots, t_m)$ is

$$F(t_1, \ldots, t_m) = \int F(t)\delta(t_1 - t) \cdots \delta(t_m - t)\mathrm{d}t,$$

that is,

$$\langle F|f_1, \ldots, f_m \rangle = \int F(t)f_1(t), \ldots, f_m(t)\mathrm{d}t.$$

20.5.15 Feynman formalism and Gaussian integration

Let $F_1(t), \ldots, F_r(t)$, $G(t)$ be as in Subsect. 20.5.11. Note that the function $t \mapsto G(t)$ takes values in $\mathrm{Pol}(\mathcal{Y})$. We will denote by G its interpretation as an element of $\mathrm{Pol}(\mathcal{V})$, using the convention in Def. 20.74.

We define the scattering operator S as in (20.49). The following theorem shows that one can reduce computations in quantum field theory to Gaussian integrals on appropriate path spaces. The theorem follows from a comparison of the formulas for the evaluation of diagrams in Defs. 20.17, 20.67, using the covariance of Defs. 20.71 or 20.72.

Theorem 20.75 *We have the following identity:*

$$(\Omega|S\Omega) = \int e^{\lambda : G:}, \tag{20.52}$$

where the right hand side is given by the formalism of Sect. 20.1.

20.6 Feynman diagrams and the scattering operator

In this section we describe how to modify the formalism of the previous section to compute the scattering operator. Diagrams for the scattering operator will have external legs of two kinds: incoming and outgoing, similarly to Friedrichs diagrams. Vertices, however, will be typical for Feynman diagrams – diagrams with a different order of time labels will not be distinguished.

Throughout this section we keep the terminology and notation of the previous section. In particular, let $\mathbb{R} \ni t \mapsto F_1(t), \ldots, F_r(t)$ be time-dependent Feynman monomials, each of a constant multi-degree with the corresponding vertices denoted by $F_1, \ldots, F_r$. $\mathfrak{V}$ denotes the set $\{F_1, \ldots, F_r\}$. We set

$$G(t) := F_1(t) + \cdots + F_r(t),$$

and perturb H_0 by $:G(t, \phi, \psi^*, \psi):$. Our aim is to compute the scattering operator

$$S = \mathrm{Texp}\left(-\mathrm{i}\lambda \int \mathrm{e}^{\mathrm{i}tH_0} :G(t, \phi, \psi^*, \psi): \mathrm{e}^{-\mathrm{i}tH_0}\, \mathrm{d}t\right). \tag{20.53}$$

20.6.1 Feynman diagrams with external legs

We assume that F_{j_i}, $i = n, \ldots, 1$, is a sequence in $\mathfrak{V}$. Let D be a Feynman diagram over $\prod_{i=n}^{1} F_{j_i}$.

Definition 20.76 *Let $i = n, \ldots, 1$. The* multi-degree *of D at the ith vertex, denoted $m_i(D)$, is defined as*

$$m_{p,i}^{\mathrm{n}}(D) := \#\mathrm{Lg}_p(D) \cap \mathrm{Lg}_p(F_{j_i}), \quad p \in \mathrm{Pr_n},$$
$$m_{q,i}^{(\pm)}(D) := \#\mathrm{Lg}_q^{(\pm)}(D) \cap \mathrm{Lg}_q^{(\pm)}(F_{j_i}), \quad q \in \mathrm{Pr_c}.$$

The detailed multi-degree *of D is the sequence $\underline{m}(D) = (m_n(D), \ldots, m_1(D))$.*

Note that the diagram has no legs iff all entries of $\underline{m}(D)$ are zero.

Recall that in Def. 20.67 we defined the evaluation of a Feynman diagram without external legs. We would like to generalize this definition to all Feynman diagrams. In the literature one can find two conventions for evaluation of such diagrams: either one includes the propagators for external legs or not. In the definition below we adopt the latter convention.

Definition 20.77 *The* amputated evaluation *of the diagram D at times $t_n, \ldots, t_1 \in \mathbb{R}$ is an element of $\overset{1}{\underset{i=n}{\otimes}} \Gamma^{m_i(D)}(\mathcal{Y})$ given by*

$$D^{\mathrm{amp}}(t_n, \ldots, t_1)$$

$$:= (-\mathrm{i}\lambda)^n \overset{1}{\underset{i=n}{\otimes}} \Theta^{m_i(D)\#} \prod_{\ell=\{l,l'\}\in\mathrm{Ln^n}(D)} \nabla_{y_l} S_\ell(t_{\mathrm{nr}(l)} - t_{\mathrm{nr}(l')})\nabla_{y_{l'}}, \tag{20.54}$$

$$\times \prod_{\kappa=\{k^{(+)},k^{(-)}\}\in\mathrm{Ln^c}(D)} \nabla_{\overline{y}_{k^{(-)}}} S_\kappa(t_{\mathrm{nr}(k^{(+)})} - t_{\mathrm{nr}(k^{(-)})})\nabla_{y_{k^{(+)}}} \prod_{i=n}^{1} F_{j_i}(t_i).$$

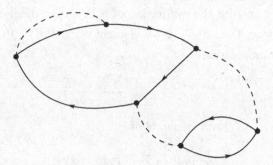

Figure 20.10 Feynman diagram without external lines.

out   in

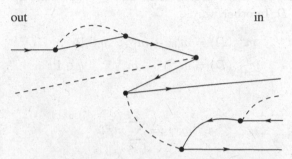

Figure 20.11 Scattering-type Feynman diagram.

Note that if D has no legs, then (20.54) coincides with $D(t_n, \ldots, t_1)$ defined in Def. 20.67.

The precise interpretation of (20.54) is similar to that of (20.12); see the discussion after Def. 20.17. The main difference is that we symmetrize or anti-symmetrize only within each vertex.

20.6.2 Feynman diagrams with incoming and outgoing external legs

Let m be a multi-degree. Recall the identification $\rho : \overline{\mathcal{Z}} \oplus \mathcal{Z} \to \mathcal{Y}$ defined in Subsect. 20.5.10. Clearly, we have the identification

$$\Gamma(\rho)^{\#} : \mathrm{Pol}^m(\mathcal{Y}) \to \mathrm{Pol}^m(\overline{\mathcal{Z}} \oplus \mathcal{Z})$$

$$\simeq \bigoplus_{m^+ + m^- = m} \mathrm{Pol}^{m^+}(\overline{\mathcal{Z}}) \otimes \mathrm{Pol}^{m^-}(\mathcal{Z}).$$

If $\underline{m} = (m_1, \ldots, m_n)$ is a sequence of multi-degrees, then this yields

$$\bigotimes_{i=1}^{n} \Gamma(\rho)^{\#} : \bigotimes_{i=1}^{n} \mathrm{Pol}^{m_i}(\mathcal{Y}) \to \bigotimes_{i=1}^{n} \mathrm{Pol}^{m_i}(\overline{\mathcal{Z}} \oplus \mathcal{Z})$$

$$\simeq \bigoplus_{\underline{m}^+ + \underline{m}^- = \underline{m}} \bigotimes_{i=n}^{1} \mathrm{Pol}^{m_i^+}(\overline{\mathcal{Z}}) \otimes \mathrm{Pol}^{m_i^-}(\mathcal{Z}). \quad (20.55)$$

Let $D^{\mathrm{amp}}(t_n,\ldots,t_1)$ be the evaluation of a Feynman diagram with external legs, which, following Def. 20.65, is an element of $\overset{n}{\underset{i=1}{\otimes}} \mathrm{Pol}^{m_i}(\mathcal{Y})$. By (20.55), we have a unique decomposition

$$\overset{n}{\underset{i=1}{\otimes}} \Gamma(\rho)^{\#} D^{\mathrm{amp}}(t_n,\ldots,t_1) = \sum_{\underline{m}^+ + \underline{m}^- = \underline{m}(D)} D^{\mathrm{amp}}_{\underline{m}^+,\underline{m}^-}(t_n,\ldots,t_1). \quad (20.56)$$

Here, $D^{\mathrm{amp}}_{\underline{m}^+,\underline{m}^-}(t_n,\ldots,t_1)$ are elements of

$$\overset{1}{\underset{i=n}{\otimes}} \mathrm{Pol}^{m_i^+}(\overline{\mathcal{Z}}) \otimes \mathrm{Pol}^{m_i^-}(\mathcal{Z}),$$

$\underline{m}(D)$ is the detailed multi-degree of D and $\underline{m}^+, \underline{m}^-$ sum up to the detailed multi-degree of D. In other words,

$$m^{\mathrm{n}}_{p,i}(D) = m^{\mathrm{n}+}_{p,i} + m^{\mathrm{n}-}_{p,i}, \quad p \in \mathrm{Pr}^{\mathrm{n}},$$
$$m^{(\pm)}_{q,i}(D) = m^{(\pm)+}_{q,i} + m^{(\pm)-}_{q,i}, \quad q \in \mathrm{Pr}^{\mathrm{c}}.$$

We also set

$$|\underline{m}^{\pm}| := \sum_{i=n}^{1} m^{\pm}_i.$$

Definition 20.78 *The* amputated evaluation *of D with $\underline{m}^-$ incoming and $\underline{m}^+$ outgoing legs at times $t_n,\ldots,t_1 \in \mathbb{R}$, denoted $D^{\mathrm{amp}}_{\underline{m}^+,\underline{m}^-}(t_n,\ldots,t_1)$, is defined by (20.56).*

Note that $D^{\mathrm{amp}}_{\underline{m}^+,\underline{m}^-}(t_n,\ldots,t_1)$ can be interpreted as an operator in

$$B\left(\overset{1}{\underset{i=n}{\otimes}} \Gamma^{m_i^-}(\mathcal{Z}), \overset{1}{\underset{i=n}{\otimes}} \Gamma^{m_i^+}(\mathcal{Z})\right).$$

In what follows we stick to this interpretation.

Definition 20.79 *The* scattering evaluation *of D with $\underline{m}^-$ incoming and $\underline{m}^+$ outgoing legs at times $t_n,\ldots,t_1 \in \mathbb{R}$ is an operator in $B(\Gamma^{|\underline{m}^-|}(\mathcal{Z}), \Gamma^{|\underline{m}^+|}(\mathcal{Z}))$ defined by*

$$D^{\mathrm{scat}}_{\underline{m}^+,\underline{m}^-}(t_n,\ldots,t_1)$$
$$= \Theta^{|\underline{m}^+|} \overset{1}{\underset{i=n}{\otimes}} \mathrm{e}^{\mathrm{i}t_i H_{m_i^+}} D^{\mathrm{amp}}_{\underline{m}^+,\underline{m}^-}(t_n,\ldots,t_1) \overset{1}{\underset{i=n}{\otimes}} \mathrm{e}^{-\mathrm{i}t_i H_{m_i^-}} \Theta^{|\underline{m}^-|}.$$

Remark 20.80 *The fact that a single Feynman diagram gives rise to many terms in the scattering operator and each of them is an analytic continuation of the others is called the* crossing symmetry.

20.6.3 Scattering operator and Feynman diagrams

Here is the analog of Thm. 20.51 in the formalism of Feynman diagrams:

Theorem 20.81

$$S = \sum_{n=0}^{\infty} \sum_{\substack{D \in \mathrm{Dg}_n\{\mathfrak{V}\} \\ \underline{m}^+ + \underline{m}^- = \underline{m}(D)}} (-i\lambda)^n \mathrm{Op}^{a^*,a}$$

$$\times \left(\int \cdots \int D^{\mathrm{scat}}_{\underline{m}^+, \underline{m}^-}(t_n, \dots, t_1) \mathrm{d}t_n \cdots \mathrm{d}t_1 \right)$$

$$= \mathrm{Op}^{a^*,a} \left(\exp \left(\sum_{n=0}^{\infty} \sum_{\substack{D \in \mathrm{Dg}_n\{\mathfrak{V}\}_{\mathrm{con}} \\ \underline{m}^+ + \underline{m}^- = \underline{m}(D)}} (-i\lambda)^n \right. \right.$$

$$\left. \left. \times \int \cdots \int D^{\mathrm{scat}}_{\underline{m}^+, \underline{m}^-}(t_n, \dots, t_1) \mathrm{d}t_n \cdots \mathrm{d}t_1 \right) \right)$$

$$\frac{S}{(\Omega|S\Omega)} = \mathrm{Op}^{a^*,a} \left(\exp \left(\sum_{n=0}^{\infty} \sum_{\substack{D \in \mathrm{Dg}_n\{\mathfrak{V}\}_{\mathrm{cnl}} \\ \underline{m}^+ + \underline{m}^- = \underline{m}(D)}} (-i\lambda)^n \right. \right.$$

$$\left. \left. \times \int \cdots \int D^{\mathrm{scat}}_{\underline{m}^+, \underline{m}^-}(t_n, \dots, t_1) \mathrm{d}t_n \cdots \mathrm{d}t_1 \right) \right).$$

Here is a reformulation of the last formula in terms of Fourier transforms of the diagrams:

$$\frac{S}{(\Omega|S\Omega)} = \mathrm{Op}^{a^*,a} \left(\exp \left(\sum_{n=0}^{\infty} \sum_{\substack{D \in \mathrm{Dg}_n\{\mathfrak{V}\}_{\mathrm{link}} \\ \underline{m}^+ + \underline{m}^- = \underline{m}(D)}} (-i\lambda)^n \int \cdots \int \mathrm{d}\xi_n^+ \cdots \mathrm{d}\xi_1^+ \mathrm{d}\xi_n^- \cdots \mathrm{d}\xi_1^- \right. \right.$$

$$\times \Theta^{|\underline{m}^+|} \overset{1}{\underset{i=n}{\otimes}} \delta(H_{m_i^+} - \xi_i^+ \mathbb{1}) \hat{D}^{\mathrm{amp}}_{\underline{m}^+, \underline{m}^-} (\xi_n^+ - \xi_n^-, \dots, \xi_1^+ - \xi_1^-)$$

$$\left. \left. \times \overset{1}{\underset{n=1}{\otimes}} \delta(H_{m_i^-} - \xi_i^- \mathbb{1}) \Theta^{|\underline{m}^-|} \right) \right).$$

Note that if the ith vertex has no incoming lines, then $H_{m_i^-} = 0$. Therefore, the delta function $\delta(H_{m_i^-} - \xi_i^- \mathbb{1})$ sets $\xi_i^- = 0$. Hence, we can drop the variable ξ_i^- altogether from the formula. A similar remark concerns outgoing lines.

20.6.4 Gell-Mann–Low scattering operator for time-independent perturbations

Assume now that $F_i(t) = F_i$ do not depend on time. Then the Fourier transform of $D^{\mathrm{amp}}(\tau_n, \dots, \tau_1)$ is supported in $\tau_n + \cdots + \tau_1 = 0$, and one can write

$$\hat{D}^{\mathrm{amp}}(\tau_n, \dots, \tau_1) = 2\pi\delta(\tau_n + \cdots + \tau_1) \hat{D}^{\mathrm{amp}}[\tau_n, \dots, \tau_1],$$

where $\hat{D}^{\mathrm{amp}}[\tau_n, \dots, \tau_1]$ is defined on $\tau_n + \cdots + \tau_1 = 0$.

Usually, the standard scattering operator does not exist; instead one can define the Gell-Mann–Low scattering operator:

$$
S_{\mathrm{GL}} = \mathrm{Op}^{a^*,a} \Bigg(\exp \Bigg(\sum_{n=0}^{\infty} \sum_{\substack{D \in \mathrm{Dg}_n \{\mathcal{V}\}_{\mathrm{link}} \\ \underline{m}^+ + \underline{m}^- = \underline{m}(D)}} (-\mathrm{i}\lambda)^n
$$

$$
\times \int \cdots \int \mathrm{d}\xi_n^+ \cdots \mathrm{d}\xi_1^+ \mathrm{d}\xi_n^- \cdots \mathrm{d}\xi_1^- \, 2\pi \delta(\xi_n^+ + \cdots + \xi_1^+ - \xi_n^+ - \cdots - \xi_1^-)
$$

$$
\times \Theta^{|\underline{m}^+|} \bigotimes_{i=n}^{1} \delta(H_{m_i^+} - \xi_i^+ \mathbb{1}) \hat{D}_{\underline{m}^+,\underline{m}^-}^{\mathrm{amp}} \left[\xi_n^+ - \xi_n^-, \dots, \xi_1^+ - \xi_1^- \right]
$$

$$
\times \bigotimes_{i=n}^{1} \delta(H_{m_i^-} - \xi_i^- \mathbb{1}) \Theta^{|\underline{m}^-|} \Bigg) \Bigg).
$$

20.6.5 Friedrichs diagrams as Feynman diagrams

It is possible to interpret Friedrichs diagrams as a kind of Feynman diagrams. In fact, suppose we use the framework of Sects. 20.3 and 20.4. All the particles from Pr we interpret as charged particles, renaming $\mathcal{Z}_q$, $q \in \mathrm{Pr}$, as $\mathcal{Y}_q$. In the bosonic case, we use the charged symplectic form equal to i times the scalar product. In the fermionic case, we just keep the scalar product. Note that there are no anti-particles: $\mathcal{Y}_p = \mathcal{Y}_p^{(+)}$. Therefore, the Feynman propagator is zero for negative times. We write the scattering operator using Wick's chronological product.

Note that the above trick, even if somewhat artificial, can be used to reduce the theory of Friedrichs diagrams to Feynman diagrams, which gives in particular a convenient way to show the linked cluster theorem in the context of Friedrichs diagrams.

20.7 Notes

A description of some elements of the diagram formalism can be found in any textbook on quantum field theory, e.g. Schweber (1962), Weinberg (1995) or Srednicki (2007).

A mathematical exposition of what we call the Friedrichs diagrams is contained in the books by Friedrichs (1963) and by Hepp (1969). Hepp describes and proves the linked cluster theorem.

The diagram formalism is also one of the basic tools of non-relativistic many-body quantum theory. Therefore, its exposition can be found in many textbooks on this subject, such as the monograph of Fetter–Walecka (1971).

A book that specializes in the topic of Feynman diagrams was written by Mattuck (1967).

21
Euclidean approach for bosons

One can distinguish two basic approaches to quantum field theory. In the more traditional approach, one views the underlying physical Hilbert space equipped with the self-adjoint generator of the dynamics – Hamiltonian or Liouvillean – as the basic object. There also exists a different philosophy, whose starting point is paths (trajectories). The physical space and the physical Hamiltonian or Liouvillean are treated as derived objects (if they can be defined at all).

The second approach is often viewed as more modern and useful by physicists active in quantum field theory. Also from the mathematical point of view, the method of paths has turned out to be in many cases more efficient than the operator-theoretic approach. This chapter is devoted to a brief description of a certain version of this method, called often the *Euclidean approach*.

Let us first explain the origin of the word *Euclidean* in the name of this approach. Originally the Euclidean approach amounted to replacing the real time variable t by the imaginary is, an operation called the *Wick rotation*. Under this transformation, the Minkowski space $\mathbb{R}^{1,d}$ becomes the Euclidean space $\mathbb{R}^{1+d}$. After the Wick rotation, the unitary group generated by the Hamiltonian e^{itH} becomes the self-adjoint group of contractions e^{-sH}. One can then study e^{-sH} from the point of view of the so-called *path space*. In particular, it is sometimes easier to construct or study interacting models of quantum field theory on the Euclidean space than on the Minkowski space.

In the literature the term "Euclidean approach" seems to have acquired a wider meaning, going beyond quantum field theory on a Euclidean space. It sometimes denotes a method for obtaining a unitary group e^{itH} by first constructing the self-adjoint semi-group e^{-sH} for $s \geq 0$. In some cases one can try to represent the integral kernel of e^{-sH} by a measure on the so-called *path space*. This allows us to use methods of measure theory, which are sometimes quite powerful. In particular, one can treat very singular perturbations with little effort, provided they fit into the framework – essentially, they need to be representable as multiplication operators.

This approach also works in ordinary quantum mechanics. For example, it can be used to construct Schrödinger Hamiltonians $H = -\frac{1}{2}\Delta_x + V(x)$ on $L^2(\mathbb{R}^d)$, where V is a real potential. In the absence of the potential, $e^{\frac{t}{2}\Delta_x}$ is simply the well-known *heat semi-group*. Its distribution kernel $K_0(t, x, y)$ can be interpreted as the probability that a Brownian path starting from y arrives at x at time t. The perturbed heat kernel $K(t, x, y)$ can now be explicitly expressed in terms of $K_0(t, x, y)$ and the integral of the potential along Brownian paths by the so-called *Feynman–Kac formula*. We will briefly describe this construction in Sect. 21.1.

In this chapter we describe the Euclidean method for bosons in an abstract framework. We describe the construction of a class of interacting Hamiltonians starting from free ones, using the Feynman–Kac(–Nelson) formula.

In the usual version of the Euclidean approach one assumes that the generator of the physical dynamics, called the *Hamiltonian*, is bounded from below. Physically, this corresponds to the zero temperature, which is typical for most applications of quantum field theory. There also exists a version of the Euclidean approach for bosonic quantum fields at positive temperatures. Its aim is to construct an interacting KMS state and a dynamics at inverse temperature β. The dynamics is now generated by a self-adjoint operator L, the *Liouvillean*, which is not bounded from below or from above. This leads to some additional technical difficulties. However, the system can be described in a way similar to zero-temperature path spaces. There is an important difference: as a consequence of the KMS condition, the path space is now β-periodic. Thus, the Euclidean space is replaced with a cylinder of circumference β.

One of the interesting features of the Euclidean approach is the use of various non-trivial tools from functional analysis. One of them is the concept of *local Hermitian semi-groups* (see Thm. 2.69). They are indispensable in the positive temperature case. They are also sometimes useful at zero temperature, which happens if the perturbation is unbounded and destroys the positivity of the generator.

To motivate the reader, let us briefly discuss *Gaussian Markov path spaces*, which are usually the starting point for applications of the Euclidean approach. Let $\mathcal{Z}$ be a Hilbert space equipped with a conjugation τ. As we have seen in Subsect. 9.3.5, in such a case the bosonic Fock space $\Gamma_s(\mathcal{Z})$ can be unitarily identified with $L^2(Q, \mathrm{d}\mu)$ for some probability space $(Q, \mathfrak{S}, \mu)$. In the Euclidean approach we study operators on $L^2(Q, \mathrm{d}\mu)$ using the space of paths, that is, functions from $\mathbb{R}$ with values in Q.

A typical situation where Euclidean methods apply arises when we consider a real (commuting with τ) self-adjoint operator $a \geq 0$ on $\mathcal{Z}$. Recall that the semi-group $\mathrm{e}^{-t\mathrm{d}\Gamma(a)}$ is then positivity improving as an operator on $L^2(Q, \mathrm{d}\mu)$. We will see that for such operators the expectation value $(F|\mathrm{e}^{-t\mathrm{d}\Gamma(a)}G)$ can be written in terms of a measure on the set of *paths*. Field operators for real (τ-invariant) arguments can be interpreted as multiplication operators on $L^2(Q, \mathrm{d}\mu)$. Therefore, operators of the form $P(\phi)$, where P is a polynomial based on $\mathcal{Z}^\tau$, the real subspace of $\mathcal{Z}$, can be interpreted as multiplication operators in the Q-space representation. The Euclidean approach gives a powerful tool to study operators of the form $\mathrm{d}\Gamma(a) + P(\phi)$.

Throughout the chapter, we will use the terminology of abstract measure theory discussed in Chap. 5. Recall, in particular, that if $\mathfrak{T}_i$, $i \in I$, is a family of subsets of a set Q, we denote by $\bigvee_{i \in I} \mathfrak{T}_i$ the σ-algebra generated by $\bigcup_{i \in J} \mathfrak{T}_i$.

Throughout the chapter, we will use t as the generic variable in $\mathbb{R}$ denoting time.

21.1 A simple example: Brownian motion

In this section we illustrate the Euclidean approach by recalling the well-known representation of the heat semi-group e^{-tH_0}, $t \geq 0$, for $H_0 = -\frac{1}{2}\Delta$ on $L^2(\mathbb{R}^d)$, using Brownian motion.

From Subsect. 4.1.8 we obtain that the distribution kernel of e^{-tH_0} is

$$e^{-tH_0}(x, y) = (2\pi t)^{-d/2} e^{-(x-y)^2/2t}. \tag{21.1}$$

Consider the real Hilbert space $\mathcal{X} = L^2([0, \infty[, \mathbb{R}^d) \simeq L^2([0, \infty[, \mathbb{R}) \otimes \mathbb{R}^d$ and the Gaussian measure on $\mathcal{X}$ with covariance $\mathbb{1}$. Let ϕ denote the generic variable in $\mathcal{X}$. The associated Gaussian $\mathbf{L}^2$ space $\mathbf{L}^2(\mathcal{X}, e^{-\frac{1}{2}\phi^2} d\phi)$ can be realized as $L^2(Q, \mathfrak{S}, d\mu)$. Following Remark 5.66, we still denote by ϕ the generic variable on Q. For a Borel subset $I \subset \mathbb{R}$, the function $\mathbb{1}_I \otimes \mathbb{1}$ is a projection in $\mathcal{X}$. The corresponding conditional expectation of a measurable function F on Q will be denoted $E_I[F]$. In particular $E_\emptyset[F] = \int F(\phi) d\mu(\phi)$.

Definition 21.1 *The Brownian motion in $\mathbb{R}^d$ is the family $\{B_t\}_{t \geq 0}$ of $\mathbb{R}^d$-valued measurable functions on Q defined by*

$$\xi \cdot B_t(\phi) := \langle \phi | \mathbb{1}_{[0,t]} \otimes \xi \rangle, \ \xi \in \mathbb{R}^d, \ t \geq 0.$$

The Wiener process in $\mathbb{R}^d$ is

$$X_t(x, \phi) := x + B_t(\phi), \ t \geq 0, \ x \in \mathbb{R}^d.$$

We will often drop ϕ from $B_t(\phi)$ and $X_t(x, \phi)$.

The following lemma expresses the Markov property of the Wiener process:

Lemma 21.2 *For $t_1, t_2 \geq 0$ and almost all (a.a.) $x \in \mathbb{R}^d$*

$$\mathrm{E}_{[0,t_1]}\Big[f\big(X_{t_2+t_1}(x) \big) \Big] = \int f\Big(X_{t_2}\big(X_{t_1}(x), \phi \big) \Big) d\mu(\phi),$$

for all bounded measurable functions $f : \mathbb{R}^d \to \mathbb{C}$.

Proof We first prove the lemma for $f(x) = e^{i\xi \cdot x}$, $\xi \in \mathbb{R}^d$. Indeed, for such a function both sides equal $e^{-\xi^2 t_2/2} e^{iX_{t_1}(x) \cdot \xi}$. By Fourier transformation, this proves the lemma for $f \in C_c^\infty(\mathbb{R}^d)$. By the usual argument, the identity extends to all bounded measurable functions f. $\qquad \square$

Proposition 21.3 *Let $f \in L^2(\mathbb{R}^d) \cap L^\infty(\mathbb{R}^d)$. Then*

$$e^{-tH_0} f(x) = \int f(X_t(x)) d\mu, \ t \geq 0, \quad \text{for a.a. } x \in \mathbb{R}^d.$$

Proof Let $f \in C_c^\infty(\mathbb{R}^d)$ and $\hat{f}$ be its Fourier transform. Then

$$\int f\big(X_t(x)\big)\mathrm{d}\mu = (2\pi)^{-d} \int \hat{f}(\xi)\mathrm{e}^{\mathrm{i}\xi \cdot x} \int \mathrm{e}^{\mathrm{i}\xi \cdot B_t}\,\mathrm{d}\mu\mathrm{d}\xi$$

$$= (2\pi)^{-d} \int \hat{f}(\xi)\mathrm{e}^{\mathrm{i}\xi \cdot x}\mathrm{e}^{-t\xi^2/2}\mathrm{d}\xi = \mathrm{e}^{-tH_0}f(x).$$

If $f \in L^2(\mathbb{R}^d) \cap L^\infty(\mathbb{R}^d)$, we choose a sequence $f_n \in C_c^\infty(\mathbb{R}^d)$ such that $f_n \to f$ in L^2, $f_n \to f$ a.e. and $\sup_n \|f_n\|_\infty < \infty$. From (21.1) we obtain that $\mathrm{e}^{-tH_0}f_n(x) \to \mathrm{e}^{-tH_0}f(x)$ for a.a. x. The convergence of the r.h.s. to $\int f\big(X_t(x)\big)\mathrm{d}\mu$ follows from the dominated convergence. $\qquad\square$

We end this section by proving the celebrated *Feynman–Kac formula* in a simple situation. We denote by $C_b(\mathbb{R}^d)$ the space of bounded continuous functions on $\mathbb{R}^d$.

Theorem 21.4 *Let $V \in C_b(\mathbb{R}^d)$ be a real potential, $f \in L^2(\mathbb{R}^d) \cap L^\infty(\mathbb{R}^d)$ and $t \geq 0$. Then, for all $x \in \mathbb{R}^d$, $\mathrm{e}^{-\int_0^t V\big(X_s(x)\big)\mathrm{d}s}f\big(X_t(x)\big)$ is a bounded measurable function on Q and*

$$\mathrm{e}^{-t(H_0+V)}f(x) = \int \mathrm{e}^{-\int_0^t V(X_s(x))\mathrm{d}s}f(X_t(x))\mathrm{d}\mu, \quad \text{for a.a. } x \in \mathbb{R}^d. \tag{21.2}$$

Lemma 21.5 *Let $g_1,\ldots,g_{n-1} \in L^\infty(\mathbb{R}^d)$, $h \in L^2(\mathbb{R}^d)$. Let $s_1,\ldots,s_n > 0$ and $t_i = t_{i-1} + s_i$, $t_1 = s_1$. Then*

$$\mathrm{e}^{-s_1 H_0}g_1\mathrm{e}^{-s_2 H_0}\cdots g_{n-1}\mathrm{e}^{-s_n H_0}h(x)$$

$$= \int \prod_{i=1}^n g_i\big(X_{t_i}(x)\big)h\big(X_{t_n}(x)\big)\mathrm{d}\mu. \tag{21.3}$$

Proof We prove (21.3) for $n = 2$; the general case follows easily by induction. We have

$$\mathrm{e}^{-s_1 H_0}g\mathrm{e}^{-s_2 H_0}h(x)$$

$$= \int g\big(X_{s_1}(x)\big)\mathrm{e}^{-s_2 H_0}h\big(X_{s_1}(x)\big)\mathrm{d}\mu$$

$$= \int g\big(X_{s_1}(x,\phi_1)\big) \int h\Big(X_{s_2}\big(X_{s_1}(x,\phi_2)\phi_1\big)\Big)\mathrm{d}\mu(\phi_2)\mathrm{d}\mu(\phi_1)$$

$$= \int g\big(X_{s_1}(x)\big)E_{[0,s_1]}\Big[h\big(X_{s_1+s_2}(x)\big)\Big]\mathrm{d}\mu$$

$$= \int g\big(X_{s_1}(x)\big)h\big(X_{s_1+s_2}(x)\big)\mathrm{d}\mu,$$

by Lemma 21.2. $\qquad\square$

Lemma 21.6 *For $V \in C_b(\mathbb{R}^d)$ and all $x \in \mathbb{R}^d$ the map*

$$[0,+\infty[\ni t \mapsto V\big(X_t(x)\big) \in L^2(Q)$$

is continuous.

Proof For $t \geq 0$, $\delta > 0$, we have

$$\int |V(X_{t+\delta}(x)) - V(X_t(x))|^2 \, d\mu$$

$$= \int V^2(X_{t+\delta}(x)) d\mu + \int V^2(X_t(x)) d\mu - 2 \int V(X_t(x)) V(X_{t+\delta}(x)) d\mu$$

$$= e^{-(t+\delta)H_0} V^2(x) + e^{-tH_0} V^2(x) - 2e^{-\delta H_0} V e^{-tH_0} V(x),$$

where in the last line we use (21.3). From (21.1) we see that e^{-tH_0} is a semi-group of contractions on $C_b(\mathbb{R}^d)$. Moreover it is easy to see that, for $G \in C_b(\mathbb{R}^d)$ and all $x \in \mathbb{R}^d$, the map

$$[0, +\infty[\ni t \mapsto e^{-tH_0} G(x) \in \mathbb{R}$$

is continuous. This proves the right continuity at all $t \geq 0$. The proof of the left continuity at all $t > 0$ is similar. □

Proof of Thm. 21.4. By Lemma 21.6, $\int_0^t V(X_s(x)) ds$ is a bounded measurable function on Q. Hence, the integrand in the r.h.s. of (21.2) is bounded measurable on Q.

Let $f \in L^2(\mathbb{R}^d) \cap L^\infty(\mathbb{R}^d)$. By Trotter's product formula (see Thm. 2.75) we have

$$e^{-t(H_0+V)} f = \lim_{n \to \infty} \left(e^{-(t/n)H_0} e^{-(t/n)V} \right)^n f, \quad \text{in } L^2(\mathbb{R}^d),$$

and after extracting a subsequence we can assume that

$$e^{-t(H_0+V)} f(x) = \lim_{n \to \infty} \left(e^{-(t/n)H_0} e^{-(t/n)V} \right)^n f(x), \quad \text{for a.a. } x.$$

Applying (21.3) to $h = e^{-(t/n)V} f$, $g_j = e^{-(t/n)V}$ for $1 \leq j \leq n-1$, we get

$$e^{-t(H_0+V)} f(x) = \int e^{-F_n(x)} f(X_t(x)) d\mu,$$

for $F_n(x) = \frac{t}{n} \sum_{j=1}^n V(X_{tj/n}(x))$. Set $F(x) = \int_0^t V(X_s(x)) ds$. We claim that

$$e^{-F_n(x)} \to e^{-F(x)} \quad \text{in } L^2(Q), \quad \text{for a.a. } x, \tag{21.4}$$

which will complete the proof of the theorem. Since $|e^{-F_n}|, |e^{-F}| \leq e^{t\|V\|_\infty}$, it suffices to prove that $F_n(x) \to F(x)$ in $L^2(Q)$ for a.a. x. Since F_n is a Riemann sum for the integral defining F, this follows from Lemma 21.6. □

21.2 Euclidean approach at zero temperature

Most of this section is devoted to a description of the Euclidean approach at zero temperature in an abstract setting. We start with the definition of an abstract version of *Markov path spaces*. We will restrict ourselves to path spaces with a finite measure, which is sufficient for most applications to quantum field theory.

Given a Markov path space there is a canonical construction of a positivity improving semi-group $\{P(t)\}_{t\in[0,\infty[}$ possessing a unique ground state. Its generator is sometimes called the *Hamiltonian*. It acts on the so-called *physical Hilbert space*. A converse construction is also possible: every contractive positivity improving semi-group with a ground state can be dilated to a Markov path space.

The concept of a Markov path space is closely related to *unitary dilations of contractive semi-groups*. Indeed, each Markov path space involves a unitary group $\{U_t\}_{t\in\mathbb{R}}$ of measure preserving transformations of the underlying space which is a dilation of the physical semi-group $\{P(t)\}_{t\in[0,\infty[}$.

The most important class of examples of Markov path spaces are *Gaussian Markov path spaces*, which can be used to describe free bosonic quantum field theories in a Euclidean setting. They can be viewed as the real-wave quantization of a dilation of a contractive semi-group.

21.2.1 Markov path spaces

Definition 21.7 *A generalized path space* $(Q, \mathfrak{S}, \mathfrak{S}_0, U_t, R, \mu)$ *consists of*

(1) *a complete probability space* $(Q, \mathfrak{S}, \mu)$;
(2) *a distinguished sub-σ-algebra* $\mathfrak{S}_0$ *of* $\mathfrak{S}$;
(3) *a one-parameter group* $\mathbb{R} \ni t \mapsto U_t$ *of measure preserving $*$-automorphisms of* $L^\infty(Q, \mathfrak{S}, \mu)$, *strongly continuous for the σ-weak topology*;
(4) *a measure preserving $*$-automorphism* R *of* $L^\infty(Q, \mathfrak{S}, \mu)$ *such that* $RU_t = U_{-t}R$, $R^2 = \mathbb{1}$.

Moreover, one assumes that

$$\mathfrak{S} = \bigvee_{t\in\mathbb{R}} U_t \mathfrak{S}_0. \tag{21.5}$$

In what follows, $(Q, \mathfrak{S}, \mathfrak{S}_0, U_t, R, \mu)$ is a generalized path space. By Prop. 5.33 (2)(iii) and (2)(iv), U_t extends to a strongly continuous group of isometries of $L^p(Q, \mathfrak{S}, \mu)$, and R extends to an isometry of $L^p(Q, \mathfrak{S}, \mu)$, for $1 \leq p < \infty$.

Definition 21.8 *We set* $\mathfrak{S}_t := U_t \mathfrak{S}_0$, $\mathfrak{S}_I := \bigvee_{t\in I} \mathfrak{S}_t$, *for* $I \subset \mathbb{R}$, *and denote by* E_I *the conditional expectation w.r.t.* $\mathfrak{S}_I$.

Definition 21.9 *The generalized path space* $(Q, \mathfrak{S}, \mathfrak{S}_0, U_t, R, \mu)$ *is a Markov path space if it satisfies*

(1) *the* reflection property: $RE_0 = E_0$,
(2) *the* Markov property: $E_{[0,+\infty[}E_{]-\infty,0]} = E_0$.

21.2.2 Reconstruction theorem

Let $(Q, \mathfrak{S}, \mathfrak{S}_0, U_t, R, \mu)$ be a Markov path space.

Definition 21.10 *The physical Hilbert space associated with* $(Q, \mathfrak{S}, \mathfrak{S}_0, U_t, R, \mu)$ *is*

$$\mathcal{H} := L^2(Q, \mathfrak{S}_0, \mu).$$

The function $1 \in \mathcal{H}$ *will be denoted by* Ω. *The Abelian* $*$-*algebra* $\mathfrak{A} := L^\infty(Q, \mathfrak{S}_0, \mu)$ *acting on* $\mathcal{H}$ *is called the* algebra of time-zero fields.

Theorem 21.11 (1) $P(t) := E_0 U_t E_0$, $t \geq 0$, *is a strongly continuous semi-group of self-adjoint contractions on* $\mathcal{H}$ *preserving* Ω.
(2) $P(t)$ *is doubly Markovian.*
(3) $P(t)$ *is a contraction semi-group on* $L^p(Q, \mathfrak{S}_0, \mu)$ *for* $1 \leq p \leq \infty$. *It is strongly continuous for* $1 \leq p < \infty$.
(4) *Let* $A_i \in L^\infty(Q, \mathfrak{S}_0, \mu)$, $i = 1, \ldots, n$ *and* $t_1 \leq \cdots \leq t_n$. *Then*

$$\left(\Omega | A_1 P(t_1 - t_2) A_2 \cdots P(t_{n-1} - t_n) A_n \Omega\right) = \int_Q \prod_{i=1}^n U_{t_i}(A_i) \mathrm{d}\mu.$$

Proof $P(t)$ is clearly a contraction. It is self-adjoint:

$$P(t)^* = E_0 U_{-t} E_0 = E_0 U_{-t} R E_0 = E_0 R U_t E_0 = E_0 U_t E_0 = P(t).$$

Let us prove the semi-group property. Note that $U_t E_0 U_{-t} = E_t$. The Markov property implies, for $t, s \geq 0$,

$$E_{-t} E_0 E_s = E_{-t} E_{]-\infty, 0]} E_{[0, +\infty[} E_s = E_{-t} E_s.$$

This yields

$$P(t)P(s) = E_0 U_t E_0 U_s E_0 = U_t E_{-t} E_0 E_s U_{-s}$$
$$= U_t E_{-t} E_s U_{-s} = E_0 U_t U_s E_0 = P(t + s).$$

Finally, since $t \mapsto U_t$ is strongly continuous, so is $t \mapsto P(t)$.

U_t, E_0 are clearly positivity preserving. Hence so is $P(t)$. U_t, E_0 preserve 1. Hence so does $P(t)$. This proves (2). (3) follows from (2) by Prop. 5.24. We leave (4) to the reader. □

Definition 21.12 *The unique positive self-adjoint operator* H *on* $\mathcal{H}$ *such that* $P(t) = \mathrm{e}^{-tH}$ *is called the* Hamiltonian.

Clearly, $H\Omega = 0$.

Remark 21.13 *Often instead of Markov path spaces one uses more general OS-positive path spaces, named after Osterwalder and Schrader, where Def. 21.9 is replaced by the condition that* $E_{[0, +\infty[} R E_{[0, +\infty[} \geq 0$. *The OS-positivity condition is one of the Osterwalder–Schrader axioms; see Osterwalder–Schrader (1973, 1975). They are Euclidean analogs of the Gårding–Wightman axioms.*

In space dimensions 2 or higher it is believed that sharp time interacting fields do not exist, hence the Markov property cannot be used. Results similar to those in this chapter can be established in the framework of OS-positive path spaces, with similar proofs.

21.2.3 Gaussian path spaces I

Let $\mathcal{X}$ be a real Hilbert space with a self-adjoint operator $\epsilon > 0$. (All the constructions of this subsection have their complex counterparts; we assume the reality to simplify the exposition and in view of the application in the next subsection.) Consider the real Hilbert space

$$L^2(\mathbb{R}, \mathcal{X}) \simeq L^2(\mathbb{R}, \mathbb{R}) \otimes \mathcal{X} \tag{21.6}$$

and the positive self-adjoint operator

$$C = (D_t^2 + \epsilon^2)^{-1}$$

on (21.6). Introduce the real Hilbert space $\mathcal{Q} := C^{-\frac{1}{2}} L^2(\mathbb{R}, \mathcal{X})$. Its dual $\mathcal{Q}^\#$ can be identified with $C^{\frac{1}{2}} L^2(\mathbb{R}, \mathcal{X})$. Note that the operator C is orthogonal from $\mathcal{Q}$ to $\mathcal{Q}^\#$.

Definition 21.14 *For $t \in \mathbb{R}$ let us define the map*

$$j_t : (2\epsilon)^{\frac{1}{2}} \mathcal{X} \ni g \mapsto \delta_t \otimes g \in \mathcal{Q}. \tag{21.7}$$

Lemma 21.15 *We have*

$$(j_{t_1} g_1 | j_{t_2} g_2)_{\mathcal{Q}} = \left(g_1 \Big| \frac{e^{-|t_1 - t_2|\epsilon}}{2\epsilon} g_2 \right)_{\mathcal{X}}.$$

In particular j_t is isometric.

Proof We use the identity

$$\int_{\mathbb{R}} e^{itk} \frac{2\epsilon}{k^2 + \epsilon^2} dk = 2\pi e^{-|t|\epsilon}, \quad t \in \mathbb{R}, \tag{21.8}$$

which follows from Fourier transform and functional calculus. □

Definition 21.16 *For $t \in \mathbb{R}$ we set $\mathcal{Q}_t := j_t (2\epsilon)^{\frac{1}{2}} \mathcal{X}$. Let e_t denote the orthogonal projection onto $\mathcal{Q}_t$.*

For $I \subset \mathbb{R}$ we set $\mathcal{Q}_I := \left(\sum_{t \in I} \mathcal{Q}_t \right)^{\mathrm{cl}}$. The orthogonal projection onto $\mathcal{Q}_I$ will be denoted e_I.

Note that $e_t = j_t j_t^\#$.

For explicit formulas, in the following proposition we prefer to use the space $\mathcal{Q}^\#$ rather than $\mathcal{Q}$, by transporting operators with the help of the operator C.

Definition 21.17 *We write e^t, resp. e^I for $Ce_t C^{-1}$, resp. $Ce_I C^{-1}$.*

Definition 21.18 *We define*

$$(rf)(s) := f(-s), \quad (u_t f)(s) = f(s - t), \quad f \in \mathcal{Q}, \ s, t \in \mathbb{R}.$$

Definition 21.19 $e^{-t\epsilon}$, *defined originally on* $\mathcal{X}$, *determines in an obvious way a contractive semi-group on* $(2\epsilon)^{\frac{1}{2}}\mathcal{X}$, *which will be denoted by the same symbol. We set* $p(t) := j_0 e^{-t\epsilon} j_0^{\#}$, *which is a contractive semi-group on* $\mathcal{Q}_0$.

Proposition 21.20 (1) *Let* t, $t_1 < t_2$, $f \in \mathcal{Q}^{\#}$. *We have*

$$e^t f(s) := e^{-\epsilon|t-s|} f(t),$$
$$e^{[t,\infty[} f(s) = \mathbb{1}_{[t,\infty[}(s) f(s) + e^{-|s-t|\epsilon} \mathbb{1}_{]-\infty,t[}(s) f(t),$$
$$e^{]-\infty,t]} f(s) = \mathbb{1}_{]-\infty,t]}(s) f(s) + e^{-|s-t|\epsilon} \mathbb{1}_{]t,\infty[}(s) f(t),$$
$$e^{[t_1,t_2]} f(s) = \mathbb{1}_{[t_1,t_2]}(s) f(s) + e^{-|s-t_1|\epsilon} \mathbb{1}_{]-\infty,t_1[}(s) f(t_1) + e^{-|s-t_2|\epsilon} \mathbb{1}_{]t_2,\infty[}(s) f(t_2).$$

(2) $C_c^{\infty}(]t_1,t_2[, \mathrm{Dom}\,\epsilon)$ *is dense in* $\mathcal{Q}_{]t_1,t_2[}$.
(3) $\mathbb{R} \ni t \mapsto u_t$ *is an orthogonal* C_0-*group on* $\mathcal{Q}$.
(4) r *is an orthogonal operator satisfying* $r u_t = u_{-t} r$ *and* $r^2 = \mathbb{1}$.
(5) $\sum_{t \in \mathbb{R}} u_t \mathcal{Q}_0$ *is dense in* $\mathcal{Q}$.
(6) $r e_0 = e_0$.
(7) $e_{[0,\infty[} e_{]-\infty,0]} = e_0$.
(8) $e_0 u_t e_0 = p(|t|)$.

Remark 21.21 *Let* $[0,\infty[\ni t \mapsto p(t)$ *be a contractive* C_0-*semi-group on a Hilbert space* $\mathcal{Q}_0$. *We say that* $(\mathcal{Q}, u_t, e_0)$ *is a* unitary dilation *of* $\{p(t)\}_{t \in [0,\infty[}$ *if* $\mathcal{Q}$ *is a Hilbert space,* e_0 *is an orthogonal projection from* $\mathcal{Q}$ *onto* $\mathcal{Q}_0$, $\{u_t\}_{t \in \mathbb{R}}$ *is a unitary* C_0-*group on* $\mathcal{Q}$ *and* $p(t) = e_0 u_t e_0$, $t \geq 0$. *We say that the dilation* $(\mathcal{Q}, u_t, e_0)$ *is* minimal *if* $\sum_{t \in \mathbb{R}} u_t \mathcal{Q}_0$ *is dense in* $\mathcal{Q}$.

Clearly, what we have constructed in this subsection is a minimal dilation of the contractive semi-group $\{p(t)\}_{t \in [0,\infty[}$.

21.2.4 Gaussian path spaces II

In this subsection we describe the main example of Markov path spaces – Gaussian path spaces. They are used to describe free quantum field theories. They are obtained by second quantizing the Markov path system constructed in the previous subsection.

Let $\mathcal{X}$ be a real Hilbert space and $\epsilon > 0$ a self-adjoint operator on $\mathcal{X}$. Let C, $\mathcal{Q}$, $\{j_t\}_{t \in \mathbb{R}}$, $\{u_t\}_{t \in \mathbb{R}}$, r be constructed as in the previous subsection. Let us consider the Gaussian $\mathbf{L}^2$ space with covariance C. According to the notation introduced in Subsect. 5.4.2, it will be denoted

$$\mathbf{L}^2\big(L^2(\mathbb{R},\mathcal{X}), e^{\phi \cdot C^{-1}\phi} d\phi\big), \tag{21.9}$$

where we use ϕ as the generic variable in $L^2(\mathbb{R},\mathcal{X})$.

As we discussed in Chap. 5, there are many ways to realize this Gaussian $\mathbf{L}^2$ space as a space $L^2(Q,\mu)$, where (Q,μ) is a probability space. (Note that the notation Q for such a measure space is traditional in a part of the literature, hence the name "Q-space representation".) A class of possible choices, which is in fact our favorite, is $Q := B^{\frac{1}{2}} L^2(\mathbb{R}, \mathcal{X})$, where $B \geq 0$ is any self-adjoint operator on $L^2(\mathbb{R}, \mathcal{X})$ such that $B^{-\frac{1}{2}} C B^{-\frac{1}{2}}$ is trace-class. Thus the Gaussian $\mathbf{L}^2$ space (21.9) becomes the concrete space $L^2(Q, \mathrm{d}\mu)$, where μ is a Borel probability measure on Q such that

$$\int_Q e^{i\phi(f)} \mathrm{d}\mu(\phi) = e^{-\frac{1}{2} f \cdot C f}, \quad f \in B^{-\frac{1}{2}} L^2(\mathbb{R}, \mathcal{X}). \tag{21.10}$$

Following Remark 5.66, we now use ϕ as the generic name for an element of $Q = B^{\frac{1}{2}} L^2(\mathbb{R}, \mathcal{X})$. $\phi(f)$ denotes the pairing of $\phi \in B^{\frac{1}{2}} L^2(\mathbb{R}, \mathcal{X})$ with $f \in B^{-\frac{1}{2}} L^2(\mathbb{R}, \mathcal{X})$.

By Prop. 5.77, we can extend the definition of

$$Q \ni \phi \mapsto \phi(f) \tag{21.11}$$

to $f \in C^{-\frac{1}{2}} L^2(\mathbb{R}, \mathcal{X})$. The function in (21.11) in general needs not to be continuous; however it still belongs to $L^p(Q, \mu)$ for all $1 \leq p < \infty$.

Definition 21.22 *Since the maps j_s defined in (21.7) are isometric, we can define for $s \in \mathbb{R}$, $g \in (2\epsilon)^{\frac{1}{2}} \mathcal{X}$, the functions*

$$\phi_s(g) := \phi(\delta_s \otimes g) \in \bigcap_{1 \leq p < \infty} L^p(Q, \mu),$$

which are called the **sharp-time fields**.

We can now define the associated path space. We lift r and $\{u_t\}_{t \in \mathbb{R}}$ to $L^2(Q, \mu)$ by setting first

$$R e^{i\phi(f)} := e^{i\phi(rf)}, \quad U_t e^{i\phi(f)} = e^{i\phi(u_{-t}f)}, \quad f \in B^{-\frac{1}{2}} L^2(\mathbb{R}, \mathcal{X}), \tag{21.12}$$

extending then R and U_t to $L^2(Q, \mu)$ by linearity and density. In particular we have

$$R\phi_s(g) = \phi_{-s}(g), \quad U_t \phi_s(g) = \phi_{s-t}(g), \quad g \in (2\epsilon)^{\frac{1}{2}} \mathcal{X}. \tag{21.13}$$

Proposition 21.23 *Let $\mathfrak{S}$ be the completion of the Borel σ-algebra on Q, $\mathfrak{S}_0$ be the σ-algebra generated by the functions $e^{i\phi_0(g)}$ for $g \in (2\epsilon)^{\frac{1}{2}} \mathcal{X}$. Let R, U_t be defined in (21.12). Then $(Q, \mathfrak{S}, \mathfrak{S}_0, U_t, R, \mu)$ is a Markov path space.*

Definition 21.24 $(Q, \mathfrak{S}, \mathfrak{S}_0, U_t, R, \mu)$ *described in Prop. 21.23 will be called the* **Gaussian path space with covariance C**.

We will later need the following lemma, which follows directly from the results on complex-wave representation in Subsect. 9.2.1.

Lemma 21.25 *Let $\mathcal{Z}$ be a Hilbert space, $\Gamma_s(\mathcal{Z})$ the associated bosonic Fock space and b a self-adjoint operator on $\mathcal{Z}$. Then*

$$\left(e^{i\phi(g_1)} \Omega | e^{-t \mathrm{d}\Gamma(b)} e^{i\phi(g_2)} \Omega \right) = e^{-\frac{1}{2} \|g_1\|^2} e^{-\frac{1}{2} \|g_2\|^2} e^{-(g_1 | e^{-tb} g_2)},$$

whenever the r.h.s. is finite.

Proof of Prop. 21.23. Using that r and u_t preserve C, formula (21.10) and the density of exponentials in $L^2(Q, \mu)$ (see Subsect. 5.2.5) we see that R is unitary on $L^2(Q, \mu)$ and that $t \mapsto U_t$ is a strongly continuous unitary group on $L^2(Q, \mu)$. R and U_t are clearly $*$-automorphisms. By Prop. 5.33, $t \mapsto U_t$ is strongly continuous on $L^\infty(Q, \mu)$ for the σ-weak topology.

From (21.13) we see that the closed vector subspace generated by $e^{i\phi_0(g)}$ for $g \in (2\epsilon)^{\frac{1}{2}}\mathcal{X}$ is invariant under R, which implies that $RE_0 = E_0$. The fact that $RU_t = U(-t)R$ is obvious.

We now check the Markov property. We unitarily identify $\mathbf{L}^2\big(L^2(\mathbb{R}, \mathcal{X}), e^{\phi \cdot C^{-1}\phi}d\phi\big)$ with $\Gamma_s(\mathbb{C}\mathcal{Q})$, as in Thm. 9.22. If $I \subset \mathbb{R}$ is a closed interval, then under this identification E_I becomes $\Gamma(e_I)$, where e_I is defined in Lemma 21.15. So the Markov property follows from the pre-Markov property proved in Prop. 21.20 (7).

It remains to check condition (21.5). We note that it is equivalent to the property that the algebra generated by $\big\{U_t f \ : \ f \in L^\infty(Q, \mathfrak{S}_0, \mu), \ t \in \mathbb{R}\big\}$ is dense in $L^2(Q, \mathfrak{S}, \mu)$. It is easy to see that finite linear combinations of $\delta_{t_i} \otimes g_i$ for $t_i \in \mathbb{R}$, $g_i \in (2\epsilon)^{\frac{1}{2}}\mathcal{X}$, are dense in $\mathcal{Q}$. It follows that if $f \in \mathcal{Q}$, the function $e^{i\phi(f)}$ can be approximated in L^2 by products of $e^{i\phi_{t_i}(g_i)}$. Since linear combinations of exponentials are dense in $L^2(Q, \mu)$, we obtain (21.5). $\qquad\square$

Theorem 21.26 *There exists a unique unitary map*

$$T_{\mathrm{eucl}} : \mathcal{H} \to \Gamma_s\big(\mathbb{C}(2\epsilon)^{\frac{1}{2}}\mathcal{X}\big)$$

such that

$$T_{\mathrm{eucl}}1 = \Omega, \tag{21.14}$$
$$T_{\mathrm{eucl}}e^{i\phi_0(g)} = e^{i(a^*(g)+a(g))}T_{\mathrm{eucl}}, \quad g \in (2\epsilon)^{\frac{1}{2}}\mathcal{X}.$$

We have

$$T_{\mathrm{eucl}}e^{-tH} = e^{-td\Gamma(\epsilon)}T_{\mathrm{eucl}}, \quad t \geq 0.$$

Proof Linear combinations of time-zero exponentials $e^{i\phi_0(g)}$, for $g \in (2\epsilon)^{\frac{1}{2}}\mathcal{X}$, are dense in $L^2(Q, \mathfrak{S}_0, \mu)$, and

$$\int_Q e^{i\phi_0(g)}d\mu = e^{-\frac{1}{2}(\delta_0 \otimes g|C\delta_0 \otimes g)} = e^{-\frac{1}{2}(g|g)},$$

by Lemma 21.15. Therefore, there exists a unique unitary map $\tilde{T}_{\mathrm{eucl}} : L^2(Q, \mathfrak{S}_0, \mu) \to \mathbf{L}^2\big((2\epsilon)^{\frac{1}{2}}\mathcal{X}, e^{-x\epsilon x}dx\big)$ such that

$$\tilde{T}_{\mathrm{eucl}}1 = 1,$$
$$\tilde{T}_{\mathrm{eucl}}e^{i\phi_0(g)} = e^{i\phi(g)}\tilde{T}_{\mathrm{eucl}}, \quad g \in (2\epsilon)^{\frac{1}{2}}\mathcal{X}.$$

Composing $\tilde{T}_{\mathrm{eucl}}$ with the map $(T^{\mathrm{rw}})^{-1}$ constructed in Thm. 9.22, we obtain the unitary map T_{eucl} with the first two properties of (21.14). To prove the third one, it suffices by density to check that, for $g_1, g_2 \in (2\epsilon)^{\frac{1}{2}} \mathcal{X}$, one has

$$\int_Q e^{-i\phi_0(g_1)} e^{i\phi_t(g_2)} d\mu = (e^{i(a^*(g_1) + a(g_1))} \Omega | e^{-td\Gamma(\epsilon)} e^{i(a^*(g_2) + a(g_2))} \Omega). \quad (21.15)$$

The l.h.s. of (21.15) equals

$$\exp\left(-\frac{1}{2}(\delta_t \otimes g_2 - \delta_0 \otimes g_1 | \delta_t \otimes g_2 - \delta_0 \otimes g_1)_{\mathcal{Q}}\right)$$
$$= \exp\left(-(g_2 | (4\epsilon)^{-1} g_2)_{\mathcal{X}} - (g_1 | (4\epsilon)^{-1} g_1)_{\mathcal{X}} + (g_1 | (2\epsilon)^{-1} e^{-t\epsilon} g_2)_{\mathcal{X}}\right),$$

by Lemma 21.15. Applying Lemma 21.25 to the Hilbert space $\mathbb{C}(2\epsilon)^{\frac{1}{2}} \mathcal{X}$, we see that this equals the r.h.s. of (21.15). □

21.2.5 *From a positivity preserving semi-group to a Markov path space*

Let (X, ν) be a measure space and $P(t) = e^{-tH}$ be positivity improving contractive semi-group on $L^2(X, \nu)$. We assume that $0 = \inf \operatorname{spec} H$ and $\inf \operatorname{spec} H$ is an eigenvalue. Recall that by the Perron–Frobenius theorem (Thm. 5.25) H has a unique positive ground state. It will be denoted by Ω.

In this subsection we present a construction converse to that of Subsect. 21.2.2.

Theorem 21.27 (1) *There exist*
  (i) *a Markov path space* $(Q, \mathfrak{S}, \mathfrak{S}_0, U_t, R, \mu)$,
  (ii) *a unitary map* $T : L^2(X, \nu) \to L^2(Q, \mathfrak{S}_0, \mu)$ *such that*

$$T\Omega = 1,$$
$$TL^\infty(X, \nu)T^{-1} = L^\infty(Q, \mathfrak{S}_0, \mu).$$

(2) *Denoting* TAT^{-1} *by* $\tilde{A}$ *for* $A \in L^\infty(X, \nu)$, *one has*

$$\int_Q \prod_{i=1}^n U_{t_i}(\tilde{A}_i) d\mu = (\Omega | A_1 e^{-(t_2 - t_1)H} A_2 \cdots e^{-(t_n - t_{n-1})H} A_n \Omega),$$

for $A_i \in L^\infty(X, \nu)$, $i = 1, \ldots, n$, $t_1 \leq \cdots \leq t_n$.

Lemma 21.28 $e^{-tH} L^\infty(X, \nu)\Omega \subset L^\infty(X, \nu)\Omega, \quad t \geq 0.$

Proof Set $\nu_\Omega = \Omega^2 \nu$ and consider the unitary map

$$T_\Omega : L^2(X, \nu) \to L^2(X, \nu_\Omega)$$
$$f \mapsto \Omega^{-1} f.$$

Setting $H_\Omega := T_\Omega H T_\Omega^{-1}$, we see that e^{-tH_Ω} is positivity preserving, with 1 as the unique strictly positive ground state. Therefore, H_Ω is doubly Markovian. Therefore, by Prop. 5.24, it is a contraction on $L^\infty(X, \nu_\Omega) = L^\infty(X, \nu)$. Now

$$e^{-tH} L^\infty(X, \nu)\Omega = T_\Omega^{-1} e^{-tH_\Omega} L^\infty(X, \nu_\Omega)1$$
$$\subset T_\Omega^{-1} L^\infty(X, \nu_\Omega)1 = L^\infty(X, \nu)\Omega. \qquad \square$$

Proof of Thm. 21.27. By the Gelfand–Naimark theorem (Sakai (1971), Thm. 1.2.1), $L^\infty(X, \nu)$ is isomorphic as a C^*-algebra to $C(Q_0)$, where Q_0 is a compact Hausdorff space. In the sequel we will denote by the same letter A an element of $L^\infty(X, \nu)$ and its image in $C(Q_0)$.

Since $L^\infty(X, \nu)$ is a W^*-algebra, we know that Q_0 is a *Stonean space*, i.e. the closure of any open set in Q_0 is open (see Sakai (1971), Prop. 1.3.2). Let Ξ be the set of characteristic functions on Q_0. By Sakai (1971), Prop. 1.3.1, the $*$-algebra generated by Ξ is dense in $C(Q_0)$.

Let $Q := Q_0^{\mathbb{R}}$ be equipped with the product topology, which is also compact by Tychonov's theorem. Note that each $q \in Q$ is a function $\mathbb{R} \ni t \mapsto q_t \in Q_0$. By the Stone–Weierstrass theorem, the $*$-algebra generated by functions f of the form $f(q) = A(q_t)$ for some $t \in \mathbb{R}$ and $A \in C(Q_0)$ is dense in $C(Q)$. By the argument above, the $*$-algebra $\mathcal{L}(Q)$ generated by the functions f of the form $f(q) = A(q_t)$ for some $t \in \mathbb{R}$ and $A \in \Xi$ is also dense in $C(Q)$.

Now let $f \in \mathcal{L}(Q)$. Clearly, f can always be written as

$$f(q) = \sum_{j=1}^{p} a_j \prod_{i=1}^{n} A_{i,j}(q_{t_i}), \ A_{i,j} \in \Xi, \ a_j \in \mathbb{C},$$

for $t_1 \leq \cdots \leq t_n$. Splitting further characteristic functions $A_{i,j}$, we can uniquely rewrite f as

$$f(q) = \sum_{j=1}^{q} b_j \prod_{i=1}^{n} B_{i,j}(q_{t_i}), \ B_{i,j} \in \Xi, \ b_j \in \mathbb{C}, \tag{21.16}$$

where $B_{i,j} B_{i,k} = 0$ for $j \neq k$. It follows that

$$\rho(f) := \sum_{j=1}^{q} b_j \big(\Omega | B_{1,j} e^{-(t_2 - t_1)H} B_{2,j} \cdots e^{-(t_n - t_{n-1})H} B_{n,j} \Omega \big), \tag{21.17}$$

defines a linear form on $\mathcal{L}(Q)$ with $\rho(1) = 1$. Now let $F \in \mathcal{L}(Q)$ with $F \geq 0$. Clearly, f can be uniquely written as in (21.16) with $b_j \geq 0$, $B_{i,j} \geq 0$. Since e^{-tH} is positivity preserving and $\Omega \geq 0$, we see that $\rho(f) \geq 0$ and ρ is a positive, hence bounded linear form on $\mathcal{L}(Q)$. We denote by $\mathfrak{S}$ the Baire σ-algebra on Q. Extending ρ to $C(Q)$ by density and using the Riesz–Markov theorem, we obtain a Baire probability measure μ such that

$$\rho(f) = \int_Q f \mathrm{d}\mu, \ f \in \mathcal{L}(Q).$$

We now set

$$rq_s := q_{-s}, \quad u_t q_s := q_{s-t}, \quad t \in \mathbb{R},$$

and

$$Rf(q) := f(rq), \quad (U_t f)(q) := f(u_{-t}q), \quad t \in \mathbb{R}.$$

Clearly, U_t and R satisfy conditions (3) and (4) of Def. 21.7. Let $\mathfrak{S}_0$ be the sub-σ-algebra of $\mathfrak{S}$ generated by the functions $q \mapsto A(q_0)$, $A \in C(Q_0) = L^\infty(X, \nu)$. Note that $\mathfrak{S} := \bigvee_{t \in \mathbb{R}} U_t \mathfrak{S}_0$. We can rewrite (21.17) as

$$\int_Q \prod_{i=1}^n A_i(q(t_i))\mathrm{d}\mu(q) = (\Omega|\bar{A}_1 \mathrm{e}^{-(t_2-t_1)H} A_2 \cdots \mathrm{e}^{-(t_n-t_{n-1})H} A_n\Omega), \qquad (21.18)$$

for $A_i \in L^\infty(X, \nu)$, $t_1 \leq \cdots \leq t_n$.

It remains to prove that $(Q, \mathfrak{S}, \mathfrak{S}_0, U_t, R, \mu)$ is a Markov path space. Property (1) of Def. 21.9 is obvious. To prove property (2) of Def. 21.9, i.e. that $E_{[0,+\infty[}E_{]-\infty,0]} = E_{\{0\}}$, it suffices by linearity and density to show that for

$$f(q) = \prod_{i=1}^n A_i(q_{t_i}), \ A_i \in L^\infty(Q, \mathfrak{S}_0), \ t_1 \leq \cdots \leq t_n \leq 0 \qquad (21.19)$$

$E_{[0,+\infty[}f$ is $\mathfrak{S}_0$-measurable. Recall that $E_{[0,+\infty[}f = g$ iff g is $\mathfrak{S}_{[0,+\infty[}$-measurable and

$$\int_Q \bar{f}h\mathrm{d}\mu = \int_Q \bar{g}h\mathrm{d}\mu, \qquad (21.20)$$

for all $\mathfrak{S}_{[0,+\infty[}$-measurable functions h. Again by linearity and density, it suffices to check (21.20) for $h(q) = \prod_{i=1}^p B_i(q_{s_i})$, $B_i \in L^\infty(Q, \mathfrak{S}_0)$ and $0 \leq s_1 \leq \cdots \leq s_p$.

For f as in (21.19), we have, using (21.18),

$$\int_Q \bar{f}h\mathrm{d}\mu$$

$$= \int_Q \prod_{i=1}^n \overline{A_i(q_{t_i})} \prod_{i=1}^p B_i(q_{s_i})\mathrm{d}\mu$$

$$= (\Omega|\overline{A_1}\mathrm{e}^{(t_1-t_2)H} \cdots \mathrm{e}^{(t_{n-1}-t_n)H} \overline{A_n}\mathrm{e}^{(t_n-s_1)H} B_1 \mathrm{e}^{(s_1-s_2)H} \cdots \mathrm{e}^{(s_{p-1}-s_p)H} B_p\Omega)$$

$$= (\mathrm{e}^{t_n H} A_n \mathrm{e}^{(t_{n-1}-t_n)H} \cdots \mathrm{e}^{(t_1-t_2)H} A_1\Omega|\mathrm{e}^{-s_1 H} B_1 \mathrm{e}^{(s_1-s_2)H} \cdots \mathrm{e}^{(s_{p-1}-s_p)H} B_p\Omega).$$

By Lemma 21.28, there exists $C \in L^\infty(X, \nu)$ such that

$$\mathrm{e}^{t_n H} A_n \mathrm{e}^{(t_{n-1}-t_n)H} A_{n-1} \cdots \mathrm{e}^{(t_1-t_2)H} A_1\Omega = C\Omega,$$

and hence

$$\int_Q \bar{f}h\mathrm{d}\mu$$

$$= (\Omega|\overline{C}\mathrm{e}^{-s_1 H} B_1 \mathrm{e}^{(s_1-s_2)H} B_2 \cdots \mathrm{e}^{(s_{p-1}-s_p)H} B_p\Omega)$$

$$= \int_Q \overline{C(q_0)} \prod_{i=1}^p B_i(q_{s_i})\mathrm{d}\mu.$$

Therefore, by (21.20) we have $(E_{[0,\infty[}f)(q) = C(q_0)$, which proves that $E_{[0,+\infty[}f$ is $\mathfrak{S}_0$-measurable and completes the proof of the Markov property.

To complete the proof of the theorem it remains to construct the unitary operator T. We first note that, since Ω is a.e. positive, $L^\infty(X, \nu)\Omega$ is dense in

$L^2(X, \nu)$. Using (21.18), it follows that the map

$$T : L^2(X, \nu) \to L^2(Q, \mathfrak{S}_0, \mu)$$

$$(TA\Omega)(q) := A(q_0), \quad A \in L^\infty(X, \nu), \quad q \in Q,$$

extends to a unitary operator. $\qquad\qquad\qquad\qquad\qquad\qquad\qquad\qquad$ □

21.3 Perturbations of Markov path spaces

We fix a Markov path space $(Q, \mathfrak{S}, \mathfrak{S}_0, U_t, R, \mu)$. Recall that this leads to a construction of a physical space $\mathcal{H}$ equipped with a Hamiltonian H. We will show how to perturb this Hamiltonian using the framework of Markov path spaces. Perturbations that can be treated by Euclidean methods are those corresponding to operators of multiplication by real $\mathfrak{S}_0$-measurable functions, i.e. by functions of time-zero fields. Sometimes the perturbation itself does not even make sense as an operator, although a perturbed Hamiltonian can be defined. These singular cases can be handled using the so-called *Feynman–Kac–Nelson kernels*.

21.3.1 Feynman–Kac–Nelson kernels

Definition 21.29 *Let* $\delta \in [0, +\infty]$. *A local Feynman–Kac–Nelson (FKN) kernel is a family* $\{F_{[a,b]}\}_{0 \leq b-a < \delta}$ *of* $\mathfrak{S}$-*measurable functions on* Q *such that*

(1) $F_{[a,b]} > 0$, $F_{[a,b]} \in L^1(Q, \mathfrak{S}_{[a,b]})$,
(2) *for* $a \in \mathbb{R}$, *the map* $[a, a+\delta[\ni b \mapsto F_{[a,b]} \in L^1(Q)$ *is continuous,*
(3) $F_{[a,b]} F_{[b,c]} = F_{[a,c]}$, *for* $a \leq b \leq c$, $c - a < \delta$,
(4) $U_s \left(F_{[a,b]} \right) = F_{[a+s,b+s]}$, *for* $s \in \mathbb{R}$,
(5) $R \left(F_{[a,b]} \right) = F_{[-b,-a]}$.

If $\delta = \infty$ *in the above definition, we will drop the word "local" and use the name "FKN kernel".*

Remark 21.30 *Let us mention a certain notational problem. Let* F *be a measurable function on* Q. $U_t(F)$ *denotes the image of* F *under the action of* U_t. *It is also a function on* Q.

The symbols F, *resp.* $U_t(F)$ *are often understood as multiplication operators. Using this meaning, we have the identity*

$$U_t(F) = U_t F U_t^*,$$

where now U_t *on the r.h.s. is understood as a unitary operator on* $L^2(Q, \mu)$. *Clearly, if we use the latter interpretation of the FKN kernel, (4) of Def. 21.29 can be rewritten as* $U_s F_{[a,b]} U_{-s}^* = F_{[a+s,b+s]}$.

Remark 21.31 *The simplest example of a FKN kernel is given by*

$$F_{[a,b]} := e^{-\int_a^b U_s(V)ds}, \tag{21.21}$$

where $V \in L^\infty(Q, \mathfrak{S}_0)$. At least formally, all FKN kernels are of this form. In fact, by (3), the operators of multiplication by $F_{[s,t]}$ form a two parameter semi-group. Their generators $V(t)$ are also operators of multiplication by $\mathfrak{S}_{\{t\}}$-measurable functions, commute with one another, and satisfy $U_s(V(t)) = V(t + s)$ by (4). Setting $V = V(0)$ we see that $F_{[a,b]}$ is formally given by (21.21).

Properties of FKN kernels obtained from formula (21.21) are described in the following lemma.

Lemma 21.32 *Let $1 \leq p < \infty$ and $V \in L^p(Q)$. Then the following hold:*

(1) $\int_a^b U_s(V)ds \in L^p(Q)$.
(2) $\|e^{-\int_a^b U_s(V)ds}\|_p \leq \|e^{-(b-a)V}\|_p = \|e^{-p(b-a)V}\|_1^{1/p}$.
(3) *Let $V \in L^p(Q)$ for some $p > 1$, and $e^{-\delta V} \in L^1(Q)$ for some $\delta > 0$. Set $F_{[a,b]} = e^{-\int_a^b U_s(V)ds}$. Then $\{F_{[a,b]}\}_{0 \leq b-a < \delta}$ is a local FKN kernel.*
(4) *Let $V \in L^1(Q)$ and $V \geq 0$. Then $\{F_{[a,b]}\}_{0 \leq b-a < \infty}$ is a FKN kernel.*

Proof (1) follows from the strong continuity of U_t on $L^p(Q)$.

To prove (2), we apply Jensen's inequality,

$$e^{-\int_a^b U_s(V)ds} \leq \frac{1}{b-a} \int_a^b e^{-(b-a)U_s(V)}ds,$$

and obtain

$$\|e^{-\int_a^b U_s(V)ds}\|_p \leq \frac{1}{b-a} \int_a^b \|e^{-(b-a)U_s(V)}\|_p ds = \|e^{-(b-a)V}\|_p,$$

since $e^{-(b-a)U_s(V)} = U_s(e^{-(b-a)V})$ and U_s is measure preserving.

To prove (3), we will use Subsect. 5.1.9. Write

$$F_{[a,b+\epsilon]} - F_{[a,b]} = (F_{[b,b+\epsilon]} - 1)F_{[a,b]}.$$

Since $F_{[a,b]} \in L^{\delta/(b-a)}(Q)$, it suffices by Hölder's inequality to prove that $F_{[b,b+\epsilon]} \to 1$ in $L^q(Q)$ for $q = \delta/(\delta - b + a)$. Since U_b is isometric on L^q, we may assume that $b = 0$. Clearly, $F_{[0,\epsilon]} \to 1$ a.e., when $\epsilon \to 0$. Hence, $F_{[0,\epsilon]} \to 1$ in measure. Using (2), we see that, for all $p' > 1$, $\|F_{[0,\epsilon]}\|_{p'} \leq C$ uniformly for $0 \leq \epsilon \leq \delta/p'$. Hence, $\{F_{[0,\epsilon]} : 0 \leq \epsilon \leq \delta/p'\}$ is an equi-integrable family. By the Lebesgue–Vitali theorem (Thm. 5.32), $F_{[0,\epsilon]} \to 1$ in $L^q(Q)$.

Finally, statement (4) is immediate, since $F_{[a,b]} \leq 1$ for all $a \leq b$. □

21.3.2 Feynman–Kac–Nelson formula

We now describe the construction of a perturbed Hamiltonian associated with a FKN kernel.

We recall that local Hermitian semi-groups were defined in Subsect. 2.3.6.

Proposition 21.33 *Let* $\{F_{[a,b]}\}_{0\leq b-a\leq\delta}$, $\delta>0$, *be a FKN kernel. For* $0\leq t<\delta/2$, *set*

$$\mathcal{D}_t := E_0\mathrm{Span}\left(\bigcup_{0\leq s<\delta/2-t}F_{[0,s]}L^\infty(Q,\mathfrak{S}_{[0,+\infty[})\right),$$

$$P_F(t) := E_0F_{[0,t]}U_t\big|_{\mathcal{D}_t}.$$

Then $\{P_F(t),\mathcal{D}_t\}_{t\in[0,\delta/2]}$ *is a local Hermitian semi-group.*

Proof We check the conditions of Def. 2.67. Since $F_{[0,s]}$ belongs only to $L^1(Q)$ it is not obvious that $\mathcal{D}_t\subset L^2(Q,\mathfrak{S}_0)=\mathcal{H}$. To prove that this is the case, we write, for $f=E_0F_{[0,s]}g\in\mathcal{D}_t$, $0\leq s<\delta/2-t$,

$$\|f\|^2 = (F_{[0,s]}g|E_0F_{[0,s]}g) = (F_{[0,s]}g|RE_0F_{[0,s]}g)$$
$$= (F_{[-s,0]}Rg|E_0F_{[0,s]}g) = (F_{[-s,0]}Rg|E_{]-\infty,0]}E_{[0,+\infty[}F_{[0,s]}g)$$
$$= (E_{]-\infty,0]}F_{[-s,0]}Rg|E_{]-\infty,0]}E_{[0,+\infty[}F_{[0,s]}g) = (F_{[-s,0]}Rg|F_{[0,s]}g) \qquad (21.22)$$
$$= (Rg|F_{[-s,0]}F_{[0,s]}g) = (Rg|F_{[-s,s]}g) \leq \|F_{[-s,s]}\|_1\|g\|_\infty^2.$$

Since $0\leq s\leq\delta/2$, $F_{[-s,s]}\in L^1(Q)$ and the r.h.s. is finite. Since $L^\infty(Q,\mathfrak{S}_0)\subset\mathcal{D}_t$, $\mathcal{D}_t$ is dense in $\mathcal{H}$. We now claim that $P_F(s)\,\mathcal{D}_t\subset\mathcal{D}_{t-s}$ for $0\leq s\leq t\leq\delta/2$. In fact, if $f=E_0F_{[0,s_1]}g\in\mathcal{D}_t$, for $0\leq s_1\leq\delta/2-t$, we have

$$P_F(s)f = E_0F_{[0,s]}U_sE_0F_{[0,s_1]}g = E_0F_{[0,s]}E_{\{s\}}F_{[s,s+s_1]}U_sg$$
$$= E_0F_{[0,s]}E_{]-\infty,s]}E_{[s,+\infty[}F_{[s,s+s_1]}U_sg = E_0F_{[0,s]}E_{]-\infty,s]}F_{[s,s+s_1]}U_sg$$
$$= E_0E_{]-\infty,s]}F_{[0,s]}F_{[s,s+s_1]}U_sg = E_0E_{]-\infty,s]}F_{[0,s+s_1]}U_sg$$
$$= E_0F_{[0,s+s_1]}U_sg \in \mathcal{D}_{t-s}, \qquad (21.23)$$

where we have used the properties of $F_{[a,b]}$ and the Markov property. The identity (21.23) also proves that if $f=E_0F_{[0,s_1]}g\in\mathcal{D}_{t+s}$ for $0\leq s_1\leq\delta/2-(t+s)$, then $P_F(t)P_F(s)f=P_F(t+s)f$.

Let us prove the weak continuity of $P_F(t)$. For $f=E_0F_{[0,s_1]}g\in\mathcal{D}_s$ and $0\leq s_1\leq\delta/2-s$ as above, we have

$$(f|P_F(t)f)_\mathcal{H} = (F_{[0,s_1]}g|E_0F_{[0,s_1+t]}U_tg) = (Rg|F_{[-s_1,s_1+t]}U_tg),$$

by the same arguments as in (21.22) and (21.23). Hence,

$$(f|P_F(t+\epsilon)f) - (f|P_F(t)f) = \left(Rg|(F_{[-s_1,s_1+t+\epsilon]} - F_{[-s_1,s_1+t]})U_{t+\epsilon}g\right)$$
$$+ \left(Rg|F_{[-s_1,s_1+t]}(U_{t+\epsilon}g - U_tg)\right).$$

The first term tends to 0 when $\epsilon\to0$ by Def. 21.29. The second term tends to 0 when $\epsilon\to0$ by the σ-weak continuity of $t\mapsto U_t$ on $L^\infty(Q)$. □

In the next two propositions we give examples of FKN kernels obtained from a real $\mathfrak{S}_0$-measurable function V as in Lemma 21.32. Note that the Hermitian operators $P_F(t)$ are now denoted by $P_V(t)$ and have slightly bigger domains. This choice will be convenient in the next subsection.

The case of positive perturbations is easier:

Proposition 21.34 *Let V be a real $\mathfrak{S}_0$-measurable function such that $V \in L^1(Q)$ and $V(x) \geq 0$ a.e. Then*

$$P_V(t) = E_0 e^{-\int_0^t U_s(V)\mathrm{d}s} U_t, \quad t \geq 0,$$

is a strongly continuous semi-group of bounded self-adjoint operators on $L^2(Q, \mathfrak{S}_0, \mu)$.

In the case of arbitrary perturbations we need to use the notion of a Hermitian semi-group.

Proposition 21.35 *Let V be a real $\mathfrak{S}_0$-measurable function such that $V \in L^{p_0}(Q)$ for some $p_0 > 2$, and $e^{-\delta V} \in L^1(Q)$ for some $\delta > 0$. Set $p(t)^{-1} := 1/2 - t/\delta$ for $0 \leq t \leq \delta/2$, and*

$$P_V(t) = E_0 e^{-\int_0^t U_s(V)\mathrm{d}s} U_t \big|_{L^{p(t)}(Q,\mathfrak{S}_0,\mu)}.$$

Then $\left\{ P_V(t), L^{p(t)}(Q, \mathfrak{S}_0, \mu) \right\}_{t \in [0, \delta/2]}$ is a local Hermitian semi-group.

Proof It follows from Lemma 21.32 that $\int_0^t U_s(V)\mathrm{d}s$ is well defined in $L^{p_0}(Q)$, and that $e^{-\int_0^t U_s(V)\mathrm{d}s} \in L^p(Q)$ for $0 \leq t \leq \delta/p$. By Hölder's inequality, $P_V(t)$ maps $L^{p(t)}(Q)$ into $L^2(Q)$, so $P_V(t)$ is well defined on $L^{p(t)}(Q, \mathfrak{S}_0)$. The fact that $P_V(t)$ maps $L^{p(t+s)}$ into $L^{p(s)}$ follows also from Hölder's inequality. The proofs of the semi-group and weak continuity properties are completely analogous to those of Prop. 21.33. $\qquad\square$

Remark 21.36 *Let us write the physical Hilbert space as $\mathcal{H} = L^2(X, \nu)$. We treat paths (elements of Q) as functions $\mathbb{R} \ni t \mapsto q_t \in X$. The expectation E_t is written as*

$$E_t G(x) =: \int G(q) \mathrm{d}\mu_{t,x}(q), \quad G \in L^1(Q, \mathrm{d}\mu), \quad x \in X.$$

Let V be a real function on X. Under some conditions on V (see for example Thms. 21.37, 21.38) one can show that $P_V(t) = E_0 e^{\int_0^t U_s(V)\mathrm{d}s} U_t = e^{-t(H+V)}$ for $t \geq 0$. This can be formally rewritten as

$$e^{-t(H_0 + V(x))} f(x) = \int \exp\left(-\int_0^t V(q_s)\mathrm{d}s\right) f(q_t) \mathrm{d}\mu_{0,x}(q). \tag{21.24}$$

Recall that in Thm. 21.4 we described the Feynman–Kac formula for the integral kernel of $e^{-t(-\frac{1}{2}\Delta + V(x))}$. The generalization (21.24) of the Feynman–Kac formula to quantum field theory was first given by Nelson. Therefore, in this context, (21.24) is usually called the Feynman–Kac–Nelson formula.

21.3.3 Perturbed Hamiltonians

We recall that H is the positive self-adjoint operator generating the group $\{P(t)\}_{t\in[0,\infty[}$ constructed in Thm. 21.11.

Let V be a real $\mathfrak{S}_0$-measurable function. The self-adjoint operator of multiplication by V on $L^2(Q, \mathfrak{S}_0, \mu)$ is also denoted by V. Under the hypotheses of Prop. 21.34 we can define a unique positive self-adjoint operator H_V such that $P_V(t) = \mathrm{e}^{-tH_V}$. Similarly, using Thm. 2.69, under the hypotheses of Prop. 21.35, we can define a unique self-adjoint operator H_V such that $P_V(t) \subset \mathrm{e}^{-tH_V}$. We now give without proof some results about the Hamiltonian H_V.

Theorem 21.37 (Positive perturbations) *Assume the hypotheses of Prop. 21.34. Then:*

(1) H_V *is bounded below.*
(2) *If $V \in L^p(Q)$ for $p > 1$, then H_V is a restriction of the form sum $H + V$.*
(3) *If $V \in L^p(Q)$ for $p \geq 2$, then H_V is the closure of $H + V$.*

Theorem 21.38 (Arbitrary perturbations) *Assume the hypotheses of Prop. 21.35. Then:*

(1) H_V *is the closure of $H + V$.*
(2) *Assume that e^{-tH} is hyper-contractive on $L^2(Q, \mathfrak{S}_0)$ and let $T > 0$ be such that e^{-TH} maps $L^2(Q)$ into $L^r(Q)$, $r > 2$. Then if $\mathrm{e}^{-\delta V} \in L^1(Q)$ for $\delta = r'T$, $1/r + 1/r' = 1/2$, H_V is bounded below.*

Remark 21.39 *The main examples of models with local interaction that can be treated by the methods of this chapter are the (space-cutoff) $P(\varphi)_2$ and $(\mathrm{e}^{\alpha\varphi})_2$ models (both at 0 and at positive temperature). The $P(\varphi)_2$ model was the first model with a local interaction to be rigorously constructed. It will be further studied in Chap. 22.*

The $(\mathrm{e}^{\alpha\varphi})_2$ model is also called the Høgh-Krohn model. Although not physical, it has the pedagogical advantage that the interaction term $\int g(x)\mathrm{e}^{\alpha\varphi(x)}\,\mathrm{d}x$ is positive, even after Wick ordering. It provides an example of where Feynman–Kac–Nelson kernels can be used even if the formal interaction does not exist. In fact, one can show that the formal interaction

$$\int g(x) :\mathrm{e}^{\alpha\varphi(x)}: \,\mathrm{d}x$$

for g a positive compactly supported function can be given a rigorous meaning iff $|\alpha| < \sqrt{2\pi}$, although the FKN kernels

$$\int_a^b \int g(x) :\mathrm{e}^{\alpha\varphi(t,x)}: \,\mathrm{d}x\mathrm{d}t$$

are well defined iff $|\alpha| < \sqrt{4\pi}$; see Simon (1974).

Another example where one has to use FKN kernels is the $P(\varphi)(0)$ model, obtained by replacing the space-cutoff $g(x)$ with the delta function δ_0; see Klein–Landau (1975).

21.4 Euclidean approach at positive temperatures

There exists a version of the Euclidean approach for bosonic fields at positive temperatures. The "Euclidean time", which at the zero temperature took values in $\mathbb{R}$, now belongs to the circle of length β. The number β has the meaning of inverse temperature. Given a β-Markov path space, we construct a von Neumann algebra equipped with a W^*-dynamics and a KMS state.

21.4.1 β-Markov path spaces

Definition 21.40 *The circle of length β, that is, $\mathbb{R}/\beta\mathbb{Z}$, is denoted by S_β, and is sometimes identified with $]-\beta/2, \beta/2]$. t will still denote the generic variable in S_β.*

Definition 21.41 *Let $(Q, \mathfrak{S}, \mathfrak{S}_0, U_t, R, \mu)$ be a generalized path space as in Def. 21.7. The path space is called*

(1) β-*periodic if $U_\beta = \mathbb{1}$, so that $S_\beta \ni t \mapsto U_t$ is a strongly continuous unitary group.*
(2) β-*Markov if in addition it satisfies*
   (i) *the β-reflection property $RE_{\{0,\beta/2\}} = E_{\{0,\beta/2\}}$,*
   (ii) *the β-Markov property $E_{[0,\beta/2]}E_{[-\beta/2,0]} = E_{\{0,\beta/2\}}$.*

It is easy to show that in a β-Markov path space we have

$$E_{\{0,\beta/2\}} = E_{[0,\beta/2]}RE_{[0,\beta/2]}. \tag{21.25}$$

21.4.2 Reconstruction theorem

We assume that we are given a β-Markov path space $(Q, \mathfrak{S}, \mathfrak{S}_0, U_t, R, \mu)$. As in Subsect. 21.2.1, we now proceed to the construction of the corresponding physical objects.

Definition 21.42 *The physical Hilbert space is*

$$\mathcal{H} := E_{\{0,\beta/2\}}L^2(Q, \mathfrak{S}, \mu) = L^2(Q, \mathfrak{S}_{\{0,\beta/2\}}, \mu),$$

and the vector $1 \in \mathcal{H}$ will be denoted by Ω. The Abelian von Neumann algebra $L^\infty(Q, \mathfrak{S}_{\{0\}}, \mu)$ acting on $\mathcal{H}$ will be denoted by $\mathfrak{A}$.

The construction of the generator of the dynamics on $\mathcal{H}$ is now more delicate than in Thm. 21.11, because U_t does not preserve $L^2(Q, \mathfrak{S}_{[0,\beta/2]}, \mu)$. In fact, U_t sends $L^2(Q, \mathfrak{S}_{[0,\beta/2]}, \mu)$ into $L^2(Q, \mathfrak{S}_{[t,t+\beta/2]}, \mu)$. In the construction the crucial role is played by the concept of a *local Hermitian semi-group* introduced in Def. 2.67.

Theorem 21.43 *Set, for $0 \le t < \beta/2$,*

$$\mathcal{M}_t := L^2(Q, \mathfrak{S}_{[0,\beta/2-t]}, \mu),$$
$$\mathcal{D}_t := E_{\{0,\beta/2\}} \mathcal{M}_t \subset \mathcal{H}.$$

Then, for any $0 \le s \le t \le \beta/2$, there exists a unique $P(s) : \mathcal{D}_t \to \mathcal{D}_{t-s}$ such that

$$P(s) E_{\{0,\beta/2\}} f = E_{\{0,\beta/2\}} U_s f, \quad f \in L^2(Q, \mu),$$

and $\{\mathcal{D}_t, P(t)\}_{t \in [0,\beta/2]}$ is a local Hermitian semi-group on $\mathcal{H}$ preserving Ω.

Proof If $s \le t$, one has $\mathcal{M}_t \subset \mathcal{M}_s$, hence $\mathcal{D}_t \subset \mathcal{D}_s$. From the definition of $\mathfrak{S}_I$ as $\bigvee_{t \in I} U_t \mathfrak{S}_0$ and from the strong continuity of U_t on $L^2(Q, \mathfrak{S}, \mu)$, we see that $\bigcup_{0 < t < \beta/2} \mathcal{M}_t$ is dense in $L^2(Q, \mathfrak{S}_{[0,\beta/2]}, \mu)$, which implies the density of $\bigcup_{0 < t < \beta/2} \mathcal{D}_t$ in $\mathcal{H}$.

We now have to check that, for $0 \le t < \beta/2$, $P(t)$ is well defined as a linear operator on $\mathcal{D}_t$.

Let us fix $0 < r \le s < t < \beta/2$ with $r + s \le t$. Let $f \in \mathcal{M}_t$. We have

$$\|E_{\{0,\beta/2\}} U_s f\|^2 = (U_s f | R U_s f)$$
$$= (U_{s-r} f | U_{-r} R U_s f) = (U_{s-r} f | R U_{s+r} f)$$
$$= (U_{s-r} f | E_{\{0,\beta/2\}} U_{s+r} f) \le \|E_{\{0,\beta/2\}} U_{s-r} f\| \|f\|. \qquad (21.26)$$

In the first line we use (21.25) and the fact that $U_s f$ is $\mathfrak{S}_{[0,\beta/2]}$-measurable. In the second line we use the unitarity of U_{-r} and $U_{-r} R = R U_r$. In the third line we apply (21.25) again, the Cauchy–Schwarz inequality and the fact that $E_{\{0,\beta/2\}}$ and U_{s+r} are contractions.

Taking $r = s$, we obtain that $\|E_{\{0,\beta/2\}} U_s f\| \le \|E_{\{0,\beta/2\}} f\|^{\frac{1}{2}} \|f\|^{\frac{1}{2}}$ for $2s \le t$. If $\frac{n+1}{n} s \le t$, for $n \in \mathbb{N}$, taking $r = s/n$ and applying recursively (21.26), we obtain

$$\|E_{\{0,\beta/2\}} U_s f\| \le \|E_{\{0,\beta/2\}} U_{s-r} f\|^{\frac{1}{2}} \|f\|^{\frac{1}{2}}$$
$$\le \|E_{\{0,\beta/2\}} U_{s-pr} f\|^{2^{-p}} \|f\|^{(2^{-1}+\cdots+2^{-p})}$$
$$\le \|E_{\{0,\beta/2\}} f\|^{2^{-n}} \|f\|^{(1-2^{-n})}.$$

This shows that $E_{\{0,\beta/2\}} f = 0$ implies $E_{\{0,\beta/2\}} U_s f = 0$ for all $0 \le s < t$. By the strong continuity of U_s, this extends to $s = t$. Thus we have proved

$$E_{\{0,\beta/2\}} f = 0, \quad f \in \mathcal{M}_t \implies E_{\{0,\beta/2\}} U_t f = 0, \qquad (21.27)$$

which means that $P(t)$ is well defined.

The semi-group property of $P(t)$ and the fact that $P(s)\mathcal{D}_t \subset \mathcal{D}_{t-s}$ are immediate. To prove that $P(t)$ is Hermitian, we write, for $f, g \in \mathcal{M}_t$,

$$\big(E_{\{0,\beta/2\}} f | P(t) E_{\{0,\beta/2\}} g\big) = (f | R U_t g) = (U_t f | R g)$$
$$= \big(E_{\{0,\beta/2\}} U_t f | E_{\{0,\beta/2\}} g\big) = \big(P(t) E_{\{0,\beta/2\}} f | E_{\{0,\beta/2\}} g\big).$$

Finally, the weak continuity of $P(t)$ follows from the strong continuity of U_t. $\qquad\qquad\qquad\qquad\qquad\qquad\qquad\qquad\qquad\qquad\qquad\qquad\square$

Definition 21.44 *The unique self-adjoint operator L on $\mathcal{H}$ such that $\mathrm{e}^{-tL}\big|_{\mathcal{D}_t} = P(t)$ is called the* Liouvillean.

Clearly, $L\Omega = 0$.

Definition 21.45 *We denote by $\mathfrak{F} \subset B(\mathcal{H})$ the von Neumann algebra defined by*

$$\mathfrak{F} := \{ \mathrm{e}^{\mathrm{i}tL} A \mathrm{e}^{-\mathrm{i}tL}, \quad A \in \mathfrak{A}, \quad t \in \mathbb{R} \}''. \tag{21.28}$$

Let

$$R_{\beta/4} := U_{\beta/2} R = U_{\beta/4} R U_{-\beta/4}$$

be the reflection around $s = \beta/4$. Clearly,

$$R_{\beta/4} E_{\{0,\beta/2\}} = E_{\{0,\beta/2\}} R_{\beta/4}. \tag{21.29}$$

Definition 21.46 *By (21.29),*

$$J E_{\{0,\beta/2\}} f := E_{\{0,\beta/2\}} R_{\beta/4} \overline{f}, \quad f \in L^2(Q, \mathfrak{S}, \mu), \tag{21.30}$$

defines an anti-unitary operator J on $\mathcal{H}$. We also introduce a state and a W^*-dynamics on $\mathfrak{F}$:

$$\omega(A) := (\Omega|A\Omega), \quad \tau_t(A) = \mathrm{e}^{\mathrm{i}tL} A \mathrm{e}^{-\mathrm{i}tL}, \quad A \in \mathfrak{F}.$$

The next theorem will be proven in the following two subsections.

Theorem 21.47 *ω is a faithful state, it satisfies the β-KMS condition for the dynamics τ, J is the modular conjugation corresponding to ω and L is the standard Liouvillean for the dynamics τ.*

21.4.3 Proof of the KMS condition

In this subsection we prove the part of Thm. 21.47 saying that ω is β-KMS. We first need to introduce additional notation. For $n \in \mathbb{N}$, we set

$$J_\beta(n) := \Big\{ (t_1, \ldots, t_n) \in \mathbb{R}^n \ : \ t_j \geq 0, \ \sum_{j=1}^n t_j \leq \beta/2 \Big\},$$

$$I_\beta(n) := \Big\{ (z_1, \ldots, z_n) \in \mathbb{C}^n \ : \ \mathrm{Re}\, z_j > 0, \ \sum_{j=1}^n \mathrm{Re}\, z_j < \beta/2 \Big\}.$$

Note that $J_\beta(n) \subset I_\beta(n)^{\mathrm{cl}}$. We denote by $\mathrm{Hol}_\beta(n)$ the space of functions (with values in $\mathcal{H}$ or in $\mathbb{C}$, depending on the context) which are holomorphic in $I_\beta(n)$ and continuous in $I_\beta(n)^{\mathrm{cl}}$.

Proposition 21.48 (1) *For* $(t_1, \ldots, t_n) \in J_\beta(n)$, $A_1, \ldots, A_n \in \mathfrak{A}$, *the vector*

$$A_n \prod_{j=n-1}^{1} (\mathrm{e}^{-t_j L} A_j)\Omega \tag{21.31}$$

belongs to $\operatorname{Dom} \mathrm{e}^{-t_n L}$.

(2) *The linear span of such vectors is dense in* $\mathcal{H}$.

(3) *Let* $(t_1, \ldots, t_n) \in J_\beta(n)$, $(s_1, \ldots, s_m) \in J_\beta(m)$ *and* $A_1, \ldots, A_n, B_1, \ldots, B_m \in \mathfrak{A}$. *Set* $t_0 := \beta/2 - (t_n + \cdots + t_1)$, $s_0 := \beta/2 - (s_n + \cdots + s_1)$. *Then one has*

$$\left(\prod_{j=n}^{1} (\mathrm{e}^{-t_j L} A_j)\Omega \Big| \prod_{i=m}^{1} (\mathrm{e}^{-s_i L} B_i)\Omega \right)$$
$$= \left(\prod_{i=0}^{m-1} (\mathrm{e}^{-s_i L} B_{i+1}^*)\Omega \Big| \prod_{j=0}^{n-1} (\mathrm{e}^{-t_j L} A_{j+1}^*)\Omega \right). \tag{21.32}$$

Proof For $A \in \mathfrak{A}$, we set $A(t) = U_t(A)$. First let us show that

$$\prod_{j=n}^{1} (\mathrm{e}^{-t_j L} A_j)\Omega = E_{\{0,\beta/2\}} \prod_{j=n}^{1} A_j(t_j + \cdots + t_n). \tag{21.33}$$

We use induction. (21.33) is clear for $n = 1$. Assume that it is true for $n - 1$, that is,

$$\prod_{j=n-1}^{1} (\mathrm{e}^{-t_j L} A_j)\Omega = E_{\{0,\beta/2\}} \prod_{j=n-1}^{1} A_j(t_j + \cdots + t_{n-1}).$$

Then

$$\mathrm{e}^{-t_n L} A_n \prod_{j=n-1}^{1} (\mathrm{e}^{-t_j L} A_j)\Omega = E_{\{0,\beta/2\}} U_{t_n} A_n \prod_{j=n-1}^{1} A_j(t_j + \cdots + t_{n-1})$$

$$= E_{\{0,\beta/2\}} \prod_{j=n}^{1} A_j(t_j + \cdots + t_n),$$

which proves (21.33) for n.

Since $A_n \prod_{j=n-1}^{1} A_j(t_j + \cdots + t_{n-1})$ belongs to $\mathcal{M}_{t_n}$, this proves that (21.31) belongs to $\operatorname{Dom} \mathrm{e}^{-t_n L}$. Hence, (1) is true.

The linear span of vectors as on the r.h.s. of (21.33) is dense in $L^2(Q, \mathfrak{S}_{[0,\beta/2]}, \mu)$, which proves (2).

We have

$$\left(\prod_{j=n}^{1} (\mathrm{e}^{-t_j L} A_j)\Omega \Big| \prod_{j=m}^{1} (\mathrm{e}^{-s_i L} B_i)\Omega \right) \tag{21.34}$$

$$= \left(\prod_{j=n}^{1} A_j(\tau_j) \Big| R \prod_{i=m}^{1} B_i(\sigma_i) \right)_{L^2(Q)}$$

$$= \left(\prod_{j=n}^{1} A_j(\tau_j) \Big| \prod_{i=m}^{1} B_i(-\sigma_i) \right)_{L^2(Q)}, \tag{21.35}$$

where

$$\tau_j = \sum_{k=j}^{n} t_k, \quad 1 \le j \le n, \qquad \sigma_i = \sum_{k=i}^{m} t_k, \quad 1 \le i \le m. \tag{21.36}$$

Since $U_\beta = \mathbb{1}$, we have

$$(f|g) = (U_{-\beta/2}f|U_{\beta/2}g)_{L^2(Q)}, \quad f, g \in L^2(Q).$$

Hence, (21.35) equals

$$\left(\prod_{j=n}^{1} A_j(-\beta/2 + \tau_j) \middle| \prod_{j=m}^{1} B_i(\beta/2 - \sigma_i) \right)_{L^2(Q)}$$

$$= \left(\prod_{i=1}^{m} \overline{B_i(\beta/2 - \sigma_i)} \middle| \prod_{j=1}^{n} \overline{A_j(-\beta/2 + \tau_j)} \right)_{L^2(Q)}$$

$$= \left(\prod_{i=0}^{m-1} (\mathrm{e}^{-s_i L} B_{i+1}^*)\Omega \middle| \prod_{j=0}^{n-1} (\mathrm{e}^{-t_j L} A_{j+1}^* \Omega) \right).$$

This proves (3). □

Proposition 21.49 *For $(z_1, \ldots, z_n) \in I_\beta(n)^{\mathrm{cl}}$ and $A_1, \ldots, A_n \in \mathfrak{A}$, the vector*

$$A_n \prod_{n-1}^{1} (\mathrm{e}^{-z_j L} A_j)\Omega \tag{21.37}$$

belongs to $\mathrm{Dom}\, \mathrm{e}^{-z_n L}$. *Furthermore, the function*

$$I_\beta(n)^{\mathrm{cl}} \ni (z_1, \ldots, z_n) \mapsto \prod_{j=n}^{1} (\mathrm{e}^{z_j L} A_j)\Omega$$

belongs to $\mathrm{Hol}_\beta(n)$ *and is bounded by* $\prod_{j=1}^{n} \|A_j\|$.

Proof We prove the result by induction in n.

By Prop. 21.48 (1), $A_1\Omega \in \mathrm{Dom}\, \mathrm{e}^{-\beta L/2}$. Therefore, Prop. 2.63 implies that the map $I_\beta(1) \ni z_1 \mapsto \mathrm{e}^{-z_1 L} A_1\Omega$ belongs to $\mathrm{Hol}_\beta(1)$. Moreover, for $z_1 \in I(1)$, we have

$$\|\mathrm{e}^{iz_1 L} A_1\Omega\| = \|\mathrm{e}^{-(\mathrm{Re}\, z_1)L} A_1\Omega\| = \|A_1^*\Omega\| \leq \|A_1\|,$$

again using Prop. 2.63. This proves the result for $n = 1$.

Assume that the result holds for $n - 1$. For $(z_1, \ldots, z_{n-1}) \in I_\beta(n-1)$, set

$$g(z_1, \ldots, z_{n-1}) = A_n \prod_{j=n-1}^{1} (\mathrm{e}^{z_j L} A_j)\Omega,$$

$$h(z_1, \ldots, z_{n-1}) = \prod_{j=1}^{n-1} (A_j^* \mathrm{e}^{z_j L}) A_n^* \Omega.$$

By the induction assumption, g, h belong to $\mathrm{Hol}_\beta(n-1)$ and are bounded by $\prod_{j=1}^{n-1} \|A_j\|$. Moreover, using (3) of Prop. 21.48 with $m = n$, $B_j = A_j$ and

$$s_n = \beta/2 - \sum_{i=1}^{n-1} s_i, \quad t_n = \beta/2 - \sum_{j=1}^{n-1} t_j,$$

we obtain that

$$g(t_1, \ldots, t_{n-1}) \in \mathrm{Dom}\, \mathrm{e}^{-t_n L}, \quad g(s_1, \ldots, s_{n-1}) \in \mathrm{Dom}\, \mathrm{e}^{-s_n L},$$

and

$$\left(e^{-t_n L} g(t_1, \dots, t_{n-1}) \middle| e^{-s_n L} g(s_1, \dots, s_{n-1})\right)$$
$$= \left(h(s_1, \dots, s_{n-1}) \middle| h(t_1, \dots, t_{n-1})\right), \tag{21.38}$$

for $(s_1, \dots, s_n)$, $(t_1, \dots, t_{n-1}) \in J_\beta(n-1)$. Denote by $\mathcal{H}_f$, resp. $\mathcal{H}_h$ the closed subspaces of $\mathcal{H}$ generated by the vectors $e^{-t_n L} g(t_1, \dots, t_{n-1})$, resp. $h(t_1, \dots, t_{n-1})$, for $(t_1, \dots, t_{n-1}) \in J_\beta(n-1)$.

Note that $h(z_1, \dots, z_{n-1})$ belongs to $\mathcal{H}_h$ for $(z_1, \dots, z_{n-1}) \in I_\beta(n-1)^{\text{cl}}$. In fact, let $\Psi \perp \mathcal{H}_h$. Then

$$\left(\Psi \middle| h(z_1, \dots, z_{n-1})\right) = 0, \quad z_1, \dots, z_{n-1} \in J_\beta(n-1). \tag{21.39}$$

Hence, by analyticity and continuity, (21.39) is true for $z_1, \dots, z_{n-1} \in I_\beta(n-1)^{\text{cl}}$.

From (21.38) we see that there exists a unique anti-unitary map $T : \mathcal{H}_f \to \mathcal{H}_h$ such that

$$T e^{-t_n L} g(t_1, \dots t_{n-1}) = h(t_1, \dots, t_{n-1}).$$

It follows that

$$f(z_1, \dots, z_{n-1}) := T^{-1} h(\overline{z_1}, \dots, \overline{z_{n-1}})$$

belongs to $\text{Hol}_\beta(n-1)$. Note that, by the definition of T, for $t_1, \dots, t_{n-1} \in J_\beta(n-1)$ one has

$$f(t_1, \dots, t_{n-1}) = e^{-t_n L} g(t_1, \dots, t_{n-1}). \tag{21.40}$$

We claim that, for $z_1, \dots, z_{n-1} \in I_\beta(n-1)$,

$$g(z_1, \dots, z_{n-1}) \in \text{Dom} \, e^{-(\beta/2 - \sum_{j=1}^{n-1} z_j)L} \tag{21.41}$$

and

$$f(z_1, \dots, z_{n-1}) = e^{-(\beta/2 - \sum_{j=1}^{n-1} z_j)L} g(z_1, \dots, z_{n-1}). \tag{21.42}$$

In fact, the scalar products of the above two functions with a fixed vector $\Psi \in \text{Dom} \, e^{-\beta L/2}$ belong to $\text{Hol}_\beta(n-1)$ and coincide on $J_\beta(n-1)$ by (21.40). By analytic continuation it follows that $g(z_1, \dots, z_{n-1})$ belongs to $\text{Dom} \, e^{-(\beta/2 - \sum_{j=1}^{n-1} z_j)L}$, and that (21.41) and (21.42) are true.

By Prop. 2.63, we obtain that the function

$$\left\{0 \leq z_n \leq \beta/2 - \sum_{j=1}^{n-1} \text{Im} \, z_j\right\} \ni z_n \mapsto e^{z_n L} g(z_1, \dots, z_{n-1})$$

is continuous and analytic in the interior of its domain. For $\text{Re} \, z_n = 0$, we have

$$\|e^{z_n L} g(z_1, \dots, z_{n-1})\| = \|g(z_1, \dots, z_{n-1})\| \leq \prod_{j=1}^{n} \|A_j\|.$$

For Re $z_n = \beta/2 - \sum\limits_{j=1}^{n-1} \mathrm{Im}\, z_j$,

$$\|\mathrm{e}^{z_n L} g(z_1, \ldots, z_{n-1})\| = \|f(z_1, \ldots, z_{n-1})\|$$
$$= \|h(\bar{z}_1, \ldots, \bar{z}_{n-1})\| \leq \prod_{j=1}^{n} \|A_j\|.$$

Therefore, by Prop. 2.63, for $0 \leq z_n \leq \beta/2 - \sum\limits_{j=1}^{n-1} \mathrm{Im}\, z_j$,

$$\|\mathrm{e}^{z_n L} g(z_1, \ldots, z_{n-1})\| \leq \prod_{j=1}^{n} \|A_j\|,$$

which ends the proof of the induction step.   □

Proof of Thm. 21.47, Part 1. By Prop. 21.49, we can analytically continue (21.42) to obtain

$$\left(\prod_{j=n}^{1} (\mathrm{e}^{-\mathrm{i}t_j L} A_j)\Omega \Big| \prod_{i=m}^{1} (\mathrm{e}^{-\mathrm{i}s_i L} B_i)\Omega \right)$$
$$= \left(\mathrm{e}^{(-\beta/2 + \mathrm{i}s_m + \cdots + \mathrm{i}s_1)L} \prod_{i=1}^{m} (B_i^* \mathrm{e}^{\mathrm{i}s_i L})\Omega \Big| \mathrm{e}^{(-\beta/2 + \mathrm{i}t_n + \cdots + \mathrm{i}t_1)L} \prod_{j=1}^{n} (A_j^* \mathrm{e}^{\mathrm{i}t_j L})\Omega \right).$$

Changing variables, this can be rewritten as

$$(A\Omega | B\Omega) = \left(\mathrm{e}^{-\beta L/2} B^* \Omega \Big| \mathrm{e}^{-\beta L/2} A^* \Omega \right),$$

$$A := \prod_{j=n}^{1} \tau_{\tau_j}(A_j),$$

$$B := \prod_{i=m}^{1} \tau_{\sigma_i}(B_i).$$

This identity implies that the (τ, β)-KMS condition (6.7) is satisfied in the ∗-algebra $\mathfrak{F}_0$ generated by $\{\tau_t(A) \,:\, A \in \mathfrak{A},\ t \in \mathbb{R}\}$. But $\mathfrak{F}_0$ is weakly dense in $\mathfrak{F}$. By Prop. 6.64, the (τ, β)-KMS condition is satisfied for all $A, B \in \mathfrak{F}$.   □

21.4.4 Identification of the modular conjugation

To complete the proof of Thm. 21.47, we need the following lemma:

Lemma 21.50 (1) $JA\Omega = \mathrm{e}^{-\beta L/2} A^* \Omega$, for all $A \in \mathfrak{F}$;
(2) $J\mathrm{e}^{\mathrm{i}tL} = \mathrm{e}^{\mathrm{i}tL} J$, for all $t \in \mathbb{R}$;
(3) $J\mathfrak{F}J \subset \mathfrak{F}'$.

Proof Let $(t_1, \ldots, t_n) \in J_\beta(n)$ and $A_1, \ldots, A_n \in \mathfrak{A}$. Then

$$J \prod_{j=n}^{1} (\mathrm{e}^{-t_j L} A_j)\Omega = E_{\{0,\beta/2\}} R_{\beta/4} \prod_{j=n}^{1} \overline{A_j(\tau_j)}$$

$$= E_{\{0,\beta/2\}} \prod_{j=n}^{1} \overline{A_j(\beta/2 - \tau_j)}$$

$$= \mathrm{e}^{-(\beta/2 - \sum_{j=1}^{n} t_j)L} \prod_{j=1}^{n-1} (A_j^* \mathrm{e}^{-t_j L}) A_n^* \Omega, \qquad (21.43)$$

where τ_j are defined in (21.36).

By Prop. 21.49, we can apply the analytic continuation to the above identity and obtain

$$J \prod_{j=n}^{1} (e^{it_j L} A_j)\Omega = e^{-\beta L/2} e^{i\sum_{j=1}^{n} t_j L} \prod_{j=1}^{n-1} (A_j^* e^{-it_j L}) A_n^* \Omega.$$

Changing variables, this can be rewritten as

$$JA\Omega = e^{-\beta L/2} A^* \Omega, \tag{21.44}$$

for

$$A = \prod_{j=n}^{1} \tau_{t_j}(A_j),$$

which proves (1) on $\mathfrak{F}_0$.

Now let $A \in \mathfrak{F}$. Since $\mathfrak{F}$ is the strong closure of $\mathfrak{F}_0$, by the Kaplansky density theorem there exists a sequence $A_n \in \mathfrak{F}_0$ such that $A_n \to A$, $A_n^* \to A^*$ strongly. Applying (21.44) to A_n, we obtain that $A_n\Omega \to A\Omega$ and $e^{-\beta L/2} A_n\Omega \to JA^*\Omega$. Since $e^{-\beta L/2}$ is closed, this implies that

$$e^{-\beta L/2} A\Omega = JA^*\Omega, \quad A \in \mathfrak{F}. \tag{21.45}$$

This proves (1) on $\mathfrak{F}$.

Let us now prove (2). Let $\Psi = E_{\{0,\beta/2\}} F$ for $F \in L^2(Q, \mathfrak{S}_{[\epsilon,\beta/2-\epsilon]}, \mu)$ and $\epsilon > 0$. For $0 \le s < \epsilon$, we have

$$Je^{-sL}\Psi = E_{\{0,\beta/2\}} R_{\beta/4} \overline{U_s F} = E_{\{0,\beta/2\}} U_{-s} R_{\beta/4} \overline{F}.$$

Since $U_{-s} R_{\beta/4} \overline{F} \in L^2(Q, \mathfrak{S}_{[\epsilon-s, \beta/2-\epsilon+s]})$, it follows that $Je^{-sL}\Psi \in \mathrm{Dom}\, e^{-sL}$ and

$$e^{-sL} Je^{-sL}\Psi = E_{\{0,\beta/2\}} R_{\beta/4}\overline{F} = J\Psi,$$

or equivalently

$$Je^{-sL}\Psi = e^{sL} J\Psi. \tag{21.46}$$

We note that $\Psi, J\Psi \in \bigcap_{|s|<\epsilon} \mathrm{Dom}\, e^{sL}$, hence they are analytic vectors for L. Therefore, we can analytically continue (21.46), using that J is anti-linear, to obtain

$$Je^{itL}\Psi = e^{itL} J\Psi.$$

Since the set of such vectors Ψ is dense in $\mathcal{H}$, this proves (2).

Let us now prove (3). Since $\mathfrak{F}$ is the strong closure of $\mathfrak{F}_0$, it suffices to show that, for $A, B \in \mathfrak{F}_0$, one has

$$[JAJ, B] = 0. \tag{21.47}$$

To prove (21.47), it suffices, using (2), to prove that

$$[JAJ, e^{itL} Be^{-itL}] = 0, \quad t \in \mathbb{R}, \quad A, B \in \mathfrak{A}. \tag{21.48}$$

Let us now prove (21.48).

First note that, for any $A_0, B_{\beta/2} \in L^\infty(Q)$, we have

$$A_0 U_{-s} B_{\beta/2} U_s = U_{-s} B_{\beta/2} U_s A_0. \qquad (21.49)$$

Assume now that $A_0 \in L^\infty(Q, \mathfrak{S}_{\{0\}}, \mu)$ and $B_{\beta/2} \in L^\infty(Q, \mathfrak{S}_{\{\beta/2\}}, \mu)$. Let $\Psi = E_{\{0,\beta/2\}} f$ for $f \in L^2(Q, \mathfrak{S}_{[\epsilon,\beta/2]}, \mu), 0 < \epsilon < \beta/2$. Since $B_{\beta/2} U_s f$ and $B_{\beta/2} U_s A_0 f$ belong to $L^2(Q, \mathfrak{S}_{[s,\beta/2]}, \mu)$, we see that

$$E_{\{0,\beta/2\}} B_{\beta/2} U_s f = B_{\beta/2} \mathrm{e}^{-sL} \Psi,$$
$$E_{\{0,\beta/2\}} B_{\beta/2} U_s A_0 f = B_{\beta/2} \mathrm{e}^{-sL} A_0 \Psi$$

belong to $\mathrm{Dom}\, \mathrm{e}^{sL}$ and (21.49) can be rewritten as

$$A_0 \mathrm{e}^{sL} B_{\beta/2} \mathrm{e}^{-sL} \Psi = \mathrm{e}^{sL} B_{\beta/2} \mathrm{e}^{-sL} A_0 \Psi. \qquad (21.50)$$

Hence, to prove (21.48) it suffices to show that

$$s \mapsto \mathrm{e}^{sL} B_{\beta/2} \mathrm{e}^{-sL} A_0 \Psi$$

can be holomorphically extended to $\{0 \leq \mathrm{Re}\, z \leq \epsilon\}$, and that its analytic extension to $s = -\mathrm{i}t$ equals $\mathrm{e}^{-\mathrm{i}tL} B_{\beta/2} \mathrm{e}^{\mathrm{i}tL} A_0 \Psi$.

Let us take a vector Ψ of the form

$$\Psi = \prod_{j=n}^{1} \mathrm{e}^{-t_j L} A_j \Omega$$

for $t_j \geq 0, t_1 + \cdots + t_n \leq \epsilon$ and $A_j \in \mathfrak{A}$. Recall from Prop. 21.48 that the linear span of such vectors is dense in $\mathcal{H}$.

Let $B_0 \in \mathfrak{A}$ such that $B_{\beta/2} = J B_0 J$. By (2), we have

$$\mathrm{e}^{sL} B_{\beta/2} \mathrm{e}^{-sL} A_0 = J \mathrm{e}^{-sL} B_0 J \mathrm{e}^{-sL} A_0.$$

Hence,

$$\mathrm{e}^{sL} B_{\beta/2} \mathrm{e}^{-sL} A_0 \Psi = J \mathrm{e}^{-sL} B_0 J \mathrm{e}^{-sL} A_0 \Psi$$

$$= J \mathrm{e}^{-sL} B_0 J \mathrm{e}^{-sL} A_0 \prod_{j=n}^{1} \mathrm{e}^{-t_j L} A_j \Omega$$

$$= J \mathrm{e}^{-sL} B_0 \mathrm{e}^{sL - (\beta/2 - \sum_{j=1}^{n} t_j)L} \prod_{j=1}^{n} (A_j^* \mathrm{e}^{-t_j L}) A_0^* \Omega,$$

using (21.43). By Prop. 21.49, this can be analytically continued to $s = \mathrm{i}t$ to give

$$J \mathrm{e}^{\mathrm{i}tL} B_0 \mathrm{e}^{-\mathrm{i}tL - (\beta/2 - \sum_{j=1}^{n} t_j)L} \prod_{j=1}^{n} (A_j^* \mathrm{e}^{-t_j L}) A_0^* \Omega$$

$$= J \mathrm{e}^{\mathrm{i}tL} B_0 \mathrm{e}^{\mathrm{i}tL} J A_0 \prod_{j=n}^{1} \mathrm{e}^{-t_j L} A_j \Omega$$

$$= \mathrm{e}^{\mathrm{i}tL} J B_0 J \mathrm{e}^{-\mathrm{i}tL} A_0 \Psi = \mathrm{e}^{\mathrm{i}tL} B_{\beta/2} \mathrm{e}^{-\mathrm{i}tL} A_0 \Psi,$$

once again using (21.43). This completes the proof of (3). $\qquad \square$

Proof of Thm. 21.47, Part 2. We will use the Tomita–Takesaki theory described in Subsect. 6.4.2. Let us check first that Ω is cyclic and separating for $\mathfrak{F}$. Let

$\Psi \in \{\mathfrak{F}_0\Omega\}^\perp$. It follows that, for all $t_1, \cdots, t_n \in \mathbb{R}$, $A_i \in \mathfrak{A}$, one has

$$\left(\Psi \mid \prod_{j=n}^{1}(A_j e^{it_j L})\Omega\right) = 0.$$

By analytic continuation and Prop. 21.49, this implies that for all $(t_1, \ldots, t_n) \in J_\beta(n)$ one has

$$\left(\Psi \mid \prod_{j=n}^{1}(A_j e^{-t_j L})\Omega\right) = 0.$$

Since the vectors of the form $\prod_{j=n}^{1}(A_j e^{-t_j L})\Omega$ span $\mathcal{H}$, this implies that $\Psi = 0$, and hence Ω is cyclic for $\mathfrak{F}$.

Since $J\Omega = \Omega$, Ω is also cyclic for $J\mathfrak{F}J$. By (3) of Lemma 21.50, this implies that Ω is separating for $\mathfrak{F}$.

By (1) of Lemma 21.50,

$$e^{-\beta L/2}B\Omega = JB^*\Omega, \quad B \in \mathfrak{F}. \tag{21.51}$$

Therefore, the operator S of the Tomita–Takesaki theory is

$$S = Je^{-\beta L/2}.$$

By the uniqueness of the polar decomposition of S, this implies that J is the modular conjugation and $e^{-\beta L/2}$ the modular operator for the state Ω. This completes the proof of the theorem. $\qquad\square$

21.4.5 Gaussian β-Markov path spaces I

We would like to describe a β-periodic version of the construction described in Subsect. 21.2.4. Let $\mathcal{X}$ be a real Hilbert space and $\epsilon > 0$ a self-adjoint operator on $\mathcal{X}$. (Again, we assume the reality just for definiteness.) Consider the real Hilbert space

$$L^2(S_\beta, \mathcal{X}) \simeq L^2(S_\beta, \mathbb{R}) \otimes \mathcal{X}$$

and the covariance

$$C = (D_t^2 + \epsilon^2)^{-1}$$

with *β-periodic boundary conditions*. (This means $-D_t^2$ is the Laplacian on the circle S_β.)

Consider also the space $Q := C^{-\frac{1}{2}}L^2(S_\beta, \mathcal{X})$ and its dual, that is, $Q^\#$, which can be identified with $C^{\frac{1}{2}}L^2(S_\beta, \mathcal{X})$.

Lemma 21.51 *Let us define for $t \in S_\beta$ the map*

$$j_t : (2\epsilon \tanh(\beta\epsilon/2))^{\frac{1}{2}}\mathcal{X} \ni g \mapsto \delta_t \otimes g \in Q. \tag{21.52}$$

Then

$$(j_{t_1} g_1 | j_{t_2} g_2)_{\mathcal{Q}} = \left(g_1 \Big| \frac{e^{-|t_1 - t_2|\epsilon} + e^{-(\beta - |t_1 - t_2|)\epsilon}}{2\epsilon(1 - e^{-\beta\epsilon})} g_2 \right)_{\mathcal{X}}.$$

In particular j_t is isometric.

Proof The proof is analogous to the proof of Lemma 21.15. In particular, we use the discrete unitary Fourier transform

$$L^2(S_\beta) \ni f \mapsto (f_n) \in l^2(\mathbb{Z}), \quad f_n = \beta^{-\frac{1}{2}} \int_{S_\beta} e^{-i2\pi nt/\beta} f(t) dt,$$

and apply

$$\frac{1}{\beta} \sum_{n \in \mathbb{Z}} \frac{e^{i2\pi nt/\beta}}{(2\pi n/\beta)^2 \mathbb{1} + \epsilon^2} = \frac{e^{-|t|\epsilon} + e^{-(\beta - |t|)\epsilon}}{2\epsilon(\mathbb{1} - e^{-\beta\epsilon})} \tag{21.53}$$

instead of (21.8). $\qquad\qquad\qquad\qquad\qquad\qquad\qquad\qquad\qquad\qquad\qquad\Box$

Definition 21.52 *For $t \in \mathbb{R}$, resp. for $I \subset \mathbb{R}$ we define $\mathcal{Q}_t$, e_t, e^t, resp. $\mathcal{Q}_I$, e_I, e^I, as in Subsect. 21.2.4.*

Definition 21.53 *We define*

$$rf(s) := f(-s), \quad u_t f(s) = f(s - t), \quad f \in \mathcal{Q}, \quad s, t \in S_\beta.$$

Proposition 21.54 (1) *Let $t, t_1, t_2 \in S_\beta$, $t_1 < t_2$ and $f \in \mathcal{Q}^\#$. We have*

$$e^t f(s) = (e^{\beta\epsilon} - e^{-\beta\epsilon})^{-1} (e^{-|t-s|\epsilon}(e^{\beta\epsilon} - \mathbb{1}) + e^{|t-s|\epsilon}(\mathbb{1} - e^{-\beta\epsilon})) f(t),$$

$$e^{[t_1,t_2]} f(s) = \mathbb{1}_{[t_1,t_2]}(s) f(s) + \left(\sinh(\beta + t_1 - t_2)\epsilon \right)^{-1}$$

$$\times \left(\mathbb{1}_{]-\beta/2,t_1[}(s) \Big(\sinh\big((s + \beta - t_2)\epsilon\big) f(t_1) - \sinh\big((s - t_1)\epsilon\big) f(t_2) \Big) \right.$$

$$\left. + \mathbb{1}_{]t_2,\beta/2[}(s) \Big(\sinh\big((s - t_2)\epsilon\big) f(t_1) \dot{-} \sinh\big((s - \beta - t_1)\epsilon\big) f(t_2) \Big) \right).$$

(2) $C_c^\infty(]t_1, t_2[, \text{Dom } \epsilon)$ *is dense in $\mathcal{Q}_{]t_1,t_2[}$.*

(3) $\mathbb{R} \ni t \mapsto u_t$ *is an orthogonal β-periodic C_0-group on $\mathcal{Q}$.*

(4) r *is an orthogonal operator satisfying $ru_t = u_{-t}r$ and $r^2 = \mathbb{1}$.*

(5) $\sum_{t \in S_\beta} u_t \mathcal{Q}_0$ *is dense in $\mathcal{Q}$.*

(6) $re_0 = e_0$.

(7) $e_{[0,\beta/2]} e_{[-\beta/2,0]} = e_{\{0,\beta/2\}}$.

21.4.6 Gaussian β-Markov path spaces II

As in Subsect. 21.2.4, we consider the Gaussian $\mathbf{L}^2$ space with covariance C. According to the notation introduced in Subsect. 5.4.2, this will be denoted

$$\mathbf{L}^2 \left(L^2(S_\beta, \mathcal{X}), e^{\phi \cdot C^{-1} \phi} d\phi \right), \tag{21.54}$$

where we use ϕ as the generic variable in $L^2(S_\beta, \mathcal{X})$. Let $L^2(Q, d\mu)$ be a concrete realization of (21.54).

Definition 21.55 *For $s \in S_\beta$ and $g \in \left(2\epsilon \tanh(\beta\epsilon/2)\right)^{\frac{1}{2}}\mathcal{X}$, we define*

$$\phi_s(g) := \phi(\delta_s \otimes g) \in \bigcap_{1 \leq p < \infty} L^p(Q),$$

called the sharp-time fields.

Set

$$Re^{i\phi(f)} := e^{i\phi(rf)}, \quad U_t e^{i\phi(f)} := e^{i\phi(u_{-t}f)}, \quad f \in \mathcal{Q}, \quad t \in S_\beta, \tag{21.55}$$

and extend R and U_t to $L^2(Q, d\mu)$ by linearity and density.

We obtain the following proposition, whose proof is completely analogous to Prop. 21.23.

Proposition 21.56 *Let $\mathfrak{S}$ be the completion of the Borel σ-algebra on Q. Let $\mathfrak{S}_0$ be the σ-algebra generated by the functions $e^{i\phi_0(g)}$ for $g \in \left(2\epsilon \tanh(\beta\epsilon/2)\right)^{\frac{1}{2}}\mathcal{X}$. Let R, U_t be defined in (21.55). Then $(Q, \mathfrak{S}, \mathfrak{S}_0, U_t, R, \mu)$ is a β-Markov path space.*

Definition 21.57 *$(Q, \mathfrak{S}, \mathfrak{S}_0, U_t, R, \mu)$ defined above will be called the* Gaussian β-Markov path space with covariance C.

The β-KMS system obtained from the Gaussian path space can be interpreted in terms of Araki–Woods CCR representations. We set

$$\rho = (e^{\beta\epsilon} - \mathbb{1})^{-1}.$$

Recall that in Subsect. 17.1.5 we defined the (left) Araki–Woods CCR representation, denoted $g \mapsto W_{\rho,\text{l}}(g)$. Recall also that J_s denoted the corresponding modular conjugation on the Araki–Woods W^*-algebra.

Theorem 21.58 *There exists a unique unitary map*

$$T_{\text{eucl}} : \mathcal{H} \to \Gamma_\text{s}\left((2\epsilon)^{\frac{1}{2}}\mathbb{C}\mathcal{X} \oplus (2\bar{\epsilon})^{\frac{1}{2}}\overline{\mathbb{C}\mathcal{X}}\right)$$

intertwining the CCR representation of the time-zero fields with the Araki–Woods CCR representation at density ρ, that is,

$$T_{\text{eucl}}1 = \Omega,$$
$$T_{\text{eucl}}e^{i\phi_0(g)}T_{\text{eucl}}^{-1} = W_{\rho,\text{l}}(y) = e^{i\phi((\mathbb{1}+2\rho)^{\frac{1}{2}}g \oplus \overline{\rho}^{\frac{1}{2}}\overline{g})}, \quad g \subset \left(2\epsilon\tanh(\beta\epsilon/2)\right)^{\frac{1}{2}}\mathcal{X}.$$

It satisfies

$$T_{\text{eucl}}L = \mathrm{d}\Gamma(\epsilon \oplus -\bar{\epsilon})T_{\text{eucl}}, \tag{21.56}$$
$$T_{\text{eucl}}J = J_\text{s}T_{\text{eucl}}. \tag{21.57}$$

Proof The proof is similar to Thm. 21.26. To construct T_{eucl}, it suffices by linearity and density to check that

$$\int_Q e^{i\phi_0(g)}\mathrm{d}\mu = e^{-\frac{1}{2}(\delta_0\otimes g|C\delta_0\otimes g)} = \exp\left(-\frac{1}{2}\big(g|(2\epsilon)^{-1}(\mathbb{1}-e^{-\beta\epsilon})^{-1}(\mathbb{1}+e^{-\beta\epsilon})g\big)_{\mathcal{X}}\right)$$

$$= \big(\Omega|e^{i\phi_{\rho,1}(g)}\Omega\big),$$

which is immediate. To check (21.56) we verify using Lemma 21.25 that

$$\int_Q e^{-i\phi_0(g_1)}e^{i\phi_t(g_2)}\mathrm{d}\mu = \big(W_{\rho,1}(g_1)\Omega|e^{-t\mathrm{d}\Gamma(\epsilon\oplus-\bar\epsilon)}W_{\rho,1}(g_2)\Omega\big), \quad 0\leq t\leq \beta/2.$$

(21.57) can be checked similarly. □

21.5 Perturbations of β-Markov path spaces

Let us fix a β-Markov path space $(Q,\mathfrak{S},\mathfrak{S}_0,U_t,R,\mu)$. In this section we describe a large class of perturbations of the measure μ that still satisfy the axioms of a β-Markov path space. We also describe the corresponding new physical space and Liouvillean.

We will restrict ourselves to perturbations given by a real $\mathfrak{S}_0$-measurable function V such that

$$V,\ e^{-\beta V}\in L^1(Q). \tag{21.58}$$

As in Sect. 21.3, it is also possible to consider more singular perturbations associated with the equivalent of a Feynman–Kac–Nelson kernel; see Klein–Landau (1981b).

21.5.1 Perturbed path spaces

By Lemma 21.32, we know that the function

$$F := e^{-\int_{S_\beta} U_s(V)\mathrm{d}s}$$

belongs to $L^1(Q)$.

Definition 21.59 *We introduce the perturbed measure*

$$\mathrm{d}\mu_V := \frac{F\mathrm{d}\mu}{\int_Q F\mathrm{d}\mu}.$$

Clearly, μ_V is a probability measure.

Note that we can write $F = F_{[-\beta/2,0]}F_{[0,\beta/2]}$ for

$$F_{[0,\beta/2]} = e^{-\int_0^{\beta/2} U_s(V)\mathrm{d}s},$$

$$F_{[-\beta/2,0]} = e^{-\int_{-\beta/2}^0 U_s(V)\mathrm{d}s} = R(F_{[0,\beta/2]}).$$

$F_{[0,\beta/2]}$, resp. $F_{[-\beta/2,0]}$ is $\mathfrak{S}_{[0,\beta/2]}$-, resp. $\mathfrak{S}_{[-\beta/2,0]}$-measurable.

Proposition 21.60 *The perturbed path space* $(Q, \mathfrak{S}, \mathfrak{S}_0, U_t, R, \mu_V)$ *is* β-*Markov.*

Proof We first check the properties of Def. 21.41. Since $F > 0$ μ–a.e., the sets of measure 0 for μ and μ_V coincide, so $\mathfrak{S}$ is complete for μ_V and $L^\infty(Q, \mu) = L^\infty(Q, \mu_V)$. The function F is clearly R and U_t invariant, hence R and U_t preserve μ_V.

Approximating F by $F_n = F\mathbb{1}_{[0,n]}(F)$ and using that U_t is strongly continuous in measure for μ, we see that U_t is also strongly continuous in measure for μ_V. By Lemma 5.33, this implies that U_t is strongly continuous on $L^2(Q, \mu_V)$.

Property (21.5) of Def. 21.7 is obvious. It remains to check the Markov property. To simplify notation we set $E_0 = E_{\{0, \beta/2\}}$, $E_+ = E_{[0, \beta/2]}$, $E_- = E_{[-\beta/2, 0]}$ and decorate with the superscript V the corresponding objects for μ_V. We also set $F_+ = F_{[0, \beta/2]}$, $F_- = F_{[-\beta/2, 0]}$, so that $F = F_+ F_-$.

Property (6) of Prop. 5.27 can be rewritten as the following operator identity, where we identify a function and the associated multiplication operator:

$$E_I^V = \left(E_I(F)\right)^{-1} E_I F, \quad I \subset S_\beta.$$

Using $R(F) = F$ and $RE_0 = E_0$, we see that $RE_0^V = E_0^V$.

Then using (2) of Prop. 5.27, we obtain that

$$E_\pm^V = \left(E_\pm(F_+ F_-)\right)^{-1} E_\pm F_+ F_- = \left(E_\pm(F_\mp)\right)^{-1} E_\pm F_\mp = \left(E_0(F_\mp)\right)^{-1} E_\pm F_\mp,$$

where in the last step we used the β-Markov property for μ. This yields

$$
\begin{aligned}
E_+^V E_-^V &= \left(E_0(F_-)\right)^{-1} E_+ F_- \left(E_0(F_+)\right)^{-1} E_- F_+ \\
&= \left(E_0(F_-)\right)^{-1} \left(E_0(F_+)\right)^{-1} E_+ F_- E_- F_+ \\
&= \left(E_0(F_-)\right)^{-1} \left(E_0(F_+)\right)^{-1} E_+ E_- F_- F_+ \\
&= \left(E_0(F_-)\right)^{-1} \left(E_0(F_+)\right)^{-1} E_+ E_- F \\
&= \left(E_0(F_-)\right)^{-1} \left(E_0(F_+)\right)^{-1} E_0 F.
\end{aligned}
\tag{21.59}
$$

Next we compute, as an identity between functions,

$$E_0(F_-) E_0(F_+) = E_+(F_-) E_-(F_+) = (E_+ E_-)(F_- F_+) = E_0(F). \tag{21.60}$$

Combining (21.59) and (21.60), we obtain that $E_+^V E_-^V = E_0^V$, which implies the Markov property for μ_V. $\qquad\square$

21.5.2 Perturbed Liouvilleans

Applying Subsect. 21.4.2, we can associate with the path space $(Q, \mathfrak{S}, \mathfrak{S}_0, U_t, R, \mu_V)$ a perturbed β-KMS system. In particular, the perturbed physical space is

$$\mathcal{H}_V^{\mathrm{int}} = L^2(Q, \mathfrak{S}_{\{0, \beta/2\}}, \mu_V).$$

It is convenient to relate this perturbed KMS system with the free KMS system obtained with the measure μ and living on the free physical space $\mathcal{H}$. We will decorate with the subscript V and the superscript int the objects obtained by Subsect. 21.4.2 for the path space involving the perturbed measure μ_V. The corresponding objects transported to $\mathcal{H}$ will be decorated with just the subscript V.

Let us first unitarily identify $\mathcal{H}$ with $\mathcal{H}_V^{\text{int}}$.

Lemma 21.61 *Let* $T_V : \mathcal{H}_V^{\text{int}} \to \mathcal{H}$ *be defined by*

$$T_V \Psi := \frac{1}{\left(\int_Q F \mathrm{d}\mu\right)^{\frac{1}{2}}} E_{\{0,\beta/2\}}(F_{[0,\beta/2]}\Psi).$$

Then T_V *is unitary.*

Proof Without loss of generality we can assume that $\int_Q F \mathrm{d}\mu = 1$. Let $\Psi, \Phi \in L^2(Q, \mathfrak{S}_{\{0,\beta/2\}}, \mu_V) = \mathcal{H}_V^{\text{int}}$. Using the reflection property and (21.60), we have

$$\begin{aligned}
(T_V \Phi | T_V \Psi)_{\mathcal{H}} &= \int_Q \overline{E_{\{0,\beta/2\}}(F_{[0,\beta/2]}\Phi)} E_{\{0,\beta/2\}}(F_{[0,\beta/2]}\Psi) \mathrm{d}\mu \\
&= \int_Q F_{[0,\beta/2]} \overline{\Phi} \Psi E_{\{0,\beta/2\}}(F_{[-\beta/2,0]}) \mathrm{d}\mu \\
&= \int_Q F_{[0,\beta/2]} \overline{\Phi} \Psi E_{[0,\beta/2]} E_{[-\beta/2,0]}(F_{[-\beta/2,0]}) \mathrm{d}\mu \\
&= \int_Q F_{[0,\beta/2]} \overline{\Phi} \Psi F_{[-\beta/2,0]} \mathrm{d}\mu \\
&= \int_Q F_{[0,\beta]} \overline{\Phi} \Psi \mathrm{d}\mu \;=\; (\Phi | \Psi)_{\mathcal{H}_V^{\text{int}}}. \qquad \Box
\end{aligned}$$

The following result is shown in Klein–Landau (1981b).

Proposition 21.62 *Let* V *be a real* $\mathfrak{S}_0$*-measurable function satisfying* (21.58). *Set*

$$F_{[0,t]} := \mathrm{e}^{-\int_0^t U_s(V) \mathrm{d}s}$$

and, for $0 < t < \beta/2$,

$$\mathcal{M}_t := \text{Span}\left(\bigcup_{0 \leq s \leq \beta/2 - t} F_{[0,s]} L^\infty(Q, \mathfrak{S}_{[0,\beta/2-t]}, \mu)\right),$$

$$\mathcal{D}_t := E_{\{0,\beta/2\}} \mathcal{M}_t.$$

Then, for any $0 \leq s \leq t \leq \beta/2$, *there exists a unique* $P_V(s) : \mathcal{D}_t \to \mathcal{D}_{t-s}$ *such that*

$$P_V(s) E_{\{0,\beta/2\}} f = E_{\{0,\beta/2\}} F_{[0,t]} U_s f, \quad f \in L^2(Q, \mu).$$

$\{\mathcal{D}_t, P_V(t)\}_{t \in [0,\beta/2]}$ *is a local Hermitian semi-group on* $\mathcal{H}$.

Definition 21.63 *The self-adjoint operator associated with* $\{\mathcal{D}_t, P_V(t)\}_{t \in [0, \beta/2]}$ *is denoted by* $\tilde{L}_V$.

The following theorem is an analog of Thms. 21.37, 21.38.

Theorem 21.64 *Assume in addition to (21.58) that either*

$$V \in L^2(Q, \mathrm{d}\mu), \quad V \geq 0, \tag{21.61}$$

or

$$V \in L^{2+\epsilon}(Q, \mathrm{d}\mu). \tag{21.62}$$

Let L be the free Liouvillean constructed in Def. 21.44. Then

$$\tilde{L}_V = (L + V)^{\mathrm{cl}}.$$

We denote by τ_V^t the dynamics on $\mathfrak{F}$ generated by $\mathrm{e}^{\mathrm{i}t\tilde{L}_V}$. We set

$$\Omega_V := \|\mathrm{e}^{-\beta \tilde{L}_V/2}\Omega\|^{-1} \mathrm{e}^{-\beta \tilde{L}_V/2}\Omega$$

and denote by ω_V the state on $\mathcal{F}$ generated by the vector Ω_V. Clearly, $(\mathfrak{F}, \tau_V, \omega_V)$ is a β-KMS system. We denote by L_V the associated standard Liouvillean (see Def. 6.55). Note that both $\tilde{L}_V$ and L_V generate the same dynamics on $\mathfrak{F}$, even though they are different operators.

We have the following result:

Theorem 21.65 *Let V be a real $\mathfrak{S}_0$-measurable function satisfying (21.58) and either (21.61) or*

$$V \in L^p(Q, \mu), \quad \mathrm{e}^{-\frac{\beta}{2}V} \in L^q(Q, \mu), \quad p^{-1} + q^{-1} = \frac{1}{2}, \quad 2 < p, q < \infty. \tag{21.63}$$

Then

$$L_V = (\tilde{L}_V - JVJ)^{\mathrm{cl}}.$$

The relationship between the two kinds of perturbed β-KMS system – $(\mathfrak{F}, \tau_V, \omega_V)$, which lives on the free space, and $(\mathfrak{F}_V^{\mathrm{int}}, \tau_V^{\mathrm{int}}, \omega_V^{\mathrm{int}})$, which lives on on the perturbed space – is described in the following theorem:

Theorem 21.66 *Let V be a real $\mathfrak{S}_0$-measurable function satisfying the assumptions of Thm. 21.64. Then*

(1) $\mathfrak{A}_V^{\mathrm{int}} = T_V^{-1} \mathfrak{A} T_V$, $\mathfrak{F}_V^{\mathrm{int}} = T_V^{-1} \mathfrak{F} T_V$;
(2) $T_V \Omega_V^{\mathrm{int}} = \Omega_V$;
(3) $T_V \tau_V^{\mathrm{int},t}(A) T_V^{-1} = \tau_V^t(T_V A T_V^{-1})$, $A \in \mathfrak{F}_V^{\mathrm{int}}$, $t \in \mathbb{R}$;
(4) $T_V J_V^{\mathrm{int}} T_V^{-1} = J$.

21.6 Notes

As explained in the introduction, the name "Euclidean approach" comes from the fact that the Minkowski space $\mathbb{R}^{1,d}$ is turned into the Euclidean space $\mathbb{R}^{1+d}$

by the Wick rotation. Hence, the Euclidean approach is usually associated with relativistic quantum field theory. As we saw, it can also be applied in other situations, as in the usual non-relativistic quantum mechanics. In constructive quantum field theory, the use of the Wick rotation was advocated by Symanzik (1965). The construction of interacting Hamiltonians through the corresponding heat semi-group appeared earlier in works of Nelson (1965) and Segal (1970). The monographs by Glimm–Jaffe (1987) and Simon (1974) contain a more detailed treatment of Euclidean methods at zero temperature, essentially in two space-time dimensions. Osterwalder–Schrader (1973, 1975) formulated a set of axioms for a Euclidean quantum theory, parallel to the Wightman axioms on Minkowski space, allowing the reconstruction of a physical theory in a way similar to the one explained here.

The treatment of this chapter follows a series of interesting papers by Klein (1978) and Klein–Landau (1975, 1981b). In particular, the proof of Thm. 21.37 can be found in Klein–Landau (1975), Thm. 3.4, and the proof of Thm. 21.38 in Klein–Landau (1981a), Sect. 2.

Our treatment of path spaces at positive temperature follows Klein–Landau (1981b) and Gérard–Jaekel (2005). In particular, Thms. 21.43 and 21.47 are due to Klein–Landau (1981b). Thms. 21.65 and 21.66 are proven in Gérard–Jaekel (2005).

22

Interacting bosonic fields

The usual formalism of interacting relativistic quantum field theory is purely perturbative and leads to formal, typically divergent expansions. It is natural to ask whether behind these expansions there exists a non-perturbative theory acting on a Hilbert space and satisfying some natural axioms (such as the Wightman or Haag–Kastler axioms). It is not difficult to give a whole list of models of increasing difficulty, well defined perturbatively, whose non-perturbative construction seems conceivable. There were times when it was hoped that by constructing them one by one we would eventually reach models in dimension 4 relevant for particle physics. The branch of mathematical physics devoted to constructing these models is called *constructive quantum field theory*.

The simplest class of non-trivial models of constructive quantum field theory is the bosonic theory in $1 + 1$ dimensions with an interaction given by an arbitrary bounded from below polynomial. It is called the $P(\varphi)_2$ *model*, where P is a polynomial, φ denotes the neutral bosonic field and $2 = 1 + 1$ stands for the space-time dimension. To our knowledge, it has no direct relevance for realistic physical systems, so its main motivation was as an intermediate step in the program of constructive quantum field theory.

The work on the $P(\varphi)_2$ model was successful and led to the development of a number of interesting and deep mathematical tools. The constructive program continued, with the construction of more difficult models, such as the Yukawa$_2$ and $\lambda\varphi_3^4$, as well as a deep analysis of Yang–Mills$_4$. Unfortunately, it seems that no models of direct physical relevance have so far been constructed within this program.

In this chapter we would like to describe some elements of the construction of the $P(\varphi)_2$ model. We will restrict ourselves to space-cutoff models and the net of local algebras associated with this model. We will not discuss the construction of the translation invariant model, which can be found in the literature. We believe that even such a limited treatment of this theory is a good illustration of many concepts of quantum field theory.

22.1 Free bosonic fields

22.1.1 Klein–Gordon equation

The simplest non-interacting relativistic model of quantum field theory describes neutral scalar bosons. It has already been discussed in Sect. 19.2. Let us recall the basic elements of its theory.

The dual phase space can be taken to be the space of real space-compact solutions of the Klein–Gordon equation

$$(\partial_t^2 - \Delta_x + m^2)\,\zeta(t,x) = 0, \quad (t,x) \in \mathbb{R}^{1,d}. \tag{22.1}$$

The model is called *massive*, resp. *massless* if $m > 0$, resp. $m = 0$. Introducing the operator $\epsilon = (-\Delta_x + m^2)^{\frac{1}{2}}$, we can rewrite (22.1) as

$$(\partial_t^2 + \epsilon^2)\zeta = 0.$$

We are in the framework of Subsect. 18.3.2 (and hence also of Subsect. 9.3.1, with $c = (2\epsilon)^{-1}$). We parametrize the dual phase space by the time-zero initial conditions

$$\vartheta(x) := \dot{\zeta}(0,x), \quad \varsigma(x) = \zeta(0,x). \tag{22.2}$$

It is natural to enlarge the dual phase space so that, in terms of the initial conditions, it is

$$\epsilon^{\frac{1}{2}} L^2(\mathbb{R}^d, \mathbb{R}) \oplus \epsilon^{-\frac{1}{2}} L^2(\mathbb{R}^d, \mathbb{R}). \tag{22.3}$$

(This is a special case of the space $\mathcal{Y}_{\mathrm{dyn}}$ defined in Subsect. 18.3.2.) The vector $\varsigma \in \epsilon^{-\frac{1}{2}} L^2(\mathbb{R}^d, \mathbb{R})$ describes the "position" and $\vartheta \in \epsilon^{\frac{1}{2}} L^2(\mathbb{R}^d, \mathbb{R})$ describes the "momentum".

The dual phase space can be treated as a Kähler space with the symplectic form

$$(\vartheta_1, \varsigma_1)\omega(\vartheta_2, \varsigma_2) = \int_{\mathbb{R}^d} (\vartheta_1(x)\varsigma_2(x) - \vartheta_2(x)\varsigma_1(x))\,dx,$$

the conjugation τ and the Kähler anti-involution j:

$$\tau = \begin{bmatrix} \mathbb{1} & 0 \\ 0 & -\mathbb{1} \end{bmatrix}, \quad \mathrm{j} = \begin{bmatrix} 0 & -\epsilon \\ \epsilon^{-1} & 0 \end{bmatrix}. \tag{22.4}$$

The dynamics is generated by the classical Hamiltonian

$$h_0(\varsigma, \vartheta) = \frac{1}{2}\int_{\mathbb{R}^d} (\vartheta^2(x) + |\nabla_x \varsigma(x)|^2 + m^2 \varsigma^2(x))\,dx.$$

One can show that the linear Klein–Gordon equation with initial conditions (22.2) possesses a unique solution, which for the Cauchy data (ϑ, ς) will be denoted $e^{ta}(\vartheta, \varsigma)$. These solutions satisfy the following basic requirements of a (classical) relativistic field theory:

Theorem 22.1 (1) *Locality. If* $\chi \in C_c^\infty(\mathbb{R}^{1,d})$, $\chi \equiv 1$ *on an open set* $\mathcal{O} \subset \mathbb{R}^{1,d}$ *and* ζ *is a solution in* $\mathbb{R}^{1,d}$, *then* $\chi\zeta$ *is a solution in* $\mathcal{O}$.

(2) *Causality. If the Cauchy data* (ϑ, ς) *are supported in a set* $K \subset \mathbb{R}^d$, *then* $e^{ta}(\vartheta, \varsigma)$ *is supported in* $J(\{0\} \times K) = \{(t,x) \in \mathbb{R}^{1,d} : \mathrm{dist}(x, K) \le |t|\}$.

(3) *Covariance. The Poincaré group acting as in Subsect. 19.2.10 preserves the space of solutions with Cauchy data in (22.3).*

22.1.2 Quantization of linear Klein–Gordon equation

Let us first briefly recall the results of Sect. 18.3 about the quantization of the linear Klein–Gordon equation. Consider the complexified dual phase space

$$\epsilon^{-\frac{1}{2}} L^2(\mathbb{R}^d) \oplus \epsilon^{\frac{1}{2}} L^2(\mathbb{R}^d). \tag{22.5}$$

The positive frequency subspace of (22.5) is taken as the one-particle space, as usual denoted by $\mathcal{Z}$. As in (8.32), and then (18.31), we parametrize it by the time-zero momenta, identifying it with $(2\epsilon)^{\frac{1}{2}} L^2(\mathbb{R}^d)$. The time reversal becomes the usual complex conjugation. One can introduce position and momentum operators for the Fock CCR representation on $\Gamma_s(\mathcal{Z})$ as in Subsect. 8.2.7. Let us recall their definition in the present context:

Definition 22.2 *The* time-zero position and momentum operators, *often called the φ and π fields, are defined as*

$$\varphi(\vartheta) := a^*(\vartheta) + a(\vartheta), \quad \vartheta \in \epsilon^{\frac{1}{2}} L^2(\mathbb{R}^d, \mathbb{R}),$$
$$\pi(\varsigma) := \tfrac{1}{2}(a^*(\epsilon\varsigma) - a(\epsilon\varsigma)), \quad \varsigma \in \epsilon^{-\frac{1}{2}} L^2(\mathbb{R}^d, \mathbb{R}).$$

Definition 22.3 *The* time-zero φ and π fields at $x \in \mathbb{R}^d$ *are defined as*

$$\varphi(x) := \varphi(\delta_x),$$
$$\pi(x) := \pi(\delta_x),$$

where δ_x is the Dirac mass at point x.

Remark 22.4 *Comparing Defs. 22.2 and 22.3, we see that the notation $\varphi(\cdot)$ and $\pi(\cdot)$ is somewhat ambiguous. We use this convention, however, and make no attempt to improve on it.*

Note that δ_x does not belong to the one-particle space. We treat the fields $\varphi(x)$ and $\pi(x)$ as "operator-valued distributions" that become well-defined closed operators only after smearing with appropriate test functions:

$$\int_{\mathbb{R}^d} \varphi(x)\vartheta(x)dx = \varphi(\vartheta),$$
$$\int_{\mathbb{R}^d} \pi(x)\varsigma(x)dx = \pi(\varsigma);$$

see Remark 3.54. Formally, $\varphi(x)$ and $\pi(x)$ satisfy the following form of the CCR:

$$[\varphi(x), \varphi(y)] = [\pi(x), \pi(y)] = 0,$$
$$[\varphi(y), \pi(x)] = i\delta(x - y)\mathbb{1}. \tag{22.6}$$

Definition 22.5 Free fields *are defined as*

$$\varphi_0(t, \vartheta) := e^{it d\Gamma(\epsilon)} \varphi(\vartheta) e^{-it d\Gamma(\epsilon)}, \quad \vartheta \in \epsilon^{\frac{1}{2}} L^2(\mathbb{R}^d, \mathbb{R}),$$
$$\varphi_0(t, x) := e^{it d\Gamma(\epsilon)} \varphi(x) e^{-it d\Gamma(\epsilon)}, \quad x \in \mathbb{R}^d.$$

(The subscript 0 indicates that we are dealing with free or, in other words, non-interacting fields.) The equations satisfied by the free fields can be expressed as an operator identity on $\Gamma_s^{\mathrm{fin}}(\mathcal{Z})$

$$\partial_t^2 \varphi_0(t, \vartheta) + \varphi_0(t, \epsilon^2 \vartheta) = 0, \tag{22.7}$$

or, equivalently, as a distributional identity

$$\left(\partial_t^2 - \Delta_{\mathrm{x}} + m^2\right) \varphi_0(t, \mathrm{x}) = 0.$$

Definition 22.6 *For* $(\vartheta, \varsigma) \in \epsilon^{-\frac{1}{2}} L^2(\mathbb{R}^d, \mathbb{R}) \oplus \epsilon^{\frac{1}{2}} L^2(\mathbb{R}^d, \mathbb{R})$, *we also introduce the corresponding* phase space fields *and* Weyl operators

$$\phi(\vartheta, \varsigma) := \varphi(\vartheta) + \pi(\varsigma), \qquad W(\vartheta, \varsigma) := \mathrm{e}^{\mathrm{i}\phi(\vartheta, \varsigma)}. \tag{22.8}$$

If ζ is the solution with the Cauchy data (ϑ, ς), we will also write $\phi(\zeta)$ and $W(\zeta)$ instead of (22.8).

22.1.3 Free dynamics and free local algebras

For concreteness, until the end of the chapter we assume that $d = 1$.

For $A \in B(\Gamma_s(\mathcal{Z}))$, we define

$$\alpha_0^t(A) := \mathrm{e}^{\mathrm{i}td\Gamma(\epsilon)} A \mathrm{e}^{-\mathrm{i}td\Gamma(\epsilon)},$$

$$\alpha_0^{\mathrm{x}}(A) := \mathrm{e}^{\mathrm{i}\mathrm{x}d\Gamma(D)} A \mathrm{e}^{-\mathrm{i}\mathrm{x}d\Gamma(D)},$$

$$\alpha_0^x(A) := \alpha_0^t \circ \alpha_0^{\mathrm{x}}(A), \qquad x = (t, \mathrm{x}) \in \mathbb{R}^{1,1},$$

where $D = D_{\mathrm{x}}$ is the momentum operator. Clearly, $\mathbb{R}^{1,1} \ni x \mapsto \alpha_0^x$ is a strongly continuous group of $*$-automorphisms of $B(\Gamma_s(\mathcal{Z}))$.

In the following definition, all the algebras are concrete and are contained in $B(\Gamma_s(\mathcal{Z}))$.

Definition 22.7 (1) *For a bounded open interval $I \subset \mathbb{R}$, the corresponding* time-zero local algebra *is defined as*

$$\mathfrak{R}(I) := \left\{ W(\vartheta, \varsigma) \; : \; \vartheta, \varsigma \in C_c^\infty(I, \mathbb{R}) \right\}''.$$

(2) *The following algebra plays the role of the algebra of all observables:*

$$\mathfrak{O} := \left(\bigcup_{I \subset \mathbb{R}} \mathfrak{R}(I) \right)^{\mathrm{cpl}}.$$

(3) *For a bounded open set $\mathcal{O} \subset \mathbb{R}^{1,1}$, the corresponding* free local algebra *is defined as*

$$\mathfrak{M}_0(\mathcal{O}) := \left\{ \alpha_0^t(A) \; : \; A \in \mathfrak{R}(I), \; \{t\} \times I \subset \mathcal{O} \right\}''.$$

As described in Subsect. 19.2.7, one can also quantize the free dynamics by abstract CCR algebras. Recall that if $\mathcal{O} \subset \mathbb{R}^{1,1}$ is an open set, then $\mathfrak{A}(\mathcal{O})$ is then the Weyl CCR C^*-algebra generated by elements $W(Gf)$ satisfying the Weyl commutation relations, with $f \in C_c^\infty(\mathcal{O})$, where $Gf(x) = \int G(x-y)f(y)\mathrm{d}y$ and G is the Pauli–Jordan function. These algebras possess a distinguished representation called the Fock representation,

$$\pi_{\mathrm{F}} : \mathfrak{A}(\mathcal{O}) \to B\big(\Gamma_{\mathrm{s}}(\mathcal{Z})\big).$$

Proposition 22.8 *The following hold:*

(1) $\mathfrak{M}_0(\mathcal{O}) = \pi_{\mathrm{F}}(\mathfrak{A}(\mathcal{O}))''$.
(2) $\alpha_0^x(\mathfrak{M}_0(\mathcal{O})) = \mathfrak{M}_0(\mathcal{O}+x), \quad x \in \mathbb{R}^{1,1}$.
(3) $\mathfrak{O} = \bigg(\bigcup_{\mathcal{O} \subset \mathbb{R}^{1,1}} \mathfrak{M}_0(\mathcal{O}) \bigg)^{\mathrm{cpl}}$.

Proof Recall first that the Klein–Gordon equation satisfies the *causality property*, i.e.

$$\mathrm{supp}(\vartheta(t), \varsigma(t)) \subset J(\{0\} \times \mathrm{supp}(\vartheta, \varsigma)), \quad t \in \mathbb{R}. \tag{22.9}$$

To prove (1), we first note that

$$\mathfrak{M}_0(\mathcal{O}) = \big\{ W\big(e^{at}(\vartheta, \varsigma)\big) : \{t\} \times (\mathrm{supp}(\vartheta, \varsigma)) \subset \mathcal{O} \big\}''.$$

We then use Thm. 19.15 and the fact that the Green's function $G(t, s)$ is $\frac{\sin((t-s)\epsilon)}{\epsilon}$. Statement (2) is obvious. Statement (3) follows from the fact that

$$\mathfrak{R}(I) \subset \mathfrak{M}(]-\epsilon_0, \epsilon_0[\times I) \subset \mathfrak{R}(I+]-\epsilon_0, \epsilon_0[), \quad \epsilon_0 > 0. \tag{22.10}$$

The first inclusion in (22.10) is obvious; the second follows from causality. $\qquad\square$

22.1.4 Q-space representation

Let τ be the canonical conjugation on $\mathcal{Z} = (2\epsilon)^{\frac{1}{2}} L^2(\mathbb{R})$ defined as $\tau\Psi(\mathrm{x}) = \overline{\Psi(\mathrm{x})}$. Recall from Subsect. 18.3.2 that τ corresponds to time reversal.

Let $T^{\mathrm{rw}} : \Gamma_{\mathrm{s}}(\mathcal{Z}) \to L^2(Q, \mathrm{d}\mu)$ be the real-wave (or Q-space) representation associated with τ, as in Subsect. 9.3.5. In the sequel we will freely identify objects on $\Gamma_{\mathrm{s}}(\mathcal{Z})$ and on $L^2(Q, \mathrm{d}\mu)$, using T^{rw}. We will also use the same symbol to denote a measurable function V on Q and the operator of multiplication by V acting on $L^2(Q, \mathrm{d}\mu)$. We are in the framework of Subsect. 9.3.1 with $c = (2\epsilon)^{-1}$.

Operators $\varphi(\vartheta)$, $\vartheta \in \epsilon^{\frac{1}{2}} L^2(\mathbb{R})$ commute with one another. In particular, polynomials in the variable φ, that is, functions of the form

$$V(\varphi) = \sum_{j=0}^n \int V_j(\mathrm{x}_1, \ldots, \mathrm{x}_j) \varphi(\mathrm{x}_1) \cdots \varphi(\mathrm{x}_j) \mathrm{d}\mathrm{x}_1 \cdots \mathrm{x}_j,$$

can be interpreted as functions on Q, and as (usually unbounded) operators on $L^2(Q, \mathrm{d}\mu) \simeq \Gamma_{\mathrm{s}}(\mathcal{Z})$.

We can also consider the Wick quantization of the polynomial V. We will use the following alternative notation:

$$\mathrm{Op}^{a^*,a}(V) = :V(\varphi): . \tag{22.11}$$

The notation on the r.h.s. of (22.11) is explained in Prop. 9.53. Clearly, $:V(\varphi):$ is a polynomial in the variable φ.

22.2 $P(\varphi)$ interaction

22.2.1 Nonlinear Klein–Gordon equation

Now let P be a real polynomial and $g : \mathbb{R} \to \mathbb{R}$ a real function. Let us consider the perturbed classical Hamiltonian

$$h(\vartheta, \varsigma) = h_0(\vartheta, \varsigma) + \int_{\mathbb{R}} g(\mathrm{x}) P\big(\varsigma(\mathrm{x})\big) \mathrm{dx}. \tag{22.12}$$

For stability reasons, we require that g be positive and that the polynomial P, and hence the Hamiltonian h, be bounded from below.

Formally, the associated field equation is the following *non-linear Klein–Gordon equation*:

$$\begin{cases} \partial_t^2 \varphi(t, \mathrm{x}) - \Delta_x \varphi(t, \mathrm{x}) + m^2 \varphi(t, \mathrm{x}) + g(\mathrm{x}) P'\big(\varphi(t, \mathrm{x})\big) = 0, \\ \varsigma(\mathrm{x}) = \zeta(0, \mathrm{x}), \quad \vartheta(\mathrm{x}) := \dot{\zeta}(0, \mathrm{x}). \end{cases} \tag{22.13}$$

22.2.2 *Formal quantization of non-linear Klein–Gordon equation*

Let us try to quantize the classical Hamiltonian (22.12). Let us assume that we can give a meaning to the formal expression

$$H = \mathrm{d}\Gamma(\epsilon) + \int_{\mathbb{R}} g(\mathrm{x}) P\big(\varphi(\mathrm{x})\big) \mathrm{dx}. \tag{22.14}$$

Set

$$\varphi(t, \mathrm{x}) := \mathrm{e}^{\mathrm{i}tH} \varphi(\mathrm{x}) \mathrm{e}^{-\mathrm{i}tH}.$$

Then formally we have

$$\partial_t^2 \varphi(t, \mathrm{x}) - \Delta_x \varphi(t, \mathrm{x}) + m^2 \varphi(t, \mathrm{x}) + g(\mathrm{x}) P'\big(\varphi(t, \mathrm{x})\big) = 0,$$

which can be rephrased as saying that we have quantized the non-linear Klein–Gordon equation (22.13).

There are two deep difficulties with the formal expression (22.14):

(1) First, $\varphi(\mathrm{x})$ does not make sense as a self-adjoint operator, so expressions like $\varphi(\mathrm{x})^p$ do not make sense (even after integration against test functions). This problem is called the *ultraviolet divergence*, and is caused by the requirement that the associated field theory should be *local*. For classical field equations it corresponds to the well-known difficulty with multiplying distributions.

(2) If $g(x) \equiv 1$, one encounters the second problem: the integral over $\mathbb{R}$ in (22.14) may not converge. This is called the *infinite-volume divergence* and is caused by the requirement that the field theory should be *translation invariant*.

One can try to tackle these problems as follows. First one modifies the Hamiltonian, introducing ultraviolet and space cutoffs. This leads to the (still formal) expression

$$H_\kappa(g) = \mathrm{d}\Gamma(\epsilon) + \int_\mathbb{R} g(x) P(\varphi_\kappa(x)) \mathrm{d}x.$$

Here, $g \in C_c^\infty(\mathbb{R})$ is a *space cutoff* and $\varphi_\kappa(x)$ is the *ultraviolet cutoff field*

$$\varphi_\kappa(x) := \int_\mathbb{R} \varphi(y) \rho_\kappa(y - x) \mathrm{d}y, \qquad (22.15)$$

where $\rho \in C_c^\infty(\mathbb{R})$ is a *cutoff function* with $\int_\mathbb{R} \rho(y) \mathrm{d}y = 1$, $\rho_\kappa(y) = \kappa\rho(\kappa y)$ and $\kappa \gg 1$ is an *ultraviolet cutoff parameter*.

Now, $\rho_\kappa \in \mathcal{Z} = (2\epsilon)^{\frac{1}{2}} L^2(\mathbb{R})$, except if $m = 0$. The case $m = 0$ is exceptional, because then

$$\|\epsilon^{-\frac{1}{2}} \rho_\kappa\|^2 = (2\pi)^{-1} \int_\mathbb{R} |k|^{-1} |\hat\rho|^2(\kappa^{-1}k) \mathrm{d}k < \infty$$

iff $\hat\rho(0) = \int_\mathbb{R} \rho(y) \mathrm{d}y = 0$. In the rest of this chapter we will always assume that $m > 0$.

Note that the interaction term

$$\int_\mathbb{R} g(x) P(\varphi_\kappa(x)) \mathrm{d}x$$

now makes sense as a self-adjoint operator on the Fock space $\Gamma_s(\mathcal{Z})$.

Next one tries to remove the ultraviolet cutoff, letting $\kappa \to \infty$ and trying to prove the existence of a (non-trivial) limit

$$H_\infty(g) = \lim_{\kappa \to \infty} \left(H_\kappa(g) - R_\kappa(g) \right) \qquad (22.16)$$

in some appropriate sense, where $R_\kappa(g)$ are the so-called *counterterms* related to the well-known need to renormalize various physical constants of the model.

In dimension $1 + 1$ one can use the counterterms

$$R_\kappa(g) := \int_\mathbb{R} g(x) \Big(P(\varphi_\kappa(x)) - {:}P(\varphi_\kappa(x)){:} \Big) \mathrm{d}x$$

obtained by the *Wick ordering* of the interaction term. It is then possible to give a meaning to the expression

$$H_\infty(g) := \mathrm{d}\Gamma(\epsilon) + \int_\mathbb{R} g(x) {:}P(\varphi(x)){:} \, \mathrm{d}x$$

as a bounded below self-adjoint operator on $\Gamma_s(\mathcal{Z})$, as we will see in Subsect. 22.2.5.

Then one tries to take the infinite-volume limit, which means putting $g = 1$. This requires a change of the representation of the CCR – it cannot be done on the original Fock space. This is related to a general argument called *Haag's theorem*.

In higher space dimensions it is no longer possible to give meaning to $H_\infty(g)$ as a self-adjoint operator on the Fock space $\Gamma_s(\mathcal{Z})$. In dimension $1 + 2$ one can quantize the classical non-linear Klein–Gordon equation if the degree of P is not greater than 4. However, even with a spatial cut-off the resulting Hamiltonian acts on a Hilbert space supporting a representation of the CCR not equivalent to the Fock representation. This is related to the so-called *wave function renormalization*.

In dimensions $1 + 3$ or higher it is believed that interacting scalar bosonic quantum fields do not exist.

22.2.3 $P(\varphi)_2$ interaction as a Wick polynomial

Until the end of this chapter we assume that $d = 1$ and $m > 0$.

Recall that

$$\varphi(\mathrm{x}) = \varphi(\delta_\mathrm{x}) = a^*(\delta_\mathrm{x}) + a(\delta_\mathrm{x}),$$
$$\varphi_\kappa(\mathrm{x}) = \varphi\big(\rho_\kappa(\cdot - \mathrm{x})\big) = a^*\big(\rho_\kappa(\cdot - \mathrm{x})\big) + a\big(\rho_\kappa(\cdot - \mathrm{x})\big)$$
$$= a^*\big(\chi(\kappa^{-1}D_\mathrm{x})\delta_\mathrm{x}\big) + a\big(\chi(\kappa^{-1}D_\mathrm{x})\delta_\mathrm{x}\big),$$

where $\chi = \hat{\rho} \in \mathcal{S}(\mathbb{R})$ satisfies $\chi(0) = 1$.

Let us fix a real bounded below polynomial

$$P(\lambda) = \sum_{p=0}^{2n} a_p \lambda^p. \tag{22.17}$$

Clearly, $\deg P = 2n$ has to be even and $a_{2n} > 0$. We also fix a space cutoff function $g \in L^2(\mathbb{R})$.

As explained in Subsect. 22.2.2, instead of the operator $P(\varphi_\kappa(\mathrm{x}))$, we prefer to use its Wick-ordered version

$$:P\big(\varphi_\kappa(\mathrm{x})\big): = \sum_{p=0}^{2n} a_p :\varphi_\kappa(\mathrm{x})^p:.$$

We refer to Prop. 9.53, where this notation is explained. In particular, we recall from (9.60) that

$$:\varphi_\kappa(\mathrm{x})^p: = \sum_{r=0}^{p} \binom{p}{r} a^*\big(\rho_\kappa(\cdot - \mathrm{x})\big)^r a\big(\rho_\kappa(\cdot - \mathrm{x})\big)^{p-r}. \tag{22.18}$$

Definition 22.9 *The operator*

$$V_\kappa := \int_\mathbb{R} g(\mathrm{x}) :P\big(\varphi_\kappa(\mathrm{x})\big): \mathrm{dx}$$

is called an ultraviolet cutoff Wick-ordered interaction.

In this subsection we investigate V_κ as a Wick polynomial.

Definition 22.10 *Let us set*

$$a(k) := (2\pi)^{-\frac{1}{2}} a(e^{ik\cdot x}), \qquad a^*(k) := (2\pi)^{-\frac{1}{2}} a^*(e^{ik\cdot x}).$$

Using the notation of Sect. 9.4, we obtain that

$$M_{p,\kappa} := \int_{\mathbb{R}} g(x) : \varphi_\kappa(x)^p : dx$$

$$= \sum_{r=0}^{p} \binom{p}{r} \int_{\mathbb{R}^p} w_{p,\kappa}(k_1, \ldots, k_r, k_{r+1}, \ldots, k_p)$$

$$\times a^*(k_1) \cdots a^*(k_r) a(-k_{r+1}) \cdots a(-k_p) dk_1 \cdots dk_p,$$

for

$$w_{p,\kappa}(k_1, \ldots, k_p) = (2\pi)^{1-2p} \hat{g}\Big(\sum_{i=1}^{p} k_i\Big) \prod_{j=1}^{p} \chi(\kappa^{-1} k_j). \qquad (22.19)$$

We denote by $w_{p,\infty}$ the function on $\mathbb{R}^p$ obtained by setting $\kappa = \infty$ in (22.19), i.e.

$$w_{p,\infty}(k_1, \ldots, k_p) = (2\pi)^{1-2p} \hat{g}\Big(\sum_{i=1}^{p} k_i\Big), \qquad (22.20)$$

which allows us to define $M_{p,\infty}$.

Lemma 22.11 *The kernels $w_{p,\kappa}$ are in $\otimes^p \epsilon^{\frac{1}{2}} L^2(\mathbb{R})$ for $0 < \kappa \leq \infty$ and, for any $\delta > 0$,*

$$\|w_{p,\kappa} - w_{p,\infty}\|_{\otimes^p \epsilon^{\frac{1}{2}} L^2(\mathbb{R})} \leq C_\delta \|g\|_{L^2(\mathbb{R})} \kappa^{-\delta}.$$

Remark 22.12 *Lemma 22.11 still holds if $g \in L^{1+\delta}(\mathbb{R})$ for some $\delta > 0$.*

Proof It clearly suffices to prove the corresponding statements with $w_{p,\kappa}$ replaced by $w_{p,\kappa} \prod_{i=1}^{p} \epsilon(k_i)^{-\frac{1}{2}}$, and $\otimes^p \epsilon^{\frac{1}{2}} L^2(\mathbb{R})$ replaced by $L^2(\mathbb{R}^p)$. We use the bound

$$\prod_{j=1}^{p} a_j \leq \sum_{i=1}^{p} \Big(\prod_{j \neq i} a_j\Big)^{p/(p-1)}, \qquad (22.21)$$

which follows from the inequality

$$\Big(\prod_{i=1}^{p} \lambda_i\Big)^{1/p} \leq \sum_{j=1}^{p} \lambda_j$$

applied to $\lambda_i = \prod_{j \neq i} a_j^{p/(p-1)}$. Applying (22.21) to $a_i = \omega(k_i)^{-\frac{1}{2}}$, we obtain that $w_{p,\infty}$, and hence $w_{p,\kappa}$ for $\kappa < \infty$, belong to $L^2(\mathbb{R}^p)$. The bound on $\|w_{p,\kappa} - w_{p,\infty}\|$ is a direct computation, using (22.21). $\qquad \square$

From Prop. 9.50 we see that

$$V_\kappa = \sum_{p=0}^{2n} a_p M_{p,\kappa}$$

is well defined as a Hermitian operator on $\mathrm{Dom}\, N^n$ for $\kappa \leq \infty$. We will use the notation

$$V := V_\infty = \int_{\mathbb{R}} g(\mathrm{x}) : P(\varphi(\mathrm{x})) : d\mathrm{x}.$$

Lemma 22.13 (1) V_κ *and V with domain $\mathrm{Dom}\, N^n$ are densely defined Hermitian operators.*
(2) *There exists $\delta > 0$ such that*

$$\|(V - V_\kappa)(N+1)^{-n}\| \leq C\|g\|_{L^2(\mathbb{R})}\kappa^{-\delta}.$$

Proof It suffices to apply the N_τ estimates of Prop. 9.50.   □

22.2.4 $P(\varphi)_2$ *interaction as a multiplication operator*

In this subsection we study the operators V_κ as multiplication operators in the Q-space representation.

Proposition 22.14 *Assume that $g \in L^2(\mathbb{R})$. Then the following are true:*

(1) V_κ *and V are multiplication operators by functions in* $\bigcap_{1 \leq p < \infty} L^p(Q, d\mu)$.

(2) *For any $\delta > 0$, there exists a constant $C_\delta > 0$ such that*

$$\|V_\kappa - V\|_{L^p(Q,d\mu)} \leq C_\delta (p-1)^n \kappa^{-\delta}, \quad p > 2.$$

(3) *Assume in addition that $g \in L^1(\mathbb{R})$ and $g \geq 0$. Then there exist constants $C > 0$, κ_0 such that, for $\kappa \geq \kappa_0$,*

$$V_\kappa \geq -C(\log \kappa)^n.$$

Proof Note that $\epsilon = (D_\mathrm{x}^2 + m^2)^{\frac{1}{2}}$ is a real operator and that $\epsilon^{-\frac{1}{2}}\rho_\kappa(\cdot - \mathrm{x})$ is real. It follows that $\varphi_\kappa(\mathrm{x})$ is a multiplication operator in the Q-space representation. Hence, by Prop. 9.53, the same is true of the operators $\int_{\mathbb{R}} g(\mathrm{x}):\varphi_\kappa(\mathrm{x})^p:d\mathrm{x}$ for $p \in \mathbb{N}$.

For $2 \leq p < \infty$, let us now consider the map $a = (p-1)^{-\frac{1}{2}}\mathbb{1}$ on $(2\epsilon)^{\frac{1}{2}}L^2(\mathbb{R})$. By Thm. 9.30, we know that $\Gamma(a) = (p-1)^{-N/2}$ is a contraction from $L^2(Q)$ into $L^p(Q)$. It follows that

$$\|\Psi\|_{L^p(Q)} \leq (p-1)^{n/2}\|\Psi\|_{L^2(Q)}, \quad \Psi \in \bigoplus_{p=0}^{n} \Gamma_\mathrm{s}^p\big((2\epsilon)^{\frac{1}{2}}L^2(\mathbb{R})\big). \tag{22.22}$$

From Lemma 22.13 we know that $(V - V_\kappa)\Omega \to 0$. So V_κ is Cauchy in $L^2(Q)$. Hence, by (22.22), it is Cauchy also in $\bigcap_{1 \leq p < \infty} L^p(Q)$. It follows that V_κ converges to a function W in $\bigcap_{1\leq p<\infty} L^p(Q)$. Now set

$$\mathcal{S} = \mathrm{Span}\{e^{i\varphi(\vartheta)} \ : \ \vartheta \in C_c^\infty(\mathbb{R},\mathbb{R})\}.$$

Clearly, $\mathcal{S}$ is dense in $L^2(Q)$ and $\mathcal{S} \subset L^\infty(Q) \cap \mathrm{Dom}\, N^p$ for all $p \in \mathbb{N}$. Using that $V_\kappa \to V$ on $\mathrm{Dom}\, N^n$, we see that $V\Psi = W\Psi$ for all $\Psi \in \mathcal{S}$. Hence, $V = W$. This completes the proof of (1).

To prove (2), we use (22.22), the fact that V_κ and V belong to $\overset{2n}{\underset{p=0}{\oplus}} \Gamma_s^p\big(L^2(\mathbb{R})\big)$, and Lemma 22.13.

It remains to prove (3). It follows from (9.26) that, for any $f \in L^2(\mathbb{R},\mathbb{R})$, one has

$$:\varphi(f)^p := \sum_{m=0}^{[p/2]} \frac{p!}{m!(p-2m)!} \varphi(f)^{p-2m} \left(-\frac{1}{4}(f|\epsilon^{-1}f)\right)^m. \tag{22.23}$$

Applying (22.23) to $f = \epsilon^{-\frac{1}{2}} \chi(\kappa^{-1}D_x)\delta_x$, we obtain that

$$:\varphi_\kappa(x)^p := \sum_{m=0}^{[p/2]} \frac{p!}{m!(p-2m)!} c(\kappa)^{2m} \varphi_\kappa(x)^{p-2m}, \tag{22.24}$$

for

$$\begin{aligned}
c(\kappa) &= \big(\delta_x|\tfrac{1}{2}\epsilon^{-1}\chi^2(\kappa^{-1}D_x)\delta_x\big)^{\frac{1}{2}} \\
&= (4\pi)^{-\frac{1}{2}}\big(\textstyle\int_{\mathbb{R}}(k^2+m^2)^{-\frac{1}{2}}\chi^2(\kappa^{-1}k)dk\big)^{\frac{1}{2}} \simeq C(\log\kappa)^{\frac{1}{2}}.
\end{aligned} \tag{22.25}$$

We will apply the bound

$$a^m b^{p-m} \leq \delta b^p + C_\delta a^p, \quad a,b \geq 0, \quad 0 \leq m \leq p, \quad \delta > 0, \tag{22.26}$$

to the terms in the r.h.s. of (22.24), setting $b = \varphi_\kappa(x)$, $a = c(\kappa)$. For $p = 2n$, we obtain, picking δ small enough,

$$:\varphi_\kappa(x)^{2n} :\geq \frac{1}{2}\varphi_\kappa(x)^{2n} - C(\log\kappa)^n.$$

For $p < 2n$, we take $\delta = 1$ and obtain

$$| :\varphi_\kappa(x)^p : | \leq C\big(|\varphi_\kappa(x)^p| + (\log\kappa)^{p/2}\big).$$

Both inequalities should be understood as inequalities between functions on the Q-space.

Since $a_{2n} > 0$, we obtain finally that

$$:P\big(\varphi_\kappa(x)\big) :\geq -C(\log\kappa)^n, \quad \text{for } \kappa \geq \kappa_0. \tag{22.27}$$

Integrating (22.27), using that $g \geq 0$ and $g \in L^1(\mathbb{R})$, we obtain (3). ⊔

Although the operators V_κ are bounded from below, this is not the case for the operator V. Nevertheless, the measure of the set $\{q \in Q \ : V(q) < 0\}$ is very small, as shown in the next proposition.

Proposition 22.15 *Assume that $g \in L^2(\mathbb{R}) \cap L^1(\mathbb{R})$ and $g \geq 0$. Then $\mathrm{e}^{-TV} \in L^1(Q, \mathrm{d}\mu)$ for all $T \geq 0$.*

Proof Let f be a positive measurable function on Q and $t \geq 0$. Set

$$m_f(t) := \mu(\{f(q) > t\}).$$

Clearly, for any $p \geq 1$,

$$m_f(t) \leq \|f\|^p_{L^p(Q)} t^{-p}. \tag{22.28}$$

Moreover, if $F : \mathbb{R}^+ \to \mathbb{R}^+$ is C^1 with $F' \geq 0$, one has

$$\int_Q F(f)\mathrm{d}\mu = \int_0^{+\infty} m_f(t) F'(t)\mathrm{d}t. \tag{22.29}$$

Let C be the constant in Prop. 22.14. We claim that there exist $c_1, c_2, \delta > 0$ such that

$$\mu\Big(\big\{q \in Q \,:\, V(q) \leq -2C(\log \kappa)^n\big\}\Big) \leq c_1 \mathrm{e}^{-c_2 \kappa^\delta}. \tag{22.30}$$

Applying (22.29) to $F(\lambda) = \mathrm{e}^{T\lambda}$ and $f = -V\mathbb{1}_{\{V \leq 0\}}$, and using (22.30), we obtain that $\mathrm{e}^{-TV} \in L^1(Q)$.

It remains to prove (22.30). Since $V_\kappa \geq -C(\log \kappa)^n$, it follows that

$$\big\{V(q) \leq -2C(\log \kappa)^n\big\} \subset \big\{|V - V_\kappa|(q) \geq C(\log \kappa)^n\big\}.$$

Hence,

$$\mu\Big(\big\{V(q) \leq -2C(\log \kappa)^n\big\}\Big) \leq m_{|V - V_\kappa|}\big(C(\log \kappa)^n\big) \leq (p-1)^{np} \kappa^{-\delta p} (\log \kappa)^{-np},$$

by (22.28). Choosing $p = \kappa^{\delta/n} + 1$ yields (22.30). $\qquad\qquad\square$

22.2.5 Space-cutoff $P(\varphi)_2$ Hamiltonian

Theorem 22.16 *Assume that $g \in L^2(\mathbb{R}) \cap L^1(\mathbb{R})$ and $g \geq 0$. Then,*

(1) $\mathrm{d}\Gamma(\epsilon) + V$ *is essentially self-adjoint on* $\mathrm{Dom}\,\mathrm{d}\Gamma(\epsilon) \cap \mathrm{Dom}\,V$;

(2) *The operator* $H = \big(\mathrm{d}\Gamma(\epsilon) + V\big)^{\mathrm{cl}}$ *is bounded from below.*

Definition 22.17 *The operator H is called a* space-cutoff $P(\varphi)_2$ *Hamiltonian.*

Proof We use the formalism of Subsect. 21.2.4. As the real Hilbert space we choose $\mathcal{X} = L^2(\mathbb{R}, \mathbb{R})$, so that $L^2(\mathbb{R}, \mathbb{R}) \otimes \mathcal{X} = L^2(\mathbb{R}^2, \mathbb{R})$. We choose the covariance

$$C = (D_t^2 + \epsilon^2)^{-1} = (D_t^2 + D_x^2 + m^2)^{-1}.$$

We consider the associated Gaussian path space introduced in Def. 21.24. By Thm. 21.26, $\Gamma_\mathrm{s}(\mathcal{Z})$ is the physical Hilbert space and $H_0 = \mathrm{d}\Gamma(\epsilon)$ is the free

Hamiltonian associated with this path space. By Props. 22.14 and 22.15, the multiplicative perturbation V satisfies the hypotheses of Prop. 21.35. By Prop. 9.29, we also see that $e^{-td\Gamma(\epsilon)} = \Gamma(e^{-t\epsilon})$ is a contraction on $L^p(Q)$ for all $1 \le p \le \infty$. Nelson's hyper-contractivity theorem (Thm. 9.30) implies that it maps $L^2(Q)$ into $L^p(Q)$ if $e^{-tm} \le (p-1)^{-\frac{1}{2}}$. Therefore, all the hypotheses of Thm. 21.38 are satisfied. This completes the proof of the theorem. □

22.2.6 Interacting dynamics and local algebras

Definition 22.18 *For $l > 0$, we set*

$$V_l := \int_{[-l,l]} \; :P(\varphi(x)): \, dx,$$

$$H_l := \left(d\Gamma(\epsilon) + V_l\right)^{cl},$$

$$\alpha_l^t(A) := e^{itH_l} A e^{-itH_l}, \quad A \in B(\Gamma_s(\mathcal{Z})),$$

which exist by Thm. 22.16.

Theorem 22.19 (Existence of interacting dynamics) *The following hold:*

(1) *For all bounded open intervals I and $A \in \mathfrak{R}(I)$, there exists the limit*

$$\alpha^t(A) := \lim_{l \to +\infty} \alpha_l^t(A).$$

(2) α^t *uniquely extends to the algebra* $\mathfrak{D}$.
(3) *Set*

$$\alpha^x := \alpha^t \circ \alpha_0^{\mathsf{x}}, \quad x = (t, \mathsf{x}).$$

Then $\mathbb{R}^{1,1} \ni x \mapsto \alpha^x$ is a group of $*$-automorphisms of $\mathfrak{D}$.

Definition 22.20 *For a bounded open set $\mathcal{O} \subset \mathbb{R}^{1,1}$, we set*

$$\mathfrak{M}(\mathcal{O}) := \left\{ \alpha^t(A) \; : \; A \in \mathfrak{R}(I), \; \{t\} \times I \subset \mathcal{O} \right\}'',$$

called the interacting local W^*-algebras.

Theorem 22.21 (Properties of interacting local algebras) *The following hold:*

(1) *One has*

$$\alpha^x(\mathfrak{M}(\mathcal{O})) = \mathfrak{M}(\mathcal{O} + x), \quad x \in \mathbb{R}^{1,1}.$$

(2) *The local interacting algebras are regular, i.e.*

$$\mathfrak{M}(\mathcal{O}) = \bigcap_{\mathcal{O}^{cl} \subset \mathcal{O}_1} \mathfrak{M}(\mathcal{O}_1) = \bigvee_{\mathcal{O}_1^{cl} \subset \mathcal{O}} \mathfrak{M}(\mathcal{O}_1).$$

(3) *If $\mathcal{O}_1$ and $\mathcal{O}_2$ are causally separated, then*

$$\mathfrak{M}(\mathcal{O}_1) \subset \mathfrak{M}(\mathcal{O}_2)'.$$

(4) If $\mathcal{O}_1$ is causally dependent on $\mathcal{O}_2$, then

$$\mathfrak{M}(\mathcal{O}_1) \subset \mathfrak{M}(\mathcal{O}_2).$$

(5)

$$\mathfrak{D} = \Big(\bigcup_{\mathcal{O} \subset \mathbb{R}^{1,1}} \mathfrak{M}(\mathcal{O}) \Big)^{\mathrm{cpl}}.$$

Proof of Thm. 22.19. Applying Trotter's product formula (Thm. 2.75), we obtain

$$e^{\mathrm{i}t H_l} = \mathrm{s} - \lim_{n\to\infty} \big(e^{\mathrm{i}t \mathrm{d}\Gamma(\epsilon)/n} e^{\mathrm{i}t V_l/n} \big)^n.$$

For $A \in \mathfrak{R}(I)$, this implies

$$\alpha_l^t(A) = \mathrm{s} - \lim_{n\to\infty} \big(\alpha_0^{t/n} \circ \gamma_l^{t/n} \big)^n (A), \qquad (22.31)$$

for

$$\gamma_l^t(A) := e^{\mathrm{i}t V_l} A e^{-\mathrm{i}t V_l}.$$

For $l' > l$, we have

$$V_{l'} - V_l = \int_{[-l',l']\setminus[-l,l]} \, :P(\varphi(\mathrm{x})): \, \mathrm{dx},$$

hence $V_{l'} - V_l$ is affiliated to $\mathfrak{R}(] - l', l'[\setminus[-l,l])$. This implies that $\gamma_{l'}^t = \gamma_l^t$ on $\mathfrak{R}(I)$ if $l, l' > |I|$. Moreover, by the causality property, we know that

$$\alpha_0^t : \mathfrak{R}(I) \to \mathfrak{R}(I + [-|t|, |t|]). \qquad (22.32)$$

Using (22.31) and (22.32), this implies that if $l, l' > |I| + |T|$ and $|t| \leq T$, then $\alpha_l^t = \alpha_{l'}^t$ on $\mathfrak{R}(I)$. This shows the existence of α^t on $\mathfrak{R}(I)$. Since $t \mapsto \alpha_t^l$ is a group of $*$-automorphisms, so is $t \mapsto \alpha^t$. This completes the proof of (1). By density, α^t uniquely extends to $\mathfrak{D}$, which proves (2).

To prove (3), we note that $\alpha_0^{\mathrm{x}} \alpha_l^t \alpha_0^{-\mathrm{x}} = \alpha_{l+\mathrm{x}}^t$, which implies (3), by letting $l \to \infty$. $\qquad\square$

Proof of Thm. 22.21. (1) follows by the definition of $\mathfrak{M}(\mathcal{O})$. (2) follows from the analogous property of the time-zero local algebras $\mathfrak{R}(I)$:

$$\mathfrak{R}(I) = \bigcap_{J \supset I^{\mathrm{cl}}} \mathfrak{R}(J) = \bigvee_{J^{\mathrm{cl}} \subset I} \mathfrak{R}(J), \qquad (22.33)$$

which is immediate.

To prove (3) and (4), instead of α^t we can use the space-cutoff dynamics α_l^t for sufficiently large l to define $\mathfrak{M}(\mathcal{O}_i)$. We note that it follows from (22.31) and the causality property that

$$\alpha_l^t\big(\mathfrak{R}(I)\big) \subset \mathfrak{R}(I +] - T, T[), \qquad |t| < T. \qquad (22.34)$$

Then (3) and (4) follow easily from (22.34). To prove (5) we again use (22.34) to get that

$$\mathfrak{R}(I) \subset \mathfrak{M}(] - \epsilon_0, \epsilon_0[\times I) \subset \mathfrak{R}(J),$$

for some $J \supset I$. This clearly implies (5). $\qquad\square$

For completeness, let us note the following theorem, which says that the interacting local algebras can also be defined only in terms of the φ fields and the interacting dynamics.

Theorem 22.22 $\mathfrak{M}(\mathcal{O}) = \left\{ \alpha^t(W(\vartheta, 0)) : \vartheta \in C_c^\infty(I, \mathbb{R}), \{t\} \times \text{supp}\,\vartheta \subset \mathcal{O} \right\}''.$

The above theorem follows easily from the following proposition.

Proposition 22.23 *For $\delta > 0$, set $\mathfrak{B}_\delta(I) = \left\{ \alpha^t(e^{i\varphi(f)}), \text{supp}\, f \subset I, |t| < \delta \right\}''.$ Then*

$$\mathfrak{R}(I) = \bigcap_{\delta > 0} \mathfrak{B}_\delta(I).$$

Proof By (22.34), we know that $\mathfrak{B}_\delta(I) \subset \mathfrak{R}(I + [-\delta, \delta])$. Hence, by (22.33), $\bigcap_{\delta > 0} \mathfrak{B}_\delta(I) \subset \mathfrak{R}(I)$.

To prove the converse inclusion, by (22.33) it suffices again to show that, for all $J^{\text{cl}} \subset I$ and small enough $\delta > 0$, one has

$$\mathfrak{R}(J) \subset \mathfrak{B}_\delta(I). \tag{22.35}$$

To prove (22.35), let us fix I and J with $J^{\text{cpl}} \subset I$, and set $\delta_0 = \frac{1}{2}\text{dist}(J, \mathbb{R} \setminus I)$. We will first prove that if $\delta < \delta_0$, $\text{supp}(\vartheta, \varsigma) \subset J$, then

$$e^{itd\Gamma(\epsilon)} W(\vartheta, \varsigma) e^{-itd\Gamma(\epsilon)} \in \mathfrak{B}_\delta(I), \quad |t| < \delta. \tag{22.36}$$

Set

$$V_I := \int_I \colon P(\varphi(\mathbf{x})) \colon d\mathbf{x}, \qquad H_I = \left(d\Gamma(\epsilon) + V_I\right)^{\text{cl}},$$

$$V_I^{(r)} := V_I \mathbb{1}_{\{|V_I| \le r\}}, \qquad H_I^{(r)} := \left(d\Gamma(\epsilon) + V_I - V_I^{(r)}\right)^{\text{cl}},$$

where $r \in \mathbb{N}$. The Hamiltonians $H_I^{(r)}$ are well defined by the methods of Sect. 22.2.5, and one has

$$H_I^{(r)} = H_I - V_I^{(r)}, \tag{22.37}$$

since $V_I^{(r)}$ is bounded.

It is easy to see that $V_I - V_I^{(r)}$ tends to 0 in $\bigcap_{1 \le p < \infty} L^p(Q)$ and for $t > 0$, $r \in \mathbb{N}$, $e^{-t(V_I - V_I^{(r)})}$ is uniformly bounded in $L^1(Q)$. Using the methods of Sect. 21.3, we prove that

$$e^{-t d\Gamma(\epsilon)} = s - \lim_{r \to +\infty} e^{-t H_I^{(r)}}, \quad t > 0,$$

hence also

$$e^{it d\Gamma(\epsilon)} = s - \lim_{r \to +\infty} e^{it H_I^{(r)}}, \quad t \in \mathbb{R}. \tag{22.38}$$

By (22.37) and Trotter's product formula, we have

$$e^{it H_I^{(r)}} = s - \lim_{p \to +\infty} \left(e^{it H_I / p} e^{-it V_I^{(r)} / p}\right)^p.$$

Noting that V_I and hence $V_I^{(r)}$ are affiliated to $\mathfrak{R}(I)$, we see that $e^{H_I^{(r)}} A e^{-it H_I^{(r)}} \in \mathfrak{B}_\delta(I)$ if $A \in \mathfrak{R}(J)$ and $|t| < \delta$. Since $\mathfrak{B}_\delta(I)$ is weakly closed, we get (22.36), using (22.38).

It follows from (22.36) that $W\left(\frac{1}{t}\left(e^{it\epsilon} h - h\right)\right) \in \mathfrak{B}_\delta(I)$ for $h = \vartheta + i\epsilon\varsigma$, $\text{supp}(\varsigma, \vartheta) \subset J$ and $|t| < \delta$. Using the strong continuity in Thm. 9.5, we obtain that $W(i\epsilon f) = e^{i\pi(f)} \in \mathfrak{B}_\delta(I)$ if $\text{supp } f \subset J$ and $f \in \text{Dom } \epsilon$. Hence, $e^{i\varphi(f)}, e^{i\pi(g)} \in \mathfrak{B}_\delta(I)$ for $\text{supp } f, \text{supp } g \subset J$. This implies (22.35) and ends the proof of the proposition. $\qquad\square$

22.3 Scattering theory for space-cutoff $P(\varphi)_2$ Hamiltonians

In this section we describe, without proof, some properties of the $P(\varphi)_2$ model. In particular, we discuss its *scattering theory*. This theory provides an interesting example of the application of the concept of CCR representations, which arise naturally as the so-called *asymptotic fields*.

In the formulation of the scattering theory we will use the symplectic space $(\mathcal{Y}, \omega)$ associated with the Klein–Gordon equation described at the beginning of Subsect. 22.1.1. Recall that it is equipped with the free dynamics $\mathbb{R} \ni t \mapsto e^{ta}$, and the free Hamiltonian H_0 implements this dynamics:

$$e^{it H_0} W(\zeta) e^{-it H_0} = W(e^{ta} \zeta), \quad \zeta \in \mathcal{Y}.$$

22.3.1 Domain of the space-cutoff $P(\varphi)_2$ Hamiltonian

Let us start with some questions about the Hamiltonian H constructed in Thm. 22.16.

The domain of H is not explicitly known, except if $\deg P = 4$, when it is known that $\text{Dom } H = \text{Dom } d\Gamma(\epsilon) \cap \text{Dom } V$. However, noting that, for all $\delta > 0$, the Hamiltonian $\delta d\Gamma(\epsilon) + V$ is also bounded below, one obtains the following bounds:

$$H_0 \leq C(H_0 + V + b\mathbb{1}), \quad \text{for some } C, b > 0. \tag{22.39}$$

These estimates are called *first-order estimates*. The following *higher-order estimates* are in practice a substitute for the lack of knowledge of the domain of H.

They are an important technical ingredient of the proof of most results described in this section.

Proposition 22.24 *Assume the hypotheses of Thm. 22.16. Then there exists $b > 0$ such that, for all $r \in \mathbb{N}$,*

$$\|N^r (H + b\mathbb{1})^{-r}\| < \infty,$$
$$\|H_0 N^r (H + b\mathbb{1})^{-n-r}\| < \infty,$$
$$\|N^r (H + b\mathbb{1})^{-1}(N + \mathbb{1})^{1-r}\| < \infty. \tag{22.40}$$

22.3.2 Spectrum of space-cutoff $P(\varphi)_2$ Hamiltonians

The following theorem about the *essential spectrum* of space-cutoff $P(\varphi)_2$ Hamiltonians was proven in Dereziński–Gérard (2000).

Theorem 22.25 *Assume the hypotheses of Thm. 22.16. Then*

$$\mathrm{spec}_{\mathrm{ess}}(H) = [\inf \mathrm{spec}\,(H) + m, +\infty[.$$

Corollary 22.26 *Therefore, H possesses a non-degenerate ground state (that is, $\inf \mathrm{spec}\, H$ is a simple eigenvalue).*

Proof Noting that $m > 0$, the existence of a ground state follows immediately from Thm. 22.25. Using the representation of e^{-tH} of Prop. 21.34, we see that e^{-tH} is positivity improving in the Q-space representation. By the Perron–Frobenius theorem (Thm. 5.25), it follows that the ground state is non-degenerate. $\qquad\square$

22.3.3 Asymptotic fields

Scattering theory of space-cutoff $P(\varphi)_2$ Hamiltonians is quite different from the usual scattering theory studied e.g. in the context of Schrödinger operators. It resembles the so-called *Haag–Ruelle scattering theory* developed in the setting of the axiomatic quantum field theory. Its main ingredients are the so-called *asymptotic fields*.

Theorem 22.27 *Assume the hypotheses of Thm. 22.16. Suppose in addition that*

$$|x|^s g(x) \in L^2(\mathbb{R}) \text{ for some } s > 1.$$

Then the following hold:

(1) For all $\zeta \in \mathcal{Y}$, the strong limits

$$W^\pm(\zeta) := \mathrm{s} - \lim_{t \to \pm\infty} e^{itH} W(e^{-ta}\zeta)e^{-itH}$$

exist. They are called the asymptotic Weyl operators.

(2) *The maps*

$$\mathcal{Y} \ni \zeta \mapsto W^{\pm}(\zeta) \in U\big(\Gamma_s(\mathcal{Z})\big)$$

are two CCR representations over the symplectic space $(\mathcal{Y}, \omega)$.

(3) *The representations* $W^{\pm}$ *are regular, so that we can define the asymptotic fields* $\phi^{\pm}(\zeta)$ *by the identity*

$$W^{\pm}(\zeta) = e^{i\phi^{\pm}(\zeta)}.$$

(4) *One has*

$$e^{itH} W^{\pm}(\zeta) e^{-itH} = W^{\pm}(e^{ta}\zeta),$$

i.e. the unitary group e^{itH} *implements the free dynamics* e^{ta} *in the CCR representations* $W^{\pm}$.

(5) *Let us equip the symplectic space* $(\mathcal{Y}, \omega)$ *with its canonical Kähler anti-involution* j *defined in (22.4). Let* $\mathcal{K}^{\pm}$ *be the corresponding space of vacua of* $W^{\pm}$ *(see Def. 11.41). Let* $\mathcal{H}_{\mathrm{pp}}(H)$ *be the point spectrum subspace for* H. *Then*

$$\mathcal{H}_{\mathrm{pp}}(H) \subset \mathcal{K}^{\pm}.$$

Proof The theorem is relatively easy to prove and can be found in Dereziński–Gérard (2000). The main step of the first statement is the so-called Cook argument: we prove that the time derivative of $t \mapsto e^{itH} W(e^{-ta}\zeta) e^{-itH}$ applied to a vector from a dense set is integrable. □

Let us note that Thm. 22.27 can be generalized to cover a much larger class of Hamiltonians. In particular, as proven in Dereziński–Gérard (1999), it holds under rather weak assumptions for the operators called sometimes *Pauli–Fierz Hamiltonians*. Operators of this form are well motivated from the physical point of view – they often appear in non-relativistic quantum physics.

It is natural to ask what type of CCR representations are defined in Thm. 22.27. Statement (5) suggests that a distinguished role is played by the Fock representation. In fact, one can prove that for space-cutoff $P(\varphi)_2$ Hamiltonians no other sectors exist.

Theorem 22.28 *Suppose that the assumptions of Thm. 22.27 hold. Then the following are true:*

(1) *The CCR representations* $W^{\pm}$ *are of Fock type for the anti-involution* j;
(2) $\mathcal{K}^{\pm} = \mathcal{H}_{\mathrm{pp}}(H)$.

Proof To prove (1) we use the number quadratic forms $n^{\pm}$ associated with the CCR representations $W^{\pm}$, defined in Subsect. 11.4.5. Let $\mathcal{V} \subset (2\epsilon)^{\frac{1}{2}} L^2(\mathbb{R})$ be a finite-dimensional space and $\Psi \in \mathrm{Dom}\,|H|^{\frac{1}{2}}$. Then, using

$$a^{\pm \sharp}(h)\Psi = \lim_{t \to \pm\infty} e^{itH} a^{\sharp}(e^{it\epsilon}h) e^{-itH}\Psi, \quad \Psi \in \mathrm{Dom}\,|H|^{\frac{1}{2}}, \tag{22.41}$$

we obtain that

$$n_{\mathcal{V}}^{\pm}(\Psi) = \sum_{i=1}^{\dim \mathcal{V}} \|a^{\pm}(e_i)\Psi\|^2$$

$$= \lim_{t \to \pm\infty} \left(e^{-itH}\Psi \mid \sum_{i=1}^{\dim \mathcal{V}} a^*(e^{it\epsilon}e_i)a(e^{it\epsilon}e_i)e^{-itH}\Psi \right)$$

$$= \lim_{t \to \pm\infty} \left(e^{-itH}\Psi \mid d\Gamma(P_t)e^{-itH}\Psi \right),$$

where P_t is the orthogonal projection on the subspace $e^{it\epsilon}\mathcal{V}$. We now note that

$$d\Gamma(P_t) \leq N \leq C(H + b\mathbb{1}),$$

by the first-order estimates (22.39). Therefore,

$$n_{\mathcal{V}}^{\pm}(\Psi) \leq C\big(\Psi, (H + b\mathbb{1})\Psi\big).$$

This implies that the number quadratic forms $n^{\pm} = \sup_{\mathcal{V}} n_{\mathcal{V}}^{\pm}$ are densely defined since $\mathrm{Dom}\,|H|^{\frac{1}{2}} \subset \mathrm{Dom}\,n^{\pm}$. By Thm. 11.52, this implies (1).

The proof of (2) is much more difficult and involves methods borrowed from N-body scattering theory; and see Dereziński–Gérard (2000) and Gérard–Panati (2008). $\qquad\square$

The two statements of Thm. 22.28 taken together are sometimes called *asymptotic completeness*, since they give a complete understanding of the asymptotic CCR representations. This form of asymptotic completeness can be proven for a much larger class of Hamiltonians. In particular, in Dereziński–Gérard (1999) it has been proven, under rather weak assumptions, for a large class of massive Pauli–Fierz Hamiltonians. The crucial assumption used in the proofs of these statements is the existence of an energy gap in the spectrum of their 1-body Hamiltonians, which is usually called the positivity of their mass.

For space-cutoff $P(\varphi)_2$ Hamiltonians the condition $m > 0$ is needed to define the model itself. On the other hand, massless Pauli–Fierz Hamiltonians are easy to define. Thm. 22.27, with minor modifications, can be proven for a large class of such Hamiltonians. An outstanding question of scattering theory is what the conditions are for asymptotic completeness to hold in the case of massless Pauli–Fierz Hamiltonians.

The central concepts of the standard formulation of scattering theory, used in quantum mechanics, are the free Hamiltonian, and the wave and scattering operators. The reader may wonder why these concepts are missing from Thms. 22.27 and 22.28.

In reality, both wave operators and the scattering operator have a natural definition, which is essentially an application of the formalism of Sect. 11.4. The role of the free Hamiltonian is to some extent played by

$$K \otimes \mathbb{1} + \mathbb{1} \otimes d\Gamma(\epsilon), \tag{22.42}$$

where $K := H\big|_{\mathcal{H}_{\mathrm{pp}}(H)}$.

Theorem 22.29 *Assume the hypotheses of Thm. 22.27. Then there exists a unique unitary operator* $S^\pm : \mathcal{H}_{\mathrm{pp}}(H) \otimes \Gamma_\mathrm{s}(\mathcal{Z}) \to \Gamma_\mathrm{s}(\mathcal{Z})$ *such that*

$$S^\pm \Psi \otimes \Omega = \Psi, \quad \Psi \in \mathcal{H}_{\mathrm{pp}}(H),$$
$$S^\pm \mathbb{1} \otimes W(\zeta) = W^\pm(\zeta) S^\pm, \quad \zeta \in \mathcal{Y}.$$

It satisfies

$$S^\pm \big(K \otimes \mathbb{1} + \mathbb{1} \otimes \mathrm{d}\Gamma(\epsilon)\big) = H S^\pm.$$

Definition 22.30 *The operators* $S^\pm$ *are called* wave *or* Møller *operators.* $S := S^{+*} S^-$ *is called the* scattering operator.

Clearly, S is a unitary operator on $\mathcal{H}_{\mathrm{pp}}(H) \otimes \Gamma_\mathrm{s}(\mathcal{Z})$ commuting with (22.42).

22.4 Notes

The first general result on existence and uniqueness of solutions for non-linear Klein–Gordon equations is due to Ginibre–Velo (1985). More recent references can be found in the book by Tao (2006). The space-cutoff $P(\varphi)_2$ model was first constructed for $P(\varphi) = \varphi^4$ by Glimm–Jaffe (1968, 1970a), for general P by Segal (1970) and Simon–Høgh-Krohn (1972), using the theory of hyper-contractive semi-groups, and by Rosen (1970). The full translation invariant model was then constructed by Glimm–Jaffe (1970b) using local algebras, as in Subsect. 22.2.6. Later a construction by purely Euclidean arguments was given by Glimm–Jaffe–Spencer (1974), Guerra–Rosen–Simon (1973a,b, 1975) and Fröhlich–Simon (1977). The higher-order estimates for the $P(\varphi)_2$ model are due to Rosen (1971).

The construction of the asymptotic fields for a large class of models is due to Høgh-Krohn (1971). The spectral and scattering theory of space-cutoff $P(\varphi)_2$ models was studied by the authors in Dereziński–Gérard (2000) and by Gérard–Panati (2008), following an earlier similar work on Pauli–Fierz Hamiltonians by Dereziński–Gérard (2004).

References

Araki, H., 1963: A lattice of von Neumann algebras associated with the quantum theory of free Bose field, J. Math. Phys. **4**, 1343–1362.

Araki, H., 1964: Type of von Neumann algebra associated with free field, Prog. Theor. Phys. **32**, 956–854.

Araki, H., 1970: On quasi-free states of CAR and Bogoliubov automorphisms, Publ. RIMS Kyoto Univ. **6**, 385–442.

Araki, H., 1971: On quasi-free states of canonical commutation relations II, Publ. RIMS Kyoto Univ. **7**, 121–152.

Araki, H., 1987: Bogoliubov automorphisms and Fock representations of canonical anti-commutation relations, Contemp. Math. **62**, 23–141.

Araki, H., Shiraishi, M., 1971: On quasi-free states of canonical commutation relations I, Publ. RIMS Kyoto Univ. **7**, 105–120.

Araki, H., Woods, E.J., 1963: Representations of the canonical commutation relations describing a non-relativistic infinite free Bose gas, J. Math. Phys. **4**, 637–662.

Araki, H., Wyss, W., 1964: Representations of canonical anti-communication relations, Helv. Phys. Acta **37**, 139–159.

Araki, H., Yamagami, S., 1982: On quasi-equivalence of quasi-free states of canonical commutation relations, Publ. RIMS Kyoto Univ. **18**, 283–338.

Bär, C., Ginoux, N., Pfäffle, F., 2007: *Wave Equations on Lorentzian Manifolds and Quantization*, ESI Lectures in Mathematics and Physics, EMS, Zurich.

Baez, J. C., Segal, I. E., Zhou, Z., 1991: *Introduction to Algebraic and Constructive Quantum Field Theory*, Princeton University Press, Princeton, NJ.

Banaszek, K., Radzewicz, C., Wódkiewicz, K., Krasiński, J. S., 1999: Direct measurement of the Wigner function by photon counting, Phys. Rev. A **60**, 674–677.

Bargmann, V., 1961: On a Hilbert space of analytic functions and an associated integral transform I, Comm. Pure Appl. Math. **14**, 187–214.

Bauer, H., 1968: *Wahrscheinlichkeitstheorie und Grundzüge der Masstheorie*, Walter de Gruyter & Co, Berlin.

Berezin, F. A., 1966: *The Method of Second Quantization*, Academic Press, New York and London.

Berezin, F. A., 1983: *Introduction to Algebra and Analysis with Anti-Commuting Variables* (Russian), Moscow State University Publ., Moscow.

Berezin, F. A., Shubin, M. A., 1991: *The Schrödinger Equation*, Kluwer Academic Publishers, Dordrecht.

Bernal, A., Sanchez, M., 2007: Globally hyperbolic space-times can be defined as "causal" instead of "strongly causal", Classical Quantum Gravity **24**, 745–749.

Birke, L., Fröhlich, J., 2002: KMS, etc, Rev. Math. Phys. **14**, 829–871.

Bloch, F., Nordsieck, A., 1937: Note on the radiation field of the electron, Phys. Rev. **52**, 54–59.

Bogoliubov, N. N., 1947a: J. Phys. (USSR) **11**, reprinted in D. Pines, ed., *The Many-Body Problem*, W. A. Benjamin, New York, 1962.

Bogoliubov, N. N., 1947b: About the theory of superfluidity, Bull. Acad. Sci. USSR **11**, 77–82.

Bogoliubov, N. N., 1958: A new method in the theory of superconductivity I, Sov. Phys. JETP **34**, 41–46.

Bratteli, O., Robinson D. W., 1987: *Operator Algebras and Quantum Statistical Mechanics*, *Volume 1*, 2nd edn., Springer, Berlin.

Bratteli, O., Robinson D. W., 1996: *Operator Algebras and Quantum Statistical Mechanics*, *Volume 2*, 2nd edn., Springer, Berlin.

Brauer, R., Weyl, H., 1935: Spinors in n dimensions. Amer. J. Math. **57**, 425–449.

Brunetti, R., Fredenhagen, K., Köhler, M., 1996: The microlocal spectrum condition and Wick polynomials of free fields on curved space-times, Comm. Math. Phys. **180**, 633–652

Brunetti, R., Fredenhagen, K., Verch, R., 2003: The generally covariant locality principle: a new paradigm for local quantum physics, Comm. Math. Phys. **237**, 31–68.

Cahill, K. E., Glauber, R. J., 1969: Ordered expansions in boson amplitude operators, Phys. Rev. **177**, 1857–1881.

Carlen, E., Lieb, E., 1993: Optimal hyper-contractivity for Fermi fields and related non-commutative integration inequalities, Comm. Math. Phys. **155**, 27–46.

Cartan, E., 1938: *Leçons sur la Théorie des Spineurs*, Actualités Scientifiques et Industrielles No 643 et 701, Hermann, Paris.

Clifford, W. K., 1878: Applications of Grassmann's extensive algebra, Amer. J. Math. **1**, 350–358.

Connes, A., 1974: Caractérisation des espaces vectoriels ordonnés sous-jacents aux algèbres de von Neumann, Ann. Inst. Fourier, **24**, 121–155.

Cook, J., 1953: The mathematics of second quantization, Trans. Amer. Math. Soc. **74**, 222–245.

Cornean, H., Dereziński, J., Ziń, P., 2009: On the infimum of the energy-momentum spectrum of a homogeneous Bose gas, J. Math. Phys. **50**, 062103.

Davies, E. B., 1980: *One-Parameter Semi-Groups*, Academic Press, New York.

Dereziński, J., 1998: Asymptotic completeness in quantum field theory: a class of Galilei covariant models, Rev. Math. Phys. **10**, 191–233.

Dereziński, J., 2003: Van Hove Hamiltonians: exactly solvable models of the infrared and ultraviolet problem, Ann. Henri Poincaré **4**, 713–738.

Dereziński, J., 2006: Introduction to representations of canonical commutation and anticommutation relations. In *Large Coulomb Systems: Lecture Notes on Mathematical Aspects of QED*, J. Dereziński and H. Siedentop, eds, Lecture Notes in Physics **695**, Springer, Berlin.

Dereziński, J., Gérard, C., 1999: Asymptotic completeness in quantum field theory: massive Pauli-Fierz Hamiltonians, Rev. Math. Phys. **11**, 383–450.

Dereziński, J., Gérard, C., 2000: Spectral and scattering theory of spatially cut-off $P(\varphi)_2$ Hamiltonians, Comm. Math. Phys. **213**, 39–125.

Dereziński, J., Gérard, C., 2004: Scattering theory of infrared divergent Pauli-Fierz Hamiltonians, Ann. Henri Poincaré **5**, 523–577.

Dereziński, J., Jakšić, V., 2001: Spectral theory of Pauli-Fierz operators, J. Funct. Anal. **180**, 241–327.

Dereziński, J., Jakšić, V., 2003: Return to equilibrium for Pauli-Fierz systems, Ann. Henri Poincaré **4**, 739–793.

Dereziński, J., Jakšić, V., Pillet, C.-A., 2003: Perturbation theory of W^*-dynamics, Liouvilleans and KMS-states, Rev. Math. Phys. **15**, 447–489.

Dimock, J., 1980: Algebras of local observables on a manifold, Comm. Math. Phys. **77**, 219–228.

Dimock, J., 1982: Dirac quantum fields on a manifold, Trans. Amer. Math. Soc. **269**, 133–147.

Dirac, P. A. M., 1927: The quantum theory of the emission and absorption of radiation, Proc. R. Soc. London A **114**, 243–265.

Dirac, P. A. M., 1928: The quantum theory of the electron, Proc. R. Soc. London A **117**, 610–624.

Dirac, P. A. M., 1930: A theory of electrons and protons, Proc. R. Soc. London A **126**, 360–365.

Dixmier, J., 1948: Position relative de deux variétés linéaires fermées dans un espace de Hilbert, Rev. Sci. **86**, 387–399.

Eckmann, J. P., Osterwalder, K., 1973: An application of Tomita's theory of modular algebras to duality for free Bose algebras, J. Funct. Anal. **13**, 1–12.

Edwards, S., Peierls, P. E., 1954: Field equations in functional form, Proc. R. Soc. A **224**, 24–33.

Emch, G., 1972: *Algebraic Methods in Statistical Mechanics and Quantum Field Theory*, Wiley-Interscience, New York.

Feldman, J., 1958: Equivalence and perpendicularity of Gaussian processes, Pacific J. Math. **8**, 699–708.

Fetter, A. L., Walecka, J. D., 1971: *Quantum Theory of Many-Particle Systems*, McGraw-Hill, New York.

Fock, V., 1932: Konfigurationsraum und zweite Quantelung, Z. Phys. **75**, 622–647.

Fock, V., 1933: Zur Theorie der Positronen, *Doklady Akad. Nauk* **6**, 267–271.

Folland, G., 1989: *Harmonic Analysis in Phase Space*, Princeton University Press, Princeton, NJ.

Friedrichs, K. O., 1953: *Mathematical Aspects of Quantum Theory of Fields*, Interscience Publishers, New York.

Friedrichs, K. O., 1963: *Perturbation of Spectra of Operators in Hilbert Spaces*, AMS, Providence, RI.

Fröhlich, J., 1980: Unbounded, symmetric semi-groups on a separable Hilbert space are essentially self-adjoint. Adv. Appl. Math. **1**, 237–256.

Fröhlich, J., Simon, B., 1977: Pure states for general $P(\phi)_2$ theories: construction, regularity and variational equality, Ann. Math. **105**, 493–526.

Fulling, S. A., 1989: *Aspects of Quantum Field Theory in Curved Space-Time*, Cambridge University Press, Cambridge.

Furry, W. H., Oppenheimer, J. R., 1934: On the theory of electrons and positrons, Phys. Rev. **45**, 245–262.

Gårding, L., Wightman, A. S., 1954: Representations of the commutation and anti-commutation relations, Proc. Nat. Acad. Sci. **40**, 617–626.

Gelfand, I. M., Vilenkin, N. Y., 1964: *Applications of Harmonic Analysis, Generalized Functions* Vol. 4, Academic Press, New York.

Gérard, C., Jaekel, C., 2005: Thermal quantum fields with spatially cut-off interactions in 1+1 space-time dimensions, J. Funct. Anal, **220**, 157–213.

Gérard, C., Panati, A., 2008: Spectral and scattering theory for space-cutoff $P(\varphi)_2$ models with variable metric, Ann. Henri Poincaré **9**, 1575–1629.

Gibbons, G. W., 1975: Vacuum polarization and the spontaneous loss of charge by black holes, Comm. Math. Phys. **44**, 245–264.

Ginibre, J., Velo, G., 1985: The global Cauchy problem for the non-linear Klein–Gordon equation, Math. Z. **189**, 487–505.

Glauber, R. J., 1963: Coherent and incoherent states, Phys. Rev. **131**, 2766–2788.

Glimm, J., Jaffe, A., 1968: A $\lambda\phi^4$ quantum field theory without cutoffs, I, Phys. Rev. **176**, 1945–1951.

Glimm, J., Jaffe, A., 1970a: The $\lambda\phi^4$ quantum field theory without cutoffs, II: the field operators and the approximate vacuum, Ann. Math. **91**, 204–267.

Glimm, J., Jaffe, A., 1970b: The $\lambda\phi^4$ quantum field theory without cutoffs, III: the physical vacuum, Acta Math. **125**, 204–267.

Glimm, J., Jaffe, A., 1985: *Collected Papers, Volume 1: Quantum Field Theory and Statistical Mechanics*, Birkhäuser, Basel.

Glimm, J., Jaffe, A., 1987: *Quantum Physics: A Functional Integral Point of View*, 2nd edn, Springer, New York.

Glimm, J., Jaffe, A., Spencer, T., 1974: The Wightman axioms and particle structure in the $P(\phi)_2$ quantum field model, Ann. Math. **100**, 585–632.

Gross, L., 1972: Existence and uniqueness of physical ground states, J. Funct. Anal. **10**, 52–109.

Grossman, M., 1976: Parity operator and quantization of δ-functions, Comm. Math. Phys. **48**, 191–194.

Guerra, F., Rosen, L., Simon, B., 1973a: Nelson's symmetry and the infinite volume behavior of the vacuum in $P(\phi)_2$, Comm. Math. Phys. **27**, 10–22.

Guerra, F., Rosen, L., Simon, B., 1973b: The vacuum energy for $P(\phi)_2$: infinite volume limit and coupling constant dependence, Comm. Math. Phys. **29**, 233–247.

Guerra, F., Rosen, L., Simon, B., 1975: The $P(\phi)_2$ Euclidean quantum field theory as classical statistical mechanics, Ann. Math. **101**, 111–259.

Guillemin, V., Sternberg, S., 1977: *Geometric Asymptotics*, Mathematical Surveys 14, AMS, Providence, RI.

Haag, R., 1992: *Local Quantum Physics*, Texts and Monographs in Physics, Springer, Berlin.

Haag, R., Kastler, D., 1964: An algebraic approach to quantum field theory, J. Math. Phys. **5**, 848–862.

Haagerup, U., 1975: The standard form of a von Neumann algebra, Math. Scand. **37**, 271–283.

Hajek, J., 1958: On a property of the normal distribution of any stochastic process, Czechoslovak Math. J. **8**, 610–618.

Halmos, P. R., 1950: *Measure Theory*, Van Nostrand Reinhold, New York.

Halmos, P. R., 1969: Two subspaces, Trans. Amer. Math. Soc. **144**, 381–389.

Hardt, V., Konstantinov, A., Mennicken, R., 2000: On the spectrum of the product of closed operators, Math. Nachr. **215**, 91–102.

Hepp, K., 1969: *Théorie de la Renormalisation*, Lecture Notes in Physics, Springer, Berlin.

Høgh-Krohn, R., 1971: On the spectrum of the space cutoff $: P(\varphi):$ Hamiltonian in two space-time dimensions, Comm. Math. Phys. **21**, 256–260.

Hörmander, L., 1985: *The Analysis of Linear Partial Differential Operators, III: Pseudo-Differential Operators*, Springer, Berlin.

Iagolnitzer, D., 1975: Microlocal essential support of a distribution and local decompositions: an introduction. In *Hyperfunctions and Theoretical Physics*, Lecture Notes in Mathematics **449**, Springer, Berlin, pp. 121–132.

Jakšić, V., Pillet, C. A., 1996: On a model for quantum friction, II: Fermi's golden rule and dynamics at positive temperature, Comm. Math. Phys. **176**, 619–644.

Jakšić, V., Pillet, C. A., 2002: Mathematical theory of non-equilibrium quantum statistical mechanics. J. Stat. Phys. **108**, 787–829.

Jauch, J. M., Röhrlich, F., 1976: *The Theory of Photons and Electrons*, 2nd edn, Springer, Berlin.

Jordan, P., Wigner, E., 1928: Über das Paulische Äquivalenzverbot, Z. Phys. **47**, 631–651.

Kallenberg, O., 1997: *Foundations of Modern Probability*, Springer Series in Statistics, Probability and Its Applications, Springer, New York.

Kato, T., 1976: *Perturbation Theory for Linear Operators*, 2nd edn, Springer, Berlin.

Kay, B. S., 1978: Linear spin-zero quantum fields in external gravitational and scalar fields, Comm. Math. Phys. **62**, 55–70.

Kay, B. S., Wald, R. M., 1991: Theorems on the uniqueness and thermal properties of stationary, non-singular, quasi-free states on space-times with a bifurcate Killing horizon, Phys. Rep. **207**, 49–136.

Kibble, T. W. B., 1968: Coherent soft-photon states and infrared divergences, I: classical currents, J. Math. Phys. **9**, 315–324.

Klein, A., 1978: The semi-group characterization of Osterwalder–Schrader path spaces and the construction of Euclidean fields, J. Funct. Anal. **27**, 277–291.

Klein, A., Landau, L., 1975: Singular perturbations of positivity preserving semi-groups, J. Funct. Anal. **20**, 44–82.

Klein, A., Landau, L., 1981a: Construction of a unique self-adjoint generator for a symmetric local semi-group, J. Funct. Anal. **44**, 121–137.

Klein, A., Landau, L., 1981b: Stochastic processes associated with KMS states, J. Funct. Anal. **42**, 368–428.

Kohn, J. J., Nirenberg, L., 1965: On the algebra of pseudo-differential operators, Comm. Pure Appl. Math. **18**, 269–305.

Kunze, R. A., 1958: L^p Fourier transforms on locally compact uni-modular groups. Trans. Amer. Math. Soc. **89**, 519–540.

Lawson, H. B., Michelson, M.-L., 1989: *Spin Geometry*, Princeton University Press, Princeton, NJ.

Leibfried, D., Meekhof, D. M., King, B. E., *et al.*, 1996: Experimental determination of the motional quantum state of a trapped atom, Phys. Rev. Lett. **77**, 4281–4285.

Leray, J., 1978: *Analyse Lagrangienne et Mécanique Quantique: Une Structure Mathématique Apparentée aux Développements Asymptotiques et à l'Indice de Maslov*, Série de Mathématiques Pures et Appliquées, IRMA, Strasbourg.

Lundberg, L. E., 1976: Quasi-free "second-quantization", Comm. Math. Phys. **50**, 103–112.

Manuceau, J., 1968: C^*-algèbres de relations de commutation, Ann. Henri Poincaré Sect. A **8**, 139–161.

Maslov, V. P., 1972: *Théorie de Perturbations et Méthodes Asymptotiques*, Dunod, Paris.

Mattuck, R., 1967: *A Guide to Feynman Diagrams in the Many-Body Problem*, McGraw-Hill, New York.

Moyal, J. E., 1949: Quantum mechanics as a statistical theory, Proc. Camb. Phil. Soc. **45**, 99–124.

Nelson, E., 1965: A quartic interaction in two dimensions. In *Mathematical Theory of Elementary Particles*, W. T. Martin and I. E. Segal, eds, MIT Press, Cambridge, MA.

Nelson, E., 1973: The free Markoff field, J. Funct. Anal. **12**, 211–227.

Neretin, Y. A., 1996: *Category of Symmetries and Infinite-Dimensional Groups*, Clarendon Press, Oxford.

Osterwalder, K., Schrader, R., 1973: Axioms for euclidean Green's functions I, Comm. Math. Phys. **31**, 83–112.

Osterwalder, K., Schrader, R., 1975: Axioms for euclidean Green's functions II, Comm. Math. Phys. **42**, 281–305.

Pauli, W., 1927: Zur Quantenmechanik des magnetischen Elektrons, Z. Phys. **43**, 601–623.

Pauli, W., Weisskopf, V., 1934: Über die Quantisierung der skalaren relativistischen Wellengleichung, Helv. Phys. Acta **7**, 709–731.

Perelomov, A. M., 1972: Coherent states for arbitrary Lie groups, Comm. Math. Phys. **26**, 222–236.

Plymen, R. J., Robinson, P. L., 1994: *Spinors in Hilbert Space*, Cambridge Tracts in Mathematics **114**, Cambridge University Press, Cambridge.

Powers, R., Stoermer, E., 1970: Free states of the canonical anti-commutation relations, Comm. Math. Phys. **16**, 1–33.

Racah, G., 1927: Symmetry between particles and anti-particles, Nuovo Cimento **14**, 322–328.

Reed, M., Simon, B., 1975: *Methods of Modern Mathematical Physics, II: Fourier Analysis, Self-Adjointness*, Academic Press, London.

Reed, M., Simon, B., 1978a: *Methods of Modern Mathematical Physics, III: Scattering Theory*, Academic Press, London.

Reed, M., Simon, B., 1978b: *Methods of Modern Mathematical Physics, IV: Analysis of Operators*, Academic Press, London.

Reed, M., Simon, B., 1980: *Methods of Modern Mathematical Physics, I: Functional Analysis*, Academic Press, London.

Rieffel, M. A., van Daele, A., 1977: A bounded operator approach to Tomita–Takesaki theory, Pacific J. Math. **69**, 187–221.

Robert, D., 1987: *Autour de l'Approximation Semiclassique*, Progress in Mathematics **68**, Birkhäuser, Basel.

Robinson, D., 1965: The ground state of the Bose gas, Comm. Math. Phys. **1**, 159–174.

Roepstorff G., 1970: Coherent photon states and spectral condition, Comm. Math. Phys. **19**, 301–314.

Rosen, L., 1970: A $\lambda\phi^{2n}$ field theory without cutoffs, Comm. Math. Phys. **16**, 157–183.

Rosen, L., 1971: The $(\phi^{2n})_2$ quantum field theory: higher order estimates, Comm. Pure Appl. Math. **24**, 417–457.

Ruijsenaars, S. N. M., 1976: On Bogoliubov transformations for systems of relativistic charged particles, J. Math. Phys. **18**, 517–526.

Ruijsenaars, S. N. M., 1978: On Bogoliubov transformations, II: the general case. Ann. Phys. **116**, 105–132.

Sakai, S., 1971: *C*-Algebras and W*-Algebras*, Ergebnisse der Mathematik und ihrer Grenzgebiete **60**, Springer, Berlin.

Sakai, T., 1996: *Riemannian Geometry*, Translations of Mathematical Monographs **149**, AMS, Providence, RI.

Schrödinger, E., 1926: Der stetige Übergang von der Mikro- zur Makromechanik, Naturwissenschaften **14**, 664–666.

Schwartz, L., 1966: *Théorie des Distributions*, Hermann, Paris.

Schweber, S. S., 1962: *Introduction to Non-Relativistic Quantum Field Theory*, Harper & Row, New York.

Segal, I. E., 1953a: A non-commutative extension of abstract integration, Ann. Math. **57**, 401–457.

Segal, I. E., 1953b: Correction to "A non-commutative extension of abstract integration", Ann. Math. **58**, 595–596.

Segal, I. E., 1956: Tensor algebras over Hilbert spaces, II, Ann. Math. **63**, 160–175.

Segal, I. E., 1959: Foundations of the theory of dynamical systems of infinitely many degrees of freedom (I), Mat. Fys. Medd. Danske Vid. Soc. **31**, 1–39.

Segal, I. E., 1963: *Mathematical Problems of Relativistic Physics*, Proceedings of summer seminar on applied mathematics, Boulder, CO, 1960, AMS, Providence, RI.

Segal, I. E., 1964: Quantum fields and analysis in the solution manifolds of differential equations. In *Analysis in Function Space*, Proceedings of a conference on the theory and applications of analysis in function space, Dedham, MA, 1963, M.I.T. Press, Cambridge, MA.

Segal, I. E, 1970: Construction of non-linear local quantum processes, I, Ann. Math. **92**, 462–481.

Segal, I. E., 1978: The complex-wave representation of the free boson field, Suppl. Studies, Adv. Math. **3**, 321–344.

Shale, D., 1962: Linear symmetries of free boson fields, Trans. Amer. Math. Soc. **103**, 149–167.

Shale, D., Stinespring, W. F., 1964: States on the Clifford algebra, Ann. Math. **80**, 365–381.

Simon, B., 1974: *The $P(\phi)_2$ Euclidean (Quantum) Field Theory*, Princeton University Press, Princeton, NJ.

Simon, B., 1979: *Trace Ideals and Their Applications*, London Math. Soc. Lect. Notes Series **35**, Cambridge University Press, Cambridge.

Simon, B., Høgh-Krohn, R., 1972: Hyper-contractive semi-groups and two dimensional self-coupled Bose fields, J. Funct. Anal. **9**, 121–180.

Skorokhod, A. V., 1974: *Integration in Hilbert Space*, Springer, Berlin.

Slawny, J., 1971: On factor representations and the C^*-algebra of canonical commutation relations, Comm. Math. Phys. **24**, 151–170.

Srednicki, M., 2007: *Quantum Field Theory*, Cambridge University Press, Cambridge.

Stratila, S., 1981: *Modular Theory in Operator Algebras*, Abacus Press, Tunbridge Wells.

Streater, R. F., Wightman, A. S., 1964: *PCT, Spin and Statistics and All That*, W. A. Benjamin, New York.

Symanzik, K., 1965: Application of functional integrals to Euclidean quantum field theory. In *Mathematical Theory of Elementary Particles*, W. T. Martin and I. E. Segal, eds, MIT Press, Cambridge, MA.

Takesaki, M., 1979: *Theory of Operator Algebras I*, Springer, Berlin.

Takesaki, M., 2003: *Theory of Operator Algebras II*, Springer, Berlin.

Tao, T., 2006: *Local and Global Analysis of Non-Linear Dispersive and Wave Equations*, CMBS Reg. Conf. Series in Mathematics **106**, AMS, Providence, RI.

Tomonaga, S., 1946: On the effect of the field reactions on the interaction of mesotrons and nuclear particles, I, Prog. Theor. Phys. **1**, 83–91.

Trautman, A., 2006: Clifford algebras and their representations. In *Encyclopedia of Mathematical Physics* 1, Elsevier, Amsterdam, pp. 518–530.

van Daele, A., 1971: Quasi-equivalence of quasi-free states on the Weyl algebra, Comm. Math. Phys. **21**, 171–191.

van Hove, L., 1952: Les difficultés de divergences pour un modèle particulier de champ quantifié, Physica **18**, 145–152.

Varilly, J. C., Gracia-Bondia, J. M., 1992: The metaplectic representation and boson fields, Mod. Phys. Lett. **A7**, 659–673.

Varilly, J. C., Gracia-Bondia, J. M., 1994: QED in external fields from the spin representation, J. Math. Phys. **35**, 3340–3367.

von Neumann, J., 1931: Die Eindeutigkeit der Schrödingerschen Operatoren, Math. Ann. **104**, 570–578.

Wald, R. M., 1994: *Quantum Field Theory in Curved Space-Time and Black Hole Thermodynamics*, University of Chicago Press, Chicago, IL.

Weil, A., 1964: Sur certains groupes d'opérateurs unitaires, Acta Math. **111**, 143–211.

Weinberg, S., 1995: *The Quantum Theory of Fields, Vol. I: Foundations*, Cambridge University Press, Cambridge.

Weinless, M., 1969: Existence and uniqueness of the vacuum for linear quantized fields, J. Funct. Anal. **4**, 350–379.

Weyl, H., 1931: *The Theory of Groups and Quantum Mechanics*, Methuen, London.

Wick, G. C., 1950: The evaluation of the collision matrix, Phys. Rev. **80**, 268–272.

Widder, D., 1934: Necessary and sufficient conditions for the representation of a function by a doubly infinite Laplace integral, Bull. AMS **40**, 321–326

Wigner, E., 1932a: Über die Operation der Zeitumkehr in der Quantenmechanik, Gött. Nachr. **31**, 546–559.

Wigner, E., 1932b: On the quantum correction for thermodynamic equilibrium, Phys. Rev. **40**, 749–759.

Wilde, I. F, 1974: The free fermion field as a Markov field. J. Funct. Anal. **15**, 12–21.

Williamson J., 1936: On an algebraic problem concerning the normal forms of linear dynamical systems, Amer. J. Math. **58**, 141–163.

Yafaev, D., 1992: *Mathematical Scattering Theory: General Theory*, Translations of Mathematical Monographs **105**, AMS, Providence, RI.

Symbols index

Subject index